Volume 1 1967

European Journal of Biochemistry

Honorary Chairman of the Editorial Board: Sir Hans Krebs (Oxford)

Editor-in-Chief: Claude Liébecq (Liège)

Editorial Board: C.Baglioni (Napoli),   J.Berthet (Leuven),   A.E.Braunstein (Moskva),
F.Chapeville (Gif-sur-Yvette),   G.N.Cohen (Gif-sur-Yvette),   L.Ernster (Stockholm),
U.Henning (Tübingen),   O.Hoffmann-Ostenhof (Wien),   A.T.James (Sharnbrook),
E.Katchalski (Rehovoth),   M.Klingenberg (Marburg/Lahn),   C.Liébecq (Liège),
U.Z.Littauer (Rehovoth),   S.Pontremoli (Ferrara),   B.R.Rabin (London),
S.M.Rapoport (Berlin),   P.Reichard (Stockholm),   J.Rudinger (Praha),   D.Shugar (Warszawa),
A.-M.Staub (Paris),   K.Wallenfels (Freiburg i. Br.),   O.Westphal (Freiburg i. Br.).

Advisory Board: H.R.V.Arnstein (London),   J.Asselineau (Toulouse),
M.Avron (Rehovoth),   G.F.Azzone (Padova),   W.Bernhard (Villejuif),
G.S.Boyd (Edinburgh),   Th.Bücher (München),   D.Cavallini (Roma),
R.M.C.Dawson (Babraham/Cambridge),   A.Ehrenberg (Stockholm),   P.Fasella (Roma),
T.W.Goodwin (Liverpool),   J.I.Harris (Cambridge),   H.Holzer (Freiburg i. Br.),
G.Ivanovics (Szeged),   L.Jaenicke (Köln), A.Kepes (Paris),   A.Klug (Cambridge),
W.Klyne (London),   S.G.Laland (Blindern/Oslo),   V.Luzzati (Gif-sur-Yvette),
F.Lynen (München),   B.G.Malmström (Göteborg),   P.Mitchell (Bodmin),   A.Pihl (Oslo),
J.Polonsky (Gif-sur-Yvette),   J.Porath (Uppsala),   D.B.Roodyn (London),
M.Schramm (Jerusalem),   M.Sela (Rehovoth),   P.Slonimski (Gif-sur-Yvette),   A.Sols (Madrid),
J.R.Tata (London),   A.Tissières (Genève),   H.H.Ussing (København),   W.J.Whelan (London),
J.-M.Wiame (Bruxelles),   Th.Wieland (Frankfurt/M.),   H.G.Zachau (München).

Special Advisor to the Editor-in-Chief: H.Mayer-Kaupp (Heidelberg)

Springer-Verlag Berlin Heidelberg GmbH

ISBN 978-3-662-23717-5   ISBN 978-3-662-25813-2 (eBook)
DOI 10.1007/978-3-662-25813-2

Table of Contents   Volume 1   1967

Nucleic Acids and Protein Synthesis — Biochemical Genetics — Morphogenesis

Protein Chemistry — Enzymology — Physical Chemistry

Author-Index

European Journal of Biochemistry

This JOURNAL perpetuates the tradition of "Biochemische Zeitschrift" founded in 1906 and edited by C. Neuberg (—Vol. 246), C. Neuberg and W. Grassmann (—Vol. 279), W. Grassmann (—Vol. 317), F. G. Fischer and K. Lang (—Vol. 325), and by Th. Bücher, F. Lynen, O. Westphal, and other leading German biochemists unto the end. Upon completion of Vol. 346 its publication was discontinued in favour of closer European cooperation.

The EUROPEAN JOURNAL OF BIOCHEMISTRY will publish papers on all aspects of biochemistry or molecular biology, provided that they describe results which make a sufficient contribution, either experimental or theoretical, to our understanding of biological problems at the chemical or physical level, or that they describe new methods applicable to biochemical problems.

The Table of Contents of the JOURNAL will be subdivided as follows:

1. Nucleic Acids and Protein Synthesis—Biochemical Genetics—Morphogenesis.
2. Protein Chemistry—Enzymology—Physical Chemistry.
3. Cellular Biochemistry and Metabolism.
4. Natural Products.
5. Immunochemistry.

The EUROPEAN JOURNAL OF BIOCHEMISTRY will accept papers from any country—European or not—whether or not the author is a member of the *Federation of European Biochemical Societies (FEBS)*.

The FEBS JOURNAL will only accept original papers concisely written and in final form. Negative results will only be accepted when they can be considered to advance our knowledge. Confirmatory information will only be published in exceptional circumstances.

Manuscripts should preferably be written in English, but papers in French or German will also be accepted (with an extended summary in English).

Submission of a paper implies that it presents the results of original research not previously published (except in the form of an abstract or preliminary note or as part of a published lecture, review or thesis), that it is not under consideration for publication elsewhere, that its contents have not been submitted in essentially the same form as a memorandum to an Information Exchange group and that, if accepted, it will not be released as such, nor be published elsewhere (in the same or in another language) without the written consent of the Editor-in-Chief.

Every author who submits any paper with a view to its publication shall be required to sign an undertaking that the Federation will become entitled to the copyright and that, when called upon to do so, he will assign to the Federation the world copyright and translation rights.

Manuscripts should conform to the "Instructions to Authors" (over-leaf) and should be submitted—in triplicate and double- or triple spaced throughout—to:

Professor Claude Liébecq, Editor-in-Chief
European Journal of Biochemistry
1 Rue des Bonnes Villes
Liège, Belgium.

For each article 40 reprints are supplied free of charge. Additional copies may be ordered at cost price; this must be done at the time when galley proofs are returned to the publisher.

Three volumes each containing four issues will be published annually.
Price per volume DM 84.—; £7.6.8; US $21.00.
Annual subscription price DM 252.—; £22; US $63.00 plus postage and handling DM 23.—; £2; US $5.75.
Subscription orders to: Springer-Verlag, 1 Berlin 31, Heidelberger Platz 3.

INSTRUCTIONS TO AUTHORS

GENERAL

1. Authors should submit *3 copies* (preferably 2 on lightweight paper) of their contribution, together with 3 copies of tables and figures to permit editorial evaluation. For figures, the originals and 2 copies are acceptable.

2. Concisely written and carefully prepared contributions are easier to edit and to read; they are always published faster than longer papers. (This does not justify the division of a large paper into several shorter papers.)

It would facilitate editorial evaluation if authors would enclose reprints of relevant preceding papers in duplicate.

3. Typescripts that are not concisely written or do not conform to the conventions of the JOURNAL will be returned to the authors for revision. A revised paper will be redated if it is not resubmitted within one month, or if it contains a significant amount of new material.

4. Manuscripts should be typewritten, *double or triple spaced throughout* (including References, Acknowledgements, Footnotes, Tables and Legends to Figures) on sheets of uniform size, preferably not larger than DIN A 4 (21×30 cm) with a margin 5 cm wide on the left, to facilitate editorial corrections.

ARRANGEMENT OF THE MANUSCRIPT

1. The *first page* should bear:
a) the title, concise but informative;
b) the initial(s) and name(s) of the author(s);
c) the name of the laboratory where the work was carried out;
d) a running title of not more than 70 letters including spaces;
e) the address to which proofs should be sent;
f) the subdivision under which the author(s) wish(es) the paper to appear in the Table of Contents (over-leaf).

2. The *second page* should list the footnotes to Page 1:
a) dedication—if any—as a footnote to the title; .
b) full postal address(es) of the author(s);
c) the list of non-standard abbreviations (see below);
d) the code numbers of enzymes mentioned in the text, preceded by the letters EC (for Enzyme Commission).

3. Each paper should be preceded by an *English Summary* of 3 to 4% of the length of the paper; this may be divided into numbered sections. The summary should be intelligible in itself.

If the paper is written in French or German, the English summary should be extended to about 10% of the length of the paper and provide as much factual information as possible.

4. The *Introduction* should be brief and state the purpose of the work in relation to other work in the same field. The introduction should not present an extensive review of the literature.

5. *Materials and Methods* should provide enough information to permit repetition of the experimental work.

6. The *Results* should be given concisely. Tables and Figures should not illustrate the same results. *Discussion* should deal with the interpretation of the results and not recapitulate them.

It may often be advantageous to combine *Results and Discussion* in one section.

7. Bibliographic *References* should be numbered sequentially—such as [13], or [17a,17b]—in order of their citation in the text.

The references will be grouped in a section to be printed at the end of the paper, and should contain the names and initials of all authors, the abbreviation of the periodical according to *Chemical Abstracts*, volume (year of publication) first page. Examples:

For *periodicals:*

Banting, F. G. and Best, C. H., *J. Lab. Clin. Med.* 7 (1922) 251.

For *books:*

Allen, M., *Studies Concerning Glycosuria and Diabetes.* Harvard University Press, Cambridge, Mass. 1913, p. 461.

For *symposia:*

Schwick, H. G., In *Immunchemie* (herausgegeben von O. Westphal und L. Ter Haak). Springer Verlag, Berlin, Heidelberg, New York 1965, S. 55.

Responsibility for the accuracy of bibliographic references rests entirely with the author.

8. Tables should be typed on separate pages and numbered sequentially with Arabic numerals. They should be understandable without reference to the text. Conditions specific to the particular experiments should be stated above the tables to which they refer and below their headings. Units in which the results are expressed should appear at the top of each column. Footnotes should be kept to a minimum.

9. *Figures* and *graphs* should be mentioned in the text and all should be numbered, using Arabic numerals also. The back of each figure should be labelled lightly in soft pencil to show the top of the figure, the author(s') name(s) and the figure number. A brief descriptive legend should be provided for each figure; these legends should be typed in sequence on manuscript paper, not on the figures.

Original *drawings* and *graphs* should be drawn with India ink in clean uniform lines on Bristol board, graph paper, blue tracing cloth or coordinate paper, printed in light blue. The labelling of all figures with letters, words, numerals, etc. should be left to the publisher. Therefore, lettering must not be placed on the figure, but instead on a cover sheet of transparent paper.

Illustrations requiring reproduction as *half-tone* plates should be avoided whenever possible. Photographs should be clean glossy prints in sharp focus and as rich in contrast as possible. They should be trimmed at precise right angles. Scales should be given.

Colour plates and half-tone illustrations may have to be charged.

ABBREVIATIONS, SYMBOLS, UNITS, ETC.

The authors should follow *internationally agreed rules* such as those adopted by the commissions of the International Union of Pure and Applied Chemistry (I.U.P.A.C.) and of the International Union of Biochemistry (I.U.B.).

The JOURNAL will essentially follow the rules defined in "Suggestions and Instructions to Authors" published in 1965 by *Biochimica et Biophysica Acta.*

A recent list of *standard abbreviations* which need not be defined has been published by various biochemical journals: see, for instance *J. Biol. Chem.* 241 (1966) 527–533.

The publication of the first issue of the EUROPEAN JOURNAL OF BIOCHEMISTRY marks an event of considerable importance for the Federation, and, hopefully, for European biochemistry at large. Science is by its nature an international cooperative endeavour. This is expressed and realized in our new journal which is based on the support and cooperation of all the constituent societies of the Federation. It is my privilege, as the present Chairman of the Federation, to express our gratitude and indebtedness to all those who have been active in the creation of the new journal and who, through their loyal support, have made it possible.

With the growth of science and the increasing flow of publications, the quality of the written presentation becomes increasingly important. If this new journal is to fulfil its purpose it must set the highest standards from the start. The Editorial Board is very conscious of this fact. The journal starts life in the best spirit and with the highest aspirations and no effort will be spared to make the journal one of the highest quality. However, this goal can only be attained if European biochemists are willing to submit their best work to the journal.

On behalf of the Federation, I would express the hope that the EUROPEAN JOURNAL OF BIOCHEMISTRY will receive the whole-hearted support of European biochemists, that it will prove to be a vital and vigorous undertaking and an important instrument for the promotion of biochemistry and scientific cooperation.

Norsk Hydro's Institute for Cancer Research
Radiumhospitalet
Montebello, Oslo
Norway

ALEXANDER PIHL
Chairman
Federation of European Biochemical
Societies

European J. Biochem. 1 (1967) 2

INTRODUCTION

The FEDERATION OF EUROPEAN BIOCHEMICAL SOCIETIES, which came into being just over three years ago, will have a profound impact on the development of biochemistry in Europe and in the rest of the world. For this reason, the Council of the INTERNATIONAL UNION OF BIOCHEMISTRY has viewed the growth of the Federation with considerable satisfaction. Closer cooperation among European biochemists was set in motion through the initiative of the BIOCHEMICAL SOCIETY several years ago to hold an annual joint meeting with some of its sister societies on the Continent. With considerable foresight the BIOCHEMICAL SOCIETY eventually proposed the setting up of a federation of biochemical societies in Europe, a proposal that was warmly endorsed by representatives of several Continental societies.

The first meeting of the Federation was held in London in the Spring of 1964. I had the privilege to attend the second (Vienna) and third (Warsaw) annual meetings and was much impressed by the rapidly growing attendance and by the quality of the scientific sessions. Upon kind invitation as president of the INTERNATIONAL UNION OF BIOCHEMISTRY I also had the privilege to attend the Federation Council meetings and to participate in the discussions that led to the decision to found a EUROPEAN JOURNAL OF BIOCHEMISTRY, a move that I enthusiastically supported and encouraged on behalf of the INTERNATIONAL UNION OF BIOCHEMISTRY. I feel certain that this journal will render an invaluable service to biochemistry for it will strengthen other endeavors of the Federation in further raising the standards of biochemistry throughout Europe and will provide an authoritative channel of wide diffusion for the publication of original contributions from European and other countries.

It is perhaps significant that the term biochemistry first came into use when Carl Neuberg founded the BIOCHEMISCHE ZEITSCHRIFT sixty years ago and that this journal, of such distinguished tradition, will now be reincarnated in the EUROPEAN JOURNAL OF BIOCHEMISTRY published for the Federation by Springer-Verlag. On behalf of the INTERNATIONAL UNION OF BIOCHEMISTRY I am happy to extend fervent good wishes to the Federation in this venture that will fill for biochemists the increased need for cooperation at all levels among the peoples of Europe.

Department of Biochemistry
New York University School of Medicine
New York, N. Y.
U.S.A.

SEVERO OCHOA
President
International Union of Biochemistry

European J. Biochem. 1 (1967) 3—11

Synthesis of Virus-Specific Proteins in *Escherichia coli* Infected with the RNA Bacteriophage MS2[1]

E. Viñuela[2], I. D. Algranati[3], and S. Ochoa

Department of Biochemistry, New York University School of Medicine, New York, N.Y.

(Received October 10, 1966)

The synthesis of virus-specific proteins following infection of *Escherichia coli* with the RNA phage MS2 was studied using spheroplasts exposed to low concentrations of actinomycin D in the presence of an amino acid mixture containing radioactive leucine or histidine. Under the conditions employed there was about $75^0/_0$ inhibition of virus production and $99^0/_0$ inhibition of host protein synthesis by actinomycin. Fractionation of the phage-induced proteins by acrylamide gel electrophoresis revealed the presence of three major peaks, I, II, and III. The largest one (peak III) was characterized as coat protein. The rate of synthesis of proteins I and II declines before that of the coat protein has reached its maximum.

In contrast to DNA T-bacteriophages, infection of susceptible cells by RNA phages [1] does not shut off the synthesis of host nucleic acids and proteins. The use of actinomycin to curtail the synthesis of host proteins poses several problems because, on the one hand, the cell wall of *Escherichia coli* is not permeable to the antibiotic and, on the other, when its penetration is enabled by the use of spheroplasts or EDTA-treated cells [2], the drug decreases to a greater or lesser extent the yield of viable phage [3—6]. However, under properly controlled conditions, infected spheroplasts or EDTA cells can be used for studying the synthesis of phage-specific proteins in the presence of actinomycin [3,4]. Under these conditions, the formation of proteins containing histidine, an amino acid that is absent from the coat of several RNA phages, signals the induction of proteins other than coat protein [3,4]. Translation of the viral RNA *in vitro* also gives rise to the appearance of coat and non-coat proteins [7—10].

Recently Darnell and collaborators [11] have used high resolution acrylamide gel electrophoresis to determine the number and the rate of synthesis of specific proteins induced upon infection of HeLa cell cultures with poliovirus in the presence of actinomycin. They found that 12—14 electrophoretically different protein chains are formed and that the rate

of synthesis of some of these proteins may vary during the infective cycle.

The small size of the genome of RNA phages provides a relatively simple system for studying the translation of a polycistronic messenger. Bacteriophage MS2, a member of the group of RNA-containing phages [1], consists of one molecule of RNA with a molecular weight of 1×10^6 daltons [12] and a coat protein made up of some 150 identical subunits each containing about 130 amino acids but lacking histidine [13]. With a coding ratio of three nucleotides per amino acid [14], MS2 RNA contains information for only a few (three to six) protein chains.

The present work was undertaken in an attempt to throw further light on the number and the nature of the proteins produced and on the factors controlling translation of the MS2 genome. In the initial phase, reported in this paper, we have used acrylamide gel electrophoresis for isolation of the phage-specific proteins synthesized by *E. coli* spheroplasts infected with MS2 phage, in the presence of actinomycin and radioactive amino acids, at various times of the viral replication cycle. Three main protein chains were synthesized one of which was characterized as coat protein. Experiments with an amber mutant of MS2, to be reported elsewhere, suggest that one of the other two proteins is a phage-specific RNA synthesizing enzyme. The total amount or radioactivity incorporated into coat protein was much greater than in the other proteins. Moreover, when the rate of coat protein synthesis had reached its maximum, the rate of synthesis of the other two proteins was already declining. The results supplement previous conclusions [4,8,11] that the translation of viral polycistronic messengers is subject to regulation.

[1] Dedicated to Professor Luis F. Leloir on the occasion of his 60th birthday.

[2] International Postdoctoral Fellow of the National Institutes of Health, U. S. Public Health Service. Permanent address: Instituto Marañón, Centro de Investigaciones Biológicas, C. S. I. C., Madrid, Spain.

[3] International Postdoctoral Fellow of the National Institutes of Health, U. S. Public Health Service. Present address: Instituto de Investigaciones Bioquímicas, Fundación Campomar, Obligado 2490, Buenos Aires, Argentina.

EXPERIMENTAL PROCEDURE

In order to achieve as much synchrony as possible of the phage replication cycle and to insure the degradation of preexisting host messenger RNA, the bacterial cells were incubated in a starvation medium (lacking glucose and amino acids) at 37° prior to infection. A few minutes after infection the cells were transferred to a suitable medium and converted to spheroplasts with lysozyme in the presence of EDTA. The spheroplast suspension was then transferred to a growth medium containing the desired labelled amino acids and actinomycin D (0.5 μg/ml), and incubated at 37°. This concentration of actinomycin, while partially inhibiting phage multiplication, caused almost complete inhibition of host protein synthesis. The subsequent isolation of the proteins was carried out under conditions favouring maximal dissociation into subunits.

Bacteria and Phage. Bacteriophage MS2 and its host *E. coli* Hfr 3000 (thiamine auxotroph) were obtained from Dr. A. J. Clark, University of California, Berkeley. The phage was grown as described by Loeb and Zinder [15]. The phage stocks (titer, 5×10^{12} p.f.u./ml) were prepared by purifying the phage from a lysate by the procedure of Strauss and Sinsheimer [12] up to the Freon step. For the preparation of radioactive capsid protein the complete purification procedure was carried out. The phage titer in *E. coli* spheroplasts was assayed by standard techniques [16] after dilution of an aliquot in 0.15 M NaCl—2.5 mM CaCl$_2$.

E. coli was grown in a synthetic medium (MTPA) containing 0.1 M Tris-HCl pH 7.5, 0.0085 M NaCl, 0.1 M KCl, 0.02 M NH$_4$Cl, 0.34 mM KH$_2$PO$_4$, 0.16 mM Na$_2$SO$_4$, 2.5 mM CaCl$_2$, 2.5 mM MgCl$_2$, 0.01 M glucose, twenty L-amino acids (each 0.1 mM) and thiamine hydrochloride (10 μg/ml). The starvation medium had the same composition except for the omission of amino acids and glucose. The spheroplast medium contained the same components as the growth medium with the further addition of sucrose (0.35 M), bovine serum albumin (0.1%, w/v) and actinomycin D (0.5 μg/ml). The specific radioactivity and concentration of any radioactive compounds added to this medium will be indicated when describing the individual experiments.

For the preparation of MS2 phage with a radioactive coat *E. coli* grown at 37° in 50 ml of MTPA medium to a concentration of $2-3 \times 10^8$ bacteria/ml, was infected with MS2 at a multiplicity of 30—50. 20 μC of [^{14}C]-arginine (specific radioactivity, 240 mC/mmole), 20 μC of [^{14}C]-lysine (specific radioactivity 240 mC/mmole) and 200 μC of [^{3}H]-leucine (specific radioactivity 1.8 C/mmole) were added immediately to the culture. After two hours at 37°, lysis was completed by incubation for 20 minutes at 37° with lysozyme (50 μg/ml) and some drops of chloroform. The phage in the lysate was purified by the method

of Strauss and Sinsheimer [12]. MS2 phage labelled with [^{3}H]-tyrosine was prepared in broth in the presence of 2.5 mC of [^{3}H]-tyrosine (specific radioactivity 38.8 C/mmole) and purified in the same way.

Spheroplasts Infected with MS2-Phage. Unless otherwise stated all operations were carried out at room temperature. In a typical experiment, 0.4 ml of an overnight culture of *E. coli* was transferred to 40 ml of MTPA medium. The culture was incubated at 37° with shaking to a density of 2×10^8 cells/ml and then filtered through a Millipore membrane (47 mm diameter; 0.45 μ pore size). The bacteria on the filter were washed three times with 10 ml of starvation medium and resuspended in 40 ml of the same medium. When using larger batches of cells centrifugation was substituted for filtration. After aeration for 1 hour at 37°, 20 ml of the culture was infected with MS2 phage at a multiplicity of 30—50. This was incubated, along with an equal volume of uninfected culture, at 37° for an additional 5 minutes. The cells were sedimented by centrifugation for 5 minutes at $10,000 \times g$, washed once with 10 ml of 0.05 M Tris-HCl, pH 8.2 containing 0.35 M sucrose, and suspended in 0.45 ml of the Tris-sucrose solution. Spheroplasts were obtained [17] by incubation with lysozyme (10 μg/ml) in the presence of EDTA (0.6 mM). At least 98% of the cells were converted to spheroplasts as followed by microscopic observation of aliquots diluted 1:10 with water.

Labelling of Phage-Specific Proteins. Aliquots of the spheroplast suspension were diluted 10-fold with spheroplast medium containing actinomycin D (0.5 μg/ml) and radioactive amino acid (time zero). The final mixture, with $5-7 \times 10^8$ spheroplasts/ml, was incubated at 37° with shaking.

To determine the rate of synthesis of the various phage-specific proteins at different times during the replicative cycle, one batch of non-infected and one of infected spheroplasts was incubated at 37° in spheroplast medium containing actinomycin (0.5 μg/ml). Aliquots of each batch were exposed to either [^{14}C]-leucine (0.01 mM, 120 mC/mmole) or [^{14}C]-histidine (0.01 mM, 240 mC/mmole) as follows: 5.0 ml aliquots for the time intervals 1—11, 10—21, and 20—31 minutes; 3.0 ml aliquots for the time intervals 30—51, 50—71, and 70—120 minutes. Portions of each of these aliquots were used for (a) assay of phage titer, (b) determination of radioactivity incorporated into acid-insoluble material, and (c) fractionation of the proteins by gel electrophoresis. A further 5.0 ml sample of each suspension was labelled throughout the 1—120 minute interval with [^{3}H]-leucine (0.01 mM, 1.8 C/mmole). After isolation of the proteins aliquots of this sample were mixed with each of the (c) samples just prior to electrophoresis. This was done to provide internal markers for the peaks of the various phage-specific proteins synthesized in the presence of ^{14}C-labelled amino acids during the specified time intervals.

The ^{14}C-radioactivity recovered from the electrophoretic fractions for each labelling period, from 1—11 to 70—120 minutes and for the 1—120 minute period, was 90 to 100% of the input radioactivity.

Isolation of Labelled Proteins. For this purpose, aliquots of the incubated mixtures were diluted to 6 ml with a solution containing Tris-HCl, pH 8.4, EDTA, and 2-mercaptoethanol to give a final concentration of 0.1, 0.01 and 0.14 M, respectively. All subsequent operations were carried out at room temperature. The suspension was shaken 3 minutes with 2 ml of redistilled phenol saturated with 0.1 M Tris-HCl, pH 8.4, 0.01 M EDTA, 0.14 M 2-mercaptoethanol. This extraction was repeated twice. The phenol phases were pooled and dialyzed, with stirring, against 80 volumes of 0.1 M acetic acid, 0.14 M 2-mercaptoethanol. After 2 hours the fluid was changed and the dialysis continued until the phenol phase at the bottom of the bag had decreased to about 1 ml. The dialysis bag was opened and the top aqueous layer, removed with a Pasteur pipette, was discarded. The bag, containing the phenol phase, was again tightly closed and dialyzed overnight against 125 volumes of 9.0 M urea, 0.05 M acetic acid 0.14 M 2-mercaptoethanol. The dialysis bag was then transferred to a beaker containing 100 volumes of 8.6 M urea, 0.01 M EDTA, 0.14 M 2-mercaptoethanol and 0.1 M Tris-HCl, pH 8.4, and the dialysis was continued for 2 hours with vigorous stirring. During this time a stream of nitrogen was bubbled through the dialyzing solution. Finally the protein was dialyzed for 24 hours against 500 volumes of 0.01 M sodium phosphate, pH 7.2, 0.1% sodium dodecylsulphate, 0.14 M 2-mercaptoethanol, changing the dialysis fluid once. For the preparation of radioactivie coat protein the purified phage was treated in the same way. The recovery of radioactivity in the final protein solutions, with respect to the acid-insoluble radioactivity present after incubation, averaged 46 and 94% for the samples from non-infected and infected spheroplasts, respectively. The low recovery in the former case may be due to the presence of a dialyzable protein (or polypeptide) the synthesis of which appears to be resistant to actinomycin [3].

Electrophoresis. The procedure was similar to that described by Summers *et al.* [11]. However, to insure accurate and efficient determination of radioactive protein in different zones of a polyacrylamide gel we have used as cross-linking agent ethylene diacrylate [18] instead of the standard *N,N'*-methylenebisacrylamide. The gel obtained in this way is soluble in Kinard's scintillation solution [19] after treatment with piperidine.

The gels contained 10% (w/v) acrylamide, 0.27% (w/v) ethylene diacrylate, 0.1% (w/v) sodium dodecylsulphate, 0.1 M sodium phosphate, pH 7.2, 0.075% (w/v) tetramethylethylenediamine (TEMED) and 0.075% (w/v) ammonium persulphate. The mixture was deaerated by evacuation before the addition of ammonium persulphate. Each column was prepared by pouring 2 ml of the solution in a glass tube (10×0.6 cm inner diameter). The polymerization was carried out at room temperature under a 1 cm layer of water. After 30—60 minutes, the columns were rinsed with water.

The sample, in 0.2 ml of 0.01 M sodium phosphate, pH 7.2, 0.1% (w/v) sodium dodecylsulphate, 0.14 M 2-mercaptoethanol and 25% (v/v) glycerol, was layered on top of the gel. The remaining space of the column and the electrode compartments were filled with a buffer containing 0.1 M sodium phosphate, pH 7.2 and 0.1% (w/v) sodium dodecylsulphate. Electrophoresis was carried out at room temperature in the analytical apparatus manufactured by Buchler Instruments, Inc., Fort Lee, New Jersey, at a constant voltage of 7 volt/cm for 4 hours or at 14 volt/cm for 1.5 hours.

After electrophoresis, the gels were removed from the tubes with gentle water pressure and placed on dry-ice. Each frozen gel was fitted in a groove in a leucite block having a transverse slit about 1 mm from the end of the groove. Slices were obtained by moving the gel along the groove and cutting with a razor blade through the slit.

Determination of Radioactivity. For measurement of the radioactivity of electrophoretic fractions, each gel slice was transferred to the bottom of a scintillation vial together with 1 ml of 1.0 M piperidine. The vial was closed and left at room temperature overnight. This was followed by the addition of 0.1 ml of 1.0 M hydroxide of hyamine and stirring in a Cyclo mixer. After adding 15 ml of Kinard's scintillation solution and stirring, followed by standing for 1 hour at room temperature, the radioactivity of the solutions was measured.

For determination of acid-insoluble radioactivity, the incubated spheroplast suspensions were treated with 5% (w/v) trichloroacetic acid. After heating for 5 minutes at 90° and cooling in ice, the precipitate was collected on Millipore filters (25 mm diameter, 0.45 μ pore size) and washed with cold 5% (w/v) trichloroacetic acid. Total radioactivity was determined an aliquots pipetted onto Millipore filters. The filters were dried either under an infrared lamp for 1 hour or by keeping overnight at 60°. The dry filters were placed in scintillation vials under 20 ml of a solution containing 4 g of 2,5-diphenyloxazole (PPO) and 50 mg of 1,4-bis-(5-phenyloxazolyl-2)-benzene (POPOP) per liter of toluene. Radioactivity was measured in a Packard TriCarb scintillation spectrometer. The counting efficiency of ^{14}C and ^{3}H was essentially the same in all cases, about 68% and 17%, respectively.

Characterization of MS2 coat protein. MS2 phage labelled with ^{14}C and ^{3}H in the coat protein was purified as described by Strauss and Sinsheimer [12]. Fig. 1A shows a coincident distribution of [14]C- and ^{3}H-radioactivity and of the virus titer as a single, symmetric peak in the last $CsCl_2$ density gradient centrifugation step of purification. The coat protein was isolated and submitted to acrylamide gel electrophoresis as described for the spheroplast proteins. The electrophoretic pattern, shown in Fig. 1 B, shows the distribution of [14]C- and ^{3}H-radioactivity as a sharp, single symmetric peak.

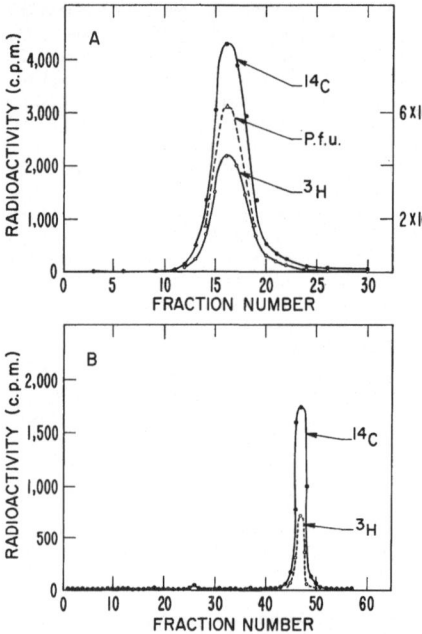

Fig. 1. *Characterization of MS2 coat protein.* A. Distribution of ^{14}C-radioactivity, ^{3}H-radioactivity and virus titer of purified MS2 phage, with doubly labelled protein, on equilibrium density centrifugation in a $CsCl_2$ gradient. B. Electrophoretic pattern of the protein isolated from the purified phage. Electrophoresis was for 4 hours at 7 V/cm. The anode is to the right in this and subsequent patterns

Reagents. ^{3}H-labelled L-leucine hydrochloride (1.8 C/mmole), ^{14}C-labelled L-leucine, L-histidine, L-lysine, and L-arginine hydrochlorides (each 240 mC/mmole), and [^{3}H]-uracil (2.8 C/mmole) were obtained from Schwarz Bio-Research, Inc.; ^{3}H-labelled L-tyrosine (38.8 C/mmole) from Nuclear-Chicago. Lysozyme was purchased from Worthington Biochemical Corporation and bovine serum albumin (sterile 30%/0 solution) from Pentex, Inc. Acrylamide, 2-mercapto- ethanol, and TEMED were from Eastman Organic Chemicals; ethylene diacrylate from the Borden Company, and ammonium persulphate from the Mallinckrodt Chemical Works. Sodium dodecyl-sulphate, from Matheson Coleman and Bell, was recrystallized from ethanol. Urea (reagent), from J. T. Baker Chemical Company was deionized and stored as described by Duesberg and Rueckert [20a]. Hydroxide of hyamine was obtained from Packard Instrument Company, PPO and POPOP from Pilot Chemicals, Inc. We are indebted to Dr. W. H. Wilkinson, Merck, Sharp and Dohme Research Laboratories, for a generous gift of actinomycin D.

RESULTS

Effect of Actinomycin on Host Protein Synthesis and Phage Yield. To select the most appropriate concentration of actinomycin for the experiments, the synthesis of host protein was determined by incubation of non-infected spheroplasts in a medium containing [^{14}C]-leucine, in the absence and in the presence of increasing concentrations of the antibiotic. The effect of the same concentrations of actinomycin on the yield of phage was also determined using infected spheroplasts.

It may be seen from the table that, at 0.5 μg/ml, actinomycin inhibited *E. coli* protein synthesis over 99%/0 but it also markedly decreased the yield of

Table. *Effect of actinomycin D on host protein synthesis and MS2 yield in E. coli spheroplasts*
Unifected and infected spheroplasts were incubated for 2 hours at 37° without and with the amounts of actinomycin indicated. After incubation, the uninfected cultures, which contained 0.1 mM [^{14}C]-leucine (12 mC/mmole), were used for assay of protein synthesis. This was determined as the radioactivity incorporated into acid-insoluble material. The infected culture was used for determination of the virus yield

Actinomycin D	Relative protein synthesis	Relative virus yield [a]
μg/ml		
0	100	100
0.25	5	45 ± 15
0.50	0.8	25 ± 10
1.0	0.6	6 ± 3
5.0	0.5	2 ± 1
10.0	0.5	2 ± 1

[a] Average of three experiments.

viable phage (64 to 85%/0 inhibition). Higher concentrations produced essentially no further inhibition of host protein synthesis while drastically depressing the yield of phage. A concentration of actinomycin D of 0.5 μg/ml was therefore used throughout this work.

Kinetics of Incorporation of Labelled Amino Acids and Uracil into Phage-Specific Protein and RNA. A study of the time course of incorporation of [^{14}C]-histidine, [^{14}C]-leucine, and [^{3}H]-uracil into protein and RNA, by actinomycin-treated non-infected and infected spheroplasts, was carried out in order to compare the kinetics of protein and RNA synthesis with that of phage multiplication and the kinetics of incorporation of histidine with that of leucine. Since histidine is absent from MS2 coat protein, the phage-induced incorporation of histidine is an index of the synthesis of virus-specific non-coat proteins.

A difference in the kinetics of incorporation into protein of histidine and other amino acids in *E. coli* infected with MS2, in the presence of actinomycin, has been reported [4]. This indicates that the transla-

[14C]-leucine (12 mC/mmole) or 0.01 mM [14C]-histidine (240 mC/mmole). After incubation, the proteins were isolated, fractionated by acrylamide gel electrophoresis, and the radioactivity of the fractions was determined as described under Methods. Typical results are shown in Fig. 4A for the leucine label and in Fig. 4B for the histidine label. It should be kept in mind, when comparing the two patterns, that the specific radioactivity of the labelled histidine was twenty times greater than that of the leucine.

Three major peaks (labelled I, II, and III) were found in the electrophoretic pattern of the proteins

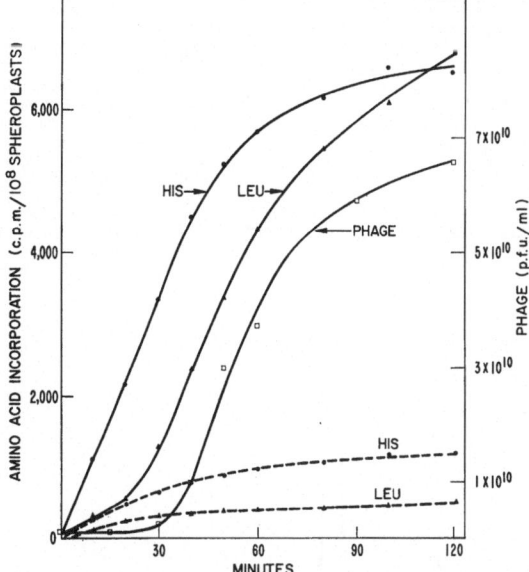

Fig. 2. *Incorporation of amino acids into phage-specific proteins as a function of time.* Concentrations and specific radio-activities of labelled amino acids were; [14C]-leucine, 0.1 mM, 12 mC/mmole; [14C]-histidine, 0.01 mM, 240 mC/mmole. Solid lines, amino acid incorporation and virus titer in infected spheroplasts. Dashed lines, amino acid incorporation in non-infected controls

tion of polycistronic viral messengers is subject to some kind of regulation. This observation was confirmed. As seen in Fig. 2, the incorporation of histidine into phage-induced protein levelled off at a time (110 minutes) when the incorporation of leucine was still going on. Fig. 3 shows, in further agreement with previous results [4], that the synthesis of viral RNA closely followed that of histidine-containing virus-specific proteins and that both preceded the formation of viable phage.

Electrophoretic Characterization of Phage-Specific Proteins. Non-infected and infected spheroplasts were incubated for 2 hours at 37° in medium containing actinomycin D (0.5 µg/ml) and either 8.1 mM

synthesized by the infected spheroplasts. Of these the largest one (peak III) was in the position of the 3H-labelled MS2 coat protein added as a marker prior to electrophoresis. This, and the virtual absence of peak III from the run with radioactive histidine, identifies the protein in this peak as coat protein.

It is apparent from Fig. 4 that, with one exception, there was little or no radioactive protein in the fractions derived from uninfected spheroplasts. The exception was the presence of a small amount of radioactivity under peak III. This may correspond to a host protein, having the same mobility as MS2 coat protein, the synthesis of which was less effec-

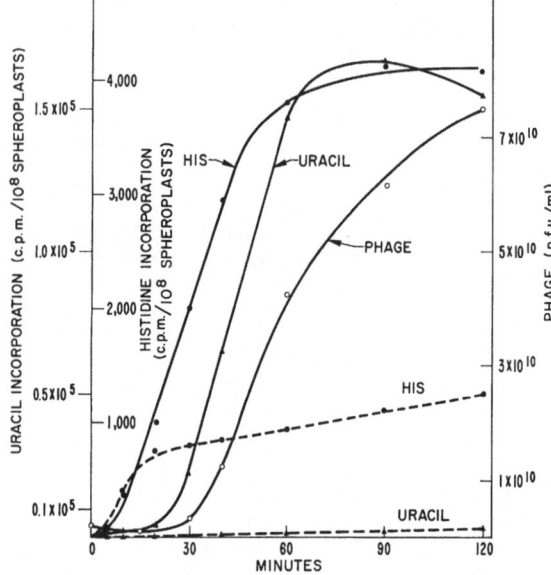

Fig. 3. *Incorporation of histidine and uracil into phage-specific protein and RNA as a function of time.* The concentrations and specific radioactivities of labelled compounds were; [14C]-histidine, 0.01 mM, 240 mC/mmole; ³H-uracil, 0.04 mM, 2.85 C/mmole. RNA radioactivity was measured as radioactivity precipitated by trichloroacetic acid in the cold. Solid lines, precursor incorporation and virus titer in infected spheroplasts. Dashed lines, precursor incorporation in non-infected controls

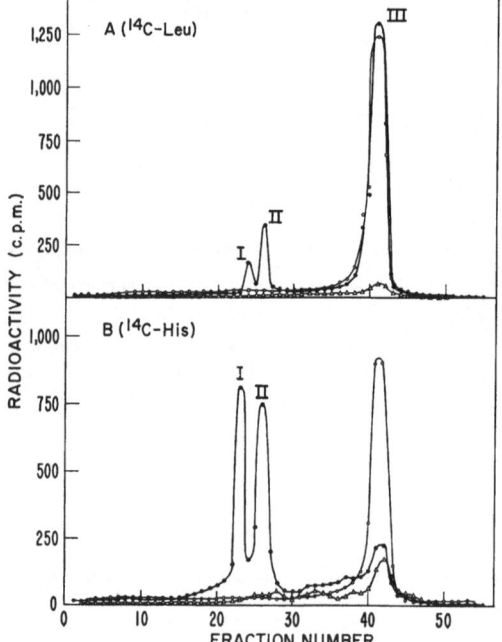

Fig. 4. *Electrophoretic patterns of proteins labelled with [14C]-leucine (A) or [14C]-histidine (B).* Electrophoresis was 90 minutes at 14 V/cm. The radioactivity is expressed in cpm/fraction. Solid circles (●—●—●) infected spheroplasts; triangles (△—△—△) non-infected spheroplasts; open circles (○—○—○) ³H-labelled coat protein marker

tively inhibited by actinomycin than that of the other cell proteins.

Time Course of Synthesis of Phage-Specific Proteins. Spheroplasts in actinomycin-containing medium, were labelled at different intervals of the phage

preceding section. It should be noted that, because of the large amounts of capsid protein synthesized between 20 and 120 minutes, the scale of the ordinate for the right half of the figure is one fifth of that for the left half.

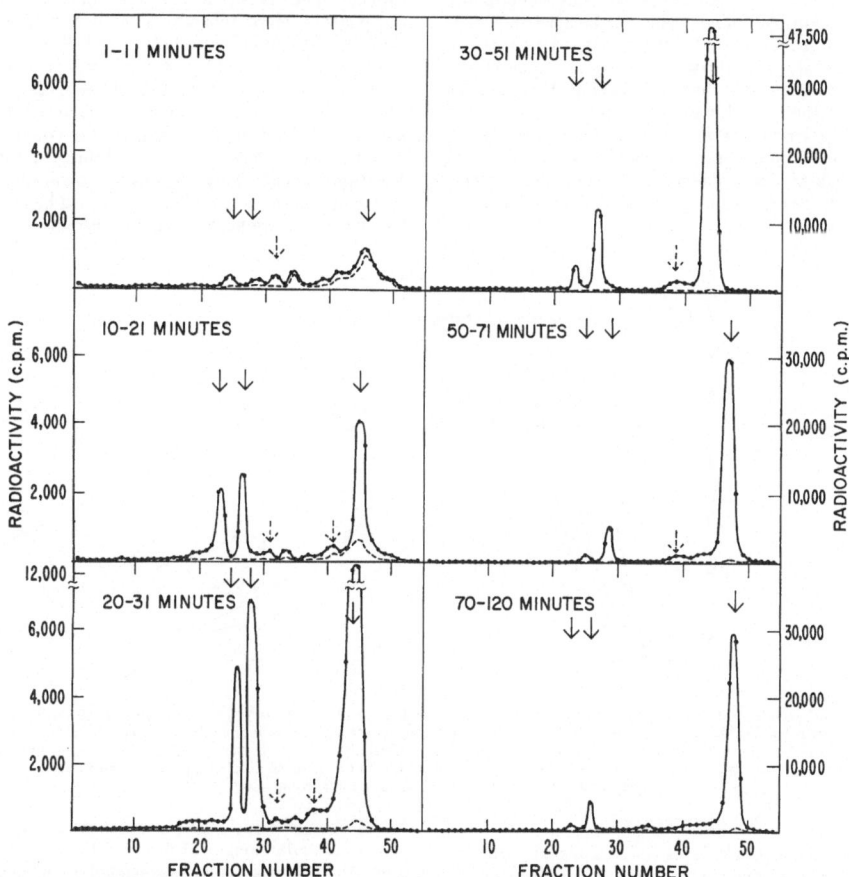

Fig. 5. *Electrophoretic patterns of proteins labelled with [¹⁴C]-leucine during different intervals of the phage replicative cycle.* The experimental procedure is described in the Methods section under "Labelling of phage-specific proteins". The solid arrows indicate the position of the three major peaks as determined by the position of the [³H]-leucine internal marker. The dashed arrows show the position of minor peaks. The ¹⁴C-radioactivity of each fraction is expressed as counts/min/ml of incubation mixture. Solid lines, infected; dashed lines, non-infected spheroplasts

replicative cycle with either [¹⁴C]-leucine or [¹⁴C]-histidine as described in the Methods section under "Labelling of phage-specific proteins". The electrophoretic patterns of the proteins labelled at various times in the presence of radioactive leucine are shown in Fig. 5. The solid arrows indicate the position of the protein peaks I, II, and III, described in the

It is apparent from Fig. 5 that the phage-specific proteins were synthesized at different rates at various times of the replicative cycle. When the relative average rate of labelling during each time interval is plotted against time, as done in Fig. 6, the times at which maximal rates of synthesis were reached were approximately 24, 28, and 40 minutes for proteins I, II, and III, respectively, with rapid subsequent decline in each case. As shown in the same figure, the rate of production of viable phage particles reached a maximum in about 40 minutes and declined after 55—60 minutes.

Further inspection of Fig. 5 suggests the presence of two additional minor peaks in the electrophoretic patterns from infected spheroplasts. These peaks, indicated by dashed arrows, are more apparent at earlier than at later times of labelling. However, because of their small magnitude and the background of radioactivity in some of the corresponding areas of the electropherograms derived from non-infected spheroplasts, the question whether these peaks belong to phage-specific proteins cannot be answered at present.

mutant in the MS2 coat protein cistron that induces the formation of large amounts of RNA-synthesizing enzyme(s) in non-permissive hosts [20b]. The amount of protein I formed in this case is increased to an extent corresponding to the amount of polymerase formed. The presence of two additional minor components, more apparent at earlier than at later time of infection, has also been noted. Whether they represent phage-specific proteins or not is uncertain.

If the entire length of MS2 RNA specifies viral proteins, with a coding ratio of three nucleotides per amino acid, the 3,000 nucleotides of the viral genome would encode peptide chains containing a total of 1,000 amino acids. Since 130 amino acids are accounted for by the capsid protein there would be information for four more peptide chains with an average of

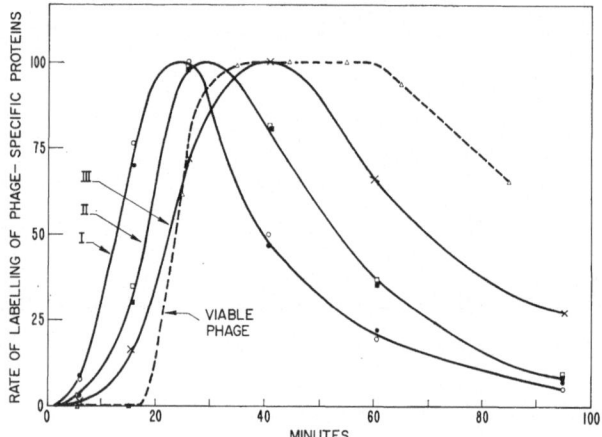

Fig. 6. *Relative rate synthesis of phage-specific proteins as a function of time after infection.* The total radioactivity in each major protein band divided by the duration of labelling was taken as the average rate of synthesis at the midpoint of each of the six labelling intervals of Fig. 5. The rate is expressed in each case as % of the maximal value. Open symbols or crosses, [14C]-leucine label; solid symbols, [14C]-histidine label. The rate of synthesis of phage particles, also expressed as per cent of the maximal rate, is given for comparison

DISCUSSION

The present work shows that three main proteins are synthesized upon translation of the MS2 genome by *E. coli* spheroplasts in the presence of actinomycin. It should be emphasized that, under our experimental conditions, these may be largely the products of translation of progeny RNA. Moreover, as already pointed out, the production of viable phage is markedly inhibited by actinomycin. To which extent, if at all, the drug impairs the fidelity and/or control of translation is unknown.

As already mentioned, the component formed in largest amounts (protein III) is capsid protein. The nature of peak II is not known but peak I probably corresponds to a viral RNA polymerase. This is suggested by experiments with MU9, an amber

220 amino acids each. In the absence of knowledge of the molecular weight of proteins I and II, and of proof that they actually correspond to protomeric subunits, all that can be said at present is that the MS2 genome may have information for a minimum of three and, possibly, for a maximum of five to six polypeptide chains. A study of amber and temperature sensitive mutants of phage f_2, of which MS2 is a close relative, by Zinder and coworkers [21] revealed the occurrence of mutations in cistrons concerned with (a) coat protein, (b) RNA polymerase, and (c) maturation of the phage. These studies indicate that f_2 RNA contains information for at least three proteins.

The mechanism of replication of the RNA of MS2 and related phages has been extensively investi-

gated in several laboratories [22]. There are indications, both from experiments with intact cells and
isolated enzymes, that the replication proceeds in
two steps: (a) formation of a replicative intermediate
by synthesis of complementary "minus" strands with
viral RNA as template, and (b) asymmetric synthesis
of progeny "plus" strands with the replicative intermediate as template. The behaviour of a temperature
sensitive mutant of phage f_2 [23], unable to carry
out step (a) while capable of performing step (b),
is in agreement with this view and indicates that
two different enzymes (whether separable or not) are
involved. Our protein fraction I may be related to the
enzyme catalyzing step (a). Whether protein fraction II is related to the enzyme involved in step (b),
to the maturation factor, or to some other phage-
specific protein, is at present a matter for speculation.

In line with the results of Ohtaka and Spiegelman [8] on translation of MS2 RNA *in vitro* and
those of Summers *et al.* [11] with poliovirus, not only
are protein chains from different MS2 cistrons formed
in different numbers but the synthesis of each protein
reaches maximal rates, to decline soon thereafter, at
different times of the replicative cycle. At 120 minutes, the total amount of leucine radioactivity in
protein II was 1.5 times and that in protein III
12.5 times greater than that in protein I. Since proteins I and II are probably larger than the capsid
protein (molecular weight, 17,000), the differences
in radioactivity incorporated into capsid protein and
into each of the other two proteins must reflect an
even greater difference in terms of actual numbers of
protein chains synthesized. Between 10 and 21 minutes after infection, twice as much radioactivity was
incorporated into coat protein as into protein I or II. It
would therefore appear that the number of coat
polypeptide chains made is greater than that of the
other major protein chains already early in infection.
An important factor in the overall production of a
greater total number of capsid chains than of the
other protein chains is undoubtedly the observed
earlier decline in the synthesis of proteins I and II.
This decline might be due to blocking of sections
of the RNA encompassing the protein I and II cistrons
by the rapidly accumulating coat protein. Since the
ribosomes are thought to start reading polycistronic
messenger RNA only from the free 5'-end [24, 25],
preferential translation of the coat protein cistron
would suggest that this cistron is close to the
5'-end of the messenger. However, our experiments
give no direct clues as to the actual order of translation
of the individual genes.

This work was aided by grants AM-01845, AM-08953,
and FR-05399 from the National Institutes of Health,
United States Public Health Service, and E. I. Du Pont de
Nemours and Co., Inc. We are indebted to Dr. Robert
C. Warner, Dr. Charles Weissmann and Dr. Wendell M. Stan-
ley, Jr. for helpful discussions. Our thanks are also due to
Miss Jana Krausova for assistance with some of the experi-
ments.

REFERENCES

1. Zinder, N. D., *Ann. Rev. Microbiol.* 19 (1965) 455.
2. Leive, L., *Proc. Natl. Acad. Sci. U. S.* 53 (1965) 745.
3. Haywood, A. M., and Sinsheimer, R. L., *J. Mol. Biol.*
   14 (1965) 305.
4. Oeschger, M. P., and Nathans, D., *Federation Proc.* 25
   (1966) 651.
5. Haywood, A. M., and Harris, J. M., *J. Mol. Biol.* 18
   (1966) 448.
6. Lunt, M. R., and Sinsheimer, R. L., *J. Mol. Biol.* 18
   (1966) 541.
7. Nathans, D., Notani, G., Schwartz, J. H., and Zinder,
   N. D., *Proc. Natl. Acad. Sci. U. S.* 48 (1962) 1424.
8. Ohtaka, Y., and Spiegelman, S., *Science* 142 (1963) 493.
9. Adams, J. M., and Capecchi, M. R., *Proc. Natl. Acad.
   Sci. U. S.* 55 (1966) 147.
10. Stanley, W. M., Jr., Salas, M., Wahba, A. J., and Ochoa,
    S., *Proc. Natl. Acad. Sci. U. S.* 56 (1966) 290.
11. Summers, D. F., Maizel, J. V., Jr., and Darnell, J. E.,
    Jr., *Proc. Natl. Acad. Sci. U. S.* 54 (1965) 505.
12. Strauss, J. H., Jr., and Sinsheimer, R. L., *J. Mol. Biol.*
    7 (1963) 43.
13. Ling, J., Tsung, C., and Fraenkel-Conrat, H., *J. Mol.
    Biol.* in press.
14. Singer, M. F., and Leder, P., *Ann. Rev. Biochem.* 35
    (1966) 195.
15. Loeb, T., and Zinder, N. D., *Proc. Natl. Acad. Sci. U. S.*
    47 (1961) 282.
16. Adams, M. H., *Bacteriophages*, Interscience, New York
    1959, p. 450.
17. Mach, B., and Tatum, E. L., *Science* 139 (1963) 1051.
18. Choules, G. L., and Zimm, B. H., *Anal. Biochem.* 13
    (1965) 336.
19. Kinard, F. E., *Rev. Sci. Instr.* 28 (1957) 293.
20a. Duesberg, P. H., and Rueckert, R. R., *Anal. Biochem.*
    11 (1965) 342.
20b. Unpublished experiments.
21. Horiuchi, K., Lodish, H. F., and Zinder, N. D., *Virology*,
    28 (1966) 438.
22. Weissmann, C., and Ochoa, S., in *Progress in Nucleic
    Acid Research and Molecular Biology* (edited by
    J. N. Davidson and W. E. Cohn), Academic Press,
    New York 1966, Vol. 6.
23. Lodish, H. F., and Zinder, N. D., *Science* 152 (1966)
    372.
24. Ames, B. N., and Martin, R. G., *Ann. Rev. Biochem.* 33
    (1964) 235.
25. Yanofsky, C., and Ito, J., *J. Mol. Biol.* 21 (1966) 313.

E. Viñuela, I. D. Algranati, and S. Ochoa
Department of Biochemistry
New York University School of Medicine
550 First Avenue, New York 16, N.Y.

European J. Biochem. 1 (1967) 12—20

Dephosphorylation of Pyrimidine Nucleotides in the Soluble Fraction of Homogenates from Normal and Regenerating Rat Liver

P. Fritzson[1]

Norsk Hydro's Institute for Cancer Research, Montebello, Oslo

(Received October 17, 1966)

The dephosphorylation of the pyrimidine nucleotides UMP, CMP, dUMP, dCMP, and dTMP was determined in the soluble fraction of homogenates from normal rat liver and from regenerating liver during an 8-day period after partial hepatectomy. The dephosphorylations were assayed at optimal pH in the presence of optimal concentrations of Mg^{++} ions and substrate. β-Glycerophosphate was included among the substrates for measurement of non-specific phosphatase activity. The intracellular origin of the enzymes was investigated by determining the rate of their appearance in the soluble fraction during homogenization and by estimating the rate at which the lysosomes were disrupted.

The results showed that the dephosphorylation of the deoxyribonucleotides was mainly due to non-specific acid phosphatase activity. This enzyme and the alkaline phosphatase activity were found to be localized in the soluble space of the liver cell. The dephosphorylation of the ribonucleotides was due largely to a specific 5'-nucleotidase. The enzyme was adsorbed or loosely bound to some sedimentable cell constituent at the time of cell disruption, but it was quantitatively released into the soluble fraction during 2 minutes homogenization.

During liver regeneration the acid and alkaline phosphatase activities showed slight variations which could not be related to the growth rate of the liver. In contrast, the 5'-nucleotidase activity showed distinct, cyclic variations which were inversely related to the growth rate. The variations in the 5'-nucleotidase activity were strikingly similar to the activity variations of uracil reductase and N-carbamoyl-β-alanine amidohydrolase observed previously in regenerating liver.

During rat liver regeneration the activities of the uracil-degrading enzymes uracil reductase and N-carbamoyl-β-alanine amidohydrolase showed strikingly similar variations which were, to a large extent, inversely related to the growth rate of the liver [1,2]. The enzymes were present entirely in the soluble fraction of the liver homogenate [1,3], indicating that they were localized in the soluble space of the intact cell. From the viewpoint of control of enzyme activity in rat liver, it was of interest to see if the activities of other pyrimidine catabolizing enzymes present in this cell compartment showed a similar variation pattern during liver regeneration. The enzymes which dephosphorylate the pyrimidine ribo- and deoxyribonucleoside-5'-monophosphates were

chosen for this study. Studies in different laboratories [4—6] had shown, however, that the 5'-nucleotidase activity of rat liver, studied with AMP as substrate, had a diffuse intracellular distribution, with a tendency to be associated preponderantly with the nuclear and microsomal fractions. There also appeared to be some doubt whether nucleotidases were present at all in the soluble fraction of rat liver homogenates [7]. Fiala et al. [8] found that the deoxynucleotidase activity of rat liver was present mainly in the mitochondria and microsomes, whereas the small activity of the soluble fraction was ascribed to an artifact due to a release of activity by mechanical disintegration of the particulate structures.

A necessary part of our study was therefore first to explore the intracellular localization of the enzymes occurring in the soluble fraction of the liver homogenate, and to devise methods for the assay of the dephosphorylation of the different nucleotides under optimal conditions. β-Glycerophosphate was included among the substrates for measurement of unspecific phosphatase activity. The present communication

---

[1] Postdoctoral Fellow of the Norwegian Cancer Society.

Non-standard Abbreviations. β-glycerophosphate, β-GP; N-carbamoyl-β-alanine, CβA.

Enzymes. Uracil reductase, or 4,5-dihydrouracil: NADP oxidoreductase (EC 1.3.1.2); N-carbamoyl-β-alanine amidohydrolase (EC 3.5.1.6); 5'-nucleotidase, or 5'-ribonucleotide phosphohydrolase (EC 3.1.3.5).

describes these experiments and presents the variation patterns of the different soluble dephosphorylating activities during liver regeneration[2].

EXPERIMENTAL PROCEDURE

Materials. All the chemicals used were of the highest purity available. The nucleotides and deoxynucleotides were purchased from the Sigma Chemical Company, St. Louis, Mo. β-Glycerophosphate was obtained from E. Merck, Darmstadt.

Animals. Female black-and-white rats from the Institute colony, about 100 days old and weighing between 170 and 200 g, were used. Partial hepatectomy was performed between 9.30 and 10.00 a.m. according to the procedure of Higgins and Anderson [9]. The feeding of the animals was as described previously [1].

Preparation of Liver Homogenates. The animals were stunned by a blow on the head, decapitated and bled. The livers were excised, placed in tared beakers containing cold 0.25 M sucrose, weighed, washed with several portions of 0.25 M sucrose, coarsely minced with scissors and homogenized for 2 minutes in 2 volumes of 0.25 M sucrose in a 190×22 mm glass tube with a motor-driven (1,600 rev./min) Teflon pestle having a clearance of 0.41 mm. The pestle was moved up and down (one stroke) with a speed of 25 strokes/min. The resulting homogenate was diluted with 0.25 M sucrose to a concentration of 1 g of liver per 4 ml of homogenate.

Preparation of Soluble Fraction. The whole homogenate was centrifuged at $105,000 \times g$ for 60 minutes in a refrigerated Spinco Model L ultracentrifuge with rotor No. 40. The supernatant was carefully separated from the sediment and fatty overlayer by withdrawing the middle layer with a capillary pipette. Part of it was used immediately for analysis. Aliquots of the remaining part were stored at $-17°$ in small centrifuge tubes so that comparable experiments could be performed on supernatant fractions frozen and thawed only once. No significant change in enzymatic activities could be detected after one week in the frozen state.

Assay of Phosphohydrolase Activity. The activity was determined by measuring the rate of liberation of inorganic phosphate. Inorganic phosphate was determined according to the method of Fiske and Subbarow [10]. The standard incubation conditions, giving optimal activity with the soluble fraction as the enzyme source, are recorded in Table 1. Each enzyme activity was tested in duplicate. About 1.2 mg of soluble protein (50 μl of soluble fraction) was used to measure dephosphorylation of β-GP at

[2] Preliminary reports were presented at the Conference on the Catabolism of Pyrimidines, Mol, Belgium, March 1965, and at the Second and Third Meetings of the Federation of the European Biochemical Societies, Vienna and Warsaw, April 1965 and 1966.

pH 8.5. The other dephosphorylating activities were measured with 100 μl of the soluble fraction. Appropriate substrate and enzyme blanks were always run. The incubations were carried out in 22×90 mm round-bottomed tubes. After 3 minutes shaking in a water bath at 37° for temperature equilibration, the reaction was started by adding the substrate. The incubation was continued for the time indicated in Table 1. The reaction was stopped by the addition of 250 μl of 25% (w/v) trichloroacetic acid. After centrifugation, an 0.4 ml aliquot of the supernatant fluid was used for phosphate determination.

RESULTS

Conditions for Measurement of Dephosphorylating Activities

It appeared that the optimal pH for the dephosphorylation of the nucleotides and β-GP varied with the experimental conditions, being dependent

Table 1. *Standard incubation conditions for measuring phosphohydrolase activities*
The total incubation volume was 0.5 ml. All concentrations given are final concentrations. The buffers used were: Glycine in incubation mixtures no. 1, 3 and 8; Tris-maleate in incubation mixtures no. 2, 4, 5 and 6; Acetate in incubation mixture no. 7. The concentration of the buffer components in the incubation mixtures was 0.05 M, except in incubation mixtures no. 7 and 8, in which the concentration was 0.10 M. pH, adjusted with sodium hydroxide, was determined at 20° in the complete incubation mixture including enzyme source but without substrate (β-GP was present during pH measurement because of its high concentration). Stock solutions of the substrates were adjusted to neutrality

Incubation mixture No.	Substrate	pH	MgCl₂	Substrate concentration	Incubation time
			mM	mM	min
1	UMP	9.2—9.6	50	10	30
2	UMP	6.8—7.3	100	10	30
3	CMP	8.0—8.6	50	10	30
4	dUMP	6.0—6.4	50	20	20
5	dTMP	5.8—6.2	50	20	20
6	dCMP	4.8—5.4	10	20	30
7	β-GP	6.1—6.4	75	100	30
8	β-GP	8.4—8.6	75	100	20

on the concentration of the substrate, the concentration of magnesium ions, and the nature of the buffer. Similarly, the optimal magnesium ion concentration and the influence of the substrate concentration on the reaction rate varied with the composition of the incubation mixture. It was therefore necessary to work out conditions which were optimal with respect to pH, Mg^{++}, and substrate at the same time. These conditions are given in Table 1. The effect of the individual factors on the enzyme activities is recorded in Fig.1—3. The shape of the pH-activity curve for UMP dephosphorylation (Fig.1) suggested that the nucleotide was attacked by two enzymes with pH optima about 7 and 9.5, respectively. Accordingly, standard assay conditions were worked out at both pH values. With β-GP as substrate, activity peaks occurred at pH 6.2 and 8.5,

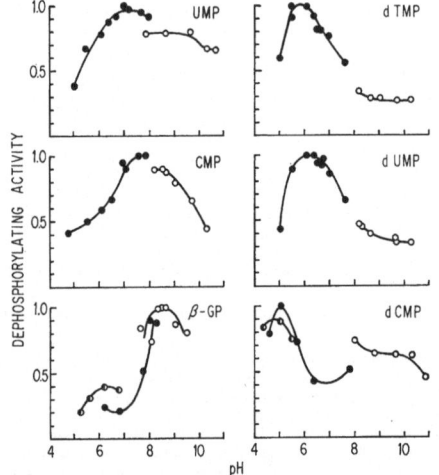

Fig. 1. *Effect of pH on the dephosphorylation of UMP, CMP, dUMP, dTMP, dCMP and β-GP.* Supernatant fluid from normal liver was used as enzyme source. The incubations were carried out under standard incubation conditions, except that pH and buffer were varied as indicated. pH was measured at 20° in the complete incubation mixture including enzyme source, but without substrate. An exception was β-GP, which was present during pH measurement. If it was omitted, the acid pH optimum occurred at pH 5. ◐——◐, Acetate buffer; ●——●, Tris-maleate buffer; ○——○, glycine buffer

indicating the presence of non-specific acid and alkaline phosphatases. The alkaline phosphatase was almost inactive in the absence of added magnesium (Fig. 2).

The validity of the methods for assay of dephosphorylating activities was verified in control experiments which showed that the phosphate released was proportional to both enzyme concentration and time. The only exception was the dephosphorylation of β-GP at pH 8.5 which showed slightly decreasing rate with increasing time and enzyme concentration. The activities obtained are taken to represent amount of enzyme.

Effect of Homogenization on the Dephosphorylating Activities of the Soluble Fraction

Information regarding the localization of enzymes in the soluble space [3] of the intact cell may be obtained by studying the rate of occurrence of enzyme activity in the soluble fraction during homogenization. Thus, enzymes localized in the soluble space may appear in the soluble fraction immediately when the cells are broken. An increase in supernatant activity during homogenization may indicate that the enzyme was associated in some way with particles. Fig. 4 shows the effect of homogenization on the protein content and dephosphorylating activity

[3] The soluble space is defined as the extraparticulate compartment of the intact cell in which proteins and metabolites are free in solution. Evidence has recently been presented [11] that the nuclear sap can be regarded as a part of the soluble space of the cell in that there is free exchange of enzymes and low molecular weight compounds between the nuclear sap and the soluble part of the cytoplasm. When homogenates are prepared in an aqueous medium such as 0.25 M sucrose, most of the nuclear sap components will occur in the soluble fraction of the homogenate.

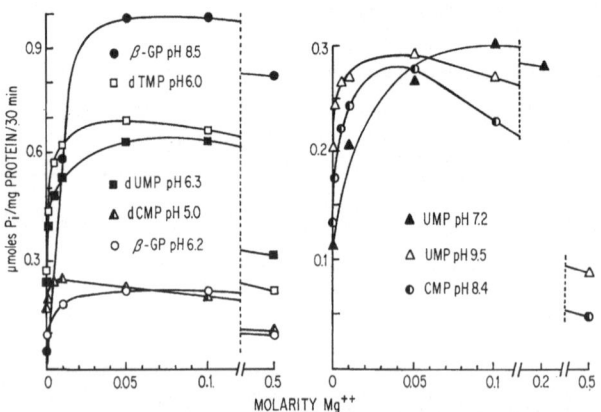

Fig. 2. *Effect of Mg^{++} ions on dephosphorylation of UMP, CMP, dUMP, dTMP, dCMP, and β-GP.* Supernatant fluid from normal rat liver was used as enzyme source. The incubations were carried out under standard incubation conditions except that the concentration of magnesium chloride was varied as indicated

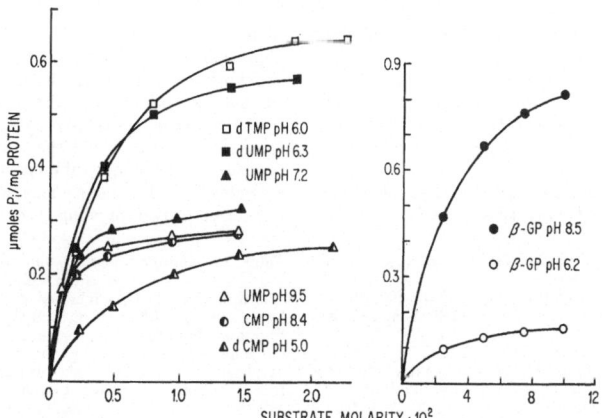

Fig. 3. *Effect of substrate concentration on the rate of dephosphorylation of UMP, CMP, dUMP, dTMP, dCMP, and β-GP.* Supernatant fluid from rat liver was used as enzyme source. The incubations were carried out under standard incubation conditions, except that the concentration of substrate was varied as indicated. The incubation time was 30 min with the nucleotides and 20 min with β-GP as substrates

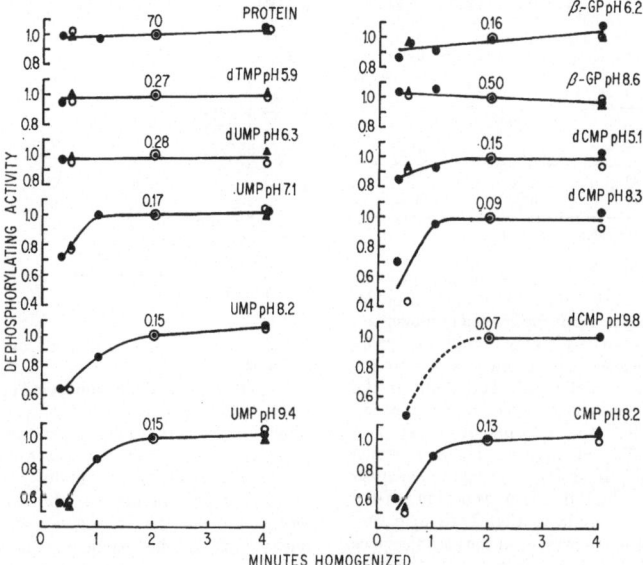

Fig. 4. *Dephosphorylation of nucleotides and β-GP in the soluble fraction of normal and regenerating rat liver as a function of time of homogenization.* The liver was disintegrated in about 20 seconds, during which time about 4 pestle strokes were made. The pestle reached the bottom of the tube at the last of these strokes. During the further homogenization the number of pestle strokes was 25 per minute. The dephosphorylating activities were determined under the standard assay conditions (Table 1). UMP dephosphorylation was measured also at pH 8.2, and dCMP dephosphorylation at pH 8.3 and 9.8. In these measurements 0.05 M glycine buffer was used. The dephosphorylating activities are activities per mg of souble protein, expressed in per cent of the activities obtained by 2 min homogenization. The activities are also a measure of the total activity of the soluble fraction, since the volume and protein content of this fraction were almost independent of the time of homogenization. Different point symbols indicate different experiments with normal (●, ○) and 2-day regenerating liver (▲). The 2-min point thus represents 3 different experiments. The figures above the 2-min points indicate μmoles phosphate released/mg soluble protein/20 min, and (top left) mg soluble protein/g liver. The values refer to normal liver

of the soluble fraction. It can be seen that the amount of soluble protein was almost independent of the time of homogenization, indicating that particulate protein was not released into the soluble fraction to a significant extent. The ratio between soluble protein and total liver protein per g of liver was about 0.33.

Fig. 4 further indicates that the enzymes which dephosphorylate dTMP and dUMP are localized in

Table 2. *Acid phosphatase activity of the soluble and lysosomal fractions*
The livers were homogenized for 2 min, and the 105,000 ×*g* supernatant was prepared from a part of the homogenate. The preparation was used to determine "Soluble activity". The other part of the homogenate was diluted with 0.25 M sucrose to give 1 g of liver in 10 ml of homogenate. A particulate fraction sedimenting between 500 ×*g* and 17,000 ×*g* was isolated. This fraction contains most of the lysosomes. The centrifugations were carried out for 10 min at 2° in a Servall centrifuge. The sediments were washed once with 0.25 M sucrose and recentrifuged. The fraction containing the lysosomes was resuspended in a known volume of 0.25 M sucrose. All acid phosphatase activities were determined according to Gianetto and de Duve [12] as follows: Aliquots of the enzyme preparations were incubated at 37° for 20 min in a total volume of 2 ml containing 0.05 M sodium acetate buffer pH 5, 0.25 M sucrose, and 0.05 M sodium β-glycerophosphate, and the phosphate released was measured. "Total lysosomal activity" was determined by adding Triton X-100 to the incubation mixture to a final concentration of 0.1% (w/v) [13]

Exp. No.	Soluble activity[a]	Lysosomal activity[a]	
		Free	Total
1	1.92	4.13	25.2
2	2.77	6.20	31.2

[a] μmoles phosphate released/g of liver/20 min.

the soluble space of the intact cell since the activities occur immediately in the soluble fraction and no increase in activity occurs during prolonged homogenization. Similarly, the curve for dCMP dephosphorylation at pH 5.1 indicates that the enzyme is situated in the soluble space of the cell. The slight initial increase in activity during homogenization may be due to a small activity at this pH of the enzyme which attacks dCMP at higher pH, and which increases considerably in activity during homogenization. The curves for β-GP dephosphorylation pH 6.2 and 8.6, respectively, indicate that both acid and alkaline phosphatases are present in the soluble space of the intact cell. The slight increase in the acid phosphatase activity during homogenization may be due to leakage of lysosomal acid phosphatase into the soluble fraction and may reflect the rate at which the lysosomes are disrupted. Table 2 indicates that at least 92% of the lysosomes were intact after 2 min homogenization[4]. It is possible that the soluble acid

[4] In the calculation, "soluble activity" is taken to reflect disrupted lysosomes, which thus amounts to 8% of the total lysosomal activity.

phosphatase activity is largely responsible for the dephosphorylation of dTMP, dUMP, and dCMP.

Our finding that acid phosphatase was present in the soluble space of the intact cell is supported by recent investigation on isoenzymes of acid phosphatase [14] which indicated that rat liver supernatant contained isoenzymes different from the lysosomal enzyme. The Mg++ dependent alkaline phosphatase has been shown previously to be present almost

Table 3. *Substrate specificity of 5'-nucleotidase solubilized during prolonged homogenization*
The activities of the solubilized enzyme were determined by taking the difference in activity between the soluble fractions obtained from livers homogenized for 2 min and 0.5 min, respectively. The difference in the liberation of phosphate from each substrate is compared with that from UMP, which is assigned an arbitrary value of 100. All substrates were tested under the standard incubation conditions for measuring UMP dephosphorylation at pH 9.5 (Table 1). Glucose 6-phosphatase activity was determined with 0.02 M substrate at 37° in 0.05 M Tris-maleate buffer pH 6.6. Incubation was carried out for 30 min

Substrate	Relative activity
5'-UMP	100
5'-CMP	75
5'-AMP	70
5'-GMP	30
Deoxy-5'-TMP	51
Deoxy-5'-UMP	57
Deoxy-5'-CMP	34
Deoxy-5'-AMP	31
Deoxy-5'-GMP	16
2',3'-UMP[a]	0
2',3'-CMP[a]	0
2',3'-AMP[a]	2
Ribose 5-phosphate	0
β-Glycerophosphate	0
Glucose 6-phosphate, pH 6.6	0

[a] Mixed isomers.

entirely in the soluble fraction of rat liver homogenate [4,15,16].

All the other dephosphorylating activities with UMP, CMP, and dCMP as substrates (Fig. 4) increased considerably during homogenization. This fact indicates that the enzyme(s) was associated with the particulate material. The finding that the activities reached a plateau after 2 minutes homogenization indicates that the enzyme(s) was quantitatively released from the particles. In terms of intracellular localization it seems clear that a distinct difference exists between the enzyme appearing in the soluble fraction and the enzyme present in the sediment[5]. For example the sedimentable activity could be due to proteins incorporated in membranous structures, whereas the soluble activity could be

[5] Unpublished experiments showed that about 90% of the dephosphorylating activity of the homogenate was still present in the sedimentable part when the soluble activity reached the plateau.

ascribed to proteins loosely bound or adsorbed to the surface of intracellular membranes.

The specificity of the soluble enzyme is shown in Table 3. It can be seen that the enzyme is specific for the hydrolysis of 5'-nucleotides although the relative activity toward these substrates is quite variable, the rate of hydrolysis of deoxy-5'-GMP being only 16% of that attained with UMP as substrate. It is

fragments in the supernatant since no glucose 6-phosphatase, which is bound entirely to microsomal membranes [18], could be detected in the 5'-nucleotidase fraction (Table 3).

Dephosphorylating Activity During Liver Regeneration

Fig. 5 shows the dephosphorylation of the pyrimidine nucleotides and β-glycerophosphate in the

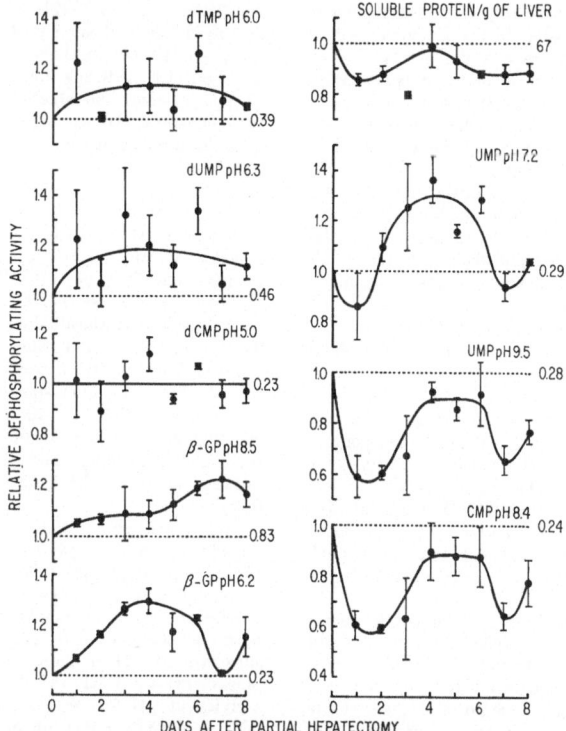

Fig. 5. *Dephosphorylation of pyrimidine nucleotides and β-GP in the soluble fraction of homogenates from regenerating rat liver during an 8-day period after 67% hepatectomy.* Homogenization of the livers was carried out for 2 min. The enzyme activities were determined at 24 h intervals under standard assay conditions (Table 1) and calculated per mg of soluble protein. The activities plotted are mean values of 2 or 3 determinations, expressed in per cent of the activities found in sham-operated rats. The bars represent standard deviations. For each determination the pooled livers from 3 partially hepatectomized and from 3 sham-operated rats were used, and 4 assays at 2 different levels of supernatant fraction were carried out on each liver preparation. The figures at the stippled lines represent mean activities (μmoles P_i released/30 min/mg soluble protein) of 17 determinations with sham-operated rats. The individual activities seemed to be independent of the time elapsed after sham-operation. Soluble protein is given in mg

also seen that the deoxy-compounds are hydrolyzed at about half the rate of the hydrolysis of the corresponding ribonucleotides.

Segal and Brenner [17] demonstrated the presence of a specific 5'-nucleotidase in rat liver microsomes. It seems clear, however, that the soluble activity cannot be ascribed to the presence of microsomal

soluble fraction of regenerating rat liver at 24-hour intervals during an 8-day period after partial hepatectomy. The amount of soluble protein per g of liver is also plotted. It can be seen that dTMP and dUMP dephosphorylation are slightly elevated, although the experimental variations were large. If we take into consideration the decreased level of soluble protein per g of liver in the regeneration period, it is apparent that the dephosphorylation of dTMP and dUMP per g of liver will be almost equal to the control activities. The activity of dCMP dephosphorylation was not significantly different from the control values.

The acid phosphatase activity determined with β-GP as substrate, shows a temporary increase during the regeneration period, the maximum activity occurring around the 4th day after partial hepatectomy. It was suggested in the foregoing section that the acid phosphatase activity was responsible for the dephosphorylation of the three deoxynucleotides. The difference between the variation patterns of β-GP dephosphorylation and the dephosphorylation of the deoxynucleotides can be explained by the presence of 2 or more different acid phosphatases [14,19—21] with different affinities for the substrates, and with slightly different variation patterns in regenerating liver. This interpretation is supported by the finding of Norberg [19] that variations in the shape of the pH-curve of acid phosphatase activity during liver regeneration was the resultant of the action of separate enzymes.

The alkaline phosphatase, determined with β-GP as substrate, shows a slightly increasing activity during the regeneration period, reaching a maximum value 20% above the control level during the 7th day. Also in this case the activity, if calculated per g of liver, is almost equal to the control activity during the regeneration period. The result agrees well with the finding of Allard et al. [22] that the alkaline phosphatase activity of the soluble fraction of homogenates from rat liver was not significantly altered during an 8-day period after partial hepatectomy.

In contrast to the dephosphorylation of the deoxynucleotides, the dephosphorylation of the ribonucleotides UMP (at pH 9.5) and CMP shows significantly decreased activity during liver regeneration. Moreover the activities undergo cyclic variations, the minimum activities occurring during the 2nd and 7th day after partial hepatectomy. The similarity of the variation patterns indicates that the same enzyme is involved. According to the findings recorded in Table 3, this enzyme is a specific 5'-nucleotidase. It seems reasonable to suggest that all the 5'-nucleotides recorded in Table 3 are dephosphorylated by the same enzyme, and thus that the dephosphorylation of these substrates at pH 9.5 would show variations similar to those of UMP dephosphorylation. In a preliminary study of dUMP dephosphorylation this was found to be true. Findings in other laboratories further support this view. Maley and Maley [23,24] using soluble rat liver preparations, found that the dephosphorylation of dUMP, dTMP, and dCMP at pH 8 was reduced by 50% or more 24—28 h after partial hepatectomy. Under similar

assay conditions Beltz [25] observed that the dephosphorylation of dAMP, dTMP, and dCMP was depressed from 30 to 38% in liver regenerating for 25 h.

The variation patterns of the alkaline phosphatase and the UMP dephosphorylating activity at pH 9.5 were entirely different, indicating that the UMP dephosphorylation could be only slightly, if at all, influenced by the alkaline phosphatase activity. That the 5'-nucleotidase (Table 3) is the predominating enzyme is also indicated by the fact that the supernatant UMP dephosphorylating activity increased by about 100% during homogenization of the liver (Fig.4) whereas the alkaline phosphatase activity was constant. If we suppose that the alkaline phosphatase contributes to the UMP dephosphorylating activity at pH 9.5 to a significant extent, the activity variations of the 5'-nucleotidase can be estimated on the basis of the curves for UMP dephosphorylation at pH 9.5 and β-GP dephosphorylation at pH 8.5 (Fig.5). For this purpose we assume that 75% of the UMP dephosphorylating activity was due to the 5'-nucleotidase, while the contribution from the alkaline phosphatase was 25%. The calculations show that the variations of the 5'-nucleotidase activity would be similar to those recorded for UMP dephosphorylation at pH 9.5 with the exception that the minimum activities during the 2nd and 7th day would both have the value 0.45 in the diagram, and the maximum activity between the 4th and 6th day would have the value 0.83.

Fig.5 also shows that the variations of UMP dephosphorylation at pH 7.2 are different from those of the other dephosphorylating activities. It is not necessary, however, to postulate the existence of a particular enzyme responsible for this activity. Thus, both the acid phosphatase measured with β-GP as substrate at pH 6.2 and the enzyme which dephosphorylates UMP at pH 9.5 may exert some activity at pH 7.2. Accordingly, the variations in UMP dephosphorylation at this pH may be explained as the resultant of the variations of the component enzyme activities. Indeed, calculations show that a curve fairly similar to that of UMP dephosphorylation at pH 7.2 can be constructed from the curves for β-GP dephosphorylation at pH 6.2 and UMP dephosphorylation at pH 9.5 by assuming that the former enzyme contributes 80% and the latter 20% to the activity at pH 7.2.

DISCUSSION

The present study has shown that the dephosphorylation of the deoxyribonucleotides dTMP, dUMP, and dCMP in the soluble fraction of rat liver homogenates is mainly due to non-specific acid phosphatase activity, whereas the dephosphorylation of the ribonucleotides UMP and CMP is due largely to a specific 5'-nucleotidase. The 5'-nucleotidase also exerted some activity on the deoxyribonucleotides. During liver regeneration the acid and alkaline phosphatase activities showed slight variations which could not be related to the growth rate (Fig. 6) of the liver. In contrast, the 5'-nucleotidase activity showed distinct, cyclic variations which were inversely related to the growth rate. This difference between the variation patterns of the enzyme activities suggests that the 5'-nucleotidase is under specific control in the cell.

Our observations do not provide any basis for a clear understanding of the factors responsible for the variations in the nucleotidase activity during liver regeneration. But some factors which could be thought to influence the activity, can be ruled out. First, the 5'-nucleotidase of the soluble fraction was adsorbed or loosely bound to particulate material in the homogenate, and homogenization for 2 min was necessary for quantitative elution of the enzyme. The decreased activity of regenerating liver could thus be explained by a more firm binding of the enzyme in the growing tissue. This possibility is, however, excluded by the finding that the relative rate of occurrence of the enzyme in the soluble fraction during homogenization was the same for normal and regenerating liver (Fig. 4). Second, variations in the level of an inhibitor of the 5'-nucleotidase activity could be responsible for the observed variations. However, the fact that these variations were strikingly similar to those of uracil reductase and $C\beta A$ amidohydrolase (Fig. 6) points to a regulatory factor common to all 3 enzymes. It seems unlikely that these enzymes, which catalyze such different reactions as hydrolysis of a phosphate ester, reduction of a double bond, and hydrolysis of a carbamoyl group, could have such a common inhibitor. Likewise it is not probable that the 3 enzymes have a common, unknown cofactor, not included in the standard incubation mixtures, which could be responsible for the activity variations observed.

One possibility that remains is thus that the variations in the nucleotidase activity occur as a result of enzyme synthesis and breakdown. The striking similarities in the variations of the 3 enzyme activities are suggestive of a regulating mechanism analogous to that underlying the coordinated control of several bacterial enzymes belonging to a common metabolic pathway [26]. Accordingly, since the 5'-nucleotidase, uracil reductase and $C\beta A$ amidohydrolase may all be regarded as belonging to the catabolic pathway which converts UMP to β alanine via uridine—uracil—dihydrouracil—$C\beta A$, the synthesis of the enzymes may be directed at the genetic level by a common operon. The similarity of the enzyme variations is most clearly expressed during the first 6 days of liver regeneration. The second drop in the 5'-nucleotidase activity during the 7th

day is less pronounced in the case of uracil reductase, and seems not at all to occur in the $C\beta A$ amidohydrolase activity. This observation points to a second factor which may also influence enzyme synthesis at some level besides that of gene transcription.

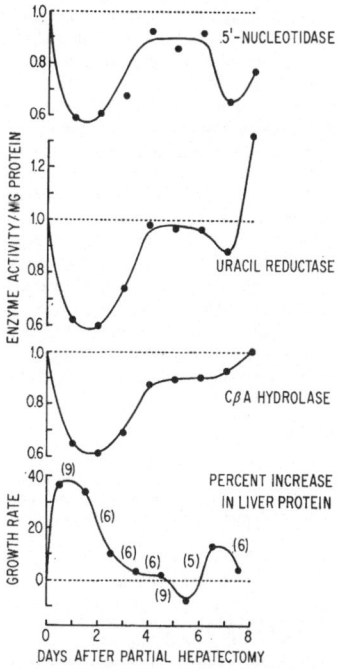

Fig. 6. *Activity variations of 5'-nucleotidase and the previously investigated enzymes* [1] *uracil reductase and N-carbamoyl-β-alanine amidohydrolase in the soluble fraction of homogenates from regenerating rat liver during an 8-day period after partial hepatectomy.* For experimental details see legend to Fig. 5. The growth rate of the liver is expressed as per cent increase in total liver protein per day (total liver protein is calculated per 100 g of body weight). The number of animals used at each time point is given in parentheses. The occurrence of two growth peaks confirms the observations published earlier [1,2]

The possible significance of the variations in the activity of the pyrimidine-catabolizing enzymes is difficult to evaluate. It seems, however, reasonable to suggest that the activity of the 5'-nucleotidase investigated in the present work could interfere with nucleic acid synthesis, particularly since enzymes necessary for the incorporation of pyrimidine nucleotides into nucleic acids may be present in the same cellular compartment [27,28]. However, in previous communications [2,29] it was calculated that the decreased activity of uracil reductase could be explained by the concept of blocked synthesis of the enzyme during the first 24 hours after partial hepatectomy. The reduction of the 5'-nucleotidase activity, which is equal to that of uracil reductase (Fig. 7), can apparently be explained in the same way. This interpretation implies that the decrease in enzyme activity is secondary to the initiation of growth, and occurs as a result of dilution of the enzyme with newly synthesized liver protein[6]. It is interesting to note also that activity variations of anabolic enzymes may be secondary to growth initiation. Thus, Eker [30] in this laboratory, studying human liver cells in tissue culture, found that variations in the activity of thymidine kinase were not involved in the initiation of deoxyribonucleic acid synthesis, but were rather directed by the rate of this synthesis, i.e., by the rate at which the nucleic acid precursor was used up.

The calculations cited above further indicated that synthesis of the catabolic enzymes starts during rapid growth of the liver and increases rapidly concomitantly with the decreasing growth rate of the liver. The possibility thus emerges that the soluble 5'-nucleotidase may be actively engaged in the mechanism which slows down and eventually stops the growth by depleting the cell of a key intermediate in nucleic acid biosynthesis.

It is a pleasure to acknowledge the expert technical assistance of Mrs. Inger Smith and Miss Unni Spaeren. The author is indebted to Miss Bergliot Bolstad Jörgensen for her skilful management of the laboratory animals.

REFERENCES

1. Fritzson, P., J. Biol. Chem. 237 (1962) 150.
2. Fritzson, P., Biochim. Biophys. Acta 91 (1964) 374.
3. Fritzson, P., J. Biol. Chem. 235 (1960) 719.
4. Novikoff, A. B., Podber, E., Ryan, J., and Noe, E., J. Histochem. Cytochem. 1 (1953) 27.
5. De Lamirande, G., Allard, C., and Cantero, A., J. Biophys. Biochem. Cytol. 4 (1958) 373.

---

[6] Turnover of the enzyme protein may contribute slightly to the decrease in activity.

6. De Lamirande, G., Allard, C., and Cantero, A., Cancer Research 18 (1958) 952.
7. Dixon, M., and Webb, E. C., Enzymes. Longmans, Green and Co., London 1964, p. 627.
8. Fiala, S., Fiala, A., Tobar, G., and McQuilla, H., J. Natl. Cancer Inst. 28 (1962) 1269.
9. Higgins, G. M., and Anderson, R. M., Arch. Pathol. 12 (1931) 186.
10. Fiske, C. H., and Subbarow, Y., J. Biol. Chem. 66 (1925) 375.
11. Siebert, G., and Humphrey, G. B., Advances in Enzymol. 27 (1965) 239.
12. Gianetto, R., and de Duve, C., Biochem. J. 59 (1955) 433.
13. Wattiaux, R., and de Duve, C., Biochem. J. 63 (1956) 606.
14. Reith, A., Schmidt, E., and Schmidt, F. W., Klin. Wochschr. 42 (1964) 915.
15. Emery, A. J., and Dounce, A. L., J. Biophys. Biochem. Cytol. 1 (1955) 315.
16. Allard, C., de Lamirande, G., Faria, H., and Cantero, A., Can. J. Biochem. Physiol. 32 (1954) 383.
17. Segal, H. L., and Brenner, B. M., J. Biol. Chem. 235 (1960) 471.
18. De Duve, C., Wattiaux, R., and Baudhuin, P., Advances in Enzymol. 24 (1962) 291.
19. Norberg, B., Acta Physiol. Scand. 19 (1950) 246.
20. Goodlad, G. A. J., and Mills, G. T., Biochem. J. 66 (1957) 346.
21. Moore, B. W., and Angeletti, P. U., Ann. N.Y. Acad. Sci. 94 (1961) 659.
22. Allard, C., de Lamirande, G., and Cantero, A., Cancer Research 17 (1957) 862.
23. Maley, F., and Maley, G. F., J. Biol. Chem. 235 (1960) 2968.
24. Maley, F., and Maley, G. F., Cancer Research 21 (1961) 1421.
25. Beltz, R. E., Arch. Biochem. Biophys. 99 (1962) 304.
26. Jacob, F., and Monod, J., Cold Spring Harbor Symp. Quant. Biol. 26 (1961) 193.
27. Bollum, F. J., and Potter, V. R., J. Biol. Chem. 233 (1958) 478.
28. Bollum, F. J., and Potter, V. R., Cancer Research 19 (1959) 561.
29. Fritzson, P., Doctoral Thesis, University of Oslo, Norway, 1964, p. 24.
30. Eker, P., J. Biol. Chem. 240 (1965) 419.

P. Fritzson
Norsk Hydro's Institute for Cancer Research
Montebello, Oslo 3, Norway

European J. Biochem. 1 (1967) 21—25

Yeast Malate Dehydrogenase: Enzyme Inactivation in Catabolite Repression

J. J. FERGUSON JR., M. BOLL, and H. HOLZER

Department of Biochemistry, University of Pennsylvania, Philadelphia and Biochemisches Institut der Universität Freiburg i. Br.

(Received October 20, 1966)

The apparent inactivation of malate dehydrogenase activity in yeast following exposure to glucose has been studied. It was found to be prevented by inhibition of protein synthesis with cycloheximide, by addition of sodium azide, and by chilling the yeast to 0°. Kinetic evidence suggests that malate dehydrogenase from acetate-induced and glucose-repressed yeast are different molecular species. Possible mechanisms for "inactivation-repression" are discussed.

Classically, the term "repression" is used to designate a process resulting in a decrease in the rate of synthesis of an enzyme following exposure of cells to a "repressor" substance [18][1]. While the biochemical basis of repression is not conclusively established, prevalent theory [8] suggests that a "co-repressor" combines with a protein "aporepressor", the resulting complex in some way modifying the expression of a specific genetically determined enzymatic capability. This modification has been thought to affect the production of enzyme-specific messenger RNA, but several studies have located the site of repressor action (see [2,15]) at the translation (i.e. polypeptide assembly) step. In a number of investigations with microorganisms unique examples of repression have been observed, in which an actual disappearance of enzyme activity occurs after exposure to a repressor [1,7,11,12,14]. Witt et al. [19] have reported that the malate dehydrogenase from a strain of Saccharomyces cerevisiae is repressed by glucose in such a unique fashion, with apparent inactivation of the enzyme. This report describes further studies of the repression of MDH by glucose in Saccharomyces cerevisiae. We have found that cycloheximide (trade name: Actidione), a potent inhibitor of protein synthesis in this yeast, completely prevents this inactivation of MDH activity by glucose.

---

[1] Lacking understanding of the molecular mechanisms involved, we use the term "repression" as defined functionally by Vogel [18], and the term "induction" to indicate merely an increase in the rate of synthesis of an enzyme after addition of "inducer." The term "inactivation-repression" is used to designate the unique type of repression described in this paper, in which specific enzymatic activity of an "induced" enzyme decreases after addition of "repressor."

Enzyme. MDH for malate dehydrogenase, or L-Malate: NAD-oxidoreductase (EC 1.1.1.37).

METHODS AND MATERIALS

The strain of S. cerevisiae var. ellipsoideus studied by Witt et al. [19] was used in these experiments. Cells were grown, harvested, washed and extracted with alumina as described by these authors. MDH activity in extracts was assayed spectrophotometrically, following either the rate of reduction of 3 mM NAD in the presence of 50 mM L-malate at pH 9.5 in glycine-hydrazine buffer [21], or, more frequently, the rate of NADH oxidation in 66 mM triethanolamine buffer at pH 7.6, 0.1 mM NADH and 3 mM oxaloacetate. Glucose-6-phosphate dehydrogenase activity in extracts was measured by the method of Kornberg and Horecker [9] to determine the activity of a reference enzyme which is neither induced nor repressed in this system. Oxygen uptake and carbon dioxide release were measured manometrically [17]. Incorporation of [1-14C] leucine into yeast protein was measured by exposing yeast cells to isotope for 3 minutes, stopping the reaction by adding the cells to an equal volume of 10% (w/v) trichloroacetic acid containing unlabeled leucine. Precipitates were collected on Millipore filters, washed, dried and counted in a Packard Tricarb Liquid Scintillation Spectrometer. Protein was measured by the biuret method [5]. Optical density of yeast suspensions was measured at 578 mμ on an Eppendorf photometer.

Unless otherwise designated, reagents were obtained from commercial sources, usually C. F. Boehringer and Sons (Mannheim). L-malate was obtained from Calbiochem. L[1-14C] leucine was purchased from the Radiochemical Centre, Amersham, England, and had a stated specific activity of 8.5 mC/mmole. Cycloheximide was purchased from Calbiochem. Puromycin was purchased from Nutritional Biochemical Co., and was neutralized prior to use. 5-methyl tryptophane and p-fluoro phenylalanine were purchased from Sigma.

In all assays, 1 unit of MDH activity is that amount of enzyme required to convert 1 μmole of substrate per minute. Specific enzyme activity is expressed in units/mg extract protein.

RESULTS
Induction of MDH by Acetate

Fig. 1 illustrates the time course of induction of MDH activity when washed, glucose-grown cells were suspended in buffered Tavlitzki medium containing 40 mM sodium acetate (hereafter called acetate medium). When initially suspended, specific enzymatic activity of MDH was typically 0.5 to 1.0 units/mg. After a lag period there occurred a marked increase in MDH specific activity above this initial

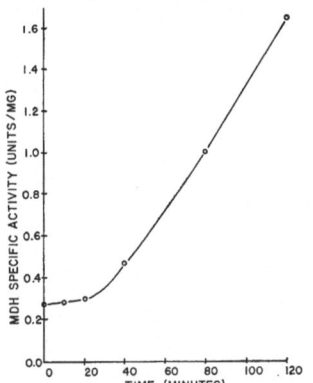

Fig. 1. *Time course of MDH induction by acetate at 30° in air.* 4 g (wet) washed glucose-grown yeast suspended per liter acetate medium. Samples iced, centrifuged, washed and ground at times indicated. Assayed in direction of oxaloacetate reduction

level. The duration of the lag and the rate and extent of the increase varied slightly from preparation to preparation. There occurred no significant change in absorbance at 578 mμ during 6 hours of such incubations. The rapid increase in MDH specific activity could be halted by several procedures. Chilling the yeast to 2° abruptly stopped the increase. Harvested chilled cells lost little MDH activity in 2 days at 2°. In a similar fashion, rapid removal of the acetate from the medium stabilized the MDH specific activity at the pre-washing level, for at least one hour of continued incubation.

Several agents known to inhibit protein synthesis in yeast were found also to prevent or diminish the increase of MDH specific activity in the presence of acetate. At a concentration of 5 μg per milliliter (17.8 μM), the antibiotic cycloheximide immediately stopped this increase. This concentration of cycloheximide was found to inhibit leucine incorporation into yeast protein more than 95% in less than

3 minutes. Other such agents were tested in a different fashion, *i.e.* by adding them at the beginning of incubation with acetate, and observing their inhibitory effect on the rise of MDH specific activity. Thus 10 mM *p*-fluoro phenylalanine, and 5 mM 5-methyl tryptophane inhibited this rise, but never fully prevented it, in incubations of 4 hours duration. These concentrations were found to inhibit [1-14C] leucine incorporation into yeast protein in the range of 80 to 90%.

Witt *et al.* [19] have reported that 2-deoxy-D-glucose prevents the rise in MDH specific activity produced by acetate, at a concentration of 15 mM. Our experiments have shown that five minutes after the addition of 2-deoxy-D-glucose at this concentration, [1-14C] leucine incorporation into yeast protein is inhibited at least 65%. It is thus suspected that 2-deoxy-D-glucose does not cause repression of MDH synthesis in the usual sense, but rather acts as a more general metabolic inhibitor of protein synthesis, thereby preventing enzyme induction.

Repression by Glucose

Between the 4th and 6th hour after suspension of yeast in acetate medium, MDH specific activity was rising in a near linear fashion. If, at this time, the medium was made 55 mM in D-glucose, the rise of MDH activity was stopped within about 2 minutes. By about 10 minutes after glucose addition a decline of MDH activity could be detected. These observations are illustrated in the representative results shown in Fig. 2. Decline of MDH specific activity was more rapid than the acetate-induced increase, returning to the pre-induction level usually within one hour. Free MDH activity could not be detected in the medium. During this hour the absorbancy at 578 mμ never increased more than 10 to 15%, indicating the occurrence of negligible growth. Specific activity never dropped to zero. Rapid chilling of the yeast stabilized the MDH activity at the level present at the time of chilling. If the inactivation by glucose was interrupted by rapidly centrifuging, washing and re-suspending the cells in medium lacking glucose but containing acetate, recovery from the inactivating effects of glucose was well established in about 30 to 40 minutes, with a return to the previous rate of induction of MDH. Thus the inhibitory effect of glucose was transient, if it was removed from the medium.

If, with glucose, cycloheximide was added at a final concentration of 5 μg/ml, there occurred a rapid cessation of the established rise of MDH activity, but *no* decrease in MDH activity such as was seen when glucose alone was added. Such an experiment is illustrated in Fig. 3. Also shown in this illustration is the effect of later addition of cycloheximide. It can be seen that if cycloheximide was added later than about 12 minutes after glucose, decrease of MDH activity was not interrupted. From other experiments we found that the decrease of MDH activity is barely detectable at 12 minutes after glucose addition (cf. Fig. 2).

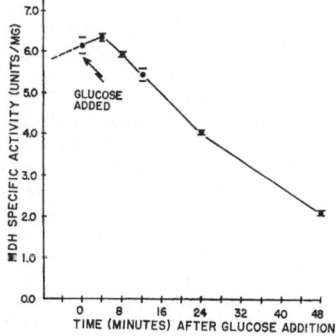

Fig. 2. *Time course of MDH repression by glucose.* 1 g (wet) yeast per 250 ml acetate medium incubated 4 hours at 30° in air. Glucose added (Time = 0) as indicated. Final concentration 55 mM. Samples iced, harvested, washed and extracted at times indicated, after glucose addition. Duplicate assays in direction of oxaloacetate reduction

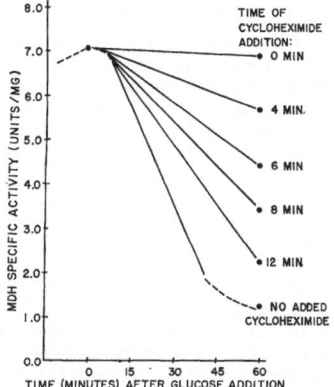

Fig. 3. *Effect of cycloheximide on inactivation-repression by glucose.* Yeast (4 g wet per liter) incubated 6 hours in acetate medium. Glucose added to 55 mM as indicated (Time = 0). Cycloheximide added either with glucose or 4, 6, 8, or 12 minutes after glucose. Samples harvested 1 hour after glucose addition. Assayed in direction of oxaloacetate reduction

These several observations on the effect of cycloheximide, an inhibitor of protein synthesis, on glucose repression of MDH in yeast, gave rise to the interesting idea that protein synthesis might be an obligatory step in this unique "inactivation-repression." To explore this possibility further, attempts were made to inhibit protein synthesis with other agents. 2-deoxy-D-glucose, p-fluoro phenyl-

alanine and 5-methyl tryptophane readily inhibited protein synthesis and blocked induction of MDH in acetate medium. However, the addition of glucose readily overcame their effectiveness in inhibiting [1-14C] leucine incorporation into protein. Thus "inactivation-repression" by glucose could be demonstrated in the presence of these compounds.

Puromycin was found to be a poor inhibitor of [1-14C] leucine incorporation into protein of intact yeast cells, except in the range of 1 mM, where inhibition was still only partial. It was not studied further. Sodium azide at a concentration of 5 mM

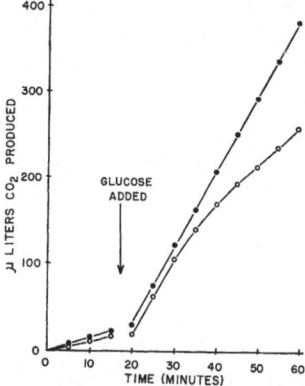

Fig. 4. *Glucose metabolism in presence of cycloheximide.* 2.5 ml 0.4% yeast (w/v) pre-incubated 3 hours in acetate medium. After manometric readings for 15 minutes, either glucose [0.2 ml of a 12.5% (w/v) solution] or glucose plus cycloheximide (1 µg/ml final) tipped from sidearm. Incubated at 30° under air at 110 cycles per minute. ●——●, glucose alone; ○——○, glucose + cycloheximide

was, like cycloheximide, found to prevent glucose "inactivation-repression," but was not studied in detail because of the known multiplicity of its effects on various enzyme systems.

Lacking another agent with which specifically to inhibit protein synthesis in yeast, the metabolic effects of cycloheximide on these cells was studied in greater detail. It has been reported [6] that cycloheximide inhibits glucose utilization by yeast. Fig. 4 shows the effect of cycloheximide on carbon dioxide release in the presence of glucose, after 3½ hours pre-incubation in acetate medium. At the completion of this preliminary incubation there was considerable basal oxygen uptake (not shown) and carbon dioxide release due to acetate utilization. Both were increased when glucose was added. If cycloheximide was added with glucose, the rate of both processes slowly and gradually decreased. One hour after addition of glucose plus cycloheximide both were decreased about 35% below the control rates. This result is typical of four separate experiments. It was thus evident that under the conditions used in glucose "inactivation-repression" of MDH, cycloheximide did significantly inhibit glucose utilization as shown manometrically. This inhibition was minimal for about 15 to 20 minutes, and gradually increased to about 30 to 40% after 1 hour. The inhibition of glucose degradation, incidentally, occurred in the presence of mM potassium cyanide, suggesting an effect on the anaerobic phase of glucose catabolism. In studies on the concentration of cycloheximide required to inhibit the rates of glucose utilization, [1-14C] leucine incorporation, and glucose "inactivation-repression," inhibition of all three rates was half maximal at a cycloheximide concentration of about 0.5 μg/ml (1.78 μM), and maximal at a concentration of just over 1.0 μg/ml.

Molecular Differences between Induced and Repressed MDH

Several miscellaneous observations deserve mention. When MDH activity of extracts was assayed

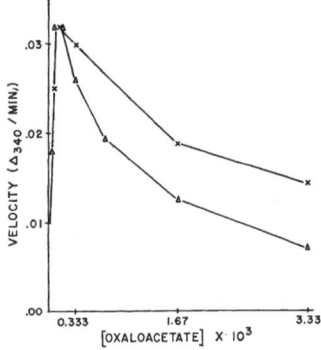

Fig.5. *MDH saturation with oxaloacetate.* Induced and repressed extracts assayed with varying oxaloacetate concentrations in presence of 0.1 mM NADH. ×——×, induced extract; △——△, repressed extract

in the direction of malate synthesis in triethanolamine buffer, the enzyme was found to exhibit marked inhibition by substrate, as has been described for MDH from other organisms [10]. This substrate inhibition was found to be greater in extracts from glucose-repressed cells than in extracts from acetate-induced cells. This is shown in Fig.5. If extracts were assayed in the presence of both high and low oxaloacetate concentrations, the extracts could be identified as coming either from induced or repressed cells by calculating the ratio of activities at the two oxaloacetate concentrations. This kinetic observation suggested that the two enzymes (*i.e.* from induced and repressed extracts) represented separate molecular species. Witt *et al.* have since

separated the two species by chromatographic procedures [20]. It should be noted that this difference in degree of inhibition by the substrate oxaloacetate magnifies the drop of MDH in extracts of repressed cells. L-malate caused only minimal inhibition of reaction rates, equal in both induced and repressed extracts.

Attempts were made to effect "inactivation-repression" in extracts of induced yeast. On the assumption that glucose might cause the synthesis of a protein which specifically inactivates MDH, extracts of induced and repressed cells were mixed and incubated for 4 hours. The combined extracts lost no activity during this period, being equal to the sum of the activities of the two extracts incubated separately. Thus a specific inactivation reaction could not be demonstrated *in vitro* by this method. Conversely, the activity of MDH from repressed cells could not be restored by dialysis, or by heating, as described for ornithine transcarbamylase by Bechet and Wiame [1]. Extracts of induced and repressed yeast neither lost nor gained MDH activity during overnight dialysis. Thus the alterations of MDH activity described are not likely to be due to the accumulation of low molecular weight activators or inhibitors.

DISCUSSION

We use the term "inactivation-repression" to describe an unusual type of catabolite repression in which enzyme activity decreases after addition of repressor. At this juncture, it is not possible to explain the inactivation of MDH observed when glucose is added to yeast. Such inactivation has been described for other enzymes in microorganisms. Spiegelman and Reiner [14] described the de-adaptation of the galactozymase system in yeast transferred from galactose-containing medium to one containing glucose. This de-adaptation was prevented by anaerobiosis or sodium azide. Robertson and Halvorson [12] found that the activity of maltose-induced maltozymase was rapidly decreased by substitution of glucose for maltose. Mandelstam [11] described the inactivation of lysine-induced lysine decarboxylase in *B. cadaveris* when cells were placed in glucose medium. This inactivation was prevented by azide and required active metabolism. However, the specific activity of lysine decarboxylase in extracts could be restored to pre-repression levels by addition of pyridoxal phosphate, suggesting an effect through alteration of cofactor binding or synthesis. The "inactivation-repression" of ornithine transcarbamylase by arginine has been described in yeast by Bechet and Wiame [1]. In this instance, arginine did not produce enzyme inactivation if added with cycloheximide, much as we have found in this present study. These authors were able to reactivate the inactivated enzyme by heating, and they propose that repression by arginine involves the synthesis of a metabolically short-lived protein which specifically binds and inactivates enzyme.

The specific fructose 1,6-diphosphate phosphatase from *Saccharomyces cerevisiae* has recently been found also to undergo "inactivation-repression" after addition of glucose [7].

A number of mechanisms can be suggested to explain the apparent disappearance of MDH activity in the presence of a repressor. The observation that cycloheximide interferes with this inactivation suggests that synthesis of a protein may be involved in this process. We must acknowledge, however, that this effect of cycloheximide may be an artifact in that any interference with glucose metabolism might lower the effective concentration of that catabolite of glucose which initiates the process of repression. Lacking, to date, an agent which selectively inhibits protein synthesis in yeast, without altering glucose utilization, this uncertainty must persist. Our studies with cycloheximide were originally undertaken to explore the possibility that induced MDH is a protein which undergoes rapid metabolic turnover, and that the observed inactivation is merely the expression of rapid catabolism of this protein following the repression of its synthesis. This possibility of course remains, in that we do not yet know if normal protein catabolism can occur in the presence of cycloheximide. More refined techniques, such as those used by Schimke *et al.* [13], must be used to evaluate this possibility.

Several reports [3,4] have provided experimental evidence that the molecular mediator of repression is a short-lived protein. If this is the case in the glucose-MDH system here described, it would seem logical that repressor formation could not occur in the absence of protein synthesis. But such a formulation does not account for the observed inactivation of MDH. The classical concept of repression offered by Jacob and Monod [8] does not readily explain this phenomenon. The studies of Sussman [16] may have relevance here, in that the "programmed" disappearance of UDP-galactose polysaccharide transferase is prevented by cycloheximide inhibition of protein synthesis in the cellular slime mold, *Dictyostelium discoideum*.

Our data indicate that in the course of repression by glucose there occurs in the induced MDH molecule an alteration which lowers its catalytic ability. This alteration could involve a degradation of MDH, by either a proteolysis of the primary structure of MDH, or a conversion of MDH into relatively inactive subunits. Either mechanism could be mediated by a specific enzyme induced by a catabolite of glucose. We have sought, and to date failed to find, evidence for such a degradation produced by extracts of repressed cells. If such a degrading enzyme exists, it is not detected in the extraction procedure we used.

Changes in conformation (and catalytic ability) of MDH could also be produced by the glucose-dependent appearance of a small molecule (eg. nucleotide), peptide or macromolecule within the yeast cell, acting at a site either remote from or at the active catalytic center(s) of MDH. If such an inactivating molecule exists, our observations suggest that protein synthesis is required for its production. It is not removed by dialysis.

With the data presented, these considerations must remain hypothetical. The several possible mechanisms described are currently under scrutiny.

Most of the work described was performed during tenure by J. J. F. of a National Science Foundation Senior Research Fellowship, at the University of Freiburg. Major support for these studies was provided by the Deutsche Forschungsgemeinschaft; part was provided by grant AM-07207 from the National Institutes of Health, U.S.A.

J.J.F. wishes to express his indebtedness to Dr. I. Witt for the benefit of many fruitful discussions of this study, as well as generous access to her data prior to its publication.

REFERENCES

1. Bechet, J., and Wiame, J. M., *Biochem. Biophys. Res. Commun.* 21 (1965) 226.
2. Bell, E., Humphreys, T., Slayton, H. S., and Hall, C. E., *Science* 148 (1965) 1739.
3. Gallant, J., and Stapleton, R., *J. Mol. Biol.* 8 (1964) 431.
4. Garen, A., and Garen, S., *J. Mol. Biol.* 6 (1963) 433.
5. Gornall, A. G., Bardawill, C. J., and David, M. M., *J. Biol. Chem.* 177 (1949) 751.
6. Greig, M. E., Walk, R. A., and Gibbons, A., *J. Bacteriol.* 75 (1958) 489.
7. Harris, W., and Ferguson, J. J., Jr., unpublished.
8. Jacob, F., and Monod, J., *Cold Spring Harbor Symp. Quant. Biol.* 26 (1961) 193.
9. Kornberg, A., and Horecker, B. L., In *Methods in Enzymology* (edited by S. P. Colowick and N. O. Kaplan), Academic Press, New York 1955, Vol. I, p. 323.
10. Kun, E., and Volfin, P., *Biochem. Biophys. Res. Commun.* 22 (1966) 187.
11. Mandelstam, J., *J. Gen. Microbiol.* 11 (1954) 426.
12. Robertson, J. J., and Halvorson, H. O., *J. Bacteriol.* 73 (1957) 186.
13. Schimke, R. T., Sweeney, E. W., and Berlin, C. M., *J. Biol. Chem.* 240 (1965) 322.
14. Spiegelman, S., and Reiner, J. M., *J. Gen. Physiol.* 31 (1947) 175.
15. Stent, G. S., *Science* 144 (1964) 816.
16. Sussman, M., *Proc. Natl. Acad. Sci. U. S.* 55 (1966) 813.
17. Umbreit, W. W., Burris, R. H., and Stauffer, J. F., *Manometric Techniques*, Burgess Publishing Co., 1957.
18. Vogel, H. J., In *The Chemical Basis of Heredity* (edited by W. D. McElroy and B. Glass). The Johns Hopkins Press, Baltimore 1957, p. 279.
19. Witt, I., Kronau, R., and Holzer, H., *Biochim. Biophys. Acta* 118 (1966) 522.
20. Witt, I., Kronau, R., and Holzer, H., *Biochim. Biophys. Acta* 128 (1965) 63.
21. Wolfe, R. G., and Neilands, J. B., *J. Biol. Chem.* 221 (1956) 61.

J. J. Ferguson, jr.
Biochemistry Department, University of Pennsylvania
Philadelphia, Pa. 19104, U.S.A.

M. Boll and H. Holzer
Biochemisches Institut der Universität
78 Freiburg i. Br., Hermann-Herder-Straße 7, Germany

European J. Biochem. 1 (1967) 26—28

Reversible Inactivation of Citrate Lyase:
Effect of Metal Ions and Adenine Nucleotides

J. McD. Blair, S. P. Datta, and S. S. Tate

Department of Biochemistry, University College London

(Received October 24, 1966)

The inactivation of citrate oxaloacetate-lyase in dilute solution is complex, at least two processes occurring, one reversible and the other irreversible. Inactivation at low magnesium concentration (0.025—0.1 mM) and reactivation at higher magnesium concentration (2.5 mM) have been investigated. The adenine nucleotides and EDTA exert effects related to the stability constants of their complexes with magnesium. Although unable to activate citrate lyase, Ca^{2+} can protect the enzyme from inactivation in dilute solution.

The substrate-induced enzyme, citrate lyase, catalyses the cleavage of citrate to oxaloacetate and acetate. This is the first reaction in the anaerobic dissimilation of citrate by various microorganisms such as *Escherichia coli*, *Aerobacter aerogenes* and Streptococci.

It was shown by Dagley and Dawes [1] that for enzymic activity, citrate lyase has an absolute requirement for a divalent metal cation such as Mg^{2+}, Mn^{2+}, Co^{2+}, etc. It was suggested by these authors and also by Harvey and Collins [2] and Tate [3] that citrate lyase acts on a metal-citrate complex. However, Ward and Srere [4] deduced from nuclear magnetic resonance studies on the enzyme from *Streptococcus diacetilactis* that a metal-enzyme acts on free citrate. The citrate lyase from *A. aerogenes* rapidly loses activity in dilute solutions and the presence of magnesium largely prevents this inactivation [5], suggesting the formation of a magnesium-enzyme complex. We have studied this protective action of magnesium in greater detail in an effort to establish the mechanism of formation of the active enzyme-metal-substrate complex.

EXPERIMENTAL

Materials

Trisodium citrate, $MgCl_2 \cdot 6H_2O$, $CaCl_2 \cdot 6H_2O$ and Na_2EDTA used were AnalaR grade. NN-bis(2-hydroxyethyl)glycine ("bicine") was "Biochemical" grade from British Drug Houses Ltd. ATP, ADP, and AMP were purchased from C. F. Boehringer & Soehne of Mannheim, Germany.

Bicine buffer ($pK_{2a} = 8.33$ at $25°$ [6]). A solution (approx. 0.2 M) of bicine was adjusted at room temperature to pH 7.6 with saturated NaOH and diluted to give a solution 0.1 M with respect to bicine.

Citrate lyase was prepared by an improved method (to be published) from extracts of *Aerobacter aero-*

Enzyme. Citrate lyase or citrate oxaloacetate-lyase (EC 4.1.3.6).

genes, NCTC strain 418. The enzyme used was in 0.05 M bicine buffer pH 7.6 containing 0.1 M KCl and mM $MgCl_2$.

Methods

The activity was assayed by the initial increase in extinction at 280 mμ caused by the oxaloacetate formed from citrate at $25°$, using a Unicam SP800 spectrophotometer. The standard assay mixture contained $MgCl_2$ (2.5 mM) and trisodium citrate (0.5 mM) in the bicine buffer. For assay, the enzyme solution (0.02 to 0.05 ml) was added to 1 ml of the assay mixture.

The time-dependent inactivation of the enzyme at low magnesium concentrations (0.02 to 0.1 mM) was followed by incubating the enzyme in bicine buffer containing the required $MgCl_2$ for periods up to 5 min, the residual activity being assayed by simultaneous addition of $MgCl_2$ and citrate (final concentrations 2.5 and 0.5 mM respectively). To follow the reactivation process, the enzyme was incubated first with low magnesium concentrations for 5 min then for periods up to 5 min with 2.5 mM $MgCl_2$, the reactions being initiated by the addition of citrate. The extent of irreversible inactivation at various low magnesium concentrations was determined by incubating the enzyme for periods of up to 5 min followed by reactivation for 5 min with 2.5 mM $MgCl_2$; citrate was added to initiate the reaction. The fall in activity under these conditions was a measure of the amount of enzyme irreversibly inactivated. An allowance was made for the extent of irreversible inactivation occurring on incubation of the enzyme with 2.5 mM $MgCl_2$ for 5 min.

The effects of adenine nucleotides and EDTA were determined by adding solutions of these compounds to the incubation mixtures to a final concentration of 0.05 mM in each case prior to the addition of the enzyme. For these experiments, the initial $MgCl_2$ concentration was 0.05 mM.

RESULTS

The time-dependent inactivation of citrate lyase in dilute solutions containing various low concentrations of magnesium (0.025 to 0.1 mM) is shown in Fig. 1. Also in Fig. 1 is shown the reactivation of the enzyme during a second preincubation with excess MgCl₂ (2.5 mM). The activity, however, could not be completely restored, showing the occurrence of more than one process—a reversible and an irreversible inactivation. The effect of increasing enzyme

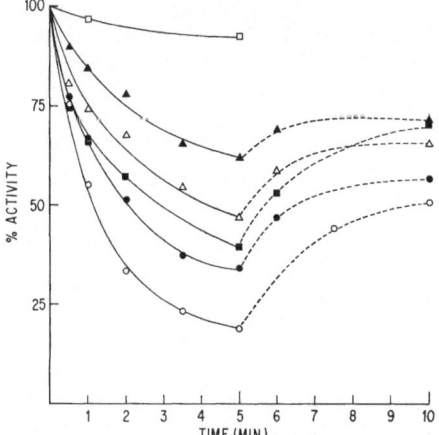

Fig. 1. *Inactivation and reactivation of citrate lyase: effect of MgCl₂ concentration.* Solid curves indicate inactivation due to the initial incubation of the enzyme (0.025 ml per ml reaction mixture) with the following concentrations of MgCl₂ (mM): ○, 0.025; ●, 0.05; △, 0.075; ▲, 0.1; and □, 2.5; in 0.1 M-bicine, adjusted to pH 7.6 with NaOH. Reactions were initiated at the time indicated by the simultaneous addition of MgCl₂- and Na₃-citrate (final concentrations, 2.5 mM and 0.5 mM respectively). ■, Initial incubation of citrate lyase (0.05 ml per ml reaction mixture) with 0.05 mM MgCl₂. Broken lines indicate reactivation by the addition of MgCl₂ (2.5 mM, final concentration) at 5 min; reactions were initiated by the addition of citrate at the times indicated. The citrate lyase was in 0.05 M bicine buffer pH 7.6 containing 0.1 M KCl and mM MgCl₂

concentration on its stability is also shown. It is evident that at higher protein concentration there is less irreversible inactivation.

The modification of the activity of citrate lyase on preincubation in dilute solution in the presence of 0.05 mM MgCl₂ and 0.05 mM of each of the adenine nucleotides, ATP, ADP and AMP, and EDTA, is shown in Fig. 2. AMP and ADP are much less effective as inhibitors than either ATP or EDTA.

The extent of irreversible inactivation with time at three low MgCl₂ concentrations is shown in Fig. 3.

Fig. 4 shows that calcium is also able to protect the enzyme against inactivation on dilution, although it is less effective than magnesium.

DISCUSSION

Our results show that the processes occurring on dilution of the enzyme, citrate lyase, in presence of low magnesium concentrations are complex. At least two inactivation processes take place—one, reversible by incubation with excess magnesium and the other apparently irreversible. Increasing concentrations of magnesium retarded both these inactivation processes showing that the active species of the enzyme is a magnesium-enzyme complex. The metal-binding

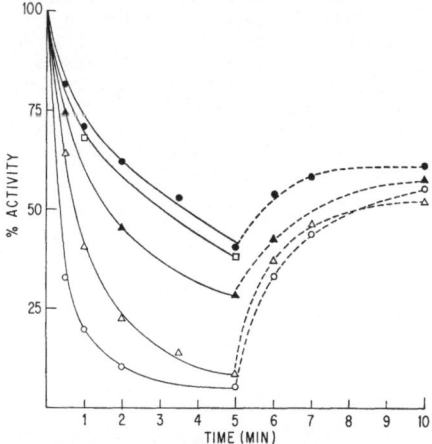

Fig. 2. *Inactivation and reactivation of citrate lyase: effect of adenine nucleotides and EDTA.* Solid curves indicate inactivation due to the initial incubation of the enzyme (0.05 ml per ml reaction mixture) with MgCl₂ (0.05 mM). In addition solutions contained 0.05 mM of: ○, EDTA; △, ATP; ▲, ADP; and □, AMP. ●, Control with MgCl₂ (0.05 mM) only. Reactions were initiated as described in Fig. 1. Broken curves indicate reactivation with MgCl₂ (2.5 mM) as described in Fig. 1

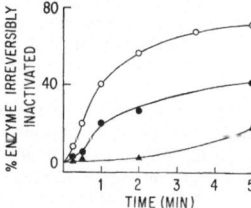

Fig. 3. *Irreversible inactivation of citrate lyase: effect of MgCl₂.* Citrate lyase (0.02 ml per ml reaction mixture) was incubated for the times indicated with the following concentrations of MgCl₂ (mM): ○, 0.02; ●, 0.03; ▲, 0.1. In each case incubation was continued for a further 5 min after the addition of MgCl₂ (2.5 mM), the residual activity being assayed by the addition of citrate (0.5 mM) final concentration. The loss in activity represents the fraction of enzyme that is irreversibly inactivated, an allowance being made for inactivation due to incubation for 5 min with MgCl₂ (2.5 mM, final concentration)

site may be identical with the active site, which would require a compulsory order mechanism for the formation of the enzyme-metal-substrate complex such as suggested by Ward and Srere [4]. However, it is possible that the metal binds at a site or sites distinct from the active site causing a conformational change to give the active species of the enzyme, in which case, the random order mechanism where the active enzyme-metal-substrate complex may form from either the enzyme-metal and citrate or from enzyme and metal-citrate complex cannot be ruled out.

The irreversible inactivation of citrate lyase may be either due to conformational changes, dissociation of the enzyme into subunits, or both. Dissociation into subunits is likely since less inactivation occurs at a higher enzyme concentration (Fig. 1). Bowen and Rogers [10] obtained evidence for the formation of a lower molecular weight species on inactivation of this enzyme on storage.

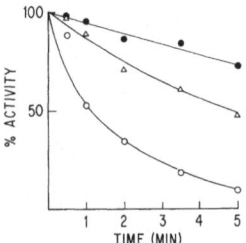

Fig. 4. *Protection of citrate lyase by* $MgCl_2$ *and* $CaCl_2$. Citrate lyase (0.02 ml perml of reaction mixture) was incubated for the times indicated in bicine buffer at the following metal concentrations (mM): O, $MgCl_2$ (0.02); ●, $MgCl_2$ (0.1); and △, $MgCl_2$ (0.02) and $CaCl_2$ (0.08). Reactions were initiated by simultaneous addition of $MgCl_2$ and Na_3-citrate (final concentrations, 2.5 mM and 0.5 mM respectively)

It would appear from the curves obtained by plotting the extent of irreversible inactivation against time at three low magnesium concentrations (Fig. 3), that dissociation of Mg^{2+} from a metal-enzyme complex precedes the irreversible inactivation.

The enhanced inactivation of citrate lyase by the adenine nucleotides appears to be due to the chelation of Mg^{2+} since EDTA has a similar effect. The table relates the per cent inhibition obtained at 30 sec in presence of the nucleotides and EDTA, the stability constants of their magnesium complexes, and the per cent reduction in Mg^{2+} concentration (calculated from stability constants (K_s^a) for experimental conditions). The percentage inhibition in each case is consistent with the reduction in Mg^{2+} concentration. The possibility arises that the ratios of the adenine nucleotides in the bacterial cell may indirectly regulate the activity of citrate lyase by controlling the free Mg^{2+} concentration. At present

it is impossible to assess whether or not this is a significant regulatory mechanism.

The protection of citrate lyase in dilute solutions by calcium is interesting since calcium does not form a catalytically active complex with the enzyme; in fact, it competitively inhibits the cleavage of citrate in presence of magnesium (Dagley and Dawes [1]). These authors attribute the non-reactivity of calcium to the fact that it has a greater ionic radius than those metals which are able to activate the enzyme. However, our results indicate that Ca^{2+} can in fact form a complex with citrate lyase, although this may not be at the catalytic site. An alternative explanation for the inability of calcium to form a catalytically

Table. *Inactivation of citrate lyase by adenine nucleotides and EDTA*
The inhibition data are obtained from Fig. 2. The control for these measurements was the experiment in which the enzyme was incubated for 30 sec with 0.05 mM $MgCl_2$

	% Inactivation after incubation for 30 sec	Log K_s^a (Stability constant) of Mg-complex at pH 7.6, approx.	% reduction in Mg^{2+} concentrations calculated from log K_s^a for experimental conditions
AMP	1.8	1.95 [7]	2
ADP	9.8	3.60 [8]	20
ATP	22.1	4.84 [8]	60
EDTA	60.1	6.01 (calc. for pH 7.6 from [9])	87

active complex with citrate lyase has been suggested based on the different structures of the magnesium and calcium chelates of citrate (Tate, in preparation).

The authors thank the Medical Research Council for a grant towards expenses and Miss A. Straker for preparing the manuscript.

REFERENCES
1. Dagley, S., and Dawes, E. A., *Biochim. Biophys. Acta* 17 (1955) 177.
2. Harvey, R. J., and Collins, E. B., *J. Biol. Chem.* 238 (1963) 2648.
3. Tate, S. S., Ph. D. Thesis, University of London, 1963.
4. Ward, R. L., and Srere, P. A., *Biochim. Biophys. Acta* 99 (1965) 270.
5. Eisenthal, R., Tate, S. S., and Datta, S. P., *Biochim. Biophys. Acta* 128 (1966) 155.
6. Datta, S. P., Grzybowski, A. K., and Bates, R. G., *J. Phys. Chem.* 68 (1964) 275.
7. Walaas, E., *Acta Chem. Scand.* 12 (1958) 528.
8. O'Sullivan, W. J., and Perrin, D. D., *Biochemistry* 3 (1964) 18.
9. Schwarzenbach, G., Gut, R., and Anderegg, G., *Helv. Chim. Acta* 37 (1954) 937.
10. Bowen, T. J., and Rogers, L. J., *Biochim. Biophys. Acta*, 67 (1963) 633.

J. McD. Blair and S. P. Datta
Dept. of Biochemistry, University College London
Gower Street, London W. C. 1, England

S. S. Tate
Dept. of Biochemistry, Tufts University School of Medicine
136, Harrison Avenue, Boston, Mass. 02111, U.S.A.

European J. Biochem. 1 (1967) 29—32

Interactions of Adenosine Tetraphosphate with Myosin and Actomyosin

J. Winand-Devigne, G. Hamoir, and C. Liébecq

Laboratoire de Biochimie de l'Institut supérieur d'Education physique et Laboratoire de Biologie générale, Université de Liège

(Received October 7, 1966)

1. Myosin-ATPase prepared from dog heart and rabbit skeletal muscle catalyzed the hydrolysis of adenosine tetraphosphate into adenosine diphosphate and orthophosphate.

2. Maximum activities were measured in the presence of Ca^{++} ions and represented 15 to 50% (according to pH and to origin) of the maximum activities measured in the presence of adenosine triphosphate (ATP). Low activities were measured in the presence of $Co^{++} > Ni^{++} > Mg^{++}$ ions.

3. Actomyosin-ATPase prepared from rabbit and from carp muscles hardly catalyzed the hydrolysis of adenosine tetraphosphate (2.5% of the rate of ATP hydrolysis) in the presence of Mg^{++} ions at low ionic strength.

4. Ultracentrifugation analysis showed that carp muscle actomyosin is split into actin and myosin by low concentrations of adenosine tetraphosphate whereas the contaminating ATP or the ATP formed from adenosine tetraphosphate during the centrifugation does not markedly influence the centrifugal pattern.

Adenosine tetraphosphate was discovered as a contaminant of various commercial preparations of ATP [1—4]. Its properties are those of a nucleotide identical to ATP but with an unbranched chain of four phosphates [3,5—7]. Its physiological significance is still unknown.

Adenosine tetraphosphate does not replace ATP in various phosphokinase reactions, such as those catalyzed by adenylate kinase [3,8—10], hexokinase [3,9] and creatine kinase [11]; adenosine tetraphosphate actually inhibits these reactions when they are measured with ATP as phosphate donor [3,9—11].

Adenosine tetraphosphate exhibits a slow reaction with glycerol and phosphoglycerate kinases, where it can replace ATP in the direct reaction [11]; it may apparently be formed very slowly from ATP in the phosphoglycerate kinase reverse reaction [10].

We had observed that homogenates of rat heart and skeletal muscles prepared in 0.5 M KCl dephosphorylated adenosine tetraphosphate whereas water homogenates of various rat tissues, as well as mitochondrial or microsomal suspensions prepared from

0.25 M sucrose homogenates of rat liver did not [11]. This suggested that the myosin-ATPase extracted by the 0.5 M solution of KCl was responsible for the hydrolysis of adenosine tetraphosphate (see also [12]).

This paper describes the hydrolysis of adenosine tetraphosphate by purified myosin prepared from heart and skeletal muscles, as well as the splitting of actomyosin by adenosine tetraphosphate during which the actomyosin-ATPase (activated by Mg^{++} ions) is transformed into myosin-ATPase (not activated by Mg^{++} ions).

A preliminary note has appeared elsewhere [13].

MATERIALS AND METHODS

Adenosine triphosphate and tetraphosphate were purchased from Sigma Chemical Company (St. Louis, Mo., U.S.A.). Adenosine tetraphosphate was rechromatographed on colums of Dowex-2 [14], replacing sodium chloride by lithium chloride. Inorganic tripolyphosphate, sodium salt, was a gift of Monsanto Chemical Company (St. Louis, Mo., U.S.A.). Analytical grade reagents were used throughout.

Myosin was prepared (by a short extraction with a neutral saline solution) from rabbit skeletal muscle, using 0.5 M KCl + 0.03 M $NaHCO_3$ [15] and from dog heart muscle, using 0.4 M KCl + 0.05 M KH_2PO_4 + 0.01 M $MgSO_4$ + 0.002 M ATP [16]. Actomyosin was prepared by a prolonged extraction with 0.5 M KCl + 0.03 M $NaHCO_3$ [17]: 15 hours at room temperature in the case of the rabbit muscle, 1 hour in the cold room in the case of the carp muscle.

Non-Standard Abbreviations. A4P and A5P for adenosine tetra- and pentaphosphate.

Enzymes. Adenosine triphosphatase (ATPase) or ATP phosphohydrolase (EC 3.6.1.3), adenylate kinase or ATP: AMP phosphotransferase (EC 2.7.4.3), hexokinase or ATP: D-hexose 6-phosphotransferase (EC 2.7.1.1), glycerol kinase or ATP: glycerol phosphotransferase (EC 2.7.1.30), creatine kinase or ATP: creatine phosphotransferase (EC 2.7.3.2), phosphoglycerate kinase or ATP: 3-phospho-D-glycerate 1-phosphotransferase (EC 2.7.2.3) and apyrase or ATP diphosphohydrolase (EC 3.6.1.5).

Myosin-ATPase activity was measured at 37° in the presence of 5 mM Ca^{++} and 50—500 mM K$^+$ ions, at pH values between 7.4 and 9.0 [12,16] and using 0.03 or 0.15 mg of protein per millilitre of enzyme assay. The ATPase activity of the actomyosin preparations was measured at 25° and pH 7.5, in the presence of Tris-Cl buffer of varying concentrations ($I = 0.04—0.80$) and of 5 mM Mg^{++} ions, using 0.5 or 1.0 mg of proteins per millilitre of enzyme assay.

Orthophosphate was determined according to Sumner [18]. Proteins were determined by a modified biuret method [19]. Nucleotides were analyzed by paper chromatography [20]; concentration, where necessary, was effected by partial evaporation in vacuo followed by precipitation with 2 volumes of methanol and 12 volumes of acetone, in the presence of approximately 0.25 M LiCl.

The splitting of actomyosin was initiated by the addition of 10 µl of a 30 mM ice-cold solution of ATP or A4P to 1 ml of a solution containing 3 mg of actomyosin in a buffer of ionic strength 0.38 composed of 0.1 M Tris-HCl (pH 7.5), 0.25 M NaCl and 0.01 M MgCl$_2$. Centrifugation at 59,780 rev./min and at 5° started 5 min later in a Spinco, model E, analytical ultracentrifuge with automatic temperature control. Two cells with aluminium centerpiece were run simultaneously; the wedge-window one contained the unsplit control. The sedimentation coefficients were corrected for water and temperature (20°), assuming the specific volumes given by Svedberg and Pedersen [21].

RESULTS AND DISCUSSION

Hydrolysis of Adenosine Tetraphosphate by Myosin

Adenosine tetraphosphate incubated in the presence of myosin from rabbit skeletal muscle was hydrolyzed into ADP without net production of ATP or AMP: 1 mole of ADP and 2 moles of orthophosphate were produced per mole of A4P. Adenosine triphosphate was probably formed first, but did not accumulate owing to its more rapid hydrolysis by myosin.

As in the case of ATP (see [22]), maximum activity was measured in the presence of Ca^{++} ions (Table 1). Other bivalent cations were less active, activity decreasing in the order Co^{++} > Ni^{++} > Mg^{++}. When inorganic tripolyphosphate was used as a substrate, activity decreased in the order Ni^{++} > Mg^{++} > Co^{++}. This was taken as evidence that our myosin preparation was essentially free of the nucleoside tetraphosphate hydrolase isolated recently from rabbit skeletal muscle [23], where A4P hydrolysis decreased in the order Co^{++} > Mg^{++} > Ni^{++} and tripolyphosphate hydrolysis in the order Ni^{++} > Co^{++} > Mg^{++}.

The hydrolysis of A4P proceeded more slowly than the hydrolysis of ATP (Table 2): 2 to 4 times more slowly at pH 7.4 in the presence of skeletal or

heart myosins. The relative activities were still further reduced at higher pH values.

Numerous natural nucleotide triphosphates [12, 24—28], synthetic analogues of ATP [29,30] as well as inorganic tripolyphosphate [31] are hydrolyzed in the presence of myosin-ATPase. Adenosine tetraphosphate is another substrate for this rather unspecific enzyme (see also [12]).

Table 1. *Phosphohydrolase activity of the myosin of rabbit skeletal muscle: activation by cations*
The experiments were performed at 37° in the presence of 0.05 mg of myosin per ml of incubation mixture buffered with 0.02 M histidine-Cl pH 7.5. The results are expressed in µmoles of orthophosphate liberated per mg of proteins and per minute

Substrate (1 mM)	Activator (5 mM)			
	Ca^{++}	Mg^{++}	Co^{++}	Ni^{++}
ATP	14.3	0.7	3.4	0.5
A4P	11.8	0.3	2.9	1.2
PPP	1.7	1.1	0.1	1.6

Table 2. *Phosphohydrolase activity of various myosin preparations in the presence of ATP or adenosine tetraphosphate (A4P)*
The experiments were performed at 37° in the presence of a uniform concentration of 5 mM Ca^{++} ions and of 0.03 to 0.05 mg of myosin per ml of incubation mixture containing histidine- or Tris-Cl buffer

Origin	Conditions		Substrate		Relative activity[a]
	pH	K$^+$	ATP	A4P	
		mM	µmoles P$_i$/mg protein/min		%
Skeletal muscle	9.0	50	1.54	0.78	26
	7.4	50	0.83[b]	0.85[b]	51
Heart muscle	8.0	500	0.25	0.08	16
	7.4	50	0.38	0.20	26

[a] A4Pase activity expressed in % of the ATPase activity, the amount of P$_i$ liberated from A4P being first divided by 2, since each mole of A4P hydrolyzed to ADP produces 2 moles of P$_i$.

[b] These figures are not comparable with those of 1.54 and 0.78 obtained at pH 9 as the enzyme preparation used (when tested again at pH 9) had lost about ¹/₃ of its initial activity.

It may be worth remembering that adenosine tetraphosphate is also a substrate for potato apyrase, another rather unspecific phosphohydrolase [32,33].

Hydrolysis of Adenosine Tetraphosphate by Actomyosin

As already observed by Hasselbach [34], the adenosine tetraphosphatase activity of actomyosin, measured in the presence of Mg^{++} ions was very low and represented only 2.5% of its ATPase activity (Fig. 1) if one assumes that actomyosin, like myosin, splits two phosphate groups per molecule of A4P. As in the case of ATP however [35], we found that this activity disappeared at high ionic strength. Similar results were obtained with carp and rabbit muscle actomyosin.

Splitting of Actomyosin by Adenosine Tetraphosphate

The viscosity of a solution of actomyosin drops after addition of ATP and this has been attributed by Szent-Györgyi [27] to the splitting of the actomyosin complex into actin and myosin. This pheno-

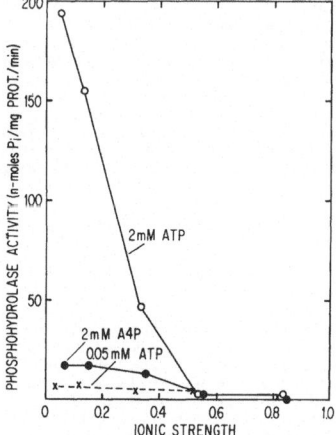

Fig.1. *Influence of the ionic strength on the phosphohydrolase activity of carp muscle actomyosin.* Ionic strength as plotted is the sum of the ionic strengths of all the constituents of the enzymatic assay, in which Tris-Cl concentration was the only variable ($I = 0.02 - 0.80$). Mg^{++} ions were added as 5 mM $MgCl_2$. The ionic strengths of ATP and of A4P were calculated on the assumption that the secondary phosphate groups were 80% ionized at pH 7.5 [40]. 0.05 mM ATP represents the ATP contamination of 2 mM A4P

addition of 0.03 mM ATP, that is at least 3 times as much as the maximum ATP contamination accompanying the addition of 0.3 mM A4P, produced only a partial splitting of actomyosin.

If actomyosin, which was present at a fairly high concentration, was able to hydrolyze adenosine tetraphosphate and hence produce ATP during the centrifugation analysis, the splitting observed in the presence of A4P could possibly be due to ATP itself.

That this is unlikely to be the case is shown by the data of Table 3 which presents the results of

Table 3. *Hydrolysis of adenosine tri- and tetraphosphates at low temperature and in the presence of actomyosin*
The experimental conditions are similar to those of a centrifugation experiment: 50 min at 5°, 0.3 mM ATP or A4P, 10 mM Mg^{++} and 3 mg of carp actomyosin/ml

	Starting material	Unincubated Control[a]	Incubated material
	mM	mM	mM
AMP	—	0.005	0.001
ADP	0.009	0.025	0.121
ATP	0.300	0.280	0.176
A4P	0.003	0.004	0.011
AMP	0.003	0.002	0.001
ADP	0.003	0.013	0.080
ATP	0.009	0.019	0.021
A4P	0.300	0.287	0.226
A5P[b]	0.033	0.026	0.020

[a] Deproteinized with $HClO_4$, neutralized with KOH, centrifuged, passed on a column of Dowex-50 [H+], neutralized with LiOH, concentrated *in vacuo*, precipitated by 2 volumes of methanol and 12 volumes of acetone in the presence of about 0.25 M LiCl.

[b] Adenosine pentaphosphate contaminates most preparations of adenosine tetraphosphate.

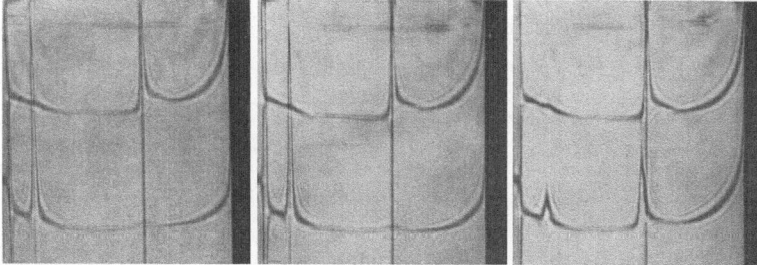

Fig.2. *Ultracentrifugation diagrams of carp muscle actomyosin (3 mg/ml) after 29 min at 59,780 rev./min at a phase plate angle of 55°.* Left: with (lower curve) or without (upper curve) 0.3 mM ATP. Middle: with (lower curve) or without (upper curve) 0.3 mM A4P. Right: with (lower curve) or without (upper curve) 0.03 mM ATP

menon can be easily studied with an ultracentrifuge [36—39]. As illustrated in Fig.2, the 28 S gradient of actomyosin from carp muscle disappeared when 0.3 mM ATP or A4P (contaminated by 0.01 mM ATP at the most) was added, and was replaced by an important 5 S gradient corresponding to myosin and a smaller 26 S gradient corresponding to F-actin. The

experiments in which ATP or A4P were incubated 50 min at 5° under the same conditions as those of a centrifugation analysis. After deproteinization with $HClO_4$, the supernatants were neutralized, concentrated and analyzed for adenine nucleotides by paper chromatography to follow their degree of hydrolysis.

The second column of Table 3 shows that deproteinization and subsequent operations altered the material added: ATP as well as A4P were moderately hydrolyzed. Incubation in the presence of 10 mM Mg^{++} ions and of actomyosin produced a more substantial degradation into ADP of about one-third of the ATP and about one-fifth of the A4P.

The degree of hydrolysis of lower concentrations of ATP (such as 0.03 mM, as used in one of the centrifugation experiments) cannot be easily determined. Even if it were twice as high (a two-third hydrolysis), the final concentration of ATP would still be about 0.01 mM.

In the case of A4P, no ATP accumulated (the increase from 0.019 to 0.021 mM for this rather minor component of the mixture cannot be considered significant) as expected from the faster hydrolysis of ATP itself. Hence, when actomyosin was split by 0.3 mM A4P, the ATP concentration must have remained essentially unchanged and close to 0.01 mM.

Since 0.3 mM A4P containing only 0.01 mM ATP is far more active than 0.03 mM ATP (see Fig.2), it may reasonably be concluded that the splitting of actomyosin was produced by A4P itself, and not by the contaminating ATP.

A firmer conclusion cannot be drawn however because it cannot be excluded that the walls of the centrifuge cells interfere with the rates of hydrolysis.

Our thanks are due to Dr. E. J. Griffith, of Monsanto Chemical Company, for a gift of inorganic tri- and tetrapolyphosphates. This work was supported by grants from the Fonds de la Recherche scientifique médicale.

REFERENCES

1. Marrian, D. H., *Biochim. Biophys. Acta*, 12 (1953) 492.
2. Liébecq, C., unpublished observations, 1953.
3. Lieberman, I., *J. Am. Chem. Soc.* 77 (1955) 3373.
4. Sacks, J., *Biochim. Biophys. Acta*, 16 (1955) 436.
5. Marrian, D. H., *Biochim. Biophys. Acta*, 13 (1954) 278.
6. Liébecq, C., *Bull. Soc. Chim. Biol.* 43 (1961) 331.
7. Liébecq, C., Jaroszewicz, K., and Lallemand, A., *Bull. Soc. Chim. Biol.* 43 (1961) 571.
8. Noda, L., *J. Biol. Chem.* 232 (1958) 237.
9. Winand-Devigne, J., and Liébecq, C., *Abstracts 2nd FEBS Meeting (Vienna)*, 1965, p. 208.
10. Small, G. D., and Cooper, C., *Biochemistry*, 5 (1966) 26.
11. Winand-Devigne, J., unpublished observations, 1965 and 1966.
12. Kielley, W. W., Kalckar, H. M., and Bradley, L. B., *J. Biol. Chem.* 219 (1956) 95.

13. Winand-Devigne, J., Hamoir, G., and Liébecq, C., *Arch. Intern. Physiol. Biochim.* 74 (1966) 948
14. Cohn, W. E., and Carter, C. E., *J. Am. Chem. Soc.* 72 (1950) 4273.
15. Hamoir, G., *Arch. Intern. Physiol. Biochim.* 63 (1955) suppl. 1.
16. Brahms, J., and Kay, C. M., *J. Biol. Chem.* 238 (1963) 198.
17. Gaspar-Godfroid, A., *Angiologica*, 1 (1964) 12.
18. Sumner, J. B., *Science*, 100 (1944) 413.
19. Beisenherz, G., Boltze, H. J., Bücher, T., Czok, R., Garbade, K. H., Meyer-Arendt, E., and Pfleiderer, G., *Z. Naturforsch.* 8b (1953) 555.
20. Devigne, J., Degueldre-Guillaume, M. J., and Liébecq, C., *Bull. Soc. Chim. Biol.* 44 (1962) 751.
21. Svedberg, T., and Pedersen, K. O., *The Ultracentrifuge.* The Clarendon Press, Oxford 1940, Appendix II.
22. Szent-Györgyi, A. G., In *Structure and Function of Muscle* (edited by G. H. Bourne). Academic Press, New York 1960, Vol. II, p. 1.
23. Small, G. D., and Cooper, C., *Biochemistry*, 5 (1966) 14.
24. Kleinzeller, A., *Biochem. J.* 36 (1942) 729.
25. Perry, S. V., *Biochem. J.* 48 (1951) 257.
26. Gergely, J., *J. Biol. Chem.* 200 (1953) 543.
27. Szent-Györgyi, A. G., *Arch. Biochem. Biophys.* 42 (1953) 305.
28. Hasselbach, W., *Biochim. Biophys. Acta*, 20 (1956) 355.
29. Ikehara, M., Ohtsuka, E., Kitagawa, S., Yagi, K., and Tonomura, Y., *J. Am. Chem. Soc.* 83 (1961) 2679.
30. Azuma, N., Ikehara, M., Ohtsuka, E., and Tonomura, Y., *Biochim. Biophys. Acta*, 60 (1962) 104.
31. Neuberg, C., and Fischer, H. A., *Enzymologia*, 2 (1937 —1938) 360.
32. Liébecq, C., Lallemand, A., and Degueldre-Guillaume, M. J., *Arch. Biochem. Biophys.* 97 (1962) 609.
33. Liébecq, C., Lallemand, A., and Degueldre-Guillaume, M. J., *Bull. Soc. Chim. Biol.* 45 (1963) 573.
34. Hasselbach, W., *Acta Biol. Med. Ger.* 2 (1959) 18.
35. Hasselbach, W., *Z. Naturforsch.* 7b (1952) 163.
36. Portzehl, H., Schramm, G., and Weber, H. H., *Z. Naturforsch.* 5b (1950) 61.
37. Johnson, P., and Landolt, H. R., *Discussions Faraday Soc.* 11 (1951) 179.
38. Hamoir, G., *Biochem. Soc. Symp. (Cambridge, Engl.)*, 6 (1951) 8.
39. Holtzer, A., Wang, T. Y., and Noelken, M. A., *Biochim. Biophys. Acta*, 42 (1960) 453.
40. Georges, P., and Rutman, R. J., *Progr. Biophys. Biophys. Chem.* 10 (1960) 1.

J. Winand-Devigne and Claude Liébecq
Laboratoire de Biochimie, Institut supérieur d'Education physique, Université de Liège
1 Rue des Bonnes Villes, Liège, Belgium

G. Hamoir
Laboratoire de Biologie générale, Université de Liège
22 Quai Edouard van Beneden, Liège, Belgium.

European J. Biochem. 1 (1967) 33—35

Kinetics of Enzyme Inactivation in Solution by Ionizing Radiation

T. Sanner and A. Pihl

Norsk Hydro's Institute for Cancer Research, The Norwegian Radium Hospital, Oslo

(Received October 3, 1966)

Equations are derived without the use of steady state assumptions for the inactivation of enzymes in dilute aqueous solution by ionizing radiation. It is shown that when the exposure times are shorter than the mean life-time of the water radicals, as is frequently the case in experiments involving pulsed radiation, the main part of the enzyme inactivation occurs after the end of the exposure. The total inactivation under such conditions can be described by the same equation as that developed earlier for protracted irradiation. The recent claim that the absolute rate constant for the inactivation of the enzyme can be determined directly from studies of the initial part of the dose inactivation curve, is not supported by the present study.

The validity of steady state approximations in the treatment of enzyme inactivation in dilute aqueous solution by ionizing radiation is examined. It is concluded that the requirements for the use of this approximation are satisfied when the exposure time exceeds greatly the lifetime of water radicals, as is the case in most enzyme experiments.

When enzymes are irradiated in dilute aqueous solution, the inactivation is due almost exclusively to the action of free radicals formed from water. In most previous treatments of the kinetics of the enzyme inactivation it has been assumed that shortly after the start of the irradiation, the water radicals exist in a steady state [1—7]. Recently, Arley [8] has presented a treatment in which kinetic equations have been solved without the use of simplifying steady state assumptions. In this treatment the steady state approximation was stated to be invalid. Furthermore, it was claimed that with the aid of an assymptotic expression derived for very short exposures, the absolute rate constant for the inactivation of enzyme by water radicals may be deduced from the initial part of experimental dose inactivation curves [8,9].

In the present investigation of the kinetics of enzyme inactivation in solution special attention is devoted to the inactivation occurring after extremely short radiation exposures, as are met with in experiments involving pulsed radiation. Furthermore, the requirements for the use of steady state approximations in treatments of radiation inactivation of enzymes are examined. It is shown that, under the experimental conditions usually employed in enzyme inactivation studies, the steady state approximation can safely be used.

THEORY

In a previous paper [7] we have treated the kinetics of inactivation of enzymes in dilute aqueous solution by ionizing radiation with the aid of steady state approximation. This general treatment included the possibility that the active and inactivated enzyme molecules may react with water radicals at different rates. In the present paper we propose to derive exact equations. In order to simplify the mathematics it is assumed for this purpose that the radicals interact with the active and the inactivated enzyme molecules at the same rate. The situation when this assumption does not hold true will be discussed later. Also, it is assumed that no contaminating substance is present in the enzyme preparation and that the inactivation is predominantly due to one type of water radical.

The following symbols will be used:

E = the concentration of active enzyme,
E_0 = the initial concentration of active enzyme,
P_0 = the concentration of added protector,
X = concentration of water radicals,
k_0 = the rate constant for inactivation of enzyme by water radicals,
k_1 = the rate constant for the disappearance of water radicals in reactions with the enzyme,
k_2 = the rate constant for disappearance of water radicals in reactions with added solute, P_0,
I = the number of radicals formed per unit of time,
t = irradiation time (length of exposure),
t' = the time from the end of the exposure,
τ = mean life time of water radicals $\left(\tau = \dfrac{1}{k_1 E_0 + k_2 P_0}\right)$.

The rate of inactivation of the enzyme is given by equation

$$\frac{dE}{dt} = -k_0 EX. \tag{1}$$

The rate of change of concentration of water radicals is given by equation (2)[1].

$$\frac{dX}{dt} = I - k_1 E_0 X - k_2 P_0 X . \qquad (2)^1$$

Equation (1) is solved for X which is substituted in equation (2), which then can be written:

$$\frac{dX}{dt} = I + \frac{1}{k_0} (k_1 E_0 + k_2 P_0) \frac{d}{dt} (\ln E). \qquad (3)$$

Upon integration of (3) from $t = 0$ to $t = t$ equation (4) is obtained:

$$\ln \frac{E_0}{E_t} = \frac{k_0 (I t - X)}{k_1 E_0 + k_2 P_0} = k_0 \tau (I t - X). \qquad (4)$$

An expression for X can be obtained by integration of equation (2) from $t = 0$ to $t = t$.

$$X = \frac{I}{k_1 E_0 + k_2 P_0} (1 - e^{-(k_1 E_0 + k_2 P_0)t}) = I \tau (1 - e^{-t/\tau}). \quad (5)$$

By substituting (5) in equation (4), equation (6) is obtained:

$$\ln \frac{E_0}{E_t} = k_0 I \tau [t - \tau (1 - e^{-t/\tau})]. \qquad (6)$$

Equation (6) describes the inactivation during the exposure. However, it should be realized that at the end of the exposure time, t, the concentration of water radicals is X (see equations (4) and (5)]. In the course of microseconds these radicals will disappear in reactions with the solutes leading to a certain inactivation. This latter inactivation must be taken into account as it is not possible to distinguish experimentally between the inactivation occurring during exposure and the total inactivation.

The extent of inactivation after exposure can be obtained from equation (1). An expression is first derived for the disappearance of X as a function of the time after exposure, t'. Since X [see equation (5)] will disappear after the irradiation according to a first order process, we have:

$$X_{t'} = I \tau (1 - e^{-t/\tau}) e^{-t'/\tau}. \qquad (7)$$

The concomitant rate of enzyme inactivation is obtained by substituting for X in equation (1), to give equation (8):

$$\frac{dE}{dt'} = - k_0 E I \tau (1 - e^{-t/\tau}) e^{-t'/\tau}. \qquad (8)$$

The inactivation occurring after exposure is obtained by integrating equation (8) from $t' = 0$ to $t' = \infty$:

$$\ln \frac{Et}{E_{t'}} = k_0 I \tau^2 (1 - e^{-t/\tau}). \qquad (9)$$

---

[1] Recombination of radicals might be expected to give rise to a term of the form $- k X^2$. A second-order recombination should increase strongly with the dose rate. Since this has not been found experimentally [6,10], second-order recombination of radicals has been disregarded in this paper.

Here E_t is the enzyme activity at the end of the exposure and $E_{t'}$ is the activity remaining when all the water radicals have disappeared.

The total inactivation is obtained by adding the inactivation during and after exposure (equations (6) and (9)] to give:

$$\ln \frac{E_0}{E_{t'}} = k_0 I \tau t = \frac{k_0 I t}{k_1 E_0 + k_2 P_0} . \qquad (10)$$

It should be noted that this equation for the total inactivation is identical with the one previously derived on the basis of steady state approximations.

The ratio between the inactivation occurring during exposure and the total inactivation is given by

$$\frac{\ln \dfrac{E_0}{E_t}}{\ln \dfrac{E_0}{E_{t'}}} = 1 - \frac{\tau}{t} (1 - e^{-t/\tau}). \qquad (11)$$

Under the conditions met in most enzyme experiments, the lifetime of the water radicals, τ, is of the order of a few microseconds. From equation (11) it follows that if the exposure time, t, is large compared to τ, the expression approaches 1. This implies that for ordinary enzyme experiments the inactivation after exposure is negligible, compared to that occurring during exposure. On the other hand if the exposure time is much smaller than the life-time of the water radicals as is usually the case when pulsed radiation is used, equation (11) reduces to:

$$\frac{\ln \dfrac{E_0}{E_t}}{\ln \dfrac{E_0}{E_{t'}}} = \frac{t}{2\tau} . \qquad (12)$$

Equation (12) is obtained from equation (11) by expanding the exponential term in series and neglecting terms of order higher than 2. It is apparent that under these conditions the predominant part of the inactivation occurs after the exposure.

It has been stated by Arley [8,9] that the extent of inactivation after short exposures is given by the equation:

$$\ln \frac{E_0}{E_t} = \frac{1}{2} k_0 I t^2, \qquad (13)$$

and that with the aid of this equation the absolute value of k_0 can be obtained directly from studies of the very beginning of the dose inactivation curve. It should be realized that equation (13), which can readily be obtained by expanding equation (6) in series, describes only the inactivation during the exposure. Since the inactivation after exposure is more important under these conditions, equation (13) is inapplicable to experimental data.

DISCUSSION

In the present paper the kinetics of the enzyme inactivation in solution by ionizing radiation has been treated without the use of the steady state approximation used in most previous studies [1—7]. In such a treatment it is necessary to take into account the fact that at the end of the exposure a certain number of water radicals are present, which will disappear in the course of microseconds and thereby cause an additional enzyme inactivation. For the exposure times usually used in enzyme inactivation studies this inactivation occurring after the end of the exposure is insignificant. However, this is not the case when the exposure times are short compared to the life-time of the water radicals as when pulsed radiation is used. Under such conditions the inactivation after the exposure proves to be more important than that occurring during the pulse. In a recent treatment [8,9] this fact was overlooked. It is shown here that when the inactivation after exposure is included, the expression derived for the total inactivation becomes identical with that previously obtained on the basis of the steady state approximations. Thus, it turns out that the same equation is valid for short and for long exposures.

Since it has been claimed [8,9] that the steady state approximation used in previous treatments is invalid, it is of some interest to examine the formal requirements for its use. This approximation involves the assumption that shortly after the start of the irradiation the rate of change in the concentration of water radicals is so small compared to the rate of radical production that it can be neglected in the mathematical derivations. This implies that dX/dt is neglected in equation (2). If this is done, equation (10) is obtained directly. Comparison of equations (10) and (4) shows that the only difference is the term X in the parenthesis in equation (4). Since $X = I\tau$ when t is large compared to τ [see equation (5)], the term in parenthesis in equation (4) then reduces to $I(t - \tau)$. As stated above, the life-time of water radicals is usually of the order of a few microseconds under conditions used in enzyme experiments.

From this it follows that the steady state approximation can safely be used when the exposure time is greater than approximately 0.1 msec.

In the present derivation the assumption was made that the active and inactivated enzyme molecules react with the water radicals at the same rate. Under conditions where this is not the case, the concentration of water radicals will change during the inactivation of the enzyme as previously pointed out [7]. An equation for the inactivation under such conditions was derived on the basis of the steady state assumption [7]. It can readily be shown that even if there is an appreciable difference in the reaction rates of the water radicals with the active and inactivated enzyme molecules respectively, the rate of change of the concentration of water radicals will be sufficiently small compared to the rate of radical production [see equation (9), ref. 7] to permit the use of the steady state approximation.

This work was supported by the Division of Radiological Health, Bureau of State Service, U.S. Public Health Service. T. Sanner is a Fellow of the Norwegian Cancer Society.

REFERENCES

1. Weiss, J., *Nature*, 153 (1944) 748.
2. Dale, W. M., Gray, L. H., and Meredith, W. J., *Phil. Trans. Roy. Soc. London*, Ser. A, 242 (1949) 33.
3. Okada, S., *Arch. Biochem. Biophys.* 67 (1957) 102.
4. Augenstine, L., *Radiation Res.* 10 (1959) 89.
5. Hutchinson, F., and Ross, D. A., *Radiation Res.* 10 (1959) 477.
6. Butler, J. A. V., Robins, A. B., and Rotblat, J., *Proc. Roy. Soc. (London)*, Ser. A, 256 (1960) 1.
7. Sanner, T., and Pihl, A., *Radiation Res.* 19 (1963) 12.
8. Arley, N., *Radiation Res.* 29 (1966) 1.
9. Arley, N., *Naturwissenschaften*, 53 (1966) 276.
10. Hutchinson, F., *Radiation Res.* 9 (1958) 13.

T. Sanner and A. Pihl
Norsk Hydro's Institute for Cancer Research
Department of Biochemistry
The Norwegian Radium Hospital,
Montebello, Oslo 3, Norway

European J. Biochem. 1 (1967) 36—45

Zum Mechanismus der Ribonuclease-Reaktion

1. Die Aufgabe der Pyrimidinbase bei der Reaktion

H. G. Gassen[1] und H. Witzel

Chemisches Institut der Universität Marburg/Lahn

(Eingegangen am 28. September 1966)

The 2',3'-cyclic phosphates VII—XXXI were prepared for further studies of the function of the base in the ribonuclease-catalysed hydrolysis of ribonucleotide diesters. As seen with purine (XXIV, XXV, XXVIII), pyridine (XXIX), and pyridazine (XXXI) derivatives there is no absolute specificity for pyrimidine bases. Minimal requirement for monomeric substrates appears to be a mesomeric ring system with a keto-group in α-position to the β-glycosidic bond. Specific binding to the enzyme by position 3, 4, 5, or 6 of the base can be excluded. Arguments that the keto-group is not involved in the binding process are derived from kinetic experiments. Alterations at the base do not affect K_m but only k_{+2}.

A relation between structural alterations at the substrates and the measured kinetic parameters k_{+2} and $K_m = (k_{-1} + k_{+2})/k_{+1}$, required an analysis of the significance of these values in connection with definite reaction steps. This is accomplished on the basis of the mechanism I—VI [3] in which the reaction occurs by a simultaneous bond exchange at three centers (see XXXV—XXXIV) initiated by a base catalysis with the 2-oxygen of the base and a polarisation at the phosphate group by two proton donating groups of the enzyme [4]. Binding to the enzyme occurs only when the dianionic intermediate state II or V has been formed. This complex breaks down after a proton catalysis by the conjugate acid of the base, which can occur alternatively in two directions indicated by k_{-1} and k_{+2}.

Arguments are given that rate determining in k_{+1} is the transfer of the proton of the 2'-OH group to the 2-oxygen of the base in I → II or by analogy in IV → V. Rate determining in k_{-1} and k_{+2} is the proton transfer from the conjugate acid to the 2'-oxygen in II → I and to the 5'-oxygen or its equivalent in II → III and by analogy in V → IV and V → VI. Both constants are coupled as shown in [32] and change by the same factor, when the catalytic function of the base is altered. If in such a case K_m remains constant, k_{+1} must have changed by the same factor too. Thus K_m cannot be treated as $K_s = k_{-1}/k_{+1}$ with k_{+2} negligibly small compared with k_{-1}.

Further results [31] indicate that the activation energy for k_{+1} contains an enthalpy term which is related to the transition XXXVII → XXXVIII and an entropy term, related to the probability of the base being in the proper position to accept the proton. Similarly k_{-1} and k_{+2} contain an enthalpy term related to the transition XXXVIII → XXXVII and an entropy term related to the probability of the base being in the form of the conjugate acid and in the proper position to transfer the proton.

Our findings that the rates depend on the polarisability of the base with K_m remaining constant and not on the acidity or basicity of the catalysing base can be explained by the assumption that increased polarisability lowers the activation energies for the transitions XXXVII ⇄ XXXVIII in both directions thus changing k_{+1} and $k_{-1} + k_{+2}$ by the same factor. Increased acidity, however, would lower the activation energy for k_{+2} in the enthalpy term, but at the same time would increase it in the entropy term; it is less probable to have the less basic base in the form of the conjugate acid. The higher rates which we found when the conjugate acid can be formed by a tautomerisation of the catalysing base can be similarly explained. Further factors influencing the rates are of sterical nature. We found differences when the catalysing α-oxygen is part of a five-membered or of a six-membered ring system or when the catalysing oxygen cannot offer its free orbital to the 2'-OH-group as discussed in the case of α-cytidylic acid (XXIII).

Another factor, the pH-dependence of k_{+2}, will be discussed in a further paper [23] on the basis of an additional catalysis of the decay by an enzyme base with a pK around 6.5.

---

[1] Liebig-Stipendiat des Fonds der Chemischen Industrie.

Die Pankreasribonuclease spaltet Ribonucleinsäure an der 5'-Diesterbindung von 3'-Pyrimidinribonucleotiden. Es entstehen zunächst cyclische 2',3'-Pyrimidinribonucleotide, die in einem zweiten Reaktionsschritt zu den entsprechenden 3'-Nucleotiden hydrolysiert werden.

Die Spezifität für die Pyrimidinnucleotide ist nicht absolut. Es kann an niedermolekularen Modellverbindungen gezeigt werden, daß die Pyrimidinbase stark variiert werden kann und daß auch eine Spaltung bei Purinribonucleotiden erfolgt.

Aus Untersuchungen an verschiedenen Derivaten von Pyrimidinnucleotiden sowie aus einem Vergleich kinetischer Daten der nichtenzymatischen und der enzymatischen Reaktion hatten wir seinerzeit geschlossen, daß die Spezifität nicht auf einer spezifischen Bindung der Base an das Enzym beruhen kann, sondern daß die Base sich spezifisch an der Katalyse

und 1964 [3, 4] an Hand eines Mechanismus versucht, bei dem die Pyrimidinbase mit ihrem C-2-Sauerstoff zunächst durch eine Basenkatalyse an der 2'-OH-Gruppe den Aufbau eines enzymstabilisierten Zwischenzustandes ermöglicht und dann in Form der conjugaten Säure den Zerfall des Zwischenzustandes katalysiert (I—III). Der zweite Reaktionsschritt (IV—VI), die Hydrolyse des 2',3'-cyclischen Diesters, verläuft völlig analog.

Als Voraussetzung für eine solche Katalyse durch die Pyrimidinbase wurde ein Ketosauerstoff angesehen, α-ständig zur β-glycosidischen Bindung der Ribose. Er muß Teil eines mesomeren Systems sein.

Diesem Mechanismus stehen andere mechanistische Vorstellungen gegenüber, die die Spezifität in einer stereospezifischen Bindung der Pyrimidinbase suchen und die die Katalyse durch Säure- und Basen-Gruppen am Enzym durchführen lassen [5, 6].

der Reaktion beteiligen muß. Für das Enzym wurde gefordert, daß es nur an der Phosphatgruppe angreift und durch eine Protonierung deren Elektrophilie erhöht [1].

Diese Vorstellung von der Funktion der Pyrimidinbase wurde durch kinetische Messungen von Witzel u. Barnard [2] gestützt. Sie zeigten, daß bei einer Änderung der Pyrimidinbase sich nicht K_m ändert, sondern nur k_{+2} beeinflußt wird.

Die Ergebnisse solcher Untersuchungen lassen sich jedoch nur dann im Hinblick auf eine Spezifität in der Bindung oder in der Katalyse verwenden, wenn eine Interpretation der gemessenen Größen K_m und k_{+2} gegeben werden kann. Dies wurde von uns 1963

Nicht allgemein gebräuchliche Abkürzungen: Uridin-2',3'-phosphat, Up; Cytidin-2',3'-phosphat, Cp; Adenosin-2',3'-phosphat, Ap.
Enzym: Pankreasribonuclease (EC 2.7.7.16).

Wir geben im folgenden eine Reihe von Untersuchungen und Ergebnissen bekannt, die wir mit dem Ziel der Aufklärung der Funktion der Pyrimidinbase durchgeführt haben. Über einige dieser Ergebnisse wurde bereits berichtet [7].

EXPERIMENTELLER TEIL

DATEN DER HERGESTELLTEN UND GEMESSENEN VERBINDUNGEN

Es wurden die 2',3'-cyclischen Phosphate der in der Tabelle aufgeführten Verbindungen hergestellt: (K_m und k_{+2}-Werte in mM bzw. sec^{-1}; die spektralen Daten gelten für Dimethylglutarsäurepuffer, pH 7; R_F-Werte in Isopropanol-1N NH$_4$OH (70:30) als Fließmittel, R_E-Werte in Boratpufferelektrophorese [1] pH 9,25 mit Up = 100).

ANGABEN ZUR PRÄPARATION

4-Thiouridin-2', (3')-phosphat

500 mg Uridin-2'(3')-phosphat (1,5 mMol) werden in 50 ml Pyridin gelöst. In die Lösung werden 2,0 g (9 mMol) P_2S_5 und 0,1 ml H_2O eingetragen. Man kocht 4 Std unter heftigem Rühren am Rückfluß. Nach dem Erkalten wird die Pyridinphase dekantiert und der harzige Rückstand zweimal mit Pyridin ausgezogen. Die vereinigten Auszüge werden im Vakuum eingeengt, zweimal mit H_2O codestilliert und über eine Säule mit Dowex 50 (H^+) gegeben. Das gelbliche Eluat wird auf ein kleines Volumen eingeengt und auf dem Papierchromatogramm gereinigt. Das Band mit einem $R_F = 11$ enthielt eine Verbindung, die in allen angegebenen Daten mit 4-Thiouridin-2'(3')-phosphat, synthetisiert nach der Methode von Ikehara et al. [10], identisch war.

1 (β-D-Ribofuranosyl)-2-pyridon

2-Trimethylsilyl-pyridon [18]:   38 g   (0,4 Mol) 2-Pyridon werden in 50 ml siedendem Pyridin gelöst. Man läßt etwas abkühlen und tropft unter Rühren 43 g (0,4 Mol) $(CH_3)_3SiCl$ zu. Dann wird 30 min unter Rückfluß gekocht und nach dem Erkalten vom Pyridinhydrochlorid unter Feuchtigkeitsausschluß abfiltriert. Das Filtrat wird über eine Kolonne destilliert.
Kp^{12}: 70—73°.
Ausbeute: 42 g (63% der Theorie).

1(β-D-2',3',5'-Tri-O-benzoyl-ribofuranosyl)-2-pyridon: 5 g (30 mMol) 2-Trimethylsilyl-pyridon werden in 50 ml trockenem Toluol gelöst und 6 g AgClO$_4$ zugesetzt. In diese Lösung läßt man 16 g (30 mMol) 1-Brom-2,3,5-tribenzoylribose [19] in 50 ml Toluol gelöst, einfließen. Es fällt sofort AgBr aus. Man läßt 1 Std bei 25° rühren, gießt die Toluolschicht von dem öligen Rückstand ab und engt sie auf die Hälfte ein. Dann läßt man sie in die vierfache Menge Petroläther tropfen, zentrifugiert den Niederschlag ab und kristallisiert aus Methanol.
Fp.: 136—138°.
Ausbeute: 7,5 g (47% der Theorie).
Spektrale Daten und Fp. stimmen überein mit der von Ukita et al. [17] nach dem Hg-Salz-Verfahren hergestellten Verbindung.

1 (β-D-Ribofuranosyl)-2-pyridon: 5,3 g (10 mMol) Tribenzoyl-ribofuranosyl-2-pyridon werden in 50 ml CH_3OH/NH_3 (bei 0° gesättigt) gelöst und bleiben 48 Std bei 25° stehen. Dann wird die Lösung im Vakuum zur Trockne eingeengt. Der Rückstand wird in 20 ml Wasser aufgenommen, mit Ameisensäure auf pH 3,0 eingestellt und dreimal mit Äther ausgeschüttelt. Die wäßrige Phase wird im Vakuum auf ein kleines Volumen eingeengt und das Nucleosid

aus wenig Wasser kristallisiert und charakterisiert nach den Daten in [17].
Fp.: 146—148°; R_F: 78; R_E: 50.
Ausbeute: 1,87 g (65% der Theorie).

1 (β-D-Ribofuranosyl)-3-methyl-2-pyridon

3-Methyl-2-trimethylsilyl-pyridon. 17 g (0,16 Mol) 3-Methyl-2-pyridon [20] werden in der Siedehitze in trockenem Pyridin gelöst. Nach dem Abkühlen tropft man 17,5 g (0,16 Mol) $(CH_3)_3SiCl$ zu und verfährt wie bei 2-Trimethylsilyl-pyridon.
Kp^{40}: 105—110°.
Ausbeute: 15 g (50% der Theorie).

1 (β-D-2',3',5'-Tri-O-benzoyl-ribofuranosyl)-3-methyl-2-pyridon. 5,1 g (30 mMol) 3-Methyl-2-trimethylsilyl-pyridon werden in 50 ml Toluol zusammen mit 6 g AgClO$_4$ gelöst. Man verfährt weiter wie bei Tribenzoyl-ribofuranosyl-2-pyridon.
Die Verbindung konnte nicht kristallisiert werden, sondern wurde als Öl weiter eingesetzt.
Ausbeute: 11,8 g (72% der Theorie).

1 (β-D-Ribofuranosyl)-3-methyl-2-pyridon: 5,5 g (10 mMol) 1 (β-D-2',3',5'-Tribenzoyl-ribofuranosyl)-3-methyl-2-pyridon werden in 50 ml CH_3OH/NH_3 gelöst und weiter behandelt wie für 1 (β-D-Ribofuranosyl)-2-pyridon. Das Produkt wird aus wenig Wasser mit Äthanol-Äther ausgefällt und durch R_F und R_E-Werte sowie die spektralen Daten charakterisiert.
R_F: 90; R_E: 50.
Ausbeute: 1,2 g (40% der Theorie).

1 (β-D-Ribofuranosyl)-3,6-dioxy-pyridazin

3,6-(Bis-O-trimethylsilyl)-dioxy-pyridazin. 11,2 g (0,1 Mol) 3,6-Dihydroxy-pyridazin werden in 50 ml trockenem Pyridin suspendiert. Man erhitzt zum Sieden und tropft 21,6 g (0,2 Mol) $(CH_3)_3SiCl$ zu. Man verfährt weiter wie für 2-Trimethylsilyl-pyridon.
Kp^{12}: 115—116°; Fp.: 69—71°.
Ausbeute: 13,5 g (53% der Theorie).

1(β-D-2',3',5'-Tri-O-benzoyl-ribofuranosyl)-3,6-di-oxy-pyridazin. 4,3 g (17 mMol) 3,6-(Bis-O-trimethylsilyl)-dioxy-pyridazin und ca. 9 g (17 mMol) 1-Chlor-tribenzoyl-ribose werden in einer Druckbirne in 250 ml Xylol gelöst. Man leitet trockenen Reinststickstoff durch die Lösung und dampft bei einer Badtemperatur von 160° die Hälfte des Xylols ab. Die Druckbirne wird verschlossen und 45 min auf 180—190° erhitzt. Nach dem Erkalten wird das Xylol am Rotationsverdampfer abgezogen und der Rückstand in 90%igem Äthanol aufgenommen. Man läßt die Lösung 5 min sieden, dampft den Alkohol ab und nimmt den Rückstand in 250 ml $CHCl_3$ auf. Die Lösung wird filtriert, mit Na_2SO_4 getrocknet und eingeengt. Der sirupartige Rückstand wird dreimal mit Äther ausgezogen und aus Äthanol kristallisiert.

$C_{30}H_{24}N_2O_9$ (556). Berechnet. $C = 64,7$; $H = 4,32$; $N = 5,04$. Gefunden: $C = 64,35$; $H = 4,43$; $N = 4,80$.
Fp.: 178—180°.
Ausbeute: 4,5 g (51% der Theorie).

1 (β-D-Ribofuranosyl)-3,6-dioxy-pyridazin. 2,2 g (3,6 mMol) 1 (β-D-Tri-O-benzoyl-ribofuranosyl)-3,6-dioxy-pyridazin werden in 50 ml CH_3OH/NH_3 gelöst und weiter behandelt wie bei β-D-Ribofuranosyl-2-pyridon. $C_9H_{12}N_2O_6$ (244). Berechnet: $C = 44,2$; $H = 4,92$; $N = 11,48$. Gefunden: $C = 44,55$; $H = 5,20$; $N = 11,28$. Fp.: 184—185°.
R_F: 50; R_E: 91.
Ausbeute: 650 mg (74% der Theorie).

Andere Präparationen

Die Tritylierung der Nucleoside folgte der Vorschrift von Levene und Tipson [21].
Die Phosphorylierung der 5'-Tritylnucleoside wurde nach Tener [22] durchgeführt.
Die Cyclisierung der Nucleotide wurde nach Michelson [12] durchgeführt.

KINETISCHE MESSUNGEN

Die Geschwindigkeiten der Umsetzungen wurden bei verschiedenen Substratkonzentrationen entweder am Cary-Spektralphotometer oder, wenn keine spektrale Änderung mit der Hydrolyse der cyclischen Nucleotide einhergeht, nach einem pH-stat-Verfahren registriert (siehe [23]). Die K_m- und k_{+2}-Werte wurden nach dem Lineweaver-Burk-Verfahren gewonnen. Temperatur 20°.

Spektrophotometrische Messungen

In einer thermostatisierten Halbmikro-Küvette ($d = 1$ cm), wurde zu 1 ml einer Lösung von 0,1 M Dimethylglutarsäure-Puffer pH 7 + 0,1 M NaCl ($I = 0,2$) mit einer Mikrometerspritze 2—15 μl der Substratlösung zugesetzt, so daß die Konzentration im Bereich von 0,3—2 mM lag. Die Reaktion wurde gestartet durch Zugabe von 10—30 μl einer Enzymlösung (Light — RNase, chromat. rein), deren Konzentration durch Messung der Extinktion bei 277 mμ bestimmt worden war. (Molekulare Extinktion der RNase = 9800.) Die Konzentration war so gewählt, daß die Halbwertszeiten im Bereich von 3—5 min lagen. Die Messungen wurden gewöhnlich bei einer Wellenlänge durchgeführt, die um 20 mμ höher lag als das Maximum der betreffenden Verbindung.

pH-stat-Messungen

In ein thermostatisiertes Titrationsgefäß wurden 1,5 ml 0,2 Mol NaCl-Lösung gegeben, dann mit einer Mikrometerspitze 3—25 μl der Substratlösung zugesetzt, deren Konzentration vorher spektrophoto-

metrisch bestimmt worden war. Unter Stickstoffspülung und Magnetrührung wurde mit 0,02 N NaOH ein pH von 7,0 eingestellt und auf Konstanz geprüft. Nach Zugabe von 10—30 μl einer Enzymlösung geeigneter Konzentration (siehe spektrophotometrische Messungen) wurde der Verbrauch an NaOH mit Hilfe eines Radiometer-Titrigraphen registriert.

DISKUSSION

Ausschluß einer spezifischen Bindung der Base an das Enzym

Wenn man argumentiert, daß die Base an das Enzym spezifisch gebunden wird [5] und dies die Voraussetzung für eine Reaktion ist, dann sollte sich das Bindungsgleichgewicht in $K_m = K_s = k_{-1}/k_{+1}$ niederschlagen. Eine Bindung an der Ribose oder der cyclischen Phosphatgruppe müßte unter diesen Umständen ausscheiden, da cyclische Ap oder N-Methyl-Up (VIII) mit den gleichen Gruppierungen am Zucker weder Substrate sind noch die Reaktion des Enzyms mit Substraten kompetitiv hemmen. Deshalb sollten Änderungen an den möglichen Bindungsstellen der Pyrimidinbase veränderte K_m-Werte liefern und k_{+2}, wenn es als vernachlässigbar klein gegenüber k_{-1} betrachtet wird [6], dürfte nicht beeinflußt werden.

Beim Vergleich von cyclischer 2',3'-Cp (XIX) mit der 4-N-Acetyl-Cp (XX) und von cyclischer 2',3'-Up (VII) mit 5,6-Dihydro-Up (IX) wurde jedoch schon festgestellt [2], daß sich nur k_{+2} ändert, nicht aber K_m. Es könnte noch argumentiert werden, daß XIX und XX gleiche Strukturmerkmale besitzen, nämlich Protonenacceptor an N-3 und Donor am 4-Substituenten und deshalb gleiches K_m besitzen könnten. Auch VII und IX sind untereinander gleich strukturiert, jedoch steht bei ihnen im Gegensatz zu den beiden ersten Verbindungen das Proton am N-3, und der 4-Sauerstoff ist Acceptor geworden. Wenn das Proton für die Bindung wesentlich wäre, dann müßte nun bei der 4-Dimethylcytidylsäure (XXII) und bei der 4-Methyl-thiouridylsäure (XIV), die beide kein Proton mehr besitzen, eine Änderung von K_m erwartet werden. Sollte nur das Elektronenpaar an N-3 zur Bindung herangezogen werden, dann dürfte keine Bindung beim Pyridon-(XXIX) oder Pyridazin-nucleotid (XXXI) erwartet werden, die beide weder das N-3 noch den C-4-Substituenten besitzen.

Die Blockierung des N-3 durch Methylierung (VIII, XVII und XVIII) verhindert zwar Bindung und Substrateigenschaft; daraus einen Rückschluß auf eine Beteiligung des N-3 an der Bindung zu ziehen ist jedoch nicht erlaubt. Einerseits ist die 2'- oder 3'-N-Methyl-uridylsäure ein kompetitiver Hemmstoff wie die 2'- oder 3'-Uridylsäure, andererseits verliert das 2-Pyridon (XXIX), das mit Sicherheit keine

Tabelle

Formel	Nr.	Substit.	Verbindung	λ_{max}	λ_{min}	R_E	R_F	K_m	k_{+2}
	VII	R=H	Uridin-2',3'-phosphat	259	232	80	30	3,6	1,4
	VIII	R=CH$_3$	3-Methyl-uridin-2',3'-phosphat[a]	259	232	60	42	—[t]	—
	IX	R=H	5,6-Dihydrouridin-2',3'-phosphat[b]	—	—	62	34	5[c]	0,5
	X	R=Cl	5-Chloruridin-2',3'-phosphat[d]	274	241	99	28	2,4	1,7
	XI	R=Br	5-Bromuridin-2',3'-phosphat[d]	276	242	99	27	1,6	1,3
	XII	R=J	5-Joduridin-2',3'-phosphat[d]	284	246	94	32	1,4	1,2
	XIII	R=SH	4-Thiouridin-2',3'-phosphat[e]	328 256	280 231	110	30	1,9	5,2
	XIV	R=SCH$_3$	4-Methylthiouridin-2',3'-phosphat[f]	301	246	55	36	3,3	1,9
	XV	R=H R'=H	Pseudouridin-2',3'-phosphat[g]	260	231	74	27	3,6	0,28
	XVI	R=H R'=CH$_3$	1-Methylpseudouridin-2',3'-phosphat[h]	266	236	74	32	1,8	1,6
	XVII	R=CH$_3$ R'=H	3-Methylpseudouridin-2',3'-phosphat[h]	262	232	77	35	—	—
	XVIII	R=CH$_3$ R'=CH$_3$	1,3-Dimethylpseudouridin-2',3'-phosphat[h]	267	236	74	42	—[t]	—
	XIX	R=H R'=H	Cytidin-2',3'-phosphat	268	242	52	35	3,3	5,5[u]
	XX	R=H R'=COCH$_3$	4-N-Acetylcytidin-2',3'-phosphat[i]	295 246	271 225			5,5	0,5[u]
	XXI	R=H R'=CH$_3$	4-N-Methylcytidin-2',3'-phosphat[k]	270	246	52	41	5,6	1,9
	XXII	R=CH$_3$ R'=CH$_3$	4-N-Dimethylcytidin-2',3'-phosphat[l]	275	240	51	47	2,5	0,2
	XXIII	R=H R'=H	α-Cytidin-2',3'-phosphat[m]	270	249	52	35	—[t]	—
	XXIV	Z=Rib-P Z'=H	N-3-Harnsäureribosid-2',3'-phosphat[n]	293	257	80	15	1,2[c]	0,5
	XXV	Z=H Z'=Rib-P	N-9-Harnsäureribosid-2',3'-phosphat[o]	292 240	268	83	15	1,0[c]	0,1

Tabelle (Fortsetzung)

Formel	Nr.	Substit.	Verbindung	λ_{max}	λ_{min}	R_E	R_F	K_m	k_{+2}
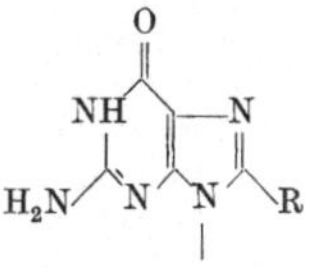	XXVI	R=Br	8-Bromguanosin-2',3'-phosphat[p]	260	239	75	15	—	—
	XXVII	R=SCH₃	8-Methylthioguanosin-2',3'-phosphat[p]	272	234	75	17	—	—
	XXVIII	R=OH	8-Oxyguanosin-2',3'-phosphat[p]	292 / 246	268 / 222	77	15	0,8[c]	1,3
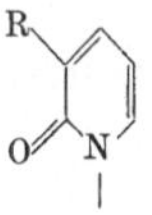	XXIX	R=H	1 (β-D-Ribofuranosyl)-2-pyridon-2',3'-phosphat[q]	300 / 225	242	62	40	1,5[c]	0,2
	XXX	R=CH₃	1 (β-D-Ribofuranosyl)-3-methyl-2-pyridon-2',3'-phosphat[r]	301 / 232	260	60	48	—	—
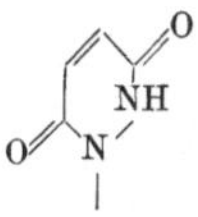	XXXI		1 (β-D-Ribofuranosyl)-3,6-dioxypyridazin-2',3'-phosphat[s]	332	261	92	29	27,8	0,2

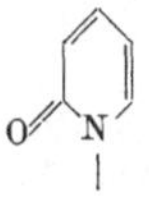

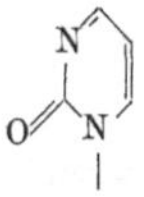

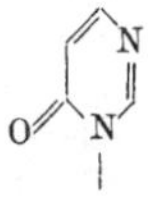

XXXII XXXIII XXXIV

Die β-Cyanoäthylester der 3'-Nucleotide von XXXII—XXXIV wurden von Ukita *et al.* [17] mit denen der Uridylsäure verglichen. Während XXXIII noch 75% der Geschwindigkeit des Uridylsäureesters erreicht, zeigten XXXII und XXXIV nur noch 15%.

[a] Methylierung mit Diazomethan in wäßriger Lösung, Hydrolyse zum Monoester mit 0,1 N HCl, 45 min, 50° nach [8].
[b] Hydrierung von 2',3'-Cp nach [2].
[c] pH-stat-Messung.
[d] Analog zur Vorschrift von Michelson [9] zur Darstellung der 5'-Phosphate.
[e] Synthese unter B, 1 beschrieben.
[f] Methylierung von XIII mit Methyljodid in 0,1 N NaOH, 30 min, 25°, neutralisiert mit Dowex 50 (H⁺) nach [10].
[g] Isoliert aus einem Hefe-RNS-Hydrolysat in 0,5 N NaOH, 72 Std, 30°, getrennt an Dowex 1×2 (Formiat).
[h] Methylierung mit Diazomethan, Hydrolyse zum Monoester mit 0,1 N HCl, 45 min, 60°, Trennung der 4 Verbindungen auf Papierchromatogramm, analog [11].

[i] Analog der Vorschrift von Michelson [12].
[k] Aus XIV, 10 min bei 5° in wäßrigem Methylamin stehen gelassen, zum Trocknen eingeengt und chromatographisch getrennt.
[l] Analog XXI mit Dimethylamin.
[m] Isoliert aus Hefe-RNS nach Gassen und Witzel [13].
[n] Das Nucleosid wurde aus Rinderblut isoliert nach [14].
[o] Nach Holmes und Robins [15].
[p] Nach Holmes und Robins [16].
[q] Synthese im exp. Teil beschrieben.
[r] Synthese im exp. Teil beschrieben.
[s] Synthese im exp. Teil beschrieben.
[t] Keine kompetitive Hemmung.
[u] Werte aus [2].

Bindung zum C-3 erwarten läßt, genau so seine Reaktivität und Bindungsfähigkeit, wenn das Proton in 3-Stellung durch eine Methylgruppe substituiert wird (XXX). Die Methylgruppe in 3-Stellung muß also auf andere Weise Einfluß nehmen.

Der C-2-Sauerstoff als Bindungsstelle zum Enzym sollte auch bei XXIII zu einer Bindung führen. Aber auch VIII oder XXX oder XVII und XVIII müßten kompetitive Hemmstoffe sein, obzwar noch gesagt werden könnte, die benachbarte CH₃-Gruppe blokkiere die Bindung am C-2-Sauerstoff. π-Wechselwirkungen sollten sich bei IX im Vergleich zu VII bemerkbar machen. Hydrophobe Bindungen sollten bei XXVI und XXVII noch besser sein als bei XXVIII. Die Purinnucleotide XXIV, XXV und

XXVIII zeigen, daß Vorstellungen von Pyrimidin-Matrizen am Enzym als Grundlage für eine spezifische Bindung unhaltbar sind.

Damit glauben wir jede Möglichkeit für eine spezifische Bindung der Base zum Enzym ausgeschlossen zu haben, wie bereits 1960 [1] und 1963 [3] aus nur wenigen Versuchen abgeleitet worden war.

Es wurde bereits darauf hingewiesen [3], daß auch bei der Bindung der 2'- und 3'-Mononucleotide als kompetitive Inhibitoren die Basen keine Rolle spielen können. Während die 2'-Cytidylsäure als dianionischer Monoester ein sehr starker Inhibitor ist, hemmen den entsprechenden Methylester oder das 2'-5'-CpA als monoanionische Diester die Reaktion nicht. Das gleiche gilt für 2'- und 3'-α-Cytidylsäure, die beide wie die β-Cytidylsäure hemmen; beim $2',3'$-cyclischen Diester wird jedoch keine Hemmung mehr beobachtet. Es mußte deshalb angenommen werden, daß nur die Phosphatgruppe, und diese nur in einem dianionischen Zustand, an das Enzym gebunden wird und die Base praktisch nichts zur Bindung beisteuern kann. Zu diesem Ergebnis führten auch die Untersuchungen von Hummel und Witzel [24] über die Protonenaufnahme und -abgabe bei der Bindung dieser Inhibitoren an das Enzym.

Es bleibt offen, ob noch eine schwache unspezifische Wechselwirkung zwischen Enzym und Base bestehen kann, die für die spektralen Änderungen bei der Bindung der Inhibitoren an das Enzym [25,26] verantwortlich wäre. Diese Wechselwirkung, vielleicht nur in Form einer Einschränkung der freien Rotation, müßte dann aber völlig unspezifisch sein, da sie bei der Adenylsäure oder beim Pyridoxalphosphat genau so auftritt wie bei den Verbindungen mit der α-Ketogruppe.

Die Aufgabe der Pyrimidinbase bei der Katalyse

Als Voraussetzung für eine Reaktion bei pH 7 gilt generell, daß entweder die Nucleophilie der am Phosphor angreifenden $2'$-OH-Gruppe (im 2. Schritt die des H_2O-Moleküls) erhöht werden muß oder aber die Elektrophilie am Phosphor oder auch beides. Eine direkte Basenkatalyse durch das Enzym an der $2'$-OH-Gruppe ist mit den früher gegebenen Argumenten auszuschließen [3]. Aus den oben diskutierten Gründen kann jetzt auch die bei der nichtenzymatischen Spaltung beobachtete Katalyse durch die Pyrimidinbase [1] nicht durch das Enzym zusätzlich verstärkt werden. So bleibt für das Enzym nur die Bindung und Aktivierung an der Phosphatgruppe mit der Konsequenz, daß ein Additionsmechanismus durchlaufen werden muß mit einem 5-bindigen dianionischen Phosphor als Zwischenzustand, wie er in I—III und IV—VI dargestellt ist.

Dieser vom Enzym stabilisierte Zwischenzustand kann, wie an anderer Stelle [3] ausführlich diskutiert worden ist, nur aufgebaut werden unter Mitwirkung der Base, die in α-Stellung zur Glycosidbindung eine Ketogruppe besitzen muß. Es sind unter den Dinucleosidphosphaten oder den cyclischen $2',3'$-Nucleotiden keine Substrate bekannt, denen diese Gruppe fehlen darf. Der Sauerstoff kann höchstens durch Schwefel ersetzt werden [27]. Auch in der Purinreihe erhält man Substrate, wenn für eine α-ständige Ketogruppe gesorgt ist wie in den Verbindungen XXIV, XXV und XXVIII, während Guanosin, 8-Bromguanosin oder 8-Thiomethylguanosin keine Substrate abgeben.

Da substratanaloge Phosphorsäurediester ohne diese Ketogruppe an der Base, z. B. $3'$-$5'$ ApC oder $2',3'$-cyclische Ap weder Substrate noch Inhibitoren sind, muß geschlossen werden, daß erst eine Basen-

katalyse an der $2'$-OH-Gruppe notwendig ist, um überhaupt die Voraussetzung zu einer Bindung an das Enzym zu schaffen, nämlich die Ausbildung der 5-bindigen dianionischen Phosphatgruppe. Die Geschwindigkeitskonstante k_{+1} kann demnach nicht die für ein vorgelagertes Bindungsgleichgewicht sein, sondern muß für die Ausbildung des Zwischenzustandes gelten und sowohl den Übergang des Protons der $2'$-OH-Gruppe auf die Base, den Angriff des Sauerstoffs am Phosphor wie auch den Übergang $P{=}O \rightarrow P{-}O^-$ unter Mitwirkung eines Protons vom Enzym enthalten. Die mechanistische Seite der Bindung einschließlich ihrer kinetischen Konsequenzen wird bei Hummel und Witzel [24] sowie bei Witzel und Barnard [23] diskutiert.

Damit läßt sich jetzt die Reaktion abstrahieren auf das Schema

$$\underset{\text{XXXV}}{\overset{\alpha \quad\quad \beta \quad\quad \gamma}{\left.B\right|\ \underset{\vphantom{|}}{H{-}O}\ \ \underset{\vphantom{|}}{P{=}O}\cdot\cdot\underset{\vphantom{|}}{H{-}E}}} \underset{k_{-1}}{\overset{k_{+1}}{\rightleftarrows}} \underset{\text{XXXVI}}{B{-}H\ \ O{-}P{-}O{-}H\cdot\cdot E}$$

mit 3 Bindungswechseln bei α, β und γ, die offensichtlich beim Erreichen der Aktivierungsenergie für den langsamsten Schritt zu gleicher Zeit erfolgen, ohne Gleichgewichte mit Zwischenprodukten. Die Herabsetzung der Aktivierungsenergien für alle Bindungswechsel geht sowohl von B wie auch von HE aus.

Geschwindigkeitsbestimmend scheint die Häufigkeit zu sein, mit der die Base bereit ist, das Proton der den Phosphor berührenden $2'$-OH-Gruppe zu übernehmen. Die Unterstützung von HE her, die bei Erfolg gleichzeitig zur Stabilisierung des sich ausbildenden stark basischen Dianions führt, scheint sehr rasch zu erfolgen. Nach den Relaxationszeit-Messungen von Hammes [28] verläuft die Bindung eines Dianions — wenigstens bei $3'$-Cp — nahezu diffusionskontrolliert. Eine weitere Rechtfertigung unserer Annahme ist die Abhängigkeit der Geschwindigkeit von der Struktur der Base und vor allem von dem Grad der Vorordnung zwischen Base, $2'$-OH-Gruppe und Phosphor, wodurch bei den Dinucleosidphosphaten die hohen Geschwindigkeiten hervorgerufen werden. Diese Zusammenhänge werden in einer folgenden Arbeit dargestellt werden [31].

Der Zerfall des Zwischenzustandes II erfordert, wie ebenfalls schon ausreichend begründet [3], eine Protonenkatalyse, die, den Produkten entsprechend, nur am $2'$- oder am $5'$-Sauerstoff ansetzen kann. Die Reversibilität aller Reaktionsschritte erfordert weiterhin, daß das auf den $2'$-Sauerstoff übertragene Proton analog herangeführt werden muß, wie es bei der Basenkatalyse weggenommen wurde, so daß offensichtlich diese Katalyse von der zur Base konjugaten Säure durchgeführt wird und deshalb ebenfalls Aufgabe der Pyrimidinbase sein muß.

Die Geschwindigkeit (XXXVI → XXXV) würde dann durch die Häufigkeit bestimmt werden, mit der die Base in der Form der konjugaten Säure das Proton auf den Sauerstoff überträgt. Dies gilt sowohl für k_{-1}, wenn die Übertragung auf den 2'-Sauerstoff erfolgt, als auch für k_{+2}, das aus der alternativen Reaktion des Protons mit dem 5'-Estersauerstoff resultiert. Beide Konstanten stehen in einem bestimmten Verhältnis zueinander und sind nach den experimentellen Ergebnissen von Wieker und Witzel [32] gekoppelt; sie ändern sich bei Änderungen an der Pyrimidinbase stets um den gleichen Faktor. $K_m = (k_{-1} + k_{+2})/k_{+1}$ wird dadurch nur zum Verhältnis der Geschwindigkeitskonstanten für den protonenkatalysierten Zerfall zu der für den basenkatalysierten Aufbau.

Die pH-Abhängigkeit für $k_{-1} + k_{+2}$, die für alle Substrate gleich ist, kommt nach Witzel und Barnard [23] durch eine Erhöhung der Nucleophilie an den beiden Sauerstoffatomen zustande. Sie ist verursacht durch eine Base am Enzym, die das stabilisierende Proton im Komplex in Anspruch nehmen kann, nicht aber durch die Nucleosidbase.

Da sich k_{-1} und k_{+2} bei einer Änderung an der Pyrimidinbase jeweils um den gleichen Faktor ändern, kann jetzt aus dem Verhalten von k_{+2} und K_m auf das Verhalten von k_{+1} geschlossen werden und somit der Einfluß von Strukturänderungen auf die Basen- und die Protonenkatalyse erkannt werden. Dadurch läßt sich jetzt die Funktion der Pyrimidinbase an Hand unserer weiteren Ergebnisse spezifizieren:

Die Basenkatalyse (k_{+1}) kann offensichtlich nicht stattfinden, wenn Base und 2'-OH-Gruppe cis-ständig sind, wie bei der cyclischen α-Cytidylsäure und α-Pseudouridylsäure [13] oder der β-Lyxouridylsäure [29]; obwohl alle notwendigen Voraussetzungen erfüllt wären, sind diese Verbindungen weder Substrate noch Inhibitoren. An Modellen läßt sich erkennen, daß bei diesen Verbindungen das Proton der 2'-OH-Gruppe eine wesentlich ungünstigere Lage zu dem freien Orbital des C-2-Sauerstoffs einnimmt. Eine höhere Aktivierungsenergie für die Übernahme des Protons muß deshalb erwartet werden. Offensichtlich muß die Base dem Proton direkt ein freies Orbital anbieten können. (Dies heißt jedoch nicht, daß eine permanente Wasserstoffbrücke zwischen Base und 2'-OH-Gruppe angenommen werden muß.) Eine zweite Interpretation wäre theoretisch möglich, wenn $k_{-1} + k_{+2}$ in die Größenordnung von k_{+1} gerückt wäre und das Verhältnis k_{-1}/k_{+2} weit zu Gunsten von k_{-1} verschoben wäre. Es müßte dann aber zusätzlich erklärt werden, warum sich diese Verhältnisse plötzlich geändert haben sollen. Wir stellen diese theoretische Möglichkeit deshalb zurück.

Der Abstand des Sauerstoffs von der 2'-OH-Gruppe ist offensichtlich kritisch. Beim N-3-Harnsäureribosid (XXIV) steht der katalysierende Sauer-

stoff an einem 6 Ring, beim N 0 Harnsäureoribosid (XXV) an einem 5-Ring. k_{+2} ist bei XXIV um einen Faktor 5 größer gegenüber XXV. Da K_m unverändert bleibt, muß auch k_{+1} um einen Faktor 5 größer sein. Dieser Faktor sollte im wesentlichen durch den unterschiedlichen Abstand des Sauerstoffs von der 2'-OH-Gruppe bedingt sein. Der nur geringe Geschwindigkeitsunterschied bei VII und XXXIII zeigt, daß der C-6-Substituent, der im 5-Ring fehlt, keine große Rolle spielen kann.

Bei der Verbindung XXV ist der α-Ketosauerstoff isoster zum N-3. Es wurde seinerzeit diskutiert, daß bei den cyclischen Purinnucleotiden die Basenkatalyse wegen des ungünstigeren Abstandes des Stickstoffs von der 2'-OH-Gruppe nicht zum Zuge kommt [1] und dadurch die Spezifität des Enzyms für die Pyrimidinnucleotide zu erklären sei. Wir sehen jetzt, daß dies nicht einen Faktor von etwa 5 übersteigt und daß wir als Ursache für die verlorene Reaktivität bei den Purinen eine wesentlich geringere Nucleophilie am N-3 annehmen müssen. Unter besonderen Bedingungen (Verlängerung der Verweilzeit im aktiven Zentrum durch zusätzliche Bindung) scheint sie offensichtlich dennoch ausreichend zu sein, wie die von Beers [30] gefundene Spaltung von Adenylsäure-Polynucleotiden beweist.

Strukturelle Änderungen an der Base, die mit einer Veränderung der meßbaren pK-Werte einhergehen, sollten bei einer kombinierten Katalyse durch die Base und die konjugate Säure den K_m-Wert ändern. Erhöhte Acidität sollte höhere $k_{-1} + k_{+2}$-Werte und verminderte k_{+1}-Werte verursachen, wodurch K_m entsprechend ansteigen müßte und umgekehrt. Witzel u. Barnard [2] hatten jedoch an den Verbindungen VII und IX sowie XIX und XX gezeigt, daß keine direkte Abhängigkeit existiert, sondern eher eine Abhängigkeit von der Polarisierbarkeit besteht. Die zusätzlichen Daten aus der Tabelle bringen kein verändertes Bild: In der Reihe XX, XXII, XXI, XIX mit ansteigenden pK-Werten verhält sich k_{+2} wie 2,5:1:10:25 und K_m wie 2:1:2:1,3. In der Reihe IX, VII, XV, XI, XIII mit fallenden pK-Werten ist das Verhältnis für k_{+2} 1,25:4:1:4:15 und für K_m 2:2:2:1:1.

Die Acidität oder Basizität im mesomeren System ist offensichtlich nicht der allein entscheidende Faktor.

Der Einfluß der „Polarisierbarkeit" geht aus der Reihe IX, VII, XIII hervor. Für k_{+2} wurde ein Verhältnis von 1,25:4:15 gefunden, für K_m 2:2:1. Auch bei dem vergleichbaren System XXII und XIV findet man für k_{+2} ein Verhältnis 1:7 und für K_m 1:1,3.

Es fällt auf, daß die Verbindungen, die durch Abgabe eines Protons an das Solvens leicht in eine tautomere Form übergehen können, höhere k_{+2}-Werte besitzen als die entsprechenden methylierten Verbindungen. In der Reihe XIX (mit 2 Protonen),

XXI (mit 1 Proton) und XXII (ohne Proton) steht k_{+2} im Verhältnis 25:10:1. Für XIII und XIV findet man ein Verhältnis für k_{+2} von 2,5:1, bei K_m 1:1,7. Bei niedriger Substratkonzentration ist XIII also viermal schneller als XIV.

Es scheint uns gerechtfertigt, aus diesen Beziehungen den Schluß zu ziehen, daß für die Übernahme eines Protons durch die Base neben dem Faktor „Häufigkeit einer richtigen Stellung", der bei den erwähnten Verbindungen etwa gleich sein sollte, ein zweiter Faktor hinzukommt, der von der Aktivierungsenergie für die Übergänge XXXVII $\rightleftarrows$ XXXVIII bestimmt wird. Begünstigt werden diese Übergänge offensichtlich, wenn das mesomere System leichter polarisierbar ist oder wenn eine Art von Tautomerisierung durch Abgabe eines Protons an das Solvens und umgekehrt möglich ist. In beiden Fällen gilt eine Herabsetzung der Aktivierungsenergie sowohl für die Ausbildung des Zwischenzustandes als auch für dessen Zerfall. Dadurch wird

XXXVII   XXXVIII

sowohl $k_{-1} + k_{+2}$ als auch k_{+1} um den gleichen Faktor geändert, während K_m praktisch konstant bleibt. Unter diesem Gesichtspunkt sind noch einige Zusammenhänge zu erkennen: Bei der Pseudouridylsäure (XV), die fünfmal langsamer als Uridylsäure (VII) ist, wird nach Methylierung an N-1 (XVI) $k_{-1} + k_{+2}$ fast sechsmal größer, während k_{+1} etwa zwölfmal größer geworden ist. Die Geschwindigkeit dieser Verbindung liegt bei den Geschwindigkeiten der 5-halogenierten Verbindungen (X—XII), bei denen k_{+1} ebenfalls größer geworden ist im Vergleich zur unsubstituierten Uridylsäure (VII). Da N-1 und C-5 bei ihnen isoster sind, scheinen diese Substituenten in p-Stellung zum 2-Sauerstoff sich auf die Basenkatalyse mehr auszuwirken als auf die Säurekatalyse.

Eine Verschiebung zu Gunsten von k_{+1} findet man auch bei den Purinnucleotiden, die alle K_m-Werte um 1 haben. Vergleicht man VII und XXIV, das als 5,6-cyclosubstituiertes Uridinnucleotid zu betrachten ist, so fällt $k_{-1} + k_{+2}$ um einen Faktor von 3, während k_{+1} unverändert geblieben sein muß. Allerdings liegt XXIV mit einem pK-Wert von 6 [14] bei pH 7 schon als Anion vor.

Bei den beiden N-9-Nucleotiden XXV und XXVIII, die ebenfalls schon Anionen sind, verhält sich k_{+2} wie 1:13 bei gleichem K_m. Hier werden also wieder $k_{-1} + k_{+2}$ und k_{+1} um den gleichen Faktor verändert, wenn die 2-Oxygruppe gegen eine 2-Aminogruppe ausgetauscht wird. 8-Hydroxyguanylsäure

wird bei niederen Substratkonzentrationen so schnell gespalten wie die Cytidylsäure, wobei ihr gegenüber aber k_{+1} etwa viermal höher ist, $k_{-1} + k_{+2}$ jedoch viermal niedriger.

Ein völlig anderes Verhältnis von $k_{-1} + k_{+2}$ zu k_{+1} findet man bei dem Pyridazinnucleotid (XXXI), das mit einem pK von 5,5 ebenfalls schon ein Anion ist. Während $k_{-1} + k_{+2}$ nur etwa siebenmal kleiner ist als bei der Uridylsäure, muß k_{+1} viermal kleiner sein. Die Verbindung war so ausgewählt worden, daß eine Tautomerisierung nicht mehr möglich ist. Die Gründe für die Verschiebung mit dem hohen K_m-Wert sind zur Zeit noch nicht durchschaubar; sie könnten, da es sich hier um ein ganz anderes System handelt, in einem anderen Gleichgewichtsverhältnis von Base zu konjugater Säure am C-6-Sauerstoff zu suchen sein.

Ebenfalls ungeklärt ist noch die Rolle, die eine Methylgruppe in der 3-Stellung spielt, ganz gleich, ob an einem Kohlenstoff XXX oder an einem Stickstoff VIII, XVIII und XVII, das entgegen früheren Angaben [3] kein Substrat ist. Die Annahme, daß durch die N-Methylierung die Resonanz im Uracilsystem eingeschränkt wird, ist bei XXX nicht ausreichend. Die Untersuchungen zu diesem Punkt sind noch nicht abgeschlossen.

Wir glauben damit ausreichend gezeigt zu haben, daß bei dieser Enzymreaktion die Spezifität für bestimmte Substrate auf einer spezifischen Katalyse beruht und nicht in einer spezifischen Bindung. K_m ist entgegen der üblichen Interpretation keine Konstante für ein vorgelagertes Bindungsgleichgewicht. k_{+1} ist hiernach die Geschwindigkeitskonstante für eine Ein-Schritt-Reaktion an drei vorgeordneten Zentren und führt direkt zum Aufbau eines enzymstabilisierten Zwischenzustandes (ES). Geschwindigkeitsbestimmend ist offensichtlich der Übergang des Protons von der 2′-OH-Gruppe auf die Base. Der Zerfall des enzymstabilisierten Zwischenzustandes wird durch eine Protonenkatalyse verursacht. Auch hier ist offensichtlich der Übergang des Protons von der konjugaten Säure auf einen Estersauerstoff geschwindigkeitsbestimmend. Die Übertragung kann auf zwei verschiedene Sauerstoffe erfolgen und damit alternativ zur Rückreaktion (k_{-1}) oder zur Produktbildung (k_{+2}) führen. Die Geschwindigkeiten hängen in einem solchen Fall von alternierender Katalyse durch Base und konjugate Säure weniger von der Basizität oder Acidität ab, als vielmehr von den Aktivierungsenergien für die Übergänge der Protonen. Mit erhöhter Polarisierbarkeit des katalysierenden Systems werden die Aktivierungsenergien in beiden Richtungen herabgesetzt.

Der für unsere Interpretation noch notwendige Nachweis, daß k_{-1} und k_{+2} gekoppelte Geschwindigkeitskonstanten sind, wird an anderer Stelle geführt; ebenso werden die hohen Geschwindigkeiten für die Dinucleosidphosphate [2] sowie die ungewöhnliche pH-Abhängigkeit von $k_1 + k_{+2}$ gesondert diskutiert werden [23].

Der Deutschen Forschungsgemeinschaft und dem Fonds der Chemischen Industrie wird für die Unterstützung gedankt.

LITERATUR

1. Witzel, H., *Ann. Chem.* 635 (1960) 182, 191.
2. Witzel, H., und Barnard, E. A., *Biochem. Biophys. Res. Commun.* 7 (1962) 289, 295.
3. Witzel, H., *Progress in Nucleic Acid Research and Molecular Biology* (Edited by T. N. Davidson and W. E. Cohn), Academic Press, New York 1963, Vol. II, S. 221.
4. Witzel, H., in *Mechanismen enzymatischer Reaktionen.* Springer-Verlag, Berlin, Göttingen, Heidelberg 1964, S. 123.
5. Findlay, D., Herries, D. G., Mathias, A. P., Rabin, B. R., und Ross, C. A., *Biochem. J.* 85 (1962) 152.
6. Rabin, B. R., und Mathias, A. P., in *Mechanismen enzymatischer Reaktionen.* Springer-Verlag, Berlin, Göttingen, Heidelberg 1964, S. 97.
7. Gassen, H. G., und Witzel, H., *Abstracts 2nd FEBS Meeting (Vienna),* 1965, S. 38.
8. Szer, W., und Shugar, D., *Acta Biochimica Polon.* 9 (1962) 131.
9. Michelson, A. M., *Biochim. Biophys. Acta,* 55 (1962) 529.
10. Ikehara, M., Ueda, T., und Ikeda, K., *Chem. Pharm. Bull. (Tokyo),* 9 (1961) 767.
11. Cohn, W. E., *J. Biol. Chem.* 235 (1960) 1488.
12. Michelson, A. M., *J. Chem. Soc.* (1959) 3655.
13. Gassen, H. G., und Witzel, H., *Biochem. Biophys. Acta,* 95 (1965) 244.
14. Forrest, H. S., Hatfield, D., und Lagowski, J. M., *J. Chem. Soc.* (1961) 963.
15. Holmes, R. E., und Robins, R. K., *J. Am. Chem. Soc.* 86 (1964) 1242.
16. Holmes, R. E., und Robins, R. K., *J. Am. Chem. Soc.* 87 (1965) 1772.
17. Ukita, T., Funakoshi, R., und Hirose, Y., *Chem. Pharm. Bull. (Tokyo),* 12 (1964) 828.
18. Birkhofer, L., Kuhltau, H. P., und Ritter, A., *Chem. Ber.* 97 (1964) 934.
19. Leonard, N. J., und Laursen, R. A., *Biochemistry,* 4 (1965) 354.
20. Seide, B., *Chem. Ber.* 57 (1924) 1805.
21. Levene, D. A., and Tipson, R. S., *J. Biol. Chem.* 104 (1937) 385.
22. Tener, G. M., *J. Am. Chem. Soc.* 83 (1961) 159.
23. Witzel, H., und Barnard, E. A., (in Vorbereitung).
24. Hummel, J. P., und Witzel, H., *J. Biol. Chem.* 244 (1966) 1123.
25. Hummel, J. P., Ver Ploeg, D. A., und Nelson, C. A., *J. Biol. Chem.* 236 (1961) 3168.
26. Ross, C. A., Mathias, A. P., und Rabin, B. R., *Biochem. J.* 85 (1962) 145.
27. Mandel, H. G., Markham, R., und Matthews, R. E. F., *Biochim. Biophys. Acta,* 24 (1957) 205.
28. Cathou, R. E., und Hammes, G. G., *J. Am. Chem. Soc.* 87 (1965) 4674.
29. Ukita, T., Hagatsu, H., und Waku, K., *J. Biochem. (Tokyo),* 50 (1961) 550.
30. Beers, R. F., *J. Biol. Chem.* 235 (1960) 2393.
31. Follmann, H., Wieker, H.-J., und Witzel, H., *European J. Biochem.* (zur Veröffentlichung eingereicht).
32. Wieker, H.-J., und Witzel, H., *European. J. Biochem.* (zur Veröffentlichung eingereicht).

H. G. Gassen und H. Witzel
Chemisches Institut der Universität
355 Marburg/Lahn, Bahnhofstraße 7, Germany

European J. Biochem. 1 (1967) 46—50

Synthesis and Use of O-Stearoyl Polysaccharides
in Passive Hemagglutination and Hemolysis

U. Hämmerling[1] and O. Westphal

Max-Planck-Institut für Immunbiologie, Freiburg-Zähringen

(Received December 5, 1966)

Lipid-free polysaccharides generally are not attached to the surface of erythrocytes. In contrast O-stearoyl derivatives of polysaccharides show a high affinity for red blood cells. By reaction of polysaccharides with stearoyl chloride in dimethyl formamide as a solvent in the presence of pyridine, derivatives are obtained which exhibit marked erythrocyte-sensitizing activity. The chemical conditions for optimal sensitization of erythrocytes were investigated for a number of polysaccharides, e.g. starch, dextran, and several lipid-free bacterial cell wall polysaccharides. A content of about 5% by weight of O-stearoyl groups has been found most suitable for producing erythrocyte-sensitizing properties without modification of the serological specificity of the polysaccharide.

Sheep erythrocytes coated with O-stearoyl polysaccharide are specifically agglutinated by anti-polysaccharide antibody (passive hemagglutination) or lysed when guinea pig complement is present at the same time (passive hemolysis).

Further applications of the method by using derivatives of O-stearoyl polysaccharides are mentioned.

In the past few years passive hemagglutination has developed into one of the most widely used methods for the detection of antibodies. There exist a large number of test antigens which can be attached to the surface of red blood cells [1,2]. Depending on the chemical properties of the test antigens this attachement can be achieved by several means, among which adsorption of the antigen onto the erythrocyte surface is the most convenient technique.

The lipopolysaccharides of Gram-negative bacteria possess erythrocyte-sensitizing properties which on mild alkaline hydrolysis are enhanced considerably [3,4]. This activation may result mainly from two alterations of the lipopolysaccharide molecule:

a) the genuine molecule with a molecular weight of 1 to 10 million undergoes depolymerisation. Treatment with dilute alkali yields a relatively homogenous product with an average molecular weight of about 200,000 [4,5].

b) the action of alkali results in a partial hydrolysis of the lipid moiety. This may impair aggregation of lipopolysaccharide molecules and may thus favour the formation of complexes between lipids of the red blood cell membrane and lipopolysaccharide molecules.

The fixation of alkali-treated lipopolysaccharide is inhibited by a number of lipids [6] such as lipoid A,

[1] Present address: Sloan-Kettering Institute for Cancer Research, New York, N.Y. 10021, U.S.A.

Non-standard Abbreviations. Dimethyl formamide, DMF; phosphate-buffered saline, PBS.

lecithin, and cholesterol. It is conceivable that these lipids react in competition with certain receptors of the erythrocyte surface or with active sites of the lipopolysaccharide molecule. These findings suggest that in the mechanism of fixation the lipid moieties of both lipopolysaccharide and erythrocyte membrane are involved; the adsorption may occur by means of hydrophobic bonds.

On complete removal of the lipid moiety of lipopolysaccharides the resulting degraded polysaccharides in general maintain their serological specificity but will not adsorb onto red blood cells. Naturally occurring lipid-free polysaccharides, with the exception of certain acidic polysaccharides [7,8], do not possess erythrocyte-sensitizing properties.

The role of the lipid moiety in the adsorption mechanism of lipopolysaccharides prompted us to investigate whether the introduction of lipoidal groups into neutral and lipid-free polysaccharides would induce erythrocyte-sensitizing activity. It has been reported previously by Tsumita and Ohashi [9] that capsular polysaccharide of *Mycobacterium tuberculosis* partially esterified with palmitic acid can be attached to red blood cells.

In the present paper the esterification of a number of polysaccharides is described. The conditions are evaluated under which O-stearoyl derivatives of polysaccharides are fixed to red blood cells and thus can be used in passive hemagglutination and passive hemolysis.

MATERIALS AND METHODS

Polysaccharides. The type-specific degraded cell wall polysaccharides of Gram-negative bacteria (*Salmonella gallinarum, Salmonella adelaide, Salmonella minnesota*, and *Escherichia coli* O 111) were extracted with dilute acetic acid according to Freeman [10]. The extracts were purified by phenol/water treatment [11]. Polysaccharides containing covalently bound peptides or lipids were subjected to treatment with hydrazine [12]. Dextran-40 was obtained from Pharmacia Co., Uppsala, starch (amylum solubile) was purchased from Merck Co., Darmstadt.

The type-specific polysaccharides from group A and group E Streptococci were donations from Professor H. D. Slade, Northwestern University, Chicago [13]. Octadecasaccharide from *E. coli* O 141: K 85(B):H4 containing the K-specificity of this strain was obtained from Dr. K. Jann, Max-Planck-Institut für Immunbiologie, Freiburg [14].

Antisera. O-antisera to *S. gallinarum, S. adelaide, S. minnesota*, and *E. coli* O 111 were prepared by immunization of rabbits with heat-killed bacteria. Antisera to "A" and "E"-streptococci and antiserum to *E. coli* O 141:K85 (OK-serum) were prepared by injection of formalin-killed cells. Rabbit-anti-dextran and rabbit-anti-starch were prepared by injection of the corresponding polysaccharide-benzylazo-edestin conjugates [15].

Complement. Fresh guinea pig complement was adsorbed with sheep red blood cells.

Reagents. Stearoyl chloride (Merck) was distilled in a nitrogen atmosphere under reduced pressure. Dimethyl formamide (DMF) was dried with calcium hydride. Pyridine was distilled repeatedly over phosphorus pentoxide.

Esterification of Polysaccharide. 50 mg of polysaccharide were dried in an ampoule (equipped with magnetic stirrer) *in vacuo* at 40° for at least 3 h. The polysaccharide was then dissolved in 4 ml anhydrous DMF at approximately 50°. 0.6 ml of anhydrous pyridine and 10 mg of stearoyl chloride dissolved in 0.1 ml of DMF were added. The ampoule was sealed and the reaction mixture was stirred at room temperature for 3 days. The reaction mixture was diluted with 0.5 ml of water and poured into 25 ml of alcohol. After standing for several hrs in the cold the precipitate was collected by centrifugation, washed with alcohol and dissolved in 5 ml of water. The solution was filtrated over a sephadex G-25 column (30 ml bed volume) and lyophilized. 40 mg of material were recovered.

Assay of Stearic Acid Ester [16,17]. About 500 μg of O-stearoyl polysaccharide were homogenized in 60 μl of pure alcohol. 125 μl of alkaline hydroxyl amine reagent were added, and the tubes were sealed with parafilm. After heating for 2 min at 65° and standing for 5 min at room temperature 315 μl of ferric-perchlorate reagent were added. The red color developed in the course of 30 min, and the absorption at 530 mμ in micro cuvets with 10 mm of path length was measured after centrifugation of the tubes. Methyl stearate was used as a standard. The determination was run in triplicates.

Sensitization of Sheep Red Blood Cells. Sheep red blood cells which were stored for at least 10 days in Alsever-solution [18] at 4° were washed three times with phosphate-buffered saline (PBS). 1 ml of a 1% suspension of red blood cells in PBS (2×10^8 cells per ml) was mixed with an appropriate amount of O-stearoyl polysaccharide dissolved in a minute volume of PBS (the amounts of O-stearoyl polysaccharides are listed in Table 2). The mixture was incubated at 37° for 30 min. The erythrocytes were washed three times with PBS and finally resuspended in 2 ml of PBS (1×10^8 cells per ml) or 2 ml of Pillemer-buffer (for the hemolysis test).

Passive Hemagglutination. Antiserum was serially diluted with PBS. To each serum dilution (0.2 ml) an equal volume (0.2 ml) of sensitized cells was added. After mixing and incubation for 30 min at 37° the cells were allowed to sediment in the cold for 3 h. The lowest serum dilution on macroscopic inspection showing agglutination was recorded as the end-point of the titration.

Passive Hemolysis Test. Twofold dilutions of antiserum in Pillemer-buffer [18] were mixed with equal volumes of a suspension of sensitized sheep red blood cells, as described above. The tubes were incubated at 37° for 1 h, and then placed in an ice bath. Guinea pig complement (5 hemolytic units [18]) was added and the tubes were incubated again for 60 min at 37°. Lysis was then stopped by cooling to 0° and the tubes were centrifuged at +4°. The absorptions of the supernatants were measured at 541 mμ in micro cuvets (path length: 10 mm). The end-point of the titration was indicated by the dilution of antiserum required to give 50% lysis.

RESULTS

Preparation of O-Stearoyl Polysaccharides

Esterification with stearoyl chloride of polysaccharides can easily be achieved in DMF in the presence of catalytic amounts of pyridine. Under the mild esterification conditions proposed it has been possible to convert the degraded cell wall polysaccharides of various Gram-negative and Gram-positive bacteria—which all are heteropolysaccharides—as well as polysaccharides of simple composition (dextran-40 and starch) into their O-stearoyl derivatives. The procedure is also useful for the esterification of acidic polysaccharides, as was demonstrated with a octadecasaccharide derived from the capsular polysaccharide of *E. coli* O 141:K85 [14].

Depending on the amount of stearoyl chloride present in the reaction mixture polysaccharides with an ester content between 1 and 15% could be obtained in a reproducible manner. In general, we endeavoured to introduce about 5% of stearoyl groups into the polysaccharide in order to maintain adequate water solubility and serological specificity.

In Table 1 the *O*-stearoyl polysaccharides prepared are listed. The analytical data represent the average ester content (measured according to [16]) obtained by reaction of 50 mg of polysaccharide with the amount of stearoyl chloride indicated.

Table 1. O-*Stearoyl polysaccharide preparations*

Preparation No	Polysaccharide	mg Stearoyl chloride per 50 mg of polysaccharide	% *O*-Stearoyl
347	*S. gallinarum*	25	6
353 A	*S. gallinarum*	35	8
353 B	*S. gallinarum*	50	15
706 B	*S. adelaide*	12	5
706 C	*E. coli* O 111	12	5
653	"A"-streptococci	12	6
698 B	"A"-streptococci	12	6
698 A	"A"-streptococci	20	9
698 C	"E"-streptococci	12	4
706 D	"E"-streptococci	12	5
706 A	Octadecasaccharide (*E. coli*)	20	6
608	Dextran-40	20	1
618	Starch	20	4

Serological Analysis of O-*Stearoyl Polysaccharides*

The antigenic specificity of *O*-stearoyl polysaccharides was analyzed by immune precipitation. The polysaccharide specificity was found not to be altered during esterification. For instance, precipitation of *S. gallinarum* polysaccharide and its 5% *O*-stearoyl derivative with specific antiserum (horse O-antiserum of Salmonella group D_1) exhibit very similar precipitation curves (see Fig.1). As had been expected, the *O*-stearoyl derivative No 347 (curve II) was less active than the genuine polysaccharide No 345 (curve I). Apparently, esterification affected only a minor part of the serologically active sites. However, the majority of the determinant groups evidently maintained their specifity producing quantitative precipitation of antibodies.

Passive Hemagglutination

All *O*-stearoyl polysaccharides prepared showed marked affinity for human, sheep, and mice red blood cells, whereas lipid-free polysaccharides did not. However, in some instances the action of *O*-stearoyl polysaccharides caused dammage to the cells:

1. In higher concentrations of *O*-stearoyl polysaccharide (above 100 µg/ml) there was swelling and enlargement of the cells, which consequently became fragile and occasionally were lysed spontaneously. (Storage at +4° for about 10 days before use rendered red blood cells more stable).

2. Deformation of cells was more marked when the ester content of the polysaccharide was increased. It appeared that an ester content higher than 5 to 6% was critical and unsuitable.

3. Occasionally spontaneous agglutination of sensitized red blood cells was observed. This was more marked in higher concentrations of *O*-stearoyl polysaccharide and with preparations of *O*-stearoyl polysaccharide with a high content of ester.

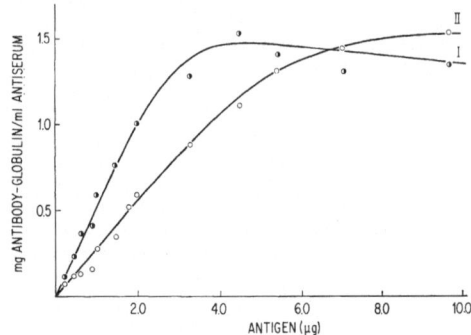

Fig.1. *Precipitation of* S. gallinarum *polysaccharide (No 345; curve I) and* S. gallinarum-O-*stearoyl polysaccharide (No 347; curve II) with horse anti-*D_1 *antiserum*

These complications could be overcome by using suitable conditions of sensitization. Besides the use of polysaccharides with an ester content of not more than 5 to 6%, the concentration of the *O*-stearoyl polysaccharide was critical and had to be determined for each preparation.

The concentrations of various *O*-stearoyl polysaccharides yielding optimal sensitization of 2×10^8 sheep red blood cells per ml varied from 0.2 to 20 µg/ml (*cf.* Table 2). *O*-stearoyl polysaccharides appeared to be more active than alkali-treated lipopolysaccharides; considerably lower concentrations of *O*-stearoyl polysaccharides were needed for optimal sensitization.

O-stearoyl polysaccharides were attached to the surface of erythrocytes thus giving rise to agglutination of the cells when anti-polysaccharide antibody was added to the system. The agglutination was specifically inhibited in the presence of soluble hapten (either polysaccharide or oligosaccharide). This finding is illustrated for the system dextran/rabbit-anti-dextran in Table 3. Dextran as well as isomaltopentaose (which represents probably the entire

determinant group of dextran [19]) completely inhibit the agglutination of O-stearoyl dextran-coated
cells with rabbit-anti-dextran.

Table 2. *Optimal concentrations of O-stearoyl polysaccharides
for sensitization of 2×10^8 sheep blood red cells/ml*

Preparation No	O-Stearoyl polysaccharide derived from	O-Stearoyl	Optimal concentration
		%	µg/ml
347	S. gallinarum	6	0.2
706 B	S. adelaide	5	4
707 C	E. coli O 111	5	4
706 A	Octadecasaccharide	6	10
653	"A"-streptococci	6	10
706 D	"E"-streptococci	5	4
608	Dextran-40	1	20
618	Starch	4	20
	Alkali-treated lipopoly-saccharide from E. coli O 111	—	50

ed by incubation with guinea pig complement, the
cells were lysed. A typical titration curve is given in
Fig. 2, where lysis of O-stearoyl dextran-coated cells
with rabbit-anti-dextran is plotted.

The hemolysis test offers two advantages over the
hemagglutination technique: (a) higher sensitivity;
(b) accurate determination of the end-point (50%
lysis) by means of photometric measurement.

DISCUSSION

The adsorption of lipopolysaccharides is probably
mediated by hydrophobic bonds between long chain
fatty acids of the lipopolysaccharide and lipid (or
protein) structures of the erythrocyte surface. The
erythrocyte-sensitizing activity of lipopolysaccharides
therefore depends largely on their content of fatty
acids. Thus the affinity of the genuine lipopolysaccharide is considerably enhanced by mild alkaline
hydrolysis, but will decrease gradually on subse-

Table 3. *Passive hemagglutination and passive hemagglutination inhibition with hapten (dextran and isomaltopentaose) in the
system: Dextran/Rabbit anti-Dextran*

Dilution of serum	2	4	8	16	32	64	128	256	512
E_c (O-stearoyl coated erythrocytes)	+++	+++	+++	+++	++	++	+	—	—
E_u (untreated erythrocytes)	—	—	—	—	—	—	—	—	—
E_c + dextran (1 µg/ml)	—	—	—	—	—	—	—	—	—
E_c + isomaltopentaose (0.1 mmole/ml)	—	—	—	—	—	—	—	—	—

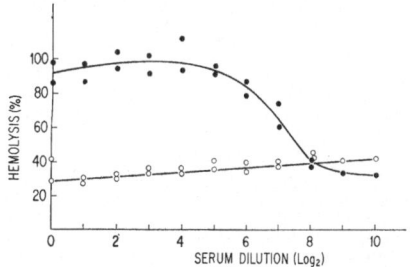

Fig. 2. *Hemolysis of O-stearoyl dextran-coated sheep red blood
cells with rabbit anti-dextran and guinea pig complement.*
○ normal rabbit serum (absorbed against dextran*); ● rabbit
antidextran antiserum

Passive Hemolysis

Since it has long been known that alkali-treated
lipopolysaccharides can be used in the passive hemolysis test [2] we attempted to adapt this technique
to O-stearoyl polysaccharides. When sheep red blood
cells sensitized with O-stearoyl polysaccharide were
incubated with anti-polysaccharide antibody, follow-

---

* The guinea pig serum contained traces of natural
dextran antibodies.

quent acid hydrolysis. Complete removal of the lipid
moiety with dilute acetic acid yields a degraded
polysaccharide which is not adsorbed onto red
blood cells.

The introduction of about 5% of stearic acid ester
groups into the degraded polysaccharide fully
restores the erythrocyte-sensitizing activity without
loss of serological polysaccharide specificity. In some
instances even much less than 5% of stearoyl groups
(about 1%) were efficient to introduce erythrocyte-
modifying properties into polysaccharides (see, for
instance, Dextran-40 prep. No 608 in Table 2). In
the present investigation anhydrous dimethyl formamide has proved to be a most effective solvent for
the esterification with stearic acid. The formation of
formyl esters according to the Vilsmeyer-reaction
occurs only in the presence of chlorinating acid chlo-
rides [20] and is negligible under the reaction conditions described in this report.

Esterification of a certain polysaccharide under
controlled conditions in our hands has proved to
yield O-stearoyl polysaccharide with a reproducible
content of ester. However, the amount of ester that
is introduced into a second polysaccharide under the
same reaction conditions may differ. This behaviour
reflects the differences in reactivity of the hydroxyl
groups available for esterification.

The introduction of stearoyl groups into the polysaccharide molecule may affect the serological specificity. A rough estimate however shows that in a polysaccharide with a content of stearoyl groups of 5% (w/w) only every 30th monosaccharide unit is substituted. As a determinant group of a polysaccharide molecule consists of an area of about a tetra to hexasaccharide[19,21] the greater part of the determinants are expected to be still accessible (cf. Fig.1).

As O-stearoyl derivatives of simple polysaccharides such as dextran and starch are attached to the surface of erythrocytes, it is assumed that O-stearoyl polysaccharides can be used generally in passive hemagglutination or hemolysis. According to several observations long chain fatty acids other than stearic acid (e.g. palmitic, oleic or lauric acid) can also produce erythrocyte-sensitizing activity.

The content of O-stearoyl groups is seen to be crucial; the affinity for the red blood cell membrane is dependant of the ester content and increases with increasing stearoyl content. O-stearoyl polysaccharides containing more than 5 to 6% of ester usually cause deformation and even lysis of red blood cells. A similar effect is observed with lower substituted polysaccharides in higher concentrations (above 100 µg/ml).

A generalized application of the method can be given according to the following directions:

a) During esterification about 5% of stearoyl groups should be introduced into the polysaccharide molecule.

b) The concentration of O-stearoyl polysaccharide for optimal sensitization of 2×10^8 red blood cells/ml ranges from 4 to 10 µg/ml. It is necessary, however, to determine the optimal concentration for each preparation.

In the presence of antibody to the polysaccharide red blood cells coated with O-stearoyl polysaccharide react specifically, giving rise to either agglutination or lysis when guinea pig complement is added. The specificity of the reaction is confirmed by hemagglutination inhibition with soluble haptens. For instance, dextran and isomaltopentaose in very low concentrations inhibit completely the agglutination of O-stearoyl dextran-coated erythrocytes with rabbit anti-dextran (Table 3), indicating the specificity of the reaction.

CONCLUSIONS

The use of O-stearoyl polysaccharides in either passive hemagglutination or passive hemolysis is a convenient method for the detection of anti-polysaccharide antibody. The method is of great advantage with respect to testing large numbers of sera, especially in the field of clinical diagnosis (cf. [22]).

O-stearoyl polysaccharides can also be used in modern plaque techniques for the identification of antibody forming cells in vitro (Jerne-technique [23]).

Moreover, erythrocytes can also be sensitized with derivatives of O-stearoyl polysaccharides. For instance p-aminobenzyl ethers of O-stearoyl polysaccharides can be coupled with protein, using Landsteiner's azo technique. O-Stearoyl polysaccharide benzylazoproteins will readily fix on the red cell surface, rendering the cells specifically agglutinable with protein antiserum and lysable in the presence of complement. We were able to detect single protein-antibody forming cells by using such sensitized erythrocytes in the Jerne technique. It is known, on the other hand, that protein-sensitized tanned erythrocytes (Boyden technique) will be agglutinated with protein antiserum, but do not lyse in the presence of complement.—Besides proteins, also smaller chemospecific determinants can be fixed on erythrocytes via their derivatives of stearoylesters of (preferably serologically inert) polysaccharides.—We will soon report on these further aspects of O-stearoyl polysaccharide derivatives [15].

The authors wish to thank Miss. Ch. Emmert for her skilful technical assistance. We are grateful to Dr. Boyse Sloan-Kettering Institute, New York, for correcting the English version of the manuscript.

REFERENCES

1. Davies, D. A. L., Crumpton, M. J., McPherson, I. A., and Hutchinson, A. M., Immunology, 1 (1964) 157.
2. Neter, E., Bacteriol Rev. 20 (1956) 166.
3. Neter, E., Westphal, O., Lüderitz, O., and Gorzinsky, E. A., Ann. N.Y. Acad. Sci. 66 (1956) 141.
4. Neter, E., Westphal, O., Gorzinsky, E. A., Lüderitz, O., and Eichenberger, E., J. Immunol. 76 (1956) 377.
5. Tauber, H., Russel, H., and Guest, W. J., Proc. Soc. exptl. Biol. Med. 107 (1961) 964.
6. Neter, E., Westphal, O., and Lüderitz, O., Proc. Soc. Exptl. Biol. Med. 88 (1955) 339.
7. Landy, M., and Lamb, E., Proc. Soc. Exptl. Biol. Med. 82 (1953) 593.
8. Jann, K., Jann, B., Ørskov, F., Ørskov, I., and Westphal, O., Biochem. Z. 342 (1965) 1.
9. Tsumita, T., and Ohashi, M., J. Exptl. Med. 119 (1964) 1017.
10. Freeman, G. G., Biochem. J. 37 (1943) 601.
11. Westphal, O., Lüderitz, O., and Bister, F., Z. Naturforsch. 7b (1952) 148.
12. Hämmerling, U., D. Sc. Thesis, Albert-Ludwigs-Universität, Freiburg 1965.
13. Slade, H. D., J. Bacteriol. 90 (1965) 667.
14. Jann, K., Jann, B., Ørskov, F., and Ørskov, I., Biochem. Z. 346 (1966) 346.
15. Hämmerling, U., Westphal, O. et al. (1966) In preparation.
16. Tauber, H., Federation Proc. 19 (1960) 245.
17. Snyder, F., and Stephens, N., Biochim. Biophys. Acta, 34 (1959) 244.
18. Kabat, E. A., and Mayer, M. M., Experimental Immunochemistry (edited by C. C. Thomas), Springfield, Illinois, 1961.
19. Kabat, E. A., J. Immunol. 84 (1960) 82.
20. Arnold, Z., Collection Czech. Chem.Commun. 26 (1961) 1723.
21. Lüderitz, O., Staub, A. M., and Westphal, O., Bacteriol. Rev. 30 (1966) 193.
22. Slade, H. D., and Hämmerling, U., Bacteriol. Proc. (1966) In press.
23. Jerne, N. K., and Nordin, A. A., Science 140, (1963) 405.

U. Hämmerling and O. Westphal
Max-Planck-Institut für Immunobiologie,
78 Freiburg-Zähringen, Postfach 1668, Germany

European J. Biochem. 1 (1967) 51—60

Physico-chemical Properties of Native and Recombined
Calf Thymus Nucleohistones

E. Fredericq and C. Houssier

Institut de Chimie physique, Université de Liège

(Received October 18, 1966)

In the preparation of calf thymus deoxyribonucleohistone a soluble component is formed when the time of storage of the glands before extraction is too high, when high speeds of homogenization are used or when dilute solutions are too strongly stirred. Under special experimental conditions, a soluble component can be obtained which is slightly degraded in comparison with the native gel-forming deoxyribonucleohistone, its protein content and its molecular weight being a little lower; electric birefringence and dichroism are however similar in both cases.

The dissociation of deoxyribonucleohistone in concentrated salt solutions was followed by centrifugation at $100,000 \times g$ under various conditions. A progressive lowering of the protein content was found in the pellet, when the concentration varies from 0.4 to 3 M NaCl; a minimum value of 0.05 was found for the protein/DNA ratio after three centrifugations in 4 M LiCl. The analysis of the residual proteins which are undissociable from DNA shows a predominance of acidic over basic amino-acids.

Deoxyribonucleohistone dissociated in salt solutions and recombined is similar to native deoxyribonucleohistone, as regards the rigidity, the intrinsic viscosity, the melting temperature and the electro-optical parameters. Combinations of histone and DNA display different properties.

Deoxyribonucleohistone with various protein contents were prepared by centrifugation and their characteristics compared with those of native deoxyribonucleohistone. It is only for low protein/DNA ratios that most of the properties are significantly altered.

The formation of gels requires a high histone content and the presence of specific proteins. The significance of gel appearance and the necessity for quantitative measurements of the modulus of rigidity have been outlined.

Extraction of thymus cells by salt solutions at ionic strength 0.15 leaves a residue which is almost entirely constituted by a combination of DNA and proteins, mostly histones. There are at present many reasons to believe that this combination is not an artefact and represents a structural unit of the chromosomes, with well defined composition and physicochemical properties.

This nucleohistone forms rigid gels in water at much lower concentrations (0.02 g/100 ml) than pure DNA. This property seems to be a characteristic of the native, undegraded state [1—7] although a soluble form can be obtained with reproducible properties [8—10]. It is well known that DNA easily combines to histones, forming artificial histone nucleates. Structural relations between soluble, gel and reconstituted deoxyribonucleohistone are still

very little known and will be examined in this paper.

Several investigators have already shown that deoxyribonucleohistone undergoes a progressive dissociation in concentrated salt solutions [10—14] reaching completion only in 5.8 M CsCl [12]. This provides a means for isolation and recombination of DNA and proteins with minimum damage and will be studied in detail below.

Finally the problem of the nature of non-histone proteins and of their structural importance in the deoxyribonucleohistone complex must be risen since it has been subject to numerous controversy. Monty and Dounce [15] isolated non-histone residual proteins from liver nuclei by rather drastic procedures and pointed out their importance for maintaining chromosomal structure. More recently, acidic proteins were extracted by milder procedures from liver or thymus nucleoproteins and completely analysed [16—19]. The total amount of non-histone protein and its amino acid analysis (in particular, the pre-

---

Non-standard abbreviations: Deoxyribonucleohistone, DNH; gel, G; soluble, S; supernatant from centrifugation in salt solution, SU; pellet from centrifugation in salt solution, P; weight ratio of content in protein and DNA, Prot/DNA.

sence of cysteine) are still a subject of great discrepancies.

We wish to point out that the deoxyribonucleoprotein studied by us is entirely soluble in M NaCl; this excludes the residual nuclear proteins not directly bound to DNA [20]. We call it here nucleohistone because of its low content in acidic proteins.

MATERIALS AND METHODS
Preparation

Calf thymus DNH was prepared in its gel form G-DNH as already described [3]. The gel is generally clear. Its turbidity can be characterized by the ratio of the absorbances at 4000 and 2600 Å. This should not exceed 0.04. However the gel always contains a few solid particles which are difficult to eliminate by centrifugation because they sediment with the gel. The greater part can be removed by the following procedure: the gel is centrifuged at 35,000 rev./min for 15 to 30 minutes at a concentration of DNA of $0.01-0.02$ g/100 ml. The supernatant is discarded. The gel fills about $^1/_3$ to $^1/_4$ of the original volume. The solid impurities are at the bottom of the tube but stick to the gel. They are removed by carefully cutting the bottom of the plastic tube with a razor blade. The clear gel is completely soluble in concentrated NaCl solutions (above 0.5 M) and will be defined here as the pure DNH. Dilutions of the gel are made in water adjusted to pH 6.5 by letting them swell for several hours in the cold and stirring without excessive shear.

Recombination of DNA and Histone

A slow mixing of 0.05 g/100 ml DNA (prepared according to Kay et al. [21]) and 0.065 g/100 ml histone [22] gives partly insoluble clusters. The soluble part has a low protein content. Consequently this method was not used further.

The same method of direct mixing is used in M NaCl solution where no precipitation occurs. The salt is dialysed out. Histone nucleates with various protein/DNA ratios can be prepared by this method. DNH (pH 6.5) is dissolved in concentrated salt solutions (from 0.4 M to 4 M), at DNA concentrations ranging from 0.005 to 0.05 g/100 ml. The solution is left overnight at 4° and centrifuged at moderate speed to eliminate insoluble clusters. The clear solution is centrifuged several hours at $105,000 \times g$ in a Spinco Model L ultracentrifuge. The supernatants constitute the SU-fractions. The pellets (P-fractions) are suspended in M NaCl, left overnight and homogenized for a few minutes with a rotating teflon pestle. After centrifugation at 10,000 rev./min, a clear solution is obtained. The two fractions can be mixed in variable amounts and dialysed versus water.

Residual Nucleoprotein after
Dounce and Hilgartner [17]

A gel DNH was extracted two times at 0° with 0.1 M HCl for 30 minutes [17]. The precipitate was dispersed in moderately alkaline water (final pH 10) with a teflon pestle homogenizer and dialysed. This residual nucleoprotein had slight gel-forming properties but was very turbid. It still contains high amounts of proteins (Prot/DNA = 0.9). The measurement of the rigidity modulus is made difficult by the high viscosity. At concentrations in DNA of 0.03 g/100 ml, it is $G' = 0.05$ dynes $\times$ cm^{-2}.

Analyses

The protein content was determined by a microbiuret method [23] or by the method of Lowry et al. [24] in M or 0.5 M NaCl. In all cases standardization was made using histone samples prepared according to Walker [22]. The DNA content is uniformly expressed as acid and not as the sodium salt, although in the complexes with a low protein content, a good part of the DNA is really in the salt form. The absorbance for a solution containing 1 g/100 ml of DNA was 215 and was used throughout for determining DNA content. This was used for defining the concentrations of the DNH. The ratio Prot/DNA was also used throughout this work for characterizing the DNH.

Rigidity of Gels

The rigidity modulus of the gel-forming solutions was measured according to the method previously described by one of us [25] for precise determinations. However for many purposes a simplified version of the method can be used as follows.

A stainless steel cylinder suspended to a torsion wire is dipped into the viscoelastic solution and set to oscillate at small angles of deviation (15°). The pseudoperiod of two oscillations is measured, using a chronometer, within 0.2 sec. The modulus G' is calculated according to the equation:

$$G' = K \left[\left(\frac{t_0}{t} \right)^2 - 1 \right]$$

where t_0 denotes the period in the solvent, t the period in the solution and K is a constant of the apparatus established from:

$$K = \frac{k}{4 \pi h} \left(\frac{1}{r^2} - \frac{1}{R^2} \right)$$

where k is the torsion constant of the wire, h is the height of the cylinder, r its radius and R the radius of the cylindrical container. In our device, $h = 4.3$ cm, $r = 0.5$, $R = 0.95$, $k = 8.42 \times g \times$ cm$^2 \times$ sec^{-2}. The DNH gel displaying rheopexy must be gently stirred with a glass rod for a few seconds just before the measurements.

Other Physical Measurements

The viscosity at zero gradient was determined by extrapolation, using a helical viscometer under various pressures. Many routine measurements however were made at finite gradient for the sake of comparison, using an Ostwald viscometer. For determining the viscosity of the DNA component, the DNH was treated by sodium dodecylsulfate (0.5 g/100 ml) and 0.1 M NaCl; the intrinsic viscosity in this case was always calculated in terms of the DNA concentration. Under such conditions we found for DNA in DNH an intrinsic viscosity of about 60 dl/g which is probably a little lower than the true value at zero gradient.

Thermal denaturation curves were determined in a Beckman DU spectrophotometer. The temperature was checked by a thermocouple dipping in an absorption cell and the absorbances were corrected for water thermal expansion. The DNH solutions had a DNA content around 0.002 g/100 ml and were in pure water pH 6.

Electro-optical measurements were made as already described [26].

Unless otherwise stated, the values of the electro-optical parameters were measured at $13-13.5$ kV/cm and at 4500 Å and 2600 Å for the birefringence (Δn) and the dichroism (D), respectively. The electric birefringence is given as $B = \Delta n/c$ (DNH), where c(DNH) is the DNH concentration in mg/100 ml, or $B' = \Delta n$/atom P. The electric dichroism is characterized by the dichroic ratio $D = -A_{\perp}/A_{\parallel}$ (since $A_{\perp} > A_{\parallel}$). In the present study, the dichroism and the birefringence were always negative; the negative sign has been omitted in the results. The relaxation time τ_{max} was determined from the slope of plots of $\log \Delta n$ versus time, at high time values, as previously described [26].

RESULTS

Factors Affecting the Rigidity and the Viscosity of DNH Preparations

Several authors [8, 27] have pointed out the necessity of controlling the speed of the mixers during homogenization of the tissue extract. They have not given however numerical values for the speed of rotation. Moreover, in the conventional commercial mixers (total volume 1 liter), the use of small volumes of liquid (200 ml) gives rise to intense turbulence which may still increase local shearing stress. We always use containers of small volumes in order to avoid turbulence. A gel preparation containing 0.05 g/100 ml DNH in water was submitted to one minute homogenization at various speeds. The modulus of rigidity and the viscosity of the DNA component are given in Table 1. One can see that the rigidity is very sensitive to shear and the intrinsic

viscosity indicates that the DNA component is partly split at moderate speeds. As a consequence, in our preparations, the tissue was subjected to homogenization at 7000 rev./min for 30 seconds and at 4000 rev./min for 3 minutes each time. (This corresponded to voltages of 60 and 45 volts respectively.)

Mazen and Champagne [6] have studied the influence of the time of storage of the thymus before extraction and found that when the gland was treated more than 30 minutes after the death of the animal, a higher yield in soluble DNH was obtained, due to a slight degradation of the native material. We confirmed and extended those findings in a systematic study of the influence of the storage time of thymus at $20°$ before extraction, on a few important

Table 1. *Effect of the speed of the homogenizer on the modulus of rigidity of native DNH and the intrinsic viscosity of its DNA*
Time of treatment : 1 minute

Speed	Modulus of rigidity	Intrinsic viscosity
rev./min	dynes/cm²	dl/g
0	8.3	60
4,200	8.3	60
5,200	5.3	46
7,500	2.2	46
12,000	0.01	36

physico-chemical properties (Table 2). Data on reduced viscosity are also given in Fig. 1. When the thymus is extracted within 15 minutes, the yield in soluble fraction is always very low (5—10%) in agreement with other observations [6, 7]; it becomes important when the thymus is stored several hours at room temperature and the turbidity of the gel tends to increase slightly. The molar absorbance of the S fraction is similar to that of G fractions. The protein content is in average slightly higher in the gel (Prot/DNA: 1.35 to 1.40 or 58% in protein) than in the soluble fraction (Prot/DNA: 1.30 to 1.35). The intrinsic viscosity is very hard to determine in the gel because of the contribution of rigidity which becomes noticeable at concentrations as low as 0.01 g/100 ml. Only estimates can be given. The values markedly decrease in the soluble fractions indicating degradation. More reliable results are obtained when viscosity is measured in 0.5 g/100 ml dodecyl sulfate. Here the viscosity is entirely due to the free DNA; the protein being denatured and complexed by dodecylsulfate does not contribute appreciably. Intrinsic viscosities around 80 dl/g indicate a high degree of polymerization of the DNA; according to the equation of Eigner and Doty [28], the molecular weight would be 16,000,000. In the soluble samples, the molecular weight falls to 9,000,000 (sample VII—S) and 4,000,000 (sample IX—S).

Table 2. *Effect of time of storage of thymus on the yield in S-fraction, the absorbance at 2600 Å, the protein content and the intrinsic viscosity of the preparation*
Values are given for the gel fraction in the case of samples I—G to VI—G and for the soluble fraction in the case of samples VII—S to IX—S

Sample	Time of storage	Yield in S-DNH	$\varepsilon_{atom\ P/l}$ at 2600 Å	Prot/DNA	$[\eta]$	
					DNH[a]	DNA[b]
	hours	%			dl/g	
I—G	0	10	6700 ± 150[c]	1.35 ± 0.02[c]	30—40	
II—G	0	10		1.35 ± 0.03[c]		
III—G to VI—G	0	5	6600 ± 200[c]	1.40 ± 0.05[c]		80 ± 10[c]
VII—S	5	77.5		1.35 ± 0.05[c]	25—30	50
VIII—S	7 1/2	100			20	50
IX—S	15	100	6650 ± 100[c]	1.33 ± 0.02[c]	12	25—30

[a] Viscosities measured in 0.7 mM phosphate pH 6.8, the concentrations being expressed as DNH.
[b] Viscosities measured in 0.5% dodecylsulfate-0.1 M NaCl, the concentrations being expressed as DNA.
[c] Mean deviation.

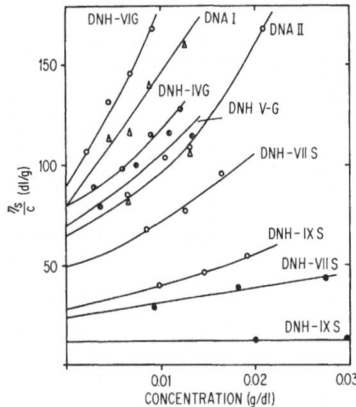

Fig. 1. *Reduced viscosities of DNH fractions as a function of concentration.* ○ ◑ ◐ in 0.5% dodecylsulfate-0.1 M NaCl, concentrations in DNA; ● in 0.7 mM phosphate, pH 6.8, concentrations in DNH; △ in 0.2 M NaCl

soluble ones 20,000,000 and 9,000,000. The characteristics of the moderately degraded sample VII—S correspond to the data given in the literature for the properties of soluble nucleohistones [8,10,29,30]. No measurements in the ultracentrifuge were possible with gel fractions which sediment very rapidly at low concentrations because of the gel structure.

From Fig. 1, it is apparent that the influence of concentration on the reduced viscosity becomes very strong for the gel samples and progressively decreases for the degraded soluble samples. Values for two DNA preparations are given for the sake of comparison. It is also interesting to compare viscosities of DNH in 0.7 mM phosphate and in dodecylsulfate since the latter value gives the contribution of the DNA component. However for making such a comparison, it is better to divide the concentration of the DNH which has been used by a factor of 2.3, in order to obtain intrinsic viscosities in terms of DNA concentrations. When those corrections are made, the intrinsic viscosities of the soluble samples in dodecylsulfate or in dilute phosphate are very

Table 3. *Electro-optical parameters of various samples of G-DNH and S-DNH*
B, B', τ_{max} and D are mean values for concentrations ranging from 5 to 10 mg/100 ml DNA (see material and methods), pH 6.5. For designation of samples, see Table 2

Sample	Solvent	$10^7\ B$	$10^3\ B'$	τ_{max}	D
		dl/mg	l/atom P	msec	
I—G	H₂O	0.89	6.5	5	1.58
	1 mM NaCl	0.45	3.25	0.60	1.28
II—G to VI—G	H₂O	0.95 ± 0.10[a]	6.7 ± 0.7[a]	1 to 5.5	1.62 ± 0.05[a]
VII—S	H₂O	1.18	8.4	5.8	1.74
	0.7 mM phosphate	0.66	4.7	1	1.39
IX—S	H₂O	1.32	9.35	4.5	1.80
	1 mM NaCl	0.80	5.7	0.5	1.43

[a] Mean deviation.

Values for the molecular weight of the DNH can be tentatively estimated by multiplying the value of DNA by a factor 2.3, taking into account the protein: this gives for the gel sample 37,000,000 and for the

close; they indicate similar extensions of the DNA in the free or in the combined state.

Table 3 compares the electro-optical parameters of gel and soluble fractions. From those data, we can see that, in all experimental conditions, the soluble fractions show greater values of D and B than the gel fractions, *i.e.* an increase in optical anisotropy. If we make a comparison with DNA [31] at the same ionic strength (1 mM NaCl), we observe that the present values of birefringence and dichroism are still far from those obtained with DNA. This would indicate some aggregation of the particles or some distortion of the DNA helix even in the soluble fractions. Data of Table 3 also show a good reproducibility of the measurements on various samples of G-DNH from distinct preparations, the values of birefringence and dichroism never differing by more than 10%; similar reproducibility has been mentioned by Itzhaki [7] on a gel forming DNH from rat thymus.

Dissociation and Centrifugation of DNH in Salt Solutions

The preparation of recombined histone nucleates makes use of the centrifugation of DNH in concentrated salt solutions (see materials and methods). The conditions affecting the centrifugation were studied also as a means for following the dissociation of DNA and protein in media of high ionic strength.

Table 4. *Influence of various factors on the ultracentrifugation of DNH in concentrated salt solutions*
The concentration in DNH at the beginning of the run is indicated as DNA (2nd column); the time of centrifugation was 4 hours except otherwise indicated

NaCl molarity	DNA	pH	Supernatant		Pellet Prot/DNA
			Protein	DNA	
	g/100 ml		g/100 ml		
0.4	0.03	6	0.0096	0.0018	0.97
0.5	0.03	6	0.0072	0.0008	1.01
0.6	0.03	6	0.0064	0.0007	0.97
1.0	0.03	6	0.0156	0.0008	0.9
1.0	0.035	6	0.0142	0.0016	0.83
1.0[a]	0.03	6			0.53
2.0	0.035	6	0.0344	0.015	0.33
2.0	0.005	6	0.004		0.18
3.0	0.018	6	0.019	0.008	0.19
3.0	0.023	6	0.02	0.003	0.21
4.0[b]	0.023	6		0.004	0.16
4.0[c]	0.02	6	0.0028	0.0026	0.055
4.0[d]	0.028	6			0.05
1.0	0.038	3.5	0.030	0.0025	0.67
1.0	0.036	5.5	0.019	0.0011	0.62
1.0	0.036	10.5	0.023	0.0017	0.42

[a] Second treatment of the dispersed pellet.
[b, c, d] Respectively first, second and third treatment in 4 M LiCl, duration of run 8 hours.

Several factors were studied: ionic strength, pH, duration of run, DNH concentration. Results are given in Table 4 in terms of analytical data for the pellet and for the supernatant. The quantity of DNA remaining in the supernatant indicates the

yield of the process and the ratio Prot/DNA in the pellet informs us about the extent of the dissociation if we assume that the dissociated protein does not sediment. The concentration of DNH was chosen in order to obtain solutions which did not display too high a viscosity. Expressed in DNA, they were between 0.02 and 0.04 g/100 ml. Best yields of sedimentation were obtained at the lower concentration.

Above times of 4 hours, there was no more important change at moderate salt concentrations ($\overline{M}$ or 2 M). When a high yield in the pellet was required, especially at high salt concentration, a time of 8 hours was used, because under such conditions, the viscosity becomes very large and slows down the sedimentation.

The supernatants were in any case composed of a high proportion of proteins. From the values of Prot/DNA, it is evident that the dissociation regularly increases with ionic strength. It may be expected that a part of the protein is taken down mechanically or by interactions, so that the pellet is richer in proteins than the dispersed particles before centrifugation. This explains why a second treatment brings about a second loss of protein in the pellet (together with a possible contribution of a shift in the equilibrium DNH $\rightleftharpoons$ DNA). LiCl was used at high concentration (4 M) because of its lower density and higher solubility. A third centrifugation did not reduce any more the protein content of the pellet. Since the ratio Prot/DNA in the pellet is 0.05, we may consider that the amount of protein which is not dissociable from the DNH in salt solution represents 4% of the total proteins originally present. The addition of 0.02 M EDTA did not change the dissociation behaviour and ultracentrifugation in 4 M guanidine chlorhydrate gives a pellet with a ratio Prot/DNA 0.3, indicating a lower dissociation than in 3 M NaCl. Consequently we may conclude that the residual protein is bound to DNA by bonds stronger than electrostatic bonds; since they are not broken by guanidine, they are likely to be covalent rather than hydrogen bonds.

The increase in pH favors the dissociation, at moderate ionic strength. In 3 M NaCl, no significant increase in dissociation occurs from pH 6 to 10.

An incomplete separation of protein and DNA by centrifugation in 2 M NaCl has been reported by other authors [10,12,13]. However Bauer and Johanson [32] claim that they achieve a complete removal of protein by adding a treatment on a sephadex column; this would indicate a total dissociation by electrostatic means. Such an assumption is however rather surprising in view of the difficulties encountered in the deproteinisation of DNA by much more drastic treatments [33,34] and may be due to a lack of sensitivity of the spectral method used by these authors for protein determination.

Amino Acid Analysis of Fractions

Table 5 gives the results of the amino acid analysis of some of the fractions obtained during the three successive centrifugations in 4 M LiCl. The fraction P_3 constituting the residual nucleoprotein is a typical acidic protein with a ratio (Lys + Arg + His)/(Glu + Asp) equal to 0.56. The results are in good agreement with those reported by Dounce and Hilgartner [17] and with the analysis of the final residue of acid extraction of Murray [18], except for the high glycine content. Since our analysis was made on the nucleoprotein complex very rich in DNA, the results of glycine content are probably too large: it has been

Table 5. *Amino acid composition of fractions obtained by centrifugation in 4 M LiCl solutions*
Method of Moore *et al.* [49]; Technicon automatic analyser. The values are percentages of total moles of recovered aminoacids

	Designation of fractions		
	1st supernatant	2nd supernatant	3rd pellet
Ala	11.1	11.8	7.3
Arg	11.2	8.7	4.2
Asp	5.1	5.2	7.2
Cys	—	—	0.6
Glu	9.4	9.4	12.6
Gly	10.4	19.4	22.2
His	1.9	2.3	2.3
Ile	4.8	3.8	3.2
Leu	9.4	7.6	6.9
Lys	9.4	9.5	4.5
Met	0.9	1.2	1.5
Phe	2.5	—	3.1
Pro	4.3	5.1	4.5
Ser	4.6	6.2	9.4
Thr	6.9	5.3	4.4
Tyr	1.9	1.4	2.0
Val	5.8	5.3	4.2

demonstrated that purine derivatives are partly converted to glycine during acid hydrolysis [35]. The supernatant S_1 is composed of histones closely corresponding to the fractions II and III [36] or F_2 and F_3 [37]. The second supernatant is very similar to the first one, if we discard again the aberrant glycine content. This shows that the proteins which are removed in the second centrifugation are histones which had been taken down in the pellet by mechanical effects during the first run.

So it appears that the dissociation in 4 M LiCl leaves the same residue as the exhaustive extraction in strong acid. The former process has the great advantage of being much less harmful both to proteins and to DNA. Our results markedly differ from those obtained on residual rat liver nucleoprotein from phenol extraction [16,19], in particular by the presence of cysteine and a very low content in basic amino acids. A thorough comparison is however

precluded by the fact that the analysis of Leveson and Peacocke [19] was made after removal of DNA. While this procedure has the advantage of eliminating possible disturbing effects due to DNA, it may, on the other hand, bring about a loss of some protein fractions.

Recombined Nucleohistones and Histone Nucleates

DNA cannot be recombined with histone in water solution to give a combination with normal proportions. It seems that random union of the components occurs and that complexes with high histone content precipitate under those conditions. On the contrary,

Table 6. *Rigidity and melting temperature of native and recombined DNH*
For the designation of samples, P and SU indicate the pellet and supernatant of centrifugation performed in salt solution, the concentration of which is given in parentheses. All measurements were made in pure water, pH 6.5; the melting temperatures were made in solutions containing 0.002 g/100 ml DNA

Sample	DNA concentration	Modulus of rigidity	Melting temperature
	g/100 ml	dynes/cm^2	°C
Native	0.03	6.0	81
Dissociated in M NaCl	0.03	6.0	81
Native	0.021	3.4	81
Dissociated in 2 M NaCl	0.021	2.8	81
Dissociated in 4 M LiCl	0.021	2.5	—
P + SU (3 M NaCl)	0.018	0.5	81
P (3 M NaCl) + histone	0.017	0.18	78
DNA + histone	0.03	0	76

a histone nucleate with the native proportions can be made in NaCl followed by dialysis; it has no gel-forming properties and shows a tendency to precipitation by heating. Recombined complexes much closer to the native DNH are formed by dissociation of DNH in M NaCl followed by dialysis; in this case they regain completely their native rigidity. After a higher degree of dissociation (in 3 or 4 M salt) the decrease in rigidity remains very slight (Table 6). The electro-optical parameters of the recombined gel-forming DNH are also very close to those of the native G-DNH as shown by the dichroism curves of Fig. 2 and by the data of Table 7. The pellet fraction from centrifugation in salt solution recombined with histone or with its supernatant has a small modulus of rigidity, probably because of being compacted in the ultracentrifugation process.

It must be concluded that the recombination of whole DNH components in concentrated NaCl takes place in a sufficiently ordered way for the reformation of the intermolecular cross links responsible for the gel structure. When pure histone is combined to DNA, part of the proteins constitutive of the DNH are missing and the gel is no more formed. Consequently, those proteins must play a special role either in the intermolecular bonds or indirectly by imposing special structural disposition. Those proteins are present in the P-fraction which explains that a significant amount of rigidity is regained in recombinations of P-fractions and histone or SU-fractions.

The extraction of a small amount of histone is sufficient to bring about almost completely the disappearance of rigidity, in DNH with Prot/DNA equal to 0.97. The intrinsic viscosity in 0.5% dodecylsulfate remains unchanged.

The thermal denaturation curves are not changed by a dissociation and recombination in agreement with Ohba's finding [38] even in the case of union

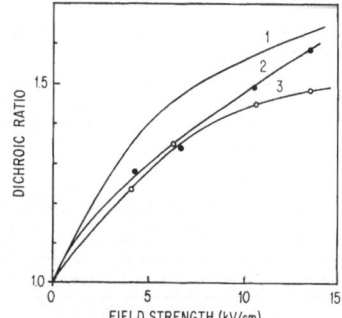

Fig. 2. *Electric dichroism of recombined DNH as a function of the field strength.* The curve 1 obtained for the native G-DNH [26] has been drawn for comparison; curve 2, recombined G-DNH after dissociation in 2 M NaCl; curve 3, recombined G-DNH after dissociation in 4 M LiCl. Solvent H_2O, pH 6.5

Table 7. *Electro-optical parameters of native, recombined and dissociated DNH, compared to DNA*
For signification of symbols, see material and methods; pH 6.5; the concentrations in DNA are around 5 mg/100 ml

Sample	Prot/DNA	Solvent	$10^2\, B'$	τ_{max}	D
			l/atom P	msec	
G-DNH	1.40	H_2O	7	4.1	1.63
		1 mM NaCl	3.25	0.65	1.30
DNH dissociated in 2 M NaCl and recombined	1.40	H_2O	6.6	3.8	1.58
DNH dissociated in 4 M LiCl and recombined	1.40	H_2O	6	3.2	1.48
Pellet from centrifugation in 0.6 M NaCl	0.97	H_2O	9.75	3.9	1.80
		1 mM NaCl	6.4	0.92	1.47
Pellet from centrifugation in 3 M NaCl	0.20	H_2O	17.3	4.5	2.41
Pellet from two centrifugations in 4 M LiCl	0.05	H_2O	18	4—5	2.5
		1 mM NaCl	13	2—2.5	2
DNA	< 0.01	1 mM NaCl	13	1.75	2

of P and SU-fractions (Fig. 3). They are slightly shifted towards lower values when pure histone is used for recombination. Here again a change has been brought about in the structure of the histone nucleate with characteristics of a less stable or less ordered structure: histone alone is not able to protect completely DNA against the loss of helical structure as well as do the total proteins from DNH. All the curves are very steep, indicating in all cases a high degree of homogeneity in helical configuration.

Data on histone nucleates with various protein contents are given in Table 8. They were obtained in the pellets of centrifugation in different salt concentrations, the Prot/DNA ratios varying from 1 to 0.05. The ratio 1.6 was obtained by direct addition of histone to DNH. Below ratios of 0.9, all samples were deprived from rigidity.

The native DNH displays the maximum value of T_m (81°); addition of histone does not increase it. With lower protein contents, the T_m decreases only to a small extent in confirmation of previous reports [38,39]; it is only at very low Prot/DNA ratio that T_m is strongly decreased. A higher degree of hetero-

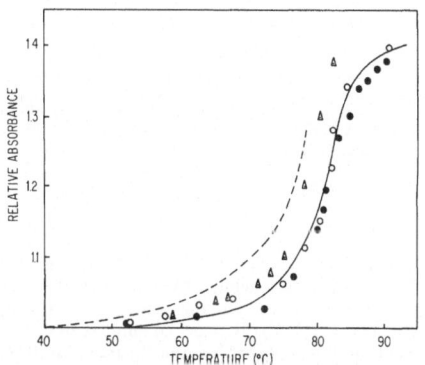

Fig. 3. *Thermal denaturation curves of native and recombined DNH.* Ordinate: ratio of the measured absorbance (at 2600 Å) to absorbance at 20°. Solvent water, pH 6, concentration in DNA 0.002 g/100 ml. Full curve, native DNH; dashed curve, DNA recombined to histone; ○ DNH dissociated in 2 M NaCl and recombined; ● pellet and supernatant recombined after centrifugation of DNH in 3 M NaCl; △ same pellet recombined to histone

geneity is evident in the sample with Prot/DNA 0.05
(Fig.4). However its intrinsic viscosity in dodecylsulfate (52 dl/g) shows that its DNA has undergone
very little degradation or denaturation by the
treatment in 4 M LiCl. The solubility at low ionic
strength is very much increased: it is quite soluble in
0.1 or 0.2 M NaCl.

Table 8. *Effect of protein content on the mean melting
temperature of DNH*
The pellets are those obtained by one (P_1) or three (P_3)
centrifugations in concentrated salt solution, the concentration of which is indicated in the parentheses

Sample	Prot/DNA	T_m
DNH + histone	1.6	81
DNH	1.4	81
Pellet (M NaCl) P_1	0.55	78.5
Pellet (3 M NaCl) P_1	0.21	78
Pellet (4 M LiCl) P_3	0.05	59

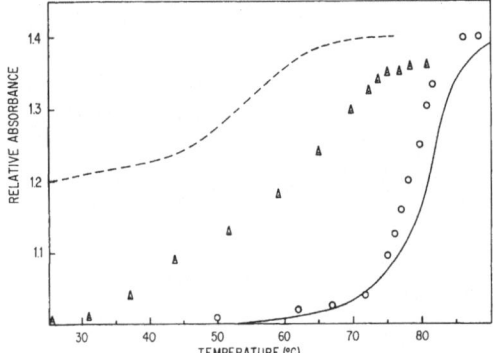

Fig.4. *Thermal denaturation curves of DNH with various
protein content.* Ordinate: see Fig.3. Solvent and concentrations, as in Fig.3. Full curve, native DNH; dashed curve,
pure DNA, the absorbance in this case being divided by the
absorbance of undenatured DNA diluted in 0.2 M NaCl at
20°; ○ pellet from centrifugation in 3 M NaCl (Prot/DNA =
0.20); △ pellet from two centrifugations in 4 M LiCl
(Prot/DNA = 0.055)

The dichroic ratios and the birefringence increase
sharply with decreasing Prot/DNA ratios, reaching
a maximum value for the lower protein content
available (Table 7). It must be emphasized that the
small protein content of the P_3-fraction (0.05) is
sufficient to prevent thermal denaturation of the
DNA in H_2O pH 6.5. This allows measurements in
these conditions of very low ionic strength where
the maximum lengthening and straightening of the
macromolecule is achieved. Making the comparison
with pure DNA at the same ionic strength (1 mM
NaCl; see ref. [31]) we observe the expected similar

optical anisotropy for the P_3-fraction (0.05). The
observations are in full agreement with the flow
birefringence and flow dichroism data of Ohba [12]
on S-DNH and dehistonized nucleohistones. Mean-
while we do not agree with the deduction of this
author that nucleohistone molecules are not as well
oriented as the DNA molecules in the gradient range
investigated. The smaller values of the rotatory
diffusion constant observed for DNA compared with
those for DNH could arise from a greater contribution to the birefringence of the smaller particles of
the DNH polydisperse system, *i.e.* to more complete
orientation of the DNH molecules. This suggestion
has been made by Zubay and Doty [8] who conclude
tentatively that "the dimensional polydispersity of
the DNH is of the same order as the DNA." Moreover,
the plots of the electric birefringence against the
field strength [26] and of the flow birefringence
against the velocity gradient [8] indicate a clear
approach to saturation in the case of DNH, but not
for DNA. So, the changes of D and B as well as that
of the flow dichroism and birefringence with the
histone content can be effectively related to the
changes in optical anisotropy of the molecules.

For the same reasons, we do not think that the
screening effect of the histone on the DNA phosphate
groups in the DNH could account for the higher
electric birefringence and dichroism of DNA compared to DNH, as it was suggested by Itzhaki [40].

DISCUSSION

When talking about rigidity of nucleoproteins and
its importance as a typical structural property, it is
necessary to make here some restrictions because of
a misuse of the term "gel-forming properties" also
pointed out by Itzhaki [7]. Although the gel state
is common in the chemistry of biopolymers, it has
been seldom studied and has not been quantitatively
characterized in most cases. The external appearance
of gel properties can be very misleading even if it is
qualitatively checked by the "bubble test" or by
thread formation [17]. When the rigidity of a gel is
measured by a physical method such as that used
in this work, one finds that many samples which
have a gel appearance, possess in fact a very low
modulus of rigidity; they are often constituted by a
heterogeneous suspension of swollen particles which
does not apparently resolve in a precipitate because
of its high viscosity but which is really in the solid
state. This state is often an indication of denaturation and aggregation. On the contrary, the DNH gel
fraction possesses a high modulus of rigidity in water,
the value of which can be measured with reproducibility, under a variety of conditions [25]. It shows
reproducible electro-optical parameters in many cases
in agreement with the observations of Itzhaki [7,
40,41]. The low turbidity of this fraction is also an

indication of homogeneity. DNH dissociated in concentrated salt solutions, in alkaline media, in 8 M urea or in dodecylsulfate has the appearance of a gel because of the very high viscosity and the tendency to produce "threads" [17]. Their modulus of rigidity is almost negligible when measured quantitatively. For instance a 0.07 g/100 ml DNH solution in M NaCl has a modulus $G' = 0.06$ dynes $\times$ cm^{-2}, in 0.5 g/100 ml dodecylsulfate, $G' = 0.13$ and in 8 M urea $G' = 0.5$ dynes $\times$ cm^{-2} (in pure water the same solution gives $G' = 18$ dynes $\times$ cm^{-2}).

In confirmation of previous statements [3], we may affirm that when the maximum caution has been taken in the preparation, the DNH is obtained almost quantitatively in the form of a true gel. The general factors which are susceptible to increase the percentage in soluble form are the time of storage of the thymus before extraction and the shearing forces exerted during homogenization or originating from excessive stirring in dilute solutions.

The exact difference between the gel and the soluble form is still not clearly established. From this work and others [6] the protein content is on the average a little lower in the soluble form and the DNA component has a somewhat lower molecular weight. However the loss of rigidity must arise from more specific transformations, although the electro-optical properties do not show any evidence for great changes in the general structure of the molecular units. We think that a special part of the proteins play an important role in the formation of the inter-molecular bonds of the gel lattice. These bonds are particularly susceptible to shear stress and proteolytic actions [4,27].

Similar conclusions can be drawn from the study of the rigidity of recombined DNH. It is clear that histones extracted from DNH are unable to form gels with pure DNA. On the contrary, the separated components in concentrated salt solution can partly do it. It appears therefore that something is missing in the extracted histones which is necessary for the gel formation. It is natural to think that this special protein fraction corresponds to the acidic ones, called residual proteins. However we cannot confirm the statement of Dounce and Hilgartner [17] according to which a complex formed of DNA and residual protein after complete removal of histones is still a rigid gel. When applying the criteria developed at the beginning of this discussion, we found that the presence of a high amount of histone is also a requisite condition for true gel formation in agreement with our previous conclusions [3]: the removal of 30% of the total proteins by acid extraction or centrifugation in 0.6 M NaCl brings about almost complete disappearance of rigidity.

Finally we conclude that gel forming properties are determined by: (a) a minimum molecular dimension of the DNA component which forms the "spinal chord" of the molecular units; (b) a high histone content which is necessary for neutralizing DNA charges; (c) a special protein component, probably constituted by the acidic "residual" protein which is determinant for the formation of intermolecular bonds.

The temperatures of denaturation are not a criterion as sensitive as rigidity. In all cases of total recombination, the values are identical to the native DNH value, except when pure histone is used. We may conclude that the T_m value is mostly dependent on the protective effect of the positive charges of histone molecules but that there are also some structural features which are necessary for obtaining a maximum value.

From all the criteria studied here, we may conclude that the reconstituted nucleohistones are almost identical with the native ones if only the easily dissociable histones have been separated and recombined. We may speak about a reversible-combination type which would mostly include the lysine-rich histones, the binding of which is the weakest [42,43]. Other physico-chemical data on native and reconstituted soluble DNH confirm these views [37,38,44—46]. However if a more complete separation of proteins from DNA has been performed, the structure of reconstituted DNH is quite transformed; and this could explain why recombinations sometimes have been reported to display higher susceptibility to enzymic degradation [4,27] or different extractivity of histones [47].

We suggest that in the DNH molecules, DNA is firmly bound to a small quantity (4%) of acidic proteins, which may also play a role of links in chromosomes [17,48]. The histones are bound to those fundamental units by electrostatic linkages, the strength of which is in proportion to the arginine content. Their removal progressively alters the structural features of the DNH, leading to a decrease in the stability and in the reversibility of the recombination.

We thank Miss Renée Hacha who performed many analyses described in this work and Mr R. Gilles who made the amino-acid analysis in the Laboratory of Biochemistry (Prof. M. Florkin).

This work was part of a programme of the Centre National de Biochimie et de Biologie Moléculaire and was supported by the Fonds de la Recherche Scientifique Fondamentale Collective.

REFERENCES

1. Shooter, K. V., Davison, P. F., and Butler, J. A. V., Biochim. Biophys. Acta, 13 (1954) 192.
2. Dounce, A. L., and O'Connell, M., J. Am. Chem. Soc. 80 (1958) 2013.
3. Fredericq, E., Biochim. Biophys. Acta, 55 (1962) 300.
4. Fredericq, E., Biochim. Biophys. Acta, 68 (1963) 167.
5. Peacocke, A. R., In E. E. Polli and S. Zanzi, Antonio Baselli Conference on Nucleic Acids and Biology, Pavia 1964, p. 60.

6. Mazen, A., and Champagne, M., *Bull. Soc. Chim. Biol.* 47 (1965) 1951.
7. Itzhaki, R. F., *Biochem. J.* 100 (1966) 211.
8. Zubay, G., and Doty, P., *J. Mol. Biol.* 1 (1959) 1.
9. Murray, K., and Peacocke, A. R., *Biochim. Biophys. Acta,* 55 (1962) 935.
10. Bayley, P. M., Preston, B. N., and Peacocke, A. R., *Biochim. Biophys. Acta,* 55 (1962) 943.
11. Oth, A., and Desreux, V., *J. Polymer Sci.* 23 (1957) 713.
12. Ohba, Y., *Biochim. Biophys. Acta,* 123 (1966) 76.
13. Robinson, M. G., Weiss, J. J., and Wheeler, C. M., *Biochim. Biophys. Acta,* 124 (1966) 176.
14. Marushige, K., and Bonner, J., *J. Mol. Biol.* 15 (1966) 160.
15. Monty, K. J., and Dounce, A. L., *J. Gen. Physiol.* 41 (1958) 595.
16. Frearson, P. M., and Kirby, K. S., *Biochem. J.* 90 (1964) 578.
17. Dounce, A. L., and Hilgartner, C. A., *Exptl. Cell Research,* 36 (1964) 228.
18. Murray, K., *J. Mol. Biol.* 15 (1966) 409.
19. Leveson, J. E., and Peacocke, A. R., *Biochim. Biophys. Acta,* 123 (1966) 329.
20. Wang, T. Y., *J. Biol. Chem.* 241 (1966) 2913.
21. Kay, E. R. M., Simmons, N. S., and Dounce, A. L., *J. Am. Chem. Soc.* 74 (1952) 1724.
22. Walker, I. O., *J. Mol. Biol.* 14 (1965) 381.
23. Itzhaki, R. F., and Gill, D. M., *Anal. Biochem.* 9 (1964) 401.
24. Lowry, O. H., Roselbrough, N. J., Farr, A. L., and Randall, R. J., *J. Biol. Chem.* 193 (1961) 265.
25. Destexhe, F., Fredericq, E., and Desreux, V., *J. Chim. Phys.* 62 (1965) 913.
26. Houssier, C., and Fredericq, E., *Biochim. Biophys. Acta,* 120 (1966) 113.
27. Sarkar, N. K., and Dounce, A. L., *Arch. Biochem. Biophys.* 92 (1961) 321.
28. Eigner, J., and Doty, P., *J. Mol. Biol.* 12 (1965) 549.
29. Giannoni, G., and Peacocke, A. R., *Biochim. Biophys. Acta,* 68 (1963) 157.
30. Lloyd, P. H., and Peacocke, A. R., *Biochim. Biophys. Acta,* 95 (1965) 522.

31. Houssier, C., and Fredericq, E., *Biochim. Biophys. Acta,* 88 (1964) 450.
32. Bauer, R. D., and Johanson, R., *Biochim. Biophys. Acta,* 119 (1966) 418.
33. Jones, A. S., and Marsh, G. E., *Biochim. Biophys. Acta,* 14 (1954) 559.
34. Champagne, M., Mazen, A., and Pouyet, J., *Biochim. Biophys. Acta,* 87 (1964) 682.
35. Lindsay, R. H., Paik, W. K., and Cohen, P. P., *Biochim. Biophys. Acta,* 58 (1962) 585.
36. Murray, K., In *The Nucleohistones* (edited by J. Bonner and P. Ts'o), Holden-Day, San Fransisco 1964, p. 21.
37. Butler, J. A. V., In *The Nucleohistones* (edited by J. Bonner and P. Ts'o), Holden-Day, San Fransisco 1964, p. 36.
38. Ohba, Y., *Biochim. Biophys. Acta,* 123 (1966) 84.
39. Huang, C. C., Bonner, J., and Murray, K., *J. Mol. Biol.* 8 (1964) 54.
40. Itzhaki, R. F., *Proc. Roy. Soc. (London) Ser. B,* 164 (1966) 411.
41. Itzhaki, R. F., *Proc. Roy. Soc. (London) Ser. B,* 164 (1966) 75.
42. Johns, E. W., and Butler, J. A. V., *Nature,* 204 (1964) 853.
43. Akinrimisi, E. O., Bonner, J., and Ts'O, P. O. P., *J. Mol. Biol.* 11 (1965) 128.
44. Zubay, G., and Wilkins, M. H. F., *J. Mol. Biol.* 9 (1964) 246.
45. Crampton, C. F., Lipshitz, R., and Chargaff, E., *J. Biol. Chem.* 206 (1954) 499.
46. Crampton, C. F., and Chargaff, E., *J. Biol. Chem.* 226 (1957) 157.
47. Crampton, C. F., and Scheer, J. F., *J. Biol. Chem.* 227 (1957) 495.
48. Ris, H., *Can. J. Genet. Cytol.* 3 (1961) 95.
49. Moore, S., Spackman, D. H., and Stein, W. H., *Anal. Chem.* 30 (1958) 1185.

E. Fredericq and C. Houssier
Institut de Chimie physique
Université de Liège
2 Rue Armand Stevart, Liège, Belgium

European J. Biochem. 1 (1967) 61—69

The Binding of Calcium at Lipid-Water Interfaces

H. HAUSER and R. M. C. DAWSON

Biochemistry Department, Agricultural Research Council, Institute of Animal Physiology, Babraham, Cambridge

(Received November 19, 1966)

Measurements have been made of the binding of radioactive calcium to unimolecular films of purified lipids equivalent to those found in cell membranes.

The adsorption of ^{45}Ca on pure phospholipid films is largely independent of the chemical nature of the phospholipid. The affinity is controlled by Coulombic forces and is directly related to the nett excess negative charge on the lipid molecule.

With pure triphosphoinositide films at collapse pressure on a subphase at pH 5.5 there is evidence that calcium adsorption is lower than that predicted assuming full ionisation of all five anionic sites. This is probably due to the high surface potential suppressing ionisation. After reducing the pressure or after dilution of the triphosphoinositide with lecithin the adsorption of calcium per molecule approaches that predicted from the other phospholipids.

The electrophoretic mobility of lecithin particles containing small amounts of anionic phospholipids indicates that at a bulk pH of pH 5.5 both negative sites on phosphatidic acid and all five negative sites on triphosphoinositide are fully ionized at the 'lecithin'-water interface.

The calcium adsorbed on anionic phospholipids and gangliosides is displaced by a large excess of Na^+ or K^+ but in no instance is one univalent cation significantly more effective than the other. The adsorbed calcium is also displaced by Mg^{2+} but with all films there is a greater preference for Ca^{2+} ranging from 4.3 fold with gangliosides to 21 fold with triphosphoinositide.

The adsorption of calcium on stearic acid films is small until the subphase is adjusted to pH values above 9. This and the electrophoretic mobility data indicate that the carboxylic acid group at an interface is only fully ionized or orientated to the aqueous phase at a bulk pH greater than 11. The calcium adsorbed on stearic acid under alkaline conditions is less readily displaced by Na^+, K^+ and Mg^{2+} than from other lipids.

The physiological effects of calcium on excitable tissues such as the nervous system [1] are well known and Douglas and his collaborators [2] have found that it is essential for secretion of the adrenal medulla, submaxillary gland and neurohypophysis. Although some of these effects might be explained by its inhibition of the 'transport' ATPase through Ca ATP competing with Mg ATP for the enzyme [3] it has often been suggested that binding of calcium to macromolecular anionic sites on cell membranes may also be involved. In this respect most of the calcium in the cell, at least in liver and kidney, is bound by the nucleus and mitochondria while probably only about 10% is present in the cytoplasm [4,5]. Even the cytoplasmic calcium is likely to be complexed with macromolecules. Thus Hodgkin and Keynes [6] have shown from electrophoresis studies that in squid nerve axoplasm only a very small percentage of the calcium present is ionized and a similar conclusion was reached by Harris [7] in his studies on the intracellular calcium of frog muscle.

Among the favourite candidates for supplying the anionic sites for calcium binding are the acidic lipids. It is known that lipids such as cephalin[8],

phosphoinositides [9,10], gangliosides [11,12], phosphatidic acid [13,14], and phosphatidyl serine [15,16] have a high affinity for both calcium and magnesium. Membranous structures of the cell such as mitochondria are rich in lipids. It has been suggested that the phospholipids of mitochondria are the initial receptor sites for the calcium and other divalent cations taken up by these organelles at least in the absence of inorganic phosphate [17—19]. Calculation suggests that phospholipids are the only contenders for receptor sites present in sufficient quantities to account for the cations imbibed.

Although it has been assumed that big quantitative differences would exist between the affinities of the anionic groups of various lipids for calcium, no attempt has been made to systematically examine this in a single investigation. In this study we have measured the adsorption of ^{45}Ca on unimolecular films of highly purified samples of nearly all the acidic phospholipids found in cell membranes. In addition the displacement of the adsorbed calcium by magnesium, sodium and potassium has been examined. While this investigation was in progress Rojas and Tobias [20] published a similar but more limited investigation of the adsorption of calcium on unimolecular films of phosphatidyl ethanolamine and phosphatidyl serine.

METHODS

Unimolecular Films and Surface Pressure

The trough was milled from a block of polytetrafluoroethylene (Fluon, I.C.I. Ltd.) and was 20 cm long, 4.5 cm wide and 0.8 cm deep (72 ml capacity). It was fixed to a rigid brass framework to prevent distortion and stirred from below using a reciprocating magnet moving a glass-sheathed steel stirrer

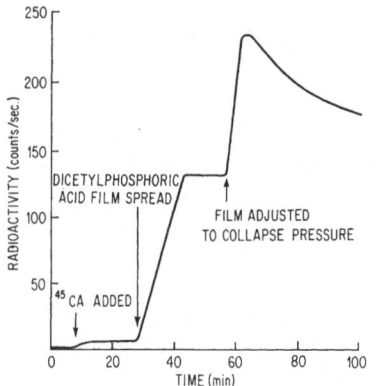

Fig. 1. *Diagram of the surface radioactivity of a dicetylphosphoric acid film with* ⁴⁵*Ca in the subphase constructed from the recorder trace.* The increase in radioactivity occurring on compressing the film to collapse pressure is mainly due to the area monitored by the counter containing more phospholipid molecules complexed with ⁴⁵Ca. Any increase in the actual affinity of ⁴⁵Ca per molecule due to the pressure change [20] is minimal

(1 mm diameter). Sufficient lipid dissolved in chloroform-ether (or chloroform-methanol with gangliosides) to form a unimolecular film was spread on the surface from a micrometer syringe (Agla). The surface pressure was adjusted by moving a Fluon barrier across the surface and measured as the force exerted on a glass dipping plate made from half a microscope cover slip suspended from the arm of a torsion balance. Attempts to use platinum (grey) as the hydrophilic dipping surface led to unsatisfactory and inconsistent results with certain lipids possibly due to the metal accelerating the autoxidation of the film.

Measurement of Calcium Adsorption

An aliquot of ⁴⁵CaCl₂ (0.0154 μmole, 3.5 μcuries, Radiochemical Centre, Amersham, England) was added to the trough which had been filled to the brim

with distilled water. The radioactivity emanating from the surface was detected by a mica-window Geiger-Müller tube located 7 mm above the surface. The pulses were integrated (over 20 sec) in an Ecko ratemeter and the output voltage monitored with a pen recorder. A know amount of lipid was then spread and the barrier moved by a calculated amount so that the film was compressed to the collapse pressure; with ganglioside the calcium adsorption was measured at 30 dynes/cm as more compressed films were unstable. The calcium adsorption on the film was measured as the increase in radioactivity above the background achieved at equilibrium (Fig.1). With most acidic lipid films there was, to a greater or lesser degree, a slow decline of the radioactivity after maximum adsorption of ⁴⁵Ca had been reached. This decrease, which can presumably be attributed to loss of film material [21] appeared to be most marked with the more acidic lipids. When appropriate a suitable correction was derived by extrapolating the radioactivity/time curve back to the time of film spreading or adjustment, it being assumed that at this point no lipid would be solubilized. When investigations were made of the displacement of adsorbed ⁴⁵Ca by other metallic cations, the radioactivity of the experimental film in the presence of the displacing cation was compared directly with a control film of the same lipid at a similar time after spreading.

To calculate the adsorption of calcium in terms of Ca ions/cm² certain experiments were performed in which the surface radioactivity increment produced on spreading a film of known area was related to the decline of radioactivity in the subphase. This then gave a direct relationship between the surface radioactivity and the amount of calcium ions/cm² adsorbed on the film.

MATERIALS

The following lipids were prepared: egg lecithin and phosphatidyl ethanolamine [22], phosphatidyl inositol [23], phosphatidic acid [24], triphosphoinositide [25], sphingomyelin [26]. Phosphatidyl serine was a gift from Mr. N. Niller, and ox brain ganglioside (28⁰/₀ N-acetylneuraminic acid) from Dr. D. B. Gammack. Dicetylphosphoric acid was a commercial sample (Albright & Wilson, Ltd., London) as was cerebroside (Koch-Light, Ltd., Colnbrook, England.) Purity of samples was checked by thin layer chromatography and alkaline degradation [27].

RESULTS

Force-area Curves

Curves relating the surface pressure to the area per molecule were constructed for each lipid. The area per lipid molecule at the collapse pressure was derived from a number of such curves and the mean values are reported in Table 1. These values were used to calculate the Ca ions adsorbed/molecule of lipid since all determinations of calcium adsorption were made at the collapse pressure except for gangliosides where because of film instability, the binding was measured at film pressures of 30 dynes/cm.

It has been reported that saturated fatty acid chains have limiting areas of 21 Å² for saturated chains and 32 Å² for unsaturated chains [28]. Thus egg lecithin, which analysis shows has roughly 50%

the head groups. The phosphatidyl serine contained predominantly saturated fatty acid residues and this may explain why the limiting area/molecule was somewhat lower than the other naturally-occurring phospholipids examined. The heart muscle sphingomyelin was a mixture of stearoyl and nervonoyltypes and its limiting area/molecule of 51 Å² is intermediate between the values of 40 Å² and 56 Å² reported for these two forms [30]. The limiting area of 36 Å²/molecule for dicetylphosphoric acid (38 Å² on 0.145 M NaCl) agress well with the experimental

Table 1. *Adsorption of ⁴⁵Ca on unimolecular films of lipids*

Lipid film (at collapse pressure unless stated)	Area/ molecule at collapse pressure	Bulk phase composition	Surface radio-activity at collapse pressure	Ca atoms adsorbed	Moles lipid	Lipid molecules per Ca atom	Number excess -ve charges on lipid	-ve charges per Ca atom
	Å²		counts/sec	cm² × 10⁻¹⁴	cm² × 10⁻¹⁴			
Phosphatidyl-choline	52	H_2O	0	0			0	—
Phosphatidyl ethanolamine	50	H_2O	7	0+			0	—
Phosphatidylserine	44	H_2O	145	0.39	2.3	5.8	1	5.8
Phosphatidyl inositol	51	H_2O	140	0.38	2.0	5.2	1	5.2
Sphingomyelin	51	H_2O	3					
Phosphatidic acid	54	H_2O	262	0.71	1.9	2.6	2	5.2
Triphosphoinositide	61 [a]	H_2O	290	0.78	1.7	2.1	5	10.5
Triphosphoinositide (2.5 dynes/cm)	183	H_2O	150	0.41	0.546	1.33	5	6.6
Lecithin + 5 mole-% Triphosphoino-sitide		H_2O	16.6	0.045	0.061	1.36	5	6.4
Ganglioside (30 dynes/cm)	78 [b]	H_2O	78	0.21	1.28	6.0	1.6 [c]	9.7
Cerebroside	36 [d]	H_2O	0	0			0	—
Stearic acid	19	H_2O	8	0+	—		0	—
		H_2O at pH 9 (NaOH)	129	0.36	5.25	14.6		
		H_2O at pH 11.5 (NaOH)	274	0.75	5.25	7.0		
Dicetylphosphoric acid	36 [e]	H_2O	184	0.50	2.8	5.5	1	5.5

[a] A value of 61 Å² was obtained by compressing the film quickly. Slow compression resulted in a lower area/molecule, presumably due to film solubility. Slow compression on 0.125 M—0.5 M NaCl also gave 61 Å²/triphosphoinositide molecule at collapse pressure.
[b] At 30 dynes/cm.

[c] Calculated from the *N*-acetylneuraminic acid content (28%) and the formula given by Ledeen [40].
[d] Assuming mol. wt. of 812.
[e] 38 Å² on 0.145 M NaCl.

of each type of acid, would be expected to have a limiting area of 53 Å/molecule which compares favourably with the experimental value (52 Å²) although it is below the value reported for synthetic lecithins [29]. All the naturally-occurring diacylphosphoglycerides examined gave values within these theoretical limits, *i.e.* 42 to 64 Å² for completely saturated and unsaturated phospholipids respectively. The limiting area per molecule for triphosphoinositide is towards the upper limit but the preparation was known to be highly unsaturated, and the large number of negative charges per molecule may cause expansion of the film by electrostatic repulsion of

values of others [31,28]. The predicted area would be 38 Å² as it is to be expected that each hydrocarbon chain would have a cross sectional area of 19 Å², *i.e.* somewhat less than a fatty acid residue, due to the absence of a ketonic acyl group. The limiting area/molecule of stearic acid was found to be 19 Å² on distilled H_2O when overcompression phenomena [32,33] were avoided. It has been reported [34] that the area per molecule of stearic acid films does not change between pH 6 and 12 and consequently this value was used for calculating the ⁴⁵Ca adsorption/mole. at higher pH values. The area per molecule occupied by the ganglioside preparation was measured at a film pressure of 30 dynes/cm due to excessive film instability at higher pressures. Its value lies between those of 66 Å² and 95 Å² (at 30 dynes/cm) obtained for mono- and di-sialo gangliosides respectively [35] and is consistent with its neuraminic acid analysis.

Adsorption of Calcium

To obtain a measurable increment in the level of surface radioactivity due to the binding of ⁴⁵Ca it was essential to add calcium of high specific radioactivity. This meant that, because of practical considerations, it was necessary to use a very low concentration of calcium in the bulk phase. However, as has been pointed out previously the level of ionized

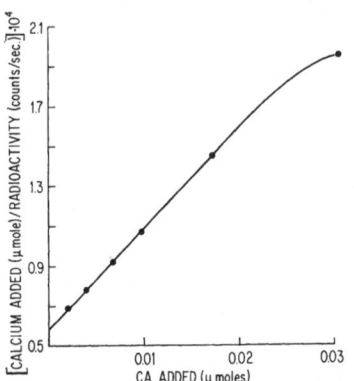

Fig.2. *Langmuir-type plot of the calcium concentration against this concentration divided by the calcium adsorbed.* Phosphatidyl inositol film—specific radioactivity of calcium was the same for all points

calcium in the cell is extremely low and the results are therefore physiologically meaningful. In addition the complications occurring through the calcium causing expansion or contraction of the unimolecular film [28] are minimized. The concentration of calcium added did not, with any lipid, saturate the available negative sites on the film and consequently a direct comparison is possible between the affinities of individual lipids for calcium at a single calcium concentration.

At low calcium concentrations, the relationship between the calcium concentration and adsorption initially conforms with the Langmuir adsorption isotherm

$$\text{adsorption (surface radioactivity)} = \frac{K_1 c}{K_2 + c}$$

where c = concentration of calcium and K_1 and K_2 are constants so that when c is plotted against c/surface radioactivity, direct proportionality is seen

(Fig.2) at least until higher concentrations of calcium are reached.

In Table 1 the adsorption of Ca ions to unimolecular films of various lipids at collapse pressure are compared. At pH 5.5 no measurable calcium adsorption occurred on films of phosphatidyl choline, phosphatidyl ethanolamine, sphingomyelin, cerebroside or stearic acid. The phospholipid films would be expected to be isoelectric at the pH of distilled water, due to their polar groups being completely ionized and forming a balanced zwitterion. The carboxylic group of stearic acid is presumably not ionised at this pH while cerebroside possesses no negative polar group in its structure.

With all phospholipid films which bind calcium at pH 5.5 the degree of adsorption was, with one exception, related to the number of excess negative charges available on the surface, assuming complete ionisation. Thus phosphatidyl inositol, phosphatidyl serine, phosphatidic acid and dicetylphosphoric acid had at this calcium concentration approximately one calcium ion distributed between 5.2—5.8 negative charges on the surface (Table 1). The exception to this picture was triphosphoinositide films at collapse pressure where the adsorption of calcium was only about 50% of that anticipated if the five negative sites available on each molecule were fully ionized, *i.e.* 1 Ca²⁺/ 10,5 -ve charges (Table 1). However, the adsorption of calcium on triphosphoinositide films at low pressures or on mixed films of lecithin + 5 mole % triphosphoinositide was much greater and approached the same value as found for the other phospholipids (Table 1). It was assumed that the lecithin was not contributing to the adsorption under such conditions.

Rather surprisingly adsorption of calcium on a stearic acid film did not become appreciable until the pH value was above 9. At pH 11.5, when presumably the carboxylic acid groups are fully ionized (see next section), the magnitude of the adsorption per negative charge is less than that on a phosphate group, and this also applies to the binding of calcium on the carboxylic groups of gangliosides at pH 5.5 (Table 1).

Electrophoresis of Lecithin Particles Mixed with Acidic Phospholipids

The above results would suggest that at the pH of distilled water (about 5.5) the interfacial anionic sites on phospholipids containing monoesterified phosphate groups such as phosphatidic acid or triphosphoinositide (admixed with lecithin) are fully ionized. Consequently experiments were made in which acidic phospholipids were added to a zwitterionic phospholipid (lecithin) and the resulting lipid mixture 'emulsified' with dilute NaCl solution at pH 5.5. Measurement of the electrophoretic mobility of such particles (Fig.3) showed that the progressive addition of acidic phospholipids produced an incre-

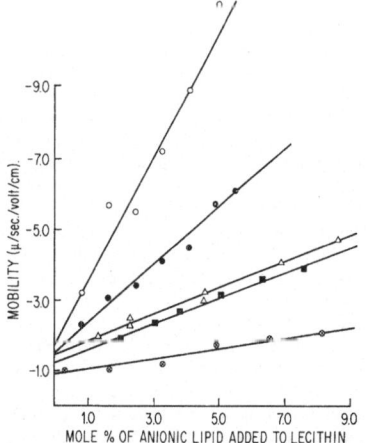

Fig. 3. *Change in electrophoretic mobility of lecithin particles when anionic lipids are added.* The lipids were mixed with ovolecithin in organic solvent solution and after removing the solvent *in vacuo* the mixed lipid samples were 'emulsified' with 0.025 M NaCl solution pH 5.5 (stearic acid, pH 8). The electrophoretic mobilities were measured in an horizontal cell [36]. ○, Triphosphoinositide; ●, Phosphatidic acid; △, Dicetylphosphoric acid; ■, Phosphatidyl inositol; ⊗, Stearic acid

Table 2. *Change in electrophoretic mobility of lecithin particles produced by adding anionic lipids*
Experimental details are described in the legend to Fig. 3. The increment in the mobility was calculated from the slope of the mobility/anionic amphipath concentration curves (Fig. 3)

Acidic lipid added	pH of bulk aqueous phase	Increment in mobility $\frac{\Delta\mu/\sec/cm/V}{\Delta \text{ mole }\%}$	Ratio of increment to that of dicetylphosphoric acid	Number of possible -ve charges
Dicetylphosphoric acid	5.5	0.367	1.0	1
Phosphatidylinositol	5.5	0.366	1.0	1
Phosphatidic acid	5.5	0.810	2.2	2
Triphosphoinositide	5.5	1.81	4.9	5
Stearic acid	5.5	0	0	0
	8.0	0.137	0.37	
	11.6	0.370	1.0	1

ment of the mobility which was, at low concentrations of anionic lipids, directly proportional to the molecular percentage of acidic phospholipid added. Moreover, the magnitude of the change in the electrophoretic mobility was directly related to the total number of possible negative charges on the acidic

phospholipid molecule (Table 2). The uniformity of these results suggests that the added anionic phospholipids are uniformly distributed throughout the lecithin particle and do not for example become concentrated at the lecithin/bulk phase interface.

Measurements of the mobility of lecithin particles containing stearic acid indicated that the carboxylic acid groups were not fully ionized, or alternatively orientated to the aqueous phase until the bulk pH rose above 11. At this point the increment in mobility per charged group (Fig. 3, Table 2) became equivalent to that of the phospholipids. At this high pH value lecithin itself would not be expected to have any effect on the increment as it still behaves as a zwitterion [36].

Displacement of Adsorbed Calcium by Sodium Potassium and Magnesium

A rapid loss of adsorbed ^{45}Ca occurred from most lipid films when either sodium, potassium or magnesium chloride was added to the subphase. A typical form of the relationship between the concentration of displacing ion added and the loss of adsorbed calcium is shown in Fig. 4. A theoretical form of the displacement curve for the univalent cations can be derived from the independent equilibrium equations (1) and (2)

$$\text{Ca}_b + \text{Li e} \rightleftharpoons \text{Ca}_s \qquad (1)$$

where Li e = number of charges/cm^2 lipid interface, Ca_s = number of calcium ions/cm^2 lipid interface, and Ca_b = concn. of calcium in bulk (μmole/ml). Then

$$\text{Li e} = \frac{\text{Ca}_s}{\text{Ca}_b} K_1 .$$

Similarly if sodium is the displacing cation

$$\text{Na}_b + \text{Li e} \rightleftharpoons \text{Na}_s \qquad (2)$$

and

$$\text{Li e} = \frac{\text{Na}_s}{\text{Na}_b} K_2$$

$$\frac{\text{Ca}_s \cdot \text{Na}_b}{\text{Na}_s \cdot \text{Ca}_b} = \frac{K_2}{K_1} = K. \qquad (3)$$

For the purpose of the calculation it is assumed that at the high Na concentrations used for displacement virtually all the available anionic sites are occupied by Na so that assuming two point electrostatic attachment of calcium

$$\text{Li e} - 2\,\text{Ca}_s = \text{Na}_s.$$

The theoretical curve is plotted in Fig. 4 and the constant K in equation (3) is given in Table 3 for a number of anionic lipids. The small variation in the constant over a range of univalent ion concentrations indicates that the relationship given in equation (3) is generally valid.

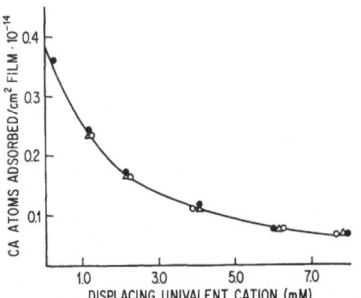

Fig. 4. *The displacement of adsorbed ⁴⁵Ca from a unimolecular film of phosphatidyl serine by sodium and potassium.* ●, NaCl added; △, KCl added; ○, Theoretical curve drawn from the mean constant K (see text) calculated for all experimental points

Table 3. *Mean K (constant) for the displacement of calcium from anionic lipids by sodium*

Lipid	K (mean ± S.E.)	Sodium concentration range
		μ atom/ml
Phosphatidyl serine	1.51 ± 0.03	1.2 – 8.0
Dicetylphosphoric acid	1.07 ± 0.05	1.3 – 10.4
Phosphatidyl inositol	0.48 ± 0.04	1.2 – 8.9
Phosphatidic acid	1.55 ± 0.11	1.3 – 7.4
Ganglioside	0.188 ± 0.014	0.8 – 2.4

Table 4. *Displacement of Ca^{2+} adsorbed on lipid/water interfaces by Na^+, K^+ and Mg^{2+}*
[All unimolecular films at collapse pressure except ganglioside (23 dynes/cm)] subphase at pH 5.5

Lipid film	Ratio of ion concentrations [a] required to reduce ⁴⁵Ca adsorption by half			Relative effeciency of displacement of Ca⁴⁵ by cations
	K^+/Ca^{2+} × 10⁻³	Na^+/Ca^{2+} × 10⁻³	Mg^{2+}/Ca^{2+}	Na : K : Mg
Phosphatidylserine	82	82	8.0	1:1 : 1050
Phosphatidylinositol	50	50	5.2	1:1 : 980
Phosphatidicacid	78	80	<16	1:1.02:>490
Triphosphoinositide	85	86	21	1:1.01: 405
Ganglioside	33	34	4.3	1:1.03: 790
Dicetylphosphoric-acid	98	98		1:1

[a] These ratios refer to the total ion concentrations present in the system.

The concentration of displacing ion required to remove half the adsorbed calcium varied with the nature of the lipid film (Table 4). Thus for example the half desorption values for K^+ varied from

3300 times the Ca^{2+} concentration for ganglioside to 9800 times with dicetylphosphoric acid while for Mg^{2+} it varied from 4.3 times the Ca^{2+} concentration with ganglioside to 21 times with triphosphoinositide. However no difference could be detected between the ability of sodium and that of potassium to displace calcium from an individual lipid film. The relative ability of magnesium to displace calcium compared with the univalent cations varied with the nature of the lipid film.

The calcium adsorbed on unimolecular films of stearic acid at a bulk pH of 11.5 was difficult to displace with other metallic cations. The adjustment to the alkaline pH meant that the calcium had already been adsorbed in competition with Na^+ added as NaOH to the system. The addition of further sodium or potassium ions (as NaCl, KCl) at 10⁵ times the concentration of the calcium ions present caused no loss of adsorbed calcium from the film. Magnesium ions added at a concentration 1.5 · 10⁴ times greater than the calcium caused a loss of only 19% of the adsorbed ⁴⁵Ca.

DISCUSSION

It must be borne in mind that the study of calcium adsorption on unimolecular lipid films at the air/water interface can only act as a model for understanding the physiological adsorption of calcium on cell membranes. Unimolecular films would in certain respects behave differently from the bimolecular lipid leaflets which may exist in cell membranes [37] and, moreover, the protein which is certainly a component of such membranes may profoundly modify the adsorption.

It is clear from the present results that at low concentrations of calcium no appreciable adsorption occurs unless the lipid constituting the unimolecular film has a nett negative charge. Thus at pH 5.5 no measurable adsorption of calcium occurred on zwitterionic phospholipids, namely phosphatidyl choline, phosphatidyl ethanolamine and sphingomyelin, which usually constitute the bulk of the polar lipids present in cell membranes. Rojas and Tobias [20] were similarly unable to detect calcium adsorption on unimolecular films of animal lecithin (ambient Ca concn. 0.1 M) although Kimizuka and Koketsu [39] reported that 'multilayer' films of an impure animal lecithin did bind calcium ions at this concentration. However the latter sample may have been contaminated with anionic phospholipid and, in fact the same authors found minimal adsorption of calcium on synthetic dipalmitoyl lecithin at 0.1 mM Ca^{2+}. It is known both from the effect of calcium on the electrophoretic mobility of lecithin particles [41] and the surface potential of lecithin films [28] that the metal is adsorbed by counter ion attraction at the lecithin/water interface but this would only be of significance at much higher concentrations of calcium. At higher bulk pH values (20) phosphatidyl ethanolamine was found to adsorb calcium and this probably corresponds with the development of a negative surface potential due to depolarisation of the amino group [22].

All of the films of pure anionic phospholipids examined gave an adsorption of calcium which was with one exception independent of the chemical nature of the phospholipid and which was in magnitude directly proportional to the total number of possible nett negative charges on the surface. At the bulk pH used (5.5) it is to be expected that the anionic phosphate sites on phosphatidyl inositol and dicetylphosphoric acid would be completely ionized. With phosphatidyl serine it is also known that the carboxyl group is fully ionized at this pH [20,42,43] so that this phospholipid would have a single nett negative charge. The calcium adsorption on phosphatidic acid can only be accounted for if it is assumed that both anionic sites on the mono-esterified phosphate group are fully ionized and then the adsorption of calcium per negative charge conforms with the other phospholipids. Films of pure phosphatidic acid would be expected to have a higher interfacial potential than the previously discussed phospholipids since phosphatidic acid would have, if fully ionized, two negative charges per two hydrocarbon chains compared with the one negative charge of the other phospholipids. Theoretically it is possible that this high interfacial negative potential would act to suppress its own ionisation by attracting H⁺ and reducing the surface pH. However there is no evidence from the calcium adsorption that this occurs at pH 5.5. This might result from the calcium on the film reducing the ζ potential and increasing the surface pH.

In other words the ionisations

$$PA \rightarrow HPA^- + H^+$$
$$HPA^- \rightarrow PA^{2-} + H^+$$

would be displaced by the reaction

$$PA^{2-} + Ca^{2+} \rightarrow CaPA.$$

The electrophoretic data obtained in the present study also indicate that phosphatidic acid at an interface, at least when diluted with the zwitterionic lecithin, is fully ionized at pH 5.5. This conclusion is difficult to reconcile with the direct titration of phosphatidic acid micelles which gives pK values of 3.8 and 8.6 [13]. However it is known that pK values obtained by the direct titration of insoluble substances must be interpreted with caution and it is likely that the present results are more relevant to the ionisation of orientated phosphatidic acid molecules present in biological membranes.

With pure triphosphoinositide films at collapse pressure it seems that the very high interfacial potential (with 5 negative charges per two hydrocarbon chains if fully ionized) is sufficient to suppress ionisation at a bulk pH of 5.5 even in the presence of the calcium added. Thus the calcium adsorption is less than that predicted from assuming complete ionisation of the anionic sites at pH 5.5. The electrophoretic studies indicate that the anionic sites on triphosphoinositide, when diluted out with lecithin are fully ionized at this pH. When the packing density of triphosphoinositide in a unimolecular film is decreased by dilution with lecithin or by reducing the surface pressure, the surface potential is reduced, the suppression of ionisation is removed and consequently the adsorption of calcium per triphosphoinositide molecule increases and per charge, approaches that predicted from the other phospholipids.

The adsorption of calcium on the carboxylic groups of gangliosides (at pH 5.5) and stearic acid (at pH 11.5) does not appear to be so strong as that on the phosphate group of phospholipids (it being assumed that the ⁴⁵Ca is adsorbed on the phosphate group of phosphatidyl serine). The adsorption of ⁴⁵Ca on ganglioside films is not increased by increasing the pH to 8.3 so it is likely that all its carboxylic groups are fully ionized at pH 5.5 in the presence of calcium. On the other hand the ⁴⁵Ca adsorption and electrophoretic studies suggest that ionisation or orientation to the aqueous phase of the carboxylic group of stearic acid takes place only at high pH values (probably above 8.5). The extent of the calcium binding may in fact be decreased by the competition of the sodium ions added as sodium hydroxide to adjust the pH although the displacement studies would not suggest that this is an important factor. The apparent ionisation of interfacial stearic acid at high pH values indicated by the present studies is in agreement with the observations of Goddard and Ackilli [44]. These workers found that above pH 9 the surface potential of stearic acid monolayers changed dramatically, indicating either a marked increase in ionisation above this pH or that additional ionized molecules are involved in a structural change which in turn involves a big change in their dipole orientation. If true ionisation is involved it is very much higher than the pK values of short-chain water-soluble fatty acids obtained by titration and the pK of 5.77 (at 35°) found by conductivity and solubility measurements on stearic acid hydrosols [45].

The ready displacement of calcium from phospholipid and ganglioside films by other metal cations indicates that in every case the bound calcium is in equilibrium with ionised calcium in the subphase and that direct competition takes place for the available anionic sites. Clearly the affinity of phospholipid anionic sites for calcium is considerably greater than for sodium and potassium by a factor of 5,000 to

10,000 fold depending on the nature of the phospholipid. With all the anionic lipids tested, identical amounts of sodium or potassium added to the bulk phase produced an equal displacement of calcium from the unimolecular film. This indicates that there is no difference between the affinity of any of the anionic sites for sodium and potassium which might explain the well known selectivity of natural membranes for these ions.

It is known that the competition of univalent cations for calcium adsorbed on phospholipid films can vary slightly with the degree of packing of the molecules [20]. However it is probably true that if the adsorption of sodium or potassium on unimolecular films of acidic lipids is influenced by steric factors these would be more likely to show at the high packing densities at which the present observations were made. Various workers have found that in biphasic solvent systems phosphatidyl serine has a higher affinity for potassium than sodium [46,16] and from isolation studies others [47,38] have concluded the converse, that phosphatidyl serine has a higher affinity for sodium than potassium. However the present results for phosphatidyl serine obtained by the unimolecular film technique which agree with those of Rojas and Tobias [20] are likely to be much more pertinent to the true physiological situation because of the complicating effect of the solvent in biphasic studies or possible ion exchanges which could occur during isolation of the phospholipid. No difference in the permeability of phosphatidyl serine bilayers to sodium and potassium can be detected except at high and unphysiological calcium concentrations when the permeability to potassium is slightly higher [48].

It is difficult to explain the difference between the present results, which show that calcium has a far greater affinity for triphosphoinositide and phosphatidyl serine than Mg^{2+}, and those of Hendrickson and Fullington [16] who found in a titration system only minor differences between the stability constants of the calcium and magnesium complexes of these phospholipids. In biphasic systems both phospholipids have a distinct preference for calcium compared with magnesium [15,49].

The difficulty in displacing of ⁴⁵Ca from stearic acid films at a high bulk pH by other metallic cations might partially be explained by assuming that the adsorbed ⁴⁵Ca was already in competition with other cations added to adjust the pH. However calculation shows that the cations added are still much less effective at displacing the adsorbed calcium than that bound to the other anionic lipids. It is likely therefore that the bound calcium is very tightly complexed with the stearic acid presumably as an insoluble calcium salt whose dissociation is negligible compared with that of ⁴⁵Ca bound to other lipid films.

One of the authors (H. H.) thanks the British Council for a research scholarship, while on leave from the University of Graz. Dr. G. H. Sloane-Stanley is thanked for carrying out an analysis of the sphingomyelin used.

REFERENCES

1. Brink, F., Pharmacol. Rev. 6 (1954) 243.
2. Douglas, W. W., and Poisner, A. M., J. Physiol. (London), 172 (1964) 1.
3. Epstein, F. H., and Whittam, R., Biochem. J. 99 (1966) 232.
4. Thiers, R. E., and Valle, B. L., J. Biol. Chem. 226 (1952) 911.
5. Hofer, M., and Kleinzeller, A., Physiol. Bohemoslov. 12 (1963) 405.
6. Hodgkin, A. L., and Keynes, R. D., J. Physiol. (London), 138 (1957) 253.
7. Harris, E. J., Biochim. Biophys. Acta, 23 (1957) 80.
8. Drinker, N., and Zinsser, H. H., J. Biol. chem. 148 (1943) 187.
9. Folch, J., J. Biol. Chem. 177 (1949) 497, 505.
10. Dawson, R. M. C., Cyclitols and phosphoinositides (edited by H. Kindl), Pergamon Press, Oxford 1966, p. 57.
11. Van Heyningen, W. E., J. Gen. Microbiol. 31 (1963) 375.
12. Quarles, R., and Folch-Pi, J., J. Neurochem. 12 (1965) 543.
13. Abramson, M. B., Katzman, R., Wilson, C. E., and Gregor, H., J. Biol. Chem. 239 (1964) 4066.
14. Abramson, M. B., Katzman, R., Gregor, H., and Curci, R., Biochemistry, 5 (1966) 2207.
15. Breyer, U., and Quadbeck, G., J. Neurochem. 13 (1966) 493.
16. Hendrickson, H. S., and Fullington, J. G., Biochemistry, 4 (1965) 1599.
17. Slater, E. C., and Cleland, K. W., Biochem. J. 55 (1953) 566.
18. Chappell, J. B., Cohn, M., and Greville, G. D., Energy-linked functions of mitochondria, Academic Press, New York 1963, p. 219.
19. Peachey, L. D., J. Cell. Biol. 20 (1964) 95.
20. Rojas, E., and Tobias, J. M., Biochim. Biophys. Acta, 94 (1965) 394.
21. Gaines, G. L., Insoluble monolayers at liquid-gas interfaces, Interscience Publ. Co., New York 1966, p. 151.
22. Dawson, R. M. C., Biochem. J. 88 (1963) 414.
23. Dawson, R. M. C., Biochem. J. 68 (1958) 352.
24. Dawson, R. M. C., and Hemington, N., Biochem. J. 102 (1967) 76.
25. Dittmer, J. C., and Dawson, R. M. C., Biochem. J. 81 (1961) 535.
26. Davenport, J. B., and Dawson, R. M. C., Biochem. J. 84 (1962) 490.
27. Dawson, R. M. C., Hemington, N., and Davenport, J. B., Biochem. J. 84 (1962) 497.
28. Shah, D. O., and Schulman, J. H., J. Lipid Res. 6 (1965) 341.
29. Van Deenen, L. L. M., Houtsmuller, U. M. T., de Haas, G. H., and Mulder, E., J. Pharm. Pharmacol. 14 (1962) 429.
30. Raper, J. H., Gammack, D. B., and Sloane-Stanley, G. H., Biochem. J. 98 (1965) 21 P.
31. Parreira, H. C., and Pethica, B. A., In Proc. Second Int. Congr. Surface Activity, Butterworths Scientific Publications, London 1957, Vol. I, p. 44.
32. Nutting, G. C., and Harkins, W. D., J. Am. Chem. Soc. 61 (1939) 1180, 2040.
33. Rabinovitch, W., Robertson, R. F., and Mason, S. G., Can. J. Chem. 38 (1960) 1881.

34. Adam, N. K., and Miller, J. G. F., *Proc. Roy. Soc. (London) Ser. A*, 142 (1933) 401.
35. Raper, T., Private communication.
36. Bangham, A. D., and Dawson, R. M. C., *Biochem. J.* 72 (1959) 486.
37. Haydon, D. A., and Taylor, J., *J. Theoret. Biol.* 4 (1963) 281.
38. Katzman, R., and Wilson, C. E., *J. Neurochem.* 7 (1961) 113.
39. Kimizuka, H., and Koketsu, K., *Nature*, 196 (1962) 995.
40. Ledeen, R., *J. Am. Oil Chemists Soc.* 43 (1966) 57.
41. Bangham, A. D., and Dawson, R. M. C., *Biochim. Biophys. Acta*, 59 (1962) 103.
42. Garvin, J. E., and Karnovsky, M. L., *J. Biol. Chem.* 221 (1956) 211.
43. Abramson, M. B., Katzman, R., and Gregor, II. P., *J. Biol. Chem.* 239 (1964) 70.
44. Goddard, E. D., and Ackilli, J. A., *J. Colloid Sci.* 18 (1963) 585.
45. Datta, N. P., *J. Indian Chem. Soc.* 16 (1939) 573.
46. Solomon, A. K., Lionetti, F., and Curran, P. F., *Nature*, 178 (1956) 582.
47. Kirschner, L. B., *J. Gen. Physiol.* 42 (1959) 231.
48. Papahadjopoulos, D., and Bangham, A. D., *Biochim. Biophys. Acta*, 126 (1966) 185.
49. Dawson, R. M. C., *Biochem. J.* 97 (1965) 134.

H. Hauser and R. M. C. Dawson
Biochemistry Department
Agricultural Research Council
Institute of Animal Physiology
Babraham, Cambridge, England

European J. Biochem. 1 (1967) 70—72

Biosynthesis of Aristolochic Acid

H. R. Schütte, U. Orban and K. Mothes

Institut für Biochemie der Pflanzen der Deutschen Akademie der Wissenschaften, Halle/Saale

(Received November 3, 1966)

Some *Aristolochia* species were investigated for the presence of aristolochic acid and magnoflorine. In all cases these two compounds were found in the roots and rhizomes of the plants, [4-^{14}C]-Tetrahydropapaverine · HCl feeding to *Aristolochia sipho* gave no radioactive aristolochic acid I. Feeding of [4-^{14}C]-norlaudanosoline-HCl yielded radioactive aristolochic acid I. The carboxyl group contained 69$^0/_0$ of the radioactivity.

Organic nitro compounds do not occur very often in nature. The most important of these substances is the antibiotic chloromycetin, which was found by Ehrlich et al. [1]. In the following years further nitro compounds were isolated from higher plants as well as fungi. The first evidence for the biosynthesis of the nitro group was given by the experiments of Birkinshaw and Dryland [2] and Gatenbeck and Forsgren [3] on the β-nitropropionic acid; the nitrogroup is formed from the aminogroup of aspartic acid.

Mixtures of nitrophenanthrene-carboxylic acids were found in the roots and rhizomes of many *Aristolochia* species. The main acid is the aristolochic acid I. Up till now we have investigated the following Aristolochia species: *Aristolochia clematitis*, *Aristolochia elegans*, *Aristolochia fimbriata*, *Aristolochia rotunda*, *Aristolochia badamae*, *Aristolochia ornithocephala*, *Aristolochia durior (sipho)*. In all cases we were able to demonstrate the presence of aristolochic acid I and of the structurally related alkaloid magnoflorine in the roots and rhizomes. Only the aerial parts are free of these two compounds, according to our investigations.

Aristolochic acid I   Magnoflorine

Both the structure of aristolochic acid and its occurrence with magnoflorine lead to suppose a biogenetic relationship with the aporphine alkaloids [4]. The aristolochic acids could originate from the aporphine skeleton by oxidative cleavage of the heterocyclic ring. An important intermediate in the biosynthetic pathway is the benzylisoquinoline norlaudanosoline, which can be formed from tyrosine or a biochemical equivalent. Norlaudanosoline yields the aporphine skeleton by phenol oxidation and dienol-benzene rearrangement [5,6].

Spenser and Tiwari [7] fed DL-[3-^{14}C]-tyrosine, DL-[2-^{14}C]-dihydroxyphenylalanine, [2-^{14}C]-dihydroxyphenylethylamine- and DL-[2-^{14}C]-noradrenaline to *A. sipho*. In all experiments they were able to isolate radioactive aristolochic acid I. The decarboxylation of the acids yielded a distribution of radioactivity in accordance with the hypothesis. According to these results the authors discuss a biogenetic pathway via norlaudanosoline → orientaline → orientalinone → orientalinol to stephanine. Based on its methoxylgroups pattern this aporphine (see page 71) alkaloid could be a precursor of aristolochic acid I. For further elucidation of the pathway we used DL-[4-^{14}C]-norlaudanosoline-HCl as precursor. The synthesis of this benzylisoquinoline was started from veratrol, which was chlormethylated with radioactive p-formaldehyde. The resultant [2-^{14}C]-3,4-dimethoxybenzylchloride was converted to the nitrile with potassium cyanide [8] and this was reduced with platinum as catalyst to [2-^{14}C]-3,4-dimethoxyphenethylamine. Reaction with 3,4-dimethoxyphenylacetylchloride gave the corresponding amide, which was ringclosed with phosphorus oxychloride to give [4-^{14}C]-3,4-dihydropapaverine. Reduction with sodium borohydride gave DL-[4-^{14}C]-tetrahydropapaverine. Demethylation with hot concentrated hydrochloric acid yielded DL-[4-^{14}C]-norlaudanosoline · HCl with a specific radioactivity of 1.06×10^9 counts/min/ mmole (measured by a flow-counter Fa. Friesecke & Hoepfner). According to the theory, norlaudanosoline should give aristolochic acid I labelled at the carboxyl group.

Norlaudanosoline   Stephanine   Aristolochic acid I

Tetrahydropapaverine

We administered DL-[4-^{14}C]-norlaudanosoline·HCl and DL-[4-^{14}C]-tetrahydropapaverine · HCl to two-year-old plants of *Aristolochia sipho*, by infusion into the stem through a cotton wick. In the experiment with tetrahydropapaverine we obtained no radioactive aristolochic acid I, no phenol oxidation can take place because of the complete etherification of the hydroxygroups. In the experiment with norlaudanosoline we isolated an aristolochic acid I with the specific radioactivity of 6.3×10^4 counts/min/mmole. This corresponds to an incorporation rate of 0.003%. Decarboxylation yielded a barium carbonate with 4.3×10^4 counts/min/mmole, corresponding to 69% of the original activity. The decarboxylated aristolochic acid I contained only 5% of the activity. These results indicate that norlaudanosoline is incorporated specifically into aristolochic acid I, the carbon atom 4 giving the carboxyl carbon atom. Whether in fact norlaudanosoline represents an intermediate in the biosynthesis of aristolochic acid has still to be elucidated. Spenser and Tiwari [7] fed DL-[2-^{14}C]-noradrenaline and found a higher incorporation than in corresponding tyrosine experiments. This could indicate, that a 4-hydroxynorlaudanosoline plays a role in the pathway of aristolochic acid biogenesis.

EXPERIMENTAL

Feeding

A solution of 6 mg DL-[4-^{14}C]-norlaudanosoline · HCl or 6 mg DL-[4-^{14}C]-tetrahydropapaverine · HCl in 4 ml water was administered to 4 plants of *Aristolochia sipho* by the cotton wick method. The plants were collected after 6 days.

Extraction and Isolation

The dried roots and rhizomes were harvested, and extracted with methanol in a soxhlet apparatus [9]. The methanolic extract was evaporated, the residue dissolved in a hot solution of Na_2CO_3, acidified and extracted with ethyl acetate. The dried organic phase was evaporated under reduced pressure, the residue dissolved in methanol and chromatographed on thin layer plates (Silica gel G, Merck) with benzene-methanol-acetic acid (85:10:5, v/v/v) as solvent. The yellow-colored zone of aristolochic acid I (R_F 0.62—0.65) was eluted with acetic acid. After rechromatography and repeated elution a further purification was carried out on a cellulose column (solvent n-butanol-methanol-water [5:5:2, v/v/v]). The evaporated eluate was recrystallised from n-butanol. Yield: 6 mg.

Decarboxylation

6 mg aristolochic acid I was refluxed 30 minutes with 3 ml fresh distilled quinoline and 20 mg copper chromite [10]. Carbon dioxide thus formed was passed into an aqueous solution of barium hydroxide with a N_2-stream. The isolated barium carbonate was washed twice with water, alcohol and ether, respectively. The decarboxylated aristolochic acid I was dissolved in ether, filtered and washed with dilute hydrochloric acid, sodium bicarbonate and water. The dried organic phase was evaporated, the residue chromatographed on Al_2O_3 with benzene and then sublimated.

DL-[4-^{14}C]-Norlaudanosoline

840 mg fresh distilled veratrol and 3 ml benzene were saturated with dry hydrogen chloride with

cooling and then 110 mg p-formaldehyde (specific activity 66 mC/g) was added in small portions with further cooling and after passing hydrogen chloride through the reaction mixture. After 60 minutes the reaction mixture was stirred and neutralized with sodium bicarbonate, evaporated and fractionated in presence of sodium bicarbonate. Veratryl chloride was distilled at 90° under vacuum (1 mm Hg), and crystallized in white needles on the cooling surface.

Yield: 258.5 mg = 38.8%.

The further synthesis of norlaudanosoline was carried out according to the procedure of Battersby et al. [8].

Yield of norlaudanosoline · HCl: 19 mg.

Specific activity: 1.06×10^9 counts/min/mmole.

Grateful acknowledgement is made to Prof. Dr. M. Pailer and Dr. P. Patt for generous gifts of aristolochic acid I and of its methyl ester.

REFERENCES

1. Ehrlich, J., Bartz, Q. R., Smith, R. M., Joslyn, D. A., and Burkholder, P. R., Science, 106 (1947) 417.
2. Birkinshaw, J. H., and Dryland, A. M. L., Biochem. J. 93 (1964) 478.
3. Gatenbeck, S., and Forsgren, B., Acta Chem. Scand. 18 (1964) 1750.
4. Pailer, M., and Pruckmayr, G., Monatsh. Chem. 90 (1959) 145.
5. Barton, D. H. R., and Cohen, T., Festschrift. Arthur Stoll, Birkhäuser Verlag, Basel 1957, p. 117.
6. Battersby, A. R., Brown, R. T., Clements, J. H., and Iverach, G. G., Chem. Comm. (1965) 230.
7. Spenser, I. D., and Tiwari, H. P., Chem. Comm. (1966) 55.
8. Battersby, A. R., Binks, R., Francis, R. J., McCaldin, D. J., and Ramuz, H., J. chem. Soc. (1964) 3600.
9. Patt, P., Arzneimittel-Forsch. 15 (1965) 90.
10. Pailer, M., Belohlav, L., and Simonitsch, E., Monatsh. Chem. 87 (1956) 249.

H. R. Schütte, U. Orban, and K. Mothes
Institut für Biochemie der Pflanzen,
Deutsche Akademie der Wissenschaften zu Berlin,
×401 Halle/Saale, Weinbergweg, Germany

European J. Biochem. 1 (1967) 73—79

Absence of β Globin Synthesis
and Excess of α Globin Synthesis in Homozygous β-Thalassemia

A. Bargellesi, S. Pontremoli, and F. Conconi

Istituto di Chimica Biologica, Università di Ferrara

(Received November 2, 1966)

Globin chain synthesis has been studied in β-thalassemia by incubating red blood cells of 10 homozygous thalassemic patients with [³H] amino acids, and by measuring the incorporation of radioactivity into the globin chains, separated by column chromatography.

Five of the ten patients were never transfused, five were polytransfused.

The following results were obtained: (a) absence of β chain synthesis in both polytransfused and never transfused patients; (b) excess of α chain synthesis in the ten cases examined (ratio α synthesis/(γ + δ) synthesis = 3.0).

In three heterozygous subjects β globin synthesis and an excess of α chain synthesis was observed (ratio α/(β + γ + δ) = 1.6). The possible interpretations of these results are presented in the discussion.

β-Thalassemia is a genetically determined anemia of man, in which the major biochemical defect is a decreased production of hemoglobin A ($\alpha_2\beta_2$) dependent on a decreased production of β chain [1]. Our studies deal with β-thalassemia which is the most common type of thalassemia in the region of Ferrara.

The problem of thalassemia should be considered in relation to the problem of the control of protein synthesis. According to Watson [2], control of protein synthesis can be achieved at various levels along the line DNA-protein:

a) at the gene level (regulatory genes, rate of transcription of the genetic message);

b) at the level of the messenger RNA (differential stability of messenger RNA, possibly genetically determined);

c) at the polyribosome level (different rates of attachment for different messenger RNAs differential translation rates). In addition, for hemoglobin it is necessary to consider the relationships between heme synthesis and globin synthesis [3,4]. A possible mechanism for the β chain defect in thalassemia could be found at each of these levels. So far, among the several possible explanations, only a few have received some experimental support.

Itano [5] has suggested that the defect in thalassemia might be a mutational change of the messenger RNA, producing a triplet coding for the same amino acid at a slower rate: a so-called modulator-mutation [6]. This modulating triplet would be transcribed at the same rate as the normally coding triplet, but its nucleotide composition would be so altered as to

fit one of minor or less available amino acid transfer RNAs [7]. Consequently, the rate of translation would be decreased. Consistent with this hypothesis are Weatherall's recent data [8] which suggest a block or a "slow point" during the assembly of the β chain.

Ingram [9] has proposed that the production of a defective or altered stable hemoglobin messenger can be the explanation of defective β globin production in thalassemia. This mechanism would operate by blocking the available ribosomal sites for β messenger RNA. The nature of the defective messenger RNA is unknown.

Possible abnormalities at the polyribosome level in thalassemia have been examined by Burka and Marks [10]. They found a normal polyribosome content in thalassemic reticulocytes compared to normal reticulocytes, associated with a lower specific activity of the polyribosomes. In further studies Bank and Marks [11], using a cell-free system prepared from human reticulocytes, found the rate of [¹⁴C] amino acid incorporation by ribosomes from the cells of 8 thalassemic subjects to be 10 times smaller than that of 9 non-thalassemic hematological patients. These authors, using polyuridilic acid as synthetic messenger, found comparable rates of phenylalanine incorporation with ribosome fractions prepared from the cells of thalassemic and non-thalassemic subjects. These data are consistent with the hypothesis that an altered or deficient messenger RNA is produced in thalassemia.

In the present studies we have investigated α, β, γ and δ globin synthesis *in vitro* in intact red blood cells from subjects homozygous and heterozygous for β-thalassemia. Evidence will be presented to show the absence of β globin synthesis in the homozygous patients examined. In addition our results indicate that in both the hetero- and homozygous state, the α chain is synthesized in a large excess over the other globin chains. This result is similar to those reported by Weatherall et al. [8], Heywood et al. [12], and Bank and Marks [13].

EXPERIMENTAL PROCEDURES

Chemicals

[^{3}H]DL-valine (137 mC/mM), [^{3}H]L-leucine (5 C/ mM) and [^{3}H]DL-lysine (1.6 PC/mM) were purchased from New England Nuclear Corp. U.S.A. [^{3}H]DL- isoleucine (100 mC/mM) was purchased from Schwarz Bio-Research Inc. U.S.A. Carboxymethylcellulose was purchased from Whatman, England.

Materials and Methods

Venous blood samples of 10 homozygous β-tha- lassemic patients, of 3 heterozygous β-thalassemic subjects, of a subject presenting a mild anemia from bleeding, and of 2 premature babies, were collected in heparinized syringes and kept at 4°. Of the 10 homozygous patients we studied, 5 had never received transfusions and were examined immedi- ately before the first one, and 5 were polytransfused (2 of this last group have had splenectomy two years before our examination; see Table). Following washing in a cold solution containing 0.15 M NaCl and 1.5 mM MgCl$_2$, 0.4 ml of packed cells were incub- ated in 2 volumes of a modified Krebs Ringer bicar- bonate buffer medium with all the amino acids required [14], of which Valine, Leucine and Lysine were tritiated. The 3 labelled amino acids were added to give for each a final concentration of 0.1 mC/ml of cell suspension.

Incubations were carried out in an atmosphere of air with continuous shaking, at 37° for 4 hours unless otherwise indicated. The cells were recovered from the incubation mixture by centrifugation at 1000×g for 10 minutes. The cells were then lysed with 2 volumes of a solution containing 0.005 M Tris-HCl buffer (pH 7.4) and 0.003 M MgCl$_2$, and after 90 sec- onds the lysate made isotonic by addition of 0.1 vol of 1.5 M KCl. Unlysed cells and cell stroma were removed by centrifugation at 25,000×g for 30 minu- tes.

Globin was prepared from the lysate by the acid- acetone method [15]. The α, β, γ and δ globin chains were separated by column chromatography using the method described by Clegg et al. [16] with minor modifications. The absorbance pattern was conti- nously recorded with a Gilford automatic recording spectrophotometer Model 2000. The proteins not held

by the column were discarded. The other proteins were collected in fractions of 2.5 ml each, and an aliquot of each fraction was counted directly in a Tri-Carb liquid scintillation counter in 10 ml of Bray's solution [17]. This direct counting technique gave better results than the usual one (precipitation in the presence of bovine albumin and filtration onto millipore disks) being more rapid and simple. Calculations were made for correction of quenching and selfabsorption when necessary.

RESULTS

Identification of α, β, γ and δ Globin Chains after CM-cellulose Chromatography

The chromatographic procedure described under "Materials and Methods" provides a separation of the four globin chains. The identification of α and β globin peaks has been previously established [16].

The position of the γ globin peak was identified in separate experiments by incubating red blood cells of a premature baby with [^{3}H]isoleucine: it is well established that human γ chain contains iso- leucine, whereas α, β and δ chains do not contain this amino acid. All the radioactivity was found under a peak eluted before the β peak.

Another way to identify the γ chain peak has been the comparison of the data obtained by starch gel electrophoresis with the data obtained by column chromatography in cases with high hemoglobin F content. By starch gel electrophoresis and densito- metric analysis, it was possible to quantitate the amount of hemoglobin F and then the percent of γ globin present in the lysate. By column chromato- graphy it was possible to separate four peaks and to determine the percent of each one. With this method, also, γ chain was identified as the peak eluted before the β chain peak. This finding is in agreement with other reports [16,13].

This last procedure was used for the identifica- tion of the δ chain peak, again comparing the percent of δ chain as determined by starch gel electrophoresis with the values obtained by column chromatography, in cases with high hemoglobin A$_2$ content.

The δ chain peak, identified through this proce- dure, preceded the γ chain peak.

The position of γ and δ chain has also been preli- minarily confirmed by NH$_2$ and COOH terminal amino acid analysis (glycine and hystidine for γ chain, valine and hystidine for δ chain).

Globin Chain Synthesis in Control Subjects

Globin chain synthesis in control subjects has been determined by incubating red blood cells of a patient presenting a mild anemia from bleed- ing with [^{3}H]amino acids. Following the procedures described under "Methods", the labelled globins

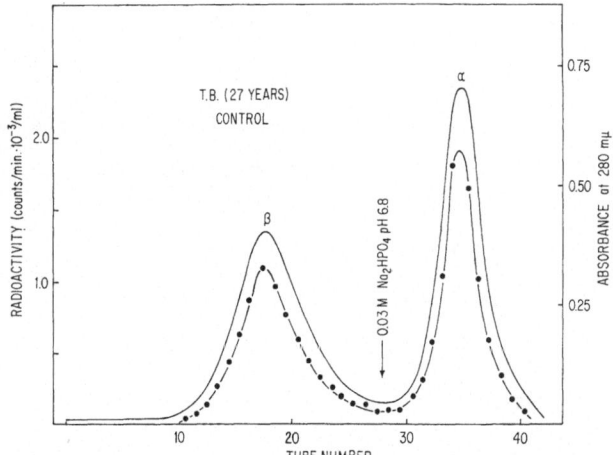

Fig. 1. *Elution pattern from CM-cellulose chromatography of the labelled globin chains, obtained by incubating the red blood cells of a subject presenting a mild anemia from bleeding, with [³H]amino acids.* The experimental procedures are reported under "Methods". The elution gradient was kept linear until the proteins of the beta chain region were completely eluted. The α chain was then eluted directly with 0.03 M Na₂HPO₄, pH 6.8. The A_{280} (——) and the radioactivity (•——•) measurements were carried out as described in the text

Table. *Amino acid incorporation into α, β, γ and δ globins in control subjects, and in heterozygous and homozygous β-thalassemic patients*

Patient	Diagnosis	Age	Radioactivity		Ratio
			$\beta + \gamma + \delta$	α	$\alpha/(\beta + \gamma + \delta)$
			counts/min [a]		
T.B.	Bleeding anemia	27 years	12,500	12,800	1.02
M.M.	Premature	7 months	68,200	70,000	1.03
S.D.	Premature	8 months	37,400	43,500	1.07
				Average	1.04
C.Ma.	Thalassemia minor	40 years	13,000	20,600	1.57
L.S.	Thalassemia minor	43 years	9,300	14,200	1.53
A.R.	Thalassemia minor	51 years	18,500	30,900	1.67
				Average	1.59

			Radioactivity			Ratio
			β	$\gamma + \delta$	α	$\alpha/(\gamma + \delta)$
			counts/min			
B.F.	Thalassemia major	10 months	0	94,000	245,000	2.7
D.L.	Thalassemia major	10 months	0	330,000	1,150,000	3.5
G.R.	Thalassemia major	15 months	0	183,000	780,000	4.1
S.R.	Thalassemia major	18 months	0	27,000	60,500	2.2
C.D.	Thalassemia major	4 years	0	35,000	102,000	2.9
I. D. [b]	Thalassemia major	4 years	0	681,000	1,362,000	2.0
N.M. [c]	Thalassemia major	6 years	0	191,000	006,000	5.1
C.Mo. [b]	Thalassemia major	7 years	0	78,000	188,000	2.4
M.V. [b]	Thalassemia major	8 years	0	54,000	160,000	2.7
C.Mass. [c]	Thalassemia major	9 years	0	121,000	280,000	2.3
					Average	3.0

[a] These counts/min represent the total amount of radioactivity incorporated into α, β, γ and δ globins by 0.4 ml of packed red blood cells after 4 hours of incubation with [³H]amino acids as described in the text.

[b] Polytransfused patients.

[c] Polytransfused and splenectomized patients.

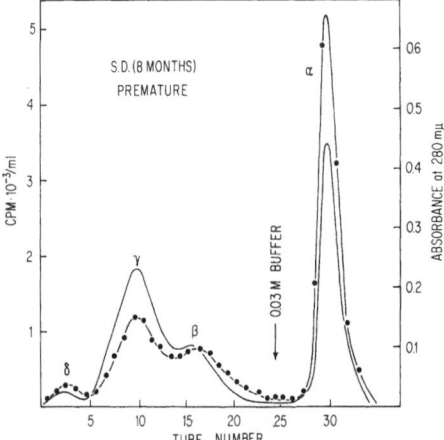

Fig. 2. *Elution pattern from CM-cellulose chromatography of the labelled globin chains obtained by incubating the red blood cells of an 8-month premature, with [³H]amino acids. The same experimental procedure described in Fig. 1, was followed. The A_{280} (——) and the radioactivity (•——•) measurements were carried out as described in the text*

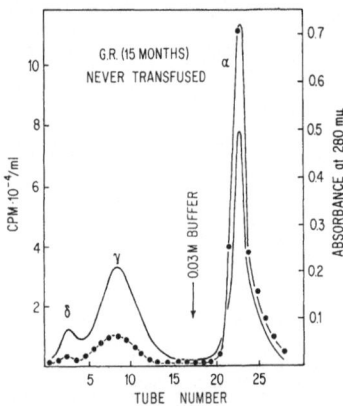

Fig. 3. *Elution pattern from CM-cellulose chromatography of the labelled globin chains obtained by incubating the red blood cells of a never transfused homozygous β-thalassemic patient, with [³H]amino acids. The same experimental procedure described in Fig. 1, was followed. The A_{230} (——) and the radioactivity (•——•) measurements were carried out as described in the text*

were submitted to column chromatography. In Fig. 1 the elution pattern reveals the presence of two peaks of protein in the α and β chain region. No γ and δ peak are detectable in this case. The same amount of protein and the same amount of incorporated radioactivity are present under the two peaks (ratio $\alpha/\beta = 1$; see Table).

We also studied the amino acid incorporation into globin by red blood cells of two premature non-thalassemic children. Four peaks of protein were separated by column chromatography, each peak corresponding respectively to α, β, γ and δ globin region. The amount of protein and the amount of incorporated radioactivity detected under the α peak were comparable to those found under the combined β, γ and δ peaks (ratio $\alpha/(\beta + \gamma + \delta) = 1$; see Table).

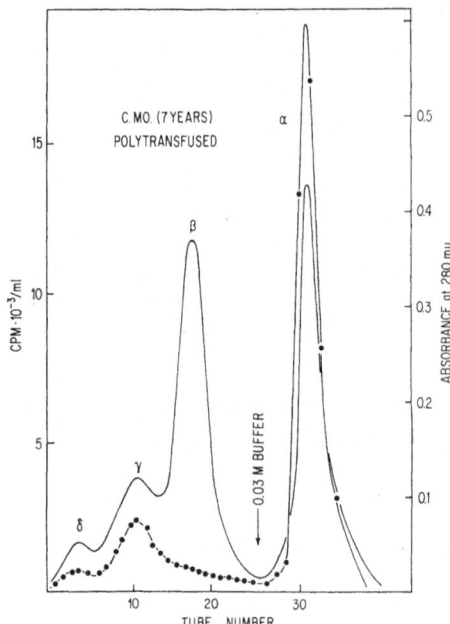

Fig. 4. *Elution pattern from CM-cellulose chromatography of the labelled globin chains obtained by incubating the red blood cells of a polytransfused homozygous β-thalassemic subject with [³H]amino acids. The same experimental procedure described in Fig. 1, was followed. The A_{280} (——) and the radioactivity (•——•) measurements were carried out as described in the text*

In the experiment carried out with red blood cells of an 8-month-old premature baby, as presented in Fig. 2, the switch from γ chain to β chain is underway as suggested by the higher specific activity of the β chain compared to that of γ chain.

Globin Chain Synthesis in Homozygous β-thalassemia

In order to follow the incorporation of radioactive amino acids into globin in homozygous β-thalassemia, red blood cells of 10 patients homozygous for the disease were incubated with [³H]amino acids as described under "Materials and Methods." The labelled globins, prepared from stroma-free lysate, were separated by column chromatography.

In Fig.3 are presented the results of an experiment carried out with red blood cells of a never transfused thalassemic.

Three protein peaks are present in the α, γ and δ chain region, while no protein peak is detectable in the β chain region. The β globin peak was absent in all the five never transfused patients examined. Furthermore, starch gel electrophoretic analysis of the lysates obtained from the red cells of the same patients never revealed the presence of hemoglobin A ($\alpha_2\beta_2$).

After chromatographic separation of labelled globin from red cells of a recently transfused homozygous β-thalassemic, the proteins present under the β globin region were eluted and submitted to a second chromatographic analysis whose pattern reported in Fig.5 indicates the absence of labelled proteins under the β peak. This provides further evidence for the absence of β globin synthesis in homozygous β-thalassemia.

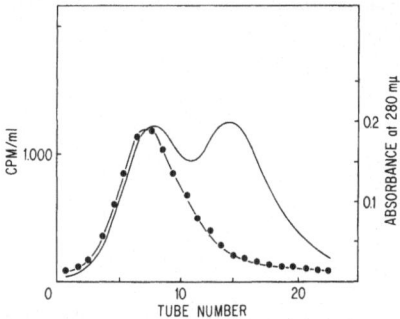

Fig.5. *Rechromatography on CM-cellulose column of the labelled proteins eluted in the β chain region.* The first chromatographic analysis was performed on the labelled globin chains obtained by incubating the red blood cells of a polytransfused homozygous β-thalassemic subject with [³H] amino acids. The peak emerging in the β chain regions was submitted to a second chromatographic analysis. The same experimental procedures described in Fig.1, were followed for both analyses. The A_{280} (——) and the radioactivity (•——•) measurements were carried out as described in the text

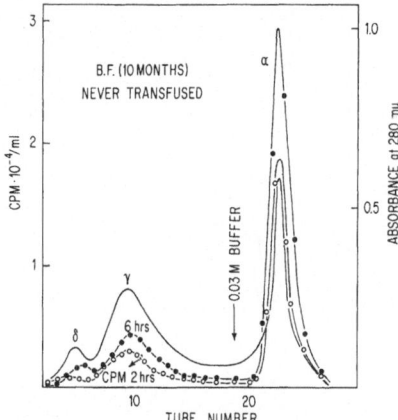

Fig.6. *Elution pattern from CM-cellulose chromatography of the labelled globin chains obtained by incubating the red blood cells of a never transfused homozygous β-thalassemic subject with [³H]amino acids.* The incubations were carried out in two separate samples for two and six hours. The same experimental procedures described in Fig.1, were followed. The A_{280} (——) and the radioactivity measurements at 2 hours (o——o) and six hours (•——•) were carried out as described in the text

In order to confirm this absence of β chain, the labelled globins obtained from the red blood cells of a never transfused patient, were mixed with unlabelled globins obtained from normal red cells. The chromatographic analysis of the 2 mixed globins demonstrated a peak of protein in the β chain region, while no peak of radioactivity was detectable under the same area. Similar results were obtained after analysis of the labelled globins from red cells of the 5 polytransfused patients; this is shown in a typical chromatographic separation illustrated in Fig.4. The amount of protein recovered under the β region in these cases was dependent on the amount of transfused blood and from the time between the transfusions and our experiments. No protein was eluted in the β chain region when the experiment was carried out 120 days after the previous transfusion. No labelled protein was ever found under the β peak.

In the 10 cases that were examined, the amount of radioactivity incorporated under the α chain region was higher than that incorporated under the combined γ and δ regions. This is clearly indicated by the ratio counts/min recovered under α/counts/min under $\gamma + \delta$ which gave an average value of 3.0 (see Table). This ratio was only slightly reduced when the incubation was prolonged from 2 to 6 hours. In fact, as observed in the experiment presented in Fig.6, the ratio $\alpha/(\gamma + \delta)$ was equal to 2.6 after 2 hours and to 2.4 after 6 hours of incubation.

In order to establish the form in which the excess of newly synthesized α chains was present in the red blood cell cytoplasm, a lysate in which the chromatographic analysis had shown a large excess of α chain synthesis was centrifuged on a sucrose density gradient (Fig.7). All the incorporated radioactivity migrated under the peak of the cold hemoglobin.

Globin Chain Synthesis in Heterozygous β-thalassemia

The red blood cells of 3 subjects with heterozygous β-thalassemia were incubated with labelled amino acids, and the globins obtained with the procedures described above were separated by column chromatography. Proteins were eluted under the regions of α, β, γ and δ chains. Four peaks of radioactivity were found in correspondence with the

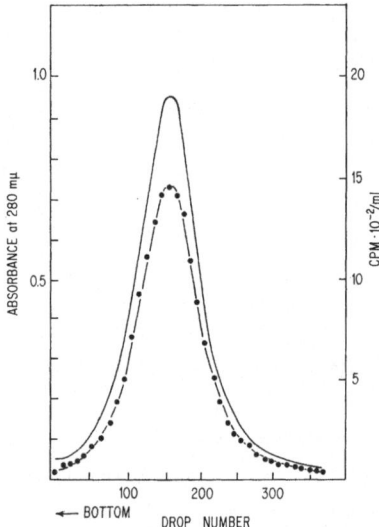

Fig. 7. *Sucrose density gradient centrifugation pattern of labelled hemoglobin in thalassemia major.* The hemoglobin was labelled by incubating the red blood cells of a never transfused homozygous β-thalassemic subject with [³H] amino acids for four hours at 37°. The hemolysate, obtained as described under "Methods" was layered on 28 ml of a 5 to 20% sucrose density gradient prepared in a solution containing 0.015 M KCl, 0.003 M MgCl₂ and 0.001 M Tris-HCl buffer pH 7.4. Centrifugation was for 48 hours at 4° in an SW 25.1 Spinco rotor at 25,000 rev./min. The absorbance at 280 mμ was continuously recorded with a Gilford recording spectrophotometer mod. 2000 (——). The radioactivity measurements (●——●) were carried out as described in the text for the labelled proteins eluted from CM-cellulose columns

same areas (Fig. 8). The amount of radioactivity incorporated under the α peak exceeded also in these cases that incorporated under the combined β, γ and δ regions. The excess of incorporation detected under the α chain region in the cases affected by heterozygous β-thalassemia was smaller than the excess found in those affected by homozygous β-thalassemia (ratio α/(β + γ + δ) = 1.6; see Table).

DISCUSSION

The present studies have shown the absence of β globin in patients with homozygous β-thalassemia from the Ferrara region. No protein has been found under the β chain region after chromatography of the globins prepared from red blood cells of 5 never transfused patients, and no hemoglobin A has been detected after analysis with starch gel electrophoresis of the lysates obtained from red cells of the same patients. Furthermore in all cases, no incorporation of labelled amino acids under the β chain region has been ever found, even after rechromatography.

The molecular mechanism causing the absence of β chain synthesis could be one of the following:

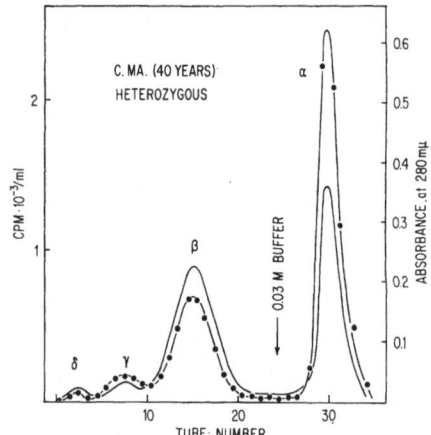

Fig. 8. *Elution pattern from CM-cellulose column chromatography of the labelled globin chains, obtained by incubating the red blood cells of an heterozygous β-thalassemic subject.* The same experimental procedures described in Fig. 1, were followed. The A_{280} (——) and the radioactivity (●——●) measurements were carried out as described in the text

(a) lack of transcription of the β gene with absence of messenger RNA for β globin; (b) synthesis of a messenger RNA for β globin abnormally sensitive to nucleases; (c) synthesis of a messenger RNA with an altered base sequence.

The altered messenger RNA for β globin either does not bind to the ribosomes, or if it can bind, the sequential addition of amino acids to the growing β chain is interrupted by the presence of a nonsense codon.

Several reports from other laboratories have indicated that in homozygous β-thalassemia the synthesis of β globin is still present, although significantly reduced as compared to normal [18—22,13].

The absence of β globin in our population of thalassemics and the presence of some hemoglobin A in other populations suggests that β-thalassemia

might be caused by a heterogenous group of mutations of the β gene [23,22], leading to the absence
or to a decreased production of β globin in patients
from different regions.

In addition to the absence of β globin synthesis,
the results presented in this paper indicate that α
chain synthesis exceeds that of $\gamma + \delta$ with a ratio
$\alpha/(\gamma + \delta)$ of 3.0 in homozygous β-thalassemia and
with a ratio of 1.6 in heterozygous β-thalassemia.
Similar results have been reported by other authors
[12,8], and were recently reported by Bank and
Marks [13]. This fact suggests that in homozygous
β-thalassemia 3 molecules of α chain are released
from the polyribosomes while only one molecule of
γ or δ chain is released per unit time. According to
the recent findings reported by Colombo and Ba-
glioni [24], in rabbit reticulocytes, the β chain of
hemoglobin combines with ribosome-associated and
completed α chain before the latter can be released.
This is not the case for the releasing of the α chain
from the polyribosome complexes of reticulocytes of
thalassemic patients, in which α globin can be
released in the absence of β globin or in the presence
of inadequate amounts of its counterparts (γ and
δ globin).

It is difficult to establish whether the excess of
α chain synthesis is a consequence of the absence of
β chain or represents an absolute increase in its
synthesis. The observation however that in hetero-
zygous β-thalassemia the ratio $\alpha/(\beta + \gamma + \delta)$ is
reduced from 3.0 to 1.6 is in favour of the first hypothesis.

The excess of newly synthesized α chain seems to
assemble in the tetramer α_4, since all the radioactivity
incorporated into protein moves in sucrose density
gradient centrifugation, under the region of cold
hemoglobin, globin monomers or dimers of new
synthesis not being detectable.

The fact that the excess of synthesis of α chain
is not accompanied by a comparable excess of α chain
detected as protein, suggests that the excess of α
chain is unstable. It is difficult to establish whether
this instability is due to a rapid destruction of the α
chain not combined with β, γ or δ chain, or to its
precipitation. If the latter interpretation is correct
this would give support to the hypothesis of Fessas [25] according to which the highly insoluble α_4
tetramer could be present in the inclusion bodies
observed in thalassemic red blood cells. This precipitation of the α globin could, according to Bank
and Marks [13], be responsible for the preferential
destruction of erythrocytes which characterizes
β-thalassemia.

We are deeply grateful to Prof. D. Gaburro, Chief of the
Department of Pediatrics of the University of Ferrara, and
to Dr. C. Menini, Director of the Blood Bank of Ferrara, and
their collegues, for providing us with blood samples which
were indispensable for this study.

This work was supported by a grant from the Italian
Consiglio Nazionale delle Ricerche, Impresa di Enzimologia,
by the Grant GM 12291-02 from the National Institutes of
Health, by the NATO Grant No. 218, and by a grant from
the Cooley's Anemia Blood and Research Foundation for
Children, Inc.

REFERENCES

1. Weatherall, D. J., *The Thalassemia Syndromes*. Blackwell Scientific Publ., Oxford 1965.
2. Watson, J. D., *Molecular Biology of the Gene*, W. A.
   Benjamin Edition, New York 1965, p. 390.
3. Bannerman, R. M., Grinstein, M., and Moore, G. C.,
   Brit. J. Haematol. 5 (1959) 102.
4. London, I. M., Bruns, G. P., and Karibian, D., *Medicine*,
   43 (1964) 798.
5. Itano, H. A., *Symposium on abnormal hemoblogins and
   enzyme deficiency*, Blackwell Scientific Publ., Oxford
   1964.
6. Ames, B. N., and Hartman, P. E., *Cold Spring Harbor
   Symp. Quant. Biol.* 28 (1963) 349.
7. Stent, G. D., *Science*, 144 (1964) 816.
8. Weatherall, D. J., Clegg, J. B., and Naughton, M. A.,
   Nature, 208 (1965) 1061.
9. Ingram, V. M., *Ann. N.Y. Acad. Sci.* 119 (1964) 485.
10. Burka, E. R., and Marks, P. A., *J. Mol. Biol.* 9 (1964)
    439.
11. Bank, A., and Marks, P. A., *Blood*, 26 (1965) 867.
12. Heywood, D., Karon, M., and Weisman, S., *J. Lab.
    Clin. Med.* 66 (1965) 476.
13. Bank, A., and Marks, P. A., *Nature*, 212 (1966) 1198.
14. Marks, P. A., Burka, E. R., and Schlessinger, D., *Proc.
    Natl. Acad. Sci. US*, 48 (1962) 2163.
15. Rossi Fanelli, A., Antonini, E., and Caputo, A., *Biochim.
    Biophys. Acta*, 30 (1958) 608.
16. Clegg, J. B., Naughton, M. A., and Weatherall, D. J.,
    Nature, 207 (1965) 945.
17. Bray, G. A., *Anal. Biochem.* 1 (1960) 279.
18. Gabuzda, T. G., Nathan, D. G., and Gardner, F. H., *J.
    Clin. Invest.* 42 (1963) 1678.
19. Guidotti, G., In *Genetics and Evolution*, (edited by
    V. M. Ingram) Columbia University Press, New York
    1963, p. 116.
20. Marks, P. A., and Burka, E. R., *Science*, 144 (1964) 522.
21. Heywood, J. D., Karon, M., and Weismann, S., *Science*,
    146 (1964) 530.
22. Motulsky, A. G., *Cold Spring Harbor Symp.* 29 (1964),
    p. 399.
23. Baglioni, C., *Correlation between genetics and chemistry of
    human hemoglobins*, In *Molecular Genetics* (edited by
    J. H. Taylor) Academic Press, New York 1963,
    p. 452.
24. Colombo, B., and Baglioni, C., *J. Mol. Biol.* 16 (1966) 51.
25. Fessas, P., and Loukopulos, D., *Science*, 143 (1964) 590.

A. Bargellesi, S. Pontremoli, and F. Conconi
Istituto di Chimica Biologica,
Università degli Studi di Ferrara,
Via Fossato di Mortara 25, Ferrara, Italy

European J. Biochem. 1 (1967) 80—91

A Protein Sequenator

P. Edman and G. Begg

St. Vincent's School of Medical Research, Melbourne, N. 6, Victoria

(Received October 5, 1966)

The protein sequenator is an instrument for the automatic determination of amino acid sequences in proteins and peptides. It operates on the principle of the phenylisothiocyanate degradation scheme. The automated process embraces the formation of the phenylthiocarbamyl derivative of the protein and the splitting off of the N-terminal amino acid as thiazolinone. The degradation proceeds at a rate of 15.4 cycles in 24 hours and with a yield in the individual cycle in excess of 98%. The material requirements are approximately 0.25 μmoles of protein. The thiazolinones are converted to the corresponding phenylthiohydantoins in a separate operation, and the latter identified by thin layer chromatography. The process has been applied to the whole molecule of apomyoglobin from the humpback whale, and it has been possible to establish the sequence of the first 60 amino acids from the N-terminal end.

Present techniques for the elucidation of protein primary structure are highly time- and labor-consuming, and quite inadequate when seen in relation to the vast amount of work waiting to be done. This contribution describes an attempt to create a more favorable situation.

The scheme is based on the phenylisothiocyanate reaction for determining amino acid sequences. In principle this procedure should allow the degradation of any sequence regardless of length, but in practice only short degradations have been feasible. The termination of a degradation is usually caused by diminishing yields, but also overlapping between consecutive steps, and nonspecific cleavage along the peptide chain tend gradually to obscure the results. For larger structures, like proteins, there is the added problem of poor solubility in the media of the degradation. However, these limitations may be made less severe by proper measures. For this an understanding of the reaction mechanism is essential [1,2]:

These reactions will be referred to as coupling (1), cleavage (2), and conversion (3, 4). The 2-anilino-5-thiazolinone derivatives (I) are too unstable to them suitable for identification purposes, and are therefore converted to the isomeric 3-phenyl-2-thiohydantoins (PTHs). Several features of the mechanism have a bearing on the degradation procedure. First, reaction (2) is fast whereas reaction (4) is slow. It is therefore advantageous to separate the thiazolinone from the shortened peptide, and convert it separately to the corresponding phenylthiohydantoin, thus minimizing the exposure of the peptide to strong acid. Second, reaction (2) proceeds readily under anhydrous conditions, and it is therefore possible to eliminate any hydrolytic action by the acid on the peptide. Finally, the phenylthiocarbamyl group has been found to be easily desulfurized by oxidation, even under the influence of oxygen dissolved in the medium [3]. This brings the degradation to a halt since obviously a thiazolinone can no longer be

$$C_6H_5 \cdot NCS + H_2N \cdot CHR' \cdot CO \cdot NH \cdot CHR'' \cdot COOH \rightarrow C_6H_5 \cdot NH \cdot CS \cdot NH \cdot CHR' \cdot CO \cdot NH \cdot CHR'' \cdot COOH \quad (1)$$

$$C_6H_5 \cdot NH \cdot CS \cdot NH \cdot CHR' \cdot CO \cdot NH \cdot CHR'' \cdot COOH \xrightarrow{H^+} C_6H_5 \cdot NH \cdot \underset{\underset{S}{\rule{2cm}{0.4pt}}}{C:N \cdot CHR' \cdot CO} + H_2N \cdot CHR'' \cdot COOH \quad (2)$$

I.

$$C_6H_5 \cdot NH \cdot \underset{\underset{S}{\rule{2cm}{0.4pt}}}{C:N \cdot CHR' \cdot CO} + H_2O \rightarrow C_6H_5 \cdot NH \cdot CS \cdot NH \cdot CHR' \cdot COOH \quad (3)$$

$$C_6H_5 \cdot NH \cdot CS \cdot NH \cdot CHR' \cdot COOH \xrightarrow{H^+} C_6H_5 \cdot \underset{\underset{\rule{4cm}{0.4pt}}{}}{N \cdot CS \cdot NH \cdot CHR' \cdot CO} + H_2O. \quad (4)$$

Non-standard abbreviations and trademarks. 3-Phenyl-2-thiohydantoin, PTH; polytetrafluoroethylene, PTFE; *N,N,N′,N′-*tetrakis-(2-hydroxypropyl)-ethylenediamine, Quadrol®.

formed. This side reaction may be eliminated simply by carrying out the procedure in an inert gas.

The various amino acids do not differ greatly in reactivity during the degradation process, and it is feasible to find one set of reaction conditions covering all cases. The degradation of a peptide may then be reduced to a repetition of a standard set of conditions. Such a degradation cycle would lend itself to programming. The theme of this contribution is the design of an instrument based on this principle.

We propose the term 'sequenator' for an instrument which determines the sequence of an ordered linear polymer by repeating a chemical process.

DESIGN

The general design of the sequenator is shown in Fig. 1. The reaction vessel is a cylindrical glass cup (A) mounted on the shaft of an electrical motor (B). The cup spins continuously, and solutions and solvents entering the cup are therefore spread as thin films on the walls of the cup. Reagents and solvents enter through the feed line (R) at the bottom of the cup. Extracting solvents climb to the groove where they are scooped off, and leave through the effluent line (S). The cup is enclosed in a bell jar (Q), and the system can be evacuated by means of a vacuum pump (P). The system is also thermostated.

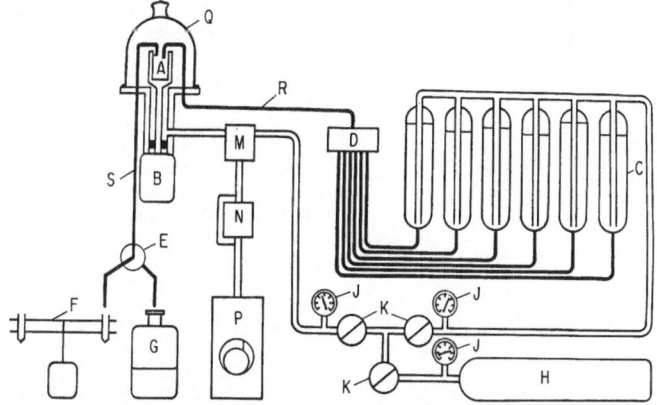

Fig. 1. *Diagram of sequenator. A*, spinning cup; *B*, electric motor; *C*, reagent (solvent) reservoir; *D*, valve assembly; *E*, outlet stopcock assembly; *F*, fraction collector; *G*, waste container; *H*, nitrogen cylinder; *J*, pressure gauges; *K*, pressure regulators; *M*, 3-way valve; *N*, 2-way valve with bypass; *P*, rotary vacuum pump; *Q*, bell jar; *R*, feed line; *S*, effluent line. Gas lines are doubly contoured, and liquid lines are filled

The design problem may be simplified by leaving the conversion reaction out of the considerations. This can be done without great sacrifice since this reaction can conveniently be carried out on a large number of samples simultaneously, and is therefore of no importance for the speed of the analysis.

The coupling and the cleavage procedure both call for a number of mechanically diverse operations, e.g. extractions, centrifugations, and dryings. In order to make the design simple it is desirable to have one mechanical operation accommodating all these processes. This has been achieved by spreading the solutions in thin films inside a rotating cylindrical cup. The spinning film is ideally suited for extraction by another film of an immiscible solvent sliding over its surface, for centrifugation and also for drying under reduced pressure, because of the large surface and the stabilizing centrifugal force.

A preliminary account of this work has appeared [4].

Reagents and solvents are stored in reservoirs (C), and are admitted to the cup through an assembly of valves (D). The reservoirs are under a constant low pressure of nitrogen supplied by a nitrogen cylinder (H) and pressure regulators (K). The contents of the bell jar are likewise held at a fixed although lower pressure of nitrogen through a similar arrangement. The pressure differerential between the reservoirs and the bell jar is constant, and the volume of reagent or solvent admitted to the cup is therefore determined by the time a valve is kept open.

The effluent line leads *via* a 3-way stopcock (E) either to a fraction collector (F) or to a waste container (G).

The valves in the assembly (D) and the gas valves (M and N) are operated by solenoids, and the 3-way outlet stopcock (E) and the fraction collector (F) by electric motors. All these functions are governed by an electronic programming unit.

The motors driving the cup and the vacuum pump run continuously.

The corrosive nature of several of the reagents used in the degradation severely limits the range of materials that can be used in the construction. Only borosilicate glass, PTFE, and gold are used in direct contact with reagents. Where the contact is solely with vapors, stainless steel (SAE 30321) is also employed.

Since highly inflammable solvents are used care has been taken to eliminate the explosion hazard by excluding sparking electric contacts. Where this is not feasible contacts are hermetically enclosed.

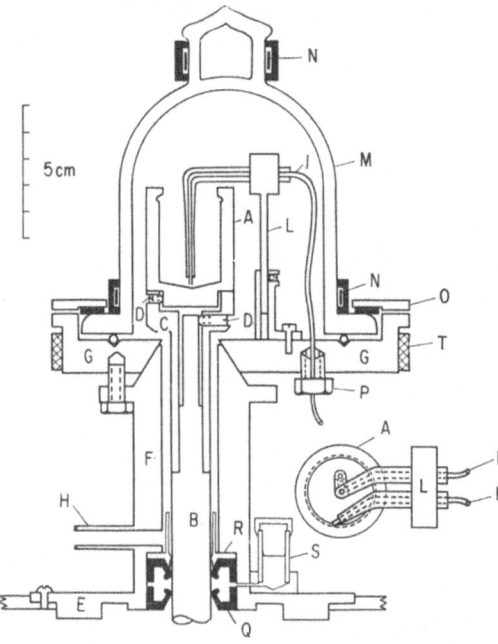

Fig. 2. *Reaction chamber.* A, glass cup; B, motor shaft extension; C, cup support; D, grub screws; E, motor support; F, column; G, base plate; H, side tube; I, feed line; K, effluent line; L, adjustable stand; M, bell jar; N, electrodes; O, rubber padded ring; P, sealing bolt; Q, oil seal; R, PTFE sleeve; S, oil reservoir; T, band heater. A top view of the cup, the feed line, and the effluent line is shown in the lower right hand corner

Rotor Assembly

It is essential for a correct operation of the cup that the inside cylindrical surface should run true. If this is not the case, disturbances in the liquid will result in mixing of the top and bottom part of the contents with diminished efficiency of the extraction as a consequence. The tolerance requirements adhered to are, that no part of the inside wall should

vary more than 10 μ in distance from the rotational axis. These requirements make considerable demands on precision in manufacturing the cup, as well as on its alignment and on motor bearings (Fig. 2).

Cup. The cup is made of Pyrex glass. The inside diameter is 26 mm, and the height of the inside cylindrical wall is 31 mm as measured from the bottom to the groove. The groove has a depth of 1 mm relative to the inside wall. The opening of the cup is slightly narrower, 24 mm in diameter, to prevent splashing during the scooping operation. The inside bottom of the cup is dished to assist withdrawal of the contents when the cup is stationary.

Motor. The requirements are a suitable and constant speed and freedom from sparking. A 3-phase, 4-pole induction motor with the ratings 1,425 rev./min and $^1/_2$ H.P. (G.M.F. Electric Motors Pty. Ltd., Arncliffe, N.S.W., Australia) is used. The power rating is far in excess of demands but the larger physical dimensions have the advantage of making the hermetic sealing of the motor shaft easier. The original ball bearings are replaced by bearings with closer tolerances (SKF Ball Bearing Co. Pty. Ltd., Melbourne, Australia, types RLS 5C2 and RLS 6C2). The top end of the motor shaft is extended by a stainless steel shaft heatshrunk on to the original shaft. The shaft extension has its diameter reduced in two steps to receive the cup support and the aligning screws respectively.

Cup Support. This is made of stainless steel. It slides freely on the extended motor shaft when not locked in position. Three PTFE-tipped grub screws hold the cup. Three other screws impinge on the top section of the extended motor shaft, and serve to align the cup within the prescribed tolerances. The alignment is made with the help of a dial gauge. A coarse alignment is first made of the outside surface of the cup support. The fine alignment is made with the feeler arm of the dial gauge resting on the inside cylindrical surface of the cup and with manual rotation of the shaft. Both the top and the bottom part of the surface conform to the tolerance limits of 10 μ.

Motor Support, Column, and Base Plate

These are the supporting structures, and are made of stainless steel. The motor support and the column are joined into one piece by silver brazing. The column and the base plate are connected in a conical joint firmly held together by three bolts. A piece of unsintered PTFE tape between the tapered surfaces ensures the vacuum seal. The column has a side tube connecting it to the 3-way vacuum-pressure valve (Fig. 1, M). The motor is firmly bolted, through its top end plate, to the motor support, and the latter is fixed to the supporting framework.

Feed Line and Effluent Line

These are PTFE tubes with inner diameters 0.5 mm and 0.9 mm respectively. They are guided into their correct positions in the cup by capillary glass tubes supported by an adjustable stand. The feed line, which is approx. 30 cm in length, terminates at the bottom of the cup somewhat off center, and with just so much clearance from the bottom that liquid flow is not obstructed. The effluent line enters the cup as nearly tangentially as possible with the tip located in the groove and facing the rotational direction of the cup. There is a small clearance all around between the tip and the groove, and the walls of the tip are thinned to reduce turbulence. The scooping is assisted by the momentum of the rotating liquid and by the higher gas pressure in the bell jar.

tion when the pressure in the bell jar is above atmospheric. It should also withstand the action of organic vapors. Two identical oil seals (Super Seals Pty. Ltd., Melbourne, Australia, type 13720N10-ABO1) facing each other and a PTFE sleeve are forced into a recess at the bottom of the column. The lips of the seals close on the shaft. The corresponding area of the shaft is polished to a mirror finish. The space between the seals and the shaft is filled with vacuum oil. The PTFE sleeve serves to reduce the contact of the synthetic rubber oil seals with organic liquids and vapors. However, the seals tend to harden with time, and should then be replaced.

Heating and Thermostating

A band heater (300 watts) is clamped around the circumference of the base plate. The base plate has

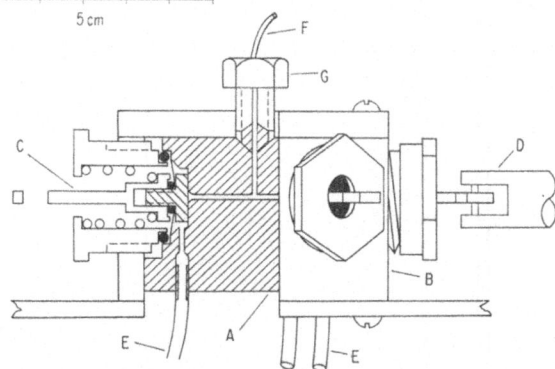

Fig.3. *Valve assembly. A*, PTFE block; *B*, stainless steel casing; *C*, S.S. shaft; *D*, solenoid core; *E*, PTFE tubes from reservoirs; *F*, feed line; *G*, sealing bolt

Bell Jar

The jar is made from a Pyrex flat flange joint (Quickfit and Quartz Ltd., Stone, Staffordshire, England, type FG 75). The volume of the bell jar is kept as small as possible in order to reduce gas space. To allow heating the outer surface has an electrically conductive coating [5], and bandshaped silver electrodes surround the bell jar at top and bottom.

Vacuum Seals

The seal between the bell jar and the base plate is provided by a PTFE O-ring. A rubber-padded stainless steel ring on top of the glass flange is bolted down to the base plate, and compresses the O-ring.

The PTFE tubes passing through the base plate are sealed by PTFE ferrules compressed by bolts.

The seal around the motor shaft is designed to ensure against leaks from the outside during vacuum stages, as well as against leaks in the opposite direc-

a thermistor heat-sensing element, and is thermostated. The bell jar has an electrically conductive coating as already described.

The temperature of the system is fixed during actual operation by first letting the base plate reach the temperature set by the thermostat. The voltage across the electrodes of the bell jar is then increased to a point where liquid no longer condenses on the walls. It has been found that a temperature setting of 50° requires approximately 35 volts (A.C.) across the electrodes.

Valve Assembly

This allows the measured admittance of reagents and solvents to the reaction cup. The design (Fig.3) should fulfil several requirements. First, an accurate dispensing by timing requires, particularly for smaller volumes, that opening and closing of the valves should be instantaneous. Second, the dead volume between the valves and the cup should be mininal. Finally, the valves should seal off volatile solvents effectively enough for the maintenance of an adequate vacuum in the bell jar during the vacuum stages.

The assembly consists of a cylindrical block of PTFE enclosed in a hexagonal stainless steel casing and six valves projecting radially at angles of 60°. The seats of the valves are formed by the PTFE block. The moving parts are cylindrical pieces of gold. The sealing surfaces are flat and polished. The

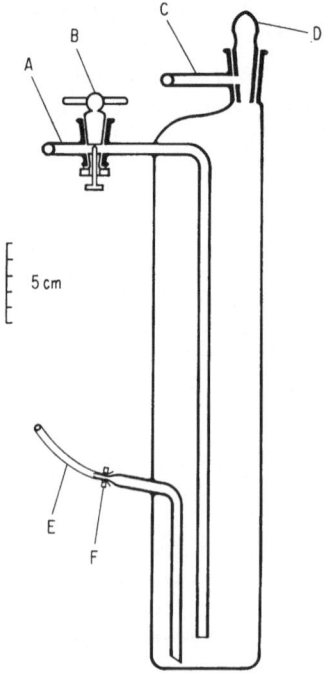

Fig.4. *Reservoir*. A, nitrogen inlet; B, stopcock; C, nitrogen outlet; D, stopper (spring loaded); E, reagent (solvent) line; F, S.S. sealing nut

pieces of gold are threaded into stainless steel shafts linking the valves to the cores of the solenoids. The valves are springloaded by stainless steel coils to a pressure of 3 kg. The valve chambers are sealed off from the outside by PTFE membranes and rubber O-rings. The assembly is connected to the reservoirs by PTFE tubes (2.7 mm inner diameter). Liquid enters through a channel at the bottom of the valve chamber, and leaves at the center of the valve seat. Capillary channels from each valve converge on the center of the PTFE block, where they merge into a common channel connected to the feed line.

The rubber O-rings of the heptafluorobutyric acid valve tend to perish because of the acid vapors, and should be replaced at intervals.

The valves are operated by solenoids (Guardian Solenoids Inc., Santurce, Puerto Rico, type 4-D.C. with high temperature winding).

Reservoirs

Nitrogen is admitted to the bottom of the reservoir (Fig.4) *via* a glass-PTFE stopcock provided with a needle valve. The stopper has an outlet, which is closed except when nitrogen is passed through a

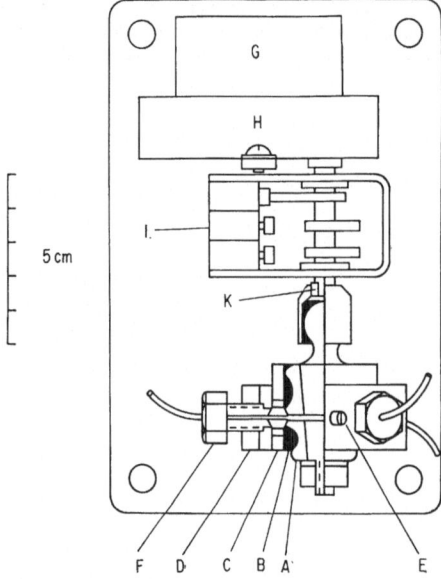

Fig.5. *Outlet stopcock assembly*. A, 3-way stopcock; B, epoxy cement; C, inner S.S. ring; D, outer S.S. ring (fixed); E, aligning screw (one of three); F, sealing bolt; G, electric motor; H, gear box; I, set of cams and microswitches; K, sliding key on drive shaft

new batch of solvent or reagent in order to remove dissolved oxygen. The solvent and reagent reservoirs hold 1,200 ml and 150 ml respectively.

Outlet Stopcock Assembly

This has three positions (Fig.5) which are termed: (a) 'vacuum', when the effluent line is closed; (b) 'collect', when the effluent line is connected to the fraction collector; and (c) 'waste', when the effluent line is connected to the waste container.

In addition to the obvious functions, two of these positions are used for other purposes. The waste position is used when a reagent is being introduced into the cup, in order to prevent a buildup of pressure due to the evaporation of the reagent. The vacuum position is also used during the reaction stages, as otherwise the passage of nitrogen through the bell jar to the outside would cause evaporation of reagents in the cup.

A standard stopcock with a glass barrel and a PTFE key (Fischer and Porter Pty. Ltd., Melbourne, Australia, type 80G2417, 2 mm bore) has been so modified that it can be operated by an electric motor. The motor is synchronous, and provided with a gear box (Philips Electrical Ltd., Eindhoven, Netherlands, type: motor AU5100/22, gear box AU5300/80DJ). It is started by a signal from the programming unit, and stopped by either of three self-operated cam and-microswitch combinations. The microswitches are hermetically sealed.

Fraction Collector

This is housed in a refrigerated stainless steel cabinet kept at $+2°$. The rack is made of stainless steel, and accomodates 50 tubes, each with a holding capacity of about 5 ml. The rack is operated by a suitably modified stepping motor of the uniselector type which is nonsparking.

The low temperature serves to minimize any decomposition of the unstable thiazolinones.

Pressure System

Two different pressures relative to the atmosphere are maintained (Fig. 1). The pressure in the bell jar, P_1, serves to assist the removal of liquid through the effluent line during extractions. P_1 is kept at 50 cm of water. A higher pressure, P_2, is maintained in the reservoirs. The latter are at a lower level than the cup, the bottom of the reservoirs being 36 cm below the bottom of the cup, and pressure is required to transfer liquid. A suitable rate of flow results when $P_2 = 150$ cm of water.

A pressure head is provided by a cylinder of compressed nitrogen equipped with a standard pressure reduction valve. On the other side of the reduction valve the line is divided in two branches, one to the bell jar and the other to the reservoirs. Each branch has a low pressure regulator and a pressure gauge. The pressure regulators are of the type commonly used for compressed butane burners (Commonwealth Industrial Gases Ltd., Melbourne, Australia, type AW51). These are well suited for their purpose since they may be accurately set, and permit a large flow of gas.

The pressure line to the bell jar joins the vacuum line at the 3-way valve (Fig. 1, M) in such a way that the bell jar may be connected either to the pressure line or to the vacuum line.

Vacuum System

Vacuum is supplied by a 2-stage rotary gas ballast pump with a pumping capacity of 30 l per minute and powered by a 3-phase induction motor (Rud Browne and Co. Pty. Ltd., Melbourne, Australia, type Dynavac 2) (Fig. 1). All lines and valves are made of stainless steel. The 3-way valve (The Skinner Chuck Co., New Britain, Conn., U.S.A., type V53DB2VAC2, coil 140 volts D.C.), connects the bell jar either to the vacuum line or to the pressure line. The former is the case when the solenoid is energized. The 2-way valve (The Skinner Chuck Co., type V52DB2077, coil 140 volts D.C.) has been so modified that it has a permanent bypass between the inlet and the outlet gates. The bypass is a coil of stainless steel tubing, length 32 cm and inner diameter 0.60 mm. When the bell jar is evacuated the 3-way valve opens immediately but the 2-way valve only after a delay period. During the delay period the system is therefore evacuated only through the bypass restriction. This arrangement serves to prevent the liquid in the cup from boiling through too sudden an evacuation.

The ultimate vacuum attainable in the system is approximately 10^{-2} mm of Hg. This degree of vacuum is ensurance that leakage from the atmosphere is sufficiently reduced. However, during actual operation this low pressure is not reached because of the contamination of the system with products of low volatility. The pressure observed at the end of a vacuum stage is usually not lower than $5 \cdot 10^{-2}$ mm of Hg.

Programming Unit

This unit (Fig. 6) is an electronic timer with 30 channels, each timing a stage in the degradation cycle. The channels are arranged in sequence so that each channel, when reaching a preset time, initiates the next, and the last channel triggers the first. The channels control, via relays, the various functions, i.e. valves and motors of the sequenator.

A set of three decade counters (X1, X10 and X100; Philips Electrical Ltd., type 88930/33), here referred to as the impulse counter, receives time pulses from a pulse generator (see below) via a pulse shaper (B; Philips Electrical Ltd., type 88930/48). Signals from the impulse counter operate sets of program stages (H; Philips Electrical Ltd., type 88930/37), which can be individually set for 1 to 999 counts.

The impulse counter is common to a number of program stages. The selection of the stage is made by a combination of a ring counter (0; Philips Electrical Ltd., type 88930/33 suitably modified), here called the stage counter, and a set of AND-gates (G), one for each program stage. One gate at a time is held open by the output from the stage counter according to its setting.

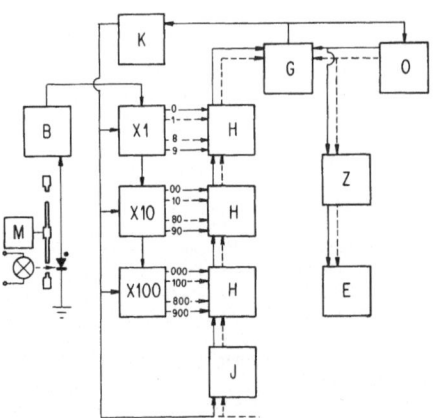

Fig.6. *Diagram of programming unit.* *X* 1, *X* 10, *X* 100, impulse counters; *M*, pulse generator; *B*, pulse shaper; *H*, program stage; *O*, stage counter; *G*, AND-gate; *J*, program initiator; *K*, reset unit; *Z*, relay amplifier; *E*, output relay

Provisions are made for the manual: (a) resetting of impulse counters and program stages; (b) resetting of stage counter; (c) selection of stage; (d) interruption of pulse generator; and (e) interruption of output commands.

Time Pulse Generator. Light from a low voltage miniature light bulb falls on a photosensitive diode (OAP 12). The light beam is interrupted by a rotating disc with two diametrically situated holes. The disc is driven by a synchronous motor (Philips Electrical Ltd., type AU 5005/22) through a gear box (Philips Electrical Ltd., type AU 5300/80 DA). The impulse frequency is 20 per minute.

The sequenator is kept in a constant temperature room at $20 \pm 2°$.

REAGENTS AND SOLVENTS

The processes performed by the sequenator are essentially the same as those in the manual procedure [6]. The protein or peptide is first exposed to a coupling medium consisting of an alkaline buffer,

Fig.7. *The sequenator*

The pulse allowed through a gate resets the impulse counter and the program initiators (J; Philips Electrical Ltd., type 88930/51), and also moves the stage counter to the next stage. The resetting pulse is modified by a reset unit (K; Philips Electrical Ltd., type 88930/54).

The signals from the stage counter operate the output relays (E) *via* relay amplifiers (Z; Philips Electrical Ltd., type 88930/57).

phenylisothiocyanate, and a solvent to form the phenylthiocarbamyl derivative. After completion of the reaction excess reagents and by-products are removed by solvent extraction, and residual solvent subsequently evaporated. The thiazolinone is then cleaved off by exposing the phenylthiocarbamyl derivative to an anhydrous acid, and is subsequently extracted by an organic solvent. After drying the shortened peptide is ready for a new degradation

Table. *Operations in a cycle of the sequenator*

Reagents 1, 2 and 3 are in reservoirs I, II and III respectively, and solvents 1, 2 and 3 occupy reservoirs IV, V and VI respectively. The valves are numbered correspondingly. Other functions are the 3-way vacuum-pressure valve (M), the 2-way vacuum valve (N), the 3-way outlet stopcock (E), and the motor driving the fraction collector (F). A + or a − sign means that the operating solenoid or motor is energized or deenergized respectively at the beginning of the stage. The + sign also means that a valve is open, in the case of valve M to the vacuum line. The three positions of the outlet stopcock are indicated by the letters c (collect), v (vacuum), and w (waste). The duration of stages refers to a reaction temperature of 50°

Stage	Duration min	(Volume) (ml)	I	II	III	IV	V	VI	M	N	E	F
1. Reagent 1	0.15	(0.40)	+								+w	+
2. Reagent 2	1.75	(0.40)	−	+								−
3. Reaction	30.00				−						+v	
4. Restric. vacuum	3.00								+			
5. Vacuum	6.00									+		
6. Delay	0.05								−	−		
7. Solvent 1	5.00	(11.5)				+					+w	
8. Solvent 2	8.00	(23.5)				−	+					
9. Delay	1.00						−					
10. Restric. vacuum	3.00								+		+v	
11. Vacuum	6.00									+		
12. Delay	0.05								−	−		
13. Reagent 3	0.70	(0.23)		+							+w	
14. Reaction	3.00				−						+v	
15. Vacuum	1.50								+	+		
16. Delay	0.05								−	−		
17. Solvent 3	1.75	(5.5)						+			+c	
18. Delay	1.00							−				
19. Restric. vacuum	3.00								+		+v	
20. Vacuum	1.00									+		
21. Delay	0.05								−	−		
22. Reagent 3	0.70	(0.23)		+							+w	
23. Reaction	3.00				−						+v	
24. Vacuum	1.50								+	+		
25. Delay	0.05								−	−		
26. Solvent 3	2.25	(7.0)						+			+w	
27. Delay	1.00							−				
28. Restric. vacuum	3.00								+		+v	
29. Vacuum	6.00									+		
30. Delay	0.05								−	−		

cycle. However, the mode of operation of the sequenator requires some modification of the reagents. The large volume of the bell jar and the high temperature (50°) require less volatile reagents. Therefore, the volatile organic buffer base has been substituted by a nonvolatile organic base, N,N,N',N'-tetrakis-(2-hydroxypropyl)-ethylenediamine (Quadrol), and the trifluoroacetic acid [7,8] has been replaced by heptafluorobutyric acid. The reagents and solvents described here are those used successfully in the degradation of myoglobin (see Applications: results).

Composition

Reagent 1 is a 5% (v/v) solution of phenylisothiocyanate in heptane. Phenylisothiocyanate is unstable in the completed coupling medium, and is therefore kept as a separate reagent. Reagent 2 is a 1.0 M Quadrol-trifluoroacetic acid buffer in n-propanol-water (3:4, v/v), pH 9.0 (glass electrode, 20°). Reagent 3 is anhydrous n-heptafluorobutyric acid. Solvent 1 is benzene. Solvent 2 is ethyl acetate

containing 0.1% (v/v) acetic acid. Solvent 3 is 1-chlorobutane.

Purification

Much emphasis is laid on the purity of reagents and solvents. Special efforts have been made to remove traces of aldehydes since these tend to react with the terminal amino group, and thus cause a progressive fall in yield. Because of their importance the purification procedures are here given in some detail.

Phenylisothiocyanate (*purissimum* grade; Fluka A.G., Buchs, Switzerland) is distilled *in vacuo* at 1 mm of Hg, and the fraction boiling at 55° collected. The reagent is somewhat unstable on storage even in the cold, and is therefore distilled only as required.

N,N,N',N'-tetrakis-(2-hydroxypropyl)-ethylenediamine (Quadrol, Wyandotte Chemicals Corp., Wyandotte, Mich., U.S.A.) is a practical grade reagent, and holds appreciable amounts of aldehydic impurities, which are difficult to remove. No procedure has been found that removes these impurities completely. The most successful method found is based on chromatography on an anion exchange resin charged with sulfite ions [9]. Five hundred grams of Amberlite IRA-500 (British Drug Houses Ltd., Poole, England, standard grade) is converted into the free base with 5 N NaOH, washed with distilled water, and made into a column (75 cm × 10 cm²). A solution of 11 g of SO_2 in 250 ml of distilled water is passed through at a flow rate of 5 ml per minute, and the column then again rinsed with water. Sixty grams of Quadrol is dissolved in distilled water to a volume of 250 ml, and passed through the column at the same flow rate. During this passage some displacement of sulfite ions down the column is apparent from a change in color of the resin. The effluent is concentrated in a rotary film evaporator, first on the water pump and finally on the rotary oil pump. This preparation is then distilled and redistilled in a molecular still (Jena Glasswork, Schott & Gen., Mainz, Germany, type 5593 mvz) at a distillation temperature of 120° and a pressure of 10^{-3} mm Hg. The resulting preparation is low in, but not entirely free of aldehydes (see Tollens' reaction).

n-Heptafluorobutyric acid (Minnesota Mining and Manufacturing Co., Saint Paul, Minn., U.S.A.) is first exhaustively oxidized at refluxing temperature with solid CrO_3, distilled off, and dried over $CaSO_4$. It is redistilled on a short Widmer column, and the fraction boiling at 119—120° collected.

Trifluoroacetic acid (Minnesota Mining & Manufacturing Co.) is purified in a similar manner.

Acetic acid (*purissimum* grade distilled from CrO_3; Fluka A.G.).

Heptane (practical grade) is shaken with several changes of concentrated H_2SO_4, washed in succession with distilled water, $10^0/_0$ (w/v) aqueous NaOH, and distilled water. This is followed by shaking in a mechanical shaker overnight with a $3^0/_0$ (w/v) aqueous solution of $KMnO_4$. The $KMnO_4$ is removed by washing with distilled water, the preparation dried over Na_2SO_4 and distilled.

Benzene (practical grade) is stirred over concentrated H_2SO_4 with several changes of acid, and then washed with distilled water. It is then shaken in a mechanical shaker overnight with a $3^0/_0$ (w/v) aqueous $KMnO_4$ solution. The $KMnO_4$ is removed by washing with distilled water, and the preparation dried over solid KOH. It is finally distilled on a 50 cm Widmer column, and the fraction boiling at 80—81° collected.

Ethyl acetate (practical grade) is stirred, first with $5^0/_0$ (w/v) aqueous Na_2CO_3 and then with a saturated aqueous $CaCl_2$ solution. The organic phase is drawn off, and shaken overnight in a mechanical shaker with solid $KMnO_4$ (1 g/l). The organic phase is drawn off, washed with distilled water, dried over $CaSO_4$ and distilled, b.p. 77°.

1-Chlorobutane is a laboratory grade reagent (Hopkin and Williams Ltd., Chadwell Heath, Essex, England), and is shaken in a mechanical shaker overnight with a $3^0/_0$ (w/v) aqueous $KMnO_4$ solution. The $KMnO_4$ is removed by repeated washing with distilled water, and the preparation then dried over $CaSO_4$. It is distilled, and the fraction boiling at 78° collected.

n-Propanol is a laboratory grade reagent (May and Baker Ltd., Dagenham, Essex, England), and is refluxed for several days with powdered Zn (5 g/l) and saturated aqueous NaOH (5 ml/l), and then dried over $CaSO_4$. The preparation is fractionated on a 50 cm Widmer column, the fractionation being followed with Tollens' reaction. Usually only the last $^1/_3$ of the distillate gives a negative reaction.

Nitrogen contains less than 10 ppm of oxygen (Commonwealth Industrial Gases, Ltd.).

Water is glass distilled.

Tollens' Reaction

Solvents and reagents are tested with this reaction for the absence of aldehydes wherever practicable (heptane, benzene, ethyl acetate, 1-chlorobutane, n-propanol, Quadrol). The test is carried out in a darkened room and in the following way. One milliliter of a $10^0/_0$ (w/v) aqueous $AgNO_3$ solution and 1 ml of a $10^0/_0$ (w/v) aqueous NaOH solution are mixed in a test tube, and a dilute aqueous NH_3 solution added dropwise until the precipitate is redissolved. To this solution is added 2 ml of the sample (Quadrol and n-propanol are first diluted 1:1 (v/v) with distilled water), and, if the solutions are immiscible, the tube is shaken at frequent intervals. No discoloration or cloudiness should appear within one hour, except in the case of Quadrol where a slight grayish discoloration indicates that some aldehydic impurities are still present. (Caution: Explosive silver fulminate is formed in the Tollens' reaction.)

OPERATION

The programming of the degradation cycle requires that it be divided in 30 stages. The stages, their durations and associated operations are shown in the Table, and only a few comments are required.

The vacuum stages 4, 5, 15, and 24 serve to concentrate or to remove the reaction medium prior to extraction. This has been found necessary since otherwise protein material tends to be carried up into the groove causing blockage of the effluent line. This is caused by the fact that at first the concentration of the extracting organic solvent is not high enough to precipitate the protein. The Quadrol offers special difficulties in this respect since it cannot be removed by evaporation. Instead, it is extracted by a solvent, benzene, with which is it miscible to a limited extent. The extraction is subsequently completed by ethyl acetate.

During the restricted vacuum stages 4, 10, 19, and 28 the 2-way valve (Fig. 1, N) is closed, and evacuation is only through the capillary bypass.

The delay stages 9, 18, and 27 are inserted between an extraction stage and a vacuum stage. Their function is to allow time for a stream of nitrogen (the outlet stopcock is in the waste or collect position) to clear the effluent line and groove of solvent. This prevents splashing of solvent when vacuum is applied.

The delay stages 6, 12, 16, 21, 25, and 30 occur between a vacuum stage and a stage where a reagent or a solvent is added. They serve to allow pressure equilibration in the bell jar.

Before use all reagents and solvents are purged of dissolved oxygen by passing a lively stream of nitrogen through the reservoir for about five minutes.

The sample (0.2—1.5 μmoles) is introduced in the following way. The programming unit is manually operated (see Programming Unit) so that the pulse generator and the outgoing commands are interrupted, and the stage counter set at stage 27. The bell jar is removed, and 0.2 ml of a solution of the sample, e.g. in water or aqueous ethanol, is deposited (it is convenient to use an all-glass syringe with a short piece of PTFE capillary tube attached to the tip) low on the walls of the spinning cup. After replacing the bell jar, the programming unit is allowed to take over the operation. It is preferable to start from stage 27 rather than from stage 1 since this ensures that the system is purged of oxygen before the degradation is started.

The volumes of buffer and heptafluorobutyric acid are so adjusted in relation to each other that the buffer rises about 3 mm higher on the wall of the cup than the acid. This helps to confine the protein to the lower part of the cup.

The dry protein can be observed at the end of the stages 11, 20, and 29. It should then have a dry, powdery appearance, and be confined to the lower $^1/_3$ of the cup.

It is convenient to use a stroboscope for observing the processes taking place in the cup.

The time required for one degradation cycle is 93.6 minutes, which is equivalent to 15.4 cycles in 24 hours. The volumes of the reservoirs allow about 50 degradation cycles to be run without attention.

<h2 style="text-align:center">APPLICATION</h2>

The sequenator has been applied to the degradation of apomyoglobin obtained from the muscles of the humpback whale *(Megaptera nodosa)*.

<h3 style="text-align:center">Materials and Methods</h3>

Myoglobin was prepared from the skeletal muscles of the humpback whale by fractionation with basic

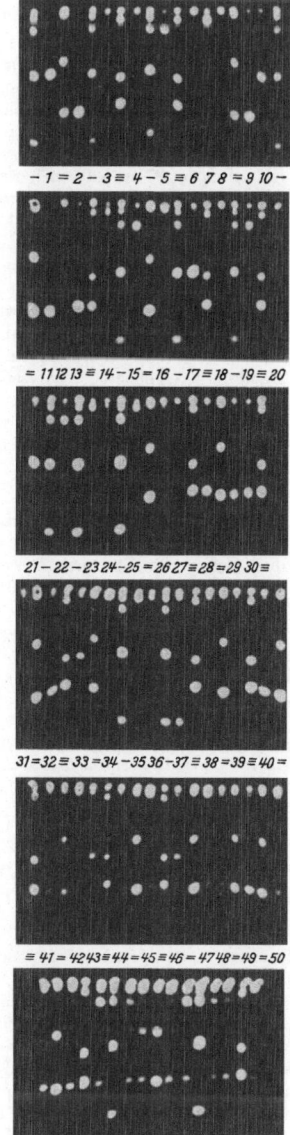

Fig. 8. *Degradation of apomyoglobin.* Thin layer chromatographic identification using solvent system D of the amino acid-PTHs obtained in first 60 degradation cycles. Reference mixtures are indicated by horizontal bars, and show in order from bottom to top the migration of (−) proline-PTH, valine-PTH, alanine-PTH, glycine-PTH; (=) leucine-PTH, methionine-PTH; (≡) isoleucine-PTH, phenyl-alanine-PTH and tryptophan-PTH

lead acetate and ammonium sulfate [10]. The filter cake was dissolved in distilled water, and potassium ferricyanide added to the solution (10 mg/g of filter cake). After removal of the salts by dialysis against distilled water and subsequent concentration of the solution, the preparation was further fractionated through chromatography on CM-cellulose [11]. A column of CM-cellulose (27 cm × 10 cm²) was equilibrated against a 0.01 M Tris phosphate buffer, pH 8.00. The column was charged with 1 g of the Fe(III)-myoglobin preparation, and developed with the same buffer at $+4°$ and with a flow rate of 50 ml per hour. The main peak, constituting about 85% of the material, appeared well resolved at an elution volume of 900 ml, but there were several smaller, partly resolved peaks earlier in the chromatogram. The main fraction was collected, dialysed against distilled water, and concentrated to approx. 10 g/l. The haem was removed by the acetone-hydrochloric acid procedure [12]. After dialysis, first against a dilute $NaHCO_3$ solution and finally against distilled water, the preparation was freeze-dried.

The conversion of the thiazolinones into PTHs was carried out somewhat differently from what has been described earlier (3). The present conversion conditions are N aqueous HCl, 80°, and 10 minutes. These conditions gave a considerably better recovery of serine-PTH without any disadvantageous effect on the other amino acid-PTHs. After evaporation of the 1-chlorobutane solution of the thiazolinone to dryness, 0.2 ml of N HCl was added, the tube

systems D and E [13] could be applied directly to thin layer chromatography. The spots were located either through the extinction of fluorescence under the low pressure mercury lamp, or through the iodine-azide-starch reaction. Arginine and histidine-PTHs were identified by thin layer electrophoresis in an apparatus and with a technique similar to that described by Honegger [14]. The electrophoresis was carried out in a 0.01 M sodium phosphate buffer, pH 8.0, for one hour at a field intensity of 10 volts per cm. The silica gel is acid in itself, and it is necessary to neutralize it beforehand. This was done by making up the slurry of silica gel used for coating the plate in the same buffer, and drying the plate in the air overnight. The electrophoretic buffer was sprayed on to the plate by the normal procedure.

Results

The degradation was performed on 5.0 mg of the apomyoglobin preparation, and was carried through 60 cycles. The yield was determined from the ultra-violet absorption at 269 mμ of the amino acid-PTHs in cycles Nos. 1 and 17 (both valines), and the average yield for a single cycle was calculated to be 98.5%. The results of the thin layer chromatography of the amino acid-PTHs using solvent system D are shown in Fig. 8 (see page 89). These results, together with those obtained by using solvent system E and thin layer electrophoresis, permitted the deduction of the following N-terminal structure for humpback whale myoglobin:

Humpback Whale: Val.Leu.Ser.*Asp.Ala*.Glu.Trp.Gln.Leu.Val.Leu.*Asn.Ile*. Trp.Ala.Lys.Val.Glu.Ala.Asp.
Sperm Whale: Val.Leu.Ser.*Glu. Gly*. Glu.Trp.Gln.Leu.Val.Leu.*His. Val*.Trp.Ala.Lys.Val.Glu.Ala.Asp.
                         1   2   3   4   5   6   7   8   9   10   11   12   13   14   15   16   17   18   19   20

                    Val.Ala.Gly.His.Gly.Gln.Asp.Ile.Leu.Ile.Arg.Leu.Phe.Lys.*Gly*.His.Pro.Glu.Thr.Leu.
                    Val.Ala.Gly.His.Gly.Gln.Asp.Ile.Leu.Ile.Arg.Leu.Phe.Lys.*Ser*.His.Pro.Glu.Thr.Leu.
                    21   22   23   24   25   26   27   28 29   30 31   32   33   34   35   36   37   38   39   40

                    Glu.Lys.Phe.Asp.*Lys*.Phe.Lys.His.Leu.Lys.Thr.Glu.Ala.Glu.Met.Lys.Ala.Ser.Glu.Asp.
                    Glu.Lys.Phe.Asp.*Arg*.Phe.Lys.His.Leu.Lys.Thr.Glu.Ala.Glu.Met.Lys.Ala.Ser.Glu.Asp.
                    41   42   43   44   45   46   47   48   49   50   51   52   53   54   55   56   57   58   59   60

flushed with nitrogen, stoppered and kept at 80° for 10 minutes. After cooling the sample was extracted with altogether 3 ml of ethyl acetate. The organic phase was evaporated to dryness, and the residue taken up with brief heating in a suitable volume of ethylene chloride (approx. 5 μmoles of PTH per ml). This solution was used for chromatography. The aqueous phase if the sample contained arginine-PTH or histidine-PTH, was evaporated to dryness *in vacuo* over pellets of KOH and P_2O_5, and the dry residue taken up in a suitable volume of distilled water for electrophoretic identification.

The amino acid-PTHs were identified by thin layer chromatography or electrophoresis on silica gel containing an ultraviolet fluorescent indicator (Fluka A.G., type D5F). The paper chromatographic

The corresponding part of the structure for sperm whale myoglobin [15] is presented for comparison, and the non-identical amino acids are italicized.

DISCUSSION

The usefulness of a sequenator process depends in the first place on the yield in the single degradation cycle. A simple calculation will illustrate this point. Let it be arbitrarily assumed that a sequence determination ceases to give useful information when the overall yield, *i.e.* the yield over *n* degradation cycles, has fallen to 30%. It may then be calculated that an average repetitive yield, *i.e.* the yield from one cycle to the next, of 97, 98 and 99% would allow 40, 60 and 120 cycles respectively. In the present work the repetitive yield has been 98% or slightly better. The cause of the loss of about 2% in each degradation cycle is at present not known. However, it should be emphasized that even a moderate reduction of this loss would allow a considerable extension of the degradation.

Another important factor in the termination of a degradation has been the appearance in the chromatograms of an increasing general background of other amino acid-PTHs, which eventually made the identification impossible. What causes this background is not certain. It may be significant that the dominating amino acids in the background are also those which occur most frequently in the structure, e.g. glycine, alanine, and leucines in the case of myoglobin (Fig. 8). A small degree of nonspecific cleavage along the peptide chain would produce this result. A mechanism is at hand to account for a nonspecific cleavage, since it has been shown that anhydrous trifluoroacetic acid may cause acidolysis of peptide bonds [16], and heptafluorobutyric acid would be expected to behave in the same way. However, in the absence of direct evidence, this explanation for the appearance of the background remains tentative.

An incomplete reaction either during the coupling or the cleavage would produce overlap between consecutive steps. Even if this occurred only to a small degree in each cycle, the overlap would become quite apparent after many cycles because of the cumulative effect. An incomplete reaction has in fact been observed during the cleavage step, and has been particularly noticable for aspartic and glutamic acid, but much less so for the amino acids with nonpolar side chains. It appears that the incomplete cleavage is due not to an insufficient reaction time, but rather to an equilibrium being reached before complete cleavage has occurred. The incompleteness of the cleavage reaction is the reason for its repetition in the present procedure. This has reduced the overlap to an insignificant level even in extended degradations.

With the limitations already discussed it is possible to identify without ambiguity every amino acid split off. However, it should be pointed out that, due to decompositions during the conversion reaction (3), asparagine, glutamine, and serine produce spots in addition to those of the expected PTH. Asparagine and glutamine-PTHs are partly hydrolysed to aspartic acid and glutamic acid-PTHs, and serine-PTH is partly converted into unidentified decomposition products. Another question is if the prolonged exposure of the protein to the conditions of the degradation could produce alterations in the amino acid residues per se. This has so far not been observed. Tryptophan is here of particular interest since it is known to be decomposed by anhydrous acids [17]. However, it should be noted that tryptophan does not occur after position 14 from the N-terminal end of the myoglobin structure.

The conditions of degradation which have been described here are those which have been found suitable for myoglobin, and also for a limited number of other proteins. However, it may be necessary to modify the conditions to suit other cases. Special difficulties are likely to be encountered with short peptides. The small difference in solubility between the thiazolinone and the short peptide makes difficult the differential extraction of the former, and yields therefore tend to fall off rapidly as the degradation approaches the C-terminal end of the peptide. This problem is at present being investigated.

We wish to acknowledge the support of this work through grants-in-aid by the National Health and Medical Research Council of Australia. Our thanks are also due to the SKF Ball Bearing Co. (Australia) for assistance with technical problems, and to the North Coast Whaling Co. and the Fisheries Division of the Commonwealth Department of Primary Industry for the supply of whale meat. We also acknowledge the skilful technical assistance of Mr. A. Pleasance and Mrs. E. Minasian.

REFERENCES

1. Edman, P., Acta Chem. Scand. 10 (1956) 761.
2. Bethell, D., Metcalfe, G. E., and Sheppard, R. C., Chem. Comm. 10 (1965) 189.
3. Ilse, D., and Edman, P., Australian J. Chem. 16 (1963) 411.
4. Edman, P., Thromb. et Diath. Haemorrhag. 17 (1963) suppl. 13.
5. Fischer, A., Z. Naturforsch. 9 a (1954) 508.
6. Blombäck, B., Blombäck, M., Edman, P., and Hessel, B., Biochim. Biophys. Acta, 115 (1966) 371.
7. Elmore, D. T., and Toseland, P. A., J. Chem. Soc. (1956) 188.
8. Edman, P., Proc. Roy. Australian Chem. Inst. (1957) 434.
9. Teremoto, S., and Ishikawa, M., Hakko Kogaku Zasshi, 32 (1954) 350.
10. Kendrew, J. C., and Parrish, R. G., Proc. Roy. Soc. (London), Ser. A, 238 (1957) 305.
11. Rumen, N. M., Acta Chem. Scand. 13 (1959) 1542.
12. Theorell, H., and Åkeson, Å., Ann. Acad. Sci. Fennicae Ser. A II, 60 (1954) 303.
13. Edman, P., and Sjöquist, J., Acta Chem. Scand. 10 (1956) 1507.
14. Honegger, C. G., Helv. Chim. Acta, 44 (1961) 173.
15. Edmundson, A. B., Nature, 205 (1965) 883.
16. Kopple, K. D., and Bächli, E., J. Org. Chem. 24 (1959) 2053.
17. Sarges, R., and Witkop, B., Biochemistry, 4 (1965) 2491.

P. Edman and G. Begg
St. Vincent's School of Medical Research
Melbourne, N. 6, Victoria, Australia

European J. Biochem. 1 (1967) 92—95

Enzymatic Synthesis of Deoxyribonucleotides

11. The Mechanism of Hydrogen Transfer
of the Ribonucleoside Diphosphate Reductase System
from *Escherichia coli* Studied with Nuclear Magnetic Resonance

L. J. Durham, A. Larsson, and P. Reichard

Department of Chemistry and Department of Biochemistry, Stanford University, Palo Alto
and Department of Chemistry 2, Karolinska Institutet, Stockholm

(Received November 11, 1966)

Deoxycytidine diphosphate was synthesized in $[^2H_2]$ water from cytidine diphosphate with the ribonucleoside diphosphate reductase system from *Escherichia coli* B. After dephosphorylation and crystallization, deoxycytidine was analyzed by nuclear magnetic resonance at 60 and 100 megaHertz. The spectra indicated that one atom of deuterium had been introduced stereospecifically into position 2'. From earlier calculations of Lemieux [10] and our own data it is tentatively concluded that the enzyme reaction occurred without change of the stereospecificity at carbon 2'.

The mechanism of hydrogen transfer during the enzymatic reduction of ribonucleotides with enzymes from *Escherichia coli* B was studied earlier in tritiated water [1]. The reaction occurs at the diphosphate level, *e.g.* CDP to deoxyCDP [2], and involves the replacement of a hydroxyl group by a hydrogen. The hydrogen donor is the reduced form of a small protein, thioredoxin-$(SH)_2$ and the over all reaction can be written as follows [3]:

$$CDP + \text{thioredoxin-}(SH)_2 \xrightarrow[\text{ATP, } Mg^{+2}]{\text{Enzymes B1 and B2}}$$

$$\text{deoxyCDP} + \text{thioredoxin-}S_2 \quad (1)$$

In tritiated water the sulfhydryl groups of reduced thioredoxin are labeled with isotope, which is then introduced into the deoxyribonucleotide during the enzymatic reduction of the ribonucleotide. After degradation of the deoxyribose obtained from deoxyCDP all tritium was found attached to carbon 2' [1]. Because of a quite large isotope effect it was not possible to distinguish whether one or both hydrogen atoms in position 2' of deoxyCDP were newly introduced. However, from other evidence it appeared likely that only one atom was incorporated and that the enzyme reaction involved a direct replacement of the hydroxyl group at carbon 2' of CDP by a hydrogen.

In the present work reaction (1) was carried out in ^2H_2O. Deoxycytidine, obtained by dephosphoryla-

tion of deoxyCDP, was analyzed by nuclear magnetic resonance. The nucleoside contained one atom of deuterium which was attached to carbon 2' and the result indicated that deuterium was introduced stereospecifically into position 2'.

Blakley *et al.* [4] independently used nmr analysis to show that the cobamide dependent ribonucleoside triphosphate reductase from *Lactobacillus leichmannii* introduced one atom of deuterium into position 2' of deoxyATP during the reduction of ATP.

EXPERIMENTAL PROCEDURES

Enzymatic Synthesis of DeoxyCDP in ^2H_2O

Ten μmoles of CDP, 20 μmoles of TPNH, 15 μmoles of ATP, 170 μmoles of MgCl₂, 50 μmoles of Tris-HCl buffer, pH 8.0, 5 μmoles of EDTA and 0.225 mg of thioredoxin were lyophilized and dissolved in 99.7% ^2H_2O. The enzymes (27 mg of Enzyme B1, 2.3 mg of Enzyme B2 and 0.15 mg of thioredoxin reductase) were equilibrated with 0.01 M Tris-HCl buffer, pH 8.0, in ^2H_2O, by passage through short columns of Sephadex G-25. The final volume of the incubation mixture was 10 ml.

CDP contained a small amount of ^3H (specific activity 21,000 counts/min/μmole) in order to make it easier to follow the purification of the product. Enzyme B1 was used after chromatography on hydroxyl apatite and Enzyme B2 after chromatography on TEAE-cellulose [2]. Thioredoxin reductase was used after chromatography on Sephadex G-100 [5]. Thioredoxin was prepared as described by Laurent *et al.* [6].

Non-standard Abbreviations. Thioredoxin-S_2 and thioredoxin-$(SH)_2$, the oxidized and reduced form, respectively, of thioredoxin [6]; Mega Hertz, MHz; nuclear magnetic resonance, nmr; nuclear magnetic double resonance, nmdr; parts per million, ppm.

Incubation was carried out at 25°. At different time intervals aliquots of the incubation mixture were diluted and their absorbance at 340 mμ was measured in a Zeiss PMQ 2 spectrophotometer. After 40 minutes, when 10 μmoles of TPNH had been consumed, another 11 μmoles of TPNH and 10 μmoles of CDP were added. After 120 minutes ^2H_2O was removed by evaporation in a vacuum. The dry residue was dissolved in 10 ml of H_2O and boiled for 10 minutes. Labile deuterium was then removed by repeated evaporation of H_2O in a vacuum.

The final residue was dissolved in 5 ml of water, the pH of the solution was adjusted to 7, and 3 mg of crude crotalus adamanteus venom were added in order to dephosphorylate all 5′—nucleotides. The reaction mixture was incubated at 37° for 2 hours and then immersed in a boiling water bath for 5 minutes. After centrifugation, the supernatant solution was adsorbed to a column of Dowex-50-H^+ (length 20 cm, diameter 1.5 cm) and cytidine and deoxycytidine were isolated by chromatography with HCl [7]. They were identified by their ultraviolet-absorbance spectra and their migration on paper chromatograms in borate medium [7]. Both had a specific activity of 21,000 counts/min/μmole.

Deoxycytidine (6.8 μmoles) was further purified by passage through a small column of Dowex-1-formate (length 10 cm, diameter 1 cm). The material was eluted with water. After neutralization of the effluent with M NH_3 and lyophilization, the dry residue was dissolved in 0.6 ml of hot absolute alcohol. A small insoluble residue was removed by centrifugation and about 0.003 ml of concentrated HCl was added to the supernatant solution. Deoxycytidine hydrochloride was allowed to crystallize at −20° over night. The crystals were washed with a small amount of cold ethanol containing HCl and dried in a vacuum over sodium hydroxide.

The residue was then dissolved in about 0.05 ml of ^2H_2O and used for the nmr spectra.

nmr Spectra

The spectra were obtained at 60 and 100 MHz using Varian A-60 and HR-100 nmr spectrometers. Chemical shifts in the ^2H_2O solutions are reported in parts per million related to external tetramethylsilane.

Field swept double resonance experiments were carried out at 100 MHz according to the method of Johnson [8] and Freeman and Whiffen [9], using the first lower sideband produced by the modulator of the V-3521 A nmr integrator for a fixed frequency and a Hewlett Packard h.p. 200 J audioscillator for a variable frequency.

RESULTS

The 100 MHz nmr spectrum of deoxycytidine (Fig. 1) is shown in Fig. 2. The assignments for the

deoxyribose moiety (primed numbers) correspond to those found for thymidine [10] and deoxyuridine [11]. Although the signals in the 60 MHz nmr spectrum of deoxycytidine (not shown here) closely resembled those of thymidine [10] and deoxyuridine [11], the 100 MHz spectrum shows that the signals for H-2′ and H-2″ are in reality much more complex than the triplet-like signal in the 60 MHz spectrum. The added complexity arises from a small non-equivalence between these protons leading to additional lines from the rather large geminal coupling between H-2′ and H-2″.

The 100 MHz nmr spectrum of the deuterated deoxycytidine hydrochloride obtained from the enzymatic reduction of CDP is shown in Fig. 3. The presence of one deuterium at carbon-2′ is confirmed by the relative area of the signal corresponding to proton(s) at C-2′ being one rather than two as observed

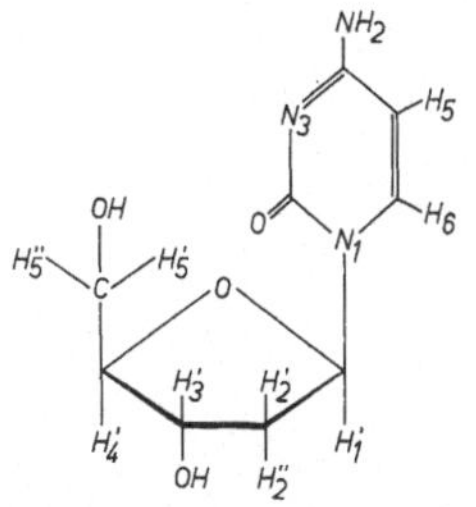

Fig. 1. *Structure of deoxycytidine.* The carbon atoms of the deoxyribosyl moiety were referred to by primed numbers

in the undeuterated material. The small difference in the position of the signals of protons at C-2′ represents the difference between the actual position of the remaining proton (H-2′, Fig. 3) and the average position of the two protons (H-2′ and H-2″, Fig. 2).

The change in the splitting patterns observed for the proton(s) at C-2′ requires that the deuterium introduced by the enzymatic reduction was placed stereospecifically. Had the deuterium entered at both sites upon removal of the hydroxyl, there would have been signals at both sites proportional to the population of protons in each position. Such a distribution would be shown by proton signals to either side of the average (2.89 ppm) position, each having their splitting patterns simplified by the removal of the large geminal (H-2′—H-2″) coupling[1]. The rather unlikely case of two atoms of deuterium at C-2′ in some molecules and none in others is also ruled out by this spectrum. In this case the signals for H-2′ and H-2″ would retain their pattern, but decrease in relative area while the splitting patterns for H-1′ and H-3′ would show the super-position of the

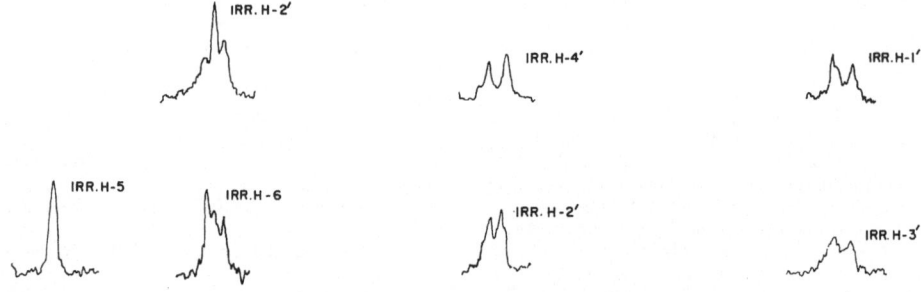

Fig. 4. *nmdr spectra of deuterated deoxycytidine at 100 MHz*

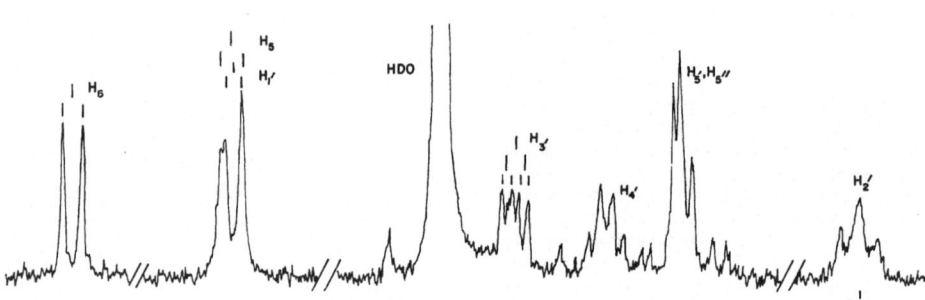

Fig. 3. *nmr spectrum of deuterated deoxycytidine at 100 MHz in* ^2H_2O

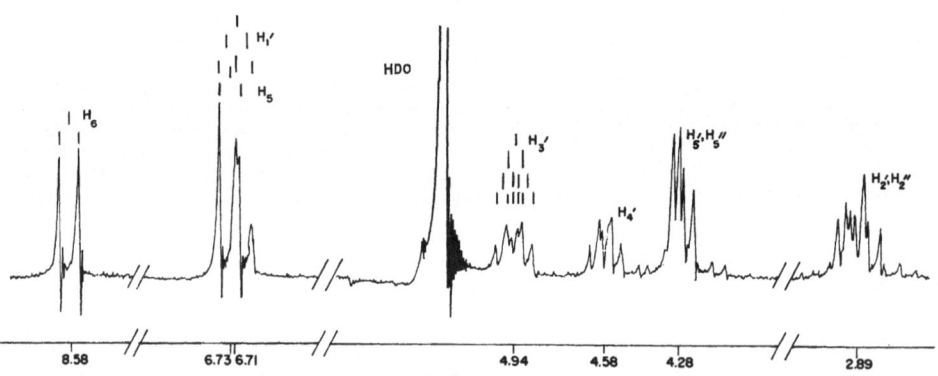

Fig. 2. *nmr spectrum of deoxycytidine at 100 MHz in 99.7%* ^2H_2O

original patterns with those in which coupling with H-2′ and H-2″ was absent[1].

Since the chemical shifts of H-2′ and H-2″ have not been identified, the disappearance of the signal to lower field (estimated position of 2.95 ppm) cannot be used to establish the location of the deuterium.

The appearance of a doublet for H-1′ in deutero-deoxycytidine (Fig. 3) with about the same spacing (6—6.5 Hz) as observed in the H-1′-triplet of the undeuterated material (Fig. 2) establishes the similarity of the couplings of H-1′ with H-2′ and H-2″ in deoxycytidine and probably in thymidine and deoxyuridine as well. With $J_{1'2'}$ and $J_{1'2''}$ equal or nearly so, the removal of one of these interactions on introduction of deuterium tells nothing about the location of the deuterium[2].

The splitting pattern of the proton at C-3′ on the other hand does give some possible clues about the location of the deuterium at C-2′. The signal for H-3′ in the non-deuterated deoxycytidine (Fig. 2) approximates a pair of triplets involving one larger and two smaller coupling interactions. After introduction of the deuterium (Fig. 3) one of the smaller couplings is lost and H-3′ appears as a pair of doublets with spacing indicative of couplings of 6 and 3.5 Hz, respectively.

Several double resonance experiments were carried out in order to establish the assignments of these coupling interactions. In Fig. 4 the "decoupled" signals are shown above the corresponding normal signals of Fig. 3, and in each case it is also indicated which signal was irradiated. Thus e.g. the singlet in Fig. 4 marked "IRR. H-5″" represents the signal of H-6 after decoupling H-6 from H-5 by irradiating the latter position.

Of particular interest are the two decoupled signals of H-3′. On irradiating H-2′ the larger coupling drops out showing that the remaining smaller coupling (3.5 Hz) represents $J_{3'4'}$. The larger coupling, remaining after irradiation of H-4′ represents $J_{2'3'}$ (= 6.5 Hz). Decoupling experiments with the proton at C-2′ gave values for $J_{1'2'}$ and $J_{2'3'}$ of about 6 and 6.5 Hz, respectively.

From the value of 6.5 Hz for $J_{2'3'}$ of deuterated deoxycytidine and the spectrum of the non-deuter-

ated material a value of 4—4.5 Hz can be calculated for the coupling of H-2″ to H-3′.

Lemieux[10] calculated values of 7.5 and 3.5 Hz for $J_{2'3'}$ and $J_{2''3'}$, respectively, by applying the Karplus relationship [12,13]. These values are in reasonable agreement with our experimental values of 6.5 and 4—4.5 Hz.

Accepting Lemieux' assignments, our results would indicate that $J_{2''3'}$ was lost in the deuterated compound and that deuterium thus had been introduced trans to the base at carbon-1′ during the enzyme reaction, without change in the stereochemistry at carbon-2′. Such a conclusion must be tentative, however, since the application of the Karplus curve to neighbouring hydrogens in furanose rings is difficult [14,15] and the assignments made by Lemieux thus cannot be considered absolutely conclusive.

Part of this research was carried out during a sabbatical leave of Peter Reichard at the Department of Biochemistry, Stanford University. This author wishes to thanks Dr. A. Kornberg and other members of the department for a very stimulating half year of work.

This investigation was supported by grants from the Damon Runyon Foundation (470), the U.S. Public Health Service (CA 7897), The Swedish Medical Research Council and The Swedish Cancer Society.

REFERENCES

1. Larsson, A., Biochemistry, 4 (1965) 1984.
2. Reichard, P., J. Biol. Chem. 237 (1962) 3513.
3. Holmgren, A., Reichard, P., and Thelander, L., Proc. Natl. Acad. Sci. US, 54 (1965) 830.
4. Blakley, R. L., Ghambeer, R. K., Batterham, T. J., and Brownson, C., Biochem. Biophys. Res. Commun. 24 (1966) 418.
5. Moore, E. C., Reichard, P., and Thelander, L., J. Biol. Chem. 239 (1964) 3445.
6. Laurent, T. C., Moore, E. C., and Reichard, P., J. Biol. Chem. 239 (1964) 3436.
7. Reichard, P., Acta Chem. Scand. 12 (1958) 2048.
8. Johnson, L. F., Varian Technical Information Bulletin, 1962, Vol. III, (3), p. 196.
9. Freeman, R., and Whiffen, D. H., Mol. Phys. 4 (1961) 321.
10. Lemieux, R. U., Can. J. Chem. 39 (1961) 116.
11. Jardetzky, C. D., J. Am. Chem. Soc. 83 (1961) 2919.
12. Karplus, M., J. Chem. Phys. 30 (1959) 11.
13. Conroy, H., Advances in Organic Chemistry, Methods and Results, (edited by R. A. Raphael, E. C. Taylor and H. Wynberg) Interscience Publ. Co., New York 1960, Vol. II, p. 265.
14. Karplus, M., J. Am. Chem. Soc. 85 (1963) 2870.
15. Lemieux, R. U., and Lineback, D. R., Ann. Rev. Biochem. 32 (1963) 155.

---

[1] Although replacement of hydrogen by deuterium actually only decreases the magnitude of the coupling, as a first approximation, it is here referred to as removing it since the remaining interaction is so much smaller. However, the geminal $^1H-^2H$ coupling is not negligible, so the splitting pattern observed for H-2′ should not be used to ascertain the coupling constants $J_{1'-2'}$ or $J_{3'-2'}$ without correcting for the $^1H-^2H$ interaction.

[2] As discussed previously [10,11], the triplet of H-1′ could indicate either a similarity of the two couplings or merely give an average of the two—particularly with H-2′ and H-2″ being nearly equivalent. Had the triplet been an average of two dissimilar couplings, learning the magnitudes of the couplings would have been helpful in locating the deuterium.

L. J. Durham
Department of Chemistry, Stanford University,
Palo Alto, Calif., U.S.A.

A. Larsson and P. Reichard,
Kemiska Institutionen II, Karolinska Institutet,
Stockholm 60, Sweden

European J. Biochem. 1 (1967) 96—101

The Mechanism of Enzyme Secretion by the Cell

4. Effects of Inducers, Substrates and Inhibitors on Amylase Secretion by Rat Parotid Slices

H. Babad, R. Ben-Zvi, A. Bdolah, and M. Schramm

Department of Biological Chemistry, The Hebrew University of Jerusalem

(Received November 23, 1966)

Preparation of rat parotid slices in cold medium caused extensive leakage of amylase during subsequent incubation. Enzyme secretion, by slices not prepared in the cold, was almost absolutely dependent on the addition of an inducer such as epinephrine. Both epinephrine and norepinephrine were more effective than other agents tested. Since epinephrine was apparently consumed during incubation, a sufficient excess of the hormone was required to achieve a maximal yield of enzyme secreted into the medium. Further experiments indicated that 3'5' cyclic AMP is an intermediate in the induction of enzyme secretion by epinephrine. The dibutyryl and monobutyryl derivatives of cyclic AMP caused amylase secretion at a rate which was even slightly higher than that obtained with epinephrine. An initial lag period was apparently due to the slow penetration of such compounds into the cell.

β-hydroxybutyrate was the only substrate which maintained a linear rate of amylase secretion up to termination of the process. The reaction stopped within 40—60 minutes when about 50% of the amylase which had been in the slice was secreted into the medium. This amount corresponded roughly to that located in the intracellular zymogen granules at zero time. Inhibitors of oxidative phosphorylation prevented enzyme secretion and caused some release of enzyme from the intracellular zymogen granules into a soluble pool within the slice. Enzyme secretion apparently required the continuous supply of energy since dinitrophenol was inhibitory even when added after onset of the process. EDTA caused discharge of amylase, presumably by releasing an endogenous inducer.

Earlier work showed that specific secretion of enzymes can be induced by epinephrine in parotid slices [1—3]. This system seemed particularly suitable for study of the mechanism of enzyme secretion because of the extremely high content of exportable α-amylase [4]. Furthermore, a large proportion of the amylase in the gland cell was located in the intracellular zymogen granules available for immediate secretion [2,4]. However, the rate of enzyme leakage from the slices in absence of epinephrine was high and the rate and extent of secretion in presence of the hormone were quite variable. Subsequent studies were therefore aimed at improving the slice system while furthering our understanding of the factors which operate in the secretion process.

It was found that enzyme leakage can be prevented by avoiding exposure of the slices to low temperatures. It was also shown that a high linear rate of secretion, up to its termination, can be assured by addition of β-hydroxybutyrate and ample amounts

of epinephrine. The present communication demonstrates these findings and reports experiments on the probable function of cyclic AMP as an inducer of secretion, on the effect of various other inducers and on the action of inhibitors of oxidative phosphorylation.

Preliminary notes on some of these findings have been published [3,5].

EXPERIMENTAL

Parotid Slice System

In order to obtain an optimally active slice system many modifications were introduced since the previously published procedure [2]. The following method is that presently used. Albino rats of 120—180 g are kept at $26 \pm 1°$. To minimize differences between glands from individual animals at least four rats are used in each experiment. About 20 h before the experiment the animals are transferred to a cage without food but with ample supply of water. The animals under heavy ether anasthesia are killed and bled by cutting through the heart. The parotid glands are removed from the animals and immediately placed in a solution of KRB [6], containing

---

Non-standard Abbreviations. Krebs Ringer bicarbonate, KRB; 2,4-dinitrophénol, DNP; carboxylcyanide p-trifluoro-phenylhydrazone, CCP; 3',5'-cyclic AMP, cyclic AMP; N^6-2-O-dibutyryl-3,'5'-cyclic AMP, dibutyryl cyclic AMP; N^6-derivative, monobutyryl cyclic AMP.

5 mM β-hydroxybutyrate at 37°. In most experiments the KCl concentration of the KRB solution was raised to 14 mM. The solution in which the glands are collected is continuously gassed with a mixture of 95% O_2, 5% CO_2.

Eight glands are cut into small pieces by a mechanical tissue slicer (Mickle Co., England), pooled and placed in a 125 ml polyethylene Erlenmeyer flask containing 40 ml KRB medium of the composition noted above, at 37°. The vessel is gassed for 20 seconds with the 95% O_2-5% CO_2 mixture, tightly closed with a rubber stopper and incubated at 37° for 15 minutes with shaking at 180 revolutions per minute (New Brunswick rotatory shaker bath, New Jersey). This period of preincubation serves to remove material from ruptured cells and blood vessels. At the end of the preincubation slices are collected on filter paper in a Buchner funnel without suction and divided into eight roughly equal portions. Each portion of slices equivalent to about one gland, containing about 5000 amylase units, is placed in a 25 ml plastic vial ($d = 23$ mm) of the type used for scintillation counting, which contains KRB medium as above, and various additions, in a final volume of 2 ml. To induce secretion, epinephrine is added to a concentration of 30 µM. The vials are gassed, tightly closed with a rubber stopper and incubated with shaking as described for the preincubation period. Since both epinephrine and oxygen are consumed it seems preferable to add epinephrine again and gas the system after 30 minutes in experiments in which the maximal extent of enzyme secretion is measured. Whenever additions are made or aliquots removed during the experiment, the vials are gassed with the O_2-CO_2 mixture and tightly stoppered before continuing the incubation. At the termination of the experiment 8 ml KRB solution are added to each vial, the contents is mixed and poured on a Buchner funnel as above to separate the medium from the slices. Amylase activity in the medium and in the homogenate prepared from the slices is determined as described elsewhere [7].

Some of the experiments reported were carried out using earlier modifications of the slice system. Such deviations are noted under the respective experiments.

Definition of per cent Secretion

The amount of amylase released into the medium plus the amount remaining in the slice, as measured in the homogenate, is defined as 100%. The amount secreted into the medium during any time interval is thus expressed as per cent of total.

Materials and Reagents

Dibutyryl cyclic AMP was a gift from Dr. Earl W. Sutherland. The compound was converted from the barium to the sodium salt by addition of Na_2SO_4.

Monobutyryl cyclic AMP was prepared by alkaline hydrolysis [8]. Epinephrine bitartarate, obtained from K and K Laboratories was dissolved in distilled water, kept at $-20°$ and used within one week after preparation. Norepinephrine was a product of Teva Co., Israel. Serotonin creatinine sulfate was purchased from Mann Research Laboratories. Oligomycin and CCP were a gift from Dr. E. Racker.

RESULTS

Leakage of Amylase and the Effect of Substrate on the Rate of Secretion

In our earlier studies glands were collected and slices prepared in ice cold KRB medium. Slices were again transferred into cold medium after preincubation. It is shown in Table 1 that preparation in cold

Table 1. *Effect of temperature during preparation of slices on subsequent leakage of amylase during incubation*
Glands were collected and slices were prepared at 4° or 37° as indicated in the table. Slices were again transferred into fresh medium at the respective temperatures after preincubation, prior to final incubation at 37°. The medium did not contain substrate

Temperature during preparation	Amylase secreted			Relative stimulation by epinephrine
	Preincubation 15 min 37°	Incubation 60 min 37°		
		No addition	Epinephrine 10 µM	
		(a)	(b)	(b/a)
°C	%	%	%	%
4	5.2	10.0	38.0	3.8
37	1.8	3.8	33.0	8.7

medium caused excessive leakage of amylase from the slices during subsequent incubation at 37°. As a result, the relative increase in the rate of enzyme release caused by epinephrine is much smaller than that achieved with slices handled at 37°. Although the problem of extensive enzyme leakage was eliminated by avoiding exposure of the slices to cold medium the rate of secretion in presence of epinephrine remained variable and often declined rapidly. Previous attempts to enhance the rate of secretion by addition of a number of substrates failed and it was therefore concluded that endogenous substrates were sufficient [2]. However, when β-hydroxybutyrate was supplied to the slices with epinephrine, a linear rate of secretion was maintained almost until termination of the process within 40—60 minutes (Fig. 1). The total amount secreted was roughly equivalent to that initially located in the intracellular zymogen granules (see also [2]). β-hydroxybutyrate did not increase the initial rate and had no effect on occasional batches of slices which maintained a linear rate without substrate. Doubling the β-hydroxybutyrate concentration did not further increase the rate or extent of secretion. The two experiments in Fig. 1 also demonstrate different extents of dependence on β-hydroxybutyrate and further show that the control without epinephrine is unaffected by the presence of this substrate. Acetoacetate, oxaloacetate, malonate, pyruvate, caprylate and the substrates which had already been tried [2] were again tested but none of these showed any consistent effect comparable to that obtained with β-hydroxybutyrate.

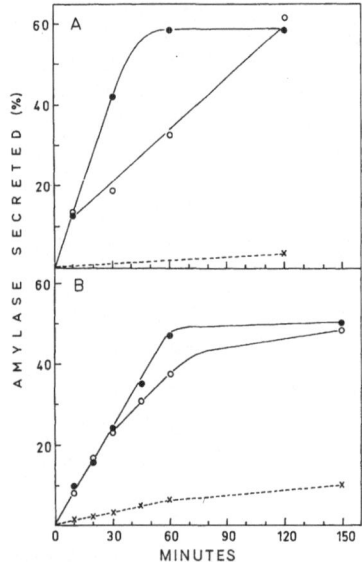

Fig. 1. *Maintenance of the secretion rate by β-hydroxybutyrate.* A and B represent two experiments carried out under the same conditions but with different batches of slices. Reagent concentrations were: epinephrine, 30 µM and β-hydroxybutyrate, 5 mM. ○, epinephrine; ●, epinephrine plus β-hydroxybutyrate; ×, no epinephrine, with and without β-hydroxybutyrate (essentially identical experimental points)

Action of Inhibitors of Oxidative Phosphorylation

Since secretion induced by epinephrine was dependent on oxidation reactions [2] the effect of oxidative phosphorylation was further studied. Table 2 demonstrates that epinephrine accelerates various inhibitors of oxygen uptake by the slices in conjunction with its induction of enzyme secretion. Oligomycin effectively inhibited the increased oxygen uptake caused by epinephrine as well as enzyme secretion. It is noted that enzyme discharge in the control without epinephrine was somewhat increased by oligomycin, and secretion in presence of the hormone was not blocked completely. The amount of enzyme released in the absence of epinephrine was high since the slices were not washed by preincubation to remove amylase released by damaged cells.

Table 2. *Oligomycin inhibition of epinephrine induced oxygen uptake and enzyme secretion*
Slices were placed in Warburg vessels in Krebs Ringer phosphate medium [6] without prior preincubation. The medium did not contain substrate. Oligomycin, 10 µg/ml was added as specified in the table and the slice systems were all kept at 25° for 30 minutes to facilitate penetration of the inhibitor into the cells. The slice systems were transferred to 37°, epinephrine, 10 µM, was added and oxygen uptake was measured. The rates of oxygen uptake were all linear throughout the incubation period of 60 minutes

Additions	Amylase secreted	Oxygen consumed
	%	µl/mg protein/hour
None	14	7
Epinephrine	48	15
Oligomycin	17	6
Oligomycin + epinephrine	22	6

Table 3. *Effect of inhibitors and absence of oxygen on the subcelllular distribution and secretion of amylase*
Reagent concentrations were: epinephrine, 10 µM; DNP, 1 mM; CCP, 1 µM. A mixture of 95% N_2, 5% CO_2 replaced the O_2—CO_2 gas phase where specified. The medium did not contain substrate. After incubation for 60 minutes the slices were removed from the medium and homogenized in 0.3 M sucrose. The homogenate was separated into a crude zymogen granule fraction and supernatant by 10 minute centrifugation at $1,300 \times g$ [2]

Additions	Relative distribution of amylase		
	Slices		Medium
	Zymogen granules	Supernatant [a]	
	%	%	%
None	50	45 (0)	5
Epinephrine	22	45 (0)	33
DNP	35	58 (+13)	7
Epinephrine + DNP	39	54 (+9)	7
CCP	40	52 (+7)	8
Epinephrine + CCP	40	49 (+4)	11
N_2	35	58 (+13)	7
Epinephrine + N_2	41	52 (+7)	7

[a] Figures in parentheses represent the increase in amylase content of the supernatant, determined by difference from the control without additions.

It is shown in Table 3 that inhibitors of oxidative phosphorylation or a nitrogen atmosphere not only prevent secretion into the medium but also cause release of amylase within the slice, from the intracellular zymogen granules into a soluble fraction. Under oxygen, in absence of inhibitors the amount of amylase in the zymogen granules remained quite constant at 50% during incubation. When epinephrine was added the amylase content of the zymogen granules dropped markedly and a roughly equivalent amount of enzyme appeared in the extracellular medium. This finding does not exclude the possibility that the enzyme is transferred from the granules into the medium *via* a soluble intracellular pool. Thus energy might also be required for the transfer from this pool into the medium. However, it seemed possible that energy might be needed only for the initiation of the secretion process and not for its maintenance. To test this possibility DNP was added 2 minutes after addition of epinephrine since it had been shown previously that this inducer activates the secretion process within ten seconds [5]. As shown in Fig. 2 the secretion rate dropped to the level of the control without epinephrine within three minutes after DNP addition. It is not known whether the three minute time interval required for complete inhibition of secretion reflects the rate of penetration of DNP into the slice cells or is due to other factors.

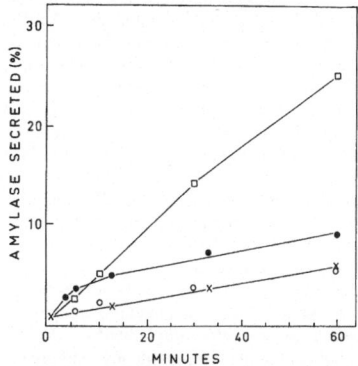

Fig.2. *Decline of enzyme secretion caused by the addition of DNP.* Reagent concentrations were: epinephrine, 10 μM and DNP, 1 mM. The medium did not contain substrate. □, epinephrine; ×, no additions; ○, DNP added five minutes before addition of epinephrine at zero time; ●, DNP added two minutes after addition of epinephrine at zero time

Inducers of Secretion and the Function of Cyclic AMP

Several agents have already been tested as inducers of enzyme secretion in the parotid slice system [2]. Table 4 shows the effect of various additional compounds. Norepinephrine was as efficient as epinephrine, while dopa which is structurally related to the catecholamines was without effect even in presence of epinephrine. Caffeine seems to act similarly to theophylline [3]. Surprisingly, serotonin also caused amylase secretion but only at a relatively high concentration.

It was observed that during incubation with EDTA in absence of calcium and magnesium ions, a relatively large amount of amylase was released into the medium (Table 5). Since DNP completely inhibited enzyme release by EDTA, this chelator probably triggered a true secretion process.

In an experiment not shown in Table 5 epinephrine further enhanced the rate of secretion, above that obtained with EDTA alone, when the hormone

was added to slices 30 minutes after EDTA. Thus, the normal function of the slices was presumably not changed drastically by the presence of the chelator. EDTA had no effect in presence of excess calcium and magnesium ions in the medium.

Table 4. *Induction of secretion by various agents*
The medium did not contain substrate. Incubation time was 45 minutes

Exp.	Additions	Amylase secreted
	M	%
I	—	5
	Epinephrine 10^{-5}	39
	Norepinephrine 10^{-5}	42
	DL-dopa 10^{-3}	7
	Epinephrine 10^{-5} + dopa 10^{-3}	43
	Caffeine 10^{-2}	23
II	—	2
	Epinephrine 10^{-5}	43
	Serotonin 10^{-3}	30
	Serotonin 10^{-4}	29
	Serotonin 10^{-5}	3

Table 5. *Discharge of amylase from the slices by EDTA and its inhibition by DNP*
Slices were prepared and incubated in the standard medium but without calcium and magnesium salts. Reagent concentrations were: epinephrine, 30 μM; DNP, 1 mM; EDTA, 5 mM. Incubation time was 60 minutes

Additions	Amylase secreted
	%
—	5
Epinephrine	61
Epinephrine + DNP	5
EDTA	23
EDTA + DNP	4

Among the various kinds of inducers tested, epinephrine was the most potent. However, using the ratio of one gland equivalent of slices per 2 ml of medium, secretion induced by 10 μM epinephrine sometimes declined after 30—40 minutes. The rate of secretion as a function of time was therefore studied at various concentrations of epinephrine (Fig.3). At twenty minutes the rate of secretion was nearly the same for the epinephrine concentrations in the range 1.8—6 μM. After 40 minutes, however, the rate of secretion declined at all epinephrine concentrations below 6 μM. It thus seems likely that the effective epinephrine concentration declines during the experiment and an ample supply must be ensured in order to achieve maximal yields of enzyme secreted into the medium.

Induction of enzyme secretion in the parotid slices by epinephrine is apparently mediated within the cell by cyclic AMP [3]. Table 6 and Fig.4 present further evidence to support this contention. Added cyclic AMP was inert in the slice system, presumably

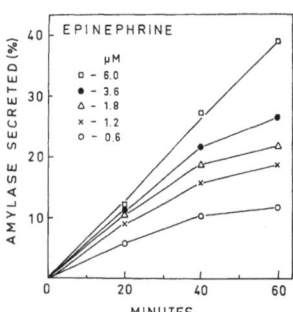

Fig.3. *Decline of enzyme secretion with suboptimal amounts of epinephrine*

Table 6. *Induction of secretion by butyryl derivatives of cyclic AMP and its inhibition by DNP*
The medium did not contain substrate. Incubation time was 60 minutes

Additions		Amylase secreted
	mM	%
—		3
Epinephrine	0.01	30
Cyclic AMP	2	3
Monobutyryl cyclic AMP	1	35
Dibutyryl cyclic AMP	1	33
Epinephrine	0.01 + DNP1	4
Dibutyryl cyclic AMP	1 + DNP1	4

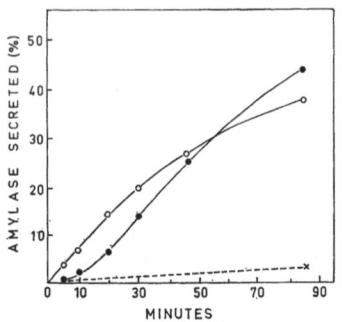

Fig.4. *Induction of enzyme secretion by dibutyryl cyclic AMP.*
The medium did not contain substrate. ●, dibutyryl cyclic AMP, 1 mM; ○, epinephrine, 10 μM; × no additions

because the cells were impermeable to this compound or because it was hydrolyzed before it reached its target in the cell. However, dibutyryl cyclic AMP as well as monobutyryl cyclic AMP were even somewhat more effective inducers of secretion than epinephrine. Table 6 also demonstrates that DNP completely blocks secretion induced by dibutyryl cyclic AMP. After an initial lag period, possibly due to

slow entry of this compound into the cells, the secretion rate becomes even higher than that obtained with epinephrine.

DISCUSSION

A considerable amount of amylase leaked out of the rat parotid slices during incubation when the tissue had been previously exposed to cold medium. Several findings indicate that low temperature causes or potentiates the release of amylase from the intracellular zymogen granules. It had been shown that zymogen granules isolated in isotonic sucrose or KCl become leaky in the cold and lose a large proportion of their amylase content [5]. The effect could be traced to temperature induced changes in the lipid structure of the granule membrane. Addition of specific lipids prevented leakage from the isolated granules in the cold. It might also be suspected that lack of energy supply in the cold causes the subsequent leakage of amylase from the slices during incubation. Indeed, as shown in the present work, inhibitors of oxidative phosphorylation or absence of oxygen caused a decline in the amylase content of the zymogen granules. However, the enzyme thus released from the granules did not find its way into the medium to any large extent but was mainly accumulated within the slice.

When exposure of the slices to low temperatures was avoided the release of enzyme into the medium became almost absolutely dependent on the addition of an inducer such as epinephrine. In presence of β-hydroxybutyrate, enzyme secretion proceeded at a linear rate and declined abruptly when 40—60% of the amylase was released into the medium. At this stage the intracellular zymogen granules are mostly depleted of amylase. Therefore it should probably be concluded that enzyme in the cell, which is not accumulated in zymogen granules is not immediately available for secretion. The very slow rate of enzyme release which still persists after the initial rapid phase is equivalent to that observed in absence of epinephrine. It was previously reported that 90% of the total amylase was sometimes released into the medium during very long incubation periods. Enzyme leakage was high and linear rates with epinephrine were not achieved [2]. It was therefore not possible at the time to distinguish sharply between the initial epinephrine dependent secretion and the subsequent independent slow enzyme release.

The importance of energy yielding oxidative processes in enzyme secretion is demonstrated in most of the experiments presented. Some supply of energy by oxidation is apparently required to maintain the amylase within the zymogen granules when the cell is not actively secreting. A further, much more stringent demand for energy arises when secretion is induced. Inhibitors of oxidative phosphorylation or replacement of oxygen by nitrogen effectively blocked the secretion process. Evidence was also presented to show that energy was required for maintaining enzyme secretion and not only for its initiation. Epinephrine caused an increase in oxygen uptake which may reflect an increase in the oxidation of fatty acids and is often referred to as its calorigenic effect [9]. Oligymycin depressed oxygen uptake in the presence of epinephrine to a value even below that of the control slices without epinephrine. Whether this is due to a strict respiratory control [10] of the mitochondria has not yet been determined.

β-hydroxybutyrate was the only substrate found which supported the secretion process. It did not increase the initial rate of secretion but maintained it, while slices incubated without this substrate often showed an early decline of the secretion rate. Because of these characteristics of β-hydroxybutyrate and the highly specific requirement for this compound it is quite possible that the endogenous substrate which drives the secretion process must also be β-hydroxybutyrate. While it is possible that DPNH generated by oxidation of β-hydroxybutyrate has some special function in the secretion process there is as yet no information which explains the specific requirement for this substrate.

Epinephrine and the closely related norepinephrine were found to be more active than all other inducers tested to date in the parotid slice system. There is evidence for the accumulation *in vivo* of these hormones in the parotid gland, apparently at nerve endings [11]. It is therefore most likely that the catecholamines serve as the natural inducers of secretion in this gland. Epinephrine is consumed during incubation in the slice system and it is also possible that part of it is taken up and segregated within remaining nerve tissue of the slices [11] so that it becomes unavailable to the secreting cells. It is therefore not only important to establish an optimal initial concentration of the hormone in the slice system, but also to ascertain that the absolute amount present will be sufficient during the entire period of the experiment. It was indeed demonstrated that epinephrine concentrations which initially caused high rates of enzyme secretion were insufficient to maintain that rate during longer incubation periods. While there is no doubt that epinephrine is most probably consumed by various processes during incubation an additional factor might contribute to the decline of secretion. It is possible that the absolute amount of enzyme secreted is dependent on the concentration of epinephrine. Either that the individual cells of the gland demonstrate different affinities for the hormone, or that within each cell the response to a given concentration of the hormone decreases progressively during secretion.

Since epinephrine is accumulated in the parotid gland *in vivo* [11] some of the inducers might act indirectly, by releasing such endogenous stores of the

hormone in the slices. The action of EDTA as noted in the present work and the effect of carbamylcholine and electric pulses reported previously [2] might fall into this category.

The evidence in this study as well as that in a previously report [3] points to cyclic AMP as the intracellular agent which activates enzyme secretion. Through the work of Sutherland and his collaborators it is known that epinephrine stimulates the synthesis of cyclic AMP in many tissues [12]. He has also shown that theophylline and caffeine inhibit the diesterase which hydrolyzes cyclic AMP [13]. Since both the mono and dibutyryl derivatives of cyclic AMP induced rapid amylase secretion and since theophylline [3] and caffeine were also active it should be concluded that cyclic AMP is the most likely intermediate in the induction of enzyme secretion by epinephrine. Induction of amylase secretion by serotonin might also be ascribed to stimulation of cyclic AMP synthesis. Mansour and Mansour showed that cyclic AMP replaced serotonin in the activation of phosphofructokinase [14].

This investigation was supported in part by a grant from the United States Public Service (No. AM 10451-01).

H. Babad thanks the United States Public Health Service for a postdoctoral fellowship.

REFERENCES

1. Bdolah, A., and Schramm, M., *Biochem. Biophys. Res. Commun.* 8 (1962) 266.
2. Bdolah, A., Ben-Zvi, R., and Schramm, M., *Arch. Biochem. Biophys.* 104 (1964) 58.
3. Bdolah, A., and Schramm, M., *Biochem. Biophys. Res. Commun.* 18 (1965) 452.
4. Schramm, M., and Danon, D., *Biochim. Biophys. Acta,* 50 (1961) 102.
5. Schramm, M., Ben-Zvi, R., and Bdolah, A., *Biochem. Biophys. Res. Commun.* 18 (1965) 446.
6. Cohen, P. P., in *Manometric Techniques* (edited by W. W. Umbreit, R. H. Burris, and J. F. Stauffer), Burgess Publishing Co., Minneapolis 1957, p. 148.
7. Schramm, M., and Loyter, A., in *Methods in Enzymology* (edited by S. P. Colowick and N. O. Kaplan), Academic Press, New York 1966, Vol. VIII, p. 533.
8. Posternak, Th., Sutherland, E. W., and Henion, W. F., *Biochim. Biophys. Acta,* 65 (1962) 558.
9. Hagen, J. H., and Hagen, P. B., in *Actions of Hormones on Molecular Processes* (edited by G. Litwack and D. Kritchevsky), John Wiley and Sons, New York 1964, p. 298.
10. Lehninger, A. L., *The Mitochondrion,* W. A. Benjamin, New York 1964, p. 132.
11. Strömblad, B. C. R., and Nickerson, K., *J. Pharmacol. Exptl. Therap.* 134 (1961) 154.
12. Sutherland, E. W., Øye, I., and Butcher, R. W., *Recent Progr. Hormone Res.* 21 (1965) 623.
13. Butcher, R. W., and Sutherland, E. W., *J. Biol. Chem.* 237 (1962) 1244.
14. Mansour, T. E., and Mansour, J. M., *J. Biol. Chem.* 237 (1962) 629.

H. Babad, R. Ben-Zvi, A. Bdolah and M. Schramm
Department of Biological Chemistry
The Hebrew University of Jerusalem
Jerusalem, Israel

European J. Biochem. 1 (1967) 102—109

Reduced Nicotinamide Dinucleotide Phosphate Diaphorase
from *Bacillus subtilis*

G. Avigad and N. Levin

From the Department of Biological Chemistry, The Hebrew University of Jerusalem

(Received November 29, 1966)

NADPH-diaphorase was extensively purified from extracts of *Bacillus subtilis* NY. FMN is most probably the prosthetic group of this enzyme. NADPH is oxidized by oxygen as the terminal acceptor only in the presence of catalytic amounts of free flavins (K_m for FMN was 1.6×10^{-6} M at pH 8.0). NADPH was directly oxidized in the presence of various electron acceptors such as quinones, ferricyanide and tetrazolium salts. Cytochrome c was a very poor acceptor in this system. K_m for NADPH was 2.0×10^{-4} M and K_i for NADP$^+$ as a competitive inhibitor was 1.1×10^{-3} M at pH 7.2. Mn^{++} or Mg^{++} at mM concentration showed some stimulation of enzyme activity. The enzyme was partially inhibited by various chelating agents and p-hydroxymercuribenzoate but not by several typical inhibitors of the respiratory chain. It is suggested that this flavoprotein can serve as a useful tool for NADP$^+$ regeneration in coupled enzyme reactions.

Several NADPH oxidizing systems of the diaphorase or cytochrome c oxidoreductase type were obtained from different biological sources and are described in the literature. Among those studied in detail were the old yellow enzyme[1], NADPH-cytochrome c reductase from yeast[2] and mammalian microsomes[3—10] and a diaphorase from chloroplasts[11]. In addition, various diaphorase type systems which oxidize both NADH and NADPH were obtained from mammalian and plant tissues [12—15] and from various microorganisms [12, 16—19]. Specific NADH oxidase systems were isolated from several bacteria [16,17,20—27]. In these cases the enzymatic activity was mostly obtained in a particulate fraction and is probably associated with the cell membrane structure and with components of the respiratory chain [28—33]. In comparison, soluble, specific NADPH-diaphorase was separated from bacterial extracts only in a very small number of cases [16,17,34] and was not studied in great detail.

While studying certain aspects of carbohydrate metabolism in a strain of *Bacillus subtilis*, the presence of a significant NADPH-diaphorase activity was noticed in extracts obtained from this organism

[35]. The enzyme catalyzing this activity was purified extensively and characterized as described in the present communication.

EXPERIMENTAL PROCEDURE
Material and Methods

Pyridine nucleotides were obtained from Boehringer & Soehne, GmbH. Flavins, purified enzyme reagents and most biochemicals used were purchased from Calbiochem and from Sigma Biochemicals Inc. Chelating agents and other fine chemicals were obtained from Fluka AG, Buchs, Switzerland. *Crotalus atrox* nucleotide phosphodiesterase was from Koch-Light Laboratories, Inc., Colnbrook, Bucks, England; and 2-n-heptyl-4-hydroxyquinoline N-oxide was a gift from Dr. J. W. Cornforth.

Absorbance was measured in a Zeiss model PMQ II spectrophotometer or with a Gilford model 2000 automatic absorbance recording spectrophotometer. Quartz cells of one cm light path were used in most experiments. Spectra were measured in a Cary Model 14 spectrophotometer. Oxidation of reduced pyridine nucleotide was followed at 340 mμ [36]. The wavelengths used to measure different components in the reaction system and their millimolar extinction coefficients ($\Delta \varepsilon$ M^{-1} cm^{-1}) employed were: 6.22 for NADPH at 340 mμ [36]; 11.3 for flavins at 450 mμ [37]; 1.0 for ferricyanide at 420 mμ [38]; 9.2 for oxidized and 27.7 for reduced cytochrome c at 550 mμ [4,23]; 20.9 for DCPIP at 600 mμ [4, 38,39]; 15.4 for INT at 520 mμ [40]. Fluorimetric

---

Non-standard Abbreviations. 2,6-dichlorophenolindophenol, DCPIP; 2-(p-iodophenyl)-3-(p-nitrophenyl)-5-phenyl tetrazolium chloride, INT.

Enzymes. Old yellow enzyme, or reduced NADP dehydrogenase, or reduced-NADP: (acceptor) oxidoreductase (EC 1.6.99.1); DT-diaphorase, or reduced NAD(P) dehydrogenase, or reduced-NAD(P): (acceptor) oxidoreductase (EC 1.6.99.2).

analysis of FMN and FAD was carried out at 535 mμ with exitation at 450 mμ in the Aminco-Bowman spectrophotofluorimeter [41]. FAD was hydrolyzed by trichloroacetic acid or by incubation with excess venom phosphodiesterase to render it sensitive for the fluorimetric analysis [41]. Paper chromatography of flavins was performed as described by Rao *et al.* [37]. Protein was determined by the phenol reagent [42] with bovine serum albumin as the standard.

Uptake of oxygen was measured manometrically by the conventional Warburg technique [43] or by the use of a polarographic oxygen electrode described by Harel *et al.* [44].

Growth of the Organism

The *Bacillus subtilis* strain employed as a source for the NADPH-diaphorase[1] was obtained in 1961 from the Department of Microbiology, New York University School of Medicine through the courtesy of Dr. B. L. Horecker. Its origin is unknown and it was designated by us as strain NY. The organism was kept on nutrient agar slants and cultivated on the following medium: Nutrient broth (Difco) 1% (w/v); NaCl, 0.2% (w/v); glycerol, 0.2% (v/v) and glucose, 0.1% (w/v). Small batches were grown in 600 ml of medium in 2 liter erlenmeyer flasks on a rotatory shaker at 30° for 24 hours. Cells were sedimented by 15 min centrifgation at 8000 ×*g* and washed twice with 0.01 M phosphate buffer, pH 7.2. The yield was 7−8 g of fresh packed cells per liter. To obtain larger quantities, cells were grown in batches of 40 liters in a stainless steel fermentor (Palbam Ltd., Ein Harod, Israel) with mechanical stirring (210 revolutions per min) and sparging of sterile air at the rate of 5 liter per min. Silicone emulsion RD (Hopkin and Williams, Ltd., Chadwell Heath, U.K.) at 0.05% (v/v) was employed for foam suppression. After 20 hours at 30° the cells were collected in the Sharpless supercentrifuge and washed twice with water.

To obtain a dry powder of bacteria, a thick suspension of cells was treated with 10 volumes of acetone at − 20°, collected by filtration, washed with cold acetone and dried in vacuum. The yield was 2.5 to 3.0 g of dry cells per liter of medium.

Preparation of Extracts

Fresh cells were suspended in 5 times their weight, and acetone dried cells in 20 times their weight of 0.01 M phosphate buffer, pH 7.2, stirred for one hour at 30° and then sonicated for 15 min in the 10 KC Raytheon oscillator. The sediment obtained after 30 min centrifugation at 10,000 ×*g* was resuspended in half the original volume of the same buffer at 30°,

---

[1] Extracts prepared from two other strains of *B. subtilis* (No. 23 and ICI) did not contain a significant NADPH oxidase activity.

then recentrifuged. The combined supernatants obtained by this procedure furnished the crude enzyme solution.

Definition of Unit of Enzyme Activity

One unit of enzyme is that amount which oxidizes one μmole NADPH per minute in a standard reaction system containing Tris-HCl, pH 7.2, 50 mM; FMN, 5 μM; MgCl$_2$, 1 mM and NADPH, 0.15 mM at 25°. NADH oxidation was assayed in a similar system but with NADH substituting NADPH as the substrate.

Catalase unit is defined as μmoles peroxide decomposed per mg protein per min in a system containing 0.01 M phosphate buffer pH 6.8 and 17 mM H$_2$O$_2$ at 25°, when recorded spectrophotometrically [47].

RESULTS

Partial Purification of the Enzyme

A solution of MnCl$_2$ was added with stirring to the crude bacterial extract up to 8 mM concentration. After 1 hour at 2°, the precipitate formed was removed by 20 min centrifugation at 10,000 ×*g*. The concentration of MnCl$_2$ in the supernatant was raised to 55 mM and the mixture was subsequently incubated for 10 hours at 2°. The precipitate collected by centrifugation contained a large portion of the NADPH-diaphorase activity and was dispersed in 0.01 M Tris buffer, pH 7.2 in about one third of the volume of the original crude extract. DNAse and RNAse were each added to a final concentration of 1 μg per ml. After 90 minutes of incubation at 30°, neutralized EDTA was added up to a concentration of 8 mM in order to clear the turbid solution (see Table 1 A).

Solid ammonium sulfate was added to the enzyme solution at 0° and the fraction obtained between 0.40 to 0.70 saturation was precipitated by centrifugation for 20 min at 12,000 ×*g*. This precipitate was washed once with 0.65 saturated ammonium sulfate solution, dissolved in 0.01 M Tris buffer, pH 7.2 in one tenth the volume of that of the original crude extract, and then dialyzed overnight against 100 volumes of 5 mM Tris buffer, pH 7.2.

The enzyme solution was applied to a DEAE-cellulose column (2.0 ×22 cm) preequilibrated with 5 mM Tris, pH 7.8. Fractions of 25 ml were collected as follows: The column was first washed with 0.1 M KCl in 0.05 M Tris, pH 8.1 (3 fractions), followed by 0.2 M KCl in the same buffer (5 fractions), then finally by 0.5 M KCl in the same buffer (4 fractions). The formation of a deep yellow colored band, which contained the enzyme activity, could be easily followed during the process of elution from the column. This component was eluted by the 0.5 M KCl buffer solutions (Fractions IX−XII, Table 1)

The combined active fractions eluted from the DEAE-cellulose were concentrated to a volume of 2 ml by overnight dialysis at 4° against an aqueous slurry of polyethylene glycol (molecular weight of 15,000). The yellow solution was then applied to a Sephadex G-200 column (1.9×25 cm) prepared in 0.5 M KCl in 0.1 M Tris buffer, pH 8.0. The column was washed with the same buffer. The yellow protein which emerged at the front contained most of the NADPH-diaphorase activity.

An alternative procedure for the purification of the enzyme was as follows: (Table 1, B). In this case the fraction obtained by precipitation between 0.4

supernatant was concentrated to 2 ml by dialysis against solid polyethylene glycol (molecular weight 15,000). The yellow suspension obtained was centrifuged for 10 min at $18,000 \times g$ and the yellow precipitate was dispesed in 0.1 M phosphate buffer, pH 7.0. This preparation (Table 1 5a, 6a) appeared as microcrystalline granules in the phase microscope.

Stability

Enzyme solutions retained more than 90% of the original activity when kept at 4° for three weeks. Ammonium sulfate fractions lost 30 to 50% of activity

Table 1. *Purification of NADPH diaphorase*
Preparation of extracts and the two purification procedures are described in the text

Preparation [a]	Fraction	ml	Protein	NADPH diaphorase units	Specific activity		Ratio $\frac{NADPH}{NADH}$	Flavin content [b]	
					NADH	NADPH		FMN	FAD
			mg					mμmole/mg protein	
	1. Crude extract	3920	10300	2180	0.19	0.21	1.1		
	2. MnCl₂ precipitate	1400	1540	940	0.30	0.61	2.0		
	3. (NH₄)₂SO₄ precipitate,								
	0.4—0.7 saturation	128	256	684	0.50	2.67	5.3	1.1	0.3
A	4. DEAE-cellulose eluate,								
	(Fractions IX—XII)	96	95	377	0.59	3.99	6.8	3.8	1.7
	5. Sephadex G-200 eluate								
	Fraction I	22	10	150	1.00	15.00	15.0	13.5	0.15
	Fraction II	10	13	100	0.70	7.70	11.0		
	1a. Crude extract	2950	5300	1330	0.21	0.25	1.2		
	2a. MnCl₂ precipitate	1000	1940	798	0.18	0.41	2.2		
	3a. (NH₄)₂SO₄ precipitate,								
	0.4—0.7 saturation	50	74	315	0.81	4.25	5.2	3.8	0.5
B	4a. (NH₄)₂SO₄ wash,								
	0.35—0.45 saturation	25	21	188	0.90	8.95	9.9	6.8	0.2
	5a. Precipitate of 4a at 100,000×g	5	6	140	0.68	23.40	34.4	12.4	0.1
	6a. Precipitate obtained from 5a								
	supernatant after dialysis and								
	concentration	2	1.4	41	0.80	29.24	36.7	18.6	0.1

[a] With 78 g acetone dried cells as the starting material in A and 70 g in B.   [b] Analyzed spectrophotometrically and fluorimetrically as indicated in Experimental Procedures.

to 0.7 saturation of ammonium sulfate as described above, was reprecipitated by addition of solid ammonium sulfate to 0.7 saturation. This precipitate was dispersed in 25 ml of 0.55 saturated (NH₄)₂SO₄ solution at 0° with stirring for one hour. The precipitate obtained after 10 min centrifugation at $8000 \times g$ was resuspended in 25 ml of 0.45 saturated (NH₄)₂SO₄ solution and treated as described above. The next extract made with 0.35 saturated (NH₄)₂SO₄ solution was yellow in color and contained most of the enzyme. After 12 hours at 4° the turbid solution was centrifuged for 30 min at $100,000 \times g$ and the deep yellow precipitate which contained a large portion of the diaphorase activity was dissolved in a small amount of 0.01 M phosphate buffer pH 7.0. The

when thawed after three months of freezing. Fractions purified on columns lost only 5 to 10% of activity upon thawing after being kept frozen for two years.

Physical Properties

Samples of purified fractions (5 and 5a, Table 1) were studied by sedimentation analysis [2] in the Spinco Model E ultracentrifuge and were found not to be homogenous. The protein concentration in the samples analyzed was 0.3% (w/v) in 0.3 M KCl and 0.01 M phosphate buffer at pH 7.5. Centrifugation

[2] Thanks are due to Mr. P. Yanai who performed these analyses.

was run at 22° and 59,780 rev./min. The samples contained a major peak (65 to 75% of the total protein) with a sedimentation constant $s_{20,w}$ of 11.0×10^{-13} sec and a lesser component with an $s_{20,w}$ of 4.7×10^{-13} sec.

Identification of the Flavin Prosthetic Group

The absorption spectrum of a purified enzyme preparation is shown in Fig. 1. The typical, absorbing peaks of flavins at 380 and 445 mμ as well as the disappearance of the 450 peak by reduction are clearly discernible. Compared to other flavoproteins, the peak at 380 mμ is relatively higher than that at 450 mμ. This could be due to the presence of a contaminating protein or a metal impurity or a

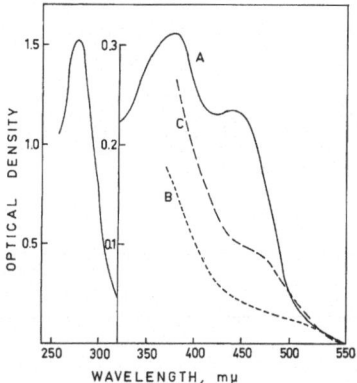

Fig. 1. *Absorption spectrum of NADPH-diaphorase.* Curve A, 1.44 mg protein per ml of a purified fraction (equivalent to step 5a, Table 1) with a specific activity of 25 units per mg, measured in a 1.0 cm light path in 0.1 M Tris-HCl, pH 7.2 buffer. Flavin concentration, 17 mμmole/ml. Reduction was accomplished by addition of excess sodium dithionite (curve B) or of 0.2 μmole NADPH (curve C)

partial reduction of the prosthetic group in these preparations [5,45,46,46a]. Analyses of total flavin in the various enzyme preparations obtained (Table 1) indicate that FMN is most probably the prosthetic group of this enzyme. The most purified preparations contained 1 mole FMN per 55,000 to 74,000 g of protein. The presence of FMN was also confirmed by paper and DEAE-cellulose chromatographic procedures performed on the flavin released from fractions of purified enzyme by acidification and heating according to the procedures described by Rao et al. [37].

Attempts were made to study the reversible removal of FMN from the enzyme. The Florisil column procedure [37] did not result in separation of apoenzyme and flavin. An enzyme fraction (Frac-

tion 3, Table 1) with an activity of 2.5 units per mg protein (when assayed with DCPIP as an acceptor) was reprecipitated in 0.7 saturated ammonium sulfate solution at pH 3.0. The precipitate which was dissolved in 0.05 Tris buffer, pH 7.5, had an activity of only 0.6 units enzyme per mg protein which increased to 1.1 units per mg protein after 2 hours incubation with 0.1 mM FMN.

Identification of the Reaction Product

It was found that even purified enzyme fractions were contaminated by catalase when assayed spectrophotometrically [47] or by the ethanol procedure [48]. Contamination in the highest purified diaphorase preparations (Nos. 5, 5a, Table 1) was between 10—25 units catalase per unit of diaphorase. This could be, in part, the reason for the difficulty in the assay of H_2O_2 formation as the possible product of NADPH oxidation. The addition of a peroxidase-chromogen system (with guaiacol or o-dianisidine as electron acceptors) as a coupling reagent, also did not indicate clearly the appearance of H_2O_2.

Reactions systems (2.2 ml) composed of 0.1 M Tris buffer, pH 7.4; NaN_3 or KCN, 5 mM; FMN, 4 mμmoles; NADPH, 0.6 to 2.0 μmoles and purified enzyme, 0.022 units were measured for oxygen uptake at 27° by the standard manometric techniques [43]. Oxygen consumption in several duplicates was found to be 0.88—0.96 moles per mole of NADPH oxidized. Similar results were obtained when equivalent reaction mixtures were measured for oxygen consumption in the oxygen electrode apparatus [44]. This result is compatible with H_2O_2 as the product of NADPH oxidation although a direct conclusive proof for its appearance in the reaction solution could not be obtained.

Accompanying Activities

The purified diaphorase fractions did not exhibit pyridine nucleotide transhydrogenase activity from NADPH to NAD^+ when assayed by two different methods [49,50]. Most of the NADH dehydrogenase activity found in the various preparations could be attributed to a separate enzyme entity as is evident from the purification data (Table 1). Oxidation of NADH by the most purified fractions was only 3 to 5% of the rate of NADPH oxidation at standard conditions of assay. Purified enzyme fractions were contaminated by catalase activity as described above.

pH Optimum

Maximal rates of NADPH oxidation occurred between pH 6.8 to 7.8. The rate of reaction in Tris-maleate and phosphate buffers was much lower than in cacodylate or Tris-HCl buffers (Fig. 2).

Kinetic Studies

The effect of substrate concentration on the rate of oxidation is illustrated in Fig. 3. The apparent K_m for NADPH at pH 7.2 was 2.0×10^{-4} M when oxygen served as the terminal electron acceptor. The K_m for FMN as a cofactor for oxidation by NADPH with O_2 as the terminal acceptor was found to be 1.6×10^{-6} M (Fig. 4).

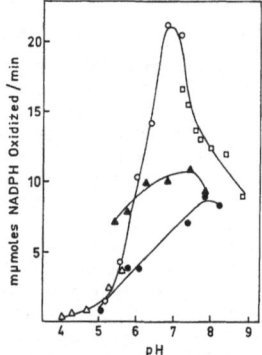

Fig. 2. *Dependence of NADPH oxidation on pH of the medium.* Reaction mixture (1.0 ml) contained, in μmoles: buffer, 50; NADPH, 0.16; FMN, 0.004, and enzyme, 0.015 units. △, citrate; ▲, phosphate; ○, cacodylate; ●, Tris-maleate; □, Tris-HCl buffers

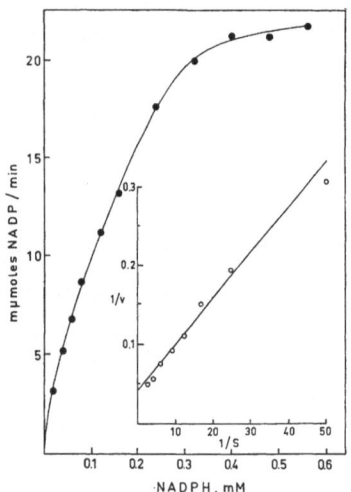

Fig. 3. *Effect of NADPH concentration on the rate of oxidation.* Reaction mixture (1.0 ml) contained, in μmoles: Tris, pH 7.2, 50; FMN, 0.002; DCPIP, 0.05; enzyme, 0.01 units and varying amounts of NADPH. Rate of NADP formation was measured at 340 mμ, using 0.5 cm quartz cells for the higher substrate concentrations. K_m (NADPH), 2.0×10^{-4} M

Electron Acceptors

Oxidation of NADPH with oxygen as the terminal acceptor occurred only when flavins in catalytic amounts were added to the reaction mixture (Fig. 4).

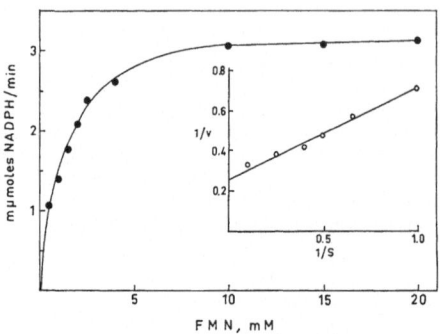

Fig. 4. *Effect of FMN concentration on the rate of NADPH oxidation.* Reaction mixture (1.0 ml) contained, in μmoles: Tris, pH 8.0, 50; $MgCl_2$, 1.0; NADPH, 0.15; FMN, as indicated and 0.004 units enzyme. K_m (FMN), 1.6×10^{-6} M

Table 2. *Rates of NADPH oxidation with various electron acceptors*

Acceptor present [a]		FMN	Relative rate of NADPH oxidation [b]	Relative rate of acceptor reduction [c]
	mM	2 μM		
O_2 (air)		—	2	—
O_2 (air)		+	100	—
DCPIP,	0.1	—	508	250
DCPIP,	0.1	+	500	266
$K_3Fe(CN)_6$,	1.0	—	212	218
$K_3Fe(CN)_6$,	1.0	+	230	265
Menadione,	0.1	—	256	—
Menadione,	0.1	+	270	—
Cytochrome c,	0.06	—	25	5
Cytochrome c,	0.06	+	262	38
INT	1.0	+or—	d	160
INT	0.4	+or—	d	105
INT	0.16	+or—	d	84

[a] Standard system under aerobic conditions consisted of Tris, 50 mM, pH 7.2; NADPH, 0.2 mM and 0.005 units enzyme per ml. Acceptor and FMN added as indicated.

[b] Read at 340 mμ.

[c] On a molar basis as compared with a value of 100 for the rate of NADPH oxidation.

[d] Absorption at 340 mμ could not be read accurately because of the very high blanks due to the formazan formed.

This was also true even for relatively crude enzyme preparations. It was found that the rates of NADPH oxidation were very similar whether riboflavin, isoriboflavin, FMN or FAD were added at 5 to 10 μM concentrations to a standard enzyme reaction mixture.

In absence of added flavins, oxidation of NADPH could proceed only if other electron acceptors were added (Table 2). Menadione, ferricyanide, DCPIP, INT as well as other tetrasolium salts [15] which were also tried, could serve as good oxidants. Cytochrome c, on the other hand, proved to be poor acceptor in this system (Table 2). It was also found that the K_m for ferricyanide as an acceptor at standard reaction conditions was 4×10^{-5} M.

Activators and Inhibitors

Millimolar concentrations of Mg^{++} or Mn^{++} had a stimulatory effect on NADPH oxidation (Fig.5), although these cations were not obligatory for the

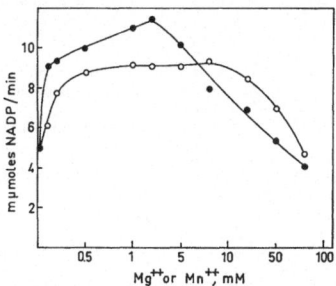

Fig.5. *Effect of Mn^{++} and Mg^{++} on the rate of NADPH oxidation.* Reaction system (1.0 ml) contained, in μmoles: Tris, pH 7.2, 100; FMN, 0.002; DCPIP, 0.1; NADPH, 0.12, enzyme 0.004 units and variable MgSO₄ or MnSO₄ concentration. Rate of NADPH oxidation was read at 340 mμ. ●, Mn^{++}; ○, Mg^{++}

reaction catalyzed by purified fractions of the dia-phorase. Various chelating agents were found to inhibit the enzyme to a certain extent (Table 3). These results are not sufficiently conclusive as proof for a metal cofactor in this enzymatic reaction. The enzyme was not inhibited by some typical inhibitors of the respiratory chain NADPH-dehydrogenase system but was partially inhibited by dicoumarol and atebrin (Table 3). Whereas N-ethylmaleimide was not an effective inhibitor, p-hydroxymercuri-benzoate did inhibit effectively as is known for other reduced pyridine nucleotide dehydrogenase flavo-proteins [4,38,46,51].

NADP⁺ inhibited the reaction competitively with NADPH (Fig.6). The apparent K_i for NADPH was 1.1×10^{-3} M at pH 7.2. NAD⁺ at 2 mM, did not exhibit an inhibitory effect on NADPH oxidation.

DISCUSSION

The NADPH-diaphorase isolated from *B. subtilis* is a flavoprotein in which FMN is most probably the prosthetic group. In this respect it is similar to the

Table 3. *Inhibition of NADPH oxidation by various reagents* Reaction system (1.0 ml) contained in μmoles: Tris, pH 7.2, 50; FMN, 0.002; DCPIP, 0.1; NADPH, 0.15; enzyme, 0.01 units and neutralized inhibitor as indicated. Rate of NADP⁺ formation was read at 340 mμ

	Concentration	Inhibition of rate of NADPH oxidation
	M.	(%)
EDTA	3×10^{-3}	25
Uramil N,N-diacetic acid[a]	3×10^{-3}	36
Quinaldic acid	3×10^{-3}	59
Nitrioltriacetic acid	1×10^{-2}	7
1,2-diaminocyclohexane N,N,N',N'-tetraacetic acid	3×10^{-2}	30
Salicyl aldoxime	1×10^{-3}	0
8-hydroxyquinoline[a]	1×10^{-2}	0
o-phenanthroline[a]	3×10^{-3}	30
2-n-heptyl-4-hydroxyquinoline N-oxide	3×10^{-5}	0[b]
Atebrin[a]	1×10^{-4}	5
Atebrin[a]	5×10^{-4}	44
Dicoumarol	5×10^{-5}	47
Dicoumarol	1×10^{-4}	55
2,4-Dinitrophenol	1×10^{-4}	0
Amytal	3×10^{-3}	0
NaCN	1×10^{-2}	0
NaN₃	1×10^{-3}	0
N-ethylmaleimide	3×10^{-3}	0
p-hydroxymercuribenzoate	3×10^{-3}	56
Ethanol	3×10^{-1}	0
Antimycin A	20 μg/ml	0

[a] System assayed without DCPIP.
[b] A slight stimulation in rates of NADPH oxidation were observed.

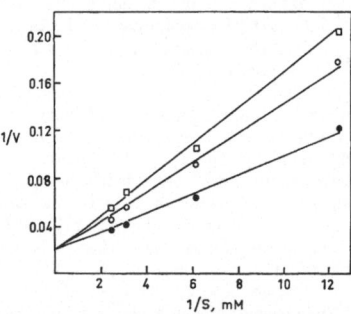

Fig.6. *Inhibition of NADPH oxidation by NADP⁺.* Reaction system (1.0 ml) contained, in μmoles: Tris, pH 7.2, 50; FMN, 0.002; DCPIP, 0.05; NADPH, varying amounts between 0.08 to 0.40 mM as in Fig.3; enzyme, 0.015 units. ●, without NADP⁺; ○, with 0.5 mM and □, with 1.25 mM NADP⁺. v = mμmoles NADP formed per minute as measured at 340 mμ. K_i (NADP⁺), 1.1×10^{-3} M

yeast NADPH-diaphorase (old yellow enzyme) and to the NADPH-cytochrome c reductase which are also FMN enzymes [1,2]. On the other hand, the plant NADPH-diaphorase [11] and probably the liver microsomal NADPH-cytochrome c reductase [4,38, 52] are FAD enzymes. In comparison it is interesting to note that both FMN and FAD were reported as the prosthetic groups of NADH-oxidoreductases isolated from various sources [20,24,37,39].

The analyses of the most purified enzyme fractions indicate the presence of one mole FMN per 55,000 to 74,000 grams of protein. The relatively high $s_{20,w}$ values obtained for these fractions as well as the fact that the enzyme emerges close after the void volume in gel filtration through Sephadex G-200 columns, suggest a molecular weight higher than 100,000 which might represent several FMN-containing subunits. An accurate physical characterization of this flavoprotein must await further studies and isolation of sufficient quantities of the purified enzyme.

Qualitatively, the *B. subtilis* NADPH-diaphorase resembles many other reduced pyridine nucleotide diaphorases in its ability to react with various electron acceptors [10—27]. In the present case, oxygen can serve as a terminal acceptor only in the presence of catalytic amounts of free flavins in the reaction system. Cytochrome *c* is a very poor electron acceptor, whereas ferricyanide, quinones and tetrazolium salts can serve as efficent oxidants in the NADPH-diaphorase reaction. The turnover number of purified enzyme fractions is about 2500 moles NADPH per 100,000 gram protein per minute when oxygen is the acceptor. This value is 2 to 5 times higher when quinones or ferricyanide are present. Thus, it is relatively a much more active protein in the catalysis of NADPH oxidation than the old yellow enzyme [1,16] or the NADPH-cytochrome *c* reductases [16] and approaches the turnover value of the plant NADPH-diaphorase [11].

As a diaphorase, like other enzymes in this category, the ultimate physiological role of the enzyme studied in the present investigation cannot be clearly defined [1,16,39]. For *in vitro* studies, however, this enzyme can conveniently be employed as an efficient regenerator of NADP$^+$ in coupled enzyme reactions. This can be accomplished by addition of the diaphorase and catalytic amounts (2 to 5 μM) of flavin to an aerobic, NADPH-producing enzyme system. The reaction can also be performed in the presence of a suitable electron acceptor, such as ferricyanide, the reduction of which can be followed colorimetrically thus providing a quantitative evaluation of the amount of NADPH oxidized in the system. The use of the *B. subtilis* NADPH-diaphorase for such purposes has been successfully achieved in our laboratory in a study of NADP$^+$-linked dehydrogenase reactions.

REFERENCES

1. Åkeson, A., Ehrenberg, A., and Theorell, H., in *The Enzymes* (edited by P. D. Boyer, H. Lardy and K. Myrbäck), 2nd edition, Academic Press, New York 1963, Vol. 7, p. 477.
2. Haas, E., Horecker, B. L., and Hogness, T. R., *J. Biol. Chem.* 136 (1940) 747.
3. Hatefi, Y., in *The Enzymes* (edited by P. D. Boyer, H. Lardy and K. Myrbäck, 2nd edition, Academic Press, New York 1963, Vol. 7, p. 495.
4. Phillips, A. H., and Langdon, G. R., *J. Biol. Chem.* 237 (1962) 2652.
5. Kamin, H., Masters, B. S. S., Gibson, Q. H., and Williams, C. H., *Federation Proc.* 24 (1965) 1164.
6. Scott, E. M., Duncan, I. W., and Ekstrand, V., *J. Biol. Chem.* 240 (1965) 481.
7. Masters, B. S. S., Bilimoria, M. H., Kamin, H., and Gibson, Q. H., *J. Biol. Chem.* 240 (1965) 4081.
8. Sato, R., Omura, T., and Nishibayashi, H., in *Oxidases and related redox systems* (edited by T. E. King, H. S. Mason and M. Morrison), John Wiley and Sons, New York 1965, Vol. 2, p. 861.
9. Schatz, G., and Klima, J., *Biochim. Biophys. Acta,* 81 (1964) 448.
10. Omura, T., Sanders, E., Cooper, D. Y., Rosenthal, O., and Estabrook, R. W., in *Non Heme Iron Proteins* (edited by A. San Pietro), The Antioch Press, Yellow Springs, Ohio 1965, p. 401.
11. Jagendorf, A. T., in *Methods in Enzymology* (edited by S. P. Colowick and N. O. Kaplan), Academic Press, New York 1963, Vol. VI, p. 430.
12. Martius, C., in *The Enzymes* (edited by P. D. Boyer, H. Lardy and K. Myrbäck), 2nd edition, Academic Press, New York 1963, Vol. 7, p. 517.
13. Ernster, L., Danielson, L., and Ljunggren, M., *Biochim. Biophys. Acta,* 58 (1962) 171.
14. Lazzarini, R. A., and San Pietro, A., *Arch. Biochem. Biophys.* 106 (1964) 6.
15. Vesco, C., and Guiditta, A., *Biochim. Biophys. Acta,* 113 (1966) 197.
16. Dolin, M. I., in *The Bacteria* (edited by I. C. Gunsalus and R. Y. Stanier), Academic Press, New York 1961, Vol. II, p. 425.
17. Robinson, D. A., and Mills, R. C., *Biochim. Biophys. Acta,* 48 (1961) 77, 85.
18. VanDemark, P. J., and Smith, P. F., *J. Bacteriol.* 88 (1964) 122.
19a.Becnofski, C., and Mills, R. C., *J. Bacteriol.* 92 (1966) 1404.
19. Bragg, P. O., *Biochim. Biophys. Acta,* 96 (1965) 263.
20. Brodie, A. F., *J. Biol. Chem.* 199 (1952) 835.
21. Doi, R. H., and Halvorson, H., *J. Bacteriol.* 81 (1961) 51.
22. Feldman, W., and O'Kane, D. J., *J. Bacteriol.* 80 (1961) 218.
23. Kogut, M., and Lightbown, J.W., *Biochem. J.* 84 (1962) 368.
24. Hoskins, D. D., Whiteley, H. R., and Mackler, B., *J. Biol. Chem.* 237 (1962) 2647.
25. Kashket, E. R., and Brodie, A. F., *J. Biol. Chem.* 238 (1963) 2564.
26. Heinen, W., Kusunose, M., Kusunose, E., Goldman, D. S., and Wagner, M. J., *Arch. Biochem. Biophys.* 104 (1964) 448.
27. Downey, R. J., *J. Bacteriol.* 88 (1964) 904.
28. Lascelles, J., in *Symposium of the Society for General Microbiology*, Cambridge University Press, London 1965, Vol. 15, p. 32.
29. Burrous, S. E., and Wood, W. A., *J. Bacteriol.* 84 (1962) 364.
30. Pollak, J. D., Razin, S., and Cleverdon, L., *J. Bacteriol.* 90 (1965) 617.
31. Boulton, A. A., *Exptl. Cell. Res.* 37 (1965) 343.
32. Fujita, M., Ishikawa, S., and Shimazono, N., *J. Biochem. (Japan),* 59 (1966) 104.
33. Pandya, K. P., and King, H. K., *Arch. Biochem. Biophys.* 114 (1966) 154.

34. Stadtman, T. C., in *Non Heme Iron Proteins* (edited by A. San Pietro), The Antioch Press, Yellow Springs, Ohio 1965, p. 439.
35. Avigad, G., Levin, N., Alroy, Y., and Englard, S., *Israel J. Chem.* 3 (1965/66) 83 p.
36. Horecker, B. L., and Kornberg, A., *J. Biol. Chem.* 175 (1948) 385.
37. Rao, N. A., Felton, S. P., Huennekens, F. M., and Mackler, B., *J. Biol. Chem.* 238 (1963) 449.
38. Williams, C. H., and Kamin, H., *J. Biol. Chem.* 237 (1962) 587.
39. Dolin, M. I., and Wood, N. P., *J. Biol. Chem.* 235 (1960) 1809.
40. Tomkins, G. M., Yielding, K. C., Curran, J. F., Summers, M. R., and Bitensky, M. W., *J. Biol. Chem.* 240 (1965) 3793.
41. Bessey, O. A., Lowry, O. H., and Love, R. H., *J. Biol. Chem.* 180 (1949) 755.
42. Lowry, O. H., Rosebrough, N. J., Farr, A. L., and Randall, R. J., *J. Biol. Chem.* 193 (1951) 265.
43. Umbreit, W. W., Burris, R. H., and Stauffer, J. F., *Manometric Techniques*, Burgess Publishing Co., Minneapolis, Minn. 1964.
44. Harel, E., Mayer, A. M., and Shain, Y., *Physiol. Plantarum*, 17 (1964) 921.
45. Nakamura, T., Yoshimura, J., and Ogura, Y., *J. Biochem. (Japan)*, 57 (1965) 554.
46. Masters, B. S., Kamin, H., Gibson, Q. H., and Williams, C. H., *J. Biol. Chem.* 240 (1965) 921.
46a. Massey, V., and Palmer, G., *Biochemistry*, 5 (1966) 3181.
47. Lück, H., in *Methods of Enzymatic Analysis* (edited by H. U. Bergmeyer), Verlag Chemie and Academic Press, New York and London 1963, p. 885.
48. Keilin, D., and Hartree, D., *Biochem. J.* 39 (1945) 293.
49. Lee, C. P., and Ernster, L., *Biochim. Biophys. Acta,* 81 (1964) 187.
50. Colowick, S. P., Kaplan, N. O., Neufeld, E. F., and Ciotti, M. M., *J. Biol. Chem.* 195 (1952) 95.
51. Singer, T., in *Non Heme Iron Proteins* (edited by A. San Pietro), The Antioch Press, Yellow Springs, Ohio 1965, p. 349.
52. Horecker, B. L., *J. Biol. Chem.* 183 (1950) 593.

G. Avigad and N. Levin
Department of Biological Chemistry
The Hebrew University of Jerusalem
Jerusalem, Israël

European J. Biochem. 1 (1967) 110—116

Studies on the Properties of (—)-2-α-Hydroxyethyl-Thiamine Pyrophosphate ("Active Acetaldehyde")

J. Ullrich and A. Mannschreck

Biochemisches Institut der Universität Freiburg
and Organisch-chemisches Institut der Universität Heidelberg

(Received November 29, 1966)

An improved enzymatic preparation of (—)-HETPP from TPP with pig heart pyruvate dehydrogenase is described, yielding an amorphous product of satisfactory elementary analysis and constant optical rotation. The optical absorption spectrum of HETPP was recorded at various pH values and briefly explained and compared with that of TPP. The proton magnetic resonance spectrum of HETPP in deuterium oxide, as compared with those of TPP, 2-α-hydroxyethyl-4-methyl-5-β-hydroxyethyl-thiazole, and 2-hydroxymethyl-3,4-dimethyl-thiazolium iodide, afforded additional and more accurate proof for the structure of HETPP. The possibility of hydrogen exchange in the α-position of the 2-side chain of a model compound, 2-hydroxymethyl-3,4-dimethyl-5-β-hydroxyethyl-thiazolium iodide, and of HETPP itself was checked under varying conditions. No proton exchange was found in the absence of enzyme at pH values below 8.6. HETPP, ³H-labelled in the 2-side chain, was prepared and treated with apo-pyruvate decarboxylase and Mg++. Unconverted HETPP from the reaction mixtures was found to have the specific radioactivity of the starting material. The experimental results are explained in terms of the 2-α-carbanion of HETPP being the real "TPP-activated acetaldehyde" rather than HETPP itself as was hitherto believed.

Since the first isolation of HETPP (I) as an intermediate in the enzymatic decarboxylation of pyruvate and the formation of subsequent products, e.g. acetoin, its low reactivity with apo-pyruvate decarboxylase as compared to that of the system pyruvate + holo-pyruvate decarboxylase has been recorded [1]. In the generally accepted reaction mechanism envisaged by Breslow [2], and Sykes [3], the formation of a carbanion (II) with the mesomeric from (III) was assumed. (II) was thought to react with different carbonyl compounds.

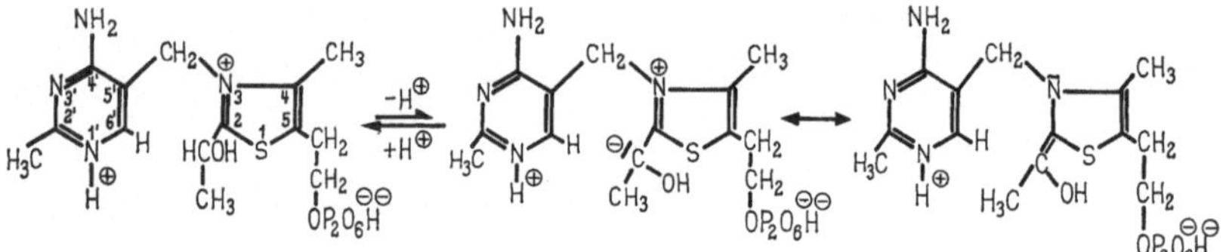

hitherto expected, and indicates that the way from (I) back to the carbanion (II), from which (I) has no doubt originally been formed, does not exist in the absence of an appropriate enzyme under otherwise physiological conditions.

EXPERIMENTAL

Materials

Thiamine pyrophosphate (cocarboxylase-tetrahydrate), sodium pyruvate, acetonitrile, deuterium

I   II   III

The work described in this paper gives evidence that the proton in the α-position of the 2-side chain of HETPP (I) is much more firmly bound than

oxide (99.75%), inorganic salts and buffer substances were commercial products from E. Merck A.G., Darmstadt, Germany. Tritiated water (5 Curie/ml) was purchased from The Radiochemical Centre, Amersham, Buckinghamshire, England. Dowex 2X8, 200—400 mesh, was used in the purified form sold as AG-2X8 by BIO-RAD Laboratories, Richmond, Calif., U.S.A. 2-α-Hydroxyethyl-4-methyl-5-β-hy-

---

Non-standard Abbreviations. 2-α-hydroxyethyl-thiamine pyrophosphate, HETPP; thiamine pyrophosphate, TPP.
Enzymes. Pyruvate dehydrogenase, or pyruvate:lipoate oxidoreductase (acceptor acetylating) (EC 1.2.4.1); pyruvate decarboxylase, or 2-oxoacid carboxy-lyase (EC 4.1.1.1).

droxyethyl-thiazole [4] and 2-hydroxymethyl-3,4-di-methyl-5-β-hydroxyethyl-thiazolium iodide were gifts from Prof. Dr. J. Kiss, Hoffmann-LaRoche A.G., Basel, Switzerland. Scintillator solution was prepared from 2,5-diphenyloxazole (PPO) and 1,4-bis-(5-phenyloxazolyl-2)-benzene (POPOP), purchased from Packard Instruments Co., LaGrange, Illinois, U.S.A., 1,4-dioxane (BASF Ludwigshafen), anisole (Farbwerke Hoechst, Frankfurt), and 1,2-dimethoxyethane (Fluka A.G., Buchs, St. Gallen, Switzerland).

Crude pyruvate dehydrogenase solution was prepared from pig heart muscle by the procedure of Scriba and Holzer [5,6]. Pyruvate decarboxylase was isolated from brewer's yeast, (a gift from Ganter-Brauerei Freiburg, Germany) by the procedure of Holzer and Beaucamp [1,7] with a slightly modified ammonium sulfate fractionation, giving an enzyme paste with a specific decarboxylation activity of 46 μmoles pyruvate/min/mg protein at pH = 6.2 and 30° under TPP-saturation. Removal of TPP by alkaline ammonium sulfate precipitation [1,8] yielded an apoenzyme which, when saturated with TPP and Mg++, had about half the specific activity of the original holoenzyme.

Preparation of
(−)-2-α-Hydroxyethyl-Thiamine Pyrophosphate

Enzymatic Reaction and Removal of Protein [5,9]. 25 mg (50 μmoles) of TPP · 4 H_2O, 60 mg of $MgSO_4$ · 7 H_2O, 50 mg of KH_2PO_4, and 20 mg of NaF were dissolved in 10 ml of crude pig heart pyruvate dehydrogenase solution [5]. The pH of the mixture was adjusted, if necessary, to 5.6 by the addition of a few droplets of dilute H_2SO_4 or $NaHCO_3$. The mixture was placed in a Warburg flask of 30—40 ml total volume, which contained 0.3 ml of 1 M Na-pyruvate in the side arm. When about 80 μmoles of CO_2 had evolved at 37° (5—10 hours), the mixture was poured into 100 ml of hot methanol. After short centrifugation, the supernate was evaporated to dryness under reduced pressure (Büchi-Rotavapor), and the residue thoroughly suspended in 20 ml of water for at least half an hour at 25°. After high speed centrifugation for removal of all turbidity, the supernate was used for chromatography. If necessary it was stored in the deepfreeze.

Preparation of Anion Exchange Column. A column of 40 cm × 2 cm was prepared from purified Dowex-2 (X8) analytical grade, 200—400 mesh (out of a large number of lots only a few were found to be suitable, the separation efficiency of a special lot being a matter of chance), washed with 2 l of 2—3 N HCl, 1 l of water, about 2 l of 2 M Na-acetate (with resuspension of the resin in order to prevent breakage of the column due to swelling of the resin), until the effluent was free of chloride, 1 l of water, 250 ml of 2 N acetic acid (with resuspension), and at least 1 l

of water. After use, the column was regenerated by application of the last two steps only (with resuspension in the acetic acid). Columns regenerated after use by this short procedure usually improved in separation efficiency during the first few runs and did not deteriorate before the tenth or fifteenth run. When a marked decrease in exchange capacity was observed, the resin was regenerated by the complete procedure.

Column Chromatography. After application of the HETPP-containing supernate, the column was washed with 100 ml of water. Subsequently a gradient was applied, consisting of 300 ml of water in the mixing chamber located about 1 m above the top of the column, and of a large storage volume of 0.0167 N acetic acid, located at the same level as the mixing chamber. The effluent from the column was collected in 25 ml fractions in a very carefully cleaned LKB-Radirac fraction collector. In order to record the ultraviolet absorption at 253.7 nm, the effluent was passed through a LKB-Uvicord, the ordinate of which was expanded by 100% in order to increase the sensitivity. An elution pattern was obtained as described earlier [9]. The HETPP-containing fractions (usually 16—20 fractions = 400—500 ml around fraction No. 60)—were pooled, and their total HETPP-content calculated from the volume and the absorbancy at 272.5 nm (the isosbestic point with ε = 7.4 cm²/μmole). Representative runs without serious failures yielded 25—35 μmoles of dissolved HETPP.

Freeze-drying. Freeze-drying of the pooled solution in 3—4 flasks of 1 l volume each yielded 9—12 mg (19—26 μmoles) of a voluminous colourless powder of low hygroscopicity, which, on rechromatography, was found to be almost pure HETPP.

Stability and Storage. Evaporation of liquid solutions of HETPP under vacuum resulted in a yellow or brownish and partially decomposed product. When kept dry, HETPP was found to be surprisingly stable, even over long periods at room temperature. In general it was stored in the deepfreeze. Solutions of HETPP with pH values between 3 and 7 were found to be stable for many weeks at −15° or below. Attempts to crystallise the product by treatment of aqueous solutions with ethanol or acetone were unsuccessful.

Elementary Analysis of HETPP

Elementary microanalysis[1] of the isolated HETPP gave values (Table 1) which—with respect to the values that could be expected from an amorphous product obtained by simple chromatography and freeze-drying—were considered to be quite satis-

---

[1] Performed by Alfred Bernhardt, Mikroanalytisches Labor im Max-Planck-Institut für Kohlenforschung, Mülheim (Ruhr), Germany.

Table 1. *Elementary analysis of HETPP*
$C_{14}H_{22}N_4O_8P_2S$; mol.wt. = 468.4

	Calculated	Found	
	%	%	%
C	35.90	35.44	35.29
H	4.74	5.71	5.07
N	11.96	11.23	10.90
S	6.85	6.21	5.95
P[a]	} 13.25	15.98	11.77
P[b]		16.7	12.4

[a] Gravimetric as ammonium phosphomolybdate.
[b] Photometric as molybdenum blue [10].

factory. However, the phosphate values found by two independent methods (gravimetric as ammonium phosphomolybdate and photometric as molybdenum blue [10] showed considerable variation between different samples. Unconverted TPP which had gone through the same isolation procedure was also sometimes found to contain excess phosphate. On the other hand, untreated commercial TPP gave normal phosphate values. The most probable explanation of this effect is the existence of traces of inorganic phosphate due to automatic washing processes using soap of a high phosphate content.

Optical Rotation of "Natural" HETPP

HETPP obtained by enzymatic reaction has long been expected to be only one of two possible optical isomers [4,11—13]. Several samples of 70—100 mg each were pooled from numerous preparations and separately measured for optical activity in a Perkin-Elmer Model 141 Photoelectric Polarimeter (Table 2). The values could be reproduced with HETPP recovered after these measurements by rechromatography and freeze-drying. The existence of optical activity and its reproducibility indicate that there is no proton exchange at the asymetric carbon of HETPP (I) under the conditions applied here.

Table 2. *Specific rotation of HETPP*
1.3—1.6% (w/v) in water; pH = 5.4—5.8; d = 10 cm. A turbidity had been centrifuged off. Average values from 3—5 measurements for each wavelength. A few solutions were yellowish and could not be measured at 365 nm

λ	$[\alpha]^{23\circ}$
nm	degrees
589 (Na)	− 10 ± 2
578 (Hg)	− 11 ± 2
546 (Hg)	− 12.5 ± 2
436 (Hg)	− 17.5 ± 2
365 (Hg)	− 25 ± 4

For chemically synthesised (−)-2-α-Hydroxy-ethyl-thiamine, resolved by dibenzoyl-D-tartaric acid, Shiobara *et al.* [14,15] found $[\alpha]_D^{22} = −13.5°$

(in 0.1 N HCl), which is in agreement with the value measured here for HETPP (Table 2). Since no appreciable changes in optical activity of these compounds can be expected from pH changes between 1 and 6, which affect only parts of the molecule very distant from the asymmetric carbon atom, the two values appear to be comparable.

Attempts to measure the optical activity of HETPP at wavelengths near the absorption bands were unsuccessful.

Optical Absorption Spectrum of HETPP

As can be expected from the structure (I), the absorption band of TPP (Fig. 1) around 270 nm, due to the pyrimidine moiety, undergoes only minor

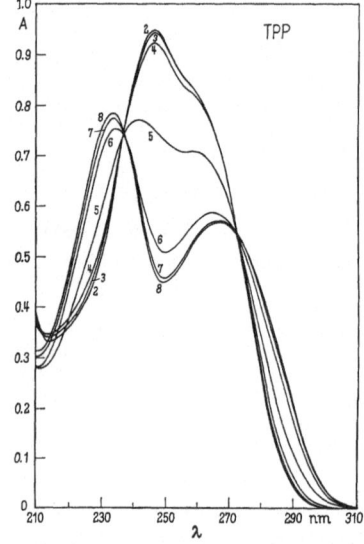

Fig. 1. *Absorbancy of 0.7×10^{-4} M TPP at different pH values.* The pH is indicated by the numbers at the lines. The spectra were recorded in 0.01 M buffers (pH = 2—3: glycine-HCl, pH = 4: Na-citrate, pH = 5—8: Na-phosphate), in 1 cm cuvettes with a Cary Model 15 Spectrophotometer, with the appropriate buffers in the reference beam

changes when the molecule is converted to HETPP [16] (Fig. 2) or to other known 2-α-hydroxyalkyl derivatives [17]. In particular, the isosbestic point at 272.5 nm with an extinction coefficient of 7.4 cm²/μmole remains unchanged within the limits of error of the measurement. This is in contrast to earlier observations exhibiting a small shift to 270.5 nm [18], but in accordance with measurements of other closely related intermediates [17]. The absorption of the thiazole moiety, however, is shifted to lower wavelengths by the substitution at C-2, and somewhat lowered in intensity. There is also a change in the pH-dependence of this part of the absorption spectrum. These spectral differences between TPP and HETPP or other known 2-substituted derivatives may be used for rough estimations of the relative content of the derivative in mixtures with TPP [18], but this method suffers from its low accuracy and its high sensitivity to impurities.

The infrared spectrum of HETPP (in KBr) exhibited no discernible difference from that of TPP [19], due to its phosphate content, and hence is not recorded here.

Proton Magnetic Resonance Spectrum of HETPP

In the proton magnetic resonance spectrum, HETPP, dissolved in deuterium oxide, exhibited the expected characteristics and changes (Fig.3) when compared with the spectra of TPP [19] and of thiamine [19,20]. The resonances measured could be correlated with the protons in the molecule (I) as shown in Table 3, by comparison with the spectra of TPP [19], 2-α-hydroxyethyl-4-methyl-5-β-hydroxyethyl-thiazole (IV) (Fig.4), and 2-hydroxymethyl-3,4-dimethyl-5-β-hydroxyethyl-thiazolium iodide (V) (Fig.5).

spectrometer: 2-hydroxymethyl-3,4-dimethyl-5-β-hydroxyethyl-thiazolium iodide (V). By repeated recording of its proton magnetic resonance spectrum (Fig.5) through one day at p^{2}H = 5.0 and at temperatures up to 80°, no change in the intensity of the crucial $-CH_2-O$ signal at $\tau = 4.90$ could be detect-

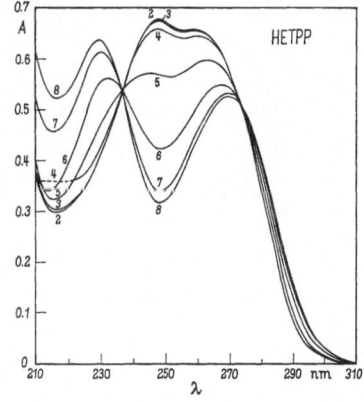

Fig.2. *Absorbancy of* 0.7×10^{-4} *M HETPP.* Conditions are the same as in Fig.1

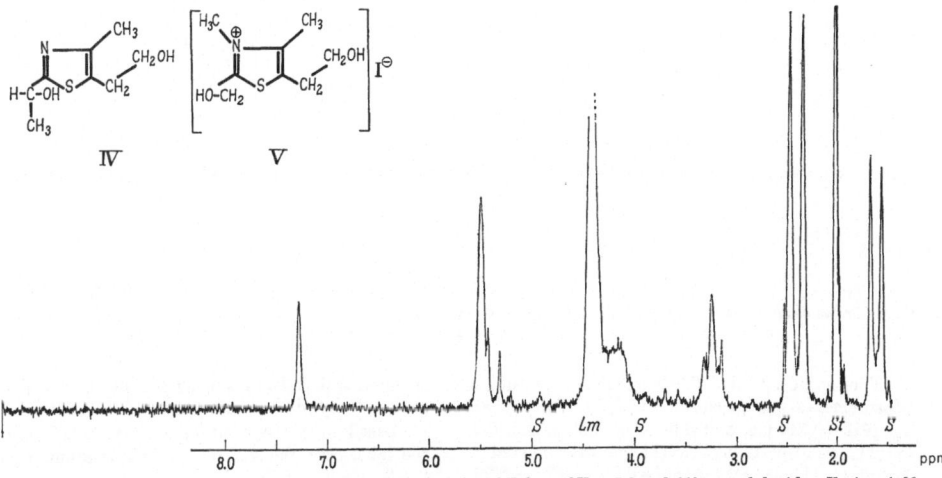

Fig.3. *Proton magnetic resonance spectrum of 0.4 M HETPP (I) in* ^2H_2O *at p^{2}H = 5.8 and 60°, recorded with a Varian A 60 NMR Spectrometer.* At lower temperatures the spectrum was not satisfactory because of very poor resolution, even after a turbidity in the sample had been centrifuged off. St = Internal Standard (acetonitrile with $\tau_{CH_3} = 7.98$ [21]); Lm = H^{2}HO in the solvent, S = Spinning side band

Check for Exchange of the α-Proton on the 2-Side Chain of HETPP

The possibility of Proton exchange in the α-position of the side chain at C-2 had first to be investigated by measuring a model compound in the NMR

ed. At p^{2}H > 9.0, however, the signal had disappeared within one day, probably due to slow alkali-catalysed ring opening of the pseudobase formed under these conditions from the thiazole ring [16].

Table 3. *Assignment of proton magnetic resonance signals in the spectrum of HETPP in ^2H_2O at $p^2H = 5.8$*
See Fig. 3, from left to right

τ	Appearance	Relative Intensity	Assignment	τ of corresponding protons		
				in TPP [19]	in (IV) (Fig. 4)	in (V) (Fig. 5)
—	singlet	1	thiazole proton at 2-C	0.32	—	—
2.72[a]	singlet	1	pyrimidine proton at 6'-C	2.06[a]	—	—
4.50	singlet	2	methylene bridge protons	4.46	—	6.17[b]
4.62	quartet (1:3:3:1) (J = 7 cps)	1	α-proton of 2-side chain	—	5.01	4.90[c]
approx. 5.8	triplet	2	outer CH_2 of 5-side chain	approx. 5.9	6.31	6.2
approx. 6.8	triplet	2	inner CH_2 of 5-side chain	approx. 6.7	7.14	6.94
7.52	singlet	3	CH_3 at 2'-C of pyrimidine	7.42	—	—
7.65	singlet	3	CH_3 at 4-C of thiazole	7.49	7.81	7.59
8.39	doublet	3	CH_3 of 2-side chain	—	8.53	—

[a] For this striking difference in τ-value of this proton between TPP and HETPP which was also found at $p^2H = 3.0$, an obvious explanation could be a conformational change of the pyridimidine ring forced out of its normal position by the bulky C-2 substituent [cf. 22, 30].
[b] Methyl group at N^+.
[c] Singlet of intensity 2.

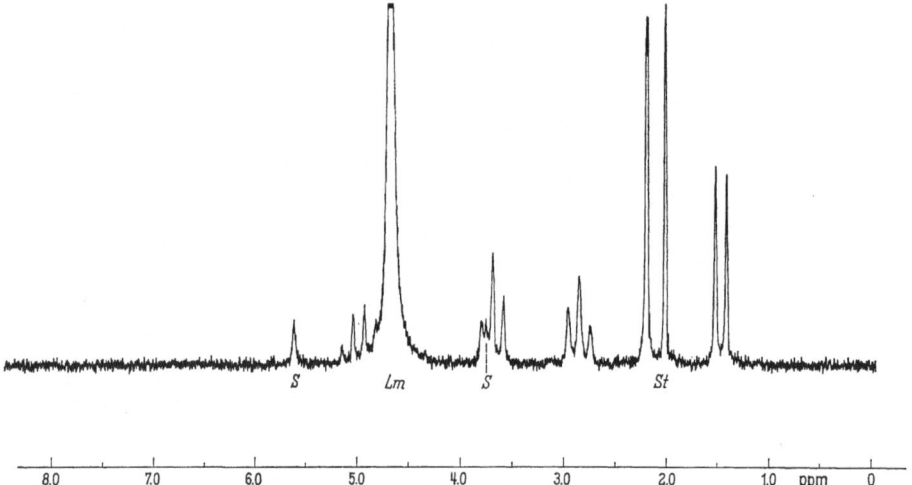

Fig. 4. *Proton magnetic resonance spectrum of 0.3 M 2-α-hydroxyethyl-4-methyl-5-β-hydroxyethyl-thiazole (IV) in ^2H_2O at 35°.*
St = Acetonitrile ($τ_{CH_3}$ = 7.98); Lm = H^2HO in the solvent; S = Spinning side band

When sufficient HETPP became available, similar measurements were performed with HETPP at 60° and $p^2H = 5.8$, 6.8, and 8.6 for periods of 3.5, 0.5, and 2 hours respectively. No change was detected in the quartet at $τ = 4.62$ and in the doublet at $τ = 8.39$, nor in any other signal of the spectrum (Fig. 3).

Even after treatment at $p^2H = 8.6$, most of the HETPP could be recovered from the sample, which had turned brown, by column chromatography, and showed the previous optical rotation.

Preparation of ^3H-labelled HETPP

HETPP was enzymatically prepared by the procedure given above in the presence of 0.1 Curie of tritiated water. The resulting product had a specific radioactivity of 30 nCurie/µmole of HETPP.

Based on the assumption of equal distribution of the label over the 4 firmly bound H-atoms at the 2-side chain during the preparation, an isotope effect of $v_1H : v_3H = 10$, which is within the normal range [23—24a] for tritium was calculated.

The label was shown to be restricted to the 2-side chain by conversion of a sample of the [^3H]-HETPP to thiochrome pyrophosphate [25], which upon repeated evaporation to dryness for the removal of water and acetic acid which had been formed, was found to contain no radioactivity. The presence of radioactive label in the CH_3 group of the 2-side chain was demonstrated qualitatively by partial isolation of the acetic acid produced in the above thiochrome reaction as Na-acetate and by detection of radioactivity.

different periods of time, and were stopped by the addition of 10 ml of hot methanol. The unconverted HETPP from the mixtures was partially recovered by anion exchange chromatography as described

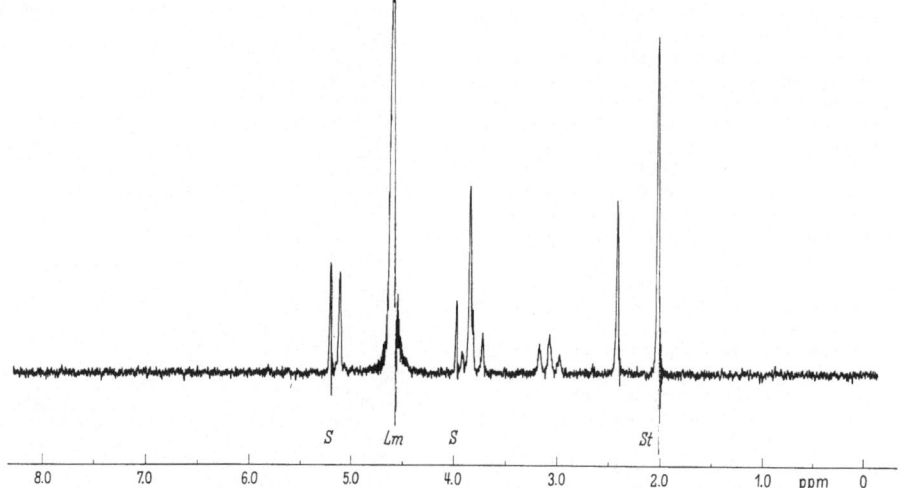

Fig. 5. *Proton magnetic resonance spectrum of 0.2 M 2-hydroxy-methyl-3,4-dimethyl-5-β-hydroxyethyl-thiazolium iodide (V) in ^2H_2O at $p^2H = 5.0$ and $35°$. St = Acetonitrile ($\tau_{CH_3} = 7.98$); Lm = H^2HO in the solvent; S = Spinning side band*

Table 4. *Specific radioactivity of ^3H-labelled HETPP after treatment with apo-pyruvate decarboxylase (PDC)*
HETPP was determined by the thiochrome assay [25]. The quantity of active apoenzyme was calculated from the protein content, the molecular weight [26], and the ratio of specific activity of the apoenzyme in the sample to the maximum specific activity of almost pure pyruvate decarboxylase [26]. ^3H was counted in a Packard Model 314E Liquid Scintillation Spectrometer, which counted 20% of the decompositions under the conditions used

apo-PDC		$\dfrac{HETPP}{apo\text{-}PDC}$	Incubation	µmoles HETPP recovered	Decompositions $\overline{min \times µmole}$
mg	nmoles		h		
9	20	150	0	1.25	32.600
9	20	150	0.5	0.82	31.500
9	20	150	1	0.72	33.800
9	20	150	2	0.52	30.000
4.5	10	300	1	1.24	33.300
0	0	—	0	(3.0)	33.000

Check for Catalysis of Proton Exchange in HETPP by Added Apo-Pyruvate Decarboxylase

Mixtures containing 50 µmoles of citrate buffer pH = 6.6, 10 µmoles of $MgSO_4$, 3 µmoles of HETPP (side chain-labelled with 45 nCuries of ^3H), and 50 nmoles of apo-pyruvate decarboxylase (9 mg of protein with a specific decarboxylation activity of 24 µmoles of pyruvate per mg protein and min) in 1.7 ml of water each were incubated at 27.5° for

above, and was analysed for specific radioactivity (Table 4). No change exceeding the limits of error was detected, which indicated that no proton exchange had taken place in the HETPP molecules not cleaved into TPP and acetaldehyde.

DISCUSSION

In accordance with earlier results [1, 22, 27], the experiments described in this paper encourage us to extend the assumption first made by Breslow [2] for the acyloin condensation mechanism, namely that carbanion (II) is the real "TPP-activated aldehyde", to the liberation of acetaldehyde by pyruvate decarboxylase as demonstrated in the following scheme:

HETPP isolated from TPP-dependent enzymatic reactions is a compound of rather low reactivity and can even be considered to be an artifact formed only when the "normal" physiological process is disturbed or interrupted. Although the mere capture of a proton from the surrounding aqueous solution, resulting in the formation of stable HETPP, must be a significantly slower reaction than the "normal" processes, e.g. liberation of free acetaldehyde, combination with a free carbonyl compound leading to an acyloin, or oxidation of the 2-side chain and its transfer to lipoic acid, it must occur while the carbanion (II) is still attached at the apoenzyme in a well defined position, or the resulting product should have no optical activity. The stabilised product, once formed, was shown not to lose its α-proton at all at the 2-side chain in the absence of enzyme at pH values below 8.6. When HETPP is combined with an appropriate apoenzyme and Mg^{++}, their influence may remove this proton and thus reproduce the "active aldehyde" in the carbanion form (II) which then can undergo all the known "normal" reactions. Some evidence has already been collected [22, 27—30] that the rate-limiting step in this "reactivation" of HETPP is its sterically hindered recombination with the apoenzyme and Mg^{++}, but the possibility has not yet been ruled out that the removal of the α-proton is an even slower reaction. If the latter were true, the K_m values previously measured for HETPP [1, 27] would be "apparent K_m values" without substantial meaning. Attempts are being made to further elucidate this problem by more decisive experiments.

Our thanks go to Prof. Dr. H. Holzer for his permanent interest in the progress of this work and for many helpful discussions, to Prof. Dr. J. Kiss, Hoffmann-LaRoche A.G., Basel, for supplying us with a number of the synthetic compounds we needed, to Frl. Ingrid Donner for valuable technical assistance, particularly for the large-scale preparation of HETPP, and to the Deutsche Forschungsgemeinschaft, Bad Godesberg, and the Bundesministerium für Wissenschaftliche Forschung, Bonn, for financial support of this work.

REFERENCES

1. Holzer, H., and Beaucamp, K., *Biochim. Biophys. Acta*, 46 (1961) 225.
2. Breslow, R., *J. Am. Chem. Soc.* 80 (1958) 3719.
3. Downes, E., and Sykes, P., *Chem. Ind.* (1957) 1095.
4. Krampitz, L. O., Greull, G., Miller, C. S., Bicking, J. B., Skeggs, H. R., and Sprague, J. M., *J. Am. Chem. Soc.* 80 (1958) 5893.
5. Scriba, P., and Holzer, H., *Biochem. Z.* 334 (1961) 473.
6. Korkes, S., Del Campillo, A., and Ochoa, S., *J. Biol. Chem.* 195 (1952) 541.
7. Holzer, H., Schultz, G., Villar-Palasi, C., and Jüntgen-Sell, J., *Biochem. Z.* 327 (1956) 331.
8. Holzer, E., Söling, H. D., Goedde, H. W., and Holzer, H., in *Methoden der enzymatischen Analyse* (herausgegeben von H. U. Bergmeyer), Verlag Chemie, Weinheim/Bergstr. 1962, p. 605.
9. Ullrich, J., and Holzer, H., *Biochem. Z.* 337 (1963) 345.
10. Bartlett, G. R., *J. Biol. Chem.* 234 (1959) 466.
11. Carlson, C. L., and Brown, G. M., *J. Biol. Chem.* 235 (1960) PC-3.
12. Holzer, H., Goedde, H. W., and Ulrich, B., *Biochem. Biophys. Res. Commun.* 5 (1961) 447.
13. Holzer, H., *Angew. Chem.* 73 (1961) 721.
14. Shiobara, Y., Sato, N., Homma, H., Hattori, R., and Murakami, M., *J. Vitaminol. (Kyoto)*, 11 (1965) 302.
15. Shiobara, Y., *Report at the Annual Meeting of The Japanese Vitamin Society*, Morioka, May 1965. Personal communication.
16. Metzler, D. E., Review in Boyer-Lardy-Myrbäck, *The Enzymes*, 2 (1960) 305.
17. Kohlhaw, G., Deus, B., and Holzer, H., *J. Biol. Chem.* 240 (1965) 2135.
18. Goedde, H. W., Blume, K. G., and Holzer, H., *Biochim. Biophys. Acta*, 62 (1962) 1.
19. Ullrich, J., and Mannschreck, A., *Biochim. Biophys. Acta*, 115 (1966) 46.
20. Sable, H. Z., and Biaglow, J. E., *Proc. Natl. Acad. Sci. U.S.* 54 (1965) 808.
21. Jones, R. A. Y., Katritzky, A. R., Murell, J. N., and Sheppard, N., *J. Chem. Soc.* (1962) 2576.
22. Schellenberger, A., Müller, V., Winter, K., and Hübner, G., *Hoppe-Seyler's Z. physiol. Chem.* 344 (1966) 244.
23. Swain, C. G., Stivers, E. C., Reuwer, J. F., jr., and Schaad, L. J., *J. Am. Chem. Soc.* 80 (1958) 5885.
24. Melander, L., *Isotope Effects on Reaction Rates*, Ronald Press, New York 1960, p. 76.
24a. Simon, H., and Palm, D., *Angew. Chem.* 78 (1966) 993.
25. Bessey, O. A., Lowry, O. H., and Davis, E. B., *J. Biol. Chem.* 195 (1952) 453.
26. Ullrich, J., Wittorf, J. H., and Gubler, C. J., *Biochim. Biophys. Acta*, 113 (1966) 595.
27. Goedde, H. W., Ulrich, B., Stahlmann, C., and Holzer, H., *Biochem. Z.* 343 (1965) 204.
28. Eyzaguirre, J., Gubler, C. J., Ullrich, J., and Holzer, H., Unpublished results.
29. Pletcher, J., and Sax, M., *Science*, 154 (1966) 1331.
30. Schellenberger, A., Winter, K., Hübner, G., Schwaiberger, R., Helbig, D., Schumacher, S., Thieme, R., Bouillon, G., and Rädler, K.-P., *Hoppe-Seilers Z. physiol. Chem.* 346 (1966) 123.

J. Ullrich
Biochemisches Institut der Universität
D-78 Freiburg i. Br.
Hermann-Herder-Straße 7, Germany

A. Mannschreck
Organisch-Chemisches Institut der Universität
D-69 Heidelberg
Tiergartenstraße, Germany

European J. Biochem. 1 (1967) 117—124

Purification and Properties of Fructose Diphosphate Aldolase from Spinach Leaves

R. Fluri[1], T. Ramasarma[2], and B. L. Horecker

Department of Molecular Biology, Albert Einstein College of Medicine, Bronx, New York

(Received December 24, 1966)

Fructose 1,6-diphosphate aldolase has been purified from spinach leaves. The preparations are homogeneous on the basis of disc gel electrophoresis and ultracentrifugal analysis. The catalytic properties closely resemble those of the mammalian muscle enzyme. The molecular weight is 120,000 and the enzyme appears to contain three active sites, estimated by reduction of the Schiff base intermediate with borohydride. Three tyrosine residues are liberated by carboxypeptidase. The protein is unusually rich in tryptophan.

The cleavage of fructose 1,6-diphosphate by an enzyme present in extracts of plant material was first reported by Tewfik and Stumpf [1]. The enzyme was partially purified from peas [2], but the best preparations were of low specific activity and as yet little is known concerning the structure and properties of the plant enzymes. Within the last few years renewed interest in this class of enzymes has derived from the work in Rutter's laboratory [3] which demonstrated that the enzyme in higher plants resembles the enzyme isolated from animal muscle or liver in that a Schiff base intermediate is formed with dihydroxyacetone phosphate. This is in contrast to the enzymes found in yeast and bacteria which are metalloproteins and which do not appear to form the Schiff base intermediate (see [3]).

Considerable information is available regarding the subunit structure [4—7] and the primary amino acid sequence at the active center [8] of the enzyme isolated from mammalian muscle. In order to provide similar information for the plant enzyme we undertook the purification of the enzyme from spinach.

It was hoped that these studies would also provide some insight into the origin and evolutionary development of this class of enzymes.

In this paper we report the isolation of fructose diphosphate aldolase from spinach. The catalytic properties of the enzyme were found to be similar to those of rabbit muscle aldolase. The enzyme appears to contain a similar number of combining sites, although the molecular weight is somewhat smaller. The amino acid composition differs considerably from that of the enzyme from rabbit muscle. In particular, it contains fewer histidine and cysteine residues, and approximately four times the complement of tryptophan.

EXPERIMENTAL PROCEDURES

Materials. Fructose 1,6-diphosphate, fructose 1-phosphate, DPNH, and TPN were purchased from the Sigma Chemical Company, St. Louis, Missouri. Glucose 6-phosphate dehydrogenase, phosphoglucose isomerase, triosephosphate isomerase, and α-glycerophosphate dehydrogenase were purchased from Boehringer und Soehne, Mannheim-Waldhof. Carboxypeptidase A (COP A-DFP) and carboxypeptidase B (COP-B-DFP) were purchased from the Worthington Biochemical Corp. Fructose 1,6-diphosphatase, prepared from *Candida utilis*, was a gift from Dr. S. Rosen of this Institution. [14C]Dihydroxyacetone phosphate was prepared from uniformly-labeled [14C]fructose (Nuclear Research Chemicals, Inc.), as previously described [9]. Unlabeled dihydroxyacetone phosphate was prepared by the same procedure from commercial fructose 1,6-diphosphate. Sephadex G-25 and DEAE-Sephadex (A-50 medium grade) were purchased from Pharmacia, Uppsala.

Aldolase Assay. The activity of fructose 1,6-diphosphate aldolase was assayed spectrophotometri-

[1] Postdoctoral fellow of the National Institutes of Health. Present address: University of Bern, Institute of General Microbiology, Bern, Switzerland.
[2] On leave of absence from the Department of Biochemistry, Indian Institute of Science, Bangalore, India; recipient of Fulbright travel grant of the U. S. Educational Foundation in India.

Enzymes. Fructosediphosphate aldolase, or fructose-1,6-diphosphate D-glyceraldehyde-3-phosphate-lyase (EC 4.1.2.13), glucosephosphate isomerase, or D-glucose-6-phosphate ketol-isomerase (EC 5.3.1.9); triosephosphate isomerase, or D-glyceraldehyde-3-phosphate ketolisomerase (EC 5.3.1.1); glucose-6-phosphate dehydrogenase, or D-glucose-6-phosphate: NADP oxidoreductase (EC 1.1.1.49); glycerol-3-phosphate dehydrogenase, or L-glycerol-3-phosphate: NAD oxidoreductase (EC 1.1.1.8); carboxypeptidase A, or peptidyl-L-amino-acid hydrolase (EC 3.4.2.1); carboxypeptidase B, or peptidyl-L-lysine hydrolase (EC 3.4.2.2).

cally with DPNH and α-glycerophosphate dehydrogenase [10]. The reaction mixtures (1 ml) contained 2 mM fructose 1,6-diphosphate, or 20 mM fructose 1-phosphate, 0.2 mM DPNH, 50 mM triethanolamine buffer, pH 7.4, 1 mM EDTA, and a mixture of α-glycerophosphate dehydrogenase and triosephosphate isomerase (10 µg). Appropriately diluted solutions of aldolase were added to start the reaction, which was then followed by measurement of the decrease in absorption at 340 mµ.

Definition of Unit of Enzyme Activity. One unit of enzyme was defined as the amount required to catalyze the formation or breakdown of one µmole of substrate per minute under the experimental conditions.

Aldolase Aassy Using Dihydroxyacetone Phosphate as the Substrate. For a study of the kinetics of the aldolase reaction in the direction of fructose 1,6-diphosphate synthesis, the amount of this substance was measured spectrophotometrically, using fructose 1,6-diphosphatase to convert the product to fructose 6-phosphate, which was then measured with phosphoglucose isomerase, glucose 6-phosphate dehydrogenase, and TPN [11]. The reaction mixture (1.0 ml) contained 50 mM triethanolamine buffer, pH 7.4, 0.5 mM EDTA, 0.5 mM TPN, 1.0 mM MgCl$_2$, 5 µg of glucose 6-phosphate dehydrogenase, 5 µg of phosphoglucose isomerase, 0.3 units of fructose 1,6-diphosphatase, 10 µg of triosephosphate isomerase, and dihydroxyacetone phosphate and spinach aldolase as specified.

Ultracentrifugation. A Spinco model E analytical ultracentrifuge equipped with standard Schleiren and Rayleigh interference optical systems was used. Sedimentation velocities were determined at various protein concentrations in 0.1 M phosphate buffer, pH 7.4. Molecular weight determinations were carried out by the sedimentation equilibrium technique of Yphantis [12]. The partial specific volume of the protein was calculated (0.74 ml/g) from the amino acid composition [13].

Sucrose Density Gradient Centrifugation. Linear density gradients of sucrose and water from 20—25% (w/v) in 10 mM Tris buffer, pH 7.4, were prepared according to Martin and Ames [14]. The sample (70 µl), containing 0.35 mg of purified spinach aldolase and 1 mg of human hemoglobin, was layered on top of the gradient (4.6 ml) and the tube centrifuged in a swinging bucket SW 39 rotor at 0—5° for 12 hours at 37,500 rev./min. After the run the tube was punctured with a fine needle at the bottom and fractions of 2 drops each were collected and analyzed for aldolase activity and absorbance at 490 mµ. The molecular weight was estimated from the relative locations of the aldolase and hemoglobin bands, as described by Martin and Ames [14].

Polyacrylamide Gel Electrophoresis. This was performed with the Canalco apparatus (Canal Industrial

Corp.) in standard 7% polyacrylamide gel, pH 9.5, at 5 mA per tube, according to the method of Davis [15].

Reduction of the Schiff Base Intermediate with Borohydride. Enzyme solutions in 50 mM phosphate buffer, pH 7.4, were mixed with an 80-fold molar excess of dihydroxyacetone phosphate (labeled with ^{14}C where indicated) and incubated at 25° for 15 minutes. The solutions were then cooled in an ice bath, adjusted to pH 5.9 with 2 N acetic acid, treated with a drop of octyl alcohol to prevent frothing, and then with 5 µl aliquots of freshly-prepared NaBH$_4$ solution (10 mg per ml) until about 50-fold molar excess had been added. During the addition the pH was maintained at 5.5—5.9. Aliquots were removed and monitored for enzyme activity. Usually over 90% inactivation was observed within a few minutes. The reduced protein was precipitated by adding trichloroacetic acid to a final concentration of 5% (v/v). The precipitate was collected by centrifugation, dissolved in 0.1 N NaOH and reprecipitated with 5% (v/v) trichloroacetic acid. This process was repeated three times. Finally, the precipitate was washed with acetone and ethyl ether and dried in air.

Amino Acid Analysis. Amino acid analyses were performed according to the method of Spackman et al. [16], using a Spinco model 120 B automatic amino acid analyzer. Before hydrolysis, the protein samples were converted to the S-carboxymethyl derivative, according to the method of Crestfield et al. [17]. The samples were then hydrolyzed with constant boiling HCl in sealed, evacuated tubes at 110° for 24 hours and 48 hours, respectively.

Determination of Tryptophan. The tryptophan content was measured by the colorimetric procedure with p-dimethylamidobenzaldehyde [18] and with N-bromosuccinamide, according to Patchornik et al. [19]. The values obtained by the two methods were in good agreement.

Determination of N-Terminal Amino Acids. The protein was treated with fluorodinitrobenzene and the dinitrophenyl (DNP)-amino acids were identified by paper chromatography in the following buffers [20]: n-butyl-acetic acid-water (4:1:5, v/v), 1.5 N sodium phosphate buffer, pH 6.0, and t-amyl alcohol saturated with phthalate buffer, pH 6.0.

Determination of Phosphorus and Protein. For phosphorus analysis the samples were ashed with 10 N H$_2$SO$_4$ and phosphate was determined by the method of Fiske and Subbarow [21]. Protein was routinely determined by the method of Bücher [22], standardized by dry weight determination of a thoroughly dialyzed sample of the purified enzyme. The results were compared with those obtained by the method of Lowry [23] and by measurement of the absorbance at 280 mµ. For determination of radioactivity, samples were plated on planchets as thin films and counted in a windowless Nuclear Chicago Corp. gas-flow counter.

RESULTS

Purification of Spinach Aldolase

Extraction of the Enzyme. One kilogram of fresh market spinach was washed and drained and homogenized for 1.5 minutes in a Waring blender with 1 liter of 1 mM EDTA solution adjusted to pH 7.4. The homogenate was centrifuged and the supernatant solution filtered through a fluted filter paper. All operations subsequent to blending were carried out at 0—5°. Unless otherwise indicated, all solutions used contained 1 mM EDTA. The results of the purification procedure are summarized in Table 1.

Table 1. *Purification of spinach aldolase*

Fraction	Total activity	Specific activity	Recovery
	units[a]	units/mg	%
Extract	2480	0.21	100
pH 5.5 Fraction	2080	1.28	84
Ammonium sulfate Fraction I	1404	3.75	58
DEAE-Sephadex Fraction	800	9.70	32
Ammonium sulfate Fraction II	720	12.80	29

[a] See Methods for definition of unit.

Acid Precipitation. The extract (1400 ml) was adjusted to pH 5.5 with 14 ml of 3.5 M acetic acid. The mixture was kept at this pH for 30 minutes and the heavy precipitate removed by centrifugation and discarded (pH 5.5 Fraction, 1300 ml).

Ammonium Sulfate Fractionation at pH 5.5. The pH 5.5 Fraction (1300 ml) was brought to 40% (v/v) saturation by the addition of 860 ml of cold saturated ammonium sulfate solution. The precipitate was collected by centrifugation and suspended in a small volume (40 ml) of 20% (v/v) saturated ammonium sulfate solution. The mixture was centrifuged and the residue again extracted with 20 ml of 20% (v/v) saturated ammonium sulfate solution. The two extracts were combined and the protein precipitated by the addition of 48 ml of saturated ammonium sulfate solution. The precipitate was dissolved in 5.0 ml of 50 mM phosphate buffer, pH 7.4, and the ammonium sulfate removed by filtering the solution through a 2×20 cm column of Sephadex G-25 equilibrated with 50 mM phosphate buffer, pH 7.4 (Ammonium sulfate Fraction I, 80 ml).

Chromatography on DEAE-Sephadex A-50. The enzyme solution was placed on a 2.5×12 cm column of DEAE-Sephadex A-50 previously equilibrated with 50 mM phosphate buffer. Elution was begun with a gradient of NaCl (0—0.2 M) in the same buffer

(Fig. 1). The enzyme fractions shown in the cross-hatched area in Fig. 1 were pooled, omitting the earlier fractions in the peak which contained a yellow protein (DEAE-Sephadex Fraction, 50 ml).

Second Ammonium Sulfate Fractionation. The DEAE-Sephadex fraction was treated with 50 ml of cold saturated ammonium sulfate solution and the mixture adjusted to pH 7.4. The precipitate was removed by centrifugation and the supernatant solution treated with 28 ml of saturated ammonium sulfate solution. The final precipitate was dissolved in 5 ml of 50 mM phosphate buffer, pH 7.4, and desalted by filtering through a small column of Sephadex G-25. The purified enzyme was stable for several weeks when stored at −20° in 50 mM phosphate buffer, pH 7.4 (Ammonium sulfate Fraction II, 5 ml).

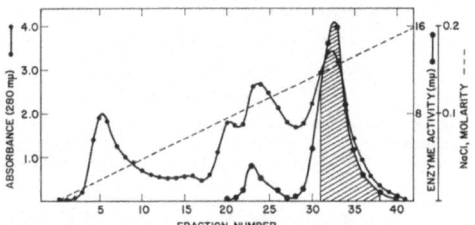

Fig. 1. *DEAE-Sephadex A-50 chromatography of spinach aldolase.* A 2.5×12 cm column equilibrated with 0.05 M phosphate buffer, pH 7.4, was used. After adsorption of the enzyme in the same buffer, elution was carried out (0.5 ml/min) with a linear gradient of NaCl (0—0.2 M). Fractions (10 ml) were collected. The protein (absorption at 280 mμ) and enzyme activity were measured in each fraction

Remarks on the Purification Procedure. Some batches of spinach yielded fractions after the first ammonium sulfate precipitation with lower specific activity; these were characteristically unstable when stored frozen and were refractory to further purification. The reason for these variations could not be determined, but were possibly related to the fact that spinach in this market is derived from many different regions. When poor results were encountered after the first ammonium sulfate fractionation the preparation was discontinued.

Precipitation with dilute acid removes most of the green pigment, yields slightly yellow solutions, and appears to be useful for the preparation of plant aldolases. A similar step was employed by Stumpf for the purification of pea aldolase [2]. After the first ammonium sulfate fractionation a number of standard procedures, including adsorption on calcium phosphate gel and fractionation by precipitation with acetone and ammonium sulfate, were unsuccessful. The key to achieving homogenous fractions appeared to lie in removing the yellow pigment during fractio-

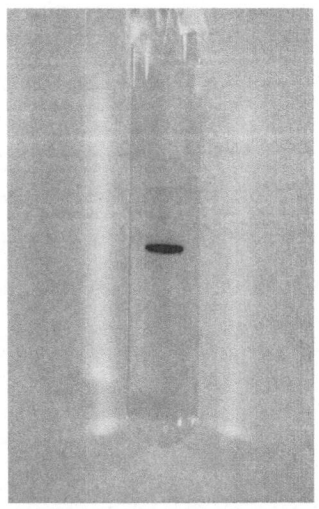

Fig. 2. *Polyacrylamide gel electrophoresis of spinach aldolase.* Electrophoresis was carried out in 7°/₀ standard polyacrylamide gel, pH 9.5, at 5 mA per tube. The sample contained 15 µg of protein

nation on DEAE-Sephadex. With some batches of spinach, where the amount of this yellow impurity appeared to be excessive, purification was unsuccessfull.

Stability of the Enzyme Preparations. The purified preparations were stable when stored in frozen solution at −20°. Solutions of the enzyme were stable at pH 3.5, but lost activity irreversibly at pH 2.8. At intermediate pH values the enzyme lost activity, recovered partially, and then on prolonged incubation was inactivated. The enzyme did not appear to dissociate into subunits at low pH and the value of $s_{20,w}$ obtained under these conditions was 6.9. The enzyme was also inactivated by high concentrations of urea (6 M), although it was stable for long periods in 3 M urea. Dilutions of the enzymes inactivated in urea did not lead to a recovery of activity.

Properties of the Purified Enzyme

Homogeneity and Molecular Weight. Although attempts to crystallize the enzyme were unsuccessful, the preparations appeared to be homogeneous on the basis of polyacrylamide gel electrophoresis (Fig. 2). Sedimentation velocity runs in the analytical ultracentrifuge showed a single symmetrical peak with

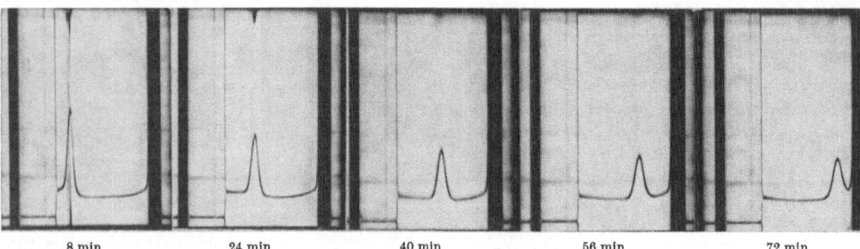

| 8 min | 24 min | 40 min | 56 min | 72 min |

Fig. 3. *Sedimentation velocity pattern of spinach aldolase.* Medium, 0.1 M phosphate buffer, pH 7.4; concentration of protein, 6 mg/ml; speed, 59,780 rev./min; temperature, 20°. In three experiments at varying dilution the following values of $s_{20,w}$ were calculated: 6.3, 6.3, 6.2

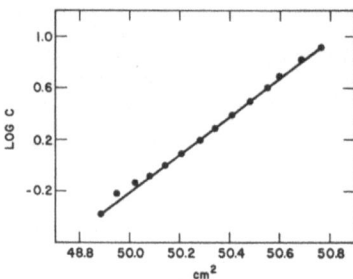

Fig. 4. *Molecular weight determination of spinach aldolase by the sedimentation equilibrium method.* Measurements were made of the net fringe displacement of an 0.2 mg/ml solution of the enzyme in 0.05 M phosphate buffer, pH 7.4, centrifuged at 21,740 rev./min for 23 h at 14°

$s_{20,w} = 6.3$, from which the molecular weight was calculated to be approximately 120,000 (Fig. 3). Determination of the molecular weight by sedimentation equilibrium yielded values of 119,000 and 121,000 (Fig. 4); analysis by the sucrose density gradient method gave a value of 117,000. For the calculations of molecular weight in the experiments to follow, the value of 120,000 was employed.

Absorption Spectrum and Amino Acid Composition. The purified preparations showed a single peak at 280 mµ. A solution in 0.05 M phosphate buffer, pH 7.4, containing 1 mg per ml gave an absorbance of 1.73 at this wave length, compared to 0.91 for rabbit muscle aldolase [24]. This indicated that the protein had a high content of aromatic amino acids, a result which was confirmed by the increased color yield compared with serum albumin in the Lowry method [23], and the composition as determined on the amino acid analyzer (Table 2). In comparison with rabbit muscle aldolase [25], the enzyme from spinach was found to contain approximately the same number of residues as phenylalanine, fewer of tyrosine, but four times as much tryptophan. The spinach enzyme also contained substantially fewer residues of cysteine and histidine, but was somewhat richer in methionine.

Table 2. *Amino acid composition of spinach aldolase*

Amino acid	Spinach Aldolase		Rabbit muscle aldolase
	24 h hydrolysate	48 h hydrolysate	
	moles/mole enzyme	moles/mole enzyme	moles/mole enzyme
Lysine	65	67	95
Histidine	9	8	39
Arginine	46	50	57
CM-Cysteine	12	11	28
Aspartic acid	92	83	106
Threonine	56	51	74
Serine	64	61	67
Glutamic acid	140	137	145
Proline	59	57	69
Glycine	101	102	108
Alanine	132	126	154
Valine	72	77	68
Methionine	20	20	11
Isoleucine	45	47	68
Leucine	107	107	127
Tyrosine	33	33	40
Phenylalanine	26	32	26
Tryptophan	25		6[a]

[a] C. Y. Lai, personal communication.

Evidence for the Formation of the Schiff Base Intermediate. The enzyme was rapidly inactivated when it was treated with sodium borohydride in the presence of dihydroxyacetone phosphate (Fig.5). It has already been indicated that this reduction is associated with the uptake of three moles of dihydroxyacetone phosphate per mole of protein. When the reduced protein was hydrolyzed with 6 N HCl and analyzed in the short column on the amino acid analyzer, a peak corresponding to authentic β-glyceryllysine was observed. This peak was found to contain all of the radioactivity present in the hydrolysate (Fig.6). This confirms the earlier report by Rutter [3] to the effect that the enzyme in higher plants is a Class 1 aldolase.

Determination of Subunits. Although the enzyme could not be dissociated into subunits with urea or at acid pH, the results of end-group analysis and evaluation of the number of combining sites by reduction of the Schiff base intermediate with sodium borohydride suggested that it is composed of subunits. To determine the number of combining sites,

the enzyme was treated with borohydride in the presence of [¹⁴C]dihydroxyacetone phosphate and the amount of radioactivity incorporated determined (Table 3). Approximately three moles of dihydroxyacetone phosphate were incorporated for each mole

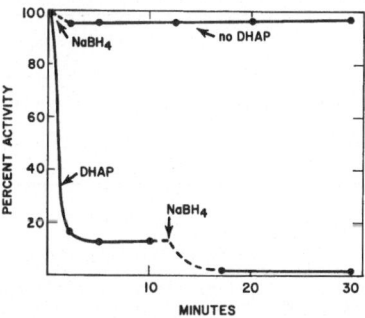

Fig. 5. *Inactivation of spinach aldolase by borohydride in the presence of dihydroxyacetone phosphate.* The reaction mixture (0.1 ml) contained 0.115 mg of spinach aldolase and 0.12 μmole of dihydroxyacetone phosphate. The pH was adjusted to 5.9 with dilute acetic acid, 1 μl of sodium borohydride (10 mg/ml) was added as indicated and aliquots taken for measurement of enzyme activity. The quantity of acetic acid required to maintain the pH between 5.9 and 6.0 was determined on a larger sample containing the same proportion of reagents

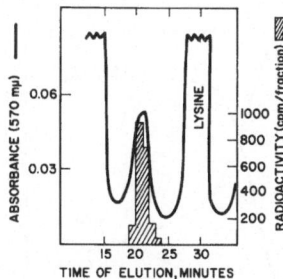

Fig. 6. *Identification of β-glyceryllysine.* Labelled aldolase (1.5 mg) was hydrolyzed with 5.7 N HCl for 24 hours at 105° in a sealed evacuated tube. Aliquots equivalent to 0.25 mg protein was processed on a short column in a Beckman-Spinco amino acid analyzer. Fractions were collected at the outlet after the colorimeter at one minute intervals and assayed for radioactivity

of enzyme inactivated, suggesting that the spinach enzyme resembles that from rabbit muscle in containing three Schiff base-forming sites [26]. The same result was obtained when the number of bound dihydroxyacetone phosphate groups was estimated from the incorporation of organic phosphorus, rather than on the basis of radioactivity. In this case, however, it was necessary to substract a rather considerable blank due to organic phosphorus present

Table 3. *Number of combining sites of spinach aldolase based on labelling with [^{14}C]-dihydroxyacetone phosphate*
Spinach aldolase (1 and 2 mg in Experiments I and II, respectively) was inactivated by adding borohydride in the presence of 80-fold molar excess of ^{14}C-labelled dihydroxyacetone phosphate (420 counts/min/mμmole). The protein was precipitated with trichloroacetic acid, washed, redissolved in 1 ml of 0.1 $\underline{N}$ NaOH and the radioactivity measured as described in the experimental section

Experiment No.	Protein recovered		Inactivation	Inactive enzyme	Radioactivity	Dihydroxyacetone phosphate	Ratio
	mg/ml	mμmoles/ml	%	mμmoles/ml	counts/min/ml	mμmoles/ml	moles/mole enzyme
I	0.86	7.2	99	7.1	8900	21.1	2.93
	0.93	7.8	99	7.7	8300	19.8	2.53
	0.80	6.7	93	6.2	6800	16.1	2.60
II	1.85	15.4	96	14.8	20,200	48.0	3.24
	1.67	13.9	99	13.8	15,800	37.8	2.84
	1.82	15.2	99	15.0	17,800	42.3	2.82

Table 4. *Number of combining sites of spinach aldolase based on determination of bound phosphorus*
Spinach aldolase (2 mg) was inactivated in the presence of 80-fold excess of DHAP by borohydride addition and processed as described in the experimental section for the determination of total phosphorus. A blank without DHAP gave a value of 27.4 mμmoles

Protein recovered		Inactivation	Inactive enzyme	Phosphate Incorporated	
mg/ml	mμmole/ml	%	mμmoles/ ml	mμmoles /ml[a]	moles/mole enzyme
1.98	16.5	99	16.3	49.3	3.03
1.94	16.2	99	16.0	50.4	3.15

[a] Corrected for the blank value.

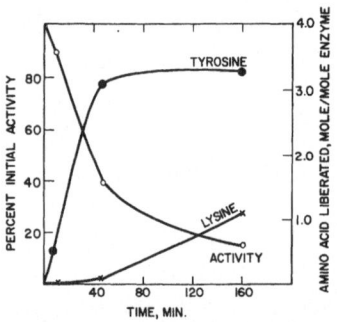

Fig. 7. *Release of amino acids on treatment with carboxypeptidase A*. The reaction mixture contained 1.04 mg of spinach aldolase in 0.2 ml of 0.05 M phosphate buffer, pH 7.4, and carboxypeptidase A at $^1/_{1000}$ molar concentration. After incubating at 25° for different time intervals, aliquots (1 μl) were withdrawn for assaying enzyme activity and 1.0 ml of citrate buffer, pH 2.2, was immediately added to stop the reaction. Aliquots of 0.5 ml from this mixture were used for amino acid analysis

in the control samples treated without dihydroxyacetone phosphate. This amounted to approximately one-third of the total organic phosphorus (Table 4).

When the enzyme was treated with carboxypeptidase A, there was a decline in activity with

fructose diphosphate associated with the release of tyrosine and lysine (Fig. 7). Approximately three equivalents of tyrosine were released rapidly, followed by about one equivalent of lysine. In other experiments with carboxypeptidase B tyrosine and lysine were released at nearly equal rates and the total number of lysine residues liberated reached approximately two. The results suggest that the enzyme is composed of three subunits, each with a COOH-terminal tyrosine residue, and that the penultimate amino acid at the COOH-terminal end may be lysine. The release of tyrosine with carboxypeptidase B was unexpected; under similar conditions the same preparation of carboxypeptidase B did not release tyrosine residues from muscle aldolase.

The only NH$_2$-terminal amino acid which could be detected following treatment of the enzyme with fluorodinitrobenzene was glycine.

In the case of fructose diphosphate aldolase from rabbit muscle, digestion with carboxypeptidase leads to a loss of activity towards fructose diphosphate, but the activity toward fructose 1-phosphate remains unchanged [27,28]. A similar pattern was obtained when fructose diphosphate aldolase from spinach was digested with either carboxypeptidase A or carboxypeptidase B, although in this case the activity towards both substrates was decreased. The loss of activity was greater when the enzyme was tested with fructose diphosphate and there was a considerable change in the ratio of activities with the two substrates (Fig. 8).

Substrate Specificity. The native enzyme is approximately twenty times as active with fructose diphosphate as with fructose 1-phosphate and the affinity for the former substrate is considerably higher (Table 5). Experiments were also carried out to determine the activity in the direction of condensation. The apparent V_{max} for dihydroxyacetone phosphate was estimated to be approximately 2×10^{-2} M, however, in these experiments the rate was probably limited by the concentration of glyceraldehyde 3-phosphate, which was twenty times lower. The maximum activity in the direction of condensation was approximately three times that observed in the direction of cleavage. This result is

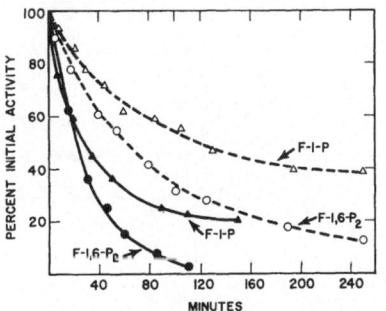

Fig. 8. *Inactivation of spinach aldolase by carboxypeptidase.* The reaction mixture contained 5 μg of spinach aldolase in 0.3 ml of 0.02 M phosphate buffer, pH 7.4, and $^1/_{1000}$ concentration (on a molar basis) of carboxypeptidase A (dashed lines) or B (solid lines). During incubation at 25° aliquots were withdrawn and the enzyme activity measured with fructose diphosphate (F-1,6-P_2) or fructose 1-phosphate (F-1-P) as described in the experimental section. The initial activities were 12.0 and 0.51 units/mg protein for FDP and F-1-P, respectively

Table 5. K_m and V_{max} values

Substrate	K_m	V_{max}
		units/mg
FDP	2×10^{-5} M	13.20
F-1-P	1×10^{-2} M	0.72
DHAP	2.2×10^{-2} M	32.00

Table 6. *Occurrence of spinach aldolase in the soluble cytoplasm*

Spinach chloroplasts were prepared according to Arnon et al. [27]. Spinach leaves (50 g), cut into small pieces, were ground in a mortar with 50 ml of 0.35 M NaCl and 10 g of glass powder. The homogenate was filtered through a double layer of cheese cloth, centrifuged at different speeds and the aldolase activity of the fractions tested

Fraction	Centrifugal force	Time	Aldolase activity	
	g	min	units	%
Homogenate			54.5	100
Cell debris, nuclei	200	2	0.7	1.2
Chloroplasts	1,000	7	1.0	2
	5,000	10	0.8	1.5
	10,000	10	0.5	1
	100,000	60	0.3	0.6
Supernatant	100,000	60	49.0	91

consistent with an important role for the spinach enzyme in the biosynthesis of hexose.

Effect of pH on Activity. The rate of the cleavage reaction falls rapidly below pH 6.0 and at pH 5.5 is

approximately one-half that observed at pH 7. There is little change in activity over the pH range from 6.0—9.0.

Occurrence of Spinach Aldolase in the Cytoplasm. Since aldolase is an important enzyme in the formation of hexoses from carbon dioxide in photosynthesis, it was of interest to establish its location in the spinach leaf. Spinach chloroplasts prepared according to Arnon et al. [29] were found to be devoid of aldolase activity; the activity was found to be confined to the supernatant solution (Table 6). This does not exclude the possibility that the enzyme was loosely bound by the chloroplasts and eluted during their preparation.

DISCUSSION

Although aldolase has been described in higher plants, this represents the first isolation of homogeneous preparations of the enzyme. It appears to be similar in many respects to that found in animal tissues; it forms a Schiff base with an active lysine residue, it contains a similar number of active sites, and the COOH-terminal groups, like those of rabbit muscle and rabbit liver aldolase [28], appear to be tyrosine. However, the molecular weight is somewhat smaller than that of the enzyme isolated from rabbit muscle or rabbit liver and the amino acid composition shows important differences, particularly with respect to the number of tryptophan and histidine residues. Like the enzyme from rabbit muscle, that from spinach is considerably more active with fructose 1,6-diphosphate than with fructose 1-phosphate, and the relative activity towards the two substrates is changed in a similar manner on removal of the COOH-terminal tyrosine residues with carboxypeptidase. Work is currently in progress in this laboratory in an effort to isolate and characterize the peptide containing the active lysine residue and to determine whether it is in any way similar to the corresponding peptides isolated from the rabbit muscle and rabbit liver enzymes.

This work was supported by grants from the National Institutes of Health (GM 11301) and the National Science Foundation (GB 1465). This is Communication No. 75 from the Joan and Lester Avnet Institute of Molecular Biology.

REFERENCES

1. Tewfik, S., and Stumpf, P. K., *Am. J. Botany*, 36 (1949) 567.
2. Stumpf, P. K., *J. Biol. Chem.* 176 (1948) 233.
3. Rutter, W. J., *Federation Proc.* 23 (1964) 1248.
4. Kowalsky, A. G., and Boyer, P. D., *J. Biol. Chem.* 235 (1960) 604.
5. Stellwagen, E., and Schachman, H. K., *Biochemistry,* 1 (1962) 1056.
6. Deal, W. C., Rutter, W. J., and Van Holde, K. E., *Biochemistry,* 2 (1963) 246.
7. Kawara, K., and Tanford, C., *Biochemistry,* 5 (1966) 1578.

8. Lai, C. Y., Hoffee, P., and Horecker, B. L., *Arch. Biochem. Biophys.* 112 (1965) 567.
9. Horecker, B. L., Rowley, P. T., Grazi, E., Cheng, T., and Tchola, O., *Biochem. Z.* 338 (1963) 36.
10. Racker, E., *J. Biol. Chem.* 167 (1947) 843.
11. Rosen, O., Rosen, S. M., and Horecker, B. L., *Arch. Biochem. Biophys.* 112 (1965) 411.
12. Yphantis, D. A., *Biochemistry*, 3 (1964) 297.
13. Schachman, H. K., in *Methods in Enzymology* (edited by S. P. Colowick and N. O. Kaplan), Academic Press, New York 1955, Vol. I, p. 70.
14. Martin, R. G., and Ames, B. N., *J. Biol. Chem.* 236 (1961) 1372.
15. Davis, B. J., *Ann. N. Y. Acad. Sci.* 212 (1964) 404.
16. Spackman, D. H., Stein, W. H., and Moore, S., *Anal. Chem.* 30 (1958) 1190.
17. Crestfield, A. M., Moore, S., and Stein, W. H., *J. Biol. Chem.* 238 (1963) 622.
18. Spies, J. R., and Chambers, D. C., *Anal. Chem.* 21 (1949) 1249.
19. Patchornik, A., Lawson, W. B., and Witkop, B., *J. Am. Chem. Soc.* 80 (1958) 4747.
20. Fraenkel-Conrat, H., Harris, J. I., and Lery, A. L., *Methods in Biochem. Analy.* 2 (1959) 359.

21. Fiske, C. H., and Subbarow, Y., *J. Biol. Chem.* 66 (1925) 375.
22. Bücher, T., *Biochim. Biophys. Acta*, 1 (1947) 292.
23. Lowry, O. H., Rosebrough, N. J., Farr, A. L., and Randall, R. J., *J. Biol. Chem.* 193 (1951) 265.
24. Taylor, J. F., Green, A. A., and Cori, G. T., *J. Biol. Chem.* 173 (1948) 591.
25. Lai, C. Y., Tchola, O., Cheng, T., and Horecker, B. L., *J. Biol. Chem.* 240 (1965) 1347.
26. Kobashi, K., Lai, C. Y., and Horecker, B. L., *Arch. Biochem. Biophys.* 117 (1966) 437.
27. Drechsler, E. R., Boyer, P. D., and Kowalsky, A. G., *J. Biol. Chem.* 234 (1959) 2627.
28. Rutter, W. J., Richards, D. C., and Woodfin, B. M., *J. Biol. Chem.* 236 (1961) 3193.
29. Arnon, D. I., Allen, M. B., and Whatley, F. R., *Biochim. Biophys. Acta*, 20 (1956) 449.

B. L. Horecker
Department of Molecular Biology
Albert Einstein College of Medicine
1300 Morris Park Avenue
Bronx, New York 10461, U.S.A.

European J. Biochem. 1 (1967) 125—134

Synthèse du poly U par la RNA polymérase avec l'acide octoadénylique (hepta adénylyl-(3',5')-adénosine) comme matrice

D. H. Hayes, R. Cukier et F. Gros

Service de Physiologie Microbienne, Institut de Biologie Physico-Chimique, Paris

(Reçu le 5 décembre 1966)

The octoadenylic acid $(Ap)_7A$ serves as a template for poly U synthesis by *Escherichia coli* RNA polymerase. Two types of polymerisation reaction can be observed by varying the divalent cation concentration in the reaction mixture. High divalent cation concentrations $(4-6\times10^{-3}$ M $MnCl_2)$ permit a limited poly U synthesis which stops when the quantity of UTP polymerised corresponds to approximately 2 moles per mole of template A. When reaction is carried out at 27° polymerisation stops completely after the plateau level of 2 U/A is reached but at 37° it continues slowly beyond this point. Addition of template to an incubation mixture in which poly U synthesis has stopped causes reaction to start again, and to continue to a new plateau corresponding to polymerisation of 2 moles of UTP per mole of additional template A. These results show that at high divalent cation concentrations poly U synthesis stops because the template is consumed, probably by its immobilisation in a complex of the type poly $A-2$ poly U. In media containing low concentrations of divalent cations $(10^{-3}$ M $MnCl_2)$ poly U synthesis is not limited by the amount of template used. Under these conditions the template functions catalytically. A biphasic incorporation curve is observed with a small reduction in the reaction rate at a point corresponding to the polymerisation of $2-3$ moles of UTP per mole of template A. Beyond this point reaction continues at a linear rate for long periods and incorporations of up to 20 moles of UTP per mole of template A have been observed.

The average chain length of the poly U synthesised in the presence of high or low divalent cation concentrations varies between 60 and 160 residues *i.e.* between 10 and 20 times that of the template. Some evidence has been obtained indicating the progressive elongation of poly U chains as a function of the duration of synthesis but the maximum variation in chain length observed was only twofold.

Sedimentation analysis of the reaction products present in incubation mixtures containing both high and low concentrations of divalent cations shows that no significant amount of poly U is free at any time. The poly U is found in fast sedimenting (>100 S) complexes dissociable by removal of divalent cations or by sodium dodecyl sulphate treatment. Both procedures liberate poly U with a sedimentation coefficient of approximately 10 S.

An interpretation of the effects of divalent cation concentration on the type of reaction catalysed by RNA polymerase is proposed.

L'activité de la RNA polymérase utilisant le DNA comme matrice peut être inhibée par l'addition de RNA tandis que sa capacité à transcrire le RNA peut être inhibée par le DNA [1—6]. Au cours d'études précédentes [6] concernant les effets inhibiteurs d'oligoribonucléotides sur la transcription du DNA par la RNA polymérase, il avait été constaté que

l'apparition du pouvoir inhibiteur dans la série $(Ap)_nA$ $(n=1 \quad 7)$ coïncide avec la capacité de l'oligonucléotide considéré à servir de matrice. Ce résultat nous a encouragés à étudier les caractéristiques des réactions catalysées par la RNA polymerase en présence du composé $(Ap)_7A$ comme matrice et d'UTP comme substrat. Ceci fait l'objet du présent mémoire. Rappelons que le rôle des oligodeoxyribonucléotides comme matrices pour la RNA polymérase a déjà été mis en évidence et que la nature du produit formé dans la réaction a déjà fait l'objet de nombreuses études [7—13]. Les résultats obtenus dans ces études antérieures sont semblables à ceux exposés ici.

Abréviations non usuelles. Hepta adénylyl-(3',5')-adénosine, $(Ap)_7A$; hepta adénylyl-(3',5')-uridine, $(Ap)_7U$; acide trichloracétique, TCA; dodécyl sulfate de sodium, SDS.

Enzymes. RNA polymérase ou nucléoside triphosphate: RNA nucléotidyltransférase (EC 2.7.7.6); RNase pancréatique ou polyribonucléotide-2-oligonucléotido-transférase (EC 2.7.7.16).

MÉTHODES

Preparations

RNA Polymerase. Plusieurs préparations de cet enzyme obtenues à partir de *Escherichia coli* W 677 par le procédé de Furth, Hurwitz et Anders [14] nous ont été gracieusement données par C. Babinet et J. M. Dubert (Institut Pasteur, Paris). D'autres préparations ont été effectuées en collaboration avec ces deux chercheurs. La teneur en RNase de l'enzyme purifié était variable. Dans certaines de nos expériences (incubations en présence de manganèse à faibles concentrations), nous n'avons employé que des préparations exemptes de RNase.

Tableau. *Caractéristiques générales de la synthèse du poly U en présence du composé (Ap)$_7$A*
Les mélanges réactionnels contenaient dans un volume de 0,1 ml les éléments suivants: RNA polymérase, 35 µg, matrice (Ap)$_7$A, 0,48—1,2 nmoles d'équivalents adénine et MnCl$_2$ et MgSO$_4$ aux concentrations indiquées. Les températures d'incubation étaient les suivantes: expérience 9, 20°, expériences 1—8, 37°, et experience 10, 52°

Expériences	UTP polymérisé en 30 min par nmole d'adénine de la matrice
	nmoles dans 0,1 ml
1 Mn 4×10⁻³ M sans enzyme	0,005
2 Mn 4×10⁻³ M sans matrice	0,05
3 sans Mn	0
4 complet Mn 10⁻³ M	7
5 complet Mn 4×10⁻³ M	1,9
6 complet Mn 6×10⁻³ M	1,8
7 complet Mn 4×10⁻³ M Mg 2×10⁻³ M	1,4
8 complet Mn 4×10⁻³ M Mg 10⁻² M	0,6
9 complet Mn 4×10⁻³ M 20°	2,2
10 complet Mn 4×10⁻³ M 52°	1,9

Poly A, acides oligoadényliques (Ap)$_n$A et [^{14}C] (Ap)$_n$U. Nous devons le poly A à l'obligeance de Mme Grunberg-Manago. Les acides oligoadényliques (Ap)$_7$A et [^{14}C](Ap)$_7$U préparés par le procédé de A. M. Michelson [15] nous ont été donnés par celui-ci.

Uridine 2'(3')-5' diphosphate. Nous avons phosphorylé l'uridine par la méthode de Michelson [16].

Autres produits. UMP 2'3', UTP et [^{14}C] UTPsont des produits de Schwarz Bioresearch Inc. Le sucrose a été fourni par Hopkin et Williams Ltd. (Grande-Bretagne).

Synthèse de poly U. Les mélanges d'incubation contiennent, dans des volumes de 0,1 à 0,5 ml, maléate de potassium, 0,05 M, pH 7,5, mercaptoéthanol, 0,02 M, [^{14}C]UTP (activité spécifique 2300, ou 27.000 coups/min/nmole) 1,5×10⁻³ M. Le MnCl$_2$, le MgSO$_4$, ainsi que la matrice et l'enzyme, sont présents comme indiqué dans les légendes des tableau et figures. L'incorporation du [^{14}C]UTP est suivie en précipitant par l'acide trichloracétique des aliquotes des mélanges réactionnels prélevés à des temps déterminés (5 ml TCA à 5%, 10 min 0°). Les précipités sont retenus sur des membranes millipores (HAWP 025, Millipore Co., Bedford, Mass., USA) qui sont ensuite lavées avec 25 ml de TCA à 5%, puis séchées à 80—100°. La radioactivité est finalement mesurée dans un compteur à « gas-flow » (Tracerlab) avec une efficacité de 15%.

Détermination de la longueur des chaines de poly U

Des aliquotes provenant des mélanges réactionnels sont refroidies à 0° et additionnées de dodécyl sulfate de sodium (SDS) à une concentration finale de 4 g/l. Le [^{14}C]UTP libre et le [^{14}C] poly U sont séparés par chromatographie sur papier DEAE (Whatman DE20) des solutions additionnées de SDS. Cette chromatographie est effectuée dans le formiate d'ammonium à 0,3 M pendant 3 heures. Les régions des chromatogrammes contenant le [^{14}C] poly U sont découpées et hydrolysées dans NaOH à 0,3 N suivant la méthode de Falaschi, Adler et Khorana [9]. Les hydrolysats sont débarrassés des débris de papier par centrifugation, et les débris lavés à l'eau (2×2 ml). Les filtrats et eaux de lavage réunis (environ 10 ml) sont amenés à pH 10 par addition d'acide formique molaire. Uridine, uridine 2'3' monophosphate, et uridine 2'(3')5' diphosphate (50 µmole) sont ajoutés et les solutions ainsi obtenues sont adsorbées sur colonne ($h = 1,8$ cm, $d = 0,8$ cm) échangeuse d'ions Dowex 1(×8) (forme formiate, 200—400 *mesh*). Celle-ci est préalablement lavée successivement à l'acide chlorhydrique 2 N, à la soude 2 N, l'acide formique 4 N et l'eau, puis mise en équilibre avec un tampon au formiate d'ammonium 0,01 M pH 9,7. Les colonnes contenant les nucléotides sont lavées au formiate d'ammonium 0,01 M pH 9,7 (100 ml) et éluées par un gradient linéaire obtenu à partir de 250 ml de formiate d'ammonium 0,01 M pH 9,7 et 250 ml de formiate d'ammonium 1,2 M pH 5. L'absorption de l'effluent dans ultraviolet (254 mµ) est enregistrée de manière continue. Les volumes approximatifs de tampons avec lesquels sont élués les différents nucléotides sont: uridine, 85 ml; urdine 5' phosphate (produit contaminant la préparation d'uridine 2'(3')5' diphosphate) 150 ml; uridine 2'(3') phosphate (200 ml); uridine 2'(3'):5' diphosphate, 340 ml. Les fractions contenant l'uridine, l'UMP 2'(3') et l'UDP sont groupées, amenées à pH 2 (HCl) et les dérivés d'uridine sont adsorbés quantitativement sur Norit A (250 mg) préalablement « désactivée » par traitement à l'hexanol [17]. La Norit A est centrifugée, lavée à l'eau (3×10 ml) et éluée par l'ammoniaque 0,3 M en solution dans l'éthanol aqueux 50% (v/v) (5×10 ml). Le lavage et l'élution se font par centrifugation. L'uridine et l'UMP 2'(3') sont récupérés quantitativement, 1' UDP 2'(3')5' avec un rendement compris entre 75 et 80%. Les éluats sont évaporés au bain marie à 45° jusqu'à faible volume (0,1—0,2 ml) et chromatographiés sur papier Whatman 3 M M pendant 16 heures dans l'éthanol-acétate d'ammonium M (75:30, v/v) [18]. Les R_F sont : uridine 0,76, uridine 2'(3') monophosphate, 0,3, uridine 2'(3')5' diphosphate, 0,07. Après séchage des chromatogrammes, les régions renfermant du matériel absorbant dans l'ultraviolet ainsi que des échantillons contrôle de R_F et taille similaires sont découpés, immergés dans une solution de 2,5-diphényloxazole (5 g) et de 1,4-bis-2-(-4-méthyl-5-phénylisoxazolyl) benzène (0,3 g) par litre de toluène, puis leur radioactivité mesurée dans un compteur à scintillation (Packard).

Centrifugation en gradients de saccharose

Nous avons décrit par ailleurs les méthodes mises en œuvre dans le fractionnement en gradients de saccharose ainsi que dans la détermination des profils de sédimentation de substances absorbant dans l'ultraviolet ou de composés radioactifs [19]. Les gradients sont préparés à partir de solutions de saccharose à 50 et 200 g/l dans les tampons Tris-HCl 0,01 M, pH 7, contenant des sels à des concentrations différentes selon les expériences. Ces concentrations sont indiquées dans les légendes des figures. Pendant leur centrifugation à 39.000 tours/min, la température des gradients, préalablement refroidis à 3°, augmente d'environ 2° par heure.

RÉSULTATS

Activité des acides oligoadényliques comme matrices pour la RNA polymérase

La Fig.1 illustre les résultats obtenus à 37° avec certains des acides oligoadényliques de la série (Ap)$_n$A. Les membres de cette famille d'oligonucléotides de longueur supérieure à six nucléotides sont actifs comme matrices dans la réaction d'incorporation de l'UTP au sein d'un produit acido précipitable. Falaschi, Adler et Khorana [9] ont obtenu des résultats similaires avec la série des acides oligothymidyliques.

Caractéristiques générales de la synthèse de poly U sur la matrice (Ap)$_7$A

Les résultats provenant d'une série d'expériences sur la synthèse de poly U sont groupés dans le tableau.

9*

Il en ressort que la réaction de polymérisation exige la présence d'UTP, ainsi que celle de l'enzyme et de la matrice, le manganèse étant cofacteur de la réaction (expériences 1, 2, 3, 4) et la réaction étant partiellement inhibée par une augmentation limitée de la concentration en ions divalents (comparer expériences 4, 5, 6, 7, 8). Ainsi que nous le verrons ci-dessous, des concentrations élevées en ions divalents n'exercent pas d'effet sur la vitesse initiale de polymérisation mais sur les propriétés (catalytique ou stœchiométrique) de la matrice dans la réaction. Les expériences 5, 9 et 10 démontrent que la RNA polymérase est active jusqu'à une température voisine de 52°.

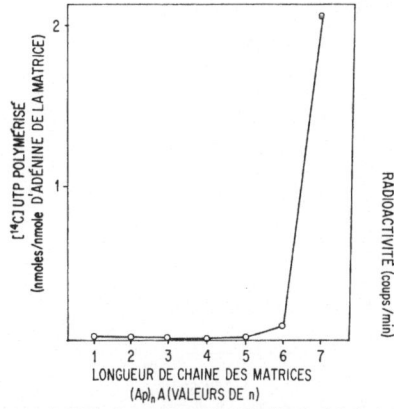

Fig.1. *Activités des acides oligoadényliques (Ap)$_n$A (n = 1—7) comme matrices pour la synthèse de poly U.* Des mélanges standard (volumes 0,1 ml) contenant 7 µg de RNA polymérase, 4×10^{-3} M MnCl$_2$, et 30 nmoles, d'adénine sous forme de matrices (Ap)$_n$A étaient incubés à 37° pendant 30 min. Les quantités de [^{14}C] UTP polymérisées sont exprimées en nmoles par nmole d'adénine des matrices

Synthèse de poly U en présence de concentrations élevées en ions divalents

La polymérisation de l'UTP dans des milieux réactionnels renfermant du MnCl$_2$ à des concentrations comprises entre 4×10^{-3} M et 6×10^{-3} M ou un mélange de MnCl$_2$ 4×10^{-3} M et de MgSO$_4$ 2×10^{-3} M se poursuit linéairement jusqu'à incorporation d'environ 2 moles d'UTP par mole de résidu adénine présent dans la matrice. Après quoi, la réaction s'arrête (incubations à basse température, Fig.2) ou continue à vitesse réduite (fortes concentrations d'enzyme et incubation à 37°, Fig.3). Dans le type de réaction qui vient d'être décrit, la vitesse de polymérisation de l'UTP (mais pas le rendement final en poly U formé) dépend de la quantité d'enzyme utilisée (Fig.2) et de la température d'incubation (Fig.3). L'addition d'enzyme après arrêt de la polymérisation n'entraîne pas une reprise de l'incorporation d'UTP (Fig. 4). En revanche, l'addition de matrice déclenche un nouveau cycle de polymérisation qui continue jusqu'à incorporation de 2 moles d'UTP par mole d'adénine présente dans la matrice ajoutée (Fig. 4). Ainsi pendant la polymérisation d'UTP dans des milieux contenant des concentrations relativement élevées de Mn^{++} ou de Mn^{++} + Mg^{++}, la matrice est consommée et ne sert pas de façon catalytique. Le rapport de la quantité de poly U fabriquée à la quantité de matrice mise en jeu

Si la concentration de Mg^{++} dans le milieu de réaction est augmentée jusqu'à 10^{-2} M, une polymérisation limitée est observée. Elle est caractérisée par un rapport de l'UTP incorporé à l'adénine initialement présente de 0,6.

Synthèse de poly U en présence de faibles concentrations de Mn^{++}

La polymérisation d'UTP en présence de faibles concentrations de Mn^{++} (10^{-3} M ou 2×10^{-3} M) se

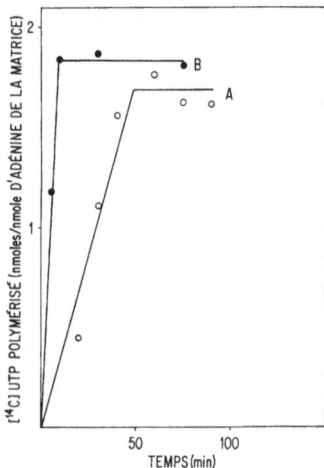

Fig. 2. *Variation de la vitesse de synthèse du poly U en fonction de la concentration en RNA polymérase (préparation manifestant une activité nucléolytique)*. Des mélanges standard (volumes 0,1 ml) contenant 4×10^{-3} M MnCl₂, 2,75 nmoles d'adénine à l'état de matrice (Ap)₇A, et: (A), 12 µg de RNA polymérase; ou (B), 60 µg de RNA polymérase étaient incubés à 27°. Les résultats sont exprimés en nmoles [¹⁴C] UTP polymérisés (TCA précipitable à 0°) par nmole de la matrice. ○ mélange A; ● mélange B

(2 moles d'UTP polymérisées par mole d'adénine présente dans la matrice) suggère que le poly U existe dans le mélange d'incubation sous la forme d'un complexe à trois brins du type poly A − 2 poly U, complexe dans lequel le brin de poly A serait remplacé par une série de molécules alignées de l'oligoadénylate. L'existence de tels complexes a déjà été mise en évidence dans des milieux renfermant du Mg^{++} 10^{-3} M et leurs propriétés ont été étudiées[1] [20,21].

---

[1] L'arrêt de la polymérisation dans une réaction du type qui vient d'être considéré pouvait éventuellement s'expliquer par la présence dans la préparation de RNA polymérase d'une certaine quantité de RNAse susceptible de dégrader tout le poly U formé au-delà de la quantité protégée [22,23] contre cette dégradation par son association avec l'(Ap)₇A. Cependant, ceci n'est pas le cas puisque nous observons également la même limitation dans la réaction d'incorpora-

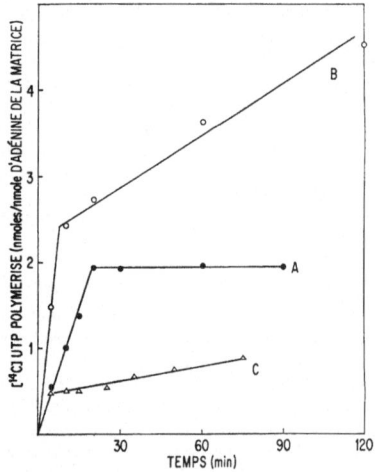

Fig. 3. *Variation de la vitesse de synthèse du poly U en fonction de la température d'incubation, et des concentrations en magnésium et enzyme (enzyme exempt d'activité nucléolytique)*. Trois mélanges standard (volumes 0,1 ml) contenaient: (A), 17,5 µg de RNA polymérase, 4×10^{-3} M MnCl₂, 2,4 nmoles d'adénine à l'état de matrice (Ap)₇A; (B), 35 µg de RNA polymérase, 4×10^{-3} M MnCl₂, 0,6 nmoles d'adénine à l'état de matrice (Ap)₇A; (C), 17,5 µg de RNA polymérase, 2×10^{-3} M MnCl₂, 10^{-2} M MgSO₄, 0,48 nmoles d'adénine à l'état de matrice (Ap)₇A. Les températures d'incubation étaient: (A), 20°, (B) et (C), 37°. Les synthèses d'UTP observées sont exprimées en nmoles de [¹⁴C] UTP polymérisés (TCA précipitable à 0°) par nmole d'adénine de la matrice. ● Mélange A; ○ Mélange B, △ Mélange C

traduit par les cinétiques décrites dans les Figs 5, 6 et 8. Contrairement à la réaction limitée qui a lieu en présence de concentrations élevées en cations divalents, on n'observe pas d'arrêt dans la cinétique d'incorporation lorsque l'enzyme employé est dépourvu d'activité RNase (Fig. 5, courbes A, B). La

tion d'UTP, que l'on mette en œuvre des préparations enzymatiques contaminées par la ribonucléase (Fig. 2 et 7) ou non (Fig. 3), l'existence éventuelle de ce contaminant étant démontrée, comme nous le verrons ci-après, en suivant grâce à l'emploi des mêmes préparations le destin du poly U dans des réactions à faible concentration de Mn^{++} (Fig. 6 et 5 respectivement).

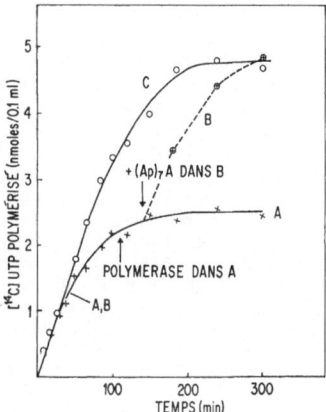

Fig. 4. *Synthèse de poly U en présence de MnCl₂ 4×10⁻³ M. Effets d'additions d'enzyme et de matrice après arrêt de la réaction.* Trois mélanges standard (volumes 0,1 ml) contenant 5 µg de RNA polymérase (préparation douée d'activité nucléolytique), 4×10^{-3} M MnCl₂ et: (A), (B), 1,2 nmoles d'adénine sous la forme (Ap)₇A ou (C) 2,4 nmoles d'adénine à l'état de matrice (Ap)₇A étaient incubés à 26°. Les additions d'enzyme (5 µg) dans (A) et de matrice (1,2 nmole d'adénine) dans (B), une fois atteint le plateau d'incorporation, étaient effectuées aux temps indiqués sur la figure. La synthèse d'UTP est exprimée en nmoles de [¹⁴C] UTP polymérisés (TCA précipitable à 0°) par 0,1 ml. + Mélange A, ⊕ Mélange B, ○ Melange C

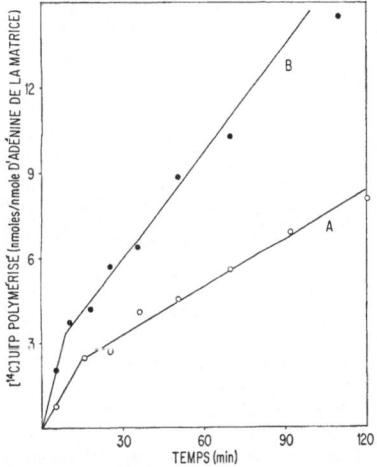

Fig. 5. *Synthèse de poly U en présence de MnCl₂ 10⁻³ M et 2×10⁻³ M (enzyme exempt d'activité nucléolytique).* Deux mélanges standard de réaction (volumes 0,1 ml), contenant 30 µg de RNA polymérase, 0,48 nmoles d'adénine sous la forme de matrice (Ap)₇A et (A), 2×10^{-3} M MnCl₂, ou (B), 10^{-3} M MnCl₂ étaient incubés à 37°. La synthèse de poly U est exprimée en nmoles de [¹⁴C] UTP polymérisés (TCA précipitable à 0°) par nmole d'adénine de la matrice. ○ Mélange A, ● Mélange B

vitesse de la réaction diminue toutefois légèrement à un stade de la polymérisation équivalent à un rapport U/A égal à 3. En revanche, si la RNA polymérase est contaminée par de la RNase (Fig. 6, courbe A), la quantité de poly U présente s'accroît, atteint un maximum, puis décroît rapidement. Néanmoins, l'effet destructeur de la RNase sur le produit de la réaction peut être inhibé par l'addition de polyvinyl sulfate de potassium (Fig. 6, courbe B). Il semble donc que la protection du poly U contre

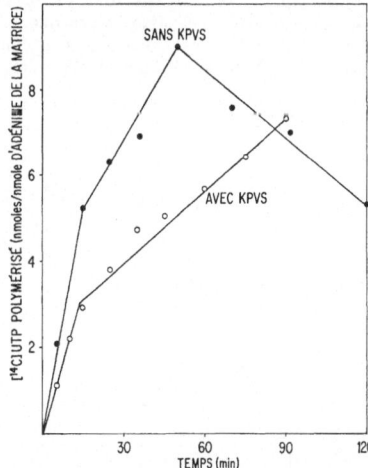

Fig. 6. *Synthèse de poly U en présence de MnCl₂ 10⁻³ M (enzyme doué d'activité nucléolytique).* Deux mélanges standard de réaction (volumes 0,1 ml), contenant 35 µg de RNA polymérase, 10^{-3} M MnCl₂, et 0,6 nmoles d'adénine à l'état de matrice (Ap)₇A étaient incubés à 35°. 1 µg de polyvinyl-sulfate de potassium était ajouté à un des mélanges au temps zéro. La synthèse de poly U est exprimée en nmoles de [¹⁴C] UTP polymérisés (TCA précipitable à 0°) par nmole d'adénine de la matrice

une action nucléasique observée quand la concentration d'ions divalents est élevée (4×10^{-3} M de Mn⁺⁺, Fig. 2 et 7) ne se produit pas si la concentration de ces ions ne dépasse pas 10^{-3} M. L'ensemble des résultats obtenus suggère que la quantité de poly U synthétisée en présence de 10^{-3} M de Mn⁺⁺ est telle que seule une faible proportion de ce produit peut être engagée dans un complexe avec la matrice. La majorité du poly U présent à un stade relativement avancé de la réaction doit donc exister à l'état de chaînes monocaténaires libres. Ceci rendrait compte en effet de leur sensibilité à la RNase.

Longueur des chaînes de poly U synthétisé en présence de concentrations variables d'ions divalents

Le mode de fractionnement des produits de polymérisation, leur hydrolyse et la séparation des produits d'hydrolyse ont été décrits dans le chapitre «Méthodes». Comme nous l'avons déjà indiqué, bien que les rendements en uridine et uridine disphosphate non radioactifs ajoutés comme entraîneurs aient atteint 100% et 75—80% respectivement, les rendements en [¹⁴C] uridine diphosphate provenant de l'hydrolyse du produit radioactif se sont avérés fort variables d'une expérience à l'autre et toujours très inférieurs (2—10 fois moindre) aux rendements en [¹⁴C]uridine. Il a par ailleurs été établi que les produits de la transcription *in vitro* de diverses matrices formées de desoxyribonucléotides telles que DNA, poly d(A—T) ou poly—dG . poly—dC comprennent des restes triphosphates sur leurs nucléotides initiaux

longueurs de chaînes dont les valeurs sont fournies dans les Figs 7 et 8 ont été calculées d'après les rendements en uridine et en uridine monophosphate.

Nos déterminations ont porté sur du poly U synthétisé soit à 25° en présence de Mn⁺⁺ 4×10⁻³ M (Fig. 7), soit à 37° en présence de Mn⁺⁺ 10⁻³ M (Fig. 8). Les valeurs obtenues montrent que le poly U a toujours une longueur moyenne 10 à 20 fois supérieure à celle de la matrice (Ap)$_7$A. Une augmentation des longueurs de chaîne en fonction de la durée de polymérisation est suggérée par les résultats, la variation maximum observée ne dépassant pas toutefois un facteur de deux. Il est par ailleurs évident que la modification du type de réaction qui accompagne les variations de la concentration en

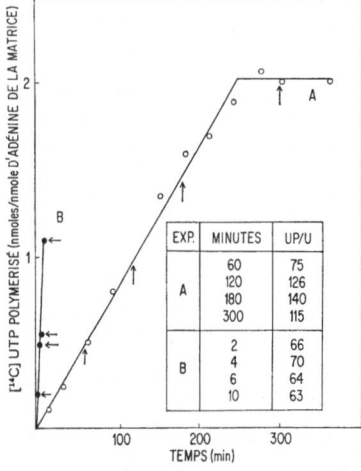

Fig. 7. *Longueurs des chaînes de poly U synthétisées en présence de MnCl₂ (4×10⁻³ M) (enzyme doué d'activité nucléolytique).* Deux mélanges standard de réaction (volumes 0,5 ml), renferment: (A), 125 μg de RNA polymérase, 4×10⁻³ M MnCl₂, et 11 nmoles d'adénine à l'état de matrice (Ap)$_7$A ou (B), 300 μg de RNA polymérase, 4×10⁻³ M MnCl₂ et 3 nmoles d'adénine à l'état de matrice (Ap)$_7$A. Incubation à 27°. Des prélèvements sont effectués soit pour mesurer la cinétique d'incorporation d'UTP (points expérimentaux), soit pour déterminer les longueurs des chaînes des produits formés (aux temps indiqués par les flèches verticales et horizontales). La synthèse de poly U est exprimée en nmoles de [¹⁴C] UTP polymérisés par nmoles d'adénine de la matrice. Le Tableau inclus dans la figure fournit les valeurs de longueurs de chaînes. ○ Mélange A, ● Mélange B

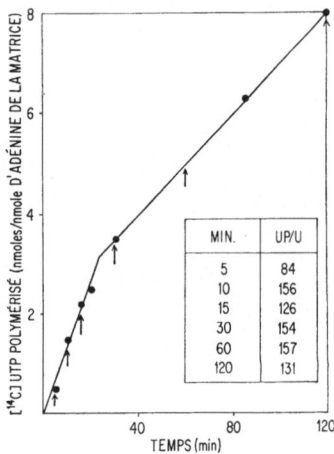

Fig. 8. *Longueurs des chaînes de poly U synthétisées en présence de MnCl₂ (10⁻³ M) (enzyme exempt d'activité nucléolytique).* Mélange standard de réaction (volume 0,3 ml), contenant 180 μg de RNA polymérase, 10⁻³ M MnCl₂, et 1,44 nmoles d'adénine à l'état de matrice (Ap)$_7$A. Incubation à 37°. Les points et flèches ont la même signification que celle indiquée dans la légende de la Fig. 7. La synthèse de poly U est exprimée dans les unités habituelles. Dans le tableau faisant partie de la figure sont portées les valeurs de longueurs de chaîne

[24, 25]. Il semble donc probable que l'[¹⁴C] uridine diphosphate provenant de l'hydrolyse du [¹⁴C] poly U synthétisé *in vitro* résulte de la conversion de ces groupements triphosphates en restes monophosphates soit pendant la synthèse elle-même (la RNA polymérase pouvant être contaminée par des phosphatases), soit au cours des manipulations d'isolement, de purification ou d'hydrolyse du poly U. Les

ions divalents (consommation ou action catalytique de la matrice) ne se réflète pas sur la longueur de chaîne du produit formé.

Etude des produits formés par sédimentation en gradient de saccharose

Les propriétés hydrodynamiques des produits formés lors des réactions de polymérisation en présence d'ions Mn⁺⁺ à concentrations variables ont été examinées par centrifugation sur gradients de saccharose. Des aliquotes provenant des mélanges réactionnels ont été déposés sur des gradients de saccharose (renferment la même concentration en ions divalents que les échantillons à analyser) soit directement, soit après addition de SDS à la concentration finale de 4 g/l.

Les Fig. 9a et 9b illustrent les résultats obtenus après 15 et 120 min d'incubation, dans des conditions où l'oligoadénylate fonctionne comme une matrice catalytique (MnCl$_2$ 10^{-3} M). Des résultats de même nature ont d'ailleurs été obtenus lorsque les prélèvements étaient réalisés après 30 et 60 min d'incubation. La Fig. 9c décrit les profils de sédimentation des produits résultant de la polymérisation limitée en présence de MnCl$_2$ 4×10^{-3} M. Il ressort de ces résultats que les produits formés dans les deux types de

et libèrent du poly U qui sédimente à environ 10 S. Pendant toute la durée des incubations, la quantité de poly U libre, c'est-à-dire de matériel sédimentant à 10 S et présent avant traitement des aliquotes au SDS, demeure négligeable. En ontre, les constantes de sédimentation du poly U libéré par traitement au SDS ne varient ni en fonction de la durée d'incubation (Figs 9a, et 9b) pour une concentration en ions Mn^{++} donnée, ni en fonction de la quantité de Mn^{++} présente dans le milieu réactionnel. En revanche, les produits de réaction non traités au SDS sédimentent de plus en plus rapidement au fur et à mesure que la durée d'incubation est plus prolongée (exemple: après 15, 30, 60 et 120 minutes d'incubation, les propor-

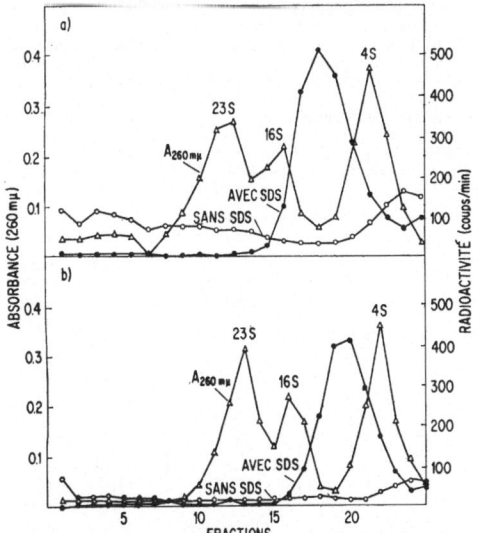

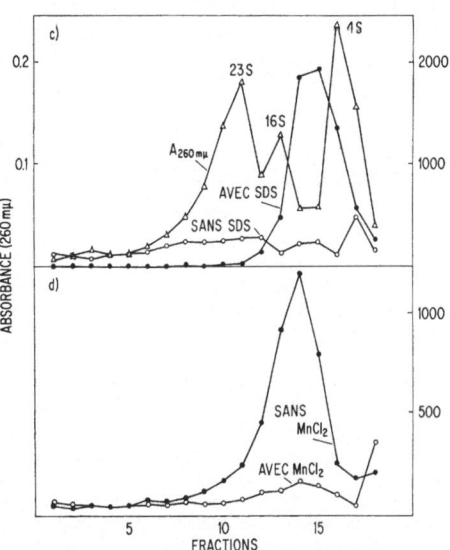

Fig. 9a—d. *Sédimentation en gradients de saccharose des produits formés au cours de la synthèse de poly U.*
a et b) Le mélange de réaction utilisé était celui dont la Fig. 8 décrit les caractéristiques. Après 15 et 120 min d'incubation, on effectuait (en double) des prélèvements de 10 µl. Chaque échantillon était dilué 20 fois dans 0,05 M K maléate pH 7,5, MnCl$_2$ 10^{-3} M. Pour chaque paire d'échantillons considérés, l'un était directement déposé après dilution sur un gradient de sucrose contenant 10^{-3} M MnCl$_2$ et l'on effectuait la même opération sur l'autre après l'avoir additionné de SDS (4 g/l). Deux gradients témoins contenant aussi 10^{-3} M MnCl$_2$ et étaient chargés chacun de 250 µg de RNA total d'*E. coli* dissous dans 200 µl de 0,05 M K maléate pH 7,5, 10^{-3} M MnCl$_2$. Les gradients étaient centrifugés par groupes de trois pendant 4 heures à 39.000 tours/min. Les résultats obtenus à partir de chaque groupe de trois gradients sont superposés dans les figures

c) Deux fractions aliquotes égales sont prélevées dans un mélange d'incubation à «haute concentration» de MnCl$_2$ (4×10^{-3} M) (1,4 nmoles de [^{14}C] UTP incorporés par nmole d'adénine de la matrice). L'une de ces fractions est additionée de SDS à 0,4% (concentration finale par poids) et l'autre non, puis elles sont centrifugées sur des gradients de sucrose contenant 4×10^{-3} M MnCl$_2$. Un gradient témoin fournit le profil de sédimentation du RNA total d'*E. coli* en présence de 4×10^{-3} M MnCl$_2$. Les résultats sont exprimés ainsi qu'il a été indiqué pour 9a, et 9b
d) Deux fractions aliquotes de volume égal sont prélevées au sein d'un mélange d'incubation à «haute concentration» de MnCl$_2$ (4×10^{-3} M) et sont analysées sur des gradients de sucrose en présence et en absence de MnCl$_2$ 4×10^{-3} M. La présentation des résultats est la même que dans les cas de 9a, 9b, et 9c

réaction précédemment décrits existent à l'état de complexes lourds renfermant vraisemblablement l'enzyme de transcription proprement dit. Ces complexes sont dissociables par traitement au SDS

tions de produit non sédimentable sont respectivement de 58%, 50%, 35% et 26%) (Fig. 9a et 9b). La présence d'une quantité suffisante d'ions Mn^{++} dans les gradients d'analyse est nécessaire à la pré-

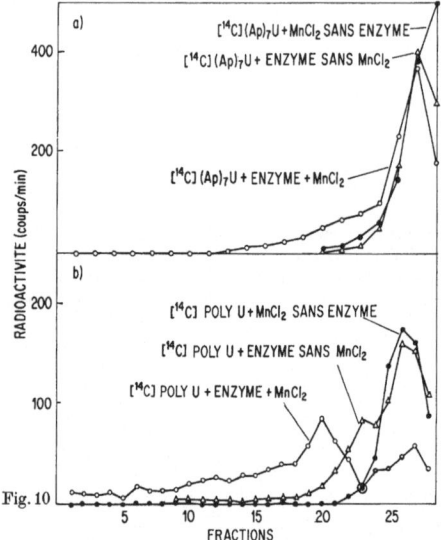

Fig. 10

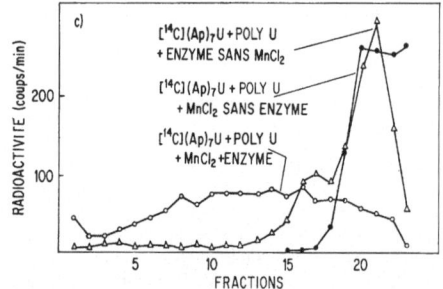

Fig. 10 a—c. *Formation des complexes entre la RNA polymérase et le poly U, l' $(Ap)_7U$ ou un mélange de poly U + $(Ap)_7U$.* On mélange dans les proportions indiquées ci-dessous les constituants suivants: RNA polymérase, poly U, [¹⁴C] poly U (activité spécifique, 340 coups/min/nmole d'uridine) [¹⁴C] $(Ap)_7U$ (activité spécifique = 300 coups/min/nmole d'adénine) et $MnCl_2$ dans 0,01 M Tris-HCl pH 7,4 (volumes finaux 200 µl). Incubation à 20° pendant 5 min. Chaque mélange est ensuite déposé sur un gradient de saccharose contenant la même concentration de $MnCl_2$ que celle du mélange déposé. Les temps de centrifugation à 39.000 tours/min étaient de 2 heures (A, B, C), 2 heures ¹/₄ (D, E, F), 3 heures (G, H, J)

servation des complexes lourds. En effet, si la sédimentation en gradient de saccharose est effectuée en absence de ces cations, tout se passe comme si du SDS avait été préalablement ajouté (Fig. 9 d).

Interactions entre la RNA polymérase d'une part, l'$(Ap)_7U$ ou l'acide polyuridylique d'autre part

Les complexes sensibles au SDS dont nous venons de décrire les propriétés ci-dessus rappellent ceux observés dans des travaux précédents avec le DNA comme matrice [26]. Rappelons d'ailleurs que la formation de tels complexes avec le DNA ou le RNA peut être observée même en absence de substrats (4,27—31]. Ceci nous a incités à étudier les associations susceptibles de s'établir entre la RNA polymérase et le poly U, l' $(Ap)_7U$ ou des mélanges de ces deux produits. Nous avons eu recours au composé $(Ap)_7U$ comme matrice au lieu de l'oligoadénylate $(Ap)_7A$, seul le premier de ces deux composés étant disponible à l'état radioactif. Sur la Fig. 10, on peut voir que l'enzyme est susceptible de former des complexes lourds soit avec l' $(Ap)_7U$ seul (Fig. 10 a), soit avec le poly U seul (Fig. 10 b), soit enfin avec le mélange de ces deux composés (Fig. 10 c). Il semble toutefois que le poly U seul (Fig. 10 b) ou l'$(Ap)_7U$ en présence de poly U (Fig. 10 c) puissent former des complexes avec l'enzyme même en l'absence d'ions divalents. Ces complexes sédimentent plus lentement que ceux formés en présence de $MnCl_2$. La formation de complexes entre la RNA polymérase et des matrices à DNA en absence de cations divalents avait déjà été signalée [4,27,29].

Mélange analysé	RNA poly- mérase	poly U	[¹⁴C] poly U	[¹⁴C] $(Ap)_7U$	M $MnCl_2$ Mélange	M $MnCl_2$ Gradient
	µg	nmoles U	nmoles U	nmoles A		
A	—	—	—	4	10^{-3}	10^{-3}
B	30	—	—	4	0	0
C	30	—	—	4	10^{-3}	10^{-3}
D	—	—	2,7	—	10^{-3}	10^{-3}
E	17,5	—	2,7	—	0	0
F	17,5	—	2,7	—	10^{-3}	10^{-3}
G	—	10	—	4	10^{-3}	10^{-3}
H	30	10	—	4	0	0
J	30	10	—	4	10^{-3}	10^{-3}

Dans chaque gradient, après centrifugation, on déterminait la distribution du matériel radioactif au sein des diverses sous-fractions recueillies. Chaque sous-fraction était placée à cet effet dans 5 ml de solution phosphorescente (naphtalène, 60 g; 2,5-diphényloxazole, 4 g; 1,4-bis-2-(-4 méthyl-5-phényl-isoxazolyl) benzène, 0,2 g; glycol d'éthylène, 20 ml; méthanol, 100 ml et dioxane à un volume final de 1000 ml). Les mesures de radioactivité étaient effectuées dans un compteur à scintillation type Packard. Les résultats obtenus pour chaque groupe de trois gradients sont superposés dans la figure

DISCUSSION

On peut rapprocher l'activité de l'$(Ap)_7A$ en tant que matrice pour la RNA polymérase de celle d'oligo-deoxyribonucléotides de longueurs de chaîne similaires [7—13]. Ces deux types de matrices courtes ont les mêmes exigences en ions divalents *etc*, et dirigent la synthèse de polymères de longueurs 10 à 20 fois supérieures à celles des matrices elles-mêmes. Deux particularités des réactions catalysées par la RNA polymérase avec l'(Ap)$_7$A comme matrice semblent toutefois mériter discussion. Il s'agit (a) de l'influence de la concentration en ions divalents sur le type de réaction obtenu, (b) du fait qu'à aucun stade de la polymérisation le produit de la réaction n'existe libre mais semble être toujours combiné à l'enzyme.

En présence de MnCl$_2$ à 10^{-3} M l'(Ap)$_7$A fonctionne comme matrice de façon catalytique. La courbe d'incorporation d'UTP est biphasique avec une discontinuité à un stade de la réaction correspondant à la polymérisation de 2—3 moles d'UTP par mole d'adénine présente dans la matrice. L'incorporation se poursuit cependant au-delà de ce stade, encore qu'à vitesse légèrement diminuée, le rapport des molécules d'UTP polymérisé aux molécules d'adénine initialement présentes dans la matrice pouvant s'élever jusqu'à 20 (Figs 5, 6, 8). Si l'on met en œuvre des concentrations plus élevées en ions divalents (MnCl$_2$ 4×10^{-3} M ou MnCl$_2$ 4×10^{-3} M + MgSO$_4$ 2×10^{-3} M), la synthèse de poly U s'arrête lorsque la quantité d'UTP polymérisé équivaut à 2 moles par mole d'adénine (Figs 2, 3, 4, 7). L'arrêt de la réaction résulte de l'épuisement de la matrice probablement du fait de son immobilisation dans un complexe du type poly A — 2 poly U. En effet, l'addition d'une nouvelle quantité d'oligoadénylate après que la réaction d'incorporation ait atteint un plateau se traduit par une reprise immédiate des synthèses de poly U jusqu'à la polymérisation de 2 moles d'UTP par mole d'adénine de la matrice additionelle. En revanche, une telle reprise des synthèses n'accompagne pas l'addition d'une nouvelle quantité d'enzyme. Des complexes oligo A — poly U du type proposé ont déjà été décrits et sont stables dans des conditions de force ionique analogues à celles réalisées dans nos mélanges d'incubation [20,21]. Nishimura, Jacob et Khorana[11] ont décrit une situation analogue à celle rapportée ici en utilisant comme matrice le composé d(TTC)$_3$ dans un milieu d'incubation renfermant du MnCl$_2$ 10^{-3} M et du MgCl$_2$ 4×10^{-3} M. Ces auteurs n'ont observé une incorporation limitée d'ATP et de GTP telle qu'en fin de réaction 1 mole de nucléoside triphosphate se trouvait polymérisée par mole de nucléotide complémentaire présent dans la matrice. Comme dans le système que nous venons de décrire, ici encore la réaction de polymérisation une fois à son terme reprenait après l'addition d'une nouvelle quantité de matrice. Hirschbein, Dubert et Babinet[32] décrivent enfin un troisième cas où la matrice (poly A dans leurs études) ne fonctionne pas catalytiquement. L'intérêt de cet ensemble d'observations réside dans leur rapport éventuel avec le mécanisme de ponctuation qui régit la transcription du DNA in vivo. A cet égard, il est intéressant de rappeler tout d'abord que Haruna et Spiegelman[33] ont observé des effets importants des ions divalents sur la replication par la RNA replicase du phage Q$_\beta$ de RNA homologues

(RNA du phage Q$_\beta$) et hétérologues (RNA de TYMV, virus de la mosaïque jaune du navet). Ces auteurs ont expliqué leurs résultats par des effets différents du Mn^{++} et du Mg^{++} sur la spécificité de la RNA replicase du phage Q$_\beta$. Il nous semble possible que la différence de concentration des ions mise en jeu dans leurs expériences (4×10^{-3} M Mg^{++}, $0,13 \times 10^{-3}$ M Mn^{++}) intervienne dans la nature de la réaction. Dans une étude de la transcription du DNA du phage λ in vitro, Naono et Gros[34] ont constaté que la concentration en ions divalents influait sur la capacité de la RNA polymérase d'E. coli à discriminer entre les moitiés «droite» et «gauche» du DNA du phage λ.

Dans les deux types de réaction qui viennent d'être décrits avec l'(Ap)$_7$A comme matrice, le produit de polymérisation n'apparaît jamais à l'état libre dans le milieu d'incubation (Fig. 9). Cette situation est analogue à celle décrite par Bremer et Conrad [26] dans le cas de la synthèse du RNA messager lors de la transcription du DNA in vitro. La vitesse de sédimentation des complexes formés après incubation enzymatique en présence de MnCl$_2$ 10^{-3} M augmente au fur et à mesure que la réaction se poursuit bien que la longueur de chaîne du poly U synthétisé demeure à peu près constante (Figs 9a, 9b, Fig. 3). En outre, les complexes dont nous observons la formation au cours de la polymérisation (Fig. 9) sédimentent beaucoup plus rapidement que ceux obtenus en l'absence d'UTP (Fig. 10). Ces résultats trouvent une explication possible dans les travaux de Sternberger et Stevens [28], de Richardson, Slayter et Hall [29] et d'Antony, Zeszotek et Goldthwait [31], lesquels démontrent que l'interaction entre RNA polymérase et matrice est renforcée par la présence de nucléosides triphosphates.

Le fait que les complexes formés pendant la polymérisation de l'UTP présentent des coefficients de sédimentation très élevés suggère que ces complexes seraient formés par l'enchaînement de plusieurs molécules d'enzyme à une molécule de poly U. Nous avons d'ailleurs proposé une interprétation [35] selon laquelle le poly U au fur et à mesure qu'il émerge du complexe qu'il forme avec l'enzyme et la matrice fixerait des molécules d'enzyme libre. Nous supposons que le poly U en cours de formation «glisse» sur l'(Ap)$_7$A selon les modalités proposées pour expliquer d'autres cas de synthèse de polymères à longue chaîne sur des matrices de faible taille [13,36] ou la synthèse de poly A pendant la transcription du DNA dénaturé en présence d'ATP comme substrat [7]. Il est évident qu'en présence de fortes concentrations de cations divalents la continuation de la polymérisation exige des molécules libres de matrice. Néanmoins, il faut noter qu'une polymérisation à vitesse réduite peut se produire au-delà du rapport de 2 U/A si la réaction a lieu à 37° en présence d'une quantité élevée d'enzyme (Fig. 3). Cette dernière situation ressemble donc opérationellement à celle observée en présence de faibles concentrations de cations divalents (Fig.5 et 8). En résumé, la réaction d'incorporation d'UTP peut se poursuivre au-delà du stade équivalent à une polymérisation stœchiométrique (2 U/1 A) soit lorsque l'on ajoute une nouvelle quantité de matrice, soit en mettant en œuvre des quantités plus élevées d'enzyme et en élevant la température d'incubation. On peut donc penser que le «glissement» du produit de polymérisation sur la matrice peut être favorisé par sa combinaison avec des molécules libres de matrice ou d'enzyme. Il semble que la stabilité de la région d'appariement entre poly U et (Ap)₇A à l'endroit où la polymérisation a lieu ainsi que l'affinité du poly U pour la RNA polymérase varient en fonction de la température et de la concentration en cations divalents. Ainsi des molécules d'enzyme libres peuvent susciter un glissement lent à 37° en présence de $MnCl_2$ 4×10^{-3} M (Fig.3, courbe B) et un glissement beaucoup plus rapide à 25° en présence de $MnCl_2$ 10^{-3} M (Figs 5, 6, 8).

Enfin un point qui peut sans doute illustrer les relations entre l'état conformationnel de la RNA polymérase et son affinité pour le complexe matrice-produit est le suivant: toutes nos expériences ont été réalisées avec une préparation enzymatique obtenue selon la méthode de Furth, Hurwitz et Anders [14]. Ainsi préparé, l'enzyme a un coefficient de sédimentation d'environ 13 S (mesuré sur gradient de saccharose en présence de 0,05 M K maléate pH 8, Babinet et Dubert, communication privée). Or, nos essais pour retrouver un effet de la concentration en cations divalents sur la cinétique de synthèse du poly U (en présence d'oligoadénylate) en utilisant la RNA polymérase préparée selon Richardson [37] se sont avérés vains. Avec l'enzyme ainsi préparé[2], enzyme dont le coefficient de sédimentation est de 24 S [36] (en gradient de saccharose en présence de 0,05 M KCl, 0,01 M Tris pH 7,9 et 0,005 M $MgCl_2$) nous n'avons pas pu observer une polymérisation limitée par la quantité de matrice mise en jeu.

BIBLIOGRAPHIE

1. Krakow, J. S., et Ochoa, S., Proc. Natl. Acad. Sci. U.S. 49 (1963) 8.
2. Fox, C. F., et Weiss, S. B., J. Biol. Chem. 239 (1964) 175.
3. Fox, C. F., Robinson, W. S., Haselkorn, R., et Weiss, S. B., J. Biol. Chem. 239 (1964) 186.
4. Fox, C. F., Gumport, R. I., et Weiss, S. B., J. Biol. Chem. 240 (1965) 2101.
5. Tissières, A., Bourgeois, S., et Gros, F., J. Mol. Biol. 7 (1963) 100.
6. Gros, F., Dubert, J. M., Tissières, A., Bourgeois, S., Michelson, M., Soffer, R., et Legault, L., Cold Spring Harbor Symp. Quant. Biol. 28 (1963) 299.

7. Chamberlain, M., et Berg, P., Proc. Natl. Acad. Sci. U.S. 48 (1962) 81.
8. Hurwitz, J., Furth, J. J., Anders, M., et Evans, A., J. Biol. Chem. 237 (1962) 3752.
9. Falaschi, A., Adler, J., et Khorana, H. G., J. Biol. Chem. 238 (1963) 3080.
10. Leder, P., Clark, B. F. C., Sly, W. S., Pestka, S., et Nirenberg, M. W., Proc. Natl. Acad. Sci. U.S. 50 (1963) 1135.
11. Nishimura, S., Jacob, J. M., et Khorana, H. G., Proc. Natl. Acad. Sci. U.S. 52 (1964) 1494.
12. Byrd, C., Ohtsuka, E., Moon, M. W., et Khorana, H. G., Proc. Natl. Acad. Sci. U.S. 53 (1965) 79.
13. Mehrotra, B. D., et Khorana, H. G., J. Biol. Chem. 240 (1965) 1750.
14. Furth, J. J., Hurwitz, J., et Anders, M., J. Biol. Chem. 237 (1962) 2611.
15. Brahms, J., Michelson, M., et van Holde, K. E., J. Mol. Biol. 15 (1966) 468.
16. Michelson, M., J. Chem. Soc. (1958) 1957.
17. Munch Petersen, A., Acta. Chem. Scand. 9 (1955) 1523.
18. Fraenkel-Conrat, H., et Singer, B., Biochemistry, 1 (1962) 120.
19. Gros, F., Gilbert, W., Hiatt, H. H., Kurland, C., Riseborough, R. W., et Watson, J. D., Nature, 190 (1961) 581.
20. Lipsett, M. N., Heppel, L. A., et Bradley, D. F., Biochim. Biophys. Acta, 41 (1960) 175.
21. Lipsett, M. N., Heppel, L. A., et Bradley, D. F., J. Biol. Chem. 236 (1961) 857.
22. Schildkraut, C. L., Marmur, J., Fresco, J. R., et Doty, P., J. Biol. Chem. 236 (1961) PC2.
23. Hayashi, M., et Spiegelman, S., Proc. Natl. Acad. Sci. U.S. 47 (1961) 1564.
24. Maitra, U., et Hurwitz, J., Proc. Natl. Acad. Sci. U.S. 54 (1965) 815.
25. Stent, G. S., Bremer, H., Conrad, M. W., et Gaines, K., J. Mol. Biol. 13 (1965) 540.
26. Bremer, H., et Conrad, M. W., Proc. Natl. Acad. Sci. U.S. 51 (1964) 801.
27. Berg, P., Kornberg, R. D., Fancher, H., et Dieckmann, M., Biochem. Biophys. Res. Comm. 18 (1965) 932.
28. Stevens, A., Emery, A. J., et Sternberger, N., Biochem. Biophys. Res. Comm. 24 (1966) 929.
29. Richardson, J. P., Slayter, H. S., et Hall, C. E., J. Mol. Biol. 21 (1966) 83.
30. Richardson, J. P., J. Mol. Biol. 21 (1966) 115.
31. Antony, D. D., Zeszotek, E., et Goldthwait, D. A., Proc. Natl. Acad. Sci. U.S. 56 (1966) 1026.
32. Hirschbein, L., Dubert, J. M., et Babinet, C., European J. Biochem. 1 (1967) 135.
33. Haruna, I., et Spiegelman, S., Proc. Natl. Acad. Sci. U.S. 54 (1965) 1189.
34. Naono, S., et Gros, F., Cold Spring Harbor Symp. Quant. Biol. 31 (1966) Sous presse.
35. Gros, F., Naono, S., Rouvière, J., Hayes, D. H., et Cukier, R., Proceedings 2nd FEBS Meeting (Vienna), 1965, Volume 4, 29 (H. Tuppy Ed., Pergamon Press, Oxford, 1966.)
36. Kornberg, A., Bertsch, L. L., Jackson, J. F., et Khorana, H. G., Proc. Natl. Acad. Sci. U.S. 51 (1964) 315.
37. Richardson, J. P., Proc. Natl. Acad. Sci. U.S. 66 (1966) 1616.

D. H. Hayes, R. Cukier et F. Gros
Service de Physiologie Microbienne
Institut de Biologie Physico-Chimique
13. Rue Pierre Curie
75 Paris-5, France

[2] Cet enzyme nous a été gracieusement fourni par le docteur J. Richardson.

European J. Biochem. 1 (1967) 135—140

Affinité différentielle de la RNA polymérase
pour divers polyribonucléotides synthétiques

L. Hirschbein, J.-M. Dubert et C. Babinet

Services de Biochimie Cellulaire et de Génétique Cellulaire, Institut Pasteur, Paris

(Reçu le 5 décembre 1966)

Using synthetic polyribonucleotides templates as models, the problems of attachment or detachment of DNA dependent-RNA polymerase to its template have been investigated. In addition, the hypothesis that sites of initiation or termination of transcription might be determined by some specific base sequences was tested with these models.

In order to allow a distinction between the transcription process and its first step, which is the binding of the enzyme to its template, the inhibition by different polyribonucletides of the transcription of DNA was examined. A wide range of relative effects was found (Fig.3). It is remarkable that the polyribonucleotides Poly G and Poly I showing the highest affinities for the enzyme, as well as the homopolymer pair Poly (G + C) of low affinity could not be transcribed into their complementary sequences when used as template.

When using the polyribonucleotides as template instead of DNA, the K_m for triphosphate binding is about 30 fold higher. It was shown in inhibition experiments that for a given system, a polyribonucleotide was more efficient as an inhibitor as the concentration of its complementary triphosphate was increased (Fig.4): this result shows that the triphosphate contributes to the stability of the polyribonucleotide-enzyme complex.

The properties of Poly I or Poly G of having very high affinity for RNA polymerase and of not being transcribed could be used to trap the enzyme not bound to its template at any time before or during the transcription process. Preliminary results are reported in Fig.5.

The fact that Poly (G + C) has barely detectable affinity for RNA polymerase can explain the kinetic results obtained when Poly C is used as a template; the G/C ratio reaches one at the end of the reaction and no further incorporation of GMP is observed when enzyme or GTP is added. Only the addition of Poly C can allow resumption of incorporation of the complementary nucleotides (Fig.2A). Consistent with these results is the observation that the enzyme and the product of the reaction migrate separately in a sucrose gradient centrifugation. Our studies show that the relative affinity of RNA polymerase for different homopolymer pairs Poly (A + U), Poly (G + C) is dependent on the nature of the bases involved in these ordered double helical structures.

Un des problèmes majeurs dans l'étude de la transcription du DNA est de déterminer la nature chimique de la ponctuation qui doit, logiquement, marquer le commencement et la fin de l'unité de transcription. L'hypothèse de plus simple est que les points d'initiation et de terminaison correspondraient à des séquences particulières de nucléotides, et donc à certaines structures caractéristiques des chaînes polynucléotidiques pour lesquelles la RNA polymérase, utilisant normalement le DNA comme matrice, aurait des affinités très différentes. La fixation et le détachement électifs de l'onzyme détermineraient ainsi respectivement le commencement et la fin de l'unité de transcription.

L'existence de sites préférentiels d'attachement de la RNA polymérase au DNA natif a été montrée par l'analyse de la stoechiométrie d'attachement de l'enzyme à divers DNA de bactériophage [1,2]. De plus, la relative abondance d'adénine à l'extrémité 5′ de la chaîne de RNA synthétisé traduit la présence préférentielle de thymine au point d'initiation [3,4]. Cependant l'identification directe des éléments de séquence constituant la ponctuation sur le DNA reste difficilement accessible.

Abréviations non usuelles. Homopolymères des ribonucléotides d'adénine, de guanine, de cytosine, d'uracile et d'inosine, Poly A, Poly G, Poly C, Poly U et Poly I, respectivement; RNA de transfert, tRNA.

Enzymes. RNase, ou ribonucléase, ou ribonucléate pyrimidine-nucléotido-2′-transférase (EC 2.7.7.16); RNA polymérase, ou RNA nucléotidyltransférase, ou nucléosidetriphosphate: RNA nucléotidyltransférase (EC 2.7.7.6).

L'utilisation comme matrice de polynucléotides synthétiques apporte dans cette étude une simplification en même temps qu'une variété considérable puisque l'on dispose de séquences connues, qu'il s'agisse d'homopolymères, d'homopolymères pairs[1] ou de séquences alternées. Divers auteurs ont montré que certains polyribonucléotides peuvent se combiner à la RNA polymérase inhibant ainsi la réaction de transcription du DNA [5,6] et peuvent également intervenir comme matrice dans la synthèse enzymatique de chaînes complémentaires [7—9]. Notre intérêt s'est porté sur les interactions entre RNA polymérase et polyribonucléotides à la suite des observations de Hirschbein et Fresco [10] et de Haselkorn et Fox [11] sur la structure des homopolymères (G + C). A la différence d'autres polymères ordonnés de ribonucléotides, le polymère (G + C) s'est révélé pratiquement dénué d'affinité pour la RNA polymérase et un argument nous semblait ainsi pouvoir être apporté en faveur du rôle supposé d'une reconnaissance directe par la RNA polymérase de certains éléments de séquence de la matrice. Le travail et son extension à d'autres polyribonucléotides est ici présenté. Les phénomènes de fixation et de détachement de l'enzyme ont été distingués de la réaction de transcription proprement dite de ces polymères utilisés comme modèles.

MATÉRIEL ET MÉTHODES

Les nucléosides triphosphates radioactifs (^{14}C et ^{3}H) proviennent de Schwartz Bioresearch, les nucléosides triphosphates non marqués de Pabst Laboratories.

Les Poly A, Poly C et Poly U proviennent de Miles Chemical Co. Le Poly G a été aimablement offert par Mme M. Grunberg-Manago. Le Poly (G + C) a été synthétisé enzymatiquement comme il est indiqué dans le texte, à partir de Poly C et de GTP. La réaction est arrêtée par addition de $2/3$ de volume de phénol fraîchement distillé. Après agitation à 4° à l'obscurité et centrifugation, la phase aqueuse est dialysée contre une solution NaCl 0,1 M, EDTA 0,2 M de pH 7, puis contre une solution de Tris 10^{-2} M pH 7,2. Le produit est soumis à l'action de la RNase pancréatique (0,22 μg par unité d'absorbance à 271 mμ) pendant 18 heures à 22°. La RNase est ensuite éliminée par passage sur une colonne de carboxy-méthyl cellulose équilibrée avec une solution Tris 5×10^{-2} M, NaCl 10^{-2} M, pH 7,3 [12]. La solution de polyribonucléotide est alors à nouveau dialysée successivement contre des solutions NaCl 0,1 M, Tris 10^{-2} M pH 7,2, puis de Tris 10^{-2} M, pH 7,2.

Les Poly (A + U) et Poly (A + 2U) ont été préparés par la méthode des mélanges continus dans du tampon Tris 10^{-3} M, pH 7, NaCl 0,12 M [13].

---

[1] Les homopolymères pairs sont formés par l'appariement des homopolymères.

Les constantes de sédimentation des polynucléotides ont été mesurées à l'aide d'une centrifugeuse Spinco modèle E, à 20°, en utilisant le système optique à absorption, la vitesse de rotation étant de 50,740 tours/min. Le solvant utilisé était du tampon tris 10^{-2} M, pH 7,2.

Le DNA de thymus de veau a été préparé selon la méthode de Kay et al. [14] suivie d'une digestion par la pronase (100 μg par ml, 16 heures à 37°), d'un nouveau traitement par le lauryl sulfate et de deux déprotéinisations selon la technique de Sevag et al. [15].

La RNA polymérase utilisant normalement le DNA comme matrice a été préparée selon la méthode de Hurwitz [15a] modifiée. L'activité spécifique de l'enzyme récemment préparé est en moyenne de 2500 à 3000 unités par mg de protéine. L'unité est définie comme la quantité d'enzyme qui catalyse, en 20 minutes et à 37°, l'incorporation d'une mμM de nucléotide, le DNA étant utilisé comme matrice.

Les concentrations des différents réactifs dans les conditions habituelles d'incubation pour la transcription du DNA sont les suivantes: tampon maléate pH 7,4 5×10^{-2} M, β-mercapto-éthanol 4×10^{-3} M. Le volume total est de 0,1 ml. La quantité saturante de DNA pour ce volume était de 5 μg. Dans le cas où des polyribonucléotides ont été utilisés comme matrices, la quantité ajoutée est indiquée dans la légende des figures ainsi que la concentration et la radioactivité spécifique du nucléoside triphosphate complémentaire.

L'incubation est effectuée à 37°. La réaction est arrêtée par addition de 3 ml de TCA à 5% refroidi dans la glace. On filtre sur millipore, on lave trois fois avec du TCA à 5% refroidi et une dernière fois avec du TCA à $0,5\%$ refroidi. La radioactivité est déterminée au compteur à scintillation Packard Tri-Carb.

RÉSULTATS

L'étude de l'incorporation de nucléotides par la RNA polymérase en présence de polynucléotides nous a conduit à mettre en évidence différents degrés d'affinité de cet enzyme en fonction des bases qui interviennent dans l'arrangement polynucléotidique.

Nous avons d'abord utilisé le Poly C comme matrice. L'analyse de ce système a révélé que le produit de la réaction est l'homopolymère pair Poly (G + C) [9,10]. Dans ce système où le Poly C est utilisé comme matrice, le K_m pour le GTP a été déterminé par la méthode graphique de Lineweaver et Burk.

Les mesures de vitesses initiales pour des concentrations de GTP comprises entre $1,3 \times 10^{-4}$ M et 13×10^{-4} M permettent de définir la valeur du K_m qui est de 5×10^{-4} M, toutefois il faut indiquer que pour des concentrations de GTP supérieures à $1,3 \times 10^{-3}$ M on observe une diminution de la vitesse de réaction. Cette valeur de K_m, 5×10^{-4} M, est élevée, environ 30 fois supérieure à celle que nous déterminons quand les quatre nucléosides triphosphates sont présents, le DNA étant la matrice.

Si l'on compare la vitesse d'incorporation des nucléotides soit avec Poly C, soit avec le DNA comme matrice, les concentrations de GTP étant respectivement de $1,3 \times 10^{-3}$ M et de $2,6 \times 10^{-4}$ M, on remarque que leur ordre de grandeur est le même: environ quatre fois moins de GMP est incorporé par unité de temps et d'activité enzymatique lorsque le DNA est utilisé comme matrice, l'incorporation portant sur les quatre types de nucléotides.

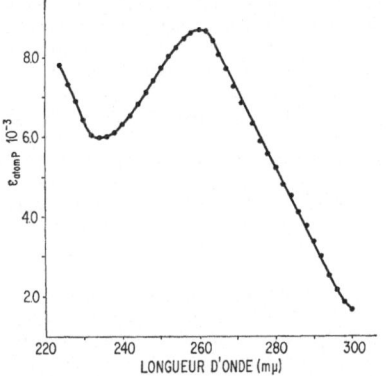

Fig. 1. *Spectre du Poly (G + C) purifié, en tampon phosphate 2×10^{-2} M, pH 7,5*

Le rapport du GMP incorporé au Poly C ajouté comme matrice tend vers 1 lorsque la réaction évolue à son terme. Cette valeur apparaît comme une limite supérieure. Dans le cas où, avec certaines préparations d'enzyme, nous avons trouvé des rapports G/C nettement inférieurs à l'unité (jusqu'à 0,6) nous avons pu mettre en évidence dans ces préparations la présence de nucléases qui dégradaient une partie du Poly C au cours de l'incubation.

Les polyribonucléotides présents à la fin de la réaction ont été isolés. Leurs propriétés sont identiques à celles de l'homopolymère pair $(G + C)$ obtenu par synthèse enzymatique catalysée par la RNA polymérase de *Micrococcus lysodeikticus* [9,10]. En particulier le spectre d'absorption dans l'ultraviolet présente un maximum à 261 mμ et un minimum à 234 mμ (Fig. 1). D'autre part cet homopolymère est très résistant à la ribonucléase: alors que 0,22 μg de RNase/ml agissant sur une solution de Poly C ayant une densité optique de 1 à 271 mμ provoque, en moins d'une minute, à 22°, une hyperchromie de 32%, aucune hyperchromie n'est observée même après 18 heures d'incubation dans les mêmes conditions et aux mêmes concentrations pour l'homopolymère pair G + C isolé, formé au terme de la réaction de transcription.

Enfin la constante de sédimentation s_{20} de $(G + C)$ varie suivant nos préparations entre 4 et 5.

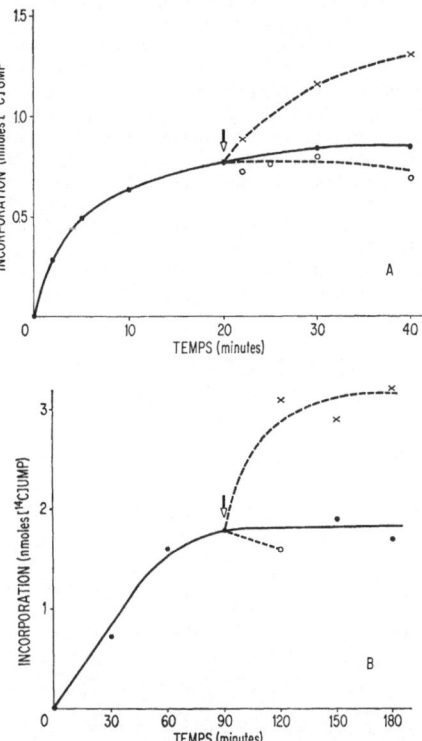

Fig. 2. *Cinétique d'incorporation du nucléotide complémentaire en présence du Poly C ou du Poly A.* (a) Poly C ajouté au temps zéro: 1 mμM. Le [^{14}C]GTP, d'activité spécifique 1700 coups/min/nM, est ajouté à la concentration finale de $5,8 \times 10^{-4}$ M. A la vingtième minute de l'incubation, on ajoute soit 1 mμM de Poly C ($\times$——$\times$), soit 0,6 unités d'enzyme (O——O). (b) Poly A ajouté au temps zéro: 0,98 mμM. Le [^{14}C]UTP d'activité spécifique 1650 coups/min/nM est ajouté à la concentration finale de 3×10^{-4} M. RNA polymérase: 2,4 unités. A 90 minutes, on ajoute soit 0,98 mμM de Poly A ($\times$——$\times$), soit 2,4 unités d'enzyme (O——O)

Le Poly C, utilisé comme matrice a une constante de sédimentation s_{20} égale à 2.

Nous avons trouvé comme Haselkorn [11] que l'homopolymère pair $G + C$ ne peut intervenir comme matrice pour l'incorporation de GMP ni de CMP. Ce résultat cependant est compatible avec deux interprétations: soit que les propriétés du polymère ne permettent pas la transcription enzymatique bien que l'association entre polymère et enzyme ait lieu, soit que la RNA polymérase n'ait que peu d'affinité pour ces structures. Les résultats des expériences que nous allons décrire sont en faveur de cette deuxième hypothèse.

Le premier argument est d'ordre cinétique. Lorsque la réaction de transcription d'une quantité donnée de Poly C est arrivée à son terme, l'incorporation de GMP peut à nouveau avoir lieu si on ajoute une nouvelle quantité de Poly C; la vitesse d'incorporation, et donc la quantité d'enzyme actif, est alors pratiquement identique à celle observée au cours de la première phase de la réaction (Fig.2A). Si par contre, dans une expérience parallèle, une nouvelle quantité d'enzyme — et non plus de Poly C — est ajoutée, aucune nouvelle incorporation de GMP n'est observée. On sait d'autre part que la transcription du Poly A en présence d'UTP se poursuit jusqu'à ce que le rapport U/A soit égal à 2

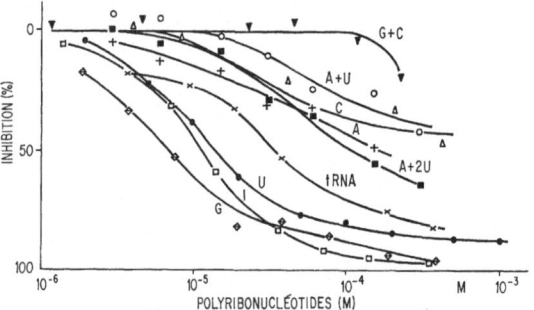

Fig.3. *Inhibition de la transcription du DNA de thymus de veau (1,6 μg) en fonction de la concentration des divers polyribonucléotides.* En présence de Poly I, Poly U, Poly A, Poly (A + U) et Poly (A + 2 U), le [14C]CTP a été utilisé (2,8 × 10⁻⁴ M). En présence de Poly G, Poly C, Poly (G + C) et du t-RNA, le [3H]UTP a été utilisé (3,2 × 10⁻⁴ M)

[6,7]. Lorsque ce rapport est atteint, une nouvelle incorporation de UMP accompagne l'addition de Poly A, mais non l'addition d'enzyme (Fig.2B). Les résultats obtenus avec le Poly A ou le Poly C comme matrice sont donc similaires. Notons que si le DNA avait été utilisé comme matrice dans des conditions par ailleurs analogues, c'est, inversement, l'addition d'enzyme au moment où la réaction approche de son terme qui eut provoqué une nouvelle synthèse de RNA alors que l'addition au même moment de DNA aurait été sans effet. Donc, contrairement à ce que l'on observe avec le DNA, l'enzyme reste disponible à la fin de la réaction de transcription de ces polyribonucléotides.

La centrifugation de zone en gradient de saccharose du milieu d'incubation où s'est effectuée la synthèse de l'homopolymère (G + C) montre que le produit formé a un coefficient de sédimentation

moyen de 5 à 6, ce qui exclut son association avec la RNA polymérase que nous avons trouvé avoir un coefficient de sédimentation de 14 S en utilisant comme repère la catalase. Le produit est détecté par la radioactivité du [14C]GMP incorporé, la RNA polymérase par son activité enzymatique. L'addition de dodécylsulfate de sodium à une concentration de 0,4% et de ClNa 0,2 M [16] ne modifie pas le profil obtenu.

Enfin des expériences d'inhibition mettent en évidence de façon directe des affinités différentielles de différents polyribonucléotides pour la RNA polymérase et tout particulièrement la trés faible affinité de l'enzyme pour l'homopolymère (G + C) (Fig.3). Le système dont on étudie l'inhibition est constitué de DNA comme matrice, à une concentration non saturante (2 × 10⁻⁵ M) et des quatre nucléosides triphosphates: le nucléoside triphosphate radioactif est choisi parmi ceux qui ne sont pas complémentaires des polyribonucléotides ajoutés comme inhibiteurs, afin que l'incorporation observée ne corresponde qu'à la transcription du DNA². Les résultats présentés dans la Fig.3 montrent que les polyribonucléotides Poly G, Poly I et Poly U ont une affinité élevée pour l'enzyme, une concentration de l'ordre de 10⁻⁵ M entraînant une inhibition de 50%.

Pour des concentrations comprises entre 10⁻⁴ et 10⁻³ M, une inhibition de 50% est obtenue avec les Poly A, Poly C, Poly (A + 2U) et Poly (A + U). De tous les polyribonucléotides que nousav ons étudiés, le Poly (G + C) se distingue donc par son absence presque totale d'activité inhibitrice aux concentrations utilisées.

A titre de comparaison, nous avons déterminé au cours de cette même série d'expériences le pouvoir inhibiteur du RNA de transfert. L'inhibition à 50% est obtenue pour une concentration de 3 × 10⁻⁵ M. L'affinité est donc légèrement inférieure à celle du Poly I, du Poly G ou du Poly U.

Les expériences d'inhibition que nous venons de rapporter ont été faites en utilisant des concentrations de triphosphates qui sont optimales pour la transcription du DNA. Mais on sait que la transcription des polyribonucléotides exige des concentrations de triphosphates plus élevées. Or les nucléosides triphosphates complémentaires s'associent au complexe enzyme matrice favorisant la formation du complexe ternaire fonctionnel ainsi que Chamberlin et Berg [17] l'ont montré dans des expériences d'inhibition où des très faibles concentrations des triphosphates autres que l'ATP arrêtaient la synthèse

---

² Il est important de préciser que c'est au mélange réactionnel contenant déjà le DNA-matrice et le polyribonucléotide qu'est ajouté, en dernier, l'enzyme.

de Poly A en présence de DNA dénaturé. On pouvait donc prévoir que l'inhibition de la transcription du DNA par les polyribonucléotides devait être fonction de la concentration des triphosphates.

Les résultats présentés dans la Fig. 4 confirment cette prévision: l'inhibition de 50% de la transcription de DNA est obtenue pour $2,5 \times 10^{-5}$ M de Poly A et pour 9×10^{-5} M de Poly A selon que la concentration de l'UTP est de $2,1 \times 10^{-3}$ M et de 10^{-4} M respectivement.

La non transcription du Poly (G + C) apparaît comme une conséquence logique de sa très faible affinité pour l'enzyme. Mais inversement l'existence d'une affinité même élevée des polyribonucléotides pour l'enzyme n'implique pas nécessairement que le processus de transcription ait lieu. En effet, parmi les homopolymères présentant une affinité importante pour l'enzyme, seul le Poly U est normalement transcrit en présence d'ATP alors que le Poly I et le Poly G ne donnent pas lieu à la synthèse de Poly C, à 37°, en présence de CTP.

Le Poly I et le Poly G ont donc deux propriétés remarquables: d'une part leur affinité élevée pour la RNA polymérase, d'autre part leur inaptitude à être transcrits. Ces propriétés doivent permettre leur utilisation au cours de la réaction de transcription pour éliminer au moment choisi l'enzyme libre, sans altérer le fonctionnement de l'enzyme qui se trouve attaché à sa matrice.

Une expérience préliminaire est présentée dans la Fig. 5. On a déterminé d'abord la quantité de Poly I qui entraîne une inhibition complète d'un système de transcription du DNA si elle est ajoutée avant l'enzyme au mélange réactionnel contenant le DNA. Si cette même quantité de Poly I est ajoutée au temps zéro de l'incubation, aussitôt après l'addition de l'enzyme au mélange complet, l'inhibition est partielle. Enfin si l'addition de Poly I est faite en cours d'incubation, par exemple après 10 minutes à 37°, on observe que la synthèse de RNA est peu modifiée pendant les 5 à 10 minutes suivantes, puis se poursuit à un taux très diminué. Si l'incorporation observée traduit bien l'activité de l'enzyme fixé au DNA au moment de l'addition du Poly I, ce type d'expérience peut apporter une simplification importante dans l'étude cinétique de la fixation de l'enzyme à sa matrice. Cependant, il nous faut préciser dans quelle mesure la formation d'un complexe entre la matrice et le Poly I peut participer à l'inhibition observée.

DISCUSSION

Qu'il s'agisse d'homopolymères ou d'homopolymères pairs, la nature des bases qui les constituent et la structure secondaire qui en résulte, déterminent les possibilités de transcription de ces matrices par la RNA polymérase. L'observation qu'un polyribonucléotide ne peut être transcrit ne renseigne pas cependant sur le degré de son affinité pour l'en-

zyme. Nos expériences ont eu pour but de déterminer les affinités différentielles des polyribonucléotides pour la polymérase, en particulier en mesurant l'inhibition de la réaction de transcription du DNA en présence de concentrations variables de divers homopolymères. Il est ainsi apparu que l'on pouvait distinguer trois groupes de polynucléotides si l'on

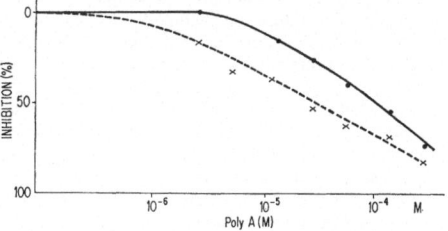

Fig. 4. *Inhibition de la transcription du DNA de thymus de veau (1,6 μg) par des concentrations variables de Poly A, en présence d'UTP 10^{-4} M (●——●) et $2,1 \times 10^{-3}$ M (×——×). Dans les deux cas, le [^{14}C]GTP était présent à la concentration de $1,7 \times 10^{-4}$ M*

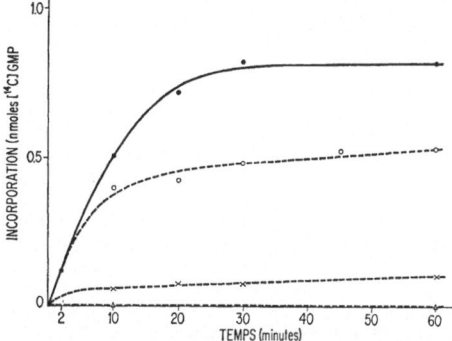

Fig. 5. *Cinétique de l'inhibition par le Poly I de la transcription du DNA de thymus de veau. DNA: 1,6 μg; [^{14}C]GTP: $1,7 \times 10^{-4}$ M. Le Poly I, à une concentration finale de 7×10^{-4} M, est ajouté au temps zéro, soit avant l'addition d'enzyme (△——△), soit aussitôt après (×——×) ou à la deuxième minute de l'incubation (○——○); incorporation en l'absence de Poly I (●——●)*

tient compte à la fois de leur affinité pour l'enzyme et de leur aptitude à être transcrits:

a) le poly G et le poly I ont une affinité très élevée et ne sont pas transcrits dans nos conditions expérimentales.

b) le poly (G + C) a une affinité très faible pour la RNA polymérase et n'est pas transcrit,

c) les autres homopolymères et homopolymères pairs que nous avons étudiés (poly U, poly C, Poly A, A + U) ont de l'affinité pour l'enzyme et sont transcrits.

Remarquons que l'inhibition observée peut ne pas résulter seulement de la fixation des polyribonucléotides à l'enzyme, mais être due également à la formation d'un complexe entre le DNA utilisé comme matrice et les polymères. S'il en était ainsi, ce complexe qui se formerait, indépendamment de la présence de l'enzyme, exercerait son effet inhibiteur de la même manière, quel que soit l'ordre d'addition du DNA, de l'enzyme et de polyribonucléotide, or ce n'est pas ce que nous observons: dans le cas du poly I, la quantité de ce polymère qui provoque une inhibition complète lorsqu'elle est ajoutée au DNA immédiatement avant l'enzyme, ne produit qu'une inhibition partielle s'il est ajouté alors que l'enzyme a déjà formé un complexe avec le DNA. D'autre part, si le Poly U est utilisé comme matrice, et non plus le DNA, des résultats similaires sont obtenus en fonction de l'ordre d'addition des constituants; nous avons vérifié par spectrophotométrie que le Poly U et le Poly I ne forment pas de complexes dans ces conditions.

Ces expériences apportent donc des arguments en faveur du rôle essentiel, sinon toujours exclusif, joué par la formation d'un complexe entre l'enzyme et les polyribonucléotides dans les inhibitions décrites.

Inversement, aucune inhibition de la réaction de transcription n'est provoquée par l'addition de polyribonucléotides (G + C). La structure de l'homopolymère pair (G + C) ne présente donc qu'une affinité très faible pour l'enzyme et l'on peut voir là la raison pour laquelle, dans nos conditions expérimentales, on n'a pas pu obtenir au cours de la transcription du Poly C un produit dont le rapport G/C soit supérieur à l'unité. On sait cependant que par la méthode des mélanges continus, il a été possible d'obtenir des complexes caractérisés par un rapport G/C supérieur à l'unité [18,19] de même que, dans certains cas, par voie enzymatique [20]. Les raisons de ces observations apparemment contradictoires peuvent être recherchées soit dans les propriétés des polyribonucléotides, en particulier dans leurs longueurs [19], soit dans les propriétés de la RNA polymérase elle-même, qui pourraient différer selon les méthodes de préparation employées, soit enfin dans les différences de concentrations en ions Mn dans le milieu d'incubation [20] dont l'importance est démontrée par Hayes et al. [21].

Des arguments d'ordre cinétique mettaient déjà en évidence que la fixation de l'enzyme à l'homopolymère pair (G + C), si elle avait lieu, était réversible et de moindre affinité que la combinaison de l'enzyme et du Poly C. En effet, dans les conditions habituelles de nos expériences, le nombre de molécules d'enzyme est de 10 à 100 fois inférieur au nombre des molécules de Poly C présentes dans le milieu d'incubation. Il est nécessaire d'admettre que l'enzyme se détache du produit après sa transcription pour se fixer à de nouvelles molécules de Poly C. Ceci n'implique aucunement une absence d'affinité de l'enzyme pour le produit de la réaction. D'ailleurs, une situation similaire est observée, par exemple pour la transcription du Poly A alors que le produit formé présente une affinité mesurable pour l'enzyme.

Seules les expériences d'inhibition montrent que, dans les conditions expérimentales où s'effectue la transcription du DNA, deux structures ordonnées, les homopolymères pairs (A + U) et (G + C) ont des affinités très différentes pour l'enzyme: la fixation de l'enzyme est donc fonction de la nature des bases participant à la structure régulière de la double hélice.

Remarquons que c'est la structure (G + C), que l'on sait présenter une stabilité extrême, qui a une très faible affinité pour l'enzyme. Cette observation serait en accord avec l'hypothèse proposée par Stent selon laquelle l'ouverture localisée de la double hélice au niveau de l'enzyme, nécessaire au processus de transcription, serait indispensable également à la fixation de l'enzym.

Ce travail a bénéficié de l'aide des National Institutes of Health des Etats-Unis, de la Délégation Générale à la Recherche Scientifique et Technique et de l'Euratom.

Nous remercions vivement Mme Grunberg-Manago d'avoir mis à notre disposition le Poly G.

BIBLIOGRAPHIE

1. Bremer, H., Konrad, M., et Bruner, R., J. Mol. Biol. 16 (1966) 104.
2. Richardson, J. P., J. Mol. Biol. sous presse.
3. Maitra, V., et Hurwitz, J., Proc. Natl. Acad. Sci. US, 54 (1965) 815.
4. Bremer, H., Konrad, M., Gaines, K., et Stent, G., J. Mol. Biol. 13 (1965) 540.
5. Gros, F., Dubert, J.-M., Tissières, A., Bourgeois, S., Soffer, R., et Legault, L., Cold Spring Harbor Symp. Quant. Biol. 28 (1963) 299.
6. Fox, C. F., Gumport, R. I., et Weiss, S. B., J. Biol. Chem. 240 (1965) 2101.
7. Nakamoto, T., et Weiss, S. B., Proc. Natl. Acad. Sci. US, 48 (1962) 880.
8. Krakow, J. S., et Ochoa, S., Proc. Natl. Acad. Sci. US, 49 (1963) 88.
9. Fox, C. F., Robinson, W. S., Haselkorn, R., et Weiss, S. B., J. Biol. Chem. 239 (1964) 186.
10. Hirschbein, L., et Fresco, J., Communication orale au Symposium sur les acides ribonucléiques, Société de Chimie Biologique, Paris 26 février 1965.
11. Haselkorn, R., et Fox, C. F., J. Mol. Biol. 13 (1965) 780.
12. Taborsky, G., J. Biol. Chem. 234 (1959) 2652.
13. Massoulié, J., Thèse Doctorat d'Etat, Paris 1966.
14. Kay, E. R. M., Simons, N. S., et Dounce, A. L., J. Am. Chem. Soc. 74 (1952) 1724.
15. Sevag, M. G., Lackman, D. B., et Smolens, J., J. Biol. Chem. 124 (1938) 425.
15a.Hurwitz, J., communication personnelle.
16. Bremer, H., et Konrad, M., Proc. Natl. Acad. Sci. US, 51 (1964) 801.
17. Chamberlin, M., et Berg, P., J. Mol. Biol. 48 (1964) 708.
18. Fresco, J., Informational Macromolecules (edited by H. J. Vogel, V. Bryson et J. O. Lampen), Academic Press, New York 1963, p. 121.
19. Pochon, F., et Michelson, A. M., Proc. Natl. Acad. Sci. US, 53 (1965) 1425.
20. Chamberlin, M. J., Federation Proc. 24 (1965) 1446.
21. Hayes, D., Cukier, R., et Gros, F., 1 (1967) 125.

L. Hirschbein, J.-M. Dubert et C. Babinet
Services de Biochimie Cellulaire et de Génétique Cellulaire
Institut Pasteur
25 rue du Docteur Roux
75 Paris-15, France

European J. Biochem. 1 (1967) 141—146

Ion Transport in Liver Mitochondria

3. Stoicheometry of Proton Release During Aerobic Calcium Ion Translocation

C. Rossi, G. F. Azzone, and A. Azzi

Unit G. Vernoni for the Study of Physiopathology,
Institute of General Pathology, University of Padova

(Received December 15, 1966)

The H^+/Ca^{++} ratio during the aerobic uptake of Ca^{++} by rat liver mitochondria, oxidizing endogenous substrates or succinate, increased with the increase of the Ca^{++}/protein ratio. At high protein concentrations an apparent "reuptake" of the H^+ ejected was observed. Mitochondrial aging also resulted in a decrease of the H^+/Ca^{++} ratio.

Succinate caused a decrease of the H^+ ejection. The effect of succinate was dependent on the pH and on the concentration of Ca^{++}. A binding of succinate to the mitochondria was observed parallel to the decrease of the H^+/Ca^{++} ratio from 2 to 1.

Addition of Ca^{++} caused a release of K^+ which was slow in fresh mitochondria and rapid in valinomycin treated mitochondria. The K^+/Ca^{++} ratio was 2 in both cases.

A H^+/Ca^{++} ratio of about 2 was usually observed when the aerobic uptake of Ca^{++} was started either by the addition of mitochondria to a Ca^{++} containing medium or by the addition of succinate to Ca^{++}- and rotenone-treated mitochondria.

In experiments with isolated rat liver mitochondria, Saris observed ejection of H^+ during aerobic Ca^{++} uptake in 1 to 1 ratio [1]. Subsequent studies conducted in many laboratories [2—5] confirmed this observation, by showing that under a great variety of experimental conditions the H^+/Ca^{++} ratio was in the vicinity of 1. A H^+/Ca^{++} ratio of 1 was therefore considered a characteristic of the process of Ca^{++} uptake. This conclusion was not modified after the observation of Rasmussen et al. [6] that permeant anions decrease the H^+/Ca^{++} ratio.

More recently however other reports appeared which suggested some reservation on the above conclusion. First, the H^+ ejection was decreased when either the pH or the ionic strength of the medium were increased [7—8]. Secondly, the H^+/Ca^{++} ratio was increased to 2 when the uptake of Ca^{++} was initiated by the addition of succinate to Ca^{++}- and rotenone-pretreated mitochondria [9]. Third, a H^+/Ca^{++} ratio of 2 was also observed when buffers were omitted from the medium [9]. A similar observation was made by Wenner with Sr^{++} [10].

The present investigation was started in the hope of finding a rationale for the variations of the H^+/Ca^{++} ratios under the various experimental conditions. It will be shown that the H^+/Ca^{++} ratio increases with the increase in the Ca^{++}/protein ratio. Data will be also reported concerning the effects of succinate, of

the release of intramitochondrial K^+ and of changing the order of addition on the H^+/Ca^{++} ratio.

The effect of the pH and of the ionic strength of the medium, in causing a decrease of the H^+ ejection is analyzed in a subsequent paper.

EXPERIMENTAL

Rat liver mitochondria prepared in 0.5 mM EGTA and then washed twice without EGTA were used in all experiments [11].

H^+ ejection was measured with a Beckman glass electrode and Beckman pH meter (Expandomatic) connected with a recorder. The uptake of Ca^{++} was measured by determining the amount of $^{45}Ca^{++}$ either bound to a mitochondrial pellet or left in the supernatant after a fast centrifugation. The mitochondrial pellet was dissolved in 1 ml of 1 N formic acid, then plated, dried and counted. Aliquots of the supernatant, taken before and after the mitochondrial uptake of Ca^{++}, were plated with a large volume of water, then dried and counted. In many experiments the determinations of Ca^{++} were carried out on both pellets and supernatants. The time for the sedimentation of the mitochondria from the incubation medium was considered to be usually less than 20 seconds.

Succinate was determined by using [1,4-^{14}C] succinate obtained from Amersham. Chloride was determined by using $^{36}Cl^-$. K^+ was measured as described previously [11].

Non-standard Abbreviation. Ethyleneglycol tetraacetate, EGTA.

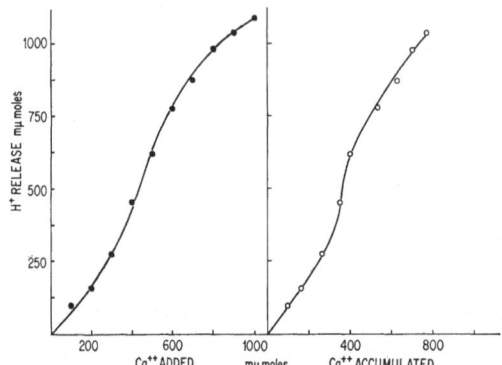

Fig.1. *H+ ejection at various Ca++ concentrations.* The medium contained, 5 mM Tris-Cl⁻ pH 7.2, 4 mM succinate Na, 1.5 μM rotenone, 0.25 M sucrose to a final volume of 2.0 ml and 7.6 mg protein. Temperature 22°

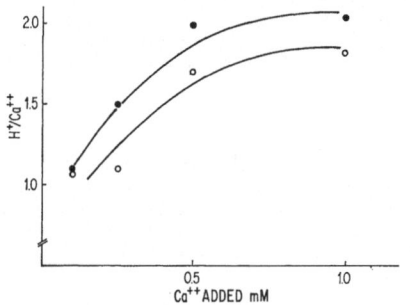

Fig.2. *Effect of Ca++ concentration on the H+/Ca++ ratio.* Experimental conditions as in Fig.1. The concentration of succinate was 250 μM in O——O and 5 mM in ●——●

RESULTS

Effect of the Ca++/Protein Ratio and of Aging on the H+/Ca++ Ratio

Addition of varying amount of Ca++ to mitochondria incubated in 0.25 M sucrose, oxidizing succinate as a substrate, resulted in a H+ ejection which did not increase parallel to the increase of the Ca++ added. A sigmoid curve was obtained when the H+ ejected was plotted against the Ca++ added or accumulated (Fig.1). At the higher Ca++ concentrations the H+/Ca++ ratio increased towards a value of 2 (Fig.2).

The H+/Ca++ ratio was dependent on the Ca++/protein ratio also when Ca++ was added to mitochondria oxidizing endogenous substrates (Fig.3). At high protein concentration (Fig.4) the rapid proton efflux was immediately followed by a partial proton influx (*cf.* Wenner [10]). At lower protein concentrations, the rate of the proton efflux was

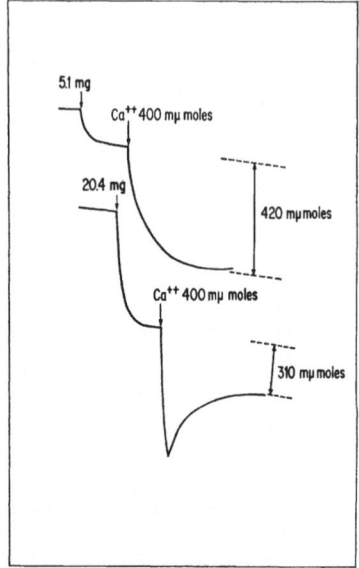

Fig. 3. *H+ ejection at various protein concentrations.* Experimental conditions as in Fig.1, pH was 7.4. The reaction was started by addition of 200 μM Ca²⁺

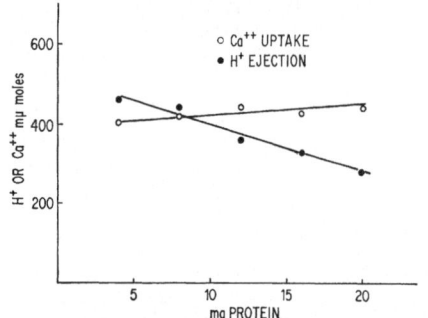

Fig.4. *H+ reuptake at high protein concentrations.* The medium contained 0.25 M sucrose and 5 mM Tris-Cl⁻, pH 7.2

slower, and was not followed by a proton reuptake.

The relation between this proton reuptake and the "oscillation" described by Carafoli *et al.* [12] will be discussed later.

Aging of the mitochondria also resulted in a decrease of the H+/Ca++ ratio. In the experiment reported in Fig.5, H+ ejection and Ca++ uptake were measured after variable periods of maintainance of the mitochondria at 18° C under nitrogen. The decrease of H+ ejection during aging was more marked than the decrease of Ca++ uptake. As a consequence the H+/Ca++ ratio became almost half after 2 hours.

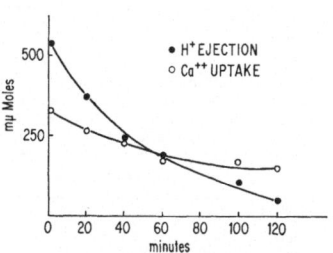

Fig. 5. *Effect of aging on the H+/Ca++ ratio.* Experimental conditions as in Fig. 1. 250 μM Ca²⁺ was added to 5.1 mg mitochondrial protein which were kept for variable periods of time at 18° under N₂ atmosphere

Effect of Succinate and of Permeant Anions on the H+ Ejection

Rasmussen *et al.* have reported that the H+/Ca++ ratio is decreased when permeant anions are present [6]. In the present section the effect of permeant anions such as acetate and P₁ is compared with that of succinate, which is the substrate most commonly used during studies on Ca++ uptake.

As shown in Table 1 addition of permeant anions to mitochondria incubated at pH 7.0 in 0.25 M sucrose caused a significant decrease, from 1.5 to 0.5 or 0.8, of the H+/Ca++ ratio. This decrease was due essentially to an apparent decrease of the H+ ejected since the amount of Ca++ accumulated was not reduced. At pH 8 the addition of permeant anion had no effect on the H+/Ca++ ratio, which on the other hand was already low, about 0.7. As will be discussed in a subsequent paper at pH 8 the permeant

Table 1. *Effect of permeant anions on the H+ ejection*
The medium contained: 0.25 M sucrose, 5 mM Tris-Anion, 1.3 μM rotenone, 1 mM succinate, 1 μg oligomycin and 10 mg mitochondrial protein. Final volume 2 ml. Amount of Ca++ added was 250 μM

	pH 7.0			pH 8.0		
	H+ ejected	Ca++ taken up	H+/Ca++	H+ ejected	Ca++ taken up	H+/Ca++
	mμMoles			mμMoles		
Tris-Cl⁻	500	238	1.54	140	192	0.75
Tris-Ac⁻	140	355	0.42	190	261	0.73
Tris-PO₄⁻	260	335	0.78	186	325	0.64

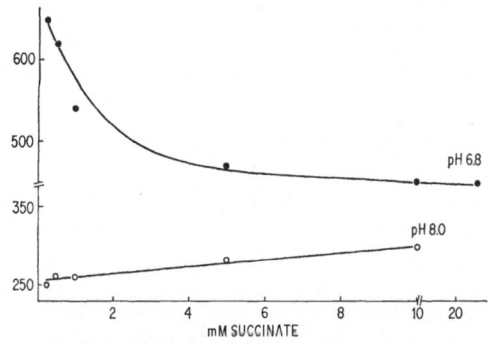

Fig. 6. *Effect of various succinate concentrations on H+ ejection.* The medium contained 0.25 M sucrose, 1.5 μM rotenone, 5 mM Tris-Cl⁻ and 6 mg mitochondrial protein; 250 μM Ca²⁺. The Ca++ accumulated varied between 390 and 450 mμmoles at pH 6.8 and between 370 and 360 mμmoles at pH 8.0

In other experiments it was tested whether uptake of chloride could explain the lowering of H+/Ca++ ratio especially at high pH. No uptake of ³⁶Cl⁻ was observed [6].

anions had a marked effect on the oxygen uptake by decreasing the apparent efficiency of the process of Ca++ uptake. In the experiment of Table 1, 1 mM succinate was present which apparently did not interfere with the effect of the permeant anions on the H+ ejection. As shown in Fig. 6 addition of increasing amounts of succinate, produced a large decrease of the H+ ejection at pH 6.8 and a very slight effect on the proton ejection at pH 8. When extrapolated to zero succinate concentration at pH 6.8 the H+/Ca++ ratio was in the vicinity of 2. The effect of succinate on the H+ ejection was also dependent on the buffering capacity of the medium. In fact at low buffering capacity of Tris, pH below 7, the lowering of the H+ ejection was dependent mainly on the concentration of succinate. Viceversa at pH 8 the lowering of the H+/Ca++ ratio was dependent on the concentration of Tris. The sum of the effects of succinate and of Tris at various pH on the H+/Ca++ ratio is reported in Fig. 7. A decrease of the H+ ejection due to Tris has already been reported by us [9] and by Wenner [10].

Experiments were also carried out with [¹⁴C]-succinate to test the possibility of a binding of succinate to the mitochondria. As shown in Table 2 a certain amount of binding of succinate to the mitochondria was observed at pH 6.5. The amount of

10*

succinate bound was much lower at pH 8.2. Ca^{++} caused an increase of the succinate bound at pH 6.5 not at pH 8.2. Parallel to the increase of succinate binding at low pH there was a decrease of the H^+ ejection and thus a decrease of the H^+/Ca^{++} ratio. The binding of succinate to the mitochondria had the following characteristics. First, it was largely inhibited by agents such as malate or malonate.

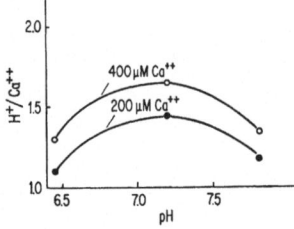

Fig. 7. H^+ ejection in presence of succinate and $Tris\text{-}Cl^-$. 200 or 400 μM Ca^{2+} were added to 9.1 mg protein incubated in 1 μM rotenone, 5 mM succinate, 0.25 M sucrose and 12.5 mM Tris-Cl⁻. The pH of the medium is indicated in the figure

Table 2. *Binding of succinate and H^+/Ca^{++} ratios at various pH*
The medium contained in 2 ml: 0.25 M sucrose, 2 mM Tris-Cl⁻, 1.5 μM rotenone, 9.4 mg protein and 1 mM [1,4-¹⁴C]-succinate

pH	Ca⁺⁺ added	Succinate binding	Ca⁺⁺ binding	H⁺ ejection	H⁺/Ca⁺⁺
	500 mμMoles	mμMoles	mμMoles	mμMoles	
6.5	—	.93			
	+	158	425	562	1.34
6.9	—	84			
	+	147	434	570	1.31
7.3	—	70			
	+	118	456	600	1.33
7.8	—	68			
	+	70	446	680	1.54
8.2	—	65.1			
	+	62.6	445	880	1.97

Second, it was not dependent on the supply of energy, as indicated by the kinetics of the binding and by the lack of effect of antimycin A. Third, it was dependent on the concentration of Ca^{++} only under very limited conditions. For example only at low pH was there an increase of succinate binding by addition of small amounts of Ca^{++}. Further addition of Ca^{++} did not increase the amount of succinate bound.

The Release of K⁺ during Ca⁺⁺ Uptake

Addition of Ca^{++} to mitochondria resulted in a release of mitochondrial K^+. The efflux of K^+ was slow, due to the low permeability of the mitochondrial membrane to K^+, while the Ca^{++} uptake was fast and completed in a few seconds. However when an excess of Ca^{++} was added to the mitochondria, after the period of fast uptake of Ca^{++} a second period of slow uptake followed. This slow uptake was parallel to the slow release of mitochondrial K^+. The K^+/Ca^{++} ratio during the period of slow Ca^{++} uptake was about 2 (Table 3).

Table 3. *K^+/Ca^{++} ratio in intact mitochondria*
The medium contained: 0.25 M sucrose, 5 mM Tris-HCl pH 7.2, and 6 mg mitochondrial protein. Ca⁺⁺ added was 250 μM

Time	Ca⁺⁺ accumulated	Δ	K⁺ released	Δ	K⁺/Ca⁺⁺
minutes	mμMoles		mμMoles		
1	260		80		
4	330	70	240	160	2.28

When the mitochondrial membrane was rendered permeable to K^+ by the addition of valinomycin, the uptake of Ca^{++} was always accompanied by a fast release of K^+ together with the ejection of H^+ [8]. The contributions of H^+ and of K^+ to the stoicheometry were dependent on the concentration of K^+ in the medium and thus on the mitochondrial K^+ content (Table 4). The release of K^+ was higher, the higher the amount of KCl in the medium. The $H^+ + K^+/Ca^{++}$ ratio was in the vicinity of 2.

Effect of the Order of Addition on the H⁺/Ca⁺⁺ Ratio

In Fig. 8A it is shown an experiment where H^+ ejection and Ca^{++} uptake were measured by addition of mitochondria to a medium containing Ca^{++}. The upper curve indicates the ejection of H^+ in the absence of Ca^{++} and the lower curve the H^+ ejection in the presence of Ca^{++}. The difference between the two is taken as the ejection of H^+ due to Ca^{++} uptake. In Fig. 8B it is shown that the H^+/Ca^{++} ratio was constant at the various pH when measured, either with Tris-HCl or with Na-glycylglycine as buffers, by the addition of mitochondria to incubation media containing Ca^{++}.

In other experiments the aerobic uptake of Ca^{++} was started by the addition of succinate. The addition of Ca^{++} to rotenone treated mitochondria resulted in a slight H^+ ejection due to the metabolism independent binding of Ca^{++} [11]. Subsequent addition of 100 μM succinate resulted in a stimulation of the H^+ ejection which rapidly slowed down. At pH 8.0 the stimulation of the H^+ ejection due to the addition of Ca^{++} was slower but slightly larger.

As seen in Table 5 the H^+/Ca^{++} ratio was affected by the amount of Ca^{++} added to the rotenone-treated mitochondria prior to the addition of succinate. At pH 7.1 the H^+/Ca^{++} ratio increased from 1.28 to 1.60 when Ca^{++} was increased from 250 to 500 μM. At pH 8 or at higher Ca^{++} concentrations

Table 4. *K⁺/Ca⁺⁺ ratios in valinomycin treated mitochondria*
The medium contained: 0.25 M sucrose, 10 mM Tris-HCl, pH 7.4, 5 mM succinate, 1 μM rotenone, 0.1 μg valinomycin. Amount of mitochondrial protein was 8 mg. Final volume 2 ml

Conditions	Ca⁺⁺ added	H⁺ ejected	K⁺ released	H⁺/Ca⁺⁺	K⁺/Ca⁺⁺	H⁺ + K⁺/Ca⁺⁺
	μM	μM	μM			
A without KCl	250	270	150	1.08	0.6	1.68
B with 2 mM KCl	125	50	160	0.4	1.28	1.69
C with 4 mM KCl	125	28	210	0.22	1.68	1.90

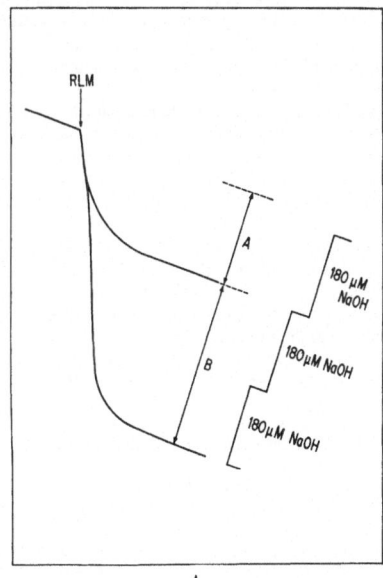

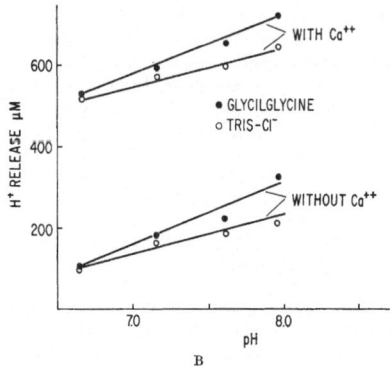

Fig.8 A and B. *H⁺ release by addition of mitochondria to Ca⁺⁺ containing media.* In A 9.0 mg protein were added to a medium containing 0.25 M sucrose, 5 mM Tris-Cl⁻. In the lower curve 200 μM Ca²⁺ was present. In B, the medium contained 5 mM Tris-Cl⁻ or glycil-glycine-Na at various pH. 10.6 mg mitochondrial protein were added in the presence or absence of 200 μM Ca⁺⁺

Table 5. *H⁺/Ca⁺⁺ ratio in Ca⁺⁺ treated mitochondria*
The medium contained: 0.25 M sucrose, 10 mM Tris-Cl⁻, 1.5 μM rotenone and 10 mg mitochondrial protein. Aerobic uptake of Ca⁺⁺ was started by the addition of 100 μM succinate. Ca⁺⁺ taken up aerobically was calculated by difference between amounts bound in the absence and presence of succinate

Ca⁺⁺ added	pH 7.1			pH 8.0		
	H⁺ ejected	Ca⁺⁺ taken up	H⁺/Ca⁺⁺	H⁺ ejected	Ca⁺⁺ taken up	H⁺/Ca⁺⁺
	mμMoles			mμMoles		
250	380	296	1.28	470	237	1.98
500	560	350	1.60	610	294	2.06

the H⁺/Ca⁺⁺ ratio was constantly in the vicinity of 2 in agreement with previous reports [9].

DISCUSSION

Knowledge of the stoicheometry accompanying the mitochondrial cation uptake is an essential requisite for establishing the molecular mechanism of this process. The following reaction mechanism has been proposed by Chance and Mela to account for the molecular event of Ca⁺⁺ transport in the mitochondria:

$$H_2X \sim I + 2\,Ca^{++} + 2\,H_2O \rightarrow 2\,Ca^{++} + 2\,OH^- + H_2XI + 2\,H^+$$

membrane   outside ← membrane ——————→ outside

According to Chance and Mela [13] the ejection of protons is an early step presumably associated with a primary binding of the cation to the membrane or to the carrier (X ∼ I).

According to Mitchell [14] the uptake of cations is due to the electrical potential established across the mitochondrial membrane by the operation of the respiratory chain loops which cause a separation of charges. Uptake of Ca++ collapses the potential thereby stimulating respiration and the translocation of protons to the mitochondrial surface.

Whichever mechanism is correct, one of the basic question to be answered concerns the neutralization of charges during the ion uptake. As we have discussed elsewhere [15], if one assumes that Ca++ goes through the mitochondrial membrane in exchange with one proton, as reported by most laboratories until now, this leads to an accumulation of positive charges inside the mitochondria and to the establishment of a large membrane potential. Uptake of 100 μMoles Ca++/g protein, in mitochondria with a capacity of 1 $\mu F/cm^2$ and a surface of 40 m^2/g protein would result in a potential of about 20 V.

From the data reported in the present paper it appears that a lowering of the H+/Ca++ ratio from 2 to 1 in mitochondria incubated in 0.25 M sucrose is observed under several conditions. For example if the mitochondrial membrane is rendered partially or completely permeable to K+, Ca++ may enter in partial or complete exchange with intramitochondrial K+. It should be noted that a certain increase of the permeability of the membrane to K+ is usually determined by Ca++ [16]. Also aging of the mitochondria results in a lower H+/Ca++ ratio, presumably due either to release of permeant anions from the mitochondria or to increased metabolism independent binding of Ca++ [11].

In fresh mitochondria the H+/Ca++ ratio is dependent on the Ca++/protein ratio both in the presence and in the absence of succinate. Uptake of Ca++ together with permeant anions released from the mitochondria could be suggested also in this case. It is to be noted however that an apparent reuptake of the H+ ejected was evident at high protein concentrations. An oscillation of the pH trace was reported by Carafoli et al. [12] but attributed to a release of Ca++ due to equilibration with extramitochondrial cations. Since in our experiments the H+ reuptake did not concern a release of the Ca++ taken up we prefer to consider it unrelated to phenomenon described by Carafoli et al. [12]. On the other hand we wish to point out that a correlation exists between rate of proton ejection during Ca++ uptake and extent of proton reuptake. Oxygen diffusion may become rate limiting at the high respiratory rates, this rendering semianaerobic the mitochondrial milieu. Thus it cannot be excluded that part of the decrease of the H+/Ca++ ratio at high protein concentrations is due to the binding of Ca++ to mitochondrial anions ionized during the preceeding pH equilibration.

The decrease of the H+ ejection observed in the presence of succinate has three peculiarities. First, it is related with the presence of Tris-HCl and thus with the buffering capacity of the medium. Second, it tends to decrease parallel to the increase of pH of the medium. Third, it is accompanied by the binding of a certain amount of succinate. Part of the binding of succinate appears to be independent of the Ca++ uptake, but another part occurs parallel to the uptake of Ca++ especially at low pH and at low Ca++ concentrations. It appears therefore reasonable to conclude that under certain conditions succinate may represent a moving anion thus causing a decrease of the H+/Ca++ ratio.

The H+ ejection was higher, both at low and at high Ca++ concentrations, either by starting the Ca++ uptake by the addition of succinate or by the addition of non equilibrated mitochondria to a Ca++ containing medium. By the first procedure, it is possible to separate the anaerobic from the aerobic Ca++ binding. In fact the Ca++ added to rotenone-treated mitochondria binds with the mitochondrial anions which are ionized during the pH equilibration. The subsequent addition of succinate induces the aerobic binding. In the second procedure the aerobic uptake of Ca++ occurs simultaneously with the metabolism independent ionization of mitochondrial anions.

The present investigation has been aided by a grant from the National Research Council. The authors wish to thank Mr. Leonardo Agosti for skilled technical assistance.

REFERENCES

1. Saris, N. E., Soc. Sci. Fennica., Commentationes Phys. Math. 28 (1963) 11.
2. Brierly, G. P., in Energy linked functions of mitochondria (edited by B. Chance), Academic Press, New York 1963, p. 237.
3. Engström, G. W., and DeLuca, H. F., Biochemistry, 3 (1964) 379.
4. Chance, B., J. Biol. Chem. 240 (1965) 2729.
5. Carafoli, E., Gamble, R. L., Rossi, C. S., and Lehninger, A. L., Biochem. Biophys. Res. Commun. 22 (1966) 431.
6. Rasmussen, M., Chance, B., and Ogata, E., Proc. Natl. Acad. Sci. U. S. 53 (1965) 1069.
7. Rossi, C., and Azzone, G. F., Biochim. Biophys. Acta, 110 (1965) 434.
8. Rossi, C., Azzi, A., and Azzone, G. F., Abstracts 3rd FEBS Meeting (Warsaw), 1966, p. 156.
9. Rossi, C., Azzi, A., and Azzone, G. F., Biochem. J. 100 (1966) 4 c.
10. Wenner, C. E., J. Biol. Chem. 241 (1966) 2810.
11. Rossi, C., Azzi, A., and Azzone, G. F., J. Biol. Chem. In press.
12. Carafoli, E., Gamble, R. L., and Lehninger, A. L., J. Biol. Chem. 241 (1966) 2644.
13. Chance, B., and Mela, L., Proc. Natl. Acad. Sci. U. S. 55 (1966) 1243.
14. Mitchell, M., and Moyle, J., Nature, 208 (1965) 147.
15. Azzone, G. F., Azzi, A., and Rossi, C., Round Table discussion on mitochondrial structure and compartmentation. Adriatica Editrice, Bari 1966.
16. Azzi, A., and Azzone, G. F., Biochim. Biophys. Acta, 113 (1966) 438.

C. Rossi, G. F. Azzone, and A. Azzi
Istituto di Patologia generale Universita di Padova
Via Loredan 16, Padova, Italia

European J. Biochem. 1 (1967) 147—151

Interactions entre pigments et acides nucléiques

4. Complexes spécifiques et non spécifiques entre la lutéoskyrine, les ions magnésium et les acides nucléiques

Y. Ohba et P. Fromageot

Service de Biochimie, Département de Biologie, Centre d'Etudes Nucléaires de Saclay, Gif-sur-Yvette

(Reçu le 16 décembre 1966)

Luteoskyrine, a bis-polyhydroxyanthraquinone, is a toxic pigment produced by *Penicillium islandicum*. Its ingestion results in hepatomas and cirrhotic degeneration of the liver in man and the rat. This pigment is able to form *in vitro* and in the presence of Mg^{2+} ions two types of complexes with DNA. The first (I) and rapidly formed complex corresponds to an interaction with the purines of the DNA. To react luteoskyrine with the DNA purines, the DNA must be denatured. The interaction is established by the following criteria:

a) In the presence of denatured DNA the insoluble luteoskyrine-Mg^{2+} complex is not formed.

b) The spectrum of luteoskyrine in aqueous solution is shifted to shorter wave lengths.

c) When submitted to ultracentrifugation and analysed at 260 mμ and at 450 mμ, the denatured DNA and the luteoskyrine sediment at the same rate, and the sedimentation coefficient of the complex is greater than that of the same sample of DNA alone.

d) After addition of luteoskyrine and magnesium ions, partially structured DNA is stabilized against further heat denaturation. Similar complexes (type I) are also formed with poly A, poly G and poly I. In each case, the spectra obtained and their stability are characteristic of the polymer utilised. Therefore the complexes of type I are said to be specific.

The second type of complex, type (II), on the contrary, is made slowly with native DNA or paired ribopolymers (poly A:U or poly G:C) irrespective of the nature of the constituent bases. Type II complexes can also be slowly formed from the type I complexes first made with denatured DNA. The spectra of the type II complexes are not characteristic of the nucleic acids involved but are very similar to those corresponding to the aggregated luteoskyrine-Mg^{2+} complex.

Un précédant travail a montré que la lutéoskyrine est susceptible de s'associer au DNA de thymus de veau en présence d'ions magnésium [1]. En outre, nous avons observé que ce pigment inhibe *in vitro* le fonctionnement de la RNA polymérase DNA dépendante extraite de *Escherichia coli* [2], et plus particulièrement lorsque le DNA a été au préalable dénaturé. Cette remarque suggère une étude des interactions entre la lutéoskyrine et le DNA dénaturé. Le présent travail montre que la lutéoskyrine peut donner deux types de complexes avec le DNA de thymus en présence d'ions Mg^{2+}. L'un est spécifique des bases puriques d'un acide nucléique dont les bases sont accessibles, l'autre, non spécifique, se forme avec tous les acides nucléiques utilisés.

MATÉRIEL ET MÉTHODES

La lutéoskyrine utilisée a été préparée au laboratoire par une technique voisine de celle de Uraguchi

Abréviations non-usuelles. Polydérivés des acides adénylique, guanylique, inosinique, cytidylique et uridylique, respectivement poly A, poly G, poly I, poly C et poly U.

Enzyme. RNA polymérase (DNA dépendante) ou nucléosidetriphosphate: RNA nucléotidyltransférase (EC 2.7.7.6).

et al. [3]. Le DNA de thymus provient du Centre de Recherches sur les Macromolécules, à Strasbourg, et nous a été offert par Mr. Pouyet. Les ribopolynucléotides proviennent de Miles Laboratories Inc., sauf le poly-G et le poly G:C (1:1), don du Dr. Michelson.

Les solutions de lutéoskyrine sont préparées en dissolvant le pigment dans 0,1 ml de NaOH 0,1 N, puis en ajoutant 9,9 ml de tampon phosphate 0,01 M, pH 7,7.

La concentration de la lutéoskyrine est mesurée en utilisant le coefficient d'extinction moléculaire 21.500 à 451 mμ et à pH 7,4. L'absorbance des solutions de lutéoskyrine croît linéairement avec la concentration au moins jusqu'à une contentration 0,1 mM en lutéoskyrine.

Les solutions de DNA sont préparées dans du tampon phosphate de sodium 0,01 M pH 7,7. Le DNA est dénaturé par chauffage à 100° pendant 10 minutes, suivi d'un refroidissement dans l'eau glacée. Le DNA apurinique a été préparé selon Chargaff [4].

Les spectres sont tracés avec le spectrophotomètre Bausch and Lomb 505, équipé d'un système de chauffage et d'enregistrement de la température des cuvettes, et avec l'appareil Cary 14.

Le mélange de DNA dénaturé 160 µM et de lutéoskyrine 42 µM, en solution dans le tampon phosphate 0,01 M pH 7,7, 0,5 mM en ions Mg²⁺, a été centrifugé dans la centrifugeuse analytique Spinco, modèle E. La sédimentation a été suivie à 260 mµ et à 440 mµ.

RESULTATS

Interaction entre la lutéoskyrine et le DNA dénaturé

Dès que l'on met en présence de DNA de thymus dénaturé par la chaleur, de la lutéoskyrine et des ions magnésium, on observe un déplacement du spectre, le

Ces résultats montrent qu'en présence d'ions Mg²⁺ la lutéoskyrine donne avec le DNA dénaturé de thymus deux types de complexes, les complexes du type I, caractérisés par des maximum d'adsorption situés à des longueurs d'onde inférieures à 451 mµ; les complexes du type II dont les maximum sont situés à des longueurs d'onde supérieures à 451 mµ.

L'association entre lutéoskyrine et DNA, dans les conditions où se forme le complexe I a aussi été mise en évidence par centrifugation. A cet effet, on a suivi la sédimentation à la fois à 260 mµ et à 440 mµ et mesuré les coefficients de sédimentation à ces deux longueurs d'onde: on trouve $s_{20} = 26,7$ à 260 mµ et

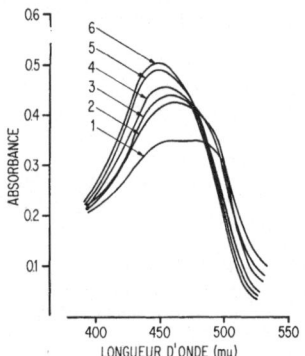

Fig. 1. *Formation du complexe I entre la lutéoskyrine, les ions Mg²⁺ et le DNA dénaturé de thymus.* DNA dénaturé, quantités variables; lutéoskyrine, 2×10^{-5} M; MgCl₂, 5×10^{-5} M; tampon phosphate, 10^{-2} M; pH 7,7. Spectres tracés après 1 heure à 37°. Rapport DNA/lutéoskyrine (nucléotides/mole de pigment): (1) 0; (2) 0,3; (3) 0,66; (4) 1,3; (5) 3,3; (6) 6,6

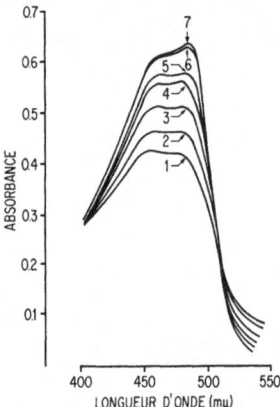

Fig. 2. *Formation du complexe II entre la lutéoskyrine, les ions Mg²⁺ et le DNA dénaturé de thymus.* DNA dénaturé, quantités variables; lutéoskyrine, $2,5 \times 10^{-5}$ M; MgCl₂, 5×10^{-5} M; tampon phosphate, 10^{-2} M; pH 7,7. Spectres tracés après 24 heures à 37°. Rapport DNA/lutéoskyrine: (1) 0; (2) 0,12; (3) 0,3; (4) 0,6; (5) 1,2; (6) 5,9; (7) 11,2

maximum dans le visible se situant à 442,5 mµ. En variant la quantité de DNA dénaturé ajoutée, on obtient un faisceau de courbes présentant un point isobestique à 476 mµ (Fig.1). On notera que ce point isobestique n'est net qu'après avoir maintenu 30 à 60 minutes le mélange à 37°. Pendant ce temps la position du maximum à 442,5 mµ reste stable. En maintenant davantage le mélange à 37°, on observe après quelques heures une modification des spectres qui est complète après 20 heures, et définitive. On voit alors deux maxima, l'un à 457,5, l'autre à 484 mµ et un point isobestique à 510 mµ (Fig.2).

Par rapport au spectre de la lutéoskyrine en solution aqueuse, à pH 7,7, on assiste donc, en ajoutant au milieu du DNA dénaturé et des ions Mg²⁺, à l'apparition de deux nouveaux spectres, le premier étant déplacé vers les courtes longueurs d'ondes, le second étant déplacé vers des longueurs d'ondes plus grandes.

27,5 à 440 mµ. Ainsi la lutéoskyrine sédimente comme le DNA. Le coefficient de sédimentation du complexe I est très supérieur à celui du DNA dénaturé seul, mesuré dans les mêmes conditions, qui a été trouvé égal à 14,1.

Le complexe I peut se transformer en complexe II. La vitesse de cette transformation dépend de plusieurs facteurs. A pH constant et égal à 7,7, interviennent la concentration en Mg²⁺, celle du DNA dénaturé et la température. L'influence de la concentration des ions Mg²⁺ est indiquée par la Fig.3. Un accroissement de la concentration de Mg²⁺ stabilise les complexes I. La Fig.3 montre ainsi côte à côte un spectre du type I (courbe 5) et un spectre du type II (courbe 1). On indiquera aussi que pour une concentration 0,5 mM en Mg²⁺, on transforme complètement le complexe I en complexe II, en 24 heures à 37°, tant que le rapport DNA/lutéoskyrine est inférieur à 0,6. Lorsque ce rapport est de 3 et supérieur, cette transformation est amorcée mais non complète en 24 heures à 37°. Elle est complète à 90° lorsqu'on atteint cette température par un chauffage progressif à raison de 2° par minute. Au contraire, si la concentration est 0,6 mM en Mg^{2+}, même avec un rapport DNA/lutéoskyrine 2,2, le complexe I reste presque inchangé, le maximum passant de 445 à 448 mμ après un cycle de chauffage qui atteint 90° en 30 minutes. En accroissant la concentration en Mg^{2+} à 1 mM, le complexe I est stable, même après le chauffage indiqué ci dessus.

Ces résultats ont conduit à examiner l'influence de la lutéoskyrine sur l'hyperchromicité que manifeste une solution de DNA de thymus, dénaturé au préalable, et réchauffée. La Fig. 4 donne la variation de

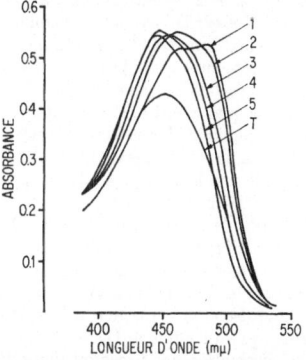

Fig. 3. *Influence de la concentration en ions* Mg^{2+} *sur la transformation du complexe I en complexe II.* DNA dénaturé, $7,3 \times 10^{-5}$ M; lutéoskyrine: $2,5 \times 10^{-5}$ M; tampon phosphate, 10^{-2} M; pH 7,7. On ajoute des quantités variables de $MgCl_2$. Concentration finale du $MgCl_2$: (1) $2,5 \times 10^{-5}$ M; (2) 10^{-4} M; (3) $2,5 \times 10^{-4}$ M; (4) 5×10^{-4} M; (5) 10^{-3} M. La courbe T correspond à une concentration en lutéoskyrine $1,7 \times 10^{-5}$ M, sans ions Mg^{2+}. Spectres tracés après 24 heures à 37°

l'absorbance à 260 mμ en fonction de la température d'une solution de DNA, préalablement dénaturée par chauffage à 100°, 0,6 mM en Mg^{2+} (tampon phosphate de sodium 1 mM), pH 7,7. Après addition de lutéoskyrine à cette solution et à la température ambiante, le chauffage ne provoque plus qu'une très faible variation à 260 mμ de l'absorbance, comme si la lutéoskyrine sous forme de complexe I était capable de stabiliser les structures secondaires du DNA dénaturé. Au contraire, en abaissant la concentration en Mg^{2+} à 0,05 mM, condition dans laquelle le chauffage provoque une transformation du complexe I en complexe II, la lutéoskyrine ne parait plus capable de protéger les structures secondaires du DNA dénaturé (courbes 3 et 6).

On a aussi chauffé du complexe II préparé avec du DNA natif. La courbe de transition obtenue est superposée à celle que l'on obtient dans les mêmes conditions en l'absence de lutéoskyrine.

Nature du complexe I

Les précédentes expériences ont montré que la lutéoskyrine forme en présence de DNA dénaturé et d'ions Mg^{2+} des complexes du type I, alors qu'en

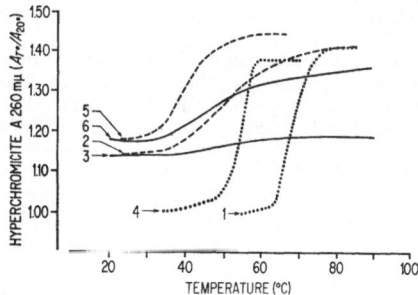

Fig. 4. *Influence du complexe I entre la lutéoskyrine et le DNA dénaturé de thymus sur l'hyperchromicité produite par chauffage.* Toutes les solutions sont faites dans du tampon phosphate 10^{-3} M, pH 7,7. (1) DNA natif, Mg^{2+}: 6×10^{-4} M; (2) DNA dénaturé, Mg^{2+}: 6×10^{-4} M; (3) comme 2), avec addition de lutéoskyrine ($2,8 \times 10^{-5}$ M) avant le chauffage; rapport DNA/lutéoskyrine: 2,2; (4) DNA natif, Mg^{2+}: 5×10^{-5} M; (5) DNA dénaturé, Mg^{2+}: 5×10^{-5} M; (6) comme 5), avec addition de lutéoskyrine ($2,8 \times 10^{-5}$ M) avant le chauffage; rapport DNA/lutéoskyrine: 2,2

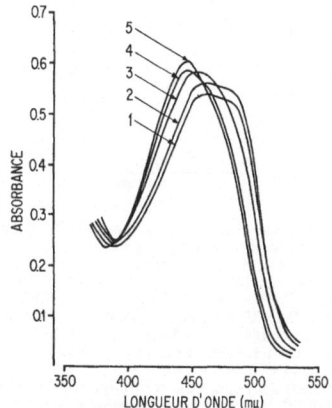

Fig. 5. *Influence de la proportion de dénaturation du DNA sur la formation des complexes avec la lutéoskyrine.* DNA, 8×10^{-5} M; lutéoskyrine, $2,4 \times 10^{-5}$ M; tampon phosphate, 2×10^{-3} M; pH 7,7; $MgCl_2$: 10^{-3} M. Spectres tracés après 18 heures à 37°. Hyperchromicité: (1) $0^0/_0$; (2) $27,5^0/_0$; (3) $59,5^0/_0$; (4) $82^0/_0$; (5) $100^0/_0$

présence de DNA natif ce sont des complexes du type II qui sont produits [1]. Aussi a-t-on examiné l'influence d'une dénaturation progressive du DNA sur la formation de ces complexes. A cet effet, on a chauffé du DNA à différentes températures. La Fig. 5 montre que, avec le DNA non chauffé, on retrouve un complexe du type II, et que plus la dénaturation progresse, plus on forme de complexes du type I. Il apparait ainsi en première analyse que les complexes du type I impliquent des interactions entre la lutéoskyrine, les ions Mg^{2+} et des constituants du DNA rendus accessibles par la séparation des chaînes du DNA.

Pour connaître la nature de ces constituants, on a utilisé en premier lieu du DNA apurinique. On constate que, en présence de lutéoskyrine (DNA apurinique/lutéoskyrine, 1,36; Mg^{2+}, 0,025 mM;

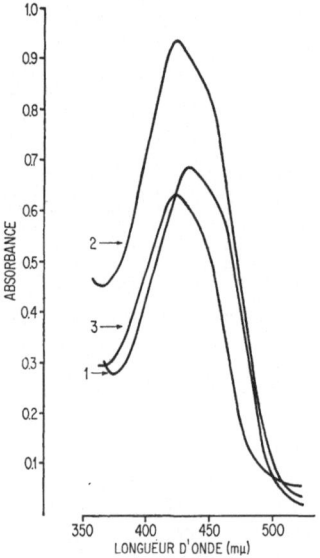

Fig. 6. *Spectres des complexes entre lutéoskyrine et ribohomopolymères.* En présence de tampon phosphate 10^{-2} M, pH 7,7, on dispose: (1) Poly A, 1×10^{-4} M; lutéoskyrine, $2,9 \times 10^{-5}$ M; $MgCl_2$, 5×10^{-4} M; (2) Poly G, $1,6 \times 10^{-4}$ M; lutéoskyrine, $3,2 \times 10^{-5}$ M; $MgCl_2$, 5×10^{-4} M; (3) Poly I, $1,08 \times 10^{-4}$ M; lutéoskyrine, $2,7 \times 10^{-5}$ M; $MgCl_2$, 5×10^{-4} M. La solution de poly G a été chauffée, préalablement à l'expérience, à 100° pendant 10 minutes. Elle est refroidie et utilisée. La solution de poly I a été dialysée contre une solution 10^{-3} M de EDTA pH 7,6, pendant 2 jours, puis contre du tampon phosphate 10^{-2} M, pH 7,7, avant d'être employée. Spectres [1] et [3] mesurés après 18 heures à la température du laboratoire. Spectre [2] mesuré après 18 heures à 37°

18 heures à 23°), on n'obtient que le complexe II, alors que du DNA dénaturé ne donne que des complexes I dans les mêmes conditions, il apparait ainsi probable que la formation des complexes I nécessite la présence de bases puriques.

Ce point de vue est étayé par l'examen du comportement de la lutéoskyrine avec les ribohomopolymères. En présence de poly A et d'une concentration en ions Mg^{2+} 0,5 mM, on obtient un complexe

de type I dont le maximum est à 446 mμ. Ce maximum est à 438 mμ avec le poly G et à 435 mμ avec le poly I. Si l'on abaisse la concentration des ions Mg^{2+} à 0,01 mM, on forme avec le poly G et le poly I les mêmes complexes dont les spectres sont identiques à ceux obtenus avec de plus grandes concentrations en Mg^{2+} et stables à température ambiante. Le complexe I avec poly I est particulièrement stable et n'est pas transformé en complexe II, même après chauffage à 100°. Au contraire, le complexe I avec poly A est particulièrement instable et se convertit en complexe II facilement. A une concentration 0,05 mM en ions Mg^{2+} le complexe I ne se forme plus. Aussi est-il vraisemblable qu'à cette concentration en ions Mg^{2+}, le complexe I qui apparait avec le DNA dénaturé corresponde essentiellement à une association de la lutéoskyrine avec la guanine présente. Cependant le maximum à 442,5 mμ du spectre obtenu avec le DNA dénaturé n'est pas celui du spectre du complexe I entre lutéoskyrine et poly G, même en présence d'une quantité de DNA dénaturée, égale à 30 fois celle du pigment. Les raisons de cette différence ne sont pas connues avec certitude mais pourraient résulter du fait que le poly G à pH 7,7 possède une structure secondaire [6] (Fig. 6).

La différence de comportement entre DNA dénaturé et DNA natif se retrouve avec les ribohomopolymères puriques seuls, ou associés à un ribohomopolymère pyrimidique: le poly A:U (1:1) ou le poly G:C (1:1) ne conduisent en présence d'ions Mg^{2+} et de lutéoskyrine, qu'au complexe II.

Ainsi la disposition des bases puriques d'un acide nucléique doit répondre à certains critères pour être capable de former un complexe de type I. Aussi doit-on se demander comment se comportent les monophosphonucléotides. A des concentrations semblables à celles des acides nucléiques, par exemple deux molécules par molécule de pigment, l'AMP ou le GMP ne préviennent pas la précipitation de la lutéoskyrine à l'état de chélate de Mg^{2+}. Des concentrations 30 fois supérieures sont nécessaires pour parvenir à ce but et le spectre obtenu est alors celui des complexes II. Un résultat semblable a été obtenu par Swanbeck [7] qui a observé la formation des complexes entre des hydroxyanthraquinones et le DNA de thymus, mais non avec les désoxymonophosphonucléotides.

DISCUSSION

On a montré précédemment [1] qu'en mettant en présence des DNA de thymus, des ions Mg^{2+} et de la lutéoskyrine, on voyait se former progressivement un complexe, dont le spectre présentait deux maxima, l'un à 457, l'autre à 484 mμ. En outre, la lutéoskyrine était maintenue en solution, alors qu'en l'absence de DNA, elle précipite sous une forme associée aux ions Mg^{2+}. En poursuivant ces recherches, on constate que le même DNA de thymus, mais dénaturé par chauffage, ainsi que divers ribohomopolymères, possèdent aussi la propriété de maintenir la lutéoskyrine en solution en présence d'ions Mg^{2+}. En ajoutant à une solution de lutéoskyrine du DNA dénaturé en quantités croissantes et des ions Mg^{2+}, on observe immédiatement une variation du spectre du mélange. Lorsque le rapport DNA/lutéoskyrine dépasse 6, le spectre obtenu présente un maximum à 442,5 mμ et le faisceau de courbes un point isobestique à 476 mμ. Ce résultat indique l'existence d'un complexe entre la lutéoskyrine, les ions Mg^{2+} et le DNA dénaturé, différent de celui antérieurement décrit, et pendant le temps qui a précédé les mesures, l'existence d'un équilibre entre le complexe et ses précurseurs (Fig. 1).

La réalité de ce complexe est étayée par d'autres arguments. Si l'on suit la sédimentation du mélange DNA dénaturé, ions Mg^{2+} et lutéoskyrine, dans la centrifugeuse analytique Spinco, on constate que la vitesse de sédimentation du complexe est la même, que l'on fasse la mesure à 260 ou à 440 mμ. En outre, cette vitesse est beaucoup plus grande ($s_{20} = 26,7$) que celle du DNA dénaturé en l'absence de lutéoskyrine ($s_{20} = 14,1$). Enfin, la lutéoskyrine s'avère capable de stabiliser les structures partiellement hélicoïdales du DNA chauffé et refroidi (Fig. 4). Ces deux dernières observations sont compatibles avec l'hypothèse que dans le complexe présentement étudié, la lutéoskyrine est distribuée sur toutes les molécules de DNA dénaturé présentes aux points où des sites sont accessibles. En outre, du fait de la nature de bis-anthraquinone substituée de la lutéoskyrine, donnant à ce pigment un charactère bifonctionnel, il peut en résulter des liaisons inter ou intra moléculaires au niveau du DNA. Que ce complexe est différent de celui qui a été décrit antérieurement résultent les remarques suivantes:

a) Il n'est pas observable, en première analyse, avec le DNA non dénaturé.

b) Lorsqu'on utilise du DNA dénaturé de façon progressive par chauffage et des températures croissantes, on voit le spectre du mélange se modifier et se déplacer vers les courtes longueurs d'ondes (Fig. 5).

c) Les spectres obtenus en présence de DNA dénaturé incubés à 37° se transforment progressivement, en se déplaçant vers les grandes longueurs d'ondes. A la limite, on retrouve les spectres obtenus avec du DNA natif (Fig. 2). La vitesse de cette transformation est fonction de la concentration en Mg^{2+}, en DNA dénaturé, et de la température. La Fig. 3 montre qu'un accroissement de la concentration en Mg^{2+} stabilise le complexe obtenu immédiatement vers le DNA dénaturé.

Pour ces motifs, nous appellerons complexe de type I celui qui se forme immédiatement entre la lutéoskyrine, les ions Mg^{2+} et le DNA dénaturé, et caractérisé par un spectre dont le maximum est à une longueur d'onde inférieure à 451 mμ, et complexe de type II celui que l'on a précédemment décrit, qui ne se forme que progressivement et dont le spectre est déplacé vers les grandes longueurs d'ondes par rapport à celui de la lutéoskyrine en solution aqueuse à pH 7,7.

Les complexes de type I constituent une famille de complexes entre la lutéoskyrine, les ions Mg^{2+} et les bases puriques. Ils ne peuvent être obtenus en présence de DNA apurinique, de poly U ou de poly C. Au contraire, ils se forment avec le DNA dénaturé, poly A, poly G et poly I. Les spectres obtenus avec chacun de ces derniers polymères possèdent des caractéristiques et une stabilité propres, indiquant un rôle spécifique de la base considérée et de la structure dans laquelle elle est engagée. La stabilité est faible avec l'adénine, grande avec la guanine et l'hypoxanthine. En outre, une structure non apairée, au sens Watson-Crick, du polymère considéré est nécessaire puisque, ni le DNA natif, ni poly A:U (1:1), ni poly G:C (1:1), ne sont capables de former de complexes de type I.

Les complexes de type II, au contraire des précédents, se forment avec le DNA natif, avec le poly C, le poly U, le poly A:U (1:1) le poly G:C (1:1), ainsi qu'avec le DNA dénaturé et le poly A lorsque la concentration en ions Mg^{2+} n'est pas suffisante pour stabiliser le complexe I. Les spectres obtenus sont toujours semblables. On peut en conclure que les complexes II ne présentent pas de spécificité de base et, comme les spectres des complexes II sont très voisins de celui du complexe associé lutéoskyrine — Mg^{2+}, il est probable que les complexes II correspondent à des associations lutéoskyrine — Mg^{2+} à la surface des acides nucléiques considérés, ces derniers se comportant comme des polyanions orientant ces associations [1].

Nous remercions Mr. Y. Arnaud, du Département de Biologie du Centre de l'Energie atomique, qui a fait pour nous les ultracentrifugations.

RÉFÉRENCES

1. Ueno, Y., Platel, A., et Fromageot, P., *Biochim. Biophys. Acta,* 134 (1967) 27.
2. Sentenac, A., Ruet, A., et Fromageot, P., *Bull. Soc. Chim. Biol.* 134 (1967) 27.
3. Uraguchi, K., Tatsuno, T., Sakai, F., Tsukioka, M., Sakai, Y., Yonemitsu, O., Ito, H., Miyake, M., Saito, M., Enomoto, E., Shikata, T., et Ishiko, T., *J. Exptl. Med. (Japan),* 31 (1961) 19.
4. Tamm, C., Hodes, M. E., et Chargaff, E., *J. Biol. Chem.* 145 (1952) 49.
5. Applequist, D., *J. Am. Chem. Soc.* 83 (1961) 3158.
6. Pochon, F., et Michelson, A. M., *Proc. Natl. Acad. Sci. U. S.* 53 (1965) 1425.
7. Swanbeck, G., *Biochim. Biophys. Acta,* 123 (1966) 630.

Y. Ohba et P. Fromageot
Service de Biochimie, Département de Biologie,
Centre d'Etudes Nucléaires de Saclay,
91 Gif-sur-Yvette, France

European J. Biochem. 1 (1967) 152—163

Sheep Kidney Nuclease

Hydrolysis of tRNA

K. Kasai and M. Grunberg-Manago

Institut de Biologie Physico-chimique, Paris

(Received January 5, 1967)

An endonuclease showing no apparent base specificity, nor any preference for the sugar moiety has been isolated from sheep kidney. The products of the digestion are 5'-P-ended oligonucleotides, and not mononucleotides. Preparation of short oligo A and oligo U (di-, tri-, tetra-) with a good yield is described. The enzyme requires Mg ions, is inhibited at high ionic strength and destroyed by temperatures of 60° and above.

The enzyme displays an extremely rigorous specificity with regard to the secondary structure of its substrate. This structural specificity is apparent from the fact that native DNA, complexes between poly A and poly U, and polyinosinic acid in high salt are completely resistant to the enzyme. Even poly A and poly C, at neutral pH, denatured DNA, ribosomal RNA, and especially tRNA, are only slowly attacked.

Preliminary experiments on the effect of addition of poly U on the hydrolysis of a copolymer A, C suggest that the enzyme will hydrolyze the unstructured loop of polyribonucleotide complexes.

The analysis of the hydrolysis of tRNAs by sheep kidney nuclease shows that the enzyme attacks all chains and produces mainly large fragments (average chain length, 40 nucleotide units), containing pGp terminal and TpΨpCpG sequences; and small oligonucleotides (average chain length 4.5 nucleotide units), containing a large amount of fragments with a CpCpA terminal.

These results are discussed in relation to the conformation of tRNA.

Purified nucleases that are highly specific, are valuable agents in the study of nucleotide sequence and secondary structure of polynucleotides. A convincing case is the brillant use of a variety of nucleases in the determination of the complete nucleotide sequence of alanine-, serine-, and tyrosine-specific transfer RNA [1—3]. In the case of polyribonucleotides, several exonucleases [4—6] have been described which only slightly attack RNA molecules that have an extensive amount of secondary structure. Such specificity has less frequently been ascribed to endonucleases [7—9]. Either T 1 or pancreatic RNase

Non-standard abbreviations. Polyuridylic, polyadenylic, polycytidylic, polyguanylic, polyxanthylic and polyinosinic acids, respectively, poly U, A, C, G, X, I; linear copolymers of adenine and cytosine in random sequence, A, C copolymers; linear copolymers of uridine and guanosine in random sequence, U, G copolymers; tobacco mosaic virus, TMV; dimethylguanylic acid, dimethyl-G; transfer RNA, tRNA; unspecified nucleotide, N; 2,5-diphenyloxazole, PPO; 1,4-bis-(5-phenyloxazolyl-2)-benzene, POPOP; trichloracetic acid, TCA.

Small oligonucleotides are designated as follows: when p (representing a phosphate) is placed to the right of a nucleoside symbol, the phosphate is esterified at 3'-C of the ribose moiety; when placed to the left of the nucleoside symbol, the phosphate is esterified at 5'-C of the ribose moiety.

Enzyme. RNase, or ribonuclease, or ribonucleate pyrimidine-nucleotido-2'-transferase (cyclizing) (EC 2.7.7.16).

—whose role was decisive in breaking the tRNA into two separate components—respond to secondary structure, but only under very special conditions [1,10—13]. Endonucleases which hydrolyze denatured DNA in preference to native DNA have been isolated from lamb brain and sheep kidney by Healey *et al.* [14].

The present paper describes a method of purification and some properties of a sheep kidney endonuclease which displays an extremely rigorous specificity (only non-structured polymers are attacked) with regard to the secondary structure of its substrate (either ribo- or deoxy-), while no specificity is observed towards internucleotide linkages. We also report some preliminary work on the use of this enzyme on unfractionated tRNA and the isolation of a digestion product some forty nucleotides long.

METHODS

Assay of Enzyme

The incubation mixture (0.1 ml) contains in mM concentrations: Tris-HCl, pH 7.5, 50; $MgCl_2$, 1.5; β-mercaptoethanol, 25; polyribonucleotide, 2; and enzyme solution. For the assay of the purified enzyme, 0.1% (w/v, final concentration) of crystalline bovine serum albumin was added. After 15 minutes incubation at 37°, 20 μl aliquots were withdrawn and added to 1 ml of 5% (w/v) trichloracetic acid containing 0.01% (w/v) uranium acetate. 50 μl of 1% (w/v) bovine serum albumin were added and the mixture was chilled in ice for 15 min, centrifuged, and the absorbance of the supernatant read at 260 mμ. One enzyme unit corresponds to an increase in A_{260} of 1 in the TCA soluble fraction after 1 hour's incubation at 37° with poly A. This in turn roughly corresponds to 0.35 μmoles of poly A hydrolyzed per incubation mixture, expressed as mononucleotides.

Amino Acid Acceptor Activity

100 μl of reaction mixture contains: Tris-HCl, pH 7.4, 100 mM, magnesium acetate, 10 mM; β-mercaptoethanol, 10; ATP, 4; CTP, 4; tRNA, about 20 A_{260}' per ml; [^{14}C] amino acid, 20—30 μM; and *Escherichia coli* amino acid activating enzyme, 700 μg/ml. Incubation was carried out at 37° for 20 minutes; 50 μg of the reaction mixture were put on DEAE-cellulose paper (Whatman DE 81). Free amino acids were removed according to the method of Ingram *et al.* [15]. The spot of tRNA was cut out, counted in a liquid scintillation counter (Tri-Carb 314 EX) in toluene containing 0.5% (w/v) PPO and 0.03% (w/v) POPOP.

Hydrolysis of DNA

The optimum Mg^{++} concentration for hydrolysis of DNA differs from that of RNA; the composition of the incubation mixture was as follows: Tris-HCl, pH 7.5, 50 mM; β-mercaptoethanol, 25 mM; MgCl$_2$, 5 mM; *E. coli* DNA, 50 μg/ml; bovine serum albumin, 1 mg/ml; and enzyme, 1.6 μg/ml. Incubation was carried out at 37°; after various intervals 0.5 ml aliquots were withdrawn and added to 0.5 ml of 10% (w/v) trichloracetic acid. The increase in acid soluble products was measured. Denatured DNA was obtained by heat treatment at 100° for 10 min, followed by rapid cooling.

Hydrolysis of tRNA

The same reaction mixture as the one described for the assay of enzyme was used, but the total volume was increased 400-fold and it contained 0.75 Mg^{++} per tRNA nucleotide. The increase in acid soluble fraction was followed in the usual manner. At the same time a 0.5 ml aliquot was taken and added to an equal volume of phenol previously saturated with water and shaken for 10 minutes. The water phase was separated by brief centrifugation and the phenol layer was extracted with 0.5 ml of water. The combined water phases were extracted five times with an equal volume of ether (free of

peroxide) and concentrated by lyophilysation. The amino acid acceptor activity was assayed in this fraction.

Elution of the Hydrolyzed Product from a Column of Sephadex G-50

Digested tRNA was separated from the enzyme by phenol treatment as described above. The concentrated digest (200—500 μl) was put on a Sephadex G-50 column (1×50 cm or 2×50 cm, see Table 9), previously equilibrated with 0.2 M ammonium formate. Elution was carried out with the same solvent. The flow rate was about 20 ml/h and aliquots of 1 ml were collected. A typical elution profile is shown in Fig. 10.

Determination of Oligonucleotides

Oligonucleotides produced by the digestion of polynucleotides with the sheep kidney enzyme were fractionated according to their chain length by descending paper chromatography on Whatman 3 MM with *n*-propanol—concentrated ammonia—water (55:10:35, v/v/v). After elution with water, their primary structure was determined by subjecting them to alkaline hydrolysis in 0.3 M KOH at 37° for 16 hours; the hydrolysate was neutralized with 5 N perchloric acid, chilled, and the KClO$_4$ precipitate removed by centrifugation. The supernatant was then analyzed by paper chromatography in *n*-propanol (for oligo U and oligo A) or paper electrophoresis in 10% (w/v) acetic acid adjusted to pH 3 with concentrated ammonia (for oligo A).

The primary structure of an oligonucleotide obtained by digestion of poly U, G was determined as follows: after separation in the *n*-propanol system, the oligonucleotide with the R_F of a pentanucleotide was eluted and the alkaline digest of this fraction was analyzed by paper electrophoresis as above. Six spots were obtained. One spot which did not move and one which moved slightly towards the cathode were identified by paper chromatography (methanol—concentrated HCl—water, 70:20:10, v/v/v) as uridine and guanosine respectively. The four other spots moved towards the anode; they were identified from their position and their behavior on paper chromatography (*n*-propanol system) as pUp, pGp, Up and Gp, starting with the fastest moving.

The hydrolysis product of tRNA, separated on Sephadex as described above, was concentrated by lyophilysation and hydrolyzed with alkali. The neutralized hydrolysate was put on Whatman 1 paper as a band and electrophoresis carried out at 40 V/cm vor 60 min in 0.45 M ammonium formate (adjusted to pH 7.5 with concentrated NH$_4$OH). Under these conditions the alkaline hydrolyzate was fractionated into three groups: nucleosides, nucleoside monophosphates and nucleoside 3',5'-diphosphates. Each band was eluted with dilute HCl (0.1 N for the nucleoside band and 0.01 N for the others).

The average chain length of peak I and II was calculated from the ratio of nucleoside monophosphate/nucleoside, or nucleoside monophosphate/nucleoside 3′,5′-diphosphate.

The mixture of nucleosides was again subjected to paper electrophoresis on Whatman 1 in 8.7% (w/v) acetic acid and 2.5% (w/v) formic acid (pH 1.8) at 40 V/cm for 100 min. Four nucleosides were well separated, each spot being eluted with 0.1 N HCl.

The mixture of nucleoside 3′,5′-diphosphates was treated with alkaline phosphatase in 0.1 M ammonium bicarbonate at 37° for 2 hours and separated as nucleosides (as described above).

The mixture of nucleotides was analyzed by two-dimensional paper chromatography, using isopropanol-water (70:30, v/v), with ammonia in the gas phase (0.35 ml/l) as the first solvent, and isobutyric acid—0.5 N NH₄OH (10:6, v/v), as the second solvent.

Materials

Sheep kidney acetone powder was purchased from Pentex (Kankakee, Ill.)

E. coli B tRNA was purchased from General Biochemical Corp. (Chargrin Falls, Ohio) lot 660870. Chromatography of this tRNA on Sephadex G-50 showed that it did not contain any small oligonucleotides. The average chain length calculated from an alkaline hydrolysis was 91. The sedimentation constant (s_{20}) was 4.2 ± 1.

Poly A (lot 15, s_{20} = 7.2), poly U (lot 61, s_{20} = 2.66) and poly U, G (lot 32) were prepared with polynucleotide phosphorylase from *E. coli*. Poly A, C (lot 33 and 34), poly I (lot 7) and poly X (lot 2) were prepared with polynucleotide phosphorylase from *Azotobacter vinelandii*. Poly 1,7-dimethyl G was a gift from Dr. Michelson, Institut de Biologie Physicochimique, Paris. *E. coli* DNA was a gift from Dr. Sach from the Laboratoire Choay, Paris.

E. coli alkaline phosphatase was a Worthington product; the Sephadex were products of Pharmacia (Uppsala) and DEAE-cellulose a product of Eastman-Kodak.

Miscellaneous

Ultracentrifugation was performed in the laboratory of Dr. R. von Rapenbusch, Hopital Saint-Louis, Paris, with a Spinco Model E, equipped with a scanner and a "demultiplex" system [16] in 0.05 M Tris-HCl, pH 7.5, 0.1 M NaCl.

Protein concentration was determined from the absorbance values at 260 and 280 mμ.

RESULTS
Purification of Enzyme[1]

All steps were carried out at 0—5°.

Crude Extract. 20 g of sheep kidney acetone powder were homogenized in a Potter Elvejem homogenizer in 200 ml of a cold solution of NaCl (0.85%, w/v). The mixture was kept in a cold room overnight and then centrifuged at 15,000 rev./min in a Servall (head SS 34) for 40 minutes. The precipitate was discarded.

Ammonium Sulfate Fractionation. Solid ammonium sulfate was added to the supernatant to give 35% saturation (205 g/l). The resulting precipitate was

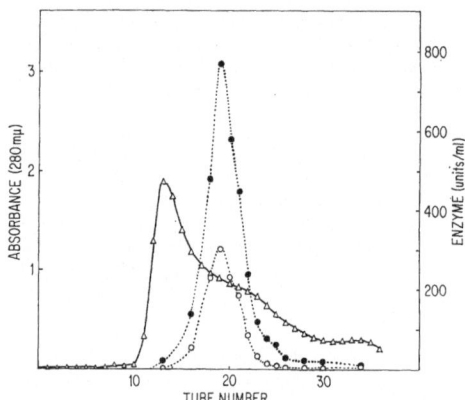

Fig. 1. *Sephadex G-100 chromatography of sheep kidney nuclease.* 10 ml of enzyme solution (9.6 mg/ml) was put onto a column of Sephadex G-100 (3×30 cm), equilibrated with 0.005 M Tris-HCl, pH 8, containing 0.01 M β-mercaptoethanol. Elution was done with the same buffer (0.5 ml/min) and aliquots of 4 ml were collected. △, A_{280}; ○, enzyme activity towards poly U; ●, enzyme activity towards poly A

collected by centrifugation (15,000 rev./min; 40 min) in a Servall (SS 34 head) and dissolved in 400 ml of 0.005 M Tris-HCl, pH 8 containing 0.01 M β-mercaptoethanol.

DEAE-cellulose. To this dilute protein solution 20 g of wet DEAE-cellulose (which had been equilibrated with the same buffer) were added; the mixture was left standing for 15 minutes with occasional stirring. The DEAE-cellulose was removed by filtration and another 20 g of DEAE-cellulose were added to the filtrate. The same procedure was repeated twice more. To concentrate this DEAE-treated solution, it was brought to 85% saturation

[1] We are grateful to Dr. G. Deutsch and Dr. Levene, Brandeis University, for their unpublished purification method which was very useful for establishing ours, and for kindly supplying the enzyme with which the work was started.

with solid ammonium sulfate (610 g/l) and left standing overnight in a cold room. The precipitate was collected by centrifugation (8,000 rev./min; 40 min) in a Servall (SS 34 head) and dissolved in 0.1 M Tris-HCl, pH 8.

Sephadex G-100. This concentrated protein solution (about 10 mg/ml) was put onto a Sephadex G-100 column (3 × 30 cm), previously equilibrated with 0.005 M Tris-HCl, pH 8, 0.01 M β-mercaptoethanol. Elution was carried out with the same buffer (0.5 ml/min) and aliquots of 4 ml were collected. A large protein peak appeared in the exclusion volume, followed by a peak of enzymatic activity (Fig. 1).

The elution profile was the same wether poly U or poly A was used as substrate in the assay of the enzyme.

DEAE-cellulose. The active fraction was put onto a small DEAE-cellulose column (1.5 × 6 cm) previously equilibrated with 0.005 M Tris-HCl pH 8, 0.01 M β-mercaptoethanol. The column was washed with the same buffer and most of the activity was

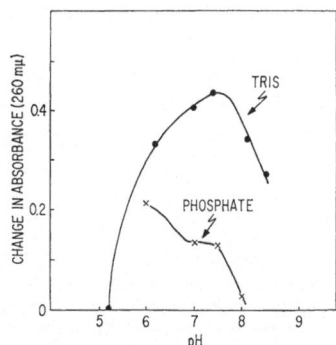

Fig. 2. *Activity pH curve of sheep kidney nuclease.* Enzyme activity was assayed against poly U in buffers of various pH. Incubation mixture contained: poly U, 2 mM; MgCl₂, 1.5 mM; β-mercaptoethanol, 25 mM; crystalline bovine serum albumin, 1 mg/ml; sheep kidney enzyme, 8 µg/ml; and buffers: acetate, Tris-HCl, phosphate, final ionic strength 0.05

Table 1. *Purification of sheep kidney nuclease*

	$\frac{A_{280}}{A_{260}}$	Protein	Enzyme activity		Total units		Activity ratio
			poly A	poly U	poly A	poly U	poly U/poly A
		mg	units/mg				
Crude extract	0.83	4,800	2.8	12	13,000	56,000	4.3
(NH₄)₂SO₄	1.0	1,500	8	29	12,200	44,000	3.6
DEAE-cellulose		176ᵃ	60	170	10,500	30,000	2.9
Sephadex G-100 chromatography	1.65	32	200	480	6,300	15,500	2.4
DEAE-cellulose		5.4ᵇ	820	1,780	4,400	9,600	2.2

ᵃ Assuming a A_{280}/A_{260} ratio of 1.
ᵇ Assuming a A_{280}/A_{260} ratio of 1.55.

recovered in this effluent. When the column was further developed with a NaCl gradient, three protein peaks were eluted, which were found to be inactive.

The enzyme thus prepared was purified about 290-fold with respect to its activity towards poly A and 150-fold with respect to poly U. The ratio of the two activities (poly U/poly A) reached 2.2 after the DEAE-cellulose step (Table 1). Some other nucleases specific for poly U (like bovine pancreatic RNase) may have been eliminated during the purification.

Properties of the Enzyme

Optimum pH. In Tris-HCl, the optimum pH is 7—7.5 (ionic strength = 0.05). In phosphate buffer of the same ionic strength, the enzyme was found to be less active at all pHs tested (Fig. 2).

Effect of Ions. The enzyme requires Mg⁺⁺. With the amount of ribonucleotide used, the enzyme is inactivated by 10^{-2} M EDTA. The optimum Mg⁺⁺ concentration is $5 × 10^{-4}$ M, higher concentrations being inhibitory: in 10^{-2} M only 50% of the activity

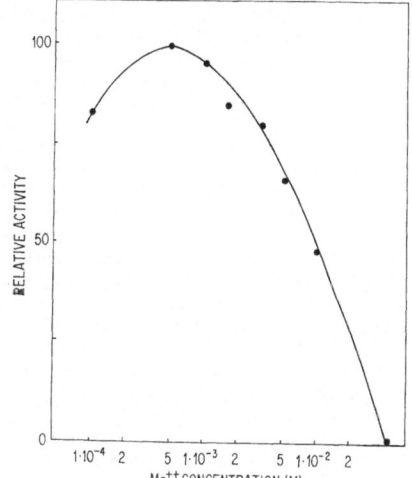

Fig. 3. *Effect of Mg⁺⁺ concentration.* Incubation mixture as described in Methods except for the concentration of MgCl₂. Substrate was poly U and enzyme concentration was 12 µg/ml

is retained and in the presence of 5×10^{-2} M Mg^{++} the enzyme is completely inhibited (Fig.3); NaCl at the same ionic strength (15×10^{-2} M) also completely inhibits the enzyme. NaCl (5×10^{-2} M) and KCl (same concentration) inhibited the enzyme 30% and 37% respectively.

Stability. Although repeated freezing and thawing of the enzyme resulted in complete inactivation, in the presence of 20% (w/v) glycerol the enzyme can be stored at $-20°$ for several months without any appreciable loss of activity. When the purified enzyme was diluted for the activity assay it became unstable: at a concentration of 0.1 mg/ml, the enzyme lost 30% of its activity when incubated for one hour at $37°$ under the conditions of the assay. When it was diluted to a concentration of 0.81 mg, 90% of the activity was lost under the same conditions. However, it is stabilized by the addition of bovine serum albumin: at the same concentration (0.01 mg/ml) only 15% of the activity is lost, in the presence of 0.1% (w/v) serum albumin.

The enzyme is heat labile: 3 minutes of heating at $60°$, resulted in complete inactivation (both in the presence and in the absence of serum albumin). Substrates (poly A or poly U) had no protective effect.

Michaelis-Menten Constant. K_m values for poly A and poly U were 2.2×10^{-3} M and 3.3×10^{-3} M, respectively (expressed as mononucleotides).

Structure of the Oligonucleotides Formed

The products of enzymatic hydrolysis of homopolymers (poly A and poly U) were fractionated by paper chromatography with the *n*-propanol system and identified by their R_F values as well as by analysis of products after alkaline hydrolysis.

The distribution of oligonucleotides depends on the amount of enzyme used and on the incubation period. After a short incubation period with a small quantity of enzyme, almost all of the products were large oligonucleotides; no mononucleotides were formed. After extensive digestion (Tables 2—4 for conditions) all products were small oligonucleotides (di-, tri-, tetra-nucleotides). From these observations it was concluded that this enzyme is an endonuclease.

The products of hydrolysis of polyribonucleotide with the sheep kindney enzyme are 5′-oligonucleotides. Alkaline hydrolysis of the oligoadenylic acids derived from poly A digestion with a relatively crude enzyme fraction (Table 2) gave rise to three components, corresponding to adenosine, AMP, and a component which moved more slowly than ADP on paper chromatography with the *n*-propanol-NH_3 system, and at the same rate as ADP on paper electrophoresis (in acetic acid). After treatment with alkaline phosphatase, this component gave only adenosine and was identified on the basis of these experiments, as pAp.

The structure of each fragment obtained by enzymatic digest was determined from the ratio of these three components. This is summarized in Table 2; fraction 7 was identified as adenosine 5′-monophosphate by paper electrophoresis in 0.015M borate buffer at pH 9.05. However, with another enzyme preparation, no mononucleotides were found.

Similar experiments were performed with poly U and the purified enzyme; they gave almost the same results (Table 3). Diuridylic acid (pUpU) and tri-uridylic acid (pUpUpU) thus obtained were incubated with a large amount of enzyme, and it was found by chromatography that these small oligonucleotides had not been attacked.

Preparation of Small Oligo A and Oligo U

One of the use of the enzyme is the preparation of a series of short of 5′-oligonucleotides. Therefore two preparations, one of oligo A and one of oligo U are described (see Table 4). It can be seen that there is quite a good yield of di-, tri- and tetranucleotides.

Specificity

The enzyme does not appear to have any specificity for the internucleotide linkage (deoxy- or ribo-). DNA, natural RNA (such as ribosomal RNA, for instance) or synthetic ribopolynucleotides are all hydrolyzed. Neither does this enzyme appear to have any specificity for the bases: poly A, poly U or poly C are all hydrolyzed at $37°$.

Poly G is not attacked, but this is probably due to the fact that it is highly structured. That it is not due to a base specificity of the enzyme was investigated in the following experiment which determined the enzymatic hydrolysis products of poly U, G (2:1). After enzymatic hydrolysis, chromatographically separated oligonucleotides of an assumed length of 5 (determined by their R_F) were hydrolyzed with alkali. Uridine, guanosine, GMP, UMP, pGp, and pUp were found in the alkaline digest and separated by paper electrophoresis (see Methods). Both pGp and G were found in much larger amounts than would be expected from the end groups of the poly U, G, since alkaline hydrolysis of an equimolar amount of untreated poly U, G gave rise to GMP and UMP only, the amounts of pGp, G, pUp and U were too small to be detectable. It can therefore be concluded that the large amounts of pGp and G found in the treated poly U, G originated from end groups of segments hydrolyzed by the enzyme. The fact that both guanosine and GpG were found indicates that the enzyme can attack either pUpG or pGpU bonds.

The enzyme hydrolyzes polymers composed of base analogs, such as polyxanthylic acid and poly dimethyl-G; at $37°$ it hydrolyzes ribosomal RNA, tRNA (see Table 5), or TMV-RNA.

Table 2. *Structure of oligoadenylic acid produced by enzymatic hydrolysis of poly A*
The incubation mixture for hydrolysis of poly A contained in mM concentrations: phosphate buffer pH 7.4, 60; $MgCl_2$, 1.8; β-mercaptoethanol, 30; poly A, 40 A_{260}/ml; and enzyme (DEAE-cellulose treated crude extract), 1.47 mg/ml. Final volume 10 ml. Incubation for 2 hours at 37°. The mixture was centrifuged and the supernatant was put onto Whatman 3 MM and developed with n-propanol system for 43 hours.
Oligonucleotides were hydrolyzed with 0.3 N KOH for 16 hours. After being neutralized with perchloric acid, each hydrolysate was put onto Whatman 3 MM and chromatographed with a solvent system of n-propanol—concentrated NH_3—water. Spots of pAp, Ap and A were eluted and A_{260} was read

Fraction	Poly A digest	R_F [a]	Ratio of components after alkaline hydrolysis			Structure
			A-3',5'-P	A-3'-P	A	
	%					
1	13	0.28				mixture of large oligo nucleotides
2	15	0.31				hexanucleatides [b]
3	— [c]	0.43				pentanucleotides [b]
4 a [d]	20	0.57	1	2.1	0.9	pApApApA
4 b [d]	13	0.69	1	1.7	0.9	pApApApA
5	8	0.77	1	1.3	0.8	pApApA
6	4.5	0.95	1	0.5	0.9	pApA
7	< 1	1.05		A-5'-P		pA

[a] ADP = 1.
[b] Assumed from R_F.
[c] Product was lost, amount between 15—20%.
[d] Some unknown substance was found in the middle of fraction 4, but with purified enzyme this substance disappeared.

Table 3. *Structure of oligouridylic acids produced by enzymatic hydrolysis of poly U*
The incubation mixture (5 ml) contains in mM concentrations: Tris-HCl, pH 7.4, 50; $MgCl_2$, 1.5; β-mercaptoethanol, 25; bovine serum albumin, 1 mg/ml; poly U, 22 A_{260}/ml; purified enzyme (specific activity, 820 units/mg protein), 7.6 µg/ml; distilled water to adjust volume. After 1 hour incubation at 37°, the mixture was chilled and put directly on Whatman paper 3 MM. It was developed by descending chromatography with n-propanol—NH_3—water (55:10:35, v/v/v) for 26 hours. Separated oligonucleotides were eluted and analyzed after alkaline hydrolysis. Incubation with more enzyme (15 µg/ml) gave predominantly di-, tri-, and tetranucleotides, a small amount of pentanucleotides and no product larger than hexanucleotides

Fraction	Poly U digest	Ratio of components after alkaline hydrolysis			Structure
		pUp	Up	U	
	%				
1	50	—	—	—	mixture of large oligonucleotids
2	13.4	1	3.6	1	pUpUpUpUpU
3	14.4	1	3.0	1.1	pUpUpUpUpU
4	12	1	1.8	1	pUpUpUpU
5	4.1	1	1.3	1	pUpUpU
6	0.9	1	0.3	1.1	pUpU

Table 4. *Preparation of oligo A and oligo U*
The incubation mixture (final volume 80 ml) contains in mMolar concentrations: Tris, pH 7.5, 50; mercaptoethanol, 25; $MgCl_2$, 1,5; bovine serum albumin, 0.1%; poly A poly U, 20 A_{260}/ml; enzyme, 140 U/ml, specific activity, 900 units/mg proteins. Incubation at 37°, 8 h for oligo A and 12 h for oligo U. The incubation mixture is diluted five times with H_2O and oligonucleotides are then separated on DEAE cellulose (column, 30×1.5 cm). Elution gradient 0.01 M [0.5 M $(NH_4)HCO_3$ + chloroform; 1/1 (w/v]; the absorbance at 260 mµ is recorded. Each peak is then concentrated to about 10 ml by lyophilysation and passed on a column of Bio-Gel P 2 (50—100 mesh; column 45×3.5 cm) to eliminate the salts; it is then eluted with H_2O adjusted to pH 0 with a few drops of NH_4OH + chloroform. The absorbance at 260 mµ is recorded and the conductivity is measured, with conductivity meter Radiometer, type CDM 2d

Oligo	A			U		
	Absorbance after DEAE	Absorbance after Bio-Gel	Recovery	Absorbance after DEAE	Absorbance after Bio-Gel	Recovery
	(mg)			(mg)		
mono	0	0	—	63 (4) [a]	57	—
di	490 (31)	480	15	184 (11.5)	170	11
tri	621 (39)	615	30	370 (23)	360	20
tetra	367 (22)	350	11	368 (23)	315	24
penta	55 (3.5)	44	2	189 (12)	177	10

[a] Values in parentheses represent percents.

Table 5. *Relative rates of hydrolysis of various polymers at 37°*
Composition of reaction mixture as described in Methods; enzyme concentration 8 µg/ml

poly U	100
poly C	45
poly A	46
poly G	0
poly UG (U/G = 2)	100
poly X	15
poly dimethyl G	2
ribosomal RNA	8.2
sRNA	4.2
denatured DNA	3.6

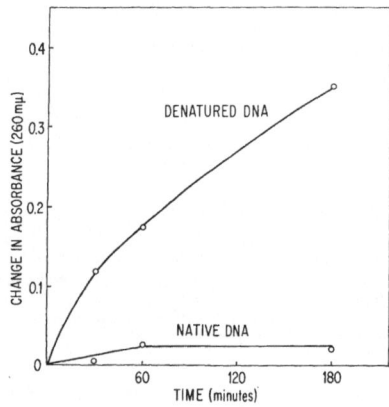

Fig.4. *Hydrolysis of DNA.* Incubation mixture as described in Methods, enzyme concentration 1.6 µg/ml

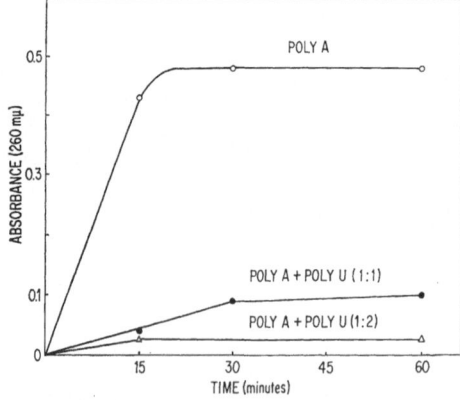

Fig.5. *Hydrolysis of poly A + poly U complex.* Incubation mixture as described in Methods. Enzyme concentration 15 µg/ml. Before addition of enzyme the mixture was pre-incubated at 37° during 1 h for formation of the complex

Effect of Secondary Structure

The enzyme exhibits a strong specificity for the secondary structure of polynucleotides. While denatured DNA is degraded, native DNA is hardly attacked (Fig.4).

For ribonucleotides there is also a strong specificity for the non-structured polymers (Fig.5). While both poly A and poly U are good substrates for the enzyme, the complex between the two polymers showed marked resistance. A poly A + poly U (1:1) mixture was poorly attacked and a (1:2) mixture, where triple strands are formed [17,18], was practically unattacked. Three-fold increase of enzyme, or prolonged incubation did not result in further hydrolysis. Poly I is not attacked either in high salt or in the presence of 5×10^{-3} M Mg^{++}; however if the Mg^{++} concentration is lowered to 0.5×10^{-3} M, the rate of hydrolysis of poly I is equivalent to that of poly C. It is known that in high salt a triple-stranded complex is formed [19,20].

Table 6. *Temperature effect on the rate of hydrolysis of polynucleotides*
Composition of reaction mixture as described in Methods. Enzyme concentration 4 µg/ml, 8 µg/ml and 240 µg/ml for poly U, poly A and tRNA, respectively

Substrate	Rate at 37°/rate at 15°
Poly U	3.1
Poly A	11.8
tRNA	11.4

The effect of temperature is much more striking on the rate of hydrolysis of poly A and tRNA than on that of poly U (Table 6). It is known that at 15° poly A has a more stacked structure than poly U [21]; it is also known that the structure of tRNA is stabilized at low temperatures [11].

The enzyme could be very useful if it would hydrolyze the non-structured part of a copolyribonucleotide selectively. In an attempt to investigate this point, the effect of the addition of poly U on the hydrolysis of A, C copolymers of different compositions was studied.

Mixing experiments (*e.g.* addition of poly U to poly A, C of different concentrations), and heating curves suggest that the stability of the complex formed between poly A, C and poly U depends on the cytidylic acid content of the copolymers (Fig.6). Experiments were made with two A,C copolymers: one which does not form a complex with poly U (A/C = 0.54), and a second which does (A/C = 1.63). Addition of poly U did not affect the hydrolysis of the former while with the latter it resulted in a considerable decrease in both the rate and extent of hydrolysis: only half of the total nucleotides was liberated after 30 minutes incubation. During that

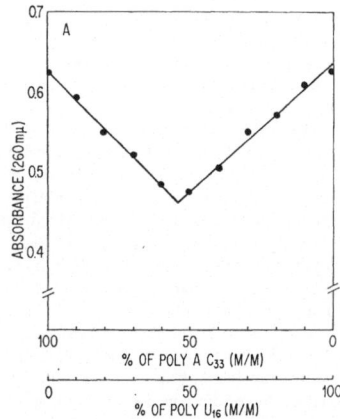 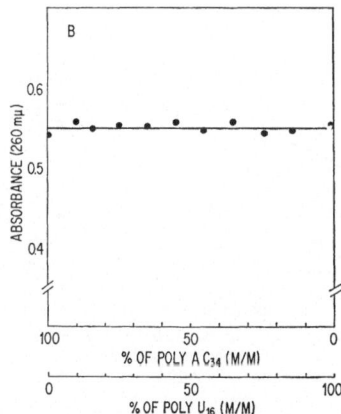

Fig. 6. *Mixing curve of poly AC + poly U.* Polymers were mixed in 0.05 M Tris, pH 7.4, and 1.5 mM MgCl₂. After keeping the solutions for 1 h at room temperature, absorbance at 260 mµ was measured. No change in the shape of the mixing curve was observed after 18 h. A, Mixing curve of poly AC_{33} (A/C = 1.63) + poly U; B, Mixing curve of poly AC_{34} (A/C = 0.54) + poly U

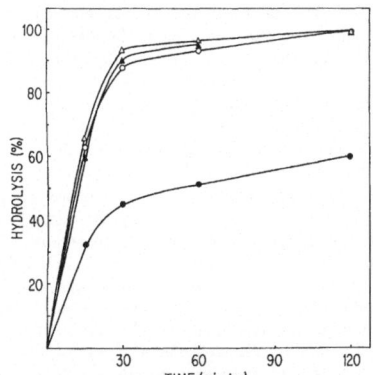

Fig. 7. *Effect of addition of poly U on hydrolysis of poly AC.* Incubation mixture as described in Methods, except that the reaction was carried out at 30°; enzyme concentration 16 µg/ml. ○, poly AC_{33} alone; ●, poly AC_{33} + poly U; △, poly AC_{34} alone; ▲, poly AC_{34} + poly U

same period all the nucleotides were liberated into the acid soluble fraction when the substrate was poly A, C (A/C = 1.63) alone (Fig. 7).

The results of the study of the hydrolysis products of a mixture of this last poly A, C + poly U is in accordance with this concept. After digestion and removal of protein, the hydrolysis products gave two peaks on Sephadex G-25: Fraction I which was excluded from the gel corresponds to the acid insoluble fraction, *i.e.* the resistant complexes of poly A, C and poly U; fraction II is a mixture of small oligonucleotides liberated from unstable parts of the

Table 7. *Base composition of fractions I and II of the hydrolysis product from poly AC + poly U*

After digestion with enzyme, 16 µg/ml at 30° for 30 min, the reaction mixture was treated with phenol, extracted with ether and concentrated by lyophilisation; it was then chromatographed on Sephadex G-25 column (1×50 cm), previously equilibrated with 0.2 M ammonium formate. Elution was carried out with the same solvent. Each fraction was hydrolyzed with alkali, followed by further treatment with *E. coli* alkaline phosphatase in 0.1 M Tris-HCl, pH 7.4, 10^{-3}MgCl₂ at 37° for two hours. Base composition was determined by paper chromatography; solvent: methanol—concentrated HCl—water (70:20:10, v/v/v)

	A	C	U
Before hydrolysis	1.63	1	2.2
Fraction I	0.9	0.2	1.8
Fraction II	0.7	0.94	0.24

complexes. Each fraction was then hydrolyzed for base identification (Table 7): fraction II was found to be much richer in cytidine and poorer in uridine than fraction I, which is consistent with the concept that regions rich in cytidine would be more susceptible to the enzyme and regions rich in uridine resistant. In fraction I the amount of uridine is double that of adenine which is probably explained by the fact that triple stranded complexes are formed between uridine and adenine regions. The excess of adenine in the acid soluble portion as compared to uridine be considered in the discussion.

Hydrolysis of E. coli tRNA

Compared with homopolymers, tRNA showed a marked resistance to the enzyme for complete hydrolysis. In order to complete the hydrolysis, the

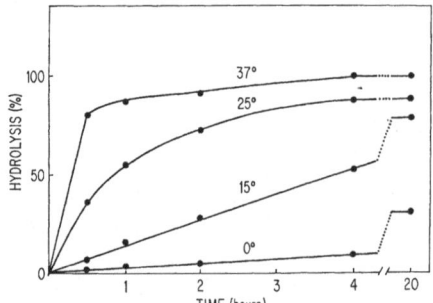

Fig. 8. *Effect of temperature on the hydrolysis of tRNA.* Incubation mixtures as described in Methods; enzyme concentration: 240 µg/ml

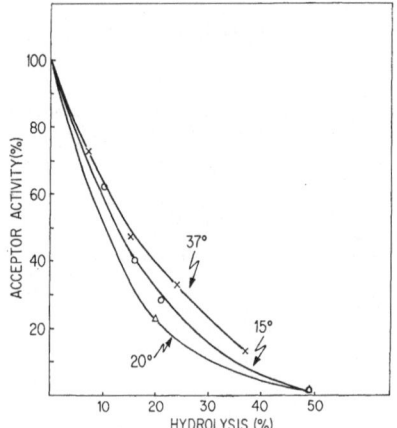

Fig. 9. *Effect of temperature on the relation between the extent of hydrolysis and remaining amino acid acceptor tRNA.* Incubation mixture as described in Methods. ○, at 15°; △, at 20°; ×, at 37°

Table 8. *Heat treatment of partially hydrolyzed tRNA with regards to amino acid acceptor activity*
Reaction was carried out at 10° (tRNA/enzyme = 8.3, w/w). Heat treatment was performed at 100° for 3 minutes, followed by rapid cooling

	Hydrolysis	Activity
	%	%
Intact tRNA	0	100
tRNA incubated 350 min without enzyme	0	96
tRNA hydrolyzed 100 min	6.6	79.4
tRNA hydrolyzed 100 min and heat-treated	6.6	77.8
tRNA hydrolyzed 200 min	13.5	53.7
tRNA hydrolyzed 200 min and heat-treated	13.5	57.8

enzyme concentration had to be increased (about ten times the amount required for poly A), or the incubation period had to be extended.

As expected, the increase in incubation temperature markedly increased the rate of digestion (Fig. 8). At a given percentage of hydrolysis, this increase in rate did not affect the relation between the extent of hydrolysis and the loss of amino acid acceptor activity (Fig. 9). Heat treatment (100° for 3 min) of partially hydrolyzed tRNA did not result in any appreciable change in the residual acceptor activity (Table 8), nor did the products of the digestion differ with increase in incubation temperature.

When incubation was carried out at 20°, after 35% hydrolysis had occured, 92% of amino acid acceptor activity was lost; a control experiment carried out under the same conditions, but in the absence of enzyme, did not show any loss of amino acid acceptor activity.

The hydrolyzed products gave two major peaks after elution on a Sephadex G-50 column, and an intermediate peak which was not studied (Fig. 10): peak I (53% of total A_{260} units eluted), which corresponds to the acid insoluble fraction, comes from the resistant part of tRNA, and peak II (30% of total A_{260} units eluted), which corresponds to the acid soluble fraction, comes from the sensitive parts of tRNA. Heat treatment (100°, 10 min) of the digest, prior to elution, resulted in a small increase near the shoulder of peak I.

The hydrolyzed products were treated so as to avoid loss of residual secondary structure which might be destroyed by phenol treatment: After digestion, the reaction mixture was diluted 2.5-fold with distilled water and put on a small DEAE-cellulose column (2×5 cm), previously equilibrated with 0.02 M Tris-HCl, pH 7.4. The column was then washed with 0.2 M NaCl in the same buffer to eliminate the enzyme and small products; larger fragments were eluted with 1.5 M NaCl in the same buffer and precipitated with ethanol. This fraction was placed on a Sephadex G-50 column and was eluted to give peak I. It had a sedimentation coefficient of 3.05. Therefore, phenol treatment did not result in any appreciable difference.

In order to determine the base composition of the two peaks, large amounts of tRNA (40 mg) were digested with the enzyme; the results of the elution were very similar to those with smaller amounts. Two major peaks were separated on G-50 Sephadex after phenol treatment (peak I, 53%; peak II, 28% of total absorbance units eluted from column). Average chain length, calculated after alkaline hydrolysis was 40.5 for peak I and 4.5 for peak II. Sedimentation coefficient of peak I was 3.01 ± 0.1, and did not change in 0.5% (w/v) formaldehyde, even after heating at 60°, 30 min in a sealed tube.

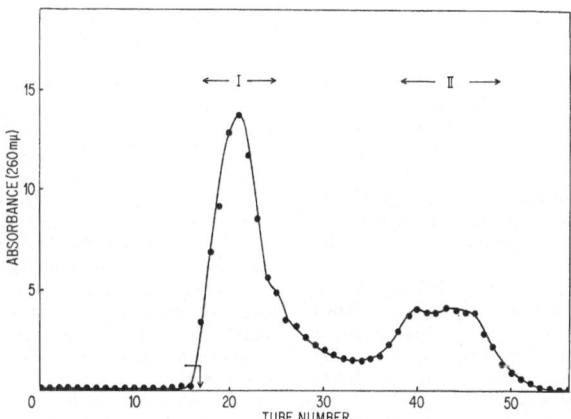

Fig. 10. *Chromatography of hydrolyzed tRNA on Sephadex G-50*. tRNA was hydrolyzed at 20° for 2 h (substrate/enzyme = 8.3, w/w). For details, see Methods

Table 9. *Composition of peak I and peak II obtained after tRNA hydrolysis*
40 mg of tRNA hydrolyzed at 20° for $5\frac{1}{2}$ hours; substrate/enzyme = 20:1 (w/w) after phenol treatment, chromatography on Sephadex G-50 column (2 × 50 cm); see methods for base analysis

| Base | Moles/80 nucleotides | | | | | | Total | Intact tRNA Np[a] |
| | Peak I | | | Peak II | | | | |
	pNp	Np	N	pNp	Np	N		
U	0.23	4.84	0.40	1.78	1.55	1.16	9.95	12.0
G	0.54	11.8	0.18	0.70	2.80	1.04	17.06	25.6
A	0.13	9.6	0.21	1.09	3.23	1.77	16.03	16.3
C	0.12	12.7	0.25	1.44	4.46	1.04	20.01	23.1
Ψ	—	0.77	—	—	0.20	—	0.97	1.7
T	—	0.61	—	—	0	—	0.61	0.88
MeA[b]	—	0.16	—	—	0.10	—	0.26	0.32
Total	1.01	40.48	1.05	5.01	12.34	5.01	64.9	80

[a] Data of Dunn *et al.* [26].
[b] Not identified, may be a mixture of methyladenylic acids.

The results of the analysis of nucleotides of peak I and peak II are summarized in Table 9. About half of the 5′-terminal of peak I was found to be guanosine while the predominant nucleoside at 3′-terminal was uridine. On the other hand, at the 3′-terminal of peak II, more adenosine was found than any other nucleoside. Less pseudouridylic acid and no thymidylic acid were found in peak II, while half of the total pseudouridylic acid and two thirds of the thymidylic acid appear in peak I.

DISCUSSION

An endonuclease showing no apparent base specificity, nor any preference for the sugar moeity has been described. The study of the hydrolysis of polynucleotides and polynucleotide complexes involv-

ing copolymers indicates that this enzyme will attack only unstructured regions. In this connection, structure in a wider sense should be understood: the stacked structure of neutral poly A and poly C are attacked very much more slowly by sheep kidney nuclease than the less structured poly U. This structural inhibition is well demonstrated by the effect of temperature on the digestion of poly A: the increase in rate of hydrolysis is the same as that observed for the highly structured tRNA (Table 6) and nearly four times the increase observed for poly U.

The most peculiar feature of this enzyme is probably the fact that its digestion products are oligonucleotides, most of which are tetramers or higher and 5′-P-ended. This is in sharp contrast to nucleases such as RNase, for instance, which all arrive at mononucleotides as the final product, but is very similar to pork liver [22] and lamb brain nucleases [14][2].

The lack of base specificity is demonstrated by the easy digestion of poly A, poly U and poly C. Although poly G is not attacked by sheep kidney nuclease, probably because of its structure, a poly U, G (2:1) copolymer is readily digested. Both guanosine and pGp are found in the alkaline digest of the degradation products, indicating that both pUpG and pGpU sequences are split. Polymers containing base analogs, like poly X or poly dimethyl-G are also attacked, though at a lower rate.

The structural inhibition of sheep kidney nuclease is apparent by the fact that native DNA is completely resistant towards the enzyme; so are the complexes between poly A and poly U, and polyinosinic acid in high salt. Even denatured DNA, ribosomal RNA, and especially tRNA are only slowly attacked.

In the experiment where a copolymer of adenine and cytidine was mixed with poly U (Table 7) attack was slow and mainly cytidine and adenine were liberated. The excess of adenine as compared uracil in the acid soluble fraction is readily explained if one considers that A,C is a random copolymer and that in a complex with poly U it will form loops of cytidine containing an occasional adenine [23]. Adenine residues will then be easily cleaved since they are located in the unstructured loop.

The complete loss of amino acid accepting activity suggests that all the tRNA molecules have been attacked. This is in complete contrast with the mode of action of some exonucleases such as polynucleotide phosphorylase which, at low temperature, only attacks some of the tRNA molecules while the others are completely resistant [24].

The analysis of hydrolyzed tRNA showed that sheep kidney nuclease produces mainly large fragments (average chain length, 40.5 nucleotide units) and small oligonucleotides (average chain length, 4.5 nucleotides). About half of the 5′-terminal of the large fragments (peak I) is guanylic acid. From the assay of amino acid acceptor activity, it appears that the contamination of intact tRNA was only a few percent. From the results of hydrolysis of poly U, G, and from analysis of the 5′-terminal of peak II, it is improbable that the enzyme attacks the bond NpGp preferentially; one can therefore assume that enzymic attack is similar for all nucleotide linkages. Thus, one would be able to consider that peak I contains either an important part of the pG half of the molecule, or that there is a guanine in all tRNAs, susceptible to preferential hydrolysis by the enzyme.

About half of all the total pseudouridylic acid and two thirds of the thymidylic acid were found in peak I. If they were derived from the sequence TpΨpCpG (which is believed to exist in all tRNAs[25]),

and if this sequence is located in the CpCpA half of the molecule (as in the case of three yeast tRNAs [1—3]) one would be able to consider that peak I contains also an important fragment of the CpCpA half of the molecule. It would be tempting to conclude therefore that peak contains fragments from both halves of the tRNA molecule, the pG and the CpCpA ends. In any event, one can conclude that by the action of sheep kidney nuclease, large fragments containing both pGp or TpΨpCpG sequences are obtained from E. coli tRNA; these sequences are not too susceptible to enzymic degradation.

On the other hand, peak I does not contain such large amounts of fragments with a CpCpA terminal which (as compared with the pGp terminal) would therefore be degraded by the enzyme. This is consistent with the excess of adenosine, as compared with other nucleosides, at the nucleotide end in peak II. These results are probably closely related to the conformation of tRNA where there is some evidence that the CpA terminal end is not involved in the structure of tRNA.

The results presented here are in very good agreement with those of Wagner et al. [13] and Armstrong et al. [11], who used pancreatic ribonuclease and RNase T 1; from experiments with purified alanine tRNA, they concluded that the central loop containing the anticodon is the most sensitive region, and that the half of the molecule containing TpΨpCpG is the most stable one.

It is apparent that sheep kidney nuclease is a useful tool in the study of nucleotide sequences. Its preference for unstructured regions makes it an ideal tool for the study of such questions as the arrangement of the loops in sRNA, or eluidating the replication mechanism of DNA. Apart from its use for studies of structure of nucleic acids, this enzyme will also be quite valuable for the preparation of series of 5′-P-oligonucleotides which are necessary for the study of coding properties and of the mechanism of action of different enzymes in vitro.

This work was started in the laboratory of Dr. Fresco while Dr. Grunberg-Manago was a visiting Professor at Princeton University. We wish to thank Dr. Fresco and his group for the hospitality extended.

The work was continued in Paris and supported by the following grants: No C-04580 of United States National Institutes of Health; Convention 6600020 of Délégation Générale à la Recherche Scientifique et Technique (Comité de Biologie Moléculaire); Comité de la Seine de la L.N.F.C.C.; French National Research Council (RCP 24); and a participation from the French Atomic Energy Commission.

We wish to thank Dr. van Rapenbusch for his help in determining all the sedimentation constants necessary for this work.

REFERENCES

1. Holley, R. W., Apagar, J., Everett, G. A., Madison, J. T., Marquisee, M., Merril, S. H., Penswick, J. R., and Zamir, A., Science, 147 (1965) 1462.
2. Zachau, H. G., Dütting, D., and Feldman, H., Angew. Chem. 78 (1966) 393.

---

[2] Lamb brain preparation of Healey et al. [14] will also hydrolyze polyribonucleotides (personal communication).

3. Madison, J. T., Everett, G. A., and Kung, H., *Science*, 153 (1966) 531.
4. Singer, M. F., and Tolbert, G., *Science*, 145 (1964) 593. Spahr, P. F., *J. Biol. Chem.* 239 (1964) 3716.
5. Grunberg-Manago, M., *J. Mol. Biol.* 1 (1959) 240.
6. Ando, T., *Biochim. Biophys. Acta*, 114 (1966) 158.
7. Linn, S., and Lehmann, I. R., *J. Biol. Chem.* 240 (1965) 1287, 1294.
8. Curtis, P. J., Burdan, M. G., and Smellie, R. M. S., *Biochem. J.* 98 (1966) 813.
9. Curtis, P. J., and Smellie, R. M. S., *Biochem. J.* 98 (1966) 818.
10. Litt, M., and Ingram, V. M., *Biochemistry*, 3 (1964) 560.
11. Armstrong, A., Hagopian, H., Ingram, V. M., and Wagner, E. K., *Biochemistry*, 5 (1966) 3027.
12. Keselev, L. L., Frovola, L. Y., Borisova, O. F., and Kukhanova, M. K., *Biokhimiya*, 29 (1964) 116.
13. Wagner, E. K., and Ingram, V. M., *Biochemistry*, 5 (1966) 3019.
14. Healey, J. W., Stollar, D., Simon, M. I., and Levine, L., *Arch. Biochem. Biophys.* 103 (1963) 461.
15. Ingram, V. M., and Pierce, J. G., *Biochemistry*, 1 (1962) 580.
16. Van Rapenbusch, R., and Deshepper, J. C., *Compt. Rend.* 262 (1966) 1365.
17. Felsenfeld, G., and Rich, A., *Biochim. Biophys. Acta*, 26 (1957) 457.
18. Blake, R. D., and Fresco, J. R., *J. Mol. Biol.* 19 (1966) 145.
   Massoulie, J., Thèse de Doctorat, Univ. Paris, Juin 1966.

19. Rich, A., *Biochim. Biophys. Acta*, 29 (1958) 502.
20. Haselkorn, R., and Doty, P., *J. Biol. Chem.* 236 (1961) 2738.
21. Holcomb, D. N., Tinoco, I., jr., *Biopolymer*, 3 (1965) 121.
22. Heppel, L. A., Ortiz, P. J., and Ochoa, S., *Science*, 123 (1956) 415.
23. Fresco, J. R., and Alberts, B. M., *Proc. Natl. Acad. Sci. U. S.* 43 (1960) 311.
24. Thang, M. N., Guschlbauer, W., Zachau, H. G., and Grunberg-Manago, M., *J. Mol. Biol.* In press.
   Monier, R., and Grunberg-Manago, M., *Acides Ribonucl. et Polyphosph.* C.N.R.S., Paris 1962, p. 163.
   Singer, M. F., Luborsky, S., Morrison, R. A., and Cantoni, G. L., *Biochim. Biophys. Acta*, 38 (1960) 568.
25. Zamir, A., Holley, R. W., and Marquisee, M., *J. Biol. Chem.* 240 (1965) 1267.
26. Dunn, D. B., Smith, J. D., and Spahr, P. F., *J. Mol. Biol.* 2 (1960) 113.

K. Kasai's present address:
Department of Biochemistry
Faculty of Sciences
Tokyo University, Tokyo, Japan

M. Grunberg-Manago
Institut de Biologie Physico Chimique
13 Rue Pierre Curie
75 Paris-5, France

European J. Biochem. 1 (1967) 164—169

Characterization of Ribonucleic Acid with Template Activity in Rat Liver Cytoplasm

A. DI GIROLAMO, M. DI GIROLAMO, S. GAETANI, and M. A. SPADONI

International Laboratory of Genetics and Biophysics, Naples and Istituto Nazionale della Nutrizione, Rome

(Received December 8, 1966)

Template activity of RNA from rat liver and the distribution of this activity along a sucrose gradient have been studied. The experiments were performed with RNA extracted from polysomes, monosomes and ribosomal subunits.

There is no correlation between template activity of RNA extracted from polysomes and early-labelled RNA, since template activity is distributed all along a sucrose gradient while the radioactivity after a 20 min pulse with [^{14}C]orotic acid is almost exclusively localized in the 18-4 S region.

Stimulatory activity in the heavy regions of the gradient is due either to unique large molecules of messenger RNA or to aggregation of ribosomal RNA with messenger RNA. Our experimental results, however, seem to exclude aggregation due to the presence of bivalent ions as well as aggregates held by hydrogen bonds.

Ribosomal RNA is itself responsible for part of the template activity found in different fractions of the gradient. Ribosomal RNA extracted from monosomes prepared from polysomes by treatment with RNase, and from subunits, showed a stimulatory activity which, although inferior to that due to messenger RNA, was consistent. Experiments were performed which excluded the possibility that residual messenger still attached to ribosomal RNA was responsible for this activity.

Treatment of ribosomal RNA with chemical agents such as EDTA and urea increased the template activity in those cases where a permanent increase in hyperchromicity could be demonstrated. Although neomycin also increased stimulatory activity, it is thought unlikely, on the basis of our experiments, that this is due to a separation of the strands of RNA.

The ability of a given RNA fraction to stimulate amino acid incorporation into ribosomal cell-free systems has been frequently adopted as one of the most valid criteria to detect mRNA. Other parameters such as early labelling, DNA-like base composition, ability to form hybrids in high proportion with homologous DNA, have been used from time to time as alternative criteria.

Since the ultimate criterion for defining mRNA is the demonstration of its ability to code for the synthesis of a specific protein *in vitro*, the above-mentioned criteria must be considered as indirect, and unless a coincidence among the various parameters is clearly demonstrated, their validity can be questioned.

In a previous paper [1], we have demonstrated that the distribution of stimulatory activity along

a sucrose density gradient of liver nuclear RNA does not coincide with that of the early-labelled RNA.

These results have led us to make a more detailed study of the significance of the stimulatory activity of RNA extracted from rat liver and of the distribution of this activity along a sucrose gradient. In the present experiments, RNA extracted from polysomes, monosomes and ribosomal subunits was studied.

MATERIALS AND METHODS

[6-^{14}C]orotic acid, specific activity varying from 29 to 42 mC/mmole, uniformly labelled L-[^{14}C]valine (161 mC/mmole), uniformly labelled L-[^{14}C]arginine (250 mC/mmole), and uniformly labelled [^{14}C]Chlorella hydrolizate (approximately 200 μC/mg), were purchased from the Radiochemical Centre, Amersham, England.

RNase was obtained from the Worthington Biochemical Corporation.

Neomycin sulphate was a gift from E. R. Squibb and Sons.

Non-standard Abbreviations. Tris-buffer containing K$^+$ ions, TK; Tris-buffer containing Mg^{++} and K$^+$ ions, TMK; messenger RNA, mRNA; ribosomal RNA, rRNA; *N*-cyclohexyl-*N''*-(methylmorpholinum)-ethyl-carboodiimide-iodide, CMEC; trichloroacetic acid, TCA; transfer RNA, tRNA.

Enzyme. RNase, or ribonuclease, or ribonucleate pyrimidine-nucleotido-2'-transferase (cyclizing) (EC 2.7.7.16).

RNA and Protein Determination

RNA was assayed by the orcinol procedure as modified by Albaum and Umbreit [2] or, when relatively pure, by the measurement of its absorption at 260 mμ, assuming that 1 μg/ml of RNA has an absorbance of 0.030 at 260 mμ [3]. Protein content was determined by the method of Lowry et al. [4].

Preparation of Polysomes, Monosomes and Subunits

Male Wistar rats, weighing approximately 200 g, were sacrificed after overnight starvation. The livers were quickly removed and polysomes prepared according to the method of Wettstein, Staehelin and Noll [5].

Polysome pellets were resuspended in a volume of TK (0.05 M Tris, pH 7.5, 0.025 M KCl) equal to one half of the original liver weight, and were fractionated in 10—40% (w/w) sucrose density gradients buffered with TMK solution (0.05 M Tris pH 7.5, 0.004 M MgCl$_2$, 0.025 M KCl).

Monosomes were produced by treatment of polysomes with RNase (0.05 μg/ml for 5 min at 37°) in TK, and were separated by centrifuging the RNase-treated particles at 24,000 rev./min for 2 hours in 10—40% (w/w) sucrose gradients. Tubes corresponding to the monosome peak were pooled and centrifuged overnight at 40,000 rev./min.

For preparation of 50 S and 30 S subunits, 3.5 to 5 mg of monosomes obtained as described above were resuspended in 1 ml of a solution mM Tris pH 7.5, 50 mM KCl, and treated with EDTA (2.5 μmoles/mg of ribosomes). The ribosomal subunits were separated on a 5—20% (w/w) sucrose gradient in the same buffer, for 8 h at 24,000 rev./min [6].

In labelling experiments, 10 to 40 μCuries of [14C]orotic acid or [14C]chlorella hydrolysate per 100 g of body weight were injected intraperitoneally 20 min prior to sacrifice.

RNA Preparation and Fractionation

RNA was isolated from polysomes, monosomes and subunits by cold phenol extraction [7] and was then fractionated on a gradient of 5—20% (w/w) sucrose in 0.1 M NaCl, 0.01 M sodium acetate at pH 5, centrifuged for 15 hours at 20,000 rev./min in a Spinco SW 25 rotor [7]. Three other types of 5—20% (w/w) sucrose gradients were employed: the first, containing 10 mM Tris pH 7.5 and mM EDTA, was centrifuged 15 h at 24,000 rev./min; the second, containing 10 mM Tris pH 7.5, 4 M urea and mM EDTA, was centrifuged 18 h at 24,000 rev./min; the third, containing 0.1 M NaCl, 0.01 M sodium acetate at pH 5, 4 M urea and mM EDTA, was centrifuged 18 h at 24,000 rev./min.

To test the stimulatory activity of various RNA fractions, the RNAs were separated by size on a sucrose density gradient. Tubes corresponding to the desired fractions were pooled and RNA precipitated as described previously [8].

Assay for Template Activity

The cell-free amino acid incorporating system from Escherichia coli of Nirenberg and Matthaei [9] was used to test the stimulatory activity of the purified RNA; the cation content of the cell-free system was modified as described previously [8]. It was found that the tRNA from E. coli which is usually added to this system did not increase the final incorporation in these experiments; it was therefore omitted from the incubation mixture. The total volume of the reaction mixture was 0.25 ml. 0.45 to 0.60 mg of incubated S-30 protein and 0.48 nmoles of [14C]-valine or 0.30 nmoles of arginine were present in each mixture. Samples were incubated at 37° for 60 min. The material precipitable with trichloroacetic acid was washed as described previously [8] and counted in a Nuclear Chicago low background counter.

Treatment with EDTA, Heat, Urea and Neomycin

In some experiments, the purified RNA extracted from monosomes was treated in various ways before being tested in an aminoacid incorporating system: (a) dissolved in mM Tris pH 7.3, 4 M urea, mM EDTA, and dialyzed before use against mM Tris pH 7.3; (b) heated to 100° for 15 min in the presence of mM EDTA, rapidly cooled, and then dialyzed against mM Tris pH 7.3; (c) treated with neomycin sulphate (8 μg/ml) either without heating, during heating or after heating and then added to the amino acid incorporating cell free system.

RESULTS

Early-labelled RNA and RNA with Template Activity

In Fig. 1 the distribution of the stimulatory activity of RNA isolated from polysomes is compared with the distribution of radioactivity after a 20 min pulse with [14C]orotic acid. It is evident that there is no correlation between these two patterns. As can be seen, template activity is found all along the gradient, whereas the early-labelled RNA is localized in the 18-4 S region. This difference in distribution is not an artifact due to overloading of the gradient since the specific stimulatory activity of the various RNA fractions was independent from the input of RNA on the gradients within the range used (0.2 to 2 mg of RNA).

The distribution of template activity shown in Fig. 1 is not in agreement with results previously reported [8,10], in which the stimulatory activity is by far the highest in the 18 S region. The possibility that the stimulatory activity in the heavy region of the gradient is due to contamination by nuclear RNA seems unlikely. In fact we find that the specific stimulatory activity of nuclear RNA is three times higher than that of ribosomal RNA. However, it should be noted that nuclear RNA accounts for only $^1/_{10}$th the total cellular RNA [11], and it can easily be calculated that contamination would not be such as to influence the pattern of stimulatory activity.

In order to study the significance of the distribution of the stimulatory activity along the sucrose gradients, the following experiment was done. Polysomes isolated from a sucrose gradient were divided

and the stimulatory activity of each was then tested in an *in vitro* system.

As can be seen from Table 1, the distribution of stimulatory activity is found to be identical in the two polysome fractions for each of the four samples tested. If there were a strict relationship between polysome size and messenger RNA size [12], it could be expected that the RNA extracted from polysomes larger than pentamers would have more stimulatory activity in the heavy regions of the gradient than would the RNA from the smaller polysomes. These results therefore might indicate that mRMA forms aggregates with ribosomal RNA

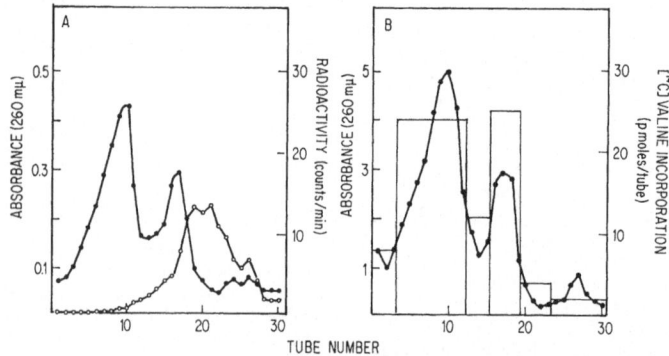

Fig.1. *Sedimentation pattern of RNA from polysomes.* The sucrose gradients [5—20°/₀ (w/w) in acetic acid buffer pH 5] were centrifuged for 15 hours at 20,000 rev./min. A, sedimentation analysis of early-labelled RNA (20 min pulse with [¹⁴C]orotic acid (29 mC/mmole; 10 µC/100 g body weight). ●, Absorbance at 260 mµ; ○, counts/min. B, capacity of RNA to stimulate amino acid incorporation into hot TCA-precipitable material; the gradient was loaded with 2 mg of RNA. The height of the columns shows pmoles of [¹⁴C]valine incorporated per tube (calculated by dividing the total incorporation per fraction by the number of tubes pooled to make up the fraction). Similar results were obtained from three different experiments. In all figures, incorporation in the absence of added RNA has been subtracted; this value never exceeded 0.6 pmoles of [¹⁴C]valine incorporated per sample

Table 1. *Template activity of RNA from polysomes of different size*
Two fractions of polysomes were isolated as described in "Results". The RNA was prepared, fractionated and assayed as described in "Methods"

	[¹⁴C] Valine incorporation			
	> 28 S	28 S	18 S	18-4 S
	pmoles/mg RNA			
polysomes > 5 n	286	310	480	425
polysomes < 5 n	280	316	458	440

into two fractions: the first fraction was composed of all polysomes larger than pentamers; the second included pentamers and smaller polysomes. The RNA in each fraction was extracted from the pellets obtained by centrifugation and was then fractionated on sucrose gradients. Samples which corresponded to RNA > 28 S, 28 S, 18 S, and 18-4 S were isolated

and that these aggregates are responsible for part of the stimulatory activity found in the heavy regions of the gradients. Formation of complexes between messenger and ribosomal RNA has been postulated also by Arnstein and Cox [13] to explain their results on stimulatory activity with rabbit reticulocytes.

In order to investigate the nature of these possible aggregates, we performed the following experiments. The first experiment was designed to test whether the aggregates were held together by divalent cationic bonds.

The RNA extracted from the polysomes was treated with mM EDTA in mM Tris pH 7.3, and then it was fractionated on sucrose gradients. The fractions of RNA obtained from each gradient were precipitated with ethanol and dialyzed overnight against mM Tris pH 7.3 and were then tested for stimulatory activity. Since the distribution of stimulatory activity does not differ from that shown in Fig.1, it can be concluded that divalent cationic bonds are not involved in the formation of these aggregates. Next we tested for the presence of hydrogen bonding by treatment of the RNA with mM EDTA and 4 M urea both before and during fractionation on sucrose gradients. Again, no change in the pattern of stimulatory activity was observed. Thus the possibility that the aggregates are stabilized by hydrogen bonds can reasonably be ruled out.

Template Activity of Ribosomal RNA

A possible explanation for the distribution of template activity all along the profile of ribosomal RNA could be the ability of ribosomal RNA itself to act as a template.

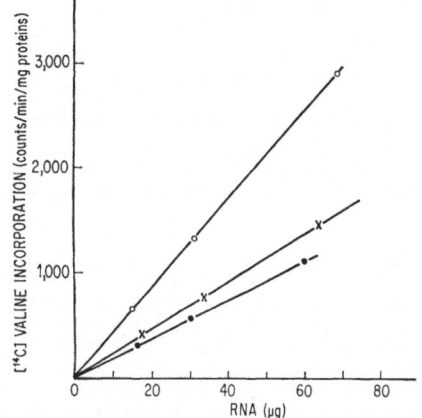

Fig. 2. *Stimulation of [¹⁴C]valine incorporation by an E. coli system following addition of RNA. RNA extracted: ×, from monosomes; ●, from 50 S subunits; ○, from 30 S subunits. In the absence of RNA, 200 counts/min/mg protein were incorporated; this number has been substracted from each value shown. Each reaction mixture of 0.25 ml contained 0.550 mg of incubated S 30 protein. The samples were incubated at 37° for 60 min*

To test this possibility RNA was extracted from polysomes, from monosomes prepared by treatment of polysomes with RNase, and from their ribosomal subunits. Fig. 2 shows that total RNA extracted from monosomes and from 50 S and 30 S subunits is still capable of stimulating the incorporation of amino acids into proteins, although to a more limited extent as compared to RNA extracted from polysomes (Table 2). Furthermore, as can be seen, there is a linear relationship between RNA added to the system and the incorporation of amino acids, at least in the range from 10 to 70 μg of RNA.

The RNA obtained from the different preparations was then fractionated on sucrose gradients. The

Table 2. *Template activity of RNA from different cellular subfractions*
The RNA was prepared, fractionated and assayed as described in "Methods". The numbers represent pmoles of [¹⁴C]valine incorporated per mg of RNA in an *E. coli* cell-free system

	[¹⁴C] Valine incorporation			
	> 28 S	28 S	18 S	18-4 S
	pmoles/mg RNA			
RNA from polysomes	325	350	529	506
RNA from monosomes	135	166	348	205
RNA from 50 S		154	200	
RNA from 30 S			395	

absorbance patterns from polysomes and from monosomes were as usual. The absorbance pattern of the RNA extracted from 30 S subunits indicated that only 18 S RNA was present. The RNA from 50 S, on the other hand, gave a large peak of 28 S RNA, with a minor peak of 18 S RNA.

On the basis of the pattern of absorbance, four fractions were isolated and tested for stimulatory activity. Each of the fractions obtained from RNA extracted from monosomes shows residual stimulatory activity, with a maximum specific activity located in the 18 S region. The RNA extracted from the subunits is also capable of stimulating amino acid incorporation; this capacity is essentially equal for the 18 S RNA obtained from the 30 S subunit and the 18 S from monosomes. In the same way, the stimulation obtained with the 28 S RNA from the 50 S subunit is equal to that obtained with the 28 S from the monosomes. Furthermore, the stimulatory activity of the 18 S RNA from the 50 S subunit is similar to that of the 28 S RNA.

The results reported in Fig. 3 indicate that the template activity reported in Table 2 is not due to fragments of messenger RNA attached to the ribosomal RNA, since, after a 20 min pulse with [¹⁴C]-orotic acid, at a time when only messenger RNA should be labelled, monosomes prepared by treatment of polysomes with RNase contain very little radioactivity (Fig. 3 B). Furthermore, when the RNA extracted from these monosomes was sedimented on a sucrose gradient, this residual activity was localized in regions of the gradient which are certainly lighter than those previously occupied by the early-labelled RNA. Likewise, when the monosomes were broken down into 50 S and 30 S subunits by treatment with EDTA, no radioactivity was found associated with either subunit. Therefore, the presence of template activity in the purified ribosomal RNA from monosomes and from subunits could be ascribed to template activity of the ribosomal RNA itself. These conclusions are based on the assumption that early-labelled RNA is representative of all messenger RNA. Otherwise, we must admit the possibility that ribosomal RNA binds slow-turnover messenger RNA with

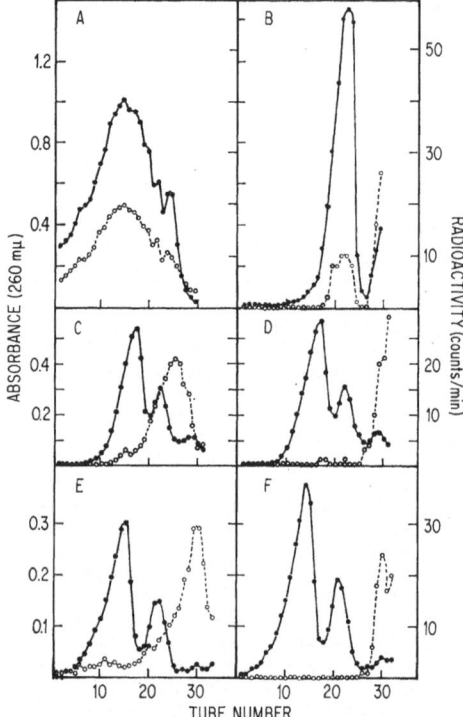

Fig. 3. Sedimentation pattern of early-labelled RNA prepared by a 20 min pulse of [14C]orotic acid (42 mC/mmole) in polysomes, monosomes and ribosomal subunits. The animals were injected with 40 μC/100 g body weight. A, polysomes; B, product of treatment of polysomes with RNase (0.05 μg/ml) for 5 min at 37°; C and D, RNA extracted from polysomes and from monosomes; E and F, 50 S and 30 S subunits obtained by treatment of polysomes and monosomes with EDTA (2.5 μmoles/mg of ribosomes). ●, absorbance at 260 mμ; ○, radioactivity

bonds different from those with which it holds rapid-turnover messenger RNA, possibily of the type postulated by Arnstein in reticulocytes [17].

The Effect of Treatment of Purified Ribosomal RNA with Urea, Heat and Neomycin on Template Activity

RNA extracted from monosomes was treated with 4 M urea and mM EDTA in mM Tris pH 7.3 and was then dialyzed overnight against mM Tris pH 7.3. In some experiments, a measurable hyperchromic effect of about 10% was found after dialysis, and in these cases it was possible to demonstrate an increase in template activity (Table 3). This hyperchromic effect cannot be ascribed to breakage of RNA since

Table 3. *Template activity of RNA from monosomes after different treatments*
The numbers represent the pmoles of [14C]valine and [14C]-arginine incorporated/mg of RNA in an *E. coli* cell-free system by RNA isolated from monosomes, and by this RNA treated with urea and EDTA, heat and EDTA and neomycin sulphate. The neomycin was also added either during or after heating of the sample

	[14C] Valine incorporated	[14C] Arginine incorporated
	pmoles/mg RNA	
RNA from monosomes	180	350
RNA from monosomes + urea		
+ EDTA	350	
RNA from monosomes heated		
+ EDTA	120	470
RNA from monosomes + neomycin	864	1505
RNA from monosomes heated in presence of neomycin	785	1550
RNA from monosomes heated, then + neomycin	800	1540

this last continues to sediment at the same velocity after treatment as before treatment with urea.

On the other hand, treatment with heat (100° for 15 min) did not cause a stable measurable hyperchromic effect and an increase in template activity. This is probably due to the fact that, under our experimental conditions, although cooling was carried out very quickly, the denaturation was completely reversed.

According to Holland, Buck and McCarthy [14], the addition of neomycin to ribosomal RNA derived from *E. coli* causes an increase in the incorporation of amino acids into protein. This effect was even greater when both neomycin and heat were used. As can be seen from Table 3, there is an increase in the template activity of rat liver rRNA after treatment with neomycin. However, simultaneous treatment with heat, under our experimental conditions, did not enhance the effect of neomycin.

The action of neomycin does not appear to be due to an increase or stabilization of the fraction of RNA present in single-stranded form. This has been demonstrated by treatment of ribosomal RNA with CMEC which preferentially binds to the uridylic and guanilic bases not linked by hydrogen bonds [15]. We found that RNA binds the same amount of CMEC in the presence of or in the absence of neomycin (unpublished results).

DISCUSSION

The principal result of these experiments has been the demonstration that rRNA itself has template activity which, in the fractions examined, accounts for more than 40% of polysomal RNA template activity. The remaining activity can be ascribed to mRNA and raises some interpretative problems which cannot be definitely resolved on the basis of the indirect evidence provided by our results.

As to the activity which is not accounted for by the template activity of rRNA itself, it can be attributed either to the actual presence of unique large molecules of mRNA or to smaller mRNA molecules linked with rRNA. The first possibility is contradicted by the finding that the template activity of RNA extracted from polysomes larger than pentamers and from smaller polysomes shows the same distribution along a sucrose gradient. However it should be noted that the existence of a strict relationship between mRNA size and polysome size has not been found [16]. On the other hand a slower turnover of the larger mRNA molecules provides a more reasonable explanation of the fact that early-labelled RNA and template activity do not coincide than does the hypothesis that mRNA with slower turnover has more affinity for rRNA. In any case, if a linkage exists between mRNA and rRNA, it does not seem to depend upon the presence of divalent cations nor on the presence of hydrogen bonds, since treatment with EDTA or with urea neither modified the distribution nor decreased the amount of template activity. This, therefore, might be a covalent type linkage, as postulated by Gould, Arnstein and Cox [17].

Experiments on the stimulatory activity of RNA extracted from monosomes and from subunits demonstrate that the rRNA itself can serve as a template. Monosomes prepared from polysomes retain only a minimal proportion of the radioactivity which had been introduced into the polysomes with a 20 min pulse with [^{14}C]orotic acid, yet the residual template activity of their RNA remains still relatively high. In addition, even in the ribosomal subunits prepared by treatment of monosomes with EDTA and in the ribosomal RNA components extracted from monosomes, no residual radioactivity could be found although the stimulatory activity remains unchanged.

That portions of single-stranded rRNA can by themselves act as a template is in accord with observations by Miura and Muto [18]. They found that single-stranded RNA from tobacco mosaic virus has stimulatory activity in vitro, whereas double-stranded RNA from rice dwarf virus has no such ability unless the structure is destroyed by heating followed by rapid cooling. More recently, Holland, Buck and McCarthy [14] have demonstrated in E. coli that RNA extracted from monosomes and from subunits has a slight template activity, and that this capacity increases notably with treatment with heat and with neomycin. According to these authors, treatment with heat and with neomycin will increase the amount of RNA with modified secondary structure, thus leading to an increase in stimulatory activity. With our preparation of ribosomal RNA from rat liver, it was not possible to obtain a stable denaturation with heat, and an increase in the stimulatory activity was never demonstrated. However, we were able, in a very few cases, to demonstrate an increase of stimulatory activity after treatment with urea. In these experiments at least, a part of the denaturation remained after removal of the urea, as was demonstrated by a hyperchromicity. Thus we can conclude that the ability of rat liver ribosomal RNA to act as a template is a result of a modification of structure of ribosomal RNA. The increase of stimulatory activity found after treatment with neomycin seems to confirm the ability of ribosomal RNA to act as a template. However, on the basis of our results with CMEC, it is not possible to ascribe this effect to an increase in, or stabilization of, portions of RNA with modified structure. At present we have no information to establish whether or not the template activity found in the rRNA is of importance in vivo for the synthesis of structural proteins of the ribosomes, as was postulated by Holland et al. for E. coli [14].

We gratefully acknowledge the skilled technical assistance of Mr. E. Busiello. This work was partially carried out under the Association Euratom-CNR-CNEN, Contract No. 012-61-12 BIAI.

REFERENCES

1. Di Girolamo, A., Di Girolamo, M., Gaetani, S., and Spadoni, M. A., Biochim. Biophys. Acta, 114 (1966) 195.
2. Albaum, H. G., and Umbreit, W. W., J. Biol. Chem. 167 (1947) 369.
3. Pardee, A. B., Paigen, K., and Prestidge, L., Biochim. Biophys. Acta, 23 (1957) 162.
4. Lowry, O. H., Rosebrough, N. J., Farr, A. L., and Randell, R. J., J. Biol. Chem. 193 (1951) 265.
5. Wettstein, F. O., Staehelin, T., and Noll, H., Nature, 197 (1963) 430.
6. Tashiro, Y., and Siekevitz, P., J. Mol. Biol. 11 (1965) 149.
7. Hiatt, H. H., J. Mol. Biol. 5 (1962) 217.
8. Di Girolamo, A., Henshaw, E. C., and Hiatt, H. H., J. Mol. Biol. 8 (1964) 479.
9. Nirenberg, M. W., and Matthaei, J. H., Proc. Natl. Acad. Sci. U. S. 47 (1961) 1588.
10. Brawerman, G., Biezunsky, N., and Eisenstadt, J., Biochim. Biophys. Acta, 103 (1965) 201.
11. Leslie, J., in The Nucleic Acids Chemistry and Biology (edited by E. Chargaff and J. N. Davidson), Academic Press, New York 1955, Vol. II, p. 12.
12. Staehelin, T., Wettstein, F. O., Oura, H., and Noll, H., Nature, 201 (1964) 264.
13. Cox, R. A., and Arnstein, H. R. V., Biochem. J. 93 (1964) 336.
14. Holland, J. J., Buck, C. A., and McCarthy, B. J., Biochemistry, 5 (1966) 358.
15. Augusti-Tocco, G., and Brown, G. L., Nature, 206 (1965) 683.
16. Latham, H., and Darnell, J. E., J. Mol. Biol. 14 (1965) 1.
17. Gould, H. J., Arnstein, H. R. V., and Cox, R. A., J. Mol. Biol. 15 (1966) 600.
18. Miura, K. J., and Muto, A., Biochim. Biophys. Acta, 108 (1965) 707.

A. and M. Di Girolamo
Laboratorio Internazionale di Genetica e Biofisica
Via G. Marconi 10
Casella postale 3061
Napoli, Italy

M. A. Spadoni and S. Gaetani
Istituto Nazionale della Nutrizione
Roma, Italy

European J. Biochem. 1 (1967) 170—178

Purification and Properties of Two Forms
of 6-Phosphogluconate Dehydrogenase from *Candida utilis*

M. Rippa, M. Signorini, and S. Pontremoli

Istituto di Chimica Biologica, Università di Ferrara

(Received November 8, 1966)

Crude extracts of *Candida utilis* contain two types of 6-phosphogluconate dehydrogenase. One can be obtained in the crystalline form and has already been described and studied. In the present paper the purification of the second one, together with the differences in the properties between the two purified proteins, is reported. The two types of enzyme have identical pH optimum, same K_m for the substrates and same specificity. They differ in amino acid composition, molecular weight, electrophoretic mobility, stability to pH and heat treatment, sensitivity to chlorodinitrobenzene and proteolytic treatment. All evidence indicates a greater instability of the crystalline enzyme in respect to the non-crystalline one and seems to exclude that the two forms of enzyme are an artefact due to the extraction or the purification procedures.

A method for the purification of the 6-phosphogluconate dehydrogenase from *Candida utilis* has been described by Horecker and Smyrniotis [1]. The enzyme has been obtained in a homogeneous crystalline form by Pontremoli *et al.* [2].

We have previously shown that, in the active center of the enzyme, there is at least a cysteine residue, a lysine residue and a phosphate attracting group. The enzyme can be inactivated by the binding of iodoacetate [3] or chlorodinitrobenzene [4] to a single cysteine residue, or by the binding of pyridoxal-5′-phosphate to a single lysine residue [5] of the enzyme molecule.

The crystalline enzyme does not account for the total 6-phosphogluconate dehydrogenase activity of the crude extract of *Candida utilis*. In the present paper we present evidence which indicates that another protein, with 6-phosphogluconate dehydrogenase activity, is present in our extracts and can be purified.

A method has been worked out for the purification of this form of enzyme. This protein does not crystallize in the conditions used for the crystallization of the other type of enzyme. The crystallizable 6-phosphogluconate dehydrogenase will be indicated as enzyme type I, the non-crystalline as enzyme type II.

From the amino acid composition, molecular weight determinations and from other parameters it appears that the enzyme type I could be derived from proteolytic digestion of the enzyme type II. The enzyme type I exhibits a greater instability to the chemical and enzymatic action of several agents.

EXPERIMENTAL PROCEDURES
Materials

Candida utilis, dried at low temperature, was supplied by the Lake States Yeast Corp., Rhinelander, Wisconsin, and kept at 4°. 6-phosphogluconate, TPN, pyridoxal-5′-phosphate, glutathione and *p*-hydroxymercuribenzoate were purchased from Sigma Chem. Co., St. Louis, Missouri. Protamine sulfate was obtained from Lilly Co. Indianapolis, Indiana. Glycerophosphate dehydrogenase, from rabbit muscle, was purchased from Boehringer, Germany. Fructose 1,6-diphosphatase was prepared as previously described [6]. This enzyme has a molecular weight of 127,000 [7]. Twice crystallized trypsin was a Worthington product. Chemicals used for disc-gel electrophoresis were obtained from Canal Industrial Co., Bethesda, Maryland. All other chemicals were reagent grade.

Methods

The ammonium sulfate concentration was measured with a Barnstead purity meter. Whatman DEAE-cellulose and phospho-cellulose were freed from fines by allowing them to settle several times after suspension in water. The resins were then suspended in 0.5 M NaOH, filtered and washed with

---

Enzymes. 6-phosphogluconic dehydrogenase, or phosphogluconate dehydrogenase, or 6-phospho-D-gluconate: NADP oxidoreductase (decarboxylating) (EC 1.1.1.44); glycerophosphate dehydrogenase, or L-glycerol-3-phosphate: (acceptor) oxidoreductase (EC 1.1.99.5); D-Fructose-1,6-diphosphate 1-phosphohydrolase, or Hexosediphosphatase (EC 3.1.3.11); ribosephosphate isomerase, or D-ribose-5-phosphate ketol-isomerase (EC 5.3.1.6).

distilled water until neutrality. The adsorption of the enzyme on the resin was carried out batchwise and was checked by analyzing, for enzymatic activity, samples of the supernatant collected after removal of the resin by centrifugation. Spectrophotometric determinations were carried out in a Zeiss PMQII spectrophotometer. All the purification procedures were carried out at 0—4°, unless otherwise stated.

Enzyme Assay

The enzymatic activity of both forms of 6-phosphogluconate dehydrogenase was determined spectrophotometrically at 22°, following the initial rate of TPN reduction in an assay mixture (1 ml) containing 0.3 mM 6-phosphogluconate, 0.3 mM TPN and 10 mM phosphate buffer, pH 7.4. The reaction was started by the addition of the enzyme and readings were taken at 5 sec intervals at 340 mμ. One unit of enzyme activity was defined as the quantity that would produce a change in absorbance of 1.0 per min.

RESULTS

Purification of the Two Forms of Enzyme

Extraction, Protamine and Heat Treatment. The dry yeast (200 g) was suspended in 1.2 liters of distilled water. After 4 hours at 37° the autolyzed suspension was centrifuged in the cold for 30 min at 20,000 × g. The supernatant (crude extract, Table 1) was adjusted to pH 6.2 with 1 M NaOH and treated with 2 g of protamine, previously dissolved in 150 ml of water. The precipitate was discarded by centrifugation. The supernatant (protamine fraction) was adjusted to pH 5.5 with 1 M CH_3COOH, heated in a water bath at 50° for 6 min, with continuous stirring, and cooled to 4°. The precipitate was discarded as above.

Column Chromatography. The supernatant obtained after centrifugation (heat fraction) was diluted 5 fold with distilled water and the pH of the solution was adjusted to 5.8. Wet phosphocellulose was added to the solution batchwise, until no more enzymatic

Table 1. *Purification procedure of the two forms of 6-phosphogluconate dehydrogenase*

Enzyme	Step	Volume	Activity		Proteins	Specific activity	Purification	Yield
		ml	units/ml	total units	mg/ml	units/mg		%
	Crude extract	900	26	23,500	29	0.9	1	100
	Protamine	1,060	21	22,500	14.2	1.6	1.6	94
	Heat	980	21	20,800	7.8	2.7	3.0	89
	Phosphocellulose	190	96	18,200	9.6	10.0	11.1	77
	First ammonium sulfate	195	90	17,500	7.0	13.1	14.4	74
Type I	First crystals	35	165	5,900	6.5	25.3	28	25
	Second crystals	36	150	5,400	2.7	56.0	62	23
	Third crystals	32	150	5,100	1.5	100	116	21
	Fourth crystals	30	160	4,800	1.0	160	176	20
Type II	Supernatant first crystals	190	59	11,250	5.9	10	11.1	48
	Second ammonium sulfate	250	34	8,500	1.6	21.5	23.2	36
	DEAE-cellulose	20	250	5,000	1.9	133	148	21

Protein concentration was determined from the absorbance at 280 mμ, based on dry weight determination. A solution containing 1.0 g per ml of pure enzyme (type I or type II) has an absorbance of 1.270 at 280 mμ. In the early steps of purification the turbidimetric method of Bucher [8] was used. The specific activity of the enzyme was defined as units per mg of protein. The specific activity of the purified crystalline 6-phosphogluconate dehydrogenase was approximatively 160. One mg of purified crystalline enzyme catalyzes the oxidation of 25.6 μmoles of 6-phosphogluconate per min.

All studies of a given kinetic or physical parameter were performed in a similar fashion with enzyme preparations of similar age and treatment in order to minimize all differences other than those specifically due to the enzyme type studied.

activity was present in the supernatant (see Methods). During this procedure the pH of the suspension was kept constant at 5.8. The resin suspension was then filtered through a Buchner funnel and the filtrate discarded. The wet cake of resin was suspended in 0.5 liters of 50 mM phosphate buffer, pH 6.2, and the suspension poured into a 4.5 × 50 cm chromatographic column. The resin was washed on the column with 2 liters of 50 mM phosphate buffer, pH 6.2, containing 0.1 mM EDTA. No enzymatic activity was detected in the washings. The average volume of the resin packed in the column was 200 ml. The enzyme was then eluted with 0.4 M phosphate buffer, pH 7.4, containing 0.1 mM EDTA and fractions of 10 ml each were collected. The fractions containing enzyme activity were pooled and adjusted to pH 6.2 (phosphocellulose column).

Ammonium Sulfate Precipitation and Crystallization. The resultant solution was treated with an amount of solid ammonium sulfate sufficient to give a 50 % saturation. During the addition of ammonium sulfate, the pH of the solution was kept at 6.2. The precipitate was discarded by centrifugation. To the supernatant (first ammonium sulfate) a cold saturated ammonium sulfate solution was added dropwise, with continuous stirring, keeping the pH at 6.2, until a slight turbidity appeared. At this point the ammonium sulfate saturation was approximately 55 %. Crystal formation began in a few minutes and was complete after 16—20 hours. The suspenrion was stored overnight at 4°.

Separation of the two Forms of Enzyme. In order to separate the two forms of 6-phosphogluconate dehydrogenase, the crystalline suspension was centrifuged and the precipitate dissolved in 20—25 ml of 10 mM phosphate buffer, pH 6.2, containing 0.1 mM EDTA. This fraction (first crystals) contained essentially only the crystallizable enzyme (6-phosphogluconate dehydrogenase, type I), while the supernatant (supernatant first crystals) contained predominantly the non-crystallizable enzyme (6-phosphogluconate dehydrogenase, type II). Each enzyme was further purified as described below.

Purification of the 6-Phosphogluconate Dehydrogenase Type I. For recrystallization, the solution containing the enzyme type I was treated with a saturated ammonium sulfate solution, added dropwise, until a slight turbidity appeared. After 4 hours at 4°, the crystalline suspension was centrifuged and the supernatant discarded. The precipitate, was again dissolved in 10 mM phosphate buffer, pH 6.2, containing 0.1 mM EDTA (second crystals). The enzyme was subjected to further crystallizations using the same procedure. Usually after 4 crystallizations the enzyme appeared to be homogeneous as judged by disc-gel and starch-gel electrophoresis, sucrose density gradient centrifugation, and no change in specific activity after column chromatography on DEAE-CM- and phosphocellulose (see below). The enzyme was kept as a crystalline suspension at 4°.

Purification of the 6-Phosphogluconate Dehydrogenase Type II. The supernatant (supernatant first crystals) was treated with a saturated ammonium sulfate solution, keeping the pH of the solution at 6.2, until a turbidity appeared. At this point the ammonium sulfate saturation was approximately 65 %. The suspension was kept at 4° for 2 days during which the turbidity increased. The suspension was then centrifuged and the inactive precipitate discarded. The supernatant was brought to pH 4.7 with 5 M CH$_3$COOH and the inactive precipitate discarded. The supernatant was adjusted to pH 6.2 with 1 M NaOH and the enzyme precipitated on raising the ammonium sulfate saturation to 85 %. The precipitate, collected by centrifugation, was

dissolved in a small volume (usually 20 ml) of 10 mM phosphate buffer, pH 7.2, containing 0.1 mM EDTA, and dialyzed 5 hours against the same buffer (second ammonium sulfate fraction). The dialyzed enzyme was then diluted to 250 ml with water and adsorbed on DEAE cellulose at pH 7.4; the wet resin was added batchwise to the solution as described before. The resin suspension was then poured into a chromatographic column (2 × 30 cm) and washed with 10 mM phosphate buffer, pH 7.4, until no more protein was detected in the effluent. The average volume of the resin packed in the column was 60 ml. The enzyme was eluted from the resin with a solution containing 0.5 % saturated ammonium sulfate in the same buffer. The fractions containing the enzyme activity were pooled (DEAE cellulose fraction) and were kept at 4° in 80 % saturated ammonium sulfate, 10 mM phosphate buffer, pH 6.2, containing 0.1 mM EDTA.

Remarks on the Purification Procedure. For the crystallization of the enzyme type I the enzymatic activity must be at least 7 units/mg.

In the case of the enzyme type II, crystalline material was not obtained at any step of the purification, under the conditions used for the type I enzyme.

Properties of the Two Forms of Enzyme

Stability. The two types of enzyme, stored as described, did not show any appreciable loss of enzymatic activity after several months.

Alternative Extraction Methods. The possibility that the two forms of enzyme represent an artefact due to the extraction procedure, was ruled out by using different methods of extraction of the enzyme from yeast. Crude extracts were prepared as follows: (a) grinding the dry yeast with silica powder in a mortar, at 4°, with water; (b) autolyzing the dry yeast for 20 min at 4°; (c) autolyzing at 37° for a period of 30 min, 4 and 8 hours; (d) autolyzing the dry yeast for 20 min at 4°, and incubating the supernatant and the precipitate (collected after centrifugation and dissolved in water) separately at 37° for 4 hours.

All extracts were submitted to the same purification procedure described in detail above. In all cases (see Table 2) the ratio of the two forms of enzyme, as calculated from the ratio of crystallization, did not vary appreciably, altough the total starting units were different, depending on the different type of extraction.

Alternative Purification Procedure. In order to eliminate the possibility of an artefactual modification of the protein, during the purification, as a cause of the occurrence of the two forms of enzyme, a modification to the present method of purification was followed, which eliminates the heat and the phospho-

Table 2. *Effect of different methods of extraction on the ratio of the levels of the two forms of 6-phosphogluconate dehydrogenase*

Extraction method	Crude extract	Before enzyme crystallization	Enzyme crystallized	$\frac{\text{Type I}}{\text{Type II}}$
	total units	total units	total units	
1. Mechanical rupture of cells	4,900	3,650	1,080	0.42
2. Autolysis for 20 min at 4°	1,600	1,120	350	0.46
3. Autolysis for 30 min at 37°	5,050	3,800	1,220	0.48
4. Autolysis for 4 hours at 37°	5,150	3,600	1,260	0.54
5. Autolysis for 8 hours at 37°	5,130	3,550	1,010	0.41
6. Incubation for 4 hours at 37° of the supernatant obtained after autolysis for 20 min at 0°	1,600	1,120	340	0.44
7. Incubation for 4 hours at 37° of the precipitate collected after autolysis for 20 min at 0°	3,400	2,050	615	0.47

Table 3. *Alternative purification procedure of the two forms of 6-phosphogluconate dehydrogenase*

Step	Volume	Activity		Protein	Specific activity	$\frac{\text{Type I}}{\text{Type II}}$
	ml	units/ml	total units	mg/ml	units/mg	
Crude extract	900	26	23,600	29.0	0.9	
Protamine	1,000	21	21,000	14.2	1.6	
First ammonium sulfate	150	126	19,000	37.0	3.4	
Phosphate gel, eluate	200	75	15,000	20.6	7.0	
Second ammonium sulfate	230	63	14,600	8.1	7.8	
First crystals	33	138	4,600	9.2	15.1	
Supernatant first crystals	210	47	9,800	7.5	6.2	0.46

cellulose column steps. The modified purification procedure was carried out as follows: the protamine fraction was treated with solid ammonium sulfate, keeping the pH at 6.2, and the fraction, precipitated at an ammonium sulfate saturation of between 50 and 75%, was collected, dissolved in 100 ml of 0.1 M phosphate buffer, pH 6.2, containing 0.1 mM EDTA, and dialyzed for 6 hours against the same buffer (ammonium sulfate fraction, Table 3). The dialysate was diluted to 400 ml with 0.1 M phosphate buffer, pH 6.2, and treated with calcium phosphate gel. This was added in an amount to allow the adsorption of inactive proteins. To the supernatant, collected after centrifugation, more calcium phosphate gel was added, sufficient to adsorb 90% of the active enzyme protein. The gel was centrifuged and the enzyme eluted with a 20% saturated ammonium sulfate solution, containing 0.1 mM EDTA. The resulting solution was submitted to further treatment as indicated under "ammonium sulfate precipitation and crystallization," and the crystallizable form separated from the non-crystallizable one.

The results obtained (Table 3) indicate that the ratio of the two forms of enzyme is identical, following the two different purification procedures.

Purity of the Two Forms of Enzyme. Both purified forms of 6-phosphogluconate dehydrogenase contained no detectable glucose-6-phosphate dehydrogenase or D-ribose-5-phosphate isomerase activities.

The disc-gel and starch-gel electrophoresis patterns of both types of enzyme, shown in Fig. 1, indi-

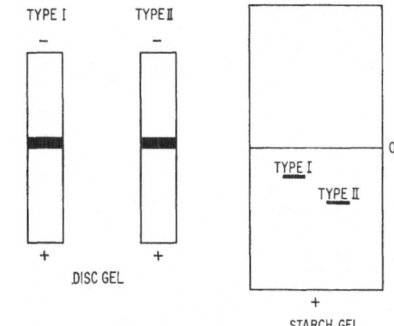

Fig. 1. *Gel electrophoresis of the two purified forms of 6-phosphogluconate dehydrogenase.* Disc-gel electrophoresis was carried out in 7% standard polycrylamide gel, pH 8.6, at 6 mA per tube. The samples contained 0.1 mg of each type of enzyme. Starch-gel electrophoresis was carried out in 0.03 M phosphate buffer, pH 7.0, at 30 mA, 200 volts for 8 hours

cate that each type of enzyme was obtained in a homogeneous form. EDTA was present in the buffer used for the electrophoresis, owing to the instability of the enzyme type I at pH 8.6. The data reported in Fig. 1 indicate that at pH 7 the two forms of enzyme have a different electrophoretic mobility as could be expected by the differences in amino acid composition.

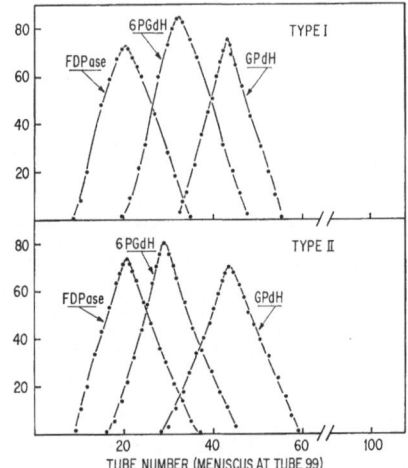

Fig. 2. *Sucrose gradient centrifugation analyses of the two forms of 6-phosphogluconate dehydrogenase (6PGdH).* The method employed was that of Martin and Ames [9]. The purified enzyme (type I or type II) was placed (0.5 mg in 0.05 ml) at the top of the sucrose gradient [5 to 20% sucrose (w/v) in 0.01 phosphate buffer, pH 6.2] together with fructose diphosphatase (FDPase) and glycerophosphate dehydrogenase (GPdH) used as internal standards. After centrifugation at 37,000 rev./min in the rotor SW 39 of the Spinco model L preparative centrifuge, for 19 hours at 4°, fractions were collected and analyzed for the three enzyme activities. Ordinate represents enzyme activities in arbitrary units

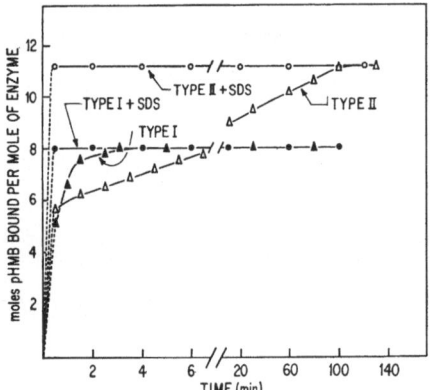

Fig. 3. *Spectrophotometric titration of the sulfhydryl groups of the two forms of 6-phosphogluconate dehydrogenase with p-hydroxymercuribenzoate (pHMB).* The ratio of moles of pHMB bound per mole of enzyme has been calculated from the increment in absorbance at 250 mμ and was standardized with a sample of glutathione. The reaction mixture contained 0.05 mg of enzyme (type I or type II), 0.1 mM pHMB, 50 mM phosphate buffer, pH 7.5 and, where indicated, 2% (w/v) sodium dodecyl sulfate. The temperature was 22°

The homogeneity of each form of 6-phosphogluconate dehydrogenase was also tested by sucrose density gradient centrifugation. Five mg of each form of enzyme, dissolved in 0.3 ml of 10 mM phosphate buffer, pH 6.2, containing 0.1 mM EDTA, were placed at the top of a 5 to 20% (w/v) sucrose gradient in the same buffer. The tubes were subjected to centrifugation at 25,000 rev./min in the rotor SW 25.1 of a Spinco L preparative centrifuge, at 4° for 39 hours. After the run the tubes were punctured and the fractions collected and analyzed for protein concentration and enzymatic activity. One symmetrical peak was obtained for each type of enzyme. The specific activity throughout the peak was equal to that of the sample before centrifugation.

No protein contamination was detected using column chromatography on DEAE- CM- and phosphocellulose.

Molecular Weight of the Two Forms of Enzyme. The approximate molecular weight of both forms of enzyme was determined by sucrose density gradient centrifugation, according to the method of Martin and Ames [9]. Fructose 1,6-diphosphatase and glycerophosphate dehydrogenase were used as markers in each tube (Fig. 2). Calculations of molecular weights, based on several runs, were in good agreement. Assuming spherical proteins, the molecular weight values were 101,000 for the enzyme type I and 111,000 for the enzyme type II.

Amino Acid Composition. The amino acid analysis of each purified form of 6-phosphogluconate dehydrogenase was carried out with the Beckman model 120 B analyzer [10]. The data reported in Table 4 indicate that the percentages of the amino acid residues present in the two proteins are different. On the basis of the molecular weight reported above it appears (Table 4, last column) that the enzyme type II contains, in addition to all the amino acid residues present in the enzyme type I, 30 basic amino acids (14 lysines, 4 histidines, 12 arginines), 46 hydrophobic amino acids (12 threonines, 10 prolines, 6 methionines, 16 leucines and 2 tryptophans) and 3 cysteine, 4 serine and 6 aspartic acid residues.

Spectrophotometric Titration of Sulfhydryl Groups. In agreement with the amino acid analysis, the number of sulfhydryl groups, determined according to the procedure of Benesch and Benesch [13], based on the work of Boyer [14], was determined to be 8 for the enzyme type I and 11 for the enzyme type II. The same values were obtained if the titrations were carried out in 2% (w/v) sodium dodecylsulfate. While p-hydroxymercuribenzoate titration was complete in 2 min in the case of the enzyme type I, two hours were required for the complete titration in the case of the enzyme type II, as it appears from Fig. 3.

Absorption Spectrum. The absorption spectra of the two forms of enzyme, in 10 mM phosphate buffer,

Table 4. *Amino acid composition of the two forms of 6-phosphogluconate dehydrogenase*[a]

Amino acid	Hydrolysis time								Average		Nearest integer per 101,000
	24 hours		24 hours[b]		48 hours		72 hours				
									μmoles	%	
6-phosphogluconate dehydrogenase type I											
Lysine	0.820	0.812	0.815	0.822	0.811	0.814	0.824	0.810	0.816	8.70	82
Histidine	0.128	0.125	0.136	0.122	0.135	0.127	0.131	0.120	0.128	1.37	13
Arginine	0.345	0.345	0.353	0.342	0.347	0.352	0.350	0.342	0.347	3.72	35
Aspartic acid	1.049	1.059	1.072	1.040	1.048	1.062	1.060	1.050	1.055	11.30	106
Threonine	0.346	0.340	0.345	0.341	0.335	0.337	0.328	0.318	0.343	3.67	34
Serine	0.429	0.409	0.413	0.426	0.405	0.410	0.390	0.400	0.419	4.47	42
Glutamic acid	1.061	1.050	1.071	1.052	1.089	1.071	1.040	1.051	1.061	12.10	106
Proline	0.407	0.418	0.410	0.401	0.418	0.412	0.420	0.410	0.412	4.41	41
Glycine	1.177	1.178	1.190	1.168	1.180	1.190	1.182	1.150	1.177	12.50	118
Alanine	0.771	0.759	0.797	0.764	0.797	0.758	0.762	0.780	0.773	8.29	77
Half cysteine[c]			0.0795	0.0791					0.0793	0.84	8
Valine	0.480	0.497	0.505	0.510	0.520	0.515	0.517	0.527	0.522	5.60	52
Methionine	0.127	0.119	0.134[d]	0.130[d]	0.122	0.135	0.124	0.120	0.132	1.41	13
Isoleucine	0.549	0.537	0.550	0.540	0.572	0.580	0.574	0.579	0.577	6.18	58
Leucine	0.749	0.756	0.768	0.760	0.765	0.759	0.551	0.760	0.756	8.09	76
Tyrosine	0.315	0.335	0.319	0.326	0.340	0.319	0.341	0.320	0.327	3.69	33
Phenylalanine	0.340	0.347	0.352	0.339	0.357	0.339	0.352	0.340	0.348	3.52	35
Tryptophan[e]									0.101	1.08	10
6-phosphogluconate dehydrogenase type II											
Lysine	0.867	0.860	0.850	0.875	0.860	0.870	0.862	0.846	0.862	9.32	96
Histidine	0.149	0.150	0.155	0.148	0.150	0.155	0.150	0.145	0.152	1.64	17
Arginine	0.425	0.429	0.430	0.421	0.418	0.427	0.431	0.416	0.425	4.60	47
Aspartic acid	1.011	0.997	1.015	0.991	1.015	1.018	0.999	1.019	1.008	10.90	112
Threonine	0.418	0.410	0.408	0.420	0.405	0.400	0.395	0.387	0.414	4.48	46
Serine	0.414	0.418	0.418	0.416	0.405	0.397	0.382	0.390	0.414	4.48	46
Glutamic acid	0.952	0.955	0.965	0.940	0.951	0.960	0.941	0.960	0.952	10.30	106
Proline	0.444	0.456	0.469	0.458	0.452	0.441	0.462	0.454	0.458	4.95	51
Glycine	1.082	1.022	1.095	1.061	1.027	1.077	1.075	1.060	1.062	11.50	118
Alanine	0.674	0.680	0.667	0.701	0.706	0.698	0.684	0.696	0.691	7.50	77
Half cysteine[c]			0.096	0.100					0.098	1.06	11
Valine	0.437	0.441	0.440	0.449	0.455	0.461	0.460	0.476	0.468	5.07	52
Methionine	0.168	0.176	0.172[d]	0.170[d]	0.168	0.174	0.170	0.168	0.171	1.85	19
Isoleucine	0.520	0.507	0.517	0.512	0.523	0.518	0.514	0.528	0.521	5.65	58
Leucine	0.819	0.830	0.810	0.815	0.847	0.824	0.847	0.822	0.826	8.95	92
Tyrosine	0.292	0.290	0.290	0.301	0.305	0.290	0.305	0.289	0.292	3.20	33
Phenylalanine	0.318	0.314	0.309	0.320	0.318	0.309	0.308	0.321	0.314	3.40	35
Tryptophan[e]									0.108	1.50	12

[a] A sample of each form of enzyme was dialyzed two days against distilled water and then evaporated to dryness. The dry material was dissolved in 5.7 N HCl, divided into three equal parts each containing 1.0 mg of protein and hydrolyzed in vacuum for 24, 48, and 72 hours. Cysteine was determined as cysteic acid and methionine as methionine sulfone in samples oxidized with performic acid [11]. Each analysis was carried out in duplicate. Corrections were made for physical loss, destruction or incomplete hydrolysis. The values reported in the table are expressed as micromoles of amino acids per 1.0 mg of protein. The amino acid composition reported in the last column was based on the molecular weight of 101,000 for the enzyme type I and 111,000 for the enzyme type II.
[b] Samples oxidized with performic acid.
[c] Determined as cysteic acid.
[d] Determined as methionine sulfone.
[e] Determined spectrophotometrically [12].

pH 7.0, are identical. There is only one peak of absorption, with a maximum at 280 mμ and no significant absorption at 340 mμ, also in the presence of 6-phosphogluconate. The ratio between the absorbancies at 280 and 260 mμ is 1.9 for both enzymes.

pH Optimum, Substrate Affinities and Specificity. The pH optimum for the activity of both forms of enzyme is the same (Fig. 4). The pH curve shows half maximal activity at pH 6.34 and 7.94 for the enzyme type I and 6.44 and 8.4 for the enzyme type II. The more rapid decrease of activity at high pH, for the enzyme type I, with respect to the type II, might be due to the greater instability of this form of enzyme at high pH (see below).

The apparent K_m values for 6-phosphogluconate were determined to be 52 μM for the enzyme type I and 31 μM for the enzyme type II. The apparent K_m value for TPN was 20 μM for both forms of enzyme.

Both forms are highly specific with respect to coenzyme, showing no detectable activity with DPN.

12*

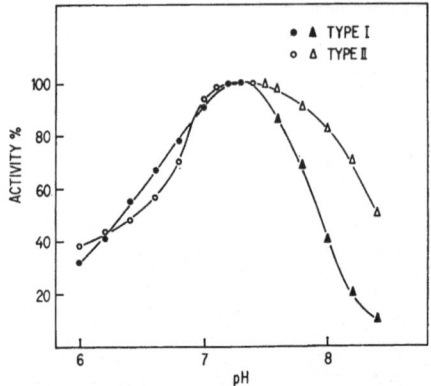

Fig. 4. *Enzymatic activities of the two forms of 6-phosphogluconate dehydrogenase as a function of pH.* 0.1 M phosphate buffer (○, ●) or 0.1 M Tris-HCl buffer (△, ▲) were used

Inhibition of Enzymatic Activity by Different Reagents. Studies on the kinetics of inhibition or of inactivation of the catalytic activity of the two forms of enzyme were undertaken to determine if differences existed between the two forms.

The data reported in Table 6 show that the treatment with chlorodinitrobenzene caused complete loss of the catalytic activity of the enzyme type I, while it had no effect on the catalytic activity of the enzyme type II.

The two forms of enzyme behaved in an identical manner with respect to the inhibitory action of pyridoxal-5′-phosphate, which was previously shown to form a Schiff base derivative with the crystalline enzyme [5].

The catalytic activity of both forms of enzyme is inhibited by incubation with *p*-hydroximercuribenzoate. The data reported in Fig. 5 show that the rate of inhibition is greater for the enzyme type I with respect to the enzyme type II. This is in agree-

Table 5. *Effect of heat and pH on the stability of the two forms of 6-phosphogluconate dehydrogenase*

Treatment	Enzyme type I		Enzyme type II	
	Duration of treatment	Inhibition	Duration of treatment	Inhibition
	min	%	min	%
Incubation at 60°ᵃ	1	100	16	50
Incubation at pH 4.1ᵇ	2	50	60	0
	12	82		
Incubation at pH 8.0ᶜ	11	50	60	20
	30	90		

ᵃ A sample containing 0.2 mg per ml of enzyme (type I or type II) in 20 mM phosphate buffer, 0.1 mM EDTA, pH 6.2, was heated at 60°. Samples were taken at intervals and analyzed for enzymatic activity.
ᵇ Solutions containing 0.2 mg per ml of enzyme in 100 mM acetate buffer, 0.1 mM EDTA, pH 4.1, were incubated at 22°. Samples were taken at intervals and analyzed.
ᶜ Solutions containing 0.2 mg per ml of enzyme in 1 M Tris-HCl buffer, pH 8.0, were incubated at 22°. Samples were taken at intervals and analyzed for enzymatic activity.

Table 6. *Effect of chlorodinitrobenzene and pyridoxal-5′-phosphate on the catalytic activity of the two forms of 6-phosphogluconate dehydrogenase*
Samples containing the type I or type II form of enzyme, at a concentration of 0.2 mg per ml, were incubated at 22° with the reagents indicated in the table under the conditions described below; at intervals samples were taken and analyzed for the enzymatic activity. Reagent 1. 0.4 mM chlorodinitrobenzene, 1 mM EDTA, 1 M Tris-HCl buffer, pH 8. Reagent 2. 0.2 mM pyridoxal phosphate, 10 mM phosphate buffer, pH 7.4

Reagent	Enzyme type I		Enzyme type II	
	Duration of treatment	Inhibition	Duration of treatment	Inhibition
	min	%	min	%
Chlorodinitrobenzene	15	50	60	0
Pyridoxal-5′-phosphate	2	50	2	50

Stability to Heat, Acid and Alkaline pH. The two forms of enzyme showed significant differences in the sensitivity to inactivation by heat treatment or by exposure to acid or alkaline medium. The data reported in Table 5 show that the catalytic activity of the enzyme type II is less sensitive than that of the enzyme type I, to the action of these agents.

ment with the titration data reported above. 6-phosphogluconate protects both forms of enzyme against the inhibition by *p*-hydroxymercuribenzoate.

Inactivation by Dodecylsulfate and Trypsin Digestion. The catalytic activity of the two forms of enzyme was also affected in a different way by the treatment with sodium dodecylsulfate and with trypsin. As shown in Fig. 6A, loss of catalytic activity of both types of enzyme was observed after incubation with dodecylsulfate, but the rate of inactivation was significantly higher for the enzyme type I. The proteolytic digestion with trypsin (Fig 6 B) caused almost complete loss of the catalytic activity of the enzyme type I in 10 min, while only 20 % of the original activity of the enzyme type II was lost in 40 min.

enzyme type II. The rate of titration of the sulfhydryl groups with p-hydroxymercuribenzoate and the consequent loss of the catalytic activity, follow different kinetics with the two enzyme proteins as indicated by the fact that the reaction, in order to be complete, requires a longer time in the case of the enzyme type II.

Thus *Candida utilis* 6-phosphogluconate dehydrogenase can be added to the increasing number of

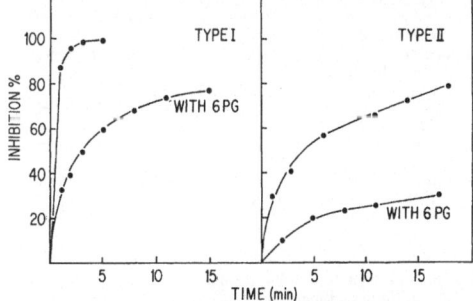

Fig. 5. *Inhibition of the enzymatic activity of the two forms of 6-phosphogluconate dehydrogenase by p-hydroxymercuribenzoate (pHMB).* 0.1 mg of enzyme (type I or type II) was incubated at 0° in a reaction mixture (0.5 ml) containing 0.01 mM pHMB, 20 mM phosphate buffer, pH 7.4, and, where indicated, 1.5 mM 6-phosphogluconate (6PG). At intervals aliquots were taken and analyzed for activity

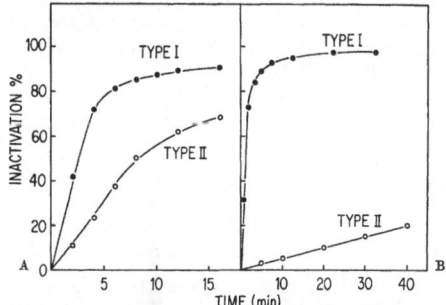

Fig. 6. *Effect of dodecylsulfate and trypsin treatment on the activity of the two forms of 6-phosphogluconate dehydrogenase.* Solutions containing 0.14 mg per ml of enzyme (type I or type II) were incubated at 22° with sodium dodecylsulfate or trypsin under the conditions described below. At intervals aliquots were taken and assayed for enzymatic activity. Fig. 6, left: 0.01% (w/v) sodium dodecylsulfate, 1 mM EDTA, 20 mM phosphate buffer, pH 6.2. Fig. 6, right: 0.01 mg per ml trypsine, 1 mM EDTA, Tris-HCl buffer, pH 8

DISCUSSION

Two types of 6-phosphogluconate dehydrogenase have been obtained in a homogeneous form from extracts of *Candida utilis*. The separation of the two forms of enzyme has been obtained taking advantage of a different property of the two forms with respect to crystallization: the enzyme type I crystallizes, while, in the same conditions, the enzyme type II does not crystallize.

The approximate value of the molecular weight is 101,000 for the enzyme type I and 111,000 for the enzyme type II. From the amino acid composition and assuming the reported molecular weight values, it appears that the enzyme type II contains, in addition to all the amino acid residues present in the enzyme type I, 89 other amino acid residues.

The two forms of enzyme differ in several respects in the response to physical, chemical and enzymatic treatments. The catalytic activity of the enzyme type II is less affected, than that of the enzyme type I, by heat treatment and exposure to acid and alkaline pH, and is more resistant to the proteolytic digestion with trypsin and to the action of dodecylsulfate.

The treatment with chlorodinitrobenzene produces a complete inactivation of the enzyme type I but has no effect on the catalytic activity of the

enzymes which have been demonstrated to exist in multiple molecular forms, although the molecular basis which could explain this phenomenon is still unknown. They would not appear to result from different polymeric states of a single enzyme, inasmuch as the molecular weight of the two forms seems to differ by only 10,000. The multiple forms would also not appear to result from extraction or purification artefacts, since the two forms are present in different conditions of autolysis and after different procedures of purification. The presence of the two forms does not depend upon the particular batch of yeast used, since two forms have been detected in 6 different batches of yeast obtained in a 2 year interval.

It is known that in *Candida utilis* two forms of fumarase [15] and at least 2 forms of transaldolase [16] exist. The presence of multiple forms of an enzyme might result from proteolytic digestion during the purification [17]. Attempts to produce interconversion of the two forms of 6-phosphogluconate dehydrogenase by the intrinsic proteolytic activity of the crude extract or by the addition of trypsin have been so far unsuccessful.

Although trypsin failed to transform one form into the other, the differences in molecular weight, amino acid composition and the properties of the two forms might suggest that the enzyme type I could be a product of a partial proteolytic digestion of the enzyme type II. The enzymatic properties of the two forms are essentially the same, and some of the different properties observed can be attributed to different structural forms.

Other explanations for the presence of two forms, based on the existence in our starting material of different strains of yeast, or on different intracellular localization of the two enzymes are still possible.

The presence of two forms of 6-phosphogluconate dehydrogenase exhibiting differences in their properties may help future studies on the structure and mechanism of action of this enzyme.

This work was supported by grants from the Italian C.N.R. (Impresa di Enzimologia), the National Institutes of Health (GM 12291-02 and TW 00227-01) and N.A.T.O. (Grant 218).

REFERENCES

1. Horecker, B. L., and Smyrniotis, P. Z., J. Biol. Chem. 193 (1951) 371.
2. Pontremoli, S., De Flora, A., Grazi, E., Mangiarotti, G., Bonsignore, A., and Horecker, B. L., J. Biol. Chem. 236 (1961) 2975.
3. Grazi, E., Rippa, M., and Pontremoli, S., J. Biol. Chem. 240 (1965) 234.
4. Rippa, M., Grazi, E., and Pontremoli, S., J. Biol. Chem. 241 (1966) 1632.
5. Rippa, M., Spanio, L., and Pontremoli, S., Arch. Biochem. Biophys. 118 (1967) 48.
6. Pontremoli, S., Traniello, S., Luppis, B., and Wood, W. A., J. Biol. Chem. 240 (1965) 3459.
7. Pontremoli, S., Luppis, B., Traniello, S., Rippa, M., and Horecker, B. L., Arch. Biochem. Biophys. 112 (1965) 7.
8. Bucher, T., Biochim. Biophys. Acta, 1 (1947) 192.
9. Martin, R. G., and Ames, B. N., J. Biol. Chem. 236 (1961) 1372.
10. Spackman, D. H., Stein, W. H., and Moore, S., Anal. Chem. 30 (1958) 1190.
11. Hirs, H. W., J. Biol. Chem. 219 (1956) 611.
12. Goodwin, T. W., and Morton, R. A., Biochem. J. 40 (1946) 628.
13. Benesch, R., and Benesch, R. E., Methods Biochem. Anal. 10 (1962) 57.
14. Boyer, P. D., J. Am. Chem. Soc. 76 (1954) 4331.
15. Hayman, S., and Alberty, R. A., Ann. N.Y. Acad. Sci. 94 (1961) 812.
16. Horecker, B. L., and Tchola, O., personal communication.
17. Kaji, A., Trayser, K. A., and Colowick, S. P., Ann. N.Y. Acad. Sci. 94 (1961) 798.

M. Rippa, M. Signorini, S. Pontremoli
Istituto di Chimica Biologica,
Università degli Studi di Ferrara
Via Fossato di Mortara 25
Ferrara, Italy

European J. Biochem. 1 (1967) 179—181

Préparation de RNA Messager de Réticulocytes de Lapin par Ultracentrifugation en Rotor Zonal

G. Huez, A. Burny, G. Marbaix et E. Schram

Laboratoires de Chimie biologique et de Morphologie animale, Faculté des Sciences, Université de Bruxelles

(Reçu le 9 février 1967)

Two methods are described for the rapid isolation of 0.6 mg to 1 mg of 9S RNA from rabbit reticulocytes. This RNA has properties expected for messenger RNA.

In the first procedure, reticulocyte polyribosomes are dissolved in 0.5% (w/v) sodium dodecylsulfate and messenger RNA is separated by sucrose gradient centrifugation in a zonal rotor (B IV-Spinco).

In the second method, messenger RNA is detached from polyribosomes by EDTA (3×10^{-2}M) treatment and separated from ribosomal sub-particles by sucrose gradient centrifugation using the same rotor.

In both cases, messenger RNA is then concentrated by adsorption on a non-ionic cellulose column.

Further purification may be obtained by sucrose gradient centrifugations using the rotor S.W. 25.2 (Spinco).

Nous avons montré précédemment [1—3] qu'on peut isoler de polyribosomes de réticulocytes une fraction ribonucléique qui représente la fibre unissant les ribosomes entre eux dans les polyribosomes; c'est le RNA messager si les conceptions actuelles sont correctes.

Ce RNA possède en effet les propriétés suivantes: (a) il représente environ 2% du RNA total; (b) sa vitesse de renouvellement est plus élevée que celle du RNA ribosomial; (c) il est beaucoup plus vite dégradé que les autres RNA cellulaires lorsqu'on traite les polyribosomes avec des quantités minimes de ribonucléase pancréatique; (d) il est détaché des polyribosomes lorsqu'on abaisse la concentration des ions bivalents [4]; (e) il se comporte comme une substance homogène à la centrifugation analytique ou en gradient de saccharose; (f) son poids moléculaire est voisin de 150.000, il peut donc contenir assez d'information pour diriger la synthèse d'une chaîne polypeptidique de la taille des chaînes α ou β de l'hémoglobine.

Dans la présente publication, nous décrivons deux méthodes de préparation rapide de quantités relativement importantes de ce RNA, méthodes qui font appel à l'ultracentrifugation en rotor zonal [5].

Abréviations non usuelles. RNA messager, mRNA; dodécylsulfate de sodium, SDS.
Enzyme. Ribonucléase ou ribonucléate pyrimidine-nucléotido-2′-transférase (EC 2.7.7.16).

MATÉRIEL ET MÉTHODES

Isolement des Polyribosomes

Les polyribosomes de réticulocytes marqués au ^{32}P sont préparés comme décrit précédemment [1], sauf que le surnageant de la centrifugation à 12.000 g est déposé sur une couche de 7 ml d'une solution de saccharose 36% (poids/volume) (Tris-HCl 10^{-2} M, KCl 5×10^{-2} M, Mg-acétate 5×10^{-3} M pH = 7,4) avant d'être centrifugé pendant 3 heures à 30.000 tours/minute dans le rotor 30 de la centrifugeuse Spinco. Ceci permet de débarrasser les polyribosomes de la majeure partie de l'hémoglobine contaminante.

Dissolution des Polyribosomes par le Dodécylsulfate de Sodium (SDS)

80 à 140 mg de polyribosomes marqués au ^{32}P sont dissous dans 20 ml de SDS 0,5% (poids/volume). Les RNA sont ainsi dissociés des protéines.

Traitement des Polyribosomes par l'Éthylène Diamine Tétraacétate de Sodium (EDTA)

80 à 140 mg de polyribosomes sont remis en suspension dans 20 ml de solution tampon (Tris-HCl 10^{-2} M, EDTA 3×10^{-2} M, pH = 7,4) à 4° au moyen d'un homogénéiseur en verre muni d'un piston en téflon.

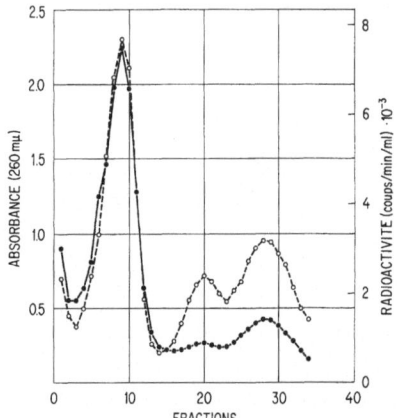

Fig. 1. *Profil de sédimentation en rotor zonal des RNA extraits au SDS de polyribosomes de réticulocytes de lapin marqués au ^{32}P. Gradient de saccharose $10\%–20\%$, Tris-HCl 5×10^{-3} M, pH = 7,4. Centrifugation de 6 heures à 40.000 tours/minute et à 4°.* ●, absorbance; ○, radioactivité

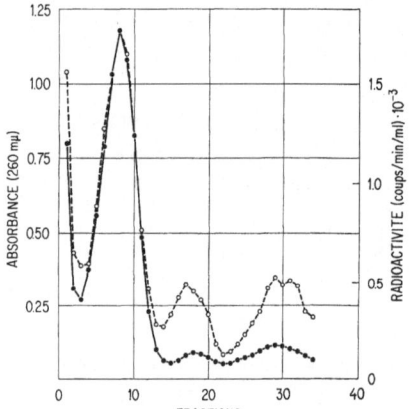

Fig. 2. *Profil de sédimentation en rotor zonal d'une suspension de polyribosomes radioactifs de réticulocytes de lapin après traitement par l'EDTA 3×10^{-2} M. Gradient de saccharose $15\%–30\%$, Tris-HCl 5×10^{-3} M, pH = 7,4. Centrifugation de 6 heures à 40.000 tours/minute et à 4°.* ●, absorbance; ○, radioactivité

Centrifugations en Gradient de Saccharose

Cas où les Polyribosomes sont dissous dans le SDS. L'échantillon est introduit à 5.000 tours/minute dans le rotor, sur un gradient préétabli de saccharose $10\%–20\%$ (poids/volume; Tris-HCl 5×10^{-3} M, pH = 7,4). Les zones à séparer se répartissant dans une fraction réduite du gradient, on diminue sensiblement la durée de centrifugation en éloignant au

maximum l'échantillon de l'axe du rotor, dès le début de la centrifugation. Ceci est réalisé en introduisant après l'échantillon 800 ml de tampon (Tris-HCl 5×10^{-3} M, pH = 7,4). La vitesse est alors amenée à 40.000 tours/minute. Après 6 heures de centrifugation, le rotor est ralenti à 5.000 tours/minute et son contenu est chassé par injection de saccharose 30% (poids/volume). Des fractions de 25 ml sont récoltées et l'absorbance à 260 mμ est mesurée. Des aliquotes sont prélevées pour la mesure de la radioactivité en milieu liquide homogène [6].

Cas où les Polyribosomes sont traités par l'EDTA. Même méthode avec les modifications suivantes: gradient linéaire de saccharose $15\%–30\%$ (poids/volume; Tris-HCl 5×10^{-3} M, pH = 7,4); injection d'une solution de saccharose 40% (poids/volume) après centrifugation.

Concentration du mRNA après Centrifugation

Les fractions contenant le mRNA sont rassemblées; à la solution globale (environ 200 ml), on ajoute du NaCl (concentration finale 0,1 M) et du SDS (concentration finale 5 mg/ml) puis 0,54 volume d'éthanol. La solution résultante est passée à travers une colonne (2×5 cm) de cellulose [7]. Le RNA fixé est élué à l'eau; il passe dans un volume de 10 à 15 ml. Cette solution est dialysée pendant 20 heures à 4° contre deux fois 100 volumes d'eau bidistillée; le RNA est alors récupéré par lyophilisation.

RÉSULTATS ET DISCUSSION

La Fig. 1 montre le profil de sédimentation en rotor zonal des RNA extraits au SDS de polyribosomes de réticulocytes de lapin. Chaque animal avait reçu une injection de 10 mC de [^{32}P]orthophosphate 15 heures avant le sacrifice. Le RNA ribosomial 26 S est au fond du gradient, le mRNA 9 S sédimente entre le RNA ribosomial 16 S et le RNA soluble 4 S. Le profil de radioactivité suit le profil d'absorption dans l'ultraviolet dans la région des RNA ribosomiaux. Il montre alors deux pics à activité spécifique élevée: celui du mRNA 9 S et celui du RNA soluble 4 S.

La Fig. 2 présente le profil de sédimentation obtenu après centrifugation d'une suspension de polyribosomes traitée par l'EDTA. La particule ribosomiale lourde 36 S est au fond du gradient; un ensemble à radioactivité spécifique élevée sédimente entre la particule ribosomiale légère 26 S et le RNA soluble 4 S. Nous avons montré précédemment que l'essentiel du matériel nucléique de cet ensemble est constitué par le RNA messager 9 S. [4].

Les polyribosomes peuvent également être dissociés par passage de la suspension à travers une colonne de CM-cellulose équilibrée avec le tampon phosphate de potassium 10^{-2} M, pH = 7,0 [4,8].

Les constantes de sédimentation des sous-particules ribosomiales sont 36 S et 26 S au lieu de 60 S et 40 S pour les particules natives. Ce phénomène est vraisemblablement dû à la désorganisation de la structure tertiaire des particules en milieu exempt d'ions Mg^{++} [9], désorganisation qui n'entraîne cependant pas de pertes de RNA, ni de protéines.

On notera que l'ensemble fortement marqué qui sédimente entre 4 S et 26 S est hétérogène, car la radioactivité spécifique du matériel qui sédimente sur le flanc léger du pic de RNA messager est plus faible que celle du matériel principal. La nature de ce contaminant à faible activité spécifique est actuellement à l'étude.

L'une ou l'autre des méthodes décrites ici permet d'obtenir rapidement 0,6 à 1 mg de RNA messager de réticulocytes. On purifie ensuite ce RNA par deux ultracentrifugations en gradient de saccharose, selon la méthode que nous avons décrite précédemment[3].

Ce travail a été effectué dans le cadre du contrat d'association Euratom-ULB n° 016-61-10-ABIB. La centrifugeuse zonale utilisée dans la présente étude a été acquise grâce à un subside du Fonds National de la Recherche Scientifique. G. Huez est boursier de l'I.R.S.I.A., G. Marbaix est aspirant du F.N.R.S. Nous remercions Mr. H. Roosens pour son excellente assistance technique.

BIBLIOGRAPHIE

1. Marbaix, G. et Burny, A., *Biochem. Biophys. Res. Commun.* 16 (1964) 522.
2. Burny, A. et Marbaix, G., *Biochim. Biophys. Acta*, 103 (1965) 409.
3. Marbaix, G., Burny, A., Huez, G. et Chantrenne, H., *Biochim. Biophys. Acta*, 114 (1966) 404.
4. Huez, G., Burny, A., Marbaix, G. et Lebleu, B., *Arch. Intern. Physiol. Biochim.* 74 (1966) 920.
5. Anderson, N. G., *J. Phys. Chem.* 66 (1962) 1984.
6. Bray, G. A., *Anal. Biochem.* 1 (1960) 270.
7. Barber, R., *Biochim. Biophys. Acta*, 114 (1966) 422.
8. Marbaix, G. et Burny, A., *Arch. Intern. Physiol. Biochim.* 72 (1964) 689.
9. Spirin, A. S., Kisselev, N. A., Shukulov, R. S. et Bogdanov, A. A., *Biokhimiya*, 28 (1963) 920.

G. Huez, A. Burny et G. Marbaix
Laboratoire de Chimie biologique
Faculté des Sciences, Université libre de Bruxelles
67 Rue des Chevaux, Rhode St-Genèse (Brabant), Belgique

E. Schram
Laboratoire de Morphologie animale
Faculté des Sciences, Université libre de Bruxelles
67 Rue des Chevaux, Rhode St-Genèse (Brabant), Belgique

European J. Biochem. 1 (1967) 182—186

Transamidinase of Hog Kidney

6. Effects of the Modification of Cysteine Residues on the Catalytic Activity

E. Grazi, V. Vigi, and N. Rossi

Istituto di Chimica Biologica, Università di Ferrara

(Received October 8, 1966)

The modification of one cysteine residue of transamidinase by reaction with 5,5'-dithiobis-2-nitrobenzoic acid alters the catalytic properties of the enzyme.

The Michaelis constant of transamidinase for arginine is increased and the capacity to utilize glycine and hydroxylamine as amidine acceptors is lost.

Arginine, the amidine donor substrate, protects the cysteine residue and prevents the changes in catalytic activity induced by the treatment with 5,5'-dithiobis-2-nitrobenzoic acid.

Transamidinase catalyzes the transfer of the amidine group from arginine to glycine with the formation of an intermediate enzyme-amidine complex as illustrated by the following scheme [1]:

arginine + enzyme = enzyme-amidine + ornithine
enzyme-amidine + glycine = enzyme + guanidino-acetate

$$\text{enzyme-amidine} + H_2O \rightarrow \text{Enzyme} + \text{urea}$$

The hydrolysis of the enzyme-amidine complex represents a side reaction occurring at about 1 % of the rate of the overall arginine-glycine transfer reaction[2].

In a previous report we have shown that dinitrophenylation of transamidinase results in the alteration of the catalytic activity. It is possible to obtain a dinitrophenylated derivative which retains completely the hydrolytic activity but in which the overall rate of the amidine transfer to glycine is greatly decreased [3]. The catalytic modifications are produced by dinitrophenylation of 3 cysteine, 1 tyrosine and 1 lysine residues[4]. Since the hydrolytic activity depends on the formation of the enzyme-amidine complex, it is clear that dinitrophenylation does not affect the group involved in the binding of the amidine group to the enzyme but only the rate at which the amidine group is transferred. It was therefore of interest to study whether the changes in the catalytic activity are linked to the modification of either one or more amino acid residues or to the introduction of the dinitrophenyl group itself in the protein molecule.

We have now found that 5,5'-dithiobis-2-nitrobenzoic acid alters the catalytic activity of transamidinase in a way similar to that of dinitrofluorobenzene through the formation of a mixed disulphide with a cysteine residue of the enzyme. Arginine, the

Non-standard Abbreviation. 5,5'-dithiobis-2-nitrobenzoic acid, DTNB.

Enzyme. Transamidinase, or L-arginine: glycine amidine transferase (EC 2.6.2.1).

amidine donor substrate, slows down the reaction of the cysteine residue with DTNB and decreases the rate of the changes in catalytic activity of transamidinase.

EXPERIMENTAL PROCEDURE

Materials

Transamidinase was prepared as previously described [5]. Glycine, L-ornithine and L-arginine were obtained from Sigma, U.S.A. Uniformly-labelled [14C]arginine was purchased from the Radiochemical Centre, Amersham, England. Hydroxylamine hydrochloride was purchased from E. Merck, Darmstadt, Germany.

5,5'-Dithiobis-2-nitrobenzoic acid and N-(4-dimethyl-amino-3,5-dinitrophenyl) maleimide were purchased from Aldrich Chemicals Co., Milwaukee, U.S.A. Cystamine hydrochloride and N-ethyl-maleimide were obtained from Nutritional Biochemical Corporation, Cleveland, Ohio, U.S.A. The sodium salt of p-hydroxymercuribenzoate was purchased from the California Foundation, Los Angeles, U.S.A.

All other reagents were analytical reagent grade and were purchased from Carlo Erba S.A., Milano, Italy.

Methods

The following reactions catalyzed by transamidinase were studied

a) L-arginine + glycine = L-ornithine + guanidinoacetate

b) L-arginine + hydroxylamine → L-ornithine + hydroxyguanidine

c) L-arginine + H_2O → L-ornithine + urea.

The reactions were followed by measuring ornithine formation with the Chinard colorimetric method [6] on the trichloroacetic acid filtrates of the incubation mixtures. The incubation mixtures contained L-arginine 0.01 M; either glycine 0.016 M or hydroxylamine 0.1 M; and phosphate buffer 0.1 M, pH 7.5. For the determination of the hydrolytic activity the acceptor substrate was omitted. Temperature was 37°; the reaction was started by the addition of the enzyme. Initial velocities were based on the early linear portion of the reaction curves involving less than 1 % of the total reaction. A unit is defined as the amount of the enzyme that catalyzes the formation of 1 μmole of ornithine or guanidino-acetate per hour at 37°.

The reaction of DTNB with transamidinase has been followed at 412 mμ using the procedure of Ellman [7].

Protein concentration was measured at 280 mμ, or with the Folin reagent [8]. The methods were calibrated by dry weight determinations on the dialyzed enzyme.

The number of sulfhydryl groups was determined spectrophotometrically at 255 mμ [9] in the presence of 8 M urea and 0.2 M acetate buffer, pH 5.0; and standardized against a sample of glutathione of known concentration.

RESULTS

Reactive Sulfhydryl Groups Determined by Titration with p-Hydroxy Mercuribenzoate

In the presence of 8 M urea, approximately 10 sulfhydryl groups are titrated. Under these conditions the reaction is complete within 5 minutes.

DTNB Treatment and Transamidinase Activity

Transamidinase treated with DTNB at pH 8.0 undergoes progressive changes in catalytic activity. The amidine transfer reactions to glycine and to hydroxylamine are affected first, while the amidine transfer to water (hydrolytic reaction) is affected much later.

The profile of the sulfhydryl titration curve with DTNB is shown in Fig.1 A. One to two sulfhydryl groups are very reactive and are titrated in less than one minute. Titration of a further three to four sulfhydryl groups is much slower and requires approximately 100 minutes. When 3.5 sulfhydryl groups per molecule of enzyme have reacted, the rate of the hydrolysis of arginine is decreased by only 5 % while the rate of transfer of the amidine group to glycine and to hydroxylamine is reduced to 20 % of the original value (Fig.1 B).

The separation of the hydrolytic reaction from the transfer reaction to glycine or hydroxylamine by DTNB is quite unique among the specific sulfhydryl reagents so far tested.

The results obtained by reacting p-hydroxymercuribenzoate with transamidinase are shown in Table 1. The inactivation of the amidine transfer to water and to glycine follows the same time course. Complete inactivation is obtained when six moles of

p-hydroxymercuribenzoate are added per mole of enzyme. Similar results have been obtained with cystamine, with N-ethylmaleimide and with N-(4-dimethylamino-3,5-dinitrophenyl)-maleimide.

Transamidinase Inactivation by DTNB and Protection by arginine

Arginine partially prevents the changes in catalytic activity of transamidinase produced by DTNB

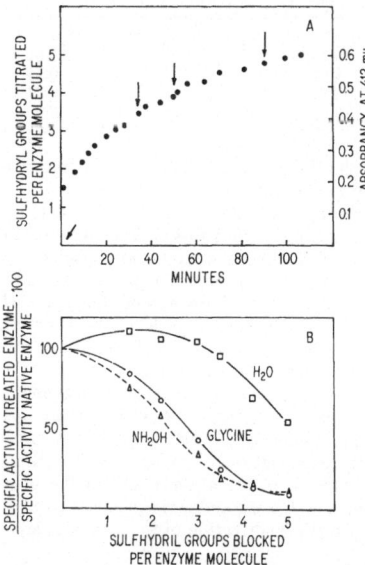

Fig.1. *Effect of DTNB treatment on transamidinase activity.* The incubation mixtures (4 ml) contained 3.6 mg of transamidinase (specific activity 65 units per mg of protein) in 0.1 M phosphate buffer, pH 8.0. At the beginning of the experiment and after 4, 33, 51 and 90 minutes 0.015 ml of 0.03 M DTNB in 0.1 M phosphate buffer, pH 7.0, were added. Temperature was 22°.

At the beginning of the experiment and after 2, 11, 23, 36, 63, and 106 minutes aliquots (0.5 ml) of the incubation mixture were taken, the protein was precipitated by addition of 250 mg of solid ammonium sulfate. The precipitate, collected by centrifugation, was dissolved in 0.05 ml of 0.02 M phosphate buffer, pH 7.5, and assayed for enzymatic activity. A control sample, treated in the same way but without addition of DTNB, retained full catalytic activity A. The reaction with DTNB was determined by the increase in absorbance at 412 mμ, in 0.1 M phosphate buffer, pH 8.0. Identical additions of DTNB were made to the sample and the blank. The readings were taken in quartz cuvettes with a light path of 1 cm. The number of sulfhydryl groups reacted per mole of enzyme was determined on the basis of a molecular weight of 100,000 [5] and of an E_M of 13,600 for the anion of 5-thio-2-nitrobenzoic acid. The arrows indicate the time of addition of the reagent

B. Transamidinase activity as a function of sulfhydryl groups titration. □———□, hydrolytic reaction; ○———○, transfer reaction to glycine; △———△, transfer reaction to hydroxylamine

Table 1. *Transamidinase inactivation by p-hydroxymercuri-benzoate*

Four incubation mixtures (0.13 ml) containing 1 mg (10 mμ-moles) of transamidinase (specific activity 68 units per mg of protein) in phosphate buffer 0.015 M, pH 7.5, were prepared. To three of the four samples were added respectively 10, 30, and 60 mμmoles of p-hydroxymercuribenzoate, the fourth received no addition. Temperature was 22°. After 20 minutes incubation the samples were analyzed for catalytic activity

p-Hydroxymercuribenzoate added per mole of enzyme	Transfer to glycine	Hydrolysis
moles	units per mg	
0	68	0.54
1	63	0.44
3	31	0.27
6	0	0.0

and affects the time course of the sulfhydryl titration curve of the enzyme.

As shown in Fig.2, one to two sulfhydryl groups react very rapidly both in the protected and in the non-protected sample with no appreciable loss of catalytic activity. From this point the titration proceeds with concomitant loss of catalytic activity in both samples. However, in the absence of arginine, after 120 minutes, 4.5 sulfhydryl groups are titrated with 82 % inactivation of the arginine—glycine transfer reaction while in the presence of arginine, after the same period of time, 3.9 sulfhydryl groups are titrated and only 35 % of the original activity is lost. The difference between the number of sulfhydryl groups titrated in the absence and in the presence of arginine is thus 0.6 and corresponds to a difference of 47 % in the inactivation of the two samples.

Properties of the Transamidinase Derivative Obtained after Reaction with DTNB

The peculiar changes in catalytic activity which result from treatment of transamidinase with DTNB have been submitted to kinetic analysis. For this study we have utilized a transamidinase derivative with 3.2 modified sulfhydryl groups which hydrolyzes arginine at almost the same rate as the unmodified protein and synthetizes guanidinoacetate and hydroxyguanidine at only 25 % of the original rate (Table 2).

The rate of the amidine transfer as a function of glycine and arginine concentrations is shown in Fig.3A and 3B. In these experiments the samples with the DTNB-treated enzyme contained twice as much protein as those with the native enzyme. The inactivation of the amidine transfer to glycine is not reversed by increasing either arginine or glycine concentration. Thus, the treatment of transamidinase with DTNB irreversibly decreases the rate of the overall transamidinase reaction between arginine and glycine but, apparently, does not change the

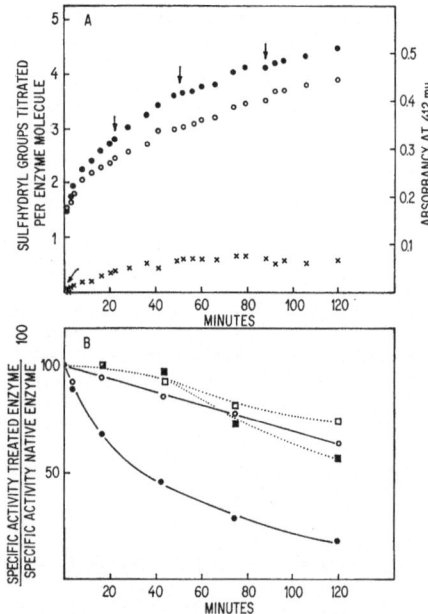

Fig.2. *Transamidinase inactivation by DTNB and protection by arginine.* The incubation mixtures (4 ml) contained 3.4 mg of transamidinase (specific activity 68 units per mg of protein) in 0.1 M phosphate buffer, pH 8.0, with or without 16 mM arginine. At the beginning of the experiment and after 22, 50, and 88 minutes 0.015 ml of 0.03 M DTNB in 0.1 M phosphate buffer, pH 7.0, were added. Temperature was 22°. At the beginning of the experiment and after 4, 16, 41, 74, and 120 minutes aliquots (0.5 ml) of the incubation mixtures were collected and the catalytic activity determined A. Titration of sulfhydryl groups. The arrows indicate the time of addition of the reagent. ●, sample without arginine; ○, sample with arginine; ×, difference in the titration induced by the presence of arginine B. Hydrolytic reaction: sample treated in the presence □----□ and in the absence ■----■ of arginine. Transfer reaction to glycine: sample treated in the presence ○——○ and in the absence ●——● of arginine

Table 2. *Properties of the DTNB treated transamidinase utilized for kinetic analysis*

The treatment with DTNB and the analysis were performed as described in Fig.1

	Transfer to glycine	Hydrolysis	Mixed S-S per molecule of enzyme
	units per mg		
Native transamidinase	65	0.56	0.0
DTNB-treated transamidinase	17	0.54	3.2

Michaelis constants of the enzyme for either of the substrates.

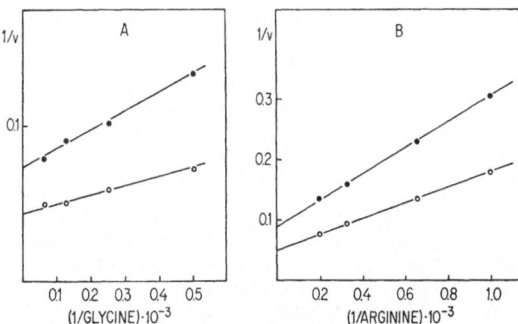

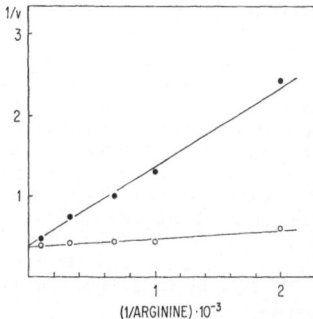

Fig. 3. *Reciprocal plot of initial velocities against arginine concentration at fixed (10 mM) glycine concentration (left) and against glycine concentration at fixed (23 mM) arginine concentration (right) in 0.1 M phosphate buffer. The temperature was 37°, the pH 7.5. The samples (0.6 ml) contained 0.02 mg of native transamidinase (specific activity 65 units per mg of protein) (O——O); or 0.04 mg of treated transamidinase (see Table 2) (●——●). Initial velocity is expressed as millimicromoles of ornithine formed per minute*

Fig. 4. *The hydrolytic reaction: reciprocal plot of initial velocity against arginine concentration. The samples (0.6 ml) contained 0.24 mg of either native transamidinase (specific activity 65 units per mg of protein) O——O; or DTNB treated transamidinase ●——● in 0.1 M phosphate buffer, pH 7.5. Arginine concentration as indicated. Temperature was 37°. Initial velocity is expressed as millimicromoles of ornithine formed per minute*

In contrast, analysis of the hydrolytic reaction shows that the apparent Michaelis constant of transamidinase for arginine is increased by treatment with DTNB. In fact, while with 10 mM arginine, the concentration usually employed for the assay of catalytic activity, the rate of the hydrolytic reaction is practically equal for the treated and for the native enzyme, with 0.5 mM arginine the treated enzyme is 75 % inhibited (Fig. 4). A similar behaviour has been observed in the exchange reaction from arginine to ornithine. In this case the apparent Michaelis constant of transamidinase for arginine is increased by the DTNB treatment but the maximal velocity is almost the same for both the treated and the native enzyme (Fig. 5).

DISCUSSION

DTNB reacts readily with the sulfhydryl groups of transamidinase to form mixed disulphides. One to two of these groups react very rapidly and their modification does not significantly affect the catalytic activity of the enzyme. A further three to four groups react more slowly; their titration decreases first the rate of the arginine-glycine and arginine-hydroxylamine transfer reactions and only later the rate of hydrolysis of arginine.

Arginine protects transamidinase against inactivation by DTNB and affects the time course of the titration curve. The analysis of the arginine-glycine transfer activity and of the titration curves of the protected and non-protected enzyme reveals that the titration of 0.6 sulfhydryl groups corresponds to the decrease of 47 % in catalytic activity. The results support the view that the modification of 1.3 sulfhydryl groups is responsible for the changes in catalytic activity. These changes only are obtained by treat-

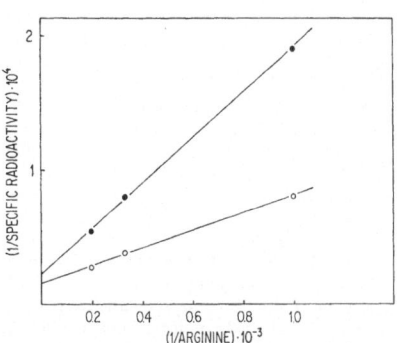

Fig. 5. *Reciprocal plot of the velocity of the arginine-ornithine exchange reaction against arginine concentration at fixed ornithine concentration. The pH was 7.5. The samples (0.1 ml) contained 0.004 mg of native transamidinase (specific activity 60 units per mg of protein) (O——O); or 0.004 mg of DTNB treated transamidinase (●——●); 5 mM ornithine and uniformly-labelled [14C]arginine (specific activity 225,000 cpm per μmole) as indicate*
The rate of the reaction is expressed as specific radioactivity of ornithine in counts/min/μmole. The samples were incubated at 37° and after 30 minutes the reaction was stopped by addition of 0.1 ml of 4 N HCl. The samples were taken to dryness, dissolved in 0.04 ml of water and aliquots chromatographed (descending) on Whatman N 1 paper with the solvent system: phenol—water—ammonia (80:20:0.5). The areas of radioactivity corresponding to ornithine were eluted. On the resulting solutions the specific radioactivity of ornithine was determined and expressed as counts/min/μmole

ment with DTNB and not with the other sulfhydryl reagents tested. This could indicate a peculiar reactivity of DTNB for the sulfhydryl group involved in the catalytic changes. Alternatively the nature of the group introduced in the enzyme could be important. This last possibility is stressed by the fact that dinitrophenylation of transamidinase results in changes in catalytic activity similar to those induced by DTNB. In that case, however, not only cysteine but also lysine and tyrosine residues have been involved.

A detailed kinetic analysis has been performed using a transamidinase derivative containing 3.2 modified sulfhydryl groups. The following results have been obtained:

a) the apparent Michaelis constant of transamidinase for arginine is increased in the hydrolytic and in the arginine-ornithine exchange reactions;

b) the apparent Michaelis constant of transamidinase for glycine is unaffected but the maximal velocity for the transfer of the amidine group to this acceptor is decreased;

c) in the arginine-glycine transfer reaction the apparent Michaelis constant for arginine is normal but the maximal velocity is decreased.

Assuming that the form of the rate equation for the inhibited enzyme is the same as for the native enzyme [10], the values for V', the apparent maximal velocity, and $K'_{m\,(gly)}$ and $K'_{m\,(arg)}$, the apparent Michaelis constants for glycine and arginine, can be calculated. The calculations from the data of Fig. 3 give: $V' = 26—23$ for the native, $8.5—7$ for the inhibited enzyme; $K'_{m\,(gly)} = 1.5 \times 10^{-3}$ M for the native, 1.7×10^{-3} M for the inhibited enzyme; $K'_{m\,(arg)} = 2.6 \times 10^{-3}$ M for the native, 2.4×10^{-3} M for the inhibited enzyme. These values are in fair agreement with those previously reported [10]. Since the effects of the inhibitor on V (which is a function of the rate limiting step for the two halves of the reaction and of the concentration of the active enzyme molecules) are great while those on the K_m (which is a function of both the enzyme-substrate affinity constants and of the rate limiting steps) are small, it seems probable that the inhibitor reduces the number of active enzyme molecules, i.e. of the molecules capable of utilizing glycine as amidine acceptor.

arginine + enzyme $\rightleftharpoons$ ornithine + enzyme-amidine (1)
glycine + enzyme-amidine $\rightleftharpoons$ guanidinoacetate + enzyme (2)

In fact, after DTNB treatment and at saturating arginine concentrations, reaction (2) is the limiting one since the rate of the arginine-ornithine exchange is higher than the rate of the arginine—glycine transfer reaction (Fig. 3 and 5).

The apparent discrepancy between the results of point (a) and (c) can be explained assuming that the reagent produces a heterogeneous enzyme population. The greater part (75%) of the enzyme has lost

completely the capacity to use glycine as amidine acceptor and has a larger Michaelis constant for arginine, while a minor fraction (25%) is still in the native form, at least for the kinetic properties, and its Michaelis constant for arginine is unaltered. Since the rate of the hydrolytic reaction is one hundred times slower than the rate of the transfer to glycine, the contribution of the ornithine produced through the hydrolytic reaction is not detected in the usual assay for the arginine—glycine transfer reaction, so that the kinetic analysis gives the picture of a pure "non-competitive" inhibition instead of the picture of a "mixed" one.

The transamidinase derivative utilized for the kinetic analysis retains full hydrolytic activity at saturating arginine levels and thus retains the full capacity to form the amidine-enzyme complex. It follows that the three sulfhydryl groups modified by DTNB are not the groups responsible for the linkage with the amidine group.

The alteration of only one of these groups, however, is sufficient to modify the kinetic properties of the enzyme, as supported by the effect of arginine on both the reaction with DTNB and the change in catalytic activity. These changes consist in the loss of the capacity of the enzyme to react with glycine and hydroxylamine and in the increase of the apparent Michaelis constant for arginine.

In previous studies we have shown that the reaction of transamidinase with glycine and hydroxylamine is dependent upon the state of dissociation of a group of the enzyme with a pK of 8.5 [10]. This could suggest that the sulfhydryl group modified by DTNB and the group with a pK of 8.5 responsible for the behaviour of the pH curve of transamidinase are identical.

Further studies are in progress in order to investigate this possibility.

This work was supported by a grant from the Italian Consiglio Nazionale delle Ricerche, Impresa di Enzimologia; by the Grant 12291-02 from the National Institutes of Health and by the Nato Grant No 218.

REFERENCES

1. Grazi, E., Conconi, F., and Vigi, V., J. Biol. Chem. 240 (1965) 2465.
2. Ratner, S., and Rochovansky, O., Arch. Biochem. Biophys. 63 (1956) 277.
3. Vigi, V., Ronca, G., and Grazi, E., Biochem. Biophys. Res. Commun. 20 (1965) 757.
4. Grazi, E., and Vigi, V., Unpublished results.
5. Conconi, F., and Grazi, E., J. Biol. Chem. 240 (1965) 2461.
6. Chinard, F. P., J. Biol. Chem. 199 (1952) 91.
7. Ellman, G. L., Arch. Biochem. Biophys. 82 (1959) 70.
8. Lowry, O. H., Rosebrough, N. J., Farr, A. L., and Randall, R. J., J. Biol. Chem. 193 (1951) 265.
9. Boyer, P. D., J. Am. Chem. Soc. 76 (1954) 4331.
10. Ronca, G., Vigi, V., and Grazi, E., J. Biol. Chem. 241 (1966) 2589.

E. Grazi, V. Vigi, N. Rossi
Istituto di Chimica Biologica, Università di Ferrara
Via Fossato di Mortara 25, Ferrara, Italy

European J. Biochem. 1 (1967) 187—192

The Localization and Function of Carnitine Acetyltransferase
in the Flight Muscles of the Locust

A. M. T. Beenakkers and P. T. Henderson

Zoological Institute, Catholic University, Nijmegen

(Received November 21, 1966)

In order to determine the extent to which carnitine acetyltransferase is restricted to mitochondria, its occurrence in fractions of the flight muscles of *Locusta migratoria*, obtained by two different procedures, was checked by simultaneous assay of known mitochondrial enzymes.

In one procedure, fractionated extraction of the muscles was achieved by gently stirring the tissue successively in isotonic sucrose and phosphate buffer after which the residue was disrupted in phosphate buffer. In the second procedure fractionation was achieved by differential centrifugation.

In both procedures the partitioning of the carnitine acetyltransferase among all fractions closely paralleled that of the intramitochondrial marker enzymes, thus strongly suggesting that carnitine acetyltransferase is exclusively localized within mitochondria—possibly with the transferase occurring in two intramitochondrial compartments separated from each other by a membrane which is impermeable to acetyl CoA.

The last-named hypothesis is furthermore supported by correlated studies on mitochondrial respiration using acetate derivatives as substrates.

Friedman and Fraenkel [1] demonstrated an enzyme in sheep and pigeon liver extracts which catalyzes the reversible acetylation of carnitine by acetyl coenzyme A, to give acetylcarnitine with an energy-rich ester linkage [2]. This enzyme, carnitine acetyltransferase, was later found in a number of tissues [3,4] and was crystallized from pigeon breast muscle by Chase *et al.* [5].

The importance of carnitine in fatty acid oxidation, first demonstrated by Fritz [6] in liver homogenate and later shown in various organs by several investigators [7—11], was thought to be connected with its functioning as a carrier of both long and short chain fatty acids to their oxidative sites within mitochondria. Moreover, Fritz and Yue [11] demonstrated that transport of acetyl groups could be mediated by carnitine. That is, according to these authors, the acetyl group associated with acetyl CoA cannot readily penetrate mitochondrial membranes unless it is first transferred to carnitine.

Non-standard Abbreviations. Carnitine acetyltransferase, CAT; citrate synthase, CS; succinate dehydrogenase, SDH; triethanolamine, TRA.

Enzymes. Carnitine acetyltransferase, or acetyl-CoA: carnitine O-acetyltransferase (EC 2.3.1.7); citrate synthase, or citrate oxaloacetate-lyase (CoA acetylating) (EC 4.1.3.7); malate dehydrogenase, or L-malate: NAD oxidoreductase (EC 1.1.1.37); succinate dehydrogenase, or succinate: (acceptor) oxidoreductase (EC 1.3.99.1); acetyl-CoA synthetase, or acetate: CoA ligase (AMP) (EC 6.2.1.1).

Bressler and Katz [13] demonstrated that the transport of acetate in the opposite direction, that is out of the mitochondria into extramitochondrial compartments, is also mediated by carnitine. According to these investigators the acetyl group derived from acetyl CoA within mitochondria is (enzymatically) transferred to carnitine in order to facilitate its transport to extramitochondrial sites. This hypothesis is supported by the observation of CAT in isolated mitochondria from several rat and pigeon organs [3].

If it may be concluded that the acetyl group in acetylcarnitine is always derived from acetyl CoA, the localization of the transferase necessary for reversible transfer of acetyl group from CoA to carnitine determines where acetylcarnitine is synthesized.

In studying several rat organs Norum [14] found most of the CAT activity in the 25,000×g fraction but also reported a significant amount in the particle-free supernatant. Marquis and Fritz [4] came to a similar conclusion using rat heart—about 20% of the CAT activity was in the supernatant above the 100,000×g residue whereas the remaining 80% was localized in the mitochondria. On the other hand, a different fractionation technique led Beenakkers and Klingenberg [3] to the conclusion that CAT activity of rat heart was localized exclusively within mitochondria.

Since, in the flight muscles of the locust, a very rapid oxidation of fatty acids occurs [15,7,8] and the CAT activity is high [3], the purpose of the present investigation was to discover the apparent site at which CAT is recovered in these muscles; moreover, the previously established relationship between CAT-catalyzed acetylcarnitine production and respiration prompted us to use respiratory measurements in an attempt to obtain correlated information on the mechanism by which carnitine, *in situ*, enhances transport of short chain fatty acids.

EXPERIMENTAL PROCEDURE

DL-acetylcarnitine was synthesized from DL-carnitine (Fluka A.G., Buchs A.G., Switzerland) according to the method of Fraenkel and Friedman [16] as modified by Bremer [17], acetylcoenzyme A according to the technique of Simon and Shemin [18] and citrate synthase from pig heart following the procedure of Srere and Kosicki [19]. Malate dehydrogenase was obtained from Boehringer GmbH, Mannheim. All other chemicals used were commercial products of the highest attainable purity. The experimental animals, *Locusta migratoria* were bred in cages at 30° at a relative humidity of about 50% with approximately 10 hours illumination *per diem*. They were fed on ditch reed, endive and rye flour. In all experiments the study material used consisted of longitudinal indirect flight muscles (nos. 81 and 112) [20] of animals sacrificed 12 days following the imaginal moult.

Assay Procedure for the Localization of CAT

The distribution of CAT in flight muscles was investigated by application of the following two methods.

First Procedure (technique of Delbrück, Zebe and Bücher [21] as modified by Pette, Brosemer and Vogell [22]). Freshly extirpated flight muscles were gently stirred for 30 minutes in 5 ml of medium consisting of 0.3 M sucrose, 10 mM TRA-HCl-buffer and 2 mM sodium-EDTA (pH = 7.3). This mixture was then centrifuged at 100,000 × g for 12 minutes after which the residue was resuspended in the same medium, extracted for 30 minutes and recentrifuged. In the third extraction the residue from the first two was stirred as before, but in 5 ml 0.1 M potassium phosphate buffer containing 2 mM sodium EDTA (pH = 7.3). After centrifugation, the remaining residue was resuspended in the same phosphate buffer and disrupted by treatment for 90 seconds with an ultraturrax disintegrator (Janke-Kunkel); this suspension was finally recentrifuged (100,000 × g for 12 minutes) and the residue was resuspended in 5 ml of fresh K-phosphate buffer.

All of these extracts and the resuspended sediment were separately assayed for CAT, CS and succinate dehydrogenase activities.

Second Procedure. Using a Potter Elvehjem apparatus, flight muscles were homogenized for a short period in sucrose medium identical with that used in the first procedure and the resulting suspension filtered through a nylon cloth. Residual fractions were obtained by means of differential centrifugation, 150 × g (10 min), 5,000 × g (15 min), 20,000 × g (15 min), 100,000 × g (40 min) and 300,000 × g (60 min). Each of these residues was then disrupted (Janke-Kunkel disintegrator) for one minute at 0° in 4 ml 0.1 M potassium phosphate buffer (pH = 7.3) containing 2 mM sodium EDTA, and each was assayed for CAT, CS and SDH. The 300,000 × g supernatant was assayed directly for the same enzymes.

Assay of Enzyme Activities

Activities were estimated spectrophotometrically at 25° using an Eppendorf photometer with automatic recording of absorption against time.

The assay mixture for each of the measured enzymes was as follows:

For Carnitine Acetyltransferase. 0.2 M Tris-HCl buffer (pH = 8.0), 5 mM sodium EDTA, 5 mM NAD+, 50 mM L-malate, malate dehydrogenase (20 units), CS (0.003 ml, specific activity 28.0); 0.3 mM CoA-SH, 2 mM acetylcarnitine; time course-absorption curves were recorded at 366 mμ.

For Citrate Synthase. 0.2 M Tris-HCl buffer (pH = 8.0), 5 mM sodium EDTA, 5 mM NAD+, 50 mM L-malate, malate dehydrogenase (20 units), 0.15 mM acetyl CoA; changes in absorption with time were automatically recorded at 366 mμ.

For Succinate Dehydrogenase. 0.1 M potassium phosphate buffer (pH = 7.4), 5 mM EDTA, 1 mM KCN, 0.1% cytochrome c, 20 mM succinate; changes in absorption were recorded at 546 mμ.

Respiratory Measurements with Flight Muscle Mitochondria

Preparation of the Mitochondria. The flight muscles of 15 locusts were ground in about 30 ml of isolation medium (0.3 M sucrose, 10 mM TRA-HCl-buffer and 2 mM sodium EDTA, pH 7.1) using a Potter Elvehjem homogenizer with teflon pestle at a low speed. The homogenate was filtered through a layer of nylon cloth and the contents remaining in the cloth washed with about 10 ml of isolation medium. The two filtrates were combined and centrifuged at 200 × g for 8 minutes and the resulting supernatant recentrifuged for 8 minutes at 7000 × g. The 7000 × g supernatant was then decanted, and the pellet resuspended in isolation medium and again centrifuged for 8 minutes at 7000 × g. This final pellet was resuspended in 1.0 ml of isolation medium. The protein content of the resulting mitochondrial suspension was 14—16 mg/ml.

Measurements of Oxygen Uptake. The oxygen uptake was measured with a conventional Warburg apparatus at 37° using 5 ml Warburg vessels containing 0.1 ml of 10% KOH in the center well and air as the gas phase. The standard incubation medium consisted of 2 mM K_2HPO_4, 1 mM ADP, 2 mM $MgCl_2$ and 3 mM ATP; 0.1 ml of mitochondrial suspension (± 1.5 mg protein) was used per vessel; the total volume was brought to 1 ml with a medium consisting of 0.3 M sucrose, 1 mM sodium EDTA and 10 mM TRA-HCl-buffer (pH = 7.1). The concentration of each substrate examined (listed in the Table and added from the side-arm at zero time) was: acetate 2 mM, succinate 1 mM, DL-carnitine 2 mM, DL-acetylcarnitine 1 mM, acetyl CoA 0.3 mM and coenzyme A 0.1 mM.

contrast with this, the two sucrose extracts were found to contains as little as 3% of the total activity. As pointed out by Pette *et al.* [22] and Beenakkers [7], these isotonic sucrose extracts contain nearly all of the extramitochondrial enzyme activity, whereas mitochondrial enzymes are present in disrupted subparticulate suspensions. That unavoidable damage of mitochondria during dissection of the tissue would give some of the mitochondrial enzyme activity in the sucrose- and first non-isotonic phosphate extracts was verified by the occurrence of the mitochondrial marker enzyme, CS, in these two fractions.

As is evident from Table 1, the sum of transferase activities in all fractions equals that in the total homogenate, an observation similarly applicable to the two marker enzymes.

Fig. 1 indicates the distribution of enzyme activities in each consecutive fraction obtained by differential centrifugation of flight muscle homogenates.

Table 1. *Distribution of carnitine acetyltransferase, citrate synthase and succinate dehydrogenase activities after repeated extraction of flight muscles*
The tissue was extracted successively in two changes of isotonic sucrose and one of phosphate buffer. The residue was disrupted and the resulting particles resuspended in phosphate buffer. This suspension was then centrifuged and the resulting pellet resuspended in phosphate buffer. Enzyme activities were measured spectrophotometrically. For comparison, the activities measured on a total homogenate are also given. Values shown are the means of five separate experiments

Treatment of the tissue	Source of enzyme activity	Enzyme activity			Total activity		
		μmoles/hour/g fresh weight			$\%$		
		CAT	CS	SDH	CAT	CS	SDH
stirring in							
isotonic sucrose I	supernatant	18	88	0	2.5	0.9	0
isotonic sucrose II	supernatant	4	108	0	0.5	1.1	0
0.1 M phosphate buffer	supernatant	46	520	0	6.4	5.3	0
disintegration in							
0.1 M phosphate buffer	supernatant	645	8815	0	88.8	89.8	0
	sediment	13	284	670	1.8	2.9	100
	Total	726	9815	670	100	100	100
Activity of total homogenate		670	9600	690			

Protein Assay

Protein was determined either by the Biuret method as adapted by Beisenherz *et al.* [23] or according to the method of Lowry *et al.* [24].

RESULTS

Table 1 summarizes data on the activities of the 3 enzymes measured in the various fractions obtained by consecutive extraction of the tissue in sucrose and phosphate. One of the most conspicuous characteristics of these data is the close correlation between the distribution of CAT and the mitochondrial marker enzyme, CS. About 90% of the activity of both enzymes was present in the $100,000 \times g$ supernatant of the mixture produced by disrupting solid material remaining after gentle extraction of the tissue in isotonic sucrose or 0.1 M phosphate buffer. In

The distribution among the various fractions was almost similar for all three enzymes. Most of the activity was recovered in the mitochondrial pellet ($5000 \times g$ residue) with the $\pm 44\%$ of the total activity in the $150 \times g$ residual fraction being undoubtedly attributable to the presence of intact muscle tissue. On the other hand, the $300,000 \times g$ supernatant fraction, containing most of the extracted protein, exhibited only about 2.7% of the CAT and CS activity. It is interesting that the $5000 \times g$ pellet showed the highest specific activity for the three investigated enzymes whereas the $300,000 \times g$ supernatant exhibited the lowest activity; no SDH activity was detectable in this fraction, probably because of its structure-bound nature. This however does not hold for CAT and CS; *i.e.* after disruption of mitochondria in phosphate buffer these enzymes easily escape into the supernatant [3]. The presence of

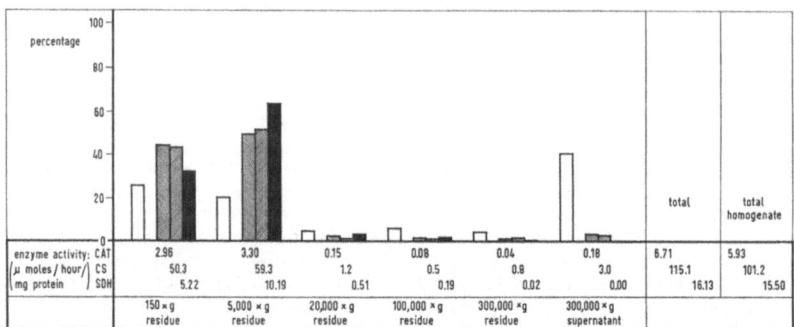

Fig. 1. *Distribution of carnitine acetyltransferase and of two marker enzyme activities after differential centrifugation of the muscles.* By differential centrifugation of a muscle homogenate the five residual fractions indicated were obtained, together with the 300,000×g supernatant fraction. The enzyme activities in each fraction were measured spectrophotometrically and the proteins content determined according to Lowry et al. [24]. The activities in a total homogenate are also given. Each value represents the mean of five separated experiments. □ total extracted protein; ▨ carnitine acetyltransferase (CAT); ▨ citrate synthase (CS); ■ succinate dehydrogenase (SDH)

Table 2. *The influence of carnitine on the oxidation of acetate by mitochondria of locust flight muscles*
The respiration of isolated mitochondria, corresponding to ± 1.5 mg mitochondrial protein, was measured by application of the conventional Warburg technique. For composition of the medium: see experimental procedure. The oxygen consumption was measured for 30 min at 37° with air as the gas phase. For each substrate the oxygen uptake was corrected by subtracting the utilization attributable to succinate. Each value expressed is the mean of at least seven experiments

Substrate	Oxygen consumption
	μatoms/min/g protein
acetate, 2 mM	3.2
acetate, 2 mM and CoA, 0.1 mM	15.0
acetyl-CoA, 0.3 mM	33.0
carnitine, 2 mM and CoA, 0.1 mM	27.0
acetyl-CoA, 0.3 mM and DL-carnitine, 2 mM	155.4
DL-acetylcarnitine, 1 mM	151.9

SDH in the 20,000, 100,000 and 300,000×g residues (3.2, 1.2 and 0.1°/₀ of the total enzyme activity respectively) suggests that these fractions contain a few light mitochondria or mitochondrial fragments produced during homogenization; if such fragments were present they might account for the CAT and SC in the three above-mentioned fractions and consequently in the 300,000×g supernatant. The data in Fig. 1 indicate that all of the enzyme activities present in a total homogenate are recovered in the various fractions prepared by this differential centrifugal procedure.

In Table 2 the influence of different substrates on respiratory rate in mitochondrial suspensions containing 1 mM succinate is summarized. Acetate alone failed to produce a significant rise in the oxygen

uptake with succinate but when accompanied by CoA, respiration increased by the value of 15 μatom oxygen/min/g protein. It is interesting to note that carnitine in the presence of CoA produced a significant increase in respiration rate. The reason for this phenomenon is not quite clear but it may be that added carnitine activates endogenous substrates or facilitates their intramitochondrial transport.

After addition of acetyl CoA the oxygen uptake was increased by 33 μatom above that observed in the controls. In the presence of both acetyl CoA and carnitine as well as acetylcarnitine, the level of respiration was significantly higher than that observed in the controls, *i.e.* acetyl CoA plus carnitine or acetylcarnitine alone, in the concentrations used, produced about the same increase in respiration rate.

DISCUSSION

The influence of carnitine on the oxidation of acetate as shown in the present experiments is in aggreement with the current hypothesis concerning the role of carnitine in fatty acid oxidation. Although-free acetate is oxidized very slowly in mitochondrial preparations from flight muscles, the oxidation of acetylcarnitine takes place very rapidly. This suggests that CAT activity inside mitochondria brings about transfer of the acetyl group which is necessary for it to enter the citric acid cycle.

The present experiments demonstrate that CAT is exclusively intramitochondrial. As already indicated the slight CAT activity in fractions normally containing extramitochondrial components of the tissue must be attributed to contamination with damaged mitochondria.

It is possible that the mitochondrial fractions in our experiments are contaminated with microsomes. But if microsomal CAT is present in the locust flight muscles, the percentage of the total activity of this enzyme in the microsomal fractions (*i.e.* 20,000 and 100,000×*g* residues in Fig. 1) would clearly differ from the percentage of the mitochondrial marker enzymes in these fractions. This is not the case and, as the sum of CAT activities in all fractions equals that in a total homogenate, no activity has been lost during fractionation-procedures.

The present results are therefore not in accordance with those of Norum [14] and Marquis and Fritz [4] who reported CAT activity in the particle-free supernatant originating from rat liver and heart respectively. In attempting to explain differences between their results and those of Beenakkers and Klingenberg [3], concerning the localization of CAT in rat heart, Marquis and Fritz [4] stated that the difference in preparation techniques could be responsible. This, of course, could be the case; and therefore in the present experiments known mitochondrial enzymes were used as markers in testing the nature of contaminants resulting from two quite different fractionation-procedures, one of which, differential centrifugation, being the same as that used by Marquis and Fritz. However, the last-named authors did not follow distribution of marker enzymes to test the purity of their fractions (*i.e.* whether their fractions were contaminated with submitochondrial fragments).

The currently popular hypothesis for explaining carnitine mediated transport of acetyl groups requires an extra- and an intramitochondrial CAT activity. As our experiments with locust flight muscles exclude the existence of extramitochondrial activity, we are led to propose a somewhat different mode of action of carnitine in acetyl group transport. In the present study on respiratory measurements using acetyl derivatives as substrates, it was found that acetyl CoA is oxidized at a relatively low rate (Table 2) suggesting a gradual diffusion of CoA derivatives into mitochondria of purified suspensions. This is probably explainable on the basis of slight damage incurred by the mitochondria during fractionation. Upon adding both acetyl CoA and carnitine to mitochondrial suspensions, the oxygen uptake is increased to values corresponding to those obtained with acetylcarnitine as a substrate. On the basis of this evidence therefore, the authors would accept the currently held hypothesis that a mitochondrial membrane exists which is impermeable to acetyl CoA but it would not appear that this is the outer membrane. It seems more in keeping with the present evidence to hypothesize that the outer mitochondrial membrane is permeable to acetyl CoA, but that between this membrane and a membrane situated more deeply within the mitochondria

(which, for convenience, is termed the "cristae membrane"), CAT activity would be present. The acetyl-carnitine generated in the space between the cristae and outer membrane could enter the deeper matrix where the acetyl group is again transferred to CoA by another CAT activity (Fig. 2). As far as the authors are aware, permeability of the outer membrane to acetyl CoA has not definitely been proved but there are a few strong indications. O'Brien and Brierley [25] stated that heart mitochondria contained at least two major compartments, "one readily penetrated by small solute molecules". Furthermore Garland *et al.* [26] recently demonstrated the existence of two intramitochondrial pools of CoA in

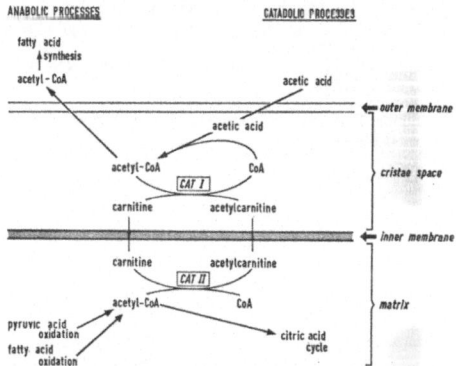

Fig. 2. *Postulated localization of carnitine acetyltransferase in the mitochondria of flight muscles and its role in transferring acetyl groups across the (acetyl CoA-impermeable) membrane.* The proposed mechanism is equally applicable to anabolic and catabolic processes

rat liver, and since exogenous CoA accelerated the oxidation rate of palmitate in these mitochondria, this cofactor presumedly had access to the site of palmityl-CoA-synthetase activity.

Similarly, in mitochondrial preparations of the authors, the acceleration of exogenous acetate oxidation by added CoA suggests that the outer membrane of the mitochondria is permeable to CoA and that acetyl-CoA synthetase is present at the intramitochondrial site. This localization is similar to that reported by Hele [27] and Webster [28] in bovine heart. In the scheme presented in Fig. 2 the synthetase activity is represented within the cristae space.

We thank Mrs W. J. Visschedijk-Niessen for her technical assistance.

REFERENCES

1. Friedman, S., and Fraenkel, G., *Arch. Biochem. Biophys.* 59 (1955) 491.
2. Fritz, I. B., Schultz, S. K., and Srere, P. A., *J. Biol. Chem.* 238 (1963) 2509.

3. Beenakkers, A. M. Th., and Klingenberg, M., *Biochim. Biophys. Acta*, 84 (1964) 205.
4. Marquis, N. R., and Fritz, I. B., *J. Biol. Chem.* 240 (1965) 2193.
5. Chase, J. F. A., Pearson, D. J., and Tubbs, P. K., *Biochim. Biophys. Acta*, 96 (1965) 162.
6. Fritz, I. B., *Acta Physiol. Scand.* 34 (1955) 367.
7. Beenakkers, A. M. Th., *Vetzuuroxidatie in de vliegspieren van* Locusta migratoria, Thesis, Thoben Offset, Nijmegen 1964.
8. Bode, C., and Klingenberg, M., *Biochim. Biophys. Acta*, 84 (1964) 93.
9. Bremer, J., *J. Biol. Chem.* 237 (1962) 3628.
10. Fritz, I. B., and McEwen, B., *Science*, 129 (1959) 334.
11. Fritz, I. B., and Yue, K. T. N., *J. Lipid Res.* 4 (1963) 279.
12. Fritz, I. B., *Am. J. Physiol.* 206 (1964) 531.
13. Bressler, R., and Katz, R. I., *J. Biol. Chem.* 240 (1965) 622.
14. Norum, K., *Acta Chem. Scand.* 17 (1963) 896.
15. Beenakkers, A. M. Th., *Acta Physiol. Pharmacol. Neerl.* 12 (1963) 332.
16. Fraenkel, G., and Friedman, S., *Vitamins and Hormones*, 15 (1957) 73.
17. Bremer, J., *J. Biol. Chem.* 237 (1962) 2228.
18. Simon, E. J., and Shemin, D., *J. Am. Chem. Soc.* 75 (1953) 2520.

19. Srere, P. A., and Kosicki, G. W., *J. Biol. Chem.* 236 (1961) 2557.
20. Albrecht, F. O., *Anatomy of the Migratory Locust*, The Athlone Press, London 1953.
21. Delbrück, A., Zebe, E., and Bücher, Th., *Biochem. Z.* 331 (1959) 273.
22. Pette, D., Brosemer, R. W., and Vogell, W., *Biochem. Z.* (In Press).
23. Beisenherz, G., Boltze, H. J., Bücher, Th., Czok, R., Garbade, K. H., Meyer-Arendt, E., and Pfleiderer, G., *Z. Naturforsch.* 8 (1953) 555.
24. Lowry, O. H., Rosebrough, N. J., Farr, A. L., and Randall, R. J., *J. Biol. Chem.* 193 (1951) 265.
25. O'Brien, R. L., and Brierley, G., *J. Biol. Chem.* 240 (1965) 4527.
26. Garland, P. B., Shepherd, D., and Yates, D. W., *Biochem. J.* 97 (1965) 587.
27. Hele, P., *J. Biol. Chem.* 206 (1954) 671.
28. Webster, L. T., jr., *J. Biol. Chem.* 240 (1965) 4158.

A. M. T. Beenakkers, and P. T. Henderson
Zoölogisch Laboratorium
Faculteit der Wiskunde en Natuurwetenschappen
Katholieke Universiteit
Driehuizerweg 200
Nijmegen, Holland

European J. Biochem. 1 (1967) 193—198

Some Properties of Alkaline Phosphatase of *Pseudomonas fluorescens*

I. Friedberg and G. Avigad

Department of Biological Chemistry, The Hebrew University, Jerusalem

(Received January 10, 1967)

The formation of alkaline phosphatase in a strain of *Pseudomonas fluorescens* was found to be induced by limiting quantities of orthophosphate in the extracellular medium. The substrate specificity of this enzyme seemed to be similar to that of *Escherichia coli*. On the other hand, it had some properties different from those of the *E. coli* enzyme. The *Pseudomonas* enzyme was several hundred fold more sensitive to inhibition by EDTA, but was not inhibited by cyanide and by various mercaptans. The apparent K_m for *p*-nitrophenylphosphate was 2.0×10^{-4} M and K_i for orthophosphate as a competitive inhibitor was 3.7×10^{-5} at pH 8.6.

Alkaline phosphatase has been detected and isolated from many biological sources [1,2]. The enzyme obtained from *Escherichia coli* cells was extensively studied in recent years. In certain *E. coli* strains its formation was found to be induced by limiting amounts of orthophosphate in the medium [3,4]. It has been purified [5—7] and its specificity and kinetics studied in great detail [8,9]. The active *E. coli* enzyme is built of two protein monomers and Zn^{++} is needed to obtain the dimer which exhibits alkaline phosphatase activity [10—18]. The primary structure of the enzyme near the active serin phosphate residue has been analysed [19—21] and mechanisms of the catalytic reaction have been proposed [22—24].

It has been shown that the *E. coli* enzyme can be completely released from the bacteria when they are converted into spheroplasts or their cell wall is damaged by chemical treatment [7,25,26]. During a study of enzyme excretion and spheroplast formation in a levan producing *Pseudomonas* [28], we noticed that the alkaline phosphatase produced by this organism had some properties different from those of the *E. coli* enzyme. The present report describes some of these observations.

EXPERIMENTAL PROCEDURES

Materials and Methods. *p*-Nitrophenylphosphate, other substrates and enzyme reagents utilized were purchased from Boehringer und Soehne GmbH, Mannheim, and Sigma Biochemical Corp., St. Louis.

Non-standard Abbreviations. *p*-Nitrophenylphosphate, NPP; *p*-nitrophenol, NP; trans-1,2-diaminocyclohexane-N,N,N',N'-tetraacetic acid, DCTA.

Enzymes. Alkaline phosphatase, or orthophosphoric monoester phosphohydrolase (EC 3.1.3.1); glucose oxidase, or β-D-glucose: oxygen oxidoreductase (EC 1.1.3.4); isocitrate dehydrogenase (EC 1.1.1.42).

Chelating agents and fine chemicals were obtained from British Drug House, Poole, England, and Fluka AG, Buchs, Switzerland.

Organisms. *Pseudomonas fluorescens*, No 11, a levan forming and nitrate denitrifying strain, was obtained from Dr. A. Fuchs [28]. *Escherichia coli* ML 308 and K 10 were from the Department of Biological Chemistry collection.

Growth of Bacteria. Cultures were kept on nutrient broth-glucose agar slants. *P. fluorescens* was grown on a synthetic medium [29] supplemented by 0.0135% Difco yeast extract. The phosphate buffer in Fraser's medium was changed to 0.1 M Tris-HCl, pH 7.4 to provide for the induction of alkaline phosphatase. The yeast extract in this case supplied the minimal concentration of 50 mμmoles orthophosphate per liter which was obligatory for growth. The phosphate-poor medium of Torriani [4] was used for growth of *E. coli* and for alkaline phosphatase derepression.

Cells were grown at 30° in a rotatory shaker at 200 rev./min, *P. fluorescens* for 48 and *E. coli* for 18 hours. After 10 min centrifugation at $10,000 \times g$, the sedimented cells were washed by 0.1 M Tris-HCl buffer, pH 7.4.

Preparation of Extracts. Bacteria were dispersed in 0.1 M Tris-HCl buffer, pH 7.4, in about one tenth of the original culture volume. After 10 min treatment in the 10 Kc Raytheon sonic oscillator, cell debris was removed by 20 min centrifugation at $10,000 \times g$ at 2°. The supernatant afforded the crude enzyme solution. In some experiments total phosphatase activity was determined in cells lysed by toluene [7].

Determination of Enzyme Activity. Standard reaction mixture (3.0 ml) contained 0.4 M Tris-HCl, pH 8.6, 1.0 mM NPP, and enzyme at room temperature. Absorbance of the released *p*-nitrophenol was read at 420 mμ in cuvettes with 1.0 cm light path using the Zeiss Model PMQII spectrophotometer. A unit of enzyme was defined as that amount which hydrolyzed one mμmole NPP per min at standard conditions of assay. *P. fluorescens* extracts contained between 40 to 67 units alkaline phosphatase per mg protein, whereas *E. coli* extracts contained 180 to 330 units enzyme per mg protein in the different batches prepared.

Stability of Enzyme. Alkaline phosphatase activity in the bacterial extracts decreased 3% per day for the *P. fluorescens* and 1% per day for the *E. coli* preparations when kept at 4°.

RESULTS

Formation of Alkaline Phosphatase in Growing Culture. Similar to what has been reported for *E. coli* [4], the enzyme appeared in *P. fluorescens*

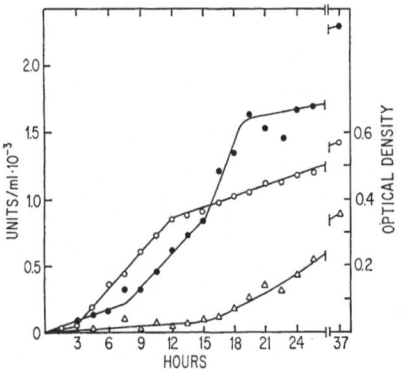

Fig. 1. *Growth of* P. fluorescens *and enzyme formation.* Cells were grown on a phosphate poor medium at 30° as described under Experimental Procedures. ○, turbidity of cell suspension; ●, total alkaline phosphatase in toluenized culture; △, extracellular enzyme

when grown on a phosphate-poor medium. The results illustrated in Fig. 1 indicate that a significant part of the enzyme leaked into the medium when growth levelled off. In another set of experiments (Table 1), it could be shown that this leakage of alkaline phosphatase is accompanied also by leakage of levansucrase, whereas a typical endocellular enzyme, isocitric dehydrogenase, did not leak into the medium under these conditions. In comparison it is evident that leakage of alkaline phosphatase from *E. coli* was very small.

In order to check whether appearance of alkaline phosphatase in the extracellular fluid did not occur as a result of mechanical damage to the cells during centrifugation and resuspension, the following experiment was performed. *P. fluorescens* cultures were grown 48 hours in Petri dishes on agar medium poor in phosphate. The plates were then sprayed with 0.5 M Tris buffer, pH 8.6, then with a 0.01 M NPP solution. The colonies, as well as a large area around them, stained yellow. When *E. coli* was similarly tested, only the colonies stained and extracellular NP liberation did not show.

Spermine was described to stabilize the cellular membrane of various bacteria [32,33]. When *P. fluorescens* cultures were treated with 1.0 mM spermine, appearance of extracellular alkaline phosphatase occurred similar to control experiments without the polyamine.

Crypticity of Alkaline Phosphatase. When enzyme activity of suspensions of whole cells was compared to that of toluenized cells, it was found that 30 to 40% of the total alkaline phosphatase activity present was exhibited by the washed bacteria. In *E. coli* ML 308 the equivalent values observed were 50 to 60% of the total enzyme activity.

Substrate Specificity. A broad range of phosphate esters was hydrolyzed by the *P. fluorescens* enzyme (Table 2). The pattern of substrates hydrolyzed resembles that of the *E. coli* enzyme [5,6].

Table 1. *Leakage of alkaline phosphatase from bacterial cells*

Cells were grown 18 hours on a phosphate poor medium to a turbidity of 60 to 70 units in the Klett colorimeter, filter 66. Total enzyme activity was determined in sonicates prepared as described in Experimental. Cells were separated from the medium by centrifugation, then washed thrice with 0.1 M Tris buffer, pH 7.4. Alkaline phosphatase was measured as described in the text. Levansucrase activity was assayed by measuring with glucose oxidase the amount of glucose liberated in a reaction system of 0.15 M sucrose in 0.05 M acetate buffer, pH 5.4 [30]. Isocitric dehydrogenase was assayed spectrophotometrically [31] in a 3.0 ml reaction system containing (in mM): Tris pH 7.45, 33; MgCl₂, 2.5; DL-isocitrate, 1.5; NADP⁺, 0.52 and a sample of enzyme

Organism	Enzyme	Total units[a]	Distribution of enzyme activity		
			Medium	Washings	Washed cells
			%	%	%
P. fluorescens	Alkaline phosphatase	1.9	25.0	20.0	49
	Levansucrase	5.0	21.0	32.0	48
	Isocitric dehydrogenase	750.0	6.5	3.2	90
E. coli ML 308	Alkaline phosphatase	11.2	0.1	trace	100
E. coli K 10	Alkaline phosphatase	52.2	2.7	1.3	98

[a] In mμmole product formed per min per ml of bacterial culture.

Table 2. *Hydrolysis of various substrates by bacterial alkaline phosphatase*

Reaction system at 30° contained: Tris, pH 8.6, 400 mM; substrate, 10 mM; *Pseudomonas* enzyme, 4.3 units or *E. coli* enzyme, 5.8 units per ml. Orthophosphate released during 30 min of incubation was measured colorimetrically [34]. When ADP was the substrate an alternative reagent was used [35]. NPP hydrolysis was measured as described under Experimental

Substrate	Relative rates of hydrolysis	
	P. fluorescens	*E. coli* ML 308
	%	%
p-Nitrophenyl phosphate	100	100
mono-Phenyl phosphate	122	115
o-Carboxyphenyl phosphate	92	62
α-Glucose-1-phosphate	85	53
Glucose-6-phosphate	98	75
Fructose-1,6-diphosphate	29	77
β-Glycerol phosphate	112	89
Adenosine monophosphate	116	111
Adenosine diphosphate	31	40
Adenosine triphosphate	60	51
Pyrophosphate	8	15

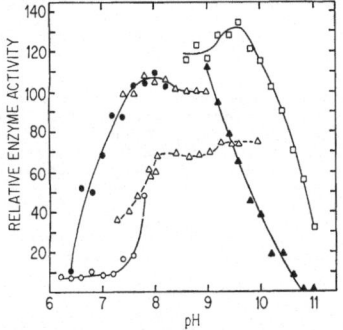

Fig. 2. *Effect of pH on enzyme activity.* Standard reaction mixture contained 0.1 M Buffer. ○, Tris-maleate-NaOH; ●, Collidine-HCl; △, Tris-HCl; ▲, Carbonate-bicarbonate; □, Glycine-NaOH; 1.7 units *Pseudomonas* enzyme per ml. After 15 min incubation at 25°, the reaction was stopped by addition of one volume of 0.2 M Na₂HPO₄ in 2 M Tris buffer, pH 8.6. Absorbance at 420 mμ was read to determine level of hydrolysis. ——, soluble enzyme; - - - -, cell suspension

Optimum pH. A broad range for maximal activity was found between pH 7.6 to 10.0 (Fig. 2). The pH curve for the *E. coli* enzyme, in comparison, has narrower peaks at pH 9.3 [4] or 8.0 [5]. The possibility that the *P. fluorescens* preparation has more than one form of alkaline phosphatase, as was detected in other organisms [36—38], cannot be excluded. A very similar broad range of maximal activity was found for whole *P. fluorescens* suspensions. It is evident that maleate and carbonate have a significant inhibitory effect on the enzyme (*cf.* ref. [1]).

Substrate Affinities. The apparent K_m values for NPP was found to be 4.6×10^{-5} M between 1×10^{-5} to 5×10^{-5} M of substrate concentration, 2.0×10^{-4} M between 5×10^{-5} to 1×10^{-2} M of NPP (Fig. 3) and 4.7×10^{-4} M between 1×10^{-2} to 5×10^{-2} M of NPP. A similar phenomenon was found also for the *E. coli* enzyme [6]. In that case, however, the K_m values were about one order of magnitude lower than those obtained for the *Pseudomonas* enzyme.

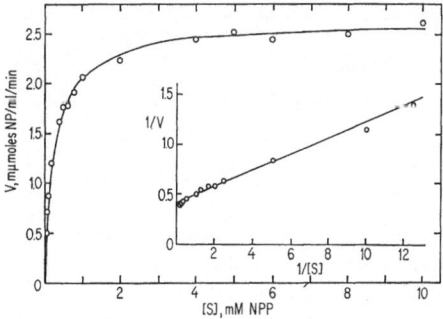

Fig. 3. *Effect of substrate concentration on rate of hydrolysis.* Standard reaction mixtures were assayed with varying NPP concentrations and 2.3 units *Pseudomonas* enzyme per ml

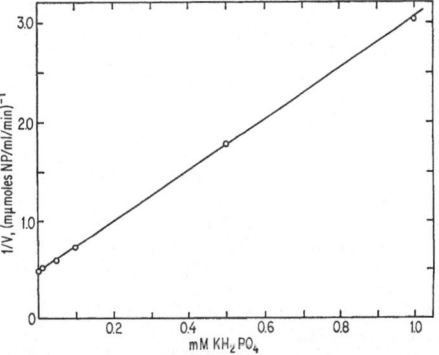

Fig. 4. *Inhibition of enzyme activity by orthophosphate.* Rate of hydrolysis of NPP in the standard reaction system was measured in presence of varying sodium orthophosphate concentrations. 2.3 units *Pseudomonas* alkaline phosphatase per ml were employed

Inhibition by Orthophosphate. Similar to alkaline phosphatase from other sources [1,2] the *Pseudomonas* enzyme was competitively inhibited by orthophosphate (Fig. 4). Apparent K_i was 3.7×10^{-5} M compared to a value of 3.7×10^{-6} M for the *E. coli* enzyme [5].

Inhibition by Various Chelating Agents. The results presented in Table 3 show a completely different pattern of inhibition of the *Pseudomonas* enzyme when compared to that of *E. coli* (Fig. 5). In the latter case, present results are similar to those reported before [5,6,11,12]. Compared to the *E. coli*

Table 3. *Effect of various chelators on the activity of bacterial alkaline phosphatase*
Standard reaction mixture containing 3.3 units enzyme per ml was preincubated 10 min with the chelator at room temperature. The reaction was started by addition of NPP

Chelator	Concentration	Enzyme activity	
		P. fluorescens	*E. coli* ML 308
	M	%	%
Ethylenediaminetetra-acetic acid (EDTA)	1×10^{-5}	100	—
	3×10^{-5}	43	—
	5×10^{-5}	0	—
	1×10^{-3}	—	78
	1×10^{-2}	—	45
	5×10^{-2}	—	19
trans-1,2-Diaminocyclo-hexane-N,N,N',N'-tetraacetic acid (DCTA)	1×10^{-5}	104	100
	5×10^{-5}	2	86
	5×10^{-2}	—	76
Nitrilotriacetic acid (NTA)	1×10^{-4}	47	65
	1×10^{-3}	0	0
Iminodiacetic acid (IDA)	1×10^{-2}	100	85
	5×10^{-2}	—	0
Uramil-N,N-diacetic acid (UDA)	1×10^{-5}	100	94
	5×10^{-5}	5	40
	1×10^{-4}	2	2
1,10-o-Phenanthroline[a]	1×10^{-3}	100	80
	5×10^{-3}	—	64
	1×10^{-2}	100	—
8-Hydroxyquinoline[a]	1×10^{-3}	100	0
	5×10^{-3}	86	—
Quinaldic acid	1×10^{-3}	100	100
	1×10^{-2}	103	102
Salicylaldoxime	1×10^{-3}	102	100
	1×10^{-2}	100	100
Diethyldithiocarbamate	1×10^{-3}	106	101
	1×10^{-2}	103	94
o-Aminothiophenol[b]	1×10^{-4}	100	103
	1×10^{-3}	117	112
Eriochrome Black T	1×10^{-4}	100	88
	5×10^{-4}	96	46
NaCN	1×10^{-3}	100	72
	1×10^{-2}	100	9
	5×10^{-2}	100	—

[a] Chelator was taken from an alcoholic stock solution. Control enzyme systems with ethanol 6% (v/v) were used as a standard.
[b] Reaction performed under nitrogen to avoid oxidation of chelator.

enzyme, the *Pseudomonas* enzyme is extremely sensitive to an immediate inhibition by EDTA and DCTA, but on the other hand was not so affected by cyanide, o-phenanthroline and 8-hydroxyquinoline.

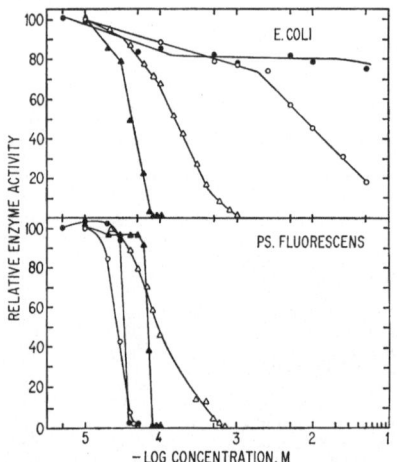

Fig. 5. *Inhibition of alkaline phosphatase from* E. coli *and* P. fluorescens *by various chelating agents.* Standard reaction mixtures were assayed as in Table 3, with varying concentrations of chelators (neutralized to pH 8.0). ○, EDTA; ●, DCTA; △, NTA; ▲, UDA (for abbreviations see Table 3)

Table 4. *Effect of mercaptans on enzyme activity*
All standard reaction mixtures with and without chelators, were flushed with nitrogen for 20 min before addition of NPP. The system contained 3.3 units phosphatase per ml, and the course of reaction measured under nitrogen

SH compound	Concentration	Enzyme activity	
		P. fluorescens	*E. coli* ML 308
	M	%	%
Cysteine	1×10^{-5}	—	99.8
	5×10^{-5}	—	82.0
	1×10^{-4}	106	20.0
	1×10^{-3}	103	0.4
	5×10^{-3}	116	—
	1×10^{-2}	116	—
Sodium thioglycolate	1×10^{-3}	—	100
	5×10^{-3}	100	86
	1×10^{-2}	100	68
2-Mercaptoethanol	5×10^{-3}	100	64
	1×10^{-2}	100	10
Sodium sulfide	1×10^{-3}	100	51
	1×10^{-2}	100	0

When *P. fluorescens* alkaline phosphatase solutions (3.3 units per ml) were inactivated by 5×10^{-5} M EDTA, for 10 minutes, addition of 1×10^{-3} M Mg^{++}, Zn^{++}, Ca^{++}, Mn^{++}, Fe^{++} or Cu^{++}, did not restore activity even after 24 hours of incubation at room temperature. When a sample of enzyme (3.8 units per ml) inactivated by 9×10^{-4} M EDTA (a concentration used for spheroplast formation [7]) was dialyzed against three changes of 100 volumes of 0.01 M Tris buffer, pH 7.4, for 24 hours, only about 1% of the original activity could be detected.

When bacterial suspensions were treated with EDTA, immediate alkaline phosphatase inactivation occurred (50% inhibition at 2.5×10^{-5} M). It was thus not possible to use EDTA for spheroplast formation [7,25,26] in order to isolate an active enzyme from this organism.

Inhibition by Mercaptans. In complete variance with *E. coli* [10,11,14,39], yeast [40] or mammalian [39,41] enzymes, the alkaline phosphatase of *P. fluorescens* was not inhibited by various thio compounds (Table 4). Sulfhydryl binding reagents such as *p*-hydroxymercuribenzoate and *N*-ethylmaleimide at 1×10^{-5} to 1×10^{-3} M did not inhibit either the *E. coli* of the *Pseudomonas* enzyme.

DISCUSSION

Alkaline phosphatases from microorganisms other than *E. coli* were studied only in a small number of cases, such as yeast [40], *Neurospora* [38] and *Bacillus subtilis* [42] enzymes. Alkaline phosphatase in *P. fluorescens* is inducible in a similar manner to the *E. coli* enzyme [3,4]. On the other hand, it has several distinct characteristics which are different from the *E. coli* enzyme, although substrate specificity of both enzymes is very similar. The *Pseudomonas* enzyme is several hundred fold more sensitive to EDTA inhibition and is also not inhibited by cyanide and mercaptans. Quantitative differences in susceptibility to inhibition by other chelating agents were also observed. It is probable that the *Pseudomonas* enzyme needs a divalent cation for activity, such as Zn^{++} which is obligatory for the *E. coli* enzyme, or Mg^{++} which activates alkaline phosphatases from other sources [1,2]. However, at present it cannot be evaluated whether the differences in the inhibition by chelating agents which exist between the *Pseudomonas* and *E. coli* enzymes, are caused by different cations or by presence of Zn^{++} in the active protein but with different ligands. Isolation and purification of *P. fluorescens* phosphatase, preferably from a constitutive strain so as to provide ample quantities of enzyme, will furnish information on the cation involved.

Other differences were also noticed between the two enzymes. The *Pseudomonas* phosphatase is less stable in dilute solution, it has a very broad pH optimum range of activity, and its apparent affinity constants for the substrate and orthophosphate as inhibitor are ten times higher than with *E. coli* enzyme.

Due to inhibition by EDTA, it was difficult to evaluate the location of alkaline phosphatase in *P. fluorescens* with similar techniques as those performed on *E. coli* [7,25,26]. However, several observations described in the present study give indirect

evidence that also in this case the enzyme is located outside the cell membrane. This is supported by the fact that a large amount of enzyme is excreted in preference to intracellular enzymes and spermine does not prevent this phenomenon. A significant leakage of alkaline phosphatase has also been noted in *B. subtilis* [43]. This has prompted the suggestion that differences in cell wall structure may determine whether peri-enzymes such as alkaline phosphatase, will be excreted [43,44]. It is hoped that further research on the properties of *P. fluorescens* alkaline phosphatase will yield more conclusive data about its location in the cell. Methods of electron microscopy like those recently employed to establish the site of alkaline phosphatase in *E. coli* [45], could be instructive when used in comparative studies with other microorganisms.

REFERENCES

1. Stadtman, T. C., in *The Enzymes* (edited by P. D. Boyer, M. Lardy, and K. Myrback), Academic Press, New York 1961, Vol. 5, p. 55.
2. Morton, R. K., in *Comprehensive Biochemistry* (edited by M. Florkin and E. M. Stotz), Elsevier Publishing Co., Amsterdam 1965, Vol. 16, p. 55.
3. Horiuchi, T., Horiuchi, S., and Mizuno, P., *Nature*, 183 (1959) 1529.
4. Torriani, A., *Biochim. Biophys. Acta*, 38 (1960) 460; also in *Procedures in Nucleic Acid Research* (edited by G. L. Cantoni and D. R. Davies), Harper and Row Publishers, New York 1966, p. 224.
5. Garen, A., and Levinthal, C., *Biochim. Biophys. Acta*, 38 (1960) 470.
6. Heppl, L. A., Harkness, D. R., and Hilmoe, R. J., *J. Biol. Chem.* 237 (1962) 841.
7. Malamy, M., and Horecker, B. L., *Biochemistry*, 3 (1964) 1889, 1893.
8. Wilson, I. B., and Dayan, J., *Biochemistry*, 4 (1965) 645.
9. Jenkins, W. T., and D'Ari, L., *J. Biol. Chem.* 241 (1966) 295.
10. Levinthal, C., Singer, R. E., and Fetherof, K., *Proc. Natl. Acad. Sci. U. S.* 48 (1962) 1230.
11. Plocke, D. J., Levinthal, C., and Valee, B. L., *Biochemistry*, 1 (1962) 373.
12. Plocke, D. J., and Vallee, B. L., *Biochemistry*, 1 (1962) 1039.
13. Garen, A., and Garen, S., *J. Mol. Biol.* 7 (1963) 13.
14. Schlesinger, M. J., and Levinthal, C., *J. Mol. Biol.* 7 (1963) 1.
15. Fan, D. P., Schlesinger, M. J., Torriani, A., Barrett, K. J., and Levinthal, C., *J. Mol. Biol.* 15 (1966) 32.
16. Cohen, S. R., and Wilson, I. B., *Biochemistry*, 5 (1966) 904.
17. Bishop, D. H. C., Roche, C., and Nisman, B., *Biochem. J.* 90 (1964) 378.
18. Schlesinger, M. J., *J. Biol. Chem.* 241 (1966) 3181.
19. Hilstein, C., *Biochim. Biophys. Acta*, 67 (1963) 171.
20. Schwartz, J. H., Crestfield, A. M., and Lipmann, F., *Proc. Natl. Acad. Sci. U. S.* 49 (1963) 722.
21. Tait, G. H., and Vallee, B. L., *Proc. Natl. Acad. Sci. U. S.* 56 (1966) 1247.
22. Schwartz, J. H., *Proc. Natl. Acad. Sci. U. S.* 49 (1963) 871.
23. Hummel, J. P., and Kalnitsky, G., *Ann. Rev. Biochem.* 33 (1964) 15.
23a. Fernley, H. N., and Walker, P. G., *Nature*, 212 (1966) 1435.

24. Lazdunski, C., and Lazdunski, M., *Biochim. Biophys. Acta*, 113 (1966) 551.
25. Malamy, M., and Horecker, B. L., *Biochem. Biophys. Res. Commun.* 5 (1961) 104.
26. Neu, H. C., and Heppel, L. A., *J. Biol. Chem.* 240 (1965) 3685.
27. Nossal, N. G., and Heppel, L. A., *J. Biol. Chem.* 241 (1966) 3055.
28. Fuchs, A., Doctoral Thesis, Rijksuniversiteit te Leiden. Waltman, Delft 1959; *Nature*, 178 (1956) 921.
29. Fraser, D., and Jerrel, E. A., *J. Biol. Chem.* 205 (1953) 291.
30. Hestrin, S., Feingold, D. S., and Avigad, G., *Biochem. J.* 64 (1956) 340.
31. Ochoa, S., in *Methods in Enzymology* (edited by S. P. Colowich and N. O. Kaplan), Academic Press, New York 1955, Vol. 1, p. 699.
32. Tabor, C. W., *J. Bacteriol.* 83 (1962) 1101.
33. Grossowicz, N., and Ariel, M., *J. Bacteriol.* 85 (1963) 293.
34. Fiske, C. H., and SubbaRow, Y., *J. Biol. Chem.* 66 (1925) 375.
35. Ernster, L., Zetterstrom, R., and Lindberg, O., *Acta Chem. Scand.* 4 (1950) 942.

36. Moss, D. W., *Biochem. J.* 94 (1965) 458.
37. Moog, F., Vire, H. R., and Grey, R. D., *Biochim. Biophys. Acta*, 113 (1966) 336.
38. Nyc, J. F., Kadner, R. J., and Crocken, B. J., *J. Biol. Chem.* 241 (1966) 1468.
39. Agus, S. G., Cox, R. P., and Griffin, M. J., *Biochim. Biophys. Acta*, 118 (1966) 363.
40. Stadtman, T. C., *Biochim. Biophys. Acta*, 32 (1959) 95.
41. Cox, R. P., and MacLeod, C. M., *Cold Spring Harbor Symp. Quant. Biol.* 29 (1964) 233.
42. Hiraga, S., *J. Bacteriol.* 91 (1966) 2192.
43. Cashel, M., and Freese, E., *Biochem. Biophys. Res. Commun.* 16 (1964) 541.
44. Trevithick, J. R., and Metzenberg, R. L., *J. Bacteriol.* 92 (1966) 1010.
45. Kushanarev, V. M., and Smirnova, T. A., *Can. J. Microbiol.* 12 (1966) 605.

I. Friedberg and G. Avigad
Department of Biological Chemistry
The Hebrew University of Jerusalem
Jerusalem, Israel

European J. Biochem. 1 (1967) 199—206

The Participation of GTP-AMP-P Transferase
in Substrate Level Phosphate Transfer of Rat Liver Mitochondria[1]

H. W. Heldt and K. Schwalbach

Physiologisch-chemisches Institut der Philipps-Universität Marburg/Lahn

(Received November 17, 1966)

From kinetic studies on the reaction sequence of substrate level phosphorylation in rat liver mitochondria, using anaerobic ketoglutarate dismutation in the presence of oligomycin and [^{32}P]phosphate, phosphohistidine appears to be the first intermediate to be labelled, followed by GTP. [^{32}P]ADP rather than [^{32}P]ATP is shown to be the main product of the reaction. The phosphorylation of AMP requires ketoglutarate and is stimulated by 2,4-dinitrophenol.

GTP-AMP-P transferase is localized in the mitochondria. This conclusion is based on enzymatic assays of fractionally ectracted rat liver and of isolated mitochondria and microsomes.

Mean values for the activities of GTP-AMP-P transferase, nucleoside diphosphate kinase and succinic thiokinase in rat liver mitochondria are given and are compared with the rate of ketoglutarate oxidation.

A possible function of GTP-AMP-P transferase for the phosphorylation of endogenous AMP is discussed with regard to the compartmentation of nucleotides in the mitochondria.

A new chromatographic assay for GTP-AMP-P transferase is reported, an assay which is not affected by nucleoside diphosphate kinase and adenylate kinase occurring in liver homogenates. An optical enzymatic assay for nucleoside diphosphate kinase is also described.

When studying substrate level phosphorylation in liver mitochondria, a phosphorylation of AMP was observed which accompanied the oxidation of ketoglutarate [1]. This was at first explained by assuming that ATP formed by substrate level phosphorylation was available to adenylate kinase [1]. Our further studies of this problem showed, however, that the direct phosphorylation of AMP by GTP was more probable. These findings recalled an earlier report on the isolation of a 6-oxypurine-nucleoside triphosphate-adenosine monophosphate transphosphorylase from liver [4,5], which catalyzed the reaction:

$$XTP + AMP \rightarrow XDP + ADP.$$

[1] Part of these results have been reported elsewhere in preliminary notes [1—3].

Non-standard Abbreviations. Triethanolamine, TRA; nucleoside diphosphate kinase, NuDiKi; hexokinase, HK; glucose-6-phosphate dehydrogenase, G6PDH; adenylate kinase, AdK; phosphohistidine, P-His; 2,4-dinitrophenol, DNP.

Enzymes. Glyceraldehydephosphate dehydrogenase, or D-glyceraldehyde-3-phosphate: NAD oxidoreductase (phosphorylating) (EC 1.2.1.12); glucose-6-phosphate dehydrogenase, or D-glucose-6-phosphate: NADP oxidoreductase (EC 1.1.1.49); hexokinase, or ATP-D-hexose-6-phosphotransferase (EC 2.7.1.1); phosphoglycerate kinase, or ATP: 3-phospho-D-glycerate 1-phosphotransferase (EC 2.7.2.3); adenylate kinase, or ATP: AMP phosphotransferase (EC 2.7.4.3); nucleosidediphosphate kinase, or ATP: nucleosidediphosphate phosphotransferase (EC 2.7.4.6).

The enzyme was shown to be highly specific for AMP as phosphate acceptor and for ITP and GTP as phosphate donors [2].

The question arose, whether this enzyme was responsible for the phosphorylation of endogenous AMP observed in liver mitochondria. Since it was not known where this enzyme occurred in the liver cell, its intracellular localization was determined. It will be shown in the following that it occurs in the mitochondria and that it is involved in substrate level phosphorylation. Since liver mitochondria do not contain ITP [6], the physiological substrate of the enzyme seems to be GTP. The enzyme will therefore be called GTP-AMP-P transferase below.

METHODS

Rat liver mitochondria were isolated according to Klingenberg and Slenczka [7]. The isolation medium contained 0.25 M sucrose, 20 mM TRA-HCl, pH 7.2, and 1 mM EDTA. Protein measurements were done according to Kröger and Klingenberg [8].

Fractional Extraction

Fractional extraction of rat liver tissue was based on the procedure of Delbrück, Zebe and Bücher [9]. The rat was killed by decapitation and the liver was immediately removed, cut into small pieces and washed in a 0.25 M sucrose medium containing 20 mM TRA, pH 7.2, and 1 mM EDTA. Care was taken to remove traces of blood. After removal of the adherent medium by filter paper, the tissue pieces were weighed and suspended in 9 times their weight of sucrose medium. The sample was homogenized with a teflon Potter-Elvehjem homogenizer and then centrifuged for 20 minutes at 80,000 $\times$ g in an ultracentrifuge Model Spinco L, rotor 40-2. Extraction with sucrose medium was repeated, and was followed by extraction with 0.1 M phosphate buffer, pH 7.2. The resulting sediment was resuspended in 0.1 M phosphate buffer, homogenized and sonicated 6 times for 5 sec at the maximal output of an ultrasonic disintegrator, Branson Model S-75. The remaining sediment was separated by centrifugation, resuspended in phosphate buffer and homogenized, and yielded the insoluble protein fraction. All steps were carried out in an ice bath, with the exception of the sonication, which was done in a NaCl-ice-bath.

Preparation of [^{32}P]GTP

The preparation of [^{32}P]GTP was based on a procedure described by Pfleiderer [10]. Since phosphoglycerate kinase is not completely ATP specific, the reaction is also suitable for labelling GTP.

Reaction Mixture. The following reaction mixture was incubated for 90 minutes at 20°: Trisbuffer pH 8, 25 μmoles; cysteine hydrochloride, 2.5 μmoles; MgCl$_2$, 3.6 μmoles; GTP, 2.0 μmoles; phosphoglyceric acid, 1.2 μmoles; [^{32}P]orthophosphate carrier free (Amersham), 2 mC; glyceroaldehydephosphate dehydrogenase (Boehringer), 25 units; phosphoglycerate kinase (Boehringer), 15 units; total volume, 1.3 ml.

The reaction was stopped by heating the mixture in a boiling water bath for 5 seconds. It was then immediately chilled in an ice bath. The protein precipitate was separated by centrifugation and washed twice with 2 ml H$_2$O. The supernatant of the first centrifugation was combined with the wash fluids. The [^{32}P]GTP was purified by two successive steps of exponential gradient ion exchange chromatography, which were followed by continuous registration of ^{32}P activity.

Step 1 Chromatography. Resin: Dowex-1 (X 8) formate, 400 mesh; column: diameter, 8 mm; length, 70 mm; reservoir: 0.67 M HCOOH + 3.00 M HCOONH$_4$ adjusted to pH 4.4; flow rate from the reservoir to the mixing chamber: 4.3 ml/hour; content of the mixing chamber at the beginning: 70 ml H$_2$O; flow rate from the mixing chamber to the column: 14 ml/hour. In this system GTP was eluted after $4^1/_2$ hours. The fraction containing the [^{32}P]GTP was evaporated by freeze-drying and was subjected to chromatography step 2.

Step 2 Chromatography. The details are the same as in step 1, except that Dowex-chloride was used

and the reservoir contained 1 N HCl. [^{32}P]GTP was eluted after 4 hours. The eluate was neutralized with KOH and stored at —20°.

By this method about 50 % of the [^{32}P]orthophosphate added was recovered as [^{32}P]GTP. A typical preparation yielded 1.1 mC in a volume of 7.6 ml with a specific activity of 0.7 C/mmole GTP. Ion exchange chromatography of this preparation (for details see assay of GTP-AMP-P transferase activity) showed that 98 % of the ^{32}P activity was found in the GTP-fraction, 1.42 % in fraction of inorganic phosphate, 0.08 % in the fraction of ADP and 0.12 % in the fraction of ATP.

Assay of GTP-AMP-P Transferase Activity

Incubation. 0.050 ml extract (equivalent to 0.25 to 0.50 mg fresh weight of tissue or 50 μg of mitochondrial protein) was incubated with 0.2 ml of the following reaction mixture for 2 minutes at 25°:

	GTP		0.15 mM
[^{32}P]	GTP	about 10	μC/ml
	AMP	10	mM
	ADP	2	mM
	ATP	10	mM
	MgCl$_2$	5	mM
	EDTA	1	mM
	TRA-HCl pH 7.2	100	mM

Incubation was stopped by addition of 0.10 ml 3 M HClO$_4$. After separation of the protein precipitate by centrifugation the supernatant was neutralized with 3 M KOH to pH 7. 0.25 ml of the extract was applied to ultra-micro scale ion exchange chromatography followed by automatic measurement of ^{32}P activity and 265 mμ absorbancy. The elution was carried out with a linear gradient. For details see Heldt and Klingenberg [6].

Conditions of Chromatography. Resin: Dowex-1 (X 8) formate; 400 mesh; column: diameter 1.1 mm, length 500 mm; temperature: 30°; reservoir: 1.27 M HCOOH + 4.80 M HCOONH$_4$, adjusted to pH 4.2; content of the mixing chamber at the beginning: 20 ml including 0.127 N HCOOH + 0.48 M HCOONH$_4$; flow rate from the mixing chamber to the column: 1.15 ml/hour. The peaks of the compounds in question appeared as follows: (values are given in hours after the start of elution) Pi: 1.5; inorganic Pyrophosphate: 3.5; ADP: 4.5; ATP: 7.0; GTP: 8.8.

For evaluation, the ^{32}P activity appearing in the ADP fraction was divided by the specific activity of [^{32}P] GTP in the reaction mixture.

Assay of Nucleoside Diphosphate Kinase Activity

For optical enzymatic assay of nucleoside diphosphate kinase activity, the following reactions were employed [11]:

$$GTP + ADP \longrightarrow ATP + GDP$$

$$ATP + glucose \xrightarrow{HK} glucose\text{-}6\text{-}P + ADP$$

$$glucose\text{-}6\text{-}P + TPN \xrightarrow{G6PDH} 6\text{-}P\text{-}gluconate + TPNH$$

The formation of TPNH was followed by recording the 340 mμ absorbancy with a Beckman spectrophotometer model DU fitted with a Gilford recorder.

The test mixture contained 1 mM EDTA, 5 mM $MgCl_2$, 20 mM glucose, 0.3 mM TPN, 0.5 mM ADP, 0.5 mM GTP, 10 mM AMP and 50 mM TRA-HCl buffer. 0.5 units of hexokinase (Boehringer) and 0.25 units of glucose-6-phosphate dehydrogenase (Boehringer) were added to a total volume of 0.5 ml. The light path was 1 cm. The reaction was started by addition of 10 μl extract, equivalent to 0.001−0.01 units of nucleoside diphosphate kinase.

The high concentration of AMP was selected to inhibit ATP formation due to the high activity of adenylate kinase in rat liver mitochondria. It may be noted, however, that this concentration also leads to slight inhibition of nucleoside diphosphate kinase activity. The hexokinase used was not fully specific for ATP. Therefore, a blank, representing the formation of glucose 6-phosphate from GTP, was measured at the beginning. This was subtracted from the activity obtained after the addition of the extract.

Assay of Adenylate Kinase

The following reactions were employed [12]:

$$2ADP \longrightarrow ATP + AMP$$

$$ATP + glucose \xrightarrow{HK} glucose\text{-}6\text{-}P + ADP$$

$$glucose\text{-}6\text{-}P + TPN \xrightarrow{G6PDH} 6\text{-}P\text{-}gluconate + TPNH$$

50 mM TRA, pH 7.2, 1 mM EDTA, 5 mM $MgCl_2$, 20 mM glucose, 1 mM ADP, 0.3 mM TPN, 0.5 units of hexokinase (Boehringer) and 0.25 units of glucose-6-phosphate-dehydrogenase (Boehringer) were added to a total volume of 0.5 ml.

For assay conditions see assay of nucleoside diphosphate kinase.

Assay of Succinic Thiokinase

The procedure of Cha and Parks [13] was followed.

RESULTS

Reaction Sequence of Substrate Level Phosphorylation

Anaerobic dismutation of α-ketoglutarate and ammonia into glutamate, succinate, and carbondioxide was employed to study the reaction sequence of substrate level phosphorylation, as shown in Fig. 1. The reaction was started by the addition of [32P]inorganic phosphate to the mitochondrial suspension. The reaction was carried out at 5° to obtain high resolution. Samples were taken at short intervals and analyzed for 32P incorporation into phosphohistidine

and into endogenous nucleotides. Phosphohistidine, which was suggested to be an intermediate of substrate level phosphorylation [14−16], is the first to be labelled. This was followed by the transfer of 32P to endogenous GTP. It appears from these data that phosphohistidine is indeed an intermediate of substrate level phosphorylation, which gives rise to the formation of the second intermediate, GTP [17]. From the time course shown, it is further evident that [32P]ADP rather than [32P]ATP is the main

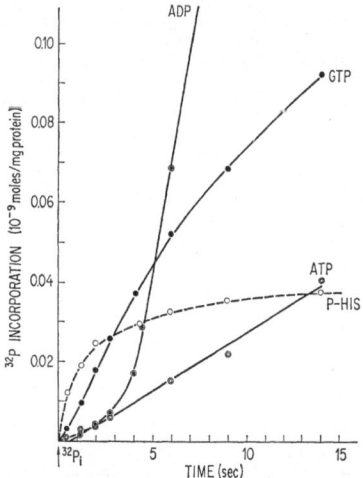

Fig. 1. *Time course of 32P incorporation into phosphohistidine and endogenous nucleotides at 5° in rat liver mitochondria.* Rat liver mitochondria (9.3 mg protein/ml) were incubated in a medium containing 0.25 M sucrose, 30 mM TRA pH 7.2, 2 mM EDTA, oligomycin (2.5 μg/ml), 2 mM malate and 1 mM KCN in a total volume of 20 ml at 5°. Ketoglutarate (2.5 mM) and NH_4Cl (5 mM) were added, and 20 sec later the reaction was started by addition of [32P]orthophosphate (0.65 mM, specific activity 12.4×10^6 counts/min/mole). Samples of about 1 ml were taken, and immediately deproteinized with 1 ml 1.5 M $HClO_4$ at 0°. From the protein precipitate, bound phosphohistidine was assayed according to Biber *et al.* [25]. Nucleotides were assayed in the neutralized extracts by ion exchange chromatography (for details see Methods: assay of GTP-AMP-P transferase). In this case the mixing chamber contained H_2O at the beginning

product of the reaction. This indicates that substrate level phosphorylation of liver mitochondria involves the phosphorylation of AMP by GTP in addition to the generally known phosphorylation of ADP by GTP.

Dependence of AMP-Phosphorylation on Substrate Level Phosphorylation

Table 1 shows the incorporation of [32P]phosphate into the endogenous nucleotides of mitochondria when succinate or ketoglutarate are substrates.

Table 1. *Dependence of the phosphorylation of endogenous AMP on ketoglutarate oxidation*

Rat liver mitochondria equivalent to 5.57 mg protein were aerobically incubated for 30 sec at 18° in the presence of 0.25 M sucrose, 20 mM TRA-HCl, 2 mM EDTA and 0.65 mM [^{32}P]phosphate (specific activity 8×10^6 counts/min/µm). Additions to the medium were made as indicated. The total volume was 1 ml. The reaction was terminated by the addition of 0.2 ml 3 M perchloric acid and the resultant soluble extract was neutralized with 3 M KOH. For analysis of [^{32}P]GTP 0.3 ml extract was subjected to ion exchange chromatography (see Methods: GTP-AMP-P transferase). The mixing chamber contained H$_2$O at the beginning. For the determination of [γ-^{32}P]ATP and [β-^{32}P]ADP + [β-^{32}P]-ATP, 0.8 ml extract was incubated 8 minutes at 25° in the presence of 4 mM MgSO$_4$, 0.1 mM ATP, 10 mM glucose, 40 mM TRA-HCl, pH 7.2 and 1.5 units of hexokinase per ml (Boehringer) in a total volume of 1 ml. The mixture was deproteinized with 0.2 ml 3 M HClO$_4$ and neutralized with 3 M KOH. 0.8 ml extract was subjected to ion exchange chromatography for determination of [^{32}P]glucose-6-phosphate and [β-^{32}P]ADP

Experimental conditions	^{32}P uptake in		
	[β-^{32}P]ADP + [β-^{32}P]ATP	[γ-^{32}P]ATP	[^{32}P]GTP
	10^{-8} moles/mg protein		
2 mM succinate as substrate:			
—	0.19	6.32	0.030
+ 0.6 mM arsenite	0.28	6.02	0.123
+ 0.6 mM arsenite + 2.5 mM Mg^{++}	0.59	6.42	0.068
2 mM ketoglutarate as substrate:			
—	1.27	6.11	0.197
+ 2.5 mM Mg^{++}	1.26	5.97	0.170

Differentiation was made between the ^{32}P incorporation into the γ-position of ATP, arising either from oxidative phosphorylation or the nucleoside diphosphate kinase reaction, and the ^{32}P incorporation into the β-position of ADP and ATP, due to the GTP-AMP-P transferase or the adenylate kinase reaction.

In all cases, the extent of the ^{32}P incorporation into the γ-position of endogenous ATP is quite similar. In contrast, the extent of the ^{32}P incorporation into the β-position of ADP and ATP (due to the phosphorylation of AMP) varies considerably. It is very low when succinate or succinate + arsenite are used as substrates. Arsenite was added to exclude the oxidation of endogenous ketoglutarate. The presence of magnesium ions produces a slight stimulation of the ^{32}P uptake into the β-position of ADP and ATP when succinate is the substrate.

In the presence of ketoglutarate, the ^{32}P incorporation into the β-position is greatly increased, indicating a close relationship between AMP phosphorylation and substrate level phosphorylation. The low ^{32}P incorporation into the β-position of ADP and ATP, which is observed in the absence of keto-

glutarate oxidation, could be due either to the adenylate kinase reaction:

$$AMP + [^{32}P]ATP \rightarrow [\beta\text{-}^{32}P]ADP + ADP$$

or to the combined reaction of nucleoside diphosphate kinase and GTP-AMP-P transferase;

$$GDP + [^{32}P]ATP \rightarrow [^{32}P]GTP + ADP$$
$$[^{32}P]GTP + AMP \rightarrow GDP + [^{32}P]ADP.$$

Since [^{32}P]GTP was always found in the mitochondria, the second possibility may be the more likely. The addition of magnesium ions, known to stimulate the adenylate kinase activity which is located in the outer compartment of the mitochondria [1,18], has only a very slight effect.

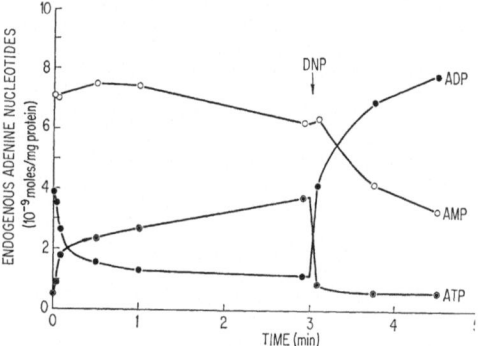

Fig. 2. *Phosphorylation and DNP-stimulated hydrolysis of endogenous adenine nucleotides.* Rat liver mitochondria (5.7 mg protein/ml) were preincubated in the absence of oxygen in a medium containing 0.25 M sucrose, 20 mM TRA-HCl pH 7.2, 2 mM EDTA, 6 mM succinate, 6 mM ketoglutarate and 0.5 mM phosphate for 3 minutes at 18°. The total volume was 12 ml. Phosphorylation was started by bubbling oxygen through the suspension. At the time indicated 0.1 mM dinitrophenol were added. Samples were taken, deproteinized with perchloric acid and analyzed by ion exchange chromatography

There is no effect of magnesium ions when ketoglutarate is used as substrate. This indicates that there is only a slight reaction between the endogenous adenine nucleotides and the adenylate kinase of the outer compartment of the mitochondria. The close relationship between the phosphorylation of AMP and substrate level phosphorylation in liver mitochondria is further demonstrated in the experiment shown in Fig. 2. The endogenous adenine nucleotides of the mitochondria are first phosphorylated by aerobic incubation in the presence of succinate + ketoglutarate. On addition of dinitrophenol there is a rapid hydrolysis of ATP and an equivalent rise in ADP. However, ADP is also formed more slowly by the phosphorylation of AMP. Since the dinitrophenol concentration used was sufficient to uncouple completely the oxidative phosphorylation, substrate level phosphorylation has to be considered for this phosphorylation of AMP. Thus the addition of dinitrophenol may be used as a tool to obtain low levels of endogenous AMP in liver mitochondria.

It may be mentioned that the addition of arsenate, which uncouples substrate level phosphorylation, results in high levels of AMP in liver mitochondria [18,19]. This effect may be explained by the reverse reaction of GTP-AMP-P transferase, following the succinic thiokinase-catalyzed arsenolysis of GTP.

Assay of GTP-AMP-P Transferase Activity

The various methods for the measurement of GTP-AMP-P transferase activity employed by other authors, are described by the following reactions (the underlined product being the one measured):

wrong measurements, due to a further reaction of the $[\beta\text{-}^{32}\text{P}]\text{ADP}$ formed:

$$[\beta^{32}\text{P}]\text{ADP} + [\gamma^{32}\text{P}]\text{GTP} \xrightarrow{\text{NuDiKi}} [\beta, \gamma\text{-}^{32}\text{P}]\text{ATP} + \text{GDP}$$

$$[^{32}\text{P}]\text{ATP} + \text{AMP} \xrightarrow{\text{AdK}} [\beta\text{-}^{32}\text{P}]\text{ADP} + \text{ADP}.$$

The first reaction would decrease the apparent GTP-AMP-P transferase measured while the second reaction would increase it. To avoid this interference by nucleoside diphosphate kinase and adenylate kinase, the assay was done in the presence of 0.15 mM $[^{32}\text{P}]\text{GTP}$, 10 mM AMP, 2 mM ADP and 10 mM ATP. The amount of extract added and the time of incubation employed were fixed so that not more than 10 % of the $[^{32}\text{P}]\text{GTP}$ reacted. Thus, the $[\beta\text{-}^{32}\text{P}]\text{ADP}$ formed by the GTP-AMP-P transferase activity is diluted more than hundredfold by unlabelled ADP. Hence, the decrease of $[\beta\text{-}^{32}\text{P}]\text{ADP}$ by the activity

1. $\text{ITP} + \text{AMP} \longrightarrow \text{IDP} + \text{ADP}$

$\text{IDP} \xrightarrow{\text{IDPase}} \text{IMP} + \underline{\text{P}_i}$ [5]

2. $[\beta\text{-}^{32}\text{P}]\text{ADP} + \text{IDP} \longrightarrow \text{AMP} + [^{32}\text{P}]\underline{\text{ITP}}$ [4]

3. $\text{IDP} + \text{ADP} \longrightarrow \text{AMP} + \text{ITP}$

$\text{ITP} + \text{ADP} \xrightarrow{\text{NuDiKi}} \text{IDP} + \text{ATP}$

$\text{ATP} + \text{Glucose} \xrightarrow{\text{HK}} \text{ADP} + \text{Glucose-6-}P$

$\text{Glucose-6-}P + \text{TPN} \xrightarrow{\text{G6PDH}} 6\text{-}P\text{-Gluconate} + \underline{\text{TPNH}}$ [4]

These methods were suitable for the assay of purified enzyme preparations. It was, however, impossible to measure GTP-AMP-P transferase activity in liver homogenates with these methods, because of the great interference by adenylate kinase, nucleoside diphosphate kinase and ATPase. Therefore, an assay procedure had to be introduced which was not affected by the presence of other enzymes in tissue homogenates. For the assay of GTP-AMP-P transferase the following reaction was employed:

$$[^{32}\text{P}]\text{GTP} + \text{AMP} \rightarrow [\beta\text{-}^{32}\text{P}]\text{ADP} + \text{GDP}.$$

$[^{32}\text{P}]\text{GTP}$ of known carrier content was added and the $[^{32}\text{P}]\text{ADP}$ formed was assayed by ion exchange chromatography with automatic measurement of ^{32}P activity.

In the presence of nucleoside diphosphate kinase and of adenylate kinase this assay may also lead to

of nucleoside diphosphate kinase becomes neglegible. $[^{32}\text{P}]\text{ATP}$ arising from nucleoside diphosphate kinase is again diluted by unlabelled ATP which prevents the formation of $[\beta\text{-}^{32}\text{P}]\text{ADP}$ by adenylate kinase activity. This assay may not yield maximal activities, but the results are not affected by other phosphate transferring enzymes. The very slight interference by adenylate kinase and nucleoside diphosphate kinase with the GTP-AMP-P transferase assayed, is illustrated by the analysis of a "pseudoextract" (Table 2). This "pseudoextract" consisted of a solution of adenylate kinase (Boehringer) and nucleoside diphosphate kinase (prepared from pig liver) in sucrose medium.

Localization of GTP-AMP-P Transferase in Rat Liver

It was necessary to determine whether the GTP-AMP-P transferase isolated from whole liver [4,5]

Table 2. *Fractional extraction of rat liver homogenate*
For details of the fractionation and the enzyme assays see Methods. The "pseudo-extract" contained isolated nucleoside diphosphate kinase from pig liver mitochondria (Schwalbach [20], 46 units) and adenylate kinase (Boehringer, 50 units) in a total volume of 10 ml sucrose medium. Experiments were performed at 25°

	GTP-AMP-P transferase	Nucleoside diphosphate kinase	Adenylate kinase
	μmoles/g fresh weight/min		
Total extract	2.59		
Sucrose extract I	0.30 ⎫ 14%	22.2 ⎫ 80%	6.20 ⎫ 22%
Sucrose extract II	0.04 ⎭	2.78 ⎭	2.50 ⎭
Phosphate extract	0.30 ⎫	3.30 ⎫	22.5 ⎫
Ultrasonic extract	1.22 ⎬ 86%	2.00 ⎬ 20%	6.20 ⎬ 78%
Insoluble protein	0.53 ⎭	1.18 ⎭	1.60 ⎭
	2.39	31.46	39.00
Pseudoextract without GTP-AMP-P transferase	0.02[a]	46	50

[a] Due to adenylate kinase and nucleotide diphosphate kinase.

Table 3. *Comparison between the content of GTP-AMP-P transferase in liver mitochondria and liver microsomes*
For details see Methods. Rat liver was homogenized in sucrose medium, centrifuged for 10 min at $500 \times g$ and the supernatant was centrifuged for 10 min at $17,000 \times g$. The resulting sediment was resuspended in sucrose medium and centrifuged for 10 min at $4,500 \times g$. This last washing was repeated and the resulting sediment taken as mitochondrial fraction. In order to obtain the microsomal fraction, the $17,000 \times g$ supernatant was centrifuged for 40 min at $80,000 \times g$, the sediment resuspended and again centrifuged for 30 min at $80,000 \times g$. The resuspended sediment was taken as the microsomal fraction. Before the assay, both the mitochondrial and the microsomal fraction were disrupted by sonication. Experiment was performed at 25°

	GTP-AMP-P transferase
	μmoles/g protein/min
Rat liver mitochondria	23
Rat liver microsomes	0.7

Table 4. *Activities of phosphorylating enzymes of rat liver mitochondria disrupted by sonication in 0.1 M phosphate buffer, pH 7.2*

Phosphorylating enzymes	Activities at 25°	
	μmoles/g protein/min	
GTP-AMP-P transferase	30 ± 8	(7)[a]
Nucleoside diphosphate kinase	83 ± 12	(16)[a]
Succinic thiokinase	24 ± 1.2	(5)[a]
Respiration with α-ketoglutarate	33[b]	(2)

[a] Means $\pm$ standard deviation (n).
[b] μmoles O/g protein/min.

actually occurs in the mitochondria. Rat liver homogenate was therefore subjected to fractional extraction [9], the results of which are shown in Table 2. The homogenized tissue was extracted twice by isotonic sucrose (see Methods) and the sediment extracted again by 0.1 M phosphate buffer. The resulting sediment was sonicated and yielded an ultrasonic extract and an insoluble fraction. The first two sucrose extracts contained the soluble extramitochondrial enzymes [9]. In the following extracts, the mitochondrial enzymes and also some microsomal enzymes are found. In contrast to the nucleoside diphosphate kinase, which is mainly localized in the soluble extramitochondrial fractions, nearly all of the GTP-AMP-P transferase activity appears in the last three fractions. The result of these assays suggested that this enzyme occurs either in the mitochondria or in the microsomal fraction. To obtain further evidence for the localization of GTP-AMP-P transferase, its activity was measured in isolated mitochondria and in microsomes. The results are shown in Table 3. From these data it is obvious that the GTP-AMP-P transferase actually occurs in the mitochondrial fraction. The low activity found in the microsomal fraction may be due to contamination by mitochondria. Table 4 shows mean values for the activities of GTP-AMP-P transferase and of related enzymes, assayed in isolated mitochondria. The activities of GTP-AMP-P transferase, succinic thiokinase and the rate of ketoglutarate oxidation are quite similar. Nucleoside diphosphate kinase activity appears to be much higher. There are indications, however, that the bulk of the measured nucleoside diphosphate kinase activity is located in the outer compartment and may not be involved in mitochondrial substrate level phosphorylation.

DISCUSSION

A survey of the functional compartmentation of phosphate transfer reactions in liver mitochondria may lead to a better understanding of the role of GTP-AMP-P transferase in mitochondrial metabolism. A working hypothesis for compartmentation has been developed in our laboratory [1,18,21,22] and is summarized in Fig.3. Substrate level phosphorylation as well as oxidative phosphorylation takes place in the inner compartment of the mitochondria. This inner compartment, as defined in functional terms, may be identical with the matrix space [18]. It is separated from the outer compartment, which may be identical with the intracristae space [18], by the translocase reaction. This reaction allows the exchanges between exogenous and endogenous ADP and ATP. AMP exchanges very slowly and guanine nucleotides not at all. The GTP being formed by substrate level phosphorylation appears to be confined to the inner compartment [1]. It follows that the nucleoside diphosphate kinase and the GTP-AMP-P transferase, which are involved in substrate level phosphorylation, are also located in this inner compartment. Adenylate kinase, on the other hand, and that part of the mitochondrial nucleoside diphosphate kinase which is not involved in substrate level phosphorylation, are located in the outer compartment.

In the course of mitochondrial phosphorylation, exogenous ADP enters the inner compartment, where

to be small compared with the GTP-AMP-P transferase activity. Thus GTP-AMP-P transferase may provide the only path for the phosphorylation of endogenous AMP in liver mitochondria.

The authors are indebted to Prof. Dr. Klingenberg for his encouragement and for many stimulating discussions concerning this work and to Dr. L. Sauer for his kind help in preparing the manuscript. The technical assistance of Miss R. Heeger is gratefully acknowledged. This research has been supported by grants from the Deutsche Forschungsgemeinschaft and from the US Public Health Service.

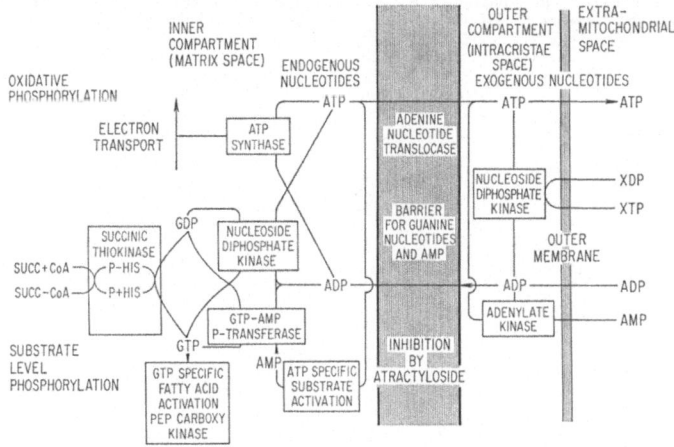

Fig. 3

phosphorylation takes place. Most of the ATP formed leaves the inner compartment again. A part of this ATP, however, may be consumed in the inner compartment by substrate activation processes, yielding AMP. Octanoate activation, for instance, may occur in the inner compartment of mitochondria [23]. This AMP is now phosphorylated by the GTP-AMP-P transferase and the ADP thus formed enters the phosphorylation cycle again.

Since AMP has been found to pass the translocase barrier in a very slow reaction [18,22], it appears to have no access to the adenylate kinase located in the outer compartment of the mitochondria. From our data, it is not possible to conclude that there is no adenylate kinase present in the inner compartment, but there is no evidence of its presence. The interconversion of endogenous AMP observed earlier [24], which indicated the presence of adenylate kinase in the inner compartment, could have resulted from the combined reaction of nucleoside diphosphate kinase and GTP-AMP-P transferase. The dependence of AMP phosphorylation on ketoglutarate oxidation suggests that, if there is any adenylate kinase activity present in the inner compartment, it is likely

REFERENCES

1. Heldt, H. W., in *Regulation of metabolic processes in mitochondria* (edited by J. M. Tager, S. Papa, E. Quagliariello, and E. C. Slater), Elsevier, Amsterdam 1966, p. 51.
2. Heldt, H. W., *Abstr. Z. Klin. Chem.* 3 (1965) 203.
3. Heldt, H. W., and Schwalbach, K., *Abstracts 3 FEBS Meeting (Warsaw)*, 1966, M 42.
4. Chiga, M., Rogers, A. E., and Plaut, G. W. E., *J. Biol. Chem.* 236 (1961) 1800.
5. Heppel, L. A., Strominger, J. L., and Maxwell, E., *Biochim. Biophys. Acta*, 32 (1959) 422.
6. Heldt, H. W., and Klingenberg, M., *Biochem. Z.* 343 (1965) 433.
7. Klingenberg, M., and Slenczka, W., *Biochem. Z.* 331 (1959) 486.
8. Kröger, A., and Klingenberg, M., *Biochem. Z.* 344 (1966) 317.
9. Delbrück, A., Zebe, E., and Bücher, Th., *Biochem. Z.* 331 (1959) 273.
10. Pfleiderer, G., *Biochim. Biophys. Acta*, 47 (1961) 389.
11. Berg, P., and Joklik, W. K., *J. Biol. Chem.* 210 (1954) 657.
12. Oliver, I. T., *Biochem. J.* 61 (1955) 116.
13. Cha, S., and Parks, R. E., *J. Biol. Chem.* 239 (1964) 1961.
14. Kreil, G., and Boyer, P. D., *Biochem. Biophys. Res. Commun.* 16 (1964) 551.

15. Mitchel, R. A., Butler, L. G., and Boyer, P. D., *Biochem. Biophys. Res. Commun.* 16 (1964) 545.
16. Slater, E. C., and Kemp, A., *Nature*, 204 (1964) 1268.
17. Heldt, H. W., Jacobs, H., and Klingenberg, M., *Biochem. Biophys. Res. Commun.* 17 (1964) 130.
18. Klingenberg, M., and Pfaff, E., in *Regulation of metabolic processes in mitochondria* (edited by J. M. Tager, S. Papa, E. Quagliariello, and E. C. Slater), Elsevier, Amsterdam 1966, vol. 7, p. 180.
19. Ernster, L., in *Round table conference on mitochondrial structure and compartmentation* (edited by J. M. Tager, S. Papa, E. Quagliariello, and E. C. Slater), Editrice Adriatica, Bari 1966, In press.
20. Schwalbach, K., Unpublished results.
21. Heldt, H. W., Jacobs, H., and Klingenberg, M., *Biochem. Biophys. Res. Commun.* 18 (1965) 174.

22. Pfaff, E., Klingenberg, M., and Heldt, H. W., *Biochim. Biophys. Acta*, 104 (1965) 312.
23. Bergh, S. G. Van den, in *Round table conference on mitochondrial structure and compartmentation* (edited by J. M. Tager, S. Papa, E. Quagliariello, and E. C. Slater), Editrice Adriatica, Bari 1966, In press.
24. Siekewitz, P., and Potter, V. R., *J. Biol. Chem.* 215 (1955) 237.
25. Biber, L. L., Lindberg, O., Duffy, J. J., and Boyer, P. D., *Nature*, 202 (1964) 1316.

H. W. Heldt, and K. Schwalbach
Physiologisch-chemisches Institut der Philipps-Universität
3550 Marburg/Lahn, Deutschhausstraße 1/2, Germany

European J. Biochem. 1 (1967) 207—215

Levels of Glycolytic Intermediates in the Musculature
of the Chick during Embryonic and Post-Embryonic Development

P. Arese, M. T. Rinaudo, and A. Bosia

Department of Biological Chemistry, University of Torino

(Received November 25, 1966)

Levels of glycogen and of glycolytic intermediates were assayed in quickly frozen embryonic mesodermic tissue (decapitated and eviscerated chick embryos) at the 5th and 7th developmental day and in the hind limb musculature in subsequent pre- and post-natal stages. Serial determination of dry weight values during embryonic development allows conversion of the levels which are reported on a wet weight basis.

While glycogen gradually increases from the 9th day onwards reaching about $^1/_3$ of the adult amount on hatching, no substantial differences between early embryonic and adult levels were noted in the case of other intermediates, such as glucose, dihydroxyacetone-phosphate, glyceraldehyde-3-phosphate, 3- and 2-phosphoglycerate, phosphoenolpyruvate, pyruvate and lactate. Glucose-1-phosphate, fructose-6-phosphate, glucose-6-phosphate and fructose-1,6-diphosphate were low during most of the embryonic development, attaining levels close to the adult ones only on hatching.

In order to study the flow rates, the equilibrium situation at each single step and the possible control points along the glycolytic pathway, the tissue was stressed by brief periods of ischaemia or by electrical stimulation at different stages of development.

Glycolytic flow rates showed a progressive increase up to hatching, where a lactate accumulation of 1.5 μmole/min/g wet weight was measured. The maximum rate of 12.9 μmoles/min/g wet weight found in electrically stimulated muscle of hatching chicks, is well below the adult value of 52.7 μmoles/min/g wet weight.

The approximate stoichiometric correspondence between glucose + glycogen breakdown and lactate formation, found to exist in the early and late embryonic stages, is absent on the 7th and incomplete on the 11th day, where the sharply reduced gluco- and glycogenolysis cannot account for the measured glycolytic rate. The existence of a vicarious anomalous glycolysis, supplying pyruvate and lactate by transamination or by other unknown mechanisms, is possible at these two stages.

The constancy of mass-action ratios under altered glycolytic fluxes and the comparison between the ratios *in vivo* and the thermodynamic equilibria show that phosphoglucomutase, phosphoglucose isomerase, aldolase, triosephosphate isomerase, phosphoglycerate mutase and enolase were maintained either at equilibrium or fairly close to equilibrium. Owing to the probable compartmentation of ATP and ADP in muscle, no calculation of kinase-catalyzed reactions was attempted.

Activation of glycolysis could not be observed on electrical stimulation of embryonic muscle at the 14th day, as the constancy of such indicator metabolites as glucose-6-phosphate, fructose-1,6-diphosphate and lactate demonstrates. An activation at the phosphorylase and phosphofructokinase steps has been indirectly shown on hatching.

Non-standard abbreviations. Dihydroxyacetone-phosphate, DAP; glyceraldehyde-3-phosphate, GAP; fructose-1,6-diphosphate, FDP.

Enzymes. Phosphoglucomutase, or α-D-glucose-1,6-diphosphate: α-D-glucose-1-phosphate phosphotransferase (EC 2.7.5.1); glucosephosphate isomerase, or D-glucose-6-phosphate ketol-isomerase (EC 5.3.1.9); fructosediphosphate aldolase, or fructose-1,6-diphosphate D-glyceraldehyde-3-phosphate-lyase (EC 4.1.2.13); triosephosphate isomerase, or D-glyceraldehyde-3-phosphate ketol-isomerase (EC 5.3.1.1); phosphoglycerate phosphomutase, or D-phosphoglycerate

2,3-phosphomutase (EC 5.4.2.1); phosphopyruvate hydratase, or 2-phospho-D-glycerate hydro-lyase (EC 4.2.1.11); hexokinase, or ATP:D hexose 6 phosphotransferase (EC 2.7.1.1); phosphofructokinase, or ATP:D-fructose-6-phosphate 1-phosphotransferase (EC 2.7.1.11); glucose-6-phosphate dehydrogenase, or D-glucose-6-phosphate:NAPD oxidoreductase (EC 1.1.1.49); malate dehydrogenase, or L-malate:NAD oxidoreductase (EC 1.1.1.37); pyruvate kinase, or ATP:pyruvate phosphotransferase (EC 2.7.1.40); alanine aminotransferase (glutamic-pyruvic transaminase) or L-alanine:2-oxoplutarate aminotransferase (EC 2.6.1.2).

Since the studies of Meyerhof and Perdigon [1] and of Novikoff *et al.* [2], Needham's view of a predominantly non-phosphorylating embryonic glycolysis [3,3a,4] has lost most of its supporting evidence. Although there is no longer any doubt as to the fundamental similarity between embryonic and adult glycolysis, recent data on the behaviour of intermediate compounds during the development of chick embryo musculature are lacking. The assay of the metabolite levels in steady-state conditions or after a short period of anaerobiosis may enable the equilibrium situation to be characterized at each single step and glycolytic rates to be measured *in vivo*. According to many writers [5—8], these levels also give an insight into the control mechanisms of intermediary metabolism.

It is the object of this paper to present a quantitative analysis of the glycolytic intermediates *in vivo* in the "muscle precursor tissue" (5-, 7- and 9-day stage), in the embryonic (11- to 21-day stage) and post-embryonic chick muscle. The effect of anaerobiosis (at all stages) and of electrical stimulation (at the 14- and 21-day stage) on most metabolites has also been taken into account. Subsequent work will deal with the reduced/oxidized state of the cytoplasmic NAD+-NADH-system and with the developmental changes of *N*-phosphoryl-creatine, creatine and adenine nucleotides[1].

EXPERIMENTAL METHODS

Fertilized chick eggs (Ledbrest & Pilch) were incubated at 38.5° and 75—80% relative humidity in a commercial incubator with forced air circulation. The eggs were rotated 6 times a day at an angle of 30°.

Early Stages

At 5, 7 and 9 days the eggs were opened, the embryos transferred to a Petri dish heated to 38.5°, care being taken to avoid damaging the membranes; with minimum manipulation they were then freed of all extraembryonic tissue, decapitated, eviscerated, gently drawn over filter paper in order to remove cutaneous layers and excess moisture, and pressed between two brass blocks previously cooled to the temperature of liquid nitrogen. Sufficient embryos to make up between 0.5 and 1 g wet weight were pooled.

Later Stages

At 11, 14, 17, and 21 days, the eggs were kept at incubation temperature (38.5°) and the shells gently peeled off over a limited zone. A hind limb was iso-

lated from the embryonic membranes and freed of skin and subcutaneous fat. The whole thigh musculature with the bone, in the case of 11- to 14-day-old embryos, or, in the case of the older embryos, muscular tissue (gastrocnemius and part of the femoral musculature), was frozen *in situ* between brass blocks cooled with liquid nitrogen, and then was cut off from the remaining tissue and dropped into liquid nitrogen. The tissue was powdered in a mortar chilled in liquid nitrogen and the powder transferred into a small glass beaker, weighed by difference and homogenized with an Ultra-Turrax-Homogenizer. After a short centrifugation (5—10 sec at $15,000 \times g$) the supernatant was neutralized with 10 N KOH. The sediment was re-extracted with 3% (w/v) perchloric acid, the collected supernatants were adjusted to pH 7.2—7.3 and filtered through a fritted glass filter (Schott G-4). The intermediates were assayed by combined enzymatic methods, using 4 to 5 cm light-path semimicro cuvettes, according to the methods of Hohorst *et al.* [9] and Bergmeyer [10]. Measurements were made with a Beckman DK-2A spectrophotometer or with an Eppendorf photometer fitted with a Philips expansion recorder. An aliquot of the powdered frozen tissue was used for glycogen assay. The polysaccharide was measured enzymatically as glucose (with NADP, glucose-6-phosphate dehydrogenase and hexokinase) after tissue digestion with 20% (w/v) NaOH, ethanol precipitation, two washings and acid hydrolysis. Anaerobiosis was obtained by maintaining the dissected embryonic tissue at 38.5° and freezing at measured time intervals with clamps cooled to a very low temperature. As no differences in lactate levels were noted on keeping the embryos in Thunberg tubes in a nitrogen atmosphere, the anaerobiosis experiments were performed in a moist air chamber. A type 104-A stimulator (American Electronics Laboratory Inc.) was used to supply square wave stimuli, which were applied directly to the leg musculature *in situ* by means of unpolarizable silver electrodes. Post-natal and 150-day-old chicks were anaesthetized by the brief administration of ether. The posterior tibial musculature was clamped, by means of metal block forceps precooled in liquid nitrogen, to a 3—4 mm thick tissue tablet. A different number of embryos or of chicks was used for each experiment, as indicated below (mean values): 5th day: 18 embryos — 7th day: 7 embryos — 9th day: 6 embryos — 11th day: 4 embryos (4 limbs) — 14th day: 3 embryos (3 limbs) — 17th day: 6 embryos (6 limbs) — 21st day: 4 embryos (4 limbs) — 5- and 10-days-old chicks: 3 chicks (3 limbs).

For the dry weight determination, embryonic tissue sampled under the same conditions as for the metabolite assay was dried to constant weight at 110°. At the 11th and 14th day, bone was dissected free of tissue and weighed separately (Table 1).

---

[1] A preliminary account of this work was presented before the Società di Biologia Sperimentale (*Boll. Soc. Ital. Biol. Sper.* 40 (1964) 1117, 40 (1964) 1890, 42 (1966) 204).

Table 1. *Dry weight in embryonic muscle of the chick* Mean values as percentage of fresh weight. Number of experiments in parentheses

Development	Dry weight	Bone
days	% of fresh weight	
5	7.27 (6)	—
7	7.71 (6)	—
11	9.53 (6)	8.9 (4)
14	11.30 (6)	8.6 (6)
17	16.54 (6)	—
21	21.40 (6)	—

RESULTS

Glycogen, Glucose and Glycolytic flow Rates

The glycogen level, which is substantially constant up to the 9th day, increases sharply between the 9th and 11th day. Thereafter, a rather continuous rise has been observed, with a net accumulation rate of about 1 μmole of glycogen (as glucose equivalent) per day per g wet weight. On hatching, glycogen is about $1/3$ of the adult levels (Table 2). Glucose is constant during the whole embryonic development. The difference between the 5th and 7th day is not significant. On hatching, a strong increase over the embryonic levels has been observed (difference between 14th and 21st day: $p < 0.001$). In the first days after hatching, glucose gradually falls to levels significantly lower than those in embryonic life. We cannot specify exactly when the still lower adult levels are attained (Table 2).

Table 2. *Glycogen and glucose levels in embryonic and post-embryonic chick muscle* Mean values in nmoles/g wet weight. N, quickly frozen tissue (*in vivo*-state); AN, anaerobiosis; TET, electrical stimulation (for details, see Experimental Methods). Glycogen expressed as glucose equivalents. Number of experiments in parentheses; number of individuals for each experiment: see Experimental Methods; n.d., not determined; H+5, H+10, etc., days after hatching

Development		Glycogen	Glucose
days		nmoles/g wet weight	
5	N	1810 (2)	2850 (5)
	8 min AN	n.d.	1116 (5)
7	N	1643 (8)	3600 (5)
	4 min AN	1990 (5)	3164 (5)
	8 min AN	n.d.	1998 (6)
9	N	1840 (3)	3656 (3)
11	N	5102 (9)	3502 (7)
	4 min AN	4821 (9)	3511 (9)
14	N	7758 (8)	3612 (7)
	6 min AN	6686 (8)	2480 (2)
	30 sec TET	n.d.	3157 (4)
17	N	10883 (6)	3400 (2)
	4 min AN	9874 (5)	2988 (5)
21	N	15800 (11)	4764 (5)
(Hatching)	4 min AN	11900 (2)	3622 (7)
	25 sec TET	11566 (3)	4290 (5)
H+5	N	28600 (4)	2990 (4)
H+10	N	26030 (3)	2590 (3)
	4 min AN	n.d.	2495 (2)
H+150	N	42033 (3)	1530 (3)
	4 min AN	n.d.	1860 (2)
	15 sec TET	n.d.	1770 (1)

Table 3. *Flux rates and stoichiometric ratios between lactate formation and carbohydrate breakdown in ischaemic and tetanized embryonic and post-embryonic chick muscle* Abbreviations used, see Table 2. The rates were calculated from the figures of Table 2 for glucose and glycogen, of Table 7 for lactate, and are expressed as nmoles/g wet weight/min

Development		Glucose breakdown	Glycogen breakdown	Lactate formation	Lactate / 2 (Glucose + Glycogen)
days		nmoles/g wet weight/min			
5	AN	216	n.d.	593	1.37
7	AN	109	(+86)[a]	542	10
11	AN	0	70	766	5.47
14	AN	189	179	866	1.17
17	AN	103	252	915	1.29
21	AN	285	975	1566	0.62
(Hatching)	TET	1134	10158	12900	0.57
H+5	AN	n.d.	n.d.	2900	n.d.
	TET	n.d.	n.d.	32300	n.d.
H+10	AN	n.d.	n.d.	3050	n.d.
H+150	AN	0	n.d.	2450	n.d.
	TET	0	n.d.	52700	n.d.

[a] At this stage, a glycogen increase after ischaemia has been observed.

Ischaemia causes both glycogenolysis (beginning at the 14-day stage) and glycolysis which roughly parallel the lactate accumulation (Table 3), except in the central sequences of embryonic development (7- and 11-day stage), where no stoichiometric correspondence between glycogen + glucose disappearance and lactate formation was observed (Table 3). At hatching, on the contrary, lactate accounts for only

62% of glycogen + glucose breakdown during anaerobiosis and 57% after electrical stimulation.

Anaerobic lactate formation rises slowly until the 17-day stage and sharply on hatching; the rate on hatching is twice that in the adult rat muscle [7], and is comparable with some experimental tumors and with dental pulp in the calf [11], but is consistently less than in mammalian liver [12] or brain [13]. Electrical stimulation in hatching and new-born chicks causes an 8-fold increase of the glycolytic flow rate. On the 5th post-natal day a peak rate of 60% of the adult value was observed (Table 3). At the 14-day stage, prolonged stimulation brings about no activation of the glycolytic flow (Table 7), although a depletion of phosphagen was observed (unpublished results).

Glucose-1-phosphate, Fructose-6-phosphate and Glucose-6-phosphate

Glucose-1-phosphate tissue level is very low up to the 17th developmental day, oscillating between 2.4 and 7.6 nmoles/g wet weight, and increases only during the last embryonic days. Owing to the small number of determinations, the data must be considered as only provisional (Table 4). Compared with the level in the adult skeletal muscle, glucose-6-phosphate is very low during the whole embryonic phase.

Though rather constant during embryogenesis it increases 5.2-fold from the 17th day up to the time of hatch. Fructose-6-phosphate shows a roughly parallel behaviour. The level ratios are not, in every developmental stage, in agreement with the thermodynamic equilibrium constant of the phosphoglucose isomerase reaction (Table 4).

A typical common feature of ischaemic tissues is the increase of glucose-6-phosphate. This has been observed in many types of musculature [14,7]. Embryonic muscle on the contrary shows no glucose-6-phosphate accumulation until the 17th day (Table 4). At the 14th developmental day, even an electrical stimulation which would otherwise involve striking acceleration of the phosphorolysis, leaves the glucose-6-phosphate unchanged. Fructose-6-phosphate and glucose-1-phosphate parallel the movements of glucose-6-phosphate at least roughly (Table 4). No correlation between glycogen, glucose-6-phosphate and fructose-6-phosphate content seems to exist. The ratio between glycogen (expressed as glucose equivalents) and glucose-6-phosphate increases up to the end of the incubation, being 73 on the 9th day (not reported in Table 3), 164 on the 11th and 262 on the 17th day, as compared with a ratio of about 40 in the adult rat skeletal [7] and 90 in the adult chick limb muscle (Table 4).

Table 4. *Levels of glucose-1-P (G-1-P), fructose-6-P (F-6-P) and glucose-6-P (G-6-P): mass-action ratios for the phosphoglucomutase and phosphoglucose isomerase reactions in normal, ischaemic and tetanized embryonic and post-embryonic chick muscle*
Mean values in nmoles/g wet weight. Abbreviations used, see Table 2. Number of experiments in parentheses; number of individuals for each experiment, see Experimental Methods. The thermodynamic equilibrium constants for the phosphoglucomutase and phosphoglucose isomerase reactions are, respectively, 0.05 and 0.44 [15]

Development		G-1-P	F-6-P	G-6-P	$\dfrac{\text{G-1-P}}{\text{G-6-P}}$	$\dfrac{\text{F-6-P}}{\text{G-6-P}}$
days			nmoles/g wet weight			
5	N	7.6 (2)	14.4 (3)	34.3 (7)	0.21 (2)	0.45 (3)
	8 min AN	n.d.	n.d.	25.3 (5)	n.d.	n.d.
7	N	2.4 (1)	8.5 (3)	25.7 (5)	0.09 (1)	0.31 (3)
	8 min AN	n.d.	16.3 (5)	27.7 (5)	n.d.	0.38 (5)
11	N	3.7 (6)	12.7 (9)	31.2 (10)	0.12 (6)	0.42 (9)
	4 min AN	3.2 (4)	10.0 (8)	29.2 (8)	0.11 (4)	0.34 (8)
14	N	2.8 (3)	6.8 (3)	35.0 (7)	0.08 (3)	0.19 (3)
	4 min AN	n.d.	4.2 (1)	34.7 (1)	n.d.	0.12 (1)
	30 sec TET	n.d.	9.9 (3)	31.6 (3)	n.d.	0.31 (3)
17	N	4.0 (3)	13.0 (4)	41.5 (5)	0.09 (3)	0.29 (4)
	4 min AN	11.1 (4)	31.8 (5)	134 (5)	0.07 (4)	0.22 (5)
21	N	22.1 (3)	36.6 (8)	215 (8)	0.06 (3)	0.17 (8)
(Hatching)	4 min AN	27.8 (6)	121 (7)	572 (7)	0.05 (6)	0.17 (7)
	25 sec TET	27.9 (3)	72.8 (5)	461 (9)	0.07 (3)	0.15 (5)
H+5	N	n.d.	42.0 (4)	204 (4)	n.d.	0.19 (4)
	4 min AN	n.d.	114 (2)	748 (2)	n.d.	0.15 (2)
H+10	N	n.d.	58.8 (3)	383 (3)	n.d.	0.15 (3)
	4 min AN	n.d.	136 (2)	962 (2)	n.d.	0.14 (2)
H+150	N	n.d.	75.0 (3)	464 (3)	n.d.	0.14 (3)
	4 min AN	n.d.	110 (2)	650 (2)	n.d.	0.16 (2)
	15 sec TET	n.d.	338 (1)	2170 (1)	n.d.	0.16 (1)

Fructose-1,6-diphosphate, Dihydroxyacetone-phosphate and Glyceraldehyde-3-phosphate

Fructose-1,6-diphosphate is constant for about the first half of incubation. Beginning with the 14th day a progressive rise was observed (Table 5): post-embryonic and adult chick levels were even higher, in contrast with other muscle types, such as pigeon *pectoralis major* [16] or rat *rectus abdominis* [7]. The products of aldolase reaction were rather constant during the whole of embryogenesis. With the exception of the 11-day-old embryo, where significantly lower levels of dihydroxyacetone-phosphate than in earlier or later stages ($p < 0.02$) were noted, at no time were we able to observe statistically significant ischaemic modifications either in fructose-1,6-diphosphate or in dihydroxyacetone-phosphate or glyceraldehyde-3-phosphate (Table 5); While tetanization of 14-day-old embryos was without significant effect on both fructose-1,6-diphosphate and dihydroxyacetone-phosphate, an evident increase of these metabolites followed electrical stimulation on hatching.

Due to the constancy, even in conditions of altered flux, of the mass-action ratio [8]: $DAP \times GAP/FDP$ within the stages where statistical evaluation was possible, we may assume a near-to-equilibrium situation for the aldolase-catalyzed reaction. A fivefold shift was observed after a strong tetanic stimulation on hatching. Such a shift compares favourably with the 4.5 fold increase observed by Hohorst [7] in tetanized rat skeletal muscle.

The triosephosphate isomerase reaction appears to be displaced from thermodynamic equilibrium by a factor varying between 2.4 and 5.5. Rise in the glycolytic flux by anaerobiosis or electrical stimulus causes no appreciable change in the ratios (Table 5).

3-Phosphoglycerate, 2-Phosphoglycerate and Phosphoenolpyruvate

The terminal intermediates of the glycolytic pathway are characterized by great steadiness. A definite and statistically significant rise in phosphoenolpyruvate, 3-phosphoglycerate and 2-phosphoglycerate was observed after anaerobiosis at the 7- and 11-day stage. These effects would seem to be unique in the whole developmental period, since they did not occur either on the 14th or the 17th day. Electrical stimulation, which had no effect on the 14-day-old embryo, was followed by a significant rise in phosphoenolpyruvate, 3- and 2-phosphoglycerate on hatching. The mass-action ratios at the enolase and phosphoglycerate mutase steps do not sensibly differ from the thermodynamic constants. They remain

Table 5. *Levels of fructose-1,6-diphosphate (FDP), dihydroxyacetone-phosphate (DAP) and glyceraldehyde-3-phosphate (GAP): mass-action ratios for the aldolase and triosephosphate isomerase reactions in normal, ischaemic and tetanized embryonic and post-embryonic chick muscle*

Mean levels in nmoles/g wet weight. Abbreviations used, see Table 2. Number of experiments in parentheses; number of individuals for each experiment: see Experimental Methods. The thermodynamic equilibrium constants for the aldolase and triosephosphate isomerase reactions are, respectively, 93 (38°, pH 7.0—7.2, Mg^{++} 5 mM) [8] and 22 (37°, pH 7.0, Mg^{++} 2 mM) [17]

Development		FDP	DAP	GAP	$\dfrac{DAP \times GAP}{FDP}$	$\dfrac{DAP}{GAP}$
days			nmoles/g wet weight			
5	N	10.9 (4)	22.4 (4)	2.8 (4)	6.23 (4)	8.1 (4)
	8 min AN	8.9 (4)	18.5 (4)	n.d.	n.d.	n.d.
7	N	8.0 (7)	19.5 (7)	1.8 (4)	8.34 (3)	11.5 (4)
	8 min AN	8.2 (6)	15.9 (6)	n.d.	n.d.	n.d.
11	N	11.7 (7)	13.1 (7)	2.2 (5)	2.88 (5)	6.5 (5)
	4 min AN	12.1 (5)	16.6 (5)	3.5 (3)	3.29 (3)	5.1 (3)
14	N	18.3 (7)	19.5 (7)	1.8 (5)	1.54 (5)	9.7 (5)
	5 min AN	20.4 (3)	12.2 (3)	2.1 (1)	1.67 (1)	7.3 (1)
	30 sec TET	21.3 (2)	27.6 (2)	n.d.	n.d.	n.d.
17	N	37.0 (2)	28.4 (3)	3.4 (3)	2.76 (2)	9.6 (3)
	4 min AN	16.7 (2)	32.6 (3)	3.8 (2)	7.32 (2)	10.1 (2)
21	N	86.0 (6)	24.7 (6)	3.8 (4)	1.52 (4)	7.9 (4)
(Hatching)	4 min AN	98.0 (6)	24.2 (6)	4.3 (5)	1.56 (5)	6.5 (5)
	25 sec TET	156 (7)	71.0 (7)	20.4 (4)	9.23 (4)	6.1 (4)
H+5	N	94.3 (4)	26.8 (4)	3.2 (4)	0.91 (4)	8.2 (4)
	4 min AN	210 (2)	33.0 (2)	n.d.	n.d.	n.d.
H+10	N	84.4 (3)	24.1 (3)	3.4 (3)	0.97 (3)	7.1 (3)
	4 min AN	463 (2)	46.0 (2)	n.d.	n.d.	n.d.
H+150	N	177 (3)	63.8 (3)	4.4 (3)	1.59 (3)	10.7 (3)
	4 min AN	96 (2)	66.0 (2)	n.d.	n.d.	n.d.
	15 sec TET	448 (1)	137 (1)	n.d.	n.d.	n.d.

Table 6. *Levels of 3-phosphoglycerate (3-PGA), 2-phosphoglycerate (2-PGA), and phosphoenolpyruvate (PEP): mass-action ratios for the phosphoglycerate mutase and enolase reactions in normal, ischaemic, and tetanized embryonic chick muscle*
Mean levels in nmoles/g wet weight. Abbreviations used, see Table 2. Number of experiments in parentheses; number of individuals for each experiment, see Experimental Methods. The thermodynamic equilibrium constants for the phosphoglycerate mutase and enolase reactions are respectively 0.17 (25°, pH 7.0) [18] and 4.5 [8]—3.0 [15]

Development		3-PGA	2-PGA	PEP	2-PGA / 3-PGA	PEP / 2-PGA
days			nmoles/g wet weight			
5	N	35.5 (5)	14.1 (5)	25.2 (6)	0.29 (5)	2.8 (5)
	8 min AN	26.5 (4)	4.9 (2)	10.1 (3)	0.17 (2)	2.3 (1)
7	N	26.6 (4)	5.9 (4)	17.9 (4)	0.17 (4)	3.6 (4)
	8 min AN	31.3 (5)	10.0 (3)	31.9 (5)	0.22 (3)	3.4 (3)
11	N	19.9 (3)	5.2 (3)	16.0 (3)	0.22 (3)	3.4 (3)
	4 min AN	47.4 (4)	9.3 (4)	24.9 (4)	0.21 (4)	2.5 (4)
14	N	34.0 (3)	13.6 (4)	24.5 (3)	0.48 (3)	2.2 (3)
	4 min AN	35.4 (1)	11.7 (1)	39.9 (1)	0.33 (1)	3.4 (1)
	30 sec TET	32.8 (4)	7.4 (3)	21.5 (4)	0.22 (3)	2.7 (3)
17	N	43.4 (3)	7.3 (3)	23.4 (4)	0.16 (3)	3.5 (3)
	4 min AN	41.7 (1)	6.9 (1)	31.1 (2)	0.17 (1)	3.9 (1)
21	N	58.2 (4)	9.7 (4)	19.7 (4)	0.16 (4)	2.1 (4)
(Hatching)	4 min AN	50.3 (4)	10.3 (4)	18.2 (4)	0.18 (4)	1.8 (4)
	25 sec TET	141 (3)	17.3 (3)	47.5 (3)	0.12 (3)	2.7 (3)

Table 7. *Levels of lactate and pyruvate: lactate/pyruvate ratios in normal, ischaemic, and tetanized embryonic and post-embryonic chick muscle*
Means levels in nmoles/g wet weight. The lactate/pyruvate ratios were calculated from the means. Abbreviations used, see Table 2. Number of experiments in parentheses; number of individuals for each experiment: see Experimental Methods

Development		Lactate	Pyruvate	Lactate / Pyruvate
days			nmoles/g wet weight	
5	N	3426 (6)	227 (8)	15.1 (6)
	8 min AN	8171 (5)	258 (4)	36.7 (4)
7	N	2200 (5)	177 (11)	12.4 (5)
	8 min AN	6536 (5)	169 (6)	38.6 (5)
9	N	2400 (3)	214 (3)	11.2 (3)
11	N	2090 (7)	188 (11)	11.1 (7)
	4 min AN	5077 (7)	155 (8)	32.7 (7)
14	N	2143 (8)	101 (10)	21.2 (8)
	6 min AN	7340 (2)	95 (2)	78.0 (2)
	30 sec TET	2010 (4)	74 (5)	27.1 (4)
17	N	1880 (5)	81 (7)	23.2 (5)
	4 min AN	6636 (3)	92 (5)	72.1 (3)
21	N	2982 (7)	164 (8)	18.2 (7)
(Hatching)	4 min AN	9246 (6)	103 (7)	89.6 (6)
	25 sec TET	7983 (10)	274 (10)	29.1 (10)
H+5	N	2380 (4)	115 (4)	20.7 (4)
	4 min AN	13950 (2)	130 (2)	107 (2)
H+10	N	1970 (3)	118 (3)	16.7 (3)
	4 min AN	14150 (2)	140 (2)	101 (2)
H+150	N	2380 (3)	104 (3)	22.8 (3)
	4 min AN	12250 (2)	181 (2)	67.6 (2)
	15 sec TET	15500 (1)	996 (1)	15.6 (1)

nearly the same during the developmental period even after ischaemic or electrical stress (Table 6).

Lactate and Pyruvate

As early as the 7-day stage lactate contents are very similar to those in the resting adult muscle.

Fluctuations during embryonic development, such as the increase observed on the 21st day (difference between the 17th and 21st day: p > 0.05) seem to be rather unimportant. Our results do not confirm the definite rise in lactate on hatching observed by Pertseva [19]. The discrepancy may be attributable to different methods of freezing. As the glycolytic flow rate caused by anaerobiosis rises very strongly at birth (Table 3), great precautions for quick freezing *in situ* must be taken. We noticed that amputation and subsequent freeze-quenching with a delay of 10—15 sec resulted in a 3-fold increase of lactate.

According to Hohorst et al. [20] lactate/pyruvate ratios of between 10 and 13 have been consistently found in a number of different tissues, with the exception of skeletal muscle, which has ratios varying between 19 and 23 [7,16,21], and tumors [11,22]. Our data show that the "muscle ratio" is actually reached at the 14-day stage together with the onset of other characteristic muscle features, such as glycogen and phosphagen accumulation. Anaerobiosis was always followed by a rise in lactate; this was taken as a basis for the flow rate calculations; peak glycolysis was estimated from lactate formation after tetanic stimulus: as shown in Table 2 the rate of maximum glycolysis is 10—20 times greater than that at rest.

Pyruvate was on the whole unaffected by ischaemia, with the exception of the 11-day-old embryo, where the ischaemic decrease was found to be significant (p < 0.001), and on hatching (p < 0.05) (Table 7). Considerable increases of the lactate/pyruvate ratio were observed in ischaemia. The opposite occurred after tetanic stimulation. The well-known pyruvate-α-glycerophosphate dismutation [22], leading to high pyruvate levels during electrical stimulation, was first observed on hatching.

Table 8. *Comparison between mass-action ratios and thermodynamic equilibrium constant at the phosphoglucose isomerase step in different tissues*
Thermodynamic equilibrium constant = 0.44 [15]

Tissue	Mass-action ratio	Reference
Yeast	0.33	[26]
Erythrocytes (man)	0.41	[27]
Brain (mouse)	0.20	[8]
Liver (rat)	0.20	[9]
Ascites tumor	0.23	[6]
Skeletal muscle (rat)	0.38	[15]
Heart (rat)	0.22	[14]
	0.18	[28]
Heart (frog)	0.20	[30]
Breast muscle (pigeon)	0.16	[16]
Sartorius (frog)	0.18	[29]

DISCUSSION

As pointed out in the experimental procedure, up to the 9-day stage, whole decapitated and eviscerated embryos, whose external layers had been removed, were used. Whereas many muscle fibres are already present on the 5th day [23,24], such material cannot be considered as musculature, but rather as a mixed population of mesenchymal cells, differentiating myoblasts and fibres. Our results show that for most substrates no great differences really exist between the "muscle precursor tissue" at 5, 7 or 9 days and the embryonic muscle at 11 days. The onset of typical muscle metabolite patterns occurs later, in the 14- to 17-day period. An exception is glycogen which sharply increases between the 9th and 11th day: it is probable that single myoblasts or undifferentiated fibres have higher glycogen levels, as our cumulative analysis indicates. Electron micrographs of heart cells a few hours after incubation has begun, show numerous glycogen granules scattered in the cytoplasm [25].

Wet weight levels may be unreliable due to water loss during development. On the other hand dry weight includes compounds such as collagen or contractile proteins which may have little influence on the control of glycolytic flow rates. For these reasons dimensionless quantities such as metabolite ratios, and changes at individual stages were preferred. Dry weight percentages were determined in a separate series of experiments using embryos of the same strain in order to permit conversion of our wet weight figures and thus allow comparison with those of other workers (see Experimental Methods and Table 1).

Generally speaking, among the two-partner reactions in embryonic muscle, the isomerase-catalyzed steps show a small but indicative displacement from equilibrium, whereas the mutase-catalyzed reactions are very close to thermodynamic equilibrium. Equilibrium at the phosphoglucose isomerase step has been studied by many authors in different cells or tissues. As seen in Table 8, where data from the literature are collected, a good agreement between mass-action ratio and thermodynamic equilibrium constant was found in yeast [26], erythrocytes [27], and rat skeletal muscle [15]. On the other hand, in line with Lowry et al. [8] for the mouse brain, with Hohorst et al. [9] for the rat liver, with Hess [6] for ascites tumor, with Williamson [28] and Herkel [14] for the rat heart, and with Özand and Narahara for the frog sartorius [29] we found a definite shift from equilibrium in other muscle types [16,30]. Data presented in this work (Table 4) show remarkable agreement with the thermodynamic constant up to the middle phase of embryonic development. The mass-action ratio then decreases twofold from the 11th to the 14th day, and attains the typical adult value on hatching. The fact that no good agreement with the thermodynamic constant has been found in brain and in muscle types such as heart or pigeon breast, which are more difficult to quench correctly, is an argument for a technical explanation of the discrepancy. On the other hand, our embryo data may partly support Özand's [29] explanation based on the assumption of a compartmentation of hexose phosphates. In fact, the onset of nonequilibrium does not occur abruptly together with the striking rise in the velocity of the ischaemic glucose-6-phosphate accumulation observed on hatching, but somewhat gradually many days before. Moreover, since glucose-6-phosphate is constant till the 17th day, the shift is mainly due to a fructose-6-phosphate decrease. The difference between the mass-action ratio at 17 and 21 days, where neither structural nor gross enzymic modifications are known to occur, may be mainly due to increased glucose-6-phosphate formation. As a matter of fact, tetanic stimulation on hatching causes a further lowering of the mass-action ratio from 0.17 to 0.15 (Table 4). Under conditions of extreme stress therefore, the isomerase could assume a bottle-neck function, as postulated by Pedersen and Sacks [31].

The reason for the divergence between mass-action ratio and thermodynamic equilibrium constant in the triosephosphate isomerase catalyzed reaction is not exactly known. According to a suggestion of Garfinkel [32] accepted by Lowry et al. [8] and Williamson [28], who had similar results working with mouse brain and perfused rat heart, this is due to the presence of two states of glyceraldehyde-3-phosphate, one possibly bound to triosephosphate isomerase and the other free.

The balance sheet between carbohydrate breakdown and lactate production is satisfying at the 5-, 14- and 17-day stages. The lack of stoichiometry at 7 and 11 days (Table 3) needs a fuller discussion. Recently, Eppenberger et al. [33] have proposed a scheme for a modified atypical glycolysis in early chick embryo. From their hypothesis, which relies upon high levels of triose phosphates, 2- and 3-phosphoglycerate and phosphoenolpyruvate, as well as low levels of ADP (all according to earlier data by Novikoff et al. [2] and very high pyruvate kinase, transaminase and malate dehydrogenase activity, one can infer that, in early embryo, the majority of lactate is formed from aminoacids and not from glucose or glycogen. According to our results, the subsidiary route proposed by Eppenberger et al. [33] should be followed only between the 7th and 11th day when glucose no longer freely permeates the cellular membrane [34,35] and the phosphorylase activity is still very low [36,37]; the strikingly reduced glucose utilization, the insufficient glycogenolysis (Table 3) and the ischaemic increase of phosphoenolpyruvate, 2- and 3-phosphoglycerate fit well into such a scheme. The close correspondence between carbohydrate breakdown and lactate accumulation and the low phosphoenolpyruvate, 2- and 3-phosphoglycerate observed on the 5th, 14th, and 17th day make the existence of a subsidiary, anomalous glycolysis in the early as well as in the late embryonic stages improbable.

An interesting question that so far can be answered only incompletely involves the development of control mechanisms at the phosphorylase and phosphofructokinase steps. Both activation effects, which are known to occur in the rest-activity transition, may be indirectly demonstrated by the behaviour of the metabolite patterns after a sudden variation in the rate of flow. So, the phosphorylase b to a conversion, which occurs rapidly in stimulated muscle [38], and the phosphofructokinase activation, occurring in anaerobic [13] as well as in stimulated tissues [7,39], are both reflected by striking increases in the levels of glucose-6-phosphate and fructose-1,6-diphosphate, respectively. It would seem plausible to suppose that phosphofructokinase activation is secondary to that of phosphorylase, in mastering the hexose monophosphate accumulation. The tetanic stimulation of the 14-day-old embryo caused a fibrillar contraction accompanied by a fall in N-phosphoryl-creatine; no activation of glycolysis, as evidenced by the constancy of lactate (Table 7), was observed. This result clearly depends upon the lack of phosphorylase activation [36]. Very small activation of the phosphofructokinase alone (moderate fructose-1,6-diphosphate accumulation) is indicated. On hatching, both control mechanisms are already fully active: in fact glucose-6-phosphate, related hexoses and fructose-1,6-diphosphate accumulate in tetanized muscle. More experimental work, with tetanus experiments spaced between the 14th and 21st day and complete kinetics of ischaemic changes, is needed in order to establish the exact onset and the characteristics of these control mechanisms.

Comparison is possible between the activity of the glycolytic enzymes as measured in vitro, taking as a basis the data of Eppenberger et al. [33], and the flow velocity in vivo. Over the whole developmental period the velocity of the slowest enzyme, phosphofructokinase, exceeds that of lactate formation by a factor of between 1.1 and 3.5. In the rat skeletal muscle Pette (quoted by [7]) found the activity of phosphofructokinase, the slowest enzymatic activity, to be 100 times greater than the glycolytic rest rate. Such narrow adjustement of the enzyme equipment to lower flow rates reflects the lack of adaptability to sudden energy needs which is proper to embryonic life. Maximum rates of lactate formation, following tetanus on hatching, are compatible with lactate dehydrogenase activity, but imply a 10-fold activation of phosphofructokinase and enolase and a 4- to 5-fold activation of glyceraldehyde-phosphate dehydrogenase and pyruvate kinase. The activation of phosphofructokinase and pyruvate kinase (for which a "feed forward" activation mechanism has been recently described by Hess [40] is plausible. At present no explanation for the activation of the other enzymes can be found.

This work was supported by the Italian Research Council (Consiglio Nazionale delle Ricerche), Roma. We are indebted to Dr. C. Giunta for valuable technical assistance and to Dr. A. Ferroni for his help in the stimulation experiments.

REFERENCES

1. Meyerhof, O., and Perdigon, E., *Enzymologia*, 8 (1940) 53.
2. Novikoff, A. B., Potter, V. R., and Le Page, G. A., *J. Biol. Chem.* 173 (1948) 239.
3. Needham, J., and Nowiński, W. W., *Biochem. J.* 31 (1937) 1165.
3a. Needham, J., Nowiński, W. W., Dixon, K. C., and Cook, R. P., *Biochem. J.* 31 (1937) 1185.
4. Needham, J., and Lehmann, H., *Biochem. J.* 31 (1937) 1210.
5. Bücher, Th., and Rüssmann, W., *Angew. Chem.* 75 (1963) 881.
6. Hess, B, In *Funktionelle und morphologische Organisation der Zelle* (herausgegeben von P. Karlson). Springer Verlag, Berlin-Göttingen-Heidelberg 1963, p. 163.
7. Hohorst, H. J., *Metabolitgehalte und Metabolitgleichgewichte in der quergestreiften Muskulatur der Ratte.* Habilitationsschrift, Marburg/Lahn 1962.
8. Lowry, O. H., and Passonneau, J. V., *J. Biol. Chem.* 239 (1964) 31.
9. Hohorst, H. J., Kreutz, F. H., and Bücher, Th., *Biochem. Z.* 332 (1959) 18.
10. Bergmeyer, H. U., *Methoden der enzymatischen Analyse.* Verlag Chemie, Weinheim/Bergstraße 1962.
11. Arese, P., and Pejrone, C. A., *Ital. J. Biochem.* In the press.
12. Hohorst, H. J., Thesis, Faculty of Philosophy, University Marburg/Lahn (1960).
13. Lowry, O. H., Passonneau, J. V., Hasselberger, F. X., and Schulz, D. W., *J. Biol. Chem.* 239 (1964) 18.
14. Herkel, L., Thesis, Faculty of Medicine. University Marburg/Lahn (1964).
15. Hohorst, H. J., Reim, M., and Bartels, H., *Biochem. Biophys. Res. Commun.* 7 (1962) 137.
16. Arese, P., Kirsten, R., and Kirsten, E., *Biochem. Z.* 341 (1965) 523.
17. Meyerhof, O., and Junowicz-Kocholaty, R., *J. Biol. Chem.* 149 (1943) 71.
18. Meyerhof, O., and Oesper, P., *J. Biol. Chem.* 170 (1947) 1.
19. Pertseva, M. N., *Biochemistry (U.S.S.R.) (Engl. Transl.)* 26 (1961) 254.
20. Hohorst, H. J., Arese, P., Bartels, H., Stratmann, D., and Talke, H., *Ann. N. Y. Acad. Sci.* 119 (1965) 974.

21. Hohorst, H. J., In *Funktionelle und morphologische Organisation der Zelle* (herausgegeben von P. Karlson), Springer Verlag, Berlin-Göttingen-Heidelberg 1963, p. 194.
22. Bücher, Th., and Klingenberg, M., *Angew. Chem.* 70 (1958) 552.
23. Allen, E. R., and Pepe, F. A., *Am. J. Anat.* 116 (1965) 115.
24. Dessouky, D. A., and Hibbs, R. G., *Am. J. Anat.* 116 (1965) 523.
25. Buffa, P., Personal communication.
26. Mann, P. F. E., Trevelyan, W. E., and Harrison, J. S., In *Recent Studies in Yeast and their Significance in Industry*, London 1958. Quoted by [5].
27. Minakami, S., Suzuki, C., Saito, T., and Yoshikawa, H., *J. Biochem. (Tokyo)* 58 (1965) 543.
28. Williamson, J. R., *J. Biol. Chem.* 240 (1965) 2308.
29. Özand, P., and Narahara, H. T., *J. Biol. Chem.* 239 (1964) 3146.
30. Arese, P, Bosia, A., and Rossini, L., *Bull. Soc. Ital. Biol. Sper.* 42 (1966) 1476.
31. Pedersen, P. L., and Sacks, J., *Arch. Biochem. Biophys.* 112 (1965) 548.
32. Garfinkel, D., *Ann. N. Y. Acad. Sci.* 108 (1963) 293.
33. Eppenberger, H. M., Fellenberg, R. von, Richterich, R., and Aebi, H., *Enzymol. Biol. Clin.* 2 (1962/63) 139.
34. Foà, P. P., Melli, M., Berger, C. K., Billinger, D., and Guidotti, G. G., *Federation Proc.* 24 (1965) 1046.
35. Guidotti, G. G., and Foà, P. P., *Am. J. Physiol.* 201 (1961) 869.
36. Cosmos, E., *Develop. Biol.* 13 (1966) 163.
37. Grillo, T. A. I., *J. Histochem. Cytochem.* 9 (1961) 386.
38. Danforth, W. H., and Lyon, J. B., *J. Biol. Chem.* 239 (1964) 4047.
39. Sacktor, B., and Wormser-Shavit, E., *J. Biol. Chem.* 241 (1966) 624.
40. Hess, B., and Brand, K., In *Control of Energy Metabolism* (edited by B. Chance, R. Estabrook, and J. R. Williamson), Academic Press, New York-London 1965, p. 111.

P. Arese, M. T. Rinaudo, A. Bosia
Istituto di Chimica Biologica dell'Università
Via Michelangelo 27, Torino, Italy

European J. Biochem. 1 (1967) 216—232

Eine neue Gruppe von Salmonella R-Mutanten

Serologische und biochemische Analyse des Heptosekerns von Lipopolysacchariden aus *Salmonella minnesota*- und *Salmonella ruiru*-Mutanten

H. J. Risse, W. Dröge, E. Ruschmann, O. Lüderitz und O. Westphal

Max-Planck-Institut für Immunbiologie, Freiburg i. Br.

und

J. Schlosshardt

Zentrallaboratorium für Bakterielle Darminfektionen beim Institut für Serum- und Impfstoffprüfung, Potsdam

(Eingegangen am 20. Dezember 1966)

From the wild types (S forms) of *Salmonella minnesota* and *Salmonella ruiru* R mutants (R forms) were isolated which produce hexose-less cell wall lipopolysaccharides of chemotype Rd. These lipopolysaccharides are composed of lipid A (glucosamine, long-chain fatty acids, phosphoric acid), 2-keto-3-deoxy-octonate (KDO), and L-glycero-D-manno-heptose (heptose), while those of the parent S forms contain additional galactosamine, glucosamine, galactose and glucose. R mutants with lipopolysaccharides of chemotype Rd could be differentiated into two groups, Rd_1 and Rd_2.

Group Rd_1 comprises the *S. minnesota* mutants mR7 and mRz. Their lipopolysaccharides contain about 14 % heptose. Partial hydrolysis of the mR7 lipopolysaccharide resulted in the formation of three main split products: lipid A, mR7A and mR7B. mR7A was identified as free 2-keto-3-deoxy-octonate, while mR7B was composed of heptose and 2-keto-3-deoxy-octonate. Depending on the method applied for the estimation of 2-keto-3-deoxy-octonate, the molar ratio of heptose/2-keto-3-deoxy-octonate was found to be 2:0.3 (on the basis of the thiobarbituric acid reaction) or 2:1 (according to the semicarbazide reaction). When permethylated mR7B was methanolyzed and the products analyzed by gas chromatography, two peaks (7.4 and 15.0 min) were observed. The 7.4 min peak was identical with one of the two peaks (7.4 and 9.6 min) obtained with permethylated authentic D-glycero-L-manno-heptose, the optical antipode of the bacterial heptose. The second peak at 15.0 min, which comprised about the same area as the 7.4 min peak, was derived from a partially methylated heptose. It is concluded that mR7B represents a heptose disaccharide linked to 2-keto-3-deoxy-octonate and that such units are present in Rd_1 lipopolysaccharides as terminal, nonreducing residues of the structure: heptosyl→heptosyl →2-keto-3-deoxy-octonate—.

Group Rd_2 comprises the *S. minnesota* mutants mR3 and mR4 and the *S. ruiru* mutant rR3. Their lipopolysaccharides contain about 7 % heptose. By partial hydrolysis of the mR3 lipopolysaccharide four main split products were obtained: lipid A, mR3A, mR3B and mR3C. Fraction mR3A was identified with free 2-keto-3-deoxy-octonate. mR3B contained heptose, phosphate and ethanolamine in a molar ratio of 1:1:1, and, in addition, 1 mole (by thiobarbituric acid reaction) or 3 moles (by semicarbacide reaction) of 2-keto-3-deoxy-octonate. mR3C contained heptose and 2-keto-3-deoxy-octonate in a ratio of 1:1 (thiobarbituric acid reaction) or 1:2 (semicarbacide reaction), but no phosphate or ethanolamine. mR3B and mR3C were methylated, the products methanolyzed and analyzed by gas chromatography. From both oligosaccharides only one heptose peak (7.4 min) was obtained which was identical with the faster peak observed with mR7B and authentic heptose. It is concluded that in the mR3 lipopolysaccharide, and generally in Rd_2 lipopolysaccharides, heptose is exclusively linked as a terminal, non-reducing monosaccharide, presumably to 2-keto-3-deoxy-octonate: heptosyl→2-keto-3-deoxy-octonate—.

The results indicate that Rd_2 lipopolysaccharides act as precursors of Rd_1 lipopolysaccharides in the biosynthesis of the wild type lipopolysaccharide and that the Rd_1 structure (Hep→Hep→ KDO—) is formed by transfer of one heptose unit to Rd_2 lipopolysaccharides (with terminal Hep-KDO).

Biosynthetic studies demonstrated that Rd_1 lipopolysaccharides, but not Rd_2 lipopolysaccharides, function as acceptor for the enzymatic incorporation of glucose from UDP-glucose. Although the Rd_1 mutants, mR7 and mRz, synthesize identical lipopolysaccharides, they are distinct regarding their enzymatic block involving the glucose anabolism. mR7 lacks glucosyl-I-transferase activity, while mRz is defective in the enzyme UDP-glucose-synthetase. — Rd_2 mutants, with the incomplete heptosyl-KDO core, are deficient with respect to the transfer of the second heptosyl residue to the first heptose; but they do contain—as expected—glucosyl-I-transferase which catalyzes the incorporation of glucose (glucose I) into the complete heptosyl-heptosyl-KDO core of Rd_1 lipopolysaccharides to form the glucosyl-heptosyl-heptosyl-KDO structure of Salmonella Rc lipopolysaccharides.

Serologically, heptose proved to be an inhibitor of precipitation in both systems, mR7/anti mR7 (Rd_1 system) and mR3/anti mR3 (Rd_2 system), which is in agreement with the concept of terminal, non-reducing heptose units occurring in both Rd_1 and Rd_2 lipopolysaccharides. However, in hemagglutination inhibition tests, the lipopolysaccharides of group Rd_1 do not show serological cross-reaction with Rd_2 lipopolysaccharides, and vice versa.

Salmonella-Bakterien und andere Enterobakteriaceen bilden ein charakteristisches Zellwand-Lipopolysaccharid, dessen prinzipielle Struktur in den letzten Jahren intensiv bearbeitet wurde (siehe [1]). Die Lipopolysaccharide bestehen einerseits aus einer Lipoid-Komponente, sog. Lipoid A, welche aus D-Glucosamin, langkettigen Fettsäuren und Phosphorsäure aufgebaut, und die für alle untersuchten Stämme gleich oder sehr ähnlich zusammengesetzt ist. Der variable Teil der Lipopolysaccharide ist die jeweilige Polysaccharid-Komponente, die ihrerseits prinzipiell aus mehreren Teilen besteht: (a) dem Heptose-Kern, welcher außer L-Glycero-D-manno-heptose [2] noch 2-Keto-3-desoxyoctonsäure [3], Äthanolamin und Phosphorsäure enthält [4]. In *Salmonella*-Polysacchariden sind an den Heptose-Kern jeweils (b) Pentasaccharid-Einheiten gebunden, die aus Glucose, Galactose und Glucosamin bestehen und die folgende Struktur besitzen [5,6]:

$$\text{GlcNAc} \rightarrow \text{Glc} \rightarrow \text{Gal} \rightarrow \text{Glc} \rightarrow (\text{Hep})$$
$$\nearrow$$
$$\text{Gal}$$

Bei den kompletten Polysacchariden der *Salmonella*-Wildformen (smooth, S-Formen) sind an dieses innere Oligosaccharid (c) lange Ketten von wiederholenden Oligosaccharid-Einheiten — im allgemeinen Tri-, Tetra- oder Pentasaccharide — gebunden. Diese langen Seitenketten, welche häufig kurze monosaccharidische Verzweigungen besitzen (siehe [1]), bestehen aus vielen verschiedenen Zuckerbausteinen, darunter Hexosen, 6-Desoxy-, 3,6-Didesoxy-hexosen und anderen, sind *species*- oder gruppenspezifisch und Träger der serologischen Spezifität (sogenannte O-Spezifität). Die Spezifität der O-Antigene ist be-

kanntlich die Basis der serologischen Klassifizierung der Salmonellen im Kauffmann-White-Schema [7].

Beim Übergang der Glattformen (S-Formen) in Rauhformen (R-Formen) — sogenannte S → R-Mutation — entstehen Verlustmutanten (R-Mutanten), die ihre serologische O-Spezifität verloren haben und deren Analyse zeigt, daß sie unvollständige Zellwand-Polysaccharide bilden. Bei der S → R-Mutation tritt ein Block in der Biosynthese des kompletten Lipopolysaccharids der Wildform ein. Es hat sich gezeigt, daß, je nach Lage des Syntheseblocks, viele R-Mutanten aus der gleichen Wildform hervorgehen können. Im Prinzip kann jedes an der Biosynthese dieser bakteriellen Heteropolysaccharide beteiligte Enzym (Synthetasen, Transferasen) durch Mutation blockiert sein, so daß (auch phänotypisch) viele R-Mutanten konzipiert werden können. So wurden schon frühzeitig RII-, RI- und M-Mutanten aus vielen *Salmonella*-Wildformen isoliert und strukturell und serologisch analysiert [6,8,10]. Es zeigte sich, daß alle ermittelten Strukturen (blockierte) Zwischenstufen in der Biosynthese des kompletten Lipopolysaccharids sind. Man kann daher durch Isolierung möglichst vieler R-Mutanten von der gleichen Wildform die Biosynthese des kompletten Lipopolysaccharids rekonstruieren und überdies jede Zwischenstufe für biosynthetische Untersuchungen einzelner Transfer-Reaktionen verwenden [5,9,11].

Nach diesen Untersuchungen bestehen die Polysaccharid-Komponenten der kompletten Zellwand-Lipopolysaccharide strukturell aus 3 Bezirken: dem Heptose-Kern (mit KDO, Heptose etc., gebunden an Lipoid A), dem inneren Oligosaccharid von R-Spezifität, und den O-spezifischen Seitenketten aus wiederholenden Oligosaccharid-Einheiten.

Die nachstehend beschriebenen Untersuchungen wurden an einer Klasse von R-Mutanten durchgeführt, welche sich von den Wildformen von *Salmonella minnesota* und *Salmonella ruiru* durch einen tiefliegenden Defekt in der Biosynthese des kompletten O-antigenen Lipopolysaccharids auszeichnen.

Nicht allgemein gebräuchliche Abkürzungen. 2-Keto-3-desoxyoctonsäure, KDO; Heptose, Hep; Dimethylsulfoxid, DMSO; Trichloressigsäure, TCE; Thiobarbitursäure-Reaktion, TBSR; Semicarbazid-Reaktion, SCR.

Diese Verlustmutanten bilden Lipopolysaccharide in der Zellwand, welche außer Lipoid A nur KDO und Heptose als Zuckerbausteine enthalten, die also hexose-frei sind und somit zum Chemotyp Rd [12] gehören. Die Wildformen von *S. minnesota* und *S. ruiru* enthalten außer Lipoid A, KDO und Heptose zusätzlich Glucose, Galactose und Glucosamin sowie Galactosamin als seitenketten-spezifischen Zuckerbaustein [1,13].

Es wird gezeigt, daß die Gruppe dieser hexoselosen Mutanten unterteilt werden kann in eine Gruppe Rd_1, deren Lipopolysaccharid etwa $14^0/_0$ Heptose enthält, und eine Gruppe Rd_2, deren Lipopolysaccharid nur etwa $7^0/_0$ Heptose enthält. Beide Gruppen unterscheiden sich auch in ihrem serologischen und biochemischen Verhalten. Die Ergebnisse der Strukturanalyse sowie der biochemischen Untersuchungen legen die Vermutung nahe, daß Rd_2 eine Vorstufe bei der Biosynthese von Rd_1 darstellt.

MATERIAL UND METHODEN

Züchtung der Bakterien und Darstellung der Lipopolysaccharide

S. minnesota Rz wurde von Dr. G. Schmidt, Max-Planck-Institut für Immunbiologie, Freiburg, als galactose-negative R-Mutante isoliert.

Die übrigen hier untersuchten R-Mutanten von *S. minnesota* und *S. ruiru*[1] wurden am Institut für Serum- und Impfstoffprüfung, Zentrallaboratorium für Bakterielle Darminfektionen, Berlin-Pankow, isoliert [12]. *S. typhimurium* TL 2 SL 848 ist eine Mutante, welche von Dr. B. A. D. Stocker, Lister Institute for Preventive Medicine, London, isoliert worden war, bei welcher das Enzym UDP-Galactose-4-Epimerase fehlt (M-Mutante). Massenkulturen der Bakterien wurden, wie kürzlich beschrieben, auf Agar gezüchtet [14]; die Lipopolysaccharide wurden mit Phenol/Wasser extrahiert und durch Ultrazentrifugation gereinigt [15]. Zur Gewinnung von Acceptor-Partikeln und Enzymüberständen für biosynthetische Versuche wurden die Bakterien in Difco-Antibiotic M3-Medium gezüchtet, bei Erreichen der halblogarithmischen Wachstumsphase geerntet und, wie weiter unten beschrieben, aufgearbeitet.

Analytische Methoden

Heptose wurde nach Dische [16] in der Abwandlung von Osborn [9] bestimmt, KDO mit Thiobarbitursäure nach Waravdekar und Saslaw [17a] in der Abwandlung von Heath [17b], sowie mit Semicarbazid nach MacGee und Doudoroff [18], Ethanolamin mit Fluordinitrobenzol nach Ghuysen und Strominger [19], Phosphor nach Lowry et al. [20], und Protein in Präzipitaten mit Folinreagens nach Lowry

---
[1] Wir benutzen das Präfix m für R-Mutanten aus *S. minnesota*, das Präfix r für R-Mutanten aus *S. ruiru*.

---

et al. [21], in bakteriellen Extrakten mit Biuret Reagens [22]. Zur Chromatographie auf Papier (Whatman No. 1) und Dünnschichtplatten (Zellulose, Macherey u. Nagel, MN 300G) wurden folgende Laufmittel benutzt: (A) Butanol-Pyridin-Wasser (6:4:3, v/v), (B) Butanol-Eisessig-Wasser (5:1:2, v/v), (C) Äthanol-1 M Ammoniumacetat, pH 7,0 (75:30, v/v), (D) $95^0/_0$ Aceton, (E) Pyridin-Eisessig-Wasser (100:40:860, v/v, pH 5,3). Die Anfärbung erfolgte mit Silbernitrat/NaOH oder Ninhydrin. Zur Erkennung von Oligosacchariden wurde das Chromatogramm vor der Silberfärbung in eine Lösung von Natriumperjodat in wäßrigem Aceton getaucht [9]. Die Papierelektrophorese wurde nach Kickhöfen und Westphal [23] durchgeführt (Laufmittel E).

L-Glycero-D-manno-heptose war aus *S. minnesota* R 7 isoliert worden [2], D-Glycero-L-manno-heptose, die als Standard bei der Heptosebestimmung benutzt wurde, haben wir von Dr. N. K. Richtmyer, N.I.H., Bethesda, Md., U.S.A., erhalten. KDO zur Präzipitationshemmung wurde aus der heptoselosen Mutante von *S. minnesota* R 595 durch Hydrolyse bei pH 3,4 und anschließende papierelektrophoretische Reinigung erhalten. Die KDO-Menge im Eluat wurde bestimmt [17] und die Lösung in eingefrorenem Zustand aufbewahrt. KDO als Standard bei der KDO-Bestimmung stammte von Dr. E. C. Heath, Johns Hopkins University, Baltimore, Md., von dem wir den Methylester der Pentaacetyl-KDO erhalten haben [3].

Methylierung und Gaschromatographie

Methylsulfinylanion [24]. 8 g NaH-Suspension ($50^0/_0$ in Öl) wurden unter Stickstoff mehrfach mit Petroläther (40—60°) gewaschen. Das erhaltene graue Pulver wurde unter Stickstoff mit 40 ml abs. Dimethylsulfoxid (DMSO) versetzt und unter gelegentlichem Schütteln 48 Std bei Zimmertemperatur stehen gelassen. Die Lösung wurde in brauner Flasche unter Stickstoff mit CaCl₂-Trockenrohr aufbewahrt.

Methylierung [25]. 30—40 mg gut getrocknetes Oligosaccharid wurden in 1,5 ml abs. DMSO gelöst und unter Stickstoff mit 1,5 ml Methylsulfinyl-Lösung versetzt. Nach 2 Std bei Zimmertemperatur wurde auf 4° gekühlt, dann vorsichtig und langsam mit 450 µl Methyljodid versetzt, unter Stickstoff geschüttelt und über Nacht bei Zimmertemperatur stehen gelassen. Dann wurden abermals 1,5 ml Methylsulfinyl-Lösung und 450 µl Methyljodid im Abstand von 2 Std zugegeben.

Schließlich wurde die DMSO-Lösung mit 15 ml Wasser versetzt und mit Chloroform mehrfach ausgeschüttelt. Die vereinigten Chloroformlösungen wurden mit H_2O gewaschen, im Vakuum eingeengt, mit ca. 5 ml Äther aufgenommen und mit Na_2SO_4 über Nacht getrocknet. Die abdekantierte Ätherlösung wurde danach im Vakuum eingeengt und 3 Tage über P_2O_5 im Ölpumpenvakuum getrocknet.

Methanolyse. Die methylierten Oligosaccharide wurden in 5 ml 1 N methanolischer HCl-Lösung gelöst und im zugeschmolzenen Glasrohr 10 Std auf 100° erhitzt. Nach dem Abkühlen wurde die Lösung mit Chloroform und Wasser versetzt und wie nach der Methylierung weiter aufgearbeitet. Das erhaltene Produkt wurde 60 Std über P_2O_5 im Ölpumpenvakuum getrocknet.

Gaschromatographie. Zur Gaschromatographie wurde ein Perkin-Elmer-Fraktometer 116-E mit Flammenionisationsdetektor (FID) benutzt. Die Säule (2 m × 4,65 mm) war gefüllt mit Neopentylglycol-succinat $0,5^0/_0$ auf Chromosorb W. Das Trägergas Stickstoff wurde auf eine Geschwindigkeit von 40 Nml/min eingestellt. Die Betriebstemperatur betrug 180°.

Serologische Methoden

Hämagglutinationshemmung [8] sowie Präzipitationshemmung [26] wurden wie früher beschrieben durchgeführt. Die Präzipitationsansätze wurden in einem Gesamtvolumen von 350 μl 48 Std bei 4° belassen, die Präzipitate dreimal gewaschen und in einem Endvolumen von 700 μl colorimetrisch bestimmt [20]. Als Bezugssubstanz diente Kaninchen γ-Globulin (15,9$^0/_0$ N). Die benutzten Antiseren waren durch Immunisierung von Kaninchen mit hitzegetöteten Bakterien gewonnen worden [12].

Herstellung von [14C]UDP-Glucose

[14C]Glucose wurde mit Hilfe gereinigter Hexokinase phosphoryliert.

Gesamtes Volumen 1,2 ml: 0,81 μMole [14C]-Glucose; 60 μMole Tris/HCl, pH 7,8; 20 μMole Mg^{++}; 30 μMole ATP; ca. 1 mg Enzym-Protein. Der Ansatz wurde nach der Inkubation (37°, 3 Std) elektrophoresiert, die mittels Autoradiographie ermittelte Bande von Glucose-6-phosphat eluiert und in folgendem Reaktionsansatz zu UDP-Glc umgesetzt: Gesamtes Volumen: 765 μl: 40 μMole Tris/HCl, pH 8,0; 7 μMole UTP; 2 μMole EDTA; 15 μMole Mg^{++}; 300 μl Hefe-Enzym [27]; 20 μl anorganische Pyrophosphatase; der Ansatz wurde 3 Std bei 37° inkubiert. Nach Elektrophorese wurde die radioaktive Bande eluiert und zweimal in dem Laufmittel C papierchromatographiert. Ausbeute an [14C]UDP-Glucose ca. 100$^0/_0$, Aktivität 13000 Imp./min/nMol.

Herstellung des Glucosyl-Acceptors (Partikelfraktion) und des (löslichen) Transferase-Rohenzyms

3 g Bakterienzellen (Feuchtgewicht) wurden in 20 ml 0,01 M Tris-Puffer, pH 7,8, unter Zugabe von 0,5 μl 2-Mercaptoäthanol suspendiert und in einem Ultraschall-Gerät (Raytheon, 9 kc) 70—80 min unter guter Kühlung beschallt (*cf.* [5,28]).

Intakte Zellen wurden durch Zentrifugieren bei 1200 × g (2 × 20 min) abgetrennt und der trübe Überstand 20 min bei ca. 20000 × g zentrifugiert. Die hierbei sedimentierende Fraktion wurde 1 mal mit dem gleichen Volumen Tris-Puffer (wie oben) gewaschen und in demselben Puffer suspendiert. Die Suspension wurde zur Zerstörung enzymatischer Aktivität 10 min auf 100° erhitzt, in dieser Form als Acceptor verwendet und bei —20° aufbewahrt. In einem aliquoten Teil der Suspension wurde KDO nach der Thiobarbitursäure-Methode [17] bestimmt. Bei den Transfer-Versuchen haben wir die Acceptormenge stets auf den so ermittelten KDO-Gehalt bezogen. Es ist möglich, daß in den Partikelpräparaten außer KDO auch andere Substanzen mit Thiobarbitursäure positiv reagieren. Die Bezugsgröße ist daher als „KDO" angegeben.

Zur Gewinnung des Roh-Enzymextraktes wurde der Überstand der 20000 × g Zentrifugation 2,5 Std bei 100000 × g ultrazentrifugiert. Der erhaltene Enzym-Überstand enthält das Transferase-Enzym (ca. 10—15 mg Protein pro ml). Die Lösung, in kleinen Portionen abgefüllt und bei —20° eingefroren, ist über mehrere Wochen aktiv.

Übertragung von [14C]Glucose auf die Partikel

Inkubationsansätze (Gesamtvolumen 250 μl) enthielten: 45 μMole Tris/HCl, pH 7,5; 10 μMole Mg^{++}; Acceptor-Partikel äquivalent ca. 100 nMol „KDO"; [14C]UDP-Glucose mit kalter UDP-Glucose verdünnt, 50—70 nMole (700 Imp./min/nMol); hochzentrifugierter Enzym-Überstand, etwa 1 mg Protein (*cf.* [5,29a]).

Die Ansätze wurden bei 37° inkubiert. Nach Abkühlen auf 0° wurden sie mit 250 μl eiskalter 10$^0/_0$iger Trichloressigsäure (TCE) versetzt und zentrifugiert (3000 U./min, 5 min). Der Niederschlag wurde 4 mal mit 500 μl 5$^0/_0$iger TCE gewaschen, anschließend in 40 μl 1 N KOH suspendiert, auf Filterpapierstücke von 1,5 × 3,5 cm aufgetragen und nach Trocknen in Vakuum im Scintillationszähler (Packard-TriCarb) gemessen. Kontrollen wurden bei 0° mit TCE versetzt und auf gleiche Weise aufgearbeitet.

Isolierung des [14C]Lipopolysaccharids aus dem TCE-Niederschlag

Die TCE-Fällung eines zehnfachen Ansatzes wurde wie oben beschrieben gewaschen, in wenig Wasser suspendiert und mit 1 N KOH neutralisiert. Zu der Suspension gaben wir als Träger ca. 3 g frische Zellen einer für den jeweiligen Versuch geeigneten Mutanten und füllten mit Wasser auf ein Endvolumen von 15 ml auf. Nach Zugabe von 15 ml 90$^0/_0$igem Phenol wurde das Gemisch 5 min bei 65° intensiv gerührt, abgekühlt und zentrifugiert (10000 × g, 5 min). Die wäßrige Schicht wurde abgehoben und die Phenolschicht zweimal mit 12 ml Wasser nachextrahiert. Die vereinigten wäßrigen Schichten wurden gegen

Wasser dialysiert (2 × 2 Liter, je 12 Std) und schließlich ultrazentrifugiert (4 Std 100000 × g). Das einmal mit Wasser gewaschene Sediment stellt das gereinigte Lipopolysaccharid dar.

Während der Aufarbeitung wurde von jeder Fraktion (TCE-Niederschlag, Phenolschicht, Dialysat usw.) eine Probe abgenommen und im Scintillationszähler gemessen.

Setzt man vor der Phenolextraktion dem neutralisierten TCE-Niederschlag keine Träger-Zellen zu, so kann das Lipopolysaccharid bzw. die Radioaktivität zum größten Teil in die Phenolschicht verschleppt werden. Das gleiche geschieht leicht, wenn man statt ganzer Zellen als Träger isoliertes (kaltes Träger-)Lipopolysaccharid zusetzt: dieses wird dann unter Umständen zu einem großen Teil im Phenol wiedergefunden.

Reagentien

ATP, Hexokinase: Boehringer, Mannheim. UTP, UDP-Glucose (nicht radioaktiv): Sigma Chemical Co., St. Louis, Mo., U.S.A. Pyrophosphatase: Worthington. [^{14}C]Glucose: Radio Chemical Center, Amersham/England.

VERSUCHE UND ERGEBNISSE

Extraktion und Analyse der Lipopolysaccharide

Die in dieser Arbeit untersuchten Bakterienstämme wurden zur Gewinnung der Lipopolysaccharide mit Phenol/Wasser (5 min, 65°) [15] extra-

saccharide der R-Formen nach Dialyse und Einengen der wäßrigen Phase spontan präzipitierten und als wasserunlösliches Präparat in einer niedertourigen Zentrifuge abgetrennt werden konnten.

Die Lipopolysaccharide von *S. minnesota* und *S. ruiru* gehören zum Chemotyp II [13] und enthalten neben den basalen Zuckern — KDO, Heptose, Glucose, Galactose und Glucosamin (im Polysaccharid und Lipoid) — zusätzlich Galactosamin (Tabelle 1). Dagegen sind die hier untersuchten R-Lipopolysaccharide lediglich aus den Zuckern KDO und Heptose aufgebaut und gehören daher zum Chemotyp Rd [12]. Ausnahmen bilden die beiden Stämme rR1 und rR541, welche eine kleine Menge Glucose bzw. Glucose und Galactose enthalten. Bestandteil aller Lipopolysaccharide ist außerdem Lipoid A, welches bei der Wildform etwa 30—50% und bei den R-Formen 60—70% ausmacht. Tabelle 1 zeigt die Ergebnisse der quantitativen Analysen der Lipopolysaccharide. Man erkennt, daß auf Grund der Heptoseanalysen die Mutanten in 2 Gruppen eingeteilt werden können: Die Lipopolysaccharide von mR7, mRz, rR541 und rR1 enthalten 10—17% Heptose, während die Lipopolysaccharide von mR3, mR4 und rR3 nur etwa 6—7% Heptose enthalten. Diese Differenzierung der Lipopolysaccharide in 2 Gruppen erwies sich auch bei den nachfolgend beschriebenen serologischen Untersuchungen als berechtigt. Wir bezeichnen die Lipopolysaccharide mit 10—17% Heptose als zur Gruppe Rd$_1$, diejenigen mit 6—7% als zur Gruppe Rd$_2$ gehörig.

Tabelle 1. *Analyse der Lipopolysaccharide von* Salmonella minnesota *S-Form und einigen R-Mutanten*
m = *S. minnesota*; r = *S. ruiru*; KDO wurde mit Thiobarbitursäure bestimmt

Chemotyp (1)	Lipopoly-saccharid	Gewichtsprozent berechnet auf Lipopolysaccharid						
		Lipoid A	KDO	Hep	Glc	Gal	GlcN (PS)	GalN
		%	%	%	%	%	%	%
II	m Wildform	50	4,5	3—5	3—4	13—16	10	14—15
Rd$_1$	mR7	58	9—10	10—13	—	—	—	—
Rd$_1$	mRz	ng^a	7,5	15	—	—	—	—
Rd$_1$(Rb)	r R541	52	10	13—17	1—2	0,3—1	—	—
Rd$_1$(Re)	r R1	ng^a	8	15	2	—	—	—
Rd$_2$	mR3	63—69	9—10	5—7	—	—	—	—
Rd$_2$	mR4	ng^a	9—10	5—7	—	—	—	—
Rd$_2$	r R3	ng^a	9	5—6	—	—	—	—

^{a} Nicht getestet.

hiert. Die Rohextrakte der wäßrigen Phase wurden in der präparativen Spinco Zentrifuge 3 mal bei 40000 U./min zentrifugiert, wobei die Lipopolysaccharide als Sediment erhalten wurden. Die Ausbeuten an Lipopolysaccharid waren bei der Wildform höher (5—6%) als bei den Mutanten (2—4%). Obwohl Züchtungs- und Aufarbeitungsbedingungen standardisiert waren, erhielten wir aus verschiedenen Chargen des gleichen Stammes unterschiedliche Ausbeuten. Gelegentlich kam es vor, daß Lipopoly-

Serologische Untersuchungen

Hämagglutinationshemmung. Die Untersuchung der R-Lipopolysaccharide als Hemmsubstanzen in den homologen, hämagglutinierenden Systemen mR7, mR3, mR4 und rR3 ergab, daß die Lipopolysaccharide des Chemotyps Rd$_1$ kreuzreagieren und das Serum mR7 hemmen, während sie die Rd$_2$-Systeme auch in hoher Konzentration nicht hemmen. Umgekehrt hemmen die Lipopolysaccharide des Chemotyps Rd$_2$ die Rd$_2$-Systeme und reagieren nicht mit

Tabelle 2. *Serologische Untersuchung der R-Lipopolysaccharide im Hämagglutinationstest*
Die Zahlen der Tabelle geben die niedrigste Konzentration an, bei welcher noch Hemmung beobachtet wurde (μg/ml). > 250 bedeutet, daß bis 250 μg/ml keine Hemmung beobachtet wurde. Die fettgedruckten Zahlen entsprechen den Hemmungen im homologen System

Chemotyp	Lipopolysaccharid	Hemmung der Systeme			
		Rd_1	Rd_2		
		mR 7/Anti mR 7	mR 3/Anti mR 3	mR 4/Anti mR 4	rR 3/Anti rR 3
		μg/ml	μg/ml	μg/ml	μg/ml
Rd_1	mR 7	2	> 250	> 250	> 250
	mRz	63	> 250	> 250	> 250
	rR 541 [a]	31	> 250	> 250	> 250
	rR 1 [a]	63	> 250	> 250	> 250
Rd_2	mR 3	> 250	8	4	4
	mR 4	> 250	16	4	4
	rR 3	> 250	8	4	4

[a] Die Lipopolysaccharide, welche geringe Mengen Hexose enthalten, zeigen zusätzliche Kreuzreaktion mit hexose-haltigen Lipopolysacchariden.

dem System mR 7/Anti mR 7 (Rd_1) (Tabelle 2). Im Hämagglutinationshemmungstest erweisen sich daher das Serum mR 7 einerseits und die Seren mR 3, mR 4 und rR 3 andererseits als spezifisch für die Chemotypen Rd_1 bzw. Rd_2. Die Tatsache, daß zwischen den beiden Gruppen keine Kreuzreaktionen beobachtet werden, bedeutet allerdings nicht, daß kreuzreagierende Antikörper in diesen Seren abwesend sind. Wie die folgenden Versuche zeigen, werden sie im Präzipitationstest sichtbar.

Präzipitation und Präzipitationshemmung. Wir haben die Präzipitation der Lipopolysaccharide mR 3 und mR 7 mit homologem Serum sowie die Kreuzreaktion mR 3-Lipopolysaccharid mit mR 7-Serum untersucht. Fig. 1 zeigt die erhaltenen Präzipitationskurven, aus denen hervorgeht, daß das mR 7-Serum kreuzreagierende Antikörper gegen mR 3-Lipopolysaccharid enthält, welche im Hämagglutinationshemmungstest nicht nachweisbar waren. Entsprechend konnten wir im mR 3-Serum kreuzreagierende Antikörper gegen mR 7-Lipopolysaccharid nachweisen. Die Vermutung lag nahe, daß mR 3 als determinanter Zucker in beiden Lipopolysacchariden, mR 3 und mR 7, fungiert, und wir haben daher untersucht, ob Heptose die Präzipitation der beiden Lipopolysaccharide zu hemmen vermag. Da KDO ein weiterer Bestandteil der R-Lipopolysaccharide ist, haben wir vergleichsweise auch KDO als Hemmsubstanz dieser Systeme getestet.

Tabelle 3 zeigt das Ergebnis der Präzipitationshemmung der homologen Systeme mR 3 und mR 7 sowie des heterologen Systems mR 7/Anti rR 3 mit L-Glycero-D-manno-heptose und KDO, welche beide aus R-Lipopolysacchariden isoliert worden waren [2]. Heptose hemmt sowohl das mR 3- wie das mR 7-System, letzteres bis zu 50% bei den angewandten Konzentrationen. Auch die Kreuzreaktion mR 7/Anti rR 3 wird durch Heptose gehemmt. Wir schließen aus diesen Befunden, daß trotz ihres verschiedenen Gehaltes an Heptose in beiden Lipopolysacchariden

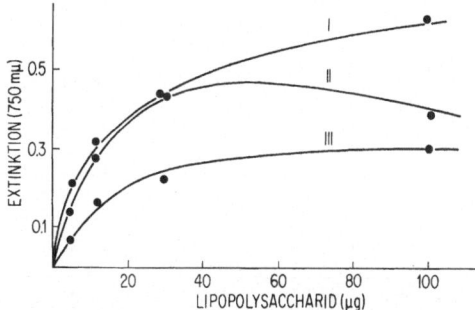

Fig. 1. *Präzipitation von R-Lipopolysacchariden durch homologes und heterologes Antiserum.* I, mR 3-Lipopolysaccharid und mR 3-Serum (30 μl). II, mR 7-Lipopolysaccharid und mR 7-Serum (15 μl). III, mR 3-Lipopolysaccharid und mR 7-Serum (15 μl)

Heptose determinanter Zucker ist und wahrscheinlich in beiden Lipopolysacchariden nicht endständig gebunden vorliegt. Die relativ geringe Hemmung der Präzipitation, die Heptose im heterologen System bewirkt, läßt vermuten, daß die beobachtete Kreuzreaktion nicht allein auf Heptose zurückgeführt werden kann.

Die Testung von KDO als Hemmsubstanz ergab bei kleinen Konzentrationen eine geringe Hemmung der Präzipitation (Tab. 3), in größeren Konzentrationen wurde jedoch mehr Protein präzipitiert als in den Kontrollversuchen ohne KDO („negative Hemmung"). Dies ist eine unspezifische Wirkung von KDO, die wir auch mit Glucuronsäure beobachtet haben. Im Gegensatz dazu wurde die homologe Präzipitation von heptoselosen, nur KDO enthaltenden Lipopolysacchariden (*S. minnesota* mR 613 und mR 595 [12]) durch KDO schon in niedrigen Konzentrationen bis zu 30% gehemmt [29b]. In diesen Lipopolysacchariden (und möglicherweise auch in

Tabelle 3. *Hemmung der Präzipitation von R-Lipopolysacchariden durch Heptose und KDO*
Um die relativ kostbaren Zucker besser auszunützen wurden die kolorimetrischen Messungen im halben Volumen (350 µl) durchgeführt

LPS	Präzipitationssystem		Hemmsubstanz	Präzipitiert. Antikörper	Präzipitations-hemmung
	Serum				
		µg/µl		µMol	µg N	%
mR3	mR3	19/10	Heptose	0	3,21	0
				0,12	2,84	13
				1,10	2,47	24
mR7	mR7	13/7	Heptose	0	1,78	0
				0,50	1,32	25
				1,70	0,89	50
mR7	rR3	15/11	Heptose	0	2,12	0
				0,55	1,92	9
				1,10	1,69	20
mR3	mR3	20/12	KDO	0	3,30	0
				0,35	2,84	14
				0,90	3,96	− 20
				1,80	3,21	3
mR7	mR7	20/12	KDO	0	3,46	0
				0,35	3,79	− 9
				0,90	5,91	− 70
				1,80	4,33	− 25
mR7	rR3	17/12	KDO	0	3,15	0
				0,32	3,03	4
				0,67	4,23	− 34

Tabelle 4. *Analyse der Oligosaccharide aus den Lipopolysacchariden von* Salmonella minnesota *R3 und R7*

Oligosaccharid aus *S. minnesota*		Papierelektrophoretische Wanderungsgeschwindigkeit (KDO = 1)	Molare Zusammensetzung der Oligosaccharide				
			Hep	KDO[a]		Phosphor	NH₂-Gruppen
				TBSR[b]	SCR[c]		
mR3	Fraktion mR3A	1,00	0	1,00		(0,1)	(0,1)
	mR3B	0,70	1,00	0,85	3,2	1,0	1,1
	mR3C	0,55	1,00	0,76	1,9	(0,07)	(0,14)
mR7	Fraktion mR7A	1,00	0	1,0		(0,1)	(0,1)
	mR7B	0,49	2,00	0,35	1,20	(0,04)	(0,02)

[a] 2-Keto-3-desoxyoctonat.
[b] Thiobarbitursäure-Reaktion.
[c] Semicarbazid-Reaktion. Es ist nicht sicher, ob mit dieser Reaktion nur KDO oder außerdem eine andere, noch nicht identifizierte Substanz erfaßt wird.
Die Werte der TBSR und SCR ohne und mit Hydrolyse waren gleich. Unter den angewandten Bedingungen (0,25 N H_2SO_4, 100°, 8 min) werden die Ketosidbindungen von KDO gespalten; bei längerer Hydrolyse wird KDO zunehmend zerstört.

mR3) scheint KDO als immunodominanter Zucker zu fungieren.

Isolierung und Analyse von Oligosacchariden aus S. minnesota R3- und R7-Lipopolysacchariden

S. minnesota *R3*. 100 mg Lipopolysaccharid mR3 wurden in 7 ml Wasser gelöst, mit etwa 0,2 ml 2 N Essigsäure auf pH 3,4 gebracht und in einer Ampulle 1 Std im siedenden Wasserbad erhitzt [4]. Nach Abkühlen wurde das ausgeflockte Lipoid A abzentrifugiert, einmal mit 1%iger Essigsäure gewaschen und im Exsikkator getrocknet (69 mg Lipoid A). Der klare Überstand und das Waschwasser wurden eingeengt und lyophilisiert, der Rückstand (30 mg) in wenig Wasser aufgenommen und der präparativen Papier-elektrophorese (3000 Volt, pH 5,3, 90 min) unterworfen. Die Anfärbung von Leitstreifen mit Silbernitrat, Perjodat/Silbernitrat und Ninhydrin zeigte die Anwesenheit von 3 Hauptfraktionen mit Wanderungsgeschwindigkeiten (bezogen auf KDO) von 1,00 = Fraktion mR3A, 0,70 = Fraktion mR3B und 0,55 = Fraktion mR3C (Tab.4). Eine Neutralfraktion, welche nur sehr schwach sichtbar war, und etwa 2% der im Lipopolysaccharid vorhandenen Gesamtheptose enthielt, wurde nicht untersucht. Die Fraktionen A, B und C wurden eluiert. Die chromatographischen und papierelektrophoretischen Eigenschaften von mR3A zeigten, daß diese Fraktion aus freier KDO besteht [ca. 30% der im Lipopolysaccharid R3 enthaltenen Menge (Thiobarbitursäure-Reaktion)]. Fraktion B und C (welche je etwa 30%

des eingesetzten KDO und je 45% der eingesetzten Heptose enthielten) wurden der wiederholten papierchromatographischen Reinigung unterworfen (Lösungsmittel B, R_{Glc} von mR 3 B : 0,25; von mR 3 C : 0,5).

In den Fraktionen mR 3 B und C wurden Heptose, Phosphor, freie NH_2-Gruppen und KDO bestimmt. Der KDO-Gehalt wurde mittels der Thiobarbitursäure- (TBSR) [17] und der Semicarbazid-Reaktion (SCR) [18] sowohl ohne wie mit Hydrolyse (0,25 N H_2SO_4, 100°, 8 min) ermittelt. Die Analysenergebnisse sind in Tabelle 4 zusammengefaßt.

mR 3 B enthält Heptose : Phosphor : NH_2-Gruppen im Verhältnis 1,0 : 1,0 : 1,1. Dazu kommen 0,8 bzw. 3,2 Mol KDO, wobei der erste Wert mit der TBSR, der letztere mit der SCR ermittelt wurde. Setzt man auch in mR 3 C die Heptose gleich 1,0, so erhält man hier 0,8 Mol (TBSR) bzw. 1,9 Mol (SCR) KDO, jedoch findet man in diesem Oligosaccharid weder Phosphor noch NH_2-Gruppen. Die in mR 3 B gefundenen NH_2-Gruppen stammen aus Äthanolamin, welches nach Totalhydrolyse des Oligosaccharids papierelektrophoretisch nachweisbar war. Der Phosphor wurde durch alkalische Phosphomonoesterase nicht freigesetzt und liegt daher wahrscheinlich in Di-esterbindung vor.

S. minnesota R 7. 100 mg Lipopolysaccharid mR 7 wurden in ähnlicher Weise wie mR 3 bei pH 3,4, 100°, hydrolysiert, jedoch nicht einmal 60 min lang, sondern dreimal je 30 min [9]. Es wurden 57 mg Lipoid A und 40 mg Oligosaccharide erhalten, welche der papierelektrophoretischen Trennung unterworfen wurden. Zwei Hauptfraktionen wurden isoliert, mR 7 A und mR 7 B. Die Fraktion mR 7 A besteht aus freiem KDO (etwa 18% des eingesetzten KDO). Die Fraktion mR 7 B wurde durch wiederholte Chromatographie auf Papier gereinigt (Lösungsmittel B: R_{Glc} = 0,125). Es fiel auf, daß mR 7 B (wie auch mR 3 C) bei wiederholter Papierchromatographie in andere Flecke überging, welche eine vergleichsweise schwächere TBSR gaben, wodurch ein immer größeres Verhältnis KDO : Heptose erhalten wurde. Das KDO : Heptose-Verhältnis im rohen mR 7 B betrug etwa 1 : 4 (TBSR), im gereinigten mR 7 B etwa 1 : 5,8. Bestimmten wir jedoch KDO mit der Semicarbazid-Reaktion, so fanden wir ein Verhältnis KDO : Heptose von 1 : 1,7 (Tab. 4).

Bei den KDO-Bestimmungen (TBSR und SCR) aller 3 Oligosaccharid-Fraktionen (mR 3 B, mR 3 C, mR 7 B) wurden mit und ohne Hydrolyse stets praktisch gleiche Werte erhalten. Im Falle der Semicarbazid-Reaktion sind wir allerdings nicht sicher, ob wir nur KDO bestimmten oder außerdem noch eine andere, bisher unbekannte Substanz erfaßten.

Methylierung und Gaschromatographie

Je 200 mg der Lipopolysaccharide von S. minnesota R3 und R7 wurden bei pH 3,4 hydrolysiert und

die nach Papierelektrophorese erhaltenen Oligosaccharidfraktionen (mR 3 B, mR 3 C, mR 7 B) nach 60 stündigem Trocknen über P_2O_5 in Anlehnung an das Verfahren von Hakomori [25] methyliert (siehe unter Methoden). Die öligen Methylierungsprodukte wurden anschließend methanolysiert und ergaben nach Isolierung folgende Ausbeuten: 32 mg aus mR 3 B, 32 mg aus mR 3 C, 47 mg aus mR 7 B. Die Präparate wurden in je 0,5 ml Chloroform gelöst und in Mengen von 30—90 μg (0,5—1,0 μl) gaschromatographisch untersucht. Vergleichsweise wurden authentische D-Glycero-L-manno-heptose und elektrophoretisch isolierte bakterielle KDO methyliert.

Fig. 2 zeigt die Gaschromatogramme der verschiedenen methylierten Präparate. Die mehr oder weniger stark ausgeprägten Peaks, welche auf dem Gaschromatogramm von methylierter KDO (Fig. 2 B) bei relativ kurzen Retentionszeiten (2,5; 3,2; 3,4 min) auftreten, sind auf Verunreinigungen zurückzuführen. Ausgeprägte Peaks werden mit KDO auch in hoher Konzentration nicht erhalten (Fig. 2 B). Das Gaschromatogramm von methylierter Heptose (Fig. 2 A) besitzt 2 Peaks mit den Retentionszeiten 7,4 und 9,5 min. Vermutlich entspricht der 9,5 min Peak einem Furanosid, da er bei Hydrolyse in 0,5 N HCl 5—10 mal rascher verschwindet als der 7,5 min Peak, der demnach einem Pyranosid entsprechen würde. Die Methanolysate der 3 methylierten Oligosaccharide zeigen im Gaschromatogramm die Peaks bei kurzen Retentionszeiten, wie sie auch auf dem Gaschromatogramm von methylierter KDO zu sehen sind (Fig. 2). Überraschend ist der stark ausgeprägte Peak, der auf dem Gaschromatogramm von mR 3 C (Fig. 2 D) nach 3,4 min auftritt. Peaks mit gleicher Retentionszeit, aber wesentlich schwächer ausgebildet, sind auch auf den Chromatogrammen der anderen Oligosaccharide und auf dem von methylierter KDO zu sehen. Sie können vorläufig noch nicht zugeordnet werden.

Neben den auch bei KDO beobachteten Peaks erscheint auf den Gaschromatogrammen der Oligosaccharide der Peak mit der Retentionszeit 7,4 min, welcher identisch ist mit dem von methylierter authentischer D-Glycero-L-manno-heptose. Dagegen fehlt der zweite Heptosepeak (9,5 min).

Das methylierte und methanolysierte Oligosaccharid mR 7 B zeigt außer dem Peak der permethylierten (endständigen) Heptose einen Peak mit der Retentionszeit 15,0 min (Fig. 2 E), welcher offensichtlich partiell methylierter Heptose zuzuschreiben ist, da nach wiederholter Hakomori-Methylierung des methylierten und methanolysierten mR 7 B-Oligosaccharids nur noch der Peak bei 7,4 min auftritt. Das Verhältnis der Flächen unter den beiden Peaks beträgt etwa 1 : 1, was zu den analytischen Ergebnissen paßt, wonach mR 7 B aus KDO und 2 Heptosen besteht (Semicarbazid-Methode).

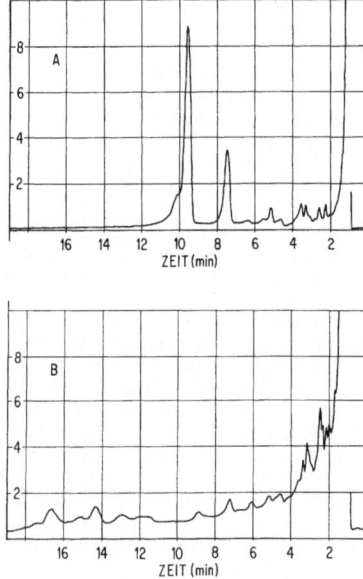

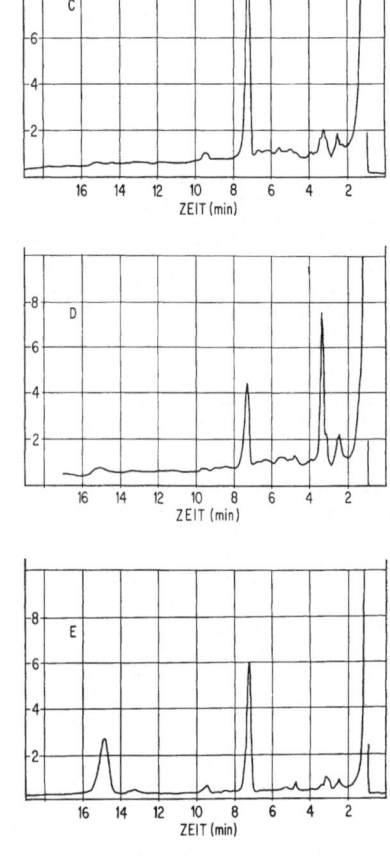

Fig. 2. *Gaschromatogramme methylierter Zucker.* Testzucker: A, Permethylierte D-Glycero-L-mannoheptose (20 µg), B, Permethylierte 3-Desoxyoctulonsäure (150 µg). Methanolysate der methylierten Oligosaccharide: C, mR3B (60 µg), D, mR3C (60 µg), E, mR7B (40 µg). Gaschromatogramme von Gemischen permethylierter authentischer Heptose mit den methanolisierten Oligosacchariden zeigten die Identität der 7,4 min-Peaks. Daß es sich bei dem 15,0 min-Peak (Fig. 2E) um partiell methylierte Heptose handelt, konnte durch Nachmethylierung bewiesen werden; das Gaschromatogramm zeigte dann nur einen einheitlichen Peak bei 7,4 min

Im Gegensatz zum mR7B-Oligosaccharid zeigen die beiden methylierten R3-Oligosaccharide keinen Peak von partiell methylierter Heptose. Die gesamte Heptose in diesen Fraktionen war permethylierbar und liegt daher in den Oligosacchariden mR3B und mR3C endständig vor.

Die Biochemie der hexoselosen Mutanten

Die Mutanten *S. minnesota* R3, R7, Rz sowie *S. ruiru* R3 und R541 enthalten neben Lipoid-Glucosamin nur Heptose und KDO als einzige Zucker in ihren Lipopolysacchariden, jedoch keine oder nur geringe Mengen von Hexosen. Nach den struktur-analytischen und biosynthetischen Befunden [1,5] ist Glucose der erste Zucker, welcher an die Heptose-einheiten des innersten Bezirks der Lipopolysaccharide geknüpft ist. Es schien daher plausibel, daß der Defekt dieser Mutanten entweder die Synthese von UDP-Glucose oder den Transfer von Glucose betreffen würde.

Wie wir in einer folgenden Arbeit zeigen [30a] besitzt die Mutante mRz einen Defekt in der Synthese von UDP-Glucose: Sie enthält, verglichen mit der Wildform, nur etwa 20% UDP-Glucose Synthetase-Aktivität. Um die Synthesedefekte der anderen Mutanten, deren UDP-Glucose-Synthese intakt ist, zu identifizieren, haben wir den Glucose-Einbau in die Lipopolysaccharide dieser Mutanten untersucht und sowohl die Glucosyltransferase-Aktivität in diesen Mutanten sowie die Fähigkeit der Rd Lipopolysaccharide, als Glucose-Acceptor zu fungieren, bestimmt.

Der Einbau von Glucose in hexoselose Lipopoly-saccharide. Zwei grundsätzliche Fragen stellten sich bei der Planung der Glucose-Inkorporationsversuche: Die meisten der von uns für den Einbau benutzten Rohenzyme enthalten UDP-Galactose-Epimerase, welche UDP-Glucose in UDP-Galactose überführt, wodurch ein Teil der Glucose für den Einbau verloren geht. Auch kann es zum nachträglichen Einbau von Galactose kommen, wodurch die Auswertung der Versuche unübersichtlich wird. Wir hatten gefunden, daß NADH bakterielle UDP-Galactose-Epimerase hemmt (*cf.* auch [30b]). Daher setzten wir unseren Inkubationsgemischen stets NADH zu, es sei denn, das Transferenzym stammte aus einer UDP-Galactose-Epimerase-losen Mutante. Die zweite Frage war, ob es möglich ist, Kreuzexperimente durchzuführen, bei welchen die Transferase aus einer und die Acceptorpartikel aus einer anderen Species stammten. Zur Standardisierung der Ansätze haben wir stets die Enzymmengen auf mg Protein und die Partikelfraktion (Acceptor) auf nMol „KDO" bezogen, was ein Maß für den Lipopolysaccharidgehalt der Partikel darstellt.

Das Ergebnis eines typischen Einbauversuchs zeigt Tabelle 5. Glucose wurde hier in die Partikelfraktion der Mutante *S. ruiru* R 1 mit Hilfe von löslichem Enzym [29a] aus der Wildform in Gegenwart und in Abwesenheit von NADH eingebaut. In einem Parallelversuch wurde das Enzym aus der epimeraselosen Mutante von *S. typhimurium* benutzt. Man sieht, daß mit dem Glattform-Enzym in Abwesenheit von NADH nur 2,8 nMol Glucose pro 100 nMol Partikel-„KDO" eingebaut wird, während dieser Wert in Gegenwart von NADH auf 12,8 ansteigt und damit dem Glucoseeinbau sehr ähnlich ist, der mit *S. typhimurium* erhalten wird. Dieser Versuch zeigt auch, daß man einen Glucose-Einbau erhält, wenn Acceptor und Enzym aus verschiedenen Species stammen.

In einem anderen Versuch haben wir den Glucose-Einbau in Abhängigkeit von der Partikelmenge, der Enzymmenge und der Zeit bestimmt. Hier wurde die Partikelfraktion aus *S. minnesota* R7 und lösliches Enzym aus *S. typhimurium* M Mutante benutzt (Fig. 3 A—C).

Um zu beweisen, daß [14C]Glucose in die Lipopolysaccharidfraktion der Partikel eingebaut wird, haben wir einen größeren Inkorporationsansatz präparativ aufgearbeitet und das Lipopolysaccharid isoliert. Die Inkubationsmischung (Partikel *S. ruiru* R 1[2], Enzym aus *S. typhimurium* M und [14C]UDP-Glucose (10⁶ Imp./min)) wurde wie unter Methoden beschrieben mit Trichloressigsäure gefällt, und das Präzipitat

[2] Wir haben die Mutante *S. ruiru* R1 für diesen Versuch gewählt, weil ihr Lipopolysaccharid etwas Glucose enthält. Dadurch erhielten wir später in den isolierten Oligosacchariden genügend Glucose und Heptose, um die Zucker mit Silbernitrat anzufärben.

Tabelle 5. *Einfluß von NADH auf den Einbau von Glucose bei Verwendung von UDPGal-4-Epimerase-haltigen Rohenzymen*

Partikel	Enzym aus	Zusatz	Glucoseeinbau
			nMol [14C]Glucose/ 100 nMol „KDO" × Std
S. ruiru R 1	*S. typhimurium* M-Mutante	—	8,5
S. ruiru	*S. ruiru* S-Form	—	2,8
S. ruiru	*S. ruiru* S-Form	10⁻³ M NADH	12,8

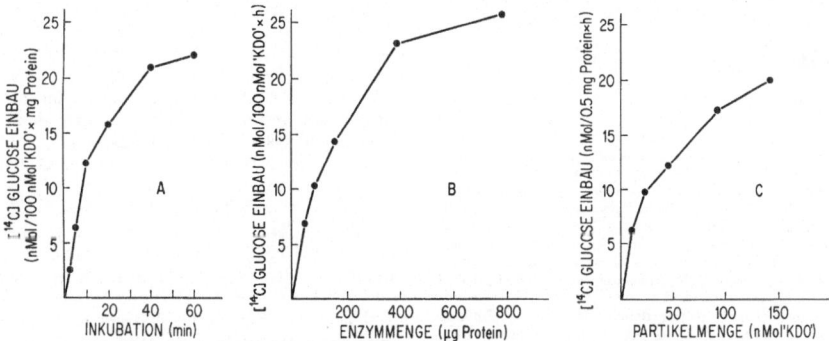

Fig. 3. *Einbau von [14C]Glucose in 20000 × g Partikel aus* Salmonella minnesota *R7 mit löslichem Enzym aus* Salmonella typhimurium *M-Mutante.* (A) Abhängigkeit von der Inkubationszeit. Der Ansatz enthielt: 360 µMol Tris/HCl, pH 7,5; 83 µMol Mg⁺⁺; 600 nMol [14C]UDP-Glucose (630 Imp/min/nMol); R7 Partikel (266 nMol „KDO"); Enzym (5 mg Protein); Gesamtvolumen 2,5 ml. Inkubation bei 37°. Nach verschiedenen Zeiten wurden Proben von 300 µl entnommen und durch Zugabe von 300 µl 10⁰/₀ Trichloressigsäure-Lösung gestoppt. (B) Abhängigkeit von der Enzymmenge. Die Ansätze enthielten: 31 µMol Tris/HCl, pH 7,5; 7,5 µMol Mg⁺⁺; 73 nMol [14C]UDP-Glucose (630 Imp./min/nMol); R7 Partikel (38 nMol „KDO"); Enzymmenge wie angegeben; Gesamtvolumen 300 µl. Inkubation 60 min bei 37°. (C) Abhängigkeit von der Partikelmenge. Die Ansätze enthielten: 46 µMol Tris/HCl, pH 7,5; 10,7 µMol Mg⁺⁺; 58 nMol [14C]-UDP-Glucose (1100 Imp./min/nMol); Enzym (0,48 mg Protein); Partikelmenge wie angegeben; Gesamtvolumen 292 µl. Inkubation 60 min bei 37°

(152000 Imp./min) nach Zusatz von 3 g feuchten *S. ruiru* R1-Bakterien als Träger mit Phenol/Wasser extrahiert und durch Ultrazentrifugation gereinigt. Das erhaltene Lipopolysaccharid besaß eine Radioaktivität von 106000 Imp./min, also 70% der Radioaktivität des TCE-Extrakts.

Wir haben dieses Lipopolysaccharid der Partialhydrolyse (1 N H_2SO_4, 30 min, 100°) unterworfen, um Oligosaccharide mit radioaktiver Glucose zu isolieren. Das Hydrolysat wurde papierchromatographisch getrennt (Laufmittel A). Die radioaktive Bande (GH auf Fig.4A, R_{Lac} = 0,38) wurde eluiert und in 1 N HCl 4 Std bei 100° hydrolysiert. Papierchromatographie (Laufmittel D) und Anfärbung mit Silbernitrat/NaOH ergab 2 Flecke, deren Laufgeschwindigkeiten identisch waren mit denen authentischer Glucose und D-Glycero-L-manno-heptose (Fig.4B).

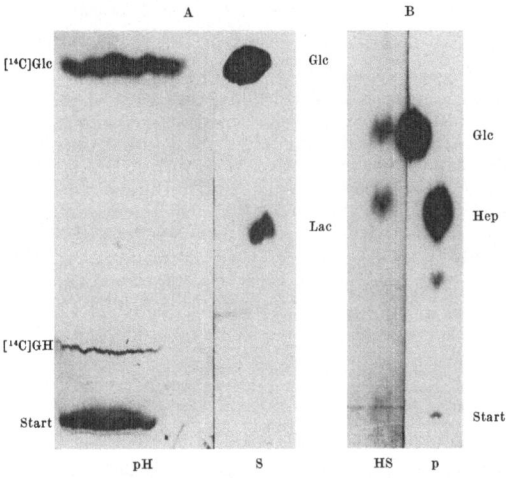

Fig.4. *Isolierung und Identifizierung von Glucose-Heptose-Oligosaccharid.* (A) Papierchromatographie des Partialhydrolysats (PH) von Lipopolysaccharid *S. ruiru* R1-[^{14}C]Glucose. Laufmittel (A). Links: Autoradiogramm, rechts: Anfärbung mit Silbernitrat/NaOH. (B) Papierchromatographie des Totalhydrolysats (TH) des radioaktiven Oligosaccharids GH. Laufmittel (D). Anfärbung mit Silbernitrat/NaOH. Die linke Seite des Chromatogramms wurde bei der photographischen Vergrößerung länger belichtet und erscheint daher dunkler als die rechte Seite. GH = radioaktives Oligosaccharid; Standardzucker (S): Glc = Glucose, Lac = Lactose, Hep = D-Glycero-L-mannoheptose

Die Ergebnisse dieser Versuche zeigen, daß Glucose aus UDP-Glucose an das Lipopolysaccharid der hexoselosen Mutante R7 gebunden wird, und aus diesem als ein Oligosaccharid aus Glucose und Heptose isoliert werden kann (siehe auch [31]).

In einem umfangreichen Versuch, bei dem alle Einzeltests unter identischen Bedingungen ausgeführt wurden, haben wir schließlich die Inkorporation von Glucose in die hexoselosen Lipopolysaccharide untersucht und dazu in homologen und heterologen Systemen die Transferase-Enzyme und die Acceptorpartikel aller Mutanten in allen Kombinationen getestet. Tabelle 6 zeigt das Ergebnis dieses Versuches. Wie zu erwarten, besitzt die UDP-Glucose-defekte Mutante mRz als einzige sowohl aktives Transfer-Enzym wie auch aktiven Rezeptor. Die übrigen Mutanten der Gruppe Rd₁ sind acceptor-aktiv, besitzen aber keine Transfer-Aktivität. Die Vertreter der Gruppe Rd₂ verhalten sich umgekehrt: Ihre Partikelfraktion ist inaktiv und inkorporiert keine Glucose, aber sie synthetisieren aktives Transfer-Enzym.

Den geringen Einbau von Glucose in rR541-Partikel mit dem Transferase-Enzym aus mR7 kann man wahrscheinlich auf die Anwesenheit von kleinen Mengen Glucose und Galactose in rR541-Lipopolysaccharid zurückführen. Wahrscheinlich wird in diesem Versuch die Glucosyl-II-Transferase-Aktivität gemessen (siehe Diskussion).

DISKUSSION

Die Lipopolysaccharide der *Salmonella* O-Antigene bestehen (siehe z.B. [1,5,32]) aus einem peripheren Bezirk, der die langen, O-spezifischen Seitenketten aus sich wiederholenden Oligosacchariden enthält, und einem zentralen Bezirk, dessen Zuckerzusammensetzung und Struktur vielen oder allen *Salmonella*-Species gemeinsam ist. *Salmonella* R-Mutanten, deren Defekt die Synthese des spezifischen Bezirks betrifft, synthetisieren das intakte, basale Kernpolysaccharid. R-Mutanten, deren Defekt die Synthese des basalen Kernpolysaccharids betrifft, synthetisieren ein mehr oder minder inkomplettes Kernpolysaccharid. Die Isolierung und das Studium zahlreicher R-Mutanten mit einem oder mehreren Blocks auf verschiedenen Stufen der Biosynthese ihres Lipopolysaccharids hat uns einen Einblick in die Biochemie, Biosynthese und Genetik von *Salmonella* O-Antigenen vermittelt, und zu einer provisorischen Vorstellung der chemischen Struktur von *Salmonella* R-Lipopolysacchariden geführt (siehe Fig.5A [1,5,12,33]). Indessen ist bislang wenig bekannt über den innersten Teil des Kernpolysacchrids, welcher Lipoid A, KDO und Heptose enthält.

Unser Arbeitskreis beschäftigt sich seit einigen Jahren mit der Analyse von R-Mutanten aus *S. minnesota* und *S. ruiru* [12]. R-Mutanten, welche die bislang zuckerärmsten Lipopolysaccharide synthetisieren, sind im Chemotyp Re zusammengefaßt und enthalten neben dem gemeinsamen Lipoid A lediglich KDO; ihm folgen Mutanten des Chemotyps Rd, deren Lipopolysaccharide KDO und Heptose enthalten, des Chemotyps Rc mit zusätzlicher Glucose

Tabelle 6. *Einbau von Glucose in die 20000×g Partikelfraktion aus Mutanten von S. minnesota und S. ruiru*
Die Werte der homologen Reaktionen sind fett gedruckt

Enzyme aus	Partikelfraktionen aus				
	m[a] R7	r[b] 541	m[a] Rz1	m[a] R3	r[b] R3
	nMol eingebaute [¹⁴C]Glucose/100 nMol „KDO" ×Std				
S. typhimurium M	12,9	13,9	8,05	0,59	0,20
S. minnesota R7	**0,02**	1,7	0,44	0,03	0
S. ruiru R541	0,30	**0,33**	0,26	0,13	0,05
S. minnesota Rz	15,3	13,6	**7,64**	0,36	0,14
S. minnesota R3	8,8	10,0	5,9	**0**	0
S. ruiru R3	14,5	11,3	6,52	0,01	**0,01**

[a] *S. minnesota.* — [b] *S. ruiru.*

Tabelle 7. *Serologische Kreuzreaktionen von Lipopolysacchariden, die zu verschiedenen R-Chemotypen gehören* [12]

Salmonella Lipopolysaccharid		Zuckerbausteine					Serologische Kreuzreaktionen[a] mit den Antiseren gegen	
Chemotyp	Mutante	KDO	Hep[b]	Glc	Gal	GlcN	GalN	
			%					
II	mR2	+	17	(+)	(+)	(+)	(+)	S-Form,	mR7
Ra	mR555	+	11	(+)	(+)	(+)	−	Ra(RII),	mR7
Rb	rR541	+	17	(+)	(+)	−	−	(Rb),	mR7
Rc	rR1	+	15	(+)	−	−	−	Rc,	mR7
Re	rR128	+	(?)					Re,	mR7

[a] Die Kreuzreaktionen wurden im Hämagglutinationshemmungstest ermittelt.
[b] % Heptose im Lipopolysaccharid.
(+): Diese Zucker wurden in kleinen Mengen nachgewiesen.
(?): Heptose wurde bislang nicht nachgewiesen, aber auf Grund der serologischen Kreuzreaktion mit mR7 ist zu erwarten, daß das Lipopolysaccharid eine kleine Menge enthält.

sowie Rb mit zusätzlicher Galactose und schließlich die Klasse der Ra-Mutanten, deren Lipopolysaccharide alle Zucker enthalten, die den basalen Kern aufbauen (vgl. Fig.5A). Darüber hinaus wurde eine weitere Klasse von Mutanten identifiziert, welche alle Zucker der Wildform im Lipopolysaccharid enthält, jedoch in geringerer Menge (sogenannte SR-Formen [34]). Aus *S. typhimurium* [5,35a] und kürzlich aus *S. godesberg* [35b] sind analoge Serien von Mutanten isoliert worden.

Die vorliegende Arbeit galt der chemischen, serologischen und biochemischen Analyse einer Reihe von *S. minnesota* und *S. ruiru* Mutanten, welche ihrer Zuckerzusammensetzung nach als Chemotyp Rd klassifiziert worden sind und deren Lipopolysaccharide Lipoid A, KDO und Heptose (L-Glycero-D-manno-heptose) enthalten [12].

Die beiden Mutanten rR1 und rR541 [12] gehören wegen des Hexosegehalts ihrer Lipopolysaccharide nicht zum Chemotyp Rd (= hexoselose Mutanten); sie zeigen aber eine starke serologische Kreuzreaktion mit dieser Gruppe. Sie verhalten sich damit ähnlich wie eine Reihe anderer R-Mutanten, die nicht zum Chemotyp Rd gehören, aber mit mR7 kreuzreagieren. Einige Beispiele solcher kreuzreagierender Mutanten verschiedenen Chemotyps sind in Tabelle 7 zusammengestellt. Die beiden Mutanten rR1 und rR541 haben wir in einige der Versuche eingeschlossen als Repräsentanten einer Gruppe, die mit dem Chemotyp Rd serologisch kreuzreagiert, ihm aber nicht angehört.

Auf Grund des Heptosegehalts ihrer Lipopolysaccharide und ihrer serologischen Spezifität haben wir bereits früher vermutet [12], daß die Gruppe Rd in 2 Untergruppen eingeteilt werden kann (Rd₁ und Rd₂), die sich strukturell unterscheiden. Um weiteren Aufschluß über Strukturunterschiede von Rd₁- und Rd₂-Lipopolysacchariden zu erhalten, haben wir je einen Vertreter, mR3 und mR7, der Partialhydrolyse unterworfen.

Die Strukturanalyse der aus den Lipopolysacchariden mR3 und mR7 erhaltenen sauren Oligosaccharide erwies sich als schwierig und ist nicht abgeschlossen. Während wir bei der Bestimmung von Heptose, Phosphor und NH₂-Gruppen zuverlässige Werte erhielten, ergab die KDO-Bestimmung verschiedene Ergebnisse je nach der angewandten Analysenmethode. Dabei lieferte die Semicarbazid-Reaktion (SCR) jeweils ein oder zwei Mole KDO mehr pro Molekül als die Thiobarbitursäure-Reaktion (TBSR). Da für einen positiven Ausfall der TBSR die Ketogruppe und die OH-Gruppen an C-4 und C-5 von KDO frei sein müssen, während für den positiven Ausfall der SCR nur die Ketogruppe zur Verfügung stehen muß, erscheint die Vermutung plausibel, daß in den Oligosacchariden KDO-Einheiten verschiedener Substitution vorliegen, welche entsprechend der Bestimmungsmethode unterschiedlich reagieren. Insbesondere erscheint es möglich, daß die erste, an Heptose gebundene KDO-Einheit den Heptosylrest an C-4 oder C-5 trägt und deshalb die TBSR nicht gibt. Daß dieses KDO auch nach Hydrolyse im TBS-Test nicht gefunden wird, liegt daran, daß die Heptosylbindung sehr säurestabil ist. Erst in der Struktur Heptosyl-KDO-KDO ist der zweite KDO-Rest mit der TBSR erfaßbar. Ein Absinken der TBSR bei wiederholter Chromatographie von mR 3 C und mR 7 B wäre so durch ein Abspalten dieses zweiten KDO-Restes zu erklären. Über die Art der KDO-Bindungen wissen wir noch nichts. Es muß auch betont werden, daß nicht sicher bewiesen ist, daß es sich bei den SC-positiven und TBS-negativen Substanzen ausschließlich um KDO handelt. Der Befund, daß die KDO-Bestimmung (TBSR) bei heptoselosen Lipopolysacchariden (Gruppe Re) etwa 16% und bei hexoselosen Lipopolysacchariden nur 10% KDO liefert, spricht jedoch dafür, daß der der Heptose folgende Zucker KDO ist.

Legt man die mit Semicarbazid erhaltenen Ergebnisse zugrunde, so handelt es sich bei mR 3 C um ein Trisaccharid, welches eine Heptose und zwei KDO-Reste enthält. Ein KDO-Rest ist vermutlich reduzierend und reagiert mit Thiobarbitursäure, mit und ohne Hydrolyse. Der andere KDO-Rest reagiert nur mit Semicarbazid.

Das zweite aus mR 3 isolierte Oligosaccharid mR3 B enthält Heptose, KDO, Phosphat und Äthanolamin im Verhältnis 1:3:1:1. Nur ein KDO-Rest reagiert mit TBS.

Das aus mR 7-Lipopolysaccharid isolierte Oligosaccharid mR 7 B ist ein Trisaccharid, welches aus 2 Molen Heptose und 1 Mol KDO besteht. Der KDO-Rest reagiert nicht in der TBSR und ist daher möglicherweise an C-4 oder C-5 mit dem einen der Heptosylreste substituiert.

Während der Status der KDO-Reste in den 3 sauren Oligosacchariden noch nicht befriedigend aufgeklärt werden konnte, erbrachten die Ergebnisse der Methylierung Aufschluß über die Stellung der Heptose. Nach Methylierung und Methanolyse von mR 3 B und mR 3 C erscheint Heptose als gaschromatographisch einheitlicher permethylierter Zucker. Permethylierte authentische Heptose gibt bei der Gaschromatographie 2 Peaks, denen wahrscheinlich die Pyranosid- bzw. Furanosid-Konfiguration zugrunde liegt. Die gesamte Heptose in den Oligosacchariden aus mR 3 liegt terminal und wahrscheinlich pyranosid vor.

Im Gegensatz hierzu führt die Methylierung von mR 7 B und die anschließende Methanolyse und Gaschromatographie zum Auftreten zweier Peaks, deren Flächen etwa gleich groß waren. Der eine Peak liegt bei der gleichen Zeit wie derjenige aus mR 3 und wie der raschere Peak von permethylierter Heptose. Der andere Peak aus mR 7 B ist wesentlich langsamer und stellt die zu erwartende partiell methylierte Heptose dar. Das Oligosaccharid mR 7 B ist daher wahrscheinlich ein Heptosyl-Heptosyl-KDO-Trisaccharid.

Osborn [9] hat erstmals KDO- und heptose-haltige Oligosaccharide aus der M-Form von *S. typhimurium* isoliert, welche außerdem Glucose enthielten. Später haben Kuriki und Kurahashi [35] KDO- und heptosehaltige Oligosaccharide aus einer hexoselosen R-Mutanten von *E. coli* erhalten. In beiden Arbeiten wurde das Verhältnis KDO:Heptose als etwa 1:10 angegeben. Kürzlich haben Cherniak und Osborn [37] diese Werte korrigiert. Unter Verwendung von Semicarbazid bei der Bestimmung von KDO fanden sie ein Verhältnis für KDO:Heptose von 1:2.

Biosynthetische Untersuchungen *in vitro* über den Einbau von Hexosen in defekte Lipopolysaccharide von *S. typhimurium*-R-Mutanten sind erstmalig von Osborn et al. [38] durchgeführt worden. In diesen Versuchen wurde eine aus der Mutanten isolierte Partikelfraktion benutzt, welche gleichzeitig den Lipopolysaccharid-Acceptor und das Transferase-Enzym enthielt; Acceptor und Enzym-Aktivität waren hier nicht getrennt. Die Partikel wurden mit [14C]UDP-Hexose inkubiert und der Einbau von [14C]Hexose in die trichloressigsäure-fällbare Fraktion (Lipopolysaccharid) gemessen. Rothfield et al. [29] fanden später, daß die Glycosyl-I-Transferase in löslicher Form erhalten werden kann, ein Befund, der Voraussetzung ist für Experimente, bei denen Acceptor und Enzym aus verschiedenen Stämmen benutzt werden sollen (heterologe Systeme). In unseren Versuchen, welche teilweise an Mutanten durchgeführt wurden, bei denen die Glucosyl-Transferase nicht synthetisiert wird, waren wir auf gekreuzte Transferversuche angewiesen; allerdings war nicht mit Sicherheit voraussehbar, ob derartige Kreuzversuche auch mit verschiedenen Species gelingen würden. Wenn jedoch, was wir vermuten, die basalen Kernpolysaccharide aller *Salmonella*-Bakterien strukturell identisch sind, wofür es viele Hinweise gibt, dann sollte die Biosynthese des Kerns bei allen Species die gleichen Zwischenstufen durchlaufen und die Transferasen sollten gleich sein. Daß der Glucose-Transfer auf hexoselose Lipopolysaccharide einer Species (z.B. *S. minnesota*) mit den Enzymen anderer Species (*S. typhimurium*, *S. ruiru*) in der Tat durchgeführt werden kann, beweist einmal mehr die Identität der Kernpolysaccharide dieser *Salmonella* Species.

Um zu beweisen, daß [14C]Glucose in das Lipopolysaccharid der Partikelfraktion eingebaut und an Heptose gebunden wird, haben wir nach Transfer von Glucose radioaktives Lipopolysaccharid aus dem Acceptorpartikeln isoliert, welches 70% der eingesetzten [14C]Glucose enthielt. Nach Partialhydrolyse erhielten wir ein Oligosaccharid, welches aus radioaktiver Glucose und Heptose bestand. Glucosyl-Heptose ist bereits früher von Nikaido [31] aus dem Lipopolysaccharid einer *Salmonella*-M-Mutante isoliert worden. Diese Ergebnisse zeigen, daß Glucose an die Heptose des zentralen Kerns gebunden wird.

Den Einbau von Glucose in die Lipopolysaccharide unserer R-Mutanten haben wir in homologen und heterologen Systemen (bezüglich Transferase und Acceptor) durchgeführt. Die Ergebnisse dieser Versuche ermöglichen uns einen Einblick in die Biosynthese der Lipopolysaccharide und bei einigen Mutanten eine unmittelbare Aussage über die Lokalisation ihres enzymatischen Defekts. Im Zusammenhang mit den Ergebnissen der chemischen und serologischen Untersuchungen können wir auch bei den übrigen Mutanten den Defekt definieren, der zur Ausbildung der entsprechenden Lipopolysaccharide führt.

nisse zeigen, daß der Defekt dieser Mutanten die Glucosyl-I-Transferase betrifft. Das von dieser Mutanten synthetisierte Lipopolysaccharid besitzt die Teilstruktur der Fig. 5 B.

Die Mutante rR 541 unterscheidet sich nur in wenigen Eigenschaften von mR 7 (Tab. 8). Ihr Lipopolysaccharid enthält außer KDO und Heptose geringe Mengen von Glucose und Galactose, und es zeigt serologische Kreuzreaktion mit Mutanten des Chemotyps Rb (Tab. 7). Wir nehmen an, daß in rR 541 einige Heptose-Endgruppen mit Glucose und Galactose besetzt sind (Fig. 5 C). R 541 besitzt, wie wir in den beschriebenen Transfer-Versuchen zeigen konnten, einen Block bezüglich der Synthese der Glucosyl-I-Transferase. Vermutlich ist dieser Defekt aber nicht vollständig, und einige Glucoseeinheiten werden transferiert und an Heptose gebunden, worauf nachfolgend auch Galactose übertragen wird. In rR 541 haben

Tabelle 8. *Chemische, serologische und biochemische Charakterisierung der untersuchten* Salmonella *R-Mutanten*

R-Mutante	Heptosegehalt[a]	Serologische Spezifität[a]	Immunodominanter[a] Zucker	Identifizierte Teilstruktur[a]	Glucosyl-Transferase-Aktivität	Acceptor Aktivität für Glucose	Defekt betreffend
	%						
mR 7	10—13	Rd₁	Heptose	Hep → Hep → KDO	—	+	Glycosyl-I-Transferase
rR 541	13—17	(Rb), Rd₁	ng[c]	ng[c]	—	+	Glucosyl-I-Transferase[b]
mRz	10	Rd₁	ng[c]	ng[c]	+	+	UDP-Glucose-Synthetase
mR 3	5—7	Rd₂	Heptose	Hep → KDO →	+	—	Synthese des Heptosekerns
rR 3	5—6	Rd₂	ng[c]	ng[c]	+	—	Synthese des Heptosekerns

[a] Im Lipopolysaccharid.
[b] Die Mutante rR 541, deren Lipopolysaccharid neben KDO und Heptose kleine Mengen an Hexosen enthält, ist eine Mutante mit mehreren Defekten.
[c] Nicht getestet.

Osborn [39] hat kürzlich über analoge Versuche an hexoselosen Mutanten von *S. typhimurium* berichtet. Eine der von dieser Autorin untersuchten Mutanten besaß einen Defekt in der UDP-Glucose-Synthese (Abwesenheit von Phosphoglucose-Isomerase). Bei den übrigen Mutanten war die UDP-Glucose-Synthese intakt. Die Vermutung, daß der Defekt die Glucosyl-I-Transferase betraf, konnte durch Testung der Transferase- und Acceptoraktivität dieser Mutanten in Kreuzexperimenten bewiesen werden. Zwei der Mutanten waren Doppelmutanten und besaßen einen zweiten Defekt, welcher die Synthese der spezifischen Seitenkette des kompletten Lipopolysaccharids der Wildform betraf.

Die wesentlichen Ergebnisse der in dieser Arbeit beschriebenen Versuche sind in Tabelle 8 zusammengefaßt. Für die Mutante mR 7 ergeben sich folgende Merkmale: Hoher Heptosegehalt des Lipopolysaccharids, welches die serologische Spezifität Rd₁ aufweist und die Struktur Hep → KDO enthält. Das Lipopolysaccharid vermag als Glucose-Acceptor zu fungieren; in der Zelle von mR 7 findet sich jedoch keine Transferase-Aktivität. Diese Ergeb-

wir einen zweiten Defekt lokalisieren können [30 a]; er betrifft das Enzym UDP-N-Acetylglucosamin-4-Epimerase, welches für die Synthese von UDP-N-Acetylgalactosamin und damit zum Aufbau der spezifischen Seitenkette der Wildform (S-Form) benötigt wird. Dieses Enzym fehlt in rR 541. Eine Mutante mit den beiden beschriebenen Defekten würde wenige, aber vollständige Ketten der Basalstruktur aufbauen und mit R II(Ra)-Mutanten serologisch kreuzreagieren. Dies ist jedoch nicht der Fall bei rR 541. Wir müssen daher noch einen weiteren Defekt postulieren, der wahrscheinlich die Verlängerung der vereinzelten Ketten über Galactose hinaus verhindert und somit die Synthese der Glucosyl-II-Transferase betrifft. Es erscheint plausibel, daß die übrigen Mutanten der Fig. 7 analoge Mutanten mit mehr als einem Block darstellen.

Beim Vergleich der Tabellen 6 und 8 fällt auf, daß die Inkubation von mR 7-Enzym mit rR 541-Partikeln zu einem Glucose-Einbau führt, der primär nicht erwartet wird. Wir glauben, daß hier die Glucosyl-II-Transferase aus mR 7 Glucose (1,7 nMol) an die Galactose des rR 541-Lipopolysaccharids transferiert.

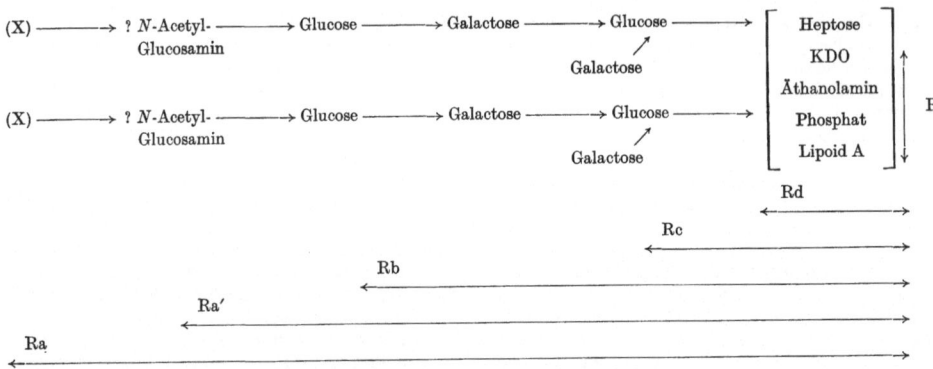

A. *Struktur des basalen Kernpolysaccharids (Chemotyp Ra) eines Salmonella-Lipopolysaccharids, von welchem sich die Strukturen der anderen R-Chemotypen Rb—Re ableiten* [1,5,6,12]

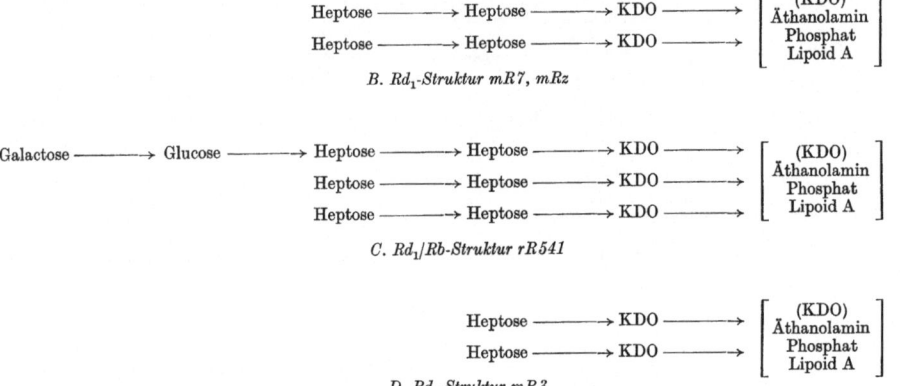

Fig. 5. *Strukturvorschläge für* Salmonella *R-Lipopolysaccharide*

Die Mutante mRz synthetisiert im Gegensatz zu der Mutanten mR7 sowohl receptor-aktives Lipopolysaccharid wie auch aktive Glucosyl-I-Transferase. Ihr Defekt betrifft die Synthese von UDP-Glucose [30a] was phänotypisch zur Bildung eines Lipopolysaccharids führt, welches die gleiche Struktur besitzt wie dasjenige aus mR7.

Die Mutanten mR3, mR4 und rR3 stellen analoge Mutanten aus *S. minnesota* und *S. ruiru* dar, die chemisch gleiche Lipopolysaccharide synthetisieren. Diese enthalten weniger Heptose als die vorher besprochenen Lipopolysaccharide, und sie besitzen die Spezifität Rd_2. Die Oligosaccharide aus mR3, welche das Disaccharid Heptosyl → KDO ... enthalten, sind vermutlich auch Bausteine der Lipopolysaccharide von mR4 und rR3 (Fig. 5D). mR3 und rR3

können nicht als Acceptor für den Einbau von Glucose fungieren. Dagegen enthalten diese Mutanten aktive Glucosyl-I-Transferase. Aus diesen Befunden schließen wir, daß die Mutanten der Gruppe Rd_2 einen Defekt besitzen, welcher einen früheren Syntheseschritt blockiert als den, der zu den Mutanten der Gruppe Rd_1 führt. Möglicherweise handelt es sich um einen Block des Heptose-II-Transfers, was — in Analogie zu den SR-Mutanten [34] — zeigen würde, daß der Transfer von Heptose-I an KDO von einem anderen Enzym katalysiert wird (Heptosyl-I-Transferase) als der Transfer von Heptose-II an Heptose-I (Heptosyl-II-Transferase).

Über den Mechanismus dieser Heptose-Transfer-Reaktionen ist nichts bekannt, auch kennen wir die aktivierten Heptosevorstufen (Heptosenucleotide?)

für den Transfer noch nicht. Es sind jedoch Mutanten bekannt, deren Block um einen Syntheseschritt tiefer liegt als der eben besprochene. In diesen Mutanten (Re-Mutanten, Fig. 5A) wird Heptose-I nicht an KDO geknüpft, sei es, weil Heptose-I-Transferase nicht synthetisiert wird, sei es, weil (aktivierte) Heptose in diesen Mutanten nicht aufgebaut werden kann. Die Lipopolysaccharide dieser Re-Mutanten enthalten nur Lipoid A und KDO. Vorläufige Ergebnisse [29 b] sprechen dafür, daß KDO in diesen Lipopolysacchariden in Form von Ketten vorliegt. Bislang ist es nicht gelungen, aus Re-Mutanten aktivierte Heptose zu isolieren.

Mit der Identifizierung von R-Mutanten, die einen Defekt in der Synthese des Heptose-Kerns ihres Lipopolysaccharids besitzen, ist die Zahl der bisher bekannten Salmonella-Mutanten erweitert. Nimmt man alle R-Mutanten zusammen, die aus S. typhimurium, S. minnesota und S. ruiru isoliert und identifiziert worden sind, und vergleicht man die Struktur ihrer Lipopolysaccharide, so erhält man ein nahezu vollständiges Bild der Zwischenstufen, die bei der Biosynthese des basalen Kern-Polysaccharids von O-Antigenen durchlaufen werden, beginnend mit Lipoid A-KDO (Re), über Rd_2- und Rd_1-Lipopolysaccharide (Fig. 5C, 5D), Rc-Lipopolysaccharid und über die Gruppe der Rb-Lipopolysaccharide schließlich zum Ra'- und Ra-Lipopolysaccharid, welches den intakten basalen Kern darstellt (Fig. 5A). Über Struktur und Biosynthese des zentralen Teils des basalen Kerns einschließlich Lipoid A sind unsere Kenntnisse noch sehr unvollständig.

Die Frage, ob die untersuchten R-Mutanten O-spezifisches Hapten synthetisieren [1, 33, 40] können wir nicht mit Sicherheit beantworten. Wenn wir die basische Fraktion des Hydrolysats einer L 1-Fraktion (siehe [40]) konzentrieren und papierchromatographisch untersuchen, so finden wir fast immer Galactosamin, den spezifischen Zucker der O-Lipopolysaccharide. Aber die Konzentration ist gering, verglichen mit der Galactosamin-Menge, die wir in der L 1-Fraktion einer typischen R I-Form von S. minnesota finden. Eine Entscheidung auf diesem indirekten Weg, ob in den hier untersuchten R-Mutanten Hapten gebildet wird, ist unzuverlässig und kann nicht getroffen werden, solange nicht Versuche zur Isolierung des Haptens durchgeführt worden sind.

Dr. N. K. Richtmyer, National Institutes of Health, Bethesda, Md., U.S.A., danken wir sehr für die Überlassung von D-Glycero-L-manno-heptose sowie Dr. E. C. Heath, the Johns Hopkins School of Medicine, Baltimore, Md., U.S.A., für den Methylester der Pentaacetyl-KDO. Den Professoren Dr. F. Sartorius und Dr. G. Starke, Zentrallaboratorium für Bakterielle Darminfektionen, Berlin-Potsdam, danken wir sehr herzlich für das stete Interesse an diesen Arbeiten.

Viele chemische und serologische Analysen wurden von den Damen I. Minner, H. Winkler und G. Stehling durchgeführt, denen wir hiermit unseren Dank aussprechen möchten.

Für Diskussionen und viele wertvolle Ratschläge sind wir vor allem Dr. A. M. Staub, Institut Pasteur, Paris, Dr. H. Mayer, Max-Planck-Institut für Immunbiologie, Freiburg und Dr. H. Nikaido, Massachusetts General Hospital, Boston, U.S.A., besonders dankbar.

Diese Arbeit wurde mit Unterstützung der Deutschen Forschungsgemeinschaft durchgeführt, der wir an dieser Stelle für die Sachbeihilfe sehr danken möchten.

LITERATUR

1. Lüderitz, O., Staub, A. M., und Westphal, O., Bacteriol. Rev. 30 (1966) 192.
2. Bagdian, G., Dröge, W., Kotelko, K., Lüderitz, O., Westphal, O., Yamakawa, T., und Ueta, N., Biochem. Z. 344 (1966) 197.
3. Ghalambor, M. A., Levine, E. M., und Heath, E. C., J. Biol. Chem. 241 (1966) 3207.
4. Grollman, A. P., und Osborn, M. J., Biochemistry, 3 (1964) 1571.
5. Osborn, M. J., Rosen, S. M., Rothfield, L., Zeleznick, L. D., und Horecker, B. L., Science, 145 (1964) 783.
6. Sutherland, I. W., Lüderitz, O., und Westphal, O., Biochem. J. 96 (1965) 439.
7. Kauffmann, F., Die Bakteriologie der Salmonella Species, Munksgaard, Copenhagen 1961.
8. Beckmann, I., Lüderitz, O., und Westphal, O., Biochem. Z. 339 (1964) 401.
9. Osborn, M. J., Proc. Natl. Acad. Sci. U. S. 50 (1963) 499.
10. Nikaido, H., Proc. Natl. Acad. Sci. U. S. 48 (1962) 1542.
11. Nikaido, H., Proc. Natl. Acad. Sci. U. S. 48 (1962) 1337.
12. Lüderitz, O., Galanos, C., Risse, H. J., Ruschmann, E., Schlecht, S., Schmidt, G., Schulte-Holthausen, H., Wheat, R., Westphal, O., und Schlosshardt, J., Ann. N. Y. Acad. Sci. 133 (1966) 349.
13. Kauffmann, F., Lüderitz, O., Stierlin, H., und Westphal, O., Zentr. Bakteriol. Parasitenk. Abt. I. Orig. 178 (1960) 442.
14. Schlecht, S., und Westphal, O., Zentr. Bakteriol. Parasitenk. Abt. I. Orig. 200 (1966) 241.
15. Westphal, O., und Jann, K., in Methods in Carbohydrate Chemistry (edited by R. L. Whistler), Academic Press, New York 1965, Vol. 5, p. 83.
16. Dische, Z., J. Biol. Chem. 204 (1953) 983.
17a. Waravdekar, V. S., und Saslaw, L. D., J. Biochem. 234 (1959) 1945.
17b. Heath, E. C., persönliche Mitteilung.
18. MacGee, J., und Doudoroff, M., J. Biol. Chem. 210 (1954) 617.
19. Ghuysen, J. M., und Strominger, J. L., Biochemistry, 2 (1963) 1110.
20. Lowry, O. H., Roberts, N. R., Leiner, K. Y., Wu, M. L., und Farr, A. L., J. Biol. Chem. 102 (1954) 1.
21. Lowry, O. H., Rosenbrough, N. J., Farr, A. L., und Randall, R. J., J. Biol. Chem. 193 (1951) 265.
22. Layne, E., in Methods in Enzymology (edited by S. P. Colowick and N. O. Kaplan), Academic Press, New York 1957, Vol. III, p. 450.
23. Kickhöfen, B., und Westphal, O., Z. Naturforsch. 7b (1952) 655.
24. Sandford, P. A., and Conrad, H. E., Biochemistry, 5 (1966) 1508.
25. Hakomori, S., J. Biochem. 55 (1964) 205.
26. Lüderitz, O., Risse, H. J., Schulte-Holthausen, H., Strominger, J. L., Sutherland, I. W., und Westphal, O., J. Bacteriol. 89 (1965) 343.
27. Lüderitz, O., Simmons, D. A. R., Strominger, J. L., und Westphal, O., Anal. Biochem. 9 (1964) 1.
28. Rothfield, L., und Takeshita, M., Biochem. Biophys. Res. Commun. 20 (1965) 521.

29a. Rothfield, L., Osborn, M. J., und Horecker, B. L., *J. Biol. Chem.* 239 (1964) 2788.
29b. Dröge,W., Dissertation, Freiburg 1967, inVorbereitung.
30a. Risse, H. J., Lüderitz, O., und Westphal, O., *European J. Biochem.* 1 (1967) 233.
30b. Robinson, E. A., Kalckar, H. M., und Troedsson, H., *J. Biol. Chem.* 241 (1966) 2737.
31. Nikaido, H., *Biochemistry,* 4 (1965) 1550.
32. Nikaido, H., Naide, Y., und Mäkelä, P. H., *Ann. N. Y. Acad. Sci.* 133 (1966) 299.
33. Lüderitz, O., und Westphal, O., *Angew. Chem.* 78 (1966) 172; *Angew. Chem. Internat. Ed. Engl.* 5 (1966) 198.
34. Naide, Y., Nikaido, H., Mäkelä, P. H., Wilkinson, R. G., und Stocker, B. A. D., *Proc. Natl. Acid. Sci. U. S.* 53 (1965) 147.
35a. Stocker, B. A. D., Wilkinson, R. G., und Mäkelä, P. H., *Ann. N. Y. Acad. Sci.* 133 (1966) 334.
35b. Schlosshardt, J., unveröffentlicht.
36. Kuriki, Y., und Kurahashi, K., *J. Biochem. (Japan),* 58 (1965) 308.

37. Cherniak, R., und Osborn, M. J., *Federation Proc.* 25 (1966) 410.
38. Osborn, M. J., Rosen, S. M., Rothfield, L., und Horecker, B. L., *Proc. Natl. Acad. Sci. U. S.* 48 (1962) 1831.
39. Osborn, M. J., *Ann. N. Y. Acad. Sci.* 133 (1966) 375.
40. Beckmann, I., Subbaiah, T. V., und Stocker, B. A. D., *Nature,* 201 (1964) 1300.

H. J. Risse, gegenwärtige Adresse:
Physiologisch-Chemisches Institut der Freien Universität
1 Berlin 33, Arnimallee 22, Deutschland (BRD)

W. Dröge, E. Ruschmann, O. Lüderitz, O. Westphal
Max-Planck-Institut für Immunbiologie
78 Freiburg, Postfach 1668, Deutschland (BRD)

J. Schlosshardt
Zentrallaboratorium für Bakterielle Darminfektionen beim
Institut für Serum- und Impfstoffprüfung
X 15 Potsdam, Deutschland (DDR)

European J. Biochem. 1 (1967) 233—242

Vergleichende Untersuchungen zur Biosynthese von Nucleotid-Zuckern in S- und R-Formen von *Salmonella minnesota* und *Salmonella ruiru*

H. J. Risse, O. Lüderitz und O. Westphal

Max-Planck-Institut für Immunbiologie, Freiburg-Zähringen

(Eingegangen am 12. Januar 1967)

About 30 R mutants derived from *Salmonella minnesota* and *Salmonella ruiru* and the corresponding wild-type strains were analysed comparatively for the presence of enzymes catalyzing the synthesis of UDP-glucose, UDP-galactose, UDP-N-acetylglucosamine and UDP-N-acetylgalactosamine.

In the biosynthesis of the corresponding O antigens these UDP-sugars function as precursors, whose sugar residues are transferred on to the growing polysaccharide by specific transferases. In cell-free extracts of the strains the following enzyme activities were determined: (quantitatively) hexokinase, phosphoglucomutase, UDP-glucose synthetase and UDP-galactose-4-epimerase; (qualitatively) glucosephosphate isomerase and UDP-N-acetylglucosamine-4-epimerase; (qualitatively, in 10 strains) glutamine:fructose-6-phosphate transaminase and (in 6 strains) UDP-N-acetylglucosamine synthetase. Most of the mutants were found to contain the enzymes in activities equal to those found in the wild-type strains. These mutants therefore probably have a block involving a transferase. Three mutants were identified to have a defect in UDP-glucose synthetase synthesis. 4 mutants lacked the enzyme UDP-N-acetylglucosamine-4-epimerase. Some of these mutants have an additional block which involves a transferase. In extracts of *S. ruiru* the enzyme system of Matsuhashi and Strominger which catalyzes the conversion of TDP-glucose to TDP-4-amino-4,6-dideoxyhexose (probably with the configuration of D-galactose), could be identified.

Hält man Wildstämme des Genus *Salmonella* längere Zeit in flüssigem Nährboden und plattiert dann auf Agar, so findet man neben glatt und abgerundet wachsenden Kolonien der Wildform (S-Form) stets

Nicht allgemein gebräuchliche Abkürzungen. Uridindiphosphoglucose, UDP-Glc; Uridindiphosphogalactose, UDP-Gal; Uridindiphospho-N-Acetylglucosamin, UDP-GlcNAc; Uridindiphospho-N-Acetylgalactosamin, UDP-GalNAc; Thymidindiphosphoglucose, TDP-Glc; Glucose-1-phosphat, Glucose-1-P; Glucose-6-phosphat, Glucose-6-P; Lipopolysaccharid, LPS; 2-Keto-3-desoxyoctonat, KDO.

Enzyme. Hexokinase, oder ATP:D-Hexose-6-Phosphotransferase (EC 2.7.1.1); Phosphoglucomutase, oder α-D-Glucose-1,6-diphosphat:α-D-Glucose-1-phosphat-Phosphotransferase (EC 2.7.5.1); UDP-Glc-Synthetase, oder UTP: α-D-Glucose-1-phosphat-Uridylyltransferase (EC 2.7.7.9); UDP-Gal-Epimerase, oder UDP-Glucose-4-Epimerase (EC 5.1.3.2); Hexosephosphat-Isomerase, oder D-Glucose-6-phosphat-ketol-Isomerase (EC 5.3.1.9); Glutamin:Fructose-6-P-Transaminase, oder L-Glutamin:D-Fructose-6-phosphat-Aminotransferase (EC 2.6.1.16); Acetylglucosamin-Phosphomutase, oder 2-Acetamido-2-desoxy-D-glucose-1,6-diphosphat:2-Acetamido-2-desoxy-D-glucose-1-phosphat-Phosphotransferase (EC 2.7.5.2); UDP-N-Acetylglucosamin-Synthetase, oder UTP:2-Acetamido-2-desoxy-α-D-glucose-1-phosphat-Uridylyltransferase (EC 2.7.7.23); UDP-N-Acetylglucosamin-Epimerase, oder UDP-2-Acetamido-2-desoxy-D-glucose-4-Epimerase (EC 5.1.3.7); Glucosamin-6-P-Acetylase, oder AcetylCoA:2-Amino-2-desoxy-D-glucose-6-phosphat-N-Acetyltransferase (EC 2.3.1.4).

glatt und abgerundet glatt und abgerundet Kolonien mutierter Formen, die eine rauhe Oberfläche und gezackte Kolonieränder aufweisen. Sie werden dem morphologischen Bild nach als Rauh-Formen (R-Formen) bezeichnet und unterscheiden sich von S-Formen durch veränderte serologische und biochemische Eigenschaften [1—6]. Sie sind nicht in der Lage, die ursprünglichen O-Antigene der Wildformen aufzubauen, sondern synthetisieren statt dessen R-Antigene, die — verglichen mit dem O-Antigen — in ihren Lipopolysacchariden unvollständige Polysaccharid-Ketten enthalten.

Im Verlauf unserer Untersuchungen über die S-R-Mutation bei Salmonellen wurden zahlreiche R-Mutanten aus den beiden Wildstämmen *S. minnesota* und *S. ruiru* (Chemotyp II, O-Antigen 21) gewonnen. Sie konnten auf Grund der Zuckerzusammensetzung ihrer Lipopolysaccharide in die Chemotypen Ra-Re klassifiziert werden, welche sich auch durch ihr serologisches Verhalten sowie hinsichtlich der Lyse durch spezifische Phagen unterscheiden [2, siehe auch 3]. Genetische Defekte, die zur Bildung von R-Mutanten führen, können prinzipiell jedes Enzym betreffen, das einen Schritt der Synthesekette vom Nährsubstrat Glucose zum fertigen Lipopolysaccharid katalysiert. Es war daher von Interesse, diese Enzyme in den R-Mutanten zu testen, um die Lage von Enzymblocks festzustellen. Die Biosynthese der Polysaccharide vollzieht sich in 2 Schritten: a) Synthese der Nucleotid-Zucker, welche die aktivierten Vorstufen für den Transfer der Monosaccharide auf die polymeren Acceptoren darstellen, b) Transfer der Monosaccharide auf das wachsende Polysaccharid.

Die vorliegende Arbeit berichtet über den Nachweis von Enzymen, welche die Synthese der Nucleotid-Derivate von Glucose, Galactose, N-Acetylglucosamin und N-Acetyl-galactosamin katalysieren, sowie über die Synthese einer TDP-4-acetamido-4,6-didesoxy-hexose [7] in *S. minnesota* und *S. ruiru*-Mutanten. Die Untersuchung von Transferase-Reaktionen wird Gegenstand einer anderen Publikation sein.

Der Biosyntheseweg von (aktivierter) Heptose, einem weiteren Baustein von S- und vielen R-Lipopolysacchariden, ist noch unbekannt und konnte daher nicht untersucht werden. KDO dagegen ist in den Lipopolysacchariden aller bislang isolierten R-Mutanten enthalten; die Biosynthese von CMP-KDO, welches die aktivierte Vorstufe von KDO darstellt [27], wurde nicht untersucht.

MATERIAL UND METHODEN

Bakterien

Bakterienstämme. Bezeichnung und Herkunft der in dieser Arbeit untersuchten R-Mutanten aus *S. minnesota* und *S. ruiru* ist in den Tab. 1 und 2 beschrieben.

Züchtung der Bakterien. S- wie R-Formen werden in 1 Liter-Ansätzen in DIFCO-Antibiotic M 3-Medium bei 37° gezüchtet. Nach Erreichen der halblogarithmischen Wachstumsphase (nach ca. 4 Std) werden die Zellen abzentrifugiert, mit 0,01 M Tris-HCl pH 7,8 gewaschen und bei −20° aufbewahrt. Aus 1 Liter Ansatz erhält man etwa 3 g Bakterien (Feuchtgewicht).

Präparationen

Darstellung der Enzymextrakte [8]. 3 g Zellen (feucht) werden in 5 ml 0,02 M Tris-HCl pH 7,5, 0,002 M EDTA suspendiert und nach Zusatz von 4,8 g Glasperlen (0,8 mm Durchmesser) unter sorgfältiger Eiskühlung 4×30 sec, mit einem Ultra-Turrax-Homogenisator (Janke und Kunkel, Staufen/Br.) desintegriert. Die Homogenate werden 10 min bei $30000 \times g$ zentrifugiert, der trübe Überstand wird 2,5 Std bei $100000 \times g$ ultrazentrifugiert. Die auf diese Weise erhaltenen Rohenzymextrakte haben einen Proteingehalt von 8—15 mg Protein/ml (Biuret-Reaktion [9]).

Darstellung von [14C]Uridindiphospho-N-acetylglucosamin. [14C]UDP-GlcNAc (an der Acetylgruppe

markiert) wird mit Hilfe von Hefe-Enzym nach bekannten Methoden synthetisiert [10,11].

Darstellung von [14C]Thymidindiphosphoglucose. [14C]Thymidindiphosphoglucose wird durch Übertragung des Thymidyl-Restes auf [14C]Glucose-1-phosphat mit Hilfe eines Enzymextraktes aus Enterokokken in Anlehnung an eine Methode von Pazur [12] hergestellt. Hierzu wurden 0,0238 mg [14C]Glucose-1-phosphat (0,01 mC) mit 45 µl Enterokokken-Extrakt (aus Stamm Nr. 12788 des Hygiene Instituts der Universität Freiburg/Br.) und 9,4 µl 0,01 M TTP versetzt und 1 Std bei 37° inkubiert. Der Reaktionsansatz wird danach in Fließmittel C papierchromatographiert. Durch Autoradiographie und Vergleich mit inaktivem TDP-Glc wird das Produkt lokalisiert. Das Eluat (125 µl) enthält 488000 Imp./min.

Bestimmung von Enzym-Aktivitäten in den Überständen von Bakterienhomogenaten

Hexokinase [13a]. Bestimmungsansatz: 50 µMol Tris-HCl pH 7,5; 10 µMol Mg++; 1 µMol EDTA; 1,3 µMol NADP; 8,2 µMol ATP; 20 µMol Glucose; 20 µl Glucose-6-phosphat-Dehydrogenase (1 mg Enzymprotein/ml). Gesamtvolumen 1,0 ml. Nach Zufügen des bakteriellen Enzymextraktes (20 µl) wird die Extinktion bei 340 mµ zu verschiedenen Zeiten (Temperatur 25°) gemessen.

Phosphoglucomutase [13b]. Bestimmungsansatz: 50 µMol Tris-HCl pH 7,5; 10 µMol Mg++; 1 µMol EDTA; 1,3 µMol NADP (Mononatriumsalz); 20 µMol Glucose-1-phosphat; 20 µl Glucose-6-phosphat-Dehydrogenase (1 mg/ml); 20 µl Enzymextrakt. Gesamtvolumen: 1,0 ml (25°). Die Messung geschieht wie bei der Hexokinase-Bestimmung.

Glucosephosphat-Isomerase [13c]. Bestimmungsansatz: 50µl Tris-HCl, pH 7,5; 10µMol Mg++; 1µMol EDTA; 1,3 µMol NADP; 20 µMol Fructose-6-phosphat; 10 µl Glucose-6-phosphat-Dehydrogenase (1 mg/ml); 10 µl Enzymextrakt. Gesamtvolumen 1,0 ml (25°). Die Messung wird wie bei der Hexokinase-Bestimmung vorgenommen.

Uridindiphosphoglucose-Synthetase [14]. Testsystem:

UDP-Glucose + Pyrophosphat → Glucose-1-P + UTP

Glucose-1-P → Glucose-6-P

Glucose-6-P + NADP → 6-P-Gluconsäure + NADPH

Bestimmungsansatz: 50 µMol Tris-HCl pH 7,5; 10 µMol Mg++; 1 µMol EDTA; 1,3 µMol NADP; 20µMol UDP-Glc; 20µl Phosphoglucomutase (2 mg Enzymprotein/ml); 20 µl Glucose-6-phosphat-Dehydrogenase (1 mg/ml); 20 µl Enzymextrakt; Gesamtvolumen 1,0 ml.

Die Extinktionsänderung bei 340 mµ (25°) wird von Minute zu Minute verfolgt. Nach 5 min werden

2,8 µMol anorganisches Pyrophosphat (neutralisiert) zugesetzt, und die Reaktion wird weiter verfolgt. Eine positive Reaktion vor Zusatz des Pyrophosphats rührt von einer Nucleotidpyrophosphatase-Aktivität her, die in den Rohextrakten stets vorhanden ist. Die Spaltungskurve nach Pyrophosphat-Zusatz beschreibt demnach die Summe der Enzymaktivitäten von Pyrophosphatase und Pyrophosphorylase (= Synthetase). Die Aktivität der Pyrophosphatase kann graphisch eliminiert werden, man erhält auf diese Weise die Umsatzrate der Synthetase allein.

D-Glucose-1-phosphat: Glucose-6-Phosphotransferase [15]. Testsystem:

Glucose + Glucose-1-*P* → Glucose-6-*P* + Glucose
Glucose-6-*P* + NADP → 6-*P*-Gluconsäure + NADPH

Bestimmungsansatz: 25 µMol Tris-Acetat pH 6,5; 0,5 µMol EDTA; 1,3 µMol NADP; 1 µMol Glucose-1-phosphat; 5 µl Glucose-6-phosphat-Dehydrogenase (1 mg/ml); 10 µl Enzymextrakt. Gesamtvolumen 1,0 ml.

Die Extinktion bei 340 mµ wird über mehrere Minuten kontrolliert. Sie sollte konstant bleiben, da Phosphoglucomutase unter diesen Bedingungen inaktiv ist (Abwesenheit von Mg++, pH 6,5). Dann werden 30 µMol Glucose zugesetzt (30 µl 1,0 M Lösung) und die Extinktionsänderung bei 340 mµ verfolgt.

Uridindiphosphogalactose-4-Epimerase [16]. Das Enzym wurde auf zweierlei Arten getestet:

a) Testsystem: UDP-Gal → UDP-Glc
UDP-Glc + 2 NAD → UDP-Glucuronsäure + 2 NADH

Bestimmungsansatz: 0,1 µMol UDP-Gal; 0,38 µMol NAD; 180 µMol Tris-HCl pH 8,7, Gesamtvolumen 1,0 ml. Nach Zusatz von 15 µl UDP-Glc-Dehydrogenase (25 mg/270 µl Wasser) wird die Extinktionsänderung bei 340 mµ kontrolliert. Sie sollte minimal sein, wenn das UDP-Galactose-Präparat wenig UDP-Glucose als Verunreinigung enthält. Dann werden 20 µl Enzymextrakt zugesetzt und die Extinktionsänderung pro Zeiteinheit gemessen.

b) Testsystem: UDP-Glc → UDP-Gal
UDP-Gal + Gal-Oxydase → UDP-Gal-Ox

Bestimmungsansätze: 9,4 µMol UDP-Glc; 94 µMol Tris-HCl, pH 8,7; 26 µMol Mg++; 1,41 µMol NAD. Gesamtvolumen 0,1 ml.

Pro Ansatz werden 35 µl Enzymextrakt zugesetzt, die Proben werden bei 37° inkubiert und nach verschiedenen Zeiten durch Erhitzen auf 100° (3 min) abgestoppt und enteiweißt. Nach Zentrifugation werden je 0,1 ml Überstand mit 0,7 ml Galactose-Oxydase-Reagens versetzt, 45 min bei 25° inkubiert, und die Extinktion bei 410 mµ gemessen.

Galactose-Oxydase-Reagens (in Anlehnung an [17]): 1 mg Peroxydase; 2 mg Galactose-Oxydase; 5,1 ml 0,1 M Phosphat pH 7,0; 1,8 µl *o*-Kresol (jeweils frisch herstellen!).

Methode b) wurde als halbquantitativer Test benutzt, um orientierend Enzymextrakte auf die Anwesenheit von Epimerase zu testen.

Glutamin: Fructose-6-phosphat-Transaminase [18]. Testsystem:

Fructose-6-*P* + Glutamin → Glucosamin-6-*P* + Glutamat

Bestimmungsansätze: 2,01 µMol Fructose-6-phosphat; 1,72 µMol Glutathion; 9,5 µMol K-Phosphat, pH 7,5; 3,16 µMol Glutamin; 90 µg Pyridoxalphosphat. Gesamtvolumen: 250 µl. Zu den Ansätzen werden je 125 µl Enzymextrakt zugefügt und bei 37° inkubiert. Zu verschiedenen Zeiten wird jeweils ein Ansatz 3 min auf 100° erhitzt und von denaturiertem Protein abzentrifugiert. Je 350 µl Überstand werden im Vakuumexsiccator getrocknet und der Morgan-Elson Reaktion [19] unterworfen.

Qualitative Bestimmung von Uridindiphospho-N-acetylglucosamin-Synthetase (siehe Diskussion). Testsystem:

UDP-GlcNAc + Pyrophosphat → GlcNAc-1-Phosphat + UTP
Glc-NAc-1-Phosphat → GlcNAc-6-Phosphat

Das Inkubationsgemisch enthält pro ml: 75 µMol Phosphat, pH 7,5; 7,5 µMol Mg++; 2,47 µMol UDP-Glc-NAc.

Bestimmungsansätze: 70 µl Inkubationsgemisch; 2 µl 0,2 M Pyrophosphat (neutral) 10 µl Enzymextrakt. Die Ansätze werden bei 37° inkubiert und die Ansätze nach verschiedenen Zeiten durch Erhitzen auf 100° abgestoppt. Pro Ansatz wird 1 % Borat, 200 µl, zugefügt, 7 min auf 100° erhitzt und Morgan-Elson-Reagens zugegeben [19]. Als Kontrolle für die Nucleotid-Pyrophosphatase-Aktivität werden jeweils Ansätze ohne Pyrophosphat mitgemessen.

Test auf Uridindiphospho-N-acetylglucosamin-4-Epimerase. Der Test wurde wie früher beschrieben durchgeführt [8].

Test auf TDP-Glucose-Oxydoreduktase [7]. Testsystem:

TDP-Glc → TDP-4-keto-6-desoxyglucose

Bestimmungsansatz: 12,4 µMol Tris-HCl, pH 7,5; 0,2 µMol TDP-Glucose; 50 µl Enzymextrakt, Gesamtvolumen: 85 µl. Die Ansätze werden bei 37° verschiedene Zeiten inkubiert und mit 600 µl 0,1 N KOH versetzt. Das Gemisch wird 20 min bei 37° gehalten, dann werden die Spektren zwischen 300 und 420 mµ aufgenommen. Die gebildete TDP-4-keto-6-desoxyglucose zeigt im alkalischen pH-Bereich ein Absorptionsmaximum bei 318 mµ.

Zur Reinigung der TDP-4-keto-6-desoxyglucose wird ein Ansatz mit 200 µl 0,01 M TDP-Glc, 173 µl 0,83 M Tris-HCl, pH 7,5, und 100 µl Enzymextrakt 4 Std bei 37° inkubiert. Nach 3 min Erhitzen auf 100° wird von denaturiertem Protein abzentrifugiert. Der Überstand wird mit 80 mg Darco-Aktivkohle versetzt und 1 min bei 0° stehen gelassen. Die Kohle wird abzentrifugiert und einmal mit demselben Volumen Wasser nachgewaschen.

Die Kohle wird mit ca. 500 µl 50 %igem Äthanol/0,1 % NH$_3$ über Nacht stehen gelassen. Dabei werden die Nucleotidzucker desorbiert.

Glutamat: TDP-4-keto-6-desoxyhexose-Transaminase; Synthese von 4-Acetamido-4,6-didesoxygalactose. Testsystem:

TDP-4-keto-6-desoxyglucose + Glutamat →
TDP-4-amino-4,6-didesoxyhexose + α-Ketoglutarat

Beim Nachweis der Transaminase haben wir uns an die von Matsuhashi und Strominger angegebenen Methoden gehalten [7].

Bestimmungsansatz: 30 µl [¹⁴C]TPD-Glucose (130000 Imp./min); 7,5 µMol Tris-HCl, pH 7,6; 0,5 µMol Mg^{++}; 1 µMol L-Glutamat; 10 µg-Pyridoxalphosphat; 400 µl *Salmonella ruiru* S-Rohextrakt. Gesamtvolumen 530 µl.

Inkubation 5 Std bei 37°. Ein Kontrollansatz enthält kein Glutamat. Nach Ende der Inkubation werden die Reaktionsgemische auf Papier (Whatman 1) aufgetragen und in Lösungsmittel C chromatographiert. Ein Autoradiogramm zeigt im Versuchsansatz 4, im Kontrollansatz 3 radioaktive Banden. Die nicht im Kontrollansatz auftretende Bande wird eluiert (60 µl) und mit einem Gemisch aus 40 µl gesättigter NaHCO$_3$ und 40 µl 5 % Acetanhydrid in Wasser acetyliert (30 min bei Zimmertemperatur) [20]. Die anschließende Elektrophorese liefert eine gleichschnell wie TDP-Glucose wandernde radioaktive Substanz. Diese wird eluiert (30 µl) und zur Spaltung in TMP und freien Acetamidozucker mit folgendem Gemisch inkubiert:

2 µMol Glycin, pH 8,6; 1,6 µMol Mg^{++}; 1 µl Phosphodiesterase aus Schlangengift (450 mg/50 µl Wasser, Worthington); 1 µl Phosphomonoesterase (aus *E. coli*, Sigma) in 70 µl Gesamtvolumen. Nach 3 Std bei 37° wird das Gemisch der Elektrophorese unterworfen; man erhält eine neutrale Substanz, die chromatographisch mit authentischen 4-Acetamido-4,6-didesoxyhexosen verglichen wird.

Papierchromatographie und Papierelektrophorese

Diese wurden wie früher beschrieben durchgeführt [21, 22]. Folgende Lösungsmittel wurden benutzt: A) Butanol-Pyridin-Wasser (6:4:3, v/v); B) Butanol-Eisessig-Wasser (5:1:2, v/v); C) Äthanol-Ammoniumacetat 1 M pH 7,0 (75:30, v/v); D) Äthylacetat-Pyridin-Eisessig-Wasser (25:25:5:15, v/v).

Für Autoradiogramme wurden Agfa Doneo Röntgenfilme verwandt.

Reagentien

Hexokinase, Phosphoglucomutase, Hexosephosphat-Isomerase, Glucose-6-phosphat-Dehydrogenase, Glucose-Oxydase, Peroxydase, ATP, NAD, NADP, Glucose-6-P; Glucose-1-P, Fructose-6-P, UDP-*N*-Acetylglucosamin und Glutathion wurden bei Boehringer, Mannheim, bezogen; UDP-Glucose, UDP-Galactose, UDP-*N*-Acetylglucosamin, UTP, TTP, UDP-Glucose-Dehydrogenase und Phosphomonoesterase aus *E. coli* von Sigma Chemical Co., St. Louis, Mo., U.S.A., Galactose-Oxydase und anorganische Pyrophosphatase von Worthington, Freehold, U.S.A., TDP-Glucose von Calbiochem, Los Angeles, Calif., U.S.A., Radioaktives Glucose-1-phosphat von Radiochemical Center, Amersham, England.

VERSUCHE UND ERGEBNISSE

Die Synthese der UDP-Zucker in S. minnesota-
und S. ruiru-*Mutanten*

Salmonella minnesota und *Salmonella ruiru* gehören zum Chemotyp II und enthalten in ihren Lipopolysacchariden neben Heptose und KDO die folgenden 4 C$_6$-Zucker: Glucose, Galactose, Glucosamin

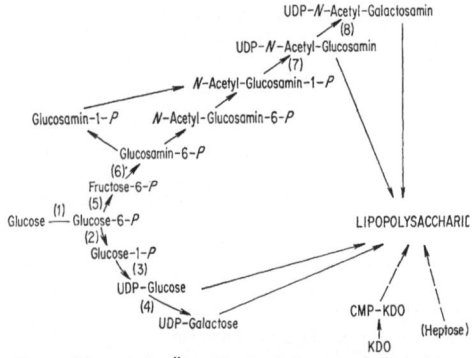

Fig. 1. *Schematische Übersicht der Polysaccharid-Biosynthese bei* S. minnesota. (1) Hexokinase; (2) Phosphoglucomutase; (3) UDP-Glc-Synthetase; (4) UDP-Gal-Epimerase; (5) Hexosephosphat-Isomerase; (6) Glutamin:Fructose-6-*P*-Transaminase; (7) UDP-*N*-Acetylglucosamin-Synthetase; (8) UDP-*N*-Acetylglucosamin-Epimerase

und Galactosamin. Diese 4 Zucker bilden UDP-Derivate als aktivierte Vorstufen für den Zucker-Transfer bei der Biosynthese der spezifischen Polysaccharide. Fig. 1 zeigt schematisch die Synthesewege, die von Glucose als Substrat zu den verschiedenen UDP-Zuckern führen [23]. Die Enzyme, die die einzelnen Syntheseschritte katalysieren, und deren genetische

Tabelle 1. *Aktivitätsbestimmungen derjenigen Enzyme, welche die Synthese von UDP-Glucose und UDP-Galactose katalysieren* m = *Salmonella minnesota*; r = *Salmonella ruiru*

Chemotyp	Mutanten	Zucker im Lipopolysaccharid	Chemotyp	Mutanten	Zucker im Lipopolysaccharid
II	mRS	KDO, Hep	Rc	mR8[b]	KDO, Hep, Glc
	r S	Glc		mR5[b]	
		Gal		r R1[b]	
II	mR1[b]	GNAc			
	mR608[b]	GalNAc			
	mR1001[b]		Rd$_1$	mR7[b]	KDO, Hep[d]
	rR2[b]			rR541[b]	
			Rd$_2$	mR3[b]	KDO, Hep[e]
Ra	mRII[a]	KDO, Hep		mR4[b]	
	mR555[b]	Glc		rR3[b]	
	mR708[b]	Gal		mR597[b]	
		GNAc			
Rb	mR1[a]	KDO, Hep	Re	mR595[b]	KDO
	mR2[b]	Glc		mR613[b]	
	mR592[b]	Gal		mR1000[b]	
				rR128[b]	

noch nicht in Chemotypen eingeordnete Mutanten:	mR27[c]	m1114 gal⁻[c]	rR562[b] m1114 gal⁻ (5)[c]
	mR43[c]	m1114 gal⁻ (2)[c]	m1114 gal⁻ (7)[c]
	mR47[c]	m1114 gal⁻ (3)[c]	mRz8[c]
	mR65[c]	m1114 gal⁻ (4)[c]	

Enzym	Hexokinase	Phosphoglucomutase	UDP-Glucose-Synthetase	UDP-Galactose-Epimerase
$\dfrac{\text{nMol Produkt}}{\text{mg Protein} \times \text{min}}$	80—120	170—250	20—40	10—20

[a] Diese R-Mutanten sind von Professor F. Kauffmann, Statens Serum Institut, Kopenhagen, isoliert worden [39].
[b] Diese R-Mutanten wurden von Frau J. Schlosshardt, Potsdam, Zentrallaboratorium für Bakt. Darminfektionen, isoliert [2].
[c] Diese R-Mutanten wurden von Dr. G. Schmidt, Max-Planck-Institut für Immunbiologie, Freiburg, nach Behandlung der Wildformen mit dem Mutagen N-Methyl-N-nitroso-N′-nitroguanidin isoliert. Rz1 und Rz2 sowie die mit gal⁻ bezeichneten Stämme sind galactose-negative Mutanten (siehe Tab. 2).
[d] Etwa 12% Heptose im Lipopolysaccharid. — [e] Etwa 6% Heptose im Lipopolysaccharid.

Blockierung zur Ausbildung von R-Mutanten führt, wurden in 33 Mutanten von *S. minnesota* und *S. ruiru* untersucht.

Das Ergebnis dieser Untersuchungen ist in den folgenden Tabellen wiedergegeben. In Tab. 1 sind etwa 30 Mutanten aufgeführt, bei denen die Enzyme Hexokinase, Phosphoglucomutase, UDP-Glucose-Synthetase und UDP-Galactose-4-Epimerase die gleiche Aktivität besitzen wie bei den Wildformen. Diese Stämme sind daher befähigt, UDP-Glucose und UDP-Galactose zu synthetisieren. Demgegenüber besitzen die Mutanten *S. minnesota* Rz 1, Rz 2 und 1111 Gal⁻ einen Block in der UDP-Glucose-Synthese (Tab. 2). Dieser betrifft die Aktivität von UDP-Glucose-Synthetase, welche bei diesen 3 Stämmen, verglichen mit der Wildform (= 100%) wesentlich vermindert (12—24%) ist. Einen ähnlichen Aktivitätsverlust von etwa 80% gegenüber der Wildform bestimmten wir bei der *E. coli* B/4$_0$-Mutanten [24], die als UDP-Glucose-Synthetase-defekter Stamm von Hattmann und Fukasawa beschrieben worden ist.

Die Ergebnisse der qualitativen Analyse einiger Enzyme, die für die Synthese der aktivierten Amino-

Tabelle 2. *Ergebnis der Synthetase-Bestimmung an UDP-Glucose-Synthetasedefekten Mutanten von* S. minnesota *und* E. coli B
Die Enzyme Hexokinase, Phosphoglucomutase, UDP-Galactose-4-Epimerase haben dieselben Aktivitäten wie die Wildform (vgl. Tab. 1)

Mutante	UDP-Glc-Synthetase-Aktivität	Aktivität (Wildstamm=100)
	nMol UDP-Glc mg Protein × min	%
S. minnesota S	10,5	100
S. minnesota Rz1[a]	2,2	21
S. minnesota Rz2[a]	2,5	24
S. minnesota 1111 gal⁻[a]	1,3	12
E. coli B	13,8	100
E. coli B/4$_0$ [24]	2,7	20

[a] Diese R-Mutanten wurden von Dr. G. Schmidt, Max-Planck-Institut für Immunbiologie, Freiburg, nach Behandlung der Wildformen mit dem Mutagen N-Methyl-N-nitroso-N′-nitroguanidin isoliert. Rz1 und Rz2 sowie die mit gal⁻ bezeichneten Stämme sind galactose-negative Mutanten.

zucker benötigt werden, sind in Tab. 3 zusammengefaßt. Wiederum enthalten die meisten Mutanten die getesteten Enzyme für die Synthese der UDP-

Tabelle 3. *Ergebnis der qualitativen Bestimmung der Enzyme, welche zur Synthese von UDP-N-Acetylglucosamin und UDP-N-Acetylgalactosamin benötigt werden*
Bestimmungsmethoden und Literatur: siehe unter Methoden

Enzym	Substrat	Untersuchte Stämme	Befund
Hexosephosphat-Isomerase	Fructose-6-P	alle Mutanten[a]	in allen Mutanten aktiv
Glutamin:Fructose-6-P-Transaminase	Fructose-6-P	*S. minnesota* S; R1; R2; R592; R595; R597; R1000; R1127; *S. ruiru* S; R128[b]	in den untersuchten Mutanten aktiv
UDP-N-Acetylglucosamin-Synthetase	UDP-GlcNAc	*S. minnesota* S; R613; R595; R1000; *S. ruiru* S; R128[b]	in allen untersuchten Mutanten aktiv
UDP-N-Acetylglucosamin-4-Epimerase	UDP-GlcNAc	alle Mutanten[a]	keine Aktivität in den Mutanten von *S. minnesota* R60; *S. ruiru* R1; R128; R541. In allen übrigen Mutanten aktiv

[a] Alle in Tab. 1 und 2 aufgeführten Mutanten.
[b] Die übrigen Mutanten wurden nicht getestet.

N-Acetylhexosamine. Lediglich in 4 Stämmen, *S. ruiru* R1, R541 und R128 sowie in *S. minnesota* R60 fehlt das Enzym UDP-N-Acetylglucosamin-4-Epimerase. Diese Stämme können daher UDP-N-Acetylgalactosamin nicht synthetisieren.

Von Fujimoto et al. [15] wurde kürzlich ein Enzym aus *Escherichia coli* beschrieben, das die Reaktion

Glucose-1-P + Glucose → Glucose-6-P + Glucose

katalysiert: Der Phosphatrest am C-Atom 1 einer Glucose wird auf das C-Atom 6 einer anderen Glucose übertragen. Da das Enzym einen Alternativweg zur Phosphoglucomutase darstellt, haben wir einige Stämme auf die Anwesenheit dieses Enzyms getestet. In allen untersuchten Stämmen (*S. minnesota* S, R1127, R1) war das Enzym vorhanden.

Die Biosynthese einer TDP-4-amino-4,6-didesoxyhexose mit S. ruiru-Enzym

Strominger u. Mitarb. [7,25] konnten in verschiedenen *E. coli*-, *Pasteurella*- und *Salmonella*-Stämmen ein Enzymsystem nachweisen, das Thymidindiphospho-glucose in eine TDP-4-acetamido-4,6-didesoxyhexose umwandelt. Je nach Herkunft der Enzyme besaß der synthetisierte Aminozucker die Glucose- oder Galactose-Konfiguration.

TDP-Glucose → TDP-4-keto-6-desoxyhexose
→ TDP-4-amino-4,6-didesoxy-
hexose
→ TDP-4-Acetamido-4,6-didesoxy-
hexose

Die erste Stufe der Reaktion wird von dem Enzym TDP-Glucose-Oxydoreduktase katalysiert, die zweite von Glutamat: 4-Keto-6-desoxy-Aminotransferase und die dritte von der entsprechenden Acetyl-Transferase.

TDP-Glucose-Oxydoreduktase konnte in den *S. minnesota*-Stämmen S, RI, Rz8 sowie in *S. ruiru* S nachgewiesen werden (andere Stämme wurden nicht untersucht). Inkubiert man TDP-Glucose mit hochtourig zentrifugiertem Bakterienextrakt (Überstand), so findet man bei 318 mµ in alkalischem Medium eine für die 4-Keto-Verbindung charakteristische Absorptionsbande (Fig. 2B). Fig. 2A zeigt den negativen Ausfall der Reaktion, wenn man UDP-Glc statt TDP-Glc als Substrat verwendet. Nach Reinigung durch Adsorption und Desorption an Aktivkohle zeigte die synthetisierte Verbindung in Alkali einen einheitlichen Absorptionspeak bei 318 mµ (Fig. 2C).

Zum Nachweis der Transaminase-Reaktion wurde [14C]TDP-Glucose in Gegenwart von Glutamat und Pyridoxylphosphat mit einem Rohextrakt aus *S. ruiru* S inkubiert. Als Kontrolle diente derselbe Ansatz ohne Glutamat. Nach der Inkubation wurden beide Ansätze im Fließmittel C papierchromatographiert. Die radioaktiven Zonen wurden autoradiographisch lokalisiert. Neben unveränderter TDP-Glc und etwas schneller wandernder TDP-Keto-Verbindung findet man im Versuchsansatz eine radioaktive Komponente, die nicht in der Kontrolle auftritt, und die noch etwas schneller als das Keto-Derivat läuft. Nach Elution aus dem Chromatogramm acetylierten wir die Substanz mit Acetanhydrid in Gegenwart von Bicarbonat. Die Elektrophorese des Reaktionsproduktes liefert eine gleichschnell wie TDP-Glucose wandernde Substanz. Da der freie Amino-Zucker außerordentlich säurelabil ist [7,26], war es nicht möglich, ihn durch Säurehydrolyse aus dem TDP-Derivat freizusetzen. Wir inkubierten das Acetylierungsprodukt deshalb bei pH 8,6 mit Phosphodiesterase und Phosphomonoesterase und unterwarfen das Reaktionsgemisch erneut der Elektrophorese. Wir erhielten eine radio-

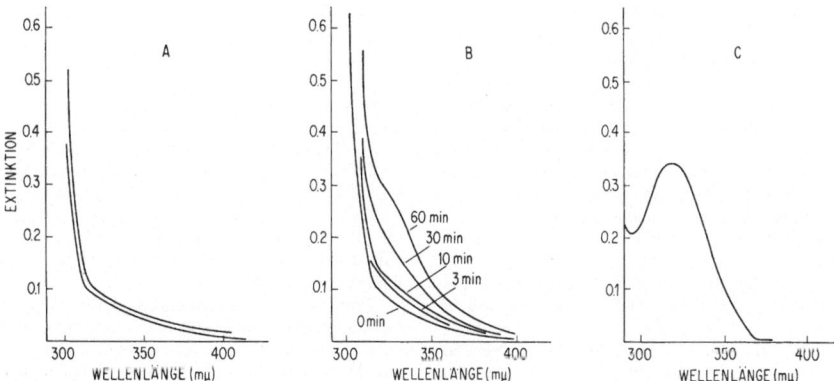

Fig. 2. *Test auf TDP-Glucose-Oxydoreduktase*. UV-Spektrum der gebildeten TDP-4-Keto-6-desoxyglucose. A, Kontrollversuch mit UDP-Glucose als Substrat. B, TDP-Glucose als Substrat, verschiedene Inkubationszeiten. C, Spektrum nach Reinigung der TDP-4-Keto-6-desoxyglucose an Tierkohle. Reaktionsbedingungen siehe Text. Enzymextrakt aus *S. minnesota* S-Form. Messung der Spektren in 0,1 N KOH

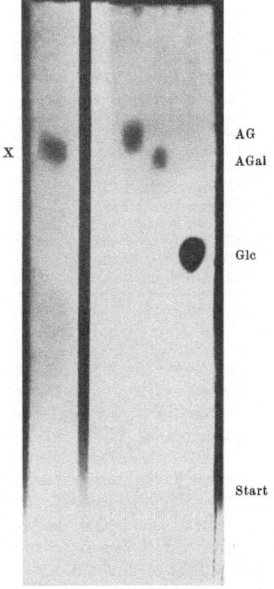

Fig. 3. *Dünnschichtchromatogramm von 4-Acetamido-4,6-didesoxyhexosen*. Die linke Seite des Chromatogramms wurde durch Autoradiographie, die rechte Seite durch Anfärbung mit Ag⁺/NaOH sichtbar gemacht. Glc = Glucose, AG = 4-Acetamido-4,6-didesoxyglucose (Testzucker), AGal = 4-Acetamido-4,6-didesoxygalactose (Testzucker), X = aus synthetisiertem TDP-Zucker isolierte 4-Acetamido-4,6-didesoxy-hexose. Fließmittel A (in Fließmittel D erhielten wir ein analoges Bild, in Fließmittel B wandern die beiden Testzucker und X mit gleicher Wanderungsgeschwindigkeit

aktive neutrale Bande, welche eluiert wurde. Dünnschichtchromatographie in 3 Lösungsmittelsystemen (A, B, D) zusammen mit authentischen 4-Acetamido-4,6-didesoxyhexosen zeigte, daß es sich bei dem synthetisierten Aminozucker vermutlich um das Derivat mit der Galactosekonfiguration handelt (Fig. 3).

DISKUSSION

S. minnesota und *S. ruiru* gehören zur *Salmonella*-Gruppe L mit dem O-Antigen 21. UDP-Glucose, UDP-Galactose, UDP-*N*-Acetylglucosamin und UDP-*N*-Acetylgalactosamin stellen in dieser Gruppe aktivierte Zuckervorstufen dar, welche für die Synthese des O-Antigens benötigt werden [8,23]. Außerdem wird KDO vermutlich in Form seines CMP-Derivates [27] verwandt, während die aktivierte Vorstufe von Heptose, einem weiteren Baustein der Polysaccharide, noch unbekannt ist. In der vorliegenden Arbeit wurden etwa 30 *S. minnesota* und *S. ruiru* R-Mutanten und vergleichsweise die entsprechenden Wildformen analysiert und die Aktivitäten einiger Enzyme bestimmt, welche die Synthese der UDP-Hexosen und UDP-Acetylhexosamine katalysieren.

Bei den weitaus meisten R-Mutanten war die Aktivität der untersuchten Enzyme gleich groß wie bei den Wildstämmen, und nur bei einer kleinen Anzahl von Mutanten konnten wir einen Defekt des UDP-Zucker Stoffwechsels identifizieren: 3 Mutanten besaßen einen Defekt der UDP-Glucose-Synthetase und 4 Mutanten einen Defekt der UDP-*N*-Acetylglucosamin-4-Epimerase. Vermutlich haben die übrigen Mutanten, bei denen ein Zucker-Synthese-Defekt nicht aufgefunden wurde, einen Transferase-Defekt. Tatsächlich haben wir kürzlich [22] einige unserer

16*

Mutanten als transferaselos identifiziert. In Tab. 4 sind alle *S. minnesota* und *S. ruiru* Mutanten zusammengefaßt, deren Defekt bekannt ist.

Wir haben allerdings nur einige Stoffwechselwege untersucht, die zu UDP-Glucose, UDP-Galactose, UDP-*N*-Acetylglucosamin und UDP-*N*-Acetylgalactosamin führen. Auch ist nicht bekannt, ob noch andere Nucleotid-Zucker eine Rolle bei der Biosynthese der Polysaccharide spielen. So hat Ginsburg [36] kürzlich in *Salmonella typhimurium* UDP-Rhamnose nachgewiesen. Die Funktion dieses Nucleotid-Zuckers ist unbekannt. Die Untersuchungen von Preiss *et al.* [37,38] über die Rolle von ADP-Glucose bei der Biosynthese von bakteriellem Glykogen zeigen jedoch, daß für Regulationen der Polysaccharid-Synthese Variationen im Nucleotidteil der aktivierten Vorstufen von großer Bedeutung sind. Es ist daher nicht ausgeschlossen, daß auch in *S. minnesota* noch unbekannte Nucleotidzucker einzelne Schritte der Polysaccharid-Synthese bestimmen.

Ob die *in vitro* gemessene Restaktivität auch in der Zelle eine Rolle spielt, erscheint unwahrscheinlich. Die kleinen Mengen Glucose (etwa 1%), welche in den bisher analysierten Lipopolysacchariden von *S. minnesota* Rz1 und Rz2 gefunden wurden, sprechen dafür, daß *in vivo* eine für die Transferase-Reaktion brauchbare Nucleotid-Glucose nicht synthetisiert wird. Wie wir kürzlich zeigen konnten [22], besitzt die Mutante Rz1 sowohl aktive Glucosyl-I-Transferase wie auch Acceptoraktivität für den Glucose-I-Transfer. Alles spricht daher dafür, daß der Defekt dieser Mutanten die Synthese von UDP-Glucose betrifft.

Trotz wiederholter Bemühungen ist es uns bisher nicht gelungen, aus *S. minnesota*- und *S. ruiru*-Stämmen eine UDP-Galactose-epimeraselose Mutante („M-Mutante") zu isolieren, wie sie bei *S. typhimurium* [29] und bei *E. coli* bekannt ist [30,31]. Bei der Untersuchung von galactose-negativen Mutanten stießen wir auf UDP-Glucose-Synthetase-lose, aber

Tabelle 4. *Zusammenstellung der* S. minnesota *und* S. ruiru R-*Mutanten, bei denen ein Enzymdefekt identifiziert wurde*
s = *Salmonella minnesota*; r = *Salmonella ruiru*

Mutanten	Zucker im Lipopolysaccharid	Art des Enzym-Defekts
mR7	KDO, Hep	Glucose-I-Transferase [22]
mR3	KDO, Hep	Synthese der Heptose-Kette [22]
rR3	KDO, Hep	Synthese der Heptose-Kette [22]
rR1	KDO, Hep (Glc)[a]	Glucose-I-Transferase [22] und UDP-GlcNAc-Epimerase
rR541	KDO, Hep (Glc, Gal)[a]	Glucose-I-Transferase [22] und UDP-GlcNAc-Epimerase
mRz1	KDO, Hep (Glc)[a]	UDP-Glucose-Synthetase
mRz2	KDO, Hep (Glc)[a]	UDP-Glucose-Synthetase
m1111 gal⁻	ng	UDP-Glucose-Synthetase
mR60	KDO, Hep, Glc, Gal, GlcNAc	UDP-GlcNAc-Epimerase
rR128	KDO	UDP-GlcNAc-Epimerase (und unbekannter Defekt)

[a] Sehr geringe Mengen von Glc bzw. Gal.

UDP-Glucose-Synthetase-defekte Mutanten waren bislang nur von *E. coli* K 12 [28] und *E. coli* B [24] isoliert worden. Die von uns aus *S. minnesota* erhaltenen Mutanten zeigen, unter den angegebenen Bedingungen in vitro getestet, eine Synthetase-Restaktivität von 12—24% verglichen mit der Wildform (= 100%). Ebenso verhielt sich die Mutante aus *E. coli* B. Das Auffinden einer gewissen Restaktivität kann folgende Ursachen haben: a) Es wird weniger Synthetase gebildet, b) der genetische Defekt führt zur Synthese eines Enzymproteins, das gegenüber dem der Wildform nur wenig verändert ist und deshalb das Substrat mit geringerer Geschwindigkeit umsetzt, c) die Synthese der UDP-Glucose-Synthetase fällt vollständig aus, es sind jedoch Synthetasen für andere Nucleotidzucker vorhanden (z. B. für ADP-Glucose), die über eine ausreichende Spezifitätsbreite verfügen, um kleine Mengen von UDP-Glucose zu synthetisieren.

nie auf epimeraselose Stämme. Es wäre möglich, daß diese M-Mutanten extrem galactosesensibel sind, oder daß Galactose in *S. minnesota* noch auf einem anderen Wege als über UDP-Gal → UDP-Glc abgebaut werden kann und die Mutante sich so der Selektionierung nach dem Merkmal galactose-negativ/galactose-sensibel entzieht. Andererseits hat man bislang vergeblich nach einer UDP-Glucose-synthetase-defekten Mutante von *S. typhimurium* gesucht.

S. minnesota R60 ist eine früher bereits beschriebene UDP-*N*-Acetylglucosamin-4-epimeraselose Mutante. Sie vermag Galactosamin nicht zu synthetisieren und kann daher die spezifischen Seitenketten des O21 Polysaccharids nicht aufbauen. Ihr Lipopolysaccharid stellt den vollständigen Polysaccharid-Kern dar (RII-Mutante [8]).

Die Mutanten *S. ruiru* R1, R541 und R128, welche ebenfalls nicht in der Lage sind, UDP-*N*-Acetylglucosamin-Epimerase zu synthetisieren, sind

Mehrfachmutanten. Der zweite Block von R 128 liegt vermutlich tiefer und betrifft möglicherweise eine Heptose-Transferase oder -Synthetase. Ihr Lipopolysaccharid enthält KDO als einzigen Zucker. Dagegen kennen wir den zusätzlichen Block in *S. ruiru* R 1 und R 541 (Chemotyp Rd₁): Er betrifft die Synthese von Glucose-I-Transferase [22]. Beide Keime enthalten KDO, Heptose und kleine Mengen Hexose im Lipopolysaccharid. In die Zellwandfraktionen dieser Mutanten kann man mit Hilfe eines löslichen Transferaseenzyms aus einer transferase-positiven Mutante Glucose einbauen [22].

Die in Tab. 4 aufgeführten Mutanten *S. minnesota* R 3 und *S. ruiru* R 3 (Chemotyp Rd₂) haben einen tiefliegenden Defekt in der Synthese des Heptose-KDO-Grundgerüstes. Ihr Lipopolysaccharid enthält KDO und, verglichen mit R 541, weniger Heptose. Sie können nicht als Acceptoren für den Einbau von Glucose dienen [22].

Während die Enzyme des UDP-Glucose- und UDP-Galactose-Weges in den Bakterienextrakten leicht bestimmt werden konnten, stießen wir bei einigen Enzymschritten des Aminozuckerstoffwechsels auf Schwierigkeiten. Bisher war es uns nicht möglich, in den rohen bakteriellen Enzympräparaten Glucosamin-6-phosphat-Acetylase zu bestimmen. Wir fanden auch, daß die mit Hefeacetylase katalysierte Glucosamin-6-*P*-Acetylase-Reaktion, die wir zur Mikrobestimmung von Glucosamin verwenden [32], durch kleine Mengen des Bakterienextrakts völlig unterdrückt wird. Die Hemmung kann durch Erhitzen des bakteriellen Extraktes auf 100°, nicht aber durch einen Überschuß an ATP oder Coenzym A aufgehoben werden. Bei dem störenden Faktor handelt es sich vielleicht um eine Acetyl-CoA-Hydrolase. Versuche von Kornfeld u. Glaser [35a] und Nikaido [35b] machen es wahrscheinlich, daß die Acetylierung des Glucosamins nicht auf der Stufe der Glucosamin-6-*P*, sondern am Glucosamin-1-*P* erfolgt (siehe Fig. 1). Die entsprechende Mutase konnte jedoch noch nicht nachgewiesen werden. In unseren Versuchen konnte nicht ausgeschlossen werden, daß die ME-positive Reaktion beim Test auf UDP-GlcNAc-Synthetase durch Bildung von freiem Acetylglucosamin verursacht wird, das durch die Einwirkung einer Phosphatase auf Acetylglucosamin-1-*P* entsteht. Die Frage nach dem tatsächlichen Ansatzpunkt der Acetylierung bei *S. minnesota* bleibt somit offen.

Die Synthese einer TDP-4-Amino-4,6-didesoxyhexose mit Hilfe von Extrakten aus *E. coli*- und *Salmonella*-Stämmen [7,25], die wir nun auch mit Enzymen aus *S. ruiru* durchführen konnten, wirft die Frage nach der Funktion eines solchen Zuckers bzw. seines Nucleotid-Derivates für das Bakterium auf. Da das vollständige Enzymsystem zur Umwandlung von TDP-Glucose in das Aminozucker-Derivat vorhanden ist, liegt es nahe, an eine Übertragung des Aminozuckers auf ein Polysaccharid zu denken. Wir haben bisher jedoch noch keinen direkten Beweis dafür, daß 4-Amino-4,6-didesoxygalactose im Lipopolysaccharid unserer *Salmonella*-Stämme vorkommt. Ein elektrophoretisch schneller als Glucosamin wandernder, ninhydrin-positiver Zucker, der auf Pherogrammen schonend hergestellter Hydrolysate von *S. minnesota*-Lipopolysaccharid gefunden wurde, konnte wegen seiner großen Labilität bisher nicht identifiziert werden.

In diesem Zusammenhang mag von Interesse sein, daß es bisher nicht gelang, in Partialhydrolysaten von *Salmonella*-Lipopolysacchariden Oligosaccharide nachzuweisen, die Zuckerbausteine sowohl aus der Basalstruktur als auch aus den spezifischen Seitenketten enthielten. Die Verknüpfung beider Strukturbereiche ist offenbar sehr labil. Man könnte daran denken, daß ein säurelabiler Zucker von der Art der 4-Amino-4,6-didesoxygalactose Zwischenglied ist und bei Hydrolyse des Lipopolysaccharids nach den üblichen Methoden stets zerstört wird. 4-Acetamido-4,6-didesoxyglucose (Viosamin) ist kürzlich im Lipopolysaccharid von *Chromobacterium violaceum* entdeckt worden [33,34].

Anmerkung bei der Korrektur: 4-Amino-4,6-didesoxygalactose und -glucose wurden kürzlich in Lipopolysacchariden aus *E. coli* nachgewiesen (diese Zeitschrift, in Vorbereitung).

Professor Dr. F. Kauffmann, Statens Seruminstitut, Kopenhagen, Frau Johanna Schlosshardt, Zentrallaboratorium für Bakterielle Darminfektionen, Potsdam, und Dr. G. Schmidt, Max-Planck-Institut für Immunbiologie, Freiburg, haben die R-Mutanten isoliert und zur Verfügung gestellt, wofür wir Ihnen bestens danken. Dr. H. Nikaido sind wir für viele theoretische und praktische Ratschläge und Hinweise sehr dankbar. Fräulein M. Rath danken wir für die Ausführung vieler Analysen und der Deutschen Forschungsgemeinschaft für die Förderung dieser Arbeiten durch eine Sachbeihilfe.

LITERATUR

1. Lüderitz, O., Staub, A. M., und Westphal, O., *Bacteriol. Rev.* 30 (1966) 192.
2. Lüderitz, O., Galanos, C., Risse, H. J., Ruschmann, E., Schlecht, S., Schmidt, G., Schulte-Holthausen, H., Wheat, R., Westphal, O., und Schlosshardt, J., *Ann. N. Y. Acad. Sci.* 133 (1966) 349.
3. Subbaiah, T. V., und Stocker, B. A. D., *Nature*, 201 (1964) 1298.
4. Beckmann, I., Subbaiah, T. V., und Stocker, B. A. D., *Nature*, 201 (1964) 1299.
5. Nikaido, H., Nikaido, K., Subbaiah, T. V., und Stocker, B. A. D., *Nature*, 201 (1964) 1301.
6. Osborn, M. S., Rosen, S. M., Rothfield, L., Zeleznick, L. D., und Horecker, B. L., *Science*, 145 (1964) 783.
7. Matsuhashi, M., und Strominger, J. L., *J. Biol. Chem.* 239 (1964) 2454.
8. Lüderitz, O., Risse, H. J., Schulte-Holthausen, H., Strominger, J. L., Sutherland, I. W., und Westphal, O., *J. Bacteriol.* 89 (1965) 343.
9. Weichselbaum, T. E., *Am. J. Clin. Pathol.* 10 (1946) 40.
10. Glaser, L., und Brown, D. H., *Proc. Natl. Acad. Sci. U. S.* 48 (1955) 2187.

11. Nathenson, S. G., und Strominger, J. L., *J. Biol. Chem.* 238 (1963) 3161.
12. Pazur, I. H., und Shuey, E. W., *J. Biol. Chem.* 236 (1961) 1780.
13. Bergmeyer, H. U., *Methoden der enzymatischen Analyse*, Verlag Chemie, Weinheim 1962, (a) S. 982; (b) S. 991; (c) S. 993.
14. Munch-Petersen, A., Kalckar, H. M., und Smith, E. E. B., *Biol. Skrifter Danske Videnskab. Selskab.* 22 (1955) 3.
15. Fujimoto, A., Ingram, P., und Smith, R. A., *Biochim. Biophys. Acta,* 96 (1965) 91.
16. Maxwell, E. S., Kurahashi, K., und Kalckar, H. M., in *Methods in Enzymology* (edited by S. P. Colowick and N. O. Kaplan), Academic Press, 1962, Vol. V, p. 174.
17. Fischer, W., und Zapf, J., *Hoppe Seyler's Z. Physiol. Chem.* 337 (1964) 186.
18. Ghosh, S., und Roseman, S., in *Methods in Enzymology* (edited by S. P. Colowick and N. O. Kaplan), Academic Press, 1962, Vol. V, p. 414.
19. Reissig, J. L., Strominger, J. L., und Leloir, L. F., *J. Biol. Chem.* 217 (1956) 959.
20. Strominger, J. L., Park, J. T., und Thompson, R. E., *J. Biol. Chem.* 234 (1959) 3263.
21. Kickhöfen, B., und Westphal, O., *Z. Naturforsch.* 7 b (1952) 655.
22. Risse, H. J., Dröge, W., Ruschmann, E., Lüderitz, O., Westphal, O., und Schlosshardt, J., *European J. Biochem.* 42 (1967) 216.
23. Ginsburg, V., *Advan. Enzymol.* 26 (1964) 35.
24. Hattman, S., und Fukasawa, T., *Proc. Natl. Acad. Sci. U. S.* 50 (1963) 279.
25. Gilbert, J., Matsuhashi, M., und Strominger, J. L., *J. Biol. Chem.* 240 (1965) 1305.
26. Stevens, S. L., Blumbergs, P., Otterbach, D. M., Strominger, J. L., Matsuhashi, M., und Dietzler, D. N., *J. Am. Chem. Soc.* 86 (1964) 2937.

27. Ghalambor, A., und Heath, E. C., *J. Biol. Chem.* 241 (1966) 3216.
28. Fukasawa, T., Jokura, K., und Kurahashi, K., *Biochem. Biophys. Res. Commun.* 7 (1962) 121.
29. Fukasawa, T., und Nikaido, H., *Biochim. Biophys. Acta,* 48 (1961) 470.
30. Sundararajan, T. A., Rapin, A. M., und Kalckar, H. M., *Proc. Natl. Acad. Sci. U. S.* 48 (1962) 2187.
31. Elbein, A. D., und Heath, E. C., *J. biol. Chem.* 240 (1965) 1919.
32. Lüderitz, O., Simmons, D. A. R., Strominger, J. L., und Westphal, O., *Anal. Biochem.* 9 (1964) 263.
33. Smith, E. J., Leatherwood, J. M., und Wheat, R., *J. Bacteriol.* 84 (1962) 1007.
34. Stevens, C. L., Blumbergs, P., Daniher, F. A., Wheat, R. W., Kujomoto, A., und Rollins, E. L., *J. Am. Chem. Soc.* 85 (1963) 3061.
35a. Kornfeld, S., und Glaser, L., *J. Biol. Chem.* 237 (1962) 3052.
35b. Nikaido, H., persönliche Mitteilung.
36. Ginsburg, V., *J. Biol. Chem.* 241 (1966) 3750.
37. Shen, L., und Preiss, J., *J. Biol. Chem.* 240 (1965) 2334.
38. Greenberg, E., und Preiss, J., *J. Biol. Chem.* 240 (1965) 2341.
39. Beckmann, I., Lüderitz, O., und Westphal, O., *Biochem. Z.* 339 (1964) 401.

H. J. Risse, gegenwärtige Adresse:
Physiologisch-Chemisches Institut der Freien Universität
1 Berlin 33, Arnimallee 22, Deutschland

O. Lüderitz, O. Westphal
Max-Planck-Institut für Immunobiologie
78 Freiburg i. Br., Postfach 1668, Deutschland

European J. Biochem. 1 (1967) 243—250

Zum Mechanismus der Ribonuclease-Reaktion

2. Die Vorordnung im Substrat als geschwindigkeitssteigernder Faktor bei Dinucleosidphosphaten und analogen Verbindungen

H. Follmann, H.-J. Wieker und H. Witzel

Chemisches Institut der Universität Marburg/Lahn

(Eingegangen am 19. Dezember 1966)

In the previous paper [1] we proposed a mechanism for the reaction of pancreatic ribonuclease with nucleotide diesters, in which the kinetic constants k_{+1} and $k_{-1} \mid k_{+2}$ are assigned to a base catalysed step for the formation, and a proton catalysed step by the conjugate acid for the breakdown, of an enzyme stabilized intermediate. In the hydrolysis reaction of the 2',3'-cyclic diesters the rates depend on the structure of the catalysing base, and in particular on its polarisability. In addition to this effect, we found that, in the dinucleoside phosphate series, the rates increased up to three orders of magnitude, depending on the nature of the base of the second nucleoside [4].

According to our suggested mechanism I—VI the accelerated rates could be due to (a) the second base influencing the polarisability of the catalysing system by π-electron interactions between the stacked bases. This effect should be reflected in the enthalpy term and should parallel the hypochromicity of the dinucleoside-phosphates exactly; (b) pre-ordering of the reacting atoms (C-2-oxygen, 2'-OH-group and phosphorus) by the base stacking, which would be reflected in a lowering of the activation energy in the entropy term and should only roughly parallel the hypochromicity.

To differentiate between these two possibilities we prepared the diesters IX—XXXIV, determined the hypochromic effect and also the values of K_m and k_{+2} at pH 7. There is only a partial correlation between k_{+2} values and the hypochromic effect. The k_{+2} values vary with the structure of the second nucleoside and show a particular dependence upon the nature of the link between base and phosphate group. Highest values are found with adenosine and deoxyadenosine. Elongating the ribose with mercaptoethanol (XIX) or substituting the relatively rigid ribofuranose ring with flexible n-hexyl-, pentyl- or butyl-chains (XX—XXII) appreciably reduces the hypochromicities and also the k_{+2} values. In the case of a propyl link (XXIII, XXIV) or in the analogously structured 3'-3'-dinucleosidephosphate XXV no hypochromic effect is found and the k_{+2} values are as low as that of 3'-cytidylic acid benzyl ester.

A second nucleoside attached to the 5' end of a substrate (XXXIII, XXXIV) does not result in an increase of the rate of diester hydrolysis at the 3' end though there is a normal hypochromicity. Therefore a pre-ordering of the reacting atoms (demonstrable on models) appears to be the accelerating factor. A specific interaction between the second nucleoside and the enzyme being responsible for the pre-ordering effect can be excluded by either blocking or varying all the potential interaction sites.

The concept of a pre-ordering in the substrate due to base stacking agrees with the results from measurements of the temperature dependence of k_{+2} and calculations of the enthalpy ($\Delta H^{\neq}$) and entropy ($\Delta S^{\neq}$) term in the activation energy of this step. When diesters with an exactly analogous structure of the intermediate II such as CpA (IX) and Cp-butyladenine (XXII) are compared, the differences are found in the entropy term only. Between the first and second step substrates such as CpA and 2',3'-Cp the differences are also found in the entropy term with a small factor in the enthalpy term.

From a comparison of Up- and Cp-diesters it can be seen that due to the differences in the enthalpy term the Up-diesters should react 5—10 times faster than Cp-diesters, but that due to the differences in the entropy term Cp-diesters react 20 times faster than Up-diesters. This is in agreement with our concept that the enthalpy term of the activation energy is derived from the transfer of a proton from the conjugate acid of the pyrimidine base to the 5'- (or 2'-)oxygen, what should be easier the more acidic the conjugate acid is. However, the probability that the base is in the form of its conjugate acid and in the right position to transfer the proton is appreciably higher in the case of the more basic cytidine than with uridine. This explains the observations made in the previous paper [1] that, in spite of an acid-base-catalysis mechanism for k_{+1} and for k_{-1} and k_{+2}, the rates do not directly parallel the acidity or basicity of the catalysing groups.

Die Reaktion der Pankreas-Ribonuclease[1] folgt, wie wir aus den Ergebnissen in der vorhergehenden Arbeit [1] annehmen müssen, dem Schema I—VI:

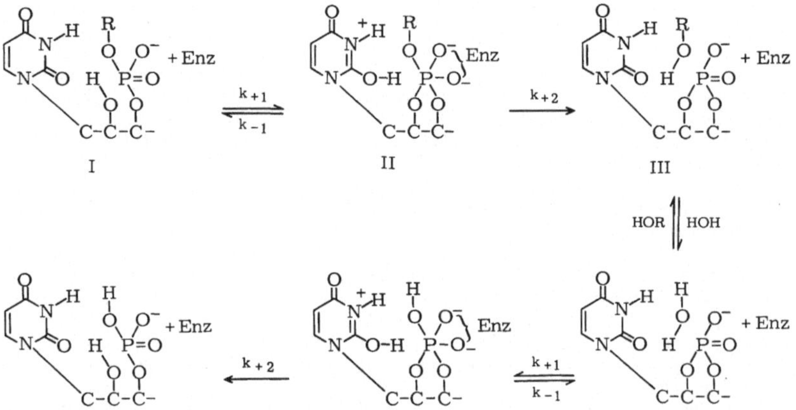

Wir erwarten aus diesem Mechanismus, daß sich die Aktivierungsenergie im k_{+2}-Wert zusammensetzt aus einem Enthalpieterm, der sich aus der Energie

Hierbei beinhaltet die Geschwindigkeitskonstante k_{+1} drei gleichzeitig ablaufende Bindungswechsel. Bei der Umesterung (I—III) sind dies die Übernahme des Protons von der 2'-OH-Gruppe durch die Pyrimidinbase (Basenkatalyse), die Reaktion des 2'-Sauerstoffs mit dem Phosphor und die Stabilisierung des pentacovalenten Zustandes durch eine Protonierung vom Enzym her; diese Protonierung wirkt sich bei dessen Ausbildung bereits als Säurekatalyse aus. Es wurde weiterhin gezeigt, daß die größte Aktivierungsenergie offensichtlich für den Übergang des Protons bei der Basenkatalyse aufzubringen ist.

k_{-1} und k_{+2} sind gekoppelte Konstanten für eine protonenkatalysierte Alternativreaktion [2]. Von der Pyrimidinbase in Form ihrer konjugaten Säure wird das Proton entweder auf den C-2'-Sauerstoff (Rückreaktion, k_{-1}) oder den C-5'-Sauerstoff (Produktbildung, k_{+2}) übertragen. Die größte Aktivierungsenergie muß wieder für diesen Protonenübergang aufgebracht werden, während die Lösung der P-O-Bindung sowie der Bindung zum Enzym offensichtlich mit geringerer Aktivierungsenergie im gleichen Zuge folgt. Für die Hydrolysereaktion (IV—VI) gelten die gleichen Zusammenhänge.

[1] *Enzym:* Pankreas-Ribonuclease (EC 2.7.7.16).

für den Übergang VIII → VII herleitet, und aus einem Entropieterm, der aus der Wahrscheinlichkeit resultiert, mit der die Base in Form der konjugaten Säure bei der Rotation um die Glykosidbindung die richtige Lage zum Acceptor-Sauerstoff einnimmt.

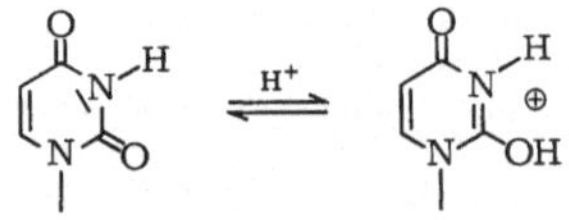

In vorhergehenden Untersuchungen an 2',3'-cyclischen Diestern mit abgewandelten Basen [1,3] hatten wir festgestellt, daß in diesen Verbindungen die Werte von k_{+2} im wesentlichen von der Polarisierbarkeit der Base abhängen (Erniedrigung der Aktivierungsenergie in VII ⇌ VIII), nicht aber von ihrer Acidität oder Basizität, obwohl eine Protonenkatalyse geschwindigkeitsbestimmend ist. Gleiches gilt für den k_{+1}-Wert des basenkatalysierten Schrittes. Es wurde beobachtet, daß bei einer Veränderung von k_{+2}, das mit k_{-1} gekoppelt ist [2], keine Veränderung von $K_m = (k_{-1} + k_{+2})/k_{+1}$ auftritt: k_{+1} muß demnach um den gleichen Faktor verändert worden sein.

Neben dieser Abhängigkeit der Geschwindigkeiten von der Struktur der Base fanden wir bei den 3'-5'-Dinucleosidphosphaten erhebliche Steigerungen in Abhängigkeit von der Base des 5'-Nucleosids [4]. Sie betragen z. B. gegenüber dem 3'-Cytidylsäurebenzylester bei CpU das 14fache, bei CpA das 1200fache, obgleich jedesmal die Esterbindung zur 5'-OH-Gruppe der Ribose gespalten wird. Da K_m wieder konstant bleibt, sind sowohl $k_{-1} + k_{+2}$ als auch k_{+1} jeweils um gleiche Faktoren erhöht.

Auf Grund einer Parallelität dieser Steigerung der Reaktionsgeschwindigkeiten mit dem Hypochromie-Effekt [5,6] der Dinucleosidphosphate hatten wir früher diskutiert, daß die Polarisierbarkeit der katalysierenden Base eventuell über die π-Elektronen-Wechselwirkungen zwischen beiden Basen erhöht sein könnte [4,7], wodurch die Aktivierungsenergie (in diesem Fall im Enthalpieterm) herabgesetzt wäre. Gestützt wurde diese Vorstellung durch die Beobachtung, daß UpA-N^1-oxid mit einer gestörten π-Wechselwirkung einen 70fach niedrigeren k_{+2}-Wert aufweist als UpA.

Der durch die neueren Untersuchungen erhaltene Einblick in die Details des Reaktionsmechanismus läßt jedoch erkennen, daß eine Erhöhung der Reaktionsgeschwindigkeiten auch über den Entropieterm erreicht werden könnte, wenn durch eine Wechselbeziehung zwischen beiden Basen die Beweglichkeit des Moleküls eingeschränkt wird. An Kalottenmodellen erkennt man, daß bei einer Übereinander-Schichtung der Basen *(base stacking)* das Aktionszentrum mit dem C-2-Sauerstoff, der 2'-OH-Gruppe und dem Phosphor fast ideal vorgeordnet wird.

Wir haben deswegen eine Reihe von Dinucleosidphosphaten synthetisiert, bei denen sowohl die zweite Base als auch die relativ starre Ribose-Verknüpfung zwischen der zweiten Base und dem Phosphor verändert wurden. In ihnen sollte, auch wenn ein *base stacking* noch möglich ist, die ideale Vorordnung des Aktionszentrums empfindlich gestört sein. Durch Untersuchung der Temperaturabhängigkeit sollte dann entschieden werden, ob sich die Veränderungen der k_{+2}-Werte im Enthalpie- oder im Entropieterm reflektieren. Wir geben im folgenden die Ergebnisse dieser Untersuchungen bekannt.

EXPERIMENTELLER TEIL

Substrate

Die Dinucleosidphosphate CpA, CpG, CpC, CpU, UpA, UpG, UpC und UpU wurden aus Partialhydrolysaten von Hefe RNS mit Bi(OH)$_3$ [8] ge wonnen, oder wie X, XII—XIV, XVII und XVIII nach einem vereinfachten Kondensationsverfahren synthetisiert, das an anderer Stelle beschrieben wird [9]. XXVII und XXX wurden hergestellt durch Oxidation von UpA bzw. UpC mit Monoperphthalsäure bei pH 7 und Raumtemperatur [10], XXXI

wurde aus UpC durch Methylierung mit Dimethylsulfat analog [11] erhalten; die dabei entstandenen Produkte wurden chromatographisch oder elektrophoretisch getrennt und durch ihre UV-Spektren vor und nach dem Abbau mit Pankreas-RNase identifiziert. Relative Wanderungsstrecken bei Papierelektrophorese in 0,02 N HCOOH (pH 2,8) mit 300 V, 12 Std: XXVI = 21, XXVII = 49, XXIX = 17, XXX = 25, Up = 100.

(3'-5')-ApC-3'-p wurde aus einem RNase-Hydrolysat von Hefe-RNS gewonnen, mit Diazomethan in Wasser/Dimethylsulfoxid bei pH 5 zum Methylester XXXIII oder mit Chlorameisensäureäthylester wie in [1] zum 2',3'-Cyclophosphat XXXIV umgesetzt und beide Verbindungen papierchromatographisch gereinigt.

Die Cytidylsäureester XIX—XXV wurden durch Kondensation von N^4, $O^{2'}$, $O^{5'}$-Triacetyl-3'-cytidylsäure [12] mit dem betreffenden nucleosidanalogen Alkohol nach folgendem allgemeinen Verfahren synthetisiert: 105 mg Triacetyl-Cp-Pyridiniumsalz (0,2 mMol), 0,3 mMol des Nucleosid-Analogen und 250 mg (1,2 mMol) Dicyclohexylcarbodiimid wurden in möglichst wenig (3—6 ml) wasserfreiem Pyridin gelöst und bis zur Trübung Wasser bei Raumtemperatur 8—10 Tage stehen gelassen. Dann wurde bis zur Trübung Wasser zugesetzt, der ausgefallene Dicyclohexylharnstoff abfiltriert, die Mischung dreimal mit je 10 ml Petroläther extrahiert und zur Trockene eingedampft. Der Rückstand wurde über Nacht mit 20 ml wasserfreiem, mit Ammoniak gesättigtem Methanol behandelt, erneut zur Trockene eingedampft und nun in Wasser aufgenommen. Nach Filtration von Ungelöstem wurde das Produkt aus der Lösung durch Papierchromatographie (absteigend in Isopropanol/Wasser/2 N NH$_3$; 7:2:1, v/v) oder Anionenaustauschchromatographie (Dowex 1 × 2, Formiatform, Elution mit einem HCOOH-Gradienten wie in [8]) in durchschnittlich 50 % Ausbeute isoliert. Die Struktur aller Diester ist gesichert durch ihre spektralen Daten und die vollständige Spaltung durch NaOH oder RNase, die nur Cytidylsäure und das betreffende Nucleosid-Analoge liefert. Die Eigenschaften der Ausgangsverbindungen und der Diester dieser Serie enthält Tab. 1.

Kinetische Messungen

Die kinetischen Messungen wurden, soweit nicht anders vermerkt, nach der spektrophotometrischen Methode wie in [1] durchgeführt. Nach Zugabe der RNase (Hormon-Chemie, München) wurde der Abfall der Extinktion im Bereich von 285—295 mμ registriert, der mit der Bildung der 2',3'-cyclischen Nucleotide einhergeht. Solange die Hydrolyse des cyclischen Diesters, die von einem Wiederanstieg der Extinktion begleitet ist, mehr als fünfmal langsamer verlief, wurde keine Korrektur vorgenommen; bei

Tabelle 1

Nucleosid-Analoges	Schmelzpunkt	λ_{max}	R_F	3'-Cp-Ester [a]	Eluiert bei pH	R_F
	°C	mμ				
2',3'-Isopropyliden-5'-deoxy-5'-(hydroxy-äthylthio)adenosin [b]	144—145	259	0,82	XIX		0,52
9-(6'-Hydroxyhexyl)-adenin [c]	188—189	261	0,78	XX	3,06	0,52
9-(5'-Hydroxypentyl)-adenin [c]	190—192	261	0,73	XXI	3,00	0,42
9-(4'-Hydroxybutyl)-adenin [14]	195—197	260	0,71	XXII	3,05	0,37
9-(3'-Hydroxypropyl)-adenin [14]	204—206	260	0,65	XXIII		0,35
2-(3'-Hydroxypropyl)-benzimidazol [d]	162—163	278 272	0,93	XXIV		0,74
5'-O-Acetyl-2'-deoxy-adenosin [15]	—	260	0,73	XXV		0,42

[a] Nummer in Tab. 2.
[b] Aus der 5'-Tosyl-Verbindung mit β-Mercaptoäthanol und Natrium in flüssigem Ammoniak analog [13].
[c] Analog zur Vorschrift in [14].
[d] Käufliches Produkt der Ega-Chemie, Heidenheim.

langsameren Substraten wurde verfahren wie in [4]. Die Anfangsgeschwindigkeiten wurden aus der Neigung bei halblogarithmischer Auftragung der Substratkonzentration gegen die Zeit errechnet, die Werte von K_m und k_{+2} nach dem Verfahren von Lineweaver-Burk ermittelt. Die Anfangskonzentration an Substrat wurde spektrophotometrisch unter Berücksichtigung der Hypochromie bestimmt.

Die Werte der Hypochromie der Diester, bezogen auf die Extinktion des Gemisches beider Komponenten, wurden aus dem Extinktionsanstieg im Absorptionsmaximum nach Zugabe von weniger als 1 % des Volumens an RNase-Lösung geeigneter Konzentration errechnet und sind auf ± 0,3 % reproduzierbar. Da das Ausmaß der Hypochromie von der Ionenstärke der Lösung abhängt, sind die Werte nicht direkt mit denen anderer Autoren [5,6] vergleichbar.

Die Konstanten K_m und k_{+2} sowie der Hypochromie-Effekt der 3'-Cytidylsäureester IX—XXV, XXXIII und XXXIV sowie der Uridylsäureester XXVI—XXXII, in 0,1 M Dimethylglutarsäure-puffer der Ionenstärke 0,2 (durch NaCl-Zusatz) bei 25° und pH 7 sind in Tab. 2 zusammengestellt.

Temperaturabhängigkeit

Fig. 1 zeigt die Auftragung von log k_{+2} gegen die reziproke absolute Temperatur $1/T$ für die angegebenen Verbindungen. Aus den Absolutwerten für k_{+2} wurden für die Aktivierungsenergie nach Gleichung (1) die $\Delta G^{\ddagger}$-Werte, aus der Neigung der Geraden in Fig. 1 und Gleichung (2) die Aktivierungsenthalpien $\Delta H^{\ddagger}$ und aus Gleichung (3) die Aktivierungsentropien $\Delta S^{\ddagger}$ ermittelt, die in Tab. 3 zusammengestellt sind.

$$\Delta G^{\ddagger} = RT \left(\ln \frac{k_B}{h} \cdot T - \ln k_{+2} \right) \qquad (1)$$

$$\frac{d(\ln k_{+2})}{d(1/T)} = -\frac{\Delta H^{\ddagger} + RT}{R} \qquad (2)$$

$$\Delta S^{\ddagger} = \frac{\Delta H^{\ddagger}}{T} - R \left(\ln \frac{k_B}{h} \cdot T - \ln k_{+2} \right) \qquad (3)$$

(k_B = Boltzmann-Konstante, R = allgemeine Gaskonstante, h = Plancksches Wirkungsquantum)

Tabelle 2

Substrat			λ_{max} (pH 7)	Hypochromie	k_{+2}	K_m
			mμ	%	sec^{-1}	mM
IX	(3'-5')Cytidylyl-adenosin	(CpA)	261	7,5	2350	1,4
X	(3'-5')Cytidylyl-2'-deoxyadenosin	(CpdA)	261	7	2350 [a]	
XI	(3'-5')Cytidylyl-guanosin	(CpG)	255, 270	5	220	1,4
XII	(3'-5')Cytidylyl-purin-9-ribosid		263	5,5	600	5,0
XIIIa	(3'-5')Cytidylyl-N^6-methyladenosin		266	8	90	1,5
XIIIb	(3'-5')Cytidylyl-N^6-dimethyladenosin		272	9,5	40 [b]	5
XIV	(3'-5')Cytidylyl-3-isoadenosin		272	10	400	5,0
XV	(3'-5')Cytidylyl-cytidin	(CpC)	269	6	160	3,3
XVI	(3'-5')Cytidylyl-uridin	(CpU)	265	4,5	27	3,7
XVII	(3'-5')Cytidylyl-N^3-methyluridin		265	6,5	18 [c]	
XVIII	(3'-5')Cytidylyl-thymidin	(CpT)	268	8	15	1,1
XIX	(3'-5')Cytidylyl-2'3'-isopropyliden-5'-deoxy-5'-thioäthyladenosin		261	4	28 [d]	0,7
XX	(3'-6')Cytidylyl-9-hexyladenin		263	2,5	60	1,6
XXI	(3'-5')Cytidylyl-9-pentyladenin		263	1,9	30	1,4
XXII	(3'-4')Cytidylyl-9-butyladenin		263	2,3	50	1,3
XXIII	(3'-3')Cytidylyl-9-propyladenin		262	0	3	
XXIV	(3'-3')Cytidylyl-2-propylbenzimidazol		272, 278	0	3	1,5
XXV	(3'-3')Cytidylyl-2'deoxyadenosin		261	0	<3 [f]	

Tabelle 2 (Fortsetzung)

3'-Cp · R

R =

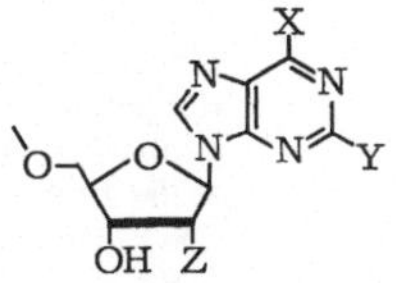

	X	Y	Z		X	Y	Z
IX	NH$_2$	H	OH	XVI	H	H	OH
X	NH$_2$	H	H	XVII	CH$_3$	H	OH
XI	OH	NH$_2$	OH	XVIII	H	CH$_3$	H
XII	H	H	OH				
XIII a	NHCH$_3$	H	OH				
XIII b	N(CH$_3$)$_2$	H	OH				

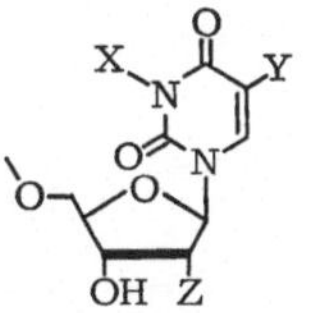

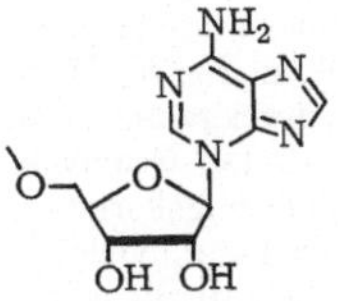

XIV

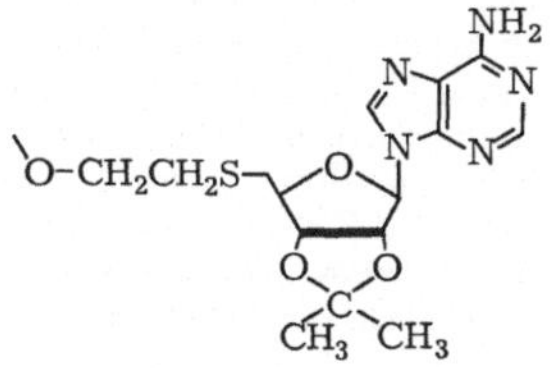

XIX

	X			
XX	n = 6			
XXI	5			
XXII	4			
XXIII	3	XXIV	XXV	

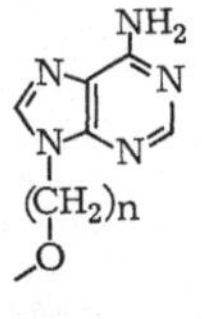

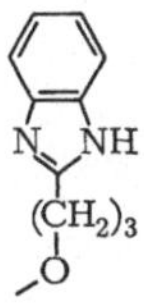

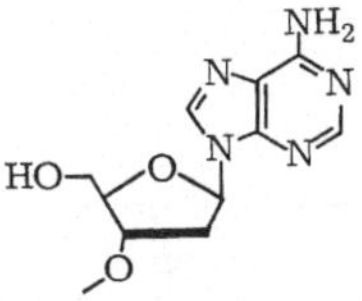

XXVI	(3'-5')Uridylyl-adenosin	(UpA)	260	3	1000	1,3
XXVII	(3'-5')Uridylyl-adenosin-N^1-oxid		261, 300	0	14	1,4
XXVIII	(3'-5')Uridylyl-guanosin	(UpG)	256	3	69	2,0
XXIX	(3'-5')Uridylyl-cytidin	(UpC)	263	1	26	1,7
XXX	(3'-5')Uridylyl-cytidin-N^3-oxid		264, 305	0	2g	
XXXI	(3'-5')Uridylyl-N^3-methylcytidin		265		20g	
XXXII	(3'-5')Uridylyl-uridin	(UpU)	261	0	11	3,7
XXXIII	(3'-5')Adenylyl-3'-cytidylsäure-methylester		261		1h	
XXXIV	(3'-5')Adenylyl-2',3'-cytidylsäure	(ApCp(c))	261	ApCp: 6,5	11	0,9

3'-Up-R

R =

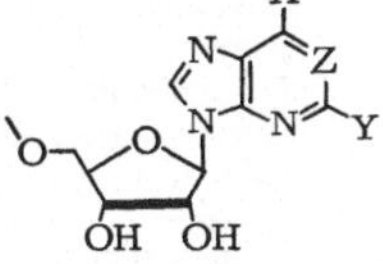

	X	Y	Z		Z
XXVI	NH$_2$	H	N	XXIX	N
XXVII	NH$_2$	H	N$^+$-O$^-$	XXX	N$^+$-O$^-$
XXVIII	OH	NH$_2$	N	XXXI	N$^+$-CH$_3$

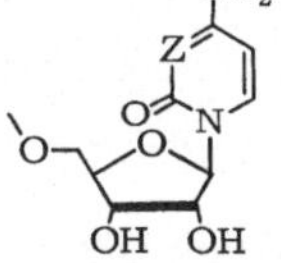

a Durch Vergleich der Anfangsgeschwindigkeit mit 3'-5'-CpA, gleicher K_m-Wert angenommen.
b Messung des Extinktionsanstieges im Absorptionsmaximum.
c Durch Vergleich mit OpU wie in a.
d Die entsprechende Verbindung ohne Isopropylidengruppe ließ sich nur in geringer Ausbeute isolieren; in der Spaltungsgeschwindigkeit besteht kein signifikanter Unterschied gegenüber XIX.
e Durch Vergleich mit XX wie in a.
f Da sich das UV-Spektrum von XXV nach RNase-Spaltung nur geringfügig ändert, wurden jeweils 0,2 µMol Substrat mit 0 2 pMol RNase in 20 µl Puffer gemischt, nach verschiedenen Inkubationszeiten die Ausgangs- und Spaltprodukte papierchromatographisch getrennt, eluiert und ihr Verhältnis bestimmt; zum Vergleich wurde mit XXIV ebenso verfahren.
g Durch Vergleich mit UpC wie in a.
h Durch Vergleich mit XXXIII wie in a.

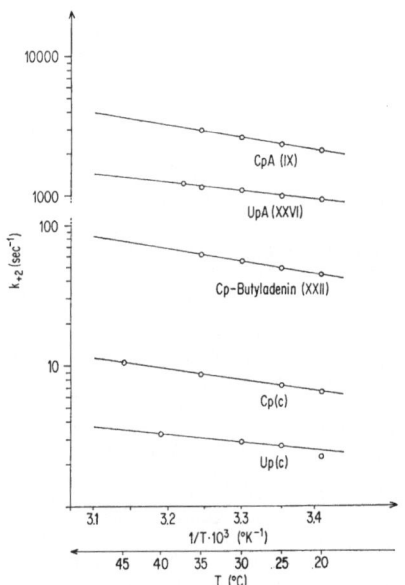

Fig.1. *Temperaturabhängigkeit von* k$_{+2}$ *für die Hydrolyse von Cytidylsäure- und Uridylsäureestern durch RNase bei pH 7,0 und der Ionenstärke 0,2*

Tabelle 3. *Thermodynamische Parameter der Aktivierungsenergie*
Spaltung durch RNase bei 25°, pH 7 (Dimethyl-glutarsäure-puffer), $I = 0,2$

Substrat	k_{+2}	$\Delta G^{\ddagger}$	$\Delta H^{\ddagger}$	$\Delta S^{\ddagger}$
	sec^{-1}	kcal/Mol		e. u.
2′,3′-Cp	7,3	16,2	2,9	−44,8
2′,3′-Up	2,7	16,8	1,8	−50,2
UpA (XXVI)	1000	13,3	2,1	−37,7
CpA (IX)	2350	12,8	3,5	−31,4
Cp-butyladenin (XXII)	50	15,1	3,5	−39,0

ERGEBNISSE UND DISKUSSION

Die Geschwindigkeit (k_{+2}) bei der Hydrolyse der 2′,3′-cyclischen Nucleotide (IV—VI) wird, wie in der vorhergehenden Arbeit begründet [1], durch die Übertragung eines Protons von der konjugaten Säure der Base auf den 2′-Sauerstoff bestimmt. Der Zwischenzustand (V) ist in bezug auf die Phosphatgruppe bei allen 2′,3′-cyclischen Substraten gleich, so daß Unterschiede in der Spaltungsgeschwindigkeit allein auf die Struktur der Base und die Aktivierungsenergie für den Übergang VIII → VII zurückzuführen sind.

Bei der Umesterungsreaktion (I—III) bleibt dieser Faktor bestehen, doch kommt mit dem Alkohol-rest -O-R, auf den das Proton übertragen wird, ein zweiter, bei den verschiedenen Substraten variabler Faktor hinzu. Unter den Alkylalkoholen wurde bei gleicher Base eine Abnahme der Spaltungsgeschwindigkeiten in der Reihe Benzyl > Methyl > Isopropyl beobachtet [4,16], in der zwar die Basizität des Sauerstoffs zunimmt, aber auch die P-O-Bindungsenergie. Die Untersuchungen zu diesem Problem sind noch nicht abgeschlossen und werden gesondert behandelt. Die in der Reihe der Dinucleosidphosphate beobachteten großen Unterschiede in den k_{+2}-Werten müssen auf einen dritten Effekt zurückgeführt werden, da alle Verbindungen die Ribose-5′-OH-Gruppe als Alkoholrest enthalten, der von der zugehörigen Base nicht wesentlich beeinflußt sein kann; vielmehr muß die Base des zweiten Nucleosids direkt oder indirekt in den Ablauf der Reaktion eingreifen.

Messungen der optischen Rotationsdispersion an Dinucleosidphosphaten lassen darauf schließen, daß diese in wäßriger Lösung eine geordnete Struktur besitzen, deren Stabilität von dem Basenpaar abhängt und weitgehend von dessen Hydrophobie bestimmt wird [6]. Als grobes Maß der Ordnung im Molekül kann schon der leicht zu messende hypochrome Effekt der Dinucleosidphosphate gelten, der ebenfalls auf der elektronischen Wechselwirkung zwischen übereinanderliegenden Basen beruht [5].

Ein Vergleich der Verbindungen IX—XXXII in Tab.2 zeigt, daß auch zwischen der Hypochromie der Substrate und ihren k_{+2}-Werten eine grobe Parallelität besteht. Beide sind offensichtlich von der Natur der zweiten Base abhängig, in Cytidyl- wie in Uridylsäureestern findet man die Reihenfolge Adenin > Guanin ~ Purin ~ Cytosin > Uracil. Die methylierten Pyrimidin- und Purinbasen verursachen höhere Hypochromie-Werte. Beim N^3-Ribosid des Adenins (in XIV) steigt die Hypochromie um 3% gegenüber dem N^9-Ribosid oder N^9-Deoxyribosid (IX, X). Trotz der höheren Hypochromie findet man jedoch gerade bei XIII und XIV einen niedrigeren k_{+2}-Wert als bei CpA. Die Wechselwirkung zwischen den Basen ist demnach nicht direkt für die hohen Geschwindigkeiten entscheidend, wie zuerst angenommen [4,7], vielmehr muß durch den Wechsel der Glykosidbindung von N-9 auf N-3 bei gleich guter Packung der Basen eine für die Spaltungsgeschwindigkeit ausschlaggebende Struktur beeinträchtigt worden sein. Das gleiche kann beim Vergleich von CpU (XVI) mit den methylsubstituierten Verbindungen XVII und XVIII sowie in XIII vermutet werden, wenn man annimmt, daß durch die größere Ausdehnung des hydrophoben Systems die Hypochromie zwar verstärkt, durch eine Verschiebung der Basen zueinander aber die Vorordnung im Ribose-Phosphor-Bereich gestört werden kann. Die langsamere Spaltung von Poly-5-methylcytidylsäure gegenüber Polycytidylsäure durch RNase weist ebenfalls in diese Richtung [17].

Für die Stärke der Wechselwirkung zwischen beiden Basen — gemessen an der Hypochromie — spielt auch die Struktur des Verbindungsstückes zwischen zweiter Base und Phosphorsäure eine wichtige Rolle. Bei allen natürlichen Dinucleosidphosphaten haben wir den relativ starren Ribofuranose-Ring. Fügt man im CpA (IX) zwischen die 5'-Methylengruppe des Adenosins und die Phosphatgruppe ein zusätzliches $-S-CH_2-CH_2$-Glied wie in XIX, so sinkt die Hypochromie von 7,5 auf 4 %, und die Geschwindigkeit geht um einen Faktor von etwa 100 herunter. Ein ähnliches Absinken findet man beim Ersatz der Ribose durch eine bewegliche n-Hexyl-, Pentyl- oder Butyl-Kette, wobei das Butyl-Zwischenglied der Struktur $-C^1-O-C^4-C^5-$ in den normalen Ribosiden entspricht.

Vorkürzt man den Abstand zwischen Base und Phosphor bis auf drei C-Atome wie in XXIII und XXIV, so sinken die Geschwindigkeiten um einen weiteren Faktor von 10 auf etwa die des Benzylesters herab. Das gleiche findet man beim 3'-3'-CpdA (XXV), wo C-1' bis C-3' des Deoxyadenosins dem Propylrest in XXIII und XXIV entsprechen; hier ist der Geschwindigkeitsabfall von drei Zehnerpotenzen gegenüber dem isomeren 3'-5'-CpdA (X) besonders auffällig. Bei allen Verbindungen mit dem Propyl-Zwischenglied ist der hypochrome Effekt verschwunden, so daß wir annehmen, daß hier ein *base stacking* nicht mehr existiert und die in den 3'-5'-Verbindungen vorhandene Ordnung im Molekül aufgehoben ist.

Bei den Uridylsäureestern ist der hypochrome Effekt allgemein geringer, doch wird auch in dieser Reihe ein Abfall in der Hypochromie von einem Absinken der k_{+2}-Werte begleitet, wie beim Vergleich von UpA und UpC (XXVI, XXIX) mit den durch N-Oxidation an der zweiten Base modifizierten Verbindungen XXVII und XXX zu beobachten ist. Bei Up-N-methyl-C (XXXI) findet man kein derartiges Absinken der Geschwindigkeit gegenüber UpC; damit kann ausgeschlossen werden, daß im N-Oxid XXX die Blockierung des N-3-Atoms für die Abnahme des k_{+2}-Wertes verantwortlich ist.

Aus diesen Vergleichen ist bereits zu schließen, daß die erhöhten Geschwindigkeiten der Dinucleosidphosphate nicht auf einer π-Wechselwirkung der Basen, sondern auf einem Vorordnungseffekt beruhen. Dieser betrifft jetzt nicht wie bei der Hypochromie den Basenteil, sondern den C-2-Sauerstoff, die 2'-OH-Gruppe und den Phosphor für k_{+1} bzw. C-2-OH, 2'-oder 5'-Sauerstoff für $k_{-1} + k_{+2}$ (in II). Geschwindigkeit und Hypochromie müssen deshalb nicht unbedingt parallel gehen, vielmehr kann durch eine festere Basenpackung die Situation am Phosphor auch verzerrt werden.

Es könnte noch diskutiert werden, ob die Vorordnung für die Reaktion allein durch eine Beziehung zwischen den beiden Basen hergestellt wird, oder ob

die zweite Base eine zusätzliche Bindung zum Enzym eingeht und so zur Fixierung des Ordnungszustandes beiträgt. Für eine derartige Bindung läßt sich aber kein allen Basen gemeinsamer Ansatzpunkt erkennen. Wie schon gezeigt, kommen N-3 der Pyrimidinbasen und aus analogem Grunde N-1 der Purinbasen nicht in Frage, beim Purinribosid (in XII) fehlen C-2- und C-6-Substituent, im Isoadenosin (XIV) sind die dem C-5 und C-6 des Cytidins entsprechenden Positionen besetzt. Auch die Ribose scheidet aus, da CpdA (X) und CpT (XVIII) mit fast gleichen Geschwindigkeiten gespalten werden wie CpA und CpU, und da eine Alkylkettte zwischen Adenin und Phosphor immer noch zu höheren Geschwindigkeiten führt als das intakte Ribose-Zwischenglied im CpU. Bei hydrophoben Beziehungen zum Enzym ist eine andere Reihenfolge der k_{+2}-Werte zu erwarten, denn sie sollten sich z. B. auf die 9-Alkyladenine stärker auswirken als auf die schon in sich fester gepackten Dinucleosidphosphate CpA und UpA. Sonderbeziehungen des zweiten Nucleosids zum Enzym [18] lassen sich daher zur Erklärung der hohen Spaltungsgeschwindigkeiten der Dinucleosidphosphate kaum heranziehen.

Eine die Reaktion beschleunigende Vorordnung kann nicht hergestellt werden, wenn die Pyrimidinbase auf der dem Aktionszentrum abgewandten Seite mit einem zweiten Nucleosid verbunden ist. ApCp zeigt eine dem CpA sehr ähnliche Hypochromie von 6,5 %, jedoch fanden wir für den 3'-Methylester (XXXIII) und das 2',3'-Cyclophosphat (XXXIV) des ApCp nur wenig höhere k_{+2}-Werte als für Cytidylsäure-3'-methylester und cyclische Cytidylsäure [4], verbunden mit einem Absinken des K_m-Wertes. (Dieser Effekt wurde an anderer Stelle [20] im Zusammenhang mit einer zusätzlichen Bindung der Nachbar-Phosphatgruppe als vierter geschwindigkeitsbeeinflussender Faktor diskutiert, auf dem z. B. auch die Spaltung von ApApA durch RNase beruht.) Die Ordnung auf der anderen Seite der Pyrimidinbase bringt also keine Herabsetzung der Aktivierungsenergie — etwa durch eine Erhöhung der Polarisierbarkeiten — mit sich, sondern nur durch eine Wechselwirkung zwischen der katalysierenden Base und den Base des abzuspaltenden Alkoholrestes wird eine wirksame Vorordnung im Aktionszentrum des Substrats hergestellt.

Legt man der Reaktion den Mechanismus I—III zugrunde und ordnet man k_{+2} dem Protonenübergang von II nach I und III zu, also der Säurekatalyse, dann darf sich infolge der Vorordnung oder stärkeren Fixierung des Aktionszentrums nur die Häufigkeit der Übergänge erhöhen. Dadurch wird die Aktivierungsenergie im Entropieterm gesenkt, nicht aber im Enthalpieterm, der im wesentlichen aus dem Übergang VIII → VII resultiert. Die Ergebnisse in Fig. 1 und Tab. 3 stimmen mit diesen Vorstellungen vollkommen überein. In Tab. 4 sind Paare von vergleichbaren Verbindungen zusammengestellt und aufgegliedert, wie sich die Unterschiede in k_{+2} auf den Entropieterm $\Delta S^{\ddagger}$ und auf den Enthalpieterm $\Delta H^{\ddagger}$ verteilen.

Tabelle 4

Substrat	Verhältnis der k_{+2}-Werte	Faktor für die Steigerung	
		aus $\Delta H^{\ddagger}$	aus $\Delta S^{\ddagger}$
CpA:Cp-butyladenin	47	1,0	48
CpA:2',3'-Cp	325	0,4	870
UpA:2',3'-Up	370	0,7	550
2',3'-Cp:2',3'-Up	2,7	0,2	15
CpA:UpA	2,3	0,1	24

Ein direkter Vergleich ist möglich für CpA und Cp-butyladenin (XXII). Hier kann sich bei gleicher katalysierender Base und gleichem Zwischenzustand an der Aktivierungsenergie im Enthalpieterm nichts geändert haben. Tatsächlich ist die gesamte Differenz im Entropieterm, also in der Vorordnung des Substrats zu finden. Beim Vergleich von CpA (1. Schritt) mit cyclischer 2',3'-Cp (2. Schritt) sowie von UpA mit cyclischer 2',3'-Up findet man ebenfalls die gesamte Beschleunigung im Entropieterm, während der Enthalpieterm nur kleine Differenzen aufweist, die wohl darauf zurückzuführen sind, daß II und V zwar analoge, aber nicht identische Zwischenzustände sind und k_{+2} im ersten Fall die Reaktion mit -O-R darstellt, im zweiten Fall die mit dem 2'-Sauerstoff.

Eine weitere wichtige Aussage liefert der Vergleich von cyclischer Cp mit cyclischer Up wie auch von CpA und UpA. Die Cytidylsäureverbindungen sind in k_{+2} um einen Faktor von etwa 2,5 schneller; im Enthalpieterm sind die Cp-Ester jedoch um ca. 5 bis 10mal langsamer, im Entropieterm dafür ca. 20mal schneller. Wenn wir in dem Enthalpieterm die Aktivierungsenergie für den Übergang VIII → VII sehen, so muß diese bei der als konjugate Säure weniger stark sauren Cytosinbase höher liegen, die Geschwindigkeit also geringer werden als bei der stärker sauren konjugaten Säure der Uracilbase. Umgekehrt ist die Häufigkeit, mit der die Cytosinbase als konjugate Säure am Übertragungsort vorliegt, wesentlich höher als beim nicht so basischen Uracil, so daß im Entropieterm Cytosin die schnellere Reaktion liefert. Dadurch wird klar, warum wir bei der Säurekatalyse keine direkte Abhängigkeit zwischen der Geschwindigkeit und der durch den pK-Wert gemessenen Acidität erwarten dürfen, wie es auch die Ergebnisse der vorhergehenden Arbeit [1] gezeigt haben.

Der aus dem Zusammenhang zwischen Geschwindigkeiten und Struktur der Substrate entwickelte Mechanismus I—VI mit der Grundvorstellung einer von der Pyrimidinbase durchgeführten Base-Säure-Katalyse [7,19] bietet nicht nur die Möglichkeit, alle

in dieser Arbeit gefundenen Ergebnisse ohne Zusatzannahmen zu interpretieren, sondern läßt andererseits, konsequent durchdacht, die hier mitgeteilten Resultate direkt erwarten. Die Untersuchungen lassen weiterhin erkennen, wie bei enzymatischen Reaktionen — unabhängig von notwendigen Ordnungszuständen im Enzym — Vorordnungen im Substrat über den Entropieterm mit Faktoren bis zu 1000 ins Gewicht fallen können.

Diese Arbeit wurde durchgeführt mit Unterstützung der Deutschen Forschungsgemeinschaft und des Fonds der Chemischen Industrie sowie der Firma Zellstofffabrik Waldhof, Mannheim. H. Follmann ist ein Stipendiat der Deutschen Forschungsgemeinschaft; H.-J. Wieker ist ein Liebig-Stipendiat des Fonds der Chemischen Industrie.

LITERATUR

1. Gassen, H. G., und Witzel, H., *European J. Biochem.* 1 (1967) 36.
2. Wieker, H.-J., und Witzel, H., *European J. Biochem.* 1 (1967) 251.
3. Witzel, H., und Barnard, E. A., *Biochem. Biophys. Res. Commun.* 7 (1962) 289.
4. Witzel, H., und Barnard, E. A., *Biochem. Biophys. Res. Commun.* 7 (1962) 295.
5. Michelson, A. M., *J. Chem. Soc. (London)*, 1959, 3655; auch in *The Chemistry of Nucleosides and Nucleotides*, Academic Press, London, New York 1963, S. 444.
6. Warshaw, M. M., und Tinoco, I., jr., *J. Mol. Biol.* 13 (1965) 54; 20 (1966) 29.
7. Witzel, H., in *Progress in Nucleic Acid Research and Molecular Biology* (edited by T. N. Davidson and W. E. Cohn), Academic Press, New York, London 1963, Bd. 2, S. 221.
8. Dimroth, K., und Witzel, H., *Liebigs Ann. Chem.* 620 (1959) 109.
9. Follmann, H., *Tetrahedron Letters*, 1967.
10. Cramer, F., Randerath, K., und Schäfer, E. A., *Biochim. Biophys. Acta*, 72 (1963) 150.
Cramer, F., und Seidel, H., *Biochim. Biophys. Acta*, 72 (1963) 157.
11. Brimacombe, R. L. C., Griffin, B. E., Haines, J. A., Haslam, W. J., und Reese, C. B., *Biochemistry*, 4 (1965) 2452.
12. Lohrmann, P., und Khorana, H. G., *J. Am. Chem. Soc.* 86 (1964) 4188.
13. Kuhn, R., und Jahn, W., *Chem. Ber.* 98 (1965) 1699.
14. Ikehara, M., Ohtsuka, E., Kitagawa, S., Yagi, K., und Tonomura, Y., *J. Am. Chem. Soc.* 83 (1961) 2679.
15. Andersen, W., Hayes, D. H., Michelson, A. M., und Todd, A. R., *J. Chem. Soc.* 1954, 1882.
16. Barker, G. R., Montague, M. D., Moss, R. J., und Parsons, M. A., *J. Chem. Soc.* 1957, 3786.
17. Szer, W., und Shugar, D., *Acta Biochim. Polon.* 13 (1966) 178.
18. Rabin, B. R., und Mathias, A. P., in *Mechanismen enzymatischer Reaktionen*, Springer-Verlag, Berlin, Göttingen, Heidelberg 1964, S. 97.
19. Witzel, H., in *Mechanismen enzymatischer Reaktionen*, Springer-Verlag, Berlin, Göttingen, Heidelberg 1964, S. 123.
20. Witzel, H., und Barnard, E. A., *Abstracts 2nd FEBS Meeting (Vienna)*, 1965, S. 39.

H. Follmann, H.-J. Wieker und H. Witzel
Chemisches Institut der Universität
355 Marburg/Lahn, Bahnhofstraße 7, Deutschland

European J. Biochem. 1 (1967) 251—258

Zum Mechanismus der Ribonuclease-Reaktion

3. Zuordnung der kinetischen Parameter k_{+1}, k_{-1}, k_{+2} und Interpretation von K_m

H.-J. Wieker und H. Witzel

Chemisches Institut der Universität Marburg/Lahn

(Eingegangen am 19. Dezember 1966)

The hydrolysis of cyclic $2',3'$-cytidylic acid (Cp) and cyclic $2',3'$-uridylic acid (Up) by pancreatic ribonuclease (E.C.2.7.7.16) at pH 7 and 5.6 is activated by adenine, adenosine, cyclic $2',3'$-adenylic acid, and adenylyl-adenosine (ApA). The activation is reflected by parallel shifts in the kinetic parameters when expressed on a $e/v - 1/s$-diagram. The measured parameters $k^\circ = \varphi \cdot k_{+2}$ and $K^\upsilon = \Phi \cdot K_m$ are increased by the same factor, i.e. $\Phi = \varphi > 1$ or $\frac{k^\circ}{k_{+2}} = \frac{K^\circ}{K_m}$ [1].

Rate equations can be derived for an interaction of the modifier with either the substrate, the enzyme, or the enzyme-substrate-complex (ES), independent of any mechanistic concept. It is then possible to calculate the conditions pertaining to the relation $\Phi = \varphi$.

The first case, an interaction between the modifier and the substrate can be excluded. In the case of an interaction between the modifier and the enzyme or ES the rate equations if $\Phi = \varphi$ require that k_{-1} and k_{+2} are equally influenced by the modifier. Thus both constants are coupled and associated with reactions of the same type.

Since the dissociation of the inhibitor $3'$-Cp from the enzyme as measured by relaxation times [21] is 10^3 times faster than the k_{+2} value in the reaction where the same $3'$-Cp is formed as the product, k_{+2} cannot be the dissociation of the product from the enzyme, but must be associated with a step involving bond exchange. Consequently k_{-1} must be associated with a step which also involves a bond exchange and is catalysed by the same principle. Both steps should therefore start from a common intermediate and lead alternatively either to the substrate (k_{-1}) or to the product (k_{+2}).

It can be further shown that k_{+1} must be associated with the complementary step to k_{-1}, also involving bond exchange, and that a simple association-dissociation-equilibrium between enzyme and substrate (not involving covalent bond exchanges in the substrate) does not appear in the measurable kinetic parameters. K_m therefore would not represent a dissociation constant in the sense of Michaelis-Menten but rather the ratio of the rate constants for the breakdown and the formation of an intermediate which is stabilized by the enzyme.

The mathematical relations do not give any information as to the nature of the bonds exchanged or the type of catalysis. On the basis of the mechanism derived from other experiments in our previous investigations [1,3] it can be concluded now that the enzyme only induces a non-rate-limiting polarisation at the monoanionic phosphodiester group of the substrates. Binding would occur in a very fast non-rate-limiting step after the rate limiting base catalysed step when the dianionic intermediate has been formed.

---

[1] *Definitionen.* K_m = Michaelis-Konstante, nach Gleichung (6); K_s = Dissoziations-Konstante (k_{-1}/k_{+1}); k_{+1}, k_{-1}, k_{+2} usw. = Geschwindigkeitskonstanten, definiert in Gleichungen (3), (9), (10) und (16); K°, k° = gemessene Konstanten [Gleichungen (1) und (2)] bei Modifizierer-Zusatz; Φ, φ = Proportionalitätsfaktoren nach Gleichungen (1) und (2); α, β, γ = Proportionalitätskonstanten nach Gleichungen (9), (10) und (16); $K' = k_{.0}/k_{+0}$ = Dissoziationskonstante für ein angenommenes, der Reaktion vorgelagertes Gleichgewicht nach Gleichung (22); K_i = Inhibitor-Konstante nach [5]; v = Reaktionsgeschwindigkeit (in Fig. 1—4 = Anfangsgeschwindigkeit). E, e, S, s, ES, P, X, x definiert vor Gleichung (8); EX, SX, ESX nach Gleichungen (9), (10) und (16); EZ = enzymstabilisierter Zwischenzustand, im Gegensatz zum allgemein gehaltenen Enzym-Substrat-Komplex (ES).

Der aus unseren bisherigen Untersuchungen abgeleitete Mechanismus der Ribonuclease-Reaktion nach dem Schema in [1,3] folgt dem Typ einer Michaelis-Menten-Kinetik.

Abweichend von der üblichen Interpretation kann jedoch bei dieser Reaktion K_m nicht eine Gleichgewichtskonstante für ein vorgelagertes Bindungsgleichgewicht ($K_m = k_{-1}/k_{+1}$, $k_{+2} \ll k_{-1}$) sein, sondern muß als ein Verhältnis der Geschwindigkeitskonstanten für den Zerfall eines enzymstabilisierten Zwischenzustands ($k_{-1} + k_{+2}$) zu der für seinen Aufbau (k_{+1}) betrachtet werden, wobei k_{-1} und k_{+2} Konstanten für Alternativreaktionen sind [4].

Den Konstanten haben wir definierte Reaktionsschritte zugrunde legen können [1]. Geschwindigkeitsbestimmend ist für die Bildung des Zwischenzustands offensichtlich die Übernahme eines Protons durch die Pyrimidinbase (im 1. Schritt bei der Umesterung von der 2'-OH-Gruppe, im 2. Schritt, bei der Hydrolyse, von einem Wassermolekül), also eine Basenkatalyse. Für den Zerfall des Zwischenzustands ist eine Protonenkatalyse durch die konjugate Säure der Pyrimidinbase geschwindigkeitsbestimmend, also die Übertragung des Protons alternativ auf den 2'-Sauerstoff (= Rückreaktion, k_{-1}) oder auf den 5'-Sauerstoff (= Produktbildung, k_{+2}). Im zweiten Schritt liegen analoge Verhältnisse vor.

Neben der in den Entropieterm eingehenden Wahrscheinlichkeit der richtigen Orientierung der Base enthält die Aktivierungsenergie für die Übertragung des Protons im Enthalpieterm einerseits einen Anteil, der mit den Übergängen der Base in die Form der konjugaten Säure und umgekehrt (I ⇌ II) zusammenhängt, andererseits einen Anteil aus der Nucleophilie der beiden Sauerstoffe, deren Verhältnis zueinander das Verhältnis von k_{-1} zu k_{+2} bestimmt. Dabei kann k_{+2} sowohl kleiner als auch größer als k_{-1} sein [2].

I   ⇌   II

Bleibt innerhalb einer Reihe von Substraten die Situation an den Sauerstoffatomen der Phosphatgruppe und am Enzym unverändert, und wird nur die Pyrimidinbase im Rahmen der aufgezeigten Bedingungen [1] verändert, so ändert sich das Verhältnis k_{-1}/k_{+2} nicht, auch wenn die als k_{+2} gemessene Geschwindigkeitskonstante verändert sein sollte. Es müssen sich dann k_{+2} und k_{-1} um denselben Faktor geändert haben, d. h. beide Konstanten sind gekoppelt.

Bleibt K_m konstant, während sich k_{+2} ändert, dann kann nicht wie üblich geschlossen werden, daß k_{+2} vernachlässigbar klein gegenüber k_{-1} sein muß, sondern nur, daß k_{+1} um denselben Faktor geändert hat wie k_{-1} und k_{+2}.

Wir haben zwei Fälle diskutiert, bei denen diese Situation auftritt:

a) Wenn durch eine Erhöhung der Polarisierbarkeit innerhalb der Pyrimidinbase die Aktivierungsenergien für die Übergänge I ⇌ II herabgesetzt werden, so werden Basen- und Säurekatalyse um den gleichen Faktor erhöht; K_m bleibt bei verändertem k_{+2}-Wert konstant [1].

b) Wenn durch eine zweite Base die freie Rotation der Pyrimidinbase um die Glykosidbindung eingeschränkt wird und dadurch der 2-Sauerstoff weitgehend fixiert wird, wie bei den Dinucleosidphospha-

ten, so wird die Häufigkeit der Protonenübergänge in beiden Richtungen erhöht. Auch in diesem Fall bleibt bei stark erhöhten k_{+2}-Werten K_m konstant [3].

In beiden Fällen sind wir bei der Interpretation der Ergebnisse bereits von der Kopplung der beiden Konstanten k_{-1} und k_{+2} ausgegangen. Ein schlüssiger Beweis für diese Kopplung war jedoch aus diesen Untersuchungen nicht abzuleiten.

Im folgenden geben wir weitere Untersuchungen bekannt, deren Ergebnisse uns zwingen, k_{-1} und k_{+2} als gekoppelte Konstanten zu betrachten und k_{+1} einem komplementären Reaktionsprozeß zuzuordnen.

KINETISCHE UNTERSUCHUNGEN UNTER ZUSATZ VON ADENINDERIVATEN ALS MODIFIZIERER

Die RNase-katalysierte Hydrolyse von cyclischer 2',3'-Cp und 2',3'-Up wurde bei 25°, $I = 0,2$ und pH 7,0 bzw. pH 5,6 untersucht. Die Messungen und ihre Auswertungen wurden durchgeführt wie in [1,3] beschrieben.

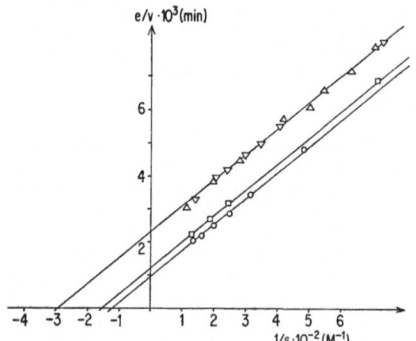

Fig.1. *Hydrolyse von 2',3'-Cp durch RNase in Gegenwart von Modifizierern.* pH 7,0, 25°, $I = 0,2$. △, 2',3'-Cp; ▽, 2',3'-Cp + 5,1×10⁻³ M Adenosin-N¹-oxid; □, 2',3'-Cp + 2,4 ×10⁻³ M ApA; ○, 2',3'-Cp + 5,9×10⁻³ M Adenin

Fig.1—4 zeigen, daß die RNase-Reaktion durch Adenin, Adenosin, cyclische 2',3'-Ap und ApA in charakteristischer Weise aktiviert wird, während Adenosin-N-1-oxid keinen Einfluß auf die Reaktion ausübt.

Im folgenden werden die Konstanten, die aus den Schnittpunkten der Geraden mit den Koordinaten im $e/v - 1/s$-Diagramm zu entnehmen sind, mit $K°$ und $k°$ bezeichnet; $K°$ ist eine Funktion von K_m, und $k°$ ist eine Funktion von k_{+2}:

$$K° = \Phi \cdot K_m. \tag{1}$$

$$k° = \varphi \cdot k_{+2}. \tag{2}$$

Im Falle der unmodifizierten Reaktion ist dann definitionsgemäß $K° = K_m$, $k° = k_{+2}$ und $\Phi = \varphi = 1$.

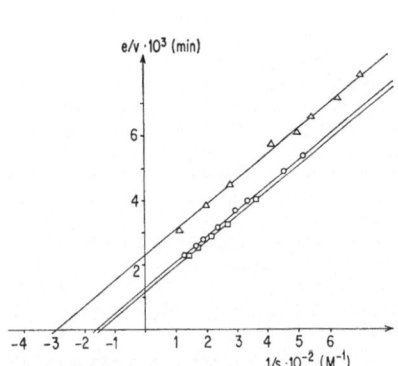

Fig. 2. *Hydrolyse von 2′,3′-Cp durch RNase in Gegenwart von Modifizierern.* pH 7,0, 25°, $I = 0,2$. △, 2′,3′-Cp; ○, 2′,3′-Cp + $5,8 \times 10^{-3}$ M Adenosin; □, 2′,3′-Cp + $5,9 \times 10^{-3}$ M 2′,3′-Ap.

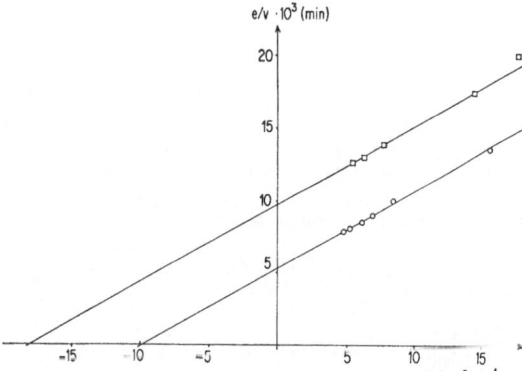

Fig. 3. *Hydrolyse von 2′,3′-Cp durch RNase in Gegenwart von Adenosin.* pH 5,6, 25°, $I = 0,2$. □, 2′,3′-Cp; ○, 2′,3′-Cp + $5,6 \times 10^{-3}$ M Adenosin

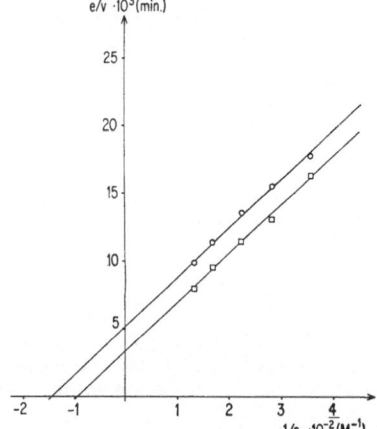

Fig. 4. *Hydrolyse von 2′,3′-Cp durch RNase in Gegenwart von Adenosin.* pH 7,0, 25°, $I = 0,2$. ○, 2′,3′-Up; □, 2′,3′-Up + $5,8 \times 10^{-3}$ M Adenosin

Tabelle

Substrat	Modifizierer	$K°$	$k°$	$\Phi = \varphi$
		mM	sec⁻¹	
cycl.	ohne	3,5	7,3	1,0
2′,3′-Cp[a]	5,9 mM Adenin	8,3	17,6	2,4
	5,8 mM Adenosin	6,3	13,1	1,8
	5,1 mM Adenosin-N-1-oxid	3,5	7,3	1,0
	5,9 mM cycl. 2′,3′-Ap	6,9	14,5	2,0
	2,4 mM ApA	6,7	14,2	1,9
cycl.	ohne	0,55	1,7	1,0
2′,3′-Cp[b]	5,6 mM Adenosin	1,0	3,2	1,9
cycl.	ohne	6,9	3,2	1,0
2′,3′-Up[a]	5,8 mM Adenosin	10,7	5,0	1,6

[a] pH 7,0; $I = 0,2$; 25°; pH-stat-Methode, in NaCl.
[b] pH 5,6; $I = 0,2$; 25°; spektralphotometrisch, in Acetat-Puffer.

In der Tabelle sind die aus den Fig. 1—4 ermittelten Werte $K°$ und $k°$ sowie die Faktoren Φ und φ zusammengestellt:

Die in Gegenwart des Modifizierers erhaltenen Geraden im e/v — $1/s$-Diagramm sind zur Geraden der unmodifizierten Reaktion parallel verschoben, d. h. $\Phi = \varphi > 1$, bzw. $K°/K_m = k°/k_{+2}$.

Diese Parallelelität wurde bisher nur für die sogenannte unkompetitive Hemmung (d. h. $\Phi = \varphi < 1$) beschrieben [5—13]. Statt der analog gebildeten Bezeichnung „unkompetitive Aktivierung" benutzen wir im folgenden die Bezeichnung „Zerfallsaktivierung" und sprechen auch von „Zerfallshemmung",

wenn dieser Kinetik-Typ dadurch hervorgerufen wird, daß der Modifizierer nur auf den Zerfall des Enzym-Substrat-Komplexes wirkt. Es wird demnächst gezeigt werden, daß in dem hier untersuchten Fall die Berechtigung dazu gegeben ist.

AUSWERTUNG DER ERGEBNISSE

In Gegenwart eines Modifizierers gilt gemäß Definition [Gleichungen (1) und (2)] für die Geschwindigkeitsgleichung bei einer Michaelis-Menten-Kinetik in Form von

$$E + S \underset{k-1}{\overset{k+1}{\rightleftharpoons}} ES \xrightarrow{k+2} E + P \qquad (3)[2]$$

[2] Da nur die Anfangsgeschwindigkeiten betrachtet werden, gehen Gleichgewichte zwischen E und P nicht in die Kinetik ein, und es ist unerheblich, ob ES → EP ⇌ E + P gesetzt wird. Wir werden später noch zeigen, daß bei der RNase-Reaktion EP → E + P nicht geschwindigkeitsbestimmend ist. Das gilt auch in Gleichung (9), Gleichung (10) und Gleichung (16).

$$\frac{v}{e} = \frac{k^\circ}{1 + \dfrac{K^\circ}{s}} = \frac{\varphi \cdot k_{+2}}{1 + \dfrac{\Phi \cdot K_m}{s}} \qquad (4)$$

und reziprok

$$\frac{e}{v} = \frac{1}{k^\circ} + \frac{K^\circ}{k^\circ} \cdot \frac{1}{s} = \frac{1}{\varphi \cdot k_{+2}} + \frac{\Phi \cdot K_m}{\varphi \cdot k_{+2}} \cdot \frac{1}{s} \qquad (5)$$

wobei K_m wie üblich definiert ist:

$$K_m = \frac{k_{-1} + k_{+2}}{k_{+1}} . \qquad (6)$$

Die experimentell gefundene Beziehung $\Phi = \varphi > 1$ ergibt sich zwar unter der Annahme, daß $k_{-1} \ll k_{+2}$ ist, wodurch $K_m = k_{+2}/k_{+1}$ wird. Jedoch liegen bis jetzt nur kinetische Befunde vor, die gegen diese Annahme sprechen. So erfolgt z. B. die Alkoholyse von cyclischen 2′,3′-Diestern zu 3′Nucleotidalkylestern etwa mit derselben Geschwindigkeit wie die Umesterung der 3′-Nucleotidalkylester zu 2′,3′-Diestern [14,15]. Somit kann die Beziehung k_{-1}

$\ll k_{+2}$ nicht zur Erklärung der hier aufgeführten Experimente herangezogen werden.

Unabhängig von jeder mechanistischen Deutung lassen sich die Geschwindigkeitsgleichungen ableiten für die Möglichkeiten einer Wechselwirkung des Modifizierers mit dem Substrat [Gleichung (9)], mit dem Enzym [Gleichung (10)] und mit dem Enzym-Substrat-Komplex ES [Gleichung (16)]. ES kann dabei ein Komplex im Sinne von Michaelis-Menten [16,17] sein oder ein enzymstabilisierter Zwischenzustand [1].

Ferner kann wegen der Analogie der entsprechenden Schritte gesetzt werden:

$$k_{+4} = \alpha \cdot k_{+2}, \quad k_{-5} = \beta \cdot k_{-1}, \quad k_{+5} = \gamma \cdot k_{+1}. \qquad (7)$$

Daraus lassen sich mit e = totale Konzentration an Enzym (E), s = totale Konzentration an Substrat (S), x = totale Konzentration an Modifizierer (X) nach der steady-state-Methode die Geschwindigkeitsgleichungen ableiten, wobei stets gilt:

$$v = k_{+2} \cdot [\text{ES}] + k_{+4} \cdot [\text{ESX}]. \qquad (8)$$

Wechselwirkung Modifizierer — Substrat

Das Reaktionsschema (9) führt zu einer Geschwindigkeitsgleichung, in der e/v nur dann proportional der 1. Potenz von 1/s ist, wenn $x \gg s$ oder $s \gg x$ gilt. Die Experimente wurden jedoch sowohl mit $x > s$ wie mit $x < s$ durchgeführt, so daß Gleichung (9) wegen der stets gefundenen Linearität zwischen e/v und 1/s hier nicht in Frage kommen kann.

Wechselwirkung Modifizierer — Enzym

Für Reaktionsschema (10) lautet die Geschwindigkeitsgleichung (4):

$$\frac{v}{e} = \frac{k^\circ}{1 + \dfrac{K^\circ}{s}} = \frac{\varphi \cdot k_{+2}}{1 + \dfrac{\Phi \cdot K_m}{s}} \qquad (4)$$

mit

$$\varphi = \frac{(\gamma k_{+3}x + k_{-3})(\alpha k_{+6}x + k_{-6}) + \alpha\gamma k_{+3}x(k_{-1} + k_{+2}) + k_{-3}(\beta k_{-1} + \alpha k_{+2})}{(\gamma k_{+3}x + k_{-3})(k_{+6}x + k_{-6}) + \gamma k_{+3}x(k_{-1} + k_{+2}) + k_{-3}(\beta k_{-1} + \alpha k_{+2})} \tag{11}$$

und

$$\Phi = \frac{(k_{+3}x + k_{-3})\cdot[(\beta k_{-1} + \alpha k_{+2})(k_{-1} + k_{+2} + k_{+6}x) + (k_{-1} + k_{+2})k_{-6}]}{[(\gamma k_{+3}x + k_{-3})(k_{+6}x + k_{-6}) + \gamma k_{+3}x(k_{-1} + k_{+2}) + k_{-3}(\beta k_{-1} + \alpha k_{+2})]\cdot(k_{-1} + k_{+2})} \ . \tag{12}$$

Um die experimentellen Befunde zu deuten, müssen die Bedingungen errechnet werden, die zu $\Phi = \varphi$ führen, wobei zu berücksichtigen ist, daß wegen $\varphi > 1$ auch $\alpha > 1$ ist. Gleichgültig, ob man $K_m = (k_{-1} + k_{+2})/k_{+1}$ [Gleichung (6)] oder $K_m = K_s = k_{-1}/k_{+1}$ setzt, folgt aus Gleichungen (11) und (12) für $\Phi = \varphi$:

$$\alpha = \beta \, , \tag{13}$$

$$\gamma = 1 \, . \tag{14}$$

Das bedeutet: der Zerfall von ESX ist in beiden Richtungen (k_{-5} und k_{+4}) um denselben Faktor modifiziert, während die Bildung von ESX nicht beeinflußt wird ($k_{+5} = k_{+1}$).

Unter diesen Bedingungen wird mit Gleichung (13) und Gleichung (14) aus Gleichung (11) und Gleichung (12)

Zur Deutung der experimentellen Befunde müssen wieder die Bedingungen errechnet werden, die zu $\Phi = \varphi$ führen. Gleichgültig, ob man $K_m = (k_{-1} + k_{+2})/k_{+1}$ oder $K_m = K_s = k_{-1}/k_{+1}$ setzt, folgt für $\Phi = \varphi$ aus Gleichung (17) und Gleichung (18):

$$\alpha = \beta \, , \tag{13}$$

$$\gamma = 0 \, . \tag{19}$$

und $$\Phi = \varphi = \frac{\alpha \cdot k_{+3}x + \alpha \cdot (k_{-1} + k_{+2}) + k_{-3}}{k_{-3}x + \alpha \cdot (k_{-1} + k_{+2}) + k_{-3}} \ . \tag{20}$$

Das bedeutet: Auch hier ist der Zerfall von ESX in beiden Richtungen (k_{-5} und k_{+4}) um denselben Faktor modifiziert, während eine Bildung von ESX aus E + S + X (Dreierstoß) nicht erfolgt. Mit Gleichung (19) wird jedoch das Schema (16) unwahrscheinlich.

$$\varphi = \Phi = \frac{[\alpha \cdot (k_{-1} + k_{+2}) + \alpha k_{+6}x + k_{-6}]\cdot[k_{+3}x + k_{-3}]}{(k_{-1} + k_{+2})\cdot(k_{+3}x + \alpha k_{-3}) + (k_{+6}x + k_{-6})(k_{+3}x + k_{-3})} \ . \tag{15}$$

Bei dieser Ableitung wurde angenommen, daß es sich bei E + X $\rightleftharpoons$ EX um ein echtes, sich sehr schnell einstellendes Gleichgewicht handelt. Eine genaue Analyse der Geschwindigkeitsgleichung, die resultiert, wenn auch für dieses Gleichgewicht eine *steady-state*-Betrachtung vorgenommen wird, ist von Ohlenbusch [12] durchgeführt worden. Unter den dort diskutierten Beziehungen zwischen den Geschwindigkeitskonstanten, die im $e/v - 1/s$-Diagramm den Typ einer unkompetiven Hemmung liefern, führt der einzige Fall, der sich auf eine Aktivierung übertragen läßt, zu denselben Bedingungen der Gleichung (13) und Gleichung (14).

Natürlich gelten diese Ableitungen allgemein für Modifizierer (Aktivator: $\varphi > 1 \rightarrow \alpha > 1$; Inhibitor $\varphi < 1 \rightarrow \alpha < 1$) und sind unabhängig von jeder Vorstellung über die Vorgänge, die der Einwirkung des Modifizierers zugrunde liegen.

DISKUSSION

Eine Entscheidung darüber, ob die Wirkung des Modifizierers auf einer Wechselwirkung mit dem Enzym [Gleichung (10)] oder mit dem Enzym-Substrat-Komplex [Gleichung (16)] beruht, braucht hier nicht gefällt zu werden; dieses Problem wird

Wechselwirkung Modifizierer — ES [3]

$$\tag{16}$$

---

[3] In Gleichung (16) würde k_{+5} die Konstante einer Reaktion dritter Ordnung sein. Setzt man in Gleichung (16) deshalb $k_{+5} \cdot x = \gamma \cdot k_{+1}$, so ergibt sich dieselbe Interpretation wie mit Gleichung (7).

17*

Für Reaktionsschema (16) sind in der Geschwindigkeitsgleichung [Gleichung (4)] zu setzen:

$$\varphi = \frac{(\beta k_{-1} + \alpha k_{+2}) + \alpha\gamma\,(k_{-1} + k_{+2})\,\mathrm{x} + (\alpha k_{+3}\mathrm{x} + k_{-3})\,(1 + \gamma\,\mathrm{x})}{(\beta k_{-1} + \alpha k_{+2}) + \gamma\,(k_{-1} + k_{+2})\,\mathrm{x} + (k_{+3}\mathrm{x} + k_{-3})\,(1 + \gamma\,\mathrm{x})} \tag{17}$$

und

$$\Phi = \frac{(\beta k_{-1} + \alpha k_{+2})\,(k_{-1} + k_{+2} + k_{+3}\mathrm{x}) + k_{-3}\,(k_{-1} + k_{+2})}{[(\beta k_{-1} + \alpha k_{+2}) + \gamma\,(k_{-1} + k_{+2})\mathrm{x} + [k_{+3}\mathrm{x} + k_{-3}]\,(1 + \gamma\,\mathrm{x})]\cdot(k_{-1} + k_{+2})} \ . \tag{18}$$

in einer folgenden Arbeit diskutiert. Aus den Gleichungen (7) und (13) folgt

$$\alpha = \frac{k_{+4}}{k_{+2}} = \frac{k_{-5}}{k_{-1}}$$

und

$$\frac{k_{+2}}{k_{-1}} = \frac{k_{+4}}{k_{-5}} \ . \tag{21}$$

Das bedeutet, daß der Modifizierer beide Zerfallsschritte (k_{-1} und k_{+2}) gleichmäßig beeinflußt, daß also der RNase-Reaktion ein Mechanismus zugrunde liegen muß, in dem eine solche gleichartige Beeinflussung möglich ist. Wären nun k_{-1} und k_{+2} Geschwindigkeitskonstanten zweier Reaktionsprozesse verschiedenartigen Typs, so würde eine gleichartige Beeinflussung durch einen Modifizierer einen Zufall darstellen. Einen solchen Zufall kann man nach Morales [18] ausschließen, wenn verschiedene Modifizierer denselben Effekt bewirken. Tatsächlich zeigen die Experimente die gleiche Zerfallsaktivierung sowohl mit verschiedenen Aktivatoren, als auch bei verschiedenen Substraten (2′,3′-Cp und 2′,3′-Up) und darüber hinaus vor allem auch bei verschiedenen pH-Werten, obgleich K_m und k_{+2} unterschiedliche pH-Abhängigkeiten aufweisen [4,19]. Wir müssen deshalb annehmen, daß k_{-1} und k_{+2} Geschwindigkeitskonstanten für Reaktionsschritte gleichen Reaktionstyps sind. Wäre nun k_{-1} die Zerfallskonstante in einem Assoziations-Dissoziations-Gleichgewicht zwischen Enzym und Substrat, so müßte k_{+2} ebenfalls nur aus einer solchen Dissoziation bestehen. Dies könnte nur die Lösung des Produkts vom Enzym bedeuten, da ausschließlich hierbei analoge Bindungskräfte und Bindungsstellen betroffen sind. Die Untersuchungen von Hammes [21] haben aber ergeben, daß die Geschwindigkeitskonstante für die Lösung des Produkts (3′-Cp) vom Enzym ca. 1000mal größer ist als der k_{+2}-Wert bei der Reaktion von cyclischer 2′,3′-Cp zu 3′-Cp. Demnach muß k_{+2} die Reaktionskonstante eines Schrittes sein, in dessen Verlauf die Bindungswechsel zum Produkt erfolgen. Analoges läßt sich auch für den ersten Schritt ableiten: Wäre k_{+2} die Konstante für die Lösung des Produkts (2′,3′-Cp) vom Enzym, so wäre dieses k_{+2} des ersten Schritts identisch mit dem k_{-1} des zweiten Schritts in dem 2′,3′-Cp das Substrat ist. Da aber die Konstanten des zweiten Schritts nicht verändert werden, wenn der cyclische Diester als Produkt eines vorhergehenden ersten Schritts oder unabhängig

vom ersten Schritt zur Reaktion gebracht wird [4,20], kann k_{+2} auch beim ersten Schritt nicht die Konstante für die Lösung des Produkts vom Enzym sein. Da nun der Schritt k_{-1} nach demselben Reaktionstyp verlaufen muß wie k_{+2}, kann k_{-1} nicht die Dissoziation des Substrats vom Enzym wiedergeben, sondern k_{-1} muß ebenfalls einen Schritt repräsentieren, in dessen Verlauf Bindungswechsel erfolgen.

Mit den Schritten k_{-1} und k_{+2} müssen demnach aus identischen oder analogen „Ausgangsverbindungen" zwei verschiedene „Endverbindungen" entstehen.

Das bedeutet also, daß im Verlauf der Reaktion mit Hilfe des Enzyms ein Zwischenzustand aufgebaut wird, der alternativ rückläufig wieder zum Substrat umgesetzt wird oder weiter zum Produkt reagiert. Dieser Zwischenzustand wird im folgenden mit EZ bezeichnet, um Verwechslungen mit dem Enzym-Substrat-Komplex ES im Sinne von Michaelis-Menten zu vermeiden; natürlich sind in den vorhergehenden Ableitungen EZ und ES gleichbedeutend. Während die Zuordnung der Konstanten k_{-1} und k_{+2} zu analogen Reaktionsschritten allein durch die mathematische Behandlung der Zerfallsaktivierung ermöglicht wurde, müssen für die Zuordnung von k_{+1} noch weitere experimentelle Befunde herangezogen werden, da Gleichung (14) bzw. Gleichung (19) keine analogen eindeutigen Rückschlüsse auf k_{+1} zulassen.

Da die k_{+2}-Werte der Dinucleosidphosphate stets im Verhältnis CpR : UpR $\simeq$ 2′,3′-Cp : 2′,3′-Up (R = 5′-Nucleosid) stehen [19] und bei der Zerfallsaktivierung das Verhältnis CpA : UpA $\simeq$ 2′,3′-Cp + A : 2′,3′-Up + A $\simeq$ 2′,3′-Cp : 2′,3′-Up gefunden wurde, müssen sich k_{-1} und k_{+2} genauso verhalten wie für den zweiten Schritt abgeleitet wurde. Die k_{+2}-Werte der Dinucleosidphosphate sind um zwei bis vier Zehnerpotenzen größer als die der entsprechenden Nucleotidalkylester, während die K_m-Werte konstant[4] und gleich denen der cyclischen Diester sind. Da sich k_{-1} ebenso wie k_{+2} geändert haben muß, kann aber K_m nur konstant bleiben, wenn sich hier auch k_{+1} um denselben Faktor verändert hat. Das kann aber wiederum nur erfolgen, wenn sich die Ge-

---

[4] Ein Parameter wird dann als „konstant" oder „unverändert" bezeichnet, wenn er sich nur wenig, ein anderer im Vergleich dazu sehr stark ändert. Variieren beide in vergleichbaren Größenordnungen, so müssen bei beiden auch kleine Änderungen berücksichtigt werden.

schwindigkeitskonstante eines Reaktionsschrittes verändert hat, in dessen Verlauf wie bei k_{+2} und k_{-1} Bindungswechsel stattfinden, d. h. k_{+1} muß die Geschwindigkeitskonstante für die Bildung des Zwischenzustands EZ sein.

Ohne Zweifel muß im Laufe der Ausbildung des Zwischenzustands EZ eine Wechselwirkung zwischen Enzym und Substrat erfolgen, so daß noch ein dem Schritt k_{+1} vorgelagertes Gleichgewicht zu diskutieren ist.

Da wir nachweisen konnten, daß an diesem Gleichgewicht seitens des Substrats nur die Phosphatgruppe beteiligt sein kann [1], sollte die zugehörige Gleichgewichtskonstante bei allen Substraten praktisch gleich groß sein.

Nun muß aber die Dissoziationskonstante eines derartigen vorgelagerten Gleichgewichts identisch sein mit der Inhibitorkonstanten von Diestern, die trotz gleicher Struktur an der Phosphatgruppe keine Substrate sind. Da aber solche Diester wie ApA und 2',3'-Ap die RNase-Reaktion nicht hemmen, während sie andererseits dieselbe Zerfallsaktivierung wie Adenosin bewirken (Tabelle), geht ein derartiges vorgelagertes Gleichgewicht nicht in die Kinetik ein, d. h. die Wechselwirkung zwischen Enzym und Phosphorsäure-diestergruppe kann keine limitierende Größe bei der Bildung von EZ sein. Damit kann nicht ausgeschlossen werden, daß sich der meßbare K_m-Wert [Gleichung (22)] von dem gemäß Gleichung (6) definierten K_m noch um einen Faktor unterscheidet:

$$K_m = \frac{k_{-1} + k_{+2}}{k_{+0}\, k_{+1}/k_{-0}} = \frac{k_{-1} + k_{+2}}{k_{+1}/K'} \qquad (22)$$

wobei $K' = k_{-0}/k_{+0}$ die Dissoziationskonstante im vorgelagerten Gleichgewicht ist. Dieser Faktor (K') sollte aber wegen der allen Diestern gemeinsamen Bindungsstelle weitgehend konstant sein und kann wegen der nicht beobachteten Hemmung nur in der Größenordnung von maximal einer Zehnerpotenz liegen. Dementsprechend stellt K_m bei der RNase-Reaktion eine Konstante dar, die das Verhältnis der Geschwindigkeitskonstanten konkreter Reaktionsschritte wiedergibt.

Über die Natur der Bindungswechsel sagen die hier behandelten kinetischen Ergebnisse nichts aus. Hierüber konnten jedoch aus Untersuchungen über den Einfluß der Struktur der Substrate auf die Reaktionsgeschwindigkeiten [1,3] exakte Aussagen gewonnen werden, wie sie einleitend schon dargestellt worden sind.

Dabei zeigt sich jetzt, daß die Funktion des Enzyms darin besteht, die Phosphatgruppe für den Aufbau des Zwischenzustands EZ zu polarisieren, wobei die Geschwindigkeiten für k_{+0} und k_{-0} im Bereich der Diffusionskontrolle liegen. Eine Bindung an das Enzym erfolgt in einem sehr schnellen, nicht geschwindigkeitsbestimmenden Schritt nur dann, wenn mit Hilfe der limitierenden Basenkatalyse eine dianionische Phosphatgruppe mit pentakovalentem Phosphor ausgebildet wird.

Für diese Form sollte wiederum ein Dissoziations-Assoziations-Gleichgewicht existieren, das mit Geschwindigkeiten sich einstellt wie das Gleichgewicht mit den dianionischen Inhibitoren. Dieses Gleichgewicht hat jedoch nichts mit der bei uns durch K_m ausgedrückten Größe zu tun.

Die aus dem vorgeschlagenen Mechanismus sich ergebenden Konsequenzen für den Bindungsprozeß bei den Substraten werden im Zusammenhang mit den Untersuchungen von Hummel und Witzel [22] über die Bindung der Inhibitoren gesondert dargestellt werden. Auch sollte dann versucht werden, die von Hammes et al. [21,23] an Inhibitoren und Substraten gemessenen Relaxationsprozesse einer mechanistischen Interpretation zuzuführen.

Wir verdanken Herrn Prof. Dr. H.-D. Ohlenbusch den Hinweis, daß neben dem Schema (10) auch noch ein Schema in Betracht gezogen werden kann, in dem ES mit einem nicht zerfallenen Isomeren (ES') im Gleichgewicht steht. Bei der nicht modifizierten Reaktion läge dann bereits eine Zerfallshemmung (unkompetitive Hemmung) vor, die durch den Modifizierer über eine Verschiebung des Isomerisierungsgleichgewichtes aufgehoben würde, ohne die Zerfallskonstanten zu verändern. Es sind jedoch keine Anhaltspunkte bekannt, daß bei der unmodifizierten RNase-Reaktion eine Zerfallshemmung vorliegt.

Diese Arbeit wurde ausgeführt mit Unterstützung der Deutschen Forschungsgemeinschaft und des Fonds der Chemischen Industrie. H.-J. Wieker ist ein Liebig-Stipendiat des Fonds der Chemischen Industrie.

LITERATUR

1. Gassen, H. G., und Witzel, H., *European J. Biochem.* 1 (1967) 36.
2. Wieker, H.-J., und Witzel, H., *Abstracts 3rd FEBS Meeting (Warsaw)*, 1966, S. 44.
3. Follmann, H., Wieker, H.-J., und Witzel, H., *European J. Biochem.* 1 (1967) 243.
4. Witzel, H., in *Progress in Nucleic Acid Research and Molecular Biology* (edited by J. N. Davidson and W. E. Cohn), Academic Press, New York 1963, Vol. II, S. 221.
5. Dixon, M., und Webb, E. C., *Enzymes*, Longmans, Green, London 1964.
6. Webb, J. L., *Enzymes and Metabolic Inhibitors*, Academic Press, New York 1963, Vol. 1.
7. Segal, H. L., in *The Enzymes* (edited by P. D. Boyer, H. Lardy, K. Myrbäck), Academic Press, New York 1959, Vol. 1, S. 1.
8. Segal, H. L., Kachmar, J. F., und Boyer, P. D., *Enzymologia*, 15 (1952) 187.
9. Laidler, K. J., *The Chemical Kinetics of Enzyme Action*, Oxford University Press, London 1958.
10. Hearon, J. Z., Bernhard, S. A., Friess, S. L., Botts, D. J., und Morales, M. F., in *The Enzymes* (edited by P. D. Boyer, H. Lardy, M. Myrbäck), Academic Press, New York 1959, Vol. 1, S. 49.
11. Bloomfield, V., und Alberty, R. A., *J. Biol. Chem.* 238 (1963) 2811.

12. Ohlenbusch, H.-D., *Die Kinetik der Wirkung von Effektoren auf stationäre Fermentsysteme*, Springer Verlag, Berlin 1962.
13. Frieden, C., *J. Biol. Chem.* 239 (1964) 3522.
14. Findlay, D., Mathias, A. P., und Rabin, B. R., *Biochem. J.* 85 (1962) 134.
15. Barker, G. R., Montague, M. D., Moss, R. J., und Parsons, M. A., *J. Chem. Soc.* 1957, 3786.
16. Rabin, B. R., und Mathias, A. P., in *Mechanismen enzymatischer Reaktionen*, Springer Verlag, Berlin, Göttingen, Heidelberg 1964, S. 97.
17. Deavin, A., Mathias, A. P., und Rabin, B. R., *Nature*, 211 (1966) 255.
18. Morales, M. F., *J. Am. Chem. Soc.* 77 (1955) 4169.
19. Herries, H. G., Mathias, A. P., und Rabin, B. R., *Biochem. J.* 85 (1962) 127.

20. Witzel, H., in *Mechanismen enzymatischer Reaktionen*, Springer Verlag, Berlin, Göttingen, Heidelberg 1964, S. 123.
21. Cathou, R. E., und Hammes, G. G., *J. Am. Chem. Soc.* 87 (1965) 4674.
22. Hummel, J. P., und Witzel, H., *J. Biol. Chem.* 241 (1966) 1023.
23. Erman, J. E., und Hammes, G. G., *J. Am. Chem. Soc.* 88 (1966) 5607 und 5614.

H.-J. Wieker und H. Witzel
Chemisches Institut der Universität
355 Marburg/Lahn, Bahnhofstraße 7, Deutschland

European J. Biochem. 1 (1967) 259—266

IUPAC-IUB Combined Commission on Biochemical Nomenclature (CBN)

Abbreviations and Symbols for Chemical Names of Special Interest in Biological Chemistry

Revised Tentative Rules (1965)[1]

The Commission on the Nomenclature of Biological Chemistry decided in 1958 that an attempt should be made to standardize the abbreviations and symbols used for chemical names of special interest in biological chemistry. A Subcommittee, consisting of L. Hellerman, W. Klyne (Chairman), and E. C. Slater, was set up early in 1959 to deal with this problem.

The original draft proposals were based on the notes given at the beginning of each number of the *Journal of Biological Chemistry* (1958 *et seq.*) and of the "Suggestions to Authors" of the *Biochemical Journal* (66 (1957) 1). These drafts were circulated to members of the Nomenclature Commission, editors of chemical and biochemical journals, and interested specialists in many fields.

The problems were discussed fully at the meeting of the Commission in Munich in September, 1959—and also in joint sessions with the Organic Nomenclature Commission and the Enzyme Commission of the International Union of Biochemistry (IUB). A third draft, incorporating the results of the Munich discussions, was widely circulated in December 1959, and many useful comments on this were received.

A fourth draft, representing the "highest common factor" of all these comments and of many personal discussions, was prepared in August 1960, and circulated to members of the Commission on the Nomenclature of Biological Chemistry and to editors of some principal journals.

The meeting of editors of biochemical journals called together by the President of IUB in Cambridge in September 1960 invited W. Klyne to attend part of their meeting. He explained the history and purpose of the memorandum on abbreviations; he emphasized that this work lies on the borderline between the provinces of the two Unions, and that agreement of both Unions in principle was therefore very desirable.

After discussion, the Secretary General of IUB, R. H. S. Thompson, proposed the following statement, which was unanimously approved. "The contents of this memorandum (*i.e.* the fourth Draft) were approved both by the Bureau of IUB and by the meeting of Editors of biochemical journals called together by the President of IUB under the Chairmanship of J. T. Edsall, at a meeting held in Cambridge on September 9, 1960."

The Tentative Rules were published in the *IUPAC Information Bulletin*, No. 13, in June 1961, and in April 1962

[1] Published by permission of IUPAC, of IUB, and the official publishers to IUPAC, Messrs. Butterworths Scientific Publications.

Comments on these re-revised proposals should be sent to the present Chairman (O. Hoffmann-Ostenhof) or Secretary (W. E. Cohn) or to any member of CBN (A. E. Braunstein, J. S. Fruton, B. Keil, W. Klyne, C. Liébecq, B. G. Malmström, R. Schwyzer, E. C. Slater), or corresponding member (N. Tamiya).

Reprints of these Tentative Rules may be obtained from Waldo E. Cohn, Director, NAS-NRC Office of Biochemical Nomenclature, Oak Ridge National Laboratory, P. O. Box Y, Oak Ridge, Tennessee 37831, U.S.A.

the Commission of Editors of IUB formally accepted the Tentative Rules thus published. (These appear in *J. Biol. Chem.*, 237 (1962) 1381.)

The IUPAC Commission has received a number of valuable suggestions for the amendment of the Tentative Rules and these were discussed at meetings in Amsterdam (April 1961), in Dobříš (near Prague) (September 1962), and in Zürich (April 1963). The *Revised Tentative Rules* (1963) were approved at the Zürich meeting.

It has been impossible to meet all the comments and criticisms that colleagues have kindly offered on the various Drafts—since so many of the comments are mutually contradictory. However, these tentative proposals represent an honest attempt to give fair weight to all the diverse opinions that have been expressed.

1. INTRODUCTION

1.1 It is sometimes convenient to use abbreviations or symbols for the names of chemical substances, particularly in equations, tables, or figures, which would otherwise require the repeated use of unwieldy terms. The limited use of abbreviations and symbols of specified meaning is therefore accepted. However, clarity and unambiguity are more important than brevity.

1.2 Some chemists deprecate the use of any abbreviations or symbols for compounds. However, in the present state of biochemistry, increasing knowledge of the structure of large molecules such as proteins, polysaccharides, and polynucleotides makes it imperative to have some "shorthand" notation in which symbols are allotted to the monomeric units (monosaccharides, amino acids, and nucleosides), which are Nature's building bricks in these complex structures. Opponents of abbreviations should consider how unwieldy the formula of insulin would appear if the "three-letter" symbols for amino acids had not been used.

1.3 Titles and summaries of papers should be generally free of abbreviations. In the body of the paper, abbreviations and symbols should be used in the text sparingly, and only if advantage to the reader results. Chemical equations, which traditionally consist of symbols, may use a shorthand expression for a term that appears in full in the neighboring text.

1.4 If, in exceptional circumstances, symbols or abbreviations are used in a summary, they should be defined in the summary, as well as in the body of the paper.

1.5 It is hoped that editors will adopt in their journals as many of the following rules as possible in the light of individual circumstances.

1.6 Even if a journal permits the use of these abbreviations without definition, nonstandard abbreviations should always be defined in each paper.

1.7 Nonstandard (*ad hoc*) abbreviations and symbols should not conflict with known ones, or with the general principles proposed in these rules (see also Section 8).

1.8 The symbols and abbreviations discussed here fall into two distinct classes.

a) *Symbols* for monomeric units in macromolecules; these symbols are used to make up abbreviated structural formulas (sometimes called "shorthand" formulas), *e.g.* Gly—Val—Thr for the tripeptide glycylvalylthreonine. These are generally used by structural organic chemists, and can be made fairly systematic.

b) *Abbreviations* for semisystematic or trivial names, *e.g.* ATP for adenosine triphosphate; FAD for flavin-adenine dinucleotide.

The abbreviations of the second kind are generally formed of three or four capital letters. They are required chiefly by biochemists and are generally introduced as required; the need is for brevity rather than for system. It is the indiscriminate coining of such abbreviations that has aroused objections to the use of abbreviations in general.

Symbols for Natural Macromolecules

1.9 There are three main series of symbols for monomeric units, *viz.* those for amino acids, monosaccharides, and mononucleosides, of which the amino acid series is the oldest. An attempt has been made here to devise a standard treatment for all the three great groups of macromolecules, which are built up from these units. The standardization of treatment will involve certain unimportant changes in the (as yet partly developed) systems for individual groups. This standardization is desirable for two reasons.

a) The work of authors, editors, and readers is made simpler if the same principles apply to polypeptides, polysaccharides, and polynucleotides.

b) Standard treatment is essential for dealing with "hybrid" compounds, built up of units of different kinds, *e.g.* the nucleotide-peptides and glycopeptides.

1.10 It is much more difficult to be completely systematic in the planning of abbreviations and "shorthand" symbols for complex substances than in the construction of organic chemical formulas and physical symbols. Experience shows that it is not only difficult, but in some cases undesirable, to be rigidly consistent with these complex symbols.

The following example will illustrate these facts. For most purposes it is convenient to use the symbol Gly—Val—Thr to represent the tripeptide glycylvalylthreonine, as solid or in solution, whatever its state of ionization. We know that at certain defined pH values, the tripeptide will exist (mainly) as cation, as anion, or as depolar ion, but it is usually unnecessary to make separate shorthand symbols to represent these different forms.

This deliberate lack of precision runs parallel with the convention by which biochemists talk about the "citric acid cycle" or "tricarboxylic acid cycle," in spite of the fact that the acids exist almost entirely as their anions at physiological pH.

In several cases it is thought desirable to recommend for the same substance two different forms of abbreviation or symbol, one or other of which is more convenient for specific purposes.

Alternative Abbreviations and Symbols

1.11 For some important compounds, it is in practice necessary to have two symbols or abbreviations. For example, most biological chemists will continue to speak of "adenosine diphosphate," or more often to abbreviate it as "ADP." Organic chemists interested in the structure and synthesis of this and related compounds will wish to call this compound "adenosine 5′-pyrophosphate," and to use a systematic symbol (Ado-5′-*P*-*P*). The abbreviation and the symbol must therefore coexist.

1.12 Abbreviations such as "ADP," which are to the organic chemist trivial, are used to form the systematic names of enzymes in the patterns proposed by the Enzyme

Commission of the International Union of Biochemistry (Elsevier, 1965).

Language Differences

1.13 It is desirable that where trivial abbreviations (such as ACTH) are necessary, they should be identical in all languages—as are chemical symbols (*e.g.* N standing for nitrogen, *azote*, and *Stickstoff*). It would be unfortunate if the substance called in English "ribonucleic acid," and abbreviated "RNA," were to retain two separate abbreviations, ARN (*acide ribonucléique*) and RNS (*Ribonukleinsäure*), in French and German, to say nothing of other languages. It is suggested that the international abbreviations should be taken from that language in which a given abbreviation first became common. Abbreviations introduced in the future may conveniently be based on Greek or Latin forms.

Structural Analogues

1.14 Structural analogues of a given compound should not generally be abbreviated as if they were derivatives of that compound.

2. POLYPEPTIDES AND PROTEINS

This system is based on the original proposals of E. Brand and J. T. Edsall (*Ann. Rev. Biochem.*, 16 (1947) 224), as developed in the monograph of J. P. Greenstein and M. Winitz (*The Chemistry of the Amino Acids*, John Wiley and Sons, Inc., New York, 1961). However, some modifications have been introduced so that it is possible to designate all amino acids found in proteins, including the acid amides and the hydroxylated compounds, by three-letter symbols-a capital followed by two lower-case letters.

2.1 The following symbols denote the common amino acids and their residues (see also 2.5):

Alanine	Ala		Hydroxylysine	Hyl
Arginine	Arg		Hydroxyproline	Hyp
Aspartic acid	Asp		Isoleucine	Ile
Asparagine	Asn		Leucine	Leu
			Lysine	Lys
			Methionine	Met
Cystine (half)	Cys or Cys		Ornithine	Orn
			Phenylalanine	Phe
			Proline	Pro
Cysteine	Cys		Serine	Ser
Glutamic acid	Glu		Threonine	Thr
Glutamine	Gln		Tryptophan	Trp
Glycine	Gly		Tyrosine	Tyr
Histidine	His		Valine	Val

Modified amino acids, such as asparagine and glutamine, may also be represented as $Asp(NH_2)$, $Glu(NH_2)$ or Asp, Glu.
$$\underset{NH_2}{\overset{|}{}} \underset{NH_2}{\overset{|}{}}$$

2.2 The abbreviations should not be used for the free amino acids in the text of papers, but only in tables, lists, and figures.

2.3 Where the sequence of residues in a peptide or protein is known, the symbols for the residues are written in order and joined by short lines (dashes, hyphens). Where the sequence is not known, the group of symbols, separated by commas, is enclosed in parentheses[2].

In the formulation of linear polypeptides or proteins, the symbol written at the left-hand end of a known sequence is that of the amino acid carrying the free amino group, and the symbol written at the right-hand end is that of the residue

---

[2] It is preferable to display the polypeptide chain as a horizontal rather than as a vertical sequence.

of the amino acid carrying the free carboxyl group. Example:
The condensed formula

Gly—Glu—Arg—Gly—Phe—(Phe, Tyr, Thr,Pro)—
Lys—Ala—Trp—Tyr—Val—Ile—Hyp—Cys—Ala

is that of a polypeptide in which the sequence of the first five
amino acids has been established, the glycine at the left carrying the free amino group. The sequence of the next four amino
acids is unknown, but the last nine amino acids are in known
order with alanine carrying the free carboxyl group.

If the direction of the link must be specfied, this may be
done with an arrow thus (→), the point of the arrow indicating the nitrogen of the peptide bond...CO→NH...

Example: Gly → Ala → Val.

The symbol → is desirable particularly for dealing with
cyclic peptides.

Unless otherwise indicated, it is assumed that poly-
functional amino acids, such as glutamic acid, aspartic acid,
and lysine, are joined by normal α-peptide bonds.

Abnormal links, *e.g.* γ-peptide bonds or links formed
through other functional groups, may be indicated by
methods such as the following:

Glu   or   Glu
└Cys—Gly   γ└Cys—Gly   = glutathione (reduced)

Comment

The links between residues have been shown previously
by peptide chemists as full points (periods, dots; ·) and by
carbohydrate chemists (generally) as short strokes (dashes,
hyphens; —). At times, special symbols have been used
(> or →) to show the direction of what is in all cases an
unsymmetrical link (peptide or glycoside).

For consistency and ease of printing, a short rule or
dash (—), which is what we normally use for a chemical bond,
should be the standard connecting symbol.

The simple usage by which Gly—Gly—Gly stands for
glycylglycylglycine appears to involve the employment of
the same three letters Gly for three different residues or
radicals—(b), (c), (d) below. However, if the dashes or
hyphens are considered as part of each symbol, we have four
distinct forms, for the free amino acid and the three residues,
viz.:

(a)   Gly   = $NH_2 \cdot CH_2 \cdot CO_2H$;[3] the free amino acid
(b)   Gly—   = $NH_2 \cdot CH_2 \cdot CO—$;   the left hand unit
(c) —Gly—   = —NH · $CH_2 \cdot CO—$;   the middle unit
(d) —Gly   = —NH · $CH_2 \cdot CO_2H$;   the right hand unit.

For peptides, a distinction may be made between the
peptide itself, *e.g.* Gly—Gly (shown without dashes at the
ends of the symbols), and the sequence, *e.g.* —Gly—Glu—
(shown with dashes at the ends of the symbols).

2.4 The amino acid symbols represent the natural L
form. Other forms are indicated by the appropriate symbols
D or DL immediately preceding the amino acid symbol and
separated from it by a hyphen. Example: Leu—D-Phe—
Gly. When it is desired to make the number of residues
appear in a more clear manner, the hyphen between the
prefix and the symbol may be omitted. Example: Leu—
DPhe—Gly.

Rare Amino Acids

2.5 The list in paragraph 2.1 is restricted to the more
common amino acids. Symbols for the more rare amino acids
are included in *Abbreviated Designation of Amino Acid
Derivatives and Polypeptides (IUPAC Information Bulletin,*
in press; *J. Biol. Chem.,* 241 (1966) 2491; *European J. Biochem.,* in preparation).

[3] Or corresponding ionized forms. Gly is ordinarily not
to be used alone in text (see 2.2).

18*

State of Ionization

2.6 As stated in paragraph 1.10, it is generally convenient
to use the same abbreviated formula for a polypeptide no
matter what its state of ionization.

In some circumstances, however, an author will wish
to show that a peptide is acting as a cation or anion; according to the convention of Greenstein and Winitz (*The Chemistry of the Amino Acids*), the amino-terminal and carboxyl-
terminal ends of the peptide are marked with H and OH,
respectively (I); these may be modified to show the appropriate state of ionization (II or III).

H—Gly—Val—Thr—OH   (I)
⁺H₂—Gly—Val—Thr—OH   (II)
H—Gly—Val—Thr—O⁻   (III)

Derivatives

2.7 Symbols for the functional groups of derivatives have
been devised by specialists in the field (*cf.* reference in paragraph 2.5).

Comment

One-letter abbreviations for amino acid residue have
been proposed. While recognizing the utility of such systems
in computer analysis of sequences in proteins, the Commission does not recommend their use in printed material or
teaching. The system proposed by Šorm *et al.* (*Collection
Czech. Chem. Commun.,* 26 (1961) 569) has found relatively
wide favor and may be recommended to those requiring a
one-letter system for computer analysis.

3. CARBOHYDRATES

A system of three-letter symbols for monosaccharide
units and their residues, similar to that already in use for
peptides, was introduced by the Carbohydrate Nomenclature Committees of the Chemical Society and the American
Chemical Society (*cf. J. Chem. Soc.,* (1952) 5121; *Chem. Eng.
News,* (1953) 1776; *J. Org. Chem.,* 28 (1963) 281). The following rules are based on this system[4].

3.1 The following symbols are used to indicate monosaccharide units and their residues in oligosaccharides and
polysaccharides.

Glucose	Glc[5]	Fructose	Fru
Galactose	Gal	Ribose	Rib
Mannose	Man		

Other monosaccharides are represented similarly by the
first three letters of their names, unless this would lead to
confusion with an existing symbol (*e.g.* Gly and Thr in the
amino acid series).

3.2 Pyranose and furanose forms are designated where
necessary by the suffixes *p* and *f*.

3.3 Configurational symbols D and L (small Roman
capital letters) and anomeric prefixes are shown where
necessary as prefixes. Examples: (a) an α-D-glucopyranose
unit, α-D-Glc*p* or Glc*p*; (b) a β-D-fructofuranose unit, β-D-
Fru*f* or Fru*f*.

3.4 Symbols thus formed are joined by short rules to
indicate the links between units. The position and nature of

[4] These proposals and those given in the Tentative Rules
of Carbohydrate Nomenclature (separate to *IUPAC Information Bulletin*, April 1963) differ in some respects. Attempts
are being made to resolve the differences by discussion between the Commissions involved.

[5] Where no ambiguity can arise, the single-letter symbol
G may be used.

the links are shown by numerals and the anomeric symbols α and β. Examples:

Maltose, Glc$p\alpha$1—4Glc
Lactose, Gal$p\beta$1—4Glc
Stachyose, Gal$p\alpha$1—6Gal$p\alpha$1—6Gal$p\alpha$1—2βFru$f$.

A branched chain tetrasaccharide:

$$Glc\beta1-3Gal\beta1-4Glc$$
$$2$$
$$|$$
$$1\,\alpha Fuc$$

Arrows may be used to indicate the direction of the glycoside link, the arrow pointing away from the hemiacetal carbon of the link; *e.g.* lactose may be represented as Gal$p\beta$1→4Glc.

3.5 A 2-deoxy sugar is designated by the symbol for its most common parent sugar with the prefix "de." Other deoxy sugars may be designated similarly with a positional numeral. Examples: 2-deoxyribose, deRib; 3-deoxyglucose, 3-deGlc.
It may sometimes be necessary to enclose such a symbol in parentheses to avoid confusion between the numeral indicating the "deoxy position" and numerals indicating the position of linkages.

3.6 Derived monosaccharide units—such as glyconic acids, glycuronic acids, 2-amino-2-deoxysaccharides, and their *N*-acetyl derivatives—may be designated by reasonable modified symbols, defined in each paper. Examples of symbols that have been used are as follows (all in the glucose, Glc, series):

Gluconic acid	GlcA
Glucuronic acid	GlcUA
Glucosamine	GlcN
N-Acetylglucosamine	GlcNAc

3.7 Symbols should not be used for the monosaccharides themselves, except in tables, lists, and figures.

4. PHOSPHORYLATED COMPOUNDS: GENERAL

4.1 Phosphorylated compounds may be designated by the name (or abbreviation) of the parent compound with a capital italic P as a prefix or suffix[6].
P is used as prefix where it symbolizes "phospho-" at the beginning of a name. P is used as a suffix where it symbolizes "phosphoric acid" or "phosphate" at the end of a name[7].
For compounds containing more than one position available for phosphorylation, the position of the phosphate group should always be indicated by number or Greek letter.
4.2 The capital P when linked to one radical indicates —PO(OH)$_2$ or any ion derived from it; when linked to two radicals in indicates —PO(OH)—, or the ion derived from it.
4.3 The pyrophosphate group (I) is represented by -P-P-:

$$
\begin{array}{cc}
O & O \\
\uparrow & \uparrow \\
-P-O-P- & \quad\text{(I)} \\
| & | \\
OH & OH
\end{array}
$$

Two separate phosphate groups, attached at different points to the same molecule, are represented by P_2.

Examples:

Glucose 6-phosphate	Glucose-6-P or Glc-6-P
Glycerol 3-phosphate or α-phosphoglycerol	Glycerol-3-P α-P-Glycerol
3-Phosphoglyceric acid Glycerate 3-phosphate	3-P-Glyceric acid Glycerate-3-P
Phosphoenolpyruvate	P-Enolpyruvate
Fructose 1,6-bisphosphate[8]	Fructose-1,6-P_2 or Fru-1,6-P_2
Creatine phosphate Phosphocreatine	Creatine-P P-Creatine

4.4 The term diphosphate (and the abbreviation DP, as in ADP) is correctly used only for the pyrophosphate group (IUPAC Inorganic Rules 7.5, 5.213, 2.251; also Organic Rules A2.5 and 67). Compounds with two or more orthophosphate residues are more properly termed bis-, tris-, tetrakis-, etc., phosphates. The older term, fructose 1,6-diphosphate, strictly interpreted, could indicate a pyrophosphate group connecting the 1- and 6-positions of fructose.

5. NUCLEOTIDES AND NUCLEIC ACIDS

5.1 Two systems are recognized, one (5.3) using three-letter symbols for the more common nucleosides (like those used for amino acids and monosaccharides in Sections 2 and 3) and a capital italic P for the phosphate residue, the other (5.4) using single capital letters for the common nucleosides and a lower-case p for the phosphate residue.

5.2 In either system, glycosyl linkages are assumed to be β and to involve only D-ribose or D-deoxyribose, and the phosphodiester linkage is assumed to be 3′-5′ from left to right unless otherwise specified by appropriate *ad hoc* symbols or numerals.

5.3 *Three-Letter Symbols*

5.3.1 The phosphate group is designated by an italic capital P (*cf.* Section 4), to distinguish it from Roman capital P for phosphorus.

5.3.2 The (ribo) nucleosides are designated by the following three-letter symbols, chosen to avoid confusion with the corresponding base:

Ado	adenosine	Thd	ribosylthymine[9]
Guo	guanosine	Cyd	cytidine
Ino	inosine	Urd	uridine
Xao	xanthosine	Ψrd	pseudouridine (5-ribosyluracil)

Ribosylnicotinamide may be designated by Nir.

5.3.3 The 2′-deoxyribonucleosides are designated by the symbols for the corresponding ribose-derivatives (5.3.2) with the prefix d. Examples are dAdo for 2′-deoxyadenosine, dThd for 2′-(deoxy)thymidine[9] (2′-deoxyribosylthimine).
The letter d may also be used as a prefix to a series (an oligonucleotide) to indicate that all the sugars in the series are 2′-deoxyribosyl units.

5.3.4 The points of attachment of phosphate residues to a sugar, if other than 3′-P-5′, are designated by the appropriate primed numerals, separated by hyphens. Examples:

---

[6] This type of partial abbreviation (*e.g.* glucose-6-P) is convenient in biochemical papers where there is much discussion of phosphorylated metabolites and intermediates. It is not commonly used in organic chemical papers.

[7] The P is italicized in order to avoid confusion with the accepted symbol for the phosphorus atom (Roman capital P).

[8] See comment on this change in 4.4.

[9] Because thymidine has traditionally been used for 2′-deoxyribosylthymine, arising at a time when the ribosyl analogue was not known, it is recommended that the prefix r (for ribo) or d (for deoxyribo) be used with Thd or with T whenever there is a possibility of misunderstanding which substance is intended.

(a) Adenosine 2′-phosphate: Ado-2′-P

    Adenosine 5′-phosphate: Ado-5′-P or P-5′-Ado

(b) 5′-O-Phosphoryldeoxyadenylyl-(3′-5′)-thymidine: P-5′-dAdo-3′-P-5′-dThd[9] or P-dAdo-P-dThd.

The positional numerals may precede a series, as in 2′-5′-(Ado-P-Guo-P-Urd-P), which specifies Ado-2′-P-5′-Guo-2′-P-5′-Urd-2′-P. When the series in the left-to-right direction is 3′-5′, as in example (b) above, they may be omitted (*cf.* 5.2).

5.3.5 A cyclic phosphate group is designated by the two positional numerals for the points of attachment of a single P, as in Cyd-2′:3′-P. (The corresponding bisphosphate would be Cyd-2′,3′-P_2 or P-2′-Cyd-3′-P.)

5.3.6 The so-called nucleoside diphosphate sugars, which are sometimes called pyrophosphates, may be represented as follows: Urd-5′-P-P-Glc for uridine diphosphate glucose [*i.e.* uridine 5′-(α-D-glucopyranosyl diphosphate)]; Urd-5′-P-P-Gal.

5.4 One-Letter Symbols [10]

5.4.1 The phosphate group is designated by a lower case p (to separate what would otherwise be a solid mass of capital letters).

5.4.2 The common (ribo)nucleosides are designated by single capital letters, thus:

A	adenosine	T	ribosylthymine[9]
G	guanosine	C	cytidine
I	inosine	U	uridine
X	xanthosine	Ψ	pseudouridine (5-ribosyluracil)

The following general symbols are also useful: Pu, unspecified purine nucleoside; Py, unspecified pyrimidine nucleoside; N, unspecified nucleoside.

5.4.3 The 2′-deoxyribonucleosides are designated by the same symbols (5.4.2) preceded by d (*cf.* 5.3.3). Thus, dA = 2′-deoxyribosyladenine, dT = 2′-deoxyribosylthymine (= thymidine)[9].

5.4.4 The points of attachment of phosphate residues, if other than 3′p5′, may be indicated as in 5.3.4. A regular 3′-5′ sequence (read left to right), as in the natural nucleic acids, need not be specified by positional numerals (5.2).

In this system, the substances in 5.3.4 become: A2′p; pA; d(pApT).

Other examples are:

ApGpUp (3′-5′ trinucleotide, ending at right in a 3′-phosphate).

ApGpU-cyclic p (the same, but ending in a 2′:3′-cyclic phosphate)[11].

pApApA (3′-5′ trinucleotide, starting with a 5′-phosphate at left, ending with unsubstituted 2′ and 3′ hydroxyl groups at right).

5.5 More Complex Structures

5.5.1 *Sequence Designation*—For more complex structures (large oligonucleotides or polynucleotides), in which known and unknown sequences may involved, the p for phosphate between two nucleosides may be replaced by a hyphen (for a known sequence) or a comma (unknown sequence) to give a system identical with that used for amino acid sequences (see Section 2.3 above). Regular 3′-5′ linking is assumed unless indicated otherwise. Thus GpApUp(CpCpUp)Gp—a

---

[10] Intended chiefly for oligonucleotides and polynucleotides. The IUPAC Commission for the Nomenclature of Organic Chemistry prefers the three-letter symbols (5.3).

[11] The symbol > for "cyclic" is useful in the one-letter system. Thus this example can be represented as ApGpU > p. Unless otherwise stated, this symbol indicates a 2′:3′-cyclic phosphate residue.

3′-ended heptanucleotide of partially known sequence—becomes G-A-U(C,C,U)Gp or G-A-U(C₂,U)Gp; d-pTpTpC-pTpTpC becomes d(pT-T-C-T-T-C).

5.5.2 *Higher Polymers.* For sequences too long or repetitive or obscure for detailed exposition, the prefix "poly" may be used in conjunction with the comma and hyphen. For example, the alternating regular sequence dA-dT-dA-dT--- becomes poly d(A-T); the random copolymer of equal amounts of U and A becomes poly (U,A). The prefix "poly" may be omitted when the number (or proportions) of nucleoside residues is specified by subscript numerals, if known, or by the subscript n (in place of the p used in the system devised by the IUPAC Commission on Macromolecules (*J. Polymer Sci.*, 8 (1952) 257) for molecules of indefinite size. Thus, poly d(A-T) may be expressed as d(A-T)$_n$, and poly (U₂,A) as (U₂,A)$_n$. Composition and size may be shown by appropriate numerical subscripts, as in (U₂,A)₅₀, which contains 100 U's and 50 A's in random sequence, and in d(A-T)₅₀, which contains 50 dA's and 50 dT's in regular alternating sequence. Multiple parentheses or brackets may be used as in organic nomenclature for blocks within polymers, side chains, etc.

5.5.3 *Nucleoside Symbols.* The more common nucleoside residues are represented by single capital letters (see 5.4.2 above). All other nucleosides should be represented by single capital letters, insofar as possible, defined as introduced (*e.g.* B for BrU). Where a nucleoside symbol must include more than one character, it should contain neither hyphens nor commas (*e.g.* −2MeG−, −6diMeA−, −BrU−). Linkages other than 3′-5′, or sugars other than ribose or deoxyribose, should be indicated by special *ad hoc* symbols. d and r may precede whole chains, groups within chains, or individual nucleoside residues, as appropriate.

5.5.4 *Association between Chains.* Associations between two or more nucleotide chains may be indicated by the center dot, as in (A)$_n$ · (U)$_n$ or (dG)$_n$ · (dC)$_n$ or [(A)$_n$ · (U)$_n$ · (U)$_n$]. The absence of association is indicated by the plus sign, as in poly (A) + poly (C). The absence of definite information on association is indicated by the comma (again meaning "unknown"), as in poly (A), poly (A-U) or (A)$_n$, (A-U)$_n$.

Special Abbreviations

5.6 The 5′-mono-, di-, and triphosphates of the common nucleosides may be designated by the customary special abbreviations, *e.g.* AMP, ADP, and ATP for the derivatives of adenosine. The corresponding derivatives of cytidine, guanosine, inosine, uridine, and pseudouridine may be designated by similar abbreviations in which the initial letters are C, G, I, U, and Ψ, respectively. Thus, for example, IMP = inosine 5′-monophosphate; UDP = uridine 5′-diphosphate. Uridine diphosphate glucose may be designated by UDPG.

These compounds should, however, be designated in more "chemical" papers by systematic symbols as indicated in paragraph 5.3.4, *e.g.* Ado-5′-P, Ado-5′-P-P, Ado-5′-P-P-P, when required for consistency with the other nucleotides.

5.7 Flavin mononucleotide (riboflavin 5′-phosphate) may be designated by the special abbreviation FMN.

5.8.1 The two types of nucleic acid are designated by their customary abbreviations:

    RNA, ribonucleic acid or ribonucleate

    DNA, deoxyribonucleic acid or deoxyribonucleate.

5.8.2 It is sometimes convenient to designate fractions or functions of RNAs by prefixes (*e.g.* mRNA for "messenger" RNA, tRNA for "transfer" RNA, rRNA for "ribosomal" RNA, nRNA for "nuclear" RNA). Such terms should be defined in each paper unless defined by the journal in which the paper is published.

5.8.3 Transfer RNAs that accept specific amino acids may be designated as, for example, tRNA^Ala for the tRNA that accepts alanine (*i.e.* "alanine tRNA"); in the case of more than one such species, they may be distinguished by subscripts, as tRNA$_1^{Ala}$, tRNA$_2^{Ala}$, etc. When such a tRNA species is bound to an amino acid, it may be designated as, for example, alanyl-tRNA^{Ala}. Specification of its source should be in parentheses before or after, as for example, alanyl-tRNA$_2^{Ala}$ (*Escherichia coli*) or (*E. coli*) analyl-tRNA$_2^{Ala}$.

6. COENZYMES

There has been much controversy about names and symbols for the nucleotide coenzymes (DPN *versus* CoI, etc.).

The Enzyme Commission of the International Union of Biochemistry decided in August 1959 to recommend the following names, for the reasons briefly stated in the comment below.

Nicotinamide-adenine dinucleotide for the compound hitherto commonly called diphosphopyridine nucleotide or Coenzyme I.

Nicotinamide-adenine dinucleotide phosphate for the compound hitherto commonly called triphosphopyridine nucleotide or Coenzyme II.

The IUPAC Commission on the Nomenclature of Biological Chemistry after discussion accepted these recommendations of the Enzyme Commission, which were formally adopted by the IUB Council at Moscow in August 1961.

Comment (cf. Dixon, M., *Nature,* 188 (1960) 464)

The two main systems of nomenclature of the nicotinamide nucleotide coenzymes (the CoI and the DPN systems) are both unsatisfactory. The first gives no indication of the chemical structure at all; the second indicates a chemical structure that is incorrect.

Since no compromise between the two systems is possible, the only satisfactory solution is to abandon both and to adopt a name that indicates the correct chemical structure. The name adopted should be consistent with the existing names of three closely related compounds, namely the corresponding mononucleotide (nicotinamide mononucleotide, NMN) and the two flavin nucleotides (flavin-adenine dinucleotide, FAD, and flavin mononucleotide, FMN).

The name the Enzyme Commission of IUB, after careful consideration of possible alternatives, has decided to recommend in place of CoI or DPN, namely nicotinamide-adenine dinucleotide (NAD), not only indicates the structure satisfactorily, but forms a logical system with the three that are already generally accepted.

CoII or TPN is a phosphorylated derivative of NAD, and may be called nicotinamide-adenine dinucleotide phosphate, conveniently abbreviated to NADP.

6.1 The dinucleotide coenzymes may be designated by the following abbreviations:

Nicotinamide-adenine dinucleotide   NAD
(formerly DPN, CoI)
Nicotinamide-adenine dinucleotide phosphate   NADP
(formerly TPN, CoII)

These abbreviations do not specify the state of oxidation of the compounds.

6.2 The oxidized and reduced forms of the coenzymes may be designated by NAD$^+$ (NADP$^+$) and NADH (NADPH), respectively[12]. They may be used in an equation as follows:

$$NAD^+ + XH_2 \rightleftarrows NADH + H^+ + X.$$

[12] The IUB Standing Committee on Enzymes has used the descriptive terminology "NAD (NADP)" and "reduced NAD (NADP)."

6.3 Other coenzymes may be designated as follows:

FAD, FADH$_2$	Flavin-adenine dinucleotide, and its reduced form
FMN, FMNH$_2$	Flavin mononucleotide, and its reduced form
GSH, GSSG	Glutathione, and its oxidized form
CoA, acetyl-CoA or CoASH, CoASAc	Coenzyme A and its acetyl derivative (alternative forms).

6.4 Systematic symbols may be built up for some of these coenzymes as shown in paragraphs 5.3.4 and 5.3.6. Examples:

$$NAD = Nir\text{-}5'\text{-}P\text{-}P\text{-}5'\text{-}Ado$$
$$NADP = Nir\text{-}5'\text{-}P\text{-}P\text{-}5'\text{-}Ado\text{-}2'\text{-}P.$$

7. MISCELLANEOUS COMPOUNDS

7.1 The following abbreviations are permitted; although they are fairly common, they should be defined in any paper if it is thought that readers might be unfamiliar with them. Some abbreviations are taken from the list published by *Annual Review of Biochemistry.*

ACTH	adrenocorticotropin, adrenocorticotropic hormone, or corticotropin
BAL	2,3-dimercaptopropanol
CM-cellulose	*O*-(carboxymethyl)cellulose
DDT	1,1,1-trichloro-2,2-bis(*p*-chloro-phenyl)-ethane
DEAE-cellulose	*O*-(diethylaminoethyl)cellulose
DFP	di-isopropyl phosphorofluoridate
DNP-	2,4-dinotrophenyl-
DOC	11-deoxycorticosterone
DOCA	11-deoxycorticosterone acetate
DOPA	3,4-dihydroxyphenylalanine
DPT	diphosphothiamine (thiamine pyrophosphate, cocarboxylase)
EDTA	ethylenediaminetetraacetic acid (or acetate)
FDNB	1-fluoro-2,4-dinitrobenzene
Hb	hemoglobin (deoxygenated)
HbCO	"carboxy" hemoglobin—*i.e.* hemoglobin plus carbon monoxide
HbO$_2$	oxyhemoglobin
MetHb	methemoglobin
Mb	deoxygenated myoglobin (may be modified in the same way as Hb)
MSH	melanocyte-stimulating hormone
P$_i$	orthophosphate (inorganic)
PP$_i$	pyrophosphate (inorganic)
TEAE-cellulose	*O*-(triethylaminoethyl)cellulose
Tris	tris(hydroxymethyl)aminomethane; 2-amino-2-hydroxymethylpropane-1,3-diol

7.2 If one-letter symbols for steroids (Compound F, Substance S) are used, the systematic name of the compound should be given at least once in each paper. Derivatives, such as "tetrahydro-E" and "11-epi-F", should also be clearly defined by systematic names.

8. STANDARDS FOR NEW ABBREVIATIONS

8.1 Abbreviations other than those listed or defined above should be constructed in accordance with the following principles.

8.2 The number should be limited; none should be introduced except where repeated use is required. Three-letter abbreviations are most convenient.

Use in another sense of an accepted abbreviation must be avoided.

Where a number of derivatives, salts, or addition compounds may be formed, the name of the common fundamental structure should be the one abbreviated, so that other symbols may be attached to it.

9. ALPHABETICAL LISTS

For convenience the symbols and abbreviations are collected in the following two alphabetical lists.

Table 1. *Symbols for monomeric units in macromolecules (or in phosphorylated compounds)*

Symbol	Monomeric unit in macromolecule
A, Ado	adenosine
Ala	alanine
Arg	arginine
Asp	aspartic acid
Asp(NH₂), Asn	asparagine
C, Cyd	cytidine
Cys[a]	cystine (half)
Cys	cysteine
de, d	(indicates "deoxy" in carbohydrates and nucleotides)
f	(suffix) furanose
Fru	fructose
Gal	galactose
G, Glc[b]	glucose
G, Guo[b]	guanosine
GlcA	gluconic acid
GlcN	glucosamine
GlcNAc	*N*-acetylglucosamine
GlcUA	glucuronic acid
Glu	glutamic acid
Glu(NH₂), Gln	glutamine
Gly	glycine
His	histidine
Hyl	hydroxylysine
Hyp	hydroxyproline
I, Ino	inosine
Ile	isoleucine
Leu	leucine
Lys	lysine
Man	mannose
Met	methionine
Nir	ribosylnicotinamide
Orn	ornithine
P, p	phosphate
p	(suffix) pyranose
Phe	phenylalanine
Pro	proline
Rib	ribose
Ser	serine
Thr	threonine
Trp	tryptophan
T, Thd	ribosylthymine[c]
Tyr	tyrosine
U, Urd	uridine
Val	valine
Xao	xanthosine

[a] With vertical bond above or below "s" (see Section 2).
[b] The one-letter symbol G must not be used if confusion between its two meanings can arise.

Table 2. *Abbreviations for semisystematic or trivial names*

ACTH	adrenocorticotropin, adrenocorticotropic hormone, or corticotropin
ADP	adenosine 5'-diphosphate (pyro)
AMP	adenosine 5'-phosphate
ATP	adenosine 5'-triphosphate (pyro)
BAL	2,3-dimercaptopropanol
CDP	cytidine 5'-diphosphate (pyro)
CM-cellulose	*O*-(carboxymethyl)cellulose
CMP	cytidine 5'-phosphate
CoA (or CoASH)	coenzyme A
CoASAc	acetyl coenzyme A
CTP	cytidine 5'-triphosphate (pyro)
DEAE-cellulose	*O*-(diethylaminoethyl) cellulose
DDT	1, 1, 1-trichloro-2,2-bis(*p*-chlorophenyl)-ethane
DFP	di-isopropyl phosphorfluoridate
DNA	deoxyribonucleic acid
DNP-	2,4-dinitrophenyl-
DOPA	3,4-dihydroxyphenylalanine
DPN[a]	diphosphopyridine nucleotide
DPT	diphosphothiamine (thiamine pyrophosphate, cocarboxylase)
EDTA	ethylenediaminetetraacetate
FAD	flavin-adenine dinucleotide
FDNB	1-fluoro-2,4-dinitrobenzene
FMN	riboflavin 5'-phosphate
GDP	guanosine 5'-diphosphate (pyro)
GMP	guanosine 5'-phosphate
GSH	glutathione
GSSG	oxidized glutathione
GTP	guanosine 5'-triphosphate (pyro)
Hb, HbCO, HbO₂	hemoglobin, carbon monoxide hemoglobin, oxyhemoglobin
IDP	inosine 5'-diphosphate (pyro)
IMP	inosine 5'-phosphate
ITP	inosine 5'-triphosphate (pyro)
Mb, MbCO, MbO₂	myoglobin, carbon monoxide myoglobin, oxymyoglobin
MetHb, MetMb	methemoglobin, metmyoglobin
MSH	melanocyte-stimulating hormone
NAD	nicotinamide-adenine dinucleotide (cozymase, Coenzyme I, diphosphopyridine nucleotide)
NADP	nicotinamide-adenine dinucleotide phosphate (Coenzyme II, triphosphopyridine nucleotide)
NMN	nicotinamide mononucleotide
P₁	inorganic orthophosphate
PP₁	inorganic pyrophosphate
RNA	ribonucleic acid
TEAE-cellulose	*O*-(trietylaminoethyl)cellulose
TPN[b]	triphosphopyridine nucleotide
Tris	tris(hydroxymethyl)aminomethane (2-amino-2-hydroxymethylpropane-1,3-diol)
UDP	uridine diphosphate (pyro)
UDPG	uridine diphosphate glucose
UMP	uridine monophosphate
UTP	uridine triphosphate (pyro)

[a] Replaced by NAD (see Section 6).
[b] Replaced by NADP (see Section 6).

These Tentative Rules have been translated into German and have appeared in *Hoppe-Seyler's Zeitschrift für physiologische Chemie* [348 (1967) 245]. They may be translated into French and would then appear in the *Bulletin de la Société de Chimie biologique*.

Other Tentative Rules of the IUPAC-IUB Commission on Biochemical Nomenclature (CBN), already published in the *Journal of Biological Chemistry* (241 (1966) 2491 and 2987) and in *Biochemistry* (6 (1967) 362) are also available from Waldo E. Cohn, Director, NAS-NRC Office of Biochemical Nomenclature, Oak Ridge National Laboratory, P. O. Box Y, Oak Ridge, Tennessee 37831, U.S.A.:

Abbreviated Designation of Amino Acid Derivatives and Polypeptides,
Trivial Names of Miscellaneous Compounds of Importance in Biochemistry,
Nomenclature of Quinones with Isoprenoid Side Chains,
Nomenclature and Symbols for Folic Acid and Related Compounds
Nomenclature of Corrinoids,
Rules for Naming Synthetic Modifications of Natural Peptides.

European J. Biochem. 1 (1967) 267—280

Synthesis, Characterization and Enzymatic Properties
of Poly-L-Lysyl Ribonuclease

A. Frensdorff and M. Sela

Section of Chemical Immunology, The Weizmann Institute of Science, Rehovoth

(Received February 8, 1967)

Poly-L-lysyl RNase was prepared by the reaction of RNase with ε,N-trifluoroacetyl-α,N-carboxy-L-lysine anhydride, either in dioxane-aqueous phosphate buffer or in dimethylformamide in the absence of phosphate. The trifluoroacetyl group was subsequently removed from poly-trifluoroacetyllysyl RNase by 1.0 M aqueous piperidine. Derivatives prepared in aqueous media and enriched with up to 70 lysine residues, and derivatives prepared in dimethylformamide enriched with up to 270 lysine residues, were soluble and retained some activity. This activity could be completely recovered in all preparations after reduction of the disulfide bonds, followed by reoxidation. Approximately four phenolic hydroxyl groups were found, by spectrophoto-metric titration, to ionise normally in polylysyl RNase, as opposed to three in the native enzyme.

The enzymic activity of polytrifluoroacetyllysyl RNase, both on RNA and on cytidine 2',3'-cyclic phosphate, under various conditions, was essentially identical to that of polyalanyl RNase. Polylysyl RNase, in which the amino groups of the attached lysine peptides are unblocked and protonated, is active on RNA as well as on cytidine 2',3'-cyclic phosphate, but optimal activity is obtained under conditions different from those of the native enzyme. Thus polylysyl RNase displays maximal activity on RNA at ionic strength 0.6—0.7. The optimum pH for activity depends on ionic strength and is shifted at low ionic strength in the same direction as that of the isoelectric point of the modified enzyme for both RNA and cytidine cyclic phosphate. The base specificity of the modified enzyme was not significantly modified.

The changes in ionic strength and pH dependence of activity are related directly to the lysine peptides attached to the enzyme and can be reversed by either reblocking the free amino groups or by removing the peptides with trypsin. Kinetic analyses suggest that the changes in activity are a result of modified substrate binding.

Chemical modification of proteins and their influence on biological properties has probably been the method most widely used in studies aimed at obtaining a better understanding of structure-function relationships. Many of the methods used and the information derived from their application have been extensively reviewed [1,2]. Additional techniques are being made available to the protein chemist continuously.

The polymerisation of N-carboxy-α-amino acid anhydrides in aqueous media, using the α- and the ε-amino groups of proteins as initiators for the polymerisation reaction, leads to the formation of poly-peptidyl proteins [3]. This reaction proceeds under

mild conditions (aqueous media, neutral pH, low temperature) which do not lead to denaturation. Owing to the great diversity of amino acids which may thus be attached to proteins, this method is useful in studying the influence of chemical groups on biological activities [4].

Poly-DL-alanyl RNase has been studied extensively [5—7]. It was shown that long poly-DL-alanyl chains can be grown on the enzyme without abolishing its solubility or activity; neither was the capacity of the molecule to regain its correct conformation after complete reduction of all the bonds stabilising its tertiary structure interfered with. These findings were extended to polytyrosyl RNase [5], polyglycyl RNase, poly-L-valyl RNase and to poly-DL-phenyl-alanyl RNase [8—10]. In all these derivatives only minimal changes were introduced into the net electro-static charge of the molecule, since a free α-amino group (that at the end of the attached polypeptide side-chain) always replaces the α- or ε-amino group which had served as the initiator. It was therefore of interest to attempt the synthesis of derivatives in

---

Non-standard abbreviations. Poly-L-lysyl RNase, pLys-RNase; poly-ε-trifluoroacetyl L lysyl RNase, pTFALys-RNase; cytidine 2',3'-cyclic phosphate, C-c-P; uridine 2',3'-cyclic phosphate, U-c-P; polyuridylic acid, poly U; poly-cytidylic acid, poly C; polyguanylic acid poly G; polyadenylic acid, poly A.

Enzymes. RNase, or ribonuclease, or ribonucleate pyri-midine-nucleotido-2'-transferase (cyclizing) (EC 2.7.7.16); trypsin (EC 3.4.4.4.).

which the polypeptide chains attached to the enzyme would carry many charged groups, and study the influence of these charges on chemical, physical and enzymic properties of the enzyme.

In the present publication we describe the synthesis, characterization and enzymic properties of poly-L-lysyl RNase; in the following one the synthesis of poly-L-ornithyl RNase, poly-L-arginyl RNase and poly-L-histidyl RNase are described and some of their properties discussed.

Poly-L-lysyl RNase was prepared by the reaction of RNase with ε,N-trifluoroacetyl-α,N-carboxy-L-lysine anhydride, with subsequent removal of the trifluoroacetyl groups by 1.0 M aqueous piperidine. Derivatives enriched thus with up to 270 lysine residues per RNase molecule were enzymically active. Quite different conditions of ionic strength and pH were required for optimal activity as compared with those of the native enzyme. Polylysyl RNase regained all its initial activity after complete reduction of all the disulfide bonds followed by reoxidation. The changes induced by attachment of the lysyl peptides to RNase are reversible, as the removal of the peptides with trypsin essentially restored the original properties of the enzyme.

MATERIALS

Ribonuclease

Bovine pancreatic RNase (Sigma, type I-A, 5 times recrystallized, lot no. 73 B 0830) was used throughout the whole study. This preparation contained, as shown by chromatography on IRC-50 according to Hirs *et al.* [11], 90% of RNase A, the remainder representing RNase B (5.5%) and aggregates of RNase A. This product, which contained approximately 9% moisture, was used without further purification for the preparation of pLysRNase and for controls in all assays of enzymic activity and will be referred to as "native RNase". Concentrations of enzyme solutions were determined by ultraviolet absorption at 280 mμ using the extinction coefficient $E_{1\ cm}^{0,1\%} = 0.695$ [12].

Poly-L-lysyl RNase

A typical preparation consisted of the following steps: Native RNase (1 g) was dissolved in 0.05 M sodium phosphate buffer, pH 7.0 (100 ml), and cooled to 0°. ε,N-Trifluoroacetyl-α,N-carboxy-L-lysine anhydride [13] (1 g in 25 ml absolute dioxane) was added and the reaction allowed to proceed with slow stirring in an ice bath for 2 hours, then left a 4° overnight. The ratio of anhydride to protein was intentionally varied in different preparations, the details concerning some representative batches being presented in Table 1. The milky white suspension

which formed during the reaction was exhaustively dialysed in the cold against water in heated dialysis casings [14] to remove salts, dioxane, lysine and small oligopeptides of lysine. The suspension was then centrifuged at 9000 rev./min for 1 hour and the supernatant fluid, representing the water-soluble fraction of poly-ε,N-trifluoroacetyl-L-lysyl RNase, and the precipitate (water-insoluble fraction) were lyophilised separately. The soluble and insoluble pTFALysRNase were then suspended separately in 1 M aqueous piperidine at 0° (7 ml of 1 M piperidine per 100 mg of protein), and left, with slow stirring for 30 hours at 4°. The clear solutions obtained were neutralised by slow addition of cold 0.5 N acetic acid, exhaustively dialysed against water and lyophilised. Native RNase was shown not to lose any of its activity by prolonged exposure to 1 N piperidine. Some batches of pLysRNase were further purified from oligopeptides of lysine by passage on a 150 cm $\times$ 0.8 cm column packed with Sephadex G-25 (fine) in 0.2 M acetic acid-0.1 M NaCl. The flow rate of this column was adjusted by a pump to 10 ml/hour, and 1.5 ml fractions were collected. The adequacy of this system was demonstrated by completely separating an artificial mixture of pLysRNase 20$_P$ and polylysine hydrobromide (n = 19). The protein emerges from this column as a sharp peak immediately after the void volume, while polylysine emerges much later as a ninhydrin positive peak which has no absorbancy at 280 mμ (Fig.1). Three preparations of pLysRNase (batches 20$_P$, 7$_8$ and 1$_8$) were thus chromatographed and found to contain free oligolysine to the extent of 1.0, 0.8 and 0.4 mg per 100 mg pLysRNase respectively. The purified pLysRNase preparations did not differ from the unpurified ones in their enzymic activity on RNA and C-c-P. Most of the studies reported were, therefore, performed on samples purified by exhaustive dialysis only.

One batch of pLysRNase was prepared in 0.05 M sodium carbonate buffer, pH 7.0 and several batches were prepared in anhydrous dimethylformamide in the following manner: 100 mg aliquots of native RNase were slowly dissolved in dimethylformamide (60 ml) and cooled to 0°. ε,N-Trifluoroacetyl-α,N-carboxy-L-lysine anhydride (100 or 200 mg dissolved in 2.5—5 ml dimethylformamide) was added and the reaction allowed to proceed at 0° with slow stirring for 6 hours. Reactions were performed in the presence as well as in the absence of phosphate ions; in the former case these were supplied as uridine 2',3'monophosphate (mixed isomers, 5 mg) which was added to the RNase solution before the reaction. The soluble and insoluble fractions of pTFALysRNase were then treated as previously described. The uridine phosphate, which was mostly insoluble in dimethylformamide, was removed during the subsequent dialyses.

Table 1. *Characterisation of some preparations of pTFALysRNase and pLysRNase*

Preparation				pTFALysRNase			pLysRNase			
NCA[a]/RNase	Solvent for NCA	Aqueous buffer pH 7.0	Designation of sample	Fluorine content	Relative activity on RNA, pH 5.0	Relative activity on C-c-P, pH 7.0	Lysine residues attached per RNase molecule	Number of reacted amino groups in RNase	Relative activity on RNA, pH 5.0	Relative activity on C-c-P, pH 7.0
w/w				%	%	%			%	%
0.5:1	Dioxane	Phosphate	7$_S$	8.2	120	200	11	7	25	150
0.5:1	Dioxane	Phosphate	7$_P$[b]	11.2	—	—	17	7	18	120
1:1	Dioxane	Phosphate	1$_S$	9.4	144	190	21	—	24	160
1:1	Dioxane	Phosphate	1$_P$	17.0	—	—	29	—	18	110
1:1	Dioxane	Phosphate	2$_S$	7.0	140	170	26	7	20	120
1:1	Dioxane	Phosphate	2$_P$	15.5	—	—	32	—	12	—
1:1	Dioxane	Phosphate	3$_S$	6.4	166	150	—	—	—	—
1:1	Dioxane	Phosphate	3$_P$	17.3	—	—	50	8	8	63
1:1	Dioxane	Phosphate	20$_S$	16.0	150	190	26	—	18	112
1:1	Dioxane	Phosphate	20$_P$	—	—	—	45	—	6	65
1:1	Dioxane	Phosphate	21$_P$	—	—	—	5	—	8	65
2:1	Dioxane	Phosphate	12$_P$	11.0	49	80	24	—	22	—
2:1	Dioxane	Phosphate	12$_S$	13.7	—	—	113	8	0	0
3:1	Dioxane	Phosphate	8$_P$	—	—	—	110	9.5	0	0
2:1	Dioxane	Carbonate	C$_P$	—	—	—	56	10.5	0	0
2:1	DMF[c]	none	14$_S$	13.0	58	105	—	—	—	—
2:1	DMF[c]	none	14$_P$	15.0	—	—	270	7.5	4	25
2:1	DMF[c]	none	15$_P$	—	—	—	245	8.5	5	35

[a] NCA = ε,N-Trifluoroacetyl-α,N-carboxy-L-lysine anhydride.

[b] The samples designated "P" were insoluble at the pTFALysRNase stage but became water-soluble after the removal of the trifluoroacetyl groups.

[c] DMF = dimethylformamide.

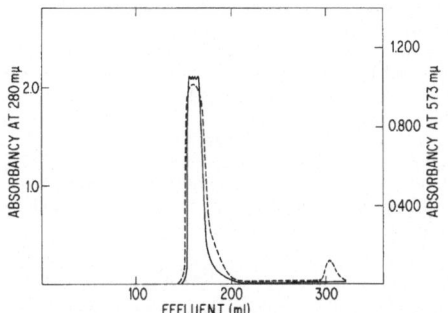

Fig. 1. *Elution pattern of an artificial mixture of pLysRNase 20$_P$ (20 mg) and poly-L-lysine HBr (2 mg, n = 19) from a 150 × 0.8 cm column of Sephadex G-25 (fine), in 0.2 M acetic acid-0.1 M NaCl. Flow rate was 10 ml/hour.* ——, absorbancy of effluent at 280 mμ; – – –, absorbancy in ninhydrin analysis at 573 mμ of one aliquot of each fraction collected

Concentrations of solutions of RNase derivatives were determined by their ultraviolet absorption at 280 mμ, using the same extinction coefficient as for the native enzyme. Thus, though the molecular weight of the derivatives differed from that of the native enzyme, solutions of equal absorbancy were equimolar in respect to the enzyme moiety.

Nucleoside 2′,3′-cyclic Phosphate

Cytidine 2′,3′-cyclic phosphate and uridine 2′,3′-cyclic phosphate were obtained from Sigma. The former was furnished as sodium salt and used as such, while the latter was supplied as barium salt and transformed to the sodium form by the reaction with 0.1 M Na_2SO_4 before use. The purity of both products was checked by chromatography according to Brown *et al.* [15]. Both cyclic phosphates gave only one spot, visualised in ultraviolet light. Uridine 2′,3′-monophosphate (mixed isomers) was equally obtained from Sigma.

RNA

High molecular weight yeast RNA was obtained from Sigma (Type IX). Low molecular weight yeast RNA was purchased from Schwartz Bioresearch, and purified by exhaustive dialysis, first against 0.1 M NaCl and then against water, followed by lyophilisation.

Polyribonucleotides

Poly A, poly C, and poly U were obtained from the Miles Co. as potassium salts and used as such. The potassium salt of poly G [16] was a gift of Dr. Y. Kimchi. Solutions of RNA, nucleoside cyclic phosphates and of polyribonucleotides were always freshly prepared before use. The latter solutions were dialysed for an hour in sterile casings in the presence of 0.001 M EDTA.

Trypsin (2 times crystallised and lyophilised) was obtained from Worthington, Poly-L-lysine HBr (average degree of polymerisation [19]) was from the departmental collection. 5,5'-Dithiobis (2-nitrobenzoic acid) (Ellman's reagent) was purchased from Aldrich Chem. Co. Ethyl thiol trifluoroacetate was prepared according to Hauptschein *et al.* [17]. Urea (AnalaR) was recrystallised twice from ethanol. Only fresh solutions were used. Desicote (Beckman) was used according to the instructions of the manufacturer to siliconise all glass surfaces in contact with enzyme solutions of less than 1 mg/ml concentration.

METHODS

Assays of Enzymic Activity

All assays were performed in triplicate. Activities were calculated by comparison to a sample of native RNase of known concentration.

Assays Using High Molecular Weight RNA as Substrate. The method described by Anfinsen *et al.* [18], measuring the formation of soluble oligonucleotides was used. Substrate concentration in the assay mixture (2.5 ml) was 3.2 mg/ml (except when otherwise stated) and enzyme concentration ranged between 0.3 and 0.6 µg of native enzyme.

Assays Using Low Molecular Weight RNA as Substrate. The assay was based on the method of Kunitz [19], following the decrease of absorbancy at 300 mµ. Changes in optical density were recorded in a Cary 14 or in a Gilford 2000 recording spectrophotometer using scale expansion, against a blank containing RNA only. Substrate concentration was 1 mg/ml except when stated otherwise. Activity was calculated from the slopes of initial velocities.

Assays Using Nucleoside 2',3'-cyclic Phosphates as Substrates. In these assays the increase in optical density accompanying hydrolysis of the substrate was recorded, essentially according to Richards [20], at 290 mµ for C-c-P and at 280 mµ for U-c-P against a blank containing substrate only. Assays were performed in a Cary 14 recording spectrophotometer using the scale expansion. Quartz inserts were used to shorten the optical pathway to 0.5 mm. The slit was kept at openings less than 0.25 mm. All reactants and the cuvette holders were kept at 26° ± 0.05 by a water circulation thermostat. The reaction mixture (1.05 ml) contained substrate (4 mg/ml), 0.1 M tris-HCl buffer at pH 7.0 (unless stated otherwise) and enzyme (50 µl). Rapid mixing of the reactants was achieved by manipulation of the insert. Specific activities were calculated from the slopes of initial velocities, which were linear for several minutes. Under these conditions the assay followed zero order kinetics in respect to substrate for enzyme concentrations up to 40 µg/ml. The determination of the pH of optimal

activity at low ionic strength, taking into consideration the contribution from the substrate, was achieved by measuring the breakdown of the sodium salt of C-c-P titrimetrically in a Radiometer TTT_1 automatic titrator at 25°. The reaction mixture (2.05 ml) contained 2 mM of the sodium salt of C-c-P and enzyme in deionised water. pH was kept constant by addition of small increments of 0.01 N NaOH. Activity was measured from the slopes of initial rates.

Assays Using Synthetic Polyribonucleotides as Substrates. These were based on measurement of the changes occurring in the ultraviolet absorption spectrum of the substrates upon enzymic hydrolysis. In analogy with the assays using nucleoside 2',3'-cyclic phosphates, these changes were recorded as a function of time at a fixed wavelength (268 mµ, 258 mµ, 228 mµ and 257 mµ for poly C, poly U, poly G and poly A respectively) in a Gilford 2000 automatic spectrophotometer. Substrate without enzyme served as blank. Substrate concentration, determined by ultraviolet absorption measurement [21,16] was approximately 80 µmoles × 10^{-3} per ml, in 0.033 M tris HCl buffer, pH 7.0. With poly C, the slopes of initial reaction rates under these conditions, were found to be proportional to enzyme concentration in the range of 10 to 200 pg RNase.

Chemical Methods

Quantitative ninhydrin estimation was performed according to Kay *et al.* [22]. Fluorine was estimated by the method of Eger and Yarden [23]. Assay of free sulfhydryl groups was carried out either with *p*-hydroxymercuribenzoate [24] or with 5',5'-dithiobis-(2-nitrobenzoic acid) [25]. Amino acid analysis was performed in a Beckman Model 120 B automatic amino acid analyser according to Spackman *et al.*[26]. Determination of the number of reacting amino groups of RNase by using the desamination method has been described previously [5].

Physical Methods

Ultraviolet Absorption. These measurements were done on a Zeiss PMQ II spectrophotometer in special quartz cells of 10 mm light path (Pyrocell) into which quartz "inserts" (Pyrocell) of varying dimensions could be inserted in order to reduce the optical pathway and reaction volume. Recording of spectra and of time-conditioned changes in optical density were performed on a Cary 14 double beam spectrophotometer equipped with a scale expanding slide wire or in a Gilford 2000 automatic cuvette changing spectrophotometer. Both instruments record changes in absorbancy linearly as function of time, and both were equipped with thermospacers controlled by a water circulation thermostat.

Spectrophotometric Titrations of the Phenolic Hydroxyl Groups of Tyrosine. Titrations were performed at 295 mμ and at 25° in a Radiometer TTT₁ pH meter linked by a nitrogen filled closed circuit to a 10 mm flow cell inserted in a Zeiss PMQ II spectrophotometer. The pH was increased by addition of small aliquots of 1 N NaOH from an Agla precision syringe. Proteins were dissolved in deionised water so that ionic strength increased during titration with addition of alkali, while protein concentration decreased. Absorbancies were corrected for the latter factor. One sample protein solution was directly brought to pH 13 with 1 N NaOH and its absorbancy was measured after 45 minutes. The value thus obtained represents the last point on each titration curve.

Analytical Ultracentrifugation. Analytical ultracentrifugation was carried out in a Spinco Model E ultracentrifuge equipped with schlieren optics. Runs were performed at 25°, either by the sedimentation velocity method at top speed (59,780 rev./min) or by the short column equilibrium approach method according to Yphantis [27].

Moving Boundary Electrophoresis. This was performed in a Spinco Model H electrophoresis diffusion instrument equipped with schlieren optics at 1° in 2 ml cells (0.125 cm² cross-section). In each experiment two cells, with native and pLysRNase, respectively, were run simultaneously. Protein solutions were dialysed against buffer before the runs. Disc electrophoresis in polyacrylamide gels was carried out according to Davies [28].

Reduction and Reoxidation of RNase and Derivatives. This was performed according to the method of Anfinsen and Haber [29,30].

It was observed in the early stages of this study that the irreversible adsorption of RNase (and, even more so, of pLysRNase) on glass surfaces [31,32] may introduce errors in determinations of RNase, when very dilute solutions are used, such as in the reoxidation experiments. In order to avoid this, all glass surfaces in contact with enzyme solutions of less than 0.1 % concentration were coated with silicone prior to the experiments. This prevented any measurable amount of adsorption.

RESULTS

A series of pTFALysRNase and pLysRNase preparations was synthesized, purified and characterized (Table 1). The trifluoroacetyl group was chosen for the reversible blocking of the ε-amino groups of lysine, as it was shown previously [33,13] to be removable under mild conditions which do not affect the enzymic activity of RNase.

Chemical Characterisation

By the use of two different chromatographic systems it could be shown that pLysRNase contains neither native RNase nor significant contamination

with free lysine peptides. Two artificial mixtures of native RNase and pLysRNase 20ₚ in the proportion of 1:10 and 1:50 could be quantitatively separated on a 30 cm × 0.9 cm column of IRC-50 (Biorex-70, -400 mesh) in 0.2 M sodium phosphate buffer, pH 6.47 (Fig. 1). The first protein to emerge from the column was eluted in two separate but overlapping peaks, and was identified as native RNase by a lysine to arginine ratio of 10/4, found upon amino acid analysis of the eluate. Even trace amounts of the protein could be detected by assaying the effluent of the column for RNase activity by the Anfinsen method. The second protein emerges only when the pH of the eluent is decreased to approximately 3.0 and the molarity increased to 0.8 NaCl. This peak was identified as pLysRNase. When 80 mg aliquots of pLysRNase (batches 20ₚ, 1ₛ and 7ₛ) were applied to this column no protein was detectable by activity assay at the place where the native enzyme was expected to emerge. Since the assay procedure is very sensitive (0.1 μg enzyme can be detected), it was concluded that no native enzyme is present in these preparations. pLysRNase preparations contained less than 1 % free oligolysine as shown by chromatography on Sephadex G-25 (fine), (see Materials). Such minor contaminations do not affect the enzymatic properties of pLysRNase or RNase.

All pLysRNase preparations were affected by ageing, *i. e.* after 2—3 months both their solubility and their specific activity decreased considerably. Neither storage in the lyophilised form nor deep freezing of solutions prevented this. Aged preparations form gels which are insoluble both in acid and alkali. Addition of crystalline trypsin to such gels results in their almost immediate solubilisation with concomitant increase of their enzymic activity. Thus, 10 mg of aged pLysRNase 7ₚ were suspended in 2 ml of 0.1 M tris-HCl buffer, pH 7.5. The absorbancy of the supernatant, measured at 280 mμ after high speed centrifugation, was 0.190 and its activity 3 % of the native enzyme (Anfinsen assay). Approximately 0.5 mg of trypsin in 50 μl buffer was added. After 3 minutes the $A_{280\,m\mu}$ of the supernatant (corrected for the absorbancy of the added trypsin) was 2.05 and its relative activity had increased to 65 %. This activity is significantly higher than that exhibited by the same pLysRNase preparation before ageing had occurred, thus suggesting that trypsin removed part of the lysine peptides from the derivative, essentially restoring the enzyme to its native form. Ageing does not seem to affect, therefore, the protein moiety of the derivative.

By varying the conditions of the polymerisation reaction, different batches of pLysRNase were prepared which varied in respect to the degree of enrichment, the number of reacted amino groups and their enzymic activity on different substrates. Some representative batches and the conditions of their preparation are listed in Table 1. It is seen from the table that the higher the N-carboxy-α-amino acid anhydride/protein ratio, the higher the degree of enrichment. However, with the increasing number of ε-TFAlysine residues being attached to RNase, the molecule tends to become insoluble in the dioxane-aqueous buffer solvent routinely used for the polymerisation reaction, until the molecules eventually drop out of solution. The maximal degree of enrichment obtainable seems, therefore, to be a function of solubility. Polymerisations leading to enzymically active derivatives could be carried out also under anhydrous conditions, in dimethylformamide. Since pTFALysRNase formed is more soluble in this solvent than in the dioxane-aqueous buffer system, higher degrees of enrichment were obtained (Table 1). Previous studies [7] have stressed that RNase, when reacted with N-carboxyamino acid anhydrides in the dioxane-aqueous buffer system, requires phosphate ions in order to stabilise some of its structure and prevent chain growing on lysine 41, the integrity of which is essential, *inter alia*, for enzymic activity [34]. The same was found to be true in the case of the reaction with ε,N-trifluoroacetyl-α,N-carboxy-L-lysine anhydride. Derivatives prepared in the presence of phosphate had generally up to 8 out of 11 amino groups reacted. A derivative prepared under identical conditions but in the absence of phosphate had 10.5 reacted amino groups and was totally inactive. It is of interest to note that when the reaction proceeds in dimethylformamide, no phosphate is required, as even in its absence enzymically active derivatives are obtained in which not more than 8 amino groups of RNase have reacted. pLysRNase preparations from which the trifluoroacetyl blocking groups have been removed by the reaction with piperidine, still contain between 0.1—1.0% fluorine. This fluorine can be removed completely by short dialysis against 0.01 M trichloroacetic acid, followed by dialysis against water, thus indicating that it is not covalently bound to the enzyme.

In order to ascertain that the changes in enzymic activity, to be described later, which occur upon removal of the trifluoroacetyl groups from pTFALysRNase are due to the unsubstituted lysine residues and not to an artefact of the preparation, we have retrifluoroacetylated pLysRNase. The modified enzyme was reacted with ethyl thiol trifluoroacetate according to the method of Goldberger and Anfinsen [33]. The reaction was carried out in the presence of 0.1 M Na_2HPO_4 in order to prevent trifluoroacetylation of those ε-amino groups in RNase on which no side chains had grown (*e.g.*, lysine 41). The final product of the reaction was partially insoluble. The supernatant fluid, obtained after centrifugation for 120 min at 17,000 rev./min possessed 125% relative activity on RNA at pH 5.0, similarly to the original pTFALysRNase preparation.

Physicochemical Properties of pLysRNase

Electrophoretic Behaviour. Bovine pancreatic RNase is electrophoretically inhomogeneous, and contains at least four active fractions of different mobilities [35]. The highly charged polylysyl derivatives of RNase were investigated by zone electrophoresis on paper, cellulose acetate, methylated or siliconised cellulose acetate, polyvinylchloride, or gels such as agar and starch. All these media led to smears rather than to banding. The only medium in which satisfactory separation could be obtained was polyacrylamide gel. Disc electrophoresis in this medium of native RNase and of purified pLysRNase 20$_P$ (enriched with 45 lysine residues per molecule) was performed either in a 0.1 M β-alanine-acetate buffer, pH 4.4 or in 0.1 M glycine-NaOH buffer, pH 9.0. A current of 4 mA per gel was applied for 20 and 60 minutes, respectively. The gels were stained with amidoblack. The inhomogeneity of the native enzyme is evident in the runs performed at pH 9.0 where it separates into 2 bands. pLysRNase behaved, as expected for a polydisperse substance, both in the acidic and in the alkaline medium. Within the stained region there are, however, 4—5 sharp bands (Fig.2). Most of the material migrates much further to the cathode than the unreacted protein. Upon moving boundary electrophoresis in 0.1 M tris-HCl buffer, pH 8.0, the mobility of the major peak of native RNase was $2.5 \times 10^{-5} \mathrm{cm^2 \times V^{-1} \times sec^{-1}}$, while that of pLysRNase 1$_s$ was $6.6 \times 10^{-5} \mathrm{cm^2 \times V^{-1} \times sec^{-1}}$ (calculated from the descending boundaries).

Molecular Weight. The molecular weight of all pLysRNase preparations was calculated from amino acid analyses. In some instances independent data were obtained by analytical ultracentrifugation, using both sedimentation velocity and short column equilibrium approach methods. The values obtained by centrifugation were considerably higher than those computed from chemical analysis. Thus a batch enriched with 21 residues with a predicted molecular weight of 16,372 whereas a molecular weight of 27,000 was calculated from a sedimentation coefficient, $s_{20} = 1.53$ S, a diffusion coefficient $D_{20} = 4.6 \times 10^{-7} \mathrm{cm^2 \times sec^{-1}}$ and a partial specific volume $\bar{v} = 0.70$ (calculated from $\bar{v}$ value of 0.695 for RNase [36] and 0.72 for lysine residue [37]. Another preparation, enriched with only 5 residues, gave values of 21,200 and 21,610 by the Yphantis and sedimentation velocity methods, respectively. Similar disagreements between predicted and found values have been reported in other polypeptidyl proteins [38,39]. They are probably due to a certain degree of aggregation.

Spectroscopic Studies. Ultraviolet absorption spectra of pLysRNase at neutral pH displayed a maximum at 276 mμ, as opposed to 277.5 mμ for native RNase (Fig.3). The spectrum of RNase in this region is determined solely by its six tyrosine residues, of

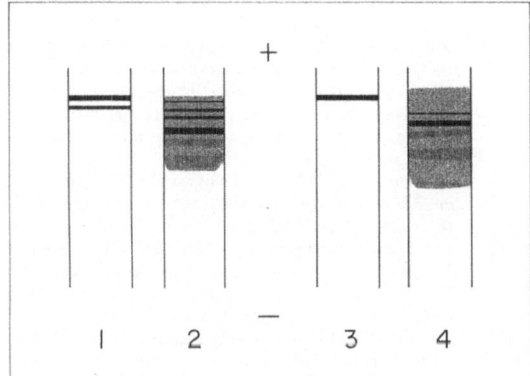

Fig. 2. *Disc electrophoresis in polyacrylamide gels of RNase (1 and 3) and of purified pLysRNase 20ₚ (2 and 4). Left, in 0.1 M glycine-NaOH buffer, pH 9.0, after 60 min; right, in 0.1 M β-alanine-acetate buffer, pH 4.4, after 20 min. Current, 4 mA/gel. Staining with amidoblack*

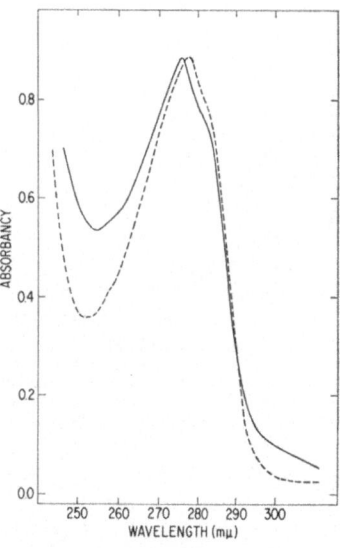

Fig. 3. *Ultraviolet absorption spectra of RNase and pLysRNase 21ₚ. Spectra were recorded on a Cary 14 recording spectrophotometer in 10 mm quartz cells at 24°. Proteins were dissolved in 0.1 M tris-HCl buffer, 7.0. Buffer served as blank. — — —, RNase; ———, pLysRNase 21ₚ*

which three have been shown to possess an abnormally high pK [40]. A shift of the wavelength of maximal absorption at neutral pH suggests a change in the availability of one or more tyrosines for ionisation. This has been shown to be the case in some modified RNases such as pepsin-inactivated RNase, performic acid-oxidised RNase, methylated RNase and RNase in 8 M urea [42,43], which all display the same blue shift of the wavelength of maximal absorption. In order to determine whether indeed polylysylation has affected the availability of some of the phenolic hydroxyl groups in the protein for ionisation, spectrophotometric titrations of three preparations of pLysRNase (batches 1_8, 3_P and 21_P; relative activity on RNA at pH 5: 20, 6 and 3 %, respectively), were performed. Preparation 1_8 became transiently turbid between pH 8.0 and 10.8, but titrated, in the remaining range as native RNase. In the two other preparations, however, approximately 4 phenolic hydroxyl groups titrated normally instead of three (Fig. 4). The spectrophotometric titration of a mixture of native RNase and polylysine HBr in a residue molar ratio of lysine to RNase similar to that in pLysRNase 3_P, did not differ from that of the native enzyme alone.

Enzymic Activity

Addition of polylysine HBr to solutions of RNA or C-c-P leads to the formation of complexes, which are mostly insoluble, so that solutions immediately become turbid. Upon reaction with RNase, the RNA moiety of the complex is degraded to oligonucleotides with concomitant clearing up of the solution [44]. No change in the enzymic activity of either native

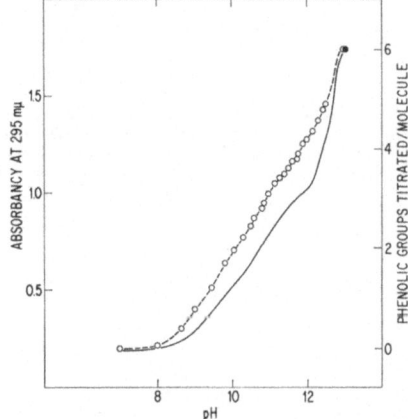

Fig. 4. *Spectrophotometric titration curves of RNase and of pLysRNase 21ₚ at 295 mμ and 24°. The proteins were dissolved in deionised water and ionic strength increased with addition of alkali. Absorbancy readings were corrected for dilution. The last point of each curve was obtained by measuring the absorbancy of an aliquot 45 min after it had been brought to pH 13. This point was assumed to represent 6 ionised phenolic groups. ———, RNase; O——O, pLysRNase 21ₚ*

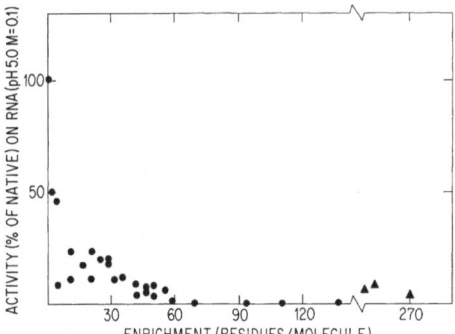

Fig.5. *Correlation between number of lysine peptides attached to pLysRNase preparations and relative activity on RNA at pH 5.0 (Anfinsen assay).* ●, batches prepared in dioxane-aqueous buffer; ▲, batches prepared in dimethylformamide. Some data not represented in Table 1 are included in this figure

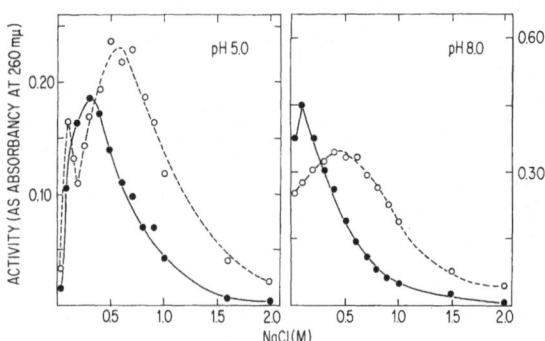

Fig.6. *The ionic strength dependence of the reaction of RNase and of pLysRNase 7₈ on RNA (Anfinsen assay) at pH 5.0 (sodium acetate buffer) and at pH 8.0 (tris HCl buffer).* Ionic strength was adjusted with NaCl. pLysRNase concentration was 5 times that of RNase. ●, RNase; ○, pLysRNase 7₈

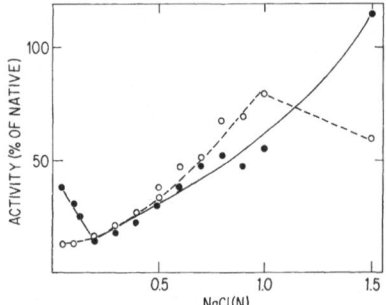

Fig.7. *Relative activity of pLysRNase 7₈ at pH 5.0 and pH 8.0 as function of ionic strength.* Conditions were as described in Fig.6. ●, pH 5.0; ○, pH 8.0

RNase or pLysRNase on RNA (Anfinsen method) was observed when polylysine HBr (0.4—200 μg/ml) was added to the reaction mixture. The influence of polylysine on the activity towards C-c-P was studied by the spectroscopic method in the range where no significant turbidity occurs (0.4—100 μg/ml), and in this range the activities of RNase and pLysRNase were similarly not affected at all.

The Activity of pLysRNase and pTFALysRNase on RNA. Different preparations were compared for their activity on RNA in 0.1 M acetate buffer, pH 5.0, according to Anfinsen *et al.* [18]. The values indicated (Table 1) are those obtained from freshly prepared batches; the same preparations were less active when tested one or two months after synthesis. All the soluble pTFALysRNase preparations are enzymically active, usually more than the native enzyme. pLysRNase, obtained by removal of the TFA group from pTFALysRNase is, on the other hand, markedly less active than the native enzyme, the activity for different preparations ranging, under the conditions of the assay, from 4—25%. There is a clear correlation between the degree of enrichment and the activity on RNA, as evident from Fig.5. Activity decreases with the increase in the number of lysine peptides attached, and approaches zero at an enrichment of 70 residues per molecule. Batches of pLysRNase prepared in dimethylformamide were, however, active even though enriched with up to 270 residues. The number of polymeric side chains attached does not influence the activity, as long as three amino groups of the enzyme remain unreacted. Three of the inactive samples showed, upon desamination, 9, 9.5 and 10.5 reacted amino groups out of a total of 11.

The ionic strength dependence of the activity of pLysRNase 7₈ on RNA was studied at pH 5.0 (in acetate) and at pH 8.0 (in tris-HCl), in buffer solutions the ionic strength of which was adjusted to the desired value by addition of NaCl in the range of 0.05—2.0. Native RNase was assayed simultaneously. pLysRNase preparations were used in concentrations 2—5 times that of the native enzyme in order to obtain comparable results. pTFALysRNase was not included in this study since it is insoluble in salt solutions of M > 0.3. From the curves of specific activity thus obtained (Fig.6) it can be seen that the activity of the native enzyme decreases rapidly when ionic strength exceeds 0.3 (at pH 5.0) and 0.1 (at pH 8.0). On the other hand, the maximum activity of pLysRNase is reached—both at pH 5.0 and 8.0— only at ionic strength of 0.5—0.6. At pH 5.0, the activity curve is bimodal and displays a first peak

at 0.2 M. The activity of pLysRNase decreases at
ionic strength above 0.6 less steeply than that of the
native enzyme, so that the relative activity (Fig.7)
at high salt concentrations increases considerably.
Thus, at $\mu = 0.05$ pLysRNase has about 15 % of the
activity of an equimolar solution of native enzyme
while at $\mu = 1.0$ its relative activity approaches
100 % (at pH 5.0) and 75 % (at pH 8.0).

The dependence of the activity upon
pH was also studied. The activity of
pLysRNase 7_P and of pTFALysRNase
on RNA was assayed at pH levels ranging
from 4.5 to 9.0 in a tris-malonic acid-
NaOH buffer system, the ionic strength of
which was calculated from the pK values
of malonic acid and adjusted to 0.1, 0.2 and
0.5, respectively, with NaCl. Substrate
concentration was increased to 6.4 mg/ml
to insure zero order reactions even at
maximal activities.

It can be seen in Fig.8 that at $\mu = 0.1$
the pH of optimal activity of pLysRNase
is shifted to 8.0 as opposed to 7—7.5 for
the native enzyme. pLysRNase has in the
pH range investigated 10—20 % of the
activity of RNase (Fig.9). At $\mu = 0.5$
pLysRNase exhibits maximum activity at
pH 7 (Fig.8), whereas relative activity
increases above pH 9.0. pTFALysRNase has, at
$\mu = 0.1$ maximal activity at pH 6.5; at this pH
value its activity is equal to that of native RNase.

*The Activity of pLysRNase and of pTFALys-
RNase on Nucleoside 2′,3′-cyclic Phosphates.* Different
preparations were compared for their activity on
C-c-P in 0.1 M tris-HCl buffer, pH 7.0. All the
pTFALysRNases assayed possessed between 1.5 and
2 times the hydrolytic activity of native RNase, as
opposed to an average of only 81 % relative activity
on RNA under identical conditions. pLysRNase preparations exhibited considerable variability in their relative activities towards C-c-P, most of them ranging
from 90—150 % of the activity of the native enzyme.
Some preparations which possessed almost no activity on RNA at pH 5.0 exhibited still 25—35 % of the
activity of the native enzyme when assayed on C-c-P.

The dependence of the activity of pLysRNase
toward C-c-P upon ionic strength was investigated
at pH 5.0 and 8.0 in buffer solutions the ionic strength
of which was adjusted with NaCl within the range
of 0.1 to 1.0 M. Changes of activity were much less
marked than when RNA served as substrate. At
pH 8.0 both RNase and pLysRNase displayed a
shallow maximum at $\mu = 0.2$. At pH 5.0 pLysRNase
activity is practically independent of ionic strength,
whereas the native enzyme has maximal activity at
$\mu = 0.2$. The activities at very low strength could not
be investigated by this method owing to the contribution of the substrate to ionic strength.

The pH dependence of this reaction was studied
by two different methods. In the spectrophotometric
method the same tris-malonic acid-NaOH buffer
system was used as described for studies on RNA.
The ionic strength (not including the contribution of
the sodium salt of C-c-P) was adjusted by addition
of NaCl to $\mu = 0.1$. Under these conditions (Fig.10)
the pH profiles of pLysRNase 20_S and pTFALys-

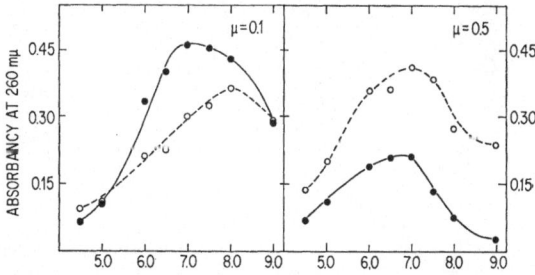

Fig. 8. *The pH dependence of the activity of RNase and pLysRNase 7_P on
RNA (Anfinsen method).* Assays were performed in a tris-malonic acid-
NaOH buffer system, the ionic strength of which was calculated from the
pK values of malonic acid and adjusted to $\mu = 0.1$ or $\mu = 0.5$ with
NaCl. pLysRNase concentration was 5 times that of RNase. ●, RNase;
○, pLysRNase 7_P

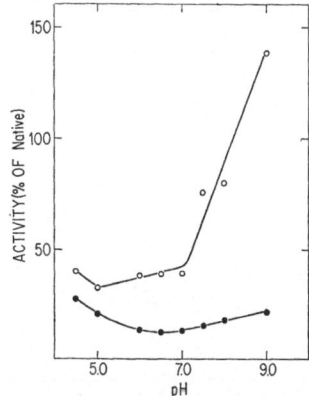

Fig. 9. *The relative activity of pLysRNase 7_P on RNA at ionic
strength $\mu = 0.1$ and $\mu = 0.5$.* ●, $\mu = 0.1$; ○, $\mu = 0.5$. Conditions were as described in Fig.8.

RNase 3_S are both not significantly displaced in
comparison with native RNase. In order to study the
pH dependence of the reaction at low ionic strength
it was necessary to reduce substrate concentrations.
The assays were performed in unbuffered aqueous
solutions in a pH-stat, after the reaction mixtures
had been brought to the desired pH value, by following the release of the acidic groups during the hydro-

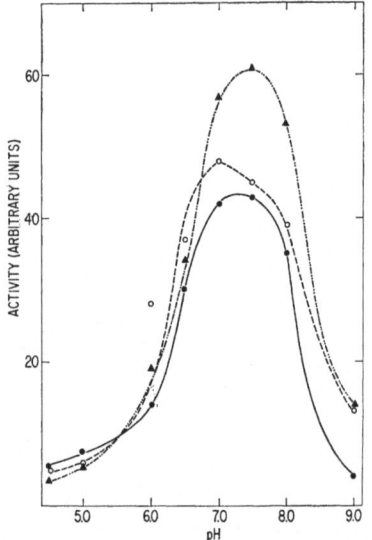

Fig. 10. *The pH dependence of the activity of RNase* (●——●), *pLysRNase 20*$_8$ (O– – –O) *and pTFALysRNase 3*$_8$ (▲–·–▲) *on C-c-P. Assay conditions are described in the text*

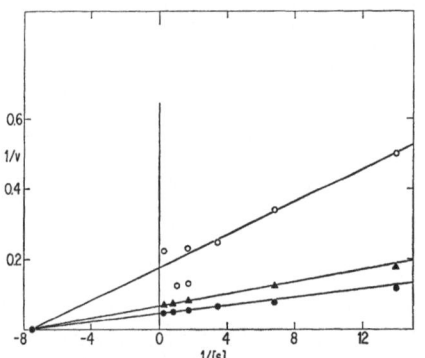

Fig. 11. *Lineweaver-Burk plots for the activity of RNase, pLysRNase 7*$_8$ *and pTFALysRNase 3*$_8$ *on RNA at pH 5.0.* Activities were measured at different substrate concentrations by the Kunitz method at 305 mµ, and at 26°. Velocities were measured from the slopes of initial reaction rates. Substrate concentration is in mg/ml. ●, RNase; O, pTFALysRNase; ▲, RNA

lysis of C-c-P. From comparison of initial velocities it was found that at a substrate concentration of 0.2 mg/ml (corresponding to 5×10^{-4} M) the maximal activity of RNase occurred at pH 7.0 whereas that of pLysRNase 20$_8$ was at pH 8.5. An increase in substrate concentration resulted in a decrease of this

shift. At a substrate concentration of 1 mg/ml there was no significant difference between pLysRNase and the native enzyme.

In view of the marked differences in the specific activities of these derivatives it was of interest to analyse the kinetic parameters of the reaction. Both the reactions of the enzyme with C-c-P and with RNA were studied, although we were aware of the fact that RNA, being heterogeneous in respect to molecular weight and being the site of three different, simultaneous reactions, is hardly a suitable substrate for kinetic studies. The reaction on RNA was studied by the Kunitz method using high molecular weight RNA as substrate. Reactions were followed at 305 mµ (instead of the customary 300 mµ) since at this wavelength the changes in absorbancy are less influenced by the hydrolytic reaction on C-c-P. The results of assays conducted at different substrate concentrations were treated by the method of Lineweaver and Burk [45] and the apparent K_m and V_{max} values were determined graphically. It is apparent from the plots shown in Fig. 11 that K_m values must be similar for RNase, pLysRNase and pTFALysRNase, and that the enzymes differ in maximal velocities. Similar results were obtained in the studies using C-c-P as substrate. The plots obtained from the experiments were treated as product inhibition plots [45] and a $K_m \times 10^{-3}$ molar value of 1.0 ± 0.2 (in 0.1 M tris-HCl buffer, pH 7.0 at 26°) calculated for the three enzymes.

Two preparations of pLysRNase (1$_8$ and 20$_8$) were assayed also on U-c-P. Wellner *et al.* [6] have reported that alanylation of RNase affected its activity differently on C-c-P and U-c-P. Table 2 illustrates the results obtained upon comparing the activity of pLysRNase, poly-DL-alanyl RNase and RNase on these two substrates at three different pH values. As seen in Table 2, U-c-P at pH 5.0 is hydrolysed faster by pLysRNase than by RNase, whereas C-c-P is hydrolysed faster by RNase than by pLysRNase. On the other hand, no significant differences between the native and the modified enzymes were observed at pH 7.0 and 8.0.

The Activity of pLysRNase and pTFALysRNase on Polycytidylic Acid. The activity of pLysRNase

Table 2. *Relative rates of hydrolysis of nucleoside 2′,3′-cyclic phosphates by pLysRNase, poly-DL-alanyl RNase and RNase* Assay conditions are described in the text. The values indicated are the rates of hydrolysis by the enzyme derivatives relative to those of RNase as 100%

Preparation	pH 5.0		pH 7.0		pH 8.0	
	C-c-P	U-c-P	C-c-P	U-c-P	C-c-P	U-c-P
pLysRNase 1$_8$	53	125	60	65	107	100
pLysRNase 20$_8$	88	150	112	95	120	108
p-DL-AlaRNase[a]	84	42	177	47	210	25

[a] From Wellner *et al.* [6].

20_s and pTFALysRNase 3_s on poly C was compared to that of the native enzyme in Tris-HCl buffer at pH 7.0; assays were performed at 0.03 M and at 0.5 M, the ionic strength of the buffer being adjusted in the latter case with NaCl. The progress of the reaction was followed in a Gilford 2000 automatic spectrophotometer, by observing changes in absorbancy of a reaction mixture vs. substrate without enzyme at 268 mμ. Both pTFALysRNase and pLysRNase were found to have about 20 % of the activity of the native enzyme at 0.03 M. At 0.5 M, (where pTFA-LysRNase is insoluble) the activities of both the native enzyme and of pLysRNase were markedly lower than at $\mu = 0.03$, but pLysRNase was found to be at this ionic strength three times more active than the native enzyme.

The Substrate Specificity of pLysRNase. Bovine pancreatic RNase cleaves bonds in RNA which follow pyrimidine bases by at least four orders of magnitude faster than bonds following purine bases [35]. In view of the fact that the attachment of polylysine or poly-DL-alanine chains to RNase affects the affinity of this enzyme to the cyclic phosphates of uridine and cytidine in different ways, it was of interest to investigate whether the base specificity of pLysRNase was not affected or modified. Both polycytidylic acid and polyuridylic acid are readily digested by RNase, though at different rates [46,47]. Polyadenylic acid is not hydrolysed unless excessive amounts of enzyme are used [46,48]. These synthetic polyribonucleotides and polyguanylic acid were used to compare the base specificity of RNase and of pLysRNase.

Solutions of the polyribonucleotides (approximately 80 μM, in 0.033 M tris-HCl buffer, pH 7.0) were incubated at 26° with RNase and with pLys-RNase. The enzyme concentrations used ranged from 10 pg/ml to 5×10^7 pg/ml. While poly C and poly U were hydrolysed in readily measurable rates by 10 and 100 pg/ml of either RNase or pLysRNase, poly A and poly G required, under similar conditions, enzyme concentrations in excess of 10^5 pg/ml. The activity of pLysRNase on poly C and poly U was 20 %, while on poly A and poly G it was approximately 50 % that of the native enzyme.

Reduction and Reoxidation of pLysRNase and of pTFALysRNase

All the preparations of pLysRNase and of pTFA-LysRNase listed in Table 1 were reduced with β-mercaptoethanol (0.32 M) in 8 molar urea solutions according to the method of Anfinsen and Haber [29]. After four hours or more the mixture was passed on a column 80×0.8 cm of Sephadex G-25 (coarse) in 0.1 M acetic acid to separate the protein from the urea, mercaptoethanol and the salts. The completeness of reduction was demonstrated by the titration of 8 free sulfhydryl groups either with PMB or, more often, with Ellman's reagent [25]. The fully reduced samples were devoid of significant enzymic activity on RNA at pH 5.0. Reoxidation was allowed to proceed by air oxygen at room temperature at protein concentrations ranging from 5 to 50 μg/ml in 0.1 M tris HCl buffer, pH 8.5 (final pH of the reoxidation mixture: 8.1—8.5). A sample of native RNase was always reduced and reoxidised simultaneously as control. Activity assays, both on RNA (by the Kunitz and the Anfinsen methods), as well as on C-c-P, were carried out on the reoxidation mixtures at 10, 20, 30, 45, 60 and 120 minutes after the onset of the experiment. All preparations regained their full initial activity on both substrates. The 50 μg/ml solutions recovered 100 % activity after 120 minutes, while at the higher dilutions full activity was recovered faster, the 5 μg/ml solutions reaching their initial activity after 45—60 minutes.

Action of Trypsin on pLysRNase

Three different preparations of pLysRNase, enriched with 17, 21 and 32 residues of lysine per molecule, respectively, and one sample of native RNase were incubated with trypsin (1 mg/8 mg RNase) in 0.1 M sodium phosphate buffer, pH 7.6, at 24° for 48 hours. Enzymic activity on RNA was assayed by both the Anfinsen and the Kunitz methods in the enzyme solutions before and after the incubation with trypsin. The trypsin preparation used in this study possessed no ribonuclease activity detectable by the above assays. As can be seen from the results (Table 3), native RNase was not affected by trypsin,

Table 3. *Effect of trypsin on pLysRNase*
Incubation was carried out at 24° for 48 hours, in 0.1 sodium phosphate buffer, pH 7.6, RNase/trypsin ratio was 8:1. Activity of RNase and derivatives was determined by the Anfinsen method

Preparation	Before incubation with trypsin			After incubation with trypsin		
	Lysine residues per RNase	Relative activity on RNA, pH 5.0	Optimal ionic strength for activity, on RNA, pH 5.0	Lysine residues per RNase	Relative activity on RNA, pH 5.0	Optimal ionic strength for activity on RNA, pH 5.0
		%			%	
RNase	10	100	0.3	10	100	0.3
pLysRNase 7_P	27	18	0.7	—	70	0.4
pLysRNase 1_s	31	22	—	—	60	—
pLysRNase 2_P	42	12	—	14	45	—

19*

while the activity of pLysRNase increased threefold. The preparation originally enriched with 17 residues had, before incubation, maximal activity on RNA (at pH 5.0) at an ionic strength of 0.7 M (opposed to 0.3 for the native enzyme); after tryptic digestion the maximum had shifted to 0.4. The sample originally enriched with 32 residues was purified after digestion by prolonged dialysis and subsequent passage on a Sephadex G-25 column, then lyophilised, hydrolysed and submitted to amino acid analysis. Of the 42 lysine residues originally present on the derivative, only 14 were detected. The electrophoretic mobility of this trypsin-treated derivative on cellulose acetate at pH 8.0 was only slightly higher than that of the native enzyme.

DISCUSSION

This paper describes the synthesis and characterisation of a series of RNase derivatives in which peptides of ε,N-trifluoroacetyl-L-lysine or of L-lysine are attached to the enzyme. The main findings are that the enrichment of RNase with as many as 270 lysine residues has led, notwithstanding the great change in the electrical charge of the molecule, to enzymically active derivatives and that all the polysyl preparations tested recovered their initial enzymic activity after reduction of all their disulfide bridges, followed by reoxidation.

The preparation of polypeptidyl derivatives carrying charged side chains requires blocking of the functional groups during the polymerisation, with subsequent removal of the blocking groups under mild conditions which do not affect the protein moiety of the derivative. In the present study the trifluoroacetyl group was used for the blocking of ε-amino groups of lysine [13,33] and subsequently removed with 1 M aqueous piperidine. pTFALys-RNase thus prepared is in itself an active enzyme, the properties of which have been studied as far as its solubility allowed. This derivative was found to behave similarly to p-DL-AlaRNase [6] with respect to its activity on RNA and C-c-P (Table 1), as well as its pH dependence.

The pLysRNase preparations obtained from pTFALysRNase have, in agreement with theoretical prediction from statistical analysis [49], a sharp molecular weight distribution. The electrophoretic pattern on polyacrylamide gel (Fig. 2) suggests, on the other hand, that pLysRNase is composed of a number of distinct molecular species. Ultracentrifugal analysis indicated that aggregation occurs, possibly progressing with time until solubility is completely lost, as is the case in aged preparations. Aggregation has been described in other polypeptidyl RNases [38,39] and for native RNase [50]. None of the molecular species in pLysRNase consists, however, of unreacted RNase, as was demonstrated by chromatography on IRC-50. This is in agreement with findings on other polypeptidyl RNases [5,7].

Removal of the trifluoroacetyl groups from pTFALysRNase causes a marked decrease in the activity of the modified enzyme, especially on RNA. Upon re-trifluoroacetylation, the initial activity of the trifluoroacetylated derivative is completely recovered, thus ruling out irreversible damage to the molecule and relating the changed activity to the presence of protonated lysine residues on the enzyme.

Polymerisation reactions on the ε-amino group of lysine 41, the integrity of which is essential for RNase function [34,7], have been prevented in the preparation of all hitherto described polypeptidyl RNases by the presence of phosphate ions during the polymerisation reaction. In the present study it was shown that polymerisation leading to an enzymically active derivative can proceed in a completely anhydrous system such as dimethylformamide, in the absence of phosphate. The fact that even extensively lysylated preparations thus prepared (Table 1) had three amino groups left unreacted, shows that lysine 41 is unavailable for the reaction with ε,N-trifluoroacetyl-α,N-carboxy-L-lysine anhydride in this medium. The advantage of the preparation of polypeptidyl RNases in anhydrous media is apparent from the successful preparation of such highly enriched soluble pLys-RNase derivatives.

Chemically modified RNases in which more than three out of the six tyrosine residues ionise normally at pK 10.6 [40,41] have been described [42,43,51], and have recently served for the identification of the three abnormal tyrosine residues in the sequence of RNase [53,54]. All these modifications are inactive on RNA. In two out of three preparations of pLys-RNase which were titrated spectrophotometrically, approximately four phenolic hydroxyl groups ionise normally (Fig. 4), and yet these preparations retain some activity towards RNA and $60-65\%$ activity towards C-c-P. Riehm and Scheraga [52] have recently reported another derivative of RNase in which four tyrosines titrate normally and which still retains some activity on RNA. The assumption that an increase in the availability of the phenolic groups for ionisation, which manifests itself also in a shift of the maximum in ultraviolet absorption spectra, is associated with loss of activity [42,43,51] has, therefore, to be reconsidered. Moreover, the fact that both pLysRNase and RNase-glycine (a derivative of RNase in which all carboxyl groups are modified by the attachment of glycine residues [55]) have, respectively, little or no activity on RNA, but retain 60 to 90% of the activity on C-c-P, raises the question of the possible role of tyrosine residues in the binding of RNA to the enzyme.

The enzymic activity of pLysRNase is quite different from that of the native enzyme, at least as far as reaction rates and conditions of optimal activity are concerned. The difference between the native and the modified enzyme is much more marked when RNA is used as a substrate than with the low molecular weight substrates. When assayed on RNA at pH 5.0, pLysRNase is much less active than the native enzyme, the loss of activity being proportional to the extent of enrichment (Fig. 5). This is, however, partly due to the marked difference in the dependence of the activity of the native and polylysylated enzyme on ionic strength. Thus, at ionic strength 0.6 to 0.7, pLysRNase is as active as the native enzyme. The increase in activity of pLysRNase at pH 5.0 upon increase in ionic strength from 0.05 to 0.6 is bimodal (Fig. 6), with a first peak at 0.2. Since no similar pattern is observed with the low molecular weight substrates, this is probably related to changes in the conformation of RNA. The ionic strength at which maximal activity occurs suggests a shielding effect of salt on the charged amino groups. Thus, at high ionic strength, pLysRNase would be expected to behave similarly to polyalanyl RNase or pTFA-LysRNase, a fact that is demonstrated by the pH dependence studies (Fig. 8). On the other hand, the activity of pLysRNase on the low molecular weight C-c-P is little affected by changes in ionic strength, similarly to the native enzyme [56].

The analysis of the pH dependence of the activity of pLysRNase supports the hypothesis that the optimum pH for activity is shifted, in modified RNases, in parallel to shifts in the isoelectric point [6]. In pLysRNase, where the isoelectric point is markedly increased, the optimum pH of activity on C-c-P is at 8.5; whereas in polyalanyl RNase, in which the isoelectric point is decreased, an optimal pH value lower than for the native enzyme was found [6]. As expected, these shifts are easily masked by salt effects (like 0.1 M buffer); they were, however, clearly demonstrated by the titrimetric assays of pLysRNase on C-c-P at low ionic strength.

Elevated pH and ionic strength form the optimal conditions for pLysRNase activity on RNA. The activity of the modified enzyme under these conditions exceeds that of native RNase (Fig. 9). Such increases of activity over that of the native enzyme are not unique to pLysRNase, and have been reported for other derivatives [6, 10, 57].

Wellner et al. [6] have reported that the rate of hydrolysis of the two low molecular weight substrates, U-c-P and C-c-P, at different pH values is affected by alanylation in a non-parallel fashion, and similar, though less marked differences, have been observed with pLysRNase (Table 3). Nevertheless, the base specificity of pLysRNase is similar to that of the native enzyme, i.e., its preference for bonds following pyrimidine bases has been preserved as evidenced from the similar low rates of hydrolysis of poly A and poly G by the native enzyme, pTFA-LysRNase and pLysRNase.

The kinetics of the reaction of pLysRNase, both on RNA and on the low molecular weight substrate, suggests that the peptidylation affects the substrate binding capacity of the enzyme, rather than its catalytical properties. These changes are more pronounced when RNA is the substrate. Apparently, this is due to the interaction of the polybase (enzyme) with the polyacid (RNA), with formation of very stable complexes located at the periphery of the enzyme and not available to the catalytic action of the active centre. This irreversible enzyme-substrate complex formation is prevented by the shielding effect of high salt concentrations, thus increasing the catalytic activity.

RNase refolds after the complete reduction of its disulfide bridges with concurrent loss of its secondary and tertiary structure to its original, active conformation [29, 30]. This refolding capacity has been conserved in various chemical modifications of the enzyme. In the derivatives studied [5, 9] there was either no change in the net charge of the molecule or only a limited one. In contradistinction, pLysRNases are derivatives in which a great number of positive charge carriers are attached to the molecule; yet, the capacity of the molecule to recover its active structure was preserved.

Reoxidation is optimal at low protein concentration, as seems to be the general rule for proteins composed of one polypeptide chain, whereas proteins consisting of several chains [58, 59] require high concentrations.

Similarly to re-trifluoroacetylation of pLysRNase, which restored to the modified enzyme the properties of pTFALysRNase, treatment of pLysRNase with trypsin, which removes most of the attached lysine peptides without affecting the protein moiety (Table 3), essentially restores the properties of the native enzyme.

This investigation was supported in part by Agreement 415100 with the National Institutes of Health, United States Public Health Service.

REFERENCES

1. Herriot, R. M., Advan. Protein Chem. 3 (1947) 170.
2. Sri Ram, J., Bier, M., and Maurer, P. H., Advan. in Enzymol. 24 (1962) 105.
3. Becker, R. R., and Stahmann, M. A., J. Biol. Chem. 204 (1953) 745.
4. Katchalski, E., Sela, M., Silman, H. I., and Berger, A., in The Proteins (edited by H. Neurath), Acad. Press, New York 1964, Vol. II, p. 405.
5. Anfinsen, C. B., Sela, M., and Cooke, J. P., J. Biol. Chem. 237 (1962) 1825.
6. Wellner, D., Silman, H. I., and Sela, M., J. Biol. Chem. 238 (1963) 1324.
7. Cooke, J. P., Sela, M., and Anfinsen, C. B., J. Biol. Chem. 238 (1963) 2034.
8. Becker, R. R., in Polyaminoacids, Polypeptides and Proteins (edited by M. A. Stahmann), University of Wisconsin Press, Madison 1962, paper 29.

9. Epstein, C. J., and Goldberger, R. F., *J. Biol. Chem.* 239 (1964) 1087.
10. Becker, R. R., and Sawada, E., *Federation Proc.* 22 (1963) 419.
11. Hirs, C. H. W., Moore, S., and Stein, W. H., *J. Biol. Chem.* 200 (1953) 493.
12. Bernfield, M. R., *J. Biol. Chem.* 240 (1965) 4753.
13. Sela, M., Arnon, R., and Jacobson, I., *Biopolymers*, 1 (1963) 517.
14. Kupke, D., *Compt. Rend. Trav. Lab. Carlsberg, Ser. Chim.* 32 (1961) 107.
15. Brown, D. M., Magrath, D. I., and Todd, A. R., *J. Chem. Soc.* (1952) 2708.
16. Thang, M. N., Graffe, M., and Grunberg-Manago, M., *Biochim. Biophys. Acta*, 108 (1965) 125.
17. Hauptschein, M., Stokes, C. S., and Nodiff, E. A., *J. Am. Chem. Soc.* 74 (1952) 4005.
18. Anfinsen, C. B., Redfield, R. R., Choate, W. L., Page, J., and Caroll, W. R., *J. Biol. Chem.* 207 (1954) 201.
19. Kunitz, M., *J. Biol. Chem.* 164 (1946) 563.
20. Richards, F. M., *Compt. Rend. Trav. Lab. Carlsberg, Ser. Chim.* 29 (1955) 315.
21. Warner, R. C., *J. Biol. Chem.* 229 (1957) 717.
22. Kay, R. E., Harris, D. C., and Entenman, C., *Arch. Biochem. Biophys.* 63 (1956) 14.
23. Eger, C., and Yarden, A., *Anal. Chem.* 28 (1956) 512.
24. Boyer, P. D., *J. Am. Chem. Soc.* 76 (1954) 4331.
25. Ellman, G. L., *Arch. Biochem. Biophys.* 82 (1959) 70.
26. Spackman, D. H., Stein, W. H., and Moore, S., *Anal. Chem.* 30 (1958) 1190.
27. Yphantis, D. A., *Ann. N. Y. Acad. Sci.* 88 (1960) 586.
28. Davies, B. J., *Ann. N. Y. Acad. Sci.* 121 (1964) 404.
29. Anfinsen, C. B., and Haber, E., *J. Biol. Chem.* 236 (1960) 1361.
30. Epstein, C. J., Goldberger, R. F., Young, D. M., and Anfinsen, C. B., *Arch. Biochem. Biophys.* Supp. 1 (1962) 223.
31. Shapira, R., *Biochem. Biophys. Res. Commun.* 1 (1959) 236.
32. Hummel, J. P., and Anderson, B. S., *Arch. Biochem. Biophys.* 112 (1965) 443.
33. Goldberger, R. F., and Anfinsen, C. B., *Biochemistry*, 1 (1962) 401.
34. Hirs, C. H. W., Halmann, M., and Kycia, J. H., in *Biological Structure and Function* (edited by T. W. Goodwin and O. Lindberg), Academic Press, New York 1961, Vol. I.
35. Scheraga, H. A., and Rupley, J. A., *Advan. Enzymol.* 24 (1962) 161.

36. Harrington, W. F., and Schellman, J. A., *Compt. Rend. Trav. Lab. Carlsberg, Ser. Chim.* 30 (1956) 21.
37. Applequist, J. B., Ph. D. Thesis, Harvard University (1959).
38. Krausz, L. M., Scaduto, F. C., Sawada, F., and Becker, R. R., *Federation Proc.* 21 (1962) 405.
39. Kettman, M. S., Nishikawa, A. H., Morita, R. Y., and Becker, R. R., *Biochem. Biophys. Res. Commun.* 22 (1966) 262.
40. Shugar, D., *Biochem. J.* 52 (1952) 142.
41. Tanford, C., Hauenstein, J. D., and Rands, D. G., *J. Am. Chem. Soc.* 77 (1955) 6409.
42. Sela, M., and Anfinsen, C. B., *Biochim. Biophys. Acta*, 24 (1957) 229.
43. Sela, M., Anfinsen, C. B., and Harrington, W. F., *Biochim. Biophys. Acta*, 26 (1957) 502.
44. Sober, H. A., Schlossman, S. F., Yaron, A., Latt, S. A., and Rushizky, G. W., *Biochemistry*, 5 (1966) 3608.
45. Lineweaver, H., and Burk, D., *J. Am. Chem. Soc.* 56 (1934) 658.
46. Michelson, A. M., *J. Chem. Soc.* (1959) 1371.
47. Zimmerman, S. B., and Sandeen, G., *Anal. Biochem.* 10 (1965) 444.
48. Michelson, A. M., *Nature* 181 (1958) 303.
49. Katchalski, E., Gehatia, M., and Sela, M., *J. Am. Chem. Soc.* 77 (1955) 6175.
50. Crestfield, A. M., Stein, W. H., and Moore, S., *J. Biol. Chem.* 238 (1963) 618.
51. Bigelow, C. C., *Compt. Rend. Trav. Lab. Carlsberg, Ser. Chim.* 31 (1960) 305.
52. Riehm, J. P., and Scheraga, H. A., *Biochemistry*, 5 (1966) 99.
53. Li, L. K., Riehm, J. P., and Scheraga, H. A., *Biochemistry*, 5 (1966) 2043.
54. Woody, R. W., Friedman, M. E., and Scheraga, H. A., *Biochemistry*, 5 (1966) 2034.
55. Wilchek, M., Frensdorff, A., and Sela, M., *Biochemistry*, 6 (1967) 247.
56. Irie, M., *J. Biochem. (Tokyo)*, 57 (1965) 355.
57. White, F. H., Jr., *Federation Proc.* 20 (1961) 221.
58. Du, Y. C., Jiang, R. Q., and Tsou, C. L., *Sci. Sinica (Peking)*, 14 (1965) 229.
59. Freedman, M. H., and Sela, M., *J. Biol. Chem.* 241 (1966) 2383.

A. Frensdorff, and M. Sela
Section of Chemical Immunology
The Weizmann Institute of Science
Rehovoth, Israel

European J. Biochem. 1 (1967) 281—288

A Comparative Study of Basic, Acidic and Neutral Polypeptidyl RNases, Including Polyarginyl and Polyhistidyl Derivatives

A. Frensdorff, M. Wilchek, and M. Sela

The Weizmann Institute of Science, Rehovoth

(Received February 8, 1967)

The syntheses of poly-L-arginyl and of poly-L-histidyl RNase are described. Poly-L-arginyl RNase was obtained by guanidisation of poly-L-ornithyl RNase, which, in turn, was prepared from native RNase by reacting it with δ,N-trifluoroacetyl-α,N-carboxy-L-ornithine anhydride, with subsequent removal of the trifluoroacetyl groups with piperidine. Derivatives enriched with up to 71 arginine residues were prepared and studied. Poly-L-histidyl RNase was prepared by the reaction of RNase with the unprotected N-carboxy-L-histidine anhydride hydrochloride. Derivatives enriched with up to 22 histidine residues per RNase molecule were obtained.

The enzymic activity of both poly-L-ornithyl RNase and poly-L-arginyl RNase towards cytidine-2′,3′-cyclic phosphate is only moderately affected by the attachment of the basic peptides. The activity on RNA is, on the other hand, strongly decreased in poly-L-ornithyl RNase and completely abolished in polyarginyl RNase, thus illustrating that the activity of the modified enzymes on these two substrates is affected independently. The pH of optimal activity of the basic polyarginyl RNase on cytidine-2′,3′-cyclic phosphate at low ionic strength was found to be increased in comparison to that of the native enzyme, while that of the acidic polyglutamyl RNase, the synthesis of which was previously described, was lowered. Polyhistidyl RNase was less active on RNA than the native enzyme, while its activity toward cytidine 2′,3′-cyclic phosphate exceeded that of unmodified RNase. All the derivatives described recovered their full initial enzymic activity after complete reduction of the disulfide bridges followed by reoxidation.

Bovine pancreatic RNase may be modified by the growing of peptide chains on some of the amino groups in its molecule [1—5]. The resulting polypeptidyl RNases are, with the exception of some preparations of polytyrosyl RNase [3], water-soluble and enzymically active, and thus may serve as useful tools in the study of the relationships between chemical structure and biological function. In the derivatives described earlier [6—8], difunctional amino acids were used for the modification of the enzyme, thus causing only limited changes in the electrostatic charges on the molecule. These studies were recently extended to derivatives carrying numerous positively [5] and negatively charged groups [4] and showed that even extensive modification of the net electrical charge on the molecule does not abolish either enzyme activity or the capacity to recover activity after complete reduction and reoxidation.

Studies of the enzymic activity of polyalanyl [7] and polylysyl [5] RNase suggested that the optimum pH is shifted in parallel to shifts in the isoelectric point. We have now extended the investigation of this hypothesis to polyglutamyl [4] and polyarginyl derivatives of RNase.

The critical role of histidine in the function of RNase, first suggested by Weil and Seibles [9], was unequivocally established by carboxymethylation studies [10—13]. These led to the conclusion that two histidine residues (His 12 and 119), one as an acid and one as a base, form part of the active site of the enzyme [14]. It was, therefore, of interest to attempt the synthesis of an RNase derivative modified by the attachment of additional histidine residues and study their effect on the properties of the enzyme.

In the present paper we describe the synthesis of poly-L-ornithyl RNase, poly-L-arginyl RNase and poly-L-histidyl RNase. Their chemical characterisation and enzymic properties, as well as those of poly-L-glutamyl RNase, the synthesis of which has been described previously [4], are also reported.

Poly-L-ornithyl RNase was obtained by reacting RNase with δ,N-trifluoroacetyl-α,N-carboxy-L-ornithine anhydride, with subsequent removal of the

---

Enzyme. RNase, or ribonuclease, or ribonucleate pyrimidine-nucleotido-2′-transferase (cyclizing) (EC 2.7.7.16).

Non-standard abbreviations. Trifluoroacetyl, TFA; poly-δ,N-trifluoroacetyl-L-ornithyl RNase, pTFAOrnRNase; poly-L-ornithyl RNase, pOrnRNase; poly-L-arginyl RNase, pArgRNase; poly-L-histidyl RNase, pHisRNase; poly-γ-phthalimidomethyl-L-glutamyl RNase, pPIMGluRNase; poly-L-glutamyl RNase, pGluRNase; cytidine 2′,3′-cyclic phosphate, C-c-P; uridine 2′,3′-cyclic phosphate, U-c-P.

trifluoroacetyl groups from the modified enzyme with 1.0 M aqueous piperidine. Poly-L-arginyl RNase was prepared by guanidisation of poly-L-ornithyl RNase with 1-guanyl-3,5-dimethylpyrazole nitrate. Poly-L-histidyl RNase was prepared by reacting RNase with the unprotected N-carboxy-L-histidine anhydride hydrochloride.

It was shown that RNase thus enriched with δ,N-trifluoroacetylornithine, ornithine, arginine, histidine or glutamic acid peptides is still enzymically active. The activity on the low molecular weight cytidine 2′,3′-cyclic phosphate is only moderately affected, often higher than that of the native enzyme, whereas that on RNA is markedly decreased. Poly-arginyl derivatives were active on the low molecular weight substrate but inactive on RNA, thus illustrating that the activity on these two substrates is affected independently. All the derivatives prepared recovered their full initial activity after complete reduction of the disulfide bridges, followed by reoxidation.

MATERIALS

Bovine pancreatic RNase, yeast RNA, cytidine 2′,3′-cyclic phosphate, urea, Ellman's reagent, Desicote, piperidine and heated dialysis tubings have been described in the preceeding paper [5]. δ,N-Tri-fluoroacetyl-α,N-carboxy-L-ornithine anhydride was prepared according to Ariely et al. [15]. 1-Guanyl-3,5-dimethylpyrazole nitrate was prepared according to Habeeb [16]. α,N-Benzyloxycarbonyl-L-histidine [17] was obtained from Yeda Co.

Preparation of Poly-L-arginyl RNase

Poly-L-arginyl RNase was prepared according to the following scheme:

RNase + δ,N-TFA-α,N-carboxy-L-ornithine an-hydride $\xrightarrow[\text{pH 7.0}]{\text{0.05 M Phosphate buffer}}$ poly-δ,N-TFA-L-ornithyl RNase $\xrightarrow[\text{30 hours}]{\text{1 M piperidine}}$ poly-L-ornithyl RNase.

Poly-L-ornithyl RNase + 1-guanyl-3,5-dimethyl-pyrazole nitrate $\xrightarrow[\text{pH 9.4}]{37°}$ poly-L-arginyl RNase.

RNase (500 mg) was dissolved in 0.05 M sodium phosphate buffer, pH 7.0 (50 ml), and cooled to 0°. δ,N-TFA-α,N-carboxy-L-ornithine anhydride (1 g in 15 ml absolute dioxane) was added and the reaction allowed to proceed for 8 hours at 0°. The milky white suspension which had formed during the reaction was exhaustively dialyzed in the cold against water in heated dialysis casings, to remove salts, dioxane and small oligopeptides of ornithine. The suspension was then centrifuged at 9,000 rev./min for one hour and the supernatant fluid, representing the water-

soluble fraction of poly-δ,N-TFA-L-ornithyl RNase, and the precipitate (water-insoluble fraction) were lyophilized separately. Poly-L-ornithyl RNase was obtained from the soluble and insoluble fractions, respectively, by suspending them separately in 1 M aqueous piperidine at 0° (7 ml 1 M piperidine/100 mg protein). The reaction was allowed to proceed for 30 hours at 4°. The clear solutions obtained were neutralized by slow addition of cold 0.5 M acetic acid, exhaustively dialysed against water and lyo-philised. The two preparations of pOrnRNase thus obtained were dissolved separately (200 mg of each) in 0,05 M Na₂HPO₄ (14 ml). To each solution 1-gua-nyl-3,5-dimethylpyrazole nitrate (400 mg) was added, and the solution adjusted to pH 9.4 by addition of 1 N NaOH. The reaction was allowed to proceed for 48 hours at 37°. The products were then exhaustively dialysed against water in the cold, concentrated by ultrafiltration to a volume of 2—3 ml and chromato-graphed on a 150×0.8 cm column of Sephadex G-25 (fine) in 0.2 M acetic acid − 0.1 M NaCl. The protein emerging immediately after the void volume of the column was dialysed against water, concentrated and stored at − 20° until needed.

Preparation of Poly-L-histidyl RNase

N-Carboxy-L-histidine Anhydride HCl. N-Benzyl-oxycarbonyl-L-histidine [17] (2.9 g) was suspended in anhydrous dioxane (10 ml) and a solution of phos-phorous pentachloride (2.2 g) in dioxane (15 ml) was added. The reaction mixture was kept at 25° by cooling with tap water. After a few minutes a cloudy solution was obtained, and shortly thereafter oil was formed. After 10 min the dioxane was decanted and ethyl acetate added. The mixture was triturated three times with ethyl acetate by decantation, and the crystals collected and dried over P₂O₅ and KOH in vacuo. The compound was then dissolved in di-methylformamide and reprecipitated with ethyl acetate. The product was very hygroscopic. Active CO₂, determined by the method of Patchornik and Shalitin [18] was 80%, based on a molecular weight of 254 (N-carboxy-L-histidine anhydride dihydro-chloride).

Poly-L-histidyl RNase. This was prepared by two alternative methods; a) RNase (500 mg) was dissolved in 0.05 M sodium phosphate buffer, pH 7.0, and cooled to 0°. N-Carboxy-L-histidine anhydride HCl (1.5 g) in powder form was suspended in this solution and the reaction allowed to proceed for 16 hours at 4°. During the reaction the pH of the solution dropped progressively to 3.8. The solution was then exhaus-tively dialysed against water, concentrated by ultra-filtration and stored at − 20° until needed. b) The reaction was performed as described above except that 0.2 M phosphate buffer was used and the pH kept at 7.0 by addition of 1 N NaOH as required.

At termination of the reaction the solution was dialysed against 6 changes of 0.2 M sodium phosphate buffer, pH 7.0. N-Carboxy-L-histidine anhydride HCl (1.5 g) was again added, and the reaction was allowed to proceed for another 16 hours. The solution was then dialysed against water, concentrated and chromatographed on Sephadex G-25 as described for pArgRNase, dialysed against water, concentrated and stored at − 20° until needed.

Preparation of Poly-L-glutamyl RNase

This preparation has been described previously[4]. The modified enzyme was purified by chromatography on Sephadex G-25 as described for pArg-RNase, then dialysed against water, concentrated and stored at − 20°.

RESULTS
Chemical Characterisation

pTFAOrnRNase, pOrnRNase and pArgRNase. RNase was reacted with δ,N-TFA-α,N-carboxy-L-ornithine anhydride [15]. The resulting poly-δ-N-TFA-L-ornithyl RNase was, in analogy to poly-ε,N-TFA-L-lysyl RNase [5], obtained as a water-soluble and a water-insoluble fraction, which were treated separately. The TFA groups were removed from the modified enzyme by treatment with 1.0 M aqueous piperidine, yielding poly-L-ornithyl RNase. Poly-L-arginyl RNase was obtained from polyornithyl RNase by reacting it with 1-guanyl-3,5-dimethyl-pyrazole nitrate [16].

From Table 1, in which the amino acid analyses and determinations of the number of reacted amino

Table 1. *Polyornithyl and polyarginyl derivatives of RNase*

Preparation and batch number		NCA[a]/RNase	Number of residues added per RNase		Number of reacted amino groups in RNase[c]	Relative activity	
			Ornithine[b]	Arginine		On RNA pH 5.0	On C-c-P pH 7.0
		w/w				%	%
pTFAOrnRNase	I[d]	2:1				94	283
pOrnRNase	I		40	—		5	152
pArgRNase	I		—	48	8	0	96
pTFAOrnRNase	II[e]	2:1				100	270
pOrnRNase	II		70	—		0	n. d.[f]
pArgRNase	II		—	76	8	0	23
pTFAOrnRNase	III	1:1				106	190
pOrnRNase	III		24	—		8	140
pArgRNase	III		—	30	6.5	0	90

[a] NCA = δ,N-TFA-α,N-carboxy-L-ornithine anhydride.
[b] Enrichment with ornithine was calculated from the single peak representing lysine and ornithine in amino acid analysis, after 10 lysine residues per molecule have been substracted.
[c] Determined by desamination [3].
[d] Preparations designated I are derived from the water-soluble fraction of pTFAOrnRNase.
[e] Preparations designated II are derived from the water-insoluble fraction of pTFAOrnRNase.
[f] Not determined.

METHODS

The methods used for determination of enzyme concentration in solutions, assays of enzymic activity, as well as the chemical and physicochemical methods used in this study have been described in the preceding paper [5].

The enrichment of derivatives with ornithine was calculated from amino-acid analyses by substracting 10 lysine residues [19] from the single peak representing the sum of lysine and ornithine.

Electrophoresis on cellulose acetate strips (Oxo, England) was carried out for 60 min, applying a potential gradient of 10 vol/cm at room temperature. Dextran (Sigma) was used as a nonmigrating marker to determine the endosmotic flow, and its location, as evidenced by the transparency of the strip, was marked before staining. Proteins were stained with amidoblack (Gurr).

groups in the modified enzyme for some of the preparations are summarised, it can be seen that consistently more arginine residues were attached to RNase than ornithine residues were found in the corresponding pOrnRNase derivatives. This is probably related to the inaccuracy of the indirect ornithine determination from the single peak representing the sum of lysine and ornithine. As expected from the experience gained with polylysyl RNase [5], the enrichment is highest in the derivative obtained from the insoluble pTFAOrnRNase, for the preparation of which a ratio of 2:1 of δ,N-TFA-α,N-carboxy-L-ornithine anhydride to RNase was used. The number of the reacted amino groups in the modified enzymes did not exceed 8 (Table 1).

Similarly to polylysyl RNase [5], pOrnRNase and pArgRNase did not lend themselves to zone electrophoresis. In free boundary electrophoresis, a mobility

Table 2. *Poly-L-histidyl RNase*

Preparation and batch number	Number of histidine residues attached per RNase	Number of reacted amino groups per RNase[b]	Relative activity	
			on RNA pH 5.0	on C-c-P pH 7.0
			%	%
pHisRNase I	4.2		65	150
pHisRNase II	6.4		42	160
pHisRNase III[a]	22	7	28	140

[a] Obtained from pHisRNase II by reacting it again with *N*-carboxy-L-histidine anhydride.
[b] Determined by desamination [3].

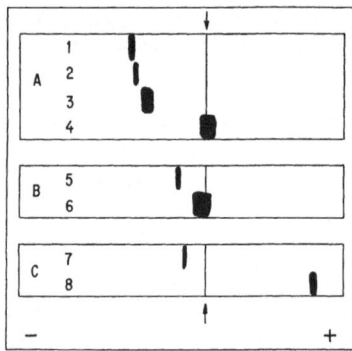

Fig. 1. *Electrophoretic patterns of RNase (bands 1, 5, 7), pHisRNase II (band 2), pHisRNase III (band 3) and pGluRNase (bands 6, 8) on cellulose acetate.* A potential gradient of 10 volt/cm was applied for 60 min at room temperature. A, in 0.1 M sodium acetate buffer, pH 5.0; B, in 0.1 M Tris-HCl buffer, pH 8.0; and C, in 0.1 M sodium veronal buffer, pH 8.6. The arrows indicate the spot at which a nonmigrating dextran marker was detected

Table 3. *Poly-L-glutamyl derivatives of RNase*

Preparation and batch number	Number of glutamyl residues attached per RNase	Number of reacted amino groups per RNase[a]	Relative activity	
			on RNA pH 5.0	on C-c-P pH 7.0
			%	%
pGluRNase I	38	8	34	35
pPIMGluRNase IV			110	160
pGluRNase IV	34	7.5	60	55
pPIMGluRNase V			100	150
pGluRNase V	71	8	16	25

[a] Determined by desamination [3].

of $8.6 \times 10^{-5} \text{ cm}^2 \times \text{V}^{-1} \times \text{sec}^{-1}$ was calculated from the descending boundary for the major fraction of pArgRNase (enriched with 30 residues) in 0.1 M tris HCl buffer, pH 8.0, as opposed to $2.5 \times 10^{-5} \text{ cm}^2 \times \text{V}^{-1} \times \text{sec}^{-1}$ for RNase run simultaneously.

pArgRNase (enriched with 30 arginine residues per molecule) was studied in the ultracentrifuge using both the sedimentation velocity and the short column equilibrium approach [20] methods. A molecular weight of 19,000 was derived from a sedimentation coefficient, $s_{20} = 2.1$ S, diffusion coefficient, $D_{20} = 9.01 \times 10^{-7} \text{ cm}^2 \times \text{sec}^{-1}$, and a partial specific volume, $\bar{v} = 0.701$ (calculated from $\bar{v}$ value of 0.695 for RNase [21] and 0.72 for arginine residues [22]). The molecular weight derived from the short column equilibrium approach method was 19,760. Both values were obtained from the preparations immediately after synthesis. Similarly to polylysyl RNase all pArgRNase preparations were subject to "ageing", and after 2—3 months progressively lost both solubility and activity. The molecular weight of the still soluble part of the same preparation, enriched with 30 arginine residues, was redetermined after approximately 4 months by the sedimentation velocity method and found to be 28,000.

Poly-L-histidyl RNase. pHisRNase was prepared by the reaction of RNase in 0.05 M aqueous sodium phosphate buffer, pH 7.0, with the unprotected *N*-carboxy-L-histidine anhydride. During the reaction, the pH dropped to 3.8. The two batches of pHisRNase thus prepared had 4.2 and 6.4 histidine residues attached to them, respectively (Table 2). A part of batch II was resubmitted to the reaction with additional *N*-carboxy-L-histidine anhydride, this time keeping the pH of the reaction mixture constant at pH 7.0. The resulting preparation (III) had 22 histidine residues attached to it.

Upon electrophoresis on cellulose acetate strips, purified pHisRNase II and III migrated as single bands, though broader than that of the native enzyme run simultaneously (Fig. 1). Both at pH 8.0 (in Tris-HCl buffer, 0.1 M) and at pH 5.0 (sodium acetate buffer, 0.1 M) they moved to the cathode with slightly lower mobility than that of the native enzyme. The ultraviolet absorption spectrum of pHisRNase at pH 7.0 did not differ from that of native RNase.

Poly-L-glutamyl RNase. The synthesis of pGluRNase via poly-γ-phthalimidomethyl-L-glutamyl RNase has been previously described [4]. Three different batches have been prepared, and the details of these preparations are given in Table 3.

Preparation V (enriched with 71 glutamyl residues) was studied in the analytical ultracentrifuge, both by sedimentation velocity and by short column equilibrium approach [20]. A molecular weight of 18,650 was derived from the sedimentation coefficient, $s_{20} = 1.61$ S, the diffusion coefficient, $D_{20} = 6.6 \times 10^{-7} \text{ cm}^2 \times \text{sec}^{-1}$, and a partial specific volume $\bar{v} = 0.68$ (calculated from $\bar{v}$ value of 0.695 for RNase [21] and 0.66 for glutamyl residues [22]). The molecular weight derived from the short column equilibrium approach was 18,950.

Electrophoresis on cellulose acetate strips in 0.1 M sodium veronal buffer at pH 8.6 showed pGluRNase to consist of a single homogeneous band migrating far to the anodal side, while RNase moved to the cathode. In acetate buffer, at pH 5.0, a broader band was obtained (Fig. 1) at proximity of the point of origin, while RNase migrated far to the cathode. At both pH values there was no evidence of unreacted RNase in the pGluRNase sample. The mobility of pGluRNase IV was determined by moving boundary electrophoresis. In 0.1 M Tris-HCl buffer, at pH 8, this preparation separated into 2 peaks with mobilities of 7.2 and 16.9×10^{-5} cm² $\times$ V⁻¹ $\times$ sec⁻¹, respectively, while the two peaks of the native enzyme had mobilities of -1.91 and -2.5×10^{-5} cm² $\times$ V⁻¹ $\times$ sec⁻¹, respectively (calculated from descending boundaries).

Enzymic Activity

Polyornithyl and Polyarginyl Derivatives. Whereas pTFAOrnRNase is as active on RNA at pH 5.0 as the native enzyme, the removal of the TFA groups by piperidine, with the resulting liberation of the δ-amino groups sharply reduces the activity on this substrate. Of the three preparations (Table 1) of pOrnRNase tested, two had 8 and 5 % activity, respectively, while the third was inactive. pArg-RNase, derived from pOrnRNase, is totally inactive on RNA in the whole pH range investigated (pH 5 to 9). Reaction mixtures became immediately turbid, suggesting complex formation between the modified enzyme and RNA. That this inactivity on RNA is not due to irreversible damage to the enzyme induced by the chemical modification is evident from the fact that these derivatives still retain activity on the low molecular weight cytidine 2′,3′-cyclic phosphate. Thus, two preparations enriched with 30 and 48 arginine residues, respectively, possessed 90 % of the activity of the native enzyme, while the one enriched with 76 residues was still 23 % active. Two of the pOrnRNase preparations were more than 100 % active toward C-c-P, and their blocked form, pTFA-OrnRNase, was 2—3 times as active on this substrate than the native enzyme. The dependence of the activity of pArgRNase toward C-c-P upon pH was investigated in the pH range between 6.5 and 9. The assays were performed at low ionic strength by the titrimetric method in a pH-stat. Activity of pArg-RNase was found to increase with increasing pH, and the highest activity was measured at pH 9.0, whereas native RNase had maximal activity at pH 7.0. The pH range investigated could not be extended to higher values, since non-enzymic hydrolysis of the substrate interfered with activity measurements. Increasing the ionic strength of the reaction mixture caused a decrease in the pH of the optimal activity of pArgRNase, which at ionic strength 0.2 was pH 7.5.

pHisRNase. The enzymic activity of the different preparations of pHisRNase, both on RNA and on the low molecular C-c-P, is indicated in Table 2. The activity of all preparations on RNA is lower than that of the native enzyme. The higher the enrichment with histidine residues, the lower the activity of the resulting modified RNase. The activity on C-c-P, on the other hand, is higher than that of the native enzyme and is little affected by the number of histidine residues attached to the derivative. The pH dependence of the activity of pHisRNase II, both on RNA and on C-c-P, was studied in a tris-malonic acid-NaOH buffer system, the ionic strength of which was adjusted to 0.1 by addition of NaCl. For both substrates, maximal activity was found at pH 7.0 as for the native enzyme.

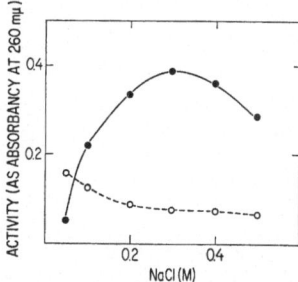

Fig. 2. *The Ionic strength dependence of the activity of RNase and of pGluRNase IV on RNA.* Anfinsen assay at pH 5.0 in 0.05 M sodium acetate buffer. Ionic strength was adjusted with NaCl. ●, RNase; ○, pGluRNase IV

pGluRNase. The enzymic activity of the different preparations of poly-γ-phthalimido-methylglutamyl RNase and of pGluRNase, both on RNA and on C-c-P, are given in Table 3. The blocked derivatives are fully active on RNA at pH 5.0, and have 1.5 times the activity of the native enzyme on C-c-P (at pH 7.0). Removal of the blocking groups resulted in a marked decrease of activity on both substrates. Preparation V, enriched with 71 glutamyl residues per molecule, was not as active as the less extensively enriched preparations. The ionic strength dependence of the reaction of pGluRNase IV on RNA was investigated in sodium acetate buffer, pH 5.0, the ionic strength of which was adjusted with NaCl in the range of 0.05 to 1.0. The highest activity was observed at the lowest ionic strength tested, even though the activity of the derivative was not markedly influenced by changes in the ionic strength (Fig. 2). At pH 4.5, and ionic strength 0.05, the relative activity of pGluRNase IV on RNA was 180 %.

The pH dependence of the reaction on RNA was studied in 0.1 M Tris-malonic acid-NAOH buffer; activity under these conditions was found to be maximal at pH 6.5 (Fig. 3). With C-c-P as substrate, the optimal activity (at low ionic strength) was found, by the titrimetric method, at pH 5.8 (as opposed to pH 7.0 for the native enzyme).

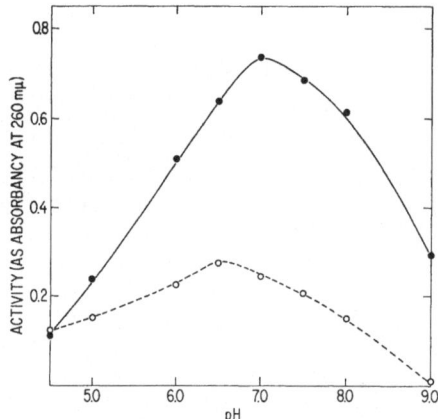

Fig. 3. *The pH dependence of the activity of RNase and of pGluRNase IV on RNA.* Assays were performed by the Anfinsen method in tris-malonic acid-NaOH buffers, the ionic strength of which was adjusted to 0.1 with NaCl. ●, RNase; ○, pGluRNase IV

Reduction and Reoxidation

pTFAOrnRNase, pOrnRNase, pArgRNase, pHisRNase, pPIMGluRNase, pGluRNase and native RNase were reduced with 0.32 M β-mercaptoethanol in 8 M urea. After 4 hours the reduced protein was passed on Sephadex G-25 in 0.1 M acetic acid, to separate the protein from salts, urea and mercaptoethanol. The completeness of the reduction was checked by titration of free sulfhydryl groups [23]. The fully reduced samples were devoid of significant enzymic activity in RNA at pH 5.0 (and in the case of pArgRNase, on C-c-P at pH 7.0). Reoxidation was allowed to proceed for 3 hours at room temperature, at a protein concentration of 20 µg/ml, in 0.1 M tris-HCl buffer, pH 8.5. All derivatives tested recovered, upon completed reoxidation, their full initial activity on both RNA and C-c-P (except for pArgRNase, which remained inactive on RNA).

DISCUSSION

The investigation of RNase modified by attachment of various peptide chains has been the subject of several recent reports [3—8]. The synthesis of some additional derivatives, in which peptides carrying side-chain functional groups were attached to the enzyme molecule is described in the present paper.

The influence of added δ-amino groups of ornithine, guanido groups of arginine, γ-carboxyl groups of glutamic acid and the imidazole functions of histidine on the properties of the modified enzyme was studied.

Two different approaches were used in the preparation of these derivatives. pGluRNase [4] and pOrnRNase were synthesised by reversible blocking of the γ-carboxyl function of glutamic acid and of the δ-amino function of ornithine, respectively, with the phthalimidomethyl group and the trifluoroacetyl group, the usefulness of which in the preparation of modified proteins has been previously described [4, 24—26,5]. In a later step, pArgRNase was prepared from pOrnRNase by a guanidisation reaction, as suggested for the preparation of polyarginyl proteins [15]. Previous studies on the guanidisation of native RNase, led to derivatives in which most of the amino groups were modified, and which still possessed enzymic activity [27]. One derivative, in which all the ε-amino groups of RNase were guanidised, was devoid of enzymic activity.

In the preparation of pHisRNase, on the other hand, RNase was reacted with the N-carboxyanhydride of L-histidine, the amino nitrogen of which was not protected. An enrichment of 22 residues per RNase molecule was obtained under these conditions. The use of unprotected amino acid anhydrides, while successful in this case, is often unadvisable. Thus, the synthesis of a polyglutamyl protein, namely, the reaction of RNase T₁ with the unprotected N-carboxyanhydride of L-glutamic acid, reported recently by Kuriyama [28] has met with only a limited success, as not more than 1—3 glutamyl residues were attached per molecule. This should be contrasted with an enrichment of RNase with up to 71 glutamyl residues residues obtained by using the γ-phthalimidomethyl-α,N-carboxy-L-glutamic acid anhydride.

RNase derivatives in which the functional groups on the attached peptides are blocked (pPIMGluRNase [4], pTFALysRNase [5], pTFAOrnRNase) are all enzymically active. Their properties are similar to polypeptidyl derivatives which possess no free functional group, e.g., poly-DL-alanyl RNase [7]. With very few exceptions they are as active or more active on RNA than native RNase, and the pH values of optimal activity are close to that of the native enzyme. They all are, on the other hand, 1.5 to 3 times more active on the low molecular weight substrate cytidine 2′,3′-cyclic phosphate than RNase. The attachment of polypeptide chains to RNase does not, therefore, *per se*, interfere with catalytic activity of the enzyme, provided that the amino group of lysine 41 in RNase remains unmodified (*cf.* Discussion of preceding paper).

The deblocking with the resulting liberation of the side-chain functional groups on the attached peptides, causes dramatic changes in the activity of the modified enzymes. These changes are not artefacts due to irreversible denaturation of the enzyme but do represent the influence of the free functional groups on the properties of the enzyme. They can be reversed either by reblocking of the functional groups, or by their removal, as demonstrated in the case of pLys-RNase [5].

In Table 4 we have compared the enzymic properties of various modified RNases, choosing as far as possible preparations enriched to a similar extent. In addition to expressing activity as percent of the activity of the native enzyme ("relative activity") which necessitates comparing, under similar conditions, enzymes having different conditions of optimal activity, we have also used the "relative maximal activity", which is the ratio between maximal activity of the native enzyme, each measured at its

100%. The above observations are based on a comparison of the derivatives and the native enzyme under conditions which do not particularly favour the activity of the native enzyme. They tend, therefore, to minimize the extent to which the activity on RNA is affected. This decrease is better expressed by the "relative maximal activity". Comparison of the few examples given in Table 4 shows clearly that activity on RNA is considerably decreased in all derivatives. The extent of the decrease was shown to be proportional to the number of the peptides attached to the enzyme.

The activity of the derivatives on the low molecular weight C-c-P is either similar to, or higher than that of the native enzyme, as reflected from both relative and "relative maximal activity" (Table 4). It is, therefore, also less influenced by the number of

Table 4. *The enzymic properties of various modified RNases*

Preparation	Number of residues attached per RNase	Relative activity		pH of optimal activity		Maximal relative activity (at optimal pH)	
		on RNA pH 5.0	on C-c-P pH 7.0	on RNA (0.1 M)	on C-c-P at low ionic strength	on RNA at ionic strength 0.1	on C-c-P at low ionic strength
		%	%			%	%
RNase	—	100	100	7—7.5	7	100	100
p-DL-AlaRNase [7]	94	50	145	6.5	6.5—7[a]	12	125[a]
pLysRNase [5]	29	18	110	8	8.5	16	120
pOrnRNase III	24	8	140	n. d.[c]	n. d.[c]	n. d.[c]	n. d.[c]
pArgRNase III	30	0	90	inactive	≥ 9	inactive	> 120
pHisRNase III	22	28	140	7	7	40	140
pGluRNase IV	34	60	55	6.5	5.8	37	50
RNase-gly [24]	11[b]	0	93	inactive	n. d.[c]	inactive	n. d.[c]

[a] Measured at ionic strength 0.1.
[b] In this case the glycine residues are attached through their amino groups to the carboxyl groups of the enzyme.
[c] Not determined.

own pH of optimal activity but at similar ionic strength. From the data on relative activity it is clear that the chemical modification affects both the activity on RNA and on the low molecular C-c-P, but in different ways. Extreme examples of this situation are to be found in pArgRNase and in RNase-glycine (an RNase derivative in which all the 11 carboxyl groups are modified by the attachment of glycine residues [24]). In both these derivatives the activity on C-c-P persists almost completely while activity on RNA is abolished. An important conclusion from these results is that lack of activity of a modified RNase on RNA does not necessarily imply complete absence of catalytic activity.

The relative activity of most derivatives on RNA at pH 5.0 is low, but some modified enzymes, e.g. polyalanyl and polyglutamyl RNase (Table 4), still retain 50—60% of the activity of the native enzyme. Moreover, the acidic pGluRNase and the basic pLys-RNase display under certain conditions (high pH and ionic strength for pLysRNase, low pH and ionic strength for pGluRNase) relative activities exceeding

residues attached to the enzyme than is the activity towards RNA, although in some very extensively peptidylated samples, activity on the small substrate was decreased. Only in the case of pGluRNase, the activity on this substrate was consistently lower than that of the native enzyme in all the preparations tested.

The shifts observed in the pH of optimal activity of the derivatives studied are in agreement with the hypothesis [7] that a correlation exists between the isoelectric point and the optimal pH for activity of the modified RNase. Thus, in the very basic modifications, pLysRNase and pArgRNase, the pH optima are higher than that of the native enzyme; pGlu-RNase, on the other hand, has a significantly lower pH optimum. Similar observations have been reported for other polypeptidyl enzymes [28,6]. The pH optimum of the activity of RNase [29] and of peptidyl RNases on RNA depends on the ionic strength. The shifts in the pH optima of the modified enzymes discussed here as compared with native RNase, are largest at very low ionic strength.

Studies of the dependence of the hydrolysis of RNA by pLysRNase on ionic strength showed that at $\mu = 0.6$ or 0.7, where the charges of the side-chain amino groups attached to the enzyme are shielded, the modified enzyme was highly active [5]. Though the rate of hydrolysis of C-c-P by either the native enzyme [30] or modified RNases is little affected by changes in ionic strength, a certain shielding effect can be attributed to salt as evident from the shifts in optimal pH, both in pLysRNase and pArgRNase (Table 4). pGluRNase, on the other hand, is most active on RNA at low ionic strength. This observation may be compared to the decrease of activity observed in polyalanyl RNase upon increasing ionic strength.

The electrostatic attraction certainly plays an important role in the interaction of RNA with the basic RNase derivatives, pLysRNase, pOrnRNase and pArgRNase. It may favour stable enzyme-substrate complexes which would lead to an apparent decrease in enzyme activity. Shielding of the charges by increasing ionic strength reverses this effect. In contrast to the above derivatives, pGluRNase is more acidic than RNase so that a negatively charged enzyme reacts with a negatively charged substrate. This may explain the low activity of this derivative both on RNA and on C-c-P. Whether the preferential effect of peptidylation on the activity of the modified enzymes towards RNA is merely due to the polyelectrolyte nature of RNA or whether RNase contains, besides the active catalytic center, additional sites [30] responsible for binding of the macromolecular substrate cannot be settled by the data presented.

Amino and guanido groups, phenolic hydroxyls of tyrosine [3], γ-carboxyl groups of glutamic acid and imidazole groups of histidine have been attached to RNase without abolishing enzymic activity. All these derivatives retained their capacity to recover their full initial activity after complete reduction and reoxidation. From the studies of polypeptidyl derivatives of RNase it appears that the hydrophilic periphery of the molecule can be modified to a great extent, and that many properties of the enzyme, including substrate binding capacity will thus be modified, but the catalytic activity and the capacity of the enzyme to recover by reoxidation its unique structure after complete reduction of the bonds stabilising tertiary structure, will not be interfered with. These basic functions of the enzyme seem to be maintained by hydrophobic and other forces stabilising the molecule in solution in its unique conformation.

This investigation was supported in part by Agreement 415100 with the National Institutes of Health, U.S. Public Health Service.

REFERENCES

1. Becker, R. R., and Stahmann, M. A., J. Biol. Chem. 204 (1953) 745.
2. Becker, R. R., in Polyaminoacids, Polypeptides and Proteins (edited by M. A. Stahmann) University of Wisconsin Press, Madison 1962, paper 29.
3. Anfinsen, C. B., Sela, M., and Cooke, J. P., J. Biol. Chem. 237 (1962) 1825.
4. Wilchek, M., Frensdorff, A., and Sela, M., Arch. Biochem. Biophys. 113 (1966) 742.
5. Frensdorff, A., and Sela, M., European J. Biochem. 1 (1967) 267.
6. Becker, R. R., and Sawada, E., Federation Proc. 22 (1963) 419.
7. Wellner, D., Silman, H. I., and Sela, M., J. Biol. Chem. 238 (1963) 1324.
8. Epstein, C. J., and Goldberger, R. F. J. Biol. Chem. 239 (1964) 1087.
9. Weil, L., and Seibles, T. S., Arch. Biochem. Biophys. 54 (1955) 368.
10. Barnard, E. A., and Stein, W. D., J. Mol. Biol. 1 (1959) 339.
11. Gundlach, H. G., Stein, W. H., and Moore, S., J. Biol. Chem. 234 (1959) 1754.
12. Crestfield, A. M., Stein, W. H., and Moore, S., J. Biol. Chem. 238 (1963) 2413; 2421.
13. Heinrikson, R. L., Stein, W. H., Crestfield, A. M., and Moore, S., J. Biol. Chem. 240 (1965) 2921.
14. Deavin, A., Mathias, A. P., and Rabin, B. R., Nature, 211 (1966) 252.
15. Ariely, S., Wilchek, M., and Patchornik, A., Biopolymers, 4 (1966) 91.
16. Habeeb, A. F. S. A., Can. J. Biochem. Physiol. 38 (1960) 493.
17. Patchornik, A., Berger, A., and Katchalski, E., J. Am. Chem. Soc. 79 (1957) 6416.
18. Patchornik, A., and Shalitin, Y., Anal. Chem. 33 (1961) 1887.
19. Hirs, C. H. W., Moore, S., and Stein, W. H., J. Biol. Chem. 219 (1956) 623.
20. Yphantis, D. A., Ann. N. Y. Acad. Sci. 88 (1960) 586.
21. Harrington, W. F., and Schellman, J. A., Compt. Rend. Trav. Lab. Carlsberg (Ser. Chim.), 30 (1956) 21.
22. Cohn, E. J., and Edsall, J. T., "Proteins, Amino Acids and Peptides". Reinhold Publ. Cor. N. Y., 1943.
23. Ellman, G. L., Arch. Biochem. Biophys. 82 (1959) 70.
24. Wilchek, M., Frensdorff, A., and Sela, M., Biochemistry, 6 (1967) 247.
25. Goldberger, R. F., and Anfinsen, C. B., Biochemistry, 1 (1962) 401.
26. Sela, M., Arnon, R., and Jacobson, I., Biopolymers, 1 (1963) 517.
27. Klee, W. A., and Richards, F. M., J. Biol. Chem. 229 (1957) 489.
28. Kuriyama, Y., J. Biochem. (Tokyo), 59 (1966) 596.
29. Kalnitsky, G., Hummel, J. P., Resnick, H., Carter, J. R., Barnett, L. B., and Dierks, O., Ann. N. Y. Acad. Sci. 81 (1959) 541.
30. Irie, M., J. Biochem. (Tokyo), 57 (1965) 355.

A. Frensdorff, M. Wilchek, and M. Sela
Section of Chemical Immunology
The Weizmann Institute of Science
Rehovoth, Israel

European J. Biochem. 1 (1967) 289—300

Regulation of the Enzymes of the β-Ketoadipate Pathway
in *Moraxella calcoacetica*

1. General Aspects

J. L. Cánovas and R. Y. Stanier

Department of Bacteriology and Immunology, University of California, Berkeley, California

(Received January 20, 1967)

The inducible enzymes responsible for the dissimilation of aromatic compounds through the β-ketoadipate pathway by *Moraxella calcoacetica* are subject to a high degree of coordinate control. The five enzymes which catalyze the conversion of protocatechuate to β-ketoadipyl-CoA are coordinately synthesized, the probable inducer being protocatechuate. In the conversion of catechol to β-ketoadipyl-CoA, there are two separate inductive events, though both are mediated by one metabolite-inducer, *cis,cis*-muconate. One is the induction of catechol oxygenase; the other is the coordinate induction of the four enzymes that catalyze conversion of *cis,cis*-muconate to β-ketoadipyl-CoA.

Although the catechol and protocatechuate branches of the β-ketoadipate pathway converge metabolically with the formation of β-ketoadipate enol-lactone, the two step-reactions required for the conversion of this intermediate to β-ketoadipyl-CoA are mediated by two isofunctional sets of inducible enzymes, one controlled coordinately with enzymes mediating steps specific to the catechol branch, the other coordinately with enzymes mediating steps specific to the protocatechuate branch.

The differing control mechanisms which govern synthesis of the enzymes of the β-ketoadipate pathway in *Moraxella calcoacetica* and in *Pseudomonas putida* are compared and discussed.

Ornston and Stanier [1] recently completed the chemical characterization of the β-ketoadipate pathway, which serves in many aerobic bacteria to transform catechol, protocatechuate and their respective aromatic precursors to succinate and acetyl-CoA. Catechol and protocatechuate are converted through three analogous but chemically distinct step-reactions to the enol-lactone of β-ketoadipate; three further common step-reactions convert this lactone to succinate and acetyl-CoA (Fig. 1). Ornston also developed assays for most of the enzymes operative in this pathway [2,3] and made a detailed study of their regulation in *Pseudomonas putida* [4]. The inductive controls which operate in *P. putida* are summarized in Fig. 2. A striking feature is the coordinate induction of the last two enzymes specific to the protocatechuate branch and of the first enzyme functional in the common terminal steps. This coordinate

block is product-induced by β-ketoadipate or, possibly, β-ketoadipyl-CoA. As a consequence, cells grown on a metabolic precursor of catechol (e.g., benzoate) synthesize full levels of two enzymes uniquely functional in the protocatechuate branch, and which are metabolically redundant under these conditions. Ornston [4] demonstrated the same regulatory linkage in two other *Pseudomonas* species *(P. aeruginosa* and *P. multivorans)*, but found that in *Moraxella calcoacetica* enzymes specific to the protocatechuate branch are not significantly induced by growth on a catechol precursor [4]. It was accordingly evident that the regulatory control of the enzymes of the β-ketoadipate pathway in *M. calcoacetica* differs significantly from that in *Pseudomonas* spp. The system of regulation operative in *M. calcoacetica* will be described in the present paper.

EXPERIMENTAL PROCEDURE

Microbiological Materials

Most experiments were conducted with strain 73 of *Moraxella calcoacetica* from the collection of this department, or mutants derived from it. In certain comparative experiments, *Pseudomonas putida* strain A.3.12 (ATCC 12633) was also used.

Enzymes. Protocatechuate oxygenase or protocatechuate:oxygen 3,4-oxidoreductase (EC 1.13.1.3); catechol oxygenase or catechol:oxygen 1,2-oxidoreductase (EC 1.13.1.1); β-ketoadipate succinyl-CoA transferase or 3-oxoadipate CoA-transferase (EC 2.8.3.6); alkaline phosphatase or orthophosphoric monoester phosphohydrolase (EC 3.1.3.1); lactic dehydrogenase or L-lactate:NAD oxidoreductase (EC 1.1.1.27); malic dehydrogenase or L-malate:NAD oxidoreductase (EC 1.1.1.37). All the other enzymes assayed in this work are not listed by the Enzyme Commission.

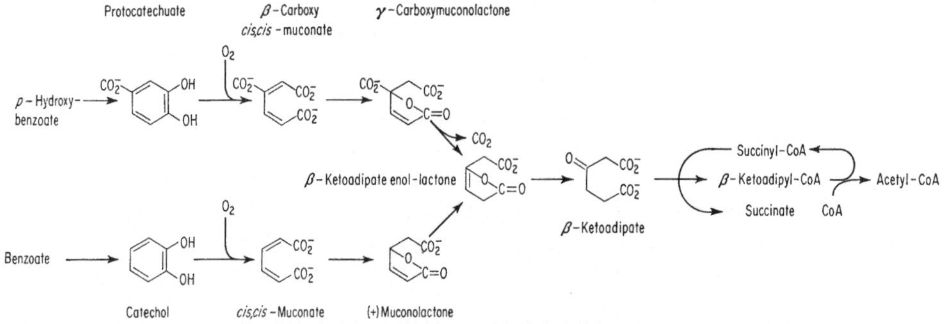

Fig. 1. *The chemistry of the β-ketoadipate pathway*

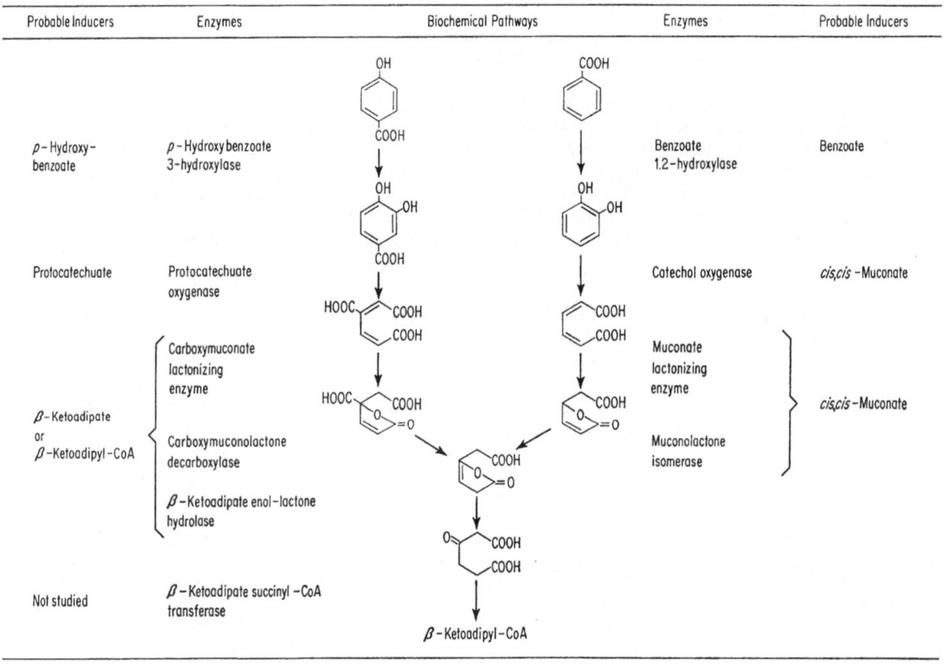

Fig. 2. *The regulation of synthesis of the enzymes catalyzing the reactions of the β-ketoadipate pathway in* Pseudomonas putida [4]

Mutants of strain 73 of *M. calcoacetica* which had lost the ability to synthesize specific enzymes operative in the β-ketoadipate pathway (blocked mutants) were obtained by nitrosoguanidine treatment [4], followed by plating on a mineral-succinate medium. Mutant colonies which had lost the ability to grow on benzoate or on *p*-hydroxybenzoate were detected by replica plating on media which contained each aromatic acid as sole source of carbon and energy. The site of the enzymatic lesion was determined in each such mutant by enzymatic analysis of an extract prepared from cells grown for about 4 generations on 20 mM succinate and 2 mM of the appropriate aromatic acid.

Permeability mutants (able to grow on *cis,cis*-muconate) were obtained without mutagenesis as described by Ornston [4].

Media and Conditions of Cultivation

Moraxella calcoacetica is an aerobic chemoheterotroph which does not require organic growth factors. The culture media used for its maintenance and for experimental purposes were identical to those described by Ornston and Stanier [1] for the growth of *Pseudomonas putida*. All cultures were incubated at 30°. Stock cultures were maintained at 4° after growth.

Cells for enzymatic analyses were grown in agitated liquid cultures in synthetic media. The course of growth was followed by turbidimetry, using a Klett-Summerson colorimeter, equipped with a No.66 filter, and cells were harvested during exponential growth at a turbidity of approximately 70 Klett units (equivalent to 10^8 cells/ml). In the preparation of synthetic media, the organic carbon and energy source was furnished at an initial concentration of 10 mM, except in the case of succinate, where the initial concentration was 20 mM. When cells were induced with an aromatic compound during growth at the expense of succinate (20 mM) the aromatic compound was furnished at an initial concentration of 2 mM.

Preparation of Cell-Free Extracts

The general procedures described by Ornston and Stanier [1] were employed, except that after sonic disintegration, the extracts were clarified by centrifugation at 35,000×*g* for 15 min. For the demonstration of low levels of activity of *p*-hydroxybenzoate hydroxylase, an additional centrifugation at 100,000×*g* for 60 min was required, in order to reduce to a minimum interfering TPNH oxidase activity.

Enzyme Assays

All assays were performed at room temperature (approximately 22°) by spectrophotometric methods, using silica cuvettes with 1.0 cm liquid path. Measurements were made with a Gilford model 2000 recording spectrophotometer.

With the exception of transferase activities, the unit of enzyme activity is defined as the amount of enzyme necessary to cause the disappearance of 1 μmole of substrate per min under the conditions of assay employed. Since the extinction coefficient of the product measured in transferase assays (β-ketoadipyl-CoA) is not known, the unit of transferase activity is defined in arbitrary terms, as the amount of enzyme necessary to cause a change of 1.0 absorbance units per min under the prescribed assay conditions. Unless otherwise indicated, assays were

initiated by addition of the enzyme to the reaction mixture.

Most enzymes of the β-ketoadipate pathway were assayed by methods already described: catechol oxygenase [2], except that Tris buffer pH 7.5 was used; *cis,cis*-muconate lactonizing enzyme [3], also using Tris buffer at pH 7.5; (+)-muconolactone isomerase [3]; protocatechuate oxygenase [5]; β-carboxy-*cis,cis*-muconate lactonizing enzyme, γ-carboxy-muconolactone isomerase and β-ketoadipate enol-lactone hydrolase [2]; and *p*-hydroxybenzoate hydroxylase [6], except that phosphate buffer at pH 7.0 was used. In this assay of the latter enzyme, the rate of TPNH oxidation by the extract in the absence of the aromatic substrate was first determined, and subtracted from the rate obtained after substrate addition.

β-Ketoadipate succinyl-CoA transferase was assayed by a modification of the method of Katagiri and Hayaishi [7]. The reaction mixture contained 100 μmoles of Tris buffer (pH 8.0), 20 μmoles of $MgCl_2$, 10 μmoles of Na β-ketoadipate and 0.5 μmoles of succinyl-CoA, in a total volume of 3.0 ml. The course of the reaction is followed by measuring the increase in absorbance at 305 mμ resulting from the formation of β-ketoadipyl-CoA. Although the reaction is readily reversible, linear rates can be obtained, provided that the assay is conducted with less than 0.03 unit of enzyme, and that the recording is amplified to a full scale of 0.100 absorbance units. With these precautions it is possible to avoid using a large amount of the expensive substrate (200 μmoles per assay), as required in the original procedure [7].

Other enzymes assayed included alkaline phosphatase [8]; lactic dehydrogenase [9]; and malic dehydrogenase [10].

Other Physical and Chemical Measurements

Cytochrome *c* and ferritin were estimated spectrophotometrically, from their absorbances at 406 and 310 mμ, respectively. Succinyl-CoA was determined by the method of Cánovas and Kornberg [11]. Protein was determined by the method of Lowry *et al.* [12]. Respirometric experiments were conducted at 30° in a Gilson differential respirometer.

Chemical and Enzymatic Reagents

The chemicals employed were obtained from commercial sources, with the following exceptions. β-carboxy-*cis,cis*-muconate and (+)-muconolactone were prepared enzymatically [1] by Dr. L. N. Ornston. *cis,cis*-Muconic acid was synthesized chemically by the method of Elvidge *et al* [13]. Succinyl-CoA was synthesized chemically by the method of Stadtman [14].

Purified preparations of β-ketoadipate enol-lactone hydrolase, (+)-muconolactone isomerase, *cis,cis*-

muconate lactonizing enzyme and β-carboxy-*cis,cis*-muconate lactonizing enzyme, required for enzyme assays, were prepared by the methods of Ornston [2,3].

Commercial enzyme preparations used were: horse heart cytochrome *c* (Sigma); pig heart malic dehydrogenase type III (Sigma); rabbit muscle lactic dehydrogenase (Boehringer); *Escherichia coli* alkaline phosphatase (Worthington) and horse spleen ferritin (Calbiochem).

Molecular Weight Determinations

Sephadex G-200 (Pharmacia) was allowed to swell for several days in 0.1 M Tris-maleate buffer (pH 7.1), made 0.1 M with respect to NaCl. The mixture was used to prepare a column 1.5 × 140 cm. A cell-free extract of either benzoate or of *p*-hydroxybenzoate-grown cells of *M. calcoacetica* (3.0 ml of extract, containing approximately 75 mg of protein) was mixed with the following "marker" proteins: cytochrome *c* (5 mg); ferritin (5 mg); lactic dehydrogenase (50 units); malic dehydrogenase (50 units); and alkaline phosphatase (20 units). This mixture was applied to the column as a layer under the above-mentioned buffer, and the column was eluted with the same buffer at a flow rate of 8—10 ml/h. Successive fractions of 1.0 ml were collected and assayed. From the observed elution volumes, molecular weights of β-ketoadipate enol-lactone hydrolases I and II were calculated by the method of Whitaker [15].

Experiments on the Kinetics of Thermal Inactivation of Enzymes

An aliquot of cell-free bacterial extract (0.5 ml, containing 2 mg of protein) was placed in a test tube, and immersed in a water-bath maintained at the desired temperature, 45 sec being allowed for the attainment of thermal equilibration. Thereafter, samples were removed at fixed intervals, immediately chilled, and subsequently assayed for the enzyme under study.

RESULTS

Patterns of Induction in Cells of the Wild Type

Levels of activity of the enzymes of the β-keto-adipate pathway in cells of *M. calcoacetica* harvested during exponential growth on succinate, benzoate and *p*-hydroxybenzoate are shown in Table 1. The three enzymes specific to the catechol branch (catechol oxygenase, muconate lactonizing enzyme and muconolactone isomerase) are undetectable in succinate-grown cells. The levels found in benzoate-grown cells represent increases over the basal level of at least several hundred-fold. All three enzymes are detectable in *p*-hydroxybenzoate-grown cells, but the specific activities are extremely low, in every case at least 30 times less than in benzoate-grown cells. The three analogous enzymes of the protocatechuate branch (protocatechuate oxygenase, carboxymuconate lactonizing enzyme and carboxymuconolactone decarboxylase) are just detectable in succinate-grown cells. The levels in *p*-hydroxybenzoate-grown cells represent increases over the basal levels of approximately 50-fold; the levels found in benzoate-grown cells probably do not represent significant increases over the basal levels. Enol-lactone hydrolase and transferase, the enzymes that mediate the first two common steps of the pathway, are present at detectable but low levels in succinate-grown cells; their specific activities increase about 50-fold over basal in *p*-hydroxybenzoate-grown cells, and slightly more in benzoate-grown cells. It is accordingly evident that the patterns of enzymatic activity induced by benzoate and *p*-hydroxybenzoate in *M. calcoacetica* show a fairly high degree of physiological specificity. The slight elevation of activities characteristic of the enzymes of the protocatechuate branch in benzoate-grown cells does not necessarily indicate non-specific induction; it may reflect non-specific catalysis by the analogous enzymes of the catechol branch. Since these enzymes have not yet been purified from *M. calcoacetica*, their absolute catalytic specificities are unknown.

Table 1. *The effect of different growth substrates on the specific activities of the enzymes of the β-ketoadipate pathway in cell-free extracts of* Moraxella calcoacetica *strain 73*

Region of pathway	Enzyme	Substrate used for growth			
		Succinate (20 mM)	Benzoate (10 mM)	*p*-Hydroxy-benzoate (10 mM)	Succinate (20 mM) + protocatechuate (2 mM)
		units/mg protein in cell-free extracts			
Catechol branch	Catechol oxygenase	< 0.002	0.70	0.004	—
	Muconate lactonizing enzyme	< 0.002	1.35	0.025	—
	Muconolactone isomerase	< 0.010	3.10	0.090	—
Protocatechuate branch	*p*-OH benzoate hydroxylase	< 0.001	—	0.07	0.003
	Protocatechuate oxygenase	0.005	0.007	0.33	0.29
	Carboxymuconate lactonizing enzyme	0.04	0.06	2.00	1.90
	Carboxymuconolactone decarboxylase	0.008	0.02	0.45	0.41
Common steps	Enol-lactone hydrolase	0.02	1.85	1.10	1.05
	Transferase	0.03	3.30	1.70	1.59

Also shown in Table 1 are the enzyme levels of wild type cells grown on succinate in the presence of protocatechuate; since protocatechuate is highly toxic for *M. calcoacetica*, it cannot be easily used as a sole carbon source for growth. The enzymatic constitution of these cells closely resembles that of *p*-hydroxybenzoate-grown cells, except for the much lower level of *p*-hydroxybenzoate hydroxylase.

Both basal and induced enzyme levels in *M. calcoacetica* are similar to the levels found in *P. putida* by Ornston [4], with the exception of the high levels of carboxymuconate lactonizing enzyme and carboxymuconolactone decarboxylase which occur in benzoate-grown cells of *P. putida*.

The Extent of Coordinacy

The enzymes of the β-ketoadipate pathway in *M. calcoacetica* are relatively insensitive to catabolite repression by succinate or acetate; however, by growing cells with a high molar ratio of catabolite repressor to inducer, it is possible to achieve a substantial reduction in the specific activities of the enzymes of the β-ketoadipate pathway, and thus to examine the extent of coordinacy. The results of a series of expriments, performed with *p*-hydroxybenzoate or protocatechuate as the inducer-substrate, are shown in Fig. 3. Over a wide range of specific activities, strict proportionality is maintained between the levels of the five enzymes responsible for the conversion of protocatechuate to β-ketoadipyl-CoA (protocatechuate oxygenase, carboxymuconate lactonizing enzyme, carboxymuconolactone decarboxylase, enol-lactone hydrolase and transferase). These five enzymes therefore constitute a coordinately-induced block, which we shall term the "protocatechuate block". The first enzyme of *p*-hydroxybenzoate dissimilation, *p*-hydroxybenzoate hydroxylase, clearly lies outside the coordinate block; as already mentioned, it is not significantly induced by growth in the presence of protocatechuate.

The results of analogous experiments performed with benzoate as the inducer-substrate are shown in Fig. 4. Over a wide range of specific activities, strict proportionality is maintained between the levels of the four enzymes responsible for the conversion of *cis,cis*-muconate to β-ketoadipyl-CoA (muconate lactonizing enzyme, muconolactone isomerase, enol-hydrolase and transferase). These four enzymes therefore constitute a coordinately-induced block, which we shall term the "muconate block". The proportionality does not extend to include catechol oxygenase, the synthesis of which is evidently regulated independently from that of the enzymes of the muconate block. The seeming paradox of assigning enol-lactone hydrolase and transferase to two separate coordinate blocks will be resolved below.

20*

The Multiplicity of Enol-lactone Hydrolase and Transferase Functions

The apparent coordinacy of enol-lactone hydrolase and transferase with enzymes of both the catechol and protocatechuate branches is at first sight incompatible with the specificities of induction demonstrated in Table 1. The only plausible way to reconcile these two sets of observations is to postulate the existence of two sets of isofunctional enzymes mediating the conversion of β-ketoadipate enol-lactone to β-ketoadipyl-CoA, each of which is subject to specific, coordinate regulatory control. If this

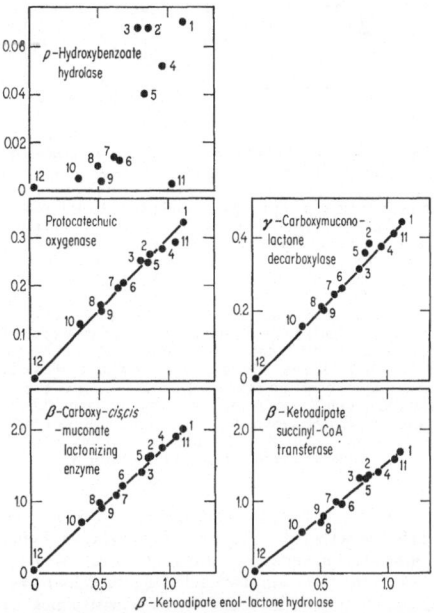

Fig. 3. *Relative specific activities of enzymes in extracts of* M. calcoacetica. The relative specific activities of *p*-hydroxybenzoate hydroxylase, protocatechuate oxygenase, β-carboxy-*cis,cis*-muconate lactonizing enzyme, γ-carboxymuconolactone decarboxylase, β-ketoadipate enol-lactone hydrolase and β-ketoadipate succinyl-CoA transferase in extracts of wild type *M. calcoacetica* prepared from cells which had been grown for at least ten generations in mineral medium containing the following carbon sources: (1) 10 mM *p*-hydroxybenzoate; (2) 5 mM *p*-hydroxybenzoate and 25 mM succinate; (3) 3 mM *p*-hydroxybenzoate and 30 mM succinate; (4) 5 mM *p*-hydroxybenzoate and 25 mM lactate; (5) 3 mM *p*-hydroxybenzoate and 30 mM lactate; (6) 10 mM *p*-hydroxybenzoate and 20 mM acetate; (7) 5 mM *p*-hydroxybenzoate and 25 mM acetate; (8) 3 mM *p*-hydroxybenzoate and 30 mM acetate; (9) 3 mM *p*-hydroxybenzoate and 15 mM succinate and 15 mM acetate; (10) 2 mM *p*-hydroxybenzoate and 20 mM succinate and 20 mM acetate; (11) 2 mM protocatechuate and 20 mM succinate; (12) 20 mM succinate. Enzymatic activities are expressed as units/mg protein

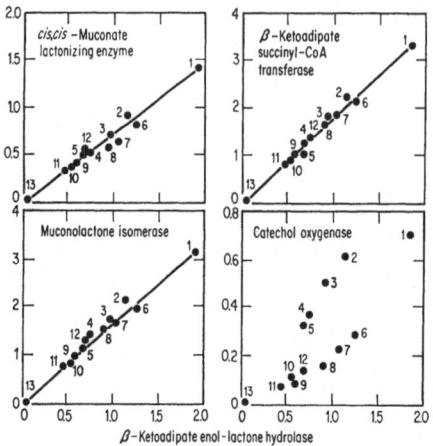

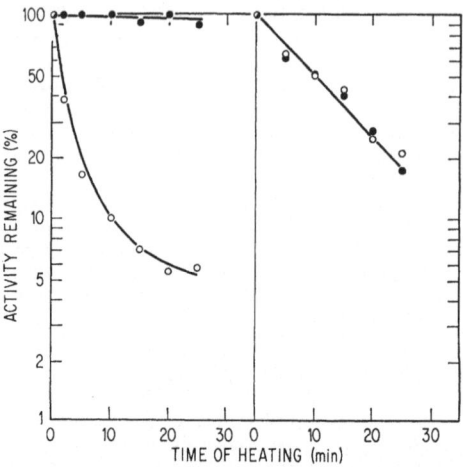

Fig. 4. *Relative specific activities of enzymes in extracts of* M. calcoacetica. The relative specific activities of catechol oxygenase, *cis-cis*-muconate lactonizing enzyme, muconolactone isomerase, β-ketoadipate enol-lactone hydrolase and β-ketoadipate succinyl-CoA-transferase, in extracts of *M. calcoacetica* (wild type unless otherwise indicated) which had been grown for at least ten generations in mineral medium with the following carbon sources: (1) 10 mM benzoate; (2) 5 mM benzoate and 25 mM succinate; (3) 3 mM benzoate and 30 mM succinate; (4) 2 mM benzoate and 20 mM succinate; (5) 1 mM benzoate and 20 mM succinate; (6) 5 mM benzoate and 25 mM lactate; (7) 3 mM benzoate and 30 mM lactate; (8) 5 mM benzoate and 25 mM acetate; (9) 3 mM benzoate and 30 mM acetate; (10) 3 mM benzoate, 15 mM acetate and 15 mM succinate; (11) 2 mM benzoate, 20 mM acetate and 20 mM succinate; (12) strain PM-2 on 10 mM *cis,cis*-muconate; (13) 20 mM succinate. Enzymatic activities are expressed as units/mg protein

Fig. 5. *The kinetics of thermal inactivation of β-ketoadipate enol-lactone hydrolase activity in crude cell-free extracts of* M. calcoacetica *and* P. putida, *grown on benzoate and p-hydroxybenzoate*. On the left, data obtained with extracts of *M. calcoacetica*, heated at 47°; on the right, data obtained with extracts of *P. putida*, heated at 49°; (●), extracts of cells grown on *p*-hydroxybenzoate; (○), extracts of cells grown on benzoate

hypothesis is correct, the enol-lactone hydrolase and transferase activities of benzoate-induced cells should be mediated by enzymes different from those responsible for the same two activities in *p*-hydroxybenzoate-induced cells. A simple and sensitive method to differentiate between two isofunctional enzymes is to demonstrate a difference between them with respect to the kinetics of their thermal inactivation. This test was accordingly performed on crude extracts of *M. calcoacetica* grown on benzoate and on *p*-hydroxybenzoate. As a control, parallel experiments were conducted with crude extracts of *Pseudomonas putida* grown on the same two aromatic substrates. Ornston [4] had established that *P. putida* synthesizes only one enol-lactone hydrolase, so that the kinetics of its thermal inactivation should be unaffected by the aromatic substrate on which the cells are grown. Figs. 5 and 6 show the results of these experiments. It is evident that the kinetics of thermal inactivation of both enol-lactone hydrolase and transferase activities of *M. calcoacetica* are different in extracts from cells grown on benzoate and on *p*-hydroxybenzoate.

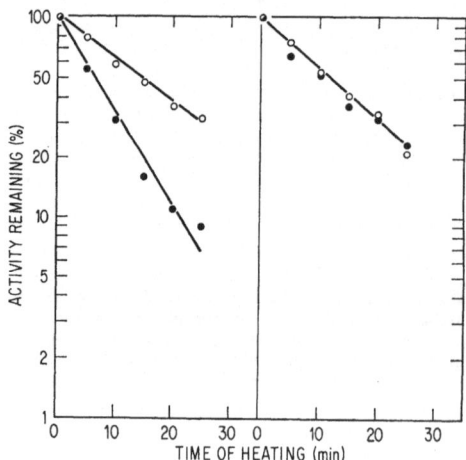

Fig. 6. *The kinetics of thermal inactivation of β-ketoadipate succinyl-CoA transferase activity in crude cell-free extracts of* M. calcoacetica *and* P. putida *grown on benzoate and on* p-hydroxybenzoate. On the left, data obtained with extracts of *M. calcoacetica*, heated at 50°; on the right, data obtained with extracts of *P. putida*, heated at 39°; (●), extracts of cells grown on *p*-hydroxybenzoate; (○), extracts of cells grown on benzoate

This indicates that two sets of enzymes able to mediate the reactions in question can be synthesized by *M. calcoacetica*; one set is induced by growth on benzoate, the other by growth on *p*-hydroxybenzoate. In *P. putida*, on the other hand, the kinetics of thermal inactivation of both enol-lactone hydrolyse and transferase activities are identical in extracts from cells grown on benzoate and on *p*-hydroxybenzoate, indicating that a single enzyme mediates each of these reactions, a fact already established by Ornston [4] for the enol-lactone hydrolase of this species.

Genetic Evidence for Two Transferases Operative in the β-Ketoadipate Pathway

The principal evidence adduced by Ornston [4] to show that *P. putida* can synthesize only one enzyme with enol-lactone hydrolase activity was the demonstration that mutants blocked in the synthesis of this enzyme lose simultaneously the ability to grow on benzoate and on *p*-hydroxybenzoate; and that reversion of such mutants restores both growth functions. Among the nitrosoguanidine-induced mutants of *M. calcoacetica* with blocks in the β-ketoadipate pathway, we isolated three which proved, on subsequent enzymatic analysis, to be defective with respect to transferase function. Two of these, 9B and 2F, were selected for inability to grow on benzoate, but could still grow normally on *p*-hydroxybenzoate. When these mutants are induced with benzoate during growth on succinate, they synthesize high levels of all the enzymes operative in catechol dissimilation, except transferase. When they are grown on *p*-hydroxybenzoate, however, they synthesize normal levels of transferase (Table 2). Mutant 9C was selected for inability to grow on *p*-hydroxybenzoate, but could still grow normally on benzoate. Table 3 presents the data which show that this mutant has lost the ability to synthesize the transferase operative in the dissimilation of *p*-hydroxybenzoate, but can still synthesize the transferase operative in benzoate dissimilation.

Additional Physical Evidence for Two Enol-lactone Hydrolases

Mutants with specific derangements of the structural genes governing enol-lactone hydrolase function have not yet been isolated. However, independent evidence for the existence of two different enzymes with hydrolase activity has been obtained by the study of the migration of enol-lactone hydrolase activity from benzoate- and *p*-hydroxybenzoate-grown cells on columns of Sephadex G-200. The activity in extracts benzoate-grown cells migrates much

Table 2. *Specific activities of certain enzymes in extracts of strains 9B and 2F, mutants unable to grow in benzoate, compared with those in extracts of the wild type of* Moraxella calcoacetica *grown under the same conditions*

Enzyme	Conditions of growth									
	Succinate (20 mM) + benzoate (2 mM)					*p*-Hydroxybenzoate (10 mM)				
	Specific activities			Percentage of wild type activitiy		Specific activities			Percentage of wild type activity	
	Wild type	9B	2F	9B	2F	Wild type	9B	2F	9B	2F
Catechol oxygenase	0.37	0.25	0.18	68	49	—	—	—	—	—
Muconate lactonizing enzyme	0.51	0.35	0.34	69	67	—	—	—	—	—
Muconolactone isomerase	1.45	0.94	1.03	65	71	—	—	—	—	—
Enol-lactone hydrolase	0.75	0.50	0.45	67	60	1.10	0.85	0.92	77	84
Transferase	1.37	0.02	0.02	1.5	1.5	1.60	1.20	1.22	75	76

Table 3. *Specific activities of certain enzymes in extracts of strain 9C, a mutant unable to grow in* p-*hydroxybenzoate, compared with the values obtained for the wild type of* Moraxella calcoacetica *grown under the same conditions*

Enzyme	Conditions of growth					
	Succinate (20 mM) + *p*-hydroxybenzoate (2 mM)			Benzoate (10 mM)		
	Specific activities		Percentage of wild type activity	Specific activities		Percentage of wild type activity
	Wild type	Mutant 9C	Mutant 9C	Wild type	mutant 9C	mutant 9C
p-Hydroxybenzoate hydroxylase	0.034	0.014	41	—	—	—
Protocatechuate oxygenase	0.31	0.18	58	—	—	—
Carboxymuconate lactonizing enzyme	1.80	2.42	134	—	—	—
Carboxymuconolactone decarboxylase	0.42	0.25	59	—	—	—
Enol-lactone hydrolase	1.00	1.14	114	1.85	1.60	86
Transferase	1.30	0.006	0.5	3.30	2.70	82

more rapidly than that in extracts of *p*-hydroxy-benzoate-grown cells, and therefore appears to be attributable to a protein of considerably higher molecular weight (Fig. 7). From the data of Fig. 7, the estimated molecular weight of the hydrolase of *p*-hydroxybenzoate-grown cells is approximately 18,500; and that of the hydrolase of benzoate-grown cells, approximately 82,500.

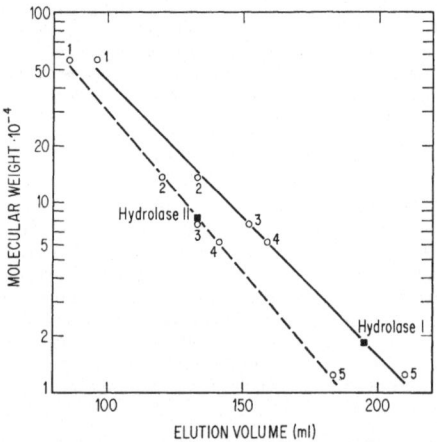

Fig. 7. *The migration of the two enol-lactone hydrolases of* M. calcoacetica *on columns of Sephadex G-200.* A crude extract of cells grown either with *p*-hydroxybenzoate (solid line) or with benzoate (dashed line) was mixed with five marker proteins of known molecular weight, and applied to the column. The points show the elution volumes at which the peak activity of each protein emerged from the column, plotted against the logarithms of the molecular weights of the following marker proteins: (1) ferritin; (2) lactic dehydrogenase; (3) alkaline phosphatase; (4) malic dehydrogenase; (5) cytochrome *c*

Several kinds of evidence accordingly demonstrate that *M. calcoacetica* can synthesize two sets of isofunctional enzymes, subject to different regulatory control, for the catalysis of the two step reactions of the β-ketoadipate pathway which follow the point of metabolic convergence. We shall designate the two enzymes belonging to the protocatechuate coordinate block as enol-lactone hydrolase I and transferase I, and the two enzymes belonging to the muconate block as enol-lactone hydrolase II and transferase II.

Demonstration of a Third Transferase in M. calcoactica

Pseudomonas aeruginosa and *P. multivorans* can use adipate as a carbon source; and growth on this substrate elicits synthesis of the coordinate block consisting of carboxymuconate lactonizing enzyme, carboxymuconolactone isomerase and enol-lactone

hydrolase [4]. Ornston [4] suggested that the physiologically non-specific induction of this coordinate block might be attributed to the formation of β-ketoadipyl-CoA as an intermediate in the dissimilation of adipate. The wild type strain of *M. calcoacetica* used in our work could grow on adipate, so that the influence of this growth substrate on the activities of enzymes of the β-ketoadipate pathway could be examined (Table 4). Growth on adipate does not induce

Table 4. *Specific activities of four enzymes of the β-ketoadipate pathway in extracts of wild type cells of* Moraxella calcoacetica *strain 73 grown on succinate, adipate, benzoate and* p-*hydroxybenzoate*

Enzyme	Substrate used for growth			
	Succinate	Adipate	Benzoate	*p*-Hydroxy-benzoate
	units/mg protein			
Carboxymuconate lactonizing enzyme	0.04	0.03	0.06	2.00
Carboxymuconolac-tone decarboxylase	0.008	0.01	0.02	0.45
Enol-lactone hydrolase	0.02	0.03	1.85	1.10
Transferase	0.03	0.35	3.30	1.70

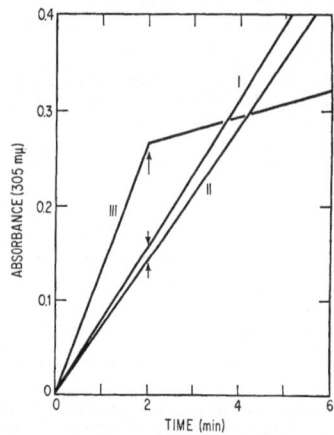

Fig. 8. *The effect of adipate on the synthesis of β-ketoadipyl-CoA by the transferases present in cell-free extracts of* M. calcoacetica *grown on benzoate (I), on* p-*hydroxybenzoate (II) and on adipate (III).* In each case, the reaction was initiated by addition of 10 μmoles of β-ketoadipate, and 1 μmole of adipate was added at the point indicated by arrows

measurable enol-lactone hydrolase activity, but increases transferase activity some ten-fold over the basal level characteristic of succinate-grown cells. Since both transferase I and transferase II are induced coordinately with enol-lactone hydrolases, the transferase activity characteristic of adipate-grown cells is evidently not attributable to either of them. We shall designate this third enzyme showing transferase activity as transferase III.

A striking functional difference between transferase III and both transferases I and II is its extreme susceptibility to inhibition by adipate (Fig. 8). This inhibition is competitive; the K_i for adipate is approximately 5.5×10^{-5}, whereas the K_m for β-ketoadipate is some two orders of magnitude higher (approximately 4×10^{-3}). Transferase III is

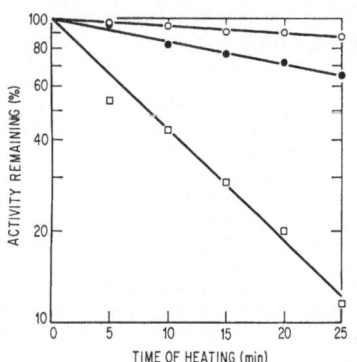

Fig. 9. *Kinetics of thermal inactivation of transferase activities at $45°$ in extracts of* M. calcoacetica *grown on p-hydroxybenzoate ($\bullet$), on benzoate ($\bigcirc$) and on adipate ($\square$)*

therefore probably an inducible adipate succinyl-CoA transferase, which has a low affinity for β-ketoadipate, and can also catalyze a non-specific transfer of CoA to this compound. We have not, however, demonstrated its putative primary function, since the assay method employed to measure transferase activity depends on the measurement of light absorption at 305 mμ, caused by absorbance of β-ketoadipyl-CoA at this wavelength, and cannot be used to detect the formation of adipyl-CoA. The formation of β-ketoadipyl-CoA by transferases I and II is wholly unaffected by the presence of adipate (Fig. 8); the

catalytic site of these two enzymes thus appears to have little or no affinity for adipate. Transferase III can also be differentiated from transferases I and II by its considerably greater thermolability (Fig. 9).

The Properties of Permeability Mutants

Intact bacteria are relatively or wholly impermeable to the non-aromatic intermediates of the β-ketoadipate pathway; but as Ornston [4] showed for *Pseudomonas putida*, the barrier can be in part surmounted by mutation. In *P. putida*, spontaneous mutants with increased permeability to *cis,cis*-muconate can be readily selected by their ability to grow on *cis,cis*-muconate. Analogous mutants of *M. calcoacetica* were obtained by the same method of selection. The levels of enzymes of the catechol pathway in one of these mutants, strain PM-2, after growth on succinate, on *cis,cis*-muconate and on benzoate are compared in Table 5. The enzymatic constitution of strain PM-2 after growth on succinate and on benzoate is virtually identical with that of the wild type grown on the same substrates. Growth of strain PM-2 on *cis,cis*-muconate elicits greatly increased levels of all the enzymes of the catechol pathway; however, the level of catechol oxygenase is only 20% of that in benzoate-grown cells and the levels of the enzymes of the muconate block, only 40%. This probably reflects the fact that the mutant is still not fully permeable to *cis,cis*-muconate; consequently, when *cis,cis*-muconate is furnished as an exogenous substrate, it never achieves an intracellular concentration that is saturating for induction. Since the oxygenative cleavage of catechol to *cis,cis*-muconate is an essentially irreversible reaction, the synthesis of catechol oxygenase during growth on *cis,cis*-muconate indicates that this enzyme is product-induced, as is also true for the catechol oxygenase of *P. putida* [4].

The effectiveness of *cis,cis*-muconate as an inducer of catechol oxygenase is confirmed by respirometric data. As shown in Fig. 10, the rate of oxidation of catechol by cells of strain PM-2 grown on *cis,cis*-muconate is almost identical with its rate of oxidation

Table 5. *Enzyme levels in cell-free extracts of a permeability mutant, strain PM-2, after growth on benzoate, cis-cis-muconate and succinate*

| Enzyme | Substrates used for growth | | | | |
| | Succinate | Benzoate | | cis,cis-Muconate | |
	Specific activity	Specific activity	Percent of Activity in benzoate-grown wild type	Specific activity	Percent of Activity in benzoate-grown wild type
Catechol oxygenase	<0.002	0.70	100	0.13	19
Muconate lactonizing enzyme	<0.002	1.20	89	0.56	42
Muconolactone isomerase	<0.010	3.10	100	1.36	43
Enol-lactone hydrolase	0.030	1.80	97	0.71	38
Transferase	0.050	3.30	100	1.28	39

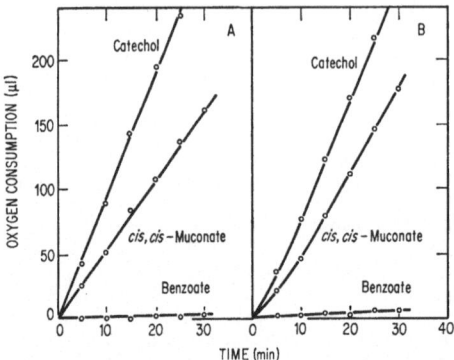

Fig. 10. *The rates of oxidation of benzoate, catechol and* cis,cis-*muconate by cells of strain PM-2 grown on 10 mM* cis,cis-*muconate (A) and on a mixture of 20 mM succinate and 2 mM catechol (B).* Cells were washed with 0.33 M disodium potassium phosphate buffer pH 6.8, and suspended in the same buffer to a density of 150 Klett units. Respirometer flasks contained 2 ml cell suspension, 0.2 ml of 1 M NaOH in the center wall and 5 μmoles of substrate in 0.5 ml of the indicated buffer in the side arm. Substrate was added at zero min and the rate of oxygen consumption followed. Values were corrected to 20°, 760 mm Hg and the endogenous respiration (0.2—0.6 μl/min) was subtracted

Table 6. *Enzyme levels in cell-free extracts of mutant 2J and of the wild type of* Moraxella calcoacetica, *grown on 20 mM succinate + 2 mM benzoate*

Enzyme	Wild type	Mutant 2J	
	Specific activity	Specific activity	Percentage of specific activity of wild type
Catechol oxygenase	0.37	0.10	27
Muconate lactonizing enzyme	0.51	< 0.0001	< 0.02
Muconolactone isomerase	1.45	0.87	60
Enol-lactone hydrolase	0.75	0.40	54
Transferase	1.37	0.79	58

by cells of the same strain induced with catechol (2 mM) during growth on succinate (20 mM). Fig. 10 also shows that cells grown with either catechol or cis,cis-muconate are incapable of oxidizing benzoate. Accordingly, the enzyme system which converts benzoate to catechol (and for which a satisfactory assay in cell-free extracts does not yet exist) appears to be substrate-induced, as it is also in *P. putida*[16].

Inductive Patterns in a Mutant Unable to Synthesize Muconate Lactonizing Enzyme

Strain 2J is a nitrosoguanidine-induced mutant, derived from the wild type, which is unable to grow on benzoate. When grown on succinate in the presence of benzoate, it synthesizes high levels of catechol oxygenase, muconolactone isomerase, enol-lactone hydrolase and transferase, but does not contain a detectable level of muconate lactonizing enzyme (Table 6). As a consequence of the absence of muconate lactonizing enzyme, it cannot synthesize (+)-muconolactone or any of the later intermediates of the β-ketoadipate pathway from benzoate. The observed induction of the three remaining enzymes of the muconate block therefore identifies cis,cis-muconate itself, rather than a later metabolic intermediate derived from it, as the direct inducer of this coordinate block.

DISCUSSION

The regulatory mechanisms governing synthesis of the enzymes of the β-ketoadipate pathway in *Moraxella calcoacetica* are schematized in Fig. 11. Although the two branches of the β-ketoadipate pathway are metabolically convergent, they do not converge enzymologically in *M. calcoacetica*. The common step-reactions are catalyzed by two sets of isofunctional enzymes, each set being linked in terms of biosynthetic control with earlier enzymes specific either to the catechol or to the protocatechuate branch.

A single inducer, cis,cis-muconate, elicits synthesis of all the enzymes operative in the conversion of catechol to β-ketoadipyl-CoA. This compound induces coordinately the four enzymes required for its own conversion to β-ketoadipyl-CoA; and also acts independently as a product-inducer of catechol oxygenase. Our data do not exclude the possibility that catechol may also be able to induce the synthesis of catechol oxygenase, but in view of its steric differences from cis,cis-muconate, such a dual inductive control does not seem likely.

The five enzymes operative in the conversion of protocatechuate to β-ketoadipyl-CoA are coordinately synthesized; and in Fig. 11, we indicate the primary substrate, protocatechuate, as the probable inducer. The data presented in this paper do not rigorously eliminate an inductive function for either of the two subsequent intermediates specific to the protocatechuate branch of the pathway, β-carboxy-cis,cis-muconate and γ-carboxymuconolactone. However, the properties of certain constitutive mutants, which will be described in a subsequent paper [17], show that the inductive effects of protocatechuate cannot be attributed to its conversion to either of these intermediates.

p-Hydroxybenzoate hydroxylase, the enzyme which converts p-hydroxybenzoate to protocatechuate, is not synthesized coordinately with the enzymes of the protocatechuate block, and is not induced by protocatechuate; it accordingly appears to be substrate-induced. The enzyme system which

Probable Inducers	Enzymes	Biochemical Pathways	Enzymes	Probable Inducers

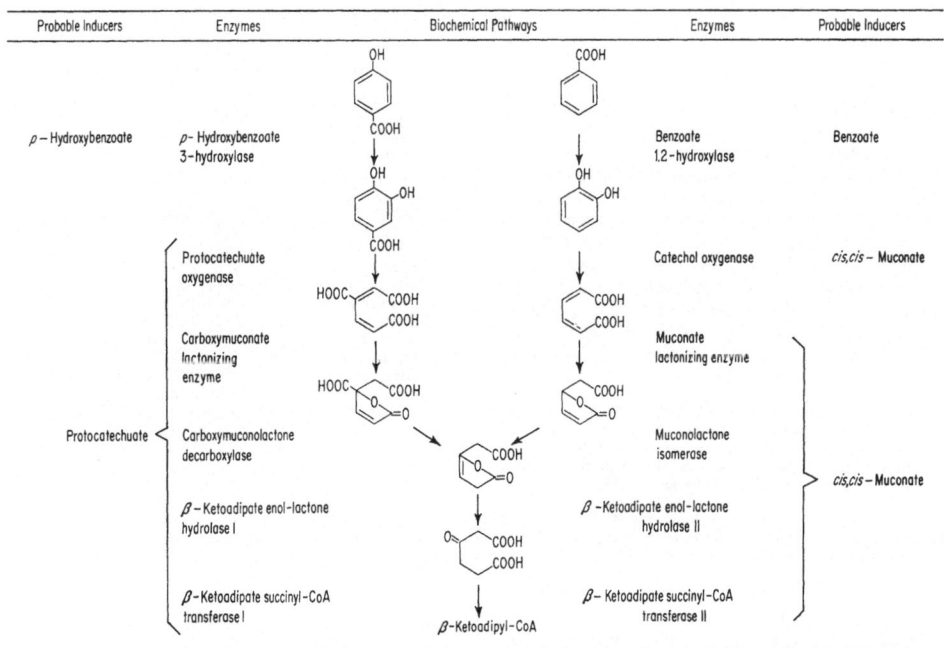

Fig. 11. *The regulation of synthesis of the enzymes catalyzing the reactions of the β-ketoadipate pathway in* Moraxella calcoacetica

converts benzoate to catechol likewise appears to be substrate-induced.

None of the later common intermediates of the β-ketoadipate pathway (β-ketoadipate enol-lactone, β-ketoadipate, β-ketoadipyl-CoA) can be ascribed an inductive function in *Moraxella calcoacetica*. If any of them were able to induce synthesis of the enzymes belonging either to the muconate or to the protocatechuate coordinate blocks, the observed physiological specificity of induction of the enzymes governing the two branches of the pathway could not exist. In a subsequent paper [17] we shall present direct evidence that β-ketoadipate is devoid of inductive function for this bacterium.

It is of some interest to compare the details of the regulatory control over the enzymes of this pathway in *M. calcoacetica* and *P. putida* (see Fig. 2). The individual step-reactions of the catechol and protocatechuate branches are chemically homologous. As Ornston [2,3] has shown, the enzymes of *P. putida* that mediate the specific homologous reactions in each branch, although catalytically specific, resemble each other in a number of properties. However, these metabolic parallelisms between the two branches of the pathway are not mirrored in parallel regulatory mechanisms. Both in *P. putida* and in *M. calcoacetica*,

protocatechuate is an inducer, whereas catechol is probably not; and *cis,cis*-muconate is an inducer, whereas its carboxylated homolog in the protocatechuate branch is not.

The two species differ most significantly with respect to the extent of coordinate control, which is much greater in *M. calcoacetica*. As a direct consequence of this difference, the synthesis by the two species of enzymes operative following the point of metabolic convergence is governed in entirely different fashions. As Ornston [4] has shown, both *P. putida* and other *Pseudomonas* spp. synthesize enol-lactone hydrolase, the enzyme which mediates the first common step-reaction, coordinately with two enzymes specific to the protocatechuate pathway. Since the inducer of this coordinate block is a metabolic product (β-ketoadipate or its CoA derivative), two enzymes specific to the protocatechuate branch are fully induced when cells are grown on precursors of the catechol branch. In discussing this regulatory linkage, Ornston pointed out that a sequential induction at the level of enol-lactone hydrolase would have obviated its physiologically awkward consequences. Following other points of metabolic convergence in the complex dissimilatory pathways characteristic of pseudomonads, sequential inductions (defined as shifts in the chemical nature of the inducer) do in fact occur, and make possible a strict specificity of inductive response to metabolic precursors in the various convergent branches. The absence of this effective regulatory device at a major point of convergence in the β-ketoadipate pathway was, therefore, somewhat unexpected.

In *M. calcoacetica*, a sequential induction at this point of metabolic convergence is again absent. The regulatory problem in *M. calcoacetica* is solved by the device of isofunctional enzymes under different regulatory control, each synthesized coordinately with a block of enzymes specific to one of the two branches, and induced by a metabolic precursor specific to the branch in question. By this means, specific inductive responses are achieved, but at the price of genetic redundancy.

Dr. L. N. Ornston very kindly provided certain of the purified enzymes necessary for the conduct of assays and the preparation of substrates. We are also grateful to the late Miss Jeannette Aagaard for assistance in the isolation of mutants, and to Mr. Mark Wheelis for help in the performance of experiments.

This work was supported by Public Health Service grant AI-1808 from the National Institute of Allergy and Infectious Diseases, and by National Science Foundation grant G-11330.

REFERENCES

1. Ornston, L. N., and Stanier, R. Y., *J. Biol. Chem.* 241 (1966) 3776.
2. Ornston, L. N., *J. Biol. Chem.* 241 (1966) 3787.
3. Ornston, L. N., *J. Biol. Chem.* 241 (1966) 3795.
4. Ornston, L. N., *J. Biol. Chem.* 241 (1966) 3800.
5. Stanier, R. Y., and Ingraham, J. L., *J. Biol. Chem.* 210 (1954) 709.
6. Hosokawa, K., and Stanier, R. Y., *J. Biol. Chem.* 241 (1966) 2453.
7. Katagiri, M., and Hayaishi, O., *J. Biol. Chem.* 226 (1957) 439.
8. Garen, A., and Levinthal, C., *Biochim. Biophys. Acta*, 38 (1960) 470.
9. Meister, A., *J. Biol. Chem.* 184 (1950) 117.
10. Ochoa, S., in *Methods in Enzymology* (edited by S. P. Colowick and N. O. Kaplan), Academic Press, New York 1955, Vol. I, p. 739.
11. Cánovas, J. L., and Kornberg, H. L., *Proc. Roy. Soc. (London) Ser. B*, 165 (1966) 189.
12. Lowry, O. H., Rosebrough, N. J., Farr, A. L., and Randall, R. J., *J. Biol. Chem.* 193 (1951) 265.
13. Elridge, J. A., Linstead, R. P., Orkin, B. A., Sims, P., Baer, H., and Pattison, D., *J. Chem. Soc.* 2228 (1950).
14. Stadtman, E. R., in *Methods in Enzymology* (edited by Academic Press), New York 1957, Vol. III, p. 931.
15. Whitaker, J. R., *Anal. Chem.* 35 (1963) 1950.
16. Sleeper, B. P., and Stanier, R. Y., *J. Bacteriol.* 59 (1950) 117.
17. Cánovas, J. L., in preparation.

J. L. Cánovas' permanent address:
Instituto de Biologia Celular,
C.S.I.C., Madrid, Spain.

R. Y. Stanier
Department of Bacteriology and Immunology
University of California
Berkeley, California 94720, U.S.A.

European J. Biochem. 1 (1967) 301—311

The Response of the Respiratory Chain and Adenine Nucleotide System to Oxidative Phosphorylation in Yeast Mitochondria

T. Onishi, A. Kröger, H. W. Heldt, E. Pfaff, and M. Klingenberg

Physiologisch-chemisches Institut der Universität Marburg

(Received November 14, 1966)

Properties and functions of the respiratory chain and of the nucleotide system were studied in mitochondria from *Saccharomyces carlsbergensis* and compared with those in mitochondria from mammalian tissues.

The composition of the respiratory chain showed a comparatively high content of DPN, ubiquinone, and cytochrome c.

The failure of DPN to become reduced in the controlled state with any substrate confirms the absence of the first phosphorylation site. The redox states of the other components are controlled by the last two phosphorylation sites as in mammalian mitochondria.

Substrates such as lactate, which feed hydrogen into the chain via cytochrome c, reduce ubiquinone in an energy-dependent reverse electron transport.

The mitochondrial content of nucleotides and of the ^{32}P incorporating components was analysed. The mitochondria were found to have a similar equipment of nucleotides etc. as mammalian mitochondria.

The phosphate transfer system in mitochondria was investigated. The phosphorylation rate of exogenous ADP is even at lower temperatures greater than that of endogenous ADP, in contrast to mammalian mitochondria.

A specific exchange between exogenous and endogenous adenine nucleotides is found with similar properties as the exchange in mammalian mitochondria.

Recently a preparation of mitochondria from *Saccharomyces carlsbergensis* has been described which exhibits an essential criterion for functional intactness, that is, the control of respiration by oxidative phosphorylation. These mitochondria were found to have some properties strikingly different from those of mammalian mitochondria: There is rapid oxidation of external DPNH coupled to oxidative phosphorylation and a lack of phosphorylation at site 1. The implication of these major findings and further properties of the mitochondria from yeast have been described in two publications [1,2].

In this paper, studies on the redox function of respiratory components in mitochondria from *S. carlsbergensis* under the influence of the oxidative phos-

phorylation are reported. Particular attention is focused on the two hydrogen carrier systems, ubiquinone and DPN. Furthermore, the function of the mitochondrial adenine nucleotide system in the phosphate transfer is investigated.

METHODS

Preparations

Cultivation of Yeast Cells. *S. carlsbergensis* cells were cultivated aerobically in a medium which contained lactate as a carbon source, as described previously [1]. In the present study, the culture medium was aerated by pumping air rapidly through large surface glass filters in bottles containing 8 liter medium. The air kept the culture in strong agitation and thus fully aerobic. In this system, cells in the logarithmic growth phase were obtained after 12 to 13 hours.

Preparation of the Protoplasts. Protoplasts were prepared according to Onishi et al. [1].

Preparation of Mitochondria. Mitochondria were prepared in essentially the same way as described [1]. Alterations: the concentration of EDTA in the preparation medium was raised to 1 mM in the present study. The composition of the standard incubation

---

Non-standard Abbreviations. Ubiquinone, UQ; reduced ubiquinone, UQ_{red}; oxidized ubiquinone, UQ_{ox}; N,N,N',N'-tetramethyl-P-phenylenediamine, TMPD; p-trifluoromethoxy-carbonyl-cyanide-phenylhydrazone, FCCP; triethanolamine, TRA; glucose-1-phosphate, G1P; adenosine tetraphosphate, APPPP; adenine nucleotides, AdN; complex of cytochrome a and a_3 (molar ratio 1:1), aa_3 (according to [22]).

Enzymes. Lactate dehydrogenase, or L-lactate:ferricytochrome c oxidoreductase (EC 1.1.2.3); D-lactate dehydrogenase, or D-lactate:ferricytochrome c oxidoreductase (EC 1.1.2.4); cytochrome oxidase, or ferrocytochrome c: oxygen oxidoreductase (EC 1.9.3.1).

medium is 0.65 M manitol, 20 mM Tris-maleate buffer (pH 6.5), 5 mM K-phosphate (pH 6.5), 10 mM KCl and 0.5 mM EDTA.

Analytical Procedures

Determination of UQ and DPN. The content and the redox states of UQ and DPN were determined in the same extract according to the method of Kröger and Klingenberg [3].

Determination of Adenine Nucleotides. The adenine nucleotide content in the mitochondria was assayed by an enzymatic analysis according to Hohorst et al. [4] and Adam [5]. The experimental methods for measuring the exchange reaction, the phosphorylation of the endogenous and exogenous adenine nucleotides, and the chromatographic assays are described in the legends of the figures and tables.

Determination of Protein. The concentration of mitochondrial protein was determined according to the Biuret method using KCN [3].

RESULTS

Content of Respiratory Components

In general the content of respiratory components in mitochondria from *S. carlsbergensis* resembles that of mitochondria from heart, when based on protein.

Table 1. *Content of respiratory components in mitochondria from* S. carlsbergensis
Cytochrome content was obtained from difference spectra: the sample cuvette containing mitochondria in the anaerobic state, and the reference cuvette aerobic mitochondria without substrate. Millimolar absorbance coefficients (E_M) used are: cyt. aa_3 (605—630 mμ), 24 [22]; cyt. ($c + c_1$) (550—540 mμ), 18; cyt. b (564—575), 19; UQ and DPN, TPN content was determined by extraction

Component	Content		Component cyt. aa_3
	μmoles/g protein		
Cyt. aa_3	0.15 ± 0.05	(5)[a]	1.0
Cyt. $c + c_1$	0.65 ± 0.15	(5)	4.3
Cyt. b	0.28 ± 0.11	(5)	1.9
UQ	5.4 ± 1.11	(8)	36.0
DPN	7.2 ± 1.71	(6)	48.0
TPN	0.9	(2)	6.0

[a] Means ± standard deviation (number of mitochondrial preparations).

However, the contents of cytochrome b, and aa_3 is considerably lower. Thus, the relative composition of the respiratory chain (ratios referred to cytochrome aa_3) gives high values for the other components. Whereas the contents of DPN and cytochrome c are comparable to heart mitochondria, the relative contents are increased several fold. The yeast mitochondria are particularly well equipped with ubiquinone with respect to both the protein and the cytochrome a content. It is to be noted that more cytochrome b is measured by the reduction with DPNH

under anaerobic conditions than with various other substrates.

The exceptional depression of the cytochrome a content as compared to the other components is typical for the cells from the logarithmic phase in lactate medium. It appears to be less pronounced in mitochondria isolated from the stationary phase in glucose medium [6].

Steady State Reduction Levels in the Respiratory Chain

The steady state reduction level of the respiratory chain in the controlled state with citrate is illustrated by a difference spectrum (minus fully oxidized state) and is compared with that of the anaerobic state (minus fully oxidized state) (Fig.1).

The α and γ bands of cytochrome c, b, and a are clearly visible indicating that the cytochromes are partially reduced in the controlled state. The absorption with a maximum at 330 mμ may be largely due to DPNH, which by enzymatic analysis was found to be formed to about 8% of the total DPN in the controlled state (Fig.2) and to about 65% in the anaerobic state with citrate. Another part of this absorption band which is connected with a trough at about 400 mμ and which is observed also in other mitochondria [7,8] may be due to the δ-band of cytochrome c.

The degree of reduction in the controlled state, as obtained by the difference spectrum for the cytochromes and by extraction and analysis for UQ and DPN, was quantitatively evaluated in redox patterns for three different substrates (Fig.2). With DPNH, the most active hydrogen donor, a comparatively high degree of reduction was observed which decreased monotonously from UQ to cytochrome aa_3. With an active DPN-linked substrate, such as citrate, a qualitatively similar redox pattern at a lower level of reduction was obtained. In contrast to the other substrates, cytochrome b was not more reduced than cytochrome ($c + c_1$) when D-lactate was the substrate. This may be due to the fact that the hydrogen from lactate enters the respiratory chain at the level of cytochrome c and therefore cytochrome b as well as UQ are reduced by reverse electron transfer. This will be discussed further below. Another component which does not follow the usual descent in the redox pattern is DPN, since it remains largely oxidized with all substrates, as shown in Fig.2. In this respect the redox pattern of mitochondria from *S. carlsbergensis* differs strikingly from animal mitochondria where DPN is usually the most reduced of all components in the controlled state, and a monotonous descent in the redox pattern from DPN to cytochrome aa_3 is observed [3].

The redox changes of the respiratory components during the transition to the various functional states under the control of oxidative phosphorylation are followed in time-dependent recordings. Examples are given in Fig.3 with two substrates, ethanol and

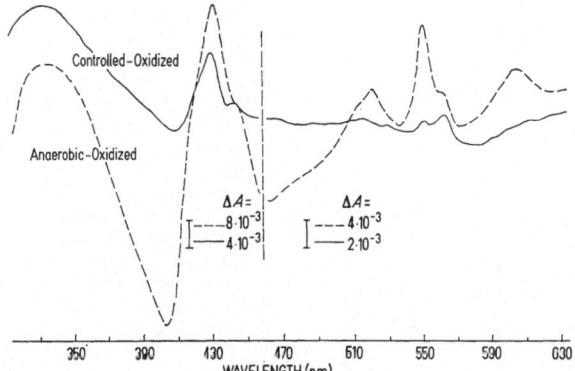

Fig.1. *Difference spectrum of mitochondria from* S. carlsbergensis *in the controlled and anaerobic state with citrate as a substrate.* Solid line: sample cuvette containing aerobic mitochondria in the controlled state with 4 mM citrate; reference cuvette containing aerobic mitochondria without addition. Dashed line: sample cuvette containing anaerobic mitochondria with citrate; reference cuvette containing mitochondria without addition. Optical pathlength = 0.5 cm; 1.7 mg mitochondrial protein/ml; standard incubation medium, oxygen saturated; 16°; composition, cf. Methods

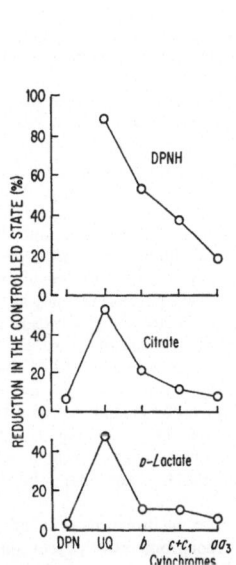

Fig.2. *Redox patterns of the respiratory chain of mitochondria from* S. carlsbergensis *in the controlled states with 3 different substrates.* Mitochondria were incubated in oxygen-saturated standard incubation medium at 16° with addition respectively of 4 mM DPNH, 4 mM citrate, or 4 mM D-lactate. The redox states of the cytochromes were determined spectrophotometrically, those of UQ and DPNH were determined by extraction

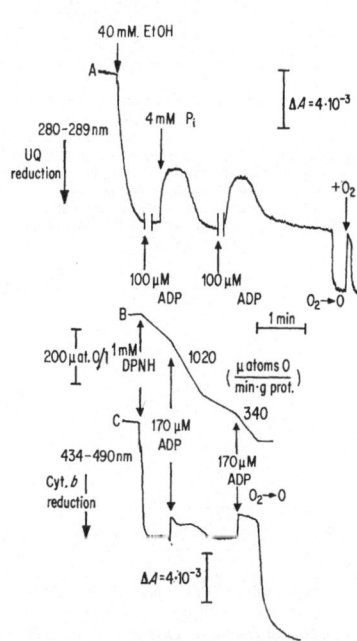

Fig.3. *Redox states of UQ and cytochrome b under the control of oxidative phosphorylation in mitochondria from* S. carlsbergensis. A. Redox changes of UQ during oxidation of ethanol. Optical pathlength = 0.2 cm. Standard medium without phosphate, 15°. 1 mg mitochondrial protein/ml. B. Amperometric recording of oxygen uptake with simultaneous recording of cytochrome b absorption as shown in part C. C. Redox changes of cytochrome b during oxidation of DPNH. Optical pathlength = 0.5 cm; standard incubation medium; 25°; 0.26 mg mitochondrial protein/ml

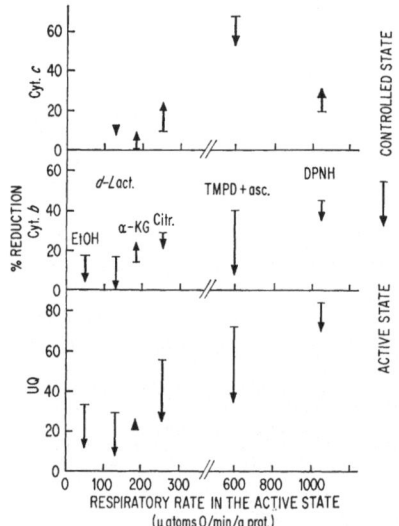

Fig. 4. *Direction of the redox changes of UQ, cytochrome* b *and* c *on the transition controlled → active state with various substrates in mitochondria from* S. carlsbergensis *correlated to the respiratory rate in the active state. The length of the arrows indicates the extent of each redox change. The redox changes of cytochrome* b *and* c *were measured spectrophotometrically, those of UQ by extraction. Standard incubation medium at 20°. Substrates: EtOH = 4 mM ethanol; d-Lact. = 4 mM D-lactate; αKG = 4 mM α-ketoglutarate; TMDP + asc. = 200 μM TMPD + 4 mM ascorbate, 1 mM DPNH*

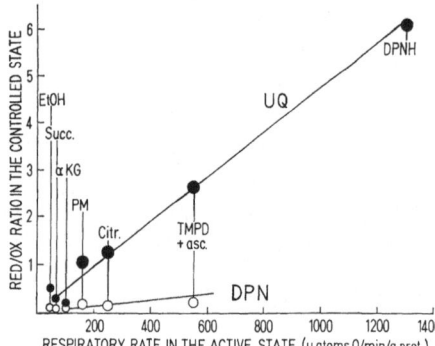

Fig. 5. *Correlation of the redox ratios of UQ and DPN in the controlled states with various substrates to the corresponding respiration rates in the active states. Oxygen uptake was measured in the active state and the same sample was extracted after the controlled state had been established. In each extract UQ and DPN were determined. Incubation medium: 0.25 M sucrose instead of mannitol, 20°. Substrates: EtOH = 4 mM ethanol, Succ. = 4 mM succinate, KG = 4 mM α-ketoglutarate, PM = 4 mM pyruvate + 4 mM malate, Citr. = 4 mM citrate, TMPD + asc. = 200 μM TMPD + 4 mM ascorbate, 1 mM DPNH*

DPNH, which are especially oxidized by yeast mitochondria. On the addition of DPNH or ethanol, UQ and cytochrome b are strongly reduced. Typical oxidation-reduction cycles are observed in the cyclic transition from the controlled to the active state, promoted by the addition of either ADP or P_i. These transitions occur simultaneously with the increase of respiration.

The steady state reduction levels of cytochrome b and c in the controlled and active states are evaluated from this type of recording. The reduction level of ubiquinone is measured by chemical analysis of extracts which were sampled simultaneously. The dependence of the redox states on the respiratory activity in the active state are reported in Fig. 4. The respiratory activity is regarded as a measure of the hydrogen donor activity of the various substrates. The degree of reduction generally increases with the hydrogen donor activity. On transition to the active state UQ and cytochrome b are, in most cases, more oxidized whereas cytochrome c is more reduced.

There are two types of exceptions: With α-ketoglutarate all components become more reduced on the transition to the active state. This might be explained by a predominant control of the respiration by the substrate level phosphorylation, *i.e.* a control occuring before the respiratory chain. With TMPD + ascorbate and D-lactate, however, all components, including cytochrome c, become more oxidized. This is due to the entrance point being at the level of cytochrome c for the electrons from these substrates. Therefore, the control point must be at the third phosphorylation site above cytochrome c. For the other substrates the control site appears to be between cytochrome b and c. With DPNH a comparatively high level of reduction of the components is observed. Upon initiation of the active state, only small changes in the degrees of reduction were seen. Apparently the respiratory chain is so nearly saturated by this extremely active substrate that in the active state the respiration is limited by the activity of the cytochrome oxidase. In contrast, with TPMD + ascorbate the redox changes of UQ and cytochrome b are relatively large since both carriers are reduced by energy-linked reverse electron transfer which is very sensitive to ADP addition [9,10]. In the active state, cytochrome c is more reduced with TMPD + accorbate than with DPNH even though the latter is oxidized about twice as actively. This may be interpreted to indicate that the oxidation of cytochrome c by cytochrome oxidase is partially inhibited in the presence of TMPD + ascorbate.

The redox behaviour of the hydrogen carrier pool UQ was further analysed by plotting its redox ratios in the controlled state, with various substrates, against the activity of these substrates in the active state (Fig. 5). The apparently linear relationship obtained over a wide range of respiratory rates demonstrates that the redox state of UQ is only dependent on the hydrogen donor activity: it is independent of the site of entrance of electrons. Thus the data for citrate (entrance at DPN), added DPNH (entrance at UQ), or TMPD + ascorbate (entrance at cytochrome c) fit the same relation. The fact that the value for α-ketoglutarate does not fit in with this rule can be explained (as discussed above) by a control at the level of substrate phosphorylation. The slope $\alpha = UQ_{red}/UQ_{ox}/(O_2)'$ —see footnote [1]— taken from this plot is $\alpha = 0.05$. This is in agreement with the result obtained from the corresponding plot for heart mitochondria, where the same slope α was found [11]. This is the more striking, since different types of substrates are active in both mitochondria, yeast mitochondria affording a particularily wide range of donor activity and types of substrate.

The dependence of the redox state of UQ on the hydrogen donor activity has been described earlier for mitochondria from rat heart muscle [12]. This relation is considered as a strong support for an energy-linked redox equilibrium regulating the redox states of the respiratory components in the controlled state.

The degree of reduction of DPN, which is also plotted in Fig. 5, remains much smaller than that of UQ for all substrates. In the redox patterns (Fig. 2) it is shown that a maximum of only about 10% of the mitochondrial DPN is reduced in the controlled state. This is in contrast to animal (heart) mitochondria where the redox ratios of DPN in the controlled states are always higher than those of UQ [11]. In summary, the relationship of the redox ratio to the donor activity is similar for UQ but differs for DPN in animal and yeast mitochondria.

The low degree of reduction of DPN in the controlled state suggests the absence of an energy-linked reduction of DPN in yeast mitochondria. On the other hand, the high degree of reduction of UQ is consistent with the presence of an energy-linked reversal of electron transfer in yeast mitochondria. Thus, energy-linked reversed electron transfer exists in the respiratory chain of yeast mitochondria as well, but does not include DPN. This suggests, in agreement with previous data [1,2], that the first phosphorylation site in the respiratory chain is missing in mitochondria from *S. carlsbergensis*.

It may be argued that the reason for the non-reducibility of the DPN in the controlled state is due to a nonparticipation of a large part of the DPN pool in hydrogen transfer. However, the high degree of reduction of DPN under anaerobic conditions (65% of total DPN as found in the presence of citrate) shows that most of the DPN is reducible. Further-

---

more, this reduction is comparatively fast. The fact that DPN is reduced only to a small extent in the controlled state with DPN-linked substrates, but to a high extent in animal mitochondria, further supports the contention that the redox state of DPN in mitochondria is energy-dependent not only with flavin-linked but also with DPN-linked substrates.

Reverse Electron Transfer

Although the reverse electron transfer to DPN cannot be demonstrated in yeast mitochondria, there is a unique opportunity to study reverse electron transfer to UQ and cytochrome b with the natural substrate lactate, from which electrons enter the respiratory chain at the cytochrome c level. The energy-linked reduction of UQ by lactate is demonstrated by a comparison with the non-energy-linked reduction of cytochrome c in Fig. 6. Lactate addition

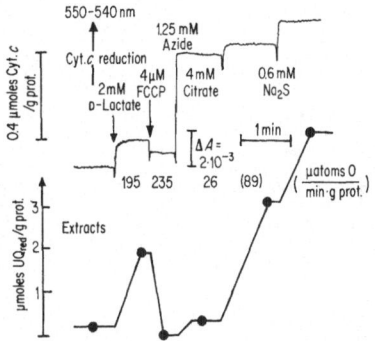

Fig. 6. *Energy-linked reduction of UQ and direct reduction of cytochrome c by D-lactate in mitochondria from S. carlsbergensis.* Recording of cytochrome c absorption and simultaneous extraction and determination of the redox state of UQ. The respiratory rates in the different states are given for control. The value in brackets is the respiratory rate with citrate in the uncoupled state. Optical pathlength = 0.5 cm. 1.28 mg mitochondrial protein/ml. Incubation medium contained 0.25 M sucrose instead of mannitol, 25°. Total amount: 6.5 μmoles UQ/g protein

leads first to the reduction of UQ which is completely reversed by the uncoupler FCCP, whereas the reduction of cytochrome c is considerably less sensitive to the uncoupler. When azide which inhibits the respiration by 90% is added, UQ remains oxidized due to the lack of energy supply, whereas at the same time cytochrome c is reduced to 78%. Only when hydrogen is supplied from a side "below" UQ, as from citrate, can UQ be largely reduced. A not fully blocking concentration of azide has been used. This allows a leakage of electrons to oxygen sufficiently large to keep UQ and cytochrome b oxidized despite the slow supply of hydrogen from pyruvate.

---

[1] $(O_2)'$ = respiratory activity in μatoms oxygen/min/g protein.

The inhibition of reverse electron transfer by antimycin A could not be demonstrated. The strong block exerted by this inhibitor allows the pyruvate

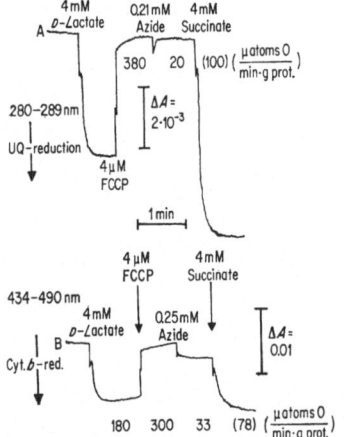

Fig. 7. *Response of UQ and cytochrome* b *in reverse electron transport in mitochondria from* S. carlsbergensis *with* D-*lactate as the substrate.* The respiratory rates in the different states are given for control. The value in brackets is the respiration rate with succinate in the uncoupled state. Standard incubation medium, 20°, 0.67 mg mitochondrial protein/ml. A. Recording of UQ absorption; optical pathlength = 0.2 cm. B. Recording of cytochrome *b* absorption; optical pathlength = 0.5 cm

formed from lactate to reduce UQ and cytochrome *b* in a forward hydrogen transfer in the uncoupled state.

Fig. 7 (A, B) illustrates the energy-linked reduction of UQ and cytochrome *b* by reverse electron transfer with lactate. The reduction of cytochrome *b* in the controlled state is also largely sensitive to uncoupler. When the respiration is inhibited to about 90% by azide, UQ and cytochrome *b* are not reduced until hydrogen is provided by a substrate of a lower potential, such as succinate.

Equipment with Mitochondrial Nucleotides

Fig. 8 shows the ion exchange chromatogram of an acid extract from *S. carlsbergensis* mitochondria incubated for 30 sec at 18° in the presence of [^{32}P]-orthophosphate with TMPD + ascorbate as substrate. The short time of incubation is selected in order to allow the ^{32}P labelling of only those compounds which show rapid ^{32}P incorporation. The peaks of the recordings of ultraviolet absorbancy and ^{32}P activity are named according to the positions these compounds occupy in the chromatogram (system I) [13] of rat liver mitochondria. Table 2 shows the quantitative evaluation of this chromatogram.

Besides adenine and pyridine nucleotides, the mitochondria also contain GTP, UTP and CTP. When compared to the adenine nucleotides, UTP occurs in *S. carlsbergensis* mitochondria in higher amounts than in mammalian mitochondria [13,14]. The unidentified compound X_6 also occurs in mamma-

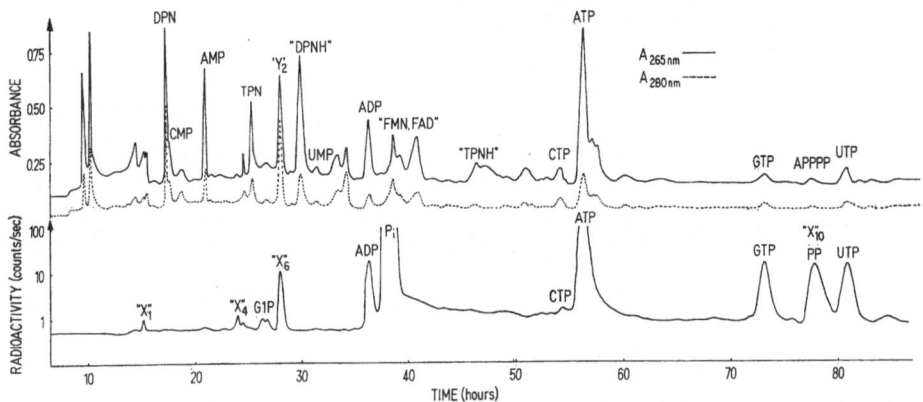

Fig. 8. *Anion exchange chromatogram of mitochondria from* S. carlsbergensis. Yeast mitochondria equivalent to 14.2 mg protein were incubated in the presence of 0.3 M sucrose, 2 mM EDTA, 20 mM TRA, pH 7.2, 3 mM ascorbate, 200 μM TMPD and 0.35 mM. [^{32}P]phosphate (specific activity = 70×10^6 counts/min/μmole) in a total volume of 3.0 ml at 18° for 30 seconds. During incubation oxygen was added by stirring. Because of the high respiration of the suspension, partial anaerobiosis might have occurred at the end, resulting in reduction of DPN. Incubation was started by addition of the mitochondria and stopped by addition of 0.525 ml 3 M perchloric acid. The supernatant acid-soluble extract was neutralized with 2 M KOH. 2.0 ml extract equivalent to 6.60 mg protein were applied to the column. For details of chromatography see Heldt and Klingenberg [13]. Chromatographic system I was employed. The recording was started from the beginning of the elution. Peaks marked with X and Y have not yet been identified

lian mitochondria though in smaller amounts. The unidentified compound X_{10}, which overlaps with pyrophosphate, is found separately between ADP and ATP, when the same extract is applied to another elution gradient (ion exchange chromatography system II [13]). From its position in both chromatographic systems it can be concluded, that X_{10} is not adenosine but might be 2,3-diphosphoglyceric acid. X_{10} is not detected in rat liver mitochondria. The pyrophosphate appearing in the chromatogram may

Table 2. *Nucleotides and other phosphate-containing substances in mitochondria from S. carlsbergensis*
The values were obtained from the chromatogram in Fig.1

Compound	Calculated from 265 mμ absorbance	³²P Incorporated	Adenine nucleotide content in stock suspension of mitochondria [b]
	μmoles/g protein [a]		
X-1		0.0005	
AMP	0.72		2.97 ± 1.78[c]
X-4		0.0042	
TPN	0.37		
G1P		0.0040	
DPNH	1.94		
X-6		0.068	
UMP	0.14		
ADP	0.86	0.118	2.64 ± 0.91[c]
TPNH	0.07		
CTP	0.20	0.0050	
ATP	4.28	2.40	0.58 ± 0.34[c]
GTP	0.26	0.153	
PP$_i$		0.048	
X-10		0.130	
APPPP	0.10		
UTP	0.43	0.137	
AdN	5.86		6.21 ± 2.57[c]

[a] Evaluation of chromatogram in Fig.8.
[b] Adenine nucleotide content was measured by enzymatic assay.
[c] Standard deviation.

be at least partially derived from an impurity in the [³²P]phosphate employed. The high level of DPNH is probably caused by a partial anaerobiosis at the end of the incubation. On the whole, the pattern of nucleotides and ³²P incorporating substances in yeast mitochondria is not very different from those of mammalian mitochondria.

The adenine nucleotide content in 10 different preparations of mitochondria was followed by enzymatic analysis in the absence of substrate respiration (last column of Table 2). It varies considerably in different preparations, as indicated by the standard deviation values, and appears to be highly dependent on the integrity of the mitochondria. The content is similar to that of rat heart mitochondria (about 9 μmoles/g protein) and much lower than that of rat liver mitochondria (about 13 μmoles/g protein [13]).

Function and Compartmentation of Mitochondrial Adenine Nucleotides

Fig.9 shows the time course for the phosphorylation of endogenous and exogenous ADP. The phosphorylation was followed by measuring ³²P uptake.

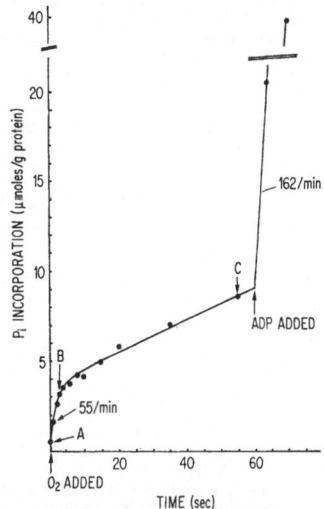

Fig.9. *Phosphorylation of endogenous and exogenous ADP.* Mitochondria (2.21 mg protein/ml), containing 9.3 μmoles of endogenous adenine nucleotides (AMP + ADP + ATP) per g of protein, as measured by enzymatic assay, were kept anaerobic for 6 min at 4° in a medium containing 0.3 M sucrose, 20 mM TRA, pH 7.2, 2 mM EDTA, 200 μM TMPD and 3 mM ascorbate. Total volume 8 ml. 3 minutes after the addition of 1 mM [³²P]phosphate (specific activity, 11 × 10⁶ counts/min/μmole) phosphorylation was started by bubbling oxygen through the suspension while stirring. After another 60 seconds 1 mM ADP was added. Samples of about 0.4 ml each were taken by a specially designed sampling device, and deproteinized with 0.6 ml 1 M $HClO_4$. After neutralization with 3 M KOH the acid soluble extract was analysed for organic phosphate by a modified Nielsen-Lehninger procedure [14]. The samples marked with A, B and C were also analyzed by ion exchange chromatography (see Table 3)

For control, the distribution of the ³²P incorporated was analysed by ion exchange chromatography (Table 3).

Most of the total ³²P uptake is due to the phosphorylation of endogenous ADP. Since the ³²P uptake in GTP and UTP is also likely to have occurred via the formation of endogenous [³²P]ATP, the initial rise of the total ³²P uptake appears to be a good measure of the rate of phosphorylation of endogenous ADP. The rate of the phosphorylation of endogenous ADP at 4° is similar to that measured in rat liver mitochondria [15]. The phosphorylation rate of exogenous ADP is almost three times as fast

Table 3. *Comparison between the assay of ^{32}P incorporation, as measured by extraction method, with the assay of ^{32}P labelled compounds by ion exchange chromatography*
From the experiment shown in Fig. 9, samples (marked A, B and C) were analysed by ion exchange chromatography [5]. The columns, filled with Dowex-1 (X 8) formate, (400 mesh) were 0.8 mm in diameter and 50 cm long. A linear concentration gradient was used. The reservoir contained 4.8 M HCOONH$_4$ + 1.27 M HCOOH, adjusted to pH 4.2, and the mixing chamber contained 10 ml distilled water at the beginning. The flow rate through the columns was 0.575 ml per hour

Compound	P_i incorporated		
	Sample A	Sample B	Sample C
	μmoles/g protein		
ADP	0	0.10	0.13
ATP	0.21	2.28	6.42
GTP	0	0.21	0.63
UTP + X-10	0	0	0.54
PP	0.35	0.38	0.34
$\Sigma\,P_i$ incorpor.	0.56	2.97	8.06
P_i incorpor. measured by extraction method	0.70	3.10	8.50

Table 4. *Measurement of exchangeability by centrifugal filtration*
Mitochondria (4.12 mg protein/ml) were suspended in a medium which contained 0.65 M mannitol, 4 mM K-phosphate (pH 6.5), 20 mM Tris-maleate buffer (pH 6.5), and 0.5 mM EDTA (standard medium). Exchange was measured with the centrifugal filtration method. Incubation conditions were: 450 μM [^{14}C]ADP or [^{14}C]ATP (0.1 μC/ml), 2 min, 4°. The exchange refers to the percentage of the endogenous adenine nucleotides labelled by the incubation

Nucleotide added	Concentration	AdN in mitochondria	[^{14}C]AdN in mitochondria	Exchange
	μM	μmoles/g protein		%
[^{14}C]ATP	469	8.30	7.53	86
[^{14}C]ADP	425	7.42	7.53	100

as that of endogenous ADP even at 4°. In contrast, in mitochondria from rat liver [15] and rat heart [14] exogenous ADP is phosphorylated more slowly than endogenous ADP at 4°.

In order to understand further the role of the endogenous adenine nucleotides in the phosphate transfer, the exchange between exogenous and endogenous adenine nucleotides was studied. The degree to which the mitochondrial adenine nucleotides can be exchanged was examined by the centrifugal filtration technique. As indicated in Table 4, after a 2 min incubation at 4° a complete exchange is observed with [^{14}C]ADP and a nearly complete exchange with [^{14}C]ATP. This result establishes the full exchangeability of the adenine nucleotides and allowed us to use the millipore-filtration method. For this method endogenous adenine nucleotides are first labelled and the exchange is measured against unlabelled

adenine nucleotides. The time course of the exchange with exogenous ADP is shown in Fig. 10. Already after 5 sec 40% of the adenine nucleotides were exchanged in a very fast reaction. The exchange thereafter, however, slows down considerably.

For the measurement of the specificity of the exchange a more extended time scale was applied. As in rat liver mitochondria ADP is exchanged faster than ATP, whereas AMP exchanges at a much slower rate. It was noted that both ADP and ATP are able to exchange 30—40% very quickly, whereas with AMP only a slow exchange was found (Fig. 11).

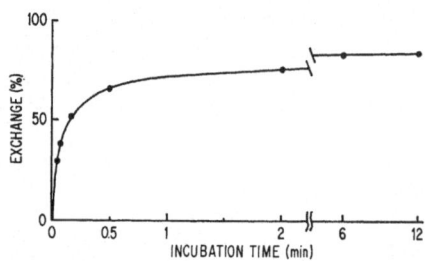

Fig. 10. *Kinetics of the exchange between endogenous and exogenous adenine nucleotides.* The mitochondrial adenine nucleotides were labelled by incubation for 50 min at 4° in a medium which contained 100 μM [^{14}C]ADP (0.5 μC/ml). The unspecifically permeated adenine nucleotides and the nucleotides in the medium were removed by two washes in 0.65 M mannitol, containing 1 mM EDTA (pH 6.3). For the exchange the labelled mitochondria were incubated in a specially designed pressure vessel for millipore filters which allows rapid mixing and filtration [23,24]. The degree of exchange was given as a % decrease of radioactivity in the mitochondria after incubation with unlabelled nucleotides against a control, in order to correct for the leakage of adenine nucleotides during incubation. The incubation conditions were: mitochondria (0.2 mg protein per ml), 400 μM ADP, Tris-maleate (pH 6.5), 4 mM K-phosphate (pH 6.5), and 0.5 mM EDTA, temperature was 8.5°. The total adenine nucleotide content of mitochondria was 7.9 μmoles/g protein

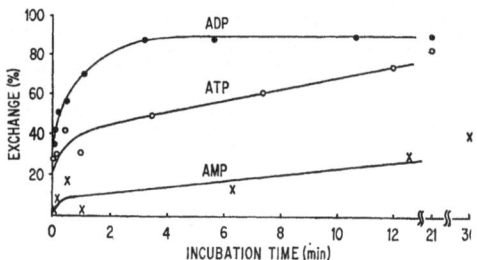

Fig. 11. *Specificity of the exchange towards exogenous ATP, ADP and AMP.* The experimental conditions were the same as described in Fig. 10 except that the mitochondrial concentration was 0.7 mg protein per ml and the temperature was 4°: 400 μM ATP, ADP or AMP were used. The total content of adenine nucleotides in the mitochondria was 3.2 μmoles/g protein

The temperature dependence of the exchange was measured at a short time (2 sec) in order to avoid the saturation region of the exchange curve at higher temperatures (Fig. 12). The temperature dependence

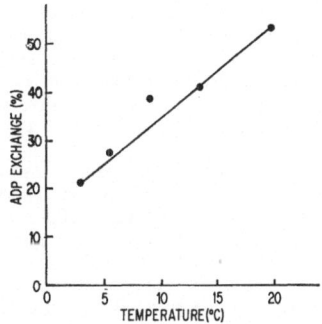

Fig. 12. *Temperature dependency of ADP exchange.* The experimental conditions were the same as described in Fig. 10. The mitochondrial concentration was 0.5 mg of protein per ml. The total adenine nucleotide content of the mitochondria was 10.2 μmoles/g protein

Table 5. *Effect of atractyloside on the exchange reaction* The incubation was carried out at 6°. Other experimental conditions are the same as shown in Fig. 12. Total of adenine nucleotide content of the mitochondria was 10.2 μmoles/g protein

Incuba-tion time	Additions			Exchange	Inhibition
	ADP	ATP	Atracyloside		
sec	μM	μM	μM	%	%
2	400	—	—	32	—
	400	—	200	2.3	93
	—	400	—	32	—
	—	400	200	0	100
10	400	—	—	49	—
	400	—	200	12	76
	—	400	—	39	—
	—	400	200	7.6	81

Table 6. *Effect of atractyloside on the phosphorylation of endogenous and exogenous ADP* See legend Fig. 9 for the methods of incubation and assay. The incubation medium contained mitochondria equal to 1.9 mg protein/ml. 25 μM and 85 μM atractyloside was added to the anaerobic mitochondrial suspension 1 min after the addition of [^{32}P]phosphate. After another 2 minutes the phosphorylation of endogenous adenine nucleotides was started by the addition of oxygen. 1 minute later 0.4 mM ADP was added for measuring the phosphorylation of exogenous ADP

Atractyloside concentration	Phosphorylation of ADP			
	endogenous		exogenous	
μM	μmoles/g protein/min			
0	40	(100%)	146	(100 %)
25	33	(83%)	25	(17 %)
85	31	(77%)	11	(7.5%)

21*

is small as compared to liver mitochondria even if one takes into account that these values do not reflect initial, linear rates.

In earlier experiments on rat liver mitochondria atractyloside was found to be a very effective inhibitor of the adenine nucleotides exchange [16,17]. The exchange in yeast mitochondria is also inhibited by atractyloside (Table 5): there is a 100% inhibition of the 2 sec exchange and 70—80% inhibition of the 10 sec exchange. However, on a prolonged time (3 min) the exchange is inhibited to a few percent in yeast mitochondria, whereas in liver mitochondria atractyloside was shown also to inhibit the 5 to 10 min exchange. This may be due to the higher exchange rate in yeast mitochondria and/or a lower effectiveness of atractyloside. In fact, considerably higher concentrations of atractyloside are required here than in liver mitochondria.

The effect of atractyloside on the phosphorylation rates of endogenous and exogenous adenine nucleotides was also studied (Table 6). Atractyloside inhibits primarily the phosphorylation of the exogenous ADP and has little effect on the phosphorylation of the endogenous ADP. This response to atractyloside is qualitatively the same as that found with liver mitochondria [15,18—20]. However, the yeast mitochondria are less sensitive since 24 μM atractyloside inhibit the phosphorylation of exogenous ADP to only 85%.

DISCUSSION

A major viewpoint of this study is the comparative one which has already proved to be a powerful means for the understanding of mitochondrial function. In this respect mitochondria from yeast are of special interest since yeast cells are among the most primitive organisms containing mitochondria. Besides two striking differences from animal mitochondria, such as the lack of one phosphorylation site and the oxidation of external DPNH, similarity in many respects is noted. The studies in this paper show furthermore that the investigated differences are more quantitative than qualitative ones.

The relative composition of the respiratory chain in yeast mitochondria as referred to cytochrome aa_3 is distinguished by the comparatively high amounts of both hydrogen transfer co-enzymes DPN and UQ, as well as of cytochrome c. All occur in several times higher relative amounts than in heart mitochondria. These differences may be understood by the function of these carriers as pools for collecting hydrogen or electrons from various dehydrogenases. In fact, the yeast mitochondria are very differently equipped with dehydrogenases linked to all three pools. For example, they contain a highly active dehydrogenase for exogenous DPNH, and D- and L-lactate dehydrogenases linked to cytochrome c.

The redox behavior of the respiratory chain follows similar laws as that of mammalian mitochondria. It differs only with respect to the DPN which, due to the lack of the first phosphorylation site, does not respond to the build-up of hydrogen in the controlled state.

Among all carriers, UQ responds most consistently to the various steady state conditions of the electron transfer, and thus acts most consistently as a member of the respiratory chain. This applies particularly to the dependency of the redox state of UQ on the hydrogen donor activity. For DPN-, flavin- and cytochrome c-linked hydrogen donors the same relation is obtained demonstrating the homogenous reactivity of almost the total UQ pool. Up to 90% of the UQ is reduced in the controlled state according to a uniform function. This is the more remarkable as the respiratory chain contains a great excess of UQ. In yeast mitochondria the same linear function between redox ratio of UQ and donor activity is found as in heart mitochondria, even with the same slope. In heart mitochondria the hydrogen donor activity available is smaller. This explains that in this case only up to 70% of the UQ can be reduced in the controlled state. These results emphasize that the second phosphorylation site and its associated energy-linked redox equilibrium are controlling the redox reactions of UQ. This is substantiated by the demonstration of a reverse electron transfer from cytochrome c-linked substrates to UQ.

The nucleotide pattern in yeast mitochondria is qualitatively similar to that in mammalian mitochondria. In a quantitative respect, the adenine nucleotide content equals that of rat heart mitochondria. When related to the content of cytochrome a, which has been taken as a measure of the respiratory capacity, the adenine nucleotide content is about 3—4 fold higher, similar to the content of DPN (Table 1).

The UTP content of yeast mitochondria is much higher than that observed in mammalian mitochondria. The participation of endogenous UTP of rat liver mitochondria in the mitochondrial ribonucleic acid synthesis has been discussed [13]. If this is the main role of UTP in mitochondria, it may not be suprising to find higher levels in yeast mitochondria, since the half life of yeast is reported to be much lower (2—3 hours) than the half life of liver mitochondria (9 days [21]). A high level of UTP may therefore reflect a higher rate of mitochondrial RNA synthesis.

The high rate of phosphorylation of exogenous ADP even at low temperature, which distinguishes the yeast from the mammalian mitochondria, raises the question as to whether the adenine nucleotides enter the yeast mitochondria by a "translocation step," first described for rat liver mitochondria [19]. This reaction was postulated to be the limiting step in the reactivity of exogenous adenine nucleotides

in rat liver mitochondria at low temperatures [15]. The result on the exchange, such as its specificity and sensitivity to atractyloside and, furthermore, the sensitivity to atractyloside of the phosphorylation of exogenous ADP, strongly indicates that there is indeed an adenine nucleotide "translocation" involved in the phosphate transfer of yeast mitochondria.

The observation, that yeast mitochondria unlike the mitochondria from rat liver and rat heart, phosphorylate exogenous ADP faster than endogenous ADP, could be explained in two alternative ways:

1. The total pool of the endogenous adenine nucleotides is involved in oxidative phosphorylation of exogenous ADP, but exogenous ADP is needed for mobilisation of the endogenous adenine nucleotides.

2. A considerable part of the endogenous adenine nucleotides is not involved in phosphorylation of the exogenous ADP, being situated on a sidepath, as defined in functional terms.

The results of the exchange experiments appear to favour the second possibility. The "initial" rate of the exchange, evaluated from a linear extrapolation, is considerably smaller than the phosphorylation of exogenous adenine nucleotides. However, the kinetics indicate that the pool is not reacting uniformly but is composed of a fast and a slower reacting part. Therefore the initial rise might be faster and even compatible with the phosphorylation rate of exogenous ADP. Taking this into account we may conclude that at least part of the endogenous adenine nucleotides has no immediate access to the phosphorylation site.

The comparatively high adenine nucleotide translocation in yeast mitochondria even at low temperatures may be explained in physiological terms. The yeast cells, unlike mammalian cells, can grow at a lower temperature. Therefore, oxidative phosphorylation may also be maintained at lower temperatures.

This work was performed during the tenure of a guest professorship at the University of Marburg by T. Onishi.

This work was supported by a grant from the "Deutsche Forschungsgemeinschaft".

REFERENCES

1. Onishi, T., Kawaguchi, K., and Hagihara, B., J. Biol. Chem. 241 (1966) 1797.
2. Onishi, T., Sottocasa, G. L., and Ernster, L., Bull. Soc. Chim. Biol. 48 (1966) 1189.
3. Kröger, A., and Klingenberg, M., Biochem. Z. 344 (1966) 317.
4. Hohorst, H. J., Kreutz, F. H., and Bücher, Th., Biochem. Z. 332 (1959) 18.
5. Adam, H., in Methoden der Enzymatischen Analyse (herausgegeben von H. U. Bergmeyer) Verlag Chemie, Weinheim/Bergstr. 1962, p. 539.
6. Hagihara, B., personal communication.
7. Chance, B., and Hagihara, B., Symposium 5th International Congress of Biochemistry (Moscow), 1961, No. V, p. 5.

8. Klingenberg, M., and Bücher, T., *Biochem. Z.* 331 (1959) 312.
9. Bücher, T., and Klingenberg, M., *Angew. Chem.* 70 (1958) 552.
10. Chance, B., and Hollunger, G., *Federation Proc.* 16 (1957) 703.
11. Kröger, A., and Klingenberg, M., In preparation.
12. Klingenberg, M., and Kröger, A., *3rd FEBS Meeting (Warsaw)*, 1966, In press.
13. Heldt, H. W., and Klingenberg, M., *Biochem. Z.* 343 (1965) 433.
14. Heldt, H. W., Unpublished results.
15. Heldt, H. W., in *Round Table Conference on Mitochondrial Structure and Compartmentation* (edited by E. Quagliariello, E. C. Slater, S. Papa, and J. M. Tager), Editrice Adriatica, Bari 1967, in press.
16. Klingenberg, M., and Pfaff, E., in *Regulation of Metabolic Processes in Mitochondria*, Elsevier, Amsterdam 1965, p. 180.

17. Pfaff, E., Klingenberg, M., and Heldt, H. W., *Biochim. Biophys. Acta*, 104 (1965) 312.
18. Chappel, J. B., and Crofts, A. R., *Biochem. J.* 95 (1965) 707.
19. Heldt, H. W., Jacobs, H., and Klingenberg, M., *Biochem. Biophys. Res. Commun.* 18 (1965) 174.
20. Kemp, A., and Slater, E. C., *Biochim. Biophys. Acta*, 104 (1965) 312.
21. Bass, R., and Neubert, D., *Abstracts 3rd FEBS Meeting (Warsaw)*, 1966, p. 168.
22. van Gelder, B. F., and Muijers, A. O., *Biochim. Biophys. Acta*, 118 (1966) 47.
23. Yang, Ch., *Rev. Sci. Instr.* 25 (1954) 807.
24. Pfaff, E., Thesis, University of Marburg (1965).
25. Pfaff, E., Unpublished results.

T. Onishi, A. Kröger, W. E. Heldt, E. Pfaff, M. Klingenberg
Physiologisch-chemisches Institut der Philipps-Universität
355 Marburg/Lahn, Deutschhausstraße 1/2, Germany

European J. Biochem. 1 (1967) 312—316

Radical Transfer Reactions and Chemical Protection
in the Dry State. Electron Spin Resonance Studies on Irradiated Mixtures
of Ribonuclease and Glycyl-Glycine

E. S. Copeland, T. Sanner, and A. Pihl

Norsk Hydro's Institute for Cancer Research, Montebello, Oslo 3

(Received January 27, 1967)

Molecular mixtures containing different proportions of ribonuclease and glycyl-glycine were irradiated with X-rays at 77° K and the electron spin resonance spectra were observed at this temperature and after heat-treatment for 10 minutes at 60°.

In mixtures freeze-dried from pH 8 the number of unpaired spins observed at 77° K corresponded to the weighted average of the yields for the components. In contrast, in mixtures freeze-dried from solutions at pH 4.4 the number of unpaired spins was much higher than expected. The qualitative spectra indicated that these "excess radicals" were located on glycyl-glycine.

The results obtained after heat-treatment indicated that in the samples freeze-dried from pH 8 a considerable transfer of spins occurred from RNase to glycyl-glycine during the heat-treatments, whereas in the samples freeze-dried from pH 4.4 little or no additional transfer seemed to occur.

The results indicate that the number of "excess radicals" at 77°K, as well as the extent of interaction in the mixtures which occurred upon heat-treatment, depended on the state of the dissociable groups of the constituents. The finding of "excess radicals" at 77° K is tentatively attributed to the action of H' atoms ejected from RNase. These H' atoms are assumed to interact with glycyl-glycine to form stable radicals, while in the absence of glycyl-glycine they are assumed to disappear primarily in combination reactions.

Recent electron spin resonance studies indicate that the radiation damage occurring when proteins are irradiated in the dry state, is partly associated with intermolecular reactions [1—5]. Thus radiation products, such as H' atoms, electrons, and molecular fragments derived from one molecule may interact with and damage neighbouring molecules. The fact that small molecular compounds may offer protection to enzymes irradiated in the dry state is probably due in part to the fact that they may scavenge such fragments, thus preventing them from attacking the enzyme molecules [2—6]. However, it is likely that the protection is also due in part to hydrogen transfer reactions, resulting in repair of damaged enzyme molecules [2, 4—8].

The intermolecular transfer of unpaired spins occurring in the dry state has been extensively studied by electron spin resonance (ESR) spectroscopy [2, 8—11]. So far, primarily mixtures of proteins with thiols have been studied. The sulfur radicals formed in such mixtures exhibit a characteristic spectrum, which can easily be identified [1—3, 12, 13]. Furthermore, thiols are among the best protective agents *in vivo* [14]. However, it has been found that in the dry state non-sulfur compounds are as protec-

Enzyme. RNase, or ribonuclease (EC 2.7.7.16).

tive as sulfur compounds [6, 15, 16]. It is therefore of interest to study the interactions in irradiated mixtures of proteins with protective non-sulfur compounds.

In the present paper the intermolecular reactions occurring in irradiated mixtures of ribonuclease (RNase) and glycyl-glycine have been studied by ESR spectroscopy. Glycyl-glycine was chosen for several reasons. Preliminary data indicate that it protects enzymes in the dry state. Furthermore, it gives a fairly well defined ESR spectrum [17, 18]. RNase was selected because this enzyme gives a pronounced sulfur resonance upon heat-treatment after irradiation [1]. It is therefore possible to assess the intermolecular transfer of unpaired spins in such mixtures by observing the effect of glycyl-glycine on the relative number of sulfur and non-sulfur radicals present.

MATERIALS AND METHODS

The ESR Spectrometer

An X-band spectrometer (9,000 MHz) with a transmission cavity and a modulation frequency of 110 kHz was used. Spectra were obtained by phase sensitive detection and recorded at 77° K as the first derivative of the actual absorption curves. The number of ESR centers was obtained by double integration of the spectra and comparison with standard samples measured under the same conditions. Anthracite carbon powder, calibrated against diphenylpicrylhydrazyl, was used as a secondary standard. The microwave frequency was measured with an ordinary wave-meter to an accuracy of about 1 MHz. The magnetic field was measured with a proton resonance field meter. The oscillation frequency for the proton resonance was measured with a Hewlett-Packard 5245 L electronic counter. The spectrometer and the procedure used for obtaining quantitative results have been previously described [1, 11]. Since a small, but in some cases significant, ESR signal can be obtained from unirradiated samples, the background signal was subtracted from each spectrum in the quantitation process.

Materials

Crystalline RNase was obtained from Armour Pharmaceutical Company (Eastbourne, England), and glycyl-glycine from Sigma Chemical Company (St. Louis, Mo., U.S.A.). These crystalline materials were used without further purification.

Experimental Procedure

"Molecular mixtures" were prepared by mixing and subsequently freeze-drying solutions of the two substances. Solutions were prepared by dissolving the substances in distilled, deionized water. In those cases where the pH of the solution was adjusted prior to lyophilization, the adjustment was carried out with minimal amounts of hydrochloric acid or sodium hydroxide.

Lyophilized samples (6 to 20 mg) were placed in glass sample tubes, evacuated and sealed off at a pressure of 10^{-4} to 10^{-5} mm Hg. The samples were irradiated at one end of the tube. Prior to ESR measurements the samples were shaken to the other end of the tube which had been shielded during irradiation. All samples were irradiated with X-rays (220 kV, 20 mA) to a dose of 4.8×10^5 R, at a dose rate of 4.6 kR/min, as measured with an ionization chamber. The irradiations were carried out at liquid nitrogen temperature. After the initial spectra had been recorded at 77° K, the samples were heated to 60° for ten minutes and then returned to 77° K for further spectroscopic investigation.

RESULTS

When proteins are irradiated at room temperature, the ESR spectra observed are not composite spectra of those from the individual amino acids, but consist mainly of two patterns [1, 3, 12, 13]. One of these is the so called "sulfur pattern", due to the radical $R—CH_2—S^·$ [2, 19, 20], while the other one is

a doublet ascribed to the radical $-NH-\overset{\cdot}{C}-CO-$
$\qquad\qquad\qquad\qquad\qquad\quad H$
[2, 18, 21]. The contribution of these to the composite patterns observed depends on the relative content of cysteine and glycine residues in the protein [1]. Since the energy absorption occurs fairly randomly, the predominance of these two patterns indicates that a migration of radiation energy to specific trapping sites occurs. This migration can be followed by irradiating the protein at 77° K and by observing the spectral changes which occur upon step-wise heat-treatment of the sample [1—4].

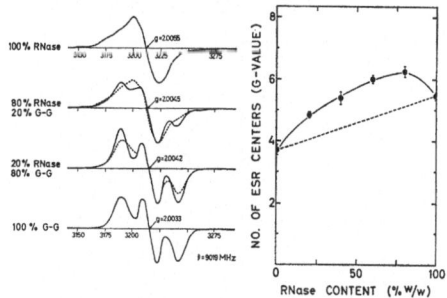

Fig. 1. *Qualitative (left) and quantitative (right) ESR data from irradiated molecular mixtures of RNase and glycyl-glycine.* RNase solutions containing different concentrations of glycyl-glycine were freeze-dried and irradiated in vacuum at 77° K to a dose of 4.8×10^5 R. The spectra were recorded at 77° K. The qualitative spectra respresent the first derivative of the absorption curves. The magnetic field strength, in gauss, is plotted on the axes. The dashed lines of the qualitative spectra represent the spectra expected on the basis of the composition of the mixture, assuming that each compound contributes in proportion to its weight fraction. The right part shows the number of ESR centers plotted as a function of the percent by weight of RNase. The dashed line represents the weighted average of the yield for the pure components. Vertical bars in this and subsequent figures represent the standard deviation of the mean, calculated from experiments on different samples

The processes occurring in irradiated mixtures can be followed by a similar approach [2, 4]. If no interaction occurs between the components, the qualitative spectra will correspond to those that can be constructed from the spectra of the constituents, using appropriate weight factors for the respective yields of radicals. Furthermore, the number of radicals observed will be the weighted average of those for the components irradiated individually.

The qualitative and quantitative ESR data of irradiated mixtures containing different proportions of RNase and glycyl-glycine are shown in Fig. 1. The mixtures were irradiated and the spectra observed at 77° K. It is apparent that the components showed distinctly different spectra (Fig. 1). Whereas RNase (top spectrum) gives a broad assymetric line, glycyl-glycine (bottom spectrum) gives a more structured, symmetric pattern. The quantitative data in Fig. 1 B demonstrate that, in contrast to previous findings on mixtures of proteins with sulfur compounds, the yields observed were in all cases appreciably greater than the weighted average of those for the pure components (the dashed line). The results could be explained if one of the substances enhanced the trapping possibilities for unpaired spins in the other component. It is more likely, however, that the presence of "excess radicals" reflects the occurrence of interactions between the components. Probably in the

RNase + GLYCYLGLYCINE

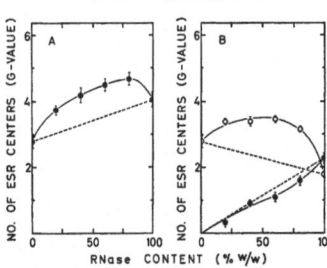

Fig. 2. *Number of ESR centers in molecular mixtures of RNase and glycyl-glycine after heat-treatment.* Samples described in Fig. 1 were heat-treated for 10 minutes at 333°K and the spectra recorded at 77°K. A, Total number of radicals. B, Results of decomposition of spectra. ●, sulfur radicals; ○, non-sulfur radicals

mixture, radiation energy that in the absence of the second component would have been dissipated without the formation of observable ESR centers, was transferred to the other component with consequent formation of unpaired spins.

In attempts to determine whether the "excess unpaired spins" were located on RNase, glycyl-glycine, or were shared between the two, theoretical qualitative spectra (dashed lines in Fig. 1 A) were constructed from the spectra of the pure components, assuming that each contributed in proportion to its weight fraction of the mixture. Comparison of the theoretical spectra with those observed indicates that the contribution of the glycyl-glycine pattern to the composite pattern is greater than expected from the composition of the mixture. Furthermore, when theoretical spectra were constructed on this latter assumption, the resulting spectra resembled closely those actually observed. These findings suggest that the "excess unpaired spins" were located on glycyl-glycine.

In order to study the reactions occurring when the radicals are mobilized by heat-treatment, the mixtures were heated for 10 minutes at 60°. This treatment resulted in an appreciable decay of radicals. It is seen (Fig. 2 A) that also after heat-treat-

ment the total number of radicals was far greater than expected from the composition of the mixture. Comparison of Fig. 1 B with Fig. 2 A indicates that the percentage "excess radicals" was about the same before and after heat-treatment. Decomposition of the spectra demonstrated (Fig. 2 B) that the "excess radicals" were of the non-sulfur type, whereas the number of sulfur radicals was only slightly lower than expected from the RNase content of the mixture. Since the spectra of non-sulfur radicals in RNase resemble closely those from glycyl-glycine radicals, it is difficult from the qualitative spectra to decide whether the excess radicals present after heat-treatment were located on RNase, glycyl-glycine or on both components. However, in view of the results obtained at 77° K it seems most likely that also in the heat-treated mixtures the "excess radicals" were located on glycyl-glycine. Altogether, the results suggest that in this case little interaction occurred between the components during the heat-treatment.

Observations are available suggesting that the yield of ESR centers in freeze-dried substances may depend on the pH of the solution prior to freeze-drying [22,23]. The possibility was therefore considered that the finding of "excess radicals" in the mixtures at 77° K might be related to the fact that the pH of the dissolved mixtures was different from that of the pure components. Thus, the solution of glycyl-glycine had a pH of 5.8 and that of RNase a pH of 4.4. Studies of the individual components were therefore carried out in which the yields of ESR centers were determined as a function of the pH of the solutions prior to freeze-drying. It is seen from Fig. 3 that the radical yield in glycyl-glycine increased strongly when the pH of the solution was lower than 5, while it was fairly independent of pH in the range 5 to 8. The results after heat-treatment indicate that the pH of the solution prior to freeze-drying had little influence on the stability of the radicals.

The results with RNase (Fig. 4) demonstrate that the number of radicals, both at 77° K and at 60°, was independent of pH prior to freeze-drying in the range 3.3 to 6.3. Above this pH the yield increased. The increase after annealing could be accounted for by an increase in the number of non-sulfur radicals (Fig. 4 B).

Since the pH of the solution prior to freeze-drying influences the yield of radicals in the components, the number of radicals expected in the mixtures at 77° K is actually somewhat higher than the dashed line in Fig. 1 B. However, even if this factor is taken into account there was still an appreciable number of "excess radicals" in the mixtures.

In attempts to reveal the mechanism underlying the formation of the "excess radicals" at 77° K experiments were carried out in which pH of the dissolved mixtures was adjusted to 4.4 and 8.0, respectively, before freeze-drying. In the mixture freeze-

GLYCYLGLYCINE

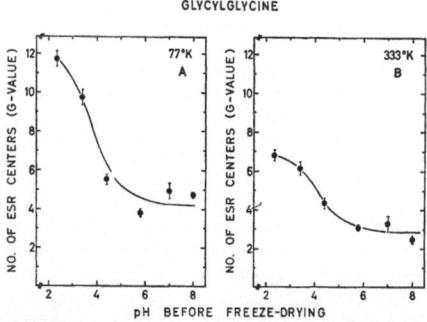

Fig. 3. *Number of ESR centers in freeze-dried glycyl-glycine as a function of pH prior to lyophilization.* The pH was adjusted by addition of hydrochloric acid or sodium hydroxide. Similar results were obtained when sulphuric or phosphoric acid was used instead of hydrochloric acid. Otherwise conditions as in Fig. 1. A, Yield at 77°K. B, Number of radicals remaining after heat-treatment for 10 minutes at 333°K

RNase

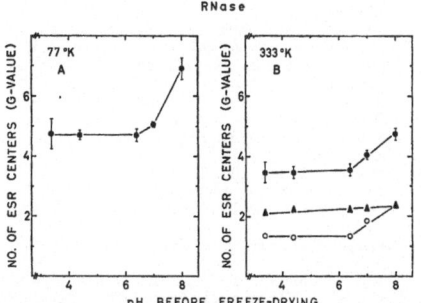

Fig. 4. *Radical yield in RNase as a function of pH before lyophilization.* A, Total yield at 77°K. B, Number of radicals remaining after heat-treatment for 10 minutes at 333°K. ●, total number of radicals; ▲, sulfur radicals; ○, non-sulfur radicals

RNase + GLYCYLGLYCINE (pH 4.4)

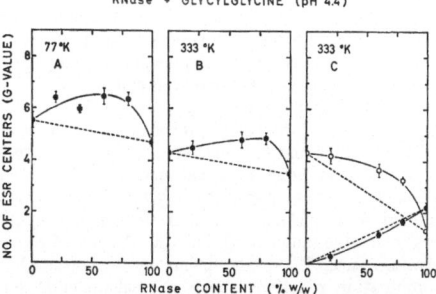

Fig. 5. *Radical yields in molecular mixtures of RNase and glycyl-glycine after freeze-drying from solutions of pH 4.4.* Conditions as in Fig. 1 and 2. The dashed lines represent the number of radicals expected from the irradiation of the pure components, assuming no interaction. A, Total yield at 77°K. B, Total number of radicals remaining after heat-treatment at 333°K for 10 minutes. C, Results of decomposition of the spectra in B. ●, sulfur radicals; ○, non-sulfur radicals

RNase + GLYCYLGLYCINE (pH 8.0)

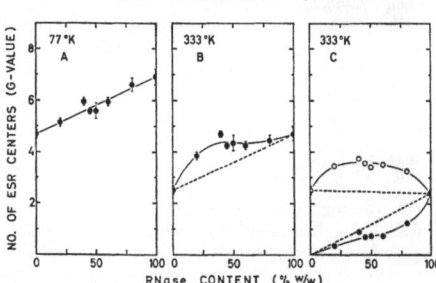

Fig. 6. *Radical yields in molecular mixtures of RNase and glycyl-glycine, freeze-dried from solutions at pH 8.* Conditions and notation as in Fig. 5

dried from the acid solution (Fig. 5) similar results were obtained as in Figs. 1 and 2. Thus, many more radicals were stabilized at 77° K than would be predicted if no interaction occurred between the irradiated components. Also, the percentage excess was again about the same after heat-treatment. Decomposition of the spectra showed that the "excess radicals" were of the non-sulfur type. The number of sulfur radicals corresponded to that expected from the RNase content of the mixture. The results suggest, in agreement with those obtained on the samples with unadjusted pH (Figs. 1 and 2), that most of the interaction occurred at 77° K.

In the experiments carried out after freeze-drying at pH 8, no "excess radicals" were found at 77° K (Fig. 6). Thus, at this temperature the yields in the

various mixtures corresponded closely to the weighted average of those obtained on the components. However, after heat-treatment at 60° the total yields of radicals were appreciably higher than predicted from the composition of the mixtures (Fig. 6B). Decomposition of the spectra demonstrated (Fig. 6C) that the yield of non-sulfur radicals was greatly increased, while the yield of sulfur radicals was lower than expected. The results indicate that at pH 8, in contrast to the results at pH 4.4, a considerable transfer of unpaired spins occurred during the heat-treatment.

DISCUSSION

The results reported in this paper demonstrate that, in mixtures of RNase and glycyl-glycine irradiated at 77° K, more radicals were observed at this temperature than could be accounted for by the yields observed for the individual components. The number of excess radicals at 77° K, as well as the extent of interaction of mixtures upon heat-treatment, was found to depend on the pH of the solutions prior to freeze-drying. The results thus indicate that the state of the dissociable groups plays an important role for the radical reactions occurring.

The finding of "excess radicals" at 77° K could be accounted for by several mechanisms. One possibility which should be considered is that the "excess radicals" may arise from the action of H· atoms. Thus, the results might be explained if H· atoms ejected from RNase interacted with glycyl-glycine to form stable radicals, while in the absence of glycyl-glycine these H· atoms primarily disappeared in combination reactions. Such a mechanism is supported by the qualitative spectra indicating that the excess radicals at 77° K were glycyl-glycine radicals.

The H· atoms formed when organic substances are irradiated in the dry state usually have a life time so short that they can not be observed by electron spin resonance spectroscopy. The H· atoms may disappear by combination with other H· atoms (equation 1), by abstracting H· atoms from neighbouring molecules (equation 2), or by adding to certain structures (e.g. by opening disulfide bonds) (equation 3). The disappearance of H· radicals according to equations 2 and 3 leads to the formation of radicals that may be detected by ESR spectroscopy.

$$H· + H· \rightarrow H_2 \tag{1}$$
$$RH + H· \rightarrow R· + H_2 \tag{2}$$
$$P + H· \rightarrow PH· \tag{3}$$

The finding that the yield of ESR centers in glycyl-glycine increased when the pH of the solution prior to freeze-drying was lowered could be explained if reaction (2) proceeds with greater ease when the carboxyl group is undissociated than when it is dissociated. If this holds true, reaction (2) may compete more efficiently with reaction (1) after freeze-drying from low pH. Accordingly, the finding of excess radicals in the mixtures irradiated at 77° K after freeze-drying at pH 4.4 could be explained by assuming that glycyl-glycine reacts with H· atoms formed from ribonuclease, thus preventing them from disappearing according to equation (1).

Intermolecular transfer of unpaired spins requires close contact between the molecules participating [24]. The fact that the intermolecular reactions occurring upon heat-treatment seem to be more extensive at alkaline than at acid pH may be related to such factors. It is conceivable that at alkaline pH, when both the protein and the glycyl-glycine have positive as well as negative charges, the components may be in closer contact than at low pH, where both components possess predominantly positive charges on the surfaces. If, as suggested above, the interaction at 77° K is due mainly to H· atoms, these will presumably be less affected by such factors, due to their high rate of diffusion.

On the basis of the mechanism suggested above, glycyl-glycine would be expected to be able to protect RNase in the dry state in several ways. In the first place it could interact with and trap H· atoms that otherwise might lead to damage and inactivation of neighbouring enzyme molecules. Secondly, in cases where damage to enzyme molecules involves the loss of a H· atom, glycyl-glycine might repair this damage by a hydrogen transfer reaction. From the present data it would be expected that the degree of protection offered will depend on the pH of the solution before freeze-drying. Experiments are now in progress to test this hypothesis.

This work was supported by Division of Radiological Health, Bureau of State Service, U.S. Public Health Service. E. S. Copeland was supported by U.S. Public Health Service Fellowship 5-F2-CA-23 from The National Cancer Institute. T. Sanner is a fellow of The Norwegian Cancer Society.

REFERENCES

1. Henriksen, T., Sanner, T., and Pihl, A., *Radiation Res.* 18 (1963) 147.
2. Henriksen, T., Sanner, T., and Pihl, A., *Radiation Res.* 18 (1963) 163.
3. Henriksen, T., in *Electron Spin Resonance and the Effects of Radiation on Biological Systems* (edited by W. Snipes), Proc. Gatlinburg Conference, Tennessee 1965. Nuclear Science Series Report Number 43, National Academy of Sciences, National Research Council, Washington D.C. 1966, p. 81.
4. Pihl, A., and Sanner, T., *Progr. Biochem. Pharmacol.* 1 (1965) 85.
5. Braams, R., *Nature*, 200 (1963) 752.
6. Sanner, T., *Radiation Res.* 26 (1965) 95.
7. Butler, J. A. V., and Robins, A. B., *Radiation Res.* 17 (1962) 63.
8. Ormerod, M. G., and Alexander, P., *Radiation Res.* 18 (1963) 495.
9. Norman, A., and Ginoza, W., *Radiation Res.* 9 (1958) 77.
10. Gordy, W., and Miyagawa, I., *Radiation Res.* 12 (1960) 211.
11. Henriksen, T., and Pihl, A., *Intern. J. Radiation Biol.* 3 (1961) 351.
12. Gordy, W., Ard, W. B., and Shields, H., *Proc. Natl. Acad. Sci. U. S.* 41 (1955) 983.
13. Gordy, W., and Shields, H., *Radiation Res.* 9 (1958) 611.
14. Bacq, Z. M., and Alexander, P., *Fundamentals of Radiobiology*, 2nd ed., Pergamon Press, London 1961.
15. Braams, R., *Radiation Res.* 12 (1960) 113.
16. Brustad, T., *Advan. Biol. Med. Phys.* 8 (1962) 161.
17. Katayama, M., and Gordy, W., *J. Chem. Phys.* 35 (1961) 117.
18. Mangiaracina, R. S., *Radiation Res.* 26 (1965) 343.
19. Kurita, Y., and Gordy, W., *J. Chem. Phys.* 34 (1962) 282.
20. Akasaka, K., Ohnishi, S. I., Suita, T., and Nitta, I., *J. Chem. Phys.* 40 (1964) 3110.
21. Gordy, W., and Shields, H., *Proc. Natl. Acad. Sci. U. S.* 46 (1960) 1124.
22. Singh, B. B., and Ormerod, M. G., *Biochim. Biophys. Acta*, 109 (1965) 204.
23. Ormerod, M. G., and Singh, B. B., *Biochim. Biophys. Acta*, 120 (1966) 413.
24. Henriksen, T., and Sanner, T., *Radiation Res.* In press.

E. S. Copeland, T. Sanner, and A. Pihl
Department of Biochemistry
Norsk Hydro's Institute for Cancer Research
Montebello, Oslo 3, Norway

European J. Biochem. 1 (1967) 317—326

An Evaluation of the Mitchell Hypothesis of Chemiosmotic Coupling in Oxidative and Photosynthetic Phosphorylation

E. C. SLATER

Laboratory of Biochemistry, B.C.P. Jansen Institute, University of Amsterdam

(Received February 17, 1967)

The Mitchell hypothesis of chemiosmotic coupling in oxidative phosphorylation is examined in the light of experimental data on oxidative phosphorylation at present available. The following objections are brought against the theory:

1. Data on the magnitude of the phosphate potential against which ATP can be synthesized by the respiratory chain require, on the basis of chemiosmotic coupling, very effective proton and/or cation extrusion and/or anion uptake by the mitochondria. Experimental evidence for this is lacking. Indeed it has been shown by Chance and Mela that mitochondria in the controlled state do not extrude protons.

2. Although reversible ATP-driven proton (chloroplasts) or cation (mitochondria, erythrocytes) pumps have been demonstrated, this is insufficient evidence for the postulate that the establishment of a proton gradient is the primary energy-conserving event of the chloroplast or mitochondrial respiratory chain. Indeed, it is general experience that cations are required for the extrusion of protons by mitochondria.

3. The stoicheiometry of proton extrusion by mitochondria and of proton uptake by chloroplasts, and the kinetics of the latter process are also difficult to reconcile with the chemiosmotic hypothesis.

4. Further experimental evidence is presented confirming that, under the conditions of the oxygen-pulse experiments of Mitchell and Moyle, the extrusion of H^+ is not associated with the oxidation of mitochondrial NADH.

5. The respiratory chain included in the chemiosmotic hypothesis is difficult to reconcile with our present knowledge of the chain.

It is concluded that the chemiosmotic theory, in its present form, is untenable.

Until recently, discussions on the mechanism of respiratory-chain phosphorylation centred on what has become known as the "chemical" or "∼" hypothesis of respiratory-chain phosphorylation, first formulated in 1953 [1]. According to this hypothesis, the initial conservation of the energy made available by the oxido-reduction reactions of the respiratory chain is by formation of a high-energy compound of an electron or hydrogen carrier in the chain. This energy is utilized for various energy-requiring reactions, including the synthesis of ATP.

This hypothesis has provided a useful framework on which to hang a large amount of information on the mechanism of respiratory-chain phosphorylation, especially on the effects of various uncouplers and inhibitors and on the utilization of the conserved energy in the absence of inorganic phosphate. However, the failure to isolate the hypothetical high-energy intermediates has stimulated consideration of other possible mechanisms. In particular, the Mitchell

hypothesis [2,3] of chemiosmotic coupling in oxidative phosphorylation has attracted much attention.

A detailed account of the Mitchell hypothesis is given in a privately distributed Research Report [3], a shortened version of which has appeared in the published literature [4]. Since several important modifications of the chemiosmotic hypothesis have been made since it was first proposed [2], the subsequent discussion will be confined to the version discussed in the Research Report [3], and in subsequent publications [5,6].

In the Introduction [3], Mitchell makes the good point that "the belief has been widely accepted that the energy-rich coupling intermediates $(C_1 \sim I_1, C_2 \sim I_2, C_3 \sim I_3$, etc.) must exist because there is no feasible alternative means of coupling electron transport to phosphorylation." His objects in proposing a working chemiosmotic hypothesis in 1961 were (a) to provide a simple rationale for the organization of the components of the oxidative phosphorylation system in the lipid membrane systems of mitochondria and chloroplasts; (b) to formulate a

---

Non-standard abbreviations. N-3,4-dichlorophenyl-N'-dimethylurea, DCMU.

type of coupling that would require no intermediates of the type of C $\sim$ I, "so that future work need no longer be so dependent upon or so circumscribed by the belief in the C $\sim$ I intermediates"; and (c) "to acknowledge the elusive character of the C $\sim$ I intermediates by admitting that they might not exist."

I believe that Mitchell has succeeded in putting forward a rational alternative to the $\sim$ hypothesis that deserves the serious consideration that it has received and is receiving from other workers in the field. This article has the limited purpose of examining the chemiosmotic hypothesis in the light of experimental data on oxidative phosphorylation at present available. Other hypotheses in which, as in Mitchell's, the production of protons by reduction of the respiratory chain plays a central role [7—9] will be left out of consideration.

MITCHELL HYPOTHESIS

The chemiosmotic system consists of four basic postulates: (a) a proton-translocating reversible ATPase system; (b) a proton-translocating oxido-reduction chain; (c) an exchange-diffusion system, coupling proton translocation to that of anions and cations; (d) an ion-impermeable coupling membrane, in which the three systems reside.

Proton-translocating Reversible ATPase System

Mitchell [ref. 3, p. 52] represents the oligomycin-sensitive proton-translocating ATPase system by the reactions

$$
\begin{aligned}
XH_R + IOH_R + ATP &\rightleftharpoons XI_R + P_1 + ADP \\
XI_R &\rightleftharpoons XI_L \\
H_2O + XI_L &\rightleftharpoons X_L^- + IO_L^- + 2H_L^+ \\
2H_R^+ + X_L^- + IO_L^- &\rightleftharpoons XH_R + IOH_R
\end{aligned}
$$

Sum:
$$2H_R^+ + H_2O + ATP \rightleftharpoons 2H_L^+ + P_1 + ADP \quad (1)$$

where the suffixes R and L refer to the right-hand and left-hand sides, respectively, of a membrane. XI, X_L^-, IO_L^-, XH and IOH are in the membrane, whereas ATP, ADP and P_1 are in the right-hand phase, which in mitochondria corresponds to the matrix or internal cristae phase. XI is considered to be a low-energy compound when it is in contact with phase L having a high electrochemical potential of H^+, and a high-energy compound ($X \sim I$) when it is in contact with phase R with a low electrochemical potential of H^+.

Proton-translocating Oxido-reduction Chain

Mitchell represents the proton-translocating oxido-reduction chain as in Fig. 1 (essentially identical with Fig. 11 of Mitchell [3]). In terms of chemical equa-

tions this may be written (representing the internal phase by the suffix R)

$$
\begin{aligned}
(SH_2)_R + NAD_R^+ &\rightarrow S_R + (NAD^+H_2)_R \\
(NAD^+H_2)_R &\rightarrow (NAD^+H_2)_L \\
(NAD^+H_2)_L + (Fe, SH)_L &\rightarrow NAD_L^+ + 2H_L^+ + \\
&\qquad (Fe, SH, 2e^-)_L \\
NAD_L^+ &\rightarrow NAD_R^+ \\
(Fe, SH, 2e^-)_L &\rightarrow (Fe, SH, 2e^-)_R \\
(Fe, SH, 2e^-)_R + FMN_R & \\
+ 2H_R^+ &\rightarrow (Fe, SH)_R + (FMNH_2)_R \\
(Fe, SH)_R &\rightarrow (Fe, SH)_L \\
(FMNH_2)_R &\rightarrow (FMNH_2)_L \\
(FMNH_2)_L + 2b_L^{3+} &\rightarrow FMN_L + 2b_L^{2+} + 2H_L^+ \\
2b_L^{2+} &\rightarrow 2b_R^{2+} \\
FMN_L &\rightarrow FMN_R \\
2b_R^{2+} + Q_R + 2H_R^+ &\rightarrow 2b_R^{3+} + (QH_2)_R \\
2b_R^{3+} &\rightarrow 2b_L^{3+} \\
(QH_2)_R &\rightarrow (QH_2)_L \\
(QH_2)_L + 2(c_1^{3+})_L &\rightarrow Q_L + 2(c_1^{2+})_L + 2H_L^+ \\
Q_L &\rightarrow Q_R \\
2(c_1^{2+})_L + \frac{1}{2}(O_2)_R + & \\
2H_R^+ &\rightarrow (2c_1^{3+})_L + H_2O_R
\end{aligned}
$$

Sum:
$$
\begin{aligned}
(SH_2)_R + \frac{1}{2}(O_2)_R & \\
+ 6H_R^+ &\rightarrow S_R + H_2O_R + 6H_L^+ \quad (2)
\end{aligned}
$$

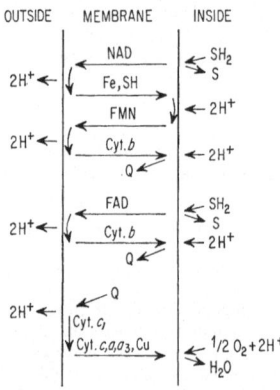

Fig. 1. *Mitchell's [3] proposal of the proton-translocating respiratory chain for oxidation of NAD-linked and FAD-linked substrates in mitochondria.* The chain is shown branching at ubiquinone (Q). Drawn from Mitchell [3], Fig. 11

If we add to this equation that representing the synthesis of ATP by the proton-translocating ATPase, written with the appropriate stoicheiometry

$$
\begin{aligned}
6H_L^+ + 3(P_1)_R + 3ADP_R & \\
&\rightarrow 6H_R^+ + 3H_2O_L + 3ATP_R
\end{aligned}
$$

we obtain

$$(SH_2)_R + \frac{1}{2}(O_2)_R + 3(P_1)_R + 3ADP_R$$
$$\rightarrow S_R + H_2O_R + 3ATP_R + 3H_2O_L$$

which correctly represents the stoicheiometry of oxidative phosphorylation.

Exchange Diffusion System

Mitchell postulates that the diffusion of ions other than protons (or OH^- ions) down the electrical gradient across the coupling membrane, and their accumulation in osmotically disruptive concentrations in the internal phase, must be counteracted by specific extrusion in exchange for protons or OH^- ions. The exchange diffusion of H^+ against cations is represented as the C^+/H^+ antiport in Fig.2 which is drawn from Fig.14 of Mitchell [3].

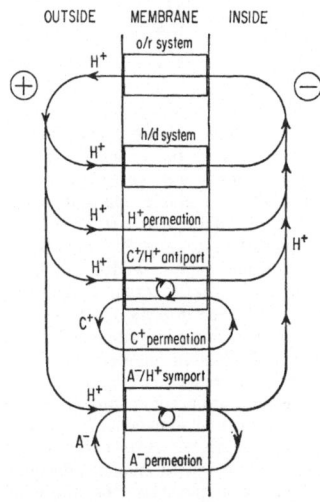

OUTSIDE   MEMBRANE   INSIDE

Fig.2. *The cation/proton antiport and the anion-proton symport proposed by Mitchell [3]. Drawn from Mitchell [3], Fig.14*

Ion-impermeable Coupling Membrane

This is identified as the inner membrane of the mitochondria, the grana membrane of chloroplasts and the plasma membrane and chromatophore membrane of bacteria.

The Mitchell theory will now be examined in the light of (a) the thermodynamics of oxidative phosphorylation; (b) experimental evidence relating to the existence and function of a proton-translocating reversible ATPase system; (c) the nature of the respiratory chain.

THERMODYNAMICS OF OXIDATIVE PHOSPHORYLATION

It is a truism to state that the thermodynamics of a chemical reaction depend only upon the initial and final states, and not on the chemical mechanism of the reaction. A calculation of the chemical potential against which ATP can be synthesized is, however, useful since it tells us the magnitude of the ΔG that can be built up by respiration in State 4 (absence of phosphate acceptor). This in turn gives us information on the magnitude of the high-energy "content" of the hypothetical $\sim$ compound of the chemical hypothesis or of the pH differential (or its equivalent) of the Mitchell hypothesis.

The chemical potential of the system

$$ATP \rightleftharpoons ADP + P_1$$

at 25° is given by

$$\Delta G' = \Delta G_0' + 1.36 \log_{10} \frac{[ADP][P_1]}{[ATP]}.$$

Cockrell et al. [10] have recently calculated, on the basis of new data on the equilibrium in State 4, that, in the absence of Mg^{2+} at pH 7.8 and 25°,

$$\Delta G' = -9.6 + 1.36 \log_{10} \frac{1 \times 10^{-4} \cdot 2 \times 10^{-3}}{5 \times 10^{-3}} \text{ kcal/mole}$$

$$= -15.6 \text{ kcal/mole.}$$

This calculated value for the phosphate potential is rather greater than that generally used in calculations of the thermodynamics of oxidative phosphorylation. From the dependence of the redox state of cytochrome *c* on the phosphate potential, Klingenberg and Schollmeyer [11] calculated a value of 12.8 kcal/mole at pH 7.2. Mitchell [3] assumes a value of 9.6 kcal. Since, however, there appears no reason to question the value calculated by Cockrell et al.[10], this will be used as the basis for subsequent calculations.

Thus, any hypothesis of oxidative phosphorylation has to provide a mechanism for the synthesis of ATP against a chemical potential of the ATP $\rightleftharpoons$ ADP + P_1 system of 15.6 kcal/mole. According to the Mitchell hypothesis, as represented by Equation (1), the synthesis is driven by a pH differential, the magnitude of which may be calculated from the equation

$$\Delta G = 1.36 \log_{10} \frac{[H_L^+]}{[H_R^+]} = \frac{15.6}{2} \text{ kcal/mole}$$

i.e. $\log_{10}[H^+]_L - \log_{10}[H^+]_R = pH_R - pH_L$

$$= \frac{15.6}{1.36 \times 2} = 5.7.$$

Since the left-hand side of the membrane, in the Mitchell hypothesis, is the space between the outer and inner mitochondrial membranes, and this space is thought to be readily permeable to ions, pH_L must be about the same as the pH of the suspension medium, *i.e.* pH 7.8 in the experiment of Cockrell *et al.* [10]. Thus, the pH of the cristae space, according to Mitchell's hypothesis, would be 13.5, which is incompatible with the operation of the intracristae enzymic systems.

In fact, Mitchell recognized this difficulty, although not its magnitude, from the beginning, and proposed the exchange-diffusion system to overcome it [2]. It was suggested that, by exchange of cations for protons, the pH differential would be replaced by a membrane potential. However, since the first version of the Mitchell theory [2], the "sidedness" of the mitochondrion has been reversed. Now [3], the inside of the mitochondrion is phase R, the outside is phase L. Thus, it is the low concentration of H^+ in phase R, rather than the high concentration of H^+ in phase L, that is supposed to drive the ATPase towards ATP synthesis.

In the most recent version of the chemiosmotic hypothesis, Mitchell has proposed two exchange-diffusion systems, a cation/proton antiport and an anion/proton symport.

Thus Equation (1) could be supplemented by the equation of the C^+/H^+ antiport, namely

$$2H_L^+ + 2C_R^+ \rightleftharpoons 2H_R^+ + 2C_L^+. \qquad (3)$$

The sum of Equations (1) and (3) is

$$H_2O + ATP + 2C_R^+ \rightleftharpoons P_i + ADP + 2C_L^+.$$

Thus, the ATP synthesis is now driven by a concentration gradient of cations across the inner membrane, the concentration being greater outside than inside. The magnitude of the concentration gradient required is given by

$$2 \times 1.36 \log_{10} \frac{[C_L^+]}{[C_R^+]} = 15.6,$$

i.e. $\log_{10} \dfrac{[C_L^+]}{[C_R^+]} = 5.7.$

Alternatively, Equation (1) could be supplemented by the equation of the A^-/H^+ symport, *e.g.*

$$2H_L^+ + A_L^{2-} \rightleftharpoons 2H_R^+ + A_R^{2-} \qquad (4)$$

assuming a dicarboxylic acid as an anion. The sum of Equations (1) and (4) is

$$H_2O + ATP + A_L^{2-} \rightleftharpoons P_i + ADP + A_R^{2-}.$$

Thus, the ATP synthesis is now driven by a concentration gradient of anions across the inner membrane, the concentration between greater inside than outside. The magnitude of the concentration gradient required is given by

$$1.36 \log_{10} \frac{[A_R^{2-}]}{[A_L^{2-}]} = 15.6,$$

i.e. $\log_{10} \dfrac{[A_R^{2-}]}{[A_L^{2-}]} = 11.5.$

The total energy available for the synthesis of ATP is the sum of the contributions made by the proton, cation and anion gradients. Thermodynamic requirements for ATP synthesis are met if

$$\Delta pH - 0.5 \, \Delta pA^{2-} + \Delta pC^+ = 5.7$$

where Δ refers to the difference, R (inside) minus L (outside), $pA^{2-} = -\log_{10} [A^{2-}]$ and $pC^+ = -\log_{10} [C^+]$. Various possible combinations yielding the required phosphate potential of 15.6 kcal are listed in Table 1.

Table 1. *Various possible combinations of proton, cation and anion gradients yielding a phosphate potential of 15.6 kcal.*

ΔpH	$-\Delta pA^{2-}$	ΔpC^+
5.7	0	0
2.0	7.5	0
2.0	5.5	1.0
2.0	3.5	2.0
2.0	1.5	3.0
2.0	0	3.7
0	11.5	0
0	7.5	2.0
0	3.5	4.0
0	0	5.7

Even if a pH differential of 2.0 is allowed (the pH of the cristae space in the experiments of Cockrell *et al.* [10] would then be 9.8), very high ionic gradients are required. Since oxidative phosphorylation proceeds actively with 0.6 M succinate in the medium, it is highly probable that $-\Delta pA^{2-}$ would be much less than 1. This would require a ΔpC^+ of 3.5, *i.e.* the external concentration of cation would have to be $10^{3.5}$ times that of the internal concentration. Since there is no evidence that the contents of mitochondria become highly alkaline during steady-state oxidative phosphorylation (see below), the requirements for cation extrusion and/or anion uptake by the mitochondria are even more stringent. In the absence of any experimental evidence of the extrusion of cations and uptake of anions of the magnitude required by the chemiosmotic hypothesis, the thermodynamic difficulties become overwhelming.

In chloroplasts, the direction of proton movements during electron transport [12] is the opposite from that found with mitochondria. The significance of this will be discussed below.

EXISTENCE AND FUNCTION OF A PROTON-TRANSLOCATING REVERSIBLE ATPASE SYSTEM

Existence

As first emphasized by SARIS [13], the energy-dependent accumulation of cations by mitochondria is accompanied by the liberation of an equivalent amount of protons. The energy may be provided either by the operation of the respiratory chain, in which case the accumulation of cations is insensitive to oligomycin, or by the hydrolysis of ATP in an oligomycin-sensitive reaction, thus

(cation uptake, proton extrusion)

where E_c represents energy conserved during the operation of the chain, without specifying the nature of the energy conservation. Thus, the existence of a proton-translocating ATPase, under certain conditions, is well-established. The question at issue is whether these conditions include all those in which oxidative phosphorylation occurs.

The reversibility of the proton-translocating ATPase has also been demonstrated, first by Jagendorf and Uribe [14] with chloroplasts, in which the direction of proton translocation is in the reverse direction from that found with mitochondria. After chloroplasts had been left for a short time at 0° and pH 3.8 in the presence of an organic acid (such as succinic acid) and DCMU (to inhibit the light reaction), the suspension was brought to pH 8.0 in the presence of $^{32}P_i$ and ADP. After a few seconds, the reaction was stopped with trichloroacetic acid and the ATP synthesis calculated from the uptake of $^{32}P_i$ into the organic phosphate fraction or the net ATP synthesis measured by the luciferase assay. This so-called "acid-bath phosphorylation", which can amount to about 40 μmoles ATP/g protein, has been confirmed by Dr. K. G. Rienits in our laboratory [15] and by McCarty and Racker [16].

It seems likely that this phosphorylation is driven by a difference of pH between the outside and inside of the chloroplasts, which, at the moment of bringing the pH of the suspension medium to 8.0, would contain a high concentration of organic acid that had diffused into the chloroplasts as uncharged acid during the acid bath. If the ADP and phosphate are added only 2 sec after adjusting the pH to 8.0, the yield of ATP is decreased by 83% [14].

Analogous experiments with rat-liver mitochondria, carried out by Dr. K. G. Rienits and Dr. M. Koivusalo in our laboratory, have failed to show any synthesis of ATP, either when the pH was raised from 5.0 to 8.4, or lowered from 10.3 to 7.0 (see Table 2 for a typical experiment). Reid et al [5], however, have reported the synthesis of a small amount of ATP when the pH was lowered from 8.8—9.0 to 4.2—4.3, in the presence of K$^+$ and valinomycin, which is known to increase the permeability of the mitochondrial membrane to K$^+$. Although the amount of ATP found is small, it is

greater than that predicted purely from reversal of the equilibrium

$$ATP^{3-} + OH^- \rightleftharpoons ADP^{2-} + P_i^{2-}.$$

After 1 sec at pH 4.2—4.3 in the experiments of Reid et al. [5], $\dfrac{[ADP][P_i]}{[ATP]} = \dfrac{(5 \times 10^{-3}) \times (1 \times 10^{-3})}{12.5 \times 10^{-6}}$

$= 0.4$ M.

At pH's between 1 and 6, $\dfrac{[ADP][P_i]}{[ATP]}$ at equilibrium equals about 10^5 M.

Although there are some features of this experiment that require further examination, we shall assume in what follows that the existence of a proton-translocating reversible ATPase system in both chloroplasts and mitochondria has been experimentally demonstrated.

Table 2. *Failure of alkali bath to effect ATP synthesis by rat-liver mitochondria*

4 ml of mitochondrial suspension (containing 72 mg protein) were mixed with 400 μmoles potassium phosphate buffer (pH 7.0) containing $^{32}P_i$, 150 μmoles MgCl$_2$, 50 μmoles EDTA and 2.5 mmoles sucrose in a total volume of 24 ml. After 4 min at 25°, 0.24 ml of 0.1 M KCN (pH 7.0) was added (final concentration, 1 mM), and 1 min later 5 ml of a solution containing 108 mmoles ethanolamine chloride (pH 11.0), 24 μmoles MgCl$_2$, 8 μmoles EDTA, 60 μmoles ADP, 6 μmoles KCN and 54 μmoles KOH. This addition brought the pH to 10.5. After 30 sec the pH was brought back to 7.0 by the addition of 5 ml of a solution containing 60 μmoles ADP, 6 μmoles KCN, 24 μmoles MgCl$_2$, 8 μmoles EDTA, 790 μmoles HCl and 720 μmoles Tris-HCl buffer (pH 7.0). Samples were withdrawn at various times and the amount of ^{32}P incorporated into esterified phosphate was measured. Control experiments showed that mitochondria could be kept at pH's between 5.0 and 10.3 for 30 sec at 25° without any loss of respiratory control or effect on the P:O ratio

Time	pH	^{32}P incorporated
sec		nmoles/mg protein
260	7.0	58
270	7.0	62
280	7.0	59
290	7.0	59
305	10.5	60
310	10.5	59
320	10.5	59
325	10.5	55
335	7.0	59
345	7.0	54
355	7.0	53
370	7.0	56

Function

The relationship between energy conservation, cation uptake and proton extrusion can be expressed, quite generally, in three possible ways.

Scheme A represents the chemiosmotic hypothesis in which the primary energy conservation is the form of translocated H$^+$ (indicated by ← H$^+$ to

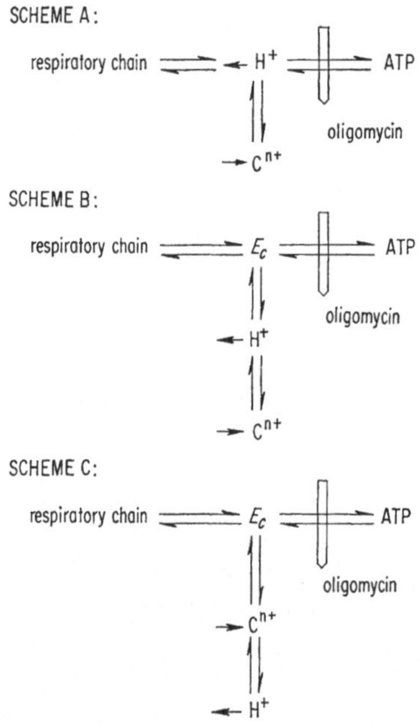

SCHEME A:

respiratory chain $\rightleftharpoons$ $\leftarrow$ H^+ $\rightleftharpoons$ ATP

oligomycin

$\rightarrow$ C^{n+}

SCHEME B:

respiratory chain $\rightleftharpoons$ E_c $\rightleftharpoons$ ATP

oligomycin

$\leftarrow$ H^+

$\rightarrow$ C^{n+}

SCHEME C:

respiratory chain $\rightleftharpoons$ E_c $\rightleftharpoons$ ATP

oligomycin

$\rightarrow$ C^{n+}

$\leftarrow$ H^+

show the translocation of H^+ from right to left). Cation accumulation ($\rightarrow C^{n+}$) takes place from left to right, as result of the membrane potential set up by the C^+/H^+ antiport and the $H^+/$anion symport of Fig. 2.

According to Scheme B, the conserved energy (E_c) may be utilized to make ATP, in an oligomycin-sensitive reaction, or to drive a proton pump, that may be linked, perhaps *via* an exchange-diffusion carrier, with cation accumulation.

According to Scheme C, the conserved energy may operate a cation pump, that is linked with the movement of H^+ in the opposite direction. Scheme C covers the more specific mechanisms proposed by Cockrell *et al.* [10] and Chance and Mela [17].

All reactions in the three schemes are written reversibly. It has been amply demonstrated that the direction of that part of the respiratory chain lying between NAD and cytochrome *c* can be reversed by ATP [18]. Reversal of the respiratory chain by a pH or cation concentration differential has not yet been demonstrated in either mitochondria or chloroplasts. However, the recently reported luminescence of chlorophyll in spinach chloroplasts [19] induced by the acid-bath treatment of Jagendorf and Uribe [14] points in this direction.

The function of the proton-translocating ATPase differs fundamentally in the three schemes. According to Scheme A (the chemiosmotic hypothesis), it is on the main path of ATP synthesis. According to Schemes B and C it is on a side path that can act as an energy store (*cf.* Jagendorf and Uribe [20]) or may be utilized for cation uptake. In Scheme B, E_c drives a proton pump, in Scheme C a cation pump.

Since all three schemes provide an explanation for the synthesis of ATP linked to a pH differential, I cannot agree with Reid *et al.* [5] that their findings "would thus seem to help to promote the status of the chemiosmotic hypothesis towards that of a theory. Conversely, the chemical hypothesis is weakened to the extent that a further *ad hoc* assumption is required to account for the new observations." ATP-driven cation pumps have for a long time featured in explanations of the Na^+ and K^+ fluxes through the cell membrane. A simple allosteric model has recently been proposed [21]. Garrahan and Glynn [22] have recently reported the synthesis of ATP by reversal of the $(Na^+ + K^+)$-stimulated ATPase of the erythrocyte membrane.

A critical test between the three schemes above is whether or not permeant cation is necessary for the extrusion of H^+ by mitochondria or the uptake of H^+ by chloroplasts, driven by respiration or by ATP. According to Schemes A and B, permeant cations need not be necessary; according to Scheme C, they must be. The careful studies of Chance and Mela [17] provide strong evidence that, although a proton concentration gradient can be clearly demonstrated when respiration is linked to the uptake of Ca^{2+} in the absence of a permeant anion (*cf.* Lehninger *et al.* [23]), no gradient is set up in the absence of a permeant cation.

By using bromthymol blue as an indicator of membrane pH and bromcresol purple as an indicator of extramitochondrial pH, Chance and Mela were able directly to demonstrate the production of H^+ outside the mitochondria and the alkalinization of the membrane on the addition of amounts of Ca^{2+} exceeding 30 μmoles per g protein to a mitochondrial suspension in State 4 (absence of ADP). Additions of Ca^{2+} below this limit caused no alkalinization of the membrane, indicating the existence of a buffer capacity, tentatively identified with phospholipid. Thus, State 4 itself, the condition of maximum conservation of energy, is not associated with the establishment of a pH gradient. Moreover, this buffer capacity would have to be neutralized before a proton gradient could be built up sufficient to drive ATP synthesis by the chemiosmotic mechanism.

Uncoupling agents, which abolished the alkalinization of the membrane by Ca^{2+}, caused no change in pH in the absence of Ca^{2+}, further supporting the conclusion that there is no detectable difference in the pH gradient in the fully activated State 4 and in the uncoupled state. Taking into account the sensitivity of the technique used, Chance and Mela concluded that no pH gradient exceeding 0.02 pH exists across the mitochondrial membrane in the energized State 4.

With rat-liver mitochondria especially prepared to minimize cation accumulation, Chance and Mela found no detectable alkalinization of the mitochondrial membrane or acidification of the medium (less than 0.36 μmole H^+ per g mitochondrial protein) on aerating an anaerobic suspension.

In contrast to this conclusion, Mitchell and Moyle [24] claimed to have demonstrated the primary production of protons. They pre-incubated a suspension of rat-liver mitochondria in a medium containing KCl, glycylglycine buffer and β-hydroxybutyrate or succinate for 20 min at 25° in the absence of oxygen. When small amounts of oxygen were introduced, H^+ was rapidly liberated into the medium, followed by a slower decline in H^+ concentration. The H^+:O ratio was 6 with β-hydroxybutyrate and 4 with succinate. When ATP was added instead of oxygen to an anaerobic mitochondrial suspension in the absence of substrate, 2 moles H^+ were produced for each mole of ATP hydrolysed, over and above the 0.8 mole H^+ expected for the hydrolysis of ATP to ADP and P_i at this pH.

Mitchell and Moyle [24] concluded that the H^+ production measured was that predicted by the chemiosmotic hypothesis. Tager et al. [15] have, however, objected to this interpretation. We pointed out that, according to Scheme A, when oxygen is added to mitochondria in the presence of substrate and the absence of P_i and ADP, it may be expected that the respiratory chain will function until a sufficient pH differential (or its equivalent) is built up to prevent the flow of reducing equivalents along the respiratory chain from substrate to oxygen. In the words of Mitchell and Moyle [25]: "under the conditions of this kind of experiment there is a 'backlash' before the translocation of protons across the coupling membrane builds up a protonmotive force sufficient to retard oxido-reduction." Thus, one must expect that, immediately after adding the oxygen, the respiration would be uncontrolled even though phosphate and phosphate acceptor are absent. With sufficient oxygen, one would proceed through the following sequences of states [26]:

State 5 → State 3 → State 4 → State 5.

The amount of oxygen consumed (and substrate oxidized) during the uncontrolled State 3 would depend upon the amount of H^+ required to build up the required protonmotive force. With smaller amounts of oxygen, State 4 would never be reached and the sequence would be:

State 5 → State 3 → State 5.

The 'backlash' limit, i.e. the amount of H^+ liberated by oxygen-pulsed mitochondria before the

steady-state oxygen consumption (State 4) sets in, was found by Mitchell and Moyle [24,25] to be about 12 μmoles per g mitochondrial protein, equivalent to 6 μmoles ATP per g mitochondrial protein. This 'backlash' limit is equivalent to the amount of energy (E_c) that can be conserved in the respiratory chain in the presence of substrate and oxygen, but absence of ADP and P_i. Van Dam [27] has shown that this is less than 0.4 μmole ATP per g mitochondrial protein. Thus, the amount of H^+ that can be liberated in an oxygen-pulse experiment is more than an order of magnitude greater than the amount of E_c that can be conserved in an ADP-pulse experiment [27]. We [15] concluded then that the H^+ liberated in the oxygen-pulse experiments does not represent the primary conservation of energy in the respiratory chain, as envisaged by Scheme A (the chemiosmotic hypothesis).

Mitchell and Moyle [3] calculate that "the translocation of only about 1 μequiv proteins per g protons in rat-liver mitochondria should bring the membrane potential to its presumed respiratory control value of some 250 mV." They [25] explain the unexpectedly large 'backlash' found in their experiments by postulating the presence of "a considerable quantity of mobile charges or dipoles." The membrane potential, which they postulate "is the main cause of respiratory control and of the re-entry of the outwardly translocated protons, would thus tend to be collapsed while the mobile charges or dipoles were redistributing in the electric field; and the major part of the backlash of some 12 μg ion H^+, or 12 μ electron equivalents, per g mitochondrial protein would be equated to the quantity of the displaceable charge." They suggest that endogenous calcium ions may be partly responsible for the 'backlash'.

Since, according to Mitchell and Moyle, 'mobile charges or dipoles' are responsible for more than 90% of the protons liberated by mitochondria in oxygen-pulse experiments, these experiments cannot be brought forward in support of Scheme A, which implies that a pH differential could be built up in the absence of 'movable ionic constituents.'

Indeed, we brought forward experimental evidence that cast doubt on the conclusion that the burst of H^+ obtained by Mitchell and Moyle [25] in their oxygen-pulse experiments is associated with uncontrolled respiration. A sensitive method of detecting a transition from controlled to uncontrolled respiration is to follow the oxidation of intramitochondrial nicotinamide nucleotide [26]. Making use of the data of Van Dam [27] on the kinetics of the oxidation of NADH on the transition from State 4 to State 3, we calculated that, if the chemiosmotic hypothesis and the interpretation of the oxygen-pulse experiments given by Mitchell and Moyle [24] were correct, at least 0.9 μmole NADH should be oxidized on the addition of oxygen [15]. However, no oxidation was found unless phosphate was added simultaneously with the oxygen.

In these experiments, NAD⁺ was determined enzymically in extracts obtained by stopping the reaction with perchloric acid. We pointed out that erroneous conclusions would be drawn if the direct spectrophotometric assay (based on changes in $A_{340-374 \text{ m}\mu}$) for reduced nicotinamide nucleotides were used, since the transition from anaerobic to aerobic conditions is accompanied by a large oxidation of cytochrome which interferes with the assay for the reduced nicotinamide nucleotides. This was demonstrated by carrying out oxygen-pulse experiments in

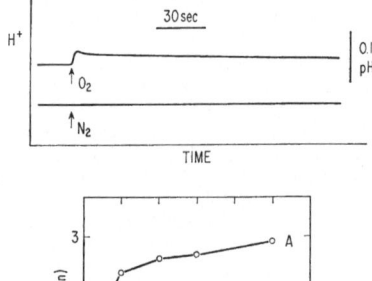

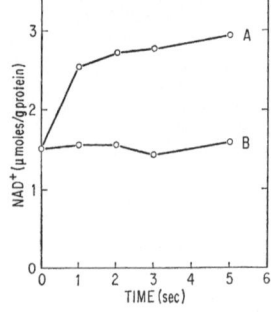

Fig. 3. *The effect of an oxygen pulse on the pH of a mitochondrial suspension and on the amount of intramitochondrial NAD⁺.* The upper figure shows traces of a recording pH meter (upward trace represents a fall in pH). In this experiment, rat-liver mitochondria (16.8 mg) were preincubated in the absence of air in 3.5 ml of a medium containing 3.3 mM glycylglycine buffer and 0.15 M KCl. The pH before addition of the mitochondria was 7.1. Nitrogen was bubbled through all solutions to free them of oxygen. Anaerobiosis was maintained by layering oxygen-free paraffin on top of the reaction mixture. The effectiveness of the procedure was tested in control experiments with an oxygen polarograph. After 20 min, 50 μl 0.15 M KCl were added, either saturated with CO_2-free air (upper trace) or with N_2 (lower trace). In the experiment shown in the lower figure, rat-liver mitochondria (5.6 mg protein) were preincubated anaerobically in 1 ml of the same medium as above. After 20 min, 0.1 ml 0.15 M KCl saturated with CO_2-free air (Curve B) or 0.1 ml 0.15 M KCl saturated with CO_2-free air and containing 0.5 μmole ADP and 5 μmoles phosphate (Curve A) was added. At the times indicated, the reaction was stopped with $HClO_4$, and NAD⁺ determined in the acid extracts as already described [15]. Each point represents a separate incubation

the presence of rotenone which inhibits the oxidation of NADH.

Mitchell and Moyle [6], while admitting the seriousness of this criticism, have very recently challenged the experimental results. Although they agree with our conclusion [15] that the direct spectrophotometric method cannot be applied without correction, they claim to have demonstrated by applying appropriate corrections to the spectrophotometric traces that NADH is oxidized. No direct chemical assays for NAD⁺ were made.

Fig. 3 shows the results of a re-examination of this question carried out by Dr. R. D. Veldsema-Currie. In order to make conditions more favourable for NADH oxidation, no substrate was added. According to Scheme A, the introduction of oxygen would be expected to lead to the oxidation of NADH at an initial rate as great as that obtained in the presence of ADP and P_1, and to the extent of more than 1 μmole NAD⁺ per g mitochondrial protein. The results clearly show the expected State 5 → State 3 transition in the presence of ADP and P_1, but in the absence of ADP and P_1, no oxidation of NADH, characteristic of the State 5 → State 3 transition, is observed. The upper trace shows the transient H⁺ production reported by Mitchell and Moyle [24]. Thus, protons are produced under these conditions without the expenditure of a detectable amount of the energy conserved by the respiratory chain in State 4.

Whether or not Mitchell and Moyle [6] have demonstrated oxidation of NADH in their experiments[1] is relatively unimportant. The fact remains that in our preparations of rat-liver mitochondria which are capable of rapid phosphorylation when ADP and P_1 are added together with the oxygen, NADH is not oxidized by oxygen in the absence of P_1 and ADP. The Mitchell chemiosmotic hypothesis and the interpretation of the experiments of Mitchell and Moyle [24] given by these authors require that the oxidation of NADH after 1 second in the experiment shown in Fig. 3 would be as great in the absence as in the presence of ADP and P_1.

We [15] showed that, under the conditions of the preincubation used by Mitchell and Moyle [24], a large part of the Mg^{2+} leaked out of the mitochondria. We suggested that the proton production observed by Mitchell and Moyle [24] was due to the reaccumulation of this Mg^{2+} and possibly K⁺ (*cf.* [10]) into the mitochondria. The accumulation of Mg^{2+} by rat-liver mitochondria requires a low rate of expenditure of energy.

In summary, Chance and Mela [17] have shown that, with some preparations of rat-liver mitochon-

---

[1] It remains possible that slowly reacting components of the chain, such as cytochrome b or ubiquinone, cause errors in the calculation by Mitchell and Moyle [6] of that part of the absorbancy change at 340 mμ—374 mμ that they attribute to NADH.

dria, aerobiosis causes no H^+ production, whereas we [15] have shown that H^+ production can be obtained in the absence of NADH oxidation. It must be concluded from these experiments that the evidence that the proton-translocating reversible ATPase system is directly concerned in oxidative phosphorylation, as indicated by Scheme A, is very weak. Indeed, the requirement of a cation for proton translocation [17] speaks strongly in favour of Scheme C over Scheme A or B.

A second method of deciding between Scheme A on one hand and Schemes B and C on the other is to measure the stoicheiometry between $\leftarrow H^+$, $\rightarrow C^{n+}$ and ATP. According to the chemiosmotic theory, ATP: $\leftarrow H^+$: $\rightarrow C^{n+}$ = $1:2:2/n$. According to Schemes B and C, there need be no strict stoicheiometry between the ATP (or $\sim P$ equivalent) and the proton or cation translocations, although one would expect stoicheiometry between the proton and cation movements. Cockrell et al. [10] have found $K^+:\sim P$ ratios of 7 (representing a thermodynamic efficiency of nearly 80%), which as they emphasize are inconsistent with the chemiosmotic hypothesis. Carafoli et al. [28] and Azzone et al. [29] have also obtained $Ca^{2+}:\sim P$ ratios for calcium uptake greater than 2 under certain circumstances.

As also pointed out by Cockrell et al. [10], the yield of ATP in the acid-bath experiment of Jagendorf and Uribe [14] is uncomfortably high to be consistent with Equation (1). According to the calculations of Jagendorf and Uribe [20], the ΔG for ATP formation under the conditions of their experiments (pH 8.4, 0°, presence of Mg^{2+}) is 10.95 kcal/mole. On the basis of the mechanism given in Equation (1), a pH difference of $\frac{10.95}{2 \times 1.25} = 4.4$ would be required to yield this amount of ATP, compared with an experimental pH jump of 4.2. In the experiments of Rienits [30], 15.7 μM ATP, 63 μM ADP and 0.83 mM P_1 were found after a jump of pH from 3.8 to 7.4 (i.e. 3.6 units). Assuming a value for $\Delta G_0'$ of 7.1 kcal/mole under these conditions, ΔG becomes 10.2 kcal/mole which would require a pH jump of 4.1, in considerable excess of the experimental 3.6. Moreover, as pointed out by Cockrell et al. [10], the initial pH differential would be expected to be rapidly dissipated. These authors conclude that, if these experiments do represent the harnessing of a pH gradient to form ATP, the $H^+:\sim P$ ratio is probably considerably greater than 2.

A third method of distinguishing between Scheme A on the one hand and Schemes B and C on the other is to compare the rate of the overall phosphorylation with that of the formation of the pH differential. The chloroplast system of Jagendorf and Uribe [20] provides the opportunity of testing this. When grana are irradiated in the presence of a redox dye but in the absence of ADP or P_1, a considerable amount of

energy (up to 20 μmoles $\sim P$ equivalent per g protein) can be conserved in a form that may be converted to ATP on the addition of ADP and P_1 in the dark. There is a good correlation between the conservation of energy in this way (X_E) and the uptake of H^+ by the grana. Jagendorf and Uribe [20] have shown that the rate of accumulation of X_E is much slower (less than one-fifth) than the rate of ATP synthesis in a complete phosphorylating system. This is most easily explained by supposing that the proton translocation is on a side-path from the main path of ATP synthesis, as in Schemes B and C.

Mitchell and Moyle [25] bring forward an additional argument in favour of the primary origin of the proton translocation during respiratory activity, based upon the fact that oxidation of ferrocyanide does not cause proton translocation, but results in the uptake of two protons in the inner mitochondrial compartment per oxygen atom reduced by cytochrome oxidase. They state: "The important point is that if the proton translocation during normal succinate or β-hydroxybutyrate oxidation were due to the utilization of an intermediate high-energy compound by a specific proton pump, or by a cation pump that incidentally translocated protons, the same energy-rich compound should have resulted in proton-translocation during ferrocyanide oxidation."

There is an obvious alternative trivial explanation of this experiment. The oxidation of ferrocyanide by oxygen

$$4 \, Fe(CN)_6^{4-} + 4 \, H^+ + O_2 \rightarrow 4 \, Fe(CN)_6^{3-} + 2 \, H_2O$$

is a sufficient explanation of the uptake of 2 protons per oxygen atom consumed. In the absence of uncouplers, there was no net production or consumption of protons during the first 30 sec after addition of ferrocyanide. If ferrocyanide was oxidized during this period, protons must have been consumed and must, therefore, have been compensated by the production of protons, possibly by the utilization of a high-energy compound for a specific proton or cation pump.

THE NATURE OF THE RESPIRATORY CHAIN

According to Mitchell (ref. [3], pp. 90—91), "The view of the respiratory chain, according to the chemical coupling hypothesis, would require us to believe that the chemical complexity of the respiratory chain is considerably greater than it seems, and that the known physical complexity has some functional significance that is, as yet, a matter of conjecture. On the other hand, the view of the respiratory chain, according to the chemiosmotic coupling hypothesis, does not require us to bias, one way or the other, the chemical and physical facts as far as they are known at present."

In fact, the chemiosmotic hypothesis takes considerable liberties with our present knowledge of the respiratory chain:

a) It would have us accept that a proton would remain attached to NADH as it moved from the right-hand side (inside) of the membrane to the left.

b) According to the respiratory chain proposed by Mitchell [3], the reduction of flavin in NADH and succinate dehydrogenases proceeds to the level of oxidation of the fully reduced leucoflavin. There is, in fact, much evidence in favour of the view that the transition is from the oxidized flavin to the semiquinone [31].

c) There is no experimental evidence that the iron-sulphur system of NADH or succinate dehydrogenase is reduced before the flavin as required by the chemiosmotic hypothesis.

d) The chemiosmotic hypothesis requires alternating hydrogen and electron carriers. Accordingly, ubiquinone is placed between cytochromes b and c_1, instead of between flavin and cytochrome b where it is usually placed [32]. The sensitivity of QH_2 oxidation to antimycin, and the close association of antimycin inhibition with cytochrome b, supports the conventional respiratory chain.

Indeed, it is very difficult, although not perhaps impossible [33], to reconcile the chemiosmotic hypothesis with what is known about the respiratory chain (cf. [17,33]).

CONCLUSIONS

Of the four basic postulates of the chemiosmotic system, experimental evidence exists for only two, the proton-translocating reversible ATPase system and the ion-impermeable membrane. There is no evidence for the existence of a proton-translocating oxidoreduction chain or for the exchange-diffusion system. Indeed, recent experimental data make it very doubtful if the respiratory chain is capable of proton translocation, in the absence of cations. Furthermore, the thermodynamic difficulties of the chemiosmotic theory are formidable.

It must be concluded that the chemiosmotic theory, in its present form, is untenable.

Our knowledge of the mechanism of oxidative phosphorylation and of cation movements is best summarized by Scheme C above. The unravelling of the nature of E_c remains a challenge. The possibility that protons, produced by the respiratory chain, raise the effective acidity of the membrane, in such a way as to favour ATP synthesis, as suggested by Williams [8,9], is not contradicted by the arguments brought forward in this article and deserves further consideration.

The experimental work reported in this paper was supported in part by grants from the Life Insurance Medical Research Fund and the U.S. Public Health Service.

REFERENCES

1. Slater, E. C., Nature, 172 (1953) 975.
2. Mitchell, P., Nature, 191 (1961) 144.
3. Mitchell, P., Chemiosmotic coupling in oxidative and photosynthetic phosphorylation, Glynn Research Ltd., Bodwin 1966.
4. Mitchell, P., Biol. Rev. 41 (1966) 445.
5. Reid, R. A., Moyle, J., and Mitchell, P., Nature, 212 (1966) 257.
6. Mitchell, P., and Moyle, J., Nature, 213 (1967) 137.
7. Robertson, R. N., Biol. Rev. 35 (1960) 231.
8. Williams, R. J. P., J. Theoret. Biol. 1 (1961) 1.
9. Williams, R. J. P., J. Theoret. Biol. 3 (1962) 209.
10. Cockrell, R. S., Harris, E. J., and Pressman, B. C., Biochemistry, 5 (1966) 2326.
11. Klingenberg, M., and Schollmeyer, P., in Intracellular Respiration: Phosphorylating and Non-Phosphorylating Oxidation Reactions, Proceedings of the Fifth International Congress of Biochemistry, Moscow 1961, Pergamon Press, Oxford 1963, Vol. 5, p. 46.
12. Jagendorf, A. T., and Hind, G., in Photosynthesis mechanisms in green plants (edited by B. Kok and A. T. Jagendorf), National Academy of Sciences, Washington 1963, p. 599.
13. Saris, N. C., Dissertation, University of Helsinki, 1963.
14. Jagendorf, A. T., and Uribe, E., Proc. Natl. Acad. Sci. U. S. 55 (1966) 170.
15. Tager, J. M., Veldsema-Currie, R. D., and Slater, E. C., Nature, 212 (1966) 376.
16. McCarty, R. E., and Racker, E., Federation Proc. 25 (1966) 226.
17. Chance, B., and Mela, L., Nature, 212 (1966) 369, 372.
18. Chance, B., and Hollunger, G., J. Biol. Chem. 236 (1961) 1534.
19. Mayne, B. C., and Clayton, R. K., Proc. Natl. Acad. Sci. U. S. 55 (1966) 494.
20. Jagendorf, A. T., and Uribe, E., in Brookhaven Symp. Biol. 19: BNL 989 (C-48), 1967, in the press.
21. Jardetzky, O., Nature, 211 (1966) 969.
22. Garrahan, P. J., and Glynn, I. M., Nature, 211 (1966) 1414.
23. Lehninger, A. L., Rossi, C. S., Bielawski, J., and Gear, A., in Mitochrondrial Structure and Compartmentation (edited by S. Papa, J. M. Tager, E. Quagliariello, and E. C. Slater), Adriatica Editrice, Bari, 1967, in the press.
24. Mitchell, P., and Moyle, J., Nature, 208 (1965) 147.
25. Mitchell, P., and Moyle, J., in Biochemistry of Mitochondria (edited by E. C. Slater, Z. Kaniuga, and L. Wojtczak), Academic Press and P.W.N., London and Warsaw, 1967, p. 53.
26. Chance, B., and Williams, G. R., Advan. Enzymol. 17 (1956) 65.
27. Dam, K. van, Biochim. Biophys. Acta, 128 (1966) 337.
28. Carafoli, E., Gamble, R. L., and Lehninger, A. L., Biochem. Biophys. Res. Commun. 21 (1965) 215.
29. Rossi, C., and Azzone, G. F., Biochim. Biophys. Acta, 110 (1965) 434.
30. Rienits, K. G., unpublished observations.
31. Slater, E. C. (Ed.), Flavins and Flavoproteins (BBA Library, Vol. 8), Elsevier, Amsterdam, 1966.
32. Slater, E. C., Colpa-Boonstra, J. P., and Links, J., in Ciba Foundation Symp. on Quinones in Electron Transport, Churchill, London, 1961, p. 161.
33. Slater, E. C., in Biochemistry of Mitochondria (edited by E. C. Slater, Z. Kaniuga, and L. Wojtczak), Academic Press and P.W.N., London and Warsaw, 1967, p. 1.

E. C. Slater
Laboratorium voor Biochemie, B.C.P.Jansen Institute
Plantage Muidergracht 12, Amsterdam-C, The Netherlands

European J. Biochem. 1 (1967) 327—333

Protection of Polynucleotide Phosphorylase
by Substrates against Urea Inactivation

R. A. Harvey, T. Godefroy, J. Lucas-Lenard, and M. Grunberg-Manago

Institut de Biologie physico-chimique, Paris

(Received February 14, 1967)

Polynucleotide phosphorylase from *Escherichia coli* is irreversibly inactivated by preincubation with urea at concentrations of 3.0 M or greater. The addition of ribonucleotide polymers, oligonucleotides or nucleoside diphosphates to the preincubation medium markedly decreases the rate of this inactivation. Both oligonucleotides terminated with an unesterified 3′-hydroxyl group or esterified with a 3′ phosphate are equally effective.

The presence of 2.0 M and 4.0 M urea in the normal reaction medium results in a complete inhibition of the polymerization of the nucleoside diphosphates UDP and ADP, respectively. Full enzymic activity can be regained either by diluting the urea or by addition of oligonucleotides terminating with an unesterified 3′-hydroxyl group. Long polynucleotides or oligonucleotides with a terminal 3′-phosphate group are ineffective in restoring enzymic activity, in contrast to their ability to protect on the enzyme against irreversible denaturation.

The phosphorolysis reaction catalyzed by polynucleotide phosphorylase is appreciably inhibited only at concentrations of urea in excess of 4.0 M.

The present results suggest that at concentrations of 4.0 M or less, urea inhibits selectively and reversibly the chain initiation reaction either by drastically decreasing the binding constant of the oligonucleotide to the enzyme, or by slowing down the first steps of the polymerization.

La polynucléotide phosphorylase extraite de *Escherichia coli* est inactivée de façon irréversible à la suite d'une pré-incubation avec de l'urée à des concentrations de 3 M ou plus. Cette inactivation est notablement diminuée par l'addition de polyribonucléotides, d'oligonucléotides ou de nucléosides diphosphates au milieu de préincubation. Les oligonucléotides sont actifs, qu'ils soient terminés par un groupe 3′-hydroxyle libre, ou qu'ils soient estérifiés en 3′ phosphate.

En présence d'urée (2 M et 4 M, respectivement pour l'UDP et l'ADP) la polymérisation des nucléosides diphosphates est complétement inhibée. L'activité enzymatiquement peut être totalement rétablie, soit par dilution de l'urée, soit par addition d'oligonucléotides terminés par un groupement 3′ hydroxyle non estérifié. Les longs polynucléotides, de même que les oligonucléotides terminés par un groupement 3′ phosphate, ne peuvent restaurer l'activité enzymatique, bien qu'ils puissent protéger l'enzyme contre une dénaturation irréversible.

La réaction de phosphorolyse catalysée par la polynucléotide phosphorylase n'est inhibée de façon appréciable par la présence d'urée qu'à des concentrations au dessus de 4 M.

Ces résultats suggèrent qu'à des concentrations de 4 M ou moins, l'urée inhibe la réaction d'initiation de la chaîne polynucléotidique de façon sélective et réversible; cette inhibition se produit soit par une forte diminution de la constante d'association entre l'oligonucléotide et l'enzyme, soit par un ralentissement des premières étapes de la polymérisation.

Polynucleotide phosphorylase catalyzes the reversible formation of long chain polymers from the corresponding nucleoside diphosphates [1,2]:

$$n\,NDP \rightleftharpoons pNp \ldots (Np)_{n-2} \ldots N + n\,P_i$$

where N represents a purine or pyrimidine base.

A detailed understanding of the mechanism of this enzyme may be useful not only for exploiting more fully its synthetic capacities, but also in providing insights into more general problems of the nature of specific protein-polynucleotide interactions.

Non-standard Abbreviations. Small oligonucleotides are designated as follows: when p, representing a phosphate, is placed at the right of a nucleoside symbol, the phosphate is esterified at C-3′ of the ribose moiety; when placed to the left of the nucleoside symbol, the phosphate is esterified at C-5′ of the ribose moiety; polyuridylic acid, poly U; polyadenylic acid, poly A.

Urea has been observed to introduce a lag in the polymerization reaction catalyzed by purified enzymes from *Escherichia coli*, and *Micrococcus lysodeikticus* [3,4]. This urea-induced inhibition has not been studied extensively and little is known concerning its mechanism. The present communication deals with the effect of protection against urea by substrates: nucleoside diphosphates, polynucleotides and oligonucleotides with a free 3'-hydroxyl group, as well as oligonucleotides with terminal 3'-phosphate residues.

MATERIALS

Nucleoside diphosphates were products of Sigma Chemical Co. and Pabst Laboratories. Polyuridylic acid and polyadenylic acid were prepared in this laboratory using polynucleotide phosphorylase from *Azotobacter vinelandii* or *Escherichia coli* [2,3]. Trinucleotides, or mixtures of oligonucleotides with an average chain length of seven residues, with free 3'-hydroxyl groups, were prepared by enzymic digestion of poly A or poly U with a sheep kidney nuclease [5]. Oligonucleotides with a 3'-phosphate group were prepared by hydrolysis of poly A in 6 M ammonia [6]. Oligonucleotides with a defined chain length were separated on DEAE-cellulose column and desalted, as described elsewhere [5]. The concentrations of all polymers and undefined mixtures of oligonucleotides are expressed in terms of their monomer units as determined by absorption at 260 mμ. Reagent grade urea was recrystallized twice from ethanol and dried in vacuum over sodium hydroxide pellets. Fresh solutions were made daily.

Polynucleotide phosphorylase was purified from *E. coli* by published procedures [3]; fractions after Sephadex G-200 step were used.

The incubation mixture for determining the rate of polymerization consisted of the following: Tris buffer, pH 8.2, 100 mM; ADP or UDP, 20 mM; MgCl$_2$, 10 mM and 4—20 units/ml of polynucleotide phosphorylase. A unit is defined as the amount of enzyme catalyzing the incorporation of 1 μmole of ^{32}P$_i$ into the nucleoside diphosphate (in the case of the phosphorolysis reaction) per hour at 37°. The reaction was stopped with 2.5% perchloric acid and the extent of the reaction measured by the release of P$_i$ [7].

The incubation for the phosphorolysis of poly A consisted of the following: Tris buffer, pH 8.2, 150 mM; poly A, 3 mM; ^{32}P$_i$ (specific activity, 1×10^6 counts/min/μmole), 10 mM; MgCl$_2$, 2.5 mM; EDTA, 0.5 mM; and 3—4 enzyme units/ml. The reaction was stopped by the addition of perchloric acid to a final concentration of 2.5% and [^{32}P]ADP was separated from ^{32}P$_i$ by the phospho-molybdate-isobutanol extraction procedure [3].

RESULTS

Effect of Urea on Polynucleotide Phosphorylase; Protection by Substrates

Experiments were performed to determine the effect of urea on polynucleotide phosphorylase. The enzyme was pre-incubated in buffered solutions containing various concentrations of urea. At suitable time intervals an aliquot of the enzyme was removed and diluted 3-fold into a reaction mixture containing no urea. The urea remaining after dilution varied from a maximum of 1.0 M (Fig. 1 A) to less than 0. 2 M (Fig. 1 B). The activity, assayed either in the phosphorolysis or the polymerization reaction, remaining after urea treatment was measured and compared to the activity of the non-treated enzyme.

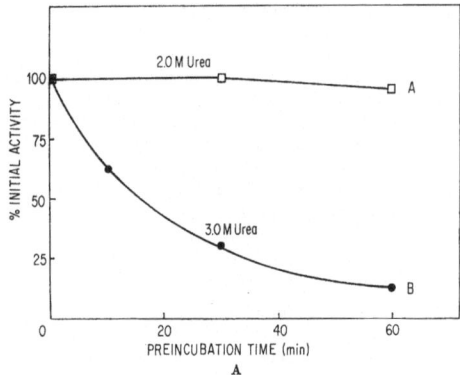

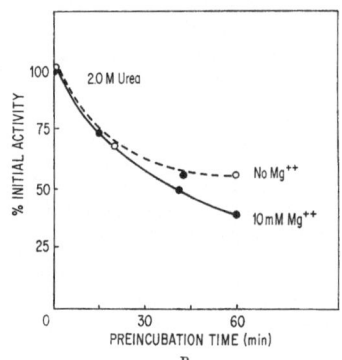

Fig. 1. *Effect of pre-incubation with urea.* (A) The reaction contained the following: Tris buffer, pH 8.2, 100 mM; EDTA, 1 mM; urea, as indicated; enzyme, 178 μg/ml. After incubation at 37° the treated enzyme was diluted two-fold with the phosphorolysis reaction medium (see Methods, poly A as substrate). (B) Same reaction mixture, except for enzyme: 42 μg/ml. After incubation at 37° for various intervals, 15 μl aliquots were removed and diluted to 45 μl with the polymerization reaction medium (see Methods, UDP assay)

Pre-incubation of polynucleotide phosphorylase in solutions containing 2.0 M urea led to little or no loss of activity, even after prolonged exposure. In contrast, similar experiments in the presence of 3.0 M urea resulted in a rapid inactivation of the enzyme assayed either in the polymerization or in the phosphorolysis reaction (Figs. 1 A, 2 A)[1]. No appreciable

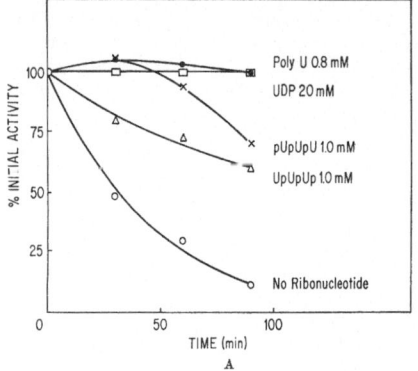

A

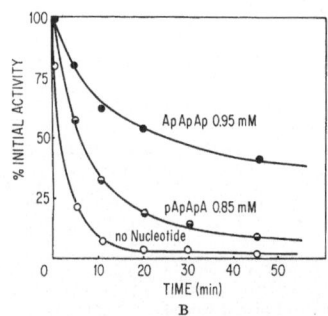

B

Fig. 2. *Effect of ribonucleotides on the pre-incubation with urea.*
(A) Same incubation mixture as in Fig. 1 B; in addition, in mMolar concentrations: urea, 3000; poly U, 0.8; UpUpUp, 1; pUpUpU, 1; and UDP, 20, were present, when indicated. (B) The enzyme, 80 µg/ml was incubated at 37° with 1.9 M urea in Tris buffer, pH 8.2, 50 µmoles/ml. 3'-OH or 3'-PO₄ triadenylic acid was added as indicated. At various times, aliquots of 0.1 ml were removed and diluted ten-fold in the phosphorolysis reaction mixture

reactivation of the enzyme occured when the urea-treated enzyme was diluted and subsequently allowed to stand for several hours at 0° or 37°, either in the presence or the absence of substrate. The observed changes in activity appear to be essentially irre-

[1] The concentration of urea irreversibly inactivating polynucleotide phosphorylase might vary with the preparations of the enzyme. Another preparation (compare Fig. 2 B with Fig. 1 A) was inhibited by 2.0 M urea.

versible under these conditions. Mg does not seem to change markedly the rate of inactivation (Fig. 1 B).

The addition of any of several nucleotides to the urea-containing pre-incubation medium markedly reduced the rate of inactivation of the enzyme (Figs. 2 A, 2 B). Substrates of the enzyme, such as poly U, 3'-hydroxyl uridine oligonucleotides and UDP at high concentration were particularly effective; oligonucleotides terminated with a 3'-phosphate group (which are inhibitors of the phosphorolysis reaction) also protect the enzyme very efficiently against the effect of urea; ApApAp appears to be more effective than pApApAp. It should be noted that the protection observed with high concentrations of UDP is probably not due to the magnesium-dependent conversion of nucleoside diphosphate to poly U by polynucleotide phosphorylase, since the incubation conditions (e.g. 10^{-3} M EDTA) were chosen so as to preclude this complication. It should also be noted that magnesium is not required for protection by substrates against urea.

Effect of Urea on the Polymerization Reaction

The addition of urea at concentrations which do not irreversibly affect the enzyme, has a marked effect on the polymerization reaction. The lag phase normally present in the polymerization reaction of UDP (and sometimes of ADP) was prolonged in the presence of 1.0 M or 1.6 M urea (Fig. 3). Furthermore,

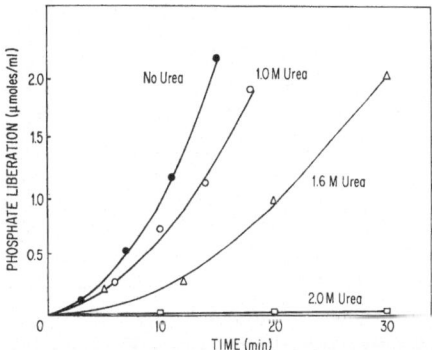

Fig. 3. *Effect of urea on the polymerization of UDP.* ●, no urea; ○, 1.0 M urea; △, 1.6 M urea; □, 2.0 M urea. The reaction mixture contained: Tris buffer, pH 8.2, 100 mM; UDP, 20 mM; MgCl₂, 10 mM; enzyme 25 µg/ml

the maximal rate obtained in the absence of urea was never observed in its presence, under the conditions of the assay (20 mM UDP). At higher concentrations of urea the lag phase became predominant until, at 2.0 M urea and above, the formation of poly U appeared to be completely inhibited. It should be noted that with the same enzyme preparation, UDP protects the enzyme completely against irreversible denaturation by 2.0 or 3.0 M urea (Fig. 2 A). The lag phase induced by urea is almost completely suppressed by the addition of 3′OH oligo U to the reaction mixture, as it is the case for the normal lag phase of UDP polymerization in the absence of urea.

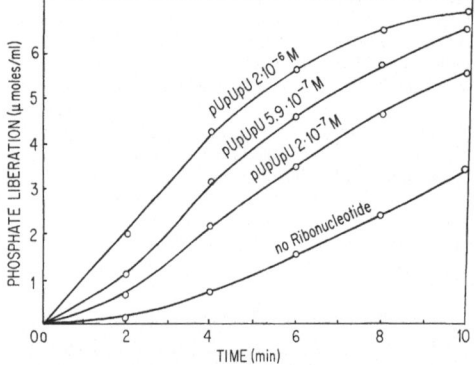

Fig. 4. *Effect of oligonucleotides on the polymerization of UDP, in the absence of urea.* Same incubation mixture as in Fig. 3, except for enzyme: 80 µg/ml; pUpUpU added as indicated

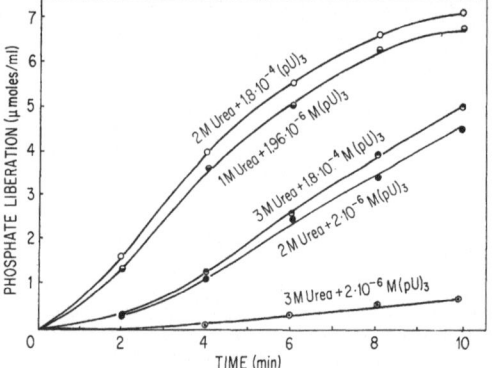

Fig. 5. *Effect of triuridylic acid on urea-inhibited polymerization of UDP.* Same reaction mixture as in Fig. 3; urea and pUpUpU added as indicated

However, the amount of pUpUpU required in the absence of urea (expressed in mononucleotides) is 2×10^{-6} M (Activator constant[2] 3×10^{-7} M) (Fig. 4) while in the presence of 2.0 M urea 100-fold more

[2] The activator constant is the concentration of activator (in our case oligonucleotide) which gives an initial rate of polymerization equal to 1/2 of the maximum obtainable with a saturating concentration.

pUpUpU is necessary (Fig. 5). Therefore in the presence of oligonucleotides 3′OH-free the action of up to 3.0 M urea appears to be reversible.

Poly U (1×10^{-3} M) of high molecular weight failed to restore the polymerization of UDP. Some preparations were active, but it was shown by Sephadex chromatography that these ribonucleotides were contaminated by small oligonucleotides. The addition of a trinucleotide UpUpUp (up to 1.2×10^{-3} M) also failed to restore enzymic activity (Fig. 6). The inability of UpUpUp to reverse the urea inhibition is a property shared by other oligonucleotides terminated with a 3′-phosphate group and does not appear

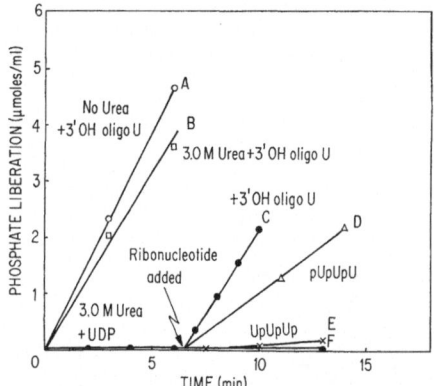

Fig. 6. *Effect of polynucleotides on urea-inhibited polymerization of UDP.* The reaction mixture contained: Tris buffer, pH 8.2, 100 µmoles/ml; UDP, 20 µmoles/ml; MgCl₂, 10 µmoles/ml; urea (except in experiments A), 3000 µmoles/ml; enzyme, 14 µg/ml. In addition, the following were added (in µmoles/ml): Experiment A: Oligo U, 0.85 (average chain length 7). Experiment B: Oligo U, 0.85. Experiment C: Oligo U, 0.85 (added at 6.5 minutes). Experiment D: pUpUpU, 0.96 (added at 6.5 minutes). Experiment E: UpUpUp, 1 (added at 6.5 minutes)

to be due to the presence of inhibitors in the preparation. Furthermore, the same UpUpUp (at the same concentration) was active in protecting polynucleotide phosphorylase against irreversible denaturation by urea (Fig. 2 A) and was also found to protect the enzyme against heat inactivation under conditions similar to those described by Lucas and Grunberg-Manago [8]. These results are in contrast with earlier observations from this laboratory [8] indicating that a 3′-phosphate oligonucleotide, GpCp, did not offer heat protection. This apparent discrepancy is probably related to the size of the oligonucleotide: dinucleotides with a 3′OH-free offer much less protection to the enzyme than trinucleotides.

The effect of urea on the rate of ADP polymerization is similar to that observed with UDP, except that higher concentrations of urea are required in the case of the purine derivative. Urea at a concentration of 3.0 M was found not only to inhibit ADP polymerization, but also to introduce a lag into the otherwise linear kinetics of the reaction (with these enzyme preparations) (Fig. 7). The presence of 4.0 M urea in

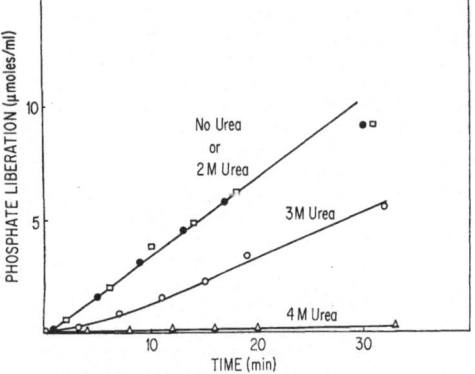

Fig. 7. *Effect of urea on ADP polymerization.* ●, no urea; ○, 3.0 M urea; △, 4.0 M urea; □, 2.0 M urea. Same reaction mixture as in Fig. 3, ADP replacing UDP

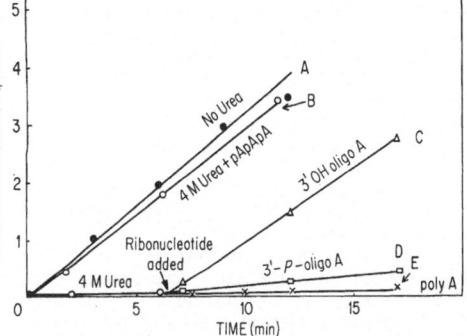

Fig. 8. *Effect of polynucleotides on urea-inhibited polymerization of ADP.* Same reaction mixture as in Fig. 3, ADP replacing UDP, except for enzyme: 14 µg/ml, urea as indicated. In addition, the following were added at 6.5 minutes in mM concentration: 3′-OH oligo A of mixed chain length (average 7), 0.53; poly A, 1; pApApA, 1; 3′-PO₄ oligo A of mixed chain length (average 7), 0.60

the assay medium led to almost complete inhibition of poly A formation. The inhibitory action of this even relatively high concentration of urea was readily reversed by the addition of oligonucleotides terminating with a 3′hydroxyl group (Fig. 8); oligonucleotides with a 3′-phosphate terminal group did

not restore enzymic activity (Fig. 8). Polyadenylic acids of high molecular weight are unable to reverse the action of urea in the polymerization reaction. These results are analogous to those obtained using UDP as substrate and again emphasize that polynucleotides must have a free 3′-hydroxyl terminal group in order to be effective in overcoming the inhibitory action of urea on the polymerization reaction.

Effect of Endogenously Produced Polymers

It is clear that neither the addition of poly U nor of poly A can efficiently reverse the inhibitory effect of urea in the polymerization of UDP or ADP, respectively. It is particularly interesting to see

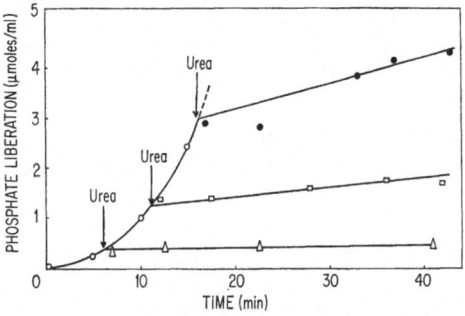

Fig. 9. *Effect of urea added at various times during the polymerization reaction.* Same reaction mixture as Fig. 3, except enzyme 14 µg/ml. Final concentration of urea added 3.0 M, at indicated times

whether the product of UDP or ADP polymerization can reverse the action of urea. It might thus be expected that the urea-induced inhibition of the polymerization reaction would disappear as the product of the reaction accumulated, that is, urea might be expected to induce an initial lag in the polymerization of UDP or ADP, but should have little or no effect on the maximal velocity of the reaction. However, is has been observed experimentally (Figs. 3 and 7) that urea does indeed affect the maximal rate of UDP or ADP polymerization. It appears that the endogenously synthesized poly U or poly A are relatively ineffective in reversing the inhibitory effect of urea. This is clearly shown in Fig. 9 where urea was added to a final concentration of 3.0 M at various times during the polymerization of UDP. The reaction was completely inhibited when the urea was introduced at zero time or after 6 minutes (0.4 µmole/ml P₁ released). A strong inhibition was produced even when urea was added after 16 minutes (3 µmoles/ml P₁ released); it should be noted that the product of UDP polymerization is present in appreciable concentrations during all but the very earliest period of the reaction. This indicated that small oligonucleotides were not produced in sufficient amounts during the synthesis to protect against urea, which is in agreement with the polymerization mechanism as it has been proposed [9].

Effect of Urea on the Phosphorolysis Reaction

The phosphorolysis reaction is less sensitive to urea than the corresponding polymerization reaction. 3.6 M urea only slightly inhibited the phosphorolysis of poly A (Table); however, the affinity of the polymer for the enzyme is changed (K_m in the absence of urea is 5×10^{-6} M, and 10^{-4} M in the presence of 3 M urea). The phosphorolysis of the pyrimidine derivative, poly U, was slightly more sensitive to the effect of urea (Table). The presence of 3 M urea did not markedly alter the apparent equilibrium of the reaction using ADP as substrate (Fig. 10); the extent of phosphorolysis of poly A was also unaffected by the presence of 4.0 M urea.

Table. *Effect of urea on the phosphorolysis reaction*
The usual phosphorolysis reaction mixture (0.1 ml) contained 76 µg of enzyme/ml

Substrate	Addition	[^{32}P]NDP[a] formed	Inhibition
		µmoles/h	%
poly A	—	1.20	
	3.6 M urea	1.03	14
poly U	—	1.18	
	3.6 M urea	0.87	26

[a] NDP: nucleoside diphosphate.

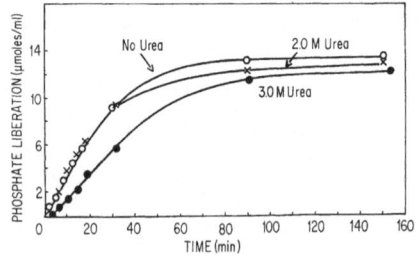

Fig. 10. *Effect of urea on the equilibrium of ADP polymerization reaction. Same as Fig. 7*

DISCUSSION

The effect of urea on polynucleotide phosphorylase appears to involve irreversible as well as reversible processes. When the enzyme is pre-incubated in a medium containing 3 M (or 2 M with some preparations) urea and is subsequently diluted into a urea-free assay mixture, at final urea concentrations which

do not affect catalytic reactions, an irreversible loss of activity is observed; neither removal of urea, nor prolonged exposure to substrate results under a return of the original enzymic activity, at least not in the conditions tried. The rate of this inactivation is markedly decreased if ribonucleotide polymers or oligonucleotides are present in the urea-containing pre-incubation medium. Oligonucleotides both with a free 3′-hydroxyl or with a terminal 3′-phosphate residue are effective. Nucleoside diphosphates at a concentration identical to that used in the polymerization reaction medium also stabilize polynucleotide phosphorylase against inactivation due to pre-incubation with urea.

The polymerization reaction with purified polynucleotide phosphorylase is very sensitive to urea. An inhibition or prolonged lag phase of the polymerization of the nucleoside diphosphates, UDP and ADP, is observed in the normal reaction medium, in the presence of 2 M and 4 M urea, respectively. From previous experiments it appears that the inhibition observed when urea is present in the reaction medium is not due to irreversible structural changes in the enzyme. The enzyme is protected against irreversible denaturation at that concentration of urea, in the presence of diphosphates.

The reversible nature of the urea-induced inhibition is emphasized by the fact that full enzymic activity can be regained not only by removing the urea by dilution, but also by the addition of a certain type of ribonucleotide. Under optimal conditions, oligonucleotides terminating with a free 3′-hydroxyl group are effective; oligonucleotides with a 3′-phosphate terminal group and polynucleotides of high molecular weight fail to restore enzymic activity, in contrast to their ability to protect the enzyme against irreversible denaturation by urea or against heat denaturation. This indicates that the nature of their binding is different during the polymerization and during the protection against urea.

This observation suggests that urea may in some way interfere with the chain initiation reaction. Under normal conditions (in the absence of urea) *E. coli* enzyme is able to synthesize polymers *de novo* [3,10], but the first steps of this synthesis are slower than the elongation process of a chain which has already been primed (see Fig. 4). Nevertheless, the presence of an oligonucleotide with a free 3′-hydroxyl end increases the rate at which the reaction starts, with a very high activator constant (for instance, for pUpUpU[3], 10^{-7} and for pApApA[3], 3×10^{-8} M were found). In view of this high activator constant it is difficult to answer the question as to whether the *de novo* synthesis requires an oligonucleotide as activator. At a concentration of 10^{-9} M an oligonucleotide significantly stimulates the synthesis and it is impossible to eliminate the

[3] Expressed as oligonucleotides.

eventuality of either the diphosphates or the enzyme preparation being contaminated by such a small amount of oligonucleotide.

This activation by oligonucleotides cannot be solely due to the fact that they are used as centers for chain elongation since the great number of chains synthesized during the reaction would be incompatible with the restricted number of oligonucleotides which are necessary to obtain maximum rate (the order of magnitude being a few molecules of activator per molecule of enzyme). For example in the presence of 10^{-2} M UDP, 3×10^{-5} M chains of an average molecular weight of 100,000, are synthesized.

The results obtained in the presence of urea could be explained if one assumes that urea strongly decreases the activator constant of oligonucleotides (the decrease of affinity for polymers in the phosphorolysis reaction would be in favor of this). One cannot decide on the basis of kinetic arguments alone, however, whether this decrease is due to a weakening of the enzyme-oligonucleotide interactions, or whether urea modifies the structure of the enzyme itself in such a way that it becomes incapable of attaching the oligonucleotide to the activating site.

On the other hand, the effect of urea could be interpreted as a deterioration of the *de novo* initiation mechanism in contrast to the situation in the absence of urea, in its presence the amount of oligonucleotides required for activation (of the order of 10^{-4} to 10^{-3} M depending on the concentration of urea) is compatible with the amount necessary to start a chain; the *E. coli* enzyme would then behave like it has been suggested that *M. lysodeikticus* enzyme does and the lag phase could be similarly explained [11]. Long polymers and oligo 3′-phosphate would be inefficient in this reaction because they do not serve as primer for chain elongation.

The phosphorolysis reaction is much less sensitive to the action of urea, which is consistent with urea inhibition in the polymerization reaction having an effect on chain initiation. The stimulation of polymer synthesis by urea, observed on crude enzyme [12] is probably due to inhibition of K-activated exonuclease present in the preparation and which is very sensitive to urea [13].

This work was supported by the following grants: No. C-O4580 of United States National Institute of Health; Convention 6600020 of Délégation Générale á la Recherche Scientifique et Technique (Comité de Biologie Moléculaire); Comité de la Seine de la L.N.F.C.C.; French National Research Council (R.C.P. No. 24); F.R.M.F.; and a participation from the French Atomic Energy Commission. Dr. Lucas-Lenard, and Dr. Harvey, were both N.I.H. fellows.

REFERENCES

1. Grunberg-Manago, M., and Ochoa, S., *J. Am. Chem. Soc.* 77 (1955) 3165.
2. Grunberg-Manago, M., Ortiz, P. J., and Ochoa, S., *Biochim. Biophys. Acta*, 20 (1956) 269.
3. Williams, F. R., and Grunberg-Manago, M., *Biochim. Biophys. Acta*, 89 (1964) 66.
4. Brennman, F. N., and Singer, M. F., *Biochem. Biophys. Res. Commun.* 17 (1964) 801.
5. Kasai, K., and Grunberg-Manago, M., *European J. Biochem.* 1 (1967) 152.
6. Nirenberg, M., and Leder, P., *Science*, 145 (1964) 1399.
7. Fiske, C. H., and Subbarow, I., *J. Biol. Chem.* 66 (1925) 375.
8. Lucas, J. M., and Grunberg-Manago, M., *Biochem. Biophys. Res. Commun.* 17 (1964) 395.
9. Thang, M. N., Harvey, R. A., and Grunberg-Manago, M., in preparation.
10. Harvey, R. A., and Grunberg-Manago, M., *Biochem. Biophys. Res. Commun.* 23 (1966) 448.
11. Singer, M. F., and Guss, J. K., *J. Biol. Chem.* 237 (1962) 182.
12. Cramer, F., and Randerath, K., *Nature*, 196 (1962) 1209.
13. Spahr, P. F., *J. Biol. Chem.* 239 (1964) 3716.

R. A. Harvey's present address:
Department of Biochemistry, Rutgers Medical School
New Brunswick, N.J., U.S.A.

T. Godefroy and M. Grunberg-Manago
Institut de Biologie physico-chimique
13 Rue Pierre Curie
75-Paris-5, France

J. Lucas-Lenard's present address:
The Rockefeller University
New York, U.S.A.

European J. Biochem. 1 (1967) 334—343

On the Mechanism of Na+- and K+-Stimulated Hydrolysis
of Adenosine Triphosphate

1. Purification and Properties of a Na+- and K+-Activated ATPase from Ox Brain

W. Schoner, C. von Ilberg, R. Kramer, and W. Seubert

Institut für vegetative Physiologie, Chemisch-physiologisches Institut der Universität, Frankfurt a. Main

(Received January 20, 1967)

The purification and properties of a Na+- and K+-activated ATPase from ox brain is de-scribed. The enzyme preparation is characterized by a high purity and a low content ($< 1\%$) of Mg++-stimulated ATPase.

It is well established that the Na+- and K+-activated ATPase system of the cell membrane is related to active cation transport [1]. A relationship between cation transport and amino acid and sugar transport through the cell membrane has also been suggested [2,3]. The enzyme was first discovered by Skou [4]. The studies of Bonting, Simon and Hawkins [5] revealed that a transport ATPase (Na+-, K+-activated ATP phosphohydrolase) is present in all tissues which are capable of transporting cations against an electrochemical gradient. The relationship between this ATPase and Na+-transport in the cell is also strengthened by the inactivation of both systems by ouabain and Ca++ [1].

From exchange studies with [^{32}P]ADP and ATP, is has been suggested that an enzyme-bound phosphorylated intermediate participates in Na+- and K+-stimulated ATP-hydrolysis [6]. Experimental evidence for the formation of a phosphorylated intermediate has been presented by several investigators on the basis of tracer studies with [^{32}P]ATP [7—12]. But differing views on the stability of the phosphorylated intermediate have appeared in the literature. Post et al., Hokin et al., Albers et al., Gibbs et al., and Nagano et al. [7—11] were able to demonstrate an intermediate stable in acid solution but labile in alkali. Heinz and Hoffman separated from erythrocyte ghosts treated with [^{32}P]ATP a small acid-labile

fraction which lost the ^{32}P-label in the presence of unlabeled ATP, presumably by exchange, whereas in the acid-stable fraction the radioactivity remained constant [12]. These findings suggested at least two different binding sites for phosphate on the ATPase, one of which is not related to Na+- and K+-stimulated ATP hydrolysis.

Clarification of the apparent discrepancies necessitates the use of a purified enzyme preparation—free of a Mg++-stimulated ATPase—for studies on the chemical nature of the phosphorylated intermediate. The present communication deals with the isolation and the properties of an active Na+- and K+-stimulated ATP-ase from ox brain.

EXPERIMENTAL PROCEDURE

Materials

Nucleotides, phospho-enolpyruvate (cyclohexyl-ammonium salt), 3-phosphoglycerate, DPNH, TPN, pyruvate kinase, lactic dehydrogenase, 3-phosphoglycerate kinase, glyceraldehyde-3-phosphate dehydrogenase, hexokinase and glucose-6-phosphate dehydrogenase were purchased from Boehringer und Soehne, Mannheim. 2-Amino-2-methyl-1,3-propanediol (AMPD) was purchased from Eastmann Organic Chemicals, Rochester, New York, 2,3-dimercapto-1-propanol (BAL) from Fluka A.G., Buchs, Switzerland, glass beads 0.25—0.30 mm diameter were from B. Braun Apparatebau, Melsungen.

All nucleotides were converted to the free acids by passage through Dowex 50 (H+-form) and neutralized with 2-amino-2-methyl-1,3-propanediol.

ATP was assayed according to Adam [13], and Lamprecht and Trautschold [14], and ADP according to Adam [15]. Hydrazine hydrochloride was neutralized with Dowex 1 and quantitated by iodometry; it contained less than 2% NH$_4$+ as assayed according to Berthelot [16].

Non-standard Abbreviations. Sodium deoxycholate, DOC.
Enzymes. ATPase, or ATP phosphohydrolase (EC 3.6.1.3); pyruvate kinase, or ATP: pyruvate phosphotransferase (EC 2.7.1.40); glycerate kinase, or ATP:D-glycerate 3-phosphotransferase (EC 2.7.1.31); glyceraldehydephosphate dehydrogenase (NAD), triosephosphate dehydrogenase (NAD), or D-glyceraldehyde-3-phosphate:NAD oxidoreductase (phosphorylating) EC 1.2.1.13); hexokinase, or ATP:D-hexose 6-phosphotransferase (EC 2.7.1.1); glucose-6-phosphate dehydrogenase, or D-glucose-6-phosphate:NADP oxidoreductase (EC 1.1.1.49); lactic dehydrogenase (EC 1.1.1.27); succinate dehydrogenase (EC 1.3.99.1); DPNH oxidase (EC 1.6.99.1); cytochrome c reductase (EC 1.6.2.1).

Optical Assay for Na^+, K^+-activated ATPase

The reaction mixture contains in 1.6 ml: 100 μmoles imidazole, pH 7.3, 5 μmoles $MgCl_2$, 0.4—0.5 μmoles DPNH, 100 μmoles NH_4Cl, 3 μmoles AMPD-ATP, 0.6 μmoles phospho-enolpyruvate, 6 units pyruvate kinase, 6 units lactic dehydrogenase, 150 μmoles NaCl or 0.15 μmoles ouabain (strophanthin G). Temperature 37°, λ = 366 mμ.

In order to compensate for the activity of the Mg^{++}-stimulated ATPase, all readings were taken against a control in which the Na^+-, K^+-stimulated ATPase was inhibited by 9.4×10^{-5} M ouabain. To avoid errors due to incorrrect pipetting of the enzyme solutions, the decrease of the optical density was controlled over a period of 1.5 minutes in both the test and the control cuvettes. The reaction was initiated by addition of 0.15 ml 1 M NaCl solution and 0.001 M ouabain to the test and the control cuvettes respectively.

For assays of the homogenate (Table 2), 0.02 ml 20% (w/v) Na-desoxycholate was added to 2 ml of the suspension and mixed in a Potter homogenizer (all operations being carried out at 0°). This treatment is no longer necessary for assays of the subsequent procedures.

Other Systems for Assays of Na^+-, K^+-activated ATPase

For a study of the effect of pH (Fig. 12) and hydrazine (Fig. 10) on the activities of the enzyme, and for specificity studies (Table 4), the enzymatic activity was calculated from inorganic phosphate and ADP liberated by the ATPase. Composition of the reaction mixture (1 ml): 100 μmoles imidazole pH 7.3, 5 μmoles $MgCl_2$, 100 μmoles NaCl, 5 μmoles KCl, 1.5 μmoles ATP, 0.02 to 0.08 units ATPase. The mixture was incubated for 10 minutes at 37°. The reaction was stopped by addition of 1 ml ice-cold 10% (w/v) perchloric acid. Inorganic phosphate was determined according to King [17]. For the enhancement of the enzymatic activity by hydrazine, the supernatant was neutralized with 5 M K_2CO_3 and after separation of the potassium perchlorate precipitate, ADP in the supernatant was assayed according to Adam [15].

Assay of DPNH-oxidase. The reaction mixture contained in 1.5 ml: 200 μmoles tris buffer, pH 8.1, 0.5 μmoles DPNH, 0.01—0.02 ml enzyme solution. The reaction was initiated by the addition of 0.1 μmole $K_3Fe(CN)_6$. The decrease of the DPNH absorption was followed at 366 mμ. Temperature 37°. A control value to correct for the autoxydation of DPNH was determined simultaneously.

Assay of DPNH-cytochrome c Reductase. 1.5 ml test volume contained 30 μmoles tris buffer, pH 7.7, 0.3—0.5 μmoles DPNH, 0.1 μmole KCN, 0.2 μmoles cytochrome c. The reaction was started by the addi-

tion of 0.01 ml enzyme solution. The absorbance at 546 mμ was a measure of the enzyme activity (ε = 2.1×10^6 cm²/mole [18]). Temperature = 37°.

Assay of Succinate Dehydrogenase. The reaction mixture contained in 2 ml: 50 μmoles potassium phosphate buffer, pH 7.5, 10 μmoles sodium succinate, 1 μmole KCN and 0.5 ml dichlorophenolindophenol (110 mg/l). The reaction was started by addition of 0.02 ml enzyme solution. The decrease of absorbance was measured at 578 mμ against control without succinate. Activity calculations are based upon the extinction coefficient for dichlorophenolindophenol (14.3×10^6 cm²/mole [19]).

One enzyme unit is defined as the amount of enzyme catalyzing the conversion of 1 μmole substrate per minute at 37°.

Ribonucleic acid was determined after extraction according to Schneider [20] with the orcine method of Lusena [21]. Protein was determined according to Lowry [22]. Na^+ and K^+ were determined with the flame-photometer Eppendorf (Netheler und Hinz, Hamburg).

Purification of the Na^+-, K^+-activated ATPase

All operations were performed at 0°. Ox brain is dissected immediately after slaughter, packed in ice and frozen at − 18°. Under these conditions, the enzymatic activity remains constant over a period of 6 months. For preparation of the enzyme the brain is slowly brought to a temperature of about − 5°. Cortex and medulla may then be easily separated by use of a scalpel on a petri dish cooled by a salt—ice mixture (− 15°).

Homogenate and 1st Sediment. 40 g of ox brain cortex are homogenized (2 minutes) in a Waring blender with 200 ml ice—cold 0.25 M sucrose containing 1 mM sodium—EDTA. Soluble material is separated by centrifugation at $40,000 \times g$ for 20 minutes in a Servall RC 2 centrifuge. The solid sediment thus obtained is taken up in 200 ml of cold 0.25 M sucrose (1 mM sodium EDTA) and again homogenized in a Potter-Elvehjem homogenizer for about 2 minutes. 20% (w/v) sodium deoxycholate is added slowly with stirring to give a final concentration of 0.2% (w/v). After stirring for an additional 15 minutes, the suspension is stored at 0—4° for 7 days.

Fractionated Centrifugation. After the enzymatic activity has reached a maximum, the suspension is centrifuged for 45 minutes at $40,000 \times g$. The pellet is then soon to be covered with a heterogenous supernatant which is decanted; all material that flows out of the tube after it has been held upside down for one minute is utilized in the next step. This is then centrifuged at $100,000 \times g$ in a Spinco L 50 centrifuge (Rotor No. 30) for 30 minutes. After this second centrifugation, the supernatant is again allowed to drain out slowly. The sediments in both cases are discarded.

Chromatography on Glass Beads. This supernatant (55 ml) is added to a glass bead column (5×42 cm) equilibrated with 0.075 M potassium phosphate buffer, pH 7.5, and 1 mM BAL. After the brownish suspension has penetrated, the column is washed with 400 ml 0.075 M phosphate buffer, pH 7.5, containing 1 mM BAL (0.4—0.6 ml/minute). The activity of the ATPase in the liquid is associated with the turbid eluates (o.s.$_{578}$ >0.05, Fig. 1). These are combined and are centrifuged for 2 h at 100,000×g. The clear yellow supernatant is then discarded.

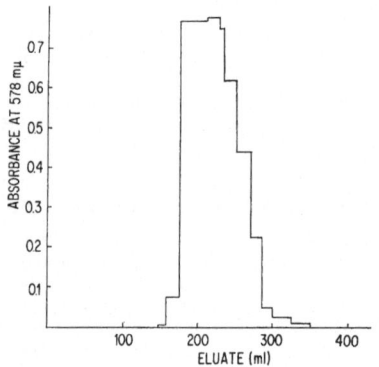

Fig. 1. *Chromatography of 55 ml supernatant (100,000×g) on glass bead column (5×42 cm)*

Washing Procedure. The pale sediment is taken up in 50 ml 0.25 M sucrose, homogenized for 2 minutes in a Potter-Elvejhem homogenizer and then diluted with 0.25 M sucrose and 1 mM BAL to a volume of 300 ml. After centrifugation for 2 hours at 100,000×g, a milky supernatant is obtained with a glassy fat layer on its surface. Both the supernatant and the fat layer are decanted. The solid pale sediment is homogenized in the same medium (50 ml), diluted again to 300 ml and centrifuged for the same time at 100,000×g. The sediment is taken up in 40 ml of the same medium, homogenized in the Potter for 1 minute and centrifuged for 20 minutes at 12,000×g. A small pale sediment is obtained which is then discarded. To the supernatant is added enough imidazole, pH 7.3, to give a final concentration of 0.02 M. Likewise a final concentration of 1 mM glutathione is contrived.

Aliquots of the enzyme are frozen quickly in liquid air and stored at −50°. The enzymatic activity remains constant during a period of two months.

Preparation and Regeneration of the Column

1 kg glass beads (diameter 0.25—0.30 mm) are suspended in 300 ml 0.075 M potassium phosphate buffer, pH 7.5, and washed 3 times with 200 ml of the same medium. For regeneration the beads are suspended for 3 to 4 hours in 33% (w/v) NaOH and subsequently washed with water until a neutral pH is obtained.

RESULTS

Assay

Because of the insolubility of the transport ATPase in aqueous solutions, the enzyme is usually assayed by determining inorganic phosphate liberated from ATP. But after treatment of the homogenate with deoxycholic acid [23], structural enzymes may

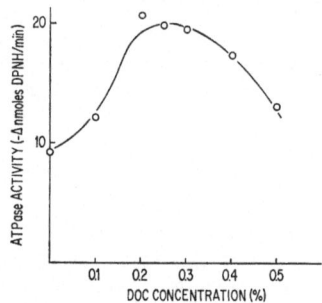

Fig. 2. *Activation of ATPase of different end concentrations of deoxycholate (0°)*

be dispersed and assayed by the more convenient and rapid optical assay according to Warburg [24]. This method was also applied to assays of the transport ATPase. As shown in Fig. 2, treatment of the ATPase suspensions with 0.2% (w/v) deoxycholate results in optimal activation of the enzyme. The activity was determined by coupling ATP hydrolysis [Eqn. (1)] with pyruvate kinase [Eqn. (2)] and lactic dehydrogenase [Eqn. (3)].

$$ATP + H_2O \xrightarrow{Mg^{++}, Na^+, NH_4^+} ADP + P_i \quad (1)$$

$$\text{phospho-enolpyruvate} + ADP \xrightarrow{Mg^{++}, NH_4^+}$$
$$\text{pyruvate} + ATP \quad (2)$$

$$\text{pyruvate} + DPNH + H^+ \to \text{lactate} + DPN^+ \quad (3)$$

$$\text{phospho-enolpyruvate} + H_2O + DPNH + H^+$$
$$\downarrow ATP, Mg^{++}, Na^+, NH_4^+$$
$$\text{lactate} + DPN^+ + P_i. \quad (4)$$

A stoichiometric relationship exists between the oxidation of DPNH and the hydrolysis of ATP

[Eqn. (4), Table 1]. Enzyme activity was calculated from the rate of DPNH-oxidation, which could be followed spectrophotometrically.

Table 1. *Balance of DPNH-oxidation and the liberation of inorganic phosphate in the optical assay for Na+-, K+-activated ATPase*
The decrease of the absorbance of DPNH at 366 mμ was followed for 3 minutes in the optical assay (enzyme protein 16.6 μg, for details see Methods). After a further 7 minutes of incubation the reaction was terminated by the addition of 1 ml 10% (w/v) perchloric acid to the test cuvettes and P_i liberated determined. The amount of DPNH oxidized/minute and P_i liberated/minute were calculated. All values were corrected by a blank with 10^{-4} M ouabain

DPNH oxidized	Inorganic phosphate liberated	$-\Delta$ DPNH : P_i
μmole/min	μmole/min	
0.057	0.056	1.008 : 1.00

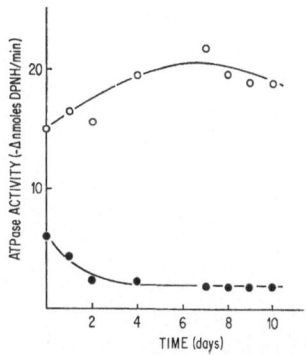

Fig. 4. *Effect of ageing in the presence of 0.2% (w/v) DOC.* O, on Na+-, K+-activated ATPase; ●, on Mg++-activated ATPase. Temperature = 0°

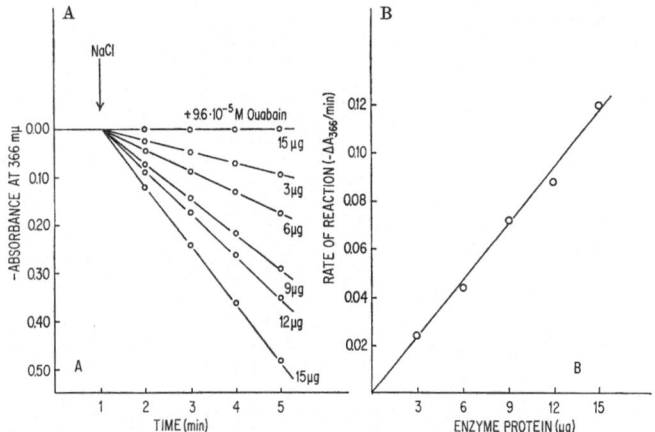

Fig. 3. *Optical assay for Na+-, K+-activated ATPase.* A, Kinetics of the assay (Figures by the curves: μg enzyme protein). B, Correlation between the rate of reaction and the amount of enzyme protein (for details, see Methods)

The auxiliary enzymes, pyruvate kinase and lactic dehydrogenase, contain considerable amounts of ammonium ions (see Methods) which, like K+, also activate the ATPase [1]. For activity determinations, potassium was therefore replaced by NH_4^+ as an activator of the ATPase [1]. In order to compensate for the activity of a Mg++-stimulated ATPase present in the crude preparations, all readings of the optical density were taken against a blank in which the transport ATPase was inhibited by 9.6×10^{-5} M ouabain.

In Fig. 3A, the kinetics of DPNH-oxidation at different enzyme concentrations are illustrated. With an excess of the auxiliary enzymes, the rate of DPNH-oxidation is proportional to the concentrations of the ATPase (Fig. 3B).

Purification of Na+- and K+-activated ATPase

The Na+-, K+-activated ATPase is mainly located in the cortex of the brain [5]. The purification procedure was therefore started with tissue dissected from the frozen brain: homogenization in isotonic medium and centrifugation at 40,000 × g results in separation of part of the soluble protein (Table 2). Subsequent ageing of the residue with 0.2% (w/v) deoxycholate for several days at 0° solubilizes the transport ATPase. The Mg++-stimulated ATPase is partially inactivated by this treatment (Fig. 4, Table 2), since the ratio of the activity of the Na+-, K+-stimulated ATPase to that of the Mg++-stimulated ATPase rises from 2 to 6.5 (Table 2, column 10). Subsequent centrifugation at 40,000 × g and 100,000 × g further removes a large part of the Mg++-activated

Table 2. *Purification of the Na+, NH4+-activated ATPase*

Step	Volume	Protein	Na+, NH4+, Mg++-stimulated ATPase			Mg++-stimulated ATPase			Ratio of the activities with $(Na^+, NH_4^+, Mg^{++}) - Mg^{++}$ / Mg^{++}
			Activity	Specific activity	Activity	Activity	Specific activity	Activity	Activity
	ml	mg	U	U/mg protein	%	U	U/mg protein	%	%
Homogenate (40 g ox brain cortex)	235	3383	1260	0.367	100.0	683	0.203	100.0	1.8
Sediment (40,000×g) + DOC (0.2%, w/v)	240	2520	910	0.370	72.3	442	0.175	64.8	2.02
7 days later	240	2520	1208	0.478	95.8	185	0.0735	27.1	6.5
Fractionated centrifugation (40,000 and 100,000×g): Supernatant (100,000×g)	165	1203	819	0.68	65.0	80.3	0.066	11.7	10.2
Chromatography on glass beads: Eluate	940	970	548	0.566	43.4	30.4	0.031	4.5	18.0
Repeated washing of the particles (100,000×g), Supernatant (12,000×g) after 2 washings	35	105.8	415	3.96a	33.0	6.8	0.064	0.99	61.0

a Optimal purification observed at this stage: 4.96 U/mg protein.

Table 3. *Relation between the activities of Na+, K+-stimulated ATPase and DPNH oxidase, succinic dehydrogenase, cytochrome c reductase and RNA-content of the individual purification steps*

Step	Na+, K+-stimulated ATPase			DPNH oxidase			Succinate dehydrogenase			Cytochrome c reductase			RNA-content
	Activity	Specific activity	Activity	Activity	Specific activity	Activity	Activity	Specific activity	Activity	Activity	Specific activity	Activity	
	U	U/mg protein	%	U	U/mg protein	%	U	mU/mg protein	%	U	mU/mg protein	%	μg/mg protein
Homogenate (40 g ox brain cortex)	1260	0.367	100.0	2560	0.762	100.0	59.2	17.7	100.0	123.5	36.8	100.0	15.5
Sediment (40,000×g) + DOC (0.2%, w/v)	910	0.37	72.3	2060	0.815	80.3	43.5	17.3	73.5	48.0	19.0	38.9	15.1
7 days later	1208	0.478	95.8	1960	0.778	75.4	16.8	6.66	28.4	8.0	3.14	6.5	—
Fractionated centrifugation (40,000 and 100,000×g): Supernatant (100,000×g)	819	0.68	65.0	1050	0.87	41.0	1.73	1.36	2.92	4.4	3.66	3.57	11.5
Chromatography on glass beads: Eluate	548	0.566	43.4	—	—	—	2.63	2.72	4.32	1.42	1.46	1.15	—
Repeated washing of the particles (100,000×g), Supernatant (12,000×g) after 2 washings	415	3.96	33.0	78.3	0.74	3.02	0.0	0.0	0.0	0.0	0.0	0.0	113.0

ATPase (Table 2, column 7—9). Mitochondrial enzymes such as succinic dehydrogenase and cytochrome c reductase also separate out at the same time (Table 3, column 8—13). The activity ratio Na^+-, K^+-stimulated ATPase/Mg^{++}-stimulated ATPase increases still further by passing the solution

Properties of the Na^+-, K^+-stimulated ATPase

In order to assay the Na^+-, K^+-stimulated ATPase under optimal conditions, the relation between reaction velocity and temperature, pH and the concentration of the various reactants and cofactors was investigated.

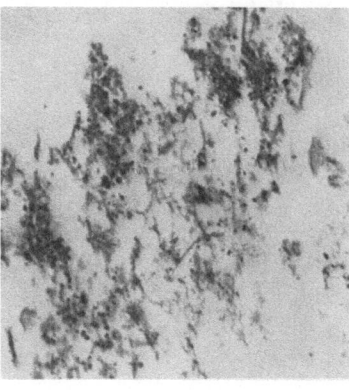

B

Fig.5. *Electron micrograph of the ATPase preparation made as described under Experimental Procedure.* The membranes were fixed with glutaric-dialdehyde/OsO_4. A, left side shows fragments of membranes partially covered with ribosomes ($\times 32,200$); B, right side: numerous free and membrane-bound ribosomes ($\times 66,000$)

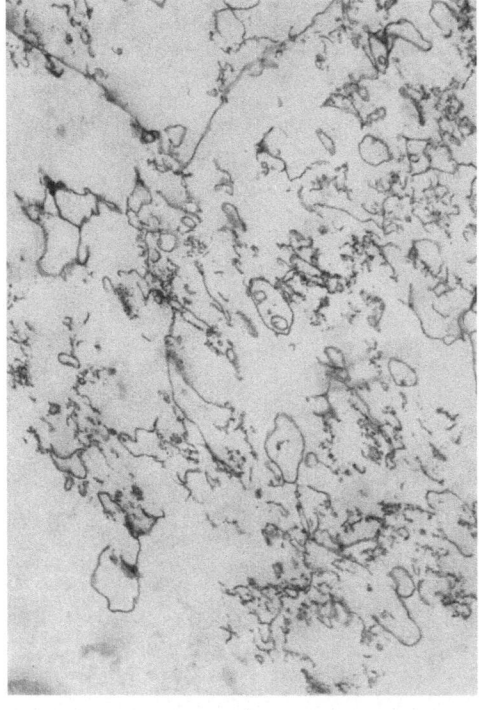

A

through a column of glass beads. The Mg^{++}-stimulated ATPase finally drops below $1^0/_0$ of the total ATPase activity after 2 subsequent washings of the insoluble material. At this step, a large amount of inactive protein including last traces of cytochrome c reductase and DPNH oxidase are also removed, since the specific activity of the ATPase rises from 0.5 to 4.0 (Table 2, column 5). During the purification of the enzyme, the proportion of RNA associated with it increases from 15.5 to 113 µg/mg protein (Table 3, column 14). This indicates that a microsomal fraction is isolated [25]. Additional evidence for this interpretation has been obtained from an electron microscope picture. Only membranes and membranous fragments could be shown to be present in this preparation (Fig.5).

In Fig.6, the relationship between the reaction velocity and Mg^{++} concentration at various levels of ATP is illustrated. The Mg^{++}-enhancement passes through a maximum, indicating an inhibitory effect of magnesium on ATPase at high concentrations of Mg^{++}. With increasing levels of ATP the activity maximum shifts towards higher values of Mg^{++} concentrations (curves II and III). In these cases, the Mg^{++}-enhancement curve is flattened If the ATP concentrations are varied at different Mg^{++}-levels (Fig.7), a graph similar to that of Fig.6 is found. These observations may be interpreted by the formation of Mg_2-ATP and/or Mg-ATP_2-complexes, which compete with the active Mg-ATP complex for the active site of the enzyme. In these cases, however, optimal enzymatic activities should be obtained at a

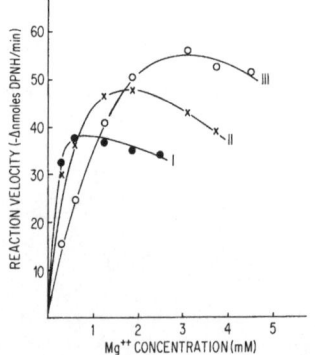

Fig. 6. *Relationship between reaction velocity and Mg++-concentration at various ATP levels.* Curve I, 0.5 mM ATP; curve II, 1.0 mM ATP; curve III, 3.0 mM ATP

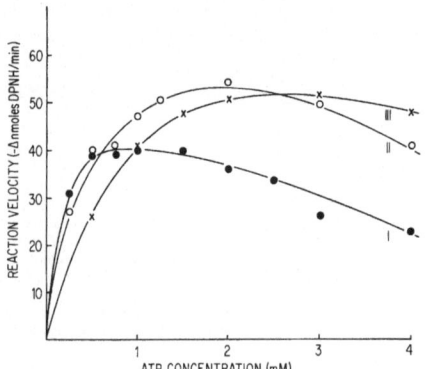

Fig. 7. *Relationship between reaction velocity and ATP-concentration at different Mg++-concentrations.* Curve I, 0.75 mM Mg++; curve II, 1.56 mM Mg++; curve III, 3.1 mM Mg++

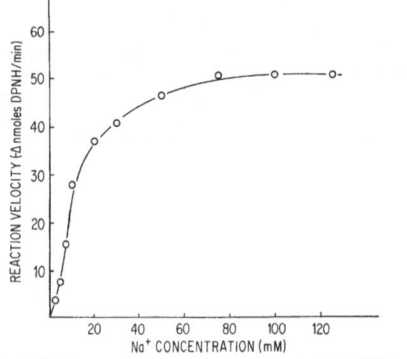

Fig. 8. *Enhancement of the reaction velocity by Na+.* 0.051 units ATPase. Additional components (see Methods)

fixed Mg/ATP ratio throughout. But as can be seen from Fig. 13, this is not the case: maximal enzymatic activities are obtained at different Mg/ATP ratios. An attempt was made to obtain further information on the type of inhibition by plotting $1/v$ against [ATP] and [Mg++] (inhibitors) with various concentrations of Mg++ and ATP respectively [26]. If the enzyme is inhibited by Mg_2-ATP or Mg-ATP_2 complexes, a competitive type of inhibition should be seen. But in the case of an inhibitory action of both reactants on intermediary steps, a mixed type of inhibition might be found. However a differentiation

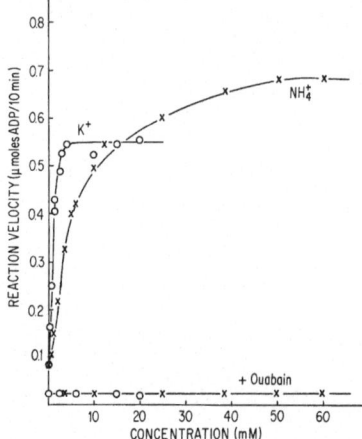

Fig. 9. *Relationship between reaction velocity and K+ or NH_4+ concentration.* 0.034 units ATPase were incubated for 10 minutes in 1.6 ml with 100 μmoles imidazole pH 7.3, 5 μmoles MgCl₂, 3 μmoles ATP and 150 μmoles NaCl at 37°. The reaction was stopped with 1 ml 10%(w/v) perchloric acid, neutralized with 5 M K₂CO₃ and the amount of ADP liberated was quantitated according to Adam [15]. The reaction was completely inhibited by 9.5×10^{-5} M ouabain (lower curve)

between a competitive and a mixed type of inhibition was not possible by this analysis: because of the enhancement of the overall reaction by both Mg++ and ATP, a V_{max} for one of the two reactants could not be obtained.

The effect of Na+ concentration on the reaction velocity is illustrated in Fig. 8. Optimal rates are observed at 10^{-1} M Na+. This value is very close to that of the extracellular Na+ concentrations.

Since pyruvate kinase depends also on K+ [Eqn. (2)], the effect of this cation on ATPase activity was determined by measuring ADP liberation from ATP (see Methods). In agreement with Skou [1] NH_4+ can substitute for K+ (Fig. 9) as can NH_2-$NH_2 \cdot$ HCl (Fig. 10). The stimulatory effect of Na+ and K+ as well as of all the other cations which substitute for K+ is inhibited by 10^{-4} M ouabain.

In Fig. 11 and 12 the effect of temperature and pH on the enzyme activity is shown. Optimal rates were observed at 50° and a pH of 7.3 respectively.

The enzyme preparation also hydrolyses ITP and GTP (Table 4). The activation by monovalent cations is different with different substrates. In presence of Mg^{++} and Na^+, potassium increases ATP-

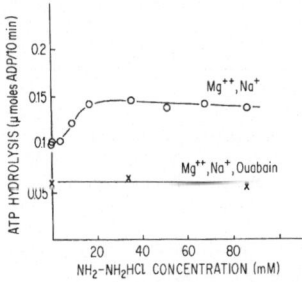

Fig. 10. *Effect of hydrazinehydrochloride on ATP-hydrolysis.* 0.064 units ATPase were assayed with varying amounts of $NH_2-NH_2 \cdot HCl$, ouabain 10^{-4} M. For details, see Methods

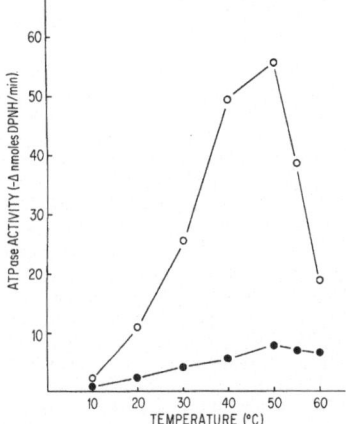

Fig. 11. *Effect of temperature on ATPase activity.* For details, see Methods

hydrolysis tenfold. ITP-hydrolysis is only doubled. The enzymatic activity with GTP remains unchanged. Ouabain inhibits the enzymatic hydrolysis of all three nucleotides.

DISCUSSION

The ATPase preparations described in this communication have a specific activity of 3—5 units/mg protein with a ratio of the activities of Na^+-, K^+-stimulated ATPase/Mg^{++}- stimulated ATPase of

23*

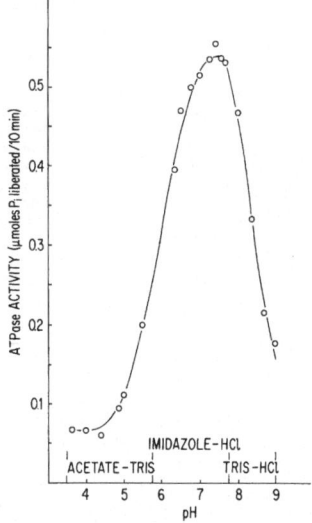

Fig. 12. *Effect of pH on ATPase activity.* For details, see Methods

Table 4. *Nucleotide specifity of Na^+-, K^+-activated ATPase* 100 mM imidazole, pH 7.3, 3 mM $MgCl_2$, 100 mM NaCl, 5 mM KCl, 1.5 mM trinucleotide; ouabain 10^{-4} M (for details see Methods)

| | P_i liberated from | | |
	ATP	ITP	GTP
	μmoles/mg protein/min		
Mg^{++}	0.27	0.065	0.035
Mg^{++}, Na^+	0.22	0.159	0.119
Mg^{++}, K^+	0.40	0.062	0.036
Mg^{++}, Na^+, K^+	2.05	0.278	0.165
Mg^{++}, Na^+, K^+,			
ouabain	0.26	0.049	0.036
	%		
Mg^{++}, Na^+, K^+	100	13.5	8

about 60. A comparison of these purity criteria with those reported in the literature (Table 5) characterizes our ATPase as one of the purest so far described. The morphologically homogeneous fraction shows only membranes (Fig. 5). The particles are free from mitochondrial fragments as indicated by the absence of succinic dehydrogenase (Table 3, column 8—10). Recent studies on the subcellular localization of the Na^+-, K^+-stimulated ATPase by Hosie [27] and Kurokawa, Sakamoto and Kato [28] revealed the activity of the enzyme to be present in the external membranes of the nerve endings, the nerve ending particles and synaptosomal limiting membranes. The high RNA content of our prepara-

Table 5. *Comparison of Na+-, K+-stimulated ATPase preparations from different sources*

Authors	Enzyme source	Specific activity	Na+-, K+-ATPase	Yield
		U/mg protein	%	U/g tissue
Schoner, von Ilberg, Kramer and Seubert	ox brain cortex	5.0	98.5	10.4
Nakao, Tashima, Nagano and Nakao [33]	rabbit brain	3.1	97.0	3.4
Post and Sen [34]	rabbit kidney guinea pig kidney	2.5	90.0	3.9
Swanson, Bradford and McIlwain [35]	guinea pig brain	1.2	83.0	10.4
Glynn, Slayman, Eichberg and Dawson [36]	electric eel electric organ	1.0[a]	100.0	2.5
Rendi and Uhr [30]	calf kidney	0.5	100.0	1.7

[a] Assayed at 25°.

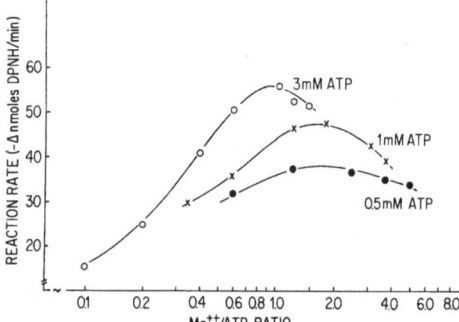

Fig. 13. *Relationship between reaction rate and Mg++/ATP ratio as a function of ATP-concentration*

tion also supports the view that the ATPase is localized in the endoplasmatic reticulum.

In the course of the purification procedure, cytochrome c reductase is completely removed and DPNH oxidase is reduced to 3% of the initial activity. These results do not support a relation between Na+-, K+-stimulated ATP hydrolysis and electron transport as suggested by Skou [29].

Attempts to solubilize the enzyme by treatment with trypsin, lipase, high concentrations of detergents, extraction of acetone powder, treatment with toluol and ultrasonic deterioration always resulted in complete inactivation of the enzyme.

The activity of the particulate enzyme is usually measured by assaying the amount of inorganic phosphate liberated from ATP. The Na+-, K+-stimulated ATPase was purified with the help of a coupled optical assay, as described by Warburg [24]. Falsifications of the rate of decrease of the optical density by sedimenting membranes were avoided by treatment of the homogenate with deoxycholate. Inhibition of the enzyme by ADP [30,31] was avoided by coupling the ATP hydrolysis with ATP regeneration catalyzed by pyruvate kinase, which also depends on K+ or NH_4^+. The optical assay is therefore unsuitable for

studies of the activating effect of these cations on the Na+-, K+-stimulated ATPase. In such cases, the enzymatic activity should be calculated from inorganic phosphate or ADP liberated by the ATPase.

The properties of the Na+-, K+-activited ATPase from ox brain are in good agreement with those of enzymes isolated from other tissue sources. In addition to the univalent cations NH_4^+, Li+, Rb+, and Cs+, hydrazine was also found to substitute for K+ (Fig. 10). The stimulatory effect of hydrazine was also abolished by ouabain. This finding indicates that hydrazine binds to the active centre of the ATPase. As reported earlier [31], this finding also contradicts the participation of an acylphosphate in Na+-, K+-stimulated ATP hydrolysis, as suggested by Post et al. [8], Hokin et al. [11] and Nagano et al. [10]: Hydrazine would cleave an acylphosphate to inorganic phosphate and a hydrazide, thus inactivating the acceptor group for phosphate.

According to Skou [6], 2 moles of Mg++ combine with 1 mole of ATP during ATP hydrolysis. This conclusion was drawn from the optimal rate of ATP hydrolysis observed at a ratio Mg++/ATP = 2:1. In contrast to these findings, Wheeler and Whittam [32] obtained optimal enzymatic activity at ratios Mg++/ATP = 1:1. As a contribution to the clarification of these apparent discrepancies, the relation between reaction velocity and the ratio Mg++/ATP was investigated as a function of ATP. As illustrated in Fig. 13, however, optimal rates of the ATP hydrolysis could be observed at different Mg++/ATP ratios depending on the ATP concentration. This suggests an interaction of one or both reactants with intermediary steps of ATP hydrolysis. No conclusion can therefore be drawn from a study of the relation between reaction velocity and Mg++/ATP ratio on the chemical configuration of the enzyme-substrate complex.

This work was supported by grants of the U.S.A. National Science Foundation (NSF-G-22 107, E. Heinz). The authors wish to thank Prof. Heinz for his encouragement. The authors are indebted to Dr. A. Nolte, Senckenbergisches Pathologisches Institut, Frankfurt, for the preparation of the electron micrograph.

REFERENCES

1. Skou, J. C., *Physiol. Rev.* 45 (1965) 596.
2. Csáky, T. Z., *Federation Proc.* 22 (1963) 3.
3. Csáky, T. Z., and Hara, Y., *Am. J. Physiol.* 209 (1965) 467.
4. Skou, J. C., *Biochim. Biophys. Acta*, 23 (1957) 394.
5. Bonting, S. L., Simon, K. A., and Hawkins, N. M., *Arch. Biochem.* 95 (1961) 416.
6. Skou, J. C., *Biochim. Biophys. Acta*, 42 (1960) 6.
7. Albers, R. W., Fahn, S., and Koval, G. J., *Proc. Natl. Acad. Sci. U. S.* 50 (1963) 474.
8. Post, R. L., Sen, A. K., and Rosenthal, A. S., *J. Biol. Chem.* 240 (1965) 1437.
9. Gibbs, R., Roddy, R. M., and Titus, E., *J. Biol. Chem.* 240 (1965) 2181.
10. Nagano, K., Kanazawa, T., Mizuno, N., Tashima, Y., Nakao, T., and Nakao, M., *Biochem. Biophys. Res. Commun.* 19 (1965) 759.
11. Hokin, L. E., Sastry, P. S., Galsworthy, P. R., and Yoda, A., *Proc. Natl. Acad. Sci. U, S.* 54 (1965) 177.
12. Heinz, E., and Hoffman, J. F., *J. Cell. Comp. Physiol.* 65 (1965) 31.
13. Adam, H., In *Methoden der enzymatischen Analyse* (herausgegeben von H. U. Bergmeyer), Verlag Chemie, Weinheim/Bergstr. 1962, p. 539.
14. Lamprecht, W., and Trautschold, I., In *Methoden der enzymatischen Analyse* (herausgegeben von H. U. Bergmeyer), Verlag Chemie, Weinheim/Bergstr. 1962, p. 543.
15. Adam, H., In *Methoden der enzymatischen Analyse* (herausgegeben von H. U. Bergmeyer), Verlag Chemie, Weinheim/Bergstr. 1962, p. 573.
16. Berthelot, M., *Repert. Chim. appl.* 1 (1854) 284.
17. King, E. J., *Biochem. J.* 26 (1932) 292.
18. Massey, V., *Biochim. Biophys. Acta*, 34 (1959) 255.
19. Seubert, W., In *Methoden der enzymatischen Analyse* (herausgegeben von H. U. Bergmeyer), Verlag Chemie, Weinheim/Bergstr. 1962, p. 433.
20. Schneider, W. C., *J. Biol. Chem.* 161 (1945) 293.
21. Lusena, C. V., *Can. J. Chem.* 29 (1951) 107.

22. Lowry, O. H., Rosebrough, N. J., Farr, A. L., and Randall, R. J., *J. Biol. Chem.* 193 (1951) 265.
23. Seubert, W., Greull, G., and Lynen, F., *Angew. Chem.* 69 (1957) 359.
24. Warburg, O., and Christian, W., *Biochem. Z.* 287 (1936) 291.
25. Schneider, W. C., In *Manometric Techniques and Tissue Metabolism* (edited by W. W. Umbreit, R. H. Burris, and J. F. Stauffer), Burgess Publishing Co., Minneapolis 1951, p. 152.
26. Dixon, M., *Biochem. J.* 55 (1953) 170.
27. Hosie, R. J. A., *Biochem. J.* 96 (1965) 404.
28. Kurokawa, M., Sakamoto, T., and Kato, M., *Biochem. J.* 97 (1965) 833.
29. Skou, J. C., *Biochem. Biophys. Res. Commun.* 10 (1963) 79.
30. Rendi, R., and Uhr, M. L., *Biochim. Biophys. Acta*, 89 (1964) 520.
31. Schoner, W., Kramer, R., and Seubert, W., *Biochem. Biophys. Res. Commun.* 23 (1966) 403.
32. Wheeler, K. P., and Whittam, R., *Biochem. J.* 85 (1962) 495.
33. Nakao, T., Tashima, Y., Nagano, K., and Nakao, M., *Biochem. Biophys. Res. Commun.* 19 (1965) 755.
34. Post, R. L., and Sen, A. K., In *Methods in Enzymology* (edited by S. P. Colowick and N. O. Kaplan), Academic Press, New York, Vol. 10, (in press).
35. Swanson, P. D., Bradford, H. F., and McIlwain, H., *Biochem. J.* 92 (1964) 235.
36. Glynn, I. M., Slayman, C. W., Eichberg, J., and Dawson, R. M. C., *Biochem. J.* 94 (1965) 692.

W. Schoner, R. Kramer, and W. Seubert
Chemisch-physiologisches Institut, Institut für Vegetative Physiologie, Johann Wolfgang Goethe-Universität
6 Frankfurt a. M.-Süd, Ludwig-Rehn-Straße 14, Germany

C. von Ilberg's present address:
Hals-Nasen-Ohrenklinik,
Johann Wolfgang Goethe-Universität
6 Frankfurt a. M.-Süd, Ludwig-Rehn-Straße 14, Germany

European J. Biochem. 1 (1967) 344—346

Differences between the Chemical Structures
of Duck and Hen Egg-white Lysozymes

J. Jollès, J. Hermann, B. Niemann, and P. Jollès

Laboratory of Biochemistry, Faculty of Sciences, Paris

(Received March 23, 1967)

Histidine-free duck egg-white lysozyme III and hen egg-white lysozyme have similar but not identical structures.

Some structural differences are reported.

The N- and C-terminal sequences are different.

The unique histidine and two phenylalanine residues of hen lysozyme are replaced by a leucine and two tyrosine residues, respectively.

The more basic character of duck lysozyme III is due to additional arginine residues.

Duck egg-white lysozymes II and III have not entirely identical chemical structures.

We recently reported the purification and the amino acid composition of three lysozymes contained in duck egg-white, two of which contained no histidine (duck lysozyme II and duck lysozyme III). The compositions of these latter two were very similar to, but not identical with that of hen egg-white lysozyme [1] (Table). This observation incited us to characterize rapidly some of the sequences in which differences between the chemical structures of one of these duck lysozymes (duck lysozyme III) and of the hen egg-white lysozyme [2a] were localized[1].

METHODS

Duck egg-white lysozyme III (100 mg) was reduced with thioglycolic acid and carboxymethylated under the conditions described by Jollès et al.[2a]. It was then digested with 2% (w/w) of trypsin at 37°; pH 7.8 (trimethylamine); 6 hours. Trypsin (Worthington) was pretreated during 16 hours with 0.0625 N HCl at 37°. The peptides obtained after tryptic hydrolysis were separated by column chromatography on Dowex 1(×2) as described in Fig. 1. 15 peaks, T 1—T 15, were characterized. The peptides obtained after chymotryptic hydrolysis [2% (w/w) of enzyme; pH 7.8; 37°; 6 hours] were separated in a similar manner. 23 peaks, C 1—C 23, were identified. When several peptides were characterized in a peak, their purification was achieved by preparative paper chromatography (Whatman No. 1 or 3; n-butanol-formic acid-water, 75-15-10, v/v) or paper electrophoresis [Whatman No. 1 or 3;

---

[1] This work is the 54th communication on lysozymes; 53rd communication: [2b].

Enzymes. Lysozyme (EC 3.2.1.17); trypsin (EC 3.4.4.4); chymotrypsin (EC 3.4.4.5); carboxypeptidase A (EC 3.4.2.1); carboxypeptidase B (EC 3.4.2.2).

---

pyridine-acetic acid-water, 100-4-4896, v/v (pH 6.5) or 2-20-1000, v/v (pH 3.5); 40 V/cm; 0°]. The indicated yields are calculated after all the purification steps. The quantitative amino acid compositions were determined after total hydrolysis (6 N HCl; 110°; 18 and/or 48 hours; sealed tubes under vacuum). In order to characterize the N- and C-terminal amino acids, the peptides were digested with the aminopeptidase described by Uhlig et al. [4] (pH 7.5; 37°; 15, 30, 60 or 120 minutes) and with the carboxypeptidases A or B (pH 7.5; 37°; DFP; 60 or 120 minutes), respectively. In order to determine the extent of the enzymatic reaction and to characterize the liberated amino acids, aliquots of the enzymatic digest were analyzed with the Technicon Auto-Analyzer.

RESULTS

Differences between Duck Egg-white Lysozyme III and Hen Egg-white Lysozyme

The N-terminal Sequence. From native duck lysozyme III, our aminopeptidase [4] was able to split off lysine and valine. Furthermore, peak T 4 contained a pentapeptide [$R_F = 0.02$; m (pH 6.5) = +0.4; yield = 7%] with two basic amino acids. Lysine was identified as the N-terminal and arginine as the C-terminal residue. This was the only peptide obtained by tryptic digestion with a N-terminal basic amino acid. After chymotryptic digestion a tripeptide: Lys-Val-Tyr and a dipeptide Gln-Arg ($R_F = 0.08$) were obtained. The structure of T 4 was: H-Lys-Val-Tyr-Gln-Arg; it constituted the *N*-terminal sequence of duck lysozyme III. The corresponding sequence of hen lysozyme was H-Lys-Val-Phe-Gly-Arg [2a]; the phenylalanine and glycine residues were replaced in the duck enzyme III by a tyrosine and a glutamine residue, respectively.

The C-terminal Sequence. From native duck lysozyme III, carboxypeptidase A was able to split off free leucine. Furthermore from peak T 5 a tetrapeptide [yield: 33%; $R_F = 0.14$; m (pH 6.5) = 0] was isolated which was the only tryptic peptide containing no C-terminal basic amino acid. Its structure: Gly-Cýs-Arg-Leu was identical with that of the C-terminal tetrapeptide of the hen enzyme [2 a].

The C-terminal octapeptide of hen lysozyme was Ala-Trp-Ileu-Arg-Gly-Cýs-Arg-Leu-OH (residues 122 to 129) with a site of chymotryptic attack between the tryptophan and isoleucine residues. In the chymotryptic digest of duck lysozyme III, we again characterized the peptide Ileu-Arg-Gly-Cýs-Arg-Leu (peak C 5; $R_F = 0.12$; yield 18%). However the tryptic digest of this same enzyme contained in peak T 5 the tripeptide Trp-Ileu-Arg [$R_F = 0.28$; m (pH 6.5) $= +0.52$; yield $= 42\%$] which was absent from the tryptic digest of hen lysozyme. This result indicated that the residue of Trp was preceded by a basic amino acid, and in fact it was possible to identify a residue of Lys at this position (peak C 2). The C-terminal octapeptide of duck lysozyme III was thus Lys-Trp-Ileu-Arg-Gly-Cýs-Arg-Leu-OH. Residue 122 (Ala) of hen lysozyme was replaced by a lysine residue in duck lysozyme III.

Tryptic Peptide from Duck Lysozyme III Corresponding to the Histidine Containing Tryptic Peptide from Hen Lysozyme. Peak T 7 contained a heptapeptide [$R_F = 0.27$; m (pH 6.5) = 0; yield $= 28\%$; composition: Asp$_2$, Gly$_1$, Leu$_2$, Tyr$_1$, Arg$_1$]. The N-terminal amino acid determined by the dansyl method [5] was leucine, the N-terminal sequence Leu-Gly-Leu and the C-terminal Tyr-Arg. The structure was deduced as Leu-Gly-Leu-Asp-Asp-Tyr-Arg, one of the Asp residues being present as Asn. This sequence was very similar to the tryptic peptide from hen lysozyme containing its unique histidine residue [2 a], *i.e.* His-Gly-Leu-Asp-Asn-Tyr-Arg. The residue of histidine was replaced by a residue of leucine in duck lysozyme III.

Peptides from Duck Lysozyme III Corresponding to the Phenylalanine Containing Sequences of Hen Lysozyme. As against the three phenylalanine residues of the hen enzyme [2 a], the duck enzyme III contains only one [1] which is found in the chymotryptic tetrapeptide C 15 b ($R_F = 0.27$; yield $= 20\%$): Glu-Ser-Gly-Phe. The corresponding sequence of hen egg-white lysozyme Glu-Ser-Asn-Phe contains residue Glu (35) which, according to Blake *et al.* [5], seems to be involved in the splitting of substrate by lysozyme.

The two other phenylalanine residues of hen lysozyme were situated one at the N-terminal side of the above mentioned peptide: Glu-Ser-Asn-Phe and the other in the N-terminal sequence of the enzyme: H-Lys-Val-Phe-Gly-Arg. Both phenylalanine residues were replaced by tyrosine residues in duck lysozyme III.

Localisation of Two Additional Basic Amino Acid Residues in Duck Lysozyme III. As indicated above, the C-terminal octapeptide of duck lysozyme III contained 3 basic amino acid residues against only 2

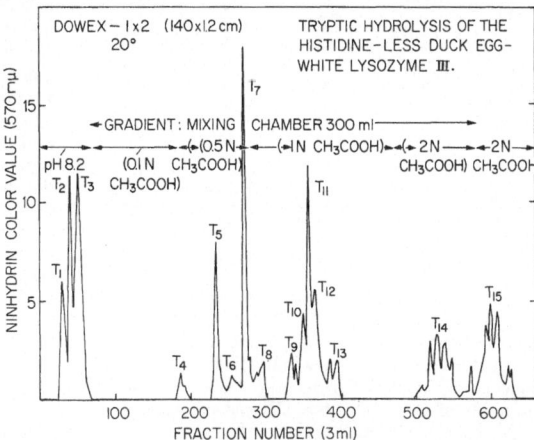

Fig. 1. *Chromatography on Dowex 1(×2) (140×1.2 cm) of the tryptic digest obtained from 100 mg reduced carboxymethylated duck egg-white lysozyme III. 1% (w/v) collidine buffer, pH 8.2, with increasing amounts of acetic acid. Ninhydrin colour value after alkaline hydrolysis determined at 570 mμ*

in the hen enzyme. Another difference between these two lysozymes was found in the tryptic peptide Thr-Pro-Gly-Ser-Arg from the hen enzyme [2 a] (residues 69—73). This sequence could not be characterized in the tryptic digest of duck lysozyme III, but peak T 2 contained the tryptic tripeptide Thr-Pro-Arg ($R_F = 0.08$; m $= +0.56$; yield $= 36\%$). Residues 71 (Gly) and 122 (Ala) of hen lysozyme which were replaced by Arg and Lys residues respectively in duck lysozyme III, were situated at the external part of the three-dimensional model of hen lysozyme described by Blake *et al.* [7].

Duck Egg-white Lysozymes II and III

Duck egg-white lysozymes II and III have very similar amino acid compositions (Table) and both enzymes contain no histidine. From the analytical data alone, it was not possible to ascertain whether their chemical structures were identical or not. A study similar to that described in the present paper

Table. *Amino acid composition of duck egg-white lysozymes II and III[a]; difference with hen egg-white lysozyme[b]*

Amino acid	Duck lysozyme II		Duck lysozyme III	
	Composition	Difference with hen lysozyme	Composition	Difference with hen lysozyme
	Residues/mole		Residues/mole	
Asp	19	−2	18−19	−2 or 3
Thr	7	0	7	0
Ser	10	0	10−11	0 or +1
Glu	5	0	5	0
Pro	2	0	2	0
Gly	12	0	12	0
Ala	11	−1	11	−1
Val	7	+1	7	+1
Cys	8	0	8	0
Met	2	0	2	0
Ileu	6	0	6	0
Leu	8	0	8	0
Tyr	5	+2	5	+2
Phe	1	−2	1	−2
Trp	6	0	6	0
Lys	6	0	6	0
His	0	−1	0	−1
Arg	12−13	+1 or 2	13−14	+2 or 3

[a] Data of Jollès et al. [1] published in 1965; Imanishi et al. published in 1966 very similar analytical data [3].
[b] Data of Jollès et al. [2a].

for duck lysozyme III is under progress for duck lysozyme II. It was possible to characterize some structural differences between these two enzymes; among them we can already indicate the N-terminal sequence H-Lys-Val-Tyr-Ser-Arg. The residue of glutamine of duck lysozyme III was replaced by a residue of serine in duck lysozyme II.

The three duck egg-white lysozymes I, II and III isolated by ion-exchange chromatography on Amberlite CG-50 were obtained from pooled eggs [1]. Actually we determine the number of lysozymes which can be prepared from a single duck egg, as Walsh et al. established that two species of bovine carboxypeptidase which were isolated from pancreatic tissue pooled from many animals may occur either separately or together in any individual animal [8].

CONCLUSION

The results reported in this paper allow the conclusion that the chemical structure of duck lysozyme III is similar to that of hen lysozyme with a limited number of changes. This latter is however higher than can be deduced from the only observation of the table. For instance both lysozymes have 5 residues of glutamic acid per molecule; however only the duck lysozyme III contains such a residue in its N-terminal sequence; on the contrary, the peptide obtained by tryptic digestion and corresponding to the sequence 117−125 of hen lysozyme has no glutamic acid.

No data concerning the position of the tryptophan residues of duck lysozyme III are reported, as these latter are found in sequences similar to those characterized in the hen enzyme. Both lysozymes also contain the same number of disulphide bonds. The sequence of duck lysozyme III corresponding to the moiety of the hen enzyme containing the two disulphide bonds Cyś (64) → Cyś (80) and Cyś (76) → Cyś (94) is actually studied in detail as several changes seem to be situated in this part.

The structural similarity of hen and duck III lysozymes is in accordance with the observation that the specific activity and the heat stability of duck lysozyme III are quite comparable to those of the hen enzyme [9]. N-acetylglucosamine was also shown to be an inhibitor. The slightly more basic behaviour of duck lysozyme III on Amberlite CG-50 may be explained by its higher Arg content (Table). As the specific activity and the heat stability of human lysozymes and of goose egg-white lysozyme are quite different from the corresponding properties of duck III and hen lysozymes, their structures probably differ more extensively from that of hen egg-white lysozyme. Jollès et al. [9] have already shown that the number of disulphide bonds and of tryptophan residues are involved in these differences.

It is worth adding that when lysozymes of different origins were purified by ion exchange chromatography on Amberlite CG-50, two biologically active peaks were usually obtained: a main one (80% to 90%) and a small one (10−20%); they were believed to differ only by their number of amide groups. The structural differences observed between duck lysozymes II and III make it necessary to investigate this question in greater detail.

This investigation was supported in part by funds from the C.N.R.S. (Equipe de recherche No 17), from the Délégation Générale à la Recherche Scientifique et Technique (Biologie moléculaire, grant 6600261) and from the Institut National de la Santé et de la Recherche Médicale (Action concertée Cancer-Leucémie, grant CR 67238). The authors wish also to express their appreciation to Misses J. Poujade and J. Feydit for technical assistance.

REFERENCES

1. Jollès, J., Spotorno, G., and Jollès, P., *Nature*, 208 (1965) 1204.
2a. Jollès, J., Jauregui-Adell, J., Bernier, I., and Jollès, P., *Biochim. Biophys. Acta*, 78 (1963) 668.
2b. Jollès, J., and Jollès, P., *Biochemistry*, 6 (1967) 411.
3. Imanishi, M., Shinka, S., Miyagawa, N., Amano, T., and Tsugita, A., *Biken's J.* 9 (1966) 107.
4. Uhlig, H., Lehmann, K., Salmon, S., Jollès, J., and Jollès, P., *Biochem. Z.* 342 (1965) 553.
5. Gray, W. H., and Hartley, B. S., *Biochem. J.* 89 (1963) 379.
6. Blake, C. C. F., Johnson, L. N., Mair, G. A., North, A. C. T., Philipps, D. C., and Sarna, V. R., *Proc. Roy. Soc. (London)* in the press.
7. Blake, C. C. F., Koenig, D. F., Mair, G. A., North, A. C. T., Phillips, D. C., and Sarna, V. R., *Nature*, 206 (1965) 757.
8. Walsh, K. A., Ericsson, L. H., and Neurath, H., *Proc. Natl. Acad. Sci. U. S.* 56 (1966) 1339.
9. Jollès, J., Dianoux, A.-C., Hermann, J., Niemann, B., and Jollès, P., *Biochim. Biophys. Acta*, 128 (1966) 568.

J. Jollès, J. Hermann, B. Niemann, and P. Jollès
Laboratoire de Chimie Biologique
Faculté des Sciences de l'Université de Paris
96 Boulevard Raspail, 75 Paris-6, France

European J. Biochem. 1 (1967) 347—352

The Role of the Allosteric Sites in the X-Ray Inactivation of Phosphorylase b

S. Damjanovich, T. Sanner, and A. Pihl

Norsk Hydro's Institute for Cancer Research, Montebello, Oslo

(Received January 24, 1967)

Crystalline rabbit muscle phosphorylase b was irradiated in dilute aqueous solution with X-rays. The enzyme was inactivated with a G-value of 0.09.

Measurements of the K_m values of the substrate, glucose-1-phosphate, and the allosteric activator, adenosine-5'-phosphate, demonstrated that these increased linearly with increasing radiation dose. The effect on the K_m for the activator was 4 times greater than that on the K_m for the substrate. The data indicate that the allosteric function is more sensitive to inactivation than the catalytic function.

The enzyme SH-groups were destroyed by X-rays with a G-value of 1.8. Comparison with data on the inactivation of the enzyme by sulfhydryl blocking agents demonstrated that the X-ray destruction of sulfhydryl groups was sufficiently large to account for the X-ray inactivation of the enzyme.

Blocking of two SH-groups with pCMB reduced the radiosensitivity of the enzyme by a factor of 2. Measurement of the K_m values showed that the pCMB blocking protected preferentially the allosteric sites.

The data indicate that the inactivation of phosphorylase b, both by sulfhydryl agents and by X-rays, involves largely an effect on the allosteric sites with loss of ability to bind the essential activator and consequent loss of ability to bind substrate.

In a previous study of aspartate transcarbamylase it was found that the regulatory function of the enzyme is far more sensitive to X-rays than is the catalytic function [1]. Since the allosteric enzymes are assumed to play an important role in the control of cellular metabolism, it was of interest to study whether a similar relationship obtains also in the case of other regulatory enzymes.

In the present paper the radiation inactivation of phosphorylase b has been studied. This enzyme, which plays a key role in glycogen metabolism consists of two subunits which are assumed to be identical[2]. Each of the subunits is believed to have one catalytic and one allosteric site [2—4]. Phosphorylase b has about 20 SH-groups which are important in maintaining the native conformation of the enzyme [5,6]. The enzyme has an absolute requirement for the allosteric activator, AMP (adenosine-5'-phosphate) which affects the binding of the substrate [3]. This implies that the enzyme activity will be lost if the ability of the allosteric sites to bind AMP is destroyed. Phosphorylase b therefore appeared to be well suited for studying the X-ray inactivation of the allosteric function as such, as well as the role of the allosteric

---

Non-standard Abbreviations. 5,5'-Dithio-bis(2-nitrobenzoic acid), DTNB; para-chloromercuribenzoate, pCMB.

Enzyme. α-Glucan phosphorylase, glycogen phosphorylase (EC 2.4.1.1).

sites in the radiation-induced loss of the catalytic activity. Part of the data have been reported in a preliminary communication [7].

MATERIALS AND METHODS

Four times recrystallized rabbit phosphorylase b was prepared according to Fischer and Krebs [8]. Cysteine and AMP were removed by passing the preparation, immediately before use, through a column of Sephadex G-100 (2.5×40 cm), equilibrated with 0.033 M glycerophosphate buffer at pH 6.8. The ratio A_{260} mμ/A_{280} mμ was 0.53, showing that the preparation was free of AMP. The homogeneity of the preparation was checked in the analytical ultracentrifuge.

The enzyme assay was carried out in the direction of glycogen synthesis. The liberation of phosphate from G-1-P (glucose-1-phosphate) was measured after incubation at 30° at pH 6.8. In most experiments the phosphate release was measured after 5 minutes, while in cases where the activity was low the incubation time was 10 minutes. In separate experiments it was shown that the phosphate release during the incubation time was proportional with the enzyme concentration and with time. The final concentrations of the constituents (in a total volume of 0.4 ml) were: G-1-P, $1.6×10^{-2}$ M; AMP, $1.0×10^{-3}$ M; glycogen,

$1 \, ^0/_0$; enzyme 1 to 5×10^{-7} M; glycerophosphate buffer 3.3×10^{-2} M. Where indicated, mercaptoethanol (final concentration 0.1 M) was added to the enzyme 30 minutes prior to the enzyme assay. After incubation for 5 or 10 minutes the reaction was stopped by adding 2.6 ml of $5 \, ^0/_0$ (w/v) trichloroacetic acid. The specific activity [9] in the presence of mercaptoethanol was 1.580 units per mg of protein. No release of phosphate was found in the absence of AMP, indicating that the preparations were free of phosphorylase a and of phosphatases.

Analytical Methods

Inorganic phosphate was determined according to Taussky and Shorr [10]. Protein was determined by the method of Appleman *et al.* [5]. For phosphorylase b a molecular weight of 242,000 was used [2]. Sulfhydryl groups were determined spectrophotometrically at 412 mμ with the use of DTNB (5,5′-dithio-bis(2-nitrobenzoic acid)) [11].

Irradiation Conditions

The enzyme, in a concentration of 3.0×10^{-6} M, in 3.3×10^{-2} M glycerophosphate buffer, pH 6.8, was irradiated in Erlenmeyer flasks in the presence of air at room temperature. The irradiation source was a Stabilipan X-ray machine. The irradiation parameters were 220 kV, 20 mA and 0.5 mm Cu filter. The dose rate was 1.970 kR/min. The dose was measured with the Fricke dosimeter [12].

RESULTS

The inactivation of different concentrations of phosphorylase b is shown in Fig. 1. The fact that linear dose-inactivation curves were found in semi-logarithmic plots (Fig. 1 A) demonstrates that the enzyme is inactivated as an exponential function of the dose. There was no after-effect, indicating that hydrogen peroxide plays a minor role in the inactivation of the enzyme. When the D_{37} dose *i.e.* the dose needed to reduce the activity to $37 \, ^0/_0$ (1/e) of the initial activity was plotted against enzyme concentration (Fig. 1 B) a straight line was found as expected. From the slope of the line the G-value for the inactivation (the number of enzyme molecules inactivated per 100 eV absorbed by the solution) was calculated to be 0.09.

Role of Sulfhydryl Groups

In our previous studies of sulfhydryl enzymes it was found [13—15] that the loss of enzyme activity could be accounted for largely by the destruction of sulfhydryl groups. Thus, blocking of SH-groups afforded a considerable protection, and simple correlations were found between loss of activity and destruction of sulfhydryl groups.

The effect of sulfhydryl group blocking on the activity of phosphorylase b is shown in Fig. 2 A. It appears that two SH-groups could be blocked without loss of activity, in agreement with previous findings

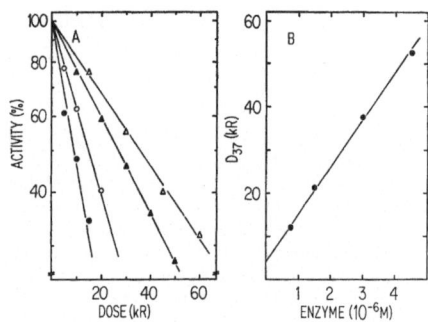

Fig. 1. *X-ray inactivation of phosphorylase b.* A, Different concentrations of enzyme were irradiated in 3.3×10^{-2} M glycerophosphate buffer, pH 6.8, in equilibrium with air at room temperature. ●, 1×10^{-6} M; ○, 1.5×10^{-6} M; ▲, 3×10^{-6} M; △, 4.5×10^{-6} M. The activity is expressed in per cent of that of the unirradiated control. B, The D_{37} dose as a function of the enzyme concentration. Data taken from Fig. 1 A

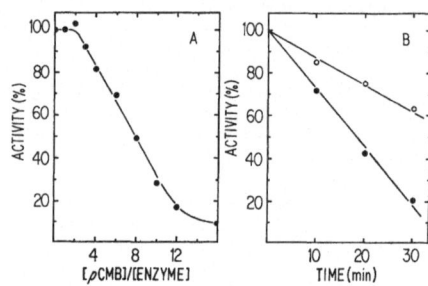

Fig. 2. *Inhibition of phosphorylase b by pCMB.* A, The enzyme, in a concentration of 3×10^{-6} M in glycerophosphate buffer, pH 6.8, was preincubated at 22° for 2 hours with increasing concentrations of pCMB. B, Effect of irradiation on rate of inhibition of phosphorylase b by pCMB. The enzyme, 3×10^{-6} M, was incubated with a 30 fold molar excess of pCMB, and the activity measured after different periods of time. ○, unirradiated enzyme; ●, enzyme irradiated with 20 kR of X-rays prior to pCMB treatment

with other sulfhydryl reagents [16]. Upon further addition of pCMB the enzyme activity decreased in a linear fashion with increasing SH-blocking. When an excess of pCMB was added, the enzyme activity disappeared initially as a linear function of time (Fig. 2 B). The inactivation occurred slowly. Thus, even after the addition of a 30 fold excess of pCMB, about $60 \, ^0/_0$ of the enzyme activity was still present after 30 minutes. After irradiation with 20 kR, which resulted in approximately 30% inactivation, the rate of inhibition of the enzyme by pCMB was considerably enhanced. This finding indicates that the irradiation caused conformation changes rendering the sulfhydryl groups more accessible to pCMB.

The effect of SH blocking on the radiation sensitivity of phosphorylase b is shown in Fig. 3. It is apparent that the D_{37} dose increased with increasing blocking of up to about 5 SH-groups. Upon further SH blocking the D_{37} dose decreased, and upon addition of 22 equivalents of pCMB a sensitization was found. The finding that the protection decreased when more than 5 SH-groups were blocked, indicates that in addition to the protection a sensitizing

30 kR reduced the total number of SH-groups from 20 to 8.

No reactivation could be obtained by treatment of the enzyme with mercaptoethanol after irradiation. The results indicate that the radiation inactivation does not involve oxidation of enzyme SH-groups to disulfide groups.

When 2 SH-groups were blocked with pCMB prior to irradiation, the rate of disappearance of the remaining SH-groups upon exposure was definitely decreased (Fig. 4A). The effect is more clearly brought out in Fig. 4B where the data are plotted on a percentage basis. It is seen that the radiation sensitivity of the SH-groups was reduced by a factor of

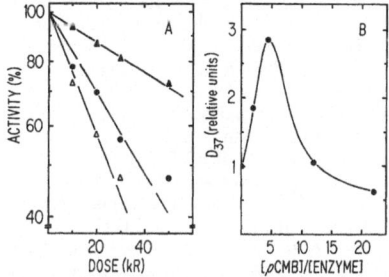

Fig. 3. *Effect of pCMB treatment on the radiation sensitivity of phosphorylase b.* The enzyme was preincubated for 2 hours with increasing concentrations of pCMB prior to irradiation. After exposure the enzyme was treated with mercaptoethanol before the enzyme assay. A, Dose inactivation curves. ●, untreated enzyme; ▲, 4.5 moles of pCMB added per mole of enzyme; △, 22 moles of pCMB added per mole of enzyme. B, D_{37} dose as a function of amount of pCMB added

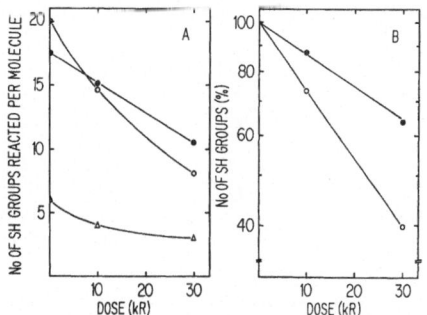

Fig. 4. *Disappearance of titratable enzyme SH-groups upon irradiation.* The enzyme, in a concentration of 3×10^{-6} M, was irradiated with increasing doses of X-rays and the remaining SH-groups were titrated with DTNB. △, SH titration in absence of lauryl sulfate; ○, SH-groups titrated in presence of 0.15% lauryl sulfate; ●, enzyme pretreated with 2.5 moles of pCMB per mole of enzyme prior to irradiation. SH-groups titrated in presence of 0.15% lauryl sulfate. A, SH-groups expressed in absolute values. B, SH-groups expressed in per cent of those of unirradiated enzyme

mechanism was concurrently operating. It is known that excess of pCMB will dissociate reversibly the enzyme into subunits [17]. It is reasonable to assume that the dissociating effect of pCMB sensitized the enzyme to the action of the water radicals.

The disappearance of the SH-groups as a function of the radiation dose is shown in Fig. 4A. When the titration was carried out in the absence of denaturing agents, only 6 SH-groups were found in the unirradiated sample. After irradiation with 10 kR, about 3 SH-groups remained, and upon further irradiation little change in the number of titratable SH-groups occurred. It seemed possible that this high resistance of the readily available SH-groups to irradiation might be apparent only and due to the fact that the radiation caused conformation changes making previously unreactive SH-groups available for reaction with DTNB. Titrations were therefore carried out in the presence of lauryl sulfate. It is seen that under these conditions the SH-groups disappeared nearly as a linear function of the dose. A dose of

2 by the pCMB blocking. The most likely explanation seems to be that the pCMB blocking protected the enzyme against radiation induced conformation changes, thereby protecting the remaining sulfhydryl groups. The data may explain the finding shown in Fig. 3 that blocking of 2 SH-groups provided protection, in spite of the fact that such blocking had no effect on the activity of the unirradiated enzyme.

Effect on Kinetic Parameters

In attempts to see whether the allosteric sites were more sensitive to X-rays than the catalytic sites, the activity after exposure was measured for high and low concentrations of the substrate, G-1-P, and for the activator, AMP, respectively. The results in Fig. 5 demonstrate that the AMP concentration affected the slope of the dose-inactivation curves more than did the G-1-P concentration, suggesting that the allosteric sites may indeed be more sensitive to irradiation than the catalytic sites.

In order to study this question in more detail, experiments were carried out in which the K_m values were measured for G-1-P and AMP, respectively. The results are summarized in Fig.6. It is apparent that both K_m values increased in a linear fashion with the radiation dose (Fig.6B). However, the

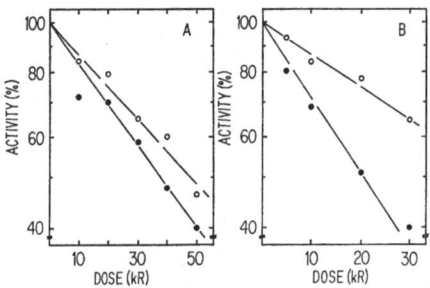

Fig.5. *Radiation sensitivity of phosphorylase b measured at different concentrations of substrate and activator.* A, ●, 2×10^{-3} M G-1-P; ○, 1.5×10^{-2} M G-1-P. In both cases the concentration of AMP was 1×10^{-3} M. B, ●, 0.1×10^{-3} M AMP; ○, 1.0×10^{-3} M AMP. In both cases the concentration of G-1-P was 1.5×10^{-2} M

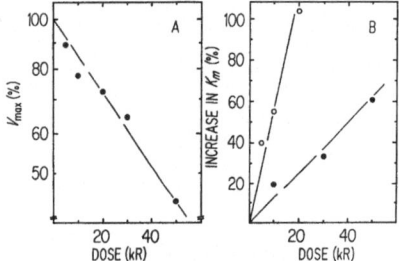

Fig.6. *Effect of irradiation on kinetic parameters of phosphorylase b.* A, V_{max} as a function of the radiation dose. B, ●, K_m for G-1-P; ○, K_m for AMP. The kinetic parameters were obtained from $1/v$ versus $1/s$ plots according to Lineweaver and Burk [19]. For the determination of the K_m of G-1-P, the AMP concentration was 1.8×10^{-3} M. Four concentrations of G-1-P were used in the range 2 to 15×10^{-3} M. For the determination of K_m of AMP, the G-1-P concentration was 1.5×10^{-2} M. Four AMP concentrations in the range 0.1 to 1.0×10^{-3} M were used. For the unirradiated enzyme the K_m values for G-1-P and AMP were 4.5×10^{-3} M and 4×10^{-5} M, respectively

effect on the K_m of AMP was much greater than the effect on the K_m for G-1-P. Thus, after a dose of 20 kR, which reduced the V_{max} by approximately 30%, the K_m for G-1-P increased by 25%, while K_m for AMP increased by more than 100%. The results indicate that the allosteric sites of phosphorylase b

are more sensitive to X-rays than the catalytic sites.

In order to obtain further information on the mechanism of inactivation of phosphorylase b, the effect of X-rays on the K_m values was compared with the effect of pCMB blocking. It is seen from Table that also when the enzyme was inhibited by pCMB, the K_m of AMP was affected to a much greater extent than the K_m of G-1-P. The ratio between the effects on the two K_m values were about the same after pCMB and X-ray treatment. However, the pCMB effect on the K_m values were much larger than that of irradiation, in spite of the fact that V_{max} was affected to about the same extent by the two different treatments. The possibility exists that the pCMB effect on the K_m values is not merely due to the loss of free SH-groups as such. The introduction of the bulky pCMB residues might interfere more with the binding of AMP than does radiation oxidation of SH-groups.

The effect of irradiation on the two K_m values was also studied on enzyme which had been pretreated with two equivalents of pCMB. In this case a larger radiation dose (50 kR) was used than for the untreated enzyme (20 kR) in order to obtain approximately the same decrease in V_{max}. It is seen (Table) that irradiation affected the K_m of G-1-P to a much greater extent in the pCMB treated enzyme than in the native enzyme. It thus appears that pCMB failed to protect the K_m of G-1-P. On the other hand, after irradiation the K_m of AMP was about the same or even somewhat less in the protected enzyme than in the native enzyme, in spite of the fact that a larger radiation dose was used in the former case. The results indicate that the binding sites for AMP were preferentially protected by the pCMB blocking.

DISCUSSION

The main findings in the present work are that in phosphorylase b, like in aspartate transcarbamylase, the allosteric function is more sensitive to X-ray inactivation than the catalytic function, and furthermore, that the allosteric function can be preferentially protected by blocking of enzyme SH-groups by pCMB.

Since damage to SH-groups plays an essential role in the X-ray inactivation of several enzymes we propose to consider first the role of these groups in phosphorylase b.

Previous studies indicate that the various sulfhydryl groups of phosphorylase b may have different significance for the enzyme activity. As pointed out above, 2 SH-groups can be blocked with either DTNB [16] or pCMB, (Fig.2A) without loss of activity. If an excess of sulfhydryl reagent is added, the effect on the activity depends on the nature of the reagent. Thus, with DTNB a 50% inhibition was

Table. *Effect of irradiation and of SH blocking on* K_m *of AMP and G-1-P*

Conditions	Radiation dose	Decrease in V_{max}	Increase in K_m[a] of	
			G-1-P	AMP
	kR	%	%	%
Unirradiated. Pretreated with 6 equivalents of pCMB	0	25	110	640
Irradiated. No pretreatment	20	30	25	110
Irradiated after pretreatment with 2 equivalents of pCMB [b]	50	30	50	80

[a] The K_m values were determined as described in legend to Fig. 6.
[b] Enzyme activity determined after treatment of enzyme with mercaptoethanol.

observed when about 4 SH-groups were blocked [16], while with pCMB 8 SH-groups had to be blocked to give the same degree of inhibition (Fig. 2A). This difference may be due to the fact that DTNB is a disulfide with a lower reactivity towards sulfhydryl groups than the mercurial pCMB. DTNB will therefore be expected to be able to seek out the most reactive sulfhydryl groups, while pCMB will react less discriminately and hence block also other sulfhydryl groups which may not be essential for the activity. This explanation is supported by the fact that in the absence of denaturing agents more sulfhydryl groups can be titrated with pCMB than with DTNB [6]. Since phosphorylase b consists of two subunits, the data indicate that each subunit contains one or two unessential SH-groups which are highly reactive. The finding that 50% inhibition was observed when one more SH-group per subunit was blocked with DTNB [16] suggests that each subunit contains at least two more reactive SH-groups, the blocking of which will abolish the enzyme activity. When the enzyme was irradiated, 50% inactivation was associated with the destruction of 16 sulfhydryl groups (compare data in Fig. 6A with those in Fig. 4B) while the same degree of inactivation was accomplished by blocking 8 SH-groups with pCMB (Fig. 2A). It thus appears that the observed X-ray destruction of SH-groups is far greater than would be necessary to account for the enzyme inactivation. The results support the view that the enzyme inactivation is associated with the destruction of specific SH-groups and they indicate that the water radicals react indiscriminately with the sulfhydryl groups.

On the basis of the data in Fig. 4 the G-value for the destruction of the SH-groups is calculated to be 1.8. This is a remarkably high value, indicating that the sulfhydryl groups of phosphorylase b are even more radiosensitive than the SH-group of papain [15]. The fact that substantial protection could be obtained by pCMB blocking further supports the view that the radiation-induced inactivation of phosphorylase b can be largely accounted for by sulfhydryl group destruction.

Since phosphorylase b has an absolute requirement for AMP, the enzyme activity can be lost either by destruction of the catalytic sites, or by destruction of the ability of the allosteric sites to bind AMP. Evidence that sulfhydryl groups are involved in the binding of AMP has been presented [16, 18]. Thus, inhibition of the enzyme by sulfhydryl blocking agents was reduced by the presence of AMP, whereas G 1 P had no such protective effect. The present finding that pCMB blocking affected the K_m of AMP much more than the K_m of G-1-P provides further support for this view. These results indicate that the inhibition of the enzyme by sulfhydryl blocking reagents can be accounted for to a large extent by an effect on the allosteric sites.

The radiation-induced inactivation of phosphorylase b likewise involved primarily the allosteric sites. Here again the K_m of AMP was more affected by X-rays than was the K_m of G-1-P, and the ratio between the effects on the two K_m values was nearly the same as found after pCMB treatment. Further support for the role of the allosteric sites in the enzyme inactivation is found in the fact that pCMB blocking protected the allosteric sites and the enzymatic activity to about the same extent, whereas the catalytic sites, as revealed by the effect on the K_m value, were not protected. It thus appears that inactivation of phosphorylase b, both by sulfhydryl agents and by X-rays, involves largely an effect on the allosteric sites with loss of ability to bind essential activator AMP and consequent loss of ability to bind substrate.

The finding that the regulatory function of allosteric enzymes may be more sensitive to ionizing radiation than the catalytic properties may prove to be of biological significance. Aspartate transcarbamylase and phosphorylases are key enzymes in the metabolism of pyrimidines and carbohydrates, respectively, and impairment of their regulatory function may have serious metabolic consequences. In cells the phosphorylase activity is influenced by numerous factors, including the concentration of the activator, AMP, and the inhibitor, ATP. Since AMP and ATP act in a competitive fashion [3], it is reasonable to assume that irradiation will decrease the binding of ATP as well. In the unirradiated enzyme variations in the concentration of these allosteric effector molecules in the range occurring in cells may influence strongly the catalytic activity of phosphorylase b. However, if the K_m of the allosteric sites is increased, such physiological variations in the concentration of nucleotides may have only slight effects on the enzyme, which may then escape the normal metabolic control.

This work was supported by the National Center for Radiological Health, U.S. Public Health Service. S. Damjanovich is a Fellow of the World Health Orginization. T. Sanner is a Fellow of The Norwegian Cancer Society.

REFERENCES

1. Kleppe, K., Sanner, T., and Pihl, A., *Biochim. Biophys. Acta*, 118 (1966) 210.
2. Krebs, E. G., and Fischer, E. H., *Advan. Enzymol.* 24 (1962) 263.
3. Madsen, N. B., *Biochem. Biophys. Res. Commun.* 15 (1964) 390.
4. Helmreich, E., and Cori, C. F., *Proc. Natl. Acad. Sci. U. S.* 52 (1964) 647.
5. Appleman, M. M., Yunis, A. A., Krebs, E. G., and Fischer, E. H., *J. Biol. Chem.* 238 (1963) 1358.
6. Damjanovich, S., and Kleppe, K., *Biochem. Biophys. Res. Commun.* 26 (1967) 65.
7. Damjanovich, S., Sanner, T., and Pihl, A., *Biochim. Biophys. Acta*. In press.
8. Fischer, E. H., and Krebs, E. G., *J. Biol. Chem.* 231 (1958) 65.
9. Illingworth, B., and Cori, G. T., *Biochem. Preparations*, 3 (1953) 1.
10. Taussky, H. H., and Shorr, E., *J. Biol. Chem.* 202 (1953) 675.
11. Ellman, G. L., *Arch. Biochem. Biophys.* 82 (1959) 70.
12. Baarli, J., and Borge, P., *Acta Radiol.* 47 (1957) 203.
13. Lange, R., and Pihl, A., *Acta Chem. Scand.* 13 (1959) 2126.
14. Lange, R., and Pihl, A., *Intern. J. Radiation Biol.* 2 (1960) 301.
15. Pihl, A., and Sanner, T., *Radiation Res.* 19 (1963) 27.
16. Damjanovich, S., and Kleppe, K., *Biochem. Biophys. Acta*, 122 (1966) 145.
17. Madsen, N. B., and Cori, C. F., *J. Biol. Chem.* 223 (1956) 1055.
18. Jókay, I., Damjanovich, S., and Tóth, S., *Arch. Biochem. Biophys.* 112 (1965) 471.
19. Lineweaver, H., and Burk, D., *J. Am. Chem. Soc.* 56 (1934) 658.

S. Damjanovich's permanent address:
Institute of Pathophysiology,
Medical University, Debrecen, Hungary

T. Sanner and A. Pihl
Norsk Hydro's Institute for Cancer Research,
Montebello, Oslo 3, Norway

European J. Biochem. 1 (1967) 353—356

Spectrométrie de masse de glycolipides

1. Structure du *"cord factor"* de *Corynebacterium diphtheriae*

M. Senn, T. Ioneda, J. Pudles et E. Lederer

Institut de Chimie des Substances Naturelles, C.N.R.S., Gif-sur-Yvette (Essonne),
et, Institut de Biochimie, Faculté des Sciences d'Orsay (Essonne)

(Reçu le 10 février 1967)

"Cord factors" are diesters of α,α-trehalose produced by Mycobacteriae and Corynebacteriae. Saponification of the "cord factor" of *Corynebacterium diphtheriae* gives an equimolecular mixture of homologues of the saturated corynomycolic and of the unsaturated corynomycolenic acids (I and II, R = H). Mass spectrometry of their methyl esters confirms their structures (see Fig. 1, R = CH_3).

The peracetylated cord factor (IV, a, b, c) gave very weak molecular ions at m/e 1630, 1632 and 1634 showing apparently that it is a mixture of the diesters having respectively two saturated, one saturated and one unsaturated, and two unsaturated acyl radicals. Strong peaks at m/e 809 and 807 are attributed to the oxonium ions (V a and V b).

These results as well as chemical degradations show that the "cord factor" of *C. diphtheriae* is a mixture of glycolipids (III a, b, c).

Ioneda, Lenz et Pudles [1] ont décrit, en 1963, un glycolipide isolé de *Corynebacterium diphtheriae*; ils ont montré que cette substance (F = 110—115°, $[\alpha]_D = +64°$) est un diester de tréhalose contenant des quantités à peu près égales d'acide corynomycolique [acide 2-tétradécyl 3-hydroxy octadécanoïque, $C_{32}H_{64}O_3$, (I) (R = H, n = 13) [2,3]] et corynomycolénique [acide 2-tétradécyl 3-hydroxy 9-octadécènoïque, $C_{32}H_{62}O_3$, (II) (R = H, n = 13) [4]].

Le présent travail, entrepris dans le cadre plus général d'une étude des possibilités d'application de la spectrométrie de masse à des glycolipides et à des dérivés d'oligosaccharides, conduit à attribuer les structures (III a, b, c) à ce mélange de glycolipides. Il s'agit donc d'homologues inférieurs du *cord factor* des Mycobactéries, qui est un 6,6'-dimycolate de tréhalose [5]. (Les acides mycoliques des Mycobactéries ont de 56 à 86 atomes de carbone [6—13]; pour une revue récente sur les glycolipides des Mycobactéries et des microorganismes apparentés, voir [14]) [1]

PARTIE EXPÉRIMENTALE

Isolement et dégradation du glycolipide

Le *cord factor* de *Corynebacterium diphtheriae* a été isolé comme décrit précédemment par Ioneda et coll. [1]; les deux fractions d'acides gras, saturé et insaturé, ont été également préparées à partir du glycolipide, comme décrit par Ioneda et coll. [1].

[1] Ce travail constitue la 19e communication sur les lipides des Corynébactéries; 18e communication, voir ref. [15]; 17e communication, voir réf. [16]; 16e communication, voir réf. [17].

La perméthylation du glycolipide, ainsi que l'identification du tri-O-méthyl glucose (effectuées par M. Yamakawa) ont été faites essentiellement selon Noll et coll. [5].

Acétylation

Environ 3 mg de substance, en solution dans 0,5 ml de pyridine sont gardés pendant une nuit après addition de 0,5 ml d'anhydride acétique. Après évaporation à sec sous vide, environ 0,1 mg du résidu a été utilisé directement pour la spectrométrie de masse.

Spectrométrie de masse

Les spectres de masse ont été mesurés avec le MS 9 (AEI); les échantillons ont été directement introduits dans la source d'ions; les spectres des esters méthyliques des acides gras ont été obtenus à 180°, ceux du *cord factor* à environ 250°. Les spectres à faibles résolution ont été effectués avec une résolution d'environ 2000, les spectres à haute résolution à environ 20.000.

ANALYSE DES RESULTATS

Structure des acides gras

Les esters méthyliques d'une fraction d'acides saturés, solide (I) (R = CH_3) et d'une fraction d'acides insaturés, liquide (II) (R = CH_3), obtenus par saponification du glycolipide et préparés selon [1], ont été analysés par spectrométrie de masse (Fig. 1). Cette analyse confirme la structure des deux acides

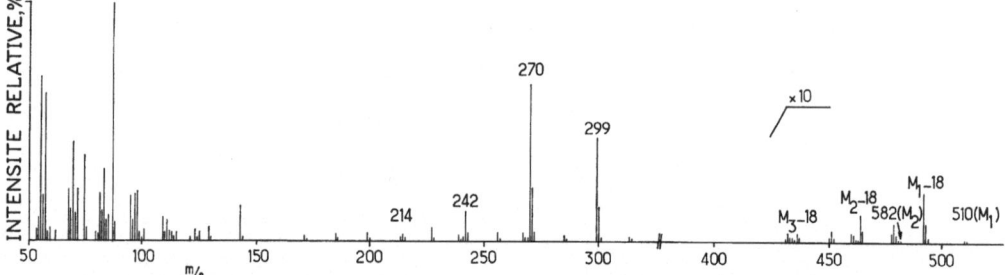

Fig. 1. Spectre de masse du corynomycolate de méthyle (I. n = 13, R = CH₃); en abscisse: unités de masse (m/e); en ordonnée: intensité relative

(I et II, n = 13) en précisant qu'ils sont accompagnés d'homologues inférieurs (n = 11 et 9). La présence d'homologues avait également été reconnue par l'analyse en phase gazeuse des esters méthyliques obtenus par pyrolyse des esters méthyliques des deux acides naturels.

masse; il est cependant évident que l'insaturation se trouve dans la chaîne principale. Le spectre de masse de l'ester insaturé (II b) est très semblable à celui de la Fig. 1 et n'est pas reproduit ici.

Ces résultats donnent les structures (I) et (II) pour les esters méthyliques saturés et insaturés,

$$CH_3(CH_2)_{14}\overset{\overset{a}{|}}{\underset{(CH_2)_nCH_3}{\overset{OH}{\underset{|}{CH}}}}\!\!\!CH\!-\!COOR \qquad n = 13,\ 11\quad 9 \qquad\qquad (I)$$

$$CH_3(CH_2)_5\!-\!\overset{H}{C}\!=\!\overset{H}{C}(CH_2)_7\overset{\overset{a}{|}}{\underset{(CH_2)_nCH_3}{\overset{OH}{\underset{|}{CH}}}}\!\!\!CH\!-\!COOR \qquad n = 13,\ 11,\ 9 \qquad (II)$$

Des ions moléculaires de très faibles intensités sont trouvés à m/e 510 (508)[2], 482 (480) et à (452). Une série de pics dûs à l'élimination d'eau est beaucoup plus intense [m/e 492 (490), 464 (462) et 436 (434)]. Le clivage en α-β du carboxyle (selon a) conduit, après transfert d'un hydrogène, aux ions à m/e 270 (270), 242 (242) et 214 (214), qui indiquent que n = 13, 11, 9. En accord avec les spectres de masse des mycolates de méthyle [8—13] des pics à 29 u. m. plus élevés sont trouvés, correspondant à la structure

$$\overset{\oplus}{HO} = CH\!-\!\underset{(CH_2)_nCH_3}{\overset{|}{CH}}\!-\!COOCH_3 \qquad \begin{array}{l} n = 13,\ m/e\ 299 \\ n = 11,\ m/e\ 271 \\ n = \ \ 9,\ m/e\ 243 \end{array}$$

Il y a une autre série de pics correspondant à la perte de CH₃OH (m* 238,4 pour la perte de méthanol à partir du fragment à m/e 299)[3]. Les résultats des mesures de masse sont indiqués dans le Tableau 1.

La position de la double liaison dans l'acide insaturé (II) ne peut pas être déduite du spectre de

respectivement; il est à remarquer que la chaîne principale paraît homogène dans les deux cas, la chaîne latérale en α présentant au moins trois homologues. Dans un travail sur les acides corynomycoliques de *Corynebacterium* 506, Etémadi et coll. [5] ont déjà décrit des séries homologues, dans les deux chaînes.

Tableau 1. *Mesure de masse des principaux pics de l'ester méthylique de l'acide solide (I) (R = CH₃)*

m/e	Mesuré	Calculé	Composition élémentaire
214	214.1929	214.1933	$C_{13}H_{26}O_2$
239	239.2009	239.2011	$C_{15}H_{27}O_2$
242	242.2244	242.2246	$C_{15}H_{30}O_2$
243	243.1957	243.1960	$C_{14}H_{27}O_3$
270	270.2557	270.2559	$C_{17}H_{34}O_2$
271	{ 271.2275	271.2273	$C_{15}H_{31}O_3$
	271.2591	271.2592	$C_{16}\,^{13}C\,H_{34}O_2$
299	299.2587	299.2586	$C_{18}H_{35}O_3$

Structure du cord factor

La spectrométrie de masse a été effectuée avec les dérivés peracétylés (IVa, b, c) du glycolipide (IIIa,

---

[2] Les chiffres entre parenthèse concernent la fraction d'acides insaturés.

[3] m* indique des pics métastables.

b, c). L'intensité des pics dans la région des ions moléculaires à m/e 1630—1634 est très faible, de sorte qu'une interprétation de cette partie de spectre est très difficile. Il existe cependant un pic à m/e 1634 (correspondant à la formule brute $C_{82}H_{162}O_{23}$ (IV a), qui pourrait être dû à l'ion moléculaire de l'octa-acétate d'un diester de tréhalose contenant deux molécules d'acide corynomycolique (I, n = 13, R = H); il existe également un pic à m/e 1632 (IV b),

qui pourrait représenter une molécule contenant une seule insaturation; un pic encore plus petit à m/e 1630 (IV c) pourrait être dû à un diester de tréhalose contenant deux molécules d'acide corynomycolénique (II, n = 13, R = H).

La fragmentation principale du glycolipide est initiée par la scission de la liaison glycosidique du tréhalose et conduit à des ions oxonium très stables. Un spectre partiel est représenté dans la Fig. 2.

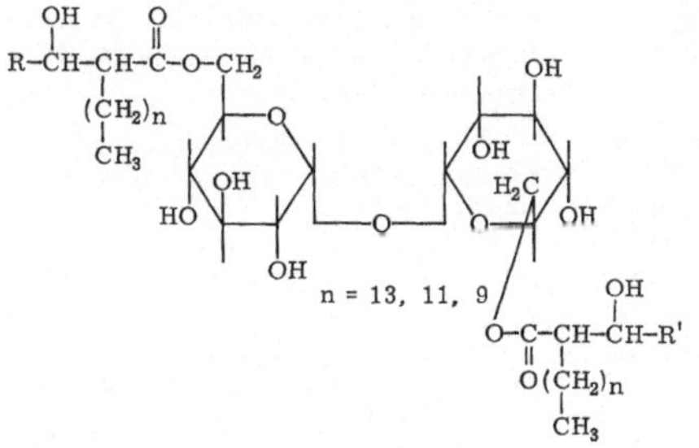

$R = R' = C_{15}H_{31}$ (III a)

$R = C_{15}H_{31}$
$R' = C_{15}H_{29}$ (III b)

$R = R' = C_{15}H_{29}$ (III c)

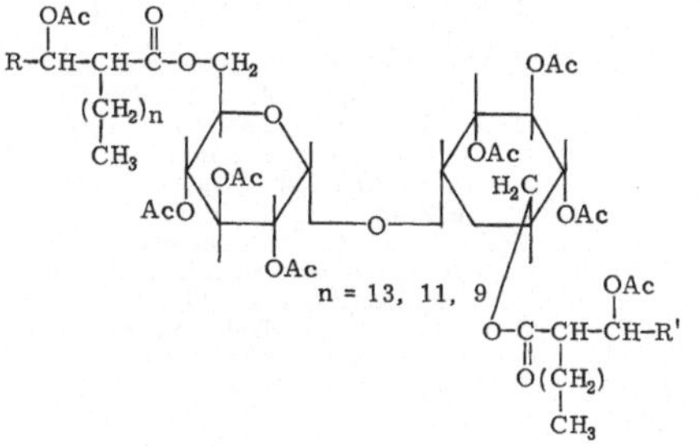

$R = R' = C_{15}H_{31}$ (IV a)
p. m. 1634 (n = 13)

$R = C_{15}H_{31}$ $R' = C_{15}H_{29}$ (IV b)
p. m. 1632 (n = 13)

$R = R' = C_{15}H_{29}$ (IV c)
p. m. 1630 (n = 13)

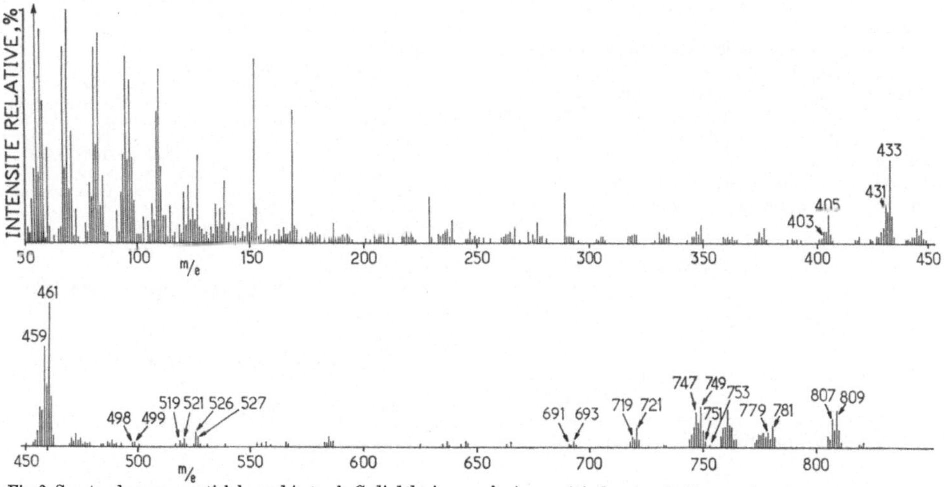

Fig. 2. Spectre de masse partiel du *cord factor* de *C. diphtheriae*; en abscisse: unités de masse (m/e); en ordonnée: intensité relativ

Tableau 2. *Mesure de masse des principaux pics du* cord
factor *peracétylé (IV a, b, c)*

m/e	Mesuré	Calculé	Composition élémentaire
461	461.4716	461.4722	$C_{32}H_{61}O$
519	519.4777	519.4777	$C_{34}H_{63}O_3$
521	521.4938	521.4933	$C_{34}H_{65}O_3$
526	526.3142	526.3142	$C_{28}H_{46}O_9$
527	527.3202	527.3219	$C_{28}H_{47}O_9$
747	747.5407	747.5411	$C_{44}H_{75}O_9$
749	749.5570	749.5567	$C_{44}H_{77}O_9$
809	809.5789	809.5778	$C_{46}H_{81}O_{11}$

Les ions oxonium (Va) à m/e 809 ($C_{46}H_{81}O_{11}$)
(Tableau 2) et (Vb) à m/e 807 perdent une molécule
d'acide acétique en donnant des pics très intenses à
m/e 749 ($C_{44}H_{77}O_9$) et à m/e 747 ($C_{44}H_{75}O_9$), respectivement. Les homologues inférieurs donnent des pics
à m/e 781 (779) et 753 (751) pour les ions oxonium,
et à m/e 721 (719) et 693 (691) après perte d'acide
acétique.

$$R = C_{15}H_{31} \quad m/e \ 809 \qquad (Va)$$
$$R = C_{15}H_{29} \quad m/e \ 807 \qquad (Vb)$$

Des pertes successives de cétène, ou d'acide acétique, comme décrites précédemment pour des sucres
peracétylés [18], sont peu prononcées; ceci nous
conduit à penser que la perte d'acide acétique
s'effectue, dans notre cas, à partir de l'acétoxyle des
groupes acyle.

Des ions dûs aux radicaux acyle sont observés à
m/e 521 ($C_{34}H_{65}O_3$) (VIa) et à m/e 519 (VIb); ils
perdent facilement de l'acide acétique en donnant
des pics intenses à m/e 461 ($C_{32}H_{61}O$) (m* 407,9) et
à m/e 459 (m* 405,9); des fragments dûs aux homologues inférieurs se trouvent à m/e 493 (491) et 465
(463), ainsi qu'à 433 (431) et 405 (403), respectivement, après perte d'acide acétique.

$$R-\overset{OAc}{\underset{(CH_2)_{13}}{CH}}-CH-C\equiv\overset{\oplus}{O} \quad (VIa) \ m/e \ 521 \xrightarrow{-AcOH} m/e \ 461 \ (R = C_{15}H_{31})$$
$$(VIb) \ m/e \ 519 \xrightarrow{-AcOH} m/e \ 459 \ (R = C_{15}H_{29})$$

Des pics à m/e 527 ($C_{28}H_{47}O_9$) et 526 ($C_{28}H_{46}O_9$)
permettent de calculer la masse de la chaîne latérale;
des mesures de masse montrent que ces fragments
sont des ions oxonium formés par clivage de (V) selon

b, avec ou sans transfert d'un hydrogène. Ces pics
indiquent également que la double liaison est localisée
dans la chaîne principale.

La spectrométrie de masse ne permet pas de
préciser la position des deux radicaux acyle sur le
tréhalose, mais elle a pu être déterminée par voie
chimique: après perméthylation du glycolipide et
hydrolyse acide, un seul sucre O-méthylé a été obtenu,
dont le comportement en phase gazeuse est identique
à celui du 2,3,4-tri-O-méthyl glucose [19].

Ces résultats prouvent que le cord factor de
C. diphtheriae est un mélange de glycolipides de
structures (IIIa, b, c).

Ce travail a bénéficié d'une subvention du National
Institute of Allergy and Infectious Diseases (U.S. Public
Health Service), Grant AI-02838.

BIBLIOGRAPHIE

1. Ioneda, T., Lenz, M., et Pudles, J., Biochem. Biophys.
   Res. Commun. 13 (1963) 110.
2. Lederer, E., et Pudles, J., Bull. Soc. Chim. Biol. 33
   (1951) 1003.
3. Lederer, E., Pudles, J., Barbezat, S., et Trillat, J. J.,
   Bull. Soc. Chim. France, (1952) p. 93.
4. Pudles, J., et Lederer, E., Biochim. Biophys. Acta, 11
   (1953) 163.
5. Noll, H., Bloch, H., Asselineau, J., et Lederer, E., Biochim. Biophys. Acta, 20 (1956) 299.
6. Asselineau, J., et Lederer, E., Chemistry and metabolism
   of bacterial lipides, in Lipide Metabolism (edited by
   K. Bloch), John Wiley and sons, New York 1960,
   p. 336.
7. Asselineau, J., Les Lipides Bactériens, Hermann, Paris
   1962; The Bacterial Lipids, Hermann, Paris, Holden-
   Day, San Fransisco, 1967.
8. Etémadi, A. H., Okuda, R., et Lederer, E., Bull. Soc.
   Chim. France, (1964) p. 868.
9. Etémadi, A. H., Miquel, A. M., Lederer, E., et Barber,
   M., Bull. Soc. Chim. France, (1964) p. 3274.
10. Etémadi, A. H., Bull. Soc. Chim. France, (1964) p. 1537.
11. Etémadi, A. H., Compt. Rend. Acad. Sci., ser. D, 263
    (1966) 1257.
12. Markovits, J., Pinte, F., et Etémadi, A. H., Compt Rend
    Acad. Sci., ser. C, 263 (1966) 960.
13. Etémadi, A. H., J. Gas Chrom. 1967, sous presse.
14. Lederer, E., Chem. and Phys. of Lipids, 1967, sous presse.
15. Etémadi, A. H., Gasche, J., et Sifferlen, J., Bull. Soc.
    Chim. Biol. 47 (1965) 631.
16. Etémadi, A. H., Bull. Soc. Chim. Biol. 45 (1963) 1423.
17. Toubiana, R., Compt. Rend. Acad. Sci. 253 (1961) 2989.
18. Biemann, K., DeJongh, D. C., et Schnoes, H. K., J. Am.
    Chem. Soc. 85 (1963) 1763.
19. Yamakawa, M. T., Communication privée.

M. Senn et E. Lederer
Institut de Chimie des Substances Naturelles du C.N.R.S.
91-Gif-sur Yvette (Essonne), France

J. Pudles et E. Lederer
Institut de Biochimie, Faculté des Sciences d'Orsay
91-Orsay (Essonne), France

Adresse actuelle de T. Ioneda:
Departamento de Bioquimica, Faculdade Medicina,
Univ. São Paulo, Brésil

European J. Biochem. 1 (1967) 357—362

Studies on Phytosterol Biosynthesis: the Sterols of *Larix decidua* Leaves

L. J. Goad and T. W. Goodwin

Department of Biochemistry and Agricultural Biochemistry, The University College of Wales, Aberystwyth

(Received January 19, 1967)

The sterols of *Larix decidua* leaves consist of β-sitosterol, campesterol, cycloartenol, 24-methylene cycloartanol, cycloeucalenol, 24-methylene lophenol, 24-ethylidene lophenol, together with a number of unidentified 4α-methyl sterols. The pattern of incorporation of [2-¹⁴C]mevalonic acid into these compounds has been examined.

Recent reports have described the isolation from plants of sterols which may be biosynthetic intermediates between squalene and the typical phytosterols[1—7] and the incorporation of [2-^{14}C] mevalonic acid, [1-^{14}C] acetate and [methyl-^{14}C] methionine into such compounds has been described [6,7]. During studies in our Department on the biosynthesis of the phytosterol side chain[8] a plant tissue was required which contained β-sitosterol as the major sterol component. Preliminary examination by gas-liquid chromatography showed that a number of conifer species contained predominantly β-sitosterol together with minor amounts of campesterol. This paper reports the results of an analysis of the sterols of the newly opened leaves of the deciduous conifer larch *(Larix decidua)*. Some observations on the incorporation of [2-^{14}C] mevalonic acid into sterols by the leaves of this tree are also included.

EXPERIMENTAL PROCEDURE

Experimental methods were generally those described previously [7]. Two batches of larch leaves were collected locally in the springtime of successive years when the leaves had been developing for about two to three weeks.

Batch 1

The non-saponifiable lipid (6.25 g) was extracted from the leaves (650 g) by homogenisation in 1.3 l of ethanol followed by addition of 200 ml of water and 200 g of potassium hydroxide and refluxing for $2^{1}/_{2}$ h. The leaf debris was removed by filtration through a pad of glass wool and the non-saponifiable lipids obtained by extraction with diethyl ether in the usual manner. The total sterol (213 mg) was obtained by digitonin precipitation [7,9], the digitonides washed with diethyl ether only and then cleaved with pyridine in the usual way. The sterols were first separated by chromatography on 20 g of alumina (Brockmann grade III) and eluting with

200 ml portions of light petroleum containing increasing amounts of diethyl ether [7]. Three sterol fractions were obtained, the 4,4'-dimethyl sterols (24 mg), the 4α-methyl sterols (7.3 mg) and the 4-desmethyl sterols (142.6 mg). These were further purified by preparative thin layer chromatography on silica gel plates incorporating Rhodamine 6G and with chloroform as developer [7]. Separation of the 4,4'-dimethyl and 4α-methyl fractions into their constituent components (see below) was achieved by acetylation followed by chromatography on thin layers of silica gel impregnated with 10% silver nitrate with 40% benzene in *n*-hexane as developer. Sterols were located by spraying with an acetone solution of Rhodamine 6G followed by examination in ultraviolet light [10].

Batch 2

2.5 kg of larch leaves were extracted exhaustively with acetone followed by diethyl ether-light petroleum (1:1, v/v) to yield 51·68 g of lipid. This was saponified by refluxing with 15% ethanolic potassium hydroxide to give 16.5 g of non-saponifiable lipid. 4.5 g of an unidentified white solid were removed from this material by precipitation from a diethyl ether solution. The remaining non-saponifiable lipid (11.88 g) was chromatographed in two portions on 200 g columns of alumina (Brockmann, grade III) and eluting with 2 l portions of light petroleum (40—60°) containing 0, 2, 6, 9, and 50% diethyl ether respectively. The 9% fractions (0.58 g) which contained the 4α-methyl sterols were rechromatographed to remove a small amount of β-sitosterol and added to the 6% fractions (2.72 g). The 4,4'-dimethyl sterols (65 mg) and the 4α-methyl sterols (12 mg) were obtained from this material by digitonin precipitation followed by preparative thin layer chromatography on silica gel. The 4,4'-dimethyl sterols were separated to give cycloartenol and 24-methylene cycloartanol by chromatography on a column of alumina impregnated with silver nitrate.

This adsorbent has been found to give good separations of sterols differing in the number and position of double bonds. The adsorbent was prepared from alumina (Merck AG., Darmstadt) which was first washed several times with distilled water and then dried overnight at 110°. 80 g of the washed alumina were stirred with 80 ml of water containing 20 g of silver nitrate and the slurry filtered through paper without suction, 15 ml of filtrate were collected. The slurry was then dried for 18—20 h at 110°. At all stages care was taken to exclude light as far as possible. For the present work a 5 g column of this material was developed sequentially with 50 ml portions of 2, 5, 10, 15, 20, 30, 40, 60, and 80% diethyl ether in light petroleum (40—60°) and 5 to 5.5 ml fractions were collected. Examination of fractions by gas liquid chromatography showed that cycloartenol (4 mg) was eluted in tubes 28—32 and 24-methylene cycloartanol (55 mg) was eluted in tubes 36—56. The 4α-methyl sterols (9 mg) were acetylated and the acetates resolved into four constituent bands, A < 1 mg, B ~ 6 mg, C ~ 2 mg, and D ~ 1 mg, by preparative thin-layer chromatography on silica gel impregnated with 10% silver nitrate. Cycloeucalenol acetate, 24-methylene lophenol acetate, and 24-ethylidene lophenol acetate were run as standards and the developing solvent was 40% benzene in *n*-hexane [10].

Gas-liquid Chromatography

A Varian-Aerograph 1522B gas chromatograph fitted with flame-ionisation detectors and on-column injection was used. Three stationary phases were employed, 1% QF-1; 1% SE-30 and 0.7% Hi-EFF8B, on 80—100 mesh silanised Chromosorb W which were prepared by the usual filtration technique [11 a]. The approximate percentage compositions of sterol mixtures were calculated from peak areas.

Incorporation of [2-14C]mevalonic Acid in Larch Sterols

A similar procedure to that previously reported for the incorporation of mevalonic acid into pea leaves [7] was found to give good incorporations with larch leaves. Three 1.0 g portions of 10—14 day old larch leaves were chopped into small pieces and moistened with 1.0 ml (1 µC/ml) portions of [2-14C] mevalonate solution (Supplied by the Radiochemical Centre, Amersham, England). The leaves were then incubated for 6 h at 25° under light. The non-saponifiable lipids were extracted and separated on alumina to give the hydrocarbon, 4,4′-dimethyl, 4α-methyl and 4-desmethyl sterol fractions as described previously [7]. Subsequent separations are described in the results section. Radioactivity was determined by liquid scintillation counting on a

Packard Tri-Carb Liquid Scintillation Spectrometer. Thin layers were scanned for radioactivity on the Desaga thin layer scanner.

RESULTS

Identification of Larch Sterols

Gas-liquid chromatography of the major sterol fraction revealed a main component (95%) with retention data on SE-30, QF-1, and Hi-EFF8B identical to β-sitosterol. A minor compound (2%) was identical to campesterol on all stationary phases. Analysis on SE-30, which will separate campesterol from stigmasterol, showed no evidence of the latter sterol. On Hi-EFF8B however, a second minor sterol was resolved with a retention time approximately the same as fucosterol but the possibility that this compound may be 29-isofucosterol [11 b] cannot be eliminated. The main sterol was crystallised from chloroform-methanol, m.p. 132°, and gave an acetate m.p. 116° (β-sitosterol 137.5—138°; β-sitosterol acetate 121—122° [12]). The infrared spectrum was identical to that of β-sitosterol and the mass spectrum had a mass peak at m/e 414 and fragmentation peaks at m/e 273 [M-side chain]; 255 [M-(side chain + water)] and 213 [M-(side chain and part of ring D + water)] consistent with this sterol being β-sitosterol [6,13].

Gas-liquid chromatography of the 4,4′-dimethyl sterols demonstrated two components with retention times corresponding to cycloartenol (8—9%) and 24-methylene cycloartanol (91—92%) respectively (Table 1). Although cyclolaudenol and 24-methylenecycloartanol show some separation on QF-1 the presence of very small amounts of cyclolaudenol in the larch preparations cannot be completely excluded. No compound could be detected corresponding in retention data to lanosterol. Separation of the cycloartenol and 24-methylene cycloartanol was achieved either by chromatography of the alcohols on silver nitrate-impregnated alumina or by thin layer chromatography of the acetates on silica gel incorporating 10% silver nitrate. The separated alcohols and acetates had retention times on GLC identical to cycloartenol and 24-methylene cycloartanol and their acetates respectively. The compound chromatographing as cycloartenol acetate had a mass peak at m/e 468 and other peaks at m/e 453, 408; 393; 365; 339; 297 (M-[side chain + acetate]) and 286 [M-ring A] which are in accord with the fragmentation pattern of cycloartenol acetate [6,14]. The compound chromatographing as 24-methylene cycloartanol acetate was crystallised from chloroform-methanol to give blades, m.p. 112—114° (24-methylene cycloartanol acetate 116—117° [15]). The infrared spectrum had peaks at 887 and 1640 cm⁻¹ characteristic of a methylene group and the mass spectrum had peaks at m/e 482; 467; 422;

407; 379; 300 [M-ring A] and 297 [M-(side chain + acetate)] which are in agreement with the identification of this compound as 24-methylene cycloartanol acetate [6,14].

Table 1. *Relative retention times*[a] *of 4,4-dimethyl and 4 α-methyl sterols*

Sterol	1% QF-1[b]	1% SE-30[c]	0.7% Hi-EFF 8B[d]
Lanosterol	4.88	3.54	11.8
Cycloartenol	5.65	3.96	15.4
24-Methylene cyclo-artanol	6.82	4.62	17.0
Cyclolaudenol	6.54	4.52	17.1
Larch 4,4-dimethyl sterols	5.68, 6.92	4.04, 4.70	—
Lophenol	3.88	2.94	10.0
24-Methylene lophenol	5.12	3.76	15.5
Cycloeucalenol	5.70	3.91	13.7
24-Ethylidene lophenol	6.41	5.35	20.6
Larch 4α-methyl sterols	4.47, 4.82, 5.12, 5.47, 5.70, 6.41	3.58, 3.76, 4.11, 5.35	10.8, 13.7, 15.5, 17.0, 20.6

a Relative to cholestane.
b 1% QF-1 on 80—100 mesh HMDS Chromosorb W. 6' × 1/8 Column 225°, Injector 250°, Detector 250°, Nitrogen 40 ml/min.
c 1% SE-30 on 80—100 mesh HMDS Chromosorb W. 6'/8'' Column 225°, Injector 245°, Detector 250°, Nitrogen 40 ml/min.
d 0.7% Hi-EFF 8B on 80—100 mesh HMDS Chromosorb W. 5' × 1/8'' Column 216°, Injector 250°, Detector 250°, Nitrogen 40/min.

The 4α-methyl sterols were examined by gas-liquid chromatography and shown to consist of a complex mixture of sterols (Table 1). The major compound (60%) had a retention time on all columns the same as 24-ethylidene lophenol; small amounts of compounds with retention times equal to 24-methylene lophenol and cycloeucalenol respectively were observed using QF-1 or Hi-EFF 8B as stationary phase. Acetylation of the 4α-methyl sterol complex followed by preparative thin-layer chromatography on silver nitrate-silica gel gave four component bands A, B, C, and D (see Fig. 3). These were analysed by gas-liquid chromatography and the results (Fig. 1, Table 2) clearly demonstrate the complexity of the 4α-methyl sterol fraction, at least fourteen compounds were observed. Fraction A had a mobility on silver nitrate—silica gel thin layers expected for sterols with a saturated side chain [10,16]. On gas-liquid chromatography five compounds were revealed, of these A2 and A4 had retention data which indicated that these compounds may be 24ζ-methyl and 24ζ-ethyl lophenol respectively. Evidence for these compounds has also been obtained in grapefruit peel [10] Fraction B, which co-chromatographed with 24-ethylidene lophenol acetate on thin-layer chromatography was the major component of the 4α-methyl sterol acetates, on gas-liquid chromatography it showed one major compound B7 with retention data identical to those of 24-ethylidene lophenol acetate. It was crystallised with difficulty from chloroform-methanol

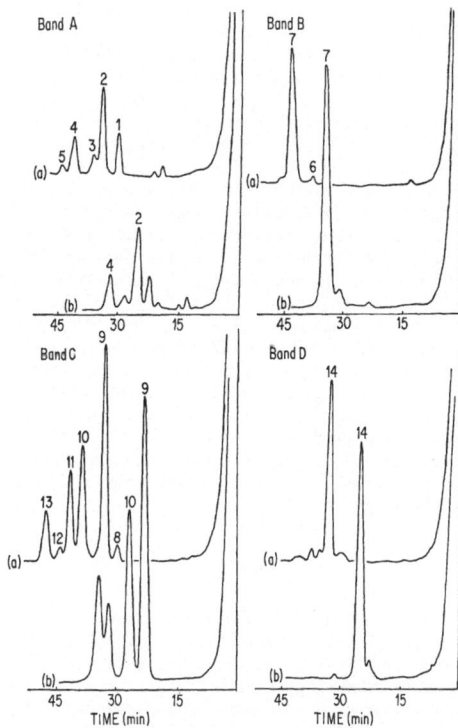

Fig. 1. *Gas-liquid chromatography.* (a) On 1% QF-1; (b) On 1% SE-30 of the larch 4α-methyl sterol acetates. Bands, A, B, C and D were separated by TLC as described in the text (see also Fig. 3)

Table 2. *Relative retention times*[a] *of 4α-methyl sterol acetates*

Sterol acetate	1% SE-30[b]	1% QF-1[c]
24-Methylene lophenol	4.91	7.30
Cycloeucalenol	5.09	8.37
24-Ethylidene lophenol	6.54	9.28
Larch 4α-methyl sterols		
A.	3.91; 4.36; 4.91 (2)[d]; 5.54; 6.27 (4)	6.53; 7.45 (2); 7.86; 8.98 (4); 9.59
B.	5.91; 6.54 (7)	8.06; 9.28 (7)
C.	4.36 (9); 5.14 (10); 6.18; 0.04	6.43; 7.17 (9); 8.37 (10); 8.98; 9.49; 10.31
D.	4.91 (14)	7.25 (14)

a Relative to Cholestane.
b 1% SE-30 in 80—100 mesh HMDS Chromosorb W. 6' × 1/8'' Column 241°, Injector 250°, Detector 237°, Nitrogen 40 ml/min.
c 1% QF-1 on 80—100 mesh HMDS Chromosorb W. 6' × 1/8'' Column 233°, Injector 245°, Detector 235°, Nitrogen 40 ml/min.
d Figures in parenthesis indicate the peak number shown in Fig. 1

to give long needles m.p. 141—143° (24-ethylidene lophenol acetate isolated from potato leaves m.p. 142—144° [3]). Mass spectrometry showed a mass peak at m/e 468 and other peaks at m/e 370 [M-part of side chain] and 327 [M-(side chain + 2H)] characteristic of 24-ethylidene lophenol acetate fragmentation [6,17]. However, a second mass peak at m/e 482 was observed indicating a higher homologue. Similar mass spectral evidence for such a sterol $(C_{31}H_{52}O)$ in the 4α-methyl sterol complex of birch wood has been reported [17,18]. Fraction C co-chromatographed with cycloeucalenol acetate on silver nitrate thin layers. Gas-liquid chromatography revealed six components, one of these, C10, had retention data identical to those of cycloeucalenol acetate. The major compound C9, however, did not correspond to any known 4α-methyl sterol but was apparently identical to one of the 4α-methyl sterols present in grapefruit [10]. Fraction D, which cochromatographed with 24-methylene lophenol acetate on thin layers, had one main component on GLC which had an identical retention time to 24-methylene lophenol acetate.

Incorporation of [2-¹⁴C]mevalonic Acid into larch Sterols

As expected 10—14 day old larch leaves incorporated [2-¹⁴C]mevalonic acid into the non-saponifiable lipid (748,300 dis./min). Chromatography on alumina showed part of this radioactivity to be distributed in the fractions which contain hydrocarbons (29,470 dis./min), 4,4'-dimethyl sterols (71,280 dis./min), 4α-methyl sterols (73,400 dis./min) and 4-desmethyl sterols (295,500 dis./min) respectively. The incorporation of [2-¹⁴C]mevalonic acid into squalene by plants is well established [6,7,19, 20] and examination of a portion of the larch hydrocarbon fraction by thin-layer chromatography confirmed that squalene was the predominantly labelled compound in this fraction. The 4,4'-dimethyl sterol fraction was purified by thin-layer chromatography on silica gel and a portion of the recovered compounds (16,900 dis./min) were acetylated. The acetates were then separated on a thin layer of silica-gel impregnated with 10% silver nitrate. Scanning for radioactivity revealed three radioactive compounds (Fig.2). The bulk of the radioactivity co-chromatographed with 24-methylene cycloartanol acetate, a lesser radioactive peak accompanied cycloartenol acetate whilst a minor radioactive peak was associated with unidentified material running ahead of cycloartenol acetate. Confirmation that the radioactive compounds were 3β-hydroxy sterols was obtained by addition of a mixture of cycloartenol and 24-methylene cycloartanol (4.8 mg) to a second portion of the radioactive 4,4'-dimethyl sterol fraction (17,800 dis./min). The sterols obtained by digi-

tonin precipitation were acetylated and separated on silica gel-silver nitrate thin layers. The cycloartenol acetate and 24-methylene cycloartanol acetate zones were eluted and contained 1,680 dis./min and 5,840 dis./min respectively. The 4α-methyl sterol fraction was

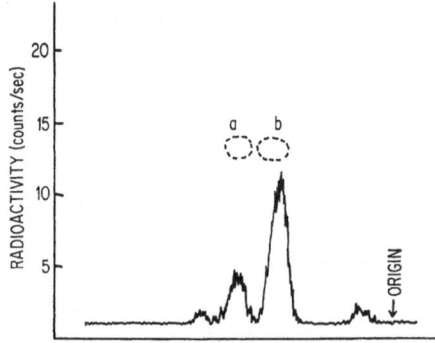

Fig.2. *Thin-layer chromatography on silica gel-silver nitrate of the radioactive 4,4'dimethyl sterol acetates obtained from larch leaves following incubation with [2-¹⁴C]mevalonic acid. Marker cycloartenol acetate (a) and 24-methylene cycloartanol acetate (b) ran in the positions indicated*

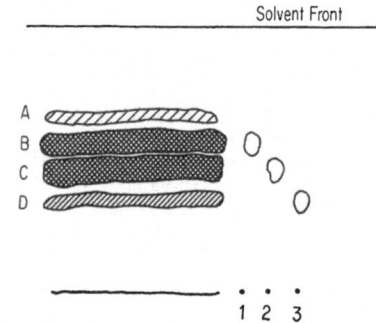

Fig.3. *Radioautograph of a silica gel-silver nitrate thin layer separation of the radioactive 4α-methyl sterol acetates biosynthesised by larch leaves in the presence of [2-¹⁴C]mevalonic acid.* (1) 24-Ethylidene lophenol acetate; (2) cycloeucalenol acetate; (3) 24-methylene lophenol acetate

further purified by thin-layer chromatography on silica gel and the sterols (33,480 dis./min) were acetylated and separated on a silica gel-silver nitrate thin layer. A radioautograph (Fig.3) showed that radioactivity was associated with the four bands A, B, C, and D. The most heavily labelled band, B, corresponded to 24-ethylidene lophenol acetate, bands C and D co-chromatographed with cycloeucalenol acetate and 24-methylene lophenol acetate respectively, band A was only weakly labelled. The bulk of the radioactivity in the 4-desmethyl sterol fraction was digitonin precipitable and co-chromatographed on thin layers with β-sitosterol.

DISCUSSION

The data presented in this report supplement and extend the results we previously obtained during a study of sterol biosynthesis in pea seedlings [7]. Similar results have also been reported following a study of sterol biosynthesis in tobacco tissue cultures[6,21].

following investigations on potato and tobacco sterols it has been suggested that cycloartenol may play a major role in phytosterol biosynthesis in these plants [5,6]. It is postulated that cycloartenol may replace lanosterol (the sterol precursor in animals and fungi) as the first cyclisation product of squalene which is subsequently modified to give the typical phytosterols. Our present demonstration of cycloartenol and 24-methylene cycloartanol, coupled with the failure to detect lanosterol, in larch leaves is compatible with such a biosynthetic route (see

Cycloartenol   24-Methylene cycloartanol

Cycloeucalenol   Intermediate

24-Methylene lophenol   24-Ethylidene lophenol   Phyto-sterols

Scheme. Possible biosynthetic sequence for the production of sterols in higher plants

The range of sterols now demonstrated in larch leaves, and in particular the complexity of the 4α-methyl sterol fraction, shows a striking similarity to the sterol composition of the manifestly different plant tissue grapefruit peel [10]. The co-occurrence of these sterols in plant tissues and the rapid incorporation of mevalonic acid into these compounds is of interest when considering possible pathways in phytosterol biosynthesis. There is increasing evidence for the widespread occurrence of cycloartenol and 24-methylene cycloartanol in higher plants [22,23] and

Scheme). However, since lanosterol has apparently been positively identified in a few plants [24—26] it is clear that this important problem requires a more detailed investigation. The present evidence for the presence of cycloeucalenol in larch and also in grapefruit [10] is compatible with a biosynthetic sequence involving 9β-19 cyclopropane triterpenes since cycloeucalenol can readily be derived from cycloartenol as indicated in the Scheme. The next step from cycloeucalenol must presumably be the opening of the 9β-19 cyclopropane ring and loss of the 14α-methyl group to give subsequently 24-methylene lophenol. By analogy with the opening of the cyclopropane ring of cycloartenol by acid catalysis [27,28] a compound such as that shown in the scheme might reasonably be expected to be produced by enzymic action upon cycloeucalenol [23], Such a compound could then lose the 14α-methyl group and this would be facilitated by the newly formed Δ^7 or Δ^8 bond [29]. By contrast the cyclopropane ring of compounds such as cycloeucalenol and cycloartenol may hinder the loss of the 14α-methyl group [30]. Although the route shown in the Scheme may ultimately prove to be the major pathway a consideration of the retention data of the minor 4α-methyl sterols in larch and also in grapefruit [10] indicates that some of the enzymes involved may be relatively non-specific, for addition of the C-29 methyl group or reduction of the side chain double bond occurs irrespective of the structure of the sterol nucleus. For example, whilst there is good evidence that C-24 methylene or ethylidene groups are reduced to produce the C-24 methyl and ethyl groups of the typical phytosterols [8,31−33] it is not clear if this reduction occurs principally at the 4α-methyl sterol level to give 24-methyl and 24-ethyl lophenol or at the 4-desmethyl level, since fucosterol or 29-isofucosterol have been reported to occur together with β-sitosterol in a number of plants [7,34,11b]. The absence of stigmasterol from larch leaves is in accord with a previous report on several pine barks where β-sitosterol was also shown to predominate [36].

We would like to thank Professor E. Lederer and Dr. H. Audier (Institut de Chimie des Substances Naturelles, Gif-sur-Yvette, France) for the determination of the mass spectra. We also wish to thank the following for samples of sterols: Professor Y. Mazur (citrostadienol) Dr. W. Lawrie (cyclolaudenol), Professor G. Ourisson (cycloartenol), Dr. W. W. Reid (cycloeucalenol), Professor K. Schreiber (24-methylene-lophenol and cycloartenol) and Dr. M. Shimizu (24-methylene cycloartanol).

Note added in proof (April 11,1967): Gaschromatography-mass spectrometry (kindly carried out by Dr. C. J. W. Brooks, Chemistry Department. The University, Glasgow, on an LKB 9000 instrument) has provided further evidence for cycloeucalenol in Band C of the 4α-methyl sterol acetate fraction. Peak C 10 had ions at m/e 468, 453, 408, 393, 365, 353, 300 and 283 (see Ref. 14).

REFERENCES

1. Mazur, Y., Weizmann, A., and Sondheimer, F., *J. Am. Chem. Soc.* 80 (1958) 1007, 6293.
2. Djerassi, C., Krakower, G. W., Lemin, A. J., Liu, L. H., Mills, J. S., and Villotti, R., *J. Am. Chem. Soc.* 80 (1958) 6284.
3. Schreiber, K., and Osske, G., *Tetrahedron*, 20 (1964) 2575.
4. Osske, G., and Schreiber, K., *Tetrahedron*, 21 (1965) 1559.
5. Von Ardenne, M., Osske, G., Schreiber, K., Steinfelder, K., and Tümmler,R., *Die Kulturpflanze*, 13 (1965) 102, 115.
6. Benveniste, P., Hirth, L., and Ourisson, G., *Phytochemistry*, 5 (1966) 31, 45.
7. Goad, L. J., and Goodwin, T. W., *Biochem. J.* 99 (1966) 735.
8. Goad, L. J., Hammam, A. S. A., Dennis, A., and Goodwin, T. W., *Nature*, 210 (1966) 1322.
9. Williams, B. L., and Goodwin, T. W., *Phytochemistry*, 4 (1965) 81.
10. Williams, B. L., Goad, L. J., and Goodwin, T. W., unpublished.
11a. Horning, E. C., Vanden Heuvel, W. J. A., and Creech, B. G., *Methods Biochem. Analy.* 11 (1963) 69.
11b. Knights, B. A., *Phytochemistry*, 4 (1965) 857.
12. Steele, J. A., and Mosettig, E., *J. Org. Chem.* 28 (1963) 571.
13. Budzikiewicz, H., Djerassi, C., and Williams, D. H., *Structure Elucidation of Natural Products by Mass Spectrometry*, Holden Day, San Francisco 1965, vol. II.
14. Audier, H. E., Beugelmans, R., and Das, B. C., *Tetrahedron Letters*, No. 36 (1966) p. 4341.
15. Ohta, G., *Chem. Pharm. Bull.* 8 (1960) 5.
16. Williams, B. L., Thesis Ph. D., University College of Wales, Aberystwyth (1966).
17. Bergman, J., Lindgren, B. O., and Svahn, C. M., *Acta Chem. Scand.* 19 (1965) 1661.
18. Paasonen, P., *On the non-volatile ethyl ether extractives of birch wood and the changes in composition effected by ageing and sulphate pulping*, Helsinki (1966).
19. Nicholas, H. J., *J. Biol. Chem.* 237 (1962) 1485.
20. Capstack, E., Baisted, D. J., Neuschwander, W. W., Blondin, G., Rosin, N. L., and Nes, W. R., *Biochemistry*, 1 (1962) 1178.
21. Reid, W. W., *Biochem. J.* 100 (1966) 13P.
22. Ourisson, G., Crabbé, P., and Rodig, O. R., *Tetracyclic Triterpenes*, Herman, Leeds 1964.
23. Goad, L. J., *Phytochemical Group Symposium on "Terpenes"* (1966), In press.
24. Gonzalez, A. G., and Mora, M. C., *Anales Real Soc. Espan. Fis. Quim. (Madrid) Ser. B*, 48 (1952) 475.
25. Gonzalez, A. G., and Toste, A. H., *Anales Real Soc. Espan. Fis. Quim (Madrid) Ser. B*, 48 (1952) 487.
26. Kaufmann, H. P., and Sen Gupta, A. K., *Fette Seifen Anstrichmittel*, 66 (1964) 461.
27. Barton, D. H. R., *J. Chem. Soc.* 1444 (1951).
28. Bentley, H. R., Henry, J. A., Irvine, D. S., and Spring, F. S., *J. Chem. Soc.* (1953) 3673.
29. Bloch, K., *Vitamins and Hormones*, 15 (1957) 119.
30. Clayton, R. B., *Quart. Rev. (London)* 19 (1965) 168.
31. Jaureguiberry, G., Law, J. H., McCloskey, J., and Lederer, E., *Biochemistry*, 4 (1965) 347.
32. Akhtar, M., Parvez, M. A., and Hunt, P. R., *Biochem. J.* 100 (1966) 38C.
33. Barton, D. H. R., Harrison, D. M., and Moss, G. P., *Chem. Comm.* No. 17 (1966) 595.
34. Anderson, B., and Krewack, B., *Acta Chem. Scand.* 11 (1957) 997.
35. Hügel, M. F., Vetter, W., Audier, H., Barbier, M., and Lederer, E., *Phytochemistry*, 3 (1964) 7.
36. Rowe, J. W., *Phytochemistry*, 4 (1965) 1.

L. J. Goad, and T. W. Goodwin's present address:
Department of Biochemistry, The University
Liverpool, 3, Great Britain

European J. Biochem. 1 (1967) 363—374

Regulation der Biosynthese der aromatischen Aminosäuren in *Saccharomyces cerevisiae*

2. Repression, Induktion und Aktivierung

F. Lingens, W. Goebel und H. Uesseler

Biochemische Abteilung, Chemisches Institut der Universität Tübingen

(Eingegangen am 25. Januar 1967)

We have studied the repression of the first enzymes of the main branch and the pathways after the branchpoint in the aromatic biosynthesis of *Saccharomyces cerevisiae*. Anthranilate synthetase is repressed by L-tryptophan and prephenate dehydratase by L-phenylalanine. 3-Deoxy-D-arabinoheptulosonic acid 7-phosphate (DAHP) synthetases are not repressed by L-phenylalanine, L-tyrosine and L-tryptophan, chorismate mutases are repressed neither by L-phenylalanine nor by L-tyrosine, and prephenate dehydrogenase shows no repression by L-tyrosine. L-Tryptophan exhibits an induction of the chorismate mutase. Besides chorismate mutase is activated by L-tryptophan. L-Phenylalanine exhibits an induction and also an activation of the prephenate dehydrogenase. By simultaneous addition of L-tryptophan, L-phenylalanine, and L-tyrosine p-aminobenzoic acid was accumulated and the production of p-aminobenzoate synthetase was induced. The tyrosine-sensitive and phenylalanine-sensitive isoenzymes of DAHP synthetase belong to the V-system of the allosteric enzymes. Chorismate mutase and prephenate dehydrogenase are members of the K-system.

In an earlier publication we have discussed the feedback inhibition of the described enzymes. In this paper we discuss the whole events of the regulation phenomena in *Saccharomyces cerevisiae*. We also considered the degradation of L-tryptophan, L-phenylalanine, and L-tyrosine to tryptophol, phenylethanol, and tyrosol respectively in connection with the regulatory events.

In der ersten Veröffentlichung über die Regulation der Biosynthese der aromatischen Aminosäuren in *Saccharomyces cerevisiae* haben wir Untersuchungen geschildert, die sich mit der Hemmung der Enzymaktivitäten an den für die Regulation wichtigen Stellen der verzweigten Biosynthesekette befaßten [1]. Da bei der Betrachtung der Regulation auch das Phänomen der Repression zu berücksichtigen ist, haben wir im weiteren Verlauf der Untersuchung geprüft, ob sich in Saccharomyces cerevisiae auch Repressionsvorgänge abspielen. Es kam auch in Betracht, daß in einer verzweigten Biosynthesekette das Endprodukt einer Kette die Bildung von Enzymen einer anderen Kette induziert oder vorhandene Enzyme aktiviert. Über die Ergebnisse dieser Untersuchungen wird im folgenden berichtet.

Ein Teil der Ergebnisse wurde von einem von uns (W. G.) am 13. Oktober 1966 auf der Tagung der Gesellschaft für physiologische Chemie in Marburg vorgetragen.

MATERIAL UND METHODEN

Verwendete Stämme von S. cerevisiae

Wildstamm S 288 C, 1071-3B trp⁻-Mutante, KP 171 phe⁻,tyr⁻-Mutante, HK 11 phe⁻-Mutante.

Die Stämme wurden entweder in einem Komplettmedium [1] im Fermentor vorgezüchtet und anschließend 4—6 Stunden zur Induktion der Enzyme in ein Minimalmedium [1] gegeben oder einen Tag in einem Saccharose-Mineralsalzmedium [1] gezüchtet. Bei den Mutanten enthielt das Medium zusätzlich $20\,\gamma/$ml der jeweils erforderlichen Endprodukte. Inkubationstemperatur: 30°.

Für die Repressionsversuche wurden die jeweiligen Endprodukte in Konzentrationen von 10^{-3} M bis 10^{-2} M zugegeben.

Herstellung der Enzymextrakte und enzymatische Tests

Die Hefezellen wurden geerntet, gewaschen und bei $-30°$ eingefroren. Die gefrorenen Hefezellen wurden mit der 1,5—2fachen Menge an Aluminium-

---

Nicht allgemein gebräuchliche Abkürzungen. 2-Keto-3-desoxy-araboheptonsäure-7 phosphat, DAHP; *p*-Hydroxyphenylbrenztraubensäure, HPB; *p*-Hydroxyphenylmilchsäure, HPM.

Enzyme. DAHP-Synthetase, oder PODH-Lyase (EC 4.1.2.15); Anthranilatsynthetase, Chorismatmutase, Prephenatdehydrogenase und Prephenatdehydratase haben noch keine EC Nummer.

oxid (Alcoa 305) verrieben und mit Phosphatpuffer (0,01 M, pH 7) oder Trispuffer (0,2 M, pH 7,2) aufgenommen. Nach Abzentrifugieren der Zellrückstände und des Aluminiumoxids (40 min bei 15000 U/ min) wurde der Rohextrakt mit der erforderlichen Menge an Protaminsulfat versetzt. Der Niederschlag wurde 10 min bei 15000 U/min abzentrifugiert. Durch Zugabe steigender Mengen Ammonsulfat wurden einzelne Fraktionen hergestellt. Für die meisten Tests wurden diese Fraktionen direkt verwendet. Sie enthielten im allgemeinen 10—15 mg Protein/ml. Für die Säulenchromatographie wurden die Ammonsulfatfraktionen vorher jeweils gegen den entsprechenden Puffer 12 Std dialysiert.

Zum Nachweis der Enzyme wurden folgende Ansätze verwendet: DAHP-Synthetase: 0,1—0,2 ml Enzymlösung (ca. 10 mg Protein/ml), 0,1 ml Phosphoenolbrenztraubensäure-Na-Salz-Lösung (1,5μMol) 0,1 ml Erythrose-4-phosphat Na-Salz-Lösung (1,5 μMol), Trispuffer (0,2 M) pH 6,4 ad 1 ml, Inkubation 5 min bei 30°. Der Nachweis des gebildeten DAHP erfolgte nach der Vorschrift von Srinivasan [2] sowie Doy und Brown [3].

Anthranilatsynthetase. 0,5 ml Ammoniumchorismatlösung (2×10⁻² M), 0,2 ml Glutaminlösung (2 mg/ml), 0,1 ml Magnesiumsulfatlösung (1 mg/ml), 0,3 ml Enzymlösung, 0,5 ml Trispuffer (0,2 M) pH 8,0, Inkubation 30 min bei 30°. Die gebildete Anthranilsäure wurde fluorimetrisch bei 435 nm sowie durch Absorption bei 330 nm nach Extraktion mit Essigester gemessen.

Chorismatmutase. 0,3 ml Ammoniumchorismatlösung (2×10⁻² M), 0,5 ml Trispuffer (0,2 M) pH 7,2, 0,3 ml Enzymlösung, Inkubation 30 min bei 30°. Die entstandene Prephensäure wurde nach der Vorschrift von Gibson [4] durch Umwandlung in Phenylbrenztraubensäure bestimmt.

Prephenatdehydratase. 0,2 ml Ammoniumprephenatlösung (10⁻² M), 0,7 ml Trispuffer (0,2 M) pH 7,2, 0,1 ml Enzymlösung, Inkubation 30 min bei 30°. Die Umsetzung zu Phenylbrenztraubensäure ließ sich durch Messung der Absorption bei 320 nm im alkalischen Milieu bestimmen.

Prephenatdehydrogenase. 0,2 ml Ammoniumprephenatlösung (10⁻² M), 0,2 ml NAD⁺-Lösung (2 mg/ ml), 0,5 ml Trispuffer (0,2 M) pH 7,2, 0,1 ml Enzymlösung. Inkubation 40 min bei 30°. Die Umsetzung wurde mit Hilfe des gebildeten NADH gemessen. Dieses wurde fluorimetrisch bei 460 nm, meist jedoch über Absorptionsmessung bei 340 nm bestimmt.

Herstellung der Substrate

Phosphoenolbrenztraubensäure wurde nach der Vorschrift von Clark und Kirby [5], Erythrose-4-phosphat nach Ballou [6] synthetisiert. Chorisminsäure wurde durch Akkumulation mit Hilfe der Hefemutante KP 171 gewonnen und mittels Essig-

ester-Extraktion aus dem Minimalmedium und anschließender Papierchromatographie in Isopropanol-Ammoniak-Wasser (8:1:1, v/v) isoliert.

Prephensäure wurde enzymatisch aus Chorisminsäure hergestellt: 4 ml Ammoniumchorismatlösung (2×10⁻² M), 2—3 ml Enzymlösung [Ammoniumsulfatfraktion: 0,65—0,8 gesätt. (NH₄)₂SO₄]. Die Umsetzung ist praktisch quantitativ.

Nachweis der Abbauprodukte der Aminosäuren

Isolierung. Die Kulturmedien wurden abzentrifugiert und der klare Überstand mehrmals mit Essigester extrahiert. Die Essigesterphase wurde zur Trockene eingeengt und der Rückstand mit wenig Methanol aufgenommen. Die Auftrennung der verschiedenen Abbauprodukte erfolgte über Papierchromatographie im System Isopropanol-Ammoniak-Wasser (8:1:1, v/v) bzw. über Dünnschichtchromatographie im System Chloroform-Tetrachlorkohlenstoff-Methanol (50:30:20, v/v).

Spektroskopie. Ultraviolett-Spektren wurden mit dem Spektralphotometer RPQ 20A der Firma Zeiss gemessen. Als Lösungsmittel diente destilliertes Methanol. Zur Messung von IR-Spektren kam der Leitz-Infrarot-Spektrograph mit Mikroeinrichtung zur Anwendung. Die Aufnahme der Massenspektren erfolgte mit dem Massenspektrographen MS 9 von AEI (Manchester).

p-Hydroxyphenylbrenztraubensäure (HPB). Die Isolierung gelang papierchromatographisch im System Benzol-Eisessig-Wasser (125:72:3, v/v) (R_F = 0,4—0,45). Mit Echtblausalz B liefert die Substanz einen intensiv blauen Kupplungsfarbstoff. Diese Reaktion diente zum Sichtbarmachen auf dem Chromatogramm. Die Identifizierung erfolgte außer durch Papierchromatographie mit authentischer HPB als Vergleich auf Grund des UV-Spektrums im neutralen und alkalischen Milieu. pH 7,0: λ_{max} = 283 nm; pH 12,0: λ_{max} = 330 nm gemessen in Methanol.

p-Hydroxyphenylmilchsäure (HPM). Durch papierchromatographischen Vergleich [System Isopropanol-Ammoniak-Wasser (8:1:1, v/v), R_F = 0,5—0,6] mit synthetischer HPM und auf Grund identischer UV-Spektren des biologisch isolierten Materials und der synthetisierten HPM konnte die Substanz identifiziert werden.

p-Hydroxyphenyläthanol-(2) (Tyrosol). Das UV-Spektrum dieser Substanz (in Methanol: neutral 276—77 nm, alk. 292 nm und 241 nm) wies dieselben Charakteristika auf wie p-Hydroxyphenylalkyl-Verbindungen. Sie zeigt keine Carboxylfunktion und reagiert mit Sprühreagentien, mit denen phenolische Substanzen nachweisbar sind (Echtblausalz B, diazot. Sulfanilsäure). Das Massenspektrum und das IR-Spektrum, das dem von Tyrosol (synthetisiert aus p-Hydroxyphenylessigsäure durch Reduktion mit LiAlH₄) übereinstimmt, führten zur Identifizierung der Substanz.

Phenylbrenztraubensäure. Der Nachweis erfolgte analog wie der von HPB. $R_F = 0,85-0,9$ im dort genannten Papierchromatographie-System. UV-Spektrum (gemessen in Methanol) neutral $\lambda_{max} = 283$ nm, alkalisch $\lambda_{max} = 323$ nm.

Phenyläthanol-(2). Die Substanz tritt als Hauptabbauprodukt des L-Phenylalanins auf. Der R_F-Wert in Isopropanol-Ammoniak-Wasser (8:1:1, v/v) ist praktisch identisch mit dem von Tyrosol. Carboxylfunktionen konnten auf Grund des IR-Spektrums und der negativen Reaktionen mit Bromkresolpurpur und Rhodamin B ausgeschlossen werden. Das sehr bandenreiche UV-Spektrum, das dem des Benzols ähnelt, ließ vermuten, daß es sich bei der Substanz um ein Alkylbenzol handelt (in Methanol neutral: $\lambda_{max} = 267,5$, 263, 258, 253, 247 nm, keine Verschiebung der Maxima im alkalischen und sauren Milieu).

p-Fluorphenyläthanol-(2). Papierchromatographisch verhielt sich die Substanz in allen Systemen wie Tyrosol und Phenyläthanol-(2) [Isopropanol-Ammoniak-Wasser (8:1:1, v/v), $R_F = 0,95$, „Partridge"-Gemisch (4:1:5, v/v), $R_F = 0,9$, Methanol/Wasser (1:1, v/v), $R_F = 0,85-0,9$]. Die Substanz trat als Hauptabbauprodukt des *p*-Fluor-DL-phenylalanins auf und ließ keine sauren Funktionen mehr erkennen. Das UV-Spektrum zeigte, daß keine Veränderungen am Aromatenring erfolgt waren (in Methanol $\lambda_{max} = 267$, 263, 258, 253 nm, ohne Verschiebung im sauren oder alkalischen Milieu).

Indolyl-3-äthanol-(2) (Tryptophol). Die Substanz wurde über Dünnschichtchromatographie in den Systemen Chloroform-Methanol-Eisessig (95:12:5, v/v) und Chloroform-Tetrachlorkohlenstoff-Methanol (50:30:20, v/v) auf Grund gleicher R_F-Werte mit synthetisiertem Tryptophol identifiziert. Auch die Farbreaktionen mit verschiedenen Indol-Sprühreagentien (Ehrlichs Reagens, Prochazka-Reagens und Salkowsky-Reagens) waren mit denen von synthetischem Tryptophol identisch.

Indolyl-3-milchsäure. Die Substanz tritt als Hauptabbauprodukt des D-Tryptophans und in geringer Menge beim Abbau des L-Tryptophans auf. Mit Bromkresolpurpur reagiert sie sauer, ein 2,4-Dinitrophenylhydrazon bildet sie nicht. Das IR-Spektrum zeigt eine C=O-Bande bei 5,9 cm^{-1}. Das UV-Spektrum entspricht dem von Indol-3-derivaten mit gesättigter Seitenkette. Die Substanz verhält sich papier- und dünnschichtchromatographisch anders als Indolyl-3-brenztraubensäure, Indolyl-3-essigsäure oder Indolyl-3-propionsäure.

p-Aminobenzoesäure. Durch Besprühen mit Ehrlichs Reagens läßt sich die Substanz auf dem Chromatogramm als gelber Fleck sichtbar machen. Das UV-Spektrum (in Methanol: neutral $\lambda_{max} = 281$ nm, alkal. $\lambda_{max} = 269$ nm) und vergleichender Papierchromatographie führten zur Identifizierung der Substanz [Isopropanol-Ammoniak-Wasser (8:1:1, v/v),

$R_F = 0,2-0,25$, Benzol-Eisessig-Wasser (25:72:3, v/v), $R_F = 0,8-0,85$].

ERGEBNISSE UND DISKUSSION

REPRESSION UND INDUKTION

DAHP-Synthetase

Wie in der ersten Arbeit [1] gezeigt wurde, läßt sich die DAHP-Synthetase (PODH-Lyase) von *S. cerevisiae* durch Ammonsulfatfraktionierung in 2 Isoenzyme auftrennen, von denen das eine fast vollständig durch L-Tyrosin und das andere fast vollständig durch L-Phenylalanin gehemmt wird. Die geringe noch verbleibende Restaktivität könnte einem Tryptophan spezifischen Isoenzym entsprechen. Beim Züchten der Hefezellen in Gegenwart von 10^{-3} M der Endprodukte L-Tyrosin, L-Phenylalanin und L-Tryptophan läßt sich bei getrenntem Zusatz der Aminosäuren noch durch Zugabe aller drei Aminosäuren eine Erniedrigung der spezifischen Aktivität der DAHP-Synthetase feststellen. Auch höhere Konzentrationen an L-Tryptophan, L-Tyrosin (jeweils 2×10^{-3} M) und L-Phenylalanin (10^{-2} M) zeigen keinen Einfluß. Verschiedene Erntezeiten (späte log-Phase oder stationäre Phase) änderten nichts am Ergebnis. In *S. cerevisiae* werden demnach die DAHP-Synthetase-Isoenzyme im Gegensatz zu denen von *Escherichia coli* [7] nicht durch die jeweiligen Endprodukte reprimiert.

Anthranilatsynthetase

Nach Wachstum der Hefezellen mit L-Tryptophan in verschiedenen Konzentrationen wurden die Enzymrohextrakte zur Vermeidung von Feedback-Effekten durch noch vorhandenes L-Tryptophan einer Ammonsulfatfällung unterworfen [0,9 gesätt. (NH$_4$)$_2$SO$_4$], da durch Dialyse stets hohe Aktivitätsverluste auftraten. Sowohl in Zellen, die in der späten log-Phase als auch in Zellen, die nach 2—4 Std in der stationären Phase geerntet wurden, waren die Repressionseffekte durch 10^{-3} M L-Tryptophan gleich (Tab. 1). Einen Hinweis auf die Repression der Anthranilatsynthetase durch L-Tryptophan in *S. cerevisiae* bietet auch die gegenüber dem Wildstamm erhöhte Aktivität der Anthranilatsynthetase bei der trp$^-$-Mutante 1071-3 B (Block hinter der Anthranilsäure), die wohl durch eine vollständigere Derepression zustande kommt. Überraschenderweise ist die Anthranilatsynthetaseaktivität auch bei der tyr$^-$, phe$^-$-Mutante KP 171 gegenüber der des Wildstamms stark erhöht. Bei der vergleichenden Überprüfung der Anthranilatsynthetase-Aktivitäten des Wildstamms S 288C, der Mutante 1071-3 B und der Mutante KP 171 war folgendes zu berücksichtigen: In den Enzymextrakten der beiden erstgenannten Stämme wird die Chorisminsäure nicht nur für die Anthranilatsynthetase sondern auch für die Chorismat-

Tabelle 1. *Repression und Derepression der Anthranilatsynthetase bei S 288 C, 1071-3 B, KP 171*
Als Enzympräparat diente eine Ammonsulfatgesamtfällung [0,9 gesätt. $(NH_4)_2SO_4$]. Die Inkubationszeit betrug jeweils 30 min bei 30°. Die spezifischen Aktivitäten wurden aus verschiedenen Versuchen gemittelt. Die Schwankungen um den Mittelwert betrugen maximal 10%

Zugesetzter Repressionsstoff	Stamm	Spezifische Aktivität der Anthranilatsynthetase
		µMol/mg Protein · min
—	S 288 C	$3,6 \times 10^{-3}$
10^{-5} M L-Trp	S 288 C	$3,8 \times 10^{-3}$
10^{-3} M L-Trp	S 288 C	$1,0 \times 10^{-3}$
—	S 288 C	$2,8 \times 10^{-3}$
10^{-2} M L-Phe + 10^{-3} M L-Tyr + 10^{-3} M L-Trp	S 288 C	$1,1 \times 10^{-3}$
—	S 288 C	$3,5 \times 10^{-3}$
—	KP 171	$6,0 \times 10^{-3}$
—	1071-3 B	$1,1 \times 10^{-2}$

mutase als Substrat benötigt. Im Extrakt der Mutante KP 171 fehlen hingegen die Chorismatmutasen. Die beobachtete Erhöhung der Anthranilatsynthetaseaktivität bei KP 171 könnte dann einfach als Folge des Fehlens der um das Substrat konkurrierenden Enzyme erklärt werden. Diese Möglichkeit ließ sich dadurch ausschließen, daß man in allen drei Versuchen für einen so hohen Überschuß an Chorisminsäure sorgte, daß der Sättigungsbereich der Anthranilatsynthetase stets erreicht wurde. Bei dieser Betrachtung kann die Umsetzung von Chorisminsäure bei der Biosynthese von *p*-Aminobenzoesäure und *p*-Hydroxybenzoesäure wegen der geringen Aktivität dieser Enzyme vernachlässigt werden. Die Erhöhung der Anthranilatsynthetaseaktivität bei der letztgenannten Mutante (KP 171) könnte auf einer Induktion der Anthranilatsynthetase durch die bei dieser Mutante überschießend gebildete Chorisminsäure beruhen.

Doy und Cooper [8] haben kürzlich ebenfalls berichtet, daß die Anthranilatsynthetase einer Repression durch L-Tryptophan unterliegt.

Chorismatmutase[1]

Während Gibson [4] bei *Acrobacter aerogenes* sowohl durch L-Tyrosin, als auch durch höhere Konzentration von L-Phenylalanin eine Reprimierung der Isoenzyme der Chorismatmutase beobachtete, konnten wir in *S. cerevisiae* weder durch L-Tyrosin, das in diesem Organismus als starker Feedback-Hemmstoff wirkt [1], in Konzentrationen von 10^{-3}M

[1] *Nachtrag bei der Korrektur* (17. 4. 1967): Eine Induktion durch Chorisminsäure scheidet aus, da in gleicher Weise bei einer auxotrophen Mutante, die unter Mangelbedingungen Shikimisäure akkumuliert, die Synthese der Chorismatmutase durch L-Tryptophan und entsprechend auch die Synthese der Prephenatdehydrogenase durch L-Phenylalanin induziert wird.

und 2×10^{-3} M, noch durch L-Phenylalanin (10^{-3} und 10^{-2} M), noch durch Kombination von L-Phenylalanin und L-Tyrosin (L-Phe: 10^{-2} M und L-Tyr: 10^{-3} M) eine Reprimierung der Chorismatmutaseaktivität feststellen. Zellen aus einer anderen Wachstumsphase (log. bzw. stationäre Phase) zeigten keine Änderung der Aktivität. Selbst in einem Komplettmedium, das noch zusätzlich mit 10^{-3} M L-Phe und 10^{-3} M L-Tyr angereichert war, konnte die volle Chorismatmutaseaktivität nachgewiesen werden. Auch hier kamen zur Vermeidung von Feedbackeffekten durch L-Tyrosin Ammonsulfatgesamtfällungen [0,9 gesätt. $(NH_4)_2SO_4$] zur Anwendung. Es wurde stets mit der unaufgetrennten Chorismatmutase gearbeitet.

Gibt man dem Kulturmedium, in dem die Zellen angezüchtet werden, L-Tryptophan zu (10^{-3} M), so beobachtet man eine Erhöhung der Mutaseaktivität auf über das Doppelte (Tab. 2). Wie später gezeigt werden soll, handelt es sich hierbei nicht um eine Aktivierung der Chorismatmutase, sondern um eine Induktion der Synthese dieses Enzyms.

Tabelle 2. *Induktion der Chorismatmutase durch L-Tryptophan*
Die Enzympräparate sind Ammonsulfatgesamtfällungen [0,9 gesätt. $(NH_4)_2SO_4$]. Die spezifische Aktivität wurde aus verschiedenen Versuchen gemittelt. Die Schwankungen um den Mittelwert betrugen maximal 8%

Zugesetzte Substanz	Stamm	Spezifische Aktivität der Chorismatmutase
		µMol/mg Protein · min
Versuche in Minimalmedium		
ohne Zusatz	S 288 C	$6,5 \times 10^{-3}$
10^{-5} M L-Trp	S 288 C	$7,2 \times 10^{-3}$
10^{-3} M L-Trp	S 288 C	$1,6 \times 10^{-2}$
2×10^{-2} M L-Phe + 10^{-3} M L-Tyr + 10^{-3} M L-Trp	S 288 C	$1,9 \times 10^{-2}$
Versuche in Komplettmedium		
ohne Zusatz	S 288 C	$7,3 \times 10^{-3}$
10^{-3} M L-Phe + 10^{-3} M L-Tyr + 10^{-4} M L-Trp	S 288 C	$9,2 \times 10^{-3}$

Prephenatdehydrogenase[1]

Einen ähnlichen Induktionseffekt, wie ihn L-Tryptophan bei der Chorismatmutase ausübt, beobachtet man auch bei der Prephenatdehydrogenase von *S. cerevisiae* bei L-Phenylalanin-Zugabe. Während in Zellen, die in Gegenwart von L-Tyrosin (10^{-3} und 2×10^{-3} M) gewachsen sind, keine Repression der Prephenatdehydrogenase festzustellen ist, steigt die spezifische Aktivität dieses Enzyms stark an, wenn man dem Medium L-Phenylalanin zusetzt. Sowohl bei Züchtung mit L-Tyrosin und L-Phenylalanin, als auch bei Züchtung in Gegenwart von L-Tyrosin, L-Phenylalanin und L-Tryptophan sowie von L-Phenylalanin allein (variierend $10^{-3}-10^{-2}$ M), beobachtet man stets ein kräftiges Ansteigen der spezifischen Aktivität der Dehydrogenase (Tab.3). Dieser Anstieg ist auf eine Induktionswirkung des L-Phenylalanins auf die Synthese der Prephenatdehydrogenase zurückzuführen, da, wie später gezeigt wird, das Enzym unter diesen Bedingungen nicht aktiviert ist.

Tabelle 3. *Spezifische Aktivität der Prephenatdehydrogenase nach Züchtung der Zellen in Gegenwart von L-Tyrosin, L-Phenylalanin und L-Tryptophan*
Die Enzympräparate wurden durch Ammonsulfatfällung gewonnen. Frakt. 0,9 gesätt. $(NH_4)_2SO_4$. Die spezifischen Aktivitäten wurden aus verschiedenen Versuchen gemittelt. Die Schwankungen um den Mittelwert betrugen maximal 5%

Zugesetzte Substanz	Stamm	Spezifische Aktivität
		μMol/mg Protein · min
Versuche in Minimalmedium		
ohne Zusatz	S 288 C	$2,0 \times 10^{-3}$
10^{-3} M L-Tyr	S 288 C	$2,2 \times 10^{-3}$
10^{-3} M L-Phe	S 288 C	$4,3 \times 10^{-3}$
10^{-2} M L-Phe	S 288 C	$4,7 \times 10^{-3}$
10^{-3} M L-Phe + 10^{-3} M L-Tyr	S 288 C	$4,0 \times 10^{-3}$
2×10^{-3} M L-Phe + 10^{-3} M L-Tyr + 10^{-3} M L-Trp	S 288 C	$4,3 \times 10^{-3}$

Prephenatdehydratase

Läßt man die Zellen in Gegenwart von 10^{-3} M L-Phenylalanin wachsen und erntet sie in der späten logarithmischen Phase, so beobachtet man eine schwache Reprimierung der Prephenatdehydratase. Durch höhere Zusätze von L-Phenylalanin (10^{-2} M) läßt sich eine etwas höhere Repression erzielen, jedoch ist das Ausmaß der Repression gering im Vergleich zu der von Gibson et al. [4] bei A. aerogenes beobachteten fast vollständigen Unterdrückung der Prephenatdehydratase-Synthese. L-Tyrosin und L-Tryptophan rufen keinen Effekt hervor. Auch die Kombination aller drei aromatischen Aminosäuren führt weder zu einer Steigerung noch zu einer Verminderung der Repressionswirkung durch L-Phenylalanin (Tab.4).

p-Aminobenzoatsynthetase

Die Aktivität dieses Enzyms ist normalerweise sehr gering[2]. Setzt man dem Medium jedoch 10^{-3} M L-Tyrosin und 10^{-2} M L-Phenylalanin zu, beobachtet man nach 3–4 Std in der frühen stationären Phase im Kulturmedium größere Mengen p-Aminobenzoesäure. Die Konzentration an p-Aminobenzoesäure im Medium läßt sich noch steigern, wenn man zusätzlich zu L-Tyrosin und L-Phenylalanin in den oben

[2] Die spezifische Aktivität kann nur ungenau angegeben werden, da die Bestimmung der gebildeten p-Aminobenzoesäure über Chromatographie und UV-Spektroskopie erfolgt: < 0,0005 μMol p-Aminobenzoesäure/mg Protein · min.

Tabelle 4. *Spezifische Aktivität der Prephenatdehydratase nach Züchtung der Zellen in Gegenwart von L-Tyrosin, L-Phenylalanin, L-Tryptophan*
Die spezifischen Aktivitäten wurden aus verschiedenen Versuchen gemittelt. Die Schwankungen um den Mittelwert betrugen maximal 10%

Zugesetzte Substanz	Stamm	Spezifische Aktivität	
		Fraktion 0,4—0,6 gesätt. $(NH_4)_2SO_4$	Gesamtfällung 0,9 gesätt. $(NH_4)_2$ SO$_4$
		μMol/mg Protein · min	
Versuche in Minimalmedium			
ohne Zusatz	S 288 C	$9,2 \times 10^{-3}$	$4,8 \times 10^{-3}$
10^{-3} M L-Phe	S 288 C	$6,8 \times 10^{-3}$	$3,1 \times 10^{-3}$
10^{-2} M L-Phe	S 288 C	$3,0 \times 10^{-3}$	
10^{-3} M L-Tyr	S 288 C	$9,1 \times 10^{-3}$	
10^{-3} M L-Trp	S 288 C	$9,1 \times 10^{-3}$	
10^{-3} M L-Phe + 10^{-3} M L-Tyr	S 288 C	$3,7 \times 10^{-3}$	
Versuche in Komplettmedium			
ohne Zusatz	S 288 C	$3,7 \times 10^{-3}$	
10^{-3} M L-Phe + 10^{-3} M L-Tyr + 10^{-3} M L-Trp	S 288 C	$2,1 \times 10^{-3}$	

angegebenen Konzentrationen 10^{-3} M L-Tryptophan dem Medium zusetzt. Die spezifische Aktivität der p-Aminobenzoatsynthetase der Hefezellen, die unter diesen Bedingungen gezüchtet wurden, ist um den Faktor 5–7 gegenüber den Bedingungen des Minimalmediums erhöht. Der Zusatz der 3 aromatischen Aminosäuren bewirkt demnach eine Induktion der p-Aminobenzoatsynthetase. Läßt man den Wildstamm S 288 C in dem beschriebenen Saccharose-Mineralsalzmedium wachsen, so kann in dem Kulturmedium nach Abzentrifugieren der Zellen praktisch keine p-Aminobenzoesäure nachgewiesen werden.

ABBAU DER AROMATISCHEN AMINOSÄUREN UNTER DEN REPRESSIONSBEDINGUNGEN

Das geringe Ausmaß der Repression der Anthranilatsynthetase durch L-Tryptophan und der Prephenatdehydratase durch L-Phenylalanin im Vergleich zu der Repressionswirkung dieser Aminosäuren bei anderen Mikroorganismen, sowie die fehlende Repression durch die aromatischen Aminosäuren im Falle der DAHP-Synthetase, Chorismatmutase und Prephenatdehydrogenase, veranlaßte uns, das Schicksal der zugesetzten Aminosäuren unter diesen Bedingungen etwas genauer zu verfolgen. Aus eigenen früheren Untersuchungen [9] wußten wir, daß ruhende Hefezellen die aromatischen Aminosäuren zu den um ein C-Atom verkürzten Alkoholen abbauen. So wird L-Tyrosin in p-Hydroxyphenyläthanol-(2) umgewandelt. Als weitere Abbauprodukte treten p-Hydroxyphenylbrenztraubensäure und p-Hydroxyphenylmilchsäure auf, wobei letztere Verbindung wohl eher als ein Reduktionsprodukt aufzufassen ist, das durch unspezifische enzymatische Hydrierung gebildet wird, und kein Zwischenprodukt des Abbaus darstellt. Der Abbau verläuft wahrscheinlich über die Stufen I—IV.

Verbindung III konnte von uns allerdings nicht nachgewiesen werden. L-Phenylalanin und L-Tryptophan werden entsprechend abgebaut. Im Falle von L-Phenylalanin werden große Mengen an Phenyläthanol-(2) gefunden, während Phenylbrenztraubensäure nur in geringer Menge nachzuweisen ist. Tryptophol tritt als Hauptabbauprodukt des L-Tryptophans auf, während im Falle von D-Tryptophan Indolylmilchsäure Hauptprodukt des in wesentlich geringerem Umfang erfolgenden Abbaus war. Entsprechende Umwandlungen von L-Tryptophan und D-Tryptophan in anderen Hefen *(Endomycopsis vernalis, Hansenula anomala* und *Torulopsis utilis)* wurden bereits von anderer Seite ausführlich beschrieben [10].

Solchen Abbaureaktionen sind auch abgewandelte aromatische Aminosäuren in quantitativ unterschiedlichem Ausmaße unterworfen. So wird *p*-Fluor-DL-phenylalanin in großem Umfang zu *p*-Fluorphenyläthanol-(2) abgebaut. Daneben ließ sich noch eine Säure als Abbauprodukt nachweisen, die nach ihren chromatographischen Eigenschaften der *p*-Fluorphenylmilchsäure entsprechen dürfte. 4-,5- und 6-Methyl-DL-tryptophan, die wir bereits als wirksame Feedback-Hemmstoffe für die Anthranilatsynthetase beschrieben haben [1], unterliegen ebenfalls einem Abbau durch *S. cerevisiae*. Neben den entsprechenden Alkoholen treten in geringem Umfang noch andere Indolderivate als Abbauprodukte auf, die nicht näher identifiziert wurden. Der Abbau von L-Tryptophan, 4-, 5-, 6- und α-Methyl-DL-tryptophan wurde unter den Züchtungsbedingungen, wie sie auch für die Repressionsansätze angewandt wurden (Ernte in der späten log-Phase), quantitativ verfolgt. Gemessen wurden jeweils die Konzentrationen der entsprechenden β-Indolyläthanole. Gleichzeitig wurde bei der trp⁻-Mutante 1071-3 B (akkumuliert Anthranilsäure) das Ausmaß der Hemmwirkung (Feedback + Repression) bei Zugabe von L-Trypto-

phan und Methyltryptophanen an der Menge der gebildeten Anthranilsäure erfaßt. Die kernsubstituierten Methyltryptophane werden in geringerem Maße abgebaut als L-Tryptophan. Die Hemmwirkung auf die Anthranilatsynthetase wird in der Tabelle 5 angegeben. α-Methyltryptophan wird nicht abgebaut, entfaltet aber auch keine Hemmwirkung. Zusatz von Methyltryptophanen zu einem Ansatz mit L-Tryptophan bewirkt keine Verringerung der Tryptopholbildung.

Tabelle 5. *Abbau von L-Tryptophan und Methyltryptophanen durch Wildstamm S 288 C. Hemmwirkung auf die Anthranilatsynthetase in Mutante 1071-3 B*
3 g Hefefeuchtmasse wurden in 500 ml Minimalmedium + jeweiligem Zusatz 2 Tage bei 30° inkubiert. Mit Essigester wurde das Gemisch der Abbauprodukte extrahiert und der Dünnschichtchromatographie unterworfen [Kieselgel HF₂₅₄, Chloroform-Tetrachlorkohlenstoff-Methanol (50:25:25, v/v)]. Die Tryptophole wurden eluiert und UV-spektroskopisch bestimmt

Zusatz	Menge an Tryptophol, bzw. Methyltryptophol	Gesamtabbau	Hemmung der Anthranilatsynthetase
	mg	%	%
ohne Zusatz	0,05		
100 mg L-Tryptophan	5,25	5,3	80
100 mg 4-Methyltryptophan	0,55	0,55	40
100 mg 5-Methyltryptophan	1,0	1,0	75
100 mg 6-Methyltryptophan	0,69	0,69	50
100 mg α-Methyltryptophan	—	—	0
100 mg L-Tryptophan + 100 mg 4-Methyltryptophan	5,9	5,9	nicht untersucht
100 mg L-Tryptophan + 100 mg 5-Methyltryptophan	5,1	5,1	nicht untersucht
100 mg L-Tryptophan + 100 mg α-Methyltryptophan	7,7	7,7	nicht untersucht

In den Kulturmedien der beschriebenen Repressionsansätze konnten nach Abzentrifugieren der Hefezellen die beschriebenen Abbauprodukte der aromatischen Aminosäuren stets nachgewiesen werden, gleichgültig ob die Zellen in der log- oder in der stationären Phase geerntet wurden.

Die Tabelle 6 gibt einen Überblick über die Ergebnisse bei den verschiedenen Repressionsansätzen. Von den Abbauprodukten wurden nur die entsprechenden Alkohole berücksichtigt und grob quantitativ nach der in Tabelle 5 angegebenen Methodik abgeschätzt. Neben den zugesetzten Aminosäuren und deren Abbauprodukten wurde auch die durch Induktion gebildete *p*-Aminobenzoesäure aufgeführt.

Überraschenderweise traten die Abbauprodukte von L-Phenylalanin, L-Tyrosin und L-Tryptophan auch im normalen Minimalmedium auf, dem diese

Tabelle 6. *Abbau der aromatischen Aminosäuren und Akkumulation der p-Aminobenzoesäure in den verschiedenen Repressionsansätzen*

Dem Minimalmedium zugesetzte Repressionsstoffe	Im Kulturmedium nachgewiesene Produkte						
	Tyrosol	Tryptophol	Phenyläthanol	*p*-Amino-benzoesäure	L-Tyr	L-Trp	L-Phe
ohne Zusatz	+	+	+	—	—	—	—
10^{-2} M L-Phe	+	+	+++	—	—	—	+++
10^{-3} M L-Trp	+	+++	+	+	—	+++	—
10^{-3} M L-Tyr	++++	+	+	(+)	++	—	—
10^{-3} M L-Tyr ⎱ + 10^{-2} M L-Phe ⎰ + 10^{-3} M L-Trp	+++	+++	+++	++	++	++	++
Dem Komplettmedium zugesetzte Repressionsstoffe							
ohne Zusatz	+	+	+	—			
10^{-3} M L-Tyr ⎱ + 10^{-3} M L-Phe ⎰ + 10^{-3} M L-Trp	+++	+++	+++	nicht untersucht	++	++	++

Aminosäuren gar nicht zugesetzt waren. Daraus kann man schließen, daß die Konzentration der entsprechenden Aminosäuren zur Induktion der erforderlichen Enzyme für den Abbau nur sehr gering zu sein braucht. Die Untersuchung des Tryptophan-Abbaus zu Tryptophol in ruhenden Zellen zeigte, daß die abbauenden Enzyme induzierbar sind. Die zur Induktion nötige [Mindestkonzentration von L-Tryptophan im Medium beträgt 1,5—2γ L-Tryptophan/ml [11]. Diese Tatsache erklärt unter Umständen das geringe Ausmaß an Repression, das bei den Enzymen der Biosynthese der aromatischen Aminosäuren beobachtet wurde. Entweder wird die Konzentration der Aminosäuren in der Zelle durch den Abbau so gering gehalten, daß noch keine Repression eintreten kann, oder der Abbauvorgang tritt überhaupt anstelle der Repression auf.

AKTIVIERUNG

Die erwähnten Enzyme der aromatischen Aminosäuren werden von den Endprodukten in ihrer Aktivität beeinflußt. Über die negative Feedback-Kontrolle haben wir bereits ausführlich berichtet [1]. Es ist eine bekannte Tatsache, daß allosterische Enzyme von Effektoren nicht nur im Sinne einer Hemmung beeinflußt werden können (negativer Effekt), sondern daß Effektoren auch einen aktivierenden Einfluß auf die Enzymaktivität ausüben können (positiver Effekt). Die theoretischen Grundlagen und Vorstellungen für diese Effekte sind von Monod, Wyman und Changeux [12] dargelegt worden.

Wir haben uns im Verlaufe unserer Untersuchungen über die Regulation der Biosynthese der aromatischen Aminosäuren mit der Frage der Aktivierung der beschriebenen Enzyme durch die aromatischen Aminosäuren beschäftigt.

DAHP-Synthetase

Die Substratsättigungskurve sowohl des Tyrosinsensitiven als auch des Phenylalanin-sensitiven

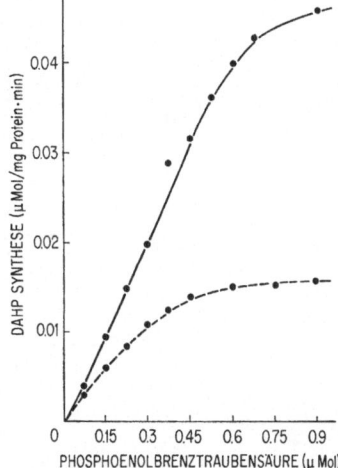

Fig. 1. *Substratsättigungskurve der Tyrosin-sensitiven DAHP-Synthetase mit Phosphoenolbrenztraubensäure.* Enzym: Ammonsulfatfraktion 0—0,5 gesätt. $(NH_4)_2SO_4$. ●——●, ohne Hemmstoffzusatz; ●——●, 5×10^{-5} M an L-Tyrosin

Isoenzyme der DAHP-Synthetase zeigt in Abwesenheit von Effektoren den Verlauf einer einfachen Michaelis-Henry-Sättigungskurve (Fig. 1 und 2). Die beiden DAHP-Synthetasen gehören zum V-System der allosterischen Enzyme, bei denen die zwei Zustände T und R dieselbe Substrataffinität haben [12]. Den gleichen Verlauf zeigen auch die Substratsättigungskurven der DAHP-Synthetasen von *E. coli* [13]. Eine Aktivierung ist in derartigen Fällen noch nicht beobachtet worden. In der Tat verliefen die Aktivierungsversuche bzw. die Versuche, die L-Tyrosin- bzw. L-Phenylalanin-Hemmung der DAHP-Synthetasen mit L-Tryptophan, *p*-Amino-

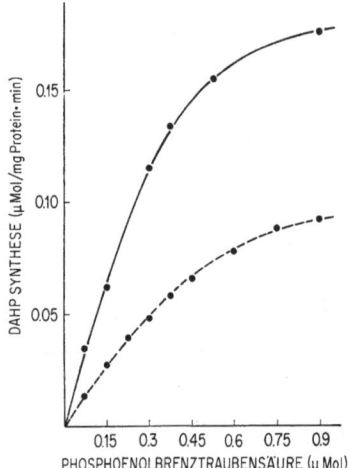

Fig. 2. *Substratsättigungskurve der Phenylalanin-sensitiven DAHP-Synthetase mit Phosphoenolbrenztraubensäure.* Enzym: Ammonsulfatfraktion 0,65—0,9 gesätt. $(NH_4)_2SO_4$. ●——● ohne Hemmstoffzusatz; ●— —● 5×10^{-5} M an L-Phenyalanin

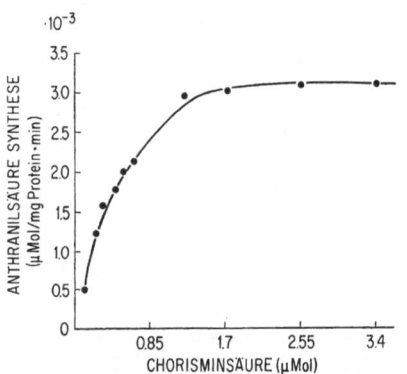

Fig. 3. *Substratsättigungskurve der Anthranilatsynthetase.* Enzym: Ammonsulfatfraktion 0,4—0,6 gesätt. $(NH_4)_2SO_4$

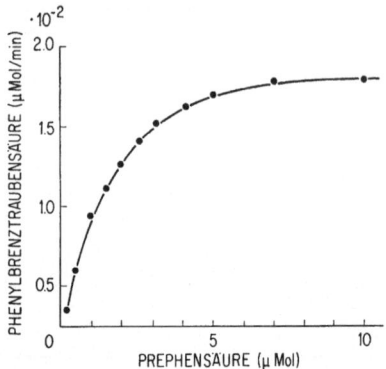

Fig. 4. *Substratsättigungskurve der Prephenatdehydratase.* Proteinkonzentration: 4 mg pro Ansatz. Enzym: Ammonsulfatfraktion 0,4—0,6 gesätt. $(NH_4)_2SO_4$

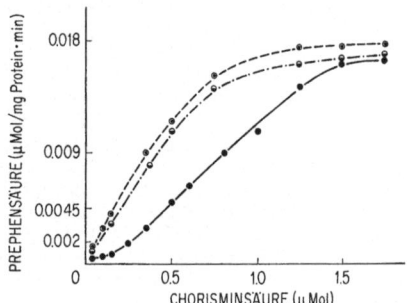

Fig. 5. *Substratsättigungskurve der Chorismatmutase.* ●——● frisches Enzym, Ammonsulfatfraktion 0,65—0,85 gesätt. $(NH_4)_2SO_4$; ⊙——⊙ Aktivierung durch Zusatz von L-Tryptophan, 2×10^{-3} M; ●— ·—● Hitzebehandelte Mutase (7 min bei 45°)

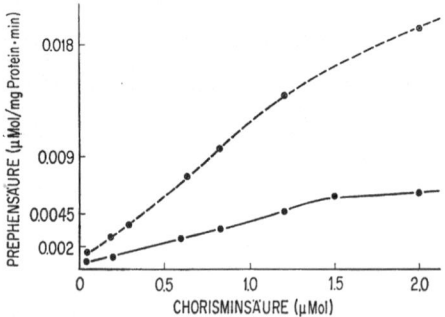

Fig. 6. *Aktivierung von gealterter Chorismatmutase durch L-Tryptophan.* Enzym: 2 Monate bei —20° aufbewahrte Ammonsulfatfraktion 0,65—0,85 gesätt. $(NH_4)_2SO_4$. ●——● ohne Zusatz; ⊙——⊙ Zusatz von L-Tryptophan, 2×10^{-3} M

benzoesäure oder *p*-Hydroxybenzoesäure aufzuheben, negativ. Hitzebehandlung (5 min, 45°) verursacht völligen Verlust der Enzymaktivität.

Anthranilsynthetase und Prephenatdehydratase

Die Substratsättigungskurven dieser beiden Enzyme zeigen denselben Verlauf wie die der DAHP-Synthetasen (Fig. 3 und 4). Aktivierung bzw. Aufhebung der Hemmung durch L-Tyrosin und L-Phenylalanin bei der Anthranilatsynthetase und durch L-Tryptophan und L-Tyrosin bei der Prephenatdehydratase konnten auch in diesen Fällen nicht beobachtet werden.

Chorismatmutase

Der Verlauf der Sättigungskurve der Chorismatmutase ist in Abwesenheit von Effektoren sigmoid (Fig. 5). Das Enzym gehört demnach dem K-System der allosterischen Proteine an [12]. Als Enzympräparat diente die Ammonsulfatfraktion 0,7—0,9 gesätt. $(NH_4)_2SO_4$, in der weder die Anthranilatsynthetase noch die Prephenatdehydratase und -dehydrogenase enthalten sind. In Gegenwart von 2×10^{-3} M L-Tryptophan erhält man dagegen eine Sättigungskurve, bei der der Sättigungsbereich nach geringeren Substratkonzentrationen verschoben ist, wie es für Aktivatoren charakteristisch ist. Denselben Verlauf erhält man auch, wenn die Chorismatmutase 7 min auf 45° erhitzt worden ist. Hierbei wird offensichtlich das allosterische Enzym in die Protomeren getrennt, denn nach dieser Hitzebehandlung wird die Chorismatmutase durch 10^{-3} M L-Tyrosin nicht mehr gehemmt. Die Aktivität der Chorismatmutase bleibt nach der Behandlung noch voll erhalten (Fig. 5). Einen sehr starken aktivierenden Einfluß hat L-Tryptophan auch auf gealterte Enzympräparate (Fig. 6).

Die Hemmwirkung des L-Tyrosins (80% bei 10^{-3} M L-Tyrosin) wird in Gegenwart von L-Tryptophan vollständig aufgehoben[3]. D-Tryptophan hat dagegen keinen Effekt (Tab. 7).

Tabelle 7. *Aufhebung der L-Tyrosinhemmung bei der Chorismatmutase durch L-Tryptophan*
Als Enzympräparat diente die Ammonsulfatfraktion 0,65 bis 0,9 gesätt. $(NH_4)_2SO_4$ des Wildstammes S 288C, in der die Chorismatmutase von Prephenatdehydratase, Prephenatdehydrogenase und von der Anthranilatsynthetase abgetrennt ist. Die spezifischen Aktivitäten wurden aus verschiedenen Versuchen gemittelt. Die Schwankungen um den Mittelwert betrugen maximal 10%

Dem Inkubationsansatz zugesetzte Effektoren	Spezifische Aktivität der Chorismatmutase
	μMol/mg Protein · min
ohne Zusatz	$8,0 \times 10^{-3}$
10^{-3} M L-Tyr	$1,0 \times 10^{-3}$
10^{-3} M L-Trp	$9,3 \times 10^{-3}$
10^{-3} M L-Tyr + $0,5 \times 10^{-3}$ M L-Trp	$8,9 \times 10^{-3}$
10^{-3} M L-Tyr + 10^{-2} M L-Trp	$9,0 \times 10^{-3}$
10^{-3} M L-Tyr + 2×10^{-3} D-Trp	$1,5 \times 10^{-3}$
Nach Hitzebehandlung:	
ohne Zusatz	$7,2 \times 10^{-3}$
2×10^{-3} M L-Tyr	
2×10^{-3} M L-Trp	$7,7 \times 10^{-3}$
2×10^{-3} M L-Tyr + 2×10^{-3} M L-Trp	$8,1 \times 10^{-3}$

Wie bereits berichtet, haben Zellen, die in Gegenwart von L-Tryptophan gezüchtet wurden, eine stark erhöhte Aktivität der Chorismatmutase gegenüber Zellen, denen dieser Zusatz beim Wachsen fehlte. Gegen die Annahme, daß im ersten Falle aktivierte

[3] *Nachtrag bei der Korrektur:* 5-Methyl-DL-Tryptophan (2×10^{-3} M) hebt ebenfalls die Hemmwirkung des L-Tyrosins auf die Chorismatmutase auf.

Chorismatmutase vorliegt, spricht der Befund, daß auch die unter diesen Bedingungen gewonnene Chorismatmutase durch L-Tryptophan noch zusätzlich aktivierbar ist (Fig. 7), durch L-Tyrosin (10^{-3} M) zu 80% gehemmt wird und daß diese Hemmung durch L-Tryptophan aufgehoben wird (Tab. 8). Der Effekt des L-Tryptophans unter den Wachstumsbedingungen muß daher einer Induktion der Chorismatmutase durch L-Tryptophan zugeschrieben werden. Bei Neurospora crassa hat kürzlich Baker [14] als erster nachgewiesen, daß L-Tryptophan als Aktivator für die Chorismatmutase wirkt.

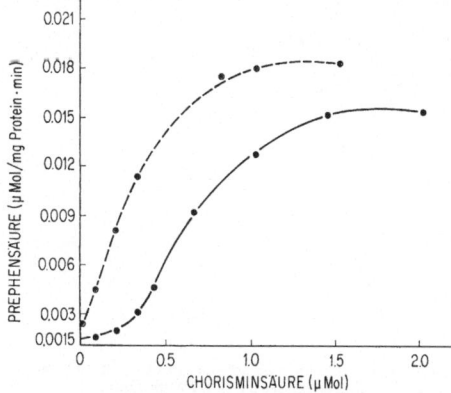

Fig. 7. *Aktivierung durch L-Tryptophan bei der Chorismatmutase aus Zellen, die in Gegenwart von L-Tryptophan gewachsen waren. Enzym: Ammonsulfatfraktion 0,6—0,9 gesätt. $(NH_4)_2SO_4$.* ●——● ohne Zusatz; ⊙——⊙ mit L-Tryptophanzusatz, 2×10^{-3} M

Tabelle 8. *Aufhebung der Tyrosinhemmung bei der Chorismatmutase, gewonnen aus Zellen, die in Gegenwart von L-Tryptophan gezüchtet wurden, durch L-Tryptophan*
Das Enzympräparat ist eine Ammonsulfatgesamtfällung [0,9 gesätt. $(NH_4)_2SO_4$] des Rohextraktes von Zellen des Stammes S 288C, der in Gegenwart von 10^{-3} M L-Tryptophan gezüchtet wurde. Die spezifischen Aktivitäten wurden aus mehreren Versuchen gemittelt. Die Schwankungen um den Mittelwert betrugen maximal 5%

Dem Inkubationsansatz zugesetzte Effektoren	Spezifische Aktivität der Chorismatmutase
	μMol/mg Protein · min
ohne Zusatz	$1,1 \times 10^{-2}$
+ 10^{-3} M L-Tyr	$0,16 \times 10^{-2}$
+ 10^{-3} M L-Trp	$1,2 \times 10^{-2}$
10^{-3} M L-Tyr + 10^{-3} M L-Trp	$1,1 \times 10^{-2}$

Prephenatdehydrogenase

Ähnlich wie bei der Chorismatmutase ist der Verlauf der Substratsättigungskurve bei der Prephenatdehydrogenase sigmoid (Fig. 8). Auch dieses Enzym gehört demnach dem K-System an. Da wir bereits bei den Repressionsbedingungen in Gegenwart von L-Phenylalanin eine Steigerung der spezifischen Aktivität dieses Enzyms beobachtet haben, lag es auf der Hand, den Einfluß von L-Phenylalanin als Aktivator für die Prephenatdehydrogenase zu untersuchen. Tatsächlich erhält man in Gegenwart von

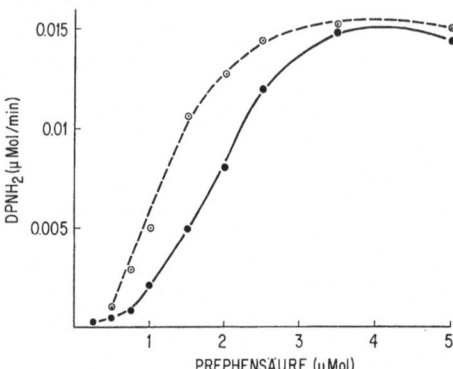

Fig. 8. *Substratsättigungskurve der Prephenatdehydrogenase.* Aktivierung durch L-Phenylalanin. Enzym: Ammonsulfatfraktion 0,4—0,6 gesätt. $(NH_4)_2SO_4$. Proteinkonzentration: 4 mg pro Ansatz. ●——●, ohne Zusatz; ⊙——⊙ Zusatz von L-Phenylalanin, 2×10^{-2} M

Tabelle 9. *Aufhebung der L-Tyrosinhemmung bei der Prephenatdehydrogenase durch L-Phenylalanin* Enzympräparat: Ammonsulfatgesamtfällung 0,9 gesätt. $(NH_4)_2SO_4$ von Wildstamm S 288 C. Die spezifischen Aktivitäten wurden aus verschiedenen Versuchen gemittelt. Die Schwankungen um den Mittelwert betrugen maximal 10%.

Dem Inkubationsansatz zugesetzte Effektoren	Spezifische Aktivität der Prephenatdehydrogenase
	μMol/mg Protein · min
ohne Zusatz	$2,1 \times 10^{-3}$
$0,5 \times 10^{-3}$ M L-Tyr	$7,2 \times 10^{-4}$
10^{-3} M L-Tyr	$4,8 \times 10^{-4}$
2×10^{-3} M L-Tyr	$3,8 \times 10^{-4}$
10^{-3} M L-Phe	$2,4 \times 10^{-3}$
10^{-3} M L-Tyr + $0,5 \times 10^{-3}$ M L-Phe	$1,9 \times 10^{-3}$
10^{-3} M L-Tyr + 10^{-3} M L-Phe	$2,1 \times 10^{-3}$
10^{-3} M L-Tyr + 2×10^{-3} M L-Phe	$2,3 \times 10^{-3}$
2×10^{-3} M p-Fluor-DL-phenylalanin	$6,1 \times 10^{-4}$
2×10^{-3} M p-Fluor-DL-phenyl-alanin + 2×10^{-3} M L-Tyr }	$2,9 \times 10^{-4}$
2×10^{-3} M p-Fluor-DL-phenyl-alanin + 2×10^{-3} M L-Phe }	$2,1 \times 10^{-3}$

2×10^{-3} M L-Phenylalanin den für positive Effectoreinflüsse typischen Verlauf der Substratsättigungskurve. Da für diese Untersuchungen Enzympräparate verwendet wurden, die durch Ammonsulfatfraktionierung hergestellt waren [0,4—0,6 gesätt. $(NH_4)_2SO_4$] und neben der Prephenatdehydrogenase auch noch die Prephenatdehydratase enthielten, die

durch L-Phenylalanin stark gehemmt wird, waren diese Ergebnisse allerdings mit einer gewissen Unsicherheit behaftet. Es wurde deshalb die Aktivierung der Prephenatdehydrogenase bei der phe⁻-Mutante HK 11, der die Prephenatdehydratase fehlt [1], zusätzlich untersucht. Hierbei konnten dieselben Ergebnisse erzielt werden (der Verlauf der Substratsättigungskurve entspricht dem von Fig. 8). Die Hemmwirkung durch L-Tyrosin und p-Fluor-DL-phenylalanin wird in Gegenwart von L-Phenylalanin aufgehoben (Tab. 9). In einer früheren Arbeit haben wir p-Fluor-DL-phenylalanin als nicht hemmend für die Prephenatdehydrogenase beschrieben [1]. Spätere Untersuchungen haben gezeigt, daß frisch gewonnenes Enzympräparat durch p-Fluor-DL-phenylalanin zu hemmen ist (Tab. 9). Ähnlich wie im Falle der Chorismatmutase läßt sich auch die Prephenatdehydrogenase von Hefezellen, die in Gegenwart von L-Phenylalanin gewachsen waren und eine erhöhte Aktivität besitzen, durch L-Phenylalanin aktivieren (der Verlauf der Substratsättigungskurve entspricht dem aus Fig. 8), durch L-Tyrosin hemmen, und diese Hemmung läßt sich durch L-Phenylalanin aufheben (Tab. 10). Damit dürfte auch der Effekt des L-Phenylalanins unter den Wachstumsbedingungen auf einer Induktion der Prephenatdehydrogenase beruhen.

Tabelle 10. *Aufhebung der L-Tyrosinhemmung bei der Prephenatdehydrogenase, gewonnen aus Zellen, die in Gegenwart von L-Phenylalanin (10^{-2} M) gezüchtet wurden, durch L-Phenylalanin* Enzympräparat: Ammonsulfatgesamtfällung 0,9 gesätt. $(NH_4)_2SO_4$ des Rohextraktes aus Zellen von S 288 C, gewachsen in Gegenwart von 10^{-2} M L-Phenylalanin. Die spezifischen Aktivitäten wurden aus verschiedenen Versuchen gemittelt. Die Schwankungen um den Mittelwert betrugen maximal 4%

Dem Inkubationsansatz zugesetzte Effektoren	Spezifische Aktivität der Prephenatdehydrogenase
	μMol/mg Protein · min
ohne Zusatz	$5,4 \times 10^{-3}$
10^{-3} M L-Tyr	$0,86 \times 10^{-3}$
10^{-3} M L-Phe	$5,1 \times 10^{-3}$
10^{-3} M L-Tyr + 10^{-3} M L-Phe	$5,5 \times 10^{-3}$

Auf Grund der früher [1] und hier beschriebenen Ergebnisse läßt sich an Hand der Fig. 9 ableiten, daß ein Überschuß an L-Tyrosin eine Hemmung des Tyrosin-sensitiven Isoenzymes der DAHP-Synthetase, der Chorismatmutase und der Prephenatdehydrogenase hervorruft. Ein Überschuß an L-Tyrosin kann durch Induktion des Abbaus zum Tyrosol beseitigt werden. L-Phenylalanin bewirkt im Überschuß eine Hemmung des Phenylalanin-sensitiven Isoenzymes der DAHP-Synthetase und der Prephenatdehydratase. Die Biosynthese der Prephenatdehydratase wird durch L-Phenylalanin reprimiert. Es erscheint sinnvoll, daß ein Überschuß an

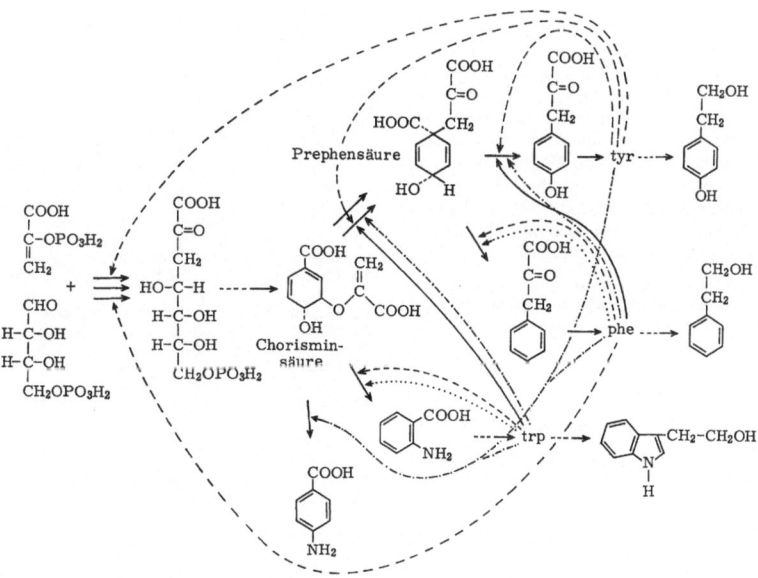

Fig. 9. *Ausschnitt aus der Biosynthese aromatischer Aminosäuren mit den Regulationsvorgängen in* S. cerevisiae. — —, Feedback-Hemmung; · · · · ·, Repression; ——, Aktivierung; — · — ·, Induktion

Tabelle 11. *Überblick über die Eigenschaften der untersuchten allosterischen Enzymsysteme*

Enzym	Substrat	Inhibitor	Aktivator	V-System	K-System
„Tyrosin"-DAHP-Synthetase	Phosphoenolbrenztrauben-säure + Erythrose-4-phosphat	L-Tyrosin	−	+	
„Phenylalanin"-DAHP-Synthetase	Phosphoenolbrenztrauben-säure + Erythrose-4-phosphat	L-Phenylalanin	−	+	
Chorismatmutase	Chorisminsäure	L-Tyrosin	L-Tryptophan		+
Prephenatdehydrogenase	Prephensäure	L-Tyrosin	L-Phenylalanin		+
Prephenatdehydratase	Prephensäure	L-Phenylalanin	−		
Anthranilatsynthetase	Chorisminsäure	L-Tryptophan	−		

Tabelle 12. *Überblick über Repression (R) und Induktion (I) bei den untersuchten Enzymsystemen*

	L-Phe	L-Tyr	L-Trp	L-Phe + L-Tyr	L-Phe + L-Tyr + L-Trp
„Tyrosin"-DAHP-Synthetase					
„Phenylalanin"-DAHP-Synthetase					
Chorismatmutase		I			I
Prephenatdehydrogenase	I			I	I
Prephenatdehydratase	R			R	R
Anthranilatsynthetase			R		R
p-Aminobenzoatsynthetase		(I)	(I)	I	I

L-Phenylalanin gleichzeitig sowohl eine Induktion wie auch eine Aktivierung der Prephenatdehydrogenase hervorruft. Der Abbau dieser Aminosäure, der zur Beseitigung überschüssigen L-Phenylalanins dienen kann, führt zum β-Phenyläthanol. Es läßt sich zeigen, daß L-Tryptophan in *S. cerevisiae* die Anthranilat-synthetase hemmt. Auch die Synthese dieses Enzyms wird durch L-Tryptophan reprimiert. In sinnvoller Weise bewirkt L-Tryptophan gleichzeitig die Aktivierung eines anderen Zweiges der verzweigten Biosynthesekette durch Induktion und Aktivierung der Chorismatmutase. Auch L-Tryptophan kann in

S. cerevisiae abgebaut werden. Dieser Abbau erfolgt bis zum Tryptophol. Setzt man die drei Aminosäuren L-Tyrosin, L-Phenylalanin und L-Tryptophan gleichzeitig im Überschuß zu, so läßt sich außer den geschilderten Phänomenen noch eine Induktion der *p*-Aminobenzoatsynthetase beobachten, die zur Abgabe von *p*-Aminobenzoesäure ins Medium führt.

Eine tabellarische Übersicht über die Eigenschaften der untersuchten Enzyme zeigt die Tabelle 11. Die Phänomene der Repression und Induktion werden in Tabelle 12 zusammengefaßt.

Fräulein J. Riekenberg danken wir für fleißige Mitarbeit. Der Deutschen Forschungsgemeinschaft und dem Fonds der Chemische Industrie danken wir für materielle Hilfe bei diesen Untersuchungen.

LITERATUR

1. Lingens, F., Goebel, W., und Uesseler, H., *Biochem. Z.* 346 (1966) 357.
2. Srinivasan, P. R., und Sprinson, D. B., *J. Biol. Chem.* 234 (1959) 731.
3. Doy, C. H., und Brown, K. D., *Biochim. Biophys. Acta,* 104 (1965) 377.
4. Cotton, R. G. H., und Gibson, F., *Biochim. Biophys. Acta,* 100 (1965) 76.
5. Clark, V. M., und Kirby, A. J., *Biochim. Biophys. Acta,* 78 (1963) 732.
6. Ballou, C. E., In *Methods in Enzymology* (edited by S. P. Colowick and N. O. Kaplan), Academic Press, New York and London 1963, vol. VI, p. 482.
7. Brown, K. D., und Doy, C. H., *Biochim. Biophys. Acta,* 118 (1966) 157.
8. Doy, C. H., und Cooper, J. M., *Biochim. Biophys. Acta,* 127 (1966) 302.
9. Goebel, W., Dissertation, Tübingen, 1965.
10. Glombitza, K.-W., und Hartmann, T., *Planta,* 69 (1966) 135.
11. Lingens, F., und Heilmann, H.-D., unveröffentlicht.
12. Monod, J., Wyman, J., und Changeux, J. P., *J. Mol. Biol.* 12 (1965) 88.
13. Smith, L. C., Ravel, J. M., Lax, S. R., und Shive, W., *J. Biol. Chem.* 237 (1962) 3566.
14. Baker, T. I., *Biochemistry,* 5 (1966) 2654.

F. Lingens, W. Goebel und H. Uesseler
Chemisches Institut der Universität
74 Tübingen, Wilhelmstraße 33, Deutschland

European J. Biochem. 1 (1967) 375—378

IUPAC-IUB Commission on Biochemical Nomenclature (CBN)

Abbreviated Designation of Amino Acid Derivatives and Peptides

Tentative Rules[1]

These Tentative Rules are an attempt to achieve a broad systematization of various types of abbreviated notation already in use [e.g. Brand and Edsall, *Ann. Rev. Biochem.* 16 (1947) 224; *Report of the Committee on Abbreviations of the American Society of Biological Chemists*, December 18, 1959; *Report of the Committee on Nomenclature of the European Peptide Symposium*, Pergamon Press, 1963, pp. 261—269; "Tentative Rules for Abbreviations and Symbols of Chemical Names of Special Interest in Biological Chemistry," IUPAC Information Bulletin No. 20, July 1963, pages 16—18; 1965 revision of the latter]. They seek to reconcile the needs of the protein chemist, *i.e.* indication of amino acid sequences, with those of persons concerned more with the chemical reactions of proteins and the synthesis of polypeptides, *i.e.* the need of conveying more detailed chemical information in abbreviated form.

1. GENERAL CONSIDERATIONS

1.1 The symbols chosen are derived from the trivial names or chemical names of the amino acids and of chemicals reacting with amino acids and polypeptides. For the sake of clarity, brevity, and listing in tables, the symbols have been, wherever possible, restricted to three letters, usually the first letters of the trivial names.

1.2 The symbols represent not only the names of the compounds but also their structural formulas.

1.3 The amino acid symbols by themselves represent the amino acids. The use of the symbols to represent the free amino acids is not recommended in textual material, but such use may occasionally be desirable in tables, diagrams, or figures. Residues of amino acids are represented by addition of hyphens in specific positions as indicated in Section 3.

1.4 Heteroatoms of amino acid residues (*e.g.* O^β and S^β of serine and cysteine, respectively, N^ε of lysine, N^α of glycine, etc.) do not explicitly appear in the symbol; such features are understood to be encompassed by the abbreviation.

---

[1] Document of the IUPAC-IUB Commission on Biochemical Nomenclature (CBN), approved by CBN in July 1965 and published by permission of the International Union of Pure and Applied Chemistry, the International Union of Biochemistry, and the official publishers to the International Union of Pure and Applied Chemistry, Messrs. Butterworths Scientific Publications. (Published in IUPAC Information Bulletin No. 25, February 1966, page 32.)

This set of Rules should be regarded as an extension, not a replacement, of Section 2 (Polypeptides and Proteins) of Abbreviations and Symbols for Chemical Names of Special Interest in Biological Chemistry [*J. Biol. Chem.* 241 (1966) 527; *European J. Biochem.* 1 (1967) 259].

Reprints of these Tentative Rules may be obtained from Waldo E. Cohn, Director, NAS-NRC Office of Biochemical Nomenclature, Oak Ridge National Laboratory, P.O.Box Y, Oak Ridge, Tennessee 37831, U.S.A.

Comments on these Tentative Rules may be sent to any member of CBN: O. Hoffmann-Ostenhof (Chairman), W. E. Cohn (Secretary), A. E. Braunstein, J. S. Fruton, B. Keil, W. Klyne, C. Liébecq, B. G. Malmström, R. Schwyzer, E. C. Slater, or corresponding member, N. Tamiya.

1.5 Amino acid symbols denote the L configuration unless otherwise indicated by D or DL appearing before the symbol and separated from it by a hyphen. When it is desired to make the number of amino acid residues appear in a clearer manner, the hyphen between the configurational prefix and the symbol may be omitted (see 5.3.1.1 *et seq.*). (Note: The designation of an amino acid residue as DL is inappropriate for compounds having another amino acid residue with an asymmetrical center.)

1.6 Structural formulas of complicated features may be used along with the abbreviated notation wherever necessary for clarity.

2. ABBREVIATIONS FOR AMINO ACIDS

2.1 Common Amino Acids

Alanine	Ala	Lysine	Lys
Arginine	Arg	Methionine	Met
Asparagine[2]	Asn[2]	Ornithine	Orn
Aspartic acid	Asp	Phenylalanine	Phe
Cysteine	Cys	Proline	Pro
Glutamic acid	Glu	Serine	Ser
Glutamine[2]	Gln[2]	Threonine	Thr
Glycine	Gly	Tryptophan	Trp
Histidine	His	Tyrosine	Tyr
Isoleucine	Ile	Valine	Val
Leucine	Leu		

2.2 Less Common Amino Acids

Abbreviations for less common amino acids should be defined in each publication in which they appear. The following principles and notations are recommended.

2.2.1 *Hydroxyamino Acids.*

Hydroxylysine	Hyl
3-Hydroxyproline	3Hyp
4-Hydroxyproline	4Hyp

2.2.2 *allo-Amino Acids.*

allo-Isoleucine	aIle
allo-Hydroxylysine	aHyl

2.2.3 *"Nor" Amino Acids.* "Nor" (*e.g.* in norvaline) is not used in its accepted sense (denoting a lower homologue) but to change the trivial name of a branched chain compound into that of a straight chain compound (compare with "iso," paragraph 2.1). "Nor" should therefore be treated as part of the trivial name without special emphasis.

Norvaline	Nva
Norleucine	Nle

2.2.4 *Higher Unbranched Amino Acids.* We suggest the following general rules for guidance in forming abbreviations: the functional prefix "amino" should be included in the symbol as the letter "A," diamino as "D."

---

[2] Asparagine and glutamine may also be denoted as Asp(NH$_2$) or Asp, and Glu(NH$_2$) or Glu, respectively.

The trivial name of the parent acid should be abbreviated to leave no more than two or three letters, as convenient and necessary for clarity. The word "acid" ("*-säure*," etc.) should be omitted from the symbol as carrying no significant information. Unless otherwise indicated (see paragraph below), single groups are in the α position, two amino groups in the α,ω (monocarboxylic acids) or α,α' positions (dicarboxylic acids). The location of amino acids in positions other than α and ω is shown by the appropriate Greek letter prefix.

Examples:

α-Aminobutyric acid	Abu
α-Aminoadipic acid	Aad
α-Aminopimelic acid	Apm
α,γ-Diaminobutyric acid	Dbu
α,β-Diaminopropionic acid	Dpr
α,α'-Diaminopimelic acid	Dpm
β-Alanine	βAla
ε-Aminocaproid acid	εAcp
β-Aminoadipic acid	βAad

2.2.5 N^α-Alkylated Amino Acids. N^α-Methylamino acids are becoming more and more common (*e.g.* in the large group of depsipeptides). This justifies special symbols.
Examples:

N-Methylglycine (sarcosine)	MeGly or Sar
N-Methylisoleucine	MeIle
N-Methylvaline, etc.	MeVal, etc.
N-Ethyglycine, etc.	EtGly, etc.

3. AMINO ACID RESIDUES

3.1 Lack of Hydrogen on the α-Amino Group

The α-amino group is understood to be at the left-hand side of the symbol when using hyphens, and—in special cases—at the point of the arrow when using arrows to indicate the direction of the peptide bond ($-CO \rightarrow NH-$, $-NH \leftarrow CO-$).

Examples:

$$-Gly: \quad -HNCH_2COOH \qquad -Ala: \quad -HN\overset{\overset{\textstyle CH_3}{|}}{C}HCOOH$$

$$>Gly \text{ or } \underline{\;}Gly: >NCH_2COOH; \quad >Ala \text{ or } \underline{\;}Ala: > N\overset{\overset{\textstyle CH_3}{|}}{C}HCOOH$$

3.2 Lack of Hydroxyl on the α-Carboxyl Group

The α-carboxyl group is always understood to be on the right-hand side of the symbol when using hyphens, and—in such special cases as 5.3.1.3—at the tail of the arrow when using arrows to indicate the direction of the peptide bond ($-CO \rightarrow NH$, $-NH \leftarrow CO-$).

Example: Gly—: H_2NCH_2CO-.

3.3 Lack of Hydrogen on Amino, Imino, Guanidino, Hydroxyl, and Thiol Functions in the Side Chain

Lys or Lys: $H_2N\overset{|}{C}HCOOH$
$(\overset{|}{C}H_2)_4$
$\overset{|}{N}H$
$|$

His or His: $H_2N\overset{|}{C}HCOOH$ or $H_2N\overset{|}{C}HCOOH$

Trp or Trp: $H_2N\overset{|}{C}HCOOH$
$\overset{|}{C}H_2$

Ser or Ser: $H_2N\overset{|}{C}HCOOH$
$\overset{|}{C}H_2$
$\overset{|}{O}$

Cys or Cys: $H_2N\overset{|}{C}HCOOH$ ("half-cystine")
$\overset{|}{C}H_2$
$-\overset{|}{S}$

(Cystine would be:
$\begin{array}{l} \overset{|}{Cys} \\ \overset{|}{Cys} \end{array}$ Cys ⌐ Cys ⌐

or Cys ⌐⌐ Cys, not Cys—Cys)

Arg or Arg: $H_2N\overset{|}{C}HCOOH$
$(\overset{|}{C}H_2)_2$
$\overset{|}{N}H$
$\overset{|}{C}$
$H_2N \diagdown \overset{\diagdown}{N}-$

Tyr or Tyr: $H_2N\overset{|}{C}HCOOH$
$\overset{|}{C}H_2$
(ring)
$\overset{|}{O}$
$|$

3.4 Lack of Hydroxyl on Carboxyl Groups in the Side Chain

Asp or Asp: $H_2N\overset{|}{C}HCOOH$
$\overset{|}{C}H_2$
$\overset{|}{C}O$
$|$

Glu or Glu: $H_2N\overset{|}{C}HCOOH$
$\overset{|}{C}H_2$
$\overset{|}{C}H_2$
$\overset{|}{C}O$
$|$

4. SUBSTITUTED AMINO ACIDS

4.1 Substitution in the α-Amino and α-Carboxyl Groups

This follows logically from 3.1 and 3.2. The following examples will make the usage clear.

N-Acetylglycine	Ac-Gly
Glycine ethyl ester	Gly-OEt
N^α-Acetyllysine	Ac-Lys
Serine methyl ester	Ser-OMe
α-Ethyl-*N*-acetylglutamate	Ac-Glu-OEt
Isoglutamine	Glu-NH$_2$

N-Ethyl-*N*-methylglycine   EtMeGly, $\underset{Me}{\overset{Et}{\diagdown}}$Gly,

$$Et\underset{\underset{\textstyle |}{\overset{\textstyle Me}{|}}}{\rule{0pt}{0pt}}Gly$$

4.2 Substitution in the Side Chain

Side chain substituents may be portrayed above or below the amino acid symbol, or by placing the symbol for the substituent in parentheses immediately after the amino acid symbol.

The use of parentheses should be reserved for a single symbol denoting a side chain substituent. Where a more complex substituent is involved, it is recommended that the vertical stroke and a two-line abbreviation be used.

Aspartic acid β-methyl ester
$$\overset{\text{OMe}}{\underset{\text{OMe}}{|}}\text{Asp}\quad\text{or}\quad\text{Asp}\quad\text{or}\quad\text{Asp(OMe)}$$

N^ε-Acetyllysine
$$\overset{\text{Ac}}{|}\text{Lys}\quad\text{or}\quad\text{Lys}\underset{\text{Ac}}{|}\quad\text{or}\quad\text{Lys(Ac)}$$

O-Acetylserine
$$\overset{\text{Ac}}{|}\text{Ser}\quad\text{or}\quad\text{Ser}\underset{\text{Ac}}{|}\quad\text{or}\quad\text{Ser(Ac)}$$

O-Methyltyrosine
$$\overset{\text{Me}}{|}\text{Tyr}\quad\text{or}\quad\text{Tyr}\underset{\text{Me}}{|}\quad\text{or}\quad\text{Tyr(Me)}$$

S-Ethylcysteine
$$\overset{\text{Et}}{|}\text{Cys}\quad\text{or}\quad\text{Cys}\underset{\text{Et}}{|}\quad\text{or}\quad\text{Cys(Et)}$$

5. POLYPEPTIDES

5.1 Polypeptide Chains

Polypeptides may be dealt with in the same manner as substituted amino acids, e.g.

Glycylglycine   Gly–Gly
α-Glutamylglycine Glu–Gly

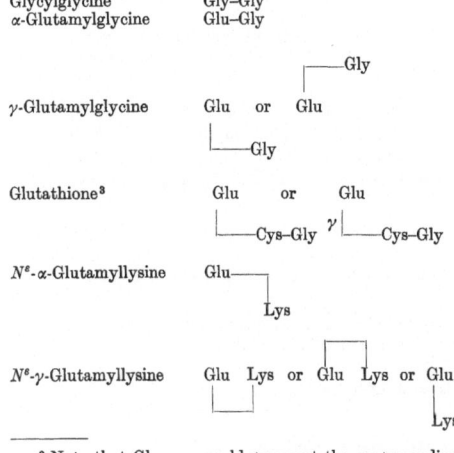

γ-Glutamylglycine

Glutathione[3]

N^ε-α-Glutamyllysine

N^ε-γ-Glutamyllysine

---

[3] Note that $\overset{\text{Glu}}{\underset{\text{Cys-Gly}}{|}}$ would represent the corresponding

thiolester with a bond between the γ-carboxyl of glutamic acid and the thiol group of cysteine.

26*

The presence of free, substituted, or ionized functional groups can be represented (or stressed) as follows:

Glycyllysylglycine
$$\text{H}-\text{Gly}-\overset{\text{H}}{\underset{}{|}}\text{Lys}-\text{Gly}-\text{OH}$$

Its dihydro-chloride
$$^+\text{H}_2-\text{Gly}-\underset{^+\text{H}_2}{|}\text{Lys}-\text{Gly}-\text{OH}\cdot2\text{Cl}^-$$

Its sodium salt Gly–Lys–Gly–O⁻ Na⁺

Its N^ε-formyl derivative
$$\text{Gly}-\underset{\text{CHO}}{|}\text{Lys}-\text{Gly}\quad\text{or}\quad\text{Gly}-\text{Lys(CHO)}-\text{Gly}$$

5.2 Peptides Substituted at N^α

Examples:

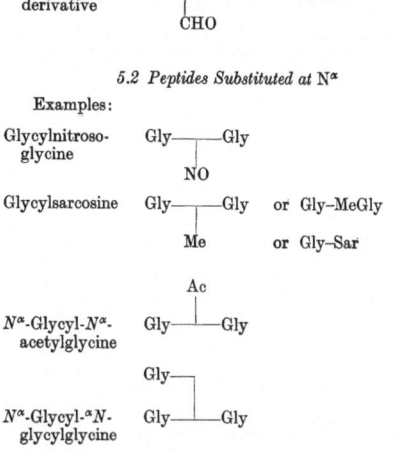

Glycylnitroso-glycine

Glycylsarcosine

N^α-Glycyl-N^α-acetylglycine

N^α-Glycyl-$^\alpha N$-glycylglycine

5.3 Cyclic Polypeptides

5.3.1 Homodetic cyclic polypeptides (the ring consists of amino acid residues in peptide linkage only). Three representations are possible:

5.3.1.1 The sequence is formulated in the usual manner but placed in parentheses and preceded by (an italic) cyclo.
Example: Gramicidin S =

cyclo-(–Val–Orn–Leu–D-Phe–Pro–Val–Orn–Leu–D-Phe–Pro–)

or (see 1.5, sentence 2)

cyclo-(Val–Orn–Leu–DPhe–Pro–Val–Orn–Leu–DPhe–Pro–)

5.3.1.2 The terminal residues may be written in one line, as in 5.3.1.1, but joined by a lengthened bond. Using the same example in the two forms (see 1.5):

$$\lfloor\text{—Val-Orn-Leu-D-Phe-Pro-Val-Orn-Leu-D-Phe-Pro—}\rfloor$$

or

$$\lceil\text{—Val-Orn-Leu-DPhe-Pro-Val-Orn-Leu-DPhe-Pro—}\rceil$$

5.3.1.3 The residues are written on more than one line, in which case the CO → NH direction must be indicated by arrows, thus (in the optimal manner of 1.5):

$$\begin{array}{l}\rightarrow\text{Val}\rightarrow\text{Orn}\rightarrow\text{Leu}\rightarrow\text{DPhe}\rightarrow\text{Pro}\rceil\\ \lfloor\text{Pro}\leftarrow\text{DPhe}\leftarrow\text{Leu}\leftarrow\text{Orn}\leftarrow\text{Val}\leftarrow\end{array}$$

5.3.2 Heterodetic cyclic polypeptides (the ring consists of other residues in addition to amino acid residues in peptide linkage): These follow logically from the formulation of substituted amino acids.

Example:

Oxytocin   Cys–Tyr–Ile–Asn–Gln–Cys–Pro–Leu–Gly–NH$_2$

Cyclic ester of threonylglycylglycylglycine

Thr–Gly–Gly–Gly⌐ or H —Thr–Gly–Gly–Gly⌐

6. ABBREVIATIONS FOR SUBSTITUENTS

Groups substituted for hydrogen or for hydroxyl may be indicated either by their structural formulas or by (accepted) abbreviations, *e.g.*

Benzoylglycine
(hippuric acid)   ⬡–CO–Gly or C$_6$H$_5$–CO–Gly or
                    Bz–Gly[4]

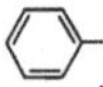

Glycine methyl   Gly–OCH$_3$ or Gly–OMe
ester

Suggestions for the abbreviations of protecting groups common in polypeptide chemistry follow. All such symbols (except those allowed by individual journals, *e.g.* Bz, Ac, Ph, Me, Et, etc.) should be defined in each paper. Although symbolization by the use of capital letters throughout would be useful for distinguishing these symbols from those of the amino acids, we propose the use of one capital letter followed by lower-case letters in order not to increase the flood of capital-letter abbreviations in biological chemistry.

---

[4] Bz– is the symbol generally used for benzoyl in organic chemistry. Its use for benzyl (which has become rather common in polypeptide chemistry) should be discouraged. We propose Bzl– for benzyl.

6.1 N-Protecting Groups of the Urethane Type

Benzyloxycarbonyl	Z–
p-Nitrobenzyloxycarbonyl	Z(NO$_2$)–
p-Bromobenzyloxycarbonyl	Z(Br)–
p-Methoxybenzyloxycarbonyl	Z(OMe)–
p-Methoxyphenylazobenzyloxycarbonyl	Mz–
p-Phenylazobenzyloxycarbonyl	Pz–
tert.-Butyloxycarbonyl	Boc–
Cyclopentyloxycarbonyl	Poc–

6.2 Other N-Protecting Groups

Acetyl	Ac–
Benzoyl	Bz–
Tosyl	Tos–
Trifluoracetyl	Tfa–
Phthalyl	Pht–
Benzyl	Bzl–
Trityl	Trt–
Tetrahydropyranyl	Thp–
Dinitrophenyl	Dnp–
Benzylthiomethyl	Btm–
o-Nitrophenylsulfenyl	Nps–

6.3 Carboxyl-protecting Groups

Methoxy (methyl ester)	–OMe
Ethoxy (ethyl ester)	–OEt
Tertiary butoxy (*tert*-butyl ester)	–OBu^t
Benzyloxy (benzyl ester)	–OBzl
Diphenylmethoxy (benzhydryl ester)	–OBzh
p-Nitrophenoxy (*p*-nitrophenyl ester)	–ONp
Phenylthio (phenylthiolester)	–SPh
p-Nitrophenylthio	–SNp
Cyanomethoxy	–OCH$_2$CN

Note: Contrary to the symbols for amino acid residues, the position of the dashes in the symbols for substituents carries no significant information.

These Tentative Rules have been translated into French and German and have appeared in the *Bulletin de la Société de Chimie biologique* [49 (1957) 121] and in *Hoppe-Seyler's Zeitschrift für physiologische Chemie* [348 (1967) 256].

All tentative Rules of the IUPAC-IUB Commission on Biochemical Nomenclature (CBN), first published in the *Journal of Biological Chemistry* [241 (1966) 527, 2491 and 2987] and in *Biochemistry* [6 (1967) 362] are available from Waldo E. Cohn, Director, NAS-NRC Office of Biochemical Nomenclature, Oak Ridge National Laboratory, P. O. Box Y, Oak Ridge, Tennessee 37831, U.S.A.:

Abbreviations and Symbols for Chemical Names of Special Interest in Biological Chemistry [see *European J. Biochem.* 1 (1967) 259],
Rules for Naming Synthetic Modifications of Natural Peptides [see *European J. Biochem.* 1 (1967) 379],
Abbreviated Designation of Amino Acid Derivatives and Polypeptides (this document),
Trivial Names of Miscellaneous Compounds of Importance in Biochemistry,
Nomenclature of Quinones with Isoprenoid Side Chains,
Nomenclature and Symbols for Folic Acid and Related Compounds,
Nomenclature of Corrinoids.

A document, OBN-1, describing the (American) NAS-NRC Office of Biochemical Nomenclature, and listing other rules affecting biochemical nomenclature is available from its Director, Dr. Waldo E. Cohn.

European J. Biochem. 1 (1967) 379—381

IUPAC-IUB Commission on Biochemical Nomenclature

Rules for Naming Synthetic Modifications of Natural Peptides

Tentative Rules[1]

During the last few years, chemists have made many compounds that are variants of naturally occurring peptides (or proteins) having trivial names. Therefore, the need has arisen for "semitrivial" names to designate these variants without the necessity of designating every residue in the chain.

After discussion with active workers in the field, the following proposals are put forward; they are based on the names used by du Vigneaud and his collaborators [cf. Bodanszky and du Vigneaud, J. Am. Chem. Soc. 81 (1959) 1258; Popenoe, Lawler, and du Vigneaud, J. Am. Chem. Soc. 74 (1952) 3713] and the symbols introduced by Schwyzer et al. [cf. Rittel, Iselin, Kappeler, Riniker, and Schwyzer, Angew. Chem. 69 (1957) 179; Riniker and Schwyzer, Helv. Chim. Acta, 44 (1961) 685; see also J. Biol. Chem. 241 (1966) 2491].

This draft has been prepared by a subcommittee consisting of J. S. Fruton, W. Klyne, and R. Schwyzer. The subcommittee is greatly indebted to many colleagues for helpful suggestions, notably to V. du Vigneaud, J. Rudinger, H. B. F. Dixon, and P. E. Verkade, chairman of the IUPAC Commission on Organic Nomenclature.

These proposals are not suitable for application to "abnormal" links in a peptide sequence; e.g., to disulfide links or γ-peptide links. They are only suitable for modifications involving normal α-peptide links.

RULES

1. Replacement

In a polypeptide of trivial name X, if the qth amino acid residue (starting from the N-terminal end of the chain) is replaced by the amino acid residue Abc, the semitrivial name of the modified polypeptide is [q-amino acid]-X and the abbreviated form, chiefly for use in tables, is [Abc^q]-X.

Examples. [8-Citrulline]-vasopressin, [Cit^8]-vasopressin [Bodanszky and Birkhimer, J. Am. Chem. Soc. 84 (1962) 4963]. [5-Isoleucine, 7-alanine]-hypertensin II, [Ile^5, Ala^7]-hypertensin II [Seu, Smeby, and Bumpus, J. Am. Chem. Soc. 84 (1962) 3883].

Comments

a) In the full name, the replacement amino acid is designated by its own full name, not the name of its radical (cf. 4 below). This name, and the position of replacement, are given in square brackets [ ], as for isotopic replacement.

---

[1] Document of the IUPAC-IUB Commission on Biochemical Nomenclature (CBN), approved by CBN in July 1966 and published by permission of the International Union of Pure and Applied Chemistry, the International Union of Biochemistry, and the official publishers to the International Union of Pure and Applied Chemistry, Messrs. Butterworths Scientific Publications.

Comments on these Tentative Rules may be sent to any member of CBN· O Hoffmann Ostenhof (Chairman), W. E. Cohn (Secretary), A. E. Braunstein, J. S. Fruton, B. Keil, W. Klyne, C. Liébecq, B. C. Malmström, R. Schwyzer, E. C. Slater, or corresponding member, N. Tamiya.

Reprints of these Tentative Rules may be obtained from Waldo E. Cohn, Director, NAS-NRC Office of Biochemical Nomenclature, Oak Ridge National Laboratory, P. O. Box Y, Oak Ridge, Tenn. 37831, U.S.A.

b) In the abbreviated form, the amino acid residues are designated by the standard three-letter symbols [J. Biol. Chem. 241 (1966) 527, 2491; European J. Biochem. 1 (1967) 259, 375], the first letter only being a capital, in square brackets [ ].

c) In the abbreviated form, the position of substitution is indicated in a special fashion, i.e., by a superior numeral q, to indicate that it is a residue, not an individual atom, that is being replaced and also for the reason indicated in comment (d).

d) The nature of the residue replaced is not designated in either the full or the abbreviated name. This is contrary to a general principle of organic nomenclature requiring that an atom (or group) that is replaced should (unless it is hydrogen) be clearly designated, as in 2-amino-2-deoxy-D-glucose. It has been decided not to insist on the designation of the residue replaced in these semitrivial names in order to keep the names as short as possible, and because the form of nomenclature in Rule 1 clearly differs from ordinary substitution nomenclature.

e) A partial analogy may be drawn with the form used for isotopic replacement, where the isotope symbol is indicated in square brackets before the name.

f) The replacement of an amino acid residue by its enantiomer may be shown logically by the application of this rule as follows: the replacement in X of L-alanine at position 7 by D-alanine results in [7-D-alanine]-X with the abbreviation [D-Ala^7]-X. An example may be found in Boissonnas, Guttman, and Pless [Experientia, 22 (1966) 526], dealing with the D-Ser^1 . . . . . . derivative of β-corticotropin; the natural compound has L-serine in position 1. Another example is the [α-D-Asp^1]-hypertensin II of Riniker and Schwyzer [Helv. Chim. Acta, 47 (1964) 2357].

2. Extension

The compounds obtained by the extension of polypeptide X at either (a) the N-terminal end or (b) the C-terminal end are designated by the kinds of names and abbreviations shown below; these are in accordance with the general principles of polypeptide nomenclature [J. Biol. Chem. 241 (1966) 2491; European J. Biochem. 1 (1967) 375].

Examples

a) Extension at N-terminal end:

Aminoacyl-X		Abc-X
e.g., Valyl-X		Val-X
or Valylglycyl-X		Val-Gly-X
(for extension by two residues).		

b) Extension at C-terminal end:

X-yl-amino acid	X-yl-Abc
e.g., X-yl-leucine	X-yl-Leu
(where X-yl is the trivial name of polypeptide X with the ending yl).	

Comment. This rule is not applicable to the extension at the C-terminal end of natural peptides having a terminal α-carboxamido group, as in the case of oxytocin or α-melanophore-stimulating hormone (α-MSH). It has been suggested that new names be given to the peptides having a free terminal α-carboxyl group (e.g., oxytocinoic acid) and that extension at the C-terminal end be denoted as in the example given above (e.g., oxytocinoyl-Abc).

3. Insertion

The compound obtained by the insertion of an additional amino acid residue Abc in the position between the qth and $(q + 1)^{th}$ residues of a polypeptide X is named qa-endo-amino acid-X (abbreviated form, endo-Abc^{qa}-X).

Example. 4a-endo-tyrosine-hypertensin II; endo-Tyr^{4a}-hypertensin II.

Comments

a) This form has analogies in other fields where endo implies the insertion of something into a structure (*e.g.*, endo-methylene). The prefix or index qa is based on analogies with the steroids where the atoms iserted in a ring after atom no. q are designated qa, qb, etc.

b) The prefix homo is not suitable for designating the insertion of a whole residue, since it is commonly used to modify the names of individual amino acids, *e.g.*, homo-serine.

c) Multiple insertions, and insertion of two or more residues together in the same place in the chain, are shown by a logical extension of this rule. For example, the insertion into the polypeptide X of threonine between residues 4 and 5, and of valine and glycine (in that order) between residues 6 and 7, is shown by the name "endo-4a-threonine, 6a-valine, 6b-glycine-X" and the abbreviation "endo-Thr^{4a}, (Val^{6a}-Gly^{6b})-X."

4. Removal

The compound obtained by the formal removal of an amino acid residue from a polypeptide X in position q is designated by the name des-q-amino acid-X, abbreviated des-Abc^{q}-X.

Example. des-7-proline-oxytocin; des-Pro^7-oxytocin [Jaquenoud and Boissonnas, *Helv. Chim. Acta*, 45 (1962) 1462].

Comments

a) Removal of a whole residue is indicated as is the removal of a ring in steroids, *e.g.*, des-A-androstane.

b) "de" is not suitable as a prefix because it is easily confused, in speaking, with D (for configuration).

5. Substitution Forming a Side Chain

The compound formed by the substitution of an additional amino acid residue as a side chain into a polypeptide X is named by applying the ordinary rules of nomenclature to the trivial name.

a) If the substitution is on a side-chain amino group of polypeptide X, the name of the additional amino residue is written (with the termination "yl") and prefixed by symbols indicating the position of substitution (residue number and atom).

Example: on imaginary compound (A)

$$\begin{array}{c} \text{Val} \\ | \\ \text{Ala-Lys-Ala} \ldots \ldots \\ 1 \quad 2 \quad 3 \\ \text{A} \end{array}$$

in which a valyl group is substituted at the ε-amino group of lysine at position 2 of the chain of a peptide X is named $N^{\varepsilon 2}$-valyl-X (abbreviated $N^{\varepsilon 2}$-Val-X).

b) If the substitution is on a side-chain carboxyl group of polypeptide X, the additional amino acid having a free α-carboxyl group, the substituted derivative is named by specifying the position of substitution (residue number, and atom) and is given the designation "X-yl-amino acid."

Example: an imaginary compound (B)

$$\begin{array}{c} \text{Val} \\ | \\ \text{Ala-Leu-Glu-Ala} \ldots \ldots \\ 1 \quad 2 \quad 3 \quad 4 \\ \text{B} \end{array}$$

in which a valine residue is substituted into the δ-carboxyl group of glutamic acid in position 3 of the chain of a peptide X would be named $C^{\delta 3}$-X-yl-valine (abbreviated $C^{\delta 3}$-X-yl-Val).

Comment. Note the importance of clear distinction from replacement as indicated in Rule 1.

6. Partial Sequences (Fragments)

Polypeptide sequences that form fragments of a longer sequence that already has a trivial name may be designated as follows. The trivial name is followed by numbers giving the positions of the first and last amino acids, and then the usual Greek designation giving the number of amino acid units in the fragment; thus: Trivial name (-X—Y-) ... peptide.

Examples: from α-MSH

Ac-Ser-Tyr-Ser-Met-Glu-His-Phe-Arg-Trp-Gly-Lys-Pro-Val-NH₂
   1   2   3   4   5   6   7   8   9 10 11 12 13

We may have:

Met-Glu-His-Phe-Arg-Trp-Gly, or α-MSH-(4—10)-heptapeptide
 4   10

or, to illustrate the nomenclature for a composition sequence of two fragments, and also for an amide-terminal group:

His-Phe-Arg-Lys-Pro-Val-NH₂, or α-MSH·(6—8)·(11—13-hexapeptide am[
 6   8   11   13

SUMMARY WITH EXAMPLES

The systematic application of these principles to the name of an imaginary pentapeptide "Iupaciubin"[2] may illustrate the symbolism.

[2] To symbolize the harmonious cooperation of IUPAC and IUB.

Table

Rule	Operation	Short name	Structure
	(Fundamental name)	Iupaciubin[a]	1 2 3 4 5 Ala-Lys-Glu-Tyr-Leu
1.	Replacement	[Phe⁴]-iupaciubin[b]	4 Ala-Lys-Glu-Phe-Leu
2a.	Extension (N terminal)	Arginyl-iupaciubin, Arg-iupaciubin	1 5 Arg-Ala-Lys-Glu-Tyr-Leu
2b.	Extension (C terminal)	Iupaciubyl-methionine, iupaciubyl-Met	1 5 Ala-Lys-Glu-Tyr-Leu-Met

Table (Continued)

Rule	Operation	Short name	Structure
3.	Insertion	Endo-Thr²ª-iupaciubin	$\overset{2\ \ \ \ 2a\ \ \ 3}{\text{Ala-Lys-Thr-Glu-Tyr-Leu}}$
4.	Removal	Des-Glu³-iupaciubin	$\overset{2\ \ \ \ \ \ 4}{\text{Ala-Lys-Tyr-Leu}}$
5a.	Side-chain substitution on amino group	$N^{\varepsilon 2}$-Val-iupaciubin	Val $\mid \varepsilon$ $\underset{2}{\text{Ala-Lys-Glu-Tyr-Leu}}$
5b.	Side-chain substitution on carboxyl group	$C\text{-}\delta^{3}$-Iupaciubyl-valine	Val $\mid \delta$ $\underset{3}{\text{Ala-Lys-Glu-Tyr-Leu}}$
6.	Partial sequence	Iupaciubin-(2—4)-tripeptide	$\overset{2\ \ \ \ 3\ \ \ \ 4}{\text{Lys-Glu-Tyr}}$

ª To symbolize the harmonious cooperation of IUPAC and IUB.
ᵇ Note that only for replacement are square brackets required.

These Tentative Rules have been translated into French and German and have appeared in the *Bulletin de la Société de Chimie biologique* [49 (1967) 325] and in *Hoppe-Seyler's Zeitschrift für physiologische Chemie* [348 (1967) 262].

All Tentative Rules of the IUPAC-IUB Commission on Biochemical Nomenclature (CBN), first published in the *Journal of Biological Chemistry* [241 (1966) 527, 2491 and 2987] and in *Biochemistry* [6 (1967) 362] are available from Waldo E. Cohn, Director, NAS-NRC Office of Biochemical Nomenclature, Oak Ridge National Laboratory, P. O. Box Y, Oak Ridge, Tennessee 37831, U.S.A.:

Abbreviations and Symbols for Chemical Names of Special Interest in Biological Chemistry [see *European J. Biochem.* 1 (1967) 259],
Abbreviated Designation of Amino Acid Derivatives and Polypeptides [see *European J. Biochem.* 1 (1967) 375],
Rules for Naming Synthetic Modifications of Natural Peptides (this document),
Trivial Names of Miscellaneous Compounds of Importance in Biochemistry,
Nomenclature of Quinones with Isoprenoid Side Chains,
Nomenclature and Symbols for Folic Acid and Related Compounds,
Nomenclature of Corrinoids.

A document, OBN-1, describing the (American) NAS-NRC Office of Biochemical Nomenclature, and listing other rules affecting biochemical nomenclature is available from his Director, Dr. Waldo E. Cohn.

European J. Biochem. 1 (1967) 382—394

Untersuchungen zur Induktion der Lac-Enzyme

1. Induktionswirkung und Permeation

W. Boos, P. Schaedel und K. Wallenfels

Chemisches Laboratorium der Universität Freiburg i. Br.

(Eingegangen am 23. Januar 1967)

The kinetics of β-galactosidase induction were investigated in two strains of *Escherichia coli*, K12 3000 ($i^+z^+y^+$) and K12 2001 ($i^+z^+y^-$). The use of a new series of β-galactoside analogues has led to a better understanding of the induction process. The unique properties of galactosyl glycerol as an inducer have been explained by the existence of a specific permease for this galactoside.

1. Previous studies of the relative efficiency of isopropyl thiogalactoside and thiogalactosyl glycerol as inducers were carried out in y^+ strains. The possibility of these inducers having differing affinities for the lac permease, made the earlier observations ambiguous. These experiments have therefore been repeated using a y^- strain. Isopropyl thiogalactoside was found to be a better inducer than thiogalactosyl glycerol.

2. An evaluation of the importance of various chemical groupings as determinants of the relative efficacy of inducers has been begun. The 6-hydroxymethyl group of the galactosyl moiety of β-galactosides appears to be of considerable importance. The substitution by a methyl group (thiofucosyl glycerol) or by a hydrogen atom (thioarabinosyl glycerol) of the 6-hydroxymethyl group (thiogalactosyl glycerol) results in the conversion of a compound which is an excellent inducer into a compound which cannot function as inducer at all. A comparison of S- and O-galactosides shows that, when an O-galactoside is an inducer, the corresponding S-galactoside is normally a better inducer. Thioallolactose is an exception to this rule. Despite the fact that it is a weak inducer of β-galactosidase, thioallolactose is an excellent inhibitor of the lac permease. When the non-inducer, thioallolactose, was reduced to thioallolactitol, it became a normal "inducteur gratuit". Since it has been observed that galactose itself can serve as an inducer of β-galactosidase in mutants lacking galactokinase, the inducer properties of the analogue 1,5-anhydro-galactitol were investigated. This compound, although not a good inducer, was superior to galactose.

3. Galactosyl galactoside exhibits similar kinetics of induction to those of lactose. It is suggested that this compound, like lactose, is not itself an inducer but that it is converted to the actual inducer by the basic level β-galactosidase, which is found in uninduced cells.

4. Galactosyl glycerol has previously been characterised as a uniquely active inducer in that (a) it can act at concentrations as low as 10^{-6} M. (b) The induction process is without an autocatalytic phase. Observations made in the present study indicate that these properties result from the presence of a seemingly constitutive permease, which is specific for galactosyl glycerol. This permease then favours the accumulation of galactosyl glycerol under conditions of low inducer concentration, when the general lac permease is not present. When the lac permease has been induced, a variety of inducers (including galactosyl glycerol) can be concentrated. The specificity of the galactosyl glycerol permease is emphasized by its lack of transport of thiogalactosyl glycerol or secondary galactosyl glycerol.

Nomenklatur und nicht allgemeingebräuchliche Abkürzungen. 1-O-β-D-Galactopyranosyl-D-glycerin = Galactosylglycerin; 2-O-β-D-Galactopyranosyl-D-glycerin = sekundäres Galactosylglycerin; 6-O-β-D-Galactopyranosyl-D-glucose = Allolactose; 6-S-β-D-Galactopyranosyl-6-thio-D-glucose = Thioallolactose; 1-S-β-D-Galactopyranosyl-1-thio-glycerin = Thiogalactosylglycerin; 1-S-β-D-Fucopyranosyl-1-thio-glyce-

rin = Thiofucosylglycerin; 1-S-β-L-Arabinopyranosyl-1-thio-glycerin = Thioarabinosylglycerin; 1-S-β-D-Galactopyranosyl-1-thio-D-sorbit = Thioallolactit; β-D-Galactopyranosyl-β'-D-galactopyranosyl-sulfid = Thiogalactosylgalactosid; β-D-Galactopyranosyl-β'-D-galactopyranosid = Galactosylgalactosid; Isopropyl-1-thio-β-D-galactopyranosid = IPTG; o-Nitrophenyl-β-D-galactopyranosid = ONPG.

Enzym. β-Galactosidase (EC 3.2.1.23).

In der Lac-Region von *Escherichia coli* ist die Information für vier Proteine niedergelegt. Drei von ihnen (β-Galactosidase [1], Galactosid-Permease [2,3] und Thiogalactosid-Transacetylase [4]) dienen der Aufnahme und Metabolisierung von β-Galactosiden. Das vierte Protein, der sogenannte Repressor [5,6], unterdrückt die Synthese dieser drei Enzyme oder gibt sie in dem Maße frei, wie der Induktor an den Repressor gebunden wird [7,8]. Die Induktion der Strukturproteine wird also durch die Bildung des Induktor-Repressor-Komplexes eingeleitet. Mit verschiedenen Substanzen wurde die Induzierbarkeit untersucht [9—12].

Die Induktorspezifität des Repressors sollte bestimmbar sein, wenn man die Bildungsgeschwindigkeit eines induzierten Enzyms in Abhängigkeit von der Konzentration eines Induktors im Medium verfolgt.

Die induzierte Enzymsynthese der Bakterienzelle als Antwort auf die Anwesenheit eines Induktors im Medium, ist zwar abhängig von der Affinität zwischen Induktor und Repressor, wird aber bestimmt durch die Induktorkonzentration im Zellinnern.

Diese Konzentration wird innerhalb der Bakterienzelle reguliert durch die Galactosidpermease und eventuell noch andere weniger spezifische Permeasen, darüber hinaus auch noch von der β-Galactosidase, sofern die zu untersuchenden Induktoren Substrate für dieses Enzym sind.

Die Thiogalactosid-Transacetylase spielt offenbar beim Induktionsvorgang keine Rolle [13].

Der Induktionsvorgang komplizert sich noch mehr, wenn die als Induktor angewendete Substanz selbst nicht induzierend wirkt, sondern erst innerhalb der Bakterienzelle durch die *basic-level*-Aktivität und in der Folge auch durch die induzierte β-Galactosidase-Aktivität in den eigentlichen Induktor umgewandelt wird. Das ist z. B. bei Lactose der Fall [10,14].

Aus allen diesen Gründen ist es schwierig, quantitative Aussagen über die Spezifität des Repressors für verschiedene Induktoren zu machen, solange dieser nicht isoliert ist.

Durch Induktionsexperimente mit strukturell variierten β-Galactosiden haben wir versucht, Gesichtspunkte für die Ermittlung der Spezifität zusammenzutragen, mit welcher der Repressor eine Verbindung als Induktor erkennt.

Von besonderem Interesse war es in diesem Zusammenhang, warum Galactosylglycerin im Induktionsexperiment in etwa 1000mal geringerer Konzentration als alle anderen untersuchten Induktoren noch volle Wirksamkeit hat. Dies konnte auf einen speziellen Permeationseffekt zurückgeführt werden.

An mehreren Beispielen wurden Verbindungspaare verglichen, welche als glycosidisches Heteroatom Sauerstoff bzw. Schwefel aufweisen.

METHODEN UND SUBSTANZEN

BAKTERIENSTÄMME UND MEDIEN

Bei den Induktionsversuchen wurden verwendet: *Escherichia coli* K 12 3000 ($i^+z^+y^+$) und K 12 2001$_c$ ($i^+z^+y^-$ thr⁻ leu⁻)[1].

Als Kulturmedium wurde das synthetische Medium 63 verwendet: 13,609 g KH_2PO_4, 2,00 g $(NH_4)_2SO_4$, 0,200 g $MgSO_4$, 0,005 g $FeSO_4$, 0,5 mg Thiamin werden in 1 l destilliertem Wasser gelöst und mit 5 N KOH auf pH 7,0 eingestellt. Als C-Quelle dient in allen Fällen 0,5% Glycerin. Zur Kultivierung von K 12 2001$_c$ werden außerdem noch 10^{-4} M Threonin und Leucin hinzugefügt. Die Stämme werden auf Schrägagarröhrchen (Standardagar II der Fa. Merck, Darmstadt) kultiviert und aufbewahrt.

SUBSTANZEN

Metabolisierbare Induktoren

Galactosylglycerin wurde synthetisiert nach [15]. Sekundäres Galactosylglycerin wurde synthetisiert nach [16]. Allolactose wurde hergestellt durch enzymatischen Galactosyltransfer von Phenylgalactosid auf Glucose [17]. Galactosylgalactosid wurde synthetisiert nach [18]. Von 3 g des Octaacetats (F = 164°) wurden 0,9 g des Galactosylgalactosids (F = 152—153°) in kristalliner Form erhalten.

Nicht metabolisierbare Induktoren

Thiogalactosylglycerin wurde synthetisiert analog der Methode von W. Schneider et al. [19] zur Darstellung von Alkylthioglucosiden: 3,72 g ($3,4 \times 10^{-2}$ Mole) 1-Thioglycerin werden in einer Lösung von 1,7 g KOH ($3,0 \times 10^{-2}$) in 50 ml Äthanol gelöst. Unter Rühren wird eine Lösung von 12,3 g ($3,0 \times 10^{-2}$ Mole) Acetobromgalactose in 50 ml Chloroform zugetropft. Dabei fällt KBr aus. Nach vollständigem Zusatz wird das Reaktionsgemisch unter Rückfluß 30 min erwärmt. Nach Abkühlen wird dreimal mit kalter Bicarbonatlösung ausgeschüttelt, die Chloroformschicht mit Na_2SO_4 getrocknet und anschließend im Vacuum zum Sirup eingeengt. Da die Substanz nicht kristallisierte, wurde der Sirup über P_2O_5 getrocknet und mit Bariummethylat bei 0° entacetyliert. Es wurde mit CO_2 neutralisiert, von dem gebildeten $BaCO_3$ abzentrifugiert und die Lösung eingeengt. Der erhaltene Sirup kristallisiert aus methanolischer Lösung. Nach viermaligem Umkristallisieren erhält man 0,8 g harte, büschelige Kristalle. F = 150—151°, Drehwert $[\alpha]_D^{20} = -7,64°$ (c 4,01 H_2O) Analyse: $C_9H_{18}O_7S$ (270,3) Ber.: 39,99% C, 6,71% H; Gef.: 39,82% C, 6,86% H.

Thioarabinosylglycerin wurde analog synthetisiert: 2,45 g ($6,3 \times 10^{-2}$ Mole) Kalium wurden in 100 ml abs. Äthylalkohol gelöst, 6,82 g ($6,3 \times 10^{-2}$

---

[1] Wir danken Prof. J. Monod, Paris, für die Überlassung der Bakterienstämme.

Mole) Thioglycerin zugesetzt und eine Lösung von 21,4 g (6,3 × 10⁻² Mole) Acetobromarabinose in 100 ml Chloroform zugetropft. Anschließend wird das Reaktionsgemisch 30 min unter Rückfluß erhitzt. Der Ansatz wurde, wie bei der Synthese des Thiogalactosylglycerins beschrieben, aufgearbeitet und mit Bariummethylat entacetyliert. Die Lösung wurde mit CO_2 neutralisiert und mit Äthanol aufgenommen. Aus der alkoholischen Lösung kristallisierten 6,0 g Thioarabinosylglycerin mit F = 73—76°. Nach mehrmaligem Umkristallisieren wurde ein konstanter Schmelzpunkt von 77—81° erhalten.

$[\alpha]_D^{20} = -15,52$ (c 2,14 H_2O)

Analyse: $C_8H_{16}O_6S$ (240,3) Ber.: 39,99% C, 6,71% H; Gef.: 40,08% C, 6,85% H.

1,5-Anhydrodulcit [37] wurde analog nach [20] synthetisiert.

Thioallolactose wurde entsprechend der Methode von Černý et al. [21] zur Darstellung von β-D-Thiogalactosiden synthetisiert: 12,7 g (3,5 × 10⁻² Mole), 2,3,4,6-Tetra-O-acetyl-1-thio-β-D-galactose und 16 g (3,5 × 10⁻² Mole) 1,2,3,4-Tetra-O-acetyl-6-desoxy-6-jodo-β-D-glucose [22] werden in 140 ml Aceton-Wasser (1:1, v/v) mit 4,82 g (3,5 × 10⁻² Mole) K_2CO_3 unter Rühren über Nacht umgesetzt. Die Lösung wurde mit Essigsäure neutralisiert, mit Wasser stark verdünnt und dreimal mit Chloroform ausgeschüttelt. Die organische Phase wurde über Na_2SO_4 getrocknet, das Lösungsmittel unter Vacuum abgezogen und der sirupöse Rückstand aus Äthylalkohol kristallisiert. Man erhält 14,1 g Octaacetylthioallolactose, F = 169°.

Analyse: $C_{28}H_{38}O_{18}S$ (694,7) Ber.: 48,41% C, 5,51% H; Gef.: 48,39% C, 5,40% H. Drehwert: $[\alpha]_D^{18} = -13,34$ (c 2.00 $CHCl_3$). 14,1 g obiger Verbindung wurden bei 0° mit Bariummethylat entacetyliert. Man erhält 5,7 g Thioallolactose, die nach dreimaligem Umkristallisieren aus Methanol/Äthanol einen Schmelzpunkt von 188—190° hat.

Analyse: $C_{12}H_{22}O_{10}S$ (358,4) Ber.: 40,22% C, 6,19% H; Gef.: 40,04% C, 6,35% H. Anfangsdrehwert: $[\alpha]_D^{20} = +63,4$ (c 2,00 H_2O), Enddrehwert: $[\alpha]_D^{20} = +46,8$ (c 2,00 H_2O).

Thiogalactosylgalactosid. 12,0 g Acetobromgalactose (2,9 × 10⁻² Mole), 8,85 g 2,3,4,6-tetra-O-acetyl-1-thio-β-D-galactose (2,4 × 10⁻² Mole) und 4,0 g K_2CO_3 (2,9 × 10⁻² Mole) werden in 60 ml Aceton-Wasser (1:1, v/v) über Nacht gerührt. Es wurden 13,5 g Octaacetylthiogalactosylgalactosid vom F = 201—203° erhalten. Die Entacetylierung mit Bariummethylat ergibt 4,8 g Thiogalactosylgalactosid F = 235°. (F nach [23] = 230°.)

Thioallolactit. Zu 3 g (8,8 × 10⁻³ Mole) Thioallolactose in 9 ml H_2O gibt man eine Lösung von 3 g (7,9 × 10⁻² Mole) $NaBH_4$ in 75 ml H_2O und rührt das Gemisch über Nacht. Die Lösung wurde mit 35 ml Ionenaustauscher (Amberlite IR 120) neutralisiert. Das Eluat wurde mit den Waschwässern vereinigt, auf

10 ml eingeengt und nach dem Ansäuern mit einigen Tropfen Eisessig auf eine Kohlesäule aufgebracht. Die anorganischen Bestandteile wurden durch Wasser ausgewaschen. Die Substanz ließ sich anschließend mit Wasser-Äthanol (4:1, v/v) eluieren. Nach Eindampfen und viermaligem Umkristallisieren aus Methanol-Äthanol wies die Substanz einen Schmelzpunkt von 142° auf. Ausbeute 2,7 g.

Analyse: $C_{12}H_{24}O_{10}S$ (360,4) Ber.: 40,00% C, 6,71% H; Gef.: 40,14% C, 6,95% H, Drehwert $[\alpha]_D^{20} = +1,41$ (c 6,04 H_2O).

Die Synthese von Thiofucosylglycerin, die über einige noch nicht beschriebene Zwischenstufen führt, soll an anderer Stelle veröffentlicht werden.

Die im Galactoseteil radioaktiv markierten Galactoside: 1-O-β-D-([1-¹⁴C]Galactosyl)D-glycerin, 2-O-β-D-([1-¹⁴C]Galactosyl)-glycerin und 6-O-β-D-([1-¹⁴C]-Galactosyl)-D-glucose wurden durch enzymatischen Transfer von radioaktiver Galactose (35 mC/mMol) auf Glycerin bzw. auf Glucose dargestellt [24].

DER INDUKTIONSVERSUCH

Die Bakterienstämme werden auf Festagar bei 37° gezogen und bei 4° höchstens 3 Monate aufbewahrt. Vom Schrägagar wird in Kulturen von 5 ml Medium 63 überimpft und über Nacht bei 37° unter Belüftung geschüttelt. Von der Vorkultur werden 1—3 ml in 100 ml frisches Medium pipettiert, die in einem Erlenmeyerkolben bei 37° in einem Reziprokschüttler (Firma New Brunswick, Typ RW-650) geschüttelt werden.

Das Wachstum der Kultur wird durch Messung der Absorption bei 578 mµ im Eppendorf-Spektralfotometer verfolgt. Die Generationszeiten für beide verwendeten Stämme betragen unter diesen Bedingungen in der logarithmischen Phase 85—100 min.

Wenn die Bakterienkultur eine optische Dichte von 0,1—0,15 erreicht hat, entsprechend einer Zelldichte von 0,25—0,375 mg/ml Feuchtgewicht, wird der zu untersuchende Induktor in 1 ml konzentrierter wäßriger Lösung zugegeben. Nach verschiedenen Zeiten werden der Zellsuspension 5 ml entnommen und in ein Reagensglas gegeben, das einen Tropfen Toluol und einen Tropfen 1%ige Natriumdesoxycholat-Lösung enthält.

Die toluolisierten Proben werden 1—1,5 Std bei 40° im Wasserbad gehalten und anschließend auf β-Galactosidase-Aktivität untersucht.

Testansatz: 0,5 ml toluolisierte Bakteriensuspension, 2,0 ml Puffer (10⁻¹ M Kaliumnatriumphosphat, pH 7,0, Mg^{2+} 10⁻³ M, 10⁻¹ M β-Mercaptoäthanol), 0,5 ml ONPG-Lösung (5 mg/ml).

Bei Ansetzen des Puffers wird β-Mercaptoäthanol vor Beginn des Tests zugesetzt; angesetzte Lösungen sind nicht lange haltbar. Die Lösungen werden zusammengegeben und die Extinktionszunahme bei 405 mµ innerhalb von 100 sec bei 40° im Eppendorf-

Spektralfotometer gemessen. Als Einheit gilt diejenige Menge Enzym, welche unter diesen Bedingungen 1 nMol o-Nitrophenol pro Minute freisetzt.

MESSUNG DER PERMEATION VON GALACTOSYLGLYCERIN

Radioaktives Galactosylglycerin (ca. 13 µg, 2 × 10^6 Imp/min) wird von Chromatographiepapier Whatman Nr. 3 mit Wasser in einen kleinen Kolben eluiert (10 ml). Die Lösung, die noch Verunreinigungen des Papiers enthält, wird mit A-Kohle behandelt, filtriert und im Vacuum zur Trockne eingedampft. Dann werden 200 µl einer nicht markierten Galactosylglycerinlösung der Konzentration $1,5 \times 10^{-3}$ M bzw. $1,5 \times 10^{-4}$ M zugegeben. Von dieser Stammlösung werden jeweils 10 µl entnommen und in kleine Gläschen gegeben. In diese Gläschen werden 1,5 ml der in Medium 63 gewachsenen Kultur gegeben, so daß dort die Konzentration an Galactosylglycerin $1,2 \times 10^{-5}$ M bzw. $2,7 \times 10^{-6}$ M beträgt. Nach 10 min wird hiervon 1 ml entnommen und durch Millipore-Filter (HA, 0,45 µ Porengröße, 13 mm Durchmesser) filtriert, mit 3 ml eiskaltem Medium 63 gewaschen und das Filter getrocknet. Die Radioaktivität der Filter wird im Scintillationszähler gezählt. Gerät: Nuclear Chicago Modell 6725; die Scintillationslösung enthält 0,4% 2,5-Diphenyloxazol und 0,005% 2-p-Phenylen-bis-(phenyloxazol) in Toluol. Gezählt wird bei einer Spannung von 940 V an Data und 1300 V an Gate. Die Spannungsniveaus im verwendeten Scaler sind 0,5 und 9,9 V.

QUANTITATIVE AUSWERTUNG
VON PAPIERCHROMATOGRAMMEN
DURCH AUTORADIOGRAPHIE

Galactosylglycerin läßt sich papierchromatographisch von Galactose, Glycerin und von seinen eigenen Isomeren am besten durch Phenol-Wasser (4:1, v/v) trennen. Für Disaccharide wie Allolactose wird Butanol-Pyridin-Wasser (6:4:3, v/v) verwendet.

Bei der enzymatischen Hydrolyse von radioaktiven Galactosiden wird dem Spaltansatz 20 µl entnommen, auf Whatman Nr. 3 gegeben und der Fleck durch einen heißen Luftstrom getrocknet, wobei gleichzeitig das Enzym denaturiert wird. Die Laufzeit des Chromatogramms beträgt bei Verwendung von Phenol/Wasser 15 Std, bei Butanol-Pyridin-Wasser 24 Std und mehr.

Das entwickelte Chromatogramm wird an der Luft getrocknet und in passenden lichtdichten Kassetten mit Röntgenfilm (Kodak-Kodirex 40 × 30 cm) belegt. Je nach Aktivität der radioaktiven Substanz wird der Film 2—14 Tage belichtet.

Die so auf dem Chromatogramm markierten Flecken werden ausgeschnitten und wie oben beschrieben im Scintillationszähler ausgezählt.

ERGEBNISSE
Induktionskinetik metabolisierbarer Induktoren

Trägt man die β-Galactosidase-Aktivität in *E. coli* K 12 3000 nach Zugabe des Induktors gegen das Zellwachstum auf, so erhält man Kurven, die den verwendeten Induktor charakterisieren. In Fig. 1 sind vier Induktionsversuche dieser Art zusammengestellt, wobei folgende β-Galactoside verwendet wurden: sekundäres Galactosylglycerin (A), Allolactose (B), Galactosylgalactosid (C) und Galactosylglycerin (D).

Bei allen Induktoren der Fig. 1 bricht die Synthese der β-Galactosidase bei der Anfangskonzentration früher oder später innerhalb eines kurzen Zeitraums ab; die dann gebildete Enzymmenge ist bei Anwendung gleicher molarer Mengen der verschiedenen Induktoren verschieden.

Allen Substanzen ist gemeinsam, daß bei Erhöhung der Induktorkonzentration eine maximale Geschwindigkeit der Enzymsynthese erreicht wird, bei welcher offenbar der Repressor gesättigt ist [12]. Diese Induktorkonzentration muß im Falle des Galactosylglycerins unter 10^{-6} M liegen; wie man aus der Fig. 1 D sehen kann, wird auch bei der niedrigsten Konzentration, bei welcher noch Induktionswirkung feststellbar ist, die maximale Synthesegeschwindigkeit erreicht, wenn auch nur für kurze Zeit.

Wesentliche Unterschiede im Induktionsverhalten ergeben sich jedoch bei den vier oben genannten Substanzen, wenn man die Anfangskinetik der Induktion betrachtet (Fig. 2).

Bei Anwendung von sekundärem Galactosylglycerin (A) und Allolactose (B) entspricht sie bekannten Vorstellungen [25—28]: eine kurze *lag*-Phase, anschließend eine mehr oder weniger lange autokatalytische Phase, in der die intracelluläre Konzentration und damit die Induktorwirksamkeit durch die selbst induzierte Lac-Permease erhöht wird; darauf folgt eine Phase konstanter Synthesegeschwindigkeit der Enzyme, die bei der entweder die Enzymsynthese maximal ist oder geringer, sofern die Induktorkonzentration unter dem Sättigungswert für die maximale Induktion liegt. Bei solchen geringen Induktorkonzentrationen ist allerdings die Phase der konstanten Synthesegeschwindigkeit kurz; es schließt sich eine Phase an, bei der die Synthesegeschwindigkeit allmählich abnimmt. Diese Abnahme tritt um so schneller ein, je geringer die angewendete Konzentration ist (Fig. 1).

Galactosylgalactosid zeigt wie Lactose eine verlängerte *lag*-Phase. Das hängt offenbar damit zusammen, daß die Substanz, wie auch die meisten isomeren Galactosidoglucosen und -galactosen, selbst kein Induktor ist, sondern erst nach Eintritt in die Zelle durch die transgalactosidierenden Eigenschaften der β-Galactosidase in einen wirksamen Induktor umgewandelt wird [10,14].

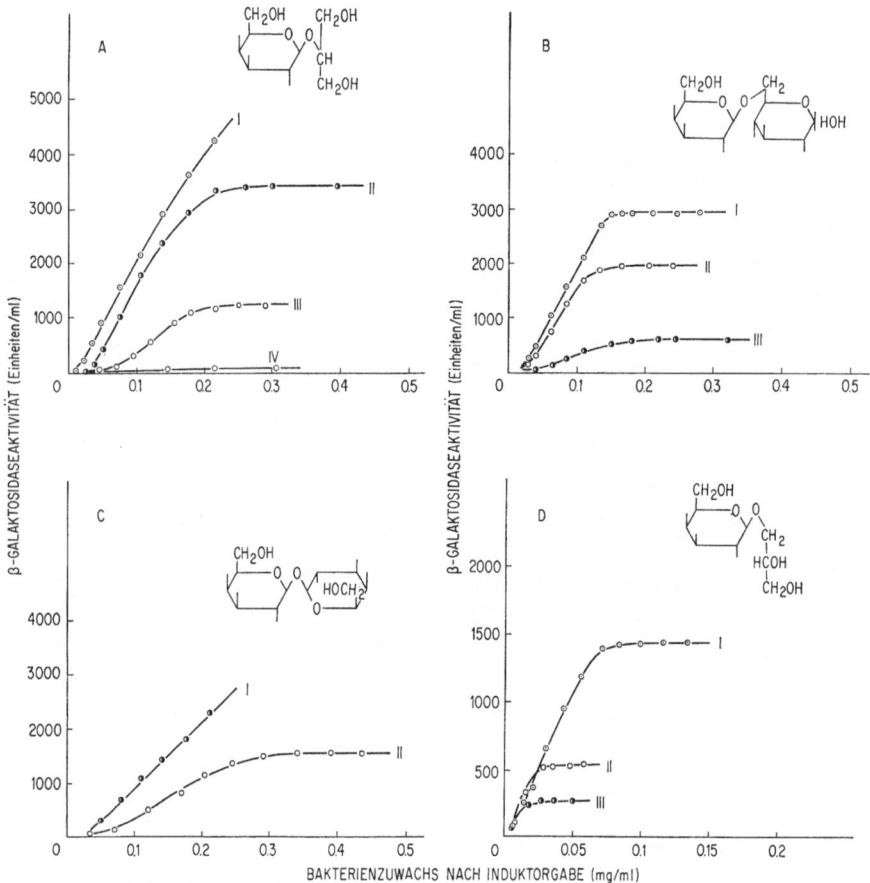

Fig. 1. *Abhängigkeit der β-Galactosidase-Aktivität in K 12 3000 während der logarithmischen Wachstumsphase von dem Zuwachs an Bakterienmasse nach Gabe des Induktors.* Als Induktoren werden verwendet: A, sek. Galactosylglycerin: I, 10^{-3} M; II, 3×10^{-4} M; III, 10^{-4} M; IV, 7×10^{-5} M. B, Allolactose: I, 10^{-4} M; II, 3×10^{-5} M; III, 10^{-5} M. C, Galactosylgalactosid: I, 10^{-3} M; II, 10^{-4} M. D, Galactosylglycerin: I, 10^{-4} M; II, 10^{-5} M; III, 10^{-6} M

Durch radioaktive Markierung des Galactosidrestes solcher Substanzen versuchen wir, diesen ,,inneren Induktor" nachzuweisen.

Galactosylglycerin wurde vor einiger Zeit als hochwirksamer Induktor erkannt [29]. Burstein [29] fand, daß diese Substanz in einem z^- Stamm noch in einer Konzentration von 10^{-7} M voll induziert; die Induktion wurde dabei durch Messung der Thiogalactosidtransacetylase-Aktivität gemessen.

Auch im Wildstamm induziert die Substanz, wie aus Fig. 1 D und 2 D ersichtlich ist, noch in sehr geringen Konzentrationen. Bemerkenswert ist, daß nach der üblichen lag-Phase die Synthese der β-Galacto-

sidase unter Wegfall einer bei anderen Induktoren stets zu beobachtenden autokatalytischen Phase sofort maximal einsetzt, um dann auch wieder rasch abzubrechen. Die Menge der gebildeten β-Galactosidase steht dabei in einem bestimmten Verhältnis zur äußeren Konzentration des Induktors. Auf dieses ungewöhnliche Induktionsverhalten wird im Anschluß noch näher eingegangen.

Bei allen Induktoren der Fig. 1 kommt die Enzymsynthese zum Stillstand, wenn die intracelluläre Induktorkonzentration unter einen bestimmten Wert gesunken ist. Diese Grenzkonzentration wird je nach Induktor verschieden rasch erreicht, da die einzelnen

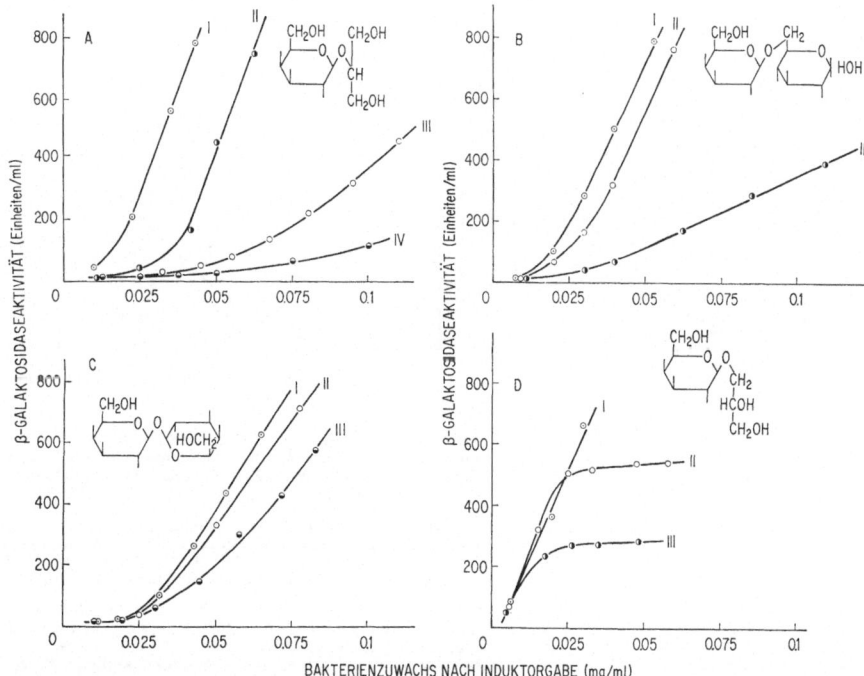

Fig. 2. *Abhängigkeit der β-Galactosidase-Aktivität in K 12 3000 während der logarithmischen Wachstumsphase von dem Zuwachs an Bakterienmasse nach Gabe des Induktors. Als Induktoren werden verwendet:* A, sek. Galactosylglycerin: I, 10^{-3} M; II, 3×10^{-4} M; III, 10^{-4} M; IV, 7×10^{-5} M. B, Allolactose: I, 10^{-4} M; II, 3×10^{-5} M; III, 10^{-5} M. C, Galactosylgalactosid: I, 5×10^{-4} M; II, 10^{-3} M; III, 10^{-4} M. D, Galactosylglycerin: I, 10^{-4} M; II, 10^{-5} M; III, 10^{-6} M

Induktoren verschieden gute Substrate der β-Galactosidase sind.

Um die Spaltungsgeschwindigkeit der metabolisierbaren Induktoren durch β-Galactosidase zu bestimmen, ließen sich die sonst üblichen Methoden nicht anwenden, und zwar aus mehreren Gründen: Einmal absorbieren die Aglyca der in Frage kommenden Galactoside nicht im sichtbaren oder ultravioletten Spektralbereich, so daß eine einfache fotometrische Bestimmung nicht möglich ist; zum anderen werden bei der enzymatischen Spaltung neben den Hydrolyseprodukten auch Transferprodukte des glyconischen Teils auf das Substrat oder dessen Spaltprodukte beobachtet. Somit liefert auch die Bestimmung der freigesetzten Galactose nach Wallenfels und Kurz [30] kein Maß für die Spaltungsgeschwindigkeit der Induktoren durch β-Galactosidase.

Wir verwendeten deshalb bei der enzymatischen Spaltung Galactoside, die im glyconischen Teil radioaktiv markiert waren und trennten die bei den Spaltansätzen nach verschiedenen Zeiten gebildeten Produkte chromatographisch auf. Durch Auszählung der Radioaktivität der gebildeten Produkte ließ sich ein einigermaßen vergleichbares Bild von der Substratqualität der einzelnen Galactoside gewinnen.

Tabelle 1. *Hydrolyse von radioaktiven Galactosiden mit β-Galactosidase aus E. coli*
Enzymatische Spaltung von radioaktiven Galactosiden mit β-Galactosidase aus *E. coli.* Spaltansatz: 200 µl Puffer (10^{-1} M Kaliumnatriumphosphat, pH [7,0, Mg^{++} 10^{-3} M) 2 mg Galactosid; 50 µl Enzymlösung, $0,54 \times 10^{-5}$ µMol β-Galactosidase. Temperatur 22°. Dem Ansatz wird nach bestimmten Zeiten 20 µl entnommen und wie in Material und Methoden weiter behandelt

	Substrat-menge	freigesetzte Galactose/ min	gebildetes Transfer-produkt	Wechselzahl
	µMol	µMol/min	µMol/min	$\dfrac{\text{Mole Produkt}}{\text{Mole Enzym} \times \text{min}}$
Allolactose	5,84	$5,18 \times 10^{-2}$	—	9590
Galactosyl-glycerin	7,80	$4,75 \times 10^{-2}$	$3,59 \times 10^{-2}$	15450
sek. Galacto-sylglycerin	7,80	$2,63 \times 10^{-2}$	$2,72 \times 10^{-2}$	9890

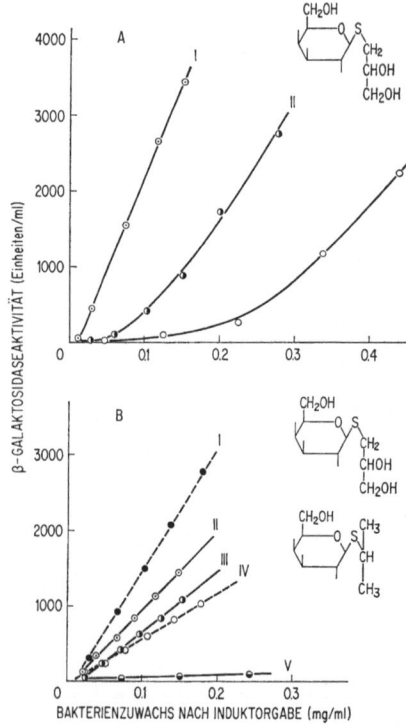

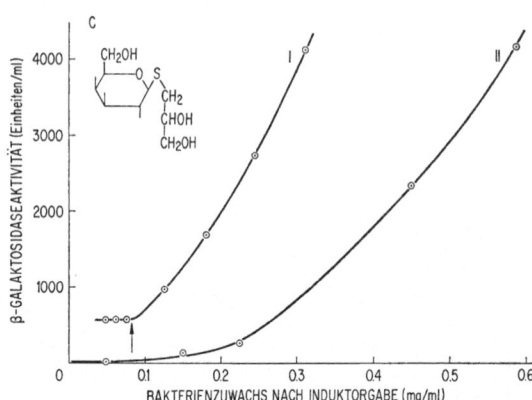

Fig. 3. A *Abhängigkeit der β-Galactosidase-Aktivität in K 12 3000 während der logarithmischen Wachstumsphase von dem Zuwachs an Bakterienmasse nach Gabe von Thiogalactosylglycerin.* Induktorkonzentrationen: I, 10^{-4} M; II, 10^{-5} M; III, 5×10^{-6} M

B *Vergleich der β-Galactosidase-Aktivität in K 12 2001$_e$ ($i^+ z^+ y^-$) während der logarithmischen Wachstumsphase nach Zugabe von IPTG bzw. Thiogalactosylglycerin.* IPTG: I, 10^{-3} M; IV, 10^{-4} M. Thiogalactosylglycerin: II, 5×10^{-3} M; III, 10^{-3} M; V, 10^{-4} M

C *Verkürzung der autokatalytischen Phase durch Vorinduktion.* I, Vorinduktion mit Galactosylglycerin 10^{-4} M bei ↑ Zugabe von 5×10^{-6} M Thiogalactosylglycerin. II, Kontrolle 5×10^{-6} M Thiogalactosylglycerin ohne Vorinduktion. Die β-Galactosidase-Aktivität ist aufgetragen gegen den Zuwachs an Bakterienmasse nach Induktorgabe

In Tabelle 1 sind die aus diesen Versuchen gemessenen Umsatzzahlen für drei Induktoren angegeben, die bei der Einwirkung von β-Galactosidase unter den Versuchsbedingungen beobachtet werden.

Induktionskinetik nicht metabolisierbarer Induktoren

Die Beteiligung der β-Galactosidase am Induktionsvorgang läßt sich eliminieren, entweder durch Induktion in einer z⁻ Mutante, oder durch Verwendung von Induktoren, die keine Substrate der β-Galactosidase sind, den sogenannten *inducteurs gratuits* [12].

Da das Galactosylglycerin noch in sehr geringen Konzentrationen als Induktor wirksam ist, war es von Interesse, wie sich das nicht metabolisierbare Thioisologe im Induktionsexperiment verhält.

Gero und Burstein [31] konnten zeigen, daß es der beste bis jetzt bekannte *inducteur gratuit* ist. Wir hatten schon früher diese Beobachtung bei der Induktion in K 12 3000 gemacht [32].

Wie man der Fig. 3 A entnehmen kann, induziert die Substanz in einer Konzentration von 5×10^{-6} M noch voll; allerdings ist die autokatalytische Phase in diesem Fall stark verlängert. Induziert man durch

Galactosylglycerin vor und gibt nach Abbruch der Enzymsynthese Thiogalactosylglycerin bis zu einer Endkonzentration von 5×10^{-6} M in das Medium, so setzt die Enzymsynthese unter Wegfall der langen autokatalytischen Phase sofort mit maximaler Synthesegeschwindigkeit ein (Fig. 3 C). Man wird wohl annehmen können, daß die durch Galactosylglycerin vorinduzierte Galactosid-Permease zum Wegfall der autokatalytischen Phase führt. In Übereinstimmung damit steht die Messung der Induktion in einem z⁺y⁻ Stamm. Hier ist Thiogalactosylglycerin ein schlechterer Induktor als IPTG (Fig. 3 B). Die hohe Empfindlichkeit im Wildstamm läßt auf eine gegenüber anderen Induktoren gesteigerte Spezifität der Lac-Permease für Thiogalactosylglycerin schließen.

Ersetzt man im Thiogalactosylglycerin die CH_2OH-Gruppe des Galactosylrestes durch CH_3 oder H und untersucht das Thiofucosyl- bzw. Thioarabinosylglycerin im Induktionsexperiment, so läßt sich keine Induktionswirkung nachweisen. Auch eine Vorinduktion der Galactosid-Permease durch Galactosylglycerin und anschließende Zugabe der beiden Thioglycoside bewirkt keine Neuinduktion der Lac-Enzyme.

Ein weiterer nicht metabolisierbarer Induktor ist 1,5-Anhydrodulcit. Diese Verbindung ist die erste im Lac-Operon induktionswirksame Substanz, die keinen intakten Galactosidrest besitzt. Auffallend ist die lange *lag*-Phase vor Beginn der Induktion (Fig. 4).

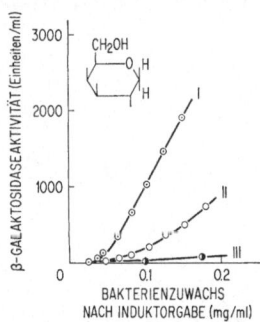

Fig. 4. *Abhängigkeit der β-Galactosidase-Aktivität in K 12 3000 während der logarithmischen Wachstumsphase von dem Zuwachs an Bakterienmasse nach Gabe von 1,5 Anhydrodulcit.* Induktorkonzentrationen: I, 10^{-2} M; II, 3×10^{-3} M; III, 10^{-3} M

Außerdem ist die Verbindung erst in relativ hoher Konzentration, über 10^{-3} M, wirksam. Wir haben die Verbindung nach Wilzbach [33] tritiiert und werden prüfen, in welchem Maß dieser Induktor in die Zelle transportiert wird.

Als weitere thioisologe Verbindung haben wir Thioallolactose hergestellt und auf ihre Induktionswirkung untersucht. Es zeigte sich, daß die Substanz selbst nur schwach induziert, jedoch ein guter Hemmstoff für die Induktion durch andere Induktoren ist. In Fig. 5 und 6 ist die Hemmwirkung der Thioallolactose auf die Induktion mit verschiedenen Induktoren dargestellt.

Die Frage, ob Thioallolactose auf der Stufe der Induktion durch Kompetition mit dem Repressor oder auf der Stufe der Permeation durch Kompetition mit der Permease wirkt, konnte durch Anwendung einer y⁻ Mutante entschieden werden. Fig. 7 A zeigt im Gegensatz zur Hemmung der Induktion im Wildstamm (Fig. 5 C), daß unter diesen Bedingungen nur bei sehr geringen Induktorkonzentrationen eine merkliche Hemmung der Induktion auftritt. Dies zeigt, daß Thioallolactose eine hohe Affinität zur Lac-Permease besitzt und die Induktion vorwiegend auf

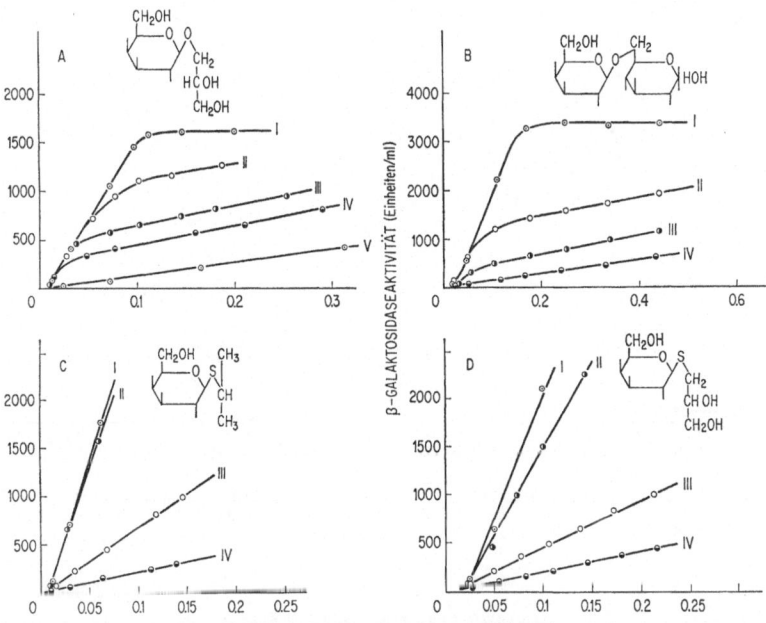

Fig. 5. *Einfluß der Thioallolactose auf die Induktion der β-Galactosidase in K 12 3000 während der logarithmischen Wachstumsphase.* Die Enzymaktivität ist aufgetragen gegen den Zuwachs an Bakterienmasse nach gleichzeitigem Hinzufügen von Thioallolactose und Induktor. A, Galactosylglycerin 10^{-4} M; B, Allolactose 10^{-4} M; C, IPTG 10^{-4} M; D, Thiogalactosylglycerin 5×10^{-5} M. Konzentrationen der Thioallolactose: I, 0 (Kontrolle ohne Thioallolactose); II, 10^{-5} M; III, 10^{-4} M; IV, 10^{-3} M. A, V: Kontrolle 10^{-3} M Thioallolactose ohne Induktor

der Stufe der Permeation hemmt. Dieser Typ von Induktionshemmung ist von Herzenberg [34] für Thiogalactosylgalactosid beschrieben worden. Wir haben daher die Hemmwirkung der Thioallolactose mit derjenigen von Thiogalactosylgalactosid verglichen. Fig. 5 D und 7 B zeigen, daß Thioallolactose bei zehnmal geringeren Konzentrationen als Thiogalactosylgalactosid in gleichem Maße die Induktionswirkung von Thiogalactosylglycerin zu reduzieren vermag. Dabei ist allerdings bemerkenswert, daß Thiogalactosylgalactosid auch im Wildstamm keine Induktion hervorruft, während Thioallolactose unter diesen Bedingungen immerhin merkbar induziert.

Auch die Induktion durch Galactosylglycerin wird durch Thioallolactose gehemmt. Es fällt aber auf, daß die Hemmung erst spät, nämlich während der Phase der konstanten Synthesegeschwindigkeit, einsetzt. Darin unterscheidet sich das Galactosylglycerin von anderen Induktoren, und sein Verhalten bei der

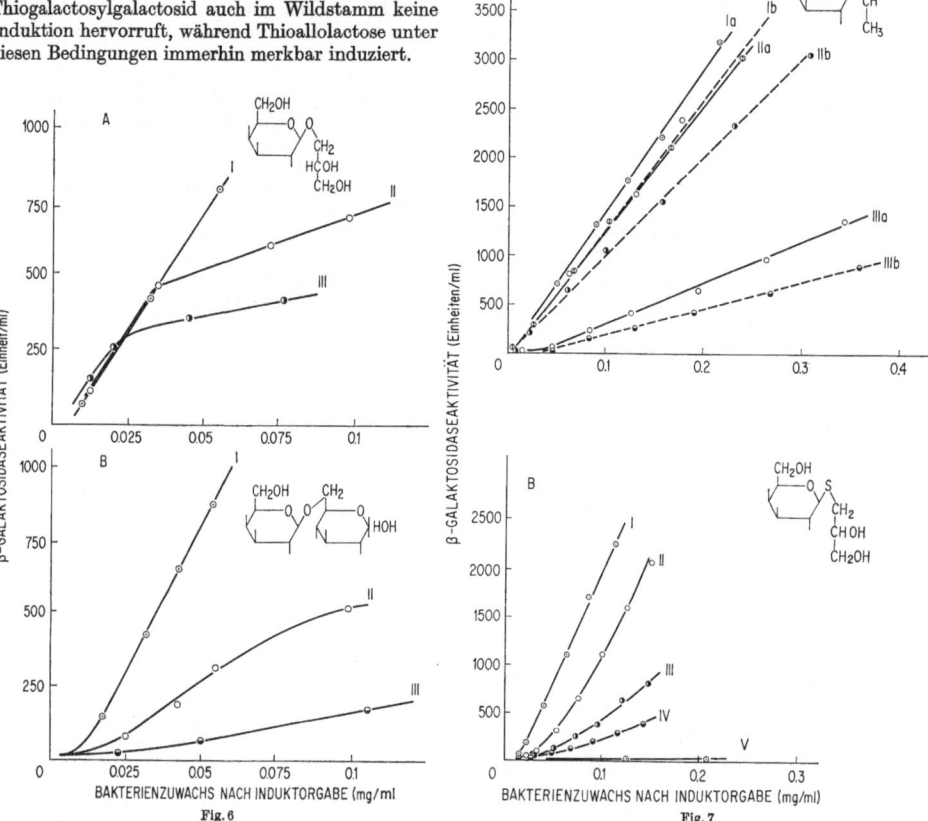

Fig. 6. *Einfluß der Thioallolactose auf die Induktion der β-Galactosidase in K 12 3000 während der logarithmischen Wachstumsphase.* Die Enzymaktivität ist aufgetragen gegen den Zuwachs an Bakterienmasse nach gleichzeitigem Hinzufügen von Thioallolactose und Induktor. A, Galactosylglycerin 10^{-4} M; B, Allolactose 10^{-4} M. Konzentration der Thioallolactose: I, 10^{-5} M; II, 10^{-4} M; III, 10^{-3} M

Fig. 7. A *Einfluß von Thioallolactose auf die Induktion der β-Galactosidase in K 12 2001$_e$ ($i^+z^+y^-$) während der logarithmischen Wachstumsphase.* Die Enzymaktivität ist aufgetragen gegen den Zuwachs an Bakterienmasse nach dem gleichzeitigen Hinzufügen von Thioallolactose und IPTG. Thioallolactosekonzentration 10^{-3} M. IPTG-Konzentration Ib, 5×10^{-4} M; IIb, 2×10^{-4} M; IIIb, 8×10^{-5} M. Kontrolle, ohne Thioallolactose: IPTG-Konzentration Ia, 5×10^{-4} M; IIa, 2×10^{-4} M; IIIa, 8×10^{-5} M. B *Einfluß von Thiogalactosylgalactosid auf die Induktion der β-Galactosidase in K 12 3000 während der logarithmischen Wachstumsphase.* Aufgetragen ist die Enzymaktivität gegen den Zuwachs an Bakterienmasse nach dem gleichzeitigen Hinzufügen von Thiogalactosylgalactosid und Thiogalactosylglycerin. Induktorkonzentration (Thiogalactosylglycerin): 5×10^{-5} M. Thiogalactosylgalactosidkonzentration: I, 0 (Kontrolle); II, 10^{-4} M; III, 10^{-3} M; IV, 10^{-2} M; V, 10^{-3} M (Kontrolle ohne Thiogalactosylglycerin)

Hemmung durch Thioallolactose erinnert an die oben beschriebene, ungewöhnliche Anfangskinetik im Induktionsexperiment (Fig.2D). Das für Galactosylglycerin charakteristische sofortige Einsetzen der maximalen Synthesegeschwindigkeit wird also durch Thioallolactose nicht gehemmt, während die Hemmung im Verlauf der späteren Phase wie bei anderen Induktoren beobachtet wird. Bei diesen setzt die Hemmung gleich bei Beginn der Induktion ein, wie die Einwirkung von Thioallolactose auf die Induktion mit Allolactose zeigt (Fig.5B, 6B). Dies läßt ver-

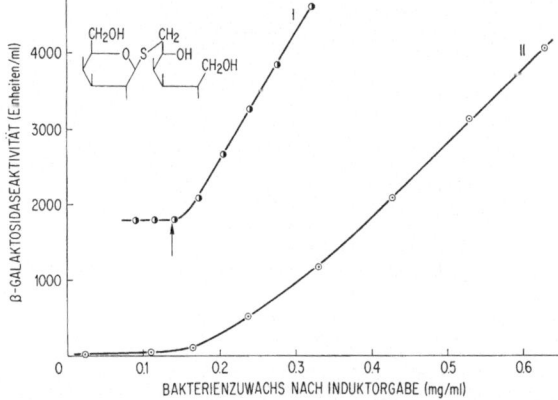

Fig.8. *Verkürzung der autokatalytischen Phase durch Vorinduktion.* I, Vorinduktion mit Galactosylglycerin 10^{-4} M, bei ↑ Zugabe von Thioallolactit 10^{-3} M. II, Kontrolle: 10^{-3} M Thioallolactit ohne Vorinduktion. Die β-Galactosidase-Aktivität ist aufgetragen gegen den Zuwachs an Bakterienmasse nach Induktorgabe

muten, daß Galactosylglycerin nicht wie die anderen Induktoren erst durch die Lac-Permease transportiert wird, deren Bildung es induziert, sondern auch von einer anderen Permease, die nicht zu den Lac-Faktoren gehört.

Durch eine einfache chemische Veränderung läßt sich Galactosylglycerin in einen normalen *inducteur gratuit* verwandeln: Reduziert man die Substanz mit NaBH$_4$, so wird durch Öffnung des Glucopyranoserings Thioallolactit gebildet. Diese Verbindung induziert in K 12 3000 mit allerdings stark verlängerter *lag*- bzw. autokatalytischer Phase (Fig.8, Kurve II). Offensichtlich wird auch diese Substanz von der Lac-Permease schlecht in die Zelle transportiert. Induziert man die Permease durch Galactosylglycerin vor, so wirkt auch Thioallolactit unter Fortfall der autokatalytischen Phase sofort maximal induzierend (Fig.8, Kurve I). Bei Verwendung von Thioallolactose läßt sich demgegenüber die beschriebene geringe Induktionswirkung durch eine solche Vorinduktion nicht steigern.

Permeation des Galactosylglycerins

Die oben erwähnte Beobachtung, daß die Hemmung der Induktion durch Galactosylglycerin mit Thioallolactose erst relativ spät einsetzt, legte die Vermutung nahe, daß Galactosylglycerin schon vor Beginn der Induktion auch ohne die Lac-Permease in die Zelle gelangt.

In Fig.9A wird die Wirkung von Galactose auf die Induktionskinetik des Galactosylglycerins in K 12 3000 gezeigt. Man sieht, daß unter diesen Bedingungen der Induktor in Konzentrationen unter 10^{-5} M nicht mehr induziert und daß bei höheren Konzentrationen die Anfangskinetik der Induktion dem Verhalten eines normalen Induktors entspricht: Die Induktion weist nun eine autokatalytische Phase auf, während sie ohne Galactosezusatz gleich mit der konstanten Synthesegeschwindigkeit beginnt (Fig.1D). Dies kann nicht auf einer Katabolitrepression durch Galactose beruhen. Die Induktion mit IPTG wird nämlich unter den gleichen Bedingungen durch Galactose nicht gehemmt (Fig.9B). Entsprechend fanden Loomis und Magasanik [35] für diesen Stamm keine Repression durch Galactose. Um die Akkumulierung des Galactosylglycerins in Gegenwart und Abwesenheit von Galactose zu messen, haben wir mit radioaktivem Galactosylglycerin in K 12 3000 induziert und die Radioaktivität der nach bestimmten Zeiten abfiltrierten Zellen gemessen. Tabelle 2 zeigt, daß in Gegenwart von 10^{-3} M Galactose die

Tabelle 2. *Einfluß der Galactose auf die Permeation des Galactosylglycerins*
Einfluß der Galactose auf die Permeation des Galactosylglycerins in K 12 3000. Von einer Bakterienkultur der angegebenen Dichte, die logarithmisch in Medium 63 mit Glycerin als C-Quelle wächst, werden 1,5 ml zu radioaktivem Galactosyl-glycerin gegeben (Endkonzentration: $1,2 \times 10^{-5}$ M und $5,9 \times 10^4$ Imp/min/ml bzw. $2,7 \times 10^{-6}$ M und $4,1 \times 10^4$ Imp/min/ml). Die Kultur wird bei 37° geschüttelt. Nach 10 min wird 1 ml entnommen und durch Millipore-Filter filtriert

Extinktion der Zellsuspension bei 578 mµ	Galactose-konzentration	Galactosylglycerin-konzentration	Radio-aktivität
	M	M	Imp./min
0,224	0	$1,2 \times 10^{-5}$	4261
0,224	10^{-3} M	$1,2 \times 10^{-5}$	323
0,111	0	$1,2 \times 10^{-5}$	939
0,111	10^{-3} M	$1,2 \times 10^{-5}$	71
0,245	0	$2,7 \times 10^{-6}$	594
0,245	10^{-3} M	$2,7 \times 10^{-6}$	53
0,284	0	$2,7 \times 10^{-6}$	907
0,284	10^{-3} M	$2,7 \times 10^{-6}$	82

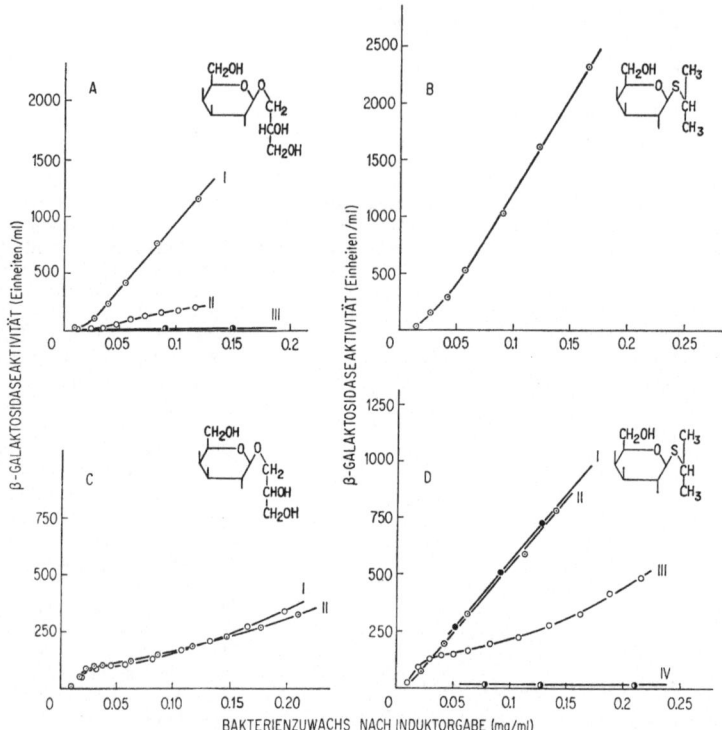

Fig. 9. A *Einfluß von Galactose auf die Induktion der β-Galactosidase in K 12 3000 während der logarithmischen Wachstumsphase.* Die Enzymaktivität ist aufgetragen gegen den Zuwachs an Bakterienmasse nach dem gleichzeitigen Hinzufügen von Galactosylglycerin und Galactose. Galactosylglycerinkonzentration: I, 5×10^{-5} M; II, 10^{-5} M; III, 10^{-6} M. Galactosekonzentration: I, 5×10^{-3} M; II, 10^{-3} M; III, 10^{-3} M

B *Einfluß von Galactose auf die Induktion der β-Galactosidase in K 12 3000 während der logarithmischen Wachstumsphase.* Die Enzymaktivität ist aufgetragen gegen den Zuwachs an Bakterienmasse nach dem gleichzeitigen Hinzufügen von IPTG 5×10^{-5} M und Galactose 5×10^{-3} M

C *Abhängigkeit der β-Galactosidase-Aktivität in K 12 2001$_e$ (i$^+$z$^+$y$^-$) während der logarithmischen Wachstumsphase vom Zuwachs der Bakterienmasse nach Hinzufügen von Galactosylglycerin.* Induktorkonzentration: I, 5×10^{-5} M; II, 5×10^{-4} M

D *Einfluß von Galactose auf die Induktion der β-Galactosidase in K 12 2001$_e$ (i$^+$z$^+$y$^-$) während der logarithmischen Wachstumsphase nach gleichzeitigem Hinzufügen von Induktor und Galactose.* Die Enzymaktivität ist aufgetragen gegen den Zuwachs an Bakterienmasse nach dem gleichzeitigen Hinzufügen von Galactose und Induktor. I, IPTG 2×10^{-4} M (Kontrolle ohne Galactose); II, IPTG 2×10^{-4} M + 5×10^{-3} M Galactose; III, Galactosylglycerin 2×10^{-4} M (Kontrolle ohne Galactose); IV, Galactosylglycerin 2×10^{-4} M + 5×10^{-3} M Galactose

Akkumulierung um mehr als 90⁰/₀ vermindert wird. Allerdings bewirkt Galactose auch eine Permeationshemmung anderer Galactoside [36], doch offenbar in weit geringerem Maße, als dies bei Galactosylglycerin der Fall ist. Auch im y⁻ Stamm wird die Induktion mit Galactosylglycerin durch Galactose völlig gehemmt, während die Induktion mit IPTG unter diesen Bedingungen durch Galactose keine Beeinflussung erfährt (Fig. 9 D).

Die Induktionskinetik des Galactosylglycerins im y⁻ Stamm zeigt Fig. 9 C. Auffallend ist hier, daß das Abbrechen der Induktion unabhängig von der Induktorkonzentration nach derselben kurzen Zeit erfolgt, also auch dann, wenn die im Medium vorhandene Konzentration an Galactosylglycerin ausreicht, um weiter induzieren zu können. Die Menge der gebildeten β-Galactosidase ist bei vergleichbaren Induktorkonzentrationen wesentlich kleiner als im Wildstamm. Es ist bemerkenswert, daß die Induktion nach ihrem Abbruch allmählich wieder einsetzt.

Die Induktionskinetik des Galactosylglycerins im y⁻ Stamm ist der im Wildstamm bei Gegenwart von

Thioallolactose ähnlich, wo durch die Blockierung der Galactosid-Permease die gleichen Permeationsverhältnisse gegeben sind (Fig.5A).

DISKUSSION

Die Spezifität der Induktion im Lac-Abschnitt des Coli-Chromosoms sollte auf die Induktor-Repressor-Wechselwirkung zurückzuführen sein. Man könnte sie dann als Spezifität des Repressorproteins für galactosidische Strukturen behandeln und mit der Spezifität der Strukturgenprodukte β-Galactosidase, Galactosidpermease und Galactosidtransacetylase vergleichen. Aus den in dieser Arbeit angeführten Beispielen läßt sich erkennen, wie schwierig es ist, eine allgemein gültige Skala der Induktionswirksamkeit aufzustellen, die für verschiedene Induktoren anwendbar ist.

Da das Ausmaß der Induktion nicht nur von der Spezifität des Repressors zum Induktor, sondern auch von seiner intracellulären Konzentration abhängt, ist es unerläßlich, die genauen Verhältnisse in der Bakterienzelle zu kennen. Stets ist im Induktionsexperiment die Permeation des Induktors durch die Zellmembran der Bindung an den Repressor vorgelagert. Bei hoher Wechselzahl der Permease kann eine gute Induktorqualität vorgetäuscht werden. So induziert Galactosylglycerin in geringen Konzentrationen nicht deshalb, weil der Repressor eine ungewöhnlich hohe Affinität zu dieser Substanz hat, sondern weil sie durch eine gute Permeation in der Zelle stark angereichert wird.

Umgekehrt läßt sich mit Thioallolactit die Enzymsynthese nur schlecht induzieren. Sorgt man aber durch Vorinduktion der Lac-Permease für einen besseren Transport dieses Induktors in die Zelle, so resultiert nach Zugabe derselben Substanz eine sofortige maximale Enzymsynthese. Man kann daher Thioallolactit allein aus seinem Induktionsverhalten nicht als schlechten Induktor klassifizieren, ohne seine wirksame Konzentration am Induktionsort zu kennen.

Eine weitere Schwierigkeit besteht darin, daß eine für den Repressor unspezifische Substanz innerhalb der Zelle in einen Induktor umgewandelt werden kann, wie man das für einige Galactoside, darunter die Lactose, nachgewiesen hat. In diesem speziellen Fall ist die β-Galactosidase für die Umwandlung verantwortlich zu machen [10,14].

Ein ähnliches Verhalten zeigt auch das Galactosylgalactosid. Nach der beobachteten Induktionskinetik ist es wahrscheinlich, daß diese Substanz wie Lactose in einen „inneren Induktor" umgewandelt wird. Die thioisologe Verbindung, die durch β-Galactosidase nicht gespalten wird, wirkt nicht induzierend, sondern ist ein Hemmstoff für die Lac-Permease [33].

Ob außer der Transgalactosidierung noch andere Reaktionen an der Umwandlung einer Substanz in

einen wirksamen Induktor beteiligt sind, ist damit noch nicht ausgeschlossen.

In diesem Zusammenhang ist die Induktion mit 1,5-Anhydrodulcit zu diskutieren. Die Wirkungsweise dieses bis jetzt einfachsten Induktors zeigt sich in einer, selbst bei hohen Induktorkonzentrationen, langen lag- wie autokatalytischen Phase. Dies kann einmal auf eine schlechte Permeation zurückzuführen sein, zum anderen besteht die Möglichkeit, daß der Induktor erst durch eine noch unbekannte Reaktion in die eigentlich induzierende Substanz umgewandelt wird.

Ein Vergleich von O- und S-isologen Verbindungen läßt erkennen, daß Thiogalactoside immer dann gute Induktoren sind, wenn auch das O-Isologe induziert. Das ist bei Methyl-, Butyl- und Glyceryl-galactosiden der Fall [11]. Merkwürdigerweise läßt sich Thioallolactose in diese Analogie nicht einbeziehen. Sie ist im Gegensatz zu Allolactose nur ein schwacher Induktor, hemmt jedoch die durch andere Galactoside hervorgerufene Induktion auf der Stufe der Permeation.

Eingehende Induktionsversuche mit Galactosylglycerin legen die Deutung nahe, daß diese Substanz bereits vor dem Erscheinen der Lac-Enzyme durch eine fremde Permease transportiert wird. Diese hat Affinität für Galactose und ist verantwortlich für die Induktionswirksamkeit des Galactosylglycerins bei Konzentrationen unter 10^{-5} M. Wird der Transport des Galactosylglycerins über diese Permease durch Galactose gehemmt, so zeigt die Substanz normales Induktionsverhalten wie z.B. die Allolactose. Im Lac-Permease-defekten Stamm wird durch diese Permease ein konstanter Transport von Galactosylglycerin in die Zelle bewirkt, der im untersuchten Bereich von der äußeren Konzentration unabhängig ist. Dieser Transport führt zu einer Erhöhung der Induktorkonzentration im Innern der Zelle. Auf diese Weise wird der Schwellenwert für die Reaktion mit dem Repressor nach wesentlich kürzerer Zeit erreicht und die induzierte Enzymsynthese beginnt sofort nach Zusatz des Induktors. Die induzierte β-Galactosidase kann nun die Konzentration des Induktors durch Spaltung verringern. Im Lac-Permease positiven Stamm wird der Transport des Induktors in die Zelle durch die neu gebildete Lac-Permease nochmals verstärkt, so daß die β-Galactosidase hier erst nach vergleichsweise längerer Zeit die Konzentration des Induktors unter den für die Induktion wirksamen Wert herabsetzen kann.

Zu erklären bleibt, weshalb beim Induktionsexperiment im y⁻ Stamm nach dem ersten Abbruch der Induktion diese allmählich wieder einsetzt (Fig.9C). Wir nehmen an, daß während der Spaltung des Galactosylglycerins ein oder mehrere Transferprodukte im Innern der Zelle gebildet werden, die für die erneute Induktion verantwortlich zu machen sind. Die chemische Natur solcher „innerer Induktoren" bleibt aufzuklären.

Das besondere Induktionsverhalten des Galactosylglycerins ist offenbar hoch spezifisch. Schon das sekundäre Galactosylglycerin wirkt wie ein normaler Induktor.

Auch der Ersatz des galactosidischen Brückensauerstoffs durch Schwefel macht die außergewöhnliche Permeation unmöglich. Thiogalactosylglycerin ist im Wildstamm zwar auch in geringen Konzentrationen ein wirksamer Induktor, zeigt jedoch bei diesen Konzentrationen eine lange autokatalytische Phase. Die Induktorwirksamkeit geht völlig verloren, wenn man im Thiogalactosylglycerin die CH_2OH Gruppe des Galactosylrestes durch CH_3 bzw. H ersetzt. Thiofucosylglycerin und Thioarabinosylglycerin sind also keine Induktoren der Lac-Enzyme.

Der Deutschen Forschungsgemeinschaft und dem Fonds der Chemischen Industrie danken wir für die Unterstützung unserer Arbeiten.

LITERATUR

1. Wallenfels, K., und Malhotra, O. P., *Advan. Carbohydrate Chem.* 16 (1961) 239.
2. Rickenberg, H. V., Cohen, G. N., Buttin, G., und Monod, J., *Ann. Inst. Pasteur*, 91 (1956) 829.
3. Fox, C. F., und Kennedy, E. P., *Proc. Natl. Acad. Sci. U. S.* 54 (1965) 891.
4. Zabin, J., Kepes, A., und Monod, J., *J. Biol. Chem.* 237 (1962) 253.
5. Bourgeois, S., Cohn, M., und Orgel, L. E., *J. Mol. Biol.* 14 (1965) 300.
6. Müller-Hill, B., *J. Mol. Biol.* 15 (1966) 374.
7. Monod, J., Changeux, J. P., und Jacob, F., *J. Mol. Biol.* 6 (1963) 306.
8. Jacob, F., und Monod, J., *Bull. Soc. Chim. Biol.* 46 (1964) 1499.
9. Jacob, F., und Monod, J., *J. Mol. Biol.* 3 (1961) 318.
10. Müller-Hill, B., Rickenberg, H. V., und Wallenfels, K., *J. Mol. Biol.* 10 (1964) 303.
11. Monod, J., Cohen-Bazire, G., und Cohn, M., *Biochim. Biophys. Acta*, 7 (1951) 585.
12. Monod, J., und Cohn, M., *Advan. Enzymol.* 13 (1952) 67,
13. Fox, C. F., Beckwith, J. R., Epstein, W., und Signer, E. R., *J. Mol. Biol.* 19 (1966) 576.

14. Burstein, C., Cohn, M., Kepes, A., und Monod, J., *Biochim. Biophys. Acta*, 95 (1965) 634.
15. Boos, W., Lehmann, J., und Wallenfels, K., *Carbohydrate Res.* 1 (1966) 419.
16. Austin, P. W., Hardy, F. E., Buchanan, J. G., und Baddiley, J., *J. Chem. Soc.* (1965) 1419.
17. Kuhn, R., Baer, H. H., und Gauhe, A., *Chem. Ber.* 88 (1955) 1713.
18. Bredereck, H., Höschele, G., und Ruck, K., *Chem. Ber.* 86 (1953) 1277.
19. Schneider, W., Sepp, J., und Stiehler, O., *Chem. Ber.* 51 (1918) 220.
20. Ness, R. K., Fletcher Jr., H. G., und Hudson, C. S., *J. Am. Chem. Soc.* 72 (1950) 4547.
21. Čzerný, M., Staněk, J., und Pacák, J., *Monatsh. Chem.* 94 (1963) 290.
22. Hardegger, E., und Montavon, R. M., *Helv. Chim. Acta*, 29 (1946) 1199.
23. Schneider, W., und Beuther, A., *Chem. Ber.* 52 (1919) 2147.
24. Boos, W., Lehmann, J., und Wallenfels, K., in Vorbereitung.
25. Pardee, A. B., und Prestidge, L. S., *Biochim. Biophys. Acta*, 49 (1961) 77.
26. Kepes, A., *Biochim. Biophys. Acta*, 76 (1963) 293.
27. Boezi, J. A., und Cowie, D. B., *Biophys. J.* 1 (1961) 639.
28. Kepes, A., und Beguin, S., *Biochim. Biophys. Acta*, 123 (1966) 546.
29. Burstein, C., Dissertation, Paris, Institut Pasteur (1964).
30. Wallenfels, K., und Kurz, G., *Biochem. Z.* 335 (1962) 559.
31. Gero, S. D., und Burstein, C., *Biochim. Biophys. Acta*, 117 (1966) 314.
32. Boos, W., Diplomarbeit, Freiburg, Chemisches Laboratorium der Universität (1965).
33. Wilzbach, K. E., *J. Am. Chem. Soc.* 79 (1957) 1013.
34. Herzenberg, L. A., *Biochim. Biophys. Acta*, 31 (1959) 525.
35. Loomis, W. F. Jr, und Magasanik, B., *J. Bacteriol.* 92 (1966) 170.
36. Koch, A. L., *Biochim. Biophys. Acta*, 79 (1964) 177.
37. Fletcher, H. G. Jr., und Hudson, C. S., *J. Am. Chem. Soc.* 70 (1948) 310.

W. Boos, P. Schaedel und K. Wallenfels
Chemisches Laboratorium der Universität Freiburg i. Br.
78 Freiburg i. Br., Albertstraße 21, Deutschland

European J. Biochem. 1 (1967) 395—399

Mechanism of Ribulose-Diphosphate Carboxydismutase Reaction

F. Fiedler, G. Müllhofer, A. Trebst, and I. A. Rose

The Institute for Cancer Research Philadelphia, Pennsylvania 19111, and
Abteilung Biochemie der Pflanze, Pflanzenphysiologisches Institut der Universität Göttingen

(Received February 11, 1967)

Ribulose-1,5-P_2 labelled with tritium in the C-3 position was converted to 2 moles of phosphoglyceric acid by the carboxydismutase. Over 98 % of the radioactivity was found in the water, less than 0.1 % being in phosphoglyceric acid, indicating that the proton lost from the 3 position of the substrate does not contribute to the formation of product. The tritiated substrate reacts at only 20 % the rate of the normal carrier species. Neither substrate nor product show significant proton exchange with the medium due to the enzyme.

Ribulose-1,5-P_2 carboxydismutase catalyzes the reaction:

$$
\begin{array}{l}
H_2COP \\
| \\
C=O \\
| \\
HCOH \\
| \\
HC-OH \\
| \\
H_2COP
\end{array}
+ CO_2 \rightarrow
\begin{array}{l}
CO_2H \\
| \\
2\ HCOH \\
| \\
H_2COP
\end{array}
$$

Recent experiments have shown that the position of C-C bond cleavage is between C_2 and C_3 of the substrate [1]. The form of the mechanism that is most often suggested [2,3] is one in which an enolization is followed by addition of CO_2 and a shift of the double bond to form a β-keto acid 6 carbon intermediate which is cleaved "hydrolytically". Whereas

the enzyme. Examples of such tight binding of reaction protons derived from an enolized substrate have been found in enzymatic isomerization reactions [6,7].

Earlier experiments have shown that a proton from the medium becomes attached to C-2 of one of the phosphoglyceric acids [1]. The existing data indicate that when the net reaction is catalyzed in ^3HOH [4,5] much less than 1 tritium is incorporated per 2 moles of 3-phosphoglyceric acid. That this discrepancy may be partly the result of an isotope effect has been suggested [4,5] and confirmed indirectly [1] when in experiments in ^2H_2O the product behaved kinetically as though it was substantially deuterated. However, the possibility that the hydrogen derived from C-3 of ribulose-P_2 might be stabilized on the enzyme as a proton and be transferred

enzymatic enolization type reactions are often distinguished by isotope exchange with medium protons, the ribulose-P_2 acquires little or no tritium when incubated in ^3HOH with enzyme in the absence of CO_2 [4,5]. On the other hand, if step 1 of the proposed sequence were occurring the proton might be prevented from exchanging by its tight binding to

Enzyme. Ribulosediphosphate carboxydismutase (EC 4.1.1.39); lactic dehydrogenase (EC 1.1.1.27); glyceraldehydephosphate dehydrogenase (EC 1.2.1.12); glycerolphosphate dehydrogenase (EC 1.1.99.5); pyruvate kinase (EC 2.7.1.40); phosphoglycerate kinase (EC 2.7.1.31); phosphoglycerate mutase (EC 5.4.2.1); phosphoglycerate enolase (EC 4.2.1.11).

in part to the developing 2 position of phosphoglyceric acid can not be ruled out by these considerations and requires direct testing with [3C-^3H]ribulose-P_2. The fact that neither the substrate nor product [5] undergoes rapid exchange of hydrogen with the medium makes a test of such a direct transfer possible.

MATERIALS AND METHODS
Preparation of [3C-^3H]ribulose-P_2

The procedure of choice for the unique tritium labelling of ribulose-1,5-P_2 at C-3 involved the labelling of D-glyceraldehyde-3-P at C-1 by reduction of 1,3-P_2-glyceric acid with DPN^{3}H enzymatically. The alternative procedure [8] in which specifically labelled dihydroxyacetone-phosphate was prepared by reaction with aldolase in ^{3}HOH was found to give rise to some non-specific exchange at other positions that would have decreased the sensitivity of the tracer experiment. The [1C-^{3}H]glyceraldehyde-P was converted to [4C-^{3}H]glucose-6-P which was reacted with the two dehydrogenases to produce [3C-^{3}H]ribulose-5-P. This was then phosphorylated in the 1 position by the specific kinase.

DPN^{3}H was prepared by reaction of glycerol kinase and glycerol-P dehydrogenase with 3-μmoles of [2C-^{3}H]glycerol (6×10^7 counts/min) and DPN in hydrazine (1 M) glycine (0.2 M) buffer, pH 9.8, and isolated by elution from a small column of DEAE-cellulose [9]. The reaction with α-glycerolphosphate dehydrogenase leads to a DPN^{3}H carrying the ^{3}H label on the side of the ring which is also attacked by glyceraldehydephosphate dehydrogenase [10].

For the preparation of [4C-^{3}H]fructose-1,6-P_2 there were incubated in a volume of 6.0 ml at room temperature in μmoles: Tris-HCl buffer, pH 8.0, 200; MgCl$_2$, 20; EDTA, 5; GSH, 10; 3-P-glyceric acid, 10; ATP, 10; dihydroxyacetone-P, 9; DPN^{3}H (appropriately diluted with unlabeled DPNH), 2.3; 3-P-glycerate kinase, 5.3 units; and glyceraldehyde-P dehydrogenase, 7 units. The reaction was followed by the decrease in absorbance at 366 mμ as a measure of DPNH oxidation. After 12 minutes there were added 1.8 units of aldolase. At 30 minutes, when the reaction had come to a stop, 0.1 ml of glacial acetic acid was added and the mixture was heated for 2 minutes on the boiling water bath in order to inactivate the enzymes. After cooling, the pH was adjusted to 5.7 with NaOH. The incubation mixture contained 1.4 μmoles of fructose-1,6-P_2 as measured in an optical test.

The fructose-1,6-P_2 was converted to ribulose-1,5-P_2 as before [1] by a series of specific enzymatic reactions.

To prove the location of tritium in the ribulose-P_2 a sample was degraded according to the specifications of Simon and Steffens [11] for the ^{14}C analysis of pentose after enzymatic dephosphorylation and dilution with 10 mmoles of unlabeled ribose. The specific activities of the (recrystallized) compounds are shown in Table 1. The somewhat lower activities of the osazone and the triazolealdehyde, respectively, may have their causes in some decomposition of the [3C-^{3}H]ribulose. There is no loss of activity in oxidizing the osazone to the aldehyde, showing the absence of tritium at C-4 and C-5 of ribulose-P_2. C-1 contains practically no tritium and practically all the radioactivity of the ribosazone must be located on C-3.

[2C-^{3}H]3-P-glyceric acid was prepared by the reaction of P-glyceric acid with enolase and P-gly-

cerate mutase in tritiated water: a mixture of 50 μmoles Tris-Cl, pH 8.0; 5 μmoles MgCl$_2$; 0.01 μmoles 2,3-P_2-glyceric acid; 0.3 unit enolase and 0.2 unit mutase; and 10 μmoles of 3-P-glyceric acid was incubated for 3 hours at 37° in tritiated water, 150,000 counts/min/μatom of H. The addition of HgNO$_3$ destroyed both the P-enolpyruvate and enzymes, and the mixture of 2- and 3-P-glyceric acids was isolated by ion exchange on Dowex-1-Cl [1]. The 2-P-glyceric acid was removed enzymatically by reaction with enolase, pyruvate kinase, and lactic dehydrogenase. The 3-P-glyceric acid remaining was found to lose 93% of its radioactivity to water upon the addition of mutase which allowed the 3-P-glyceric acid to be converted to P-enolpyruvate. The recovery was 1.5 μmoles of 3-P-glyceric acid with a specific activity of 100,000.

Table 1. *Degradation of the [3 C-^{3}H]ribulose-P_2*

Compound	Positions measured for tritium	Specific activities	Measured position
		counts/min/mmole	%
ribulose + ribose carrier	C-1 to C-5	155,000	100.0
ribosazone	C-3 to C-5, and ½ on C-1	143,000	92.0
2-phenyl-triazole-aldehyde	C-3, and ½ on C-1	147,000	95.0
2-phenyl-triazole-carboxylic acid	½ on C-1	550	0.35

The enzymes used in the preparation were obtained commercially except the ribulose-5-P kinase [12] and the carboxydismutase [13]. All enzyme activities are given in units which correspond to micromoles of product formed per minute.

In determining the volatile tritium radioactivity present in a sample of [3C-^{3}H]ribulose-P_2, the ribulose-P_2 was reduced at pH 8 with a large excess of NaBH$_4$. This procedure was necessary to prevent the erratic results produced by the exchange of tritium during the sublimation of the water.

RESULTS

Reaction of [3C-^{3}H]Ribulose-P_2 with CO$_2$ and Ribulose-P_2-carboxydismutase

Ribulose-P_2-carboxydismutase (1.5 mg protein; 1.5 units) was incubated at room temperature for 2 minutes in a volume of 1.6 ml containing in μmoles: Tris-HCl buffer, pH 8.0, 100; MgCl$_2$, 80; EDTA, 8; GSH, 8; and NaHCO$_3$, 80. Then were added 6.5 ml of a solution containing 4 μmoles of [3C-^{3}H]ribulose-P_2 (137,000 counts/min/μmole) with 980 μmoles Tris-HCl buffer, pH 8.0. After 2 hours reaction time at room temperature, a sample of 0.5 ml was taken to dryness. The residue was dissolved in 1 ml of water and again lyophilized, and this procedure was repeated once. The sublimed water was collected and it accounted for 98.3% of the original radioactivity. Thus the maximum amount of tritium from the ribulose-P_2 that could be present in unused substrate, starting impurity, or possibly in 3-P-glyceric acid was only 1.7%. The phosphoglyceric acid was recovered from a column of Dowex-1-Cl$^-$ (0.75 × 15 cm) by elution with 0.05 N HCl following treatment of the column with 0.0125 and 0.025 N HCl (30 ml each). Phosphoglyceric acid (6 μmoles) containing 360 counts/min was obtained. In order to determine how much radioactivity was present in the C-2 position, a part of this phosphoglyceric acid (0.8 μmole) was converted to lactic acid by the combined reaction of phosphoglycerate mutase, enolase, pyruvate kinase and lactic dehydrogenase in the presence of added ADP, DPNH, 2,3-P_2-glyceric acid, and Mg^{++}. The progress of the formation of lactate was followed to completion spectrophotometrically. The solution was taken to dryness and the volatile counts were taken as those present in the C-2 position of 3-P-glyceric acid which would be labilized in the enolase reaction. It was found that only one-third of the counts occupied this position. Thus the specific activity of phosphoglyceric acid would be 40 counts/min per 2 μmoles of 3-P-glyceric acid at C-2, or 0.03% of that of the starting substrate. This value, while higher than could be expected from incorporation of a proton via the medium, which attained a specific activity of about 0.6 counts/min/μg atom of hydrogen at the end of the reaction, indicates that the proton removed from C-3 and the one added to the developing C-2 position of phosphoglyceric acid are not mixing in a common pool to any significant extent. Since all other hydrogens of ribulose-P_2 would be represented in the products in stable positions, the ribulose-P_2 must not have been labelled significantly in any position other than C-3.

Concerning Exchange Between Phosphoglyceric Acid and ^3HOH

The above conclusion is valid only if 3-P-glyceric acid containing ^3H at C-2 is not subject to enzyme-catalyzed exchange with the medium during the reaction. Earlier experiments imply that such an exchange is not significant [5]. To examine this possibility [2C-^{3}H]P-glyceric acid (0.4 mM at 100,000 counts/min/μmole) was incubated for 45 minutes in 1 ml of solution containing triethanolamine Cl, pH 8, 0.1 M; MgCl$_2$, 0.01 M; EDTA, 1 mM; glutathione, 0.3 mM; Na$_2$CO$_3$, 0.03 M; and 0.4 unit of carboxydismutase. Samples (0.15 ml) of solution were taken at 0 min and 45 min and the water recovered by sublimation was found to be without significant activity in both cases. A sample taken at

45 min was assayed for enzymatic activity by the addition of [3C-^{3}H]ribulose-P_2, and the rate of appearance of tritium in water used as a measure of reaction rate. The rate measured in this way was significantly lower than expected. That this discrepancy might be due to an isotope effect was next examined.

Discrimination Against [3C-^{3}H]Ribulose-P_2 in the Dismutase Reaction

In order to determine the effect of tritium substitution at C-3 of ribulose-diphosphate on the rate of its reaction, the rate of tritium appearance in the medium was compared with the rate of $^{14}CO_2$ fixation in a single incubation. In addition, the experiments were done in either H$_2$O or ^{2}H$_2$O to determine whether an effect on net rate could be observed. This last possibility is suggested by the earlier observation of discrimination between medium hydrogen isotopes [4,5].

Two solutions of 2.67 ml volume in H$_2$O or ^{2}H$_2$O contained: [3C-^{3}H]ribulose-P_2 (3.7 μmoles at 44,500 counts/min/μmole), Tris-Cl, pH 8 (250 μmoles), MgCl$_2$ (20 μmoles), EDTA (5 μmoles), glutathione (10 μmoles), and NaH^{14}CO$_3$ (60 μmoles at 1.25 × 10$^-$ counts/min/μmole). The final pH's of both incubations were 8.1 by adjustment with the glass electrode. The reactions at 25° were begun by the addition of 0.02 ml of pre-activated enzyme containing 0.036 unit. Samples of 0.3 ml were combined with 0.1 ml of 1 N HCl to stop the reaction and to eliminate unreacted carbonate. A sample of 0.1 ml of this was removed for ^{14}C determination and the remainder was treated with NaBH$_4$ in Tris buffer, pH 8, in order to reduce the keto group of the remaining substrate. This was necessary to prevent the labilization of ^{3}H during the subsequent sublimation of water. The results obtained for the two incubations are given in Table 2 as ^{14}C fixed or ^{3}H labilized in the entire incubation at the times of sampling. The value for tritium obtained at the time ^{14}C fixation was no longer changing was taken to give the 100% value for ^{3}H labilization and the isotope effect was calculated by using ^{14}C fixation as a measure of the extent of the reaction of the II species of ribulose-P_2 and ^{3}H labilization as a measure of the reaction of the ^{3}H species. It is evident from these data that the tritiated species of ribulose-P_2 reacts much less rapidly, by a factor of 4—6, than the normal species. This value decreases with time unexpectedly. It is possible that the slow spontaneous exchange of tritium from the ribulose-P_2 which occurs at pH 8 over the prolonged period of incubation would have resulted in a lowering of the apparent isotope effect. A comparison of the $^{14}CO_2$-fixation in the two incubations provided an estimate of the effect of ^{2}H$_2$O on the reaction rate. Since this ratio must

Table 2. *Discrimination against [3C-³H]ribulose-P_2 in H_2O and ²H_2O*
The values in parentheses are the fractions (f) of the reaction that occurred as measured with ¹⁴C or ³H

Time	¹⁴CO₂-fixed (fraction reacted)		³H-released (fraction reacted)		$k_H/k_{³H}$ [a]		$v_{H_2O}/v_{²H_2O}$
	in H_2O	in ²H_2O	in H_2O	in ²H_2O	in H_2O	in ²H_2O	
min	$10^{-3} \times$ counts/min		$10^{-3} \times$ counts/min				
0	3.2 (0)	3.9 (0)	0.426 (0)	2.32 (0)			
5	23.0 (0.036)	14.5 (0.020)	1.17 (0.0062)	2.84 (0.005)	5.9	4.0	1.8
10	57.4 (0.090)	34.2 (0.057)	2.48 (0.017)	3.36 (0.010)	5.5	5.8	1.6
15	92.4 (0.15)	57.0 (0.107)	4.23 (0.031)	5.04 (0.027)	5.0	4.1	1.4
20	126.0 (0.20)	77.5 (0.145)	6.78 (0.052)	7.29 (0.049)	4.1	3.1	1.4
105	473.0 (0.75)	343.0 (0.64)	44.0 (0.36)	35.4 (0.328)	3.1	2.6	1.2
245	610.0 (0.97)	524.0 (0.98)	86.5 (0.72)	81.0 (0.780)			
—	635.0 (1.0)	531.0 (1.0)	121.0 (1.0)	104.0 (1.0)			

[a] k_H/k_T determined by $\dfrac{\log(1-f_H)}{\log(1-f_{³H})}$, [13], where f_H, $f_{³H}$ are the fractions of the H species and ³H species that have reacted as shown by ¹⁴C-fixation and ³H-release.

approach 1.0 as both reactions approach completion, the later values are not to be considered. A small decrease in rate is evident in the ²H_2O medium.

DISCUSSION

In unpublished experiments from this laboratory using [3C-³H]ribulose-P_2, attempts have been made to demonstrate an isotope exchange catalyzed by the enzyme in the absence of added CO_2. The rate of appearance of radioactivity in the water was only 2.5% of the rate observed when CO_2 was later added to the incubation, and this value appeared to be largely explained by P-glyceric acid formation due to contaminating CO_2. The comparison of the two tritium rates makes it unnecessary to correct for an isotope effect and so indicates that the rate of exchange of proton in the absence of CO_2 is negligible. The conclusion that the exchange in the absence of CO_2 is very slow agrees with published work [4,5]. Although the proposed mechanism would not predict a CO_2 requirement for ³H exchange at C-3, it is possible that CO_2 acting in some way other than as a reactant might be required to activate the protein

to carry out the enolization step, so that such results cannot rule out this mechanism. As an example, it is found that enolization of pyruvate by pyruvate kinase requires ATP, but it is clear that the ATP is not acting as a reactant since inactive analogues will replace it in allowing the enolization reaction [14]. Similarly the enolization of acetyl CoA is catalyzed by the citrate synthetizing enzyme only if oxalo-acetate or an analogue such as malate is present [15]. Spectral evidence for an interaction between ribulose-P_2 and enzyme in the absence of Mg^{++} or CO_2 has been reported and interpreted as due to an enolization reaction involving an -SH group on the enzyme [3]. To conform with this interpretation it would be necessary to suppose that in the absence of CO_2 the enolization step occurs, but that the proton is not free to exchange from its conjugate acid group on the enzyme in the absence of a subsequent reaction that depends on CO_2. The occurrence of a large isotope effect in which the substrate tritiated at C-3 is used at 20% the rate of the 3C-¹H form indicates that in the presence of CO_2 the enolization step is followed by a fast and irreversible step.

The experiment with [3C-^{3}H]ribulose-P_2-^{3}H indicates that the proton lost from C-3 of the substrate is not the one that enters at what was the C-2 position in the formation of phosphoglyceric acid. This may mean that two independent enzyme sites are functioning in the two protolytic steps (mechanism A). Alternatively, there could be one site alone which undergoes complete exchange with the medium protons, H_M^+, in the conjugate acid form prior to proton transfer. Evidently there is sufficient time for this exchange to occur since the significant discrimination [4,5] against tritium derived from ^{3}HOH in the subsequent transfer of hydrogen to the developing phosphoglyceric acid indicates that this step (step 3) is slow.

Under the same experimental conditions used for Table 2, the tritium content per two molecules of phosphoglyceric acid formed was 4.3 to 4.6 times lower than the specific activity of the medium. The occurrence of such an isotopic discrimination does not necessarily signify that this step is rate-determining for the over-all sequence [16], but does indicate that it is slow relative to subsequent steps. The effect of ^{2}H$_2$O on the net reaction rate was only 1.6 (Table 2) which is smaller than would be expected from the discrimination against tritium in the ^{3}HOH experiment if step 3 were rate-determining. A small effect is to be expected since in adjusting both incubations to 8.1 in the pH meter, the measurement in ^{2}H$_2$O is 0.4 unit low [17] and, according to Hurwitz [12] the rate at pH 8.5 is somewhat lower than at 8.1. The discrimination against [3C-^{3}H]ribulose-P_2 may signify that the enolization is ratedetermining for the net reaction.

This research was supported by grants CA-07818 and CA-06927 from the National Institutes of Health, U.S.P.H.S. G. Müllhofer was supported by a stipend from the German Academic Exchange Organization.

REFERENCES

1. Müllhofer, G., and Rose, I. A., J. Biol. Chem. 240 (1965) 1341.
2. Calvin, M., J. Chem. Soc. (1956) 1895.
3. Rabin, B. R., and Trown, P. W., Nature, 202 (1964) 1290.
4. Simon, H., Dorrer, H. D., and Trebst, A., Z. Naturforsch. 19b (1964) 734.
5. Hurwitz, J., Jakoby, W. B., and Horecker, B. L., Biochim. Biophys. Acta, 22 (1956) 194.
6. Wang, S. F., Kawahara, F. S., and Talalay, P., J. Biol. Chem. 238 (1963) 576.
7. Rose, I. A., and O'Connell, E. L., J. Biol. Chem. 236 (1961) 3086.
8. Rognstad, R., Kemp, R. G., and Kratz, J., Arch. Biochem. Biophys. 109 (1965) 372.
9. Pastore, E. J., and Friedkin, M., J. Biol. Chem. 236 (1961) 2314.
10. Levy, H. R., and Vennesland, B., J. Biol. Chem. 228 (1957) 85.
11. Simon, H., and Steffens, J., Chem. Ber. 95 (1962) 358.
12. Hurwitz, J., In Methods in Enzymology (edited by S. P. Colowick and O. Kaplan), Academic Press, New York 1956, Vol. V, p. 258.
13. Racker, E., In Methods in Enzymology (edited by S. P. Colowick and O. Kaplan), Academic Press, New York 1956, Vol. V, p. 266.
14. Rose, I. A., J. Biol. Chem. 235 (1960) 1170.
15. Eggerer, H., Biochem. Z. 343 (1965) 111.
16. Kaplan, L., J. Am. Chem. Soc. 76 (1954) 4645.
17. Glasoe, P. K., and Long, F. A., J. Phys. Chem. 64 (1960) 188.

F. Fiedler's present address:
Klinisch-chemisches Institut an der Chirurgischen Klinik der Universität,
8 München, Germany

G. Müllhofer's present adress:
Physiologisch-chemisches Institut der Universität
8 München, Germany

A. Trebst
Abteilung Biochemie der Pflanze, Pflanzenphysiologisches Institut der Universität, 34 Göttingen, Germany

I. A. Rose
Institute for Cancer Research, 7701 Burholme Avenue, Fox Chase, Philadelphia, Pa. 19111, U.S.A.

European J. Biochem. 1 (1967) 400—410

Dissoziation der Rinderleber-Katalase in ihre Untereinheiten

H. Sund, K. Weber und E. Mölbert

Chemisches Laboratorium und Pathologisches Institut der Universität Freiburg i. Br.

(Eingegangen am 7. Dezember 1966)

Earlier results from other laboratories have shown that beef liver catalase is built up of subunits. However there was no integral relationship between the number of subunits and the number of active sites (heme groups). To clarify the question how the heme groups are distributed among the subunits, the quaternary structure of catalase was studied under different denaturing conditions.

1. 5 M guanidine-HCl, in the presence of 0.1 M β-mercaptoethanol, caused dissociation of catalase into subunits each with a molecular weight of 59,000 ($s_{20,w}^{\circ} = 2.10$ S; $D_{20,w}^{\circ} = 3.20$ F). In the absence of β-mercaptoethanol the molecular weight in 5 M guanidine-HCl was found to be about 140,000 ($s_{20,w}^{\circ} = 3.50$; $D_{20,w}^{\circ} = 2.17$ F).

2. Succinylation of native catalase with succinic anhydride at neutral pH yielded subunits each with a molecular weight of 65,000 ($s_{20,w}^{\circ} = 3.22$ S; $D_{20,w}^{\circ} = 4.43$ F).

3. After short incubation (up to six hours) at pH 12.5—12.8 catalase dissociated into subunits each with a sedimentation coefficient ($s_{20,w}^{\circ}$) of 2.89 S. The molecular weight of these subunits was estimated to be 54,000 and 67,000 respectively. The first value was obtained by combining sedimentation and viscosity measurements ($[\eta] = 17.78$ ml/g) from the relation deduced by Scheraga and Mandelkern using a β-value of 2.5×10^6. The principle of Archibald was used for the estimation of the second value.

Longer incubation (e.g. twenty hours) caused splitting into several components, the main component after sixty hours possesses a sedimentation coefficient of 1.7 S. Probably this splitting is due to hydrolysis of peptide bonds. For this reason it was not possible to estimate the diffusion coefficient of the $S_{2.9}$-component which needs a dialyzed solution.

4. During the incubation of native catalase in the presence of 7.3—14.6 mM sodium dodecyl sulfate an incomplete dissociation into a component with $s_{20,w}^{\circ} = 3.4$ S was observed. Separation of the dissociation product by means of a separation cell yielded a single component with a molecular weight of about 80,000 ($s_{20,w}^{\circ} = 3.3$ S; $D_{20,w}^{\circ} = 3.3$ F). The significance of this component is not clear, because the molecular weight disagrees with the value of 61,000 obtained by all other methods.

Treatment of catalase or apocatalase with formic acid followed by incubation with sodium dodecyl sulfate caused dissociation into subunits with a sedimentation coefficient of 3.8 S ($D_{20,w}^{\circ} = 5.1$ F) and 4 S respectively. The molecular weight of the $S_{3.8}$-component was found to be 60,000.

5. 8 M urea resulted in a dissociation into subunits each with a molecular weight of 133,000 ($s_{20,w}^{\circ} = 3.3$ S; $D_{20,w}^{\circ} = 2.23$ F). About the same molecular weight (120,000) was obtained by Samejima and Yang at pH 3.

By comparison with the molecular weight of the native enzyme (240,000—250,000) it was concluded that beef liver catalase consists of four subunits of equal size (molecular weight 60,000—65,000). This conclusion was confirmed by electron microscopic studies. In 5 M guanidine-HCl, or in 8 M urea, or at pH 3 the catalase molecule dissociated into two subunits only, whereas all other denaturing conditions used gave rise to four subunits. Together with the reported re-naturation experiments the dissociation scheme of beef liver catalase is shown in the following equation:

$$240,000 \rightleftharpoons 2 \times 120,000 \rightleftharpoons 4 \times 60,000.$$

As mentioned above guanidine-HCl (and probably urea), in the presence of β-mercaptoethanol, caused dissociation into four subunits, whereas in the absence of β-mercaptoethanol only two subunits were obtained. It is assumed that the reason for this difference might be the oxidation of SH groups to disulfide bridges during the denaturation linking subunits together. This process is prevented in the presence of β-mercaptoethanol. In the native enzyme molecule the four subunits are held together only by noncovalent bonds. It was concluded that each of the four subunits contains one active center.

Enzym. Katalase oder $H_2O_2 : H_2O_2$-Oxidoreductase (EC 1.11.1.6).

Physikalisch-chemische und chemisch-analytische Methoden haben gezeigt, daß zahlreiche Enzymmoleküle, besonders wenn ihre Molekulargewichte größer als 100000 sind, aus mehreren Untereinheiten oder Polypeptidketten bestehen und mehrere aktive Zentren besitzen. Häufig kommen innerhalb eines Moleküls gleich viele Untereinheiten wie aktive Zentren vor, so daß man jeder Untereinheit ein aktives Zentrum zuordnen kann [2].

Katalase aus Rinderleber enthält vier Hämingruppen [3] und dissoziiert unter dem Einfluß verschiedener Denaturierungsmittel in Untereinheiten. Je nach den Versuchsbedingungen wurden Dissoziationsprodukte beschrieben, deren Molekulargewichte $1/6$, $1/3$, $1/2$ oder $2/3$ des Wertes nativer Katalase (Molekulargewicht 250000 [4]) besitzen sollen [5—9], Entsprechend dem Hämingehalt sollte man jedoch eine Dissoziation in 4 oder 4 n Untereinheiten erwarten. Ursprünglich nahm man auch an [8], daß Säuredenaturierung zu vier Untereinheiten führt, später konnte jedoch gezeigt werden [9], daß unter diesen Bedingungen nur zwei Untereinheiten vom Molekulargewicht 120000 entstehen. Nach den bisherigen Ergebnissen war es daher schwer verständlich, wie die vier Hämingruppen innerhalb des Katalase-Moleküls auf die Untereinheiten verteilt sein sollten.

Es war das Ziel dieser Arbeit[1], die Quartärstruktur der Rinderleber-Katalase eingehend zu untersuchen. Insbesondere lag uns daran zu prüfen, ob nicht doch eine symmetrische Verteilung von Hämingruppen und Polypeptidketten vorliegt. Ein Aufbau aus drei oder sechs Untereinheiten [6] müßte eine unsymmetrische Verteilung der aktiven Zentren auf die Untereinheiten voraussetzen. Dies wäre ein bei Enzymen bisher unbekanntes Bauprinzip. In der vorliegenden Arbeit können wir nun zeigen, daß ein derartiges unsymmetrisches Bauprinzip bei Katalase aus Rinderleber offenbar nicht vorliegt. Sowohl die Succinylierung als auch die Denaturierungen durch Guanidin-HCl, NaOH oder Natriumdodecylsulfat führen zu einer Dissoziation des Katalase-Moleküls in vier gleich große Untereinheiten. Dieses Ergebnis wird gestützt durch elektronenmikroskopische Untersuchungen.

EXPERIMENTELLER TEIL

Reagentien

KH_2PO_4 und Na_2HPO_4 nach Sörensen, Natriumcitrat p.a., H_2SO_4 (95—97%ig, p.a.), NaOH p.a., Ameisensäure (98—100%ig, p.a.), Phosphorwolframsäure (p.a.), 1,10-Phenanthrolin p.a., Hydrochinon, reinst (alle E. Merck, Darmstadt), H_2O_2 (30%ig, p.a., Riedel-de Haën, Seelze-Hannover), Glutardialdehyd (25%ige Lösung in H_2O, Th. Schuchardt, München)

_____
[1] Vorläufige Mitteilung: Lit. [1].

sowie Merkaptoäthanol (reinst, Serva, Heidelberg) wurden ohne weitere Reinigung verwendet. Guanidin-HCl (,,100%ig", Th. Schuchardt, München): Aus Äthanol-Wasser umkristallisiert. Natriumdodecylsulfat (rein, Henkel u. Cie., Düsseldorf): Einmal aus Wasser umkristallisiert. Bernsteinsäureanhydrid (Riedel-de Haën, Seelze-Hannover): Zweimal aus Methanol umkristallisiert, Fp 120°. Für alle Versuche verwendeten wir quarzdestilliertes Wasser. Der verwendete Phosphat-Puffer ist M/15 Kaliumnatriumphosphat-Puffer pH 7,6.

Katalase

Katalase aus Rinderleber wurde als Kristallsuspension von der Firma C. F. Boehringer und Soehne GmbH, Mannheim, erhalten. Die spezifische Aktivität der beiden verwendeten Katalase-Präparate betrug im Test nach Bergmeyer [10] 4500 bzw. 4730 Einheiten/mg, das entspricht 58500 bzw. 61500 I.E./mg (vgl. auch [11]). Die Katalase-Präparate waren in der analytischen Ultrazentrifuge einheitlich (Fig.2, A).

Die Konzentrationsbestimmung erfolgte beim nativen Enzymprotein auf Grund der Absorption bei 405 mµ mit $E_{1\ cm}^{1\ \%} = 13,5$ entsprechend einem molaren Extinktionskoeffizienten von $3,24 \times 10^5$ (bezogen auf ein Molekulargewicht von 240000) [9]. Bei den nach Denaturierung erhaltenen Präparaten (mit Ausnahme der in Tabelle 3 angegebenen und bei pH 12,5 durchgeführten Versuche) haben wir die Konzentration aus der Fläche unter den Konzentrationsgradienten der Ultrazentrifugen-Aufnahmen (Phasenwinkel 60°, 12 mm-Zelle, 20°), die durch Planimetrieren nach zehnfacher Vergrößerung und gegebenenfalls nach Berücksichtigung der radialen Verdünnung erhalten wurde, berechnet. Die Umrechnung erfolgte unter der Annahme, daß sich der Brechungsindex der Katalase durch die Denaturierung nicht verändert hat. Vor den Diffusions- und Sedimentationsmessungen wurde das Präparate (mit Ausnahme der in Tabelle 3 angegebenen Versuche sowie bei der Inkubation in Gegenwart von NaOH) 48 Std dialysiert.

Auf Grund des Katalase-Spektrums [9,20] errechnet sich der Hämingehalt der verwendeten Präparate zu 3,8 Hämingruppen pro Katalase-Molekül. Nach Veraschen der Katalase mit conc. Schwefelsäure und H_2O_2 ergab die optische Bestimmung des Eisengehaltes nach der Phenanthrolin-Methode [30] bei Einwaagen von 10, 24 und 35 mg einen Gehalt von 4,1 Mol Eisen pro Mol Enzym. Diese Werte stimmen mit dem erwarteten Wert für vier Hämingruppen pro Katalase-Molekül gut überein.

Die pH-Messungen wurden mit dem pH-Meter PHM 4 der Firma Radiometer, Kopenhagen, vorgenommen.

Diffusion und Sedimentation

Die Diffusions- und Sedimentationsmessungen wurden wie früher [12,23] beschrieben in der analytischen Ultrazentrifuge Spinco, Modell E (Beckman Instruments, München) vorgenommen. Die extrapolierten t_0-Zeiten bei den Diffusionsversuchen lagen zwischen -1 und -5 min (vgl. auch Fig. 4). Die Überschichtungen wurden in Ventiltyp- oder Capillartyp-Zellen, die Sedimentationen je nach Proteinkonzentration in 3, 12 oder 30 mm-Zellen vorgenommen. Die Berechnung der Sedimentations- und Diffusionskoeffizienten erfolgte nach zehnfacher zeichnerischer Vergrößerung auf Millimeterpapier, die Archibald-Versuche in der Auftragung nach Trautman und Crampton [13] wurden nach zwanzigfacher Vergrößerung ausgewertet. Aus der Konzentrationsabhängigkeit der Sedimentationskoeffizienten wurde $s°$ nach der Methode der kleinsten Quadrate berechnet. Die für die Umrechnung der Diffusions- und Sedimentationskoeffizienten auf Normalbedingungen ($D_{20,w}$ und $s_{20,w}$ [14,15]) benötigten Angaben über Dichte und Viskosität der Lösungsmittel wurden wie früher [12] erhalten. Die Diffusionskoeffizienten (D_A, nach der Flächenmethode berechnet) werden in Fick-Einheiten ($F = 10^{-7}$ cm$^2 \times$ sec^{-1}), die Sedimentationskoeffizienten in Svedberg-Einheiten ($S = 10^{-13}$ sec) angegeben.

Die Bestimmung der Molekulargewichte (M) erfolgte nach der Svedberg-Formel [14] bzw. nach der Archibald-Methode [13].

Aus Materialmangel war es nicht möglich, das partielle spezifische Volumen $\bar{v}$ unter den verschiedenen Lösungsbedingungen zu bestimmen. Wir legten deshalb für die verschiedenen Berechnungen den für das native Enzymprotein bestimmten Wert von 0,73 ml/g [4] zugrunde.

Viskosität

Die Bestimmung der Viskositätszahl [η] erfolgte wie früher beschrieben [12] mit einem Ostwald-Viskosimeter für 3 ml Lösung, das eine Durchlaufzeit für Wasser bei 20° von 257 sec besaß (Zeitfehler $<0,1°/_0$). Zur Entfernung der Staubteilchen wurde die zu untersuchende Lösung durch eine Glasfritte G 3 gepreßt. Die gemessene kinematische Viskositätszahl [ν] lieferte nach der von Tanford [16] eingeführten Beziehung die Viskositätszahl [η]. Messungen bei 20,0 $\pm$ 0,1°.

Denaturierungen

Einwirkung von Guanidin-HCl. 1 ml einer etwa 10 M Guanidin-HCl-Lösung (in Phosphat-Puffer pH 7,6) wurde mit 1 ml Katalase-Lösung (15—20 mg/ml, gelöst in Phosphat-Puffer pH 7,6) überschichtet, anschließend rasch durchmischt und dann gegen 100 ml 5 M Guanidin-HCl-Lösung (in Phosphat-Puffer pH 7,6) 48 Std bei Zimmertemperatur dialysiert. Bei den Versuchen in Gegenwart von Merkaptoäthanol wurde genau so verfahren, die Guanidin-Lösungen enthielten zusätzlich jeweils 0,2 M (bei 10 M Guanidin-HCl) bzw. 0,1 M Merkaptoäthanol (Dialyse-Außenflüssigkeit).

Succinylierung [17]. 15—18 mg Katalase, gelöst in 1 ml Phosphat-Puffer pH 7,6, wurden bei Zimmertemperatur dreimal mit 10 mg Bernsteinsäureanhydrid versetzt (2. und 3. Zugabe nach 10 und 30 min). Während der Inkubation (insgesamt 2 Std) wurde der pH-Wert der Reaktionslösung mit NaOH (0,5 N oder 5,0 N) auf etwa pH 7,6 gehalten, anschließend dialysierten wir 48 Std bei 4° gegen Phosphat-Puffer pH 7,6. Bei unvollständiger Spaltung wurde die Succinylierung wiederholt.

Einwirkung von NaOH. Katalase, gelöst in Phosphat-Puffer pH 7,6 (6—14 mg/ml), wurde mit 0,5 N NaOH auf pH 12,5 bis 12,8 gebracht, die angegebene Zeit inkubiert und dann untersucht.

Einwirkung von Natriumdodecylsulfat. 5—10 mg lufttrockener Katalase oder Apokatalase (entsprechend der Vorschrift für Globin [18] dargestellt) wurden mit 1—2 ml Ameisensäure (98°/_0ig) versetzt, im Exsiccator über NaOH eingetrocknet, anschließend mit gepufferter (0,1 N Tris) Natriumdodecylsulfat-Lösung (pH 9,5) aufgenommen und dann gegen eine Natriumdodecylsulfat-Lösung gleicher Konzentration dialysiert. Bei den Versuchen ohne vorherige Ameisensäure-Behandlung wurden die Katalase-Lösungen mit Natriumdodecylsulfat-Lösung auf die angegebenen Konzentrationen gebracht.

Elektronenmikroskopische Untersuchungen

Die Katalase-Lösungen wurden nach Dialyse gegen Phosphat-Puffer pH 7,6 mit gepuffertem Glutardialdehyd (1°/_0ig; pH 7,3) versetzt. 3 min später wurde das gleiche Volumen einer 2°/_0igen Phosphorwolframsäure-Lösung (mit KOH auf pH 7,3 gebracht) zugegeben und die erhaltene Lösung dann auf kohlebeschichtete elektronenmikroskopische Objektträger aufgebracht oder aufgesprüht. Für die Untersuchungen verwendeten wir das Elektronenmikroskop Elmiskop I bzw. I A der Firma Siemens mit Spitzenkathode und Objektraumkühlung. Die Aufnahmen wurden bei Vergrößerungen von 80000, 100000 bzw. 160000 aufgenommen.

ERGEBNISSE

Native Katalase

Native Katalase ist durch einen Sedimentationskoeffizienten $s_{20,w}°$ von 11,29 S („S_{11}-Komponente") und einer Konzentrationsabhängigkeit des Sedimentationskoeffizienten entsprechend Gleichung (1) mit k_s von $6,2 \times 10^{-3}$ ml/mg charakterisiert (Fig. 1).

$$s_{20,w} = s_{20,w}° \, (1 - k_s c) \qquad (1)$$

c = Proteinkonzentration in mg/ml

k_s = Konstante (ml/mg)

Den Diffusionskoeffizienten, gemessen bei vier Konzentrationen zwischen 1,6 und 3,4 mg/ml, erhielten wir zu 4,17 ± 0,05 F (vgl. auch Fig. 4). Aus diesen

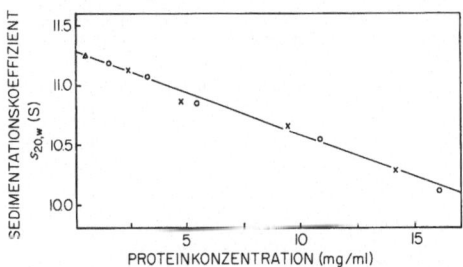

Fig. 1. *Abhängigkeit des Sedimentationskoeffizienten von der Proteinkonzentration bei Katalase aus Rinderleber.* Messungen in Phosphat-Puffer pH 7,6 bei 20,0 ± 0,1° und 50740 (△) bzw. 59780 (×, ○) U/min mit zwei verschiedenen Präparaten (△, × bzw. ○)

Einwirkung von Guanidin-HCl

In Abwesenheit von Merkaptoäthanol. Inkubiert man Katalase in Gegenwart von 5 M Guanidin-HCl, so erhält man ein offenbar monodisperses Dissoziationsprodukt (Fig. 2 B), das durch einen Sedimentationskoeffizienten (s_{20}^0) von 1,60 S charakterisiert ist (Fig. 3 A). Nach Umrechnung erhält man für $s_{20,w}^0$ einen Wert von 3,50 S. Die Ergebnisse der Diffusionsmessungen sind der Tabelle 1 zu entnehmen. $D_{20,w}^0$ ergibt sich zu 2,17 F. In Fig. 4 ist die Auftragung, nach der die Berechnung von D_A vorgenommen wurde, angegeben. Aus $s_{20,w}^0$ und $D_{20,w}^0$ errechnet sich das Molekulargewicht zu 144000.

In Gegenwart von Merkaptoäthanol. Inkubiert man Katalase in Gegenwart von 5 M Guanidin-HCl außerdem noch unter Zusatz von 0,1 M Merkaptoäthanol, dann erhält man ebenfalls ein monodisperses Dissoziationsprodukt (Fig. 2 C), dessen Sedimentationskoeffizient aber wesentlich niedriger ist als derjenige, den man in Abwesenheit von Merkaptoäthanol beobachtet (Fig. 3). s_{20}^0 wird zu 0,96 S und $s_{20,w}^0$ zu 2,10 S erhalten. Die Ergebnisse der Diffusionsmes-

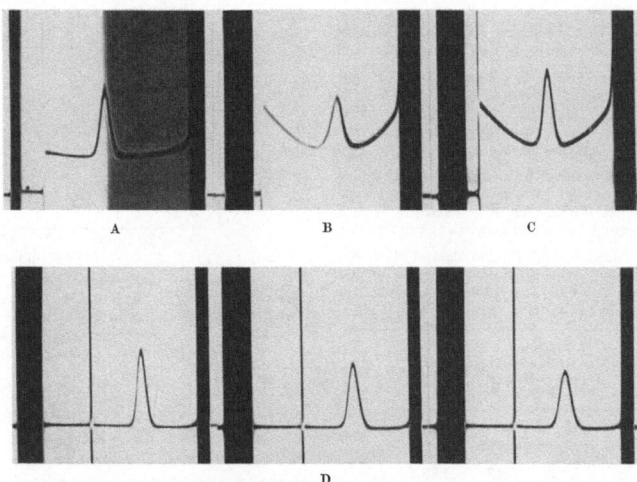

Fig. 2. *Sedimentation von Rinderleber-Katalase.* Versuche bei 59780 U/min bei A—C bzw. 4059 U/min bei D in der 12 mm-Zelle (A) bzw. nach Überschichtung in der Ventiltyp-Zelle (B—D). Temperatur jeweils 20,0 ± 0,1°; Winkel 60°. A, Kontrolle: Katalase, 4,8 mg/ml, in Phosphat-Puffer pH 7,6, Aufnahme nach 32 min. B, Katalase (c = 4,63 mg/ml) in 5 M Guanidin-HCl (gelöst in Phosphat-Puffer pH 7,6), Aufnahme nach 76 min. Messung im Anschluß an die Bestimmung von D_A. C, Katalase (c = 4,41 mg/ml) in 5 M Guanidin-HCl/0,1 M Merkaptoäthanol (gelöst in Phosphat-Puffer pH 7,6), Aufnahme nach 78 min. D, Versuch zur Messung von D_A: Katalase (c = 3,84 mg/ml) in 5 M Guanidin-HCl/0,1 M Merkaptoäthanol (gelöst in Phosphat-Puffer pH 7,6), Aufnahmeabstand 16 min (Versuch 3 der Tab. 1)

Werten errechnet sich das Molekulargewicht der Katalase zu 243000, was mit den früher gemessenen Werten von 240000 bis 250000 [4,5,9] gut übereinstimmt.

sungen sind der Tabelle 1 sowie den Fig. 2 D und 4 zu entnehmen. $D_{20,w}^0$ ergibt sich zu 3,20 F. Aus $s_{20,w}^0$ und $D_{20,w}^0$ errechnet sich das Molekulargewicht zu 59000.

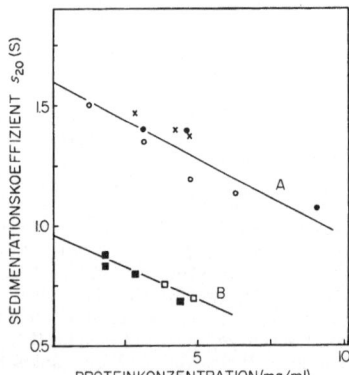

Fig. 3. *Abhängigkeit des Sedimentationskoeffizienten von der Proteinkonzentration bei Katalase in 5 M Guanidin-HCl* (gelöst in Phosphat-Puffer pH 7,6). Messungen bei 59780 U/min und 20,0 ± 0,1° nach Überschichtung in der Ventiltyp- bzw. Capillartyp-Zelle. A, in Abwesenheit von Merkaptoäthanol (drei Denaturierungsansätze); B, in Gegenwart von 0,1 M Merkaptoäthanol (zwei Denaturierungsansätze)

Tabelle 1. *Diffusionskoeffizienten von Rinderleber-Katalase nach Einwirkung verschiedener Denaturierungsmittel*

A. Versuche in Phosphat-Puffer pH 7,6 bei 19,7—20,0° und 4059 U/min in der Capillartyp- (Versuch Nr. 3) bzw. Ventiltyp-Überschichtungszelle (alle anderen Versuche). Versuche 3 bzw. 1, 2 und 4 stellen zwei verschiedene Succinylierungsansätze dar

B. Versuche in 5 M Guanidin-HCl (in Phosphat-Puffer) — 0,1 M Merkaptoäthanol bei pH 7,6 in der Ventiltyp-Überschichtungszelle bei 4059 U/min. Die Versuche 1, 3, 5 bzw. 2, 4 stellen zwei verschiedene Denaturierungsansätze dar

C. Versuche in 5 M Guanidin-HCl (in Phosphat-Puffer) bei pH 7,6 und 20° in der Ventiltyp-Überschichtungszelle bei 4059 U/min. Die Versuche 2 bzw. 1, 4, 6 bzw. 3, 5 stellen drei verschiedene Denaturierungsansätze dar

Versuch Nr.	A. Succinylierte Katalase		B. Versuche in 5 M Guanidin-HCl + 0,1 M Merkaptoäthanol		C. Versuche in 5 M Guanidin-HCl	
	c (mg/ml)	D_A	c (mg/ml)	D_A	c (mg/ml)	D_A
1	2,04	4,45	2,36	3,03	2,41	2,04
2	4,09	4,47	2,82	3,43	2,89	2,25
3	5,83	4,30	3,84	3,19	4,63	1,92
4	6,63	4,51	4,41	3,12	4,70	2,09
5			7,71	3,24	4,73	2,14
6					9,10	1,99
$D^\circ_{20,w}$ (in F)	4,43		3,20		2,17	

Die Einwirkung von Guanidin-HCl führt demnach in Gegenwart von Merkaptoäthanol zur Dissoziation in vier Untereinheiten, während in Abwesenheit von Merkaptoäthanol wahrscheinlich nur zwei Untereinheiten gebildet werden.

Succinylierte Katalase

Bei Proteinen mit relativ hohem Lysin-Gehalt führt die Umsetzung der ε-Aminogruppen mit Bernsteinsäureanhydrid zu einer starken Änderung der Gesamtladung und damit zu einer Änderung der Konformation des Proteinmoleküls. Bei Hemerythrin [17] und β-Galaktosidase [19] hatte die Succinylierung eine Dissoziation in Untereinheiten zur

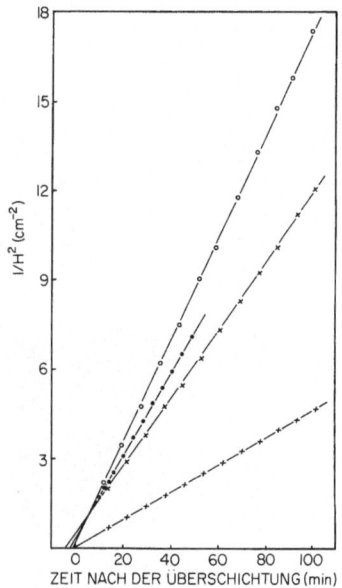

Fig. 4. *Auftragung zur Berechnung von* D_A. ●, Native Katalase (3,3 mg/ml); +, Succinylierte Katalase (Versuch 4 der Tabelle 1); ×, Katalase in 5 M Guanidin-HCl (Versuch 1 der Tabelle 1); ○, Katalase in 5 M Guanidin-HCl/0,1 M Merkaptoäthanol (Versuch 1 der Tabelle 1)

Folge. Führt die Succinylierung zu einer Dissoziation, dann bietet diese Methode den Vorteil, daß die Molekulargewichtsbestimmung in wäßriger Pufferlösung vorgenommen werden kann. Hierbei entfallen dann die für konzentrierte Guanidin-HCl- oder Harnstoff-Lösungen notwendigen erheblichen Korrekturen auf Grund des Faktors $(1 - \bar{V}\varrho)$, da die Dichte nur wenig von der des Wassers abweicht.

Katalase enthält 116 Lysin-Reste pro Molekül [20]. Durch die Einwirkung von Bernsteinsäureanhydrid wird Katalase vollständig in ein monodisperses Reaktionsprodukt übergeführt (Fig. 5), das durch einen Sedimentationskoeffizienten $s_{20,w} = 3,22$ S charakterisiert ist. Die Abhängigkeit des Sedimentationskoeffizienten von der Proteinkonzentration ist der Fig. 6A zu entnehmen. Der Diffusionskoeffizient ist in dem untersuchten Konzentrationsbereich unabhängig von der Proteinkonzentration (Tabelle 1). Er wurde zu 4,43 F gefunden. Die Auftragung zur Berechnung des Diffusionskoeffizienten ist der Fig. 4 zu entnehmen. Das Molekulargewicht des nach Succinylierung erhaltenen Dissoziationsproduktes errechnet sich zu 65000. Die Succinylierung führt demnach zu einer Dissoziation in vier nach Sedimentation und Diffusion gleich große Untereinheiten.

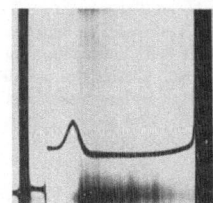

Fig. 5. *Sedimentation von succinylierter Katalase.* c = 2,73 mg/ml. Messung in Phosphat-Puffer pH 7,6 bei 59780 U/min und 20,0 ± 0,1°, Winkel 60°, Aufnahme nach 48 min

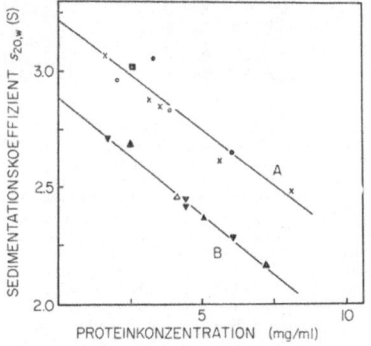

Fig. 6. *Abhängigkeit des Sedimentationskoeffizienten von der Proteinkonzentration bei Katalase nach Succinylierung* (A: ■, ×, ○, ●) *bzw. nach ein- bis fünfstündiger Inkubation bei pH 12,5—12,8 und 20°* (B: △, ▲, ▼). Messungen bei 59780 U/min und 19,2—20,0° nach Überschichtung (×, ○, ●) bzw. ohne Überschichtung (alle anderen). Succinylierte Katalase: Versuche in Phosphat-Puffer pH 7,6; drei Succinylierungsansätze. Versuche bei pH 12,5—12,8 mit lyophilisierter (△) bzw. gelöster (▲, ▼) Katalase

Nach Succinylierung bleibt die Hämingruppe noch an das Protein gebunden. Dies geht hervor aus der gleichen Wanderungsgeschwindigkeit des Konzentrationsgradienten und der Absorption während der Sedimentation.

Einwirkung von NaOH

Inkubiert man gelöste oder lyophilisierte Katalase kurzzeitig (bis zu 6 Std) bei pH 12,5—12,8 und 20°, so wird sie in eine offenbar monodisperse Komponente übergeführt (Fig. 7, A). Diese Komponente ist durch einen Sedimentationskoeffizienten $s_{20,w}^0$ = 2,89 S („$S_{2,9}$-Komponente") und eine Konzentrationsabhängigkeit des Sedimentationskoeffizienten entsprechend Gleichung (1) mit k_s von $3,5 \times 10^{-2}$ ml/mg charakterisiert (Fig. 6 B).

Inkubiert man längere Zeit, z. B. 20 Std, so tritt neben der $S_{2,9}$-Komponente eine weitere langsamer sedimentierende Komponente auf (Fig. 7 B), nach 60 stündiger Inkubation erhält man ein Präparat, dessen Hauptkomponente mit einem Sedimentationskoeffizienten von 1,7 S sedimentiert. Möglicherweise ist die durch längere Inkubation hervorgerufene Spaltung durch die Hydrolyse von Peptidbindungen hervorgerufen. Während der für die Bestimmung von Diffusionskoeffizienten erforderlichen ausgedehnten Dialyse wird die $S_{2,9}$-Komponente in

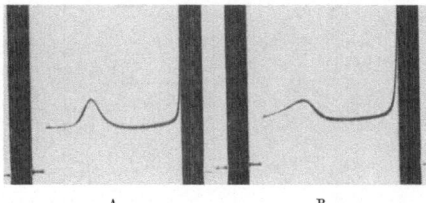

Fig. 7. *Sedimentation von Katalase* (c = 4,1 mg/ml) *nach Inkubation bei pH 12,5.* Inkubation 2 (A) bzw. 20 (B) Std. Messungen bei 59780 U/min und 20,0 ± 0,1° in der Kel-F-Zelle, Winkel 60°, Aufnahmen jeweils nach 79 min

die $S_{1,7}$-Komponente übergeführt. Aus diesem Grunde war der Diffusionskoeffizient der $S_{2,9}$-Komponente und damit das Molekulargewicht auf Grund der Svedberg-Gleichung nicht zu bestimmen. Das Molekulargewicht erhielten wir deshalb nach dem Archibald-Verfahren, und zwar in der von Trautman und Crampton angegebenen Modifikation [13, 15].

In vier Versuchen bei Konzentrationen zwischen 4,02 und 6,03 mg/ml erhielten wir das Molekulargewicht zu 67000 (Tabelle 2 und Fig. 8).

Zur weiteren Bestätigung der Ergebnisse haben wir außerdem die Viskosität gemessen. Unter der Annahme, daß die Einwirkung von Denaturierungsmitteln die mehr oder weniger vollständige Entfaltung eines Proteinmoleküls zur Folge hat, läßt sich auf Grund der von Scheraga und Mandelkern [21] entwickelten Gleichung (2) mit einem für statistische Knäuel charakteristischem β-Wert von $2,5 \times 10^6$ [22] durch Kombination von Sedimentations- und Viskositäts-Daten das Molekulargewicht bestimmen.

$$\beta = \frac{N_L s^0 [\eta]^{1/3} \eta_0}{M^{2/3} (1 - \overline{V}\varrho)} \quad N_L = \text{Loschmidt-Zahl} \quad (2)$$

Aus der Abhängigkeit der reduzierten Viskosität η_{sp}/c von der Proteinkonzentration erhält man für

Tabelle 2. *Molekulargewicht der nach Inkubation bei pH 12,5 bis 12,8 und 20° entstehenden Untereinheiten von Rinderleber-Katalase*
Inkubation bis zu 6 Std. Versuchsbedingungen siehe Legende zu Fig. 8

Proteinkonzentration	Molekulargewicht
mg/ml	
6,03	74 000
5,03	61 000
4,92	63 000
4,02	69 000
Mittel	67 000 ± 7 000

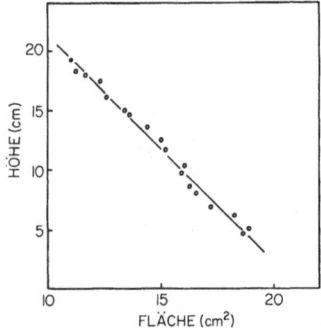

Fig. 8. *Bestimmung des Molekulargewichtes der bei pH 12,6 entstehenden Untereinheit der Katalase nach dem Archibald-Verfahren.* Auftragung nach Trautman und Crampton [13]. Proteinkonzentration 5,03 mg/ml, Temperatur 20,0 ± 0,1°, 29500 U/min. Berechnetes Molekulargewicht: 61000

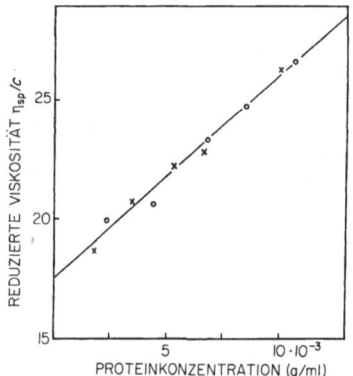

Fig. 9. *Abhängigkeit der reduzierten Viskosität (η_{sp}/c mit c in g/ml) bei pH 12,5—12,7 von der Proteinkonzentration bei Katalase.* Messungen bei 20,0 ± 0,1°, Inkubation bis zu 5 Std, zwei Denaturierungsansätze

die kinematische Viskositätszahl [ν] einen Wert von 17,51 ml/g (Fig. 9). Umrechnung entsprechend [16] liefert für die Viskositätszahl [η] einen Wert von 17,78 ml/g. Dieser relativ hohe Wert für [η] (für native Katalase wurde [η] zu 3,9 ml/g bestimmt [9]) zeigt an, daß das Katalase-Molekül unter dem Einfluß von NaOH weitgehend entfaltet wird und es somit gerechtfertigt ist, die Bestimmung des Molekulargewichtes entsprechend Gleichung (2) vorzunehmen. Mit $s_{20,w}^{0} = 2,89\ S$ und [η] = 17,78 ml/g erhält man das Molekulargewicht zu 54000.

Sowohl die Archibald-Versuche als auch die Kombination von Sedimentations- und Viskositätsmessungen ergeben übereinstimmende Ergebnisse. Danach dissoziiert Katalase nach kurzzeitiger Inkubation bei pH 12,5—12,8 in vier Untereinheiten vom Molekulargewicht etwa 60000.

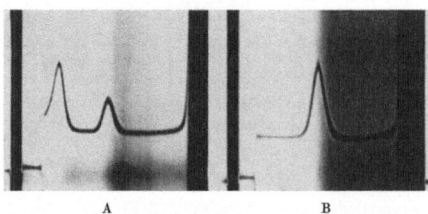

Fig. 10. *Sedimentation von Katalase nach Inkubation in Gegenwart von Natriumdodecylsulfat.* Versuche bei 59780 U/min und 20,0 ± 0,1° in der 12 mm-Zelle, Winkel 60°. A, 21 Std in Gegenwart von 7,3 mM Natriumdodecylsulfat, anschließend 2 Std in Gegenwart von 14,6 mM Natriumdodecylsulfat (gelöst in M/15 Kaliumnatriumphosphat-Puffer pH 7,6) bei 20° inkubiert, Aufnahme nach 35 min, Konzentration 6 mg/ml. B, Dissoziationsprodukt nach Inkubation in Gegenwart von 26,4 mM Natriumdodecylsulfat (gelöst in Phosphat-Puffer pH 7,6) und Abtrennung von der S_{11}-Komponente in einer Separationszelle. Aufnahme nach 126 min, c = 6,74 mg/ml

Einwirkung von Natriumdodecylsulfat

Während der Inkubation von Katalase in Gegenwart von Natriumdodecylsulfat bildet sich aus dem nativen Enzymprotein (S_{11}-Komponente) ein langsamer sedimentierendes Dissoziationsprodukt (Fig. 10), das je nach Proteinkonzentration einen Sedimentationskoeffizienten von 2,6 bis 3,1 S aufweist ($s_{20,w}^{0} = 3,4\ S$). Das Ausmaß der Bildung dieser Komponente ist von der Inkubationszeit abhängig (Tabelle 3). Nach etwa 40stündiger Inkubation wird ein Endwert erreicht, ohne daß eine vollständige Umwandlung der nativen Katalase in die $S_{3,4}$-Komponente stattfindet.

Trennt man die $S_{3,4}$-Komponente in einer Separationszelle (mit Siebplatte, vgl. [23]) von ungespaltener Katalase ab, dann erhält man das Dissoziationsprodukt in offenbar einheitlicher Form (Fig. 10 B). Die Abhängigkeit des Sedimentationskoeffizienten von

Tabelle 3. *Einwirkung von Natriumdodecylsulfat auf Rinderleber-Katalase*
Inkubation in Phosphat-Puffer pH 7,6 bei 20° in Gegenwart von $7,3 \times 10^{-3}$ M (bis zu 21 Std) bzw. von $14,6 \times 10^{-3}$ M Natriumdodecylsulfat (ab 21 Std). Katalase: 6 mg/ml

Inkubationszeit	$S_{3,4}$-Komponente
Std	%
4	50
21	55
23	65
27	72
42	76

der Proteinkonzentration ergibt für $s_{20,w}^0$ einen Wert von 3,33 S. Den Diffusionskoeffizienten bestimmten wir zu 3,3 F, so daß sich aus diesem und dem $s_{20,w}$-Wert ein Molekulargewicht von 91000 errechnet.

Auch die Behandlung von Katalase oder Apokatalase mit Ameisensäure, anschließender Lyophilisierung und Inkubation in Gegenwart von 1,7 bis $2,6 \times 10^{-2}$ M Natriumdodecylsulfat führt zur Aufspaltung in Untereinheiten. Geht man von Katalase aus, so erhält man $s_{20,w}^0$ zu 3,8 S und $D_{20,w}$ zu 5,1 F und damit das Molekulargewicht zu 67000. Im Falle der Apokatalase erhält man neben einem Dissoziationsprodukt mit einem Sedimentationskoeffizienten ($s_{20,w}^0$) von 4 S eine schneller sedimentierende Komponente mit $s_{20,w}^0 = 4,8$ S.

Das durch Natriumdodecylsulfat-Behandlung erhaltene Dissoziationsprodukt setzt sich aus einem Proteinanteil und dem an das Protein gebundenen Natriumdodecylsulfat zusammen. Letzteres läßt sich z. B. auf Grund des Brechungsindexinkrementes quantitativ bestimmen [23,24], so daß das Molekulargewicht des Proteinanteils berechnet werden kann. Unterläßt man diese Bestimmung und vernachlässigt bei der Berechnung des Molekulargewichtes die Natriumdodecylsulfat-Bindung, dann liegt das gefundene bis zu 10% höher als das wirkliche Molekulargewicht [25]. Stellt man 10% in Rechnung, dann erhält man das Molekulargewicht des nach Natriumdodecylsulfat-Einwirkung erhaltenen Dissoziationsproduktes nach Vorbehandlung mit Ameisensäure zu etwa 60000. Das entspricht einer Dissoziation in vier Untereinheiten. Die Komponente mit einem Sedimentationskoeffizienten von 4 S, die aus Apokatalase erhalten wurde, entspricht offenbar ebenfalls der Viertel.

Für das Dissoziationsprodukt, das ohne Vorbehandlung mit Ameisensäure durch Einwirkung von Natriumdodecylsulfat erhalten wurde, ergibt sich das Molekulargewicht nach Korrektur zu etwa 80000. Das würde etwa der Dissoziation in drei Untereinheiten entsprechen. Auch im Falle der β-Galaktosidase hatten wir schon früher gefunden [23], daß Natriumdodecylsulfat je nach den Versuchsbedingungen zu Dissoziationsprodukten unterschied-

lichen Molekulargewichtes führt. Auf Grund der beschriebenen Dissoziation in vier Untereinheiten durch Einwirkung der anderen Denaturierungsmittel zweifeln wir daran, daß dem Dissoziationsprodukt vom Molekulargewicht etwa 80000 reelle Bedeutung zukommt. Möglicherweise ist es ein Sekundärprodukt, das während der relativ langen Inkubationszeit (Separation, Dialyse) entstanden ist. Weitere Versuche müssen diese Frage klären.

Untersuchungen mit dem Elektronenmikroskop

Elektronenmikroskopisch ist das Katalase-Molekül nach der negative-staining-Methode als flacher Quader mit einer Kantenlänge von 70—90 Å und einer Höhe von 30—40 Å zu beschreiben. Daraus errechnet sich für das Molekülvolumen ein Mittelwert von $2,2 \times 10^5$ Å^3 und ein Molekulargewicht von etwa 200000. Dieser Wert erlaubt uns anzunehmen, daß die im Elektronenmikroskop sichtbaren Partikel dem Katalasemolekül zugeordnet werden dürfen. Die Partikel zeigen Substruktur, eine symmetrische Vierteilung ist deutlich festzustellen (Fig. 11).

DISKUSSION

Die Ergebnisse der vorliegenden Untersuchung, zusammengestellt in Tabelle 4, zeigen, daß die Einwirkung von Guanidin-HCl in Gegenwart von Merkaptoäthanol, NaOH oder Natriumdodecylsulfat sowie die Succinylierung zur Dissoziation der Katalase in vier Untereinheiten führt. Unter den angewendeten Versuchsbedingungen zur Bestimmung der Molekulargewichte treten weitere Dissoziationsprodukte nicht auf.

Obwohl die erhaltenen Molekulargewichte der Untereinheiten zwischen 54000 und 67000 gefunden wurden (Mittelwert 61000), gibt das Sedimentationsverhalten keinen Hinweis auf eine Uneinheitlichkeit

Tabelle 4. *Die hydrodynamischen Eigenschaften der Rinderleber-Katalase und ihrer Untereinheiten*

	$s_{20,w}^0$	$D_{20,w}^0$	Molekular-gewicht	f/f_0
	S	F		
Native Katalase (vgl. auch [4,5,9])	11,29	4,17	243000	1,24
in 8 M Harnstoff	3,30	2,23	133000	2,83
bei pH 3 [9]	4,50	—	120000 [a]	1,94
in 5 M Guanidin-HCl	3,50	2,17	144000	2,83
in 5 M Guanidin-HCl 0,1 M Merkapto-äthanol	2,10	3,20	59000	2,59
Succinylierte Katalase	3,22	4,43	65000	1,81
in 20 mM Natriumdode-cylsulfat (nach Einwirkung von HCOOH)	3,8	5,1	60000	1,56
in 0,1 N NaOH	2,89	$[\eta] =$ 17,78 ml/g	54000 [b] 67000 [a]	1,90

[a] Bestimmung nach dem Archibald-Verfahren.
[b] Aus $s_{20,w}^0$ und $[\eta]$ berechnet.

der Dissoziationsprodukte. Die erhaltenen Unterschiede sind wahrscheinlich auf Grund der Fehlerbreite, die bei der Molekulargewichtsbestimmung unter denaturierenden Bedingungen größer ist als bei der Molekularbestimmung nativer Proteine, zu erklären und nicht darauf zurückzuführen, daß die nach Dissoziation erhaltenen Untereinheiten unterschiedliche Molekulargewichte besitzen.

In Abwesenheit von Merkaptoäthanol führt Guanidin-HCl zur Dissoziation in eine offenbar einheitliche Komponente vom Molekulargewicht etwa 140000. Diese ist wahrscheinlich — wie die in Gegen-

zeitige Einwirkung von NaOH (vgl. Ergebnisse). Diese Dissoziation der Katalase in vier Untereinheiten ist daher die Folge einer Veränderung der durch Nebenvalenzbindungen stabilisierten Konformation des Proteinmoleküls, d. h. im nativen Enzymmolekül sind die Untereinheiten ausschließlich durch Nebenvalenzbindungen miteinander verbunden.

In Gegenwart von Guanidin-HCl, Harnstoff oder bei pH 3 werden dagegen offenbar Konformationen stabilisiert, die nur eine Dissoziation in das halbe Molekül gestatten.

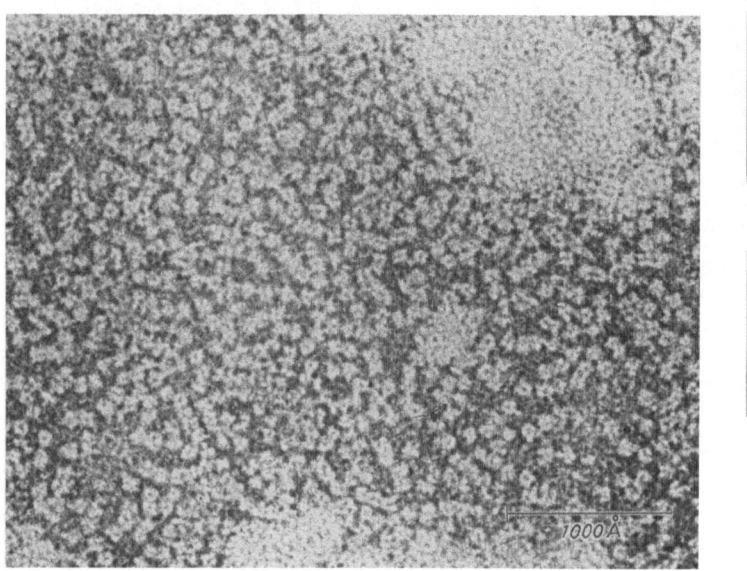
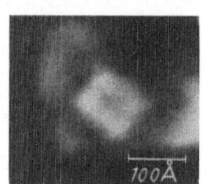

B

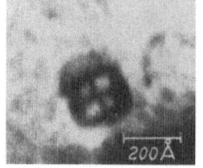

C

A

Fig. 11. *Elektronenmikroskopische Aufnahmen von Rinderleber-Katalase nach der negative-staining-Methode.* Vergrößerung 280000× (A), 1000000× (B) bzw. 500000× (C). Die Aufnahmen B und C zeigen eine deutliche Aufspaltung in vier Untereinheiten

wart von Harnstoff[2] oder bei pH 3 (Molekulargewicht 120000 [9]) erhaltenen Dissoziationsprodukte — durch Dissoziation des Katalase-Moleküls in zwei Untereinheiten entstanden.

Es ist berechtigt anzunehmen, daß unter unseren Versuchsbedingungen durch die Einwirkung von Natriumdodecylsulfat oder Bernsteinsäureanhydrid weder Peptid- noch Disulfidbindungen gespalten wurden. Dasselbe gilt offenbar auch für die kurz-

Das Dissoziationsschema der Rinderleber-Katalase ist nach diesen Ergebnissen folgendermaßen zu formulieren[3]:

$$240000 \rightleftharpoons 2 \times 120000 \rightleftharpoons 4 \times 60000. \qquad (3)$$

Mehrere Gründe sind für die Differenzen zwischen diesem Dissoziationsschema der Katalase und den früher aufgestellten [5—7] anzuführen. Erstens wurde für die Bestimmung der Molekulargewichte in den früheren Untersuchungen nicht in allen Fällen die Konzentrationsabhängigkeit der hydrodynamischen Eigenschaften untersucht. Wie die Fig. 3 und 6

---

[2] Nach Einwirkung von 7 M Harnstoff wurde ein Dissoziationsprodukt vom Molekulargewicht 151000 erhalten [6]. Wir können dieses Ergebnis prinzipiell bestätigen. Nach Inkubation in Gegenwart von 8 M Harnstoff erhalten wir ein Dissoziationsprodukt vom Molekulargewicht 133000 ($s_{20,w}^{\circ} = 3,30$ S; $D_{20,w}^{\circ} = 2,23$ F).

[3] Nach Inkubation bei pH 12 konnte eine 50%ige Reaktivierung erreicht werden [5]. Die Säuredenaturierung ist ebenfalls reversibel [9].

zeigen, beobachtet man unter denaturierenden Bedingungen eine starke Abhängigkeit der Sedimentationskoeffizienten von der Proteinkonzentration. Berücksichtigt man diese nicht, dann erhält man für die Molekulargewichte keine korrekten Werte. Das gilt ganz allgemein für denaturierte Proteine, die nicht mehr als globuläre, kompakte Moleküle zu beschreiben sind, sondern im entfalteten Zustand vorliegen (vgl. Tabelle 4, letzte Spalte). Zweitens wurden die Diffusionskoeffizienten offenbar häufig mit Lösungen bestimmt, die vorher nicht dialysiert wurden. Nach Überschichten solcher Lösungen sind Salzgradienten unvermeidlich, die Folge davon sind im allgemeinen zu hohe Diffusionskoeffizienten. Als dritter Grund ist möglicherweise anzuführen, daß unter den angewendeten Bedingungen die Denaturierungen nicht zu monodispersen Systemen führten[4].

Besonders interessant ist das Ergebnis der Dissoziation durch Guanidin-HCl in Gegenwart und Abwesenheit von Merkaptoäthanol. Nur in Anwesenheit von Merkaptoäthanol führt Guanidin-HCl zur Dissoziation in vier Untereinheiten, ohne Merkaptoäthanol entsteht eine Komponente vom Molekulargewicht etwa 140000. Die weitergehende Dissoziation in Gegenwart von Merkaptoäthanol ist möglicherweise darauf zurückzuführen, daß sich während der Denaturierung in Abwesenheit von Merkaptoäthanol durch Oxydation der freien SH-Gruppen Disulfidbrücken zwischen den Untereinheiten ausbilden, so daß hierdurch die Dissoziation in die vier Untereinheiten verhindert wird[5]. Vermutlich gelten diese Überlegungen auch für die durch Harnstoff hervorgerufene Dissoziation. Weitere Versuche müssen dieses Problem des Einflusses von Merkaptoäthanol auf die Dissoziation der Katalase klären[6].

Das Ergebnis des Aufbaus der Katalase aus vier Untereinheiten wird durch die elektronenmikroskopischen Untersuchungen gestützt. Das Auftreten von bis zu fünf Isozymen in Mäuse-, Ratten- und Schafsleber sowie in Mais [27,28] läßt sich mit dem Aufbau des Katalase-Moleküls aus vier Untereinheiten erklären, sofern man die für Rinderleber-Katalse erhaltenen Ergebnisse auf die Katalasen anderen Ursprungs übertragen darf.

Nimmt man an, daß sich jede der vier Hämingruppen [3] in einem aktiven Zentrum befindet, dann ist jeder Untereinheit vom Molekulargewicht 60000 ein aktives Zentrum zuzuordnen. Die Frage, ob diese Untereinheit schon die Polypeptidkette darstellt oder ob sie noch aus mehreren Polypeptidketten besteht, kann noch nicht beantwortet werden[7]. Die partielle Aufklärung der Aminosäuresequenz der Katalase spricht für den Aufbau aus vier identischen Untereinheiten [26]. Freie N-terminale Aminogruppen konnten bisher nicht gefunden werden. Der Nachweis von 4,3 bis 6,4 (im Mittel 5,3) Acetylgruppen pro Katalase-Molekül läßt vermuten, daß die Untereinheiten vom Molekulargewicht 60000 aus mehr als einer Polypeptidkette bestehen [26]. Diese können dann aber nicht identisch sein.

Frau H. Fröhlich danken wir für ihre ausgezeichnete Mitarbeit bei der Durchführung der Ultrazentrifugen-Versuche. Der Deutschen Forschungsgemeinschaft und dem Fonds der Chemischen Industrie danken wir für finanzielle Unterstützung. Herrn Dr. Bergmeyer (C. F. Boehringer und Soehne GmbH) danken wir für die großzügige Überlassung der verwendeten Katalase-Präparate. Der Henkel u. Cie GmbH danken wir für die Überlassung eines reinen Natriumdodecylsulfat-Präparates.

LITERATUR

1. Weber, K., und Sund, H., Angew. Chem. 77 (1965) 621; Int. Ed. 4 (1965) 597.
2. Sund, H., und Weber, K., Angew. Chem. 78 (1966) 217; Int. Ed. 5 (1966) 231.
3. Nicholls, P., und Schonbaum, G. R., in The Enzymes (edited by P. D. Boyer, H. Lardy und K. Myrbäck), Academic Press, 2nd ed., New York-London 1963, Vol. 8, S. 147.
4. Sumner, J. B., und Gralén, N., J. Biol. Chem. 125 (1938) 33.
5. Samejima, T., J. Biochem. (Tokyo), 46 (1959) 155.
6. Samejima, T., und Shibata, K., Arch. Biochem. Biophys. 93 (1961) 407.
7. Itoh, M., Nakamura, Y., und Shibata, K., Can. J. Biochem. Physiol. 40 (1962) 1327.
8. Tanford, C., und Lovrien, R., J. Am. Chem. Soc. 84 (1962) 1892.
9. Samejima, T., und Yang, J. T., J. Biol. Chem. 238 (1963) 3256.
10. Bergmeyer, H. U., Biochem. Z. 327 (1955) 255.
11. Boehringer-Information „KAT", August 1962.
12. Sund, H., und Weber, K., Biochem. Z. 337 (1963) 24.
13. Trautman, R., und Crampton, C. F., J. Am. chem. Soc. 81 (1959) 4036.

[4] Ein Hinweis für einen Aufbau der Katalase entsprechend Gleichung (3) wurde schon früher gegeben [8]. Lyophilisierte, partiell denaturierte und inaktivierte Katalase-Präparate enthalten neben der S_{11}-Komponente, die der nativen, enzymatisch aktiven Katalase entspricht, noch zwei weitere Komponenten mit Sedimentationskoeffizienten von 7,6 S und 4,2 S [8]. Auf Grund des Vergleiches der Sedimentationskoeffizienten wurde angenommen, daß der Komponente mit den Sedimentationskoeffizienten von 7,6 das halbe Molekulargewicht, der Komponente mit 4,2 S ein Viertel des Molekulargewichtes nativer Katalase zukommt [8]. Eine vollständige Dissoziation in zwei bzw. vier Untereinheiten konnte jedoch nicht beobachtet werden, stets lag ein Gemisch von mindestens drei Komponenten vor. Notwendigerweise war daher auf Grund dieser Ergebnisse die Frage nach der Quartärstruktur der Katalase nicht eindeutig zu beantworten.

[5] Im nativen Proteinmolekül sind 6 freie SH-Gruppen, nach Denaturierung 16 freie SH-Gruppen nachweisbar [9]. Da das Katalase-Molekül etwa 16 Cys/2-Reste enthält [26], sind Disulfidbrücken offenbar nicht anwesend.

[6] In diesem Zusammenhang ist es interessant zu erwähnen, daß nach Oxydation mit Perameisensäure und anschließender Einwirkung von Guanidin-HCl ein Dissoziationsprodukt vom Molekulargewicht etwa 60000 erhalten wurde. Im Falle der Succinylierung mit anschließender Inkubation in Gegenwart von Guanidin-HCl wurde das Molekulargewicht dagegen zu etwa 170000 gefunden.

[7] Versuche in Ameisensäure-Essigsäure geben Hinweise darauf, daß das Katalase-Molekül möglicherweise aus acht Polypeptidketten aufgebaut ist.

28*

14. Svedberg, T., und Pedersen, K. O., *Die Ultrazentrifuge.* Theodor Steinkopff, Dresden-Leipzig 1940.
15. Elias, H. G., *Ultrazentrifugen — Methoden.* (2. Aufl.) Beckman Instruments GmbH, München 1961.
16. Tanford, C., *J. Phys. Chem.* 59 (1955) 798.
17. Klotz, I. M., und Keresztes-Nagy, S., *Biochemistry*, 2 (1963) 445.
18. Schroeder, W. A., Shelton, J. R., Shelton, J. B., Cormick, J., und Jones, R. T., *Biochemistry*, 2 (1963) 992.
19. Weber, K., Dissertation, Freiburg 1964.
20. Schroeder, W. A., Saha, A., Fenninger, W. D., und Cua, J. T., *Biochim. Biophys. Acta*, 58 (1962) 611.
21. Scheraga, H. A., und Mandelkern, L., *J. Am. Chem. Soc.* 75 (1953) 179.
22. Kawahara, K., Kirshner, A. G., und Tanford, C., *Biochemistry*, 4 (1965) 1203.
23. Wallenfels, K., Sund, H., und Weber, K., *Biochem. Z.* 338 (1963) 714.
24. Hofmann, T., und Harrison, P. M., *J. Mol. Biol.* 6 (1963) 256.
25. Hersh, R. T., und Schachman, H. K., *Virology*, 6 (1958) 234.
26. Schroeder, W. A., Shelton, J. R., Shelton, J. B., und Olson, B. M., *Biochim. Biophys. Acta*, 89 (1964) 47.
27. Scandalios, J. G., *Proc. Natl. Acad. Sci. U. S.* 53 (1965) 1035.
28. Holmes, R. S., und Masters, C. J., *Arch. Biochem. Biophys.* 109 (1965) 196.
29. Osbahr, A. J., und Eichhorn, G. L., *J. Biol. Chem.* 237 (1962) 1820.
30. Sandell, E. B., *Colorimetric Determination of Traces of Metals* (3rd edition), Interscience Publishers, New York-London 1959, S. 537.

H. Sund's jetzige Adresse:
Fachbereich Biologie der Universität,
775 Konstanz, Deutschland

K. Weber's jetzige Adresse:
Biological Laboratories, Harvard University,
Cambridge, Mass., Vereinigte Staaten von Amerika

E. Mölbert's jetzige Adresse:
Department of Physics, Physical Science Building,
Kansas State University, Manhattan, Kansas,
Vereinigte Staaten von Amerika

European J. Biochem. 1 (1967) 411—418

Isolement de l'autoantigène responsable de la formation d'autoanticorps chez les malades atteints de cancer gastrique

D. Karitzky et P. Burtin

Institut de Recherches Scientifiques sur le Cancer, Villejuif

(Reçu le 1 mars 1967)

Several studies have already originated from this laboratory dealing with the presence of autoantibodies in patients suffering from carcinomas of the digestive tract. The antigen inducing these autoantibodies has been shown to be present in the tumor cell as well as in the normal one. The following work is concerned with the isolation of the antigen found in normal and cancerous gastric mucosa.

The active products have been found to be soluble in both perchloric and phytic acids and to be precipitated by ammonium sulfate. Electrophoretically the fractions have been found to be still impure (Table).

After chromatography on DEAE-cellulose the graphs of the normal and cancerous extracts were found to be similar (Fig. 1 and 2). The activity of the autoantigen was found mainly in the first, and to a lesser degree in the third of the seven peaks (P_1, P_3).

P_1 was therefore subjected to Sephadex G 100 filtration, after which three peaks were generally seen: A, B and C. Slight differences were observed within these peaks: C was lower after the passage of tumor extract than after that of normal extracts. Antigenic activity was seen only in some tubes of peak C (hachured part of Fig. 3 and 4). 10 μg of isolated peak C material, coming from a fraction of phytic acid extracted cancerous mucosa were sufficient to absorb out all antibodies of the patients' sera. After migration in acrylamide agarose this product showed one single band of α_2-globulin mobility (Fig. 5). Immunoelectrophoresis revealed only one precipitin line (Fig. 7). When P_3 was filtered on Sephadex G 100, an active antigen was detected in the A region. Its relationship with the P_1 antigen is still under discussion.

The physico-chemical analysis gave the following results: the purest isolated fraction contained 6% hexoses, its activity was destroyed by chymotrypsin but left unaltered by trypsin, ribonuclease and heating. Ultra-centrifugation indicated a sedimentation constant of 1,83 S.

A highly concentrated, highly active product which represents most probably the autoantigen in highly purified state, has been isolated from the cancerous gastric mucosa. The finding of one band after electrophoretic migration, of a single line after immunoelectrophoresis, and of one peak in analytical ultracentrifugation, are strong arguments in favour of the purity of the component. The substance is of small molecular weight and almost entirely protein in nature (only 6% hexoses were detected) which makes it comparable to the antigen characterized in experimental encephalomyelitis in guinea pigs.

Gel filtration of non-lyophilized unfractionated extracts of normal gastric mucosa yields a product having the same immunological specificity as the pure product. This is strong evidence against the possibility of denaturation during fractionation.

Comparison between the fractions obtained from normal and tumorous mucosa showed complete identity as demonstrated by electrophoretic and immunoelectrophoretic studies. It was not possible to obtain with material from normal mucosa the level of purity of the tumor material.

By chemical studies, it could be shown that the antigen isolated from normal gastric mucosa is different from the microsomial antigen reacting with autoantibodies of pernicious anemia sera, the properties of which have been studied by Baur et al.

As a hypothetical mechanism for the autoantigenicity of a normal component it was assumed that substances normally sequestered within the cell are liberated into the blood stream, thus giving rise to the formation of autoantibodies.

Plusieurs travaux de ce laboratoire ont été effectués sur les autoanticorps présents dans le sérum des malades atteints de cancer digestif [1—4].

Ces autoanticorps ne sont pas spécifiques de la maladie cancéreuse et n'ont donc pas une grande valeur diagnostique. Ils sont en fait dirigés contre un constituant présent dans la cellule normale comme dans la cellule cancéreuse. Ce travail décrit l'isolement et l'étude de l'autoantigène présent dans la muqueuse gastrique normale et tumorale.

MATERIEL

Extraits tissulaires

Nous avons utilisé des estomacs prélevés chirurgicalement soit pour un cancer gastrique, soit pour un ulcère (48 tumeurs, 68 estomacs non cancéreux). Dans tous les cas, le diagnostic clinique fut contrôlé par un examen histologique. Les muqueuses gastriques normales et tumorales ont été découpées, lavées et homogénéisées par l'Ultra-Turrax (Jahnke & Kunkel, K.G. Staufer i. Br., Allemagne) dans un tampon phosphate 0,1 M pH 8,0 [5]. Après centrifugation à 30.000 × g pendant 30 min (MSE High Speed 18) le surnageant a été lyophilisé. En général, pour les stades ultérieurs de la purification nous avons mélangé plusieurs extraits provenant d'estomacs atteints de modifications pathologiques semblables.

Sérums humains

Nous avons utilisé huit sérums de malades donnant un titre élevé ($^1/_{80}$, $^1/_{160}$) en hémagglutination passive [6,7]. Quatre étaient atteints d'un cancer d'estomac, un d'un cancer du sein, un d'un cancer de la parotide, un d'un cancer de l'oesophage et une malade présentait une carcinomatose pelvienne.

Immunsérums de lapin

Les immunsérums ont été préparés chez le lapin par injection d'extraits de muqueuse gastrique tumorale partiellement purifiés par traitement avec l'acide perchlorique et le sulfate d'ammonium (anti TSPP). L'immunisation intradermique avec trois mg de protéines additionnées d'adjuvant de Freund complet, dans la pulpe de la patte, a été suivie de quatre séries d'injections de rappel à un mois d'intervalle. Chaque série de rappel a consisté en une injection sous-cutanée et deux injections intraveineuses, chacune de 3 mg de protéines additionnées d'une solution d'alun d'ammonium à 1%.

METHODES

Précipitation par l'acide perchlorique et le sulfate d'ammonium

Le surnageant obtenu après centrifugation de l'extrait tissulaire, contenant 1 à 2 g de protéines

mesurées par le réactif du biuret dans 250 ml de tampon phosphate, ou le lyophilisat redissous dans l'eau distillée (mêmes quantités environ), a été mélangé avec une quantité égale d'acide perchlorique à 10% refroidi à +4°C. La concentration d'acide perchlorique dans le mélange était de 0,5 M. La précipitation a été immédiatement suivie d'une centrifugation à +4°C (15.500 × g pendant 15 min), le surnageant a été ajusté à un pH de 6,8 avec NaOH 0,5 M, puis précipité par le sulfate d'ammonium à saturation. Le précipité obtenu après centrifugation a été redissous dans 10—20 ml d'une solution isotonique de NaCl et dialysé pendant 5 jours contre de l'eau distillée renouvelée deux fois par jour. Le dialysat a été finalement lyophilisé.

Précipitation par l'acide phytique et le sulfate d'ammonium

Comme autre réactif d'extraction nous avons utilisé l'acide inositohexaphosphorique ou phytique [8]. Deux solutions ont été préparées: (a) eau distillée 600 ml, phytate de sodium 1,3 g, HCl pur 2 ml, pH ajusté à 2,1 avec HCl 1 M; (b) extrait tissulaire (5 à 8 g en protéines) dans 400 ml d'une solution de NaCl isotonique.

Après mélange de la totalité des solutions A et B et ajustement du pH à 2,1 avec HCl 1 M, on a laissé la préparation pour une heure à la température du laboratoire. Après centrifugation (5.500 × g pendant 15 min) 1 ml du surnageant a été agité avec 1 ml de la solution (a) pour vérifier qu'il ne contenait plus de protéines précipitables avec l'acide phytique. Le surnageant a été précipité par le sulfate d'ammonium à saturation, le précipité redissous dans une solution de NaCl isotonique, dialysé et lyophilisé.

En traitant la muqueuse gastrique normale et les tumeurs gastriques selon les deux méthodes précédentes, nous avons obtenu les quatre produits suivants: (a) extrait perchlorique de la muqueuse gastrique normale précipité par le sulfate d'ammonium à saturation (NSPP); (b) extrait perchlorique de la muqueuse gastrique tumorale précipité par le sulfate d'ammonium à saturation (TSPP); (c) extrait phytique de la muqueuse gastrique normale précipité par le sulfate d'ammonium à saturation (NSPhyP); (d) extrait phytique de la muqueuse gastrique tumorale précipité par le sulfate d'ammonium à saturation (TSPhyP).

C'est le TSPP qui a servi à l'immunisation des lapins mentionnée au chapitre Matériel.

Chromatographie sur DEAE-cellulose

Nous avons utilisé, pour séparer les différents produits semi purifiés, des colonnes de DEAE cellulose (Whatman Ion-Exchange cellulose, DEAE-cellulose, 1 mEq/g, W & R Balston Ltd) de 20 cm de hauteur et de 1,5 cm de diamètre. Les protéines ont été éluées au départ avec un tampon phosphate 0,005 M, pH 6,8. Puis la molarité a été augmentée par paliers grâce à l'utilisation de tampons phosphates de concentrations croissantes (0,05 M, 0,1 M, 0,2 M, 0,3 M, 0,5 M, 1 M). Les éluats ont été mesurés au spectrophotomètre à 280 et 260 mμ.

Filtration sur Sephadex G 100

Les éluats réunis ont été dialysés contre l'eau distillée pour 24 h, lyophilisés et redissous dans une quantité minimale de NaCl en solution à 8,5⁰/₀₀ (pH 7,3). Cinquante à deux cents mg dans un volume de 3 ml environ ont été filtrés sur une colonne de Sephadex G 100 de 60 cm/2 cm (Pharmacia, Uppsala, Suède) également en solution isotonique de NaCl. Le contenu des différentes fractions a été mesuré au spectrophotomètre à 280 et 260 mμ et les tubes groupés en fonction de la courbe obtenue. Après une dialyse pendant 36 h contre de l'eau distillée trois fois renouvelée les fractions ont été lyophilisées.

Hémagglutination passive

L'hémagglutination passive a été employée selon la technique de Boyden [6], modifiée par Stavitsky [7]. Les globules rouges de mouton (Institut Pasteur) ont été tannés par l'acide tannique à $1/_{80.000}$ puis sensibilisés par les fractions des différents stades de purification. Ils ont été mis en contact avec les sérums humains précédemment sélectionnés, inactivés par chauffage à 56° pendant 30 min. Les sérums ont été dilués de $1/_2$ en $1/_2$ à partir du $1/_{10}$ avec du tampon phosphate à pH 7,4, 0,15 M additionné de EDTA (10 mg par litre) et de sérum de lapin normal inactivé à 1⁰/₀.

L'hémagglutination passive a été aussi effectuée après absorption des autoanticorps par ces fractions plus ou moins purifiées. Pour l'absorption, les sérums de malades ont été incubés pendant deux heures à 37° avec différentes quantités de ces fractions. Dans ces expériences, le NSPP a été utilisé pour sensibiliser les hématies, à la concentration de 0,2 mg de protéines par ml.

Electrophorèse simple en acrylamide agarose

La purification de l'autoantigène a été suivie par électrophorèse simple en acrylamide-agarose [9] à pH 8,2 (tampon véronal 0,025 M, temps de migration 2 h 30 min à 10 V/cm). La coloration des protéines après la migration a été effectuée avec le Coomassie Brillant Blue (R. E. Gurr Ltd. London S. W. 14) [10].

Electrophorèse simple en agarose

Elle a été effectuée en agarose 1⁰/₀ (Industrie Biologique Française, Gennevilliers, France) dans un tampon véronal 0,05 M à pH 8,2 (voltage 5 V/cm). Après un temps de migration de 1 h 30 min les plaques ont été colorées selon la méthode d'Uriel utilisant la réaction acide périodique-NADI (α-naphtol et para-phénylène diamine) pour détecter les glucides [11].

Immunodiffusion

La méthode de double diffusion en gélose a été employée selon Ouchterlony [12] et l'analyse immunoélectrophorétique selon Grabar et Williams [13]. La migration a été faite en gélose à 1,5⁰/₀ (tampon véronal 0,05 M à pH 8,2), 5 V/cm.

Dosage des glucides

Les dosage des hexoses, a été fait avec le réactif Orcinol + H_2SO_4 (témoin glucose) celle des pentoses par la réaction de Bial (Orcinol + $FeCl_3$ en solution chlorhydrique) avec le désoxy-D-ribose comme témoin.

Ultracentrifugation

Des ultracentrifugations analytiques ont été effectuées dans une ultracentrifugeuse Spinco Modèle E (Rotor type: Cand, cellule à frontière préformée en aluminium).

RESULTATS

Les résultats des fractionnements ont été suivis deux manières:

a) d'une part en déterminant la concentration de l'autoantigène, jugée d'après la quantité minimale de substance absorbant les autoanticorps des sérums étudiés, donc inhibant l'hémagglutination passive.

b) d'autre part en jugeant les résultats des méthodes physicochimiques parmi lesquelles avant tout l'électrophorèse en acrylamide-agarose et l'analyse immunoélectrophorétique.

RESULTATS DES FRACTIONNEMENTS

La précipitation par l'acide perchlorique ou l'acide phytique et le sulfate d'ammonium

Le produit actif s'est révélé soluble dans l'acide perchlorique et précipitable par le sulfate d'ammonium à saturation. Il en a été de même lorsque l'acide perchlorique a été remplacé par l'acide phytique, considéré comme moins dénaturant.

Les produits ainsi obtenus, que ce soient à partir de la muqueuse gastrique normale (NSPP et NSPhyP) ou tumorale (TSPP et TSPhyP), étaient très impurs. (La Fig. 5 montre que TSPhyP donne de nombreuses bandes après électrophorèse en acrylamide agarose). Cependant l'autoantigène y est déjà assez concentré, puisqu'il faut environ 50 fois moins de ces fractions que des extraits bruts pour inhiber l'hémagglutination passive (Tableau).

Tableau. *Quantités* (en mg) *de protéines nécessaires pour l'absorption des auto-anticorps contenus dans 1 ml de sérum*

	Muqueuse gastrique normale				Muqueuse gastrique tumorale			
Extraction aqueuse	50				20			
	Précipitation[a] préalable à l'acide							
	perchlorique		phytique		perchlorique		phytique	
Précipitation par le sulfate d'ammonium saturé	(NSPP) 0,5—1,0		(NSPhyP) 0,5—1,0		(TSPP) 0,5		(TSPhyP) 0,5	
Chromatographie sur DEAE-cellulose	(P_1) 0,3	(P_3) 1,0	(P_1) 0,4	(P_3) 1,0	(P_1) 0,2	(P_3) 0,4	(P_1) 0,2	(P_3) 0,3
Filtration sur Sephadex G 100	0,05—0,1		0,05—0,1		≤ 0,03		0,01	0,1

[a] Le précipité est éliminé; le surnageant est précipité par le sulfate d'ammonium à saturation.

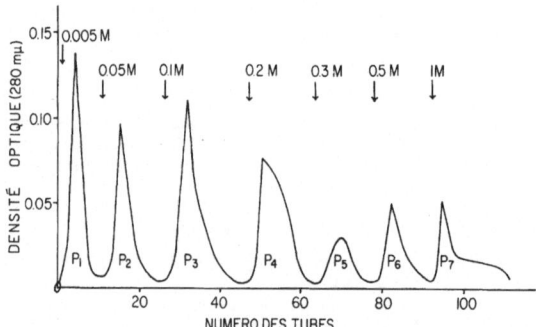

Fig. 1. *Chromatographie sur DEAE-Cellulose en tampon phosphate à pH 6,8 de l'extrait TSPP.* Changement de la molarité du tampon phosphate par paliers. Lecture des éluats à 280 mμ

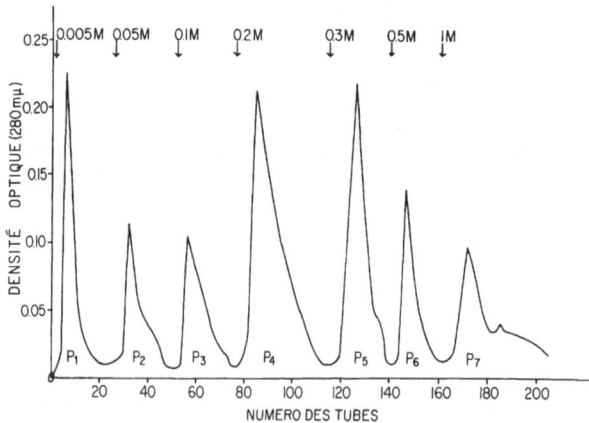

Fig. 2. *Chromatographie sur DEAE-Cellulose en tampon phosphate à pH 6,8 de l'extrait NSPP.* Changement de la molarité du tampon phosphate par paliers. Lecture des éluats à 280 mμ

La chromatographie sur DEAE-cellulose

En chromatographiant les quatre extraits (NSPP, NSPhyP, TSPP, TSPhyP) les diagrammes obtenus ont été comparables avec cependant des variations dans les hauteurs des pics. La Fig. 1 montre une chromatographie de l'extrait TSPP, la Fig. 2 une de l'extrait NSPP. L'activité autoantigénique a été retrouvée dans le premier pic (P_1) et dans une moindre mesure dans le troisième (P_3). Il est à noter, que P_1 inhibe l'agglutination par les hématies sensibilisées par P_3, et inversement. La concentration de l'antigène est seulement 2 à 3 fois plus grande que dans les extraits non chromatographiés (Tableau). Par contre, le nombre des bandes après électrophorèse en acrylamide-agarose est moins grand (Fig. 5).

La filtration sur Sephadex G 100

Premier pic chromatographique. C'est presque toujours les substances contenues dans le premier pic (P_1) recueilli après chromatographie sur DEAE-cellulose, qui ont été soumises à la filtration sur gel. Trois pics ont été le plus souvent obtenus, et nommés A, B et C. Parfois A et B étaient fusionnés en un seul pic assez large. Le pic C provenant de NSPP était moins haut et plus étalé que celui obtenu à partir de NSPhyP. Dans l'ensemble le rendement en pic C a été plus faible pour les extraits tumoraux que pour les extraits normaux. Le produit actif a été trouvé seulement dans les tubes correspondant à une partie du pic C (partie indiquée par une zone noire ou hachurée sur les Fig. 3 et 4). Pour inhiber totalement l'hémagglutination passive, il as uffi de 10 µg/ml du produit isolé à partir de TSPhyP (TSPhyP-P_1-C) et de 30 µg/ml du produit obtenu à partir de TSPP. A poids égal les substances provenant des extraits normaux NSPP et NSPhyP étaient beaucoup moins

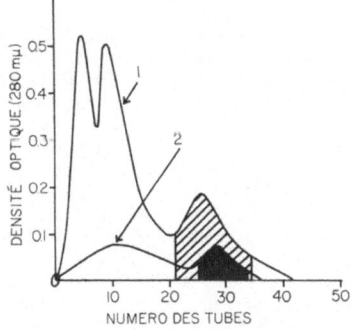

Fig. 3. *Filtration sur Sephadex G 100 du TSPhyP-P_1 (1) et TSPP-P_1 (2).* Lecture des éluats à 280 mµ. Les régions noires et hachurées correspondent aux fractions contenant l'autoantigène (TSPhyP-P_1-C et TSPP-P_1-C)

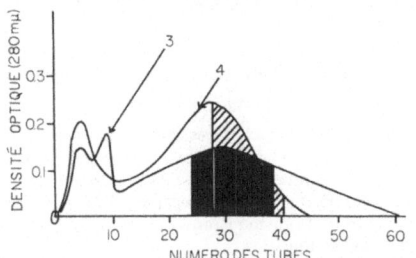

Fig. 4. *Filtration sur Sephadex G 100 du NSPP-P_1 (3) et NSPhyP-P_1 (4).* Lecture des éluats à 280 mµ. Les régions noires et hachurées correspondent aux fractions contenant l'autoantigène (NSPP-P_1P-C et NSPhyP-P_1-C)

actives. Le rendement final de l'extraction est faible: à partir d'un extrait brut contenant 5 à 10 g de protéines, nous avons recueilli quelques milligrammes d'autoantigène purifié. Le produit obtenu à partir des extraits normaux était un peu plus abondant, mais moins pur.

L'électrophorèse en acrylamide-agarose a montré:

a) dans le cas du produit obtenu à partir de TSPhyP (= TSPhyP-P_1-C) une seule bande de mobilité assez rapide, correspondant à celle des α2-globulines sériques, et une traînée atteignant le réservoir de départ et le dépassant en arrière. La

bande nette a la même mobilité que l'une de celles données par TSPhyP et sa fraction de chromatographie P_1 (Fig. 5).

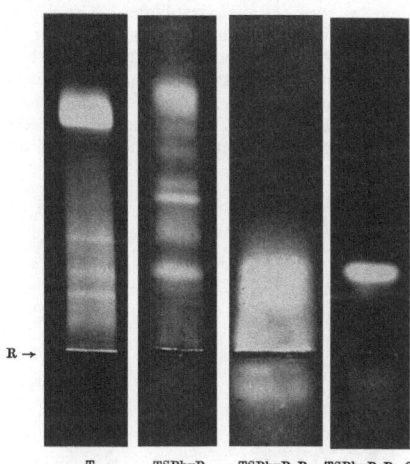

Fig. 5. *Electrophorèse simple en acrylamide-agarose montrant la purification successive de l'autoantigène à partir de TSPhyP.* A gauche, comme témoin: l'extrait aqueux brut de la muqueuse gastrique tumorale (T). Pôle positif en haut, R = réservoir de départ

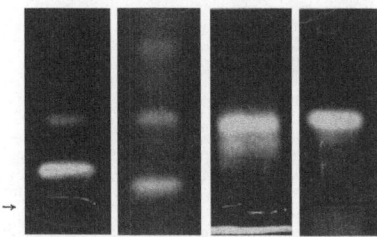

Fig. 6. *Electrophorèse simple en acrylamide-agarose des fractions purifiées par filtration sur Sephadex, à partir des différents extraits normaux et tumoraux.* Pôle positif en haut, R = réservoir de départ

b) dans le cas des produits obtenus à partir des autres fractions (NSPP, NSPhyP, TSPP) plusieurs bandes dont l'une de mobilité comparable à celle donnée par TSPhyP-P_1-C (Fig. 6).

L'analyse immunoélectrophorétique de TSPhyP-P_1-C, avec un immunsérum anti TSPP, a donné une seule ligne de précipitation à double courbure, la principale de mobilité rapide, l'autre entourant le réservoir de départ (Fig. 7). Avec le même antisérum, NSPhyP-P_1-C a donné plusieurs lignes de précipitation, dont une seule était forte: cette ligne avait la même mobilité que celle donnée par TSPhyP-P₁-C, et à peu près la même forme, avec cependant une courbure lente plus faible (Fig. 8).

Troisième pic chromatographique. En ce qui concerne les pics chromatographiques 3, seul celui obtenu à partir de TSPhyP a été fractionné sur Sephadex G 100, 3 pics ont été obtenus. L'activité autoantigénique a été retrouvée seulement dans le premier pic P₃A et seulement trois fois plus concentrée que dans le produit P₃ non fractionné. L'électrophorèse en acrylamide-agarose de P₃A a montré une

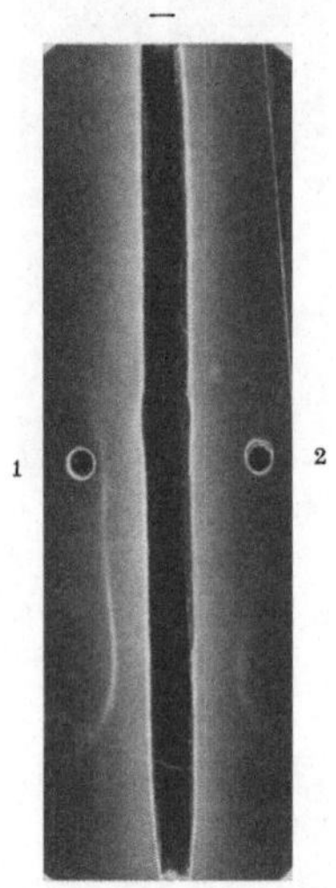

Fig. 7 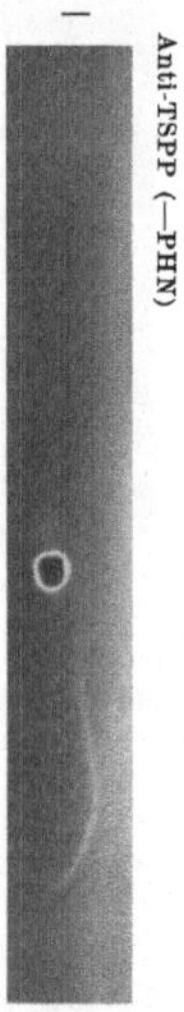 **Fig. 8**

Fig. 7. *Analyse immunoélectrophorétique de TSPhyP-P₁-C.* (1) TSPhyP-P₁-C (4 mg/ml); (2) TSPhyP-P₁-C, traité par la chymotrypsine. Concentration initiale 4 mg/ml. Gouttière: antisérum de lapin anti TSPP, absorbé par le plasma humain normal. Pôle positif en bas

Fig. 8. *Analyse immunoélectrophorétique de NSPhyP-P₁-C avec l'immunsérum anti TSPP* (il existe outre les lignes de précipitation principales une faible ligne de mobilité β non visible sur la photographie)

faible tache aux limites peu nettes dans une zone assez rapide, et une forte et étroite bande juste en avant du réservoir de départ.

L'analyse immunoélectrophorétique de P₃A avec l'immunsérum anti TSPP a donné une ligne de précipitation entourant le réservoir de départ. Cette ligne n'a pas été modifiée après absorption de l'antisérum par TSPhyP-P₁-C. De plus en double diffusion elle croisait celle donnée par TSPhyP-P₁-C. Elle correspond donc à un antigène différent de celui responsable de l'hémagglutination passive.

Extrait total. La filtration sur Sephadex G 100 de l'extrait total de la muqueuse gastrique normale a

été faite deux fois: (a) sur un extrait lyophilisé; (b) sur un extrait non lyophilisé, concentré par pervaporation.

La courbe d'élution de l'extrait lyophilisé (900 mg) a comporté de nombreux pics. L'autoantigène a été élué tardivement dans un volume comparable à celui qui était nécessaire pour obtenir les produits actifs au cours de la filtration des produits partiellement purifiés (NSPP-P₁ et NSPhyP-P₁); la quantité de fraction active nécessaire pour absorber les autoanticorps était de 2 à 5 mg (en protéines) par ml de sérum.

Pour l'autre filtration sur Sephadex G 100, nous avons utilisé un extrait aqueux de la muqueuse gastrique normale concentré par pervaporation, contenant 300 mg de protéines. La fraction capable d'inhiber l'hémagglutination passive a été trouvée principalement dans les éluats qui passaient à la colonne assez tard (les premiers présentant néanmoins aussi une faible activité). En AIE, cette fraction active a donné avec l'anti TSPP beaucoup de lignes de précipitation dont l'une à double courbure d'une mobilité assez rapide était comparable à celle obtenue avec le produit TSPhyP-P₁-C. Une analyse immunoélectrophorétique comparative entre cette fraction et le TSPhyP-P₁-C, en utilisant la méthode des gouttières interrompues [14], a donné une identité complète. En électrophorèse en acrylamide-agarose, la grande quantité et la superposition des bandes n'ont pas permis de retrouver celle due à l'autoantigène.

ETUDES PHYSICO-CHIMIQUES SUR L'AUTOANTIGÈNE

L'autoantigène purifié

Dosage des hexoses et pentoses. L'analyse du produit TSPhyP-P₁-C donnait les valeurs suivantes: le produit contenait 6% de hexoses (substances positives avec l'orcinol). Le réactif de Bial n'a donné aucune coloration, montrant l'absence de pentoses, et d'acide sialique.

Action de la chymotrypsine. Le TSPhyP-P₁-C, dissous dans un tampon phosphate à pH 8 a été soumis pendant 24 h à une protéolyse à 20° par la chymotrypsine (α-Chymotrypsine crist., A grade, Calbiochem, Los Angeles/USA).

La solution contenait 2 mg par ml de TSPhyP-P₁-C et 60 μg de chymotrypsine. La réaction a été arrêtée par l'addition d'une goutte de diisopropyl fluorophosphate 10⁻³ M. Après cette digestion, le produit a perdu son activité autoantigénique. La ligne de précipitation obtenue avec l'immunsérum anti TSPP a été très modifiée: il restait une ligne très faible de mobilité proche de celle de la courbure rapide de la ligne de l'autoantigène non traité par l'enzyme (Fig. 7).

Ultracentrifugation analytique. Le TSPhyP-P₁-C a donné un seul pic. La constante de sédimentation (non extrapolée à zéro) était $s = 1,83$, ce qui correspond approximativement à un poids moléculaire compris entre 10000 et 20000 (Fig. 9).

↓ Ménisque

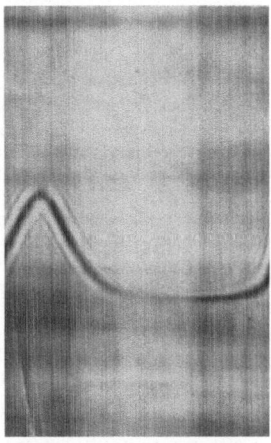

Fig. 9. *Sédimentation d'un échantillon de TSPhyP-P_1-C juste après formation de la frontière.* Centrifugé en cellule à frontière synthétique, vitesse de rotation 50.740 tours/min, temps d'exposition 40 secondes; température 20°, angle de phase 45°. Concentration du produit 7,5 mg de protéines/ 0,5 ml de NaCl (8,5 g/l), pH 7,3

L'extrait brut

Action des enzymes. Un extrait brut de la muqueuse gastrique tumorale a été traité par la trypsine aux concentrations de 1 et 10 mg/100 mg et par la ribonucléase à 1 mg/100 mg. Aucune modification n'a été trouvée quant à la faculté du produit d'absorber les autoanticorps dans les sérums étudiés.

Le même extrait brut, sous l'action de la chymotrypsine à 1%, a perdu tout pouvoir inhibiteur de l'hémagglutination passive.

Chauffage. Un chauffage du produit à 90° pendant 5 min n'a pas modifié sa capacité d'absorber les autoanticorps.

DISCUSSION

L'autoantigène purifié

A partir de la muqueuse gastrique tumorale, nous avons donc obtenu un autoantigène hautement concentré et purifié (TSPhyP-P_1-C). Avec les critères de pureté que nous avons utilisés cette substance donne : (a) une seule bande en électrophorèse simple en acrylamide-agarose (avec une faible trainée entre la bande et le réservoir de départ); (b) une seule ligne en AIE avec deux courbures, dont la deuxième a la même mobilité que la traînée observée en électrophorèse simple; (c) un seul pic en ultracentrifugation analytique.

C'est seulement grâce à l'extraction par l'acide phytique que ce résultat a été obtenu. L'extraction par l'acide perchlorique a donné une fraction moins pure.

Cet autoantigène est une protéine de poids moléculaire faible. Ceci rappelle le poids moléculaire trouvé pour l'autoantigène de l'encéphalomyélite allergique du cobaye isolé par Laatsch *et al.* [15]. Notre préparation contient un peu d'hexoses, cependant, ceux-ci pourraient être dus à une contamination par d'autres constituants. En effet, dans la fraction TSPhyP-P_1-B qui précède la fraction C dans la filtration sur Sephadex G 100, on trouve peu de protéines et une quantité importante de glucides, donnant une forte coloration après électrophorèse simple en agarose. Une contamination du TSPhyP-P_1-C par des glucides provenant de la fraction précédente ne peut donc être exclue.

Malgré sa teneur faible en glucides, cette substance est soluble dans l'acide perchlorique et l'acide phytique, c'est sans doute son faible poids moléculaire qui est responsable de cette propriété. L'hétérogénéité de notre préparation purifiée a été observée en électrophorèse et donnait lieu à une double courbure de la ligne de précipitation en AIE. On peut se demander toutefois si ces aspects ne sont pas dus à une certaine dénaturation survenue au cours du fractionnement. Contre cette hypothèse il y a deux arguments:

a) en filtrant sur le Sephadex G 100 un extrait frais de la muqueuse gastrique normale, ni lyophilisé, ni extrait en milieu acide, on obtient un produit actif en hémagglutination passive qui donne une ligne de précipitation à double courbure, identique à la ligne obtenue avec TSPhyP-P_1-C;

b) en ultracentrifugation analytique, cette fraction active donne un pic (entre d'autres), qui a aproximativement la même constante de sédimentation que le TSPhyP-P_1-C.

Il est donc vraisemblable que la protéine autoantigénique, homogène en ultracentrifugation, est électrophorétiquement hétérogène.

La substance dans le pic 3 de chromatographie sur DEAE-cellulose

Cette substance a les mêmes propriétés immunologiques que le TSPhyP-P_1-C, puisqu'elle absorbe complètement les autoanticorps. Mais outre son comportement chromatographique différent, elle doit avoir un poids moléculaire nettement plus élevé, puisqu'elle passe avec le premier pic après filtration sur Sephadex G 100. Plusieurs hypothèses peuvent être faites sur les rapports entre cette substance et celle contenue dans le TSPhyP-P_1-C. Les propriétés physico-chimiques de l'autoantigène trouvé dans P_3 peuvent être expliquées par: (a) une dénaturation au cours de la préparation; (b) un complexe avec une autre molécule d'où un changement de la charge électrique et du poids moléculaire; (c) l'existence d'un polymère de la substance contenue dans le premier pic de chromatographie.

La très faible quantité de cette substance que nous avons obtenue rend son étude difficile. Elle est, en outre fortement impure, puisqu'elle contient un grand pourcentage d'un constituant électrophorétique plus lent, immunochimiquement différent de TSPhyP-P$_1$-C. Aussi est-il difficile de discuter les hypothèses précédemment émises. Il existe quand même un argument contre l'hypothèse de la dénaturation: en filtrant un extrait total aqueux de la muqueuse gastrique normale non lyophilisée sur une colonne de Sephadex G 100, il passe dans les premiers tubes une substance active en hémagglutination passive. Il est donc possible qu'une substance de poids moléculaire relativement élevé et à activité autoantigénique soit présente dans la muqueuse gastrique normale.

Rapport entre les substances actives obtenues à partir de la muqueuse gastrique tumorale et normale

Nous avons étudié parallèlement les autoantigènes de la muqueuse gastrique normale et tumorale. Ils ont les mêmes comportements sur colonne de DEAE-cellulose et de Sephadex G 100 et la même spécificité immunologique. Mais on ne peut prouver leur identité physicochimique, puisque nous n'avons pas pu obtenir à l'état pur l'autoantigène de la muqueuse gastrique normale. Cette identité est cependant très vraisemblable, pour les raisons suivantes: les extraits NSPP-P$_1$-C et NSPhyP-P$_1$-C ont donné plusieurs bandes en acrylamide-agarose (Fig. 6), dont une analogue à celle de l'autoantigène extrait de la muqueuse tumorale (TSPhyP-P$_1$-C). La principale ligne de précipitation donnée par NSPhyP-P$_1$-C est d'aspect presque identique à celle obtenue avec TSPhyP-P$_1$-C (Fig. 7 et 8). Il y a dans NSPP-P$_1$ un constituant qui à l'ultracentrifugation se révèle avoir une constante de sédimentation d'environ 1,5 S. Une valeur comprise entre 1,5 et 2 S a été trouvée pour un constituant de la fraction autoantigéniquement active de l'extrait de muqueuse gastrique tumorale filtré directement sur Sephadex G 100.

Ainsi y a-t-il des arguments nombreux et concordants pour une très grande parenté et probablement une identité, entre les autoantigènes normal et tumoral.

Comparaison avec l'autoantigène de la muqueuse gastrique normale impliqué dans les anémies pernicieuses

Parmi les autoanticorps présents dans le sérum des malades atteints d'anémie pernicieuse, Irvine *et al.* [16] ont décrit des autoanticorps dirigés contre un constituant cytoplasmique des cellules gastriques pariétales. Ce constituant est en fait situé sur les microsomes. Ses propriétés ont été étudiées par Baur

et al. [17]. Il s'agit d'une lipoprotéine insoluble dont l'activité est détruite par les détergents, donc très différente de l'autoantigène que nous avons isolé. D'ailleurs chez les mêmes malades on trouve des titres très différents pour les anticorps dirigés contre l'antigène de l'anémie pernicieuse et ceux réagissant avec l'antigène des cancers gastriques [3].

Mécanisme de l'autoantigénicité

L'autoantigène que nous avons isolé est un constituant de la cellule gastrique normale. Il est présent aussi dans l'estomac embryonnaire [3]. Il n'a cependant pas produit de tolérance immunologique. La raison en est peut-être que c'est un antigène normalement «séquestré» à l'intérieur de la cellule. Lorsqu'il est libéré dans la circulation générale, en raison d'une lyse cellulaire accrue, il donne lieu à la production d'autoanticorps. Un trouble du mécanisme de reconnaissance des autoantigènes n'est pas nécessairement à envisager.

Nous sommes reconnaissants au Docteur P. May et à Monsieur Paoletti d'avoir bien voulu effectuer les ultracentrifugations analytiques mentionnées dans ce travail. D. Karitzky est boursier du Deutscher Akademischer Austauschdienst.

BIBLIOGRAPHIE

1. Bonatti, A., Rapp, W., et Burtin, P., *Protides of the Biological Fluids*, 11th Colloquium 1964, Elsevier, Amsterdam 1965, p. 219.
2. Kleist, S. v., et Burtin, P., *Protides of the Biological Fluids*, 11th Colloquium 1964, Elsevier, Amsterdam 1965, p. 222.
3. Burtin, P., Kleist, S. v., Rapp, W., Loisillier, F., Bonatti, A., et Grabar, P., *IVth International Symposium Immunopathology*, Monte Carlo 1965, Schwabe & Co., Basel, Stuttgart 1966, p. 91.
4. Kleist, S. v., et Burtin, P., *Immunology*, 10 (1966) 507.
5. Rapp, W., Aronson, S. B., Burtin, P., et Grabar, P., *J. Immunol.* 92 (1964) 579.
6. Boyden, S. V., *J. Exptl. Med.* 93 (1951) 107.
7. Stavitsky, A. B., *J. Immunol.* 72 (1954) 360.
8. Agneray, J., *Thèse de Doctorat ès Sciences*, Lille, 1965.
9. Uriel, J., *Bull. Soc. Chim. Biol.* 48 (1966) 969.
10. Fazekas de Saint Groth, S., Webster, R. G., et Datyner, A., *Biochim. Biophys. Acta*, 71 (1963) 377.
11. Grabar, P., et Burtin, P., *Analyse immunoélectrophorétique*, Masson, Paris 1960.
12. Ouchterlony, O., *Progr. Allergy*, 5 (1958) 1.
13. Grabar, P., et Williams, C. A., *Biochim. Biophys. Acta*, 10 (1953) 193.
14. Heremans, J., *Les globulines sériques du système gamma, leur nature et leur pathologie*, Arscia, Bruxelles et Masson, Paris 1960.
15. Laatsch, R. H., Kies, M. W., Gordon, S., et Alvord, E. C., jr., *J. Exptl. Med.* 115 (1962) 777.
16. Irvine, W. J., Davies, S. A., Delamore, J. W., et Williams, A. W., *Brit. Med. J.* 1 (1962) 454.
17. Baur, S., Roitt, J. M., et Doniach, D., *Immunology*, 8 (1965) 62.

Adresse actuelle de D. Karitzky:
Universitäts-Kinderklinik
78 Freiburg i. Br., Mathildenstr. 1, Allemagne
P. Burtin
Institut de Recherches Scientifiques sur le Cancer
Boîte postale 8, 94-Villejuif, France

European J. Biochem. 1 (1967) 419—426

Conversion of P-450 to P-420 by Neutral Salts and some other Reagents

Y. Imai and R. Sato

Institute for Protein Research, Osaka University, Osaka

(Received February 9, 1967)

Exposure of liver microsomes to high concentrations of neutral salts resulted in the conversion of the microsomal hemoprotein, P-450, to a modified form called P-420. The efficiency of various salts in causing the conversion obeyed the order known as Hofmeister's lyotropic series of ions, i.e. $SCN^- > I^- > Br^- > Cl^- > SO_4^{--}$, CH_3COO^- for anions, and $Li^+ > K^+$, Na^+ for cations. The conversion induced by salts proceeded more rapidly in dithionite-treated microsomes, in which the hemoprotein was kept reduced, than in aerobic microsomes. Even in the absence of added salts the conversion took place to some extent when pH was moved far from neutrality. The addition of salts narrowed the pH region where the hemoprotein was resistant to the conversion. Guanidine hydrochloride, lysolecithin, and high concentrations of aniline also caused the conversion of P-450. Although iodine was an effective conversion agent, it caused considerable destruction of the hemoprotein. Based on these and other findings, the structure around the heme group of P-450 as well as the mechanism of its conversion to P-420 are discussed.

The presence of a CO-binding pigment in liver microsomes was first described by Klingenberg [1] and by Garfinkel [2]. Omura and Sato [3,4] have later established that this pigment, called P-450, is a hemoprotein possessing protoheme as the prosthetic group. Its spectral properties are, however, anomalous for a hemoprotein, as manifested by the following observations: (a) the CO-difference spectrum of reduced P-450 has only one peak at 450 mμ [1—3]; (b) the binding of ethyl isocyanide to the reduced pigment causes a spectral change having two Soret bands [3,5]; (c) these two Soret bands in the ethyl isocyanide-difference spectrum are affected by pH in a characteristic way [6]; and (d) the spectral difference between the reduced and oxidized forms of P-450 is small and of ambiguous shape [3,7].

Following the earlier observation by Klingenberg [1] that a CO-binding protohemochromogen was formed by the action of cholate on microsomes, Omura and Sato [3,4] have also found that treatment of liver microsomes with deoxycholate, snake-venom phospholipase A, or steapsin results in the conversion of P-450 to a modified form called P-420. This latter form of the pigment, in contrast to P-450, has been shown to possess typical hemoprotein spectra characteristics of a b-type cytochrome [3,4]. It therefore seems that the anomalous spectra of P-450 arise from its abnormal state of existence in the microsomal membrane and that the normal hemoprotein spectra are restored when it is liberated from this state and converted to P-420.

Enzymes. Phospholipase C (EC 3.1.4.3); phospholipase A (EC 3.1.4.).

On the other hand, it has been established that P-450 is a key component of the NADPH-requiring hydroxylase systems in liver microsomes as well as adrenal cortex microsomes and mitochondria [8—15]; in the hydroxylation reactions the hemoprotein acts not only as the oxygen-activating enzyme [8—12] but also as the substrate-binding site [13—15]. Since the conversion of P-450 to P-420 is always accompanied by the inactivation of the hydroxylase systems and P-420 is incapable of reacting with hydroxylatable substrates in the same way as P-450 [14], it is likely that the integrity of the unusual state of P-450 is essential for its functions. Clearer understanding of the anomaly of P-450 is, therefore, required for elucidation of the molecular mechanism of the hydroxylation reactions. An approach to the anomalous state of P-450 would be through a detailed analysis of the conversion of P-450 to P-420. This conversion has been shown to be caused not only by the three reagents mentioned above but also by a wide variety of reagents such as trypsin [16], p-chloromercuribenzoate [17], bathocuproine sulfonate [18], and urea [16]. Such diversity of effective reagents, though highly suggestive, appears to have caused a confusion in the interpretation of the state of P-450.

In this paper, we report that high concentrations of certain neutral salts such as KSCN and NaI could cause the conversion of P-450 to P-420. The conversion by some other reagents is also reported. The results obtained suggest that a specific conformation of the protein of P-450 or certain hydrophobic interactions of the heme region with protein or phospholipids are responsible for the unique spectral properties of the hemoprotein.

EXPERIMENTAL PROCEDURE

Rabbit-liver microsomes, free from hemoglobin contamination, were prepared in isotonic KCl as described previously [3]. The washed microsomes were suspended in deionized water, usually 20 mg of protein per ml, and stored at −20° under nitrogen. The frozen microsomes were rehomogenized before use.

Cytochrome b_5 was partially purified from rabbit-liver microsomes by the procedure of Strittmatter and Velick [19]. Lysolecithin prepared from egg lecithin by the method of Hanahan et al. [20] and phospholipase C purified from culture filtrates of *Clostridium perfringens* by the method of Ikezawa et al. [21] were kindly supplied by Mr. A. Ito of this laboratory. Guanidine hydrochloride was purified by crystallization from 90% methanol. CO was prepared from formic acid and concentrated sulfric acid by the conventional method and purified by passing through an NaOH solution. Ethyl isocyanide synthesized by the method of Jackson and McKusick [22] was a generous gift from Dr. Y. Izumi and Mr. N. Oshino of this Institute.

P-450 and P-420 were determined from the CO-difference spectrum of dithionite-treated microsomes as described by Omura and Sato [3], assuming values of 91 and −11 cm^{-1} mM^{-1} for molar extinction increments between 450 and 490 mμ for P-450 and P-420, respectively, and −41 and 110 cm^{-1} mM^{-1} between 420 and 490 mμ for P-450 and P-420, respectively. Difference spectra were recorded at room temperature in a Cary model 14 spectrophotometer, using cuvettes of 1 cm optical path; the scanning speed was 1 mμ per second.

Unless otherwise stated, the conversion of P-450 to P-420 by neutral salts or other reagents was carried out by incubating the mixture containing microsomes (2 to 3 mg of protein per ml), 0.2 M potassium phosphate or Tris acetate buffer (pH 7.25), and a suitable concentration of the reagent for about 10 minutes at room temperature under aerobic conditions. When lysolecithin and phospholipase C were used as the reagents, the mixture was incubated anaerobically at 37° for 30 minutes. After the incubation, the mixture was treated with a small amount of solid sodium dithionite and the CO-difference spectrum was measured as rapidly as possible.

Protein was determined by the method of Lowry et al. [23], using bovine serum albumin as the standard.

RESULTS

Conversion of P-450 to P-420 by Neutral Salts

The addition of CO to dithionite-treated liver microsomes, suspended in 0.2 M phosphate buffer,

pH 7.25, caused a difference spectrum having a peak at 450 mμ, due to the formation of the CO compound of reduced P-450 [1—3]. When the CO-difference spectrum was measured in the presence of KSCN, a new peak was observable at 420 to 422 mμ. As shown in Fig. 1, this new peak was intensified as the KSCN concentration was increased and this intensification was accompanied by a decrease in the height of the peak at 450 mμ. At a KSCN concentration of 1.2 M, the peak at 450 mμ disappeared completely and only the peak at 420 mμ was visible. It was thus concluded

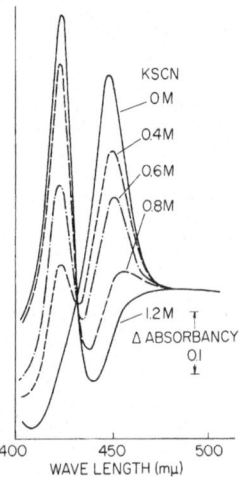

Fig. 1. *CO-difference spectra of reduced microsomes incubated with KSCN.* Microsomes (final concentration, 2.6 mg of protein per ml) in 0.2 M K-phosphate buffer, pH 7.25, were placed in both the sample and reference cuvettes of the spectrophotometer and incubated for 10 min aerobically in the presence of indicated final concentrations of KSCN. CO was then added to the sample cuvette, and the contents of both cuvettes were reduced by minimal amounts of solid Na₂S₂O₄. Scanning was started from 500 mμ, 2 min after the addition of Na₂S₂O₄

that KSCN, like phospholipase A and deoxycholate [3], induced the conversion of P-450 to P-420. This conversion could be visualized more clearly in Fig. 2, in which the concentrations of both P-450 and P-420 are plotted against the KSCN concentration. It will be seen that the sum of the amounts of P-450 and P-420 was preserved in spite of the profound effect induced.

The conversion could be induced not only by KSCN but also by some other neutral salts. The relative efficiency of various neutral salts in causing the conversion was studied and the results are summarized in Fig. 3. As will be seen, KSCN, causing 50% conversion at about 0.5 M, was most effective among the salts examined. NaBr and LiCl, on the other hand, were much less effective; half maximal conversion by these salts at pH 7.25 took place at as high a concentration as about 3 M. Although not shown in Fig.3, KBr was almost as effective as NaBr. NaI, KI and LiBr were more effective than NaBr, LiCl and KBr, but were less effective than

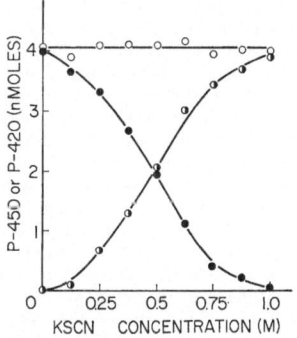

Fig.2. *Conversion of P-450 to P-420 at varying concentrations of KSCN.* Conditions were the same as those described in Fig.1. Concentrations of both forms of the hemoprotein were calculated from the CO-difference spectrum as described in Experimental Procedure. ●, P-450; ◑, P-420; ○, P-450 + P-420

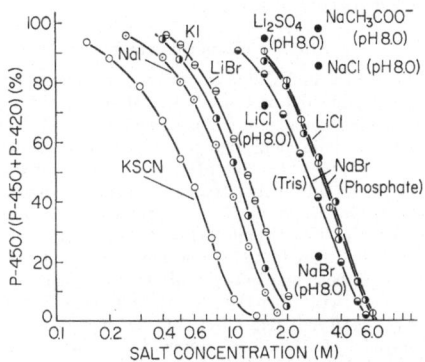

Fig.3. *Conversion of P-450 to P-420 as a function of the concentration of various salts added.* Microsomes (2.0 mg of protein per ml) in either 0.2 M K-phosphate or Tris-acetate buffer were used. Conditions of treatments and of measurements were the same as those described in Fig.1. The extents of the conversion were determined from concentrations of P-450 and P-420 calculated from the CO-difference spectrum as described in Experimental Procedure

KSCN. Fig.3 also shows that the conversion-inducing effect of a salt was more pronounced at pH 8.0 than at pH 7.25. Salts such as sodium acetate induced little conversion even at pH 8.0 and at a concentration of 3 M. From the data shown in Fig.3 it was

evident that Li^+ was more effective than Na^+ and K^+. A comparison of the efficiency of lithium salts indicated an order of $Br^- > Cl^- > SO_4^{--}$ for anions. On the other hand, the data obtained with sodium and potassium salts revealed the following order of efficiency; $SCN^- > I^- > Br^-$. It was thus clear the order of efficiency of various salts in causing the conversion agreed well with Hofmeister's lyotropic series of ions; *i.e.* $SCN^- > I^- > Br^- > SO_4^{--}$, CH_3COO^- for anions, and $Li^+ > Na^+$, K^+ for cations.

Time Course of Salt-Induced Conversion

In the experiments described above, microsomes were incubated with salts for 10 minutes under aerobic conditions and the extent of conversion effected during this period of time was estimated by

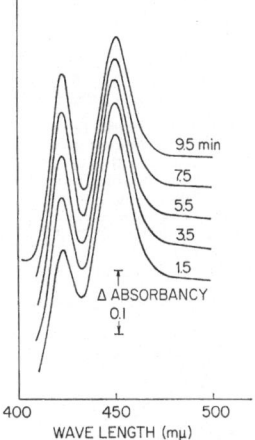

Fig.4. *Changes in CO-difference spectrum of microsomes with time in the presence of KSCN.* Microsomes (2.6 mg of protein per ml) in 0.2 M K-phosphate buffer, pH 7.25, were treated with 0.4 M KSCN for 10 min, treated with CO, and then reduced with a minimal amount of dithionite. Scanning was started from 500 mμ at indicated time after the addition of dithionite.

measuring the CO difference spectrum immediately after the addition of dithionite. It was, however, noticed that the addition of dithionite greatly accelerated the conversion of the hemoprotein. Fig.4 shows the time-dependent conversion of P-450 to P-420 observed in an experiment, in which the CO-difference spectrum was measured at different time intervals after the addition of dithionite to aerobic microsomes that had been incubated for 10 minutes with 0.6 M KSCN. The progress of the conversion under these conditions is illustrated in a quantitative manner in Fig.5, which also includes the results obtained in the presence of 0.4 M KSCN. It may be seen that the conversion reaction under these conditions was of first order and proceeded faster in the presence of higher concentrations of the salt. In contrast to the conversion in the presence of dithionite, the formation of P-450 was much slower when aerobic microsomes were incubated with salts. Fig. 6 shows the rates of the conversion in aerobic microsomes in the presence of 0.6 M KSCN or 0.6 M NaI.

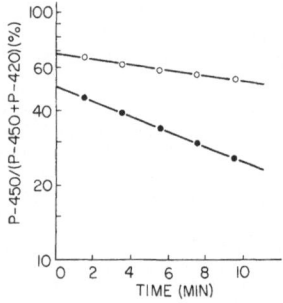

Fig. 5. *Time course of the conversion of reduced P-450 to P-420 in the presence of KSCN.* Conditions were the same as those described in Fig. 4, except that the concentrations of KSCN were varied. The extents of conversion were determined from concentrations of P-450 and P-420 calculated from the CO-difference spectrum as desribed in Experimental Procedure. ●, 0.6 M KSCN; ○, 0.4 M KSCN

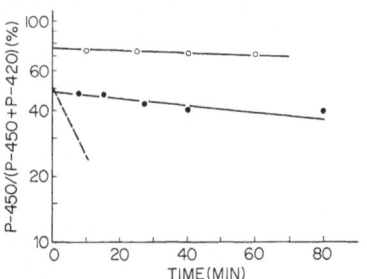

Fig. 6. *Time course of the conversion of oxidized P-450 to P-420.* Conditions were the same as those described in Fig. 5, except that the time of treatment in the reduced state was fixed at 2 min and that in the oxidized state was changed as indicated. The extents of the conversion were determined as in Fig. 5. ○, 0.6 M KSCN; ●, 0.6 M NaI; - - - - was reproduced from Fig. 5 (lower curve) for comparison

In these experiments, dithionite was added to the mixture that had been incubated aerobically for a desired period of time and the CO-difference spectrum was measured exactly 2 minutes after the addition of dithionite. As is shown, the conversion measured under these conditions was also a first-order reaction and was slower by a factor of more than 10 as compared with that in the presence of dithionite (*cf.* Curve C, included in Fig. 6 for comparison).

Since P-450 is in the fully oxidized state in aerobic microsomes and dithionite reduces the hemoprotein completely [3], the data described above suggested that reduced P-450 was more susceptible to the action of salts than the oxidized hemoprotein. Higher rates of the conversion in the presence of dithionite were also observed on treatment with other reagents such as *p*-chloromercuribenzoate.

Prolonged incubation of microsomes with very high concentrations of strongly effective salts, *e.g.* 1.5 M KSCN, resulted in an apparent extent of the conversion exceeding $100^0/_0$, as estimated from the CO-difference spectrum. The peak at 420 mμ in the CO-difference spectrum was also shifted slightly towards shorter wavelengths under these conditions. These observations could be accounted for by the denaturation of another microsomal hemoprotein, cytochrome b_5, to a form capable of binding CO. Actually, partially purified cytochrome b_5, when treated with 1.5 M KSCN, showed a peak at 417 mμ in the CO-difference spectrum as compared with that at 420 to 422 mμ due to the CO binding to P-420. The peak due to the CO compound of denatured cytochrome b_5 was, however, much lower than that of P-450, and the contribution of the former to the absorbance at 420 mμ was very small. Moreover, no appreciable denaturation of cytochrome b_5 took place under the conditions in which the P-450 was usually studied.

Effect of pH on Conversion

As shown in Fig. 7, the conversion of P-450 to P-420 was affected profoundly by pH of the medium. Even in the absence of salts, except for 0.2 M Tris-acetate or potassium phosphate added as buffer, P-450 was stable only at pH 6 to 8. At pH values lower than 6 and higher than 8, it was converted to P-420 to some extent. In the presence of neutral salts such as KSCN and NaI, the hemoprotein was most stable at pH 7.2 to 7.5, and the stability declined rapidly on both sides of this pH region. Such steep pH curves with a maximum around pH 7.5 were always observed regardless of the neutral salts employed. It therefore seems that this profile reflects the effect of pH on the stability of the ligand state of P-450, either due to a direct action on the ligand itself or to an indirect effect through the protein and/or phospholipid moieties.

Properties of P-420 Produced by Neutral Salts

Although it has been reported that P-420 produced by the action of deoxycholate or snake-venom phospholipase A is in a soluble state [3], the conversion of P-450 induced by neutral salts was not accompanied by its solubilization. Thus, P-420 formed in the presence of 1.5 M KSCN was recovered almost quantitatively in the microsomal pellet when the conversion mixture was diluted and centrifuged at 105,000 ×g. Non-solubilization of P-420 has also been observed in the steapsin-induced conversion [4]. However, the spectra of microsomal bound P-420 produced by KSCN were identical with those of partly purified P-420 that had been solubilized by deoxycholate and phospholipase A [4]. It has been reported that ethyl isocyanide [5] and aniline [14] combine with oxidized P-450 to give characteristic spectra. The addition of 3 mM ethyl isocyanide or 20 mM aniline (saturating concentrations for P-450) to KSCN-produced oxidized P-420 caused only slight spectral changes in the Soret region as in the case of purified P-420. The spectral interaction of ethyl

somes contain large amounts of lecithin, it was likely that the conversion induced by phospholipase A was actually due to the formation by this enzyme of lysolecithin which has a powerful detergent activity. As shown in Fig. 8, it was in fact found that anaerobic incubation of microsomes with 0.2% lysolecithin caused complete conversion of P-450 to P-420. On the other hand, phospholipase C, which splits phosphorylcholine from lecithin, caused only 20% of the conversion, although this treatment removed 50% of the total phospholipid phosphorous from microsomes [15a]. As reported preliminarily [14], guanidine hydrochloride also caused complete conversion of P-450 to P-420 at 2.0 M, whereas the con-

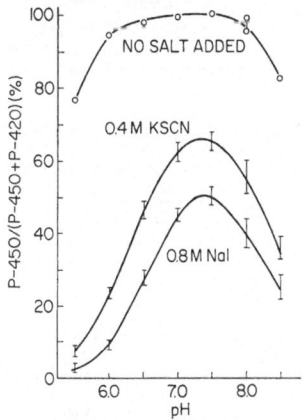

Fig. 7. *Effect of pH on the conversion.* Microsomes (2.1 mg of protein per ml) in either 0.2 M K-phosphate (pH 5.5 to 8.0) or Tris-acetate (pH 8.0 to 8.5) buffer were treated with salts for 10 min and then reduced with a minimal amount of dithionite. CO-difference spectra were recorded 2 min after the addition of dithionite. The extents of the conversion were determined as described in Fig. 5

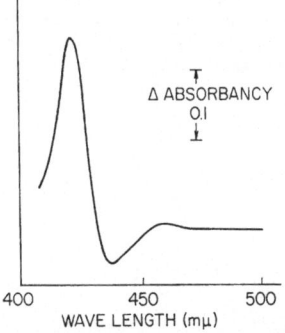

Fig. 8. *CO-difference spectrum of lysolecithin-treated microsomes.* Microsomes (1.6 mg of protein per ml) in 0.2 M Tris-acetate buffer, pH 8.0, were treated with 1.5% of lysolecithin at 37° for 30 min under an atmosphere of nitrogen, and then the CO-difference spectrum of treated microsomes was recorded

isocyanide with the dithionite-reduced form of KSCN-produced P-420 was also identical with that with purified P-420. Anomalous effects of the ethyl isocyanide concentration on this interaction will be reported in a later publication. From spectral properties it was inferred that KSCN-produced P-420, though still bound to the microsomal membrane, was identical with or closely similar to solubilized P-420 in the molecular conformation at least around the heme group.

Conversion by Other Reagents

As mentioned above, snake-venom phospholipase A has been shown to induce an effective conversion of P-450 to P-420 [3]. Since phospholipase A produces lysolecithin from lecithin and liver micro-

centration of urea, another protein denaturant, required for the complete conversion has been reported to be 4.0 M [16].

When 1 mM iodine was added to microsomal suspensions containing 1 to 2 mg protein per ml, P-450 was extensively destroyed probably owing to the oxidative decomposition of the heme. At lower microsomal concentrations, however, the conversion of P-450 to P-420 could be observed without extensive destruction of hemoprotein; about 10 moles of iodine per mole of P-450 were required for the conversion at pH 7.25 when the microsomal concentration was 0.1 mg of protein per ml.

A characteristic spectral change having a Soret peak at 427 mμ has been observed on interaction of aniline with oxidized P-450 in aerobic microsomes and this interaction has been shown to be an essential step for the hydroxylation of aniline [14,15b]. As shown in Fig. 9 (Curve A), incubation of microsomes with 20 mM aniline, causing the fully developed spectral change mentioned above, did not induce appreciable conversion of P-450 to P-420. However, when the aniline concentration was raised to 100 mM, considerable formation of P-420 was observed (Fig.9, Curve B). As in the case of treatment with neutral salts, the conversion by high concentrations of aniline was more extensive in dithionite-treated microsomes than in untreated preparations (Fig.9, Curve C). Spectral changes similar to that produced by 20 mM aniline have also been observed on addition of a few percent of organic solvents, especially alcohols, to aerobic microsomes [14]. However, in agreement with previous work [16], the same solvents, at high concentrations, caused the conversion of P-450 to P-420. The relative efficiency of alcohols in causing the conversion was isoamyl > isobutyl > isopropyl > ethyl > methyl. For instance, isopropanol induced the conversion at a concentration of 15%, whereas 20% ethanol was required to cause the same effect.

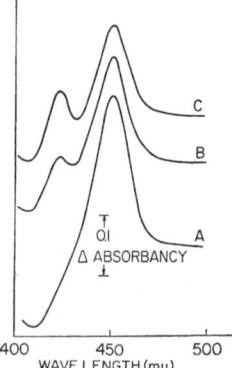

Fig.9. *CO-difference spectra of aniline-treated microsomes.* Microsomes (1.7 mg of protein per ml) in 0.2 M Tris-acetate buffer, pH 8.0, were treated aerobically with aniline for 10 min, and then reduced with a minimal amount of dithionite in case of Curves A and B but 4 min in case of Curve C. Concentration of aniline present was as follows: Curve A; 20 mM; Curves B and C, 100 mM

DISCUSSION

The microsomal bound hemoprotein, P-450, has unusual spectral properties and can be converted by various treatments to a form called P-420 which shows normal spectra for a hemoprotein. The data reported here indicate that this conversion can also be effected by high concentrations of neutral salts. The relative efficiency of various salts in causing the conversion agrees with Hofmeister's lyotropic series of ions, which was originally determined for the capability of salts to precipitate proteins from aqueous solutions [24]. With minor exceptions, the same order of activity of ions can also be applied for

destabilization of marcomolecules such as ribonuclease [25,26], collagen [27], myosin [28], and DNA [29], as well as for salting in of nonelectrolytes such as benzene and aniline [30]. Since the concentrations of salts required for the conversion of the hemoprotein are in the same range as for destabilization of macromolecules, it is suggested that the salt-induced conversion of P-450 to P-420 also involves the destabilization of certain macromolecular structures.

It should, however, be noticed that neutral salts are not the only agent causing the conversion. Previous work as well as the present study have shown that the conversion can be induced by a wide variety of reagents, including detergents [3], proteases [16], phospholipase A [3], chelating agents [18], mercurials [17], iodine, organic solvents [16], and high concentrations of hydroxylatable substrates such as aniline, as summarized in the Table. Even extreme

Table. *Conversion of P-450 to P-420 by a variety of reagents* The abbreviations used are: PCMB; *p*-chloromercuribenzoate, PCMS; *p*-chloromercuribenzene sulfonate, AmOH; amyl alcohol, ProOH; propanol, BuOH; butanol, EtOH; ethanol, and MeOH; methanol

Reagents or treatments	Concentration	References
Heated venom of the snake	0.1%	[3,4]
Trimeresurus flavoviridis (Phospholipase A)		
Sodium deoxycholate	0.5%	[3,4]
Lysolecithin	0.2%	this paper
Bathocuproine sulfonate	1 mM	[18]
Trypsin		[16]
Urea	4 M	[16]
Guanidine hydrochloride	2 M	this paper
Neutral salts (consistent with Hofmeister's lyotropic series)	1~6 M	this paper
PCMB, PCMS	0.1 mM	[17]
Iodine	10 μM[a]	this paper
Aniline	100 mM	this paper
Organic solvents (isoAmOH > isoBuOH > isoProOH > EtOH > MeOH)	for example EtOH 20% isoProOH 15%	[16] and this paper
Acidic	pH 6	
Alkaline	pH 8	

[a] About 10 moles of iodine per mole of P-450. It is necessary to lower the concentration of microsomal protein below about 0.1 mg of protein per ml.

pH values can cause appreciable conversion of the hemoprotein. Therefore, such diversity of the conversion reagents has to be taken into consideration in interpreting the mechanism of the conversion and the structure of P-450. Based on the nature of the conversion reagents, Mason *et al.* [16] have proposed a hypothetical structure for P-450, in which a sulfhydryl group and phospholipids play important roles. Yamano *et al.* [31] have suggested from electron spin resonance studies that a lipid substance acts as a ligand for the heme of P-450. Although these suggestions are instructive, they can explain the observed facts only partially. A more general interpretation of the structure of P-450 and its conversion to P-420 is therefore desired. In what follows, we will attempt to look for such an interpretation based on the observations described in this and previous reports.

P-450 in liver microsomes acts as a site of both oxygen and substrate activation for drug hydroxylations [14, 15b]. Since only those drugs possessing high solubility in lipid solvents are hydroxylated by the microsomal system [32], it seems reasonable to assume that the reactive area of P-450, i.e. the vicinity of the heme, is in contact with or burried in a highly hydrophobic part of P-450 protein or of the lipids of the microsomal membrane. The specific structural relationship of the heme to the hydrophobic region seems to be maintained by the special ternary conformation of the P-450 protein or by the high lipid content in the membrane. The unusual spectral properties of P-450 may thus be ascribed to a hydrophobic interaction of the heme with some components which are situated nearby. These assumptions seem to be supported by the fact that the spectral properties of chlorophyll in lipid solvents are markedly different from those in aqueous solution [33]. The observation that P-450 can be solubilized from microsomes only by the action of detergents, but not by proteases which solubilize about half the microsomal protein [4], may also be favorably explained by the existence of the strong hydrophobic interaction. It seems, therefore, likely that the conversion of P-450 to P-420 results from the disturbance of the hydrophobic environment around the heme either by the primary action of the conversion reagents or by the secondary effects through conformational change in the hemoprotein.

According to the above interpretation, the conversion of P-450 to P-420 must be accompanied by destruction of the hydrophobic environment around the heme or by profound changes in the conformation of the hemoprotein. Detergents are known to disrupt hydrophobic bonds in general and denature proteins. As mentioned above, the action of phospholipase A on microsomes produces lysolecithin, which is a powerful detergent. Certain chelating agents such as bathocuproine sulfonate are soluble in both aqueous and lipid solvents and may also act as detergents. Proteases primarily cause modifications in the native conformation of the hemoprotein, breaking down secondarily the hydrophobic interaction of the heme. Protein denaturants such as urea and guanidine hydrochloride can also destroy hydrophobic bonds. The effect of high concentrations of neutral salts may also be interpreted either by their action to destabilize the protein conformation or by their ability to destroy

the interaction of the heme with its surroundings in an analogous way as in the salting in phenomenon. The addition of moderate concentrations of alcohols as well as aniline and some other substrates of the microsomal hydroxylase system has been shown to produce characteristic spectral changes [14], probably reflecting certain alterations in the conformation of the hemoprotein. These changes are, however, not sufficiently extensive to modify the molecular feature of P-450. On further increase in the concentrations of these reagents, it is expected that the changes become increasingly more drastic up to a point where the integrity of the specific structure of P-450 can no longer be maintained, leading to the conversion to P-420.

According to recent studies by Mason et al. [16], Murakami and Mason [34], and Yamano et al. [31], P-420 can exist in three states, designated as β, γ and δ, which show the same spectra but differ from one another in their spin states. The action of detergents, organic solvents or proteases on P-450 (the α state of the hemoprotein) produces the β and then γ states. The δ or high-spin state of the ferrihemoprotein, where the natural ligand is replaced by water, can be obtained by the action of mercurials on either the α or γ state. It is, therefore, likely that in the case of mercurials and perhaps iodine as the reagents, a conversion mechanism different from those induced by the other reagents is operative. However, it is also true that mercurials and iodine can cause modifications of protein residues, especially sulfhydryl groups. In view of Mason's proposal that sulfide is serving as a ligand in P-450 [16], direct attack of these reagents on the ligand itself may also cause the conversion to P-420.

The conversion of P-450 to P-420 induced by neutral salts proceeds more rapidly in the reduced form of the hemoprotein than in the oxidized form. The possibility that the accelerated conversion in the reduced form is due to an artefact resulting from the use of dithionite as the reductant could be excluded by an experiment in which the hemoprotein reduced by NADPH under CO atmosphere was treated with neutral salts. The time course of the conversion under these conditions was identical with that obtained for the dithionite-reduced hemoprotein. Differences in stability between the oxidized and reduced forms have also been noticed for several hemoproteins. For instance, Okunuki [35] has reported the higher stability of reduced cytochrome c than the oxidized form towards various treatments, and physicochemical studies have recently been published to indicate that the reduced and oxidized forms of cytochrome c posses different secondary and tertiary structures [36]. Furthermore, Nakazima et al. [37] have shown that the conversion of horse-radish peroxidase from the low- to high-spin states occurs more readily in the reduced form.

It seems that the pH curves for the salt-induced conversion of P-450 to P-420 (Fig. 7) are the reflection of the pH stability of the environment around the heme of P-450. The decline on the acid side of these curves appears to be due to the labilization of the ligand itself, in view of recent findings by Murakami and Mason [34] that the conversion of the hemoprotein from the low- to high-spin states occurs at acidic pH values. On the other hand, the instability of P-450 in the alkaline region seems to be related to that of the structural integrity of the microsomal membrane. The labilization of the membrane structure at high pH values is suggested by the fact that solubilization of microsomal protein by phospholipase A is more efficient in the alkaline region than at neutral pH's which are optimal for the phospholipase action [38]. At neutral pH's some purified proteins exist in the polymerized form because of hydrophobic interactions, but dissociate to the monomeric form at alkaline pH's [39]. Therefore, it is also likely that the instability of P-450 at high pH values may be related to the hydrophobic nature of the environment of the heme.

REFERENCES

1. Klingenberg, M., *Arch. Biochem. Biophys.* 75 (1958) 376.
2. Garfinkel, D., *Arch. Biochem. Biophys.* 77 (1958) 493.
3. Omura, T., and Sato, R., *J. Biol. Chem.* 239 (1964) 2370.
4. Omura, T., and Sato, R., *J. Biol. Chem.* 239 (1964) 2379.
5. Nishibayashi, H., Omura, T., and Sato, R., *Biochim. Biophys. Acta*, 118 (1966) 651.
6. Imai, Y., and Sato, R., *Biochem. Biophys. Res. Commun.* 23 (1966) 5.
7. Nishibayashi, H., and Sato, R., *J. Biochem. (Tokyo)*, In press.
8. Estabrook, R. W., Cooper, D. Y., and Rosenthal, O., *Biochem. Z.* 338 (1963) 741.
9. Sato, R., Omura, T., and Nishibayashi, H., in *Oxidases and Related Redox Systems* (edited by T. King, M. Morrison, and H. S. Mason), John Wiley and Sons, New York 1965, p. 861.
10. Orrenius, S., Ericsson, J., and Ernster, L., *J. Cell Biol.* 25 (1965) 627.
11. Omura, T., Sato, R., Cooper, D. Y., Rosenthal, O., and Estabrook, R. W., *Federation Proc.* 24 (1965) 1181.
12. Cooper, D. Y., Levin, S., Narasimhulu, S., Rosenthal, O., and Estabrook, R. W., *Science*, 147 (1965) 400.
13. Narasimhulu, S., Cooper, D. Y., and Rosenthal, O., *Life Sci.* 4 (1965) 2101.
14. Imai, Y., and Sato, R., *Biochem. Biophys. Res. Commun.* 22 (1966) 620.
15a. Ito, A., and Sato, R., Unpublished observation.
15b. Remmer, H., Schenkman, J., Estabrook, R. W., Sasame, H., Gillette, J., Narasimhulu, S., Cooper, D. Y., and Rosenthal, O., *Mol. Pharmacol.* 2 (1966) 187.
16. Mason, H. S., North, J. C., and Vanneste, M., *Federation Proc.* 24 (1965) 1172.
17. Cooper, D. Y., Narasimhulu, S., Rosenthal, O., and Estabrook, R. W., in *Oxidase and Related Redox Systems* (edited by T. King, M. Morrison, and H. S. Mason), John Wiley and Sons, New York 1965, p. 838.
18. Mason, H. S., Yamano, T., North, J. C., Hashimoto, Y., and Sakagishi, P., in *Oxidase and Related Redox Systems* (edited by T. King, M. Morrison, and H. S. Mason), John Wiley and Sons, New York 1965, p. 879.
19. Strittmatter, P., and Velick, S. F., *J. Biol. Chem.* 221 (1956) 253.
20. Hanahan, D. J., Rodbell, M., and Turner, L. D., *J. Biol. Chem.* 206 (1954) 431.
21. Ikezawa, H., Yamamoto, A., and Murata, R., *J. Biochem. (Tokyo)*, 56 (1964) 480.
22. Jackson, H. L., and McKusick, B. C., in *Organic Syntheses* (edited by T. L. Cairns), John Wiley and Sons, New York 1955, Vol. 35, p. 62.
23. Lowry, O. H., Rosenbrough, N. J., Farr, N. J., and Randall, R. J., *J. Biol. Chem.* 193 (1951) 265.
24. Hofmeister, L., *Arch. Exptl. Pathol. Pharmacol.* 24 (1888) 247.
25. von Hippel, P. H., and Wong, K.-Y., *Science*, 145 (1964) 577.
26. von Hippel, P. H., and Wong, K.-Y., *J. Biol. Chem.* 240 (1965) 3909.
27. von Hippel, P. H., and Wong, K.-Y., *Biochemistry*, 1 (1962) 664.
28. Tonomura, Y., Sekiya, K., and Imamura, K., *J. Biol. Chem.* 237 (1962) 3110.
29. Hamaguchi, K., and Geiduschek, E. P., *J. Am. Chem. Soc.* 84 (1962) 1329.
30. Long, F. A., and McDevit, M. F., *Chem. Rev.* 51 (1952) 119.
31. Yamano, T., Ichikawa, Y., and Mori, K., Preprint of *18th Symposium on Enzyme Chemistry*, Sapporo 1966, p. 40.
32. Gaudette, L. E., and Brodie, B. B., *Biochem. Pharmacol.* 2 (1959) 89.
33. Izawa, S., Itoh, M., Ogawa, T., and Shibata, K., Microalgae and Photosynthetic Bacteria (Special issue of *Plant and Cell Physiol.*), (1963) 413.
34. Murakami, K., and Mason, H. S., Preprint of *18th Symposium on Enzyme Chemistry*, Sapporo 1966, p. 36.
35. Okunuki, K., *Advances in Enzymol.* 23 (1961) 30.
36. Ulmer, D. D., *Biochemistry*, 4 (1965) 902.
37. Nakazima, T., Tamura, M., Yokota, T., and Yamazaki, I., Preprint of *18th Symposium on Enzyme Chemistry*, Sapporo 1966, p. 1.
38. Hanahan, D. J., *Progr. Chem. Fats Lipids*, 4 (1957) 141.
39. Criddle, R. S., Bock, R. M., Green, D. E., and Tisdale, H., *Biochemistry*, 1 (1962) 827.

Y. Imai and R. Sato
Institute for Protein Research
Osaka University
Joancho 36, Kita-Ku, Osaka, Japan

European J. Biochem. 1 (1967) 427—433

Palmityl-CoA: carnitine O-palmityltransferase in the Mitochondrial Oxidation of Palmityl-CoA

J. Bremer and K. R. Norum

Institute of Clinical Biochemistry, University of Oslo, Rikshospitalet, Oslo

(Received January 30, 1967)

1. The kinetics of the carnitine palmityltransferase and of the carnitine-stimulated oxidation of palmityl-CoA to acetoacetate in isolated rat liver mitochondria have been studied.

2. The particle-bound enzyme shows the same kinetic properties as the solubilized enzyme.

3. The rate of respiration is found to depend on the concentration of palmitylcarnitine formed in the reaction mixture, as long as this concentration is less than 2×10^{-6} M.

4. The capacity of isolated mitochondria for palmitylcarnitine formation exceeds severalfold their capacity for oxidation of palmitylcarnitine to acetoacetate. It is likely therefore that the long-chain acylcarnitine/carnitine substrate pair is nearly in equilibrium with the extramitochondrial palmityl-CoA/CoA in the intact cell.

5. Relatively high concentrations of CoA and carnitine inhibit the oxidation of palmityl-CoA when the concentration of this substrate is low. These effects are in accordance with the kinetic properties of the carnitine palmityltransferase and may contribute to the slow rate of fatty acid oxidation when the concentration of long-chain acyl-CoA in the cell is low.

6. High concentrations of palmityl-CoA are oxidized by isolated mitochondria even in the absence of carnitine, possibly because of a detergent effect. Albumin inhibits this carnitine-independent oxidation, presumably because palmityl-CoA is bound to the protein.

7. The regulation of fatty acid oxidation is discussed. It is suggested that both the cellular concentration of long-chain acylcarnitines and the availability of mitochondrial free CoA are of importance. The existence of such a system permits regulation of fatty acid oxidation without interference with extramitochondrial CoA-dependent processes.

Carnitine accelerates the mitochondrial oxidation of added long-chain acyl-CoA [1]. This effect of carnitine is mediated by the enzyme palmityl-CoA: carnitine palmityltransferase.

Recently we have shown that the carnitine palmityltransferase is localized exclusively in the mitochondria [2], most likely in the inner membrane of these particles [3]. Studies on the kinetic properties of the solubilized enzyme showed that palmityl-CoA is both a substrate for the enzyme and a strong competitive inhibitor to the second substrate (carnitine). Palmityl-CoA therefore shows a pronounced substrate inhibition which varies with the concentration of carnitine. At low palmityl-CoA concentrations high concentrations of carnitine are also inhibitory [4].

In the present communication results are presented which show that these kinetic properties of the carnitine palmityltransferase can be demonstrated also in intact mitochondria.

Non-standard Abbreviation. N-tris(hydroxymethyl)methyl-2-aminoethansulfonic acid, TES.

Enzymes. Palmityl-CoA: carnitine O-palmityltransferase carnitine palmityltransferase); Palmityl-CoA hydrolase (EC 3.1.2.2).

MATERIALS AND METHODS

Animals

Rats of Wistar strain (approximately 200 g) were used. They had free access to food and water until they were killed.

Chemicals

[Me-^{3}H]DL-carnitine (approximately 75 µC/µmole) was prepared as previously described [4]. L-Carnitine was isolated from Difco bovin extracts [5]. L-Palmitylcarnitine was prepared according to Bremer [6], palmityl-CoA according to Seubert [7] and carnitine palmityltransferase according to Norum [8]. CoA was purchased from Boehringer und Soehne, Mannheim, Germany; N-Tris (hydroxymethyl) methyl-2-aminoethan-sulfonic acid (TES) from Calbiochem, Los Angeles, California, U.S.A.; and crystalline bovine albumin from Sigma Chemical Co., St. Louis, Missouri, U.S.A.

Rat Liver Mitochondria

Rat liver mitochondria were prepared by homogenizing rat liver in a Potter-Elvehjem homogenizer with a tight fitting teflon pestle in approximately

10 parts of 10% sucrose containing 2×10^{-3} M neutralized ethylenediaminetetraacetic acid. Nuclei and cell debris were sedimented at $800 \times g$ for 5 min. The mitochondria were sedimented at $12000 \times g$ for 5 min and washed twice in the same volume of homogenizing medium. The particles were finally suspended in homogenizing medium in a concentration corresponding to particles from approximately 2 g of liver per ml. In the experiments 0.2 ml of this suspension (5—7 mg of protein) was used per incubation (2.5 ml).

In some experiments the mitochondrial suspension was treated with ultrasonic vibration in a glass vessel for 1 min (five periods of 12 sec) at 20000 cycles/sec and 6 A with a Branson S-75 Sonifier (standard tip). After ultrasonic treatment an aliquot was centrifuged at $50,000 \times g$ for 2 hours in order to separate the mitochondrial membranes from the soluble protein.

Assay Procedures

Oxygen disappearance was measured with a Clark oxygen electrode (Yellow Springs Instrument Co., Yellow Springs, Ohio, U.S.A.) at $25°$ essentially as described by Chappell [9]. All experiments were done in the presence of malonate (4×10^{-4} M) and dinitrophenol (2×10^{-4} M). Previously we have shown that acylcarnitines are oxidized almost exclusively to acetoacetate by liver mitochondria in the presence of malonate [10]. Thus, the increase in oxygen uptake upon addition of palmityl-CoA or palmitylcarnitine in the reported experiments is a measure of the β-oxidation of fatty acids and therefore also of the rate of transfer of palmityl groups to the mitochondrial site of β-oxidation. The rate of oxygen uptake was the same both when the respiration was stimulated with dinitrophenol and when it was stimulated with ADP and phosphate.

The formation of [Me-^{3}H]L-palmitylcarnitine from [Me-^{3}H]carnitine and- palmityl-CoA was assayed as previously described [4]. When both oxygen uptake rates and the formation of palmitylcarnitine were measured, the course of the oxygen uptake was first followed with the Clark electrode. In an identical experiment with radioactive carnitine samples of 0.25 ml of the reaction mixture were taken at intervals and diluted to 2 ml with N hydrochloric acid. The content of palmitylcarnitine was then assayed by butanol extraction as described [4].

Protein content was calculated from the nitrogen content determined with a micro-Kjeldahl procedure.

RESULTS

Fig. 1 shows that when mitochondria oxidize palmityl-CoA in the presence of carnitine, the rate of oxygen uptake depends on the concentration of palmitylcarnitine formed in the reaction mixture.

The rate of oxygen uptake was nearly maximum with a palmitylcarnitine concentration of 2×10^{-5} M, while there was a significantly lower rate of uptake with a palmitylcarnitine concentration of 10^{-6} M. Addition of relatively great amount of palmityl-CoA increased the accumulation of palmitylcarnitine in the reaction mixture without any significant increase in respiration rate. If albumin was left out of the reaction mixture the accumulation of palmitylcarnitine was more rapid (not shown), and as expected, more palmitylcarnitine accumulated also

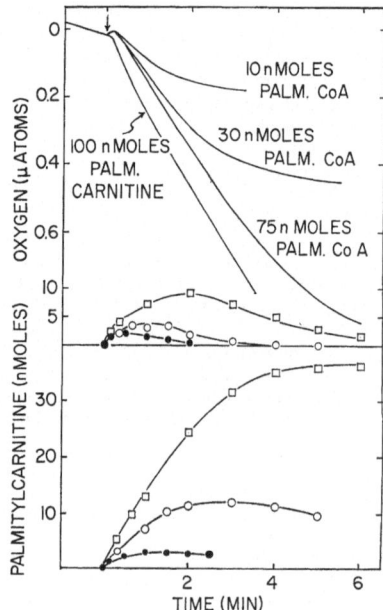

Fig. 1. *Carnitine stimulated oxidation of palmityl-CoA in rat liver mitochondria*. Mitochondria (6 mg protein) were incubated with TES buffer (pH 7.3), 25 μmoles; malonate, 10 μmoles; albumin, 10 mg; 2,4-dinitrophenol, 0.5 μmoles; L-carnitine, 1 μmole. ●——●, palmityl-CoA, 10 nmoles; O——O, palmityl-CoA, 30 nmoles; □——□, palmityl-CoA, 75 nmoles. Upper curves, oxygraph tracings. Middle curves, palmitylcarnitine concentration in experiments identical with the oxygraph experiments. Lower curves, palmitylcarnitine concentration in experiments identical with the oxygraph experiments except that respiration was blocked with 2.5 μmoles of cyanide. Total volume, 2.5 ml, temperature $25°$

when the respiration was blocked with cyanide (Fig. 1, lower curves).

Both Fig. 1 and other experiments showed that with palmityl-CoA plus carnitine as substrate some time (15—30 sec) was required before maximum rate of oxygen uptake was reached, while the maximum rate of oxygen uptake was reached almost immediately with palmitylcarnitine as substrate. These observations confirm that the rate of oxygen uptake depends on the accumulated concentration of palmitylcarnitine as long as this concentration is low.

Previously we have shown that palmityl-CoA is both a substrate for the carnitine palmityltransferase and a competitive inhibitor for the second substrate, carnitine [4]. It might be expected therefore that high concentrations of palmityl-CoA would inhibit both its own oxidation and the oxidation of palmitylcarnitine in isolated mitochondria. No such phenomena could be demonstrated in isolated mitochondria. On the contrary, relatively high concentrations of palmityl-CoA (30—150 μM) were oxidized

presence of albumin. Nearly maximum effect was obtained with 10^{-4} M L-carnitine, and there was no significant difference in the effect of carnitine with the two concentrations of palmityl-CoA used.

In the experiments with palmityl-CoA as substrate it was observed that the total oxygen uptakes obtained always amounted to far less than expected from a complete oxidation of the added palmityl groups, while with palmitylcarnitine as substrate nearly theoretical uptakes were obtained. These observations suggested that palmityl-CoA is rapidly hydrolyzed by isolated mitochondria while palmitylcarnitine is not hydrolyzed to the same extent. Fig.3 confirms this assumption. When the added palmityl-CoA and carnitine were preincubated with carnitine palmityltransferase (free from palmityl-

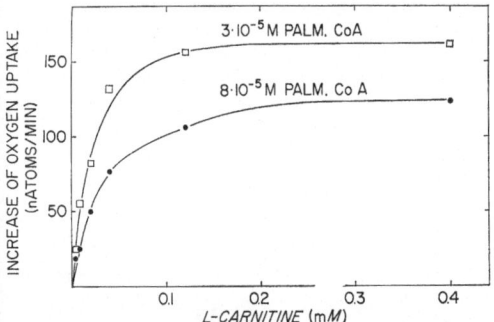

Fig.2. *Effect of* L-*carnitine concentration on the oxidation of palmityl-CoA by mitochondria (5 mg of protein) in the presence of albumin (10 mg).* Palmityl-CoA in the concentrations shown was added followed by L-carnitine approximately 2 min later. The curves show the increase in oxygen uptake rate after the addition of carnitine. Other additions as stated in legend to Fig.1

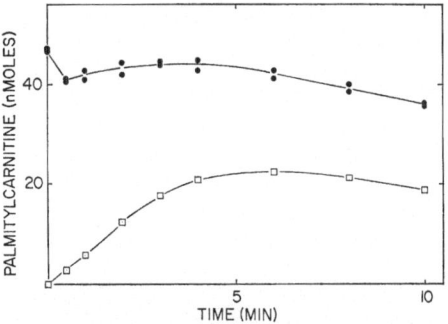

Fig.3. *Deacylation of palmityl-CoA in mitochondria (6 mg of protein).* Additions: TES buffer (pH 7.3), 25 μmoles; potassium cyanide, 5 μmoles; palmityl-CoA, 50 nmoles; L-carnitine, 0.5 μmoles; Total volume, 2.5 ml, temperature 25°. ●———●, partially purified carnitine palmityltransferase (free from palmityl-CoA hydrolase) was added to the incubation mixture 15 min before the mitochondria to allow the reaction to reach equilibrium; □———□, no carnitine palmityltransferase added

even in the absence of carnitine, possibly because of the detergent properties of palmityl-CoA. With the highest concentrations of palmityl-CoA no effect of carnitine could be observed. The carnitine-independent oxidation was inhibited by albumin, and in the presence of this protein carnitine again stimulated the oxidation.

Addition of palmityl-CoA apparently did not affect the oxidation of palmitylcarnitine when estimated with the oxygen uptake rate. However, in these experiments part of the oxygen uptake was most likely due to the carnitine-independent oxidation of palmityl-CoA. Thus, palmityl-CoA probably gave a certain inhibition of palmitylcarnitine oxidation, but this effect was not necessarily due to an inhibition of the carnitine palmityltransferase.

Fig.2 shows how the oxidation of palmityl-CoA depends on the concentration of carnitine in the

CoA hydrolase) before the addition of mitochondria (and cyanide), about twice as much palmitylcarnitine was obtained as when the transferase in the mitochondria catalyzed the reaction. When equilibrium had been reached, only a slow deacylation took place in both cases. It is thus evident that palmityl-CoA hydrolase (and possibly other enzymes) competes efficiently with the carnitine palmityltransferase at high palmityl-CoA concentrations, while it is relatively inactive at low palmityl-CoA concentrations. These results fit well with the relatively high K_m (about 2×10^{-5} M) reported for palmityl-CoA hydrolase [11]. The initial rapid drop in palmitylcarnitine concentration after the addition of mitochondria seen in Fig.3 is probably due to a relatively rapid transfer of palmityl groups to mitochondrial CoA. This is followed by a slower reformation of palmitylcarnitine from the added palmityl-CoA and carnitine, thus establishing a new equilibrium.

Recently it has been suggested that mitochondria contain two carnitine palmityltransferases with different kinetic properties [12]. Fig. 4 shows that there were no significant differences in the behaviour of the carnitine palmityltransferase in the soluble fraction of sonicated mitochondria, in the particulate fraction of sonicated mitochondria, and in a soluble buffer extract of the lyophilized particulate

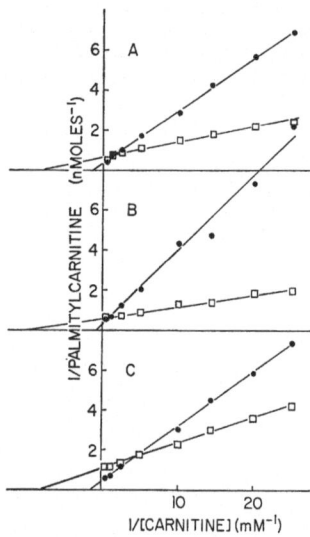

We have also tested the effect of palmityl-CoA concentration on the K_m for carnitine of the carnitine palmityltransferase extracted from the liver of the primitive fish *myxine glutinosa*. Also in this case the K_m for carnitine was found to increase with the concentration of palmityl-CoA. This property of the enzyme therefore is likely to be present in all species.

Fig. 5 shows similar kinetic experiments performed with intact mitochondria in the presence and in the absence of albumin. As with the subfractions of mitochondria the K_m for carnitine increased with the concentration of palmityl-CoA. The values of K_m for carnitine however, were lowered with a factor of about two in comparision with those obtained with subfractions of mitochondria (Fig. 5). A further lowering of the K_m was obtained when albumin was added to the medium. Concomitantly there was a lowering of the V_{max} of the reaction. These results indicate that the concentration of palmityl-CoA

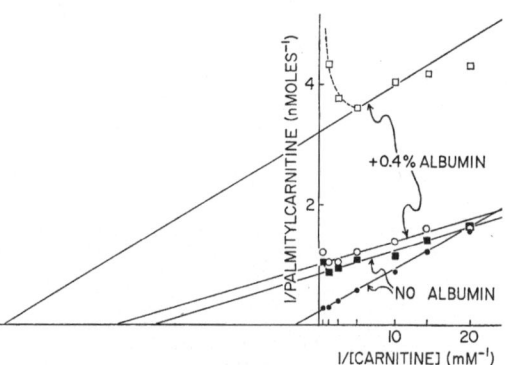

Fig. 4. *The effect of palmityl-CoA concentration on the values of K_m for L-carnitine in subfractions of rat liver mitochondria.* Sonicated mitochondria were separated into a soluble and a particulate fraction by centrifugation at $50,000 \times g$ for 2 hours. Part of the sediment was lyophilized and extracted with TRIS buffer according to Norum [8]. Incubation mixtures: TRIS buffer, (pH 7.4), 50 µmoles; glutathione, 2.5 µmoles; L-carnitine, 0.025—1.0 µmoles; palmityl-CoA, 5 or 30 nmoles. Total volume, 0.5 ml. Incubation time 5 min at 30°. (A) Soluble fraction. 0.035 mg of protein/vessel; (B) Particulate fraction. 0.06 mg of protein/vessel; (C) Extract of lyophilized particulate fraction. 0.015 mg of protein/vessel. □———□, 10^{-5} M palmityl-CoA; ●———●, 6×10^{-5} M palmityl-CoA

Fig. 5. *The effects of palmityl-CoA concentration and of albumin on the value of K_m for L-carnitine in intact mitochondria.* Incubation mixtures: Tris buffer (pH 7.4), 10 µmoles; potassium cyanide, 1 µmole; potassium chloride, approximately 50 µmoles; mitochondria (in 0.05 ml in $10^0/_0$ (w/v) sucrose), 0.1 mg of protein; L-carnitine, 0.025—1.0 µmoles; palmityl-CoA, 5 or 30 nmoles. Total volume 0.5 ml. Incubation time 2 min at 30°. □———□, 10^{-5} M palmityl-CoA + 2 mg of albumin; ○———○, 6×10^{-5} M palmityl-CoA + 2 mg of albumin; ■———■, 10^{-5} M palmityl-CoA; ●———●, 6×10^{-5} M palmityl-CoA

fraction. These experiments gave values of K_m for carnitine of approximately 10^{-4} M and 7×10^{-4} M at palmityl-CoA concentrations of 10^{-5} M and 6×10^{-5} M respectively. Although these few experiments do not permit any accurate calculation of the "true" K_m for carnitine of the rat liver enzyme, they do show that the K_m of the rat liver enzyme is probably several times lower than the "true" K_m for carnitine of the calf liver enzyme [4].

available to the enzyme is lower in the experiments with intact mitochondria and that this concentration is lowered further by albumin. As the carnitine palmityltransferase is probably localized on the inner membrane of the mitochondria [3], the observed effects indicate that both the mitochondrial outer membrane and albumin bind palmityl-CoA, thus creating a "depot" effect. This effect presumably explains how albumin prevents the carnitine-independent stimulation of respiration by high concentrations of palmityl-CoA. Binding of free fatty acids liberated by the palmityl-CoA hydrolase may also be an important effect of albumin, as Yates *et al.* [12] have shown that free fatty acids can provoke oxidation of palmityl-CoA in the absence of carnitine, presumably by a detergent effect.

The total oxygen uptake obtained with a certain amount of palmityl-CoA was greater in the presence of albumin than in its absence. This effect is easily explained by the "depot" effect of albumin which will lower the activity of the hydrolase because of the relatively high K_m for palmityl-CoA of this enzyme [11].

Fig. 5 shows that as the concentration of palmityl-CoA available to the transferase is lowered, high concentrations of carnitine have an increasing inhibiting effect. This effect is also visible in the experiment with subfractions of mitochondria (Fig 4), and it has previously been observed with the soluble calf liver enzyme [4]. This property of the carnitine palmityltransferase may explain the inhibiting effect of high carnitine concentrations on the oxidation of low concentrations of palmityl-CoA (Table 1). This

Table 1. *Effect of L-carnitine concentration on the rate of palmityl-CoA oxidation*

Mitochondria (approximately 6 mg of protein) were incubated with TES buffer (pH 7.3), 25 μmoles; malonate, 10 μmoles; 2,4-dinitrophenol, 0.5 μmoles; albumin, 10 mg; and L-carnitine as shown. Palmityl-CoA, 20 nmoles, was added last, and the increase in oxygen uptake rate was measured. Total volume, 2.5 ml, temperature 25°

L-Carnitine concentration	Increase in oxygen uptake rate
mM	n atoms/min
0	3
0.12	54
0.2	60 67
0.8	59
1.6	42

effect may also be explained by an inhibiting effect of carnitine on the oxidation of the palmitylcarnitine formed as an intermediate, as palmitylcarnitine and carnitine can be assumed to have a common attachment site on the carnitine palmityltransferase [4]. Table 2 shows that carnitine can indeed inhibit the oxidation of palmitylcarnitine and that this inhibition is relieved by an increased concentration of palmitylcarnitine. Relatively high concentrations of carnitine are required to obtain this effect, but it must also be noted that the palmitylcarnitine concentration used in the experiments of Table 2 is much higher than the palmitylcarnitine concentration accumulated in the experiments of Table 1 (see also Fig. 1). Thus, the inhibition of palmityl-CoA oxidation by high concentrations of carnitine may be due to both an inhibited formation of palmitylcarnitine and an inhibited formation of intramitochondrial palmityl-CoA from the palmitylcarnitine formed.

Table 2. *The effect of L-carnitine on the oxidation of L-palmitylcarnitine*

Mitochondria (approximately 7 mg of protein) were incubated with TES buffer (pH 7.3), 25 μmoles; malonate, 10 μmoles; 2,4-dinitrophenol, 0.5 μmoles; and other additions as shown. Palmitylcarnitine was added last and the increase in oxygen uptake rate was measured. Total volume, 2.5 ml, temperature 25°

Additions	Increase in oxygen uptake rate
	n atoms/min
L-palmitylcarnitine,　40 nmoles	342
L-palmitylcarnitine,　40 nmoles L-carnitine,　　　　　3 μmoles	342
L-palmitylcarnitine,　40 nmoles L-carnitine,　　　　30 μmoles	194
L-palmitylcarnitine, 200 nmoles L-carnitine,　　　　30 μmoles	268

Table 3. *The effect of CoA on the carnitine-stimulated oxidation of palmityl-CoA*

Mitochondria (approximately 7 mg of protein) were incubated with TES buffer (pH 7.3), 25 μmoles; malonate, 10 μmoles; 2,4-dinitrophenol, 0.5 μmoles; palmityl-CoA, 50 nmoles; and other additions as shown. L-Carnitine was added last (one min after the palmityl-CoA), and the increase in the oxygen uptake rate was measured. (A substantial part of the palmityl-CoA is hydrolyzed by palmityl-CoA hydrolase before the addition of carnitine in these experiments). Total volume, 2.5 ml, temperature 25°

Additions	Increase in oxygen uptake rate
	n atoms/min
L-carnitine, 0.3 μmoles	108
L-carnitine, 0.3 μmoles CoA, 0.5 μmoles	85
L-carnitine, 0.3 μmoles albumin, 10 mg	70
L-carnitine, 0.3 μmoles, albumin, 10 mg, CoA, 0.5 μmoles	34
L-carnitine, 3 μmoles albumin, 10 mg	113
L-carnitine, 3 μmoles albumin, 10 mg, CoA, 0.5 μmoles	80

Table 3 shows that high concentrations of free CoA can slow down the oxidation of palmityl-CoA in the presence of carnitine. An increased concentration of carnitine counteracts this effect of CoA.

Fig. 6 shows that the inhibiting effect of CoA may be due to both a lowered equilibrium concentration of palmitylcarnitine and to a slowed down formation of this intermediate. Both the equilibrium concentration and the rate of formation of palmitylcarnitine is increased by a high carnitine concentration under the conditions of this experiment.

Fig. 6 also shows that the presence of CoA from the beginning of the incubation gave a slightly higher equilibrium concentration of palmitylcarnitine than when CoA was added after a high concentration of palmitylcarnitine had been formed beforehand in the absence of CoA. As a low equilibrium concentration of palmitylcarnitine necessarily corresponds to a high concentration of palmityl-CoA, a higher activity of the palmityl-CoA hydrolase should be expected with CoA present from the beginning of the incubation. As the opposite is evidently the case, free CoA seems to inhibit the hydrolase.

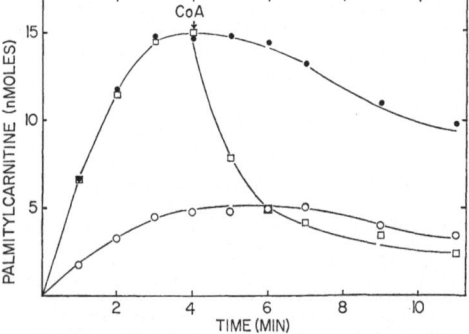

Fig. 6. *Effect of CoA on the formation of palmitylcarnitine from palmityl-CoA and L-carnitine in mitochondria.* (7 mg of protein.) TES buffer, malonate, albumin, dinitrophenol and potassium cyanide were added as stated in legend to Fig. 1. Other additions: Palmityl-CoA, 55 nmoles; L-carnitine, 0.5 μmoles. Total volume, 2.5 ml. Temperature 25°. ●——●, No CoA; ○——○, 0.5 μmoles of CoA; □——□, 0.5 μmoles of CoA after 4 min

DISCUSSION

The results presented in the present communication show that in the carnitine-dependent β-oxidation of palmityl-CoA by rat liver mitochondria, the level of palmitylcarnitine formed as an intermediate is determining for the oxidation rate as long as the concentration of palmitylcarnitine is low (less than 2×10^{-6} M) (Fig. 1). At higher palmitylcarnitine concentrations the acetoacetate-forming system seems to be rate limiting. A saturating concentration of palmitylcarnitine from palmityl-CoA and carnitine is easily obtained. In the experiments shown in Fig. 5 the maximum rate of palmitylcarnitine formation obtained was approximately 20 nmoles/min/mg mitochondrial protein in the absence of albumin and about 5 nmoles/min/mg mitochondrial protein in the presence of 0.4°/₀ (w/v) albumin. The maximum rates of oxygen uptake obtained with palmitylcarnitine or with palmityl-CoA plus carnitine corresponded to β-oxidation of 2.5—3 nmoles palmitate/min/mg mitochondrial protein.

This rate of acetoacetate formation may not correspond to the maximum capacity of the β-oxidiz-

ing enzyme system as such, as we have previously shown that the acetoacetate formation from acetyl-CoA may be rate limiting in the allover reaction in rat liver mitochondria when the citric acid cycle is not operating [10]. It is unlikely though that the capacity of the mitochondria to β-oxidize fatty acids exceeds the capacity of acetoacetate formation to any great extent [10]. In the intact cell the respiratory control will probably also prevent a maximum rate of fatty acid oxidation, even in a starving or diabetic condition. Liver mitochondria therefore appear to contain a carnitine palmityltransferase capacity which will exceed the capacity for fatty acid oxidation, even under extreme conditions.

The relatively great reserve capacity for palmitylcarnitine formation in the mitochondria suggests that the carnitine/palmitylcarnitine substrate pair is in close equilibrium with palmityl-CoA/CoA *in vivo*. In agreement with this view it is found that the level of long-chain acylcarnitines in the tissues fluctuates widely with the nutritional state of the animal [13,14]. It is all the same possible that the inhibitory effect of high carnitine concentrations on the formation of palmitylcarnitine from low concentrations of palmityl-CoA will effect a certain deviation from equilibrium to the advantage of palmityl-CoA, *e.g.* in carbohydrate feeding when the level of both palmityl-CoA [13,15] and long-chain acylcarnitines [14] is low. The inhibitory effect of high palmityl-CoA concentrations is more difficult to understand. The observation that protein (albumin) counteracts this effect of palmityl-CoA makes it possible that inhibitory concentrations are not reached in the intact cell and that this property of the carnitine palmityltransferase is of no physiological significance.

Altogether it is likely that the level of extra-mitochondrial long-chain acyl-CoA and of long-chain acylcarnitines is very important in determining the rate of fatty acid oxidation in the tissues, but it is unlikely that this level alone is regulating. Previously we have obtained results which suggest that the level of free CoA in the mitochondria may be of importance and that the availability of other substrates, especially succinate, may reduce the rate of acylcarnitine and pyruvate oxidation [10,16]. Succinate might reduce the level of free CoA by succinyl-CoA formation.

If availability of mitochondrial CoA is an important factor in this system, the rate of fatty acid oxidation (and the oxidation of other CoA-dependent substrates) can be regulated in the mitochondria without affecting CoA-requiring processes in the extramitochondrial compartment(s) of the cell.

The function of acyl-CoA hydrolase in the tissues is obscure. If its function in the intact cell is to hydrolyze CoA esters, the activity of this enzyme evidently will vary with the level of long-chain acyl-CoA because of its relatively high K_m for acyl-CoA

[11]. It may thus contribute to the hypermetabolism in conditions with a high rate of fatty acid oxidation [17]. The deacylase may under such conditions represent a "security valve" preventing all CoA from being bound as acyl-CoA.

This work has been supported by the Norwegian Council of Cardiovascular Diseases.

REFERENCES

1. Fritz, I. B., and Yue, K. T. N., *J. Lipid Res.* 4 (1963) 279.
2. Norum, K. R., and Bremer, J., *J. Biol. Chem.* 242 (1967) in press.
3. Norum, K. R., Farstad, M., and Bremer, J., *Biochem. Biophys. Res. Commun.* 24 (1966) 797.
4. Bremer, J., and Norum, K. R., *J. Biol. Chem.* 242 (1967) 1744.
5. Friedman, S., McFarlane, J. E., Bhattacharyya, P. K., and Fraenkel, G., *Biochem. Prep.* 7 (1960) 26.
6. Bremer, J., *Biochem. Prep.* in press.
7. Seubert, W., *Biochem. Prep.* 7 (1960) 80.
8. Norum, K. R., *Biochim. Biophys. Acta*, 89 (1964) 95.
9. Chappell, J. B., *Biochem. J.* 90 (1964) 225.
10. Bremer, J., *Biochim. Biophys. Acta*, 116 (1966) 1.
11. Srere, P. A., Seubert, W., and Lynen, F., *Biochim. Biophys. Acta*, 33 (1959) 313.
12. Yates, D., Shepherd, D., and Garland, P., *Nature*, 209 (1966) 1213.
13. Pearson, D. J., and Tubbs, P. K., *Biochim. Biophys. Acta*, 84 (1964) 772.
14. Bøhmer, T., Norum, K. R., and Bremer, J., *Biochim. Biophys. Acta*, 125 (1966) 244.
15. Bortz, W. M., and Lynen, F., *Biochem. Z.* 339 (1963) 77.
16. Bremer, J., *Biochim. Biophys. Acta*, 104 (1965) 581.
17. Challoner, D. R., *Lancet* II (1966) 681.

J. Bremer and K. R. Norum
Institute of Clinical Biochemistry
Rikshospitalet, Oslo, Norway

European J. Biochem. 1 (1967) 434—438

Role of Sulphydryl Groups in Adenosine Deaminase

G. RONCA, C. BAUER, and C. A. ROSSI

Istituto di Chimica Biologica, University of Pisa

(Received April 8, 1967)

The reactivity and the role of sulphydryl groups in adenosine deaminase have been investigated. In the sodium dodecylsulphate-denatured enzyme 2.1 sulphydryl groups are titratable with p-mercuribenzoate. The native enzyme is inactivated by p-mercuribenzoate or phenylmercuric acetate and a stoichiometric relationship exists between the equivalents of mercaptide formed and the inactivation of the enzyme. Substrate analogs protect the enzyme against p-mercuribenzoate and phenylmercuric acetate inactivation. Iodoacetamide, iodoacetate, and N-ethylmaleimide, which do not affect the enzyme activity, do not react with sulphydryl groups. The results show that a sulphydryl group unreactive towards the alkylating reagents is essential in some way for the enzymatic activity. The lack of reactivity of the essential sulphydryl group with alkylating reagents suggests that the sulphydryl group may be contained in a hydrophobic region of the enzyme molecule.

Adenosine deaminase catalyzes the hydrolytic breakdown of adenosine to inosine and free ammonia. The reaction is irreversible and does not require activators [1]. The enzyme is distributed widely in animal tissues and in microorganisms and has been obtained in highly purified form from calf duodenal mucosa and epiphyseal bone marrow [2,3].

The enzyme from bovine intestinal mucosa is relatively specific in deaminating, in addition to adenosine, 2'-deoxyadenosine, 2-fluoroadenosine, 2,6-diaminopurine riboside and 3'-deoxyadenosine [4], and could also convert 6-chloropurine riboside to inosine [5]. Adenine, adenylic acid, nicotinamide-adenine dinucleotide, and cytidine are not substrates for the enzyme.

The nature of the amino acid residues involved in the catalytic reaction has not been studied so far. We have found that the enzyme is inactivated by PMB and PhHgAc and that a stoichiometric relationship exists between the equivalents of mercaptide formed and the inactivation of the enzyme. Substrate analogs protect the adenosine deaminase against PMB and PhHgAc inactivation.

Some other reagents of —SH groups such as N-ethylmaleimide, iodoacetamide and iodoacetate do not affect the enzyme activity and do not react with the sulphydryl groups of adenosine deaminase.

EXPERIMENTAL PROCEDURE

Materials

Na p-chloromercuribenzoate was obtained from Sigma Chemical Co., PhHgAc, iodoacetamide, and

Non-standard Abbreviations. p-mercuribenzoate, PMB; phenylmercuric acetate, PhHgAc.
Enzyme. Adenosine deaminase (EC 3.5.4.4).

iodoacetate from BDH, all these chemicals were recrystallized before use. All other chemicals were of Analytical Reagent grade and were used as obtained.

Adenosine deaminase was purified from calf intestinal mucosa [2]. The final enzyme preparation was devoid of any phosphatase activity and showed a specific activity of 450 units per mg protein. A value of 430 units per mg protein has been reported to correspond to homogeneous adenosine deaminase preparations [2]. From Sephadex G-75 and DEAE-cellulose columns the enzyme was eluted as a single symmetric peak with constant specific activity.

Methods

Protein concentration was measured by the micro-biuret method, using lyophilized adenosine deaminase as a standard. The molecular weight of the enzyme has been determined by sucrose-gradient centrifugation [6] and a value corresponding to a molecular weight of 35,000 was found. This value is in close agreement with that reported by Brady and O'Sullivan [7].

Enzymatic activity was determined spectrophotometrically by following the decrease in absorbance at 265 mμ according to Kalckar [8]. One enzyme unit is the amount of enzyme which will convert 1 μmole of adenosine to inosine per minute at pH 7 and at 37°.

The number of —SH groups was determined using PMB according to a procedure of Benesch and Benesch [9] based on the work of Boyer [10]. In a typical experiment 11 nmoles of adenosine deaminase were dissolved in 1.0 ml of 0.05 M sodium phosphate buffer, pH 7, and 1% sodium dodecyl-sulphate in 1.0 ml silica cuvette. PMB $(3 \times 10^{-4}$ M) was added in increments of 10 µl. Absorbance at 250 mµ was measured after each addition of PMB.

When the loss of enzymatic activity and the mercaptide formation as a function of time of incubation with PMB were followed, the incubation mixtures contained, in a final volume of 1 ml: 10 to 14 nmoles of adenosine deaminase, 10 to 200 nmoles of PMB and 0.1 M sodium acetate buffer, pH 5.2 or 5.6, or 0.05 M sodium phosphate buffer, pH 7. Samples of 10 µl were removed at intervals, diluted with 0.1 M sodium phosphate buffer, pH 7, to a concentration of approximately 4 µg adenosine deaminase per ml. For the determination of enzymatic activity 10 µl of the diluted enzyme solutions were added to a cuvette containing 3 ml of 60 µM adenosine in 0.1 M phosphate buffer, pH 7.

The equivalents of the mercaptide formed have been estimated at pH 5.2 and at pH 5.6 from the increment of the absorption at 255 mµ, on the basis of a molar extinction increment of 6,200 [11]. At pH 7 a molar extinction increment at 250 mµ of 8,000 obtained from the titration of the sulphydryl groups of sodium dodecylsulphate-denatured adenosine deaminase has been used for calculation.

The same experimental procedure was used when the enzyme was incubated with PhHgAc; in this case, however, the mercaptide formation was not determined.

When the protection by substrate analogs against PhHgAc and PMB inactivation was tested, the incubation mixtures contained in a final volume of 1 ml: 3 mM 2,6-diaminopurine or 3 mM 8-azaguanine or 3 mM 8-azaguanine plus 0.65 mM adenosine, 14 nmoles adenosine deaminase, 35 nmoles PhHgAc or 420 nmoles PMB and 0.05 phosphate buffer, pH 7. The determinations of enzyme activity were performed as previously indicated; the final concentrations of 2,6-diaminopurine and 8-azaguanine in the enzyme assay were not inhibitory.

RESULTS

Effect of PMB and PhHgAc on the Enzymatic Activity

In the sodium dodecylsulphate-denatured adenosine deaminase 2.1 sulphydryl groups are titratable with PMB. In the native enzyme one —SH group reacts with PMB, its modification being followed by loss of enzymatic activity.

Fig.1 shows the mercaptide formation and the percent inhibition of the enzyme activity as a function of time. The reaction was carried out in sodium acetate buffer, pH 5.2 or 5.6, and in sodium phosphate buffer, pH 7. At pH 5.2 and at pH 5.6 with a molar ratio PMB/enzyme of 1.2 and 1.10 a strong inhibition (70 to 85%) of the enzymatic activity has been obtained. The equivalents of the mercaptide formed parallel the percent inhibition of the enzy-

matic activity. At pH 7 the reactivity of the sulphydryl group with PMB is lower than at pH 5.2 and 5.6, but also in this condition the equivalents of mercaptide formed parallel the inhibition; this experiment was carried out with a molar ratio PMB/enzyme of 14.

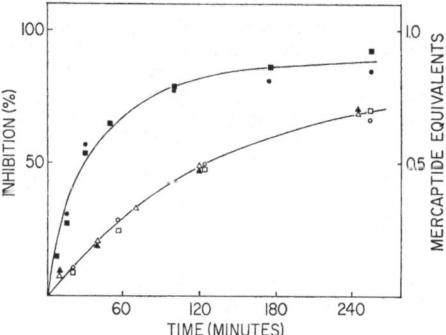

Fig.1. *Inactivation of adenosine deaminase by PMB at pH 5.2, 5.6, and 7.* The reaction mixtures contained: 11.9 µM adenosine deaminase, 13.9 µM PMB and 0.1 M acetate buffer, pH 5.2; 10.2 µM adenosine deaminase, 10.9 µM PMB and 0.1 M acetate buffer, pH 5.6; 10 µM adenosine deaminase, 140 µM PMB and 0.05 M phosphate buffer, pH 7. The ratio PMB/enzyme was 1.2, 1.1, and 14 at 5.2, 5.6, and 7, respectively. The increase in absorbance at 250 and 255 mµ was followed with time; at intervals aliquots were taken for the determination of enzymatic activity. The controls without PMB retained full activity. Temperature 24°. For other details see Methods. Percent inhibition at pH 5.2 (●), 5.6 (○), 7 (▲); equivalents of mercaptide formed at pH 5.2 (■), 5.6 (□), 7 (△)

Table 1. *Second-order velocity constants of the reaction of adenosine deaminase with PMB and PhHgAc*

Buffer	pH	Reagent	K[a]
			$M^{-1} sec^{-1}$
0.1 M sodium acetate	5.2	PMB	35
0.1 M sodium acetate	5.6	PMB	18
0.05 M sodium phosphate	7.0	PMB	0.75
0.1 M sodium acetate	5.0	PhHgAc	40
0.05 M sodium phosphate	7.0	PhHgAc	8
0.1 M triethanolamine-HCl	8.6	PhHgAc	18

[a] The second-order velocity constants reported are the mean of two or three determinations; K has been calculated from the mercaptide formed for the reaction with PMB and from the percent inhibition of enzyme activity for the reaction with PhHgAc.

The reaction of the sulphydryl group with PMB follows a second-order course at the three pH values tested. The second-order velocity constants, calculated according to Boyer [10] from the amount of mercaptide formed, are reported in Table 1.

PhHgAc also inactivates adenosine deaminase; Fig.2 shows the percent inhibition of the enzymatic activity as a function of incubation time with three different PhHgAc concentrations. The reaction was carried out in sodium phosphate buffer, pH 7; with a molar ratio PhHgAc/enzyme of 1.3 an inhibition of 85% has been obtained in 380 min; with 2.6 moles of PhHgAc added per mole of enzyme the rate of inactivation is higher than that obtained at pH 7 with 14×molar excess of PMB.

The reaction was also carried out at three different pH values. The rate of inhibition at pH 5 in acetate buffer and at pH 8.6 in triethanolamine buffer is higher than at pH 7 in phosphate buffer. On the basis of the observed inhibition of enzymatic activity and assuming that one mole of PhHgAc reacted with

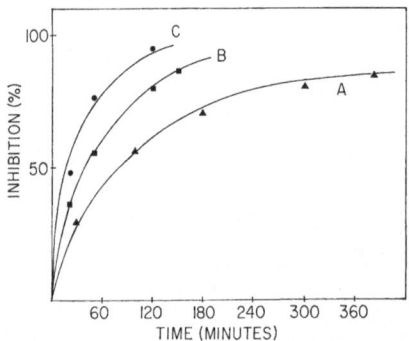

Fig.2. *Inactivation of adenosine deaminase by PhHgAc at pH 7.0.* The reaction mixtures contained 14 μM adenosine deaminase, 18.2 μM (Curve A), 36.4 μM (Curve B), and 72.8 μM (Curve C) PhHgAc and 0.05 M phosphate buffer, pH 7. The controls without PhHgAc retained full activity. Temperature 24°

one mole of inactivated enzyme the velocity constants of the reaction of sulphydryl group with PhHgAc at ·the three pH values have been calculated. With PhHgAc also the reaction follows a second-order course. The values of the second-order velocity constants are reported in Table 1.

Effect of Alkylating Reagents on Enzymatic Activity

The enzyme has been treated with other −SH group reagents such as N-ethylmaleimide, iodoacetamide, and iodoacetate. The incubation mixtures contained 140 μM adenosine deaminase, 0.05 M N-ethylmaleimide or 0.1 M iodoacetamide or 0.1 M iodoacetate in 0.05 M sodium phosphate buffer at pH 6.5, 7, 7.5 or 8; prolonged incubation (10 hours) of the enzyme at 20° with these reagents does not affect the enzymatic activity.

Titration of Sulphydryl Groups of Adenosine Deaminase Treated with Alkylating Reagents

In order to determine whether the alkylating reagents, which do not affect the enzyme activity, react with the sulphydryl groups of adenosine deaminase, the enzyme was treated with 0.1 M iodoacetamide in 0.05 M sodium phosphate buffer, pH 7.5, for 10 hours at 20°. The enzyme was dialysed at 4° for 2 days against 0.1 M phosphate buffer, pH 7. A control sample without iodoacetamide was subjected to the same procedure. In no case the dialysis procedure appreciably affected the enzyme activity. The iodoacetamide-treated adenosine deaminase and the control were denatured with 1% (w/v) sodium dodecylsulphate and the sulphydryl groups were determined with PMB. Table 2 reports the sulphydryl group determination of the iodoacetamide-treated enzyme and the control: it can be seen that no variation in the number of sulphydryl groups titratable with PMB after enzyme denaturation has been observed.

Table 2. *Titration of sulphydryl groups of iodoacetamide-treated adenosine deaminase*
The incubation mixture contained 140 μM adenosine deaminase, 0.1 M iodoacetamide and 0.05 M sodium phosphate buffer, pH 7.5; after 10 hours at 20° the enzyme was dialysed for 2 days against 0.1 M phosphate buffer, pH 7. A control sample without iodoacetamide was subjected to the same procedure. After denaturation with sodium dodecylsulphate sulphydryl groups were determined with PMB as described under Methods. The values reported are the mean of two experiments

	Specific activity before dialysis	Specific activity after dialysis	−SH per mole of denatured enzyme
	units/mg	units/mg	
Control	445	425	1.98
Iodoacetamide-treated adenosine deaminase	448	430	2.05

This was further confirmed by an additional experiment in which the iodoacetamide-treated adenosine deaminase and the control were incubated with 30×molar excess of PMB at pH 7. The mercaptide formation and the inactivation of the two samples were followed as a function of incubation time. No differences in the rate of inactivation and in the mercaptide formation could be observed (Fig.3).

Similar results have been obtained with adenosine deaminase treated with iodoacetate and N-ethylmaleimide.

Protection by Substrate Analogs

Owing to its rapid deamination it was not possible to use adenosine for the study of the protection by substrate against the PMB and PhHgAc inactivation. We have used 2,6-diaminopurine which is a competitive inhibitor of adenosine deaminase [12] and 8-azaguanine which is an uncompetitive inhibitor [13a].

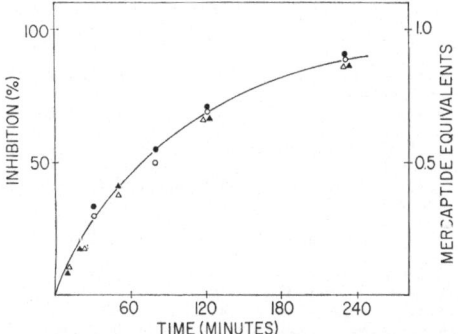

Fig. 3. *Mercaptide formation and inactivation of iodoacetamide-treated adenosine deaminase by PMB.* The incubation mixtures contained 10 μM adenosine deaminase or iodoacetamide-treated adenosine deaminase, 300 μM PMB and 0.05 M phosphate buffer, pH 7. The controls without PMB retained full activity. Percent inhibition of adenosine deaminase (●) and of iodoacetamide-treated adenosine deaminase (○). Equivalents of mercaptide formed: adenosine deaminase (▲) and iodoacetamide-treated adenosine deaminase (△)

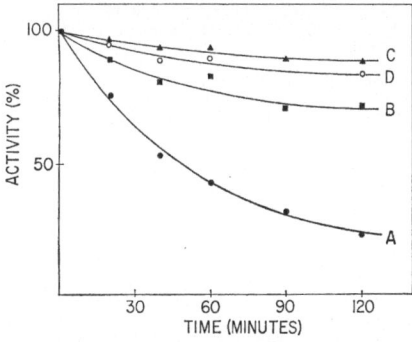

Fig. 4. *Inactivation of adenosine deaminase by PhHgAc and protection by substrate analogs.* For details see Methods. The control samples without PhHgAc retained full activity. The curves show the activity in the presence of PhHgAc alone (Curve A) or plus 3 mM 8-azaguanine (Curve B) or plus 3 mM 8-azaguanine and 0.63 mM adenosine (Curve C), or plus 3 mM 2,6-diaminopurine (Curve D)

The reaction of sulphydryl group of adenosine deaminase with PhHgAc and PMB was performed in the presence of 2,6-diaminopurine and 8-azaguanine with and without adenosine as described under "Methods." Fig. 4 shows that the presence of substrate analogs strongly protects the enzyme against

PhHgAc inactivation. Protection by substrate analogs against PMB inactivation has also been observed.

DISCUSSION

The amino acid composition of adenosine deaminase (mol. wt. 35,000) has been determined [13b] and 9 half-cystine residues per molecule of enzyme have been found, while in the sodium dodecylsulphate-denatured enzyme only 2.1 sulphydryl groups are titratable with PMB. These findings might be attributed to the presence of —S—S— bonds in the enzyme molecule and/or to the inaccessibility of —SH groups to PMB in the sodium dodecylsulphate-denatured adenosine deaminase.

Concentrations of PMB and PhHgAc equimolar to the enzyme produce an almost complete inactivation of native adenosine deaminase. Moreover with PMB the equivalents of the mercaptide formed parallel the percent inhibition of the enzyme activity. These data suggest that one of the two sulphydryl groups titratable with PMB in the denatured enzyme is essential in some way for the enzymatic activity. As in all such experiments, an uncertainty exists as to whether the sulphydryl group in question is functionally or merely structurally essential. The strong protection that 2,6-diaminopurine and 8-azaguanine exert against PMB and PhHgAc inactivation would suggest that the essential sulphydryl group is in or near the catalytic site. This kind of evidence however cannot be taken as proof for the direct participation of the —SH group in the catalysis.

The values of the second-order velocity constants (Table 1) show that at pH 7 the rate of the reaction of the essential sulphydryl group with PhHgAc is tenfold higher than with PMB. At decreasing pH values an increase of the second-order velocity constants has been observed; at pH 5 the increase is 45-fold for PMB and 5-fold for PhHgAc with respect to pH 7. For several proteins the rate of the reaction of —SH groups with PMB is greater at pH 4.6 than at pH 7. This is at least partly due to the high affinity of PMB for hydroxyl ions that would increase its reactivity as the pH is lowered. Moreover the reactivity of PMB can be affected by the anions used in the buffer system [10]. These factors seem to affect to a minor degree the reaction of the essential —SH group with PhHgAc.

The second-order velocity constants of the reaction of the —SH group with PMB and PhHgAc are comparable to those observed with egg albumin and β-lactoglobulin [10]; however the reactivity is low in comparison with that of the essential sulphydryl groups of many enzymes. The essential —SH group in native adenosine deaminase shows a marked lack of reactivity with the alkylating reagents; iodoacetamide, iodoacetate, and N-ethylmaleimide do not affect the enzyme activity. In adenosine deaminase treated with the alkylating reagents the rate of inactivation by PMB, the amount of the mercaptide formed and the number of —SH groups titratable with PMB after sodium dodecylsulphate denaturation are unchanged.

However the low reactivity of the sulphydryl group towards —SH reagents does not rule out its possible role in the enzyme catalysis, in view of the fact that non-polar residues in the vicinity of the —SH group may affect its reactivity [14]. In lactic dehydrogenase from beef heart, Gold and Segal [15] have determined the amino acid sequence in the vicinity of an —SH group essential for the enzyme activity showing a lack of reactivity with alkylating reagents. They found a high proportion of non-polar residues around the essential —SH group. Adenosine deaminase is competitively inhibited by aliphatic alcohols in the decreasing order of n-butanol, n-propanol, iso-propanol, ethanol, and methanol [16]. This phenomenon has also been observed with leucine aminopeptidase, α-chymotrypsin and pepsin [17,18, 19], and has been interpreted by assuming that these enzymes contain regions which combine with the non-polar moiety of the substrates through a hydrophobic type of interaction. On the basis of various kinetic studies [18], Hartley [20] proposed that a "hydrophobic slit" may be present in α-chymotrypsin. It is conceivable that the active site of adenosine deaminase might contain regions with high proportion of non-polar residues in view of the hydrophobic character of the purine moiety of adenosine.

We wish to thank Dr. T. G. Brady for sending us his data of the amino acid composition of adenosine deaminase

from calf intestinal mucosa before publication. This investigation was supported by the Italian Consiglio Nazionale delle Ricerche, Impresa di Enzimologia.

REFERENCES

1. Brady, T. G., Biochem. J. 36 (1942) 478.
2. Brady, T. G., and O'Connell, W., Biochim. Biophys. Acta, 62 (1962) 216.
3. Ipata, P. L., and Rossi, C. A., Boll. Soc. It. Biol. Sper. 38 (1962) 1117.
4. Cory, J. G., and Suhadolnik, R. J., Biochemistry, 4 (1965) 1729.
5. Cory, J. G., and Suhadolnik, R. J., Biochemistry, 4 (1965) 1733.
6. Martin, R. G., and Ames, N. B., J. Biol. Chem. 236 (1961) 1372.
7. Brady, T. G., and O'Sullivan, M., Biochim. Biophys. Acta, 132 (1967) 127.
8. Kalckar, H. M., J. Biol. Chem. 167 (1947) 445.
9. Benesch, R., and Benesch, R. E., Methods Biochem. Analy. 10 (1962) 43.
10. Boyer, P. D., J. Am. Chem. Soc. 76 (1954) 4331.
11. Fraenkel-Conrat, H., Methods Enzymol. 4 (1956) 247.
12. Ronca, G., Biochim. Biophys. Acta, 132 (1967) 214.
13a. Feigelson, P., and Davidson, J. D., J. Biol. Chem. 223 (1956) 65.
13b. Brady, T. G., (1966) personal communication.
14. Cecil, R., The Proteins, 1 (1963) 379.
15. Gold, A. H., and Segal, H. L., Biochemistry, 4 (1965) 1506.
16. Bauer, C., Ronca, G., and Rossi, C. A., Boll. Soc. It. Biol. Sper. 41 No. 20 bis (1965) 22.
17. Hill, R. L., and Smith, E. L., J. Biol. Chem. 224 (1957) 209.
18. Neurath, H., and Hartley, B. S., J. Cell. Comp. Physiol. 54 (1959) 179.
19. Tang, J., J. Biol. Chem. 240 (1965) 3810.
20. Hartley, B. S., Brookhaven Symposia in Biology, 15 (1962) 85.

G. Ronca, C. Bauer, and C. A. Rossi
Istituto di Chimica Biologica
Via Roma 55, Pisa, Italy

European J. Biochem. 1 (1967) 439—446

Inhibition of Respiration under the Control of Azide Uptake by Mitochondria

F. PALMIERI and M. KLINGENBERG

Physiologisch-Chemisches Institut, Universität Marburg/Lahn

(Received January 3, 1967)

The inhibition of mitochondrial respiration by azide under various steady states was analyzed and found to be controlled by an active accumulation of azide.

The azide inhibition is found to be uncompetitive with the respiratory rate. Thus the $50^0/_0$ inhibiting concentration of azide decreases inversely to the respiratory rate, independent of the substrate.

The respiration activated by ion uptake is about 10 times more sensitive to azide than the uncoupled respiration.

The respiration of particles is about 20-fold less sensitive to azide when compared to mitochondria under identical conditions (ion pumping state) and at equal respiratory rates.

Azide can be demonstrated to be a permeant anion under conditions of passive swelling.

An active accumulation of azide, as deduced from these results, is demonstrated by respiratory assays in sedimented mitochondria.

Azide does not inhibit electron transfer to ferricyanide but rather stimulates it by activation of ion uptake.

It is concluded that azide on mitochondria acts also by means of its accumulation and that its interference with energy transfer is a secondary result of its accumulation as a permeant anion.

Since Keilin [1] has shown that azide inhibits cytochrome oxidase, its effect on respiration has been widely studied in microorganisms and animal systems. In particular mitochondrial reactions involved in respiration and oxidative phosphorylation have been tested for their sensitivity to azide. The inhibition of respiration by azide has been further studied at the level of the isolated cytochrome oxidase [2].

The studies on mitochondria suggested that azide can have further points of attack. Thus Loomis and Lipman [3], Judah [4] and Slater [5] showed that azide can uncouple oxidative phosphorylation. Furthermore, the Mg^{2+}-activated ATPase was found to be stimulated by low concentrations of azide [6,7] and inhibited by higher concentrations [6,8,9]. The inhibitory effect of azide on the mitochondrial ATPase is relieved by dithionite [10], whereas that on the microsomal ATPase is not [11]. Also other reactions of oxidative phosphorylation are affected by azide, such as the ATP-ADP exchange [12—14] (but not in digitonine particles [15]), the ATP-P$_1$ exchange [16,17] and the ATP-induced contraction of swollen mitochondria [18]. Wilson and Chance [19] reported a competitive release by uncouplers of

azide inhibition of the ADP-stimulated respiration and stated that this is evidence for "a direct relationship between the effects of azide on oxidative phosphorylation and its inhibition of electron transfer." Bogucka and Wojtczak [20] concluded that azide has a dual effect in energy transfer: it acts as an uncoupler of oxidative phosphorylation at the first phosphorylation site and as an oligomycin-like inhibitor at the second and third site (cf. scheme I of ref. [20]).

In this paper evidence has been obtained that the effects of azide on mitochondria are controlled by its uptake as a permeant anion under the influence of energy-linked ion transport. On this basis the interference of azide with energy transfer, as widely reported in the literature, is explained as a secondary effect. The only site of action of azide is postulated to be on cytochrome a, once azide is present within the mitochondria.

METHODS

Rat-liver mitochondria were isolated as described by Klingenberg and Slenczka [21]. For the preparation of particles, mitochondria were sonicated at $0°$ in the presence of 5 mM Mg^{2+} using a Branson sonifier. After removal of the unbroken mitochondria, the sonicated particles were isolated by centrifugation ($70,000 \times g$ for 20 min). The oxygen uptake was measured polarographically, using 2—3 mg mito-

---

Non-standard abbreviations. Tetramethyl-p-phenylenediamine, TMPD; carbonylcyanide-phenylhydrazone, CCP; bovine serum albumin, BSA; dinitrophenol, DNP.

Enzyme. Cytochrome oxidase (EC 1.9.3.1).

chondrial protein in a standard reaction mixture containing: 0.25 M sucrose, 20 mM triethanolamine buffer, pH 7.2, 1 mM EDTA, 10 mM KCl in a final volume of 1 ml at 25°.

The turbidity changes of the mitochondrial suspension were recorded by following the decrease of absorbance at 546 mμ.

Ferricyanide reduction was monitored in the mitochondrial suspension by using the double-beam spectrophotometer at a wavelength pair of 440 and 490 mμ. The extinction coefficient of this wavelength pair is $\Delta\varepsilon = 0.48 \ mM^{-1} \times cm^{-1}$. This wavelength pair has approximately half of the maximum extinction coefficient. It has been chosen in order to keep the absorption changes comparatively small and thus to avoid a nonlinearity in the recordings, since the applied instrument records only absorption changes.

Protein was determined by the Biuret method [22].

RESULTS

Interference of Azide with Electron Transport

The effect of azide on the mitochondrial respiration has been studied using three substrates with widely different respiratory activities, namely glutamate plus malate, succinate and ascorbate plus TMPD, from which electrons enter at different sites to the respiratory chain. The respiration in the active state with all these substrates is inhibited by azide. Usually the inhibition of respiration with different substrates is compared as % inhibition versus the inhibitor concentration (Fig. 1 A). Here the inhibitory power of azide decreases going from ascorbate plus TMPD to succinate and glutamate plus malate. In other words $I_{1/2}$[1] increases inversely with the respiratory activity of the various electron donors. The inhibition by azide is more adequately analyzed by plotting the reciprocal of respiratory activities versus azide concentration. In this way straight lines are obtained (Fig. 1 B), which are displaced parallel for each substrate in the sequence of their respiratory activity. This indicates an uncompetitive inhibition of respiration by azide (e.g. at the level of cytochrome a).

This relation can be derived from the common equation for uncompetitive inhibition by substituting the substrate dependent term with the uninhibited respiratory rate:

$$\frac{1}{v} = \frac{K_m}{V_{max} S} + \frac{I}{V_{max} K_i} + \frac{1}{V_{max}} \quad (1)$$

v = specific respiratory activity, I = azide concentration.

For uninhibited respiration v_o holds:

$$\frac{1}{v_o} = \frac{K_m}{V_{max} S} + \frac{1}{V_{max}} \quad (2)$$

which is substituted in equation (1)

$$\frac{1}{v} = \frac{1}{v_o} + \frac{I}{K_i V_{max}} . \quad (3)$$

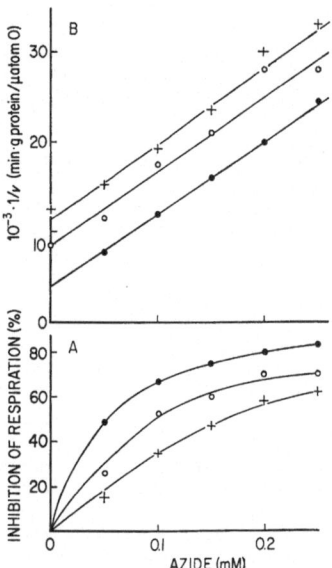

Fig. 1. A, Dependence on the concentration of azide of the inhibition of liver mitochondrial respiration in the active state. As substrates were used ascorbate plus TMPD, succinate, or glutamate plus malate. B, Kinetic analysis of the inhibition by a 1/v versus azide concentration plot. The reaction mixture contained the basic components with the following additions: 2.64 mg mitochondrial protein, 2.5 mM phosphate, 1 mM ADP and either 2 mM ascorbate plus 0.2 mM TMPD, 2 mM succinate, or 2 mM glutamate plus 2 mM malate as substrate. With ascorbate plus TMPD, rotenone (2 μg) was present. Azide was added to the ADP-activated respiration in the concentrations indicated. The uninhibited respiratory activities in the absence of azide were (in μatoms O/min/g protein) with ascorbate plus TMPD 250, with succinate 118, and with glutamate plus malate 80. ●, ascorbate + TMPD; ○, succinate; +, glutamate + malate

On the basis of equation (3) a plot of $1/v$ versus $1/v_o$ i.e. versus the reciprocal of the uninhibited respiratory rate with the various substrates, should give a straight line with a slope of 1 at a given concentration of azide. This relation is fulfilled as shown in Fig. 2, where the points approximately obey the straight lines with a slope of 1. This also further supports the uncompetitive type of the inhibition by azide. The uncompetitive nature can be explained on the basis of a point of attack of azide at cytochrome *a* only, to which azide binds only in the reduced form [1,2,19,23]. Reduced cytochrome *a* would represent the enzyme-substrate complex (ES), which in the uncompetitive mechanism is the binding form of the enzyme for the inhibitor. Therefore, the inhibition by azide is independent of the type of electron donating substrate as well as of the phosphorylation sites involved, as suggested in the literature [20], but rather dependent on the respiratory activity, *i.e.* the turnover of cytochrome oxidase in the absence of inhibitor.

Further evidence for the conclusion that azide interferes with electron transport only at the cytochrome oxidase is obtained using ferricyanide as electron acceptor in cyanide-inhibited mitochondria. In this system the span of electron transport to cytochrome *c* is separated from cytochrome oxidase. As shown below in Fig.8, no inhibition by azide on electron transfer to ferricyanide is observed.

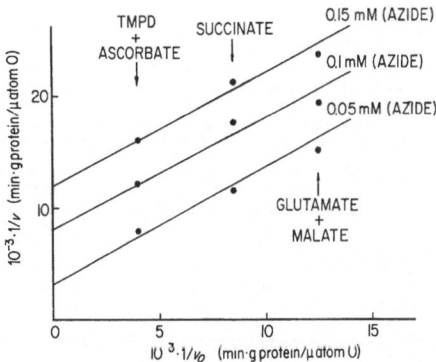

Fig.2. *Kinetic analysis of the inhibition of respiration by azide.* Data from the experiment in Fig.1 are given in a double reciprocal plot of the respiration $(1/v)$ versus the uninhibited respiration $(1/v_o)$, demonstrating the uncompetitive type of the azide inhibition according to equation (3)

Interference of Azide with Energy Transfer

In the course of kinetic studies on the respiratory chain, a particular interference of low concentrations of azide with electron transport stimulated by ion uptake, such as with Ca^{2+} or K^+ plus valinomycin was observed, both on the level of oxygen uptake and of the redox state of respiratory components. This inhibition was found to be abolished by uncouplers. As shown in the following studies the effect of azide on ion transport gives a key to understanding its interference with energy transfer.

For a closer analysis the sensitivity to azide of respiration in the ion pumping state (*i.e.* stimulated by induced ion uptake) is compared to that in the active (in the presence of ADP and P_i) and uncoupled state with succinate as substrate. In the $1/v$ versus azide concentration plot, a linear function is obtained in all three conditions (Fig.3).The slope of the line is much higher for the ion pumping state (with valinomycin) than for the active state. For the uncoupled state the slope is still smaller. The same large difference in the sensitivity to azide holds between the Ca^{2+}-stimulated and uncoupled respiration (not shown). Respiration in the ion pumping state is always much more sensitive to azide than in the active and particularly in the uncoupled state. This shows that the difference of the inhibition

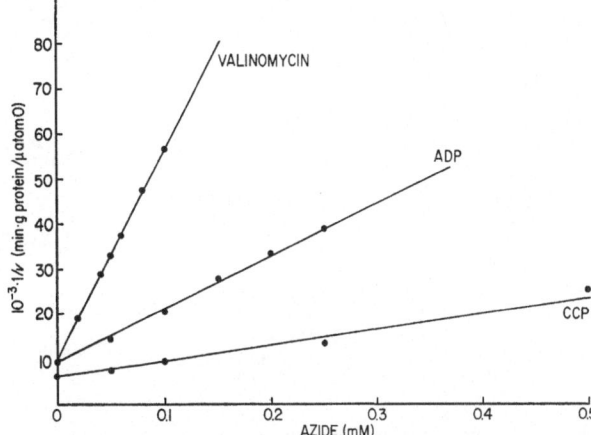

Fig.3. *Sensitivity of respiration to azide of mitochondria in various steady states, such as in the ion pumping, active and uncoupled state.* In addition to the basic components plus 2 mM succinate and 2 μg rotenone, the reaction mixtures contained 10 mM acetate and 0.3 μg valinomycin (ion pumping state), or 2.5 mM phosphate and 1 mM ADP (active state), or 12.5 μM carbonylcyanide-phenylhydrazone (CCP) (uncoupled state). Azide was added to the actively respiring system in the concentrations indicated

between various functional states is due to another mechanism (see below) than the uncompetitive inhibition seen with various substrates in a given state.

The azide concentrations required for half maximal inhibition, as evaluated from Fig.3 and from other experiments, are reported in Table 1. The $I_{1/2}$ is lowest for the ion transport-linked respiration. In the active state the $I_{1/2}$ is 5 times higher and in the uncoupled state even 10 times increased. This holds

Table 1. *Azide concentrations required for half maximal inhibition ($I^{1}/_{2}$) of respiration in different states and at different respiratory activities*

Conditions as in Fig.3, using as substrate either 2 mM succinate plus 2 µg rotenone, 2 mM ascorbate plus 0.2 mM TMPD plus rotenone, or 2 mM glutamate plus 2 mM malate. The values are the mean of different experiments

Substrates	$I^{1}/_{2}$ of azide for respiration in		
	Ion pumping state	Active state	Uncoupled state
	mM	mM	mM
Ascorbate + TMPD	0.01	0.05	0.11
Succinate	0.02	0.10	0.18
Glutamate + malate	0.04	0.15	—

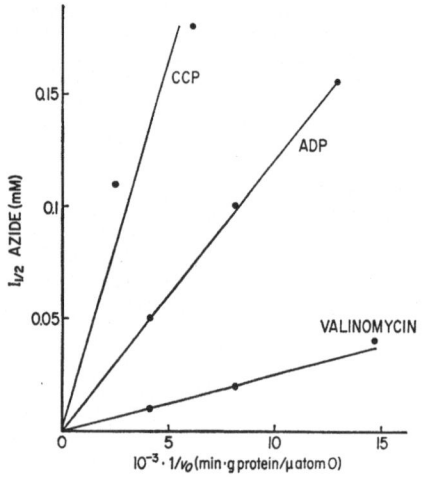

Fig.4. *The linear relation between $I^{1}/_{2}$ and the reciprocal of the uninhibited respiration. $I^{1}/_{2}$ values from Table 1*

tration of azide will also be only somewhat higher than the external concentration. With uncouplers, the ion accumulation is inhibited and therefore the azide concentration within the mitochondria cannot exceed that present outside. Therefore, the $I^{1}/_{2}$ does not necessarily represent the actual half-maximum concentration effective at the level of cytochrome a inside the mitochondria.

In order to verify this interpretation, the inhibitory effect of azide on respiration of submitochondrial particles was studied, because here it can be assumed that the concentration of azide at the site of cytochrome a is the same as in the incubation medium. Therefore, the inhibition of respiration by

Table 2. *Azide concentrations required for half maximal inhibition ($I^{1}/_{2}$) of respiration in mitochondria and isolated particles, at three different respiratory rates*

In addition to the basic components, the reaction mixtures contained 10 mM acetate and 0.3 µg valinomycin. The uninhibited respiratory rates indicated in the Table were adjusted to be approximately equal for mitochondria and particles by using different substrates, for allowing the direct comparison of the $I^{1}/_{2}$ in the various preparations. The substrate concentrations were: 2 mM ascorbate plus 0.2 mM TMPD, 2 mM (in mitochondria) and 4 mM (in particles) succinate, 2 mM glutamate plus 2 mM malate, 2 mM DPNH. Azide was added last

Substrates added to		Respiratory activity	$I^{1}/_{2}$ of azide in	
Mitochondria	Particles		Mitochondria	Particles
		µ atoms O/min/g protein	mM	mM
Ascorbate + TMPD	Succinate + DPNH + ascorbate + TMPD	200—230	0.01	0.18
Succinate	Succinate + DPNH	110—120	0.02	0.41
Glutamate + malate	Succinate	65—80	0.04	0.86

for all three substrates used. The differences of $I^{1}/_{2}$ between the different substrates at a given state can be understood on the basis of the uncompetitive type inhibition by plotting $I^{1}/_{2}$ versus $1/v_{o}$ (Fig.4), since from equation (3) it follows that $I^{1}/_{2} = V_{max} \times K_{i}/v_{o}$.

The results reported so far lead to the working hypothesis that azide can be accumulated within the mitochondria if conditions allow active ion transport. In this way the apparent $I^{1}/_{2}$ of azide for mitochondrial respiration can be largely diminished. Thus, particularly when, in the presence of valinomycin, K^{+} is accumulated, azide can accompany the cation uptake similarly to other permeant anions. The resulting higher internal concentration of azide exerts a correspondingly stronger inhibition of cytochrome oxidase. When the ion accumulation is diminished in the active state, the internal concen-

azide is compared in mitochondria, sonicated mitochondria and isolated particles under conditions which support maximum ion uptake by mitochondria. Valinomycin is added not only to the mitochondria but also to the fragmented preparations for control purposes. On the basis of the results and kinetic principles of azide inhibition, as discussed in context with Fig.1 and 2, for comparative purposes the uninhibited respiration of the different preparations was adjusted to approximately equal activity using different substrates (*cf.* legend to Fig.5 and Table 2). The analysis of such an experiment (Fig.5) in a $1/v$ versus azide concentration plot shows decrease of the linear slope from mitochondria to particles. In comparing this to Fig.3, the inhibition function for the particles is similar to that for the uncoupled mitochondria. The inhibition functions of the different preparations are evaluated in terms of $I_{1/2}$ and presented in Table 2. Here also the dependence of the $I_{1/2}$ on the respiratory activities, obtained using different substrates, is considered. The $I_{1/2}$ increases from intact to sonicated mitochondria 5-fold (not shown) and in the isolated sonic particles even 20-fold. This holds for a 4-fold range of the uninhibited respiratory activity.

The properties of azide permeation to mitochondria can also be demonstrated on the basis of swelling of mitochondria. We noted that azide does not promote a swelling sustained by active, *i.e.* respiration linked, ion uptake since an uptake of azide leads to an inhibition of respiration. However, swelling induced by azide can be studied using the method described by Chappell and Crofts [24], in which permeation can be followed without active ion uptake. Here iso-osmotic solutions of different salts leads to swelling of mitochondria when both anion and cation can penetrate the mitochondria, such as with NH_4^+-phosphate and NH_4^+-acetate. As illustrated in Fig. 6, azide induces mitochondrial swelling similar to phosphate, although at a lower rate, whereas there is no swelling with an impermeant anion such as Cl^-. In controls it is verified that the swelling induced by azide requires presence of the permeant cation NH_4^+. It can be concluded that azide behaves as a permeant anion. Therefore, in the presence of an active ion pump, an accumulation of azide within the mitochondria can be expected.

In view of the lack of a sensitive chemical assay for azide, a biological test was applied to determine an accumulation of azide in mitochondria. Evidence that the concentration of azide within the mitochondria can vary in different states is obtained by allowing the mitochondria to accumulate azide, separate them and test their respiratory activity, which should reflect the amount of the azide retained. For this purpose the mitochondria have been allowed to respire in the active and in the uncoupled state, with and without the addition of azide. After removal of the mitochondria by centrifugation, the response of their respiration to ADP was examined in an azide free medium. Mitochondria still have a good respiratory control after this procedure, as shown in the control experiment (Fig. 7A). No difference was observed if the mitochondria were incubated in the active or in the uncoupled state prior to the

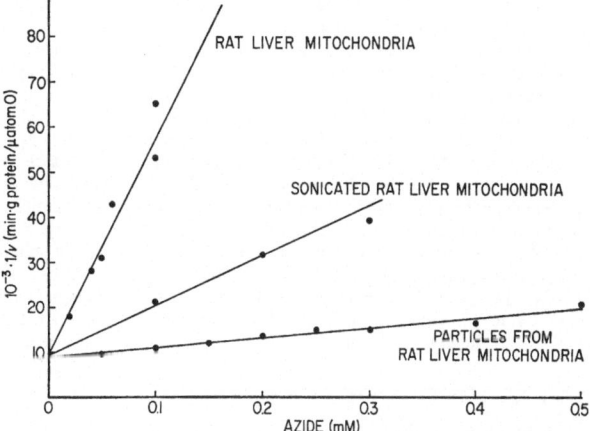

Fig. 5. *Comparison of azide inhibition of respiration in mitochondria, sonic mitochondria and isolated sonic particles.* The reaction mixtures contained the basic components plus 10 mM acetate and 0.3 µg valinomycin with the following additions: a) 2 mM succinate plus 2 µg rotenone with intact mitochondria, or b) 4 mM succinate plus 2 mM DPNH with sonicated mitochondria and isolated particles. In all cases 2 mg protein were present. Azide was added to the actively respiring system in the concentrations indicated

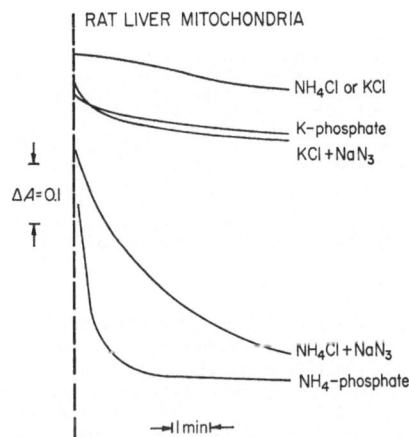

Fig. 6. *Turbidity changes corresponding to the swelling of rat-liver mitochondria induced by azide and phosphate in the presence of NH_4^+.* 0.44 mg protein were added to iso-osmotic (0.13 M) solutions of different inorganic salts, containing 20 mM tris-HCl, pH 7.2, 2 µg rotenone and 1 mM EDTA in a final volume of 1 ml at 25°. In the experiments with azide, 80 mM NaN₃ plus 50 mM KCl or 80 mM NaN₃ plus 50 mM NH₄Cl were present. The turbidity changes were followed as indicated under methods. The downward deflection of the trace indicates decrease of absorbance

polarographic assay, provided that sufficient amount of BSA is present in the assay mixture in agreement with Weinbach and Garbus [25]. In mitochondria treated in the active state with azide and then reincubated without azide, the respiration remains inhibited despite the addition of ADP, but is released on the addition of uncoupler (Fig.7B). In contrast, in mitochondria treated in the uncoupled state with azide and then reincubated without azide, the respiration is activated by ADP and then inhibited by the addition of 200 μM azide (Fig.7C). This shows that the concentration of azide within the mitochondria is higher in the active state than in the uncoupled state. Valinomycin cannot be used in this system since it cannot be removed from the mitochondria by the available procedures.

azide does not increase the rate of ferricyanide reduction in the absence of both valinomycin and ADP. This clearly indicates the ability of azide as an anion to accompany active cation accumulation.

DISCUSSION

Azide has been useful as an inhibitor of respiration since, by varying the concentration of azide, respiration can be adjusted over a wide range of activity. In this application a shift of the cross-over in the redox changes of respiratory carriers has been obtained by Chance and Williams [27] and Klingenberg and Bücher [28]. Generally, low concentrations, which only lead to small inhibition of respiration, have been used to increase the redox changes on

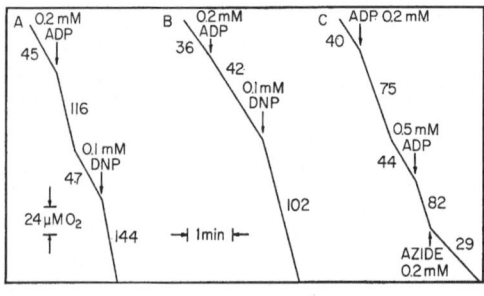

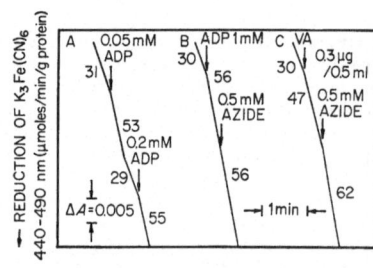

Fig. 7   Fig. 8

Fig. 7. *Demonstration of azide accumulation by a respiratory assay of azide loaded mitochondria in azide free medium.* Mitochondria were first preincubated in the presence or in the absence of azide: the reaction mixtures contained the basic components plus 2 mM succinate and 2 μg rotenone with the following additions: 0.5 mM azide (B and C), 1 mM phosphate and 1 mM ADP (A and B), 0.1 mM DNP (C). After 40 seconds at 25°, the mitochondria were removed by centrifugation and suspended in small volume. For the polarographic assay of the preincubated mitochondria in azide-free medium, the reaction mixtures (A, B and C) contained the basic components with the following additions: 3 mg mitochondrial protein, 2 mM succinate plus 2 μg rotenone, 1 mM phosphate and in C 1 mM BSA. The other additions are indicated in the Fig.

Fig. 8. *The activation of ferricyanide reduction by azide.* The reaction mixtures contained the basic components with the following additions: 3.4 mg/ml mitochondrial protein, 4 mM MgCl₂, 2 mM malonate, 4 mM glutamate, 4 mM malate, 1 mM KCN, 1 mM K₃Fe(CN)₆ and in A and in B 2.5 mM phosphate in a cuvette of 0.5 cm light path. Final volume 0.5 ml. Temperature: 25°. The other additions are indicated in the Fig. VA = valinomycin. The ferricyanide reduction was followed as indicated under methods

The energy dependence of the uptake of azide can be demonstrated by following the reduction of ferricyanide, since in this system a counteraction by azide inhibition of respiration is circumvented. As demonstrated by Estabrook [26], the reduction of ferricyanide shows the respiratory control by ADP (Fig.8A). Azide added to ferricyanide reduction in the active state (Fig.8B) does not increase the rate of electron transport. However, in the case where, despite the presence of valinomycin but in the absence of phosphate, electron transfer linked to cation uptake is limited by the lack of permeant anion, the addition of azide stimulates ferricyanide reduction (Fig.8C). In controls, the same concentration of

transition between various steady states of cytochromes *a* and *c*, which in the absence of azide are largely oxidized and show changes which are difficult to follow [29]. Furthermore, azide has been used in adjusting electron acceptor activity at the oxygen side in studying the dependence of the redox state of the respiratory carriers on the electron donor and acceptor activities (Kröger and Klingenberg [30]). In all these studies an interference of azide with cytochrome oxidase only has been considered.

The interference of azide with energy transfer, as suggested by a variety of phenomena mentioned in the introduction, is explained here as a secondary consequence of azide being a permeant anion to the mitochondria and thus it can be accumulated by energy-linked processes. The main indication for the control of azide inhibition by ion accumulation is the large difference for the inhibitory affinity of azide between the respiration in the ion pumping state and the uncoupled state. This cannot be explained by assuming that azide interferes with the energy transfer sequence on the basis of any of the proposed hypothetical reaction sequences. All the subsequent experimental data reported, *e.g.* the comparison of mitochondria with particles, give supporting evidence for ion accumulation of azide as controlling its inhibition of respiration.

On this basis the various phenomena of interference of azide with energy transfer, reported in the literature, might be explained. One argument has been that azide blocks energy transfer, either at the site of oligomycin action [20] *i.e.* on an energy rich intermediate "I ~ X," or directly at the primary energy rich intermediate of a respiratory carrier such as cytochrome *a* [19]. This is based in the first case on the release of the azide inhibition by uncouplers and in the second case on the competitive nature of this release. In the present study the releasing action by uncouplers has been an important argument in favour of the ion accumulation control of azide inhibition. Furthermore in agreement with this interpretation, the competition of this release between uncoupler and azide is much less pronounced in sonic particles than in mitochondria (unpublished results). It is suggested that the increasing concentration of uncouplers serves to inhibit fully the ion accumulation. In that case ion accumulation needs a higher concentration of uncoupler than ATP formation. An analogy to energy linked transhydrogenation may be mentioned, which for its inhibition also requires higher concentration of uncoupler than oxidative phosphorylation [31]. In both cases, for the transhydrogenation and for the ion transport, only low potential energy is required. This interpretation makes the assumption of Wilson and Chance [19] of an energy rich intermediate of reduced cytochrome *a*, which is stabilized by azide, superfluous.

The often reported uncoupling effect of azide [3—5,20] can also be explained in the light of the present theory. Thus the ion accumulation of azide can lead to stimulation of electron transport, as seen with ferricyanide as acceptor in Fig. 8C. A small stimulation of respiration, as that reported by Bogucka and Wojtczak [20], will also be seen when oxygen is the terminal acceptor, because of the uncompetitive type of azide inhibition which requires high electron transfer rate for effective inhibition.

Therefore the only direct action of azide within the mitochondria is concluded to be at the level of cytochrome *a*. Among the reported results the following supporting evidence may be quoted. Azide inhibits electron transfer to oxygen but not to cytochrome *c*-

ferricyanide. The uncompetitive type of inhibition of mitochondrial respiration by azide, observed by varying the electron flow rate, is also in agreement with the concept of binding of azide to the reduced cytochrome *a*, as established in studies with isolated cytochrome oxidase [2] and earlier indicated by the studies of Keilin [1] with non-phosphorylating heart preparations. This is further supported by Wilson and Chance [19] who observe a shift of the α-band of reduced cytochrome *a* in the presence of azide.

The ability of azide to penetrate the mitochondria as a permeant anion appears to be plausible by analogy with the highly permeable acetate on the basis of the similar pK ($pK_{HN_3} = 4.67$, $pK_{acetic\ acid} = 4.75$). Therefore, it could enter the mitochondria in the undissociated or anionic form as it has been suggested for acetic acid [24]. On the other hand a certain chemical similarity between N_3^- and Cl^-, which is based on the similar size of the ions, is in contrast to the different permeabilities of the two anions to the mitochondria. This suggests that it is the high pK of HN_3, as compared to that of HCl, which facilitates its permeation probably in the undissociated form. As a permeant anion azide has the ability to accumulate concomitantly with cation uptake which is elicited in particular by valinomycin plus K^+ or by Ca^{2+}.

Although the binding of cytochrome *a* to the cristae membrane is widely acknowledged, there is no evidence on which side of the membrane it is located. The present study implies that cytochrome *a* is located on the inner surface of the cristae. Here, in direct contact with the osmotically active space, its activity can be controlled by the ion accumulation of azide. This agrees with the identification of the cristae membrane as the site of ion translocation with the osmotically active matrix space [32].

The present study was performed during the tenure by F. Palmieri of a NATO fellowship. Financial support was obtained from the Deutsche Forschungsgemeinschaft.

REFERENCES

1. Keilin, D., *Proc. Roy. Soc. London Ser. B*, 121 (1936) 165.
2. Yonetani, T., and Ray, G. S., *J. Biol. Chem.* 240 (1965) 3392.
3. Loomis, W. F., and Lipmann, F., *J. Biol. Chem.* 179 (1949) 503.
4. Judah, J. D., *Biochem. J.* 49 (1951) 271.
5. Slater, E. C., *Biochem. J.* 59 (1955) 392.
6. Robertson, H. E., and Boyer, P. D., *J. Biol. Chem.* 214 (1955) 295.
7. Myers, D. K., and Slater, E. C., *Biochem. J.* 67 (1957) 572.
8. Novikoff, A. B., Hecht, L., Podber, E., and Ryan, J., *J. Biol. Chem.* 194 (1952) 153.
9. Siekevitz, P., Löw, H., Ernster, L., and Lindberg, O., *Biochim. Biophys. Acta*, 29 (1958) 378.
10. Lindberg, O., Löw, H., Conover, T. E., and Ernster, L., *Biological Structure and Function* (edited by T. W. Goodwin and O. Lindberg), Academic Press, London 1961, Vol. 2, p. 3.

446 F. PALMIERI and M. KLINGENBERG: Azide Uptake and Respiratory Inhibition European J. Biochem.

11. Ernster, L., and Jones, L. C., *J. Cell Biol.* 15 (1962) 563.
12. Wadkins, C. L., and Lehninger, A. L., *J. Biol. Chem.* 233 (1958) 1589.
13. Wadkins, C. L., *J. Biol. Chem.* 236 (1961) 221.
14. Wadkins, C. L., and Lehninger, A. L., *J. Biol. Chem.* 238 (1963) 2555.
15. Lehninger, A. L., Wadkins, C. L., Cooper, C., Devlin, T. M., and Gamble, J. L., jr., *Science*, 128 (1958) 450.
16. Swanson, M. A., *Biochim. Biophys. Acta*, 20 (1956) 85.
17. Löw, H., Siekevitz, P., Ernster, L., and Lindberg, O., *Biochim. Biophys. Acta*, 29 (1958) 392.
18. Lehninger, A. L., *J. Biol. Chem.* 234 (1959) 2187.
19. Wilson, D. F., and Chance, B., *Biochim. Biophys. Res. Commun.* 23 (1966) 751.
20. Bogucka, K., and Wojtczak, L., *Biochim. Biophys. Acta*, 122 (1966) 381.
21. Klingenberg, M., and Slenczka, W., *Biochem. Z.* 331 (1959) 486.
22. Szarkowska, L., and Klingenberg, M., *Biochem. Z.* 338 (1963) 674.
23. Minnaert, K., *Biochim. Biophys. Acta*, 50 (1961) 23.
24. Chappell, J. B., and Crofts, A. R., in *Regulation of Metabolic Processes in Mitochondria* (edited by J. M. Tager, S. Papa, E. Quagliariello, and E. C. Slater), Elsevier, Amsterdam 1966, p. 293.
25. Weinbach, E. C., and Garbus, J., *J. Biol. Chem.* 241 (1966) 3708.
26. Estabrook, R. W., *J. Biol. Chem.* 236 (1961) 3051.
27. Chance, B., and Williams, J. R., *J. Biol. Chem.* 221 (1956) 477.
28. Klingenberg, M., and Bücher, T., *Ann. Rev. Biochem.* 29 (1960) 669.
29. Klingenberg, M., and Schollmeyer, P., *Biochem. Z.* 335 (1961) 231.
30. Kröger, A., and Klingenberg, M., unpublished.
31. Danielson, L., and Ernster, L., in *Energy-linked Functions of Mitochondria* (edited by B. Chance), Academic Press, New York 1963, p. 157.
32. Klingenberg, M., and Pfaff, E., in *Regulation of Metabolic processes in Mitochondria* (edited by J. M. Tager, S. Papa, E. Quagliariello, and E. C. Slater), Elsevier, Amsterdam 1966, p. 180.

F. Palmieri
Istituto di Chimica Biologica
Università di Bari
Via Crisanzio 1, Bari, Italy

M. Klingenberg
Physiologisch-Chemisches Institut der Universität
355 Marburg, Deutschhausstraße 1/2, Germany

European J. Biochem. 1 (1967) 447—475

Über das Katalyseprinzip der Malat-synthase

H. Eggerer und A. Klette

Chemisches Laboratorium, Institut für Biochemie der Universität München

Eingegangen am 19. Dezember 1966/3. Mai 1967

Malate-synthase catalyzes an aldol condensation: The enzyme enolizes acetyl-CoA and hydrolyzes malyl-CoA.

I. *Acetyl-CoA-enolase*, *"monofunctional"*. Enolization of acetyl-CoA by malate-synthase was demonstrated by isotopic-exchange between the methyl hydrogen of the acetyl group and tritiated water. [³H]Acetyl-CoA was determined as [³H] p-nitroacetanilide. Under optimal conditions, the rate of enolization was approximately a thousand times slower than that of the synthesis of malate. The isotopic exchange was dependent on high Mg^{++} and enzyme concentrations, pH and on long incubation times. Mg^{++} ions could be partially replaced by other divalent metal ions. The action of the metal ions is considered an acid catalysis, *i.e.* the thioester carbonyl when bound to the Lewis acid becomes polarized, the methyl hydrogen becomes acidic. The fact, that the related enzyme citrate-synthase required base-catalysis for the enolization of acetyl-CoA, suggested that a partial mechanism was demonstrated with each enzyme (acid- and base-catalysis) both of which are active cooperatively in both the enzymes. The chemical model of this cooperative action is the bifunctional catalyst of Swain and Brown. With malate-synthase it was therefore attempted to demonstrate the participation of the carboxylate anion of glyoxylate as a base-catalyst in the enolization of acetyl-CoA, using substrate-analogues of glyoxylate.

II. *Acetyl-CoA-enolase*, *"bifunctional"*. α-Ketoacids stimulated the rate of enolization in the order pyruvate > oxalacetate ≫ α-ketobutyrate > α-ketoglutarate ≅ α-ketovalerate, *i.e.* the more, the greater the structural similarity to glyoxylate. Other carboxylic acids were inactive. Tritio-acetyl-CoA-yielding side reactions were excluded. Pyruvate inhibited the synthesis of malate competitively ($K_i = 10^{-3}$ M). The affinity of pyruvate for the enzyme yielded ordinary kinetics ($K_m = 10^{-3}$ M) as determined by the isotopic-exchange. No sign of aggregation or dissociation of the enzyme was detectable, as judged from sedimentation studies with and without acetyl-CoA and pyruvate. Taken together these results exclude an allosteric action of the α-keto-acids and agree with their action as base-catalysts. The fact, that citramalyl-CoA, the condensation product of acetyl-CoA and pyruvate, was not attacked by the enzyme, suggested that only the carboxylate anion of the α-ketoacids participated in the enolization. Due to steric hindrance at the active site, the ketocarbonyl of these acids is in a position too far removed for reaction with the acetyl-CoA carbanion formed. The α-ketoacids therefore induce the isotopic exchange. The rate of enolization in the presence of pyruvate was stimulated a thousand times and was nearly equal to that of the synthesis of malate. As in the chemical model both the nucleophilic and the electrophilic groups act cooperatively: removal of either Mg^{++} or pyruvate abolished the enolization. Taken all together, these results provide proof for the participation of the carboxylate anion of glyoxylate in the enolization of acetyl-CoA in the natural system.

III. *Malyl-CoA-hydrolase*. A coupled optical test was used for the enzymatic hydrolysis of malyl-CoA, in which the hydrolysis of the substrate with the subsequent oxidation of CoA-SH by ferricyanide was determined from the decrease of absorption. Both the affinity and turnover number of malyl-CoA when used as a substrate, were approximately 10^{-3} times less than those of the natural reactants. This is in agreement with the existence of enzyme bound malyl-CoA in the natural system. The enzyme required Mg^{++} and catalyzed the hydrolysis of (S)-malyl-CoA faster than that of the diastereomeric mixture and of the (R)-diastereomer.

IV. *Rate Determining Step*. Malate, when synthesized in tritiated water in the absence of pyruvate contained no tritium; acetyl-CoA after partial reaction with glyoxylate remained unlabelled: the enolization of acetyl-CoA is the rate-determining step in the synthesis of malate.

V. *Equivalence of the Methyl Hydrogens in Enolization and Synthesis*. In the equilibium of the isotopic exchange, the specific (sA) acitivity of p-nitroacetanilide, corrected for losses during its isolation, corresponded to sA (acetyl-CoA) = 0.94 × sA(H$_2$O). Theoretically it should be sA (acetyl-CoA) = 3/2 sA(H$_2$O) without mass effect, since the methyl hydrogens are chemically equivalent.

The value obtained experimentally may reflect a thermodynamic isotopic effect and may also indicate stereospecific exchange of two methyl hydrogens with tritium. The latter was excluded by using chemically prepared [³H-2C]acetyl-CoA, which, in the presence of pyruvate and malate-synthase in water lost its tritium content completely. [³H]malate, synthesized enzymatically from either chemically or enzymatically prepared tritio-acetyl-CoA was recognized as (S)-2-hydroxy-3,3-ditritio-succinate by the isotopic exchange catalyzed by fumarase. As expected, the methyl hydrogens of the acetyl group are thus equivalent in the reaction with either pyruvate for the enolization, or glyoxylate for the enolization and subsequent synthesis of malate.

VI. *Principle of Catalysis.* The following results illustrate that the principle of catalysis is the approximation of the reactants:

a) No acetyl-enzyme is formed between acetyl-CoA and malate-synthase.

b) Cooperative catalysis with Mg^{++} and the carboxylate-anion of glyoxylate generates enolic acetyl-CoA.

c) In the natural system this immediately reacts with the aldehyde-carbonyl of glyoxylate to form (S)-malyl-CoA.

d) The enzyme catalyzes the hydrolysis of (S)-malyl-CoA used as a substrate.

e) The chemical hydrolysis of substituted succinyl-monothioesters in neutral aqueous medium is facilitated by neighboring group participation and proceeds *via* substituted succinic-anhydride.

f) The bifunctional enzymatic catalysis corresponds in every respect with the bifunctional chemical catalysis. Proper orientation of the acidic and basic groups in the chemical model can be achieved by their insertion into one molecule, and in the enzymatic catalysis by their corresponding orientation on the enzyme.

The facts, that Mg^{++} was required for the enolization of acetyl-CoA as well as for the hydrolysis of malyl-CoA, and that the enolization was abolished when either acid- or base-catalyst was removed, indicate the formation of a complex between Mg^{++} and the substrates on the enzyme. In this complex the reactants with both their carbonyls are coordinatively bound to the Lewis acid and then by a forced intramolecular reaction are converted to the products.

VII. *Related Enzyme Catalyses.* In the related catalyses the Mg^{++} may be replaced by a Lewis acid constituitively present in the protein, and the ketoacid may be arranged sterically opposite to the position of either glyoxylate on malate-synthase or oxalacetate on citrate-synthase. The carbonyl group is then attacked from the R-side by the acetyl-CoA-carbanion and the (R)-hydroxyacid is formed as the product.

Malat-synthase wurde von Wong u. Ajl in Extrakten aus *Escherichia coli* entdeckt [1] und seither in vielen Organismen nachgewiesen (vgl. [4]). Das Enzym katalysiert die Synthese des (S)-Malats[1] aus Acetyl-CoA und Glyoxylat (Gl.1) und hat im anaplerotischen Glyoxylsäurecyclus Kornbergs [3] eine Schlüsselstellung.

$$HO_2C-CHO + CH_3COSCoA + H_2O \rightleftharpoons$$
$$HO_2CCH(OH)CH_2CO_2H + CoA-SH \qquad (1)$$

In sorgfältigen Untersuchungen über die Eigenschaften der gereinigten Malat-synthase aus Hefe

---

Nicht allgemein gebräuchliche Abkürzungen. [1—¹⁴C]-Acetyl-CoA, ¹⁴A; Malat-synthase, MS; *p*-Nitroacetanilid, *p*NAA; [³H₂—3C](2S)Ditritiomalat, MTA; [³H₁—3C] (2S, 3R)Monotritiomalat, MTM; spezifische Aktivität, sA; Äthylendiamintetraacetat, EDTA; 3-Hydroxy-3-Methylglutarsäure, HMG.

Enzyme. Aldolase (EC 4.1.2.7); Arylamin-transacetylase (EC 2.3.1.5); Citrat-synthase (EC 4.1.3.7); Fumarase (EC 4.2.1.2); HMG-CoA-Synthase (EC 4.1.3.5); α-Ketoglutaratdehydrogenase (EC 1.2.4.2); Lactat-dehydrogenase (EC 1.1.1.27); Malat-Synthase (EC 4.1.3.2); Thiolase (EC 2.3.1.9).

[1] Das Nomenklatur-System von Cahn, Ingold u. Prelog [2] wurde verwendet.

und *Pseudomonas ovalis* erzielten Dixon, Kornberg u. Lund folgende Ergebnisse [4]: Das Enzym benötigt Mg^{++}-Ionen als Cofaktoren und unterscheidet sich darin von der eng verwandten Citrat-synthase. Es arbeitet substratspezifisch; Acetyl-CoA kann nicht von anderen Acyl-coenzymen A, Glyoxylat nicht von α-Ketosäuren ersetzt werden. Im praktisch irreversiblen Syntheseverlauf entsteht kein Acetyl-Enzym. Ein Malyl-CoA ungeklärter Struktur wird von Malat-synthase nicht angegriffen.

Formal gehört die Biosynthese des Malats zu den nucleophilen Reaktionen des Acetyl-CoA [5], deren Mechanismus in einem Fall aufgeklärt werden konnte: Citrat-synthase enolisiert Acetyl-CoA [6][2], spaltet Citryl-CoA zu Acetyl-CoA und Oxalacetat [7], hydrolysiert Citryl-CoA [7,45], katalysiert also zur Synthese des Citrats eine aldolartige Reaktion zwischen Acetyl-CoA und Oxalacetat [6]. Wie Citrat-synthase, muß auch Malat-synthase bei aldolartigem Syntheseverlauf (Gl.1) formal drei Schritte katalysieren: Enolisierung des Acetyl-CoA (Gl.2), Addition des

---

[2] Im Text wird zwischen Acetyl-CoA-Carbanion und -Enol nicht unterschieden.

$$CH_3-COSCoA \quad\rightleftharpoons\quad \overset{\ominus}{C}H_2-COSCoA + H^{\oplus} \quad\rightleftharpoons\quad CH_2=C\overset{\displaystyle OH}{\underset{\displaystyle SCoA}{}} \tag{2}$$

$$HO_2C-CHO + CH_2=C\overset{\displaystyle OH}{\underset{\displaystyle SCoA}{}} \quad\rightleftharpoons\quad HO_2C\overset{H\;\;OH}{\underset{CH_2COSCoA}{\diagdown\!C}} \tag{3}$$

$$HO_2C\overset{H\;\;OH}{\underset{CH_2COSCoA}{\diagdown\!C}} + H_2O \quad\rightleftharpoons\quad HO_2C\overset{H\;\;OH}{\underset{CH_2CO_2H}{\diagdown\!C}} + CoA-SH \tag{4}$$

$$
\begin{array}{c}
R-CO_2^{\ominus} + H-CH_2COSCoA + E \\
\Updownarrow \\
E(R-CO_2^{\ominus} + H-CH_2COSCoA \rightleftharpoons R-CO_2H + \overset{\ominus}{C}H_2COSCoA \xrightleftharpoons[-H^{\oplus}]{+^3H^{\oplus}} R-CO_2^{\ominus} + {}^3H-CH_2COSCoA) \\
\Updownarrow \\
R-CO_2^{\ominus} + {}^3H-CH_2COSCoA + E
\end{array}
\tag{5}
$$

E = Citrat-synthase: R = CH(OH)CH$_2$CO$_2$H;   E = Malat-synthase: R = COCH$_3$

enolischen Acetyl-CoA an Glyoxylat zu (S)-Malyl-CoA (Gl. 3) und dessen Hydrolyse zu (S)-Malat und CoA-SH (Gl. 4).

Da (S)-Malyl-CoA synthetisch zugänglich war [8], schien es möglich, den Mechanismus der Biosynthese des Malats durch enzymatische C—C-Spaltung und Hydrolyse der synthetisierten Verbindung als Aldolkondensation zu kennzeichnen. Auch dieses (S)-Malyl-CoA ist aber unter den Bedingungen des optischen Tests zur Synthese des Malats [4] kein Substrat der Malat-synthase [9]. Es wird statt dessen von der substratspezifischen Citrat-synthase hydrolysiert [9]. Damit war es möglich, den Mechanismus der Biosynthese des Citrats als Aldolkondensation zu stützen und die Konfiguration des aus Acetyl-CoA und Oxalacetat am Enzym entstehenden Zwischenproduktes als (S)-Cityl-CoA festzulegen [9]. Sie stimmt mit der absoluten Konfiguration biologisch entstandenen Citrats überein, die Martius u. Schorre abgeleitet [10], Hanson u. Rose [11] und unabhängig davon Weber u Arigoni [12] bewiesen haben.

Citrat- und Malat-synthase katalysieren die Carboxymethylierung einer α-Oxosäure mit Acetyl-CoA, weshalb man für beide Enzyme ein ähnlich aufgebautes aktives Zentrum annehmen darf. Da im (S)-Malyl-CoA nur die aus Oxalacetat stammende Carboxymethylgruppe des (S)-Cityl-CoA durch Wasserstoff ersetzt ist, ist die Hydrolyse des (S)-Malyl-CoA durch Citrat-synthase chemisch sinnvoll. Da aber Citrat-synthase eine aldolartige Reaktion katalysiert, muß aus dem gleichen Grund gefordert werden, daß auch Malat-synthase die Aldolreaktion

zwischen Acetyl-CoA und Glyoxylat erleichtert, und (S)-Malyl-CoA folglich auch ein Substrat der Malat-synthase ist. Die nachfolgend beschriebenen Versuchsergebnisse beweisen die aldolartige Synthese des Malats: Malat-synthase[3] ist eine Acetyl-CoA-Enolase und eine Malyl-CoA-Hydrolase.

I. MALAT-SYNTHASE WIRKT MIT LEWIS-SÄUREN ALS ACETYL-COA-ENOLASE

Aldolreaktionen erleichternde Enzyme des Embden-Meyerhof Schemas und des Pentosephosphat-Cyclus katalysieren die Bildung des Carbanions der entsprechenden Substrate [13,14]. Zur Enolisierung des Acetyl-CoA dagegen ist das Enzym allein ungenügend [15,16]. Citrat-synthase benötigt dafür zusätzlich den Reaktionspartner Oxalacetat, dessen Carboxylat-Anion als basischer Katalysator wirkt (vgl. [16]). Der Nachweis gelang mit (S)-Malat, das wie Oxalacetat ein Proton von der Methylgruppe des Acetyl-CoA löseu, sich aber nicht an das entstandene Enol addieren kann. (S)-Malat induziert daher die Enolisierung und bewirkt im tritiumhaltigen Lösungsmittel den Isotopenaustausch an Acetyl-CoA (Gl. 5) [6].

Zu dessen Bestimmung wurde Arylamin-transacetylase als Hilfsenzym verwendet und markiertes Acetyl-CoA als p-Nitroacetanilid gemessen [6] (Gl. 6).

$$CH_2{}^3HCOSCoA + H_2N-C_6H_4-NO_2 \rightleftharpoons$$
$$CH_2{}^3HCONH-C_6H_4-NO_2 + CoA-SH \tag{6}$$

[3] Alle nachfolgend beschriebenen Versuche wurden mit ca. tausendfach angereichertem Enzym aus Bäckerhefe (sA = 10—15) durchgeführt. Über die Reinigung soll an anderer Stelle berichtet werden.

Diese Technik anwendend, wurde versucht nachzuweisen, daß auch Malat-synthase eine Acetyl-CoA-Enolase ist: Glyoxylat-ähnliche Carbonsäuren (Acetat, Glycolat und Oxalat) wurden auf ihre Eignung untersucht, die Enolisierung nach Gl. (5) zu induzieren. Um enzymatische Aktivität zu entfalten, benötigt Malat-synthase entgegen Citrat-synthase Mg^{++}-Ionen [4]. Wir fanden, daß nicht die eingesetzten Carbonsäuren, sondern die Mg^{++}-Ionen den Isotopenaustausch des Acetyl-CoA (Gl. 7) katalysierten. Darüber unterrichten die Ergebnisse folgender Abhängigkeiten der enzymatischen Aktivität:

$$CH_3COSCoA + H^3HO \rightleftharpoons CH_2{}^3HCOSCoA + H_2O \quad (7)$$

Mg^{++}-Ionen

Wie die Synthese des Malats, ist auch die Enolisierung des Acetyl-CoA von der Konzentration der Mg^{++}-Ionen abhängig (Tab. 1). Während aber bei Synthese des Malats die 1 mM $MgCl_2$-Lösung genügt, um das Enzym zu sättigen [4], erfordert das vorliegende System zur Sättigung eine 10 mM $MgCl_2$-Lösung. In Gegenwart des komplexbildenden EDTA werden die Mg^{++}-Ionen davon gebunden und wird die Enolisierung gehemmt, aber auch bei großem Überschuß des EDTA nicht vollständig aufgehoben. Dieses Ergebnis deutet auf eine verhältnismäßig feste Bindung zwischen Protein und Metallion (vgl. Fig. 7 und 10). Die Enolisierung mit dem Mg^{++}-Ion als Cofaktor erfolgt langsam. In 10 mM $MgCl_2$-Lösung ist ihre Anfangsgeschwindigkeit (Fig. 1) rund tausendmal geringer als die Geschwindigkeit der Synthese des Malats.

Enzym

Die Enzymabhängigkeit der Enolisierung ist aus Tab. 2 ersichtlich. Die Markierung des Acetyl-CoA nimmt proportional zur Konzentration an Malat-Synthase zu. Die geringe Markierung, die in den Kontrollversuchen (ohne $MgCl_2$ aber mit EDTA) auftritt, wird von der $MgCl_2$-haltigen Enzymlösung selbst verursacht. Mit dem Enzym gelangt unvermeidlich $MgCl_2$ in die Reaktionslösung, worin es vom EDTA nur unvollständig gebunden wird (vgl. Tab. 1).

Zeit

Auch nach langer Inkubationszeit gelangte man nicht bis in den Bereich des Isotopenaustausch-

Tabelle 1. *Mg^{++}-Abhängigkeit der Enolisierung*
Standardsätze mit je 9,6 E Malat-synthase (je 40 µl Lösung des Enzyms in 1 mM $MgCl_2$) und Änderungen wie angegeben. Inkubationsdauer: 15 Std; $sA(H_2O) = 8,6 \times 10^4$ Imp./min $\times$ µMol

Zusätze	$-\Delta E_{465}$	pNAA	Aktivität	
µMol		µMol	Imp. / 3 min	Imp./min / µMol pNAA
1,0 Na$_2$H$_2$-EDTA	0,150	0,54	690	430
1,0 Na$_2$H$_2$-EDTA + 0,5 K$_2$Mg-EDTA	0,140	0,50	1740	1160
0,01 MgCl$_2$	0,150	0,54	7340	4530
0,10 MgCl$_2$	0,130	0,47	8040	5700
1,0 MgCl$_2$	0,130	0,47	14230	10100
10,0 MgCl$_2$	0,150	0,54	15930	9850
1,0 MgCl$_2$ + 0,5 K$_2$Mg-EDTA	0,120	0,43	13700	10090
1,0 MgCl$_2$ + 1,0 Na$_2$H$_2$-EDTA	0,140	0,50	5860	3900
1,0 MgCl$_2$ + 2,0 Na$_2$H$_2$-EDTA	0,140	0,50	1050	700
1,0 MgCl$_2$ + 10,0 Na$_2$H$_2$-EDTA	0,140	0,50	320	210

Tabelle 2. *Enzymabhängigkeit der Enolisierung*
Standardansätze mit Änderungen wie angegeben. Inkubationsdauer: 15 Std; $sA(H_2O) = 8,6 \times 10^4$ Imp./min $\times$ µMol

Zusätze	$-\Delta E_{465}$	pNAA	Aktivität	
µMol		µMol	Imp. / 3 min	Imp./min / µMol pNAA
0 Einheiten MS + 1,0 MgCl$_2$ + 0,5 K$_2$Mg-EDTA	0,120	0,43	0	0
3,3 Einheiten MS + 1,0 MgCl$_2$ + 0,5 K$_2$Mg-EDTA	0,120	0,43	5170	4010
6,6 Einheiten MS + 1,0 MgCl$_2$ + 0,5 K$_2$Mg-EDTA	0,120	0,43	11210	8200
9,9 Einheiten MS + 1,0 MgCl$_2$ + 0,5 K$_2$Mg-EDTA	0,110	0,40	17650	14900
13,2 Einheiten MS + 1,0 MgCl$_2$ + 0,5 K$_2$Mg-EDTA	0,115	0,41	19810	16100
3,3 Einheiten MS + 1,5 Na$_2$H$_2$-EDTA	0,120	0,43	500	390
6,6 Einheiten MS + 1,5 Na$_2$H$_2$-EDTA	0,120	0,43	900	700
9,9 Einheiten MS + 1,5 Na$_2$H$_2$-EDTA	0,120	0,43	1700	1320
13,2 Einheiten MS + 1,5 Na$_2$H$_2$-EDTA	0,120	0,43	2470	1920

Gleichgewichtes. Die Enolisierung ist zwar erwartungsgemäß zeitabhängig (Fig.1; Kurve I; linke Ordinate), wird aber nach dem Kurvenverlauf zu schließen, mit der Zeit immer stärker gehemmt. Die spezifische Aktivität der letzten Meßpunkte entspricht erst der Inkorporation von ca. 0,6 µAtom³H/ µMol Acetyl-CoA (vgl. V.). Da Malat-synthase im gleichen Zeitraum stabil ist (Kurve II; rechte Ordinate), Acetyl-CoA aber während der Inkubation hydrolysiert (Kurve III; rechte Ordinate), liegt es nahe, die Hemmung des Isotopenaustausches auf eine Hemmung des Enzyms mit hydrolytisch entstandenem CoA−SH zurückzuführen. Übereinstimmend damit hemmt CoA−SH die Enolisierung (Tab.3).

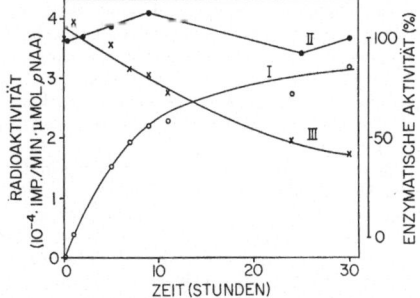

Fig.1. *Zeitabhängigkeit der Bildung des Tritio-Acetyl-CoA.* Zehnfacher Standardansatz mit 150 E Malat-synthase. Start der Reaktion mit dem Enzym. Je 0,10 ml des Reaktionsansatzes wurden bei t [Std] (siehe Fig.) abgestoppt und aufgearbeitet (siehe Methoden). $sA(H_2O) = 8{,}6 \times 10^4$ Imp./min × µMol.
Kurve I: Bildung des Tritio-Acetyl-CoA (linke Ordinate). Kurve II: Prozent der enzymatischen Aktivität (rechte Ordinate). Optische Bestimmung bei 232 mµ (siehe Methoden) mit je 10 µl des Reaktionsansatzes nach geeigneter Verdünnung. Kurve III: Hydrolyse des Acetyl-CoA (rechte Ordinate). Bestimmung des Acetyl-CoA als p-Nitroacetanilid (siehe Methoden)

Tabelle 3. *Hemmung der Enolisierung mit CoA−SH*
Standardansätze mit je 5 Einheiten Malat-synthase und CoA−SH, wie angegeben. Inkubationsdauer: 7 Std; $sA(H_2O) = 8{,}6 \times 10^4$ Imp./min × µMol

CoA-SH	$-\Delta E_{405}$	pNAA	Aktivität	
µMol		µMol	Imp. 3 min	Imp./min µMol pNAA
0	0,160	0,575	18350	10600
0,25	0,160	0,575	15680	9100
0,50	0,150	0,54	7680	4750
1,00	0,155	0,555	5990	3600

pH

Bei Verwendung verschiedener Puffer-Anionen entstand kein regelmäßiges pH-Profil der Enolisierung (Tab.4). Statt der pH-Abhängigkeit wird mit den verschiedenen Puffersubstanzen tatsächlich ein die enzymatische Aktivität unterschiedlich beeinflussender Salzeffekt gemessen, der auch im optischen Test zur Synthese des Malats zu beobachten war.

Tabelle 4. *pH- und Pufferabhängigkeit der Enolisierung*
Standardansätze mit je 11,1 Einheiten Malat-synthase und Puffersubstanz (je 30 µMol; Borat, Collidin-Lutidin und Pyrophosphat je 6 µMol) wie angegeben. Inkubationsdauer: 15 Std; $sA(H_2O) = 8{,}6 \times 10^4$ Imp./min/µMol

Pufferlösung	pH	$-\Delta E_{405}$	pNAA	Aktivität	
			µMol	Imp. 3 min	Imp./min µMol pNAA
β,β-Dimethylglutarat	6,0	0,175	0,63	29880	15850
β,β-Dimethylglutarat	6,5	0,155	0,56	42620	25600
Phosphat	7,0	0,155	0,56	28160	16800
Phosphat	7,4	0,127	0,46	23640	17300
Tris	7,4	0,095	0,34	10830	10600
Tris	8,1	0,047	0,17	5900	11630
Borat	8,1	0,177	0,64	41560	21800
Collidin-Lutidin	8,1	0,170	0,61	40990	22400
Pyrophosphat	8,1	0,152	0,55	3070	1900
Tris	8,5	0,040	0,14	5440	12500

Tabelle 5. *Acetyl-CoA-Abhängigkeit der Enolisierung*
Standardansätze mit je 7,5 Einheiten Malat-synthase (0 Einheit in den Kontrollansätzen) und Acetyl-CoA wie angegeben. Inkubationsdauer: 15 Std; $sA(H_2O) = 8{,}6 \times 10^4$ Imp./min × µMol

Acetyl-CoA	$-\Delta E_{405}$	pNAA	Aktivität	
µMol		µMol	Imp. 3 min	Imp./min µMol pNAA
0,4	0,070	0,25	18910	24600
0,8	0,140	0,50	26420	17300
1,2	0,210	0,75	32370	14200
1,6	0,255	0,915	27860	10000

Parallelansätze ohne Malat-synthase:

0,4	0,090	0,32	0	—
0,8	0,170	0,61	0	—
1,2	0,225	0,81	0	—
1,6	0,250	0,91	0	—

Tabelle 6. *Eignung von Metallionen als Cofaktoren der Enolisierung*
Standardansätze mit je 8,4 Einheiten Malat-synthase, 0,5 µMol Na_2H_2-EDTA und je 1,5 µMol Metallsalz wie angegeben. Inkubationsdauer: 15 Std; $sA(H_2O) = 8{,}6 \times 10^4$ Imp./min × µMol

Salz	$-\Delta E_{405}$	pNAA	Aktivität	
		µMol	Imp. 3 min	Imp./min pMol pNAA
	0,150	0,54	120	450
$ZnCl_2$	0,055	0,21	3100	5050
$Hg(Ac)_2$	0	0	0	—
$CuCl_2$	0,105	0,38	0	0
$CoCl_2$	0,085	0,32	29280	30500
$MnCl_2$	0,145	0,54	5210	3220
$MgCl_2$	0,130	0,49	14060	9450

Acetyl-CoA

Bei Bestimmung der Enolisierung mit zunehmender Konzentration an Acetyl-CoA wurde die erwartete Enzymkinetik erhalten (Tab.5). Die enzymatische Aktivität (Imp./3 min) ergibt gegen die zugehörige Konzentration des pNAA aufgetragen eine Sättigungskurve. Die spezifische Aktivität des pNAA fällt daher mit steigender Konzentration. Mit diesem Ergebnis übereinstimmend sind die Mg++-Ionen allein wirkungslos, erfolgt in den enzymfreien Kontrollansätzen kein Isotopenaustausch.

Zweiwertige Metallionen

Das Mg++-Ion als Cofaktor für die Synthese des Malats kann teilweise durch Mn++-, nicht aber durch Zn++-, Cu++, Hg++- und Co++-Ionen ersetzt werden [4]. Ähnlich dazu sind Cu++- und Hg++-Ionen auch als Cofaktoren der Enolisierung ungeeignet, wirken Zn++- und Mn++- schlechter, Co++- aber besser als Mg++-Ionen (Tab.6). Diese unterschiedliche Wirkung der Metallionen in beiden Testsystemen ist mit deren verschiedener Zusammensetzung und Empfindlichkeit leicht erklärbar.

Hitzeinaktivierung des Enzyms

Für die voranstehend beschriebenen Versuche wurden gereinigte, nicht aber reine Enzymlösungen verwendet (vgl. Fig.4). Um zu beweisen, daß Malat-Synthase für die Enolisierung verantwortlich ist,

wurde die Enzymlösung auf 60° erhitzt und der Verlust der katalytischen Aktivität mit dem Test zur Enolisierung und mit dem Test zur Synthese des Malats zeitlich verfolgt. Da mit beiden Methoden das gleiche Ergebnis erzielt wurde (Tab.7), sind Malat-synthase und Acetyl-CoA-Enolase identisch.

Die Fähigkeit des Acetyl-CoA bifunktionell zu wirken, hat einen gemeinsamen Grund [5,22,86]: Die freien Elektronen des Schwefels in Thioestern nehmen wahrscheinlich weniger an der C—S Bindung teil als die des Sauerstoffes in Carbonsäureestern an der C—O Bindung [17,18]. Das Thioestercarbonyl ist daher stärker polarisiert als das Carbonsäureestercarbonyl und ist deshalb ein besseres elektrophiles Agens. In die Derivate der Carboxylgruppe mit steigender Resonanzstabilisierung [19] eingereiht, liegt der Thioester entsprechend seiner Reaktivität zwischen Carbonsäure-chlorid und -ester: Thioester sind schwächer elektrophil als Carbonsäurehalogenide und stärker elektrophil als Carbonsäureester. Diese Zuordnung ist in vergleichenden Untersuchungen über die Reaktivität der Thioester gut gesichert [20,21].

Die verminderte Resonanzstabilisierung in Thioestern bewirkt aber auch ihre Eignung zu nucleophilen Reaktionen: Im Vergleich zum Carbonsäureester ist der Carbonylkohlenstoff im Thioester stärker positiviert, ist die Abspaltung des Protons vom benachbarten Methylkohlenstoff erleichtert (Gl.8). Der α-ständige Wasserstoff in Thioestern ist daher acider als in entsprechenden Carbonsäureestern.

R = Alkyl

Tabelle 7. *Kinetik der Hitzeinaktivierung der Malat-synthase*

Je 0,07 ml Lösung der Malat-synthase (300 Einheiten/ml; sA = 12) wurden im Zentrifugenbecher bei $t = 0$ min im Bad von 60° inkubiert und bei t min im Eisbad gekühlt. Denaturiertes Protein wurde abzentrifugiert (15 min, 20000 g), der Überstand zur Bestimmung der enzymatischen Aktivität verwendet: Je 10 μl wurden verdünnt (siehe Tabelle) und die Synthese des Malats damit (μl wie angegeben) bestimmt. Je 30 μl der unverdünnten Lösung wurden im Standardansatz für den Radioaktivitäts-Test verwendet. Inkubationsdauer: 15 Std; $sA(H_2O) = 8,6 \times 10^4$ Imp./min × μMol

			Synthese des Malats [a]			Bildung des Tritio-Acetyl-CoA [b]			
t	Verdünnung	Zusatz	$\dfrac{-\Delta E_{233}}{min}$	Relative Aktivität	$-\Delta E_{405}$	pNAA	Aktivität		Relative Aktivität
							$\dfrac{Imp.}{3\ min}$	$\dfrac{Imp./min}{\mu Mol\ pNAA}$	
min		μl		%		μMol			%
0	1:100	10	0,045	100	0,148	0,53	23270	14600	100
2	1:100	10	0,034	75	0,165	0,59	18340	10350	71
5	1:100	10	0,025	56	0,167	0,60	13440	7460	51
7	1:100	10	0,015	33	0,175	0,63	8810	4660	34
10	1:100	10	0,013	29	0,175	0,63	7040	3730	26
15	1:100	50	0,044	20	0,175	0,63	5450	2880	20
60	1:10	10	0,017	4	0,185	0,66	1140	575	4

[a] Nach Gleichung (1).
[b] Nach Gleichung (7).

Den experimentellen Hinweis dafür erbrachten Lynen u. Wessely bereits 1953 [22,23]: Der pK' des N-Acetyl-S-acetacetylcysteamins (R = CH$_3$CO, R' = CH$_2$CH$_2$NHCOCH$_3$) ist um 2,2 Einheiten nied-

$$R-C\begin{matrix}H\\|\\|\\H\end{matrix}C\begin{matrix}O\\\diagup\\\diagdown SR'\end{matrix} \rightleftharpoons R-CH=C\begin{matrix}OH\\\diagup\\\diagdown SR'\end{matrix} \qquad (8)$$

Acetyl–CoA: R = H, R' = CoA

riger als der pK' des Acetessigsäureäthylesters. Die Neigung der Thioester zur Carbanionbildung wurde in jüngerer Zeit durch folgende Befunde gestützt: Wie im Fall der Acetacetylester ist auch der pK' des Cyanessigsäure-äthylthioesters (R = CN, R' = C$_2$H$_5$) um 2,5 Einheiten niedriger als der pK' des Sauerstoff-analogen, Cyanessigsäureäthylester [21]. N-Carbo-benzoxy-β-cyano(S)-alanylthiophenylester racemi-siert bei basischer Katalyse 45mal rascher als der entsprechende O-Phenylester [24]. In den angeführ-ten Beispielen wird die C—H Acidität von den be-nachbarten Substituenten bewirkt; sie ist schon bei den Sauerstoffanalogen vorhanden und wird im ent-sprechenden Thioester nur erhöht. Die Thioester-gruppierung allein als Substituent an der Methyl-gruppe ist zur Erzeugung der C—H Acidität un-genügend: Acetyl-CoA enolisiert nicht [15,16]. Statt durch Einführung eines geeigneten Substituenten in die Methylgruppe (die Carboxylierung des Acetyl-CoA z. B. ergibt das C—H acide Malonyl-CoA) ließe sich die Ablösung des Protons von der Methylgruppe auch durch eine noch stärkere Polarisierung des Thioestercarbonyls mit geeigneten Katalysatoren erzielen. Diese Bedingung erfüllt die Säurekatalyse (Gl. 9), deren Wirkung mit der Erzeugung einer posi-

$$CH_2\begin{matrix}\diagup C\\|\\H\end{matrix}\begin{matrix}OH^{\oplus}\\\diagup\\\diagdown SCoA\end{matrix} \rightleftharpoons CH_2=C\begin{matrix}OH\\\diagup\\\diagdown SCoA\end{matrix}+H^{\oplus} \qquad (9)$$

tiven Ladung im Substrat beschrieben werden kann. Da ein Metallion diese positive Ladung noch stärker als das Proton erzeugen kann, ist es bei geeigneter Bindung an das Substrat auch ein besserer Säure-katalysator [25]. Die Wirkung der enzymgebundenen Lewis-Säuren zur Enolisierung des Acetyl-CoA (Tab. 6) besteht folglich in der Polarisierung des Thioestercarbonyls (Gl. 10).

$$CH_2\begin{matrix}\diagup C\\|\\H\end{matrix}\begin{matrix}OMg^{++}\\\diagup\\\diagdown SCoA\end{matrix} \rightleftharpoons CH_2=C\begin{matrix}OMg^+\\\diagup\\\diagdown SCoA\end{matrix}+H^{\oplus} \qquad (10)$$

Acetyl-CoA und die Lewis-Säure werden vom Enzym genau so fixiert, daß Thioestercarbonyl und Metallion eine Bindung eingehen müssen. Der damit ausgelöste Elektronensog (Gl. 10) positiviert den Carbonylkohlenstoff stark genug, um die Ablösung des Protons vom benachbarten Methylkohlenstoff

und damit den Isotopenaustausch zu ermöglichen. Übereinstimmend damit acylieren Thioester in Gegenwart von Metallionen Amine [26] und kataly-siert MgCl$_2$ die Hydrolyse des Malyl-CoA, während andere Salze wirkungslos sind (Tab. 8).

Tabelle 8. *Salzbewirkte Hydrolyse des (R.S)-Malyl-CoA*
Die Reaktionsansätze enthielten bei 25° im Vol. 0,20 ml je 20 µMol Trispuffer, pH 8,1; 1 µMol K$_3$Fe(CN)$_6$; 0,13 µMol (R,S)-Malyl-CoA (Bestimmung als Hydroxamsäure) und Salze wie angegeben. Messung gegen eine Kontrollküvette, die alle Zusätze, aber kein Malyl-CoA enthielt. $\lambda = 436$ mµ; $d = 1$ cm; Photometer Eppendorf. Bestimmung der Hydro-lyse nach Gl. (17)

Salzlösung	MgCl$_2$	KCl	BaCl$_2$	NaCl
mM		$-\Delta E_{436}/10$ min		
0	0,000	0,000	0,000	0,000
10	0,008	—	—	—
20	0,024	—	—	—
40	0,028	0,000	0,000	0,000
80	0,040	—	—	—
100	0,050	0,000	0,000	0,000
300	0,077	0,005	0,002	0,010

Aus einer Sicht war die enzymatische Enolisierung des Acetyl-CoA mit den Lewis-Säuren sehr zufrieden-stellend: Dabei können keine, die Enolisierung vor-täuschende Nebenreaktionen erfolgen. Die voran-stehend angeführten Untersuchungen bieten daher einen strengen Beweis für die mit Citrat-synthase und dem Induktor (S)-Malat erzielten Ergebnisse. Für den Mechanismus dieser Reaktion konnte aber die basische Katalyse, Zug des Carboxylat-Anions am Methylwasserstoff (Gl. 11) wahrscheinlich gemacht werden [6], während er sich bei Malat-synthase als Säurekatalyse (Gl. 10) abzeichnet. Wie bei Citrat-synthase erfolgt auch die Enolisierung des Acetyl-

$$CH_2\begin{matrix}\diagup C\\|\\\diagdown H\end{matrix}\begin{matrix}\diagup O\\\diagdown SCoA\end{matrix} \rightleftharpoons CH_2=C\begin{matrix}\diagup O^{\ominus}\\\diagdown SCoA\end{matrix} \qquad (11)$$

Citrat-synthase: R = COCH$_2$CO$_2$H; Malat-synthase: R = CHO

coenzym A mit Malat-synthase und Mg^{++}-Ionen viel langsamer als die Gesamtreaktion (Gl. 1). Während die geringe Reaktionsgeschwindigkeit bei Citrat-synthase mit der geringen Affinität des „unnatür-lichen" Induktors (S)-Malat zum Enzym zufrieden-stellend erklärt werden kann, sind die Mg^{++}-Ionen bei Malat-synthase die natürlichen Cofaktoren des Enzyms. Wären sie allein für die Enolisierung ver-antwortlich, so müßte der Teilschritt mit einer der Gesamtreaktion ähnlichen Geschwindigkeit ablaufen. Umgekehrt kann das Metallion in Citrat-synthase so fest vom Protein gebunden sein, daß keine Abhängig-keit dazu meßbar ist, oder es kann durch eine andere, konstitutiv im Protein enthaltene Lewis-Säure ersetzt sein. Damit wäre auch besser verständlich, warum das schwach basische Carboxylat-Anion des Oxalacetats bzw. des (S)-Malats bei Citrat-synthase zur Ablösung des Protons von der Methylgruppe des Acetyl-CoA genügt. Dazu paßt auch, daß Citrat-synthase allein Acetyl-CoA nicht enolisiert: Auch mit Malat-synthase ist mit wenig, zur Synthese (Gl.1) genügenden Mg^{++}-Ionen, praktisch keine Enolisierung nachweisbar. Schließlich ist das Carboxylat-Anion dann ein wirksameres nucleophiles Agens, wenn es sterisch richtig angeordnet in Nachbarschaft des elektrophilen Reaktanten gebunden ist [25,61,62].

(12)

Citrat-synthase: R = COCH$_2$CO$_2$H; Malat-synthase: R = CHO

Es schien daher wahrscheinlich, daß mit Malat- und Citrat-synthase je ein Teilmechanismus zur Enolisierung des Acetyl-CoA — Säurekatalyse (Gl.10), Basenkatalyse (Gl.11) — nachgewiesen wurde, die in beiden Enzymen kooperativ arbeiten (Gl.12).

Das chemische Modell dafür ist der bifunktionelle Katalysator von Swain u. Brown [27]: Die geeignete, in α-Hydroxypyridin verwirklichte sterische Anordnung der sauren und basischen Gruppe, erhöht die Geschwindigkeit der säure- und basen-katalysierten Mutarotation der Tetramethylglucose (Gl.13) außerordentlich.

(13)

Es wurde daher versucht, an Malat-synthase auch die basische Katalyse nachzuweisen, und zwar wie bei Citrat-synthase mit dem Salz einer substratähnlichen Carbonsäure.

II. MALAT-SYNTHASE WIRKT BIFUNKTIONELL

Der Schlüsselversuch für den Nachweis der „bifunktionellen enzymatischen Katalyse" bestand in der systematischen Suche nach einem geeigneten Induktor (Gl.5). Dafür wurden zwei verschieden empfindliche Versuchsreihen gewählt: In einer Reihe (A, Fig.2) enthält das Inkubationsgemisch Tritium-

wasser, Malat-synthase, Acetyl-CoA, das zu untersuchende Salz der organischen Säure und MgCl$_2$. Die zweite Reihe (B, Fig.2) ist genau gleich zusammengesetzt, enthält aber statt MgCl$_2$ das komplexbildende EDTA. Damit wird die Fähigkeit des Enzyms Acetyl-CoA zu enolisieren stark vermindert, aber nicht vollständig aufgehoben (vgl. Tab.1). Da der Induktor aktiv an der Enolisierung teilnimmt, sollte er sie auch in Gegenwart des EDTA in Reihe B stimulieren und sich darin von der salzabhängig andersartig veränderten enzymatischen Aktivität (vgl. Tab.4) der Reihe A unterscheiden. Das Versuchsergebnis (Fig.2) stimmt damit überein und deutet auf α-Ketosäuren als Induktoren: Gegenüber dem Bezugswert (vollständiger Reaktionsansatz mit MgCl$_2$, aber ohne carbonsaures Salz) wird die enzymatische Aktivität in Reihe A mit z. B. Acetat, Succinat stimuliert, und mit Glycolat, Malonat gehemmt. Keines der in Reihe A stimulierenden Salze ist aber in Reihe B aktiv und keines dieser Salze kann daher induzierend wirken. Dieses Ergebnis wird mit α-ketosauren Salzen grundlegend geändert: Pyruvat, Oxalacetat, α-Ketobutyrat, α-Ketovalerat und α-Ketoglutarat sind auch in Reihe B aktiv und induzieren daher wahrscheinlich die Enolisierung. Da sich Reihe B durch ungünstigere Bedingungen für die enzymatische Aktivität auszeichnet, sollte die Enolisierung darin, wie z. B. mit α-Ketobutyrat verwirklicht, langsamer erfolgen als in den Parallelversuchen der Reihe A. Mit Pyruvat und Oxalacetat werden aber in beiden Reihen gleiche Werte gemessen. Das deutet auf Sättigung, d. h. auf die in diesem System höchstmögliche Tritierung des Acetyl-CoA. Bei Oxalacetat muß berücksichtigt werden, daß es während der Inkubation decarboxyliert und genügend Pyruvat liefern kann, um dessen Wirkung vorzutäuschen.

Eine Stütze dafür, daß die α-Ketosäuren nach Gl. (5) induzierend in die Enzymkatalyse eingreifen, bietet die mit anorganischen Salzen beeinflußte enzymatische Aktivität (Tab.9). Wie zu erwarten, verhalten sich diese Salze in den Versuchsreihen A und B ähnlich den gewöhnlichen carbonsauren Salzen.

Kontrollen der Wirkung des Pyruvats

Die Teilnahme des Carboxylat-Anions an der enzymatischen Enolisierung kann durch Nebenreaktionen vorgetäuscht werden:

a) Bei Decarboxylierung des Pyruvats entsteht im tritiumhaltigen Lösungsmittel Tritio-Acetaldehyd, der bei Bestimmung des Acetyl-CoA als p-Nitroacetanilid mit p-Nitroanilin ein Azomethin oder ein Aldehyd-Aminat liefern könnte. Diese Neutralsubstanzen würden bei Aufarbeitung des p-Nitroacetanilids nicht entfernt werden und daher die

---

[4] Nach Zusatz von Trägersubstanz bleibt die spezifische Aktivität bei Umkristallisation unverändert.

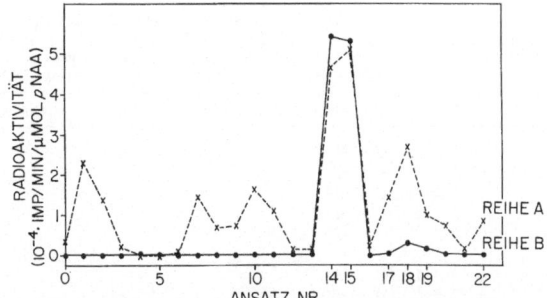

Fig. 2. α-*Ketosäuren induzieren die Enolisierung des Acetyl-CoA.* Bezugswert Reihe A: Standardansatz mit 3,8 E Malat-synthase. Reihe A: ×, Wie Bezugswert, aber mit je 10 µMol Salz Nr. 1—22. Bezugswt Reihe B: Standardansatz mit 3,8 E Malat-synthase und 2 µMol Na₂H₂-EDTA statt MgCl₂ und K₂Mg-EDTA. Reihe B: ●, Wie Bezugswert, aber mit je 10 µMol Salz Nr. 1—22. Inkubationsdauer: 15 Std; sA(H₂O) = 8,6×10⁴ Imp./min×µMol; Meßdaten: siehe Methoden
0 —; 1 Formiat; 2 Acetat; 3 Propionat; 4 Oxalat; 5 Glycolat; 6 Malonat; 7 Succinat; 8 Glutarat; 9 β,β-Dimethyl-glutarat; 10 (R.S)-Citramalat; 11 (R)-Malat; 12 (R.S)-Malat; 13 (S)-Malat; 14 Pyruvat; 15 Oxalacetat; 16 γ-Keto-valerat; 17 α-Ketovalerat; 18 α-Ketobutyrat; 19 α-Ketoglutarat; 20 Citrat; 21 (2R.3S)-Isocitrat; 22 trans-Aconitat

Tabelle 9. *Beeinflussung der enzymatischen Aktivität mit anorganischen Salzen*
Standardansätze, aber mit je 30 µMol Trispuffer, pH 8,1 (statt Phosphatpuffer, pH 7,0) und je 6,8 Einheiten Malat-synthase. Reihe A: mit je 10 µMol Salz wie angegeben. Reihe B: Standardansätze wie voranstehend, aber mit je 2 µMol Na₂H₂-EDTA (statt MgCl₂ und K₂Mg-EDTA) und je 10 µMol Salz, wie angegeben. Inkubationsdauer: 15 Std; sA(H₂O) = 8,6×10⁴ Imp./min×µMol

Salz	Reihe A				Reihe B			
	$-\Delta E_{405}$	pNAA	Aktivität		$-\Delta E_{405}$	pNAA	Aktivität	
		µMol	Imp. / 3 min	Imp./min / µMol pNAA		µMol	Imp. / 3 min	Imp./min / µMol pNAA
—	0,090	0,32	5540	5770	0,075	0,27	840	1040
KCl	0,090	0,32	4070	4240	0,070	0,25	300	400
NaCl	0,090	0,32	3600	3750	0,070	0,25	350	470
K₂HPO₄	0,075	0,27	4170	5150	0,070	0,25	100	130
Na₂H₂P₂O₇	0,070	0,25	2020	2690	0,070	0,25	420	560
Na₂SO₄	0,075	0,27	3680	4540	0,060	0,22	0	0
Na₂HAsO₄	0,070	0,25	3640	4850	0,050	0,18	200	370
K₃Fe(CN)₆	0,070	0,25	6100	8140	0,020	0,07	0	0

Tabelle 10. *Kontrolle der Bildung des Tritioacetaldehyds*
Standardansätze mit je 2 µMol Na₂H₂-EDTA (statt MgCl₂ und K₂Mg-EDTA); Pyruvat, Acetyl-CoA (A) und Malat-synthase (MS) wie angegeben. Inkubationsdauer: 15 Std; sA(H₂O) = 8,6×10⁴ Imp./min×µMol

Versuch	Änderungen im Reaktionsansatz	$-\Delta E_{405}$	pNAA	Aktivität	
			µMol	Imp. / 3 min	Imp./min / µMol pNAA
1	+5 µMol Pyruvat + A + 3,6 Einheiten MS	0,170	0,61	84300	46000
2	+5 µMol Pyruvat − A + 3,6 Einheiten MS	0	0	0	0
3	+5 µMol Pyruvat − A − MS	0	0	0	0
4	+5 µMol Pyruvat + A − MS	0,180	0,65	0	0

Tritierung des Acetyl-CoA vortäuschen. Die Kontrollversuche (Tab. 10) zeigen aber, daß Radioaktivität (des Tritio-pNAA⁴) nur im vollständigen Ansatz zur Enolisierung des Acetyl-CoA gemessen wird (Versuch 1). Das Ergebnis der Versuche 3 und 4 beweist, daß keine chemische, das von Versuch 2,

daß keine enzymatische Bildung des Tritio-Acetaldehyds nachweisbar ist.
b) Da Pyruvat an der Methylgruppe durch Keto-Enol-Tautomerie markiert wird, könnte in der Reaktionslösung entstandenes Tritio-Pyruvat in enzymatischer CO₂-Austauschreaktion markiertes Acetyl-

Tabelle 11. *Kontrolle des CO_2-Austausches zwischen Acetyl-CoA und Pyruvat*
Standardansätze ohne $MgCl_2$, mit je 1,0 μMol Pyruvat, 4,8 Einheiten Malat-synthase (MS) und Änderungen wie angegeben. Inkubationsdauer: 15 Std; $sA(H_2O) = 8,6 \times 10^4$ Imp./min × μMol (Versuch 3—5; K-1). Pyruvat wurde als 2,4-Dinitrophenylhydrazon (DNPH) isoliert und bestimmt. Acetyl-CoA, A; [1—^{14}C] Acetyl-CoA, ^{14}A

Versuch	Änderungen im Reaktionsansatz	Imp./min × μMol Pyruvat-2,4-DNPH
1	$+ ^{14}A - H^3HO + MS$	0
2	$+ ^{14}A - H^3HO - MS$	0
3	$- A + H^3HO - MS$	2400
4	$- A + H^3HO + MS$	2900
5	$- A + H^3HO - MS + 10$ μMol NaOH	30000
		Imp./min × μMol pNAA
K-1	$+ A + H^3HO + MS$	66000
K-2	$+ ^{14}A - H^3HO + MS$	152000

Tabelle 12. *Zeitabhängigkeit der Enolisierung im „bifunktionellen System" ohne und mit Ferricyanid*
Standardansätze ohne K_2Mg-EDTA, mit je 1 Einheit Malat-synthase, 1 μMol Pyruvat, und wo angegeben mit je 10 μMol $K_3Fe(CN)_6$; $sA(H_2O) = 8,6 \times 10^4$ Imp./min × μMol

	Ohne $K_3Fe(CN)_6$				Mit $K_3Fe(CN)_6$			
t	$-\Delta E_{405}$	pNAA	Aktivität		$-\Delta E_{405}$	pNAA	Aktivität	
min		μMol	$\frac{\text{Imp.}}{3 \text{ min}}$	$\frac{\text{Imp./min}}{\mu\text{Mol } p\text{NAA}}$		μMol	$\frac{\text{Imp.}}{3 \text{ min}}$	$\frac{\text{Imp./min}}{\mu\text{Mol } p\text{NAA}}$
0	0,195	0,72	0	0	—	—	—	—
5	0,195	0,72	6710	3100	0,090	0,33	3500	3540
10	0,200	0,73	13660	6220	0,075	0,29	7810	8950
15	0,190	0,70	17780	8450	0,085	0,31	10500	11300
20	0,190	0,70	25300	12040	0,090	0,33	14350	14500
25	0,190	0,70	29000	13800	0,090	0,33	17020	17200
30	0,190	0,70	34150	16300	0,090	0,33	19510	19700

$$\underset{HO_2C}{\overset{HO}{\diagdown}}\overset{CH_3}{\diagup}CH_2CO_2H + CoA-SH \rightleftharpoons \underset{HO_2C}{\overset{O}{\diagdown}}\overset{CH_3}{\diagup} + CH_2{}^3HCOSCoA + H_2O + H^{\oplus} \tag{14}$$

CoA liefern. Nach Inkubation mit [1-^{14}C]Acetyl-CoA und Enzym isoliertes Pyruvat ist aber nicht radioaktiv (Tab. 11; Versuch 1), während es bei äquilibriertem CO_2-Austausch mit $6,5 \times 10^4$ Imp./min × μMol markiert sein müßte. Versuch 2 kontrolliert Versuch 1 und zeigt, daß bei Bereitung des 2,4-Dinitrophenylhydrazons kein [1-^{14}C]Acetyl-2,4-dinitrophenylhydrazid entsteht. An Versuch 3 ist erkenntlich, daß Pyruvat chemisch viel langsamer markiert wird, als daß der hohe Tritiumgehalt des Acetyl-CoA (Versuch K-1) damit erklärt werden könnte. Versuch 4 zeigt, daß die Markierung des Pyruvats enzymatisch nicht beschleunigt wird. Eine Kontrolle dazu bietet Versuch 5: Mit NaOH als Katalysator wird erwartungsgemäß mehr Tritio-Pyruvat gebildet. Versuch K-2 kontrolliert die Beständigkeit des [1-^{14}C]Acetyl-CoA unter den Versuchsbedingungen und dient der Bestimmung seiner spezifischen Aktivität.

c) Statt in enzymatischer Austauschreaktion mit Tritio-Pyruvat, kann markiertes Acetyl-CoA auch durch teilweises Abreagieren des Acetyl-CoA mit unmarkiertem Pyruvat gebildet werden: Bei reversibler Synthese des Citramalats (Gl. 14) [28—30] würde markiertes Acetyl-CoA durch Äquilibrierung entstehen. Wie Citrat-synthase [6] ist aber auch Malat-synthase in 0,1 M Ferricyanid-Lösung stabil, während CoA—SH darin oxydiert und die Äquilibrierung daher vollständig gehemmt wird (vgl. [6]). Da die Kinetik des Isotopenaustausches von Ferricyanid-Lösung nicht beeinflußt wird (Tab. 12), können alle CoA—SH abhängigen Nebenreaktionen ausgeschlossen werden.

d) Um zu beweisen, daß Malat-synthase für die Enolisierung im System mit Pyruvat verantwortlich ist, wurde die Hitzeinaktivierung des Enzyms mit dem Test zur Synthese des Malats und mit dem pyruvathaltigen Test zur Enolisierung zeitlich verfolgt: Beide Methoden ergaben ein übereinstimmendes Ergebnis (Tab. 13).

e) Statt als basischer Reaktant die Rolle des Glyoxylats zu übernehmen, könnte Pyruvat als allosterischer Effektor [31] die Konformationsänderung des Enzyms von der katalytisch inaktiven zur aktiven Form bewirken. Die nachfolgend beschrie-

Tabelle 13. *Kinetik der Hitzeinaktivierung der Malat-synthase*

Die Hitzeinaktivierung wurde mit je 0,10 ml Lösung der Malat-synthase (105 Einheiten/ml, sA = 13) wie bei Tab. 7 angegeben ausgeführt und bestimmt: Je 10 µl der Überstände wurden 1:100 verdünnt und die Synthese des Malats damit (µl wie angegeben) bestimmt, und je 10 µl unverdünnte Lösung wurden im 1 µMol Pyruvat enthaltenden Standardansatz zur Enolisierung eingesetzt. Inkubationszeit: 30 min; sA(H₂O) = 9,1 × 10⁴ Imp./min × µMol

	Synthese des Malats [a]			Bildung des Tritio-Acetyl-CoA [b]			
t	Zusatz (1:100 Verdünnung)	$\dfrac{-\varDelta E_{232}}{\text{min}}$	Relative Aktivität	$-\varDelta E_{405}$	pNAA	Aktivität	Relative Aktivität
min	µl		%		µMol	Imp. / 3 min · Imp./min / µMol pNAA	%
0	20	0,033	100	0,180	0,66	44170 · 22400	100
2	20	0,031	94	0,180	0,66	37890 · 19100	86
5	20	0,026	77	0,180	0,66	28100 · 14200	64
7	20	0,023	70	0,180	0,66	29470 · 14900	67
10	20	0,019	58	0,180	0,66	27240 · 13750	61
15	20	0,014	42	0,180	0,66	23170 · 11700	52
30	20	0,009	27	0,180	0,66	13690 · 6900	31
60	40	0,013	20	0,190	0,70	10700 · 5100	23

[a] Nach Gleichung (1).
[b] Nach Gleichung (7).

benen Versuchsergebnisse lassen aber dafür keinen experimentellen Anhalt erkennen.

Die rund tausendfach angereicherte Lösung der Malat-synthase wird im Rohrzuckergradienten zentrifugiert in zwei Proteinzonen aufgetrennt, wovon eine enzymatische Aktivität parallel zum Protein enthält (Fig. 3). Bei Sedimentation in der Ultrazentrifuge wird eine vergleichbare Auftrennung beobachtet (Fig. 4 A), die im Acetyl-CoA und Pyruvat enthaltenden Parallelansatz nicht geändert wird (Fig. 4 B). Da Acetyl-CoA unter diesen Bedingungen ständig enolisiert (vgl. Tab. 14—19), muß Pyruvat anders als das Protein spaltend oder zusammenlagernd wirksam sein.

Würde Pyruvat die Enolisierung durch einen allosterischen Vorgang am Enzym auslösen, so müßte im natürlichen System auch Glyoxylat ein allosterischer Effektor sein: Acetyl-CoA wird von Malatsynthase allein nicht enolisiert (vgl. Fig. 7), aber mit

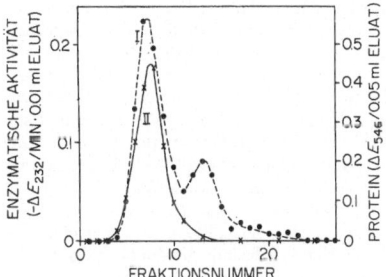

Fig. 3. *Verhalten der gereinigten Malat-synthase bei Sedimentation im Rohrzucker-Dichtegradienten*
Kurve I: ●, Protein (mit je 0,05 ml Eluat nach Lowry et al. [85] abgeändert bestimmt); rechte Ordinate. Kurve II: ×, Enzymatische Aktivität (mit 5 oder 10 µl Eluat bei 232 mµ optisch bestimmt; siehe Methoden); linke Ordinate

Glyoxylat und Enzym rasch in Malat umgewandelt. Die Aldehydsäure müßte daher sowohl am allosterischen wie auch am aktiven Zentrum gebunden werden. Würde Pyruvat statt Glyoxylat nur am allosterischen Zentrum reagieren, so sollte es die Synthese des Malats gar nicht oder stimulierend beeinflussen, nicht aber hemmend. Tatsächlich hemmt Pyruvat die Synthese kompetitiv (Fig. 5). Reagiert Pyruvat im Vorgang der Enolisierung mit beiden Zentren, so sollte es zwar die Synthese des Malats kompetitiv hemmen, aber statt der parabolischen Sättigungskurve eine sigmoide Pyruvatabhängigkeit des Isotopenaustausches liefern. Tatsächlich besteht dafür eine normale Michaelis-Menten Kinetik (Fig. 6).

In ähnlicher Weise könnte Pyruvat das im pyruvatfreien Zustand inaktive, flexible aktive Zentrum des Enzyms nur in die zur Enolisierung geeignete Paßform bringen. Eine Voraussetzung der *induced fit theory* [32] ist die präzise Bindung der konformationsändernd wirkenden Reaktanten [33]. Im vorliegenden Fall würde man daher nicht erwarten, daß Pyruvat mit der raumerfüllenden Methylgruppe das Glyoxylat so zur „Induktion der Paßform" ersetzen kann, daß die Enolisierung wie im natürlichen System müßte: Ihre Geschwindigkeit müßte weit unterhalb der Synthesegeschwindigkeit des Malats liegen. Tatsächlich verlaufen aber beide Prozesse mit ähnlicher Geschwindigkeit (vgl. V.).

Ein flexibles Protein im Katalysevorgang kann und soll mit diesen Ergebnissen nicht ausgeschlossen werden. Sie zeigen aber, daß kein Grund dazu besteht, eine andere Wirkung des Pyruvats und des Glyoxylats anzunehmen, als die in Gl. (11) angegebene. Übereinstimmend damit wird Malyl-CoA als Substrat von Malat-synthase hydrolysiert ohne dafür Glyoxylat oder Pyruvat zu benötigen (vgl. III.).

f) Statt nur den ersten Schritt der Synthese des Malats zu katalysieren, könnte Pyruvat das Glyoxylat

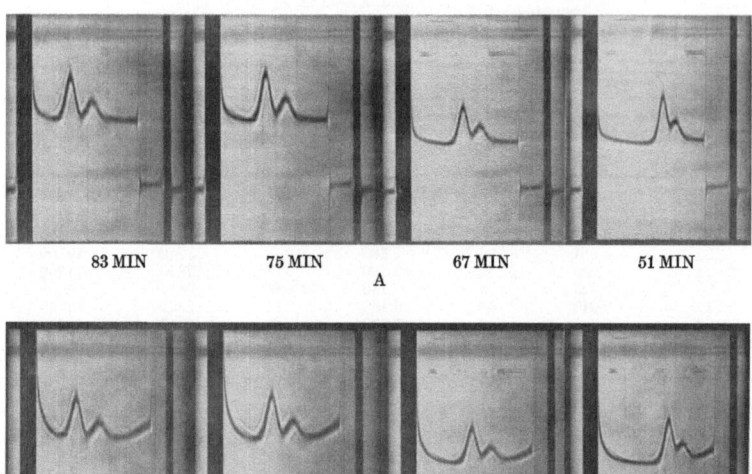

B

Fig. 4. *Verhalten der gereinigten Malat-synthase bei Sedimentation in der Ultrazentrifuge ohne und mit den Cofaktoren zur Enolisierung.* A: 6,7 mg Protein (sA 10) in 1,0 ml Mg-Trispuffer, pH 8,0. B: Wie I, aber mit 2,4 μMol Acetyl-CoA und 10 μMol Pyruvat. Analytische Ultrazentrifuge, Modell E, Beckmann Instruments Co. Zelle: *Centerpiece Al single sector*, 12 mm, 4°, Temperatur = 6°C, 59,780 U/min bei 9,6°C

auch im zweiten Schritt, der Bildung des β-Hydroxy-acyl-CoA ersetzen. Anstatt den dritten Schritt des natürlichen Systems, die Hydrolyse einzugehen, müßte dabei entstandenes Citramalyl-CoA [Gl.(15)] zurückreagieren, da während der Inkubation mit

$$CH_3-\overset{\overset{\displaystyle O}{\|}}{C}\overset{\underset{\displaystyle O}{\|}}{\underset{\displaystyle O^\ominus}{C}} \quad \overset{\overset{\displaystyle O}{\|}}{\underset{\displaystyle H}{CH_2-C}} \overset{\displaystyle O}{\underset{\displaystyle SCoA}{}} \rightleftharpoons \quad \overset{\displaystyle HO}{\underset{\displaystyle HO_2C}{\overset{\displaystyle }{C}}} \overset{\displaystyle CH_3}{\underset{\displaystyle CH_2COSCoA}{}} \quad (15)$$

Pyruvat kein Acetyl-CoA verbraucht wird. Die Frage, warum dann Citramalyl-CoA nicht auch den dritten Schritt der Enzymkatalyse eingeht, warum Malat-synthase nicht auch die Synthese des Citramalats oder des Citrats katalysiert, kann nur unbefriedigend beantwortet werden. Es erscheint daher wahrscheinlicher, daß Pyruvat zwar das aktive Zentrum des Enzyms erreicht, aber nur das Carboxylat-Anion, nicht das Ketocarbonyl am Katalysegeschehen aktiv beteiligt ist. Eine Möglichkeit, die unterbleibende Addition [Gl.(15)] zu diskutieren, bietet die größere Reaktivität des Aldehydcarbonyls gegenüber dem Ketocarbonyl [34]. Da aber an anderer Enzym-katalyse Pyruvat mit Acetyl-CoA zu Citramalat reagiert [28—30] und angenommen werden muß,

daß der Katalysemechanismus dieser Reaktion prinzipiell gleichartig abläuft, ist dieses Argument hinfällig. Die Fähigkeit des jeweiligen Enzyms ein bestimmtes Carbonyl zu erkennen, muß eine andere Ursache haben. Die beste Erklärung für das Reak-tionsgeschehen zwischen Pyruvat und Acetyl-CoA an Malat-synthase stützt sich auf die sterische An-ordnung der Reaktanten: Das Carboxylat-Anion des Pyruvats gelangt zwar in die zur Enolisierung des Acetyl-CoA geeignete Bindung, die sperrige Methyl-gruppe des Pyruvats kann aber den Platz des Wasser-stoffes im Glyoxylat nicht einnehmen. Die Carbonyl-gruppe des Pyruvats wird daher in eine Lage ge-bracht, die zur Addition an das enolische Acetyl-CoA zu weit entfernt ist (vgl. [38]). Übereinstimmend damit wird Citramalyl-CoA [35], das wie voran-stehend diskutiert, rasch zu Pyruvat und Acetyl-CoA gespalten werden sollte, auch von großen Mengen Malat-synthase nicht angegriffen (siehe Methoden).

Das relative Induktorvermögen verschiedener α-Ketosäuren bietet eine weitere Stütze dafür, daß die Raumerfüllung der Substrate bei der enzyma-tischen Katalyse eine entscheidende Rolle spielt. Je kurzkettiger und damit glyoxylatähnlicher die α-Ketosäuren sind, um so geeigneter sind sie als Co-

Tabelle 14. *Relative Induktorwirkung der α-Ketosäuren*
Standardansätze ohne K_2Mg-EDTA, mit je 1 μMol der angegebenen α-Ketosäuren und je 2 Einheiten Malat-synthase.
Start der Reaktion mit Acetyl-CoA. Inkubationsdauer:
15 min; $sA(H_2O) = 9,6 \times 10^4$ Imp./min×μMol

α-Ketosäure	$-\Delta E_{405}$	pNAA	Aktivität	
			Imp.	Imp./min
	μMol		3 min	μMol pNAA
—	0,191	0,696	0	0
Pyruvat	0,194	0,706	80 270	37 800
Oxalacetat	0,186	0,676	14 720	7 260
α-Ketobutyrat	0,193	0,704	870	410
α-Ketoglutarat	0,184	0,640	110	60
α-Ketovalerat	0,176	0,640	80	40
γ-Ketovalerat	0,194	0,706	0	0

Tabelle 15. *Zeitabhängigkeit im „bifunktionellen System"*
Zwei zehnfach vergrößerte Standardansätze ohne K_2Mg-
EDTA, die je 10 mC H^3HO, 100 μMol $K_3Fe(CN)_6$, 12,5 μMol
Acetyl-CoA, 50 μMol Pyruvat und verschieden 80 bzw.
40 Einheiten Malat-synthase enthielten. Je 0,10 ml der An-
sätze wurden zur Zeit t abgestoppt und aufgearbeitet.
$sA(H_2O) = 4,8 \times 10^4$ Imp./min×μMol

t	$-\Delta E_{405}$	pNAA	Aktivität	
			Imp.	Imp./min
Std	μMol		3 min	μMol pNAA
Mit 8 Einheiten Malat-synthase:				
0	0,223	0,82	0	0
0,25	0,238	0,88	79 510	30 000
0,5	0,225	0,83	77 860	31 100
0,75	0,236	0,87	85 780	32 800
1	0,233	0,86	80 170	31 000
2	0,233	0,86	84 850	32 800
3	0,225	0,83	84 370	33 800
5	0,237	0,88	90 100	34 000
7	0,250	0,92	94 280	34 000
Mit 4 Einheiten Malat-synthase:				
0	0,247	0,91	0	0
0,5	0,250	0,92	74 000	26 800
1	0,256	0,94	82 900	29 400
2	0,254	0,93	89 140	31 900
3	0,241	0,89	88 150	33 000
5	0,250	0,92	92 330	34 400
7	0,249	0,92	92 900	34 700
9	0,255	0,94	88 400	31 400

Tabelle 16. *Enzymabhängigkeit im „bifunktionellen System"*
Standardansätze ohne K_2Mg-EDTA, mit je 1 μMol Pyruvat
und Malat-synthase wie angegeben. Inkubationsdauer:
15 min; $sA(H_2O) = 9,6 \times 10^4$ Imp./min×μMol

Malat-synthase	$-\Delta E_{405}$	pNAA	Aktivität	
			Imp.	Imp./min
Einheiten	μMol		3 min	μMol pNNA
0	0,211	0,769	0	0
0,5	0,193	0,701	17 500	8 400
1	0,202	0,736	34 080	15 500
2	0,190	0,692	60 630	29 300
3	0,198	0,722	81 530	37 800
4	0,207	0,756	92 090	40 600
6	0,194	0,707	108 510	51 200
8	0,203	0,704	120 730	54 400

faktoren der Enolisierung (Tab.14). Glyoxylat paßt
genau in das aktive Zentrum: Sein Carboxylat-Anion
löst das Proton von der Methylgruppe des Acetyl-
CoA, sein Aldehyd-Carbonyl reagiert mit dem Enol,
der intermediäre Thioester hydrolysiert zu den Produkten. Pyruvat hat beschränkt Platz: Sein Carboxylat-Anion katalysiert noch die Enolisierung, sein
Keto-Carbonyl kann aber mit dem Enol nicht mehr
reagieren. Das größere Oxalacetat paßt schlechter in
das aktive Zentrum und ist folglich weniger aktiv,
während das noch größere α-Ketoglutarat darin
keinen Platz mehr findet. Wie das etwa gleich große
α-Ketovalerat ist es daher zur Induktion praktisch
ungeeignet. Diese Abhängigkeit von der enzymatischen
Aktivität von der Kettenlänge des Substrates
(Tab.14) zeigt auch wie treffend Fischers Vergleich
von Enzym und Substrat mit Schlüssel und Schloß
ist [36].

Zusammengefaßt beweisen diese und die mit
Citrat-synthase erzielten Ergebnisse [6] die Teilnahme
des Carboxylat-Anions an der Enolisierung. Diese
Katalyse wird insbesondere durch Arbeiten von Bell
gestützt. Bell u. Mitarb. haben die katalytische Wirkung des Carboxylat-Anions verschiedener Carbonsäuren zur Enolisierung zahlreicher Carbonylverbin-
dungen klar nachgewiesen [39].

Eigenschaften des „bifunktionellen enzymatischen Systems"

Qualitativ gleichen die mit Pyruvat gemessenen
enzymatischen Abhängigkeiten denen des unter I.
beschriebenen „monofunktionellen Systems". Abweichend davon ist das Mg^{++}-Bedürfnis geringer,
stören Salze weniger und entsteht kein hemmendes
CoA—SH. Quantitativ unterscheiden sie sich aber
durch außerordentlich erhöhte Enolisierungsgeschwindigkeit. Aus der Zeitabhängigkeit (Tab.15) ist ersichtlich, daß sich der Isotopenaustausch mit 4 und 8 Einheiten Malat-synthase bereits nach 30 bzw. 15 minutiger Inkubation im Gleichgewicht befindet. Mit einer
Enzymeinheit und 1 μMol Pyruvat im Reaktionsansatz wird nach 15 min ein Maß der Enolisierung
erreicht (Tab.16), für das im „monofunktionellen
System" bei 15 stündiger Inkubation 10 Einheiten
erforderlich waren (Tab.2). Ähnlich dazu wurde das
Gleichgewicht des Isotopenaustausches unter den
Bedingungen der Tab.17 in allen Proben, und bei
Bestimmung der Abhängigkeit der Enolisierung von
der Konzentration des Enzyms (Tab.16) nach
15 minutiger Inkubation mit 6—8 Einheiten erreicht.
Das pH-Profil der Enolisierung (Tab.18) ist mit
seinem breiten Optimum dem der Synthese des
Malats [4] ähnlich und unterscheidet sich von dem
unter I. ermittelten (Tab.4): Die Salzwirkung ist im
„bifunktionellen System" praktisch belanglos. Der
Versuchsanordnung kann entnommen werden, daß
das hohe Mg^{++}-Erfordernis des „monofunktionellen

Tabelle 17. *Gleichgewicht des Isotopenaustausches* [Gl.(7)] Standardansätze ohne $MgCl_2$, mit je 1 µMol Pyruvat und Malat-synthase wie angegeben. Inkubationsdauer: 7 Std; $sA(H_2O) = 8{,}6 \times 10^4$ Imp./min × µMol

Malat-synthase	$-\Delta E_{405}$	pNAA	Aktivität	
Einheiten		µMol	Imp.	Imp./min
			3 min	µMol pNAA
0	0,135	0,495	0	0
1,8	0,130	0,475	72740	51000
3,6	0,135	0,495	73150	49200
7,2	0,140	0,51	80430	52500
10,8	0,135	0,495	77640	52200

Tabelle 18. *pH-Abhängigkeit im „bifunktionellen System"* Standardansätze ohne $MgCl_2$, mit je 10 µMol $K_3Fe(CN)_6$, 1 µMol Pyruvat, 2 Einheiten Malat-synthase und 30 µMol Puffersubstanz wie angegeben. Inkubationsdauer: 15 min; $sA(H_2O) = 8{,}6 \times 10^4$ Imp./min × µMol

Pufferlösung	pH	$-\Delta E_{405}$	pNAA	Aktivität	
			µMol	Imp.	Imp./min
				3 min	µMol pNAA
β,β-Dimethylglutarat	6,0	0,115	0,49	12340	8400
β,β-Dimethylglutarat	6,5	0,110	0,47	21940	15600
Phosphat	7,0	0,110	0,47	29630	21000
Phosphat	7,4	0,105	0,45	30480	22800
Tris	7,8	0,100	0,42	28350	22500
Tris	8,1	0,100	0,42	28060	22300
Tris	9,0	0,080	0,34	25930	22700

Tabelle 19. *Acetyl-CoA-Abhängigkeit im „bifunktionellen System"* Standardansätze ohne $MgCl_2$, mit je 10 µMol $K_3Fe(CN)_6$, 1 µMol Pyruvat, 2 Einheiten Malat-synthase und Acetyl-CoA wie angegeben. Inkubationsdauer: 15 min. $sA(H_2O) = 8{,}6 \times 10^4$ Imp./min × µMol

Acetyl-CoA	$-\Delta E_{405}$	pNAA	Aktivität	
µMol		µMol	Imp.	Imp./min
			3 min	µMol pNAA
0,4	0	0	0	0
0,4	0,055	0,20	19260	32100
0,8	0,085	0,31	22610	24400
1,2	0,115	0,41	19690	16000
1,6	0,145	0,53	17450	11000
2,4	0,225	0,82	16390	6660

Standardansätze ohne $MgCl_2$, mit je 1 µMol Pyruvat, 2 Einheiten Malat-synthase und Acetyl-CoA wie angegeben. Inkubationsdauer: 15 min; $sA(H_2O) = 8{,}6 \times 10^4$ Imp./min × µMol

0	0	0	0	0
0,4	0,090	0,325	24170	24800
0,8	0,185	0,68	33920	16600
1,2	0,280	1,01	29780	9850
1,6	0,360	1,30	36000	9220
2,4	0,480	1,74	30000	5700

Systems" (Tab. 1) mit Pyruvat stark erniedrigt wird: Um das System zu sättigen, genügt das in der Enzym-lösung enthaltene $MgCl_2$.

Die Abhängigkeit der Enolisierung von der Konzentration des Acetyl-CoA (Tab. 19) gleicht derjenigen des „monofunktionellen Systems" (Tab. 5): Im ferricyanidfreien Ansatz ergibt die enzymatische Aktivität (Imp./3 min) gegen die Substratkonzentration (µMol pNAA) aufgetragen eine Sättigungskurve. Im ferricyanidhaltigen Ansatz wird mit mehr als ca. 10 mM Acetyl-CoA wahrscheinlich das Enzym gehemmt, da ab dieser Konzentration eine Hemmkurve erhalten wurde. Mit zunehmender Konzentration des Acetyl-CoA wird in beiden Fällen die spezifische Aktivität des pNAA erniedrigt. Aus analytischen Gründen (vgl. Methoden) kann sie nicht mit geringeren Mengen Acetyl-CoA bestimmt werden, womit sie der Konzentration annähernd proportional wäre. Die als Maß der Enolisierungsgeschwindigkeit dienende spezifische Aktivität des pNAA liefert daher mit dem 0,8 µMol Acetyl-CoA enthaltenden Standardansatz unvermeidlich zu geringe Werte.

Wenn Pyruvat das Glyoxylat am aktiven Zentrum des Enzyms zur Enolisierung ersetzt, dann muß es die Synthese des Malats kompetitiv hemmen. Zum

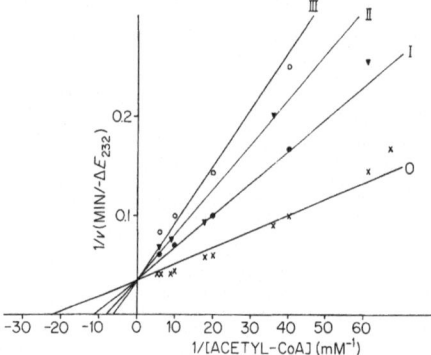

Fig. 5. *Pyruvat hemmt die Synthese des Malats kompetitiv.* Lineweaver-Burk-Darstellung [37] der Meßpunkte (siehe Methoden). Die Anfangsgeschwindigkeit der Synthese des Malats wurde abhängig von der Konzentration an Acetyl-CoA bei 232 mµ bestimmt: Gerade 0 (ohne Pyruvat); Gerade I, II und III (mit 1, 2 und 3 mM Pyruvat)

Nachweis dafür wurde die Anfangsgeschwindigkeit der Synthese des Malats abhängig von der Konzentration des Acetyl-CoA pyruvatfrei, und in 1 bis 3 mM Pyruvatlösung gemessen. Die graphisch [37] ausgewerteten Meßpunkte lassen die kompetitive Hemmung klar erkennen (Fig.5): K_m (Acetyl-CoA) $= 4 \times 10^{-5}$ M; K_i (Pyruvat) $= 10^{-3}$ M.

Übereinstimmend damit kann die Affinität des Pyruvats zum Enzym wie die eines gewöhnlichen Cofaktors bestimmt werden: Die Abhängigkeit des Isotopenaustausches von der Pyruvatkonzentration ergibt eine typische Michaelis-Menten Kinetik (Fig. 6), die graphisch ausgewertet [37] $K_m = 1,25 \times 10^{-3}$ M und $V_{max} = 1,8 \times 10^3$ Imp./min$^2 \times \mu$Mol lieferte. Auf 1 Einheit bezogen (die Enzymansätze enthielten je 2 Einheiten Malat-synthase) beträgt $V_{max} = 0,9 \times 10^3$ Imp./min$^2 \times \mu$Mol. Da der Enolisierung von 1 μMol Acetyl-CoA die berechnete sA(pNAA korr) = 4,5 $\times 10^3$ Imp./min $\times \mu$Mol entspricht (vgl. V), beträgt das mit dem Test zur Enolisierung bestimmte V_{max}

der vorinkubierten Enzymlösung zur Zeit t min je 2 Einheiten entnommen und die enzymatische Aktivität mit dem Test zur Enolisierung bestimmt (0,8 μMol Acetyl-CoA, Inkubationsdauer: 15 min):

Bei $t = 0$ min der Vorinkubation wurde der Enzymlösung MgCl$_2$ zugesetzt (Pfeil 1) und die enzymatische Aktivität sofort anschließend und nach 10 minutiger Vorinkubation bestimmt. Mit beiden

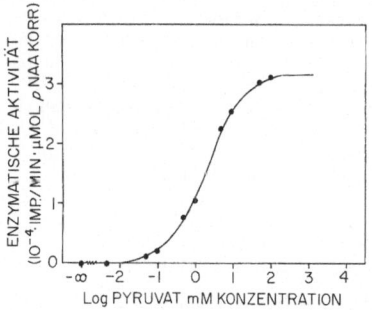

Fig. 6. *Abhängigkeit der Enolisierung des Acetyl-CoA von der Konzentration an Pyruvat.* Standardansätze ohne K$_2$Mg-EDTA, mit je 0,5 μMol MgCl$_2$, 10 μMol K$_3$[Fe(CN)$_6$], 2 Einheiten Malat-synthase und Pyruvat wie angegeben. Inkubationsdauer: 15 min; sA(H$_2$O) = 8,6 $\times 10^4$ Imp./min $\times \mu$Mol. Bezüglich Meßdaten und sA(pNAA korr) siehe Methoden

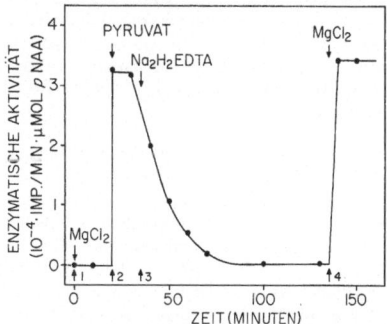

Fig. 7. *Kooperative Wirkung des Mg^{++}-Ions mit Pyruvat.* Standardansätze; Änderungen dazu vgl. Text. Über Meßanordnung und -daten siehe Methoden. Inkubationsdauer: 15 min; sA(H$_2$O) = 9,6 $\times 10^4$ Imp./min $\times \mu$Mol

20 % des mit dem Test zur Synthese des Malats bestimmten. Berücksichtigt man, daß die angewandte Acetyl-CoA Konzentration die spezifische Aktivität des pNAA erniedrigt (Tab. 19) und tritiertes Acetyl-CoA die Markierung rasch mit dem Lösungsmittel austauscht (Tab. 26), so sind die Geschwindigkeiten von Enolisierung des Acetyl-CoA und Synthese des Malats einander praktisch gleich. Wie unter VI angeführt, soll die prinzipielle Aufgabe des Enzyms darin bestehen, die Reaktanten in der zur intramolekularen Säure-Basenkatalyse erforderlichen Anordnung zusammenzuhalten. Wenn diese Interpretation der Versuchsergebnisse richtig ist, darf das Zustandekommen der hohen Synthesegeschwindigkeit des Malats unter physiologischen Bedingungen als im wesentlichen geklärt betrachtet werden. Dafür genügt offenbar die bifunktionelle Katalyse (vgl. Fig. 7)

2-Hydroxypyridin katalysiert in 0,001 M Lösung die Mutarotation [Gl. (13)] 7000 mal wirksamer als die Mischung von 0,001 M Pyridin und 0,001 M Phenol, obschon es 10000 mal weniger basisch als Pyridin und 100 mal schwächer sauer als Phenol ist [27]. Ähnlich dazu enolisiert Acetyl-CoA im „bifunktionellen System" ca. 1000 mal rascher als im „monofunktionellen" (Fig. 7). In diesem Versuch wurde die Enzymlösung mit den jeweiligen Reagentien vorinkubiert (Pfeile 1—4; Fig. 7) und die davon bewirkte Änderung der katalytischen Aktivität gemessen. Dafür wurden

Proben des „monofunktionellen Systems" ist unter den angewandten Bedingungen praktisch keine Enolisierung nachweisbar. Nach 20 minutiger Vorinkubation wurde dem MgCl$_2$-haltigen Ansatz Pyruvat zugesetzt (Pfeil 2) und die enzymatische Aktivität sofort anschließend und nach 10 minutiger Inkubation bestimmt ($t = 20$ und 30 min). Im jetzt „bifunktionellen System" wird die Geschwindigkeit der Enolisierung ungemein erhöht; die spezifische Aktivität des pNAA dieser Meßpunkte entspricht der Enolisierung von ca. 10 μMol Acetyl-CoA (vgl. V.).

Danach wurde eine neue Meßreihe begonnen, bei $t = 0$ min der Vorinkubation Pyruvat statt MgCl$_2$ zugesetzt wurde, die aber noch das MgCl$_2$ der Enzymlösung selbst enthielt. Bei $t = 32$ min wurde das komplexbildende EDTA zugesetzt (Pfeil 3). In langsamer Reaktion (32—130 min) wird enzymgebundenes Mg^{++} Ion entfernt und die Enolisierung daher vermindert. Bei 130 min hört sie praktisch auf: Das System ist wieder „monofunktionell", und zwar mit Pyruvat statt MgCl$_2$. Erwartungsgemäß kann die kooperative Wirkung beider Cofaktoren auch von hier aus wiederhergestellt werden: Zusatz von MgCl$_2$ (Pfeil 4) restauriert das „bifunktionelle System" und bewirkt erneut die hohe Geschwindigkeit der Enolisierung.

Die kooperative Wirkung von Pyruvat und Mg^{++}-Ion stimmt mit Gl. (12) überein und ist der säureund basenkatalysierten Mutarotation der Tetramethylglucose [Gl. (13)] analog: Das elektrophile

Phenol wird im enzymatischen Prozeß vom Mg^{++}-Ion, der nucleophile Pyridin-Stickstoff vom Carboxylat-Anion des Pyruvats ersetzt. Die geeignete sterische Anordnung der sauren und basischen Gruppen wird bei chemischer Katalyse mit deren Einfügen in eine Molekel, bei enzymatischer Katalyse mit entsprechender Bindung an das Protein erreicht. Unterschiedlich zum chemischen Modell wird aber im natürlichen enzymatischen Vorgang eine zusätzliche Variante eingeführt, kommt dem reagierenden Glyoxylat eine zweifache Rolle zu: Sein Carboxylat-Anion ist nucleophiler Katalysator zur Enolisierung des Acetyl-CoA, sein Aldehyd-Carbonyl elektrophiler Reaktant mit diesem Enol.

III. MALYL-COENZYM A IST EIN SUBSTRAT DER MALAT-SYNTHASE

Der unter I und II angeführte Nachweis des ersten Schrittes der Enzymkatalyse kennzeichnet den Mechanismus der Synthese des Malats als aldolartig. Man muß daher auf dem Syntheseweg intermediäres (S)-Malyl-CoA annehmen und erwarten, daß es als Substrat eingesetzt von Malat-synthase hydrolysiert wird. (S)-Malyl-CoA ist aber unter den Bedingungen des optischen Tests zur Synthese des Malats kein Substrat der Malat-synthase [4,9]. Da es die Synthese [Gl.(1)] nicht hemmt, ist auch seine Affinität zum Enzym gering. Wie die langsamen Umsetzungen des Citryl-CoA mit Citrat-synthase [7], können diese Ergebnisse damit erklärt werden, daß der intermediäre Thioester nicht frei, sondern enzymgebunden auftritt (Schema 1). In der Enzymkatalyse unterscheidet sich (S)-Malyl-CoA als Substrat vom natürlich entstandenen wie folgt:

a) Enzymgebundenes Malyl-CoA entsteht im natürlichen System auf einem anderen Weg als mit Malyl-CoA und Enzym, und daher mit verschiedener Geschwindigkeit. Sie wird bei Synthese des Malats von der Enolisierung des Acetyl-CoA bestimmt (vgl. IV). Der langsamste Schritt der enzymatischen Hydrolyse des Malyl-CoA als Substrat kann seine Bindung an das Enzym sein. Damit übereinstimmend ist Malat-synthase als „einheitliches Enzym" wirksam, das bei Reinigung nicht in eine Enolase, Ligase und Hydrolase aufgetrennt werden kann.

b) Das aktive Zentrum des Enzyms muß so konstruiert sein, daß es Acetyl-CoA und Glyoxylat zur Enolisierung des Acetyl-CoA und für die nachfolgende Reaktion zu (S)-Malyl-CoA rasch und sterisch richtig angeordnet bindet. Dafür ist bei Glyoxylat das Aldehydcarbonyl bedeutungsvoll (vgl. Tab.14). Mit dem Katalyseprinzip als „Annäherung der Reak-

tanten" übereinstimmend (vgl. VI), ist es daher verständlich, daß das für Acetyl-CoA und Glyoxylat codierte aktive Zentrum, Malyl-CoA als Substrat nur schlecht zu binden vermag. Da im optischen Test zur Synthese des Malats nur wenig Enzym eingesetzt werden kann, ist damit eine langsam erfolgende Hydrolyse des Malyl-CoA nicht erfaßbar.

Mit viel Enzym im Reaktionsansatz ist die optische Messung der Thioesterhydrolyse bei 234 mµ wegen der zu hohen Absorption des Proteins undurchführbar. In orientierenden Versuchen wurde daher die enzymatische Hydrolyse des Malyl-CoA im diskontinuierlichen Test am freigesetzten CoA—SH colorimetrisch [84] verfolgt und beobachtet, daß sie nach kurzer Zeit praktisch stillstand. Da CoA—SH die Enolisierung des Acetyl-CoA hemmt (vgl. Tab.3), schien es wahrscheinlich, daß es auch die Hydrolyse des Malyl-CoA verhindert. Mit Ferricyanid-Lösung, worin Malat-synthase stabil ist (vgl. Tab.12), bot sich ein kontinuierlicher optischer Test an, bei dem CoA—SH oxydiert wird [Gl.(17)].

$$Malyl\text{-}CoA + H_2O \rightleftharpoons Malat + CoA\text{-}SH \qquad (4)$$

$$CoA\text{-}SH + [Fe(CN)_6]^{-3} \rightleftharpoons$$
$$1/2(CoA\text{-}S)_2 + [Fe(CN)_6]^{-4} + H^+ \qquad (16)$$

$$Malyl\text{-}CoA + H_2O + [Fe(CN)_6]^{-3} \rightleftharpoons$$
$$Malat + 1/2\ (CoAS)_2 + [Fe(CN)_6]^{-4} + H^+ \quad (17)$$

Dabei wird die enzymatische Hydrolyse des Malyl-CoA [Gl.(4)] mit der nachgeschalteten Indikatorreaktion [40] [Gl.(16)] bei 436 mµ an der Entfärbung des Ferricyanids bestimmt. Unter den angewandten Bedingungen (Vol. 0,20 ml; $d = 1$ cm) ist die Bestimmung sehr empfindlich. Da $\Delta\varepsilon_{436}$ $= 7,25 \times 10^2$ $M^{-1} \times cm^{-1}$, wird die optische Dichte unter den Versuchsbedingungen um $\Delta E_{436} = -0,36/$

(A-Enol + G)MS $\rightleftharpoons$ (Malyl-CoA)MS $\xrightarrow{+H_2O}$ Malat + CoA-SH + MS

$\Updownarrow$ $\Updownarrow$

A + G + MS   Malyl-CoA + MS

A = Acetyl-CoA

G = Glyoxylat   ,   (Schema 1)

MS = Malat-synthase

Tabelle 20. *Enzymabhängigkeit der Hydrolyse des (S)-Malyl-CoA*
Die Reaktionsansätze, Vol. 0,20 ml, enthielten bei 25° je 20 µMol Trispuffer, pH 8,1; 1 µMol $K_3Fe(CN)_6$; 1 µMol $MgCl_2$; 0,31 (Meßreihe I), 0,50 (Meßreihe II); 1,0 (Meßreihe III) µMol (S)-Malyl-CoA, und Malat-synthase wie angegeben. Messung gegen eine Kontrollküvette, die alle Zusätze, aber kein Enzym enthielt. $\lambda = 436$ mµ; $d = 1$ cm; Photometer Eppendorf

Malat-synthase	$-\Delta E_{436}$/min		
	I	II	III
Einheiten			
0	0	0	0
2,5	0,014	0,015	0,022
5	0,025	0,026	0,031
10	0,036	0,041	0,048
20	0,055	0,061	0,074

Tabelle 21. *pH-Profil der enzymatischen Hydrolyse des (S)-Malyl-CoA*
Reaktionsansätze und Meßanordnung wie bei Tab. 20 angegeben, aber mit je 0,1 μMol (S)-Malyl-CoA; 10 Einheiten Malatsynthase und je 20 μMol der angegebenen Puffersubstanzen

Pufferlösung	pH	$-\Delta E_{436}$/min
β,β-Dimethylglutarat (K⁺)	6,0	0,006
Phosphat (K⁺)	7,0	0,008
Phosphat (K⁺)	7,4	0,016
Tris (HCl)	7,8	0,026
Tris (HCl)	8,1	0,067
Tris (HCl)	9,0	0,084
Tris (HCl)	9,5	0,089
Tris (HCl)	10,0	0,074

bestimmte Kinetik der Hitzeinaktivierung des Enzyms ist gleich (Tab. 22), die Malyl-CoA-Hydrolase benötigt Mg⁺⁺-Ionen (Fig. 10). Ähnlich der Wirkung als Acetyl-CoA-Enolase (vgl. Fig. 7) wird Malat-synthase auch als Malyl-CoA-Hydrolase bei Entzug der Lewis-Säure mit EDTA in zeitabhängiger Reaktion gehemmt (Fig. 10, Pfeil 1) und bei ihrem Zusatz reaktiviert (Fig 10, Pfeil 2). Die Beteiligung des Mg⁺⁺-Ions zur Hydrolyse des β-Hydroxyacyl-thioesters ist ver-

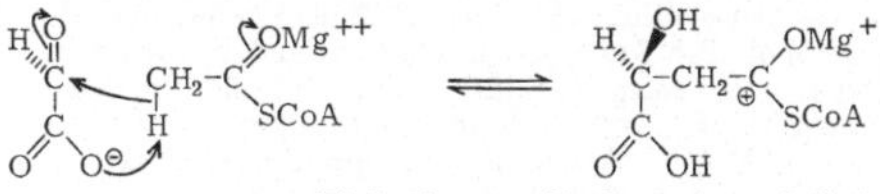

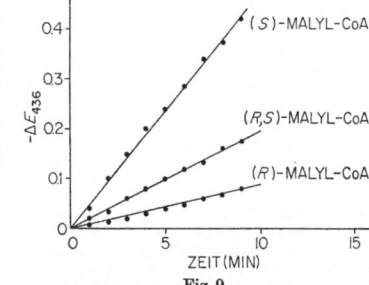

(18)

ständlich: Aus enolischem Acetyl-CoA und Glyoxylat entstandenes (S)-Malyl-CoA ist am Thioestercarbonyl positiviert Gl. (18).

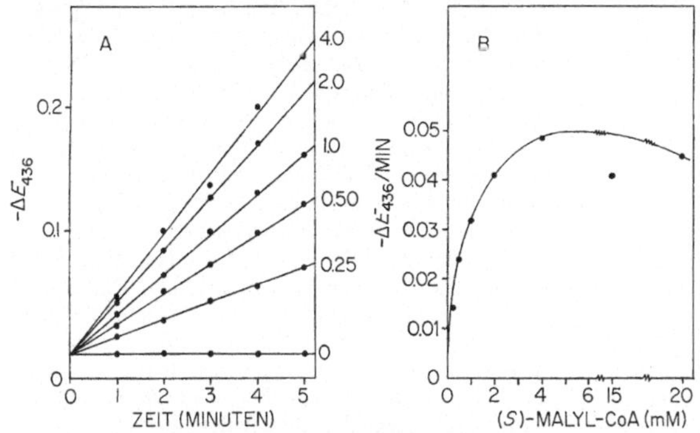

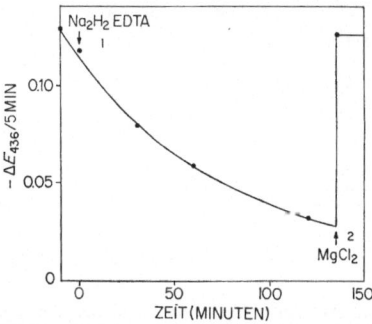

Fig. 8

Fig. 9

Fig. 8. *Optischer Test für die Malyl-CoA-Hydrolase.* A: Kinetik des Tests (Zahlen am Rand: mM Substratkonzentration) bei 25° mit je 10 Einheiten Malat-synthase. B: Substratabhängigkeit (Ordinate: ΔE_{436}/min aus Fig. 8 A; Abszisse: mM Substratkonzentration)
Fig. 9. *Stereospezifität der Malyl-CoA-Hydrolase.* Bei 25° mit je 10 Einheiten Malat-synthase und je 0,13 μMol (Bestimmung als Hydroxamsäure) der Malyl-CoA-Diastereomeren nach Gl. (17) bestimmt. Citrat-synthase hydrolysierte von je 0,13 μMol (Bestimmung als Hydroxamsäure) 0,10 μMol (S)-, 0,05 μMol (R.S)- und 0 μMol (R)-Malyl-CoA; vgl. [9]

0,1 μMol CoA—SH geändert. Die nachfolgenden, Malyl-CoA als Substrat der Malat-synthase kennzeichnenden Versuche, wurden mit diesem Test erzielt.

Die Hydrolyse des Malyl-CoA ist enzymabhängig (Tab. 20) und erfolgt am raschesten bei pH 9—10 (Tab. 21). Sie ist von der Konzentration des Malyl-CoA abhängig (Fig. 8 A) und zeigt dabei in allen Proben innerhalb 5 min konstante Anfangsgeschwindigkeit. Das Enzym wird mit Substratkonzentrationen über 5 mM schwach gehemmt und mit 10⁻³ M (S)-Malyl-CoA halbgesättigt; $V_{max} = -\Delta E_{436} = 0,049$/min × 10 E $= 1,36 × 10^{-3}$ μMol/min × 1 E (Fig. 8 B): K_m und V_{max} weichen von den entsprechenden Daten der natürlichen Substrate [4] um rund drei Zehnerpotenzen ab.

Citrat-synthase hydrolysiert (S)-Malyl-CoA [9]. Ähnlich dazu, aber weniger stereospezifisch hydrolysiert Malat-synthase (S)-Malyl-CoA rascher als das Diastereomerengemisch und die unnatürliche (R)-Form (Fig. 9).

Folgende Ergebnisse weisen die Malyl-CoA-Hydrolase als Malat-synthase aus: Die mit Malyl-CoA [Gl. (17)] und mit Acetyl-CoA und Glyoxylat [Gl. (1)]

Fig. 10. *Mg⁺⁺-Abhängigkeit der Malyl-CoA-Hydrolase.* 0,12 ml (60 Einheiten) Lösung der Malat-synthase wurden bei $t = 0$ min mit 0,12 ml 0,1 M Na₂H₂-EDTA-Lösung gemischt (Pfeil 1) und bei 25° gehalten. Die enzymatische Aktivität wurde mit je 0,04 ml davon und je 0,2 μMol (S)-Malyl-CoA nach Gl. (17) bestimmt (Zeiten wie angegeben). Der Wert vor $t = 0$ min wurde mit 20 μl der EDTA-freien Enzymlösung in gleicher Weise ermittelt. Bei $t = 135$ min wurden 0,10 ml des Inkubationsgemisches mit 0,10 ml 0,1 M MgCl₂-Lösung versetzt (Pfeil 2) und die enzymatische Aktivität mit 0,08 ml bestimmt

Erwartungsgemäß wird die enzymatische Hydrolyse des Malyl-CoA statt durch Entzug der Mg^{++}-Ionen auch mit den natürlichen Substraten Acetyl-CoA und Glyoxylat, und mit dem „unnatürlichen" Induktor Pyruvat gehemmt (Tab. 23): Die Substrate konkurrieren mit Malyl-CoA um das aktive Zentrum des Enzyms.

Tabelle 22. *Kinetik der Hitzeinaktivierung der Malat-synthase*
Je 0,1 ml Malat-synthase-Lösung (110 Einheiten/ml; sA=12) wurden wie bei Tab. 7 angegeben behandelt und die Aktivität des Überstandes bestimmt: A: Mit Acetyl-CoA und Glyoxylat bei 232 mμ [Gl. (1)] nach Verdünnung wie angegeben. B: Mit je 0,20 μMol Malyl-CoA und je 50 μl des Überstandes bei Tab. 20 angegeben

Zeit um 60°	Bestimmung nach			
	A		B	
	Zusatz (1:100 Verdünnung)	$-\Delta E_{232}$ min	Relative Aktivität	$-\Delta E_{436}$ 5 min	Relative Aktivität
min	μl		%		%
0	0,02	0,034	100	0,108	100
5	0,02	0,012	36	0,038	35
10	0,06	0,016	16	0,015	14
60	0,24	0,002	0,5	0	0

Tabelle 23. *Hemmung der enzymatischen Hydrolyse des (S)-Malyl-CoA mit Reaktanten der Malat-synthase*
Die Reaktionsansätze enthielten bei 25° im Vol. 0,20 ml je 20 μMol Carbonatpuffer, pH 8,5; 1 μMol $K_3Fe(CN)_6$; 1 μMol $MgCl_2$; 0,1 μMol (S)-Malyl-CoA und 10 Einheiten Malat-synthase. Messung gegen eine Kontrollküvette, die alle Zusätze, aber keine Malat-synthase enthielt. $\lambda = 436$ mμ; $d = 1$ cm; Photometer Eppendorf. Bestimmung der enzymatischen Aktivität nach [Gl. (17)]

Hemmstoff	Enzymatische Aktivität		
	Pyruvat	Acetyl-CoA	Glyoxylat
mM	%	%	%
0	100	100	100
0,30	—	83	—
0,50	—	—	75
0,60	—	65	—
1,00	92	48	—
2,00	84	—	51
4,00	—	—	37
5,00	69	—	—

IV. REVERSIBILITÄT UND SCHRITTMACHER DER SYNTHESE DES MALATS

Bei reversibler Synthese des Malats muß im tritiumhaltigen Lösungsmittel (S)-2-Hydroxy-3,3-ditritiosuccinat entstehen: C—C Spaltung des Malats bei Syntheseumkehr liefert Glyoxylat und Tritio-Acetyl-CoA [Gl. (19)] ([16], vgl. [6]).

Wegen des praktisch vollständig auf Seite der Synthese liegenden Gleichgewichtes [Gl. 1)] entsteht folglich an der Carboxymethylgruppe zweifach tritiertes (S)-Malat (vgl. V). Bei 15 stündiger Inkubation des (S)-Malats (0,1—1,0 μMol) mit CoA—SH (0,1 bis 0,5 μMol) und Malat-synthase (2—10 Einheiten) im Standardansatz wird aber kein Tritiomalat gebildet. Wie auf anderem Wege schon nachgewiesen [4], ist die Synthese des Malats praktisch irreversibel.

Ebensowenig ist die Aldolspaltung des (S)-Malyl-CoA nachweisbar: In Versuchen dazu wurden je 0,16 μMol (S)-Malyl-CoA im Vol. 0,25 ml mit Malat-synthase, p-Nitroanilin, Arylamin-transacetylase und tritiumhaltigem Wasser inkubiert. Bei Aldolspaltung des Malyl-CoA entsteht Tritio-Acetyl-CoA [analog Gl. (19)], das im zusammengesetzten Test mit p-Nitroanilin und dem Hilfsenzym irreversibel Tritio-p-nitroacetanilid liefert [Gl. (6)]. Zwischen pH 6,5 und 8,1 konnte aber keine, die Acetylierung des Amins anzeigende, abnehmende optische Dichte bei 405 mμ gemessen, und nach Zusatz von Trägersubstanz im isolierten p-Nitroacetanilid keine Radioaktivität nachgewiesen werden. Dabei erwies sich auch der Zug an den Produkten der Hydrolyse und Aldolspaltung (CoA—SH und Glyoxylat) mit Hilfsenzymen (α-Ketoglutarat-dehydrogenase; Lactat-dehydrogenase) als wirkungslos. Da Arylamintransacetylase aus praktischen Gründen in geringerer Konzentration eingesetzt werden muß als Malat-synthase, schien es möglich, daß zwar die Aldolspaltung des Malyl-CoA, nicht aber die Acetylierung des Amins abließ. In diesem Fall würde die Synthese des Malats damit erfolgreich konkurrieren und wie voranstehend diskutiert, markiertes Malat entstehen. Nach Zusatz von Trägersubstanz isoliertes (S)-Malat war aber ebenfalls frei von Radioaktivität: Die enzymatische Aldolspaltung ist weder mit Malat und CoA—SH noch mit Malyl-CoA nachweisbar.

Zwei Versuchsergebnisse kennzeichnen die Enolisierung bei der Synthese des Malats als geschwindigkeitsbestimmend: In Abwesenheit von Pyruvat im tritiumhaltigen Wasser synthetisiertes Malat ist tritiumfrei (siehe Methoden); Acetyl-CoA wird bei teilweisem Abreagieren mit Glyoxylat nicht markiert (Tab. 24). Wäre ein Schritt der Synthese langsamer als die Enolisierung, so würde die Äquilibrierung mit dem entsprechenden Zwischenprodukt ein markiertes Acetyl-CoA bzw. Malat liefern.

V. GLEICHWERTIGKEIT DER METHYLWASSERSTOFFE GEGENÜBER PYRUVAT UND GLYOXYLAT

Die unter I und II angegebenen spezifischen Aktivitäten des p-Nitroacetanilids bestimmen die Ge-

$$\underset{HO_2C}{\overset{HO}{\diagdown}}\hspace{-0.3em}\underset{CH_2CO_2H}{\overset{H}{C}} + CoA-SH \underset{}{\overset{^3H^\oplus}{\rightleftarrows}} \underset{HO_2C}{\overset{O}{\diagdown}}\hspace{-0.3em}\overset{H}{C} + CH_2{}^3HCOSCoA + H_2O + H^\oplus \tag{19}$$

Tabelle 24. *Bildung des Tritio-Acetyl-CoA durch Äquilibrierung mit Intermediaten der Synthese des Malats* Standardansätze ohne MgCl$_2$, mit je 10 μMol K$_3$Fe(CN)$_6$, 20 μMol Puffersubstanz und Glyoxylat wie angegeben. Start der Reaktion mit je 1,8 Einheiten Malat-synthase. Inkubationsdauer: 30 min; sA(H$_2$O) = 8,6 × 10^4 Imp./min × μMol

Pufferlösung	Glyoxylat	$-\Delta E_{405}$	pNAA	Aktivität
	μMol		μMol	$\frac{\text{Imp./min}}{\text{μMol } p\text{NAA}}$
β,β-Dimethyl-	0	0,140	0,51	0
glutarat, pH 6,0	0,2	0,120	0,44	0
	0,4	0,110	0,40	0
Phosphat, pH 7,0	0	0,140	0,51	0
	0,2	0,105	0,38	0
	0,4	0,090	0,33	0
Pyrophosphat,	0	0,120	0,44	0
pH 8,1	0,2	0,100	0,37	0
	0,4	0,090	0,33	0

schwindigkeit der Enolisierung fehlerhaft. Sie liefern qualitativ richtige Abhängigkeiten, da in allen Versuchen „übereinstimmende Zustände" herrschen: Die Reaktionsansätze sind gleichartig zusammengesetzt und werden genau gleich aufgearbeitet. Das Ergebnis vorhandener Bestimmungsfehler entspricht daher nur der geänderten spezifischen Aktivität des tritiumhaltigen Lösungsmittels, die im „monofunktionellen System" wegen der langsamen Reaktionsgeschwindigkeit belanglos war. Im raschen „bifunktionellen System" dagegen wird das Gleichgewicht des Isotopenaustausches in kurzer Zeit erreicht, erfolgt der Teilschritt [Gl. (2)] mit einer der Gesamtreaktion [Gl. (1)] ähnlichen Geschwindigkeit. Sie wird in der enzymatischen Katalyse nach der Enzymeinheit charakterisiert: 1 Einheit katalysiert die Enolisierung von 1 μMol Acetyl-CoA pro Minute. Sie wird bei Reaktion von Glyoxylat optisch an der abnehmenden Thioesterabsorption [Gl. (1)], und bei Reaktion mit Pyruvat an der Radioaktivität des pNAA [Gl. (6)] gemessen. Um die katalytische Wirkung des Enzyms in beiden Vorgängen vergleichen zu können, muß die spezifische Aktivität des pNAA mehrfach korrigiert werden:

a) Der Isotopenaustausch ist reversibel (Tab. 26), die Synthese des Malats praktisch irreversibel. Mit gleichen Enzymmengen bestimmt, ist daher die „Anfangsgeschwindigkeit" des Isotopenaustausches geringer als die Synthesegeschwindigkeit des Malats.

b) Die spezifische Aktivität des pNAA fällt mit zunehmender Konzentration an Acetyl-CoA (Tab. 19). Da aus analytischen Gründen eine hohe Konzentration gewählt werden muß (im Standardansatz 0,8 μMol; vgl. Methoden), ist die tatsächliche enzymatische Aktivität unter optimalen Bedingungen höher als die mit dem Radioaktivitäts-Test bestimmte.

c) Die spezifische Aktivität wird aus der optisch bestimmten Konzentration des pNAA vor seiner Isolierung und der danach vorhandenen Radio-

aktivität bestimmt. Die bei Aufarbeitung unvermeidlich erfolgenden Substanzverluste betragen 25% (vgl. Methoden): sA(pNAA)/0,75 = sA(pNAA korr).

d) Der kinetische Isotopeneffekt bei Tritierung des Acetyl-CoA beträgt $k_{^1H^+}/k_{^3H^+}$ = 6,7⁵[16]. Die Inkorporation von 1 μAtom³H/μMol entsprechende sA(pNAA korr) wird daher erst bei Enolisierung von 6,7 μMol Acetyl-CoA erreicht.

e) Der thermodynamische Isotopeneffekt [41] vermindert die spezifische Aktivität des Acetyl-CoA bzw. des pNAA gegenüber derjenigen des Wassers. Da sich die Markierung im Wasser auf zwei O—H Bindungen und im Acetyl-CoA auf drei C—H Bindungen verteilt, müßte die spezifische Aktivität des Acetyl-CoA im Gleichgewicht des Isotopenaustausches ohne Masseneffekt sA(Acetyl-CoA) = 3/2 sA(H$_2$O) betragen. Experimentell wurde statt dessen sA(pNAA korr) = 0,94 sA(H$_2$O) gefunden (vgl. Methoden).

Mit diesen Daten kann die spezifische Aktivität des pNAA berechnet werden, die bei Enolisierung von 1 μMol Acetyl-CoA entstehen muß: Im Standardansatz beträgt 3/2 sA(H$_2$O) = 14,4 × 10^4 Imp./min × μMol (vgl. Methoden). Unter Berücksichtigung des thermodynamischen Isotopeneffektes entspricht ihr bei erfolgter Äquilibrierung der drei Methylwasserstoffe sA(pNAA korr) = 0,94 sA(H$_2$O) = 9,0 × 10^4 Imp./min × μMol, und folglich bei Reaktion nur einer C—H Gruppe sA(pNAA korr) = 3,0 × 10^4 Imp./min × μMol. Wegen des kinetischen Isotopeneffektes ($k_{^1H}/k_{^3H}$ = 6,7) erst bei Enolisierung von 6,7 μMol erreicht. Der Enolisierung von 1 μMol Acetyl-CoA entspricht somit die sA(pNAA korr) = 4,5 × 10^3 Imp./min × μMol. Verwendet man diese spezifische Aktivität zur Auswertung der Meßdaten und berücksichtigt man die beiden ersten Fehlerquellen, so erfolgt die Enolisierung des Acetyl-CoA im „bifunktionellen System" so rasch wie die Synthese des Malats (vgl. II, Fig. 6). Dieses Ergebnis wird mit den nachfolgend beschriebenen Kontrollversuchen gestützt, mit denen die Gleichwertigkeit der Methylwasserstoffe des Acetyl-CoA bei Reaktion mit Pyruvat und mit Glyoxylat nachgewiesen und die Zuverlässigkeit der Bestimmung überprüft wurde.

Im Standardansatz zur Enolisierung beträgt das Verhältnis Acetyl-CoA: ^{1}H : ^{3}H $\cong$ 1 : 10^4 : 0,1. Im System mit Pyruvat wird deshalb nur ein geringer Anteil der drei C—H Gruppen durch reversiblen Isotopenaustausch gleichmäßig markiert. Bei Reaktion dieses markierten Acetyl-CoA mit Glyoxylat wird im geschwindigkeitsbestimmenden Schritt der Synthese des Malats nur eine der drei markierten C—H Bindungen irreversibel gelöst. Wegen des kinetischen Isotopeneffektes ($k_{^1H}/k_{^3H}$ = 6,7) werden aber statt 33,3% aller C—^{3}H Bindungen nur 5% davon gelöst,

---

⁵ Dieser Wert wurde mit Citrat-synthase bestimmt. Der genaue Wert für Malat-synthase kann davon abweichen (vgl. [87]).

so daß ein Ditritiomalat entstehen muß, das $95^0/_0$ der gesamten Markierung des Acetyl-CoA enthält.

Der im Gleichgewicht des Isotopenaustausches experimentell bestimmte Wert sA(pNAA korr) = 0,94 sA(H_2O) würde ohne Masseneffekt auf die stereospezifische Reaktion von zwei C—H Bindungen hinweisen, wofür sA(pNAA korr) = sA(H_2O) zu erwarten wäre. Ein Vergleich mit Citrat läßt aber erkennen, daß der stereospezifische Austausch an zwei C—H Bindungen nicht auf die Bildung eines meso-Kohlenstoffatoms [42—44] zurückgehen kann. Im Citrat (I) liegt das meso-Kohlenstoffatom strukturell vor und erfolgen die enzymatischen Umwandlungen daher asymmetrisch [10,45—47]. Im Acetyl-CoA (II) wird

das meso-Kohlenstoffatom erst bei Monotritierung erzeugt. Da aber im Standardansatz weit mehr Protonen als Tritonen enthalten sind, liegt im Gleichgewicht des Isotopenaustausches neben unmarkiertem Acetyl-CoA nur die Molekelart $CH_2{}^3HCOSCoA$

Dieses im „bifunktionellen System" mit Pyruvat entstandene und nur an zwei C—H Positionen markierte Acetyl-CoA würde bei stereospezifischer Reaktion mit Glyoxylat Malat A [sA = 1/2 sA (Acetyl-CoA)], Malat B oder C [sA = 3/4 sA(Acetyl-CoA)] liefern [Gl.(2)]. Wird dagegen im System mit Pyruvat erwartungsgemäß jede der drei C—H Bindungen markiert und bei Reaktion mit Glyoxylat eine davon irreversibel gelöst, so muß wie voranstehend diskutiert, Malat D [sA = 0,95 sA(Acetyl-CoA)] entstehen.

Um den Verlauf der erwarteten Reaktion zu beweisen, wurde im „bifunktionellen System" gebildetes, markiertes Acetyl-CoA isoliert [sA(pNAA korr) = 1×10^5 Imp./min $\times \mu$Mol] und nachfolgend mit Glyoxylat und Malat-synthase in Malat (MTA) umgewandelt (sA = $9,3 \times 10^4$ Imp./min $\times \mu$Mol]. Da Malat MTA nahezu die gesamte Radioaktivität des Acetyl-CoA enthält, wird Malat A als Syntheseprodukt [Gl.(21)] ausgeschlossen und B bzw. C unwahrscheinlich. Deren Vorliegen wurde mit Fumarase überprüft, die an (S)-Malat trans-Abspaltung und -Addition von Wasser katalysiert [48,49] und es folglich in tritiumhaltigem Wasser stereospezifisch markiert [Gl.(22)] oder in wäßriger Lösung die Markierung des (S)-Malats entsprechend löst [50]. Malat B

$$\tag{20}$$

$$\tag{21}$$

$$\tag{22}$$

vor, an deren Bildung jeder der drei Methylwasserstoffe gleichwertig beteiligt sein muß. Der stereospezifische Austausch zweier Methylwasserstoffe ist daher unwahrscheinlich und wäre nur dann möglich, wenn die Methylgruppe nicht frei rotieren würde oder wenn das Enzym nach einem unvorstellbaren Mechanismus eine Auswahl treffen könnte. Die Reaktion ließe sich dann wie folgt formulieren [Gl.(20)].

muß daher in wäßriger Lösung 2/3, Malat C 1/3 und Malat D 1/2 des Tritiumgehaltes an das Lösungsmittel abgeben. Als Kontrollsäure für diesen Versuch wurde markiertes Malat MTM nach Gl.(22) dargestellt: sA = $1,01 \times 10^5$ Imp./min $\times \mu$Mol. Beide Tritiomalate wurden in wäßriger Lösung mit Fumarase behandelt und die Kinetik des Isotopenaustausches [Gl.(22)] bestimmt. Das Ergebnis zeigt (Tab.25), daß MTM die Markierung vollständig ver-

Tabelle 25. *Kinetik des Isotopenaustausches tritierter Malate bei Einwirkung der Fumarase*
Über Reaktionsansätze und Ausführung der Messung, siehe Methoden

Zeit	$\Delta E_{366}/0,2$ ml Eluat	(S)-Malat	Imp. 3 min	Imp./min 11,6 µMol (S)-Malat
Std		µMol		
Ansatz MTA:				
0	0,391	11,72	20150	6650
0,5	0,310	9,30	11740	4880
1	0,308	9,24	10340	4320
2	0,311	9,32	9800	4060
4	0,309	9,27	9860	4120
Ansatz MTM:				
0	0,388	11,62	24960	8320
0,5	0,306	9,16	5230	2170
1	0,301	9,03	1210	540
2	0,305	9,15	340	140
4	0,313	9,40	160	70
Ansatz MTA + MTM:				
0	0,387	11,60	44160	14750
0,5	0,309	9,27	18220	7600
1	0,308	9,24	12170	5060
2	0,316	9,48	9520	3880
4	0,334	10,02	10220	4030

Tabelle 26. *Reversibilität des Isotopenaustausches* [Gl. (7)]
Über Reaktionsansätze und Ausführung der Messung, siehe Methoden

Zeit	$-\Delta E_{405}$	pNAA		Aktivität	
Std	µMol		Imp. 3 min	korr.	Imp./min µMol pNAA (korr.)
0	0,235	0,865	103800	138000	53200
0,25	0,220	0,81	13900	18500	7600
0,50	0,230	0,845	11280	15000	6500
1	0,220	0,81	9360	12450	5100
2	0,220	0,81	8530	11400	4700
3	0,225	0,825	8200	10900	4400
4	0,225	0,825	8300	11000	4460
5	0,225	0,825	8040	10700	4320
7	0,235	0,865	8010	10700	4120

liert, während sie in MTA zur Hälfte erhalten bleibt. Da im Mischversuch MTA + MTM gleiches erfolgt, wird aus enzymatisch gebildetem Tritio-Acetyl-CoA wie erwartet Malat D synthetisiert.

Die Zuverlässigkeit der Meßanordnung zur Bestimmung des Isotopenaustausches mit sA(pNAA korr) und das erwartete gleichartige Verhalten von enzymatisch und chemisch erhaltenem Tritio-Acetyl-CoA gegenüber Pyruvat und Glyoxylat wird mit folgender Meßreihe nachgewiesen: Mit [³H]Acetanhydrid (sA = 1,02 × 10⁵ Imp./min × µMol) wurde [³H]pNAA (sA = 4,9×10⁴ Imp./min × µMol) und [³H]Acetyl-CoA dargestellt, dessen Bestimmung sA(pNAA korr) = 5,2×10⁴ Imp./min ×µMol ergab [sA(Acetanhydrid : pNAA : Acetyl-CoA) Ber. = 2:1:1, Gef. = 2:0,97:1,02]. Aus diesem [³H]Acetyl-CoA mit Glyoxylat und Malat-Synthase entstandenes [³H]-Malat (sA = 4,4×10⁴ Imp./min ×µMol) verlor bei Einwirkung der Fumarase die Hälfte des Tritiumgehaltes (siehe Methoden). Erwartungsgemäß gab das chemisch dargestellte [³H]Acetyl-CoA in Gegenwart von Pyruvat und Malat-synthase seinen Tritiumgehalt rasch an das Lösungsmittel ab (Tab. 26), womit die Reversibilität des Isotopenaustausches gezeigt wird. Die Restaktivität (ca. 8 %), die in diesem Versuch scheinbar im Acetyl-CoA erhalten bleibt, geht auf die chemische Darstellung des [³H]Acetyl-CoA und auf die Spezifität der verwendeten Enzyme zurück: Bei der chemischen Acetylierung des CoA—SH werden auch darin enthaltene Begleitmercaptane verestert. Während Malat-synthase substratspezifisch ist [4], setzt das Hilfsenzym Arylamin-transacetylase auch andere Acetyl-CoA-ähnliche Acetyl-thioester

um [51], weshalb von Malat-synthase nicht angreifbare [³H]Acetyl-thioester als [³H]pNAA erfaßt werden.

Aldolase katalysiert die stereospezifische Ablösung eines Protons vom meso-Kohlenstoffatom der Hydroxymethylgruppe des Dihydroxyacetonphosphats [52—54]. In Gegenwart von 3-Phosphoglycerinaldehyd wird Fructose-1,6-diphosphat synthetisiert und genau dieses Proton des Dihydroxyacetonphosphats an das Lösungsmittel abgegeben. Das gleichartige Verhalten des Wasserstoffes im Teilschritt und in der Gesamtreaktion beweist deren zweistufigen Mechanismus und schließt aus, daß das Carbonyl des Reaktionspartners, 3-Phosphoglycerinaldehyd, an der Carbanionbildung beteiligt ist [52,54]. In diesem Fall wird die C—H Bindung vom Enzym gelöst: Der optisch nachgewiesene Enzym-Substrat-Komplex zwischen Aldolase und Dihydroxyacetonphosphat [55] ist dessen Azomethin mit der ε-Aminogruppe eines enzymgebundenen Lysins [56]. Citrat- und Malat-synthase allein sind zur Enolisierung des Acetyl-CoA ungenügend und benötigen dafür das Carboxylat-Anion des jeweiligen Reaktionspartners als Base. Das natürliche Substrat kann bei Citrat-synthase mit (S)-Malat, bei Malat-synthase mit Pyruvat ersetzt werden. Das gleichartige Verhalten der Methylwasserstoffe des Acetyl-CoA gegen Glyoxylat und Pyruvat stimmt mit Gl. (12) überein.

VI. DAS KATALYSEPRINZIP DER MALAT-SYNTHASE IST DIE „ANNÄHERUNG DER REAKTANTEN"

An der aldolartigen enzymatischen Synthese des Malats sind folgende Schritte beteiligt: Der Methylwasserstoff des Acetyl-CoA wird in Mg⁺⁺-Ion bewirkter Säurekatalyse acid (vgl. I). Kooperativ dazu wird dieser Schritt in basischer Katalyse vom Carboxylat-Anion des Glyoxylats katalysiert (vgl. II). Dabei entstandenes enolisches Acetyl-CoA reagiert sofort und praktisch irreversibel mit dem Aldehydcarbonyl des Glyoxylats (vgl. IV, V) zu (S)-Malyl-CoA, das nachfolgend hydrolysiert (vgl. III). Tatsächlich beweisen die erzielten Ergebnisse nur den aldolartigen Mechanismus der Synthese und nicht wie die Reaktanten am Enzym aktiviert und verwandelt werden. Da aber die Teilnahme des Glyoxylats an der Enolisierung des Acetyl-CoA als gesichert erscheint, und das Aldehydcarbonyl des Glyoxylats sofort mit dem enolischen Acetyl-CoA reagiert, müssen die Reaktanten in der dafür geeigneten sterischen Anordnung vorliegen. Die prinzipielle Aufgabe des Enzyms besteht folglich darin, die Substrate anzunähern, wofür die sich Komplexbildung mit dem Mg^{++}-Ion [57,58], Koordinationszahl = 6, anbietet. Übereinstimmend damit ist die Lewis-Säure auch zur Hydrolyse des Malyl-CoA unentbehrlich (vgl. III) und hört die enzymatische Aktivität im „bifunktionellen System" auf, wenn das Mg^{++}-Ion entfernt wird (Fig. 7). Wir nehmen an, daß es das Thioestercarbonyl des Acetyl-CoA und das Aldehydcarbonyl des Glyoxylats bindet. Bei deren Komplexbildung erfolgt die Synthese des Malats intramolekular und wird als zusätzlicher Gewinn das Aldehydcarbonyl des Glyoxylats polarisiert, die Addition des enolischen Acetyl-CoA erleichtert. Mit den experimentellen Befunden übereinstimmend zeichnet sich folgender Reaktionsverlauf ab [Gl. (23)]:

Malat-synthase: R = H

Die Enzym-Substratverbindung ist ein Chelat, in dem die Carbonylgruppen des Acetyl-CoA und Glyoxylats koordinativ an das Mg-Protein gebunden sind. Das Thioestercarbonyl wird polarisiert und der jetzt acide Methylwasserstoff im langsamsten Schritt der Synthese vom kooperativ arbeitenden Carboxylat-Anion des Glyoxylats als Proton abgelöst. In rascher Folgereaktion reagiert das entstandene enolische Acetyl-CoA mit dem polarisierten und daher reaktionsbereiten Carbonyl des Glyoxylats und schließt die C—C Bindung. Dabei wird das Carbonyl an der S-Seite angegriffen, während bei Reduktion mit DPNH und Lactat-dehydrogenase, wie Johnson et al. und unabhängig davon Weber u. Arigoni bewiesen haben [59,12] die R-Seite den Wasserstoff aufnimmt. Das danach vorliegende (S)-Malyl-CoA ist am Thioestercarbonyl positiviert. Seine aus Glyoxylat stammende Carboxylgruppe liegt wegen der rasch erfolgenden Deprotonierung [60] als Anion vor. Es be-

findet sich in idealer sterischer Position und reagiert daher mit dem Thioestercarbonyl unter Bildung des 2-Hydroxysuccinanhydrids. CoA—SH wird eliminiert, die Komplexbildung mit dem Protein unter Aufnahme einer Molekel Wasser gelöscht, und (S)-Malat an das Lösungsmittel abgegeben. Das Enzym bindet also die Substrate so angeordnet, daß sie in zwangsläufig erfolgender Reaktion intramolekular in die Produkte umgewandelt werden.

Zusammenfassend weisen folgende Befunde das Katalyseprinzip der Malat-synthase als die „Annäherung der Reaktanten" [61] aus:

a) Im Syntheseverlauf [Gl.(1)] entsteht kein Acetyl-Enzym [4].

b) Enolisches Acetyl-CoA wird in kooperativer Katalyse mit Mg^{++}-Ion und dem Carboxylat-Anion des Glyoxylats gebildet.

c) Enolisches Acetyl-CoA entsteht im langsamsten Schritt der Synthesefolge und reagiert sofort und praktisch irreversibel mit dem Aldehydcarbonyl des Glyoxylats.

d) Mit seinem enzymgebundenen intermediären Auftreten übereinstimmend, wird als Substrat eingesetztes (S)-Malyl-CoA von Malat-synthase hydrolysiert, und zwar langsam und mit geringer Affinität zum Enzym.

e) Die chemische Hydrolyse substituierter Succinylmonothioester in neutraler wäßriger Lösung erfolgt über substituiertes Succinanhydrid (vgl. [25,61,62a, 62b]).

f) Das „bifunktionelle System" zur Enolisierung des Acetyl-CoA ist dem bifunktionellen Katalysator von Swain u. Brown [27] in jeder Hinsicht vergleichbar.

VII. NUCLEOPHILE ENZYMATISCHE REAKTIONEN DES ACETYL-COENZYM A

Zur Carboxymethylierung mit Acetyl-CoA können drei Verbindungstypen das Acceptorcarbonyl liefern: CO_2 als Biotin-CO_2, das Thioestercarbonyl des Acetyl-CoA, das Ketocarbonyl des Acetacetyl-CoA und das der α-Ketosäuren. Aus dieser Einteilung ist bereits erkenntlich, daß enolisches Acetyl-CoA als reaktives Agens der Synthesen nicht nach einem einheitlichen Mechanismus entstehen kann. Dementsprechend sind bei Thiolase und HMG-CoA-Synthase Acyl-Enzyme am Katalysevorgang beteiligt [63—65]. Da aber die Synthese des Citrats und des Malats mit Oxalacetat bzw. Glyoxylat als Acceptorcarbonyl in gleichartigen Schritten erfolgt, nehmen wir an, daß alle entsprechenden α-Ketosäuren, wie unter VI angeführt, carboxymethyliert werden. Dabei kann das Metallion [Gl. (23)] in den verwandten Enzymkatalysen von einer konstitutiv im Protein enthaltenen Lewis-Säure ersetzt und kann der Rest R der α-Ketosäure [Gl. (23)] von der „Paßform" des Enzyms bestimmt auch sterisch entgegengesetzt angeordnet sein. Statt der (S)-

Hydroxysäure Citrat [Gl. (23); $R = CH_2CO_2H$] oder Malat ($R = H$) entsteht dann, wie kürzlich nachgewiesen [66,67] die, (R)-Hydroxysäure α-Isopropylmalat [$R = CH(CH_3)_2$] oder Citrat ($R = CH_2CO_2H$).

Ließe sich zeigen, daß das Carboxylat-Anion des N-Carboxybiotins an der Enolisierung der entsprechenden Substrate beteiligt ist, so dürfte auch den enzymatischen Carboxylierungen ein Gl. (23) analoges Reaktionsgeschehen zugrunde liegen.

Die Frage, ob diese Vorstellungen und Gl. (23) gültig sind, wie das Enzym die Substrate sterisch richtig angeordnet bindet, wie es in die Hydrolyse des intermediären β-Hydroxyacyl-CoA eingreift, kann erst mit weiteren Versuchen beantwortet werden. Mit Sicherheit dürfen wir aber die nucleophilen enzymatischen Reaktionen des Acetyl-CoA mit den vorliegenden Ergebnissen als aldolartige Synthesen einteilen. Die Ergebnisse bestätigen daher einen von Cornforth geprägten Satz [68]: *„Nature, it seems, is an organic chemist having some predilection for the aldol condensation."*

METHODEN UND MESSDATEN

SUBSTRATE UND ENZYME

CoA—SH, Fumarase und Lactat-dehydrogenase der Fa. C. F. Boehringer und Soehne, Mannheim; Serumalbumin der Behringwerke, Marburg; [^{14}C]Acetanhydrid, [^{3}H—2C]-Acetanhydrid und H^{3}HO der Fa. Buchler, Braunschweig; Glyoxylat der Fa. C. Roth, Karlsruhe; Toluol und Aceton (p.a.) der Fa. Merck, Darmstadt. Unmarkiertes, [^{14}C]- und [^{3}H]-markiertes Acetyl-CoA (80 mM) wurde durch Acetylierung mit dem entsprechenden Anhydrid bereitet [69], mit Dowex-50 (H$^+$) entsalzt, danach gefriergetrocknet, in Wasser gelöst und mit KOH neutralisiert. Citramalyl-CoA nach [35], (R)-, (R,S)- und (S)-Malyl-CoA nach [8] synthetisiert, Arylamin-transacetylase nach [70], Malat-synthase aus Bäckerhefe gereinigt (für alle Versuche wurde in Mg-Trispuffer^6 gelöstes Enzym verwendet). Angegebene Konzentrationen der Substrate sind enzymatisch bestimmt, wenn nicht anders vermerkt. Zur Entsalzung von Lösungen wurde stets Dowex-50 (H$^+$; 200—400 *mesh*; Vern.-Grad 8) verwendet.

ENZYMATISCHE BESTIMMUNGSMETHODEN

Proteinbestimmungen erfolgten nach [71], wenn nicht anders vermerkt. Enzymaktivitäten sind in internationalen Einheiten ausgedrückt. Arylamin-transacetylase wurde nach [35], Malyl-CoA nach [9], Acetyl-CoA nach [72], Glyoxylat und Pyruvat nach [73] und (S)-Malat nach [74] bestimmt.

Optischer Test zur Synthese des Malats

Die Küvette ($d = 0.5$ cm) enthielt in 1,5 ml 40 μMol Pyrophosphatpuffer^7, pH 8,1; 0,2 μMol Acetyl-CoA; 5 μMol MgCl$_2$; 1 μMol K$_2$Mg-EDTA; 0,5 μMol Glyoxylat und Malatsynthase nach [4]. Start der Reaktion mit Glyoxylat. $\lambda = 232$ mμ (Spektralphotometer Zeiss). Messung gegen eine

---

^{6} Mg-Trispuffer = 1 mM MgCl$_2$ in 5 mM Trispuffer pH 8,0 (vgl. [4]).

^{7} Trispuffer [4] reagiert mit Glyoxylat zu einem Aldehyd-Aminat, vermindert daher die Substratkonzentration und stört die Bestimmung.

Küvette, die ATP statt Acetyl-CoA enthielt. Zur Berechnung einer Einheit aus der Anfangsgeschwindigkeit wurde $\Delta \varepsilon$ (Acetyl-CoA) = 4.5×10^3 M^{-1}cm^{-1} verwendet.

Optischer Test zur Hydrolyse des Malyl-CoA

Die auf 25° temperierte Küvette ($d = 1$ cm) enthielt im Vol. 0,20 ml 20 μMol Trispuffer, pH 8,1; 1 μMol K$_3$Fe(CN)$_6$; 1 μMol MgCl$_2$; Malyl-CoA und Malat-synthase [Gl. (17)]. Start der Reaktion mit dem Enzym. $\lambda = 436$ mμ (Photometer Eppendorf mit Mikrospalt). Messung gegen eine Küvette, die alle Zusätze aber keine Malat-synthase enthielt. $\Delta \varepsilon_{436} = 7.25 \times 10^2$ M^{-1}cm^{-1} wurde mit reinen Ferro- und Ferricyanid-Lösungen bestimmt.

Radioaktivitäts-Test zur Enolisierung des Acetyl-CoA

Standardansatz („monofunktionelles System"). Das während der Inkubation verschlossene Analysenglas enthielt bei 25° im Vol. 0,10 ml 20 μMol Phosphatpuffer, pH 7,0; 1 μMol MgCl$_2$; 0,5 μMol K$_2$Mg-EDTA; 1 mg Serumalbumin; 0,8 μMol Acetyl-CoA; 2 mC H^{3}HO und Malat-synthase. Inkubationsdauer: 15 Std.

Standardansatz („bifunktionelles System"). Wie voranstehend, aber ohne K$_2$Mg-EDTA und Serumalbumin, mit 0,5 μMol MgCl$_2$, 1 μMol Pyruvat und wo zutreffend mit 10 μMol K$_3$Fe(CN)$_6$. Inkubationsdauer: 15 min. Veränderungen beider Standardansätze sind an den Tabellen und Figuren vermerkt.

Bestimmung des Acetyl-CoA als pNAA [Gl. (6); vgl. [6]]; *ferricyanidfreie Ansätze.* Jeder Ansatz wurde nach beendeter Inkubation mit 10 μl 3 M Trichloressigsäure abgestoppt, danach mit 0,20 ml M Phosphatpuffer, pH 7,0 (oder M Trispuffer, pH 8,1) neutralisiert und 10 μl (1 μMol) 0,1 M Na$_2$H$_2$-EDTA, nachfolgend 1,0 ml p-Nitroanilin-Lösung (ca. 2,5mM) zugesetzt. Die Acetylierung des Amins wurde mit 0,02 bis 0,05 ml (5 mE) Arylamin-transacetylase gestartet und der Ansatz zur vollständigen Umsetzung 3 Std (oder über Nacht) im verschlossenen Gefäß bei 25° inkubiert. Denaturiertes Protein wurde danach abzentrifugiert und der Gehalt an p-Nitroanilin des Überstandes gemessen: Je 0,10 ml davon wurden in die 1,40 ml Wasser enthaltende Küvette ($d = 0.5$ cm) pipettiert und die optische Dichte der Lösung bei 405 mμ bestimmt (Photometer Eppendorf). Der Wert der reinen p-Nitroanilin-Lösung wurde in gleicher Weise mit 0,10 ml eines Parallelansatzes, ohne Acetyl-CoA, bestimmt und der Gehalt an p-Nitroacetanilid der Restlösung (Vol. 1,24 bis 1,27 ml) mit $\Delta \varepsilon_{405} = 10.2 \times 10^3$ M^{-1}cm^{-1} aus der Differenz beider Bestimmungen ermittelt. Bei Serienansätzen wurde für alle Bestimmungen eine Pipette benutzt. Um die Verschleppung der Radioaktivität und Bestimmungsfehler zu vermeiden, wurde sie vor Verwendung jeweils in der Lösung von 10 ml p-Nitroanilin-Lösung und 3,3—3,5 ml Wasser gespült.

Metallionenhaltige Ansätze wurden in gleicher Weise, aber nach Zusatz von überschüssigem Na$_2$H$_2$-EDTA bestimmt.

Ferricyanidhaltige Ansätze wurden nach Neutralisation der abgestoppten Ansätze mit 15 μl (15 μMol) M Thioäthanol reduziert und dann wie voranstehend beschrieben behandelt. Da überschüssiges Mercaptan chemisch zu enzymatisch inaktivem Acetyl-thioester [70] austauscht [75,76], wird unter diesen Bedingungen weniger p-Nitroacetanilid gefunden, als Acetyl-CoA vorliegt.

Aufarbeitung. Dem nach Gehaltsbestimmung vorliegenden Reaktionsansatz, Vol. 1,24—1,27 ml, wurden 0,02 ml (2 μMol) 0,1 M alkoholisch-acetonische p-Nitroacetanilid Lösung als Träger zugesetzt. Danach wurde mit 0,5 ml 5 N HCl angesäuert, zur Isolierung des p-Nitroacetanilids mit 1 ml Essigester ausgeschüttelt und zentrifugiert. Nach dem Abpipettieren der organischen Phase in einen 5 ml

Scheidetrichter wurde die Extraktion weitere zweimal mit je 0,5 ml Essigester wiederholt und die Überstände vereinigt. Zur Reinigung wurden sie mit je 0,5 ml wie folgt gewaschen: Zweimal mit N NaOH, viermal mit 5 N HCl, einmal mit Wasser und gesättigter KHCO$_3$-Lösung, dreimal mit Wasser. Die gewaschenen Extrakte wurden am Rotationsverdampfer im Vakuum eingedampft, der Rückstand zweimal in je 5 ml Äthanol, einmal in 5 ml Aceton gelöst und jeweils zur Trockene gebracht, zuletzt in 1 ml Aceton gelöst und mit insgesamt 14 ml Scintillationslösung in das Zählgefäß gespült.

Scintillationslösung. 12 g 2,5-Diphenyloxazol und 100 mg 1,4-Bis-[4-methyl-5-phenyloxazolyl-(2)]-benzol [77] wurden mit Toluol zu 1 Liter gelöst.

Zählgerät. Tri-Carb Liquid Scintillation-Spectrometer, Modell 314 EX, Packard Instrument Company, La Grange, Illinois, U.S.A.

Meßanordnung. Die Radioaktivität der Probe wurde bei 1150 V 3 min gemessen. Die Meßwerte (Imp./3 min) der Tabellen und Figuren sind um den Leerwert (400—600 Imp./3 min; Bestimmung in Parallelansätzen ohne Acetyl-CoA oder ohne Enzym) korrigiert.

MESSDATEN

Daten zu Fig.1

	Bildung des Tritio-Acetyl-CoA				Bestimmung der enzymatischen Aktivität		
Zeit	$-\Delta E_{405}$ [a]	pNAA	Aktivität		Zeit	$\dfrac{-\Delta E_{233} \,[b]}{\text{min}}$	Aktivität
Std		μMol	Imp. 3 min	Imp./min μMol pNAA	Std		Einheiten ml
0	0,165	0,59	0	siehe Fig.1	0	0,022	siehe Fig.1
1	0,180	0,645	7800		2	0,023	
5	0,160	0,575	25110		5	0,024	
7	0,140	0,50	28930		9	0,025	
9	0,135	0,485	31660		25	0,021	
11	0,120	0,43	29150		30	0,023	
24	0,080	0,29	23420				
30	0,070	0,25	24150				

[a] Wie voranstehend beschrieben bestimmt.
[b] Je 10 μl des Reaktionsansatzes wurden bei t (Std) 1:100 verdünnt und je 10 μl der Verdünnung für den optischen Test der Synthese des Malats verwendet (siehe Enzymatische Methoden).

Daten zu Fig.2

	Reihe B				Reihe A		
Salz Nr [a]	$-\Delta E_{405}$	pNAA	Aktivität		$-\Delta E_{405}$	pNAA	Aktivität
		μMol	Imp. 3 min	Imp./min μMol pNAA		μMol	Imp. 3 min	Imp./min μMol pNAA
0	0,165	0,59	1220	siehe Fig.2	0,165	0,59	5720	siehe Fig.2
1	0,170	0,61	390		0,175	0,63	44180	
2	0,170	0,61	1060		0,165	0,59	24760	
3	0,160	0,58	1220		0,170	0,61	1750	
4	0,160	0,58	1170		0,180	0,65	750	
5	0,070	0,25	80		0,145	0,52	70	
6	0,155	0,56	1110		0,170	0,61	900	
7	0,160	0,58	1250		0,155	0,56	23950	
8	0,160	0,58	1130		0,150	0,54	11570	
9	0,165	0,59	1150		0,170	0,61	13330	
10	0,160	0,58	1160		0,155	0,56	27720	
11	0,165	0,59	1120		0,155	0,56	17660	
12	0,160	0,58	1200		0,160	0,58	2100	
13	0,155	0,56	1190		0,170	0,61	2600	
14	0,180	0,65	102860		0,170	0,61	85450	
15	0,155	0,65	89190		0,150	0,54	83540	
16	0,165	0,59	620		0,180	0,65	4920	
17	0,180	0,65	1400		0,170	0,61	26500	
18	0,180	0,65	5900		0,180	0,65	52400	
19	0,165	0,59	3360		0,170	0,61	18690	
20	0,160	0,58	1520		0,180	0,65	14380	
21	0,160	0,58	1250		0,175	0,63	1120	
22	0,160	0,58	1380		0,145	0,52	12800	

[a] Siehe Fig.2; mit Ausnahme von Pyruvat, α-Ketobutyrat und -Valerat (Na[+]) wurden die K[+]-Salze der Säuren eingesetzt.

Daten zu Fig. 3

Der Rohrzucker-Dichtegradient (20–5%) [78], Vol. 4,6 ml, wurde mit 0,15 ml Lösung der Malat-synthase (0,85 mg Protein; 8,3 E), nachfolgend mit ca. 0,3 ml Paraffinöl überschichtet und 22 Std zentrifugiert (Spinco, 38000 U/min, ausschwingender Rotor, ca. 5° Rotortemperatur). Die zentrifugierte Lösung wurde mit je 20 Tropfen (ca. 0,2 ml) in 27 getrennte Fraktionen eluiert und wie folgt bestimmt:

a) 0,05 ml des jeweiligen Eluats wurden mit 0,29 ml Wasser verdünnt, das Protein mit je 0,06 ml 3 M Trichloressigsäure gefällt, zentrifugiert (6 min 4000 U/min) und nach dem Dekantieren des Überstandes in 0,129 ml N NaOH gelöst. Nach 30minutigem Stehen wurden 1,25 ml Biuret-Reagens [79a], nach weiterem 10minutigem Stehen 0,125 ml Folin-Ciocalteu-Reagens (Merck; mit Wasser auf das Eineinhalbfache verdünnt) zugesetzt und die optische Dichte der Lösung 30 min danach bestimmt [79b]: $\lambda =$ 546 mµ, $d = 2$ cm; Photometer Eppendorf. Messung gegen eine Küvette, die alle Zusätze, aber kein Protein enthielt.

b) Die enzymatische Aktivität wurde mit je 5–10 µl des Eluats optisch bei 232 mµ gemessen (siehe enzymatische Methoden).

Daten zu Fig. 5

Zur Bestimmung (siehe enzymatische Methoden) wurden je 0,015 E Malat-synthase eingesetzt. Start der Reaktion mit je 0,03 ml 0,01 M Glyoxylat.

Acetyl-CoA	Pyruvat			
	—	1 mM	2 mM	3 mM
mM		$-10^2 \times \Delta E_{232}$/min		
0,016 (0,015)	7 (6)	5	4	—
0,027 (0,025)	11 (10)	6	5	4
0,054 (0,050)	17 (16)	10	11	7
0,109 (0,10)	24 (22)	14	13	10
0,163 (0,15)	24 (23)	16	15	12

Daten zu Fig. 6

Pyruvat	$-\Delta E_{405}$	pNAA		Aktivität (korr.[a])
		µMol	Imp. 3 min	Imp./min µMol pNAA
0	0,155	0,56	0	siehe
0,005 mM	0,130	0,47	265	Fig. 5
0,05 mM	0,115	0,42	1380	
0,10 mM	0,130	0,47	2800	
0,50 mM	0,110	0,40	9250	
1,0 mM	0,055	0,20	6270	
5,0 mM	0,140	0,51	35000	
10,0 mM	0,145	0,52	39700	
50,0 mM	0,140	0,51	46500	
100,0 mM	0,115	0,42	39400	

[a] Um den Verlust bei Aufarbeitung des pNAA korrigiert (+ 25%; nachfolgend beschrieben).

Daten zu Fig. 7

Zeit 0—30 min. Vier Reaktionsansätze enthielten bei 25° je 20 µMol Phosphatpuffer, pH 7,0; 2 mC H³HO und 1 µMol MgCl₂. Das Vol. betrug zur Vorinkubation 0,08—0,09 ml, zur Enolisierung des Acetyl-CoA 0,10 ml. $t = 0$ min: Der Ansatz enthält zusätzlich 0,8 µMol Acetyl-CoA; die Enolisierung wird mit 2 E Malat-synthase gestartet. $t = 10$ min: Der Ansatz enthält zusätzlich 2 E Malat-synthase und wird nach 10minutiger Vorinkubation mit 0,8 µMol Acetyl-CoA gestartet. $t = 20$ und 30 min: Der Ansatz enthält zusätzlich 2 E Malat-synthase, 1 µMol Pyruvat und wird nach 20 bzw. 30minutiger Vorinkubation mit je 0,8 µMol Acetyl-CoA gestartet.

Zeit 40—150 min. Der Inkubationsansatz (25°) enthielt bei $t = 0$ min 20 µMol Phosphatpuffer, pH 7,0; 20 mC H³HO, 10 µMol Pyruvat und 20 E Malat-synthase. Start der Mg⁺⁺-Entfernung vom Enzym bei 32 min mit 20 µMol Na₂H₂-EDTA. Inkubationsvol. 0,80 ml. Je 0,08 ml des Ansatzes wurden bei t min (Fig. 5) in 20 µl (0,8 µMol) Acetyl-CoA-Lösung pipettiert. Bei $t = 135$ min wurden zu 0,24 ml des Ansatzes 0,03 ml M MgCl₂-Lösung pipettiert, damit vorinkubiert, und bei 140 und 150 min je 0,09 ml davon in je 10 µl Acetyl-CoA-Lösung (0,8 µMol) pipettiert. Die kompletten Enolisierungsansätze beider Versuchsreihen wurden nach 15minutiger Inkubation bei 25° mit je 10 µl 3 M Trichloressigsäure abgestoppt und p-Nitroacetanilid wie voranstehend beschrieben, gebildet und bestimmt.

Zeit	Änderungen im Reaktionsansatz	$-\Delta E_{405}$	pNAA	Aktivität	
min			µMol	Imp. 3 min	Imp./min µMol pNAA
0	+ 1 µMol MgCl₂	0,191	0,696	100	siehe Fig. 4
10	+ 1 µMol MgCl₂	0,206	0,750	60	
20	+ 1 µMol MgCl₂ + 1 µMol Pyruvat	0,192	0,696	68400	
30	+ 1 µMol MgCl₂ + 1 µMol Pyruvat	0,192	0,696	66170	
40	− MgCl₂ + 1 µMol Pyruvat + 2 µMol Na₂H₂-EDTA	0,204	0,744	44810	
50	− MgCl₂ + 1 µMol Pyruvat + 2 µMol Na₂H₂-EDTA	0,194	0,705	24320	
60	− MgCl₂ + 1 µMol Pyruvat + 2 µMol Na₂H₂-EDTA	0,181	0,695	10800	
70	− MgCl₂ + 1 µMol Pyruvat + 2 µMol Na₂H₂-EDTA	0,187	0,684	4960	
100	− MgCl₂ + 1 µMol Pyruvat + 2 µMol Na₂H₂-EDTA	0,195	0,709	810	
130	− MgCl₂ + 1 µMol Pyruvat + 2µMol Na₂H₂-EDTA	0,187	0,684	330	
140	+ 1 µMol Pyruvat + 2 µMol Na₂H₂-EDTA + 10 µMol MgCl₂	0,185	0,673	69960	
150	+ 1 µMol Pyruvat + 2 µMol Na₂H₂-EDTA + 10 µMol MgCl₂	0,189	0,690	70820	

ISOLIERUNG DES [³H]pNAA AUS ANSÄTZEN DES
„BIFUNKTIONELLEN SYSTEMS" UND VERHALTEN DER
SPEZIFISCHEN AKTIVITÄT BEI UMKRISTALLISATION

Vereinigte Meßproben des „bifunktionellen Systems" (vgl. II), Vol. 250 ml (ca. 35 μMol pNAA), 620 Imp./min × ml wurden mit 54 mg (300 μMol) pNAA als Träger versetzt, bis zur Lösung zum Rückfluß erhitzt und danach im Vakuum zur Trockene eingeengt. Der Rückstand wurde in 5 ml absolutem Toluol digeriert, der unlösliche Anteil (Rohprodukt) isoliert, wiederholt aus abs. Toluol umkristallisiert und die spezifische Aktivität des jeweiligen Kristallisats gemessen.

Substanz	Schmelz-punkt	Einwaage	Aktivität	
	°C	mg	Imp. 3min	Imp./min [a] μMol
Rohprodukt	—	1,50	9870	395
1. Umkristallisation	209—211	1,18	8480	430
2. Umkristallisation	209—211	1,07	7980	450
3. Umkristallisation	209—211	1,42	10090	430
4. Umkristallisation	209—211	1,01	7800	460

[a] Berechnet: 460 Imp./min × μMol.

VERSUCHE ZUR ALDOLSPALTUNG DES CITRAMALYL-COA

Die Küvette (d = 1 cm) enthielt im Vol. 1,0 ml 100 μMol Trispuffer (in Parallelversuchen 40 μMol Pyrophosphatpuffer) pH 8,1; 0,05—0,15 μMol Citramalyl-CoA (Bestimmung nach [35]); 0,2 μMol DPNH; 18 Einheiten Lactatdehydrogenase und 6—12 Einheiten Malat-synthase (damit wurde die Reaktion gestartet). λ = 366 mμ (Photometer Eppendorf) (vgl. II; Kontrollen).

Ergebnis. Keine Änderung der optischen Dichte; nachträglich zugesetztes Pyruvat wurde rasch reduziert. Statt Malyl-CoA im optischen Test zur enzymatischen Hydrolyse [Gl.(17); siehe enzymatische Methoden] eingesetztes Citramalyl-CoA wird von Malat-synthase (10 Einheiten) auch nicht hydrolysiert.

DIE SPEZIFISCHE AKTIVITÄT DES pNAA ALS
MASS DER ENOLISIERUNGSGESCHWINDIGKEIT

Tritiumgehalt des verwendeten H³HO

0,10 ml H³HO (200 mC/ml) wurden mit Wasser zu 100 ml verdünnt und die verdünnte Lösung (200 μC/ml) zur Umkristallisation von 180 mg (1 mMol) p-Nitroacetanilid verwendet. Nach dem Trocknen über P_2O_5 wurde die Selbstabsorption des tritierten p-Nitroacetanilids bestimmt und die Werte auf unendliche Verdünnung extrapoliert. Gef. 520 Imp./min × μMol; Ber.: 480 Imp./min × μMol (mit 12,0 % Zählausbeute). Zur Berechnung spezif. Aktivitäten wurden 200 mC/ml verwendet.

Zählausbeute

Bestimmung mit Hexadecan-³H-Standard: 1,98 μC/g. Reparaturen am Zählgerät veränderten die Zählausbeute während der Untersuchungen. Gef.: 475 (10,8 %), 500 (11,4 %) und 530 Imp./min × μMol (12,0 %). Die zugehörige, mit der jeweiligen Zählausbeute berechnete sA(H_2O) ist an den Tabellen und Figuren angegeben.

Substanzverlust bei Aufarbeitung des pNAA

20 Ansätze mit je 2 μMol [³H]pNAA, spezifische Aktivität = 5,15 × 10⁴ Imp./min × μMol wurden im Standardansatz (ohne H³HO und Acetyl-CoA) gelöst und, wie voranstehend beschrieben, aufgearbeitet. Der aufarbeitungsfrei gemessenen Lösung als Bezugswert, wurde der Substanzverlust bei Aufarbeitung im Mittel zu 25 % bestimmt. Die um diesen Verlust korrigierte spezifische Aktivität ist als sA(pNAA korr) angegeben.

Überprüfung des sek. Isotopeneffektes

12 Reaktionsansätze, Vol. je 1,20 ml, enthielten bei 25° je 100 μMol Trispuffer, pH 8,1; 1,0 ml p-Nitroanilin (ca. 2,5 mM); 2 μMol Na_2H_2-EDTA; 5 mE Arylamin-transacetylase und 1,08 μMol [³H]Acetyl-CoA (ca. 0,2 μC/μMol). Start der Reaktion mit dem Enzym. An je einem Ansatz wurde die optische Dichte mit je 0,10 ml Lösung bei t = 0, 15, 30, 60, 150, 180, 240, 300, 420 min und nach 24 Std bestimmt, die Restlösung, Vol. 1,10 ml, sofort mit 0,5 ml 5 N HCl abgestoppt und pNAA wie voranstehend beschrieben bestimmt und gemessen.

Ergebnis. Die Acetylierung war nach 2 Std beendet; die spezifische Aktivität des pNAA ergab gegen μMol pNAA aufgetragen eine Gerade: Die Bildung des [³H]pNAA aus [³H]Acetyl-CoA erfolgt ohne meßbaren sekundären Isotopeneffekt [41,80].

Bestimmung des Isotopenaustausch-Gleichgewichtes

a) sA(pNAA korr) im Gleichgewicht des Isotopenaustausches

Die Meßdaten der Tab. 15 (Kurve I, II) wurden um den Substanzverlust bei Aufarbeitung (25 %) korrigiert, in Fig. 11 als Kurve I—, II— korr eingetragen. Ein Vergleich mit den als C gekennzeichneten, experimentell bestimmten Grenzen des Substanzverlustes bei Aufarbeitung (25 ± 5 %) läßt erkennen, daß ihre Berücksichtigung an der Lage des Gleichgewichtes, sA(pNAA korr) = 0,94 sA(H_2O), nur wenig ändert.

Auch bei Berücksichtigung des Fehlers der optischen Bestimmung von pNAA ändert sich die Gleichgewichtslage kaum: Wie voranstehend beschrieben, werden dem Acetylierungsansatz, Vol. 1,34—1,37 ml, je 0,10 ml zur Bestimmung entnommen und wird der Gehalt an pNAA aus der gemessenen Differenz der optischen Dichte zu einem Kontrollansatz (ohne Acetyl-CoA) berechnet. Diese Bestimmung ist bei hoher Differenz der optischen Dichte, wie im vorliegenden Versuch zutreffend, praktisch fehlerfrei. Nimmt man an, daß jede Bestimmung (Leerwert und Meßwert) einen Ablesefehler $E_{405} = ± 0,010$ enthält, der Bestimmungsfehler also $ΔE_{405} ± 0,020$ beträgt, so ändert sich sA(pNAA korr) nur um den Betrag B in Fig. 11.

Dieser Fehler wird um so größer, je weniger pNAA vorliegt. Mit 0,10 μMol Acetyl-CoA im Ansatz zur Enolisierung beträgt die berechnete Änderung der optischen Dichte bei vollständiger Umsetzung zu pNAA nur $ΔE_{405} = —0,025$ (Bestimmung wie voranstehend beschrieben). Um den Fehler praktisch auszuschließen, wurden im Standardansatz zur Enolisierung 0,8 μMol Acetyl-CoA eingesetzt.

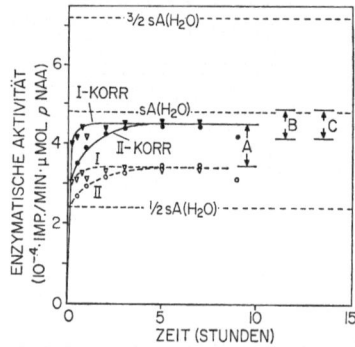

Fig. 11. Fehler bei der Bestimmung des pNAA-³H. Meßdaten der Tab. 15. Beachte: Die Ansätze der Tab. 15 enthalten nur den halben H³HO-Gehalt (1 mC) des Standardansatzes. Dessen sA(H_2O) = 9,6 × 10⁴ Imp./min × μMol

b) Kontrollen dazu mit [³H]Acetanhydrid

[³H]Acetanhydrid. 25 mC (50 µMol) [³H]Acetanhydrid wurden mit 5,9 ml Träger-Acetanhydrid zu ca. 0,4 µC/µMol verdünnt und destilliert: Kp = 134—136° 10 µl (106 µMol) des Destillates wurden mit Aceton zu 1,06 mM verdünnt, die Radioaktivität der Lösung bestimmt: sA = 1,02×10⁵ Imp./min × µMol.

[³H]p-Nitroacetanilid. 138,1 mg (1 mMol) p-Nitroanilin wurden in 5 ml absoluten Benzol mit 0,15 ml (1,6 mMol) [³H]Acetanhydrid acetyliert [81], das Reaktionsprodukt aus absolutem Toluol und Essigester-Petroläther mehrfach umkristallisiert: F = 209—211°; sA = 4,9×10⁴ Imp./min × µMol.

[³H]Acetyl-CoA. 79 mg CoA—SH wurden in 1 ml Wasser gelöst und mit 12 µl (127 µMol) [³H]Acetanhydrid acetyliert. Die Lösung wurde mit Dowex-50 angesäuert, zur Entfernung der markierten Essigsäure 3 Std kontinuierlich mit Äther extrahiert (NaOH in der Vorlage), durch Passieren einer Dowex-50-Säule (0,8×5 cm) vollständig entsalzt und das Eluat gefriergetrocknet. Der Rückstand wurde in 1 ml Wasser gelöst und mit verdünntem KOH neutralisiert: 54 mM; sA(pNAA korr) = 5,2×10⁴ Imp./min × µMol.

Acetyl-CoA-Lösung	—ΔE_{405}	pNAA	Aktivität	
µl		µMol	Imp. 3 min	Imp./min µMol pNAA (korr.)
10	0,130	0,447	54010	53800
20	0,265	0,912	105770	51400
20	0,270	0,93	104670	50000
30	0,385	1,325	157120	52500

Reversibilität der enzymatischen Tritierung des Acetyl-CoA. Der Reaktionsansatz, Vol. 1,0 ml, enthielt bei 25° 200 µMol Phosphatpuffer, pH 7,0; 10 mg Serumalbumin; 10 µMol MgCl₂; 10 µMol K₃Fe(CN)₆; 50 µMol Pyruvat; 10,8 µMol [³H]Acetyl-CoA und 43 E Malat-synthase. Start der Reaktion mit dem Enzym. Je 0,10 ml des Ansatzes wurden bei t Std mit je 10 µl 3 M Trichloressigsäure abgestoppt und Acetyl-CoA als p-Nitroacetanilid bestimmt. Ergebnis: Tab. 26.

Tritiomalat aus [³H]Acetyl-CoA. Der Reaktionsansatz, Vol. 1,0 ml, enthielt bei 25° 100 µMol KHCO₃—K₂CO₃-Puffer, pH 8,7; 10 µMol MgCl₂; 14 µMol Glyoxylat; 10,8 µMol [³H]Acetyl-CoA und 10,5 E Malat-synthase. Start der Reaktion mit Glyoxylat. Nach 60 minutiger Inkubation wurde das Enzym durch Ansäuern mit ca. 50 mg Dowex-50 denaturiert, Malat durch Behandlung mit Kohle und Dowex-50 gereinigt (siehe Isotopenaustausch tritierter Malate) und nach dem Gefriertrocknen des Eluats in 1 ml Wasser gelöst: sA = 4,4×10⁴ Imp./min × µMol.

Isotopenaustausch an Tritiomalat bei Einwirkung der Fumarase. Der Reaktionsansatz, Vol. 0,30 ml, enthielt bei 25° 1,4 µMol (1×10⁴ Imp./min) Tritiomalat, 40 µMol (S)-Malat als Träger und 50 µMol Trispuffer, pH 7,8. Start der Reaktion mit 10 µl (7 E) Fumarase. Je 0,05 ml (10⁴ Imp./min, 1,02 µMol Malat bei t = 0 Std) des Reaktionsansatzes wurden bei t Std abgestoppt und aufgearbeitet (siehe Isotopenaustausch tritierter Malate). Ergebnis:

Zeit	(S)-Malat	Aktivität	
Std	µMol	Imp. 3 min	Imp./min 10,2 µMol Malat
0	10,0	27310	9300
0,5	8,66	18470	7280
1,0	8,60	15300	6050
2,5	8,60	13180	5200
4,0	8,95	12570	4780

32*

Zusammenfassung. Meßwerte (Fig. 11) und Kontrollen dazu liefern ein übereinstimmendes Ergebnis: Im Gleichgewicht des Isotopenaustausches [Gl. (7)] beträgt sA(pNAA korr) = 0,94×sA(H₂O). Die Gleichgewichtskonstante beträgt daher

$$K = \frac{[CH_2{}^3HCOSCoA][H_2O]}{[CH_3COSCoA][H^3HO]} = \frac{0,94}{3} \times 2 = 0,63.$$

AUS ACETYL-COA UND GLYOXYLAT ENTSTEHT IM TRITIUMHALTIGEN LÖSUNGSMITTEL UNMARKIERTES MALAT

Drei Reaktionsansätze, Vol. 0,10 ml, enthielten bei 25° 0,0, 0,8 bzw. 1,6 µMol Acetyl-CoA; 0,0, 2,0 bzw. 4,0 µMol Glyoxylat; je 20 µMol Trispuffer, pH 8,1; 2 mC H³HO; 0,5 µMol MgCl₂ und 4,2 E Malat-synthase. Start der Reaktion mit Glyoxylat. sA(H₂O) = 9,6×10⁴ Imp./min × µMol. Nach 15 minutiger Inkubation wurde durch 3 minutiges Erhitzen (100°) abgestoppt, vom denaturierten Protein zentrifugiert, Malat nach Zusatz von je 10 µMol (S)-Malat als Trägersubstanz (Endvol. 0,30 ml) mit je 30 µl M Bleiacetat als Bleisalz gefällt und viermal mit je 0,3 ml Wasser gewaschen. Nach Suspension in je 0,5 ml Wasser wurde durch Ansäuern mit je ca. 30 mg Dowex-50 (H⁺) gelöst, vorhandenes Nucleotid an Kohle (0,5×3 cm) adsorbiert und das bei 260 mµ nicht absorbierende Eluat an Dowex-50 (H⁺; 0,8×5 cm) vollständig entsalzt: Vol. je 25 ml; 7,95, 7,0 bzw. 8,15 µMol (S)-Malat (Bestimmung mit 0,4 ml; siehe Enzymatische Methoden). Die Lösungen wurden im Vakuum eingedampft, mit 5 ml HCl, nachfolgend zweimal mit 5 ml Wasser, zuletzt mit Äthanol und Aceton gleichartig behandelt, die Rückstände in 1 ml Aceton gelöst und mit insgesamt je 14 ml Scintillationslösung in das Zählgefäß gespült (vgl. IV).

Ergebnis: Keine vom Kontrollwert (ohne Acetyl-CoA; 400 Imp./3 min) verschiedene Aktivität im synthetisierten Malat.

Kontrolle dazu: Zwei Parallelansätze mit 0,8 und 1,6 µMol Acetyl-CoA wurden wie voranstehend, aber ohne H³HO, mit Glyoxylat und Malat-synthase umgesetzt und nach der 15 minütlichen Inkubation mit je 10 µl 3 M Trichloressigsäure abgestoppt. Die Bestimmung des Acetyl-CoA als p-Nitroacetanilid gab ΔE_{405} = 0, entspr. vollständiger Umsetzung des Acetyl-CoA.

In Versuchen, die Reversibilität der Synthese des Malats [Gl. (1)] nachzuweisen (vgl. IV), wurde Malat nach erfolgter Inkubation, wie voranstehend beschrieben, nach Zusatz von je 10 µMol (S)-Malat als Trägersubstanz aufgearbeitet und bestimmt. Ergebnis: Kein Nachweis der Tritiuminkorporation in (S)-Malat.

ISOLIERUNG VON ENZYMATISCH GEBILDETEM TRITIO-ACETYL-COA

Der Reaktionsansatz, Vol. 2,0 ml, enthielt bei 25° 200 µMol Phosphatpuffer, pH 7,0; 200 µMol K₃Fe(CN)₆; 20 µMol Pyruvat; 10 µMol MgCl₂; 60 mC H³HO; 24,8 µMol Acetyl-CoA und 41 E Malat-synthase. Start der Reaktion mit dem Enzym. Inkubationsdauer 120 min sA(H₂O) = 1,44×10⁵ Imp./min × µMol.

Zur Kontrolle des Reaktionsverlaufes wurden bei t min je 0,10 ml entnommen und pNAA bestimmt: Nach 10, 30, 60 und 120 min betrug pNAA korr. 20200, 56400, 78500 und 98000 Imp./min × µMol.

Nach Entnahme der letzten Probe wurde der Ansatz, Vol. 1,60 ml, mit 0,2 ml 5 N HCl abgestoppt, Acetyl-CoA nach Zusatz von 2 ml gesättigter Ammoniumsulfat-Lösung durch Phenolextraktion [82] isoliert, gefriergetrocknet, danach in insgesamt 1 ml Wasser gelöst und mit verd. KOH neutralisiert: 13 mM; sA = 1,0×10⁵ Imp./min × µMol (Bestimmung als pNAA korr).

ISOLIERUNG DER TRITIERTEN MALATE MTM UND MTA

(2S,3R)-[³H−3C]Monotritiomalat: (MTM)

20 µMol (S)-Malat wurden im Vol. 0,10 ml mit 30 µMol Trispuffer, pH 7,8; 6 mC H³HC und 14 E Fumarase 20 Std bei 25° gehalten. sA(H₂O) = 2,86 × 10⁵ Imp./min × µMol. Das Enzym wurde danach durch dreiminutiges Erhitzen (100°) denaturiert, die Lösung mit 0,20 ml Wasser verdünnt, Malat mit 0,10 ml M Bleiacetat als Bleisalz gefällt und abzentrifugiert. Der Niederschlag wurde durch dreimaliges Suspendieren in je 0,3 ml Wasser gewaschen, jeweils zentrifugiert und die Überstände verworfen. Der Rückstand wurde in 1 ml Wasser suspendiert, mit ca. 50 mg Dowex-50 angesäuert und gelöst, und nach Zusatz von 1 ml gesättigter Ammonsulfat-Lösung 8 Std kontinuierlich mit Äther extrahiert. Das Lösungsmittel wurde im Vakuum verdampft, der Rückstand zur Entfernung austauschbaren Tritiums in 10 ml Wasser gelöst und gefriergetrocknet⁸, nachfolgend mit 5 ml Wasser gleichartig behandelt und zuletzt in insgesamt 1 ml Wasser aufgenommen: 5,9 mM; sA = 1,01 × 10⁵ Imp./min × µMol (zur Bestimmung wurden den Meßproben je 5 µMol (S)-Äpfelsäure als Träger zugesetzt, die Lösung im Vakuum zur Trockene gebracht, der Rückstand in 1 ml Aceton gelöst und mit 14 ml Scintillations-Lösung in das Zählgefäß gespült).

(2S)-[³H₂−3C]Ditritiomalat: (MTA)

Der Reaktionsansatz, Vol. 0,89 ml, enthielt bei 25° 40 µMol Trispuffer, pH 8,1; 10 µMol [³H]Acetyl-CoA (sA = 1,0 × 10⁵ Imp./min × µMol, Isolierung voranstehend beschrieben, und 10 E Malat-synthase. Die Synthese des Malats wurde mit 14 µl (14 µMol) M Glyoxylat gestartet und 10 min danach weitere 4 µMol Glyoxylat zugesetzt. Nach insgesamt 90minutiger Inkubation wurde das Enzym durch 3 minutiges Erhitzen (100°) denaturiert, die Lösung zur Abtrennung von CoA−SH und Protein durch eine Kohlesäule (0,5 × 5 cm) geschickt und mit Wasser nachgewaschen. Das bei 260 mµ (d = 1 cm) nicht absorbierende Eluat, Vol. 10 ml, wurde an Dowex-50 (0,8 × 7 cm) entsalzt und mit Wasser nachgewaschen: Vol. 25 ml; 5 µMol (S)-Malat (Bestimmung mit 0,40 ml). Das Eluat wurde gefriergetrocknet und der Rückstand in insgesamt 1 ml Wasser gelöst: 4,8 mM; sA = 9,3 × 10⁴ Imp./min × µMol.

KINETIK DES ISOTOPENAUSTAUSCHES
DER TRITIERTEN MALATE MTA UND MTM
BEI EINWIRKUNG DER FUMARASE

Der Standardansatz, Vol. 0,30 ml, enthielt bei 25° 50 µMol Trispuffer, pH 7,8; 69 µMol (S)-Malat als Träger und das zu untersuchende Tritiomalat wie folgt:
Ansatz MTA: mit 0,44 µMol (4,1 × 10⁴ Imp./min) Malat MTA.
Ansatz MTM: mit 0,50 µMol (5,05 × 10⁴ Imp./min) Malat MTM.
Ansatz MTA + MTM: mit 0,44 µMol MTA und 0,50 µMol MTM (4,1 × 10⁴ + 5,05 × 10⁴ Imp./min).
Je 0,05 ml Lösung der drei Reaktionsansätze [je 11,6µMol (S)-Malat] wurden zur Bestimmung bei t = 0 Std des Isotopenaustausches entnommen, der Isotopenaustausch danach mit je 10 µl (14 E) Fumarase gestartet und je 0,05 ml des Ansatzes bei t Std durch dreiminutiges Erhitzen (100°) abgestoppt. Malat wurde wie folgt isoliert und bestimmt: Die abgestoppten Proben wurden mit je 0,2 ml Wasser verdünnt, an Dowex-50 (0,8 × 5 cm) entsalzt und zu 20 ml eluiert. Der Malatgehalt wurde mit 0,20 ml des jeweiligen Eluats enzymatisch bestimmt und die Restlösung, Vol. 19,8 ml,

---

⁸ Bei gewöhnlicher Aufarbeitung entstand Malomalsäure [83], die enzymatisch (Malat-dehydrogenase, Fumarase) inaktiv ist und daher falsche Ergebnisse liefert.

im Vakuum zur Trockene gebracht. Der Rückstand wurde zweimal in je 5 ml Äthanol, einmal in 5 ml Aceton gelöst und im Vakuum jeweils zur Trockene gebracht, zuletzt in 1 ml Aceton gelöst und mit insgesamt 14 ml Scintillations-Lösung in das Zählgefäß gespült. Ergebnis: Tab. 25.

Der Deutschen Forschungsgemeinschaft danken wir für großzügige Unterstützung, der Fa. C. F. Boehringer & Soehne, Mannheim, für die Überlassung von Coenzym A. Herrn Dr. U. Gehring und Fräulein C. Riepertinger danken wir für die Messungen mit der Ultrazentrifuge, Herrn M. Krella für die Anfertigung der Illustrationen und Herrn Dr. G. Lust für Korrekturen an der englischen Kurzfassung.

LITERATUR

1. Wong, D. T. O., und Ajl, S. J., J. Am. Chem. Soc. 78 (1956) 3230.
2. Cahn, R. S., Ingold, C. K., und Prelog, V., Experientia, 12 (1956) 81.
3. Kornberg, H. L., Angew. Chem. 77 (1965) 601.
4. Dixon, G. H., Kornberg, H. L., und Lund, P., Biochem. Biophys. Acta, 41 (1960) 217.
5. Jaenicke, L., und Lynen, F., In The Enzymes (edited by P. D. Boyer, H. Lardy, and K. Myrbäck), Academic Press, New York and London 1960, Vol. 3, p. 3.
6. Eggerer, H., Biochem. Z. 343 (1965) 111.
7. Eggerer, H., und Remberger, U., Biochem. Z. 337 (1963) 202.
8. Eggerer, H., und Grünewälder, C., Liebigs Ann. Chem. 677 (1964) 200.
9. Eggerer, H., Remberger, U., und Grünewälder, C., Biochem. Z. 339 (1964) 436.
10. Martius, C., und Schorre, G., Liebigs Ann. Chem. 570 (1950) 140, 143.
11. Hanson, K. H., and Rose, I. A., Proc. natl. Acad. Sci. U.S. 50 (1963) 981.
12. Weber, H., Promotionsarbeit, Eidgenössische Techn. Hochschule Zürich (Prom. Nr. 3591), 1965.
13. Topper, Y. J., in The Enzymes (edited by P. D. Boyer, H. Lardy and K. Myrbäck), Academic Press, New York and London 1961, Vol. 5, p. 429.
14. Rose, I. A., In Brookhaven Symposia in Biology, Associated Universities, Brookhaven National Laboratory, Upton (N. Y.) 1962, No. 15, p. 293.
15. Markus, A., und Vennesland, B., J. Biol. Chem. 233 (1958) 726.
16. Bové, J., Martin, R. O., Ingraham, L. L., und Stumpf, P. K., J. Biol. Chem. 234 (1959) 999.
17. Chronyn, M. W., Chang, M. P., and Wall, R. A., J. Am. Chem. Soc. 77 (1955) 3031.
18. Bruice, T. C., In Organic Sulfur Compounds (edited by N. Kharasch), Pergamon Press, New York 1961, Vol. 1, p. 423.
19. Ingold, C. K., Structure and Mechanism in Organic Chemistry, Cornell University Press, Ithaca (N. Y.) 1953.
20. Cornforth, J. W., J. Lipid Res. 1 (1959) 1.
21. Lienhard, G. E., und Jencks, W. P., J. Am. Chem. Soc. 87 (1965) 3863.
22. Lynen, F., Federation Proc. 12 (1953) 683.
23. Wessely, L., Promotionsarbeit, Universität München 1955.
24. Liberek, B., Bull. Acad. Polon. Sci., Ser. Sci. Chim. 11 (1963) 677.
25. Bender, M., Chem. Rev. 60 (1960) 53.
26. Schwyzer, R., und Hürlimann, C., Helv. Chim. Acta, 37 (1954) 155.
27. Swain, C. G., und Brown, F. J., J. Am. Chem. Soc. 74 (1952) 2538.

28. Losada, M. L., Trebst, A. V., Ogata, S., und Arnon, D. I., *Nature*, 186 (1960) 753.
29. Gray, C. T., und Kornberg, H. L., *Biochem. Biophys. Acta*, 42 (1960) 371.
30. Losada, M., Cánovas, J. L., und Ruitz-Amil, M., *Biochem. Z.* 340 (1964) 60.
31. Monod, J., Changeux, J. P., und Jacob, F., *J. Mol. Biol.* 6 (1963) 306.
32. Koshland, D. E., *Proc. natl. Acad. Sci. U. S.* 44 (1958) 98.
33. Koshland, D. E., In *The Enzymes* (edited by P. D. Boyer, H. Lardy and K. Myrbäck), Academic Press, New York and London 1959, Vol. 1, p. 332.
34. Frost, A. A., und Pearson, R. G., *Kinetics and Mechanism*, John Wiley and Sons, New York and London 1962.
35. Eggerer, H., und Grumm, I., *Biochem. Z.* 342 (1965) 40.
36. Fischer, E., *Ber. Deut. Chem. Ges.* 27 (1894) 2985.
37. Lineweaver, H., und Burk, D., *J. Am. Chem. Soc.* 56 (1934) 658.
38. Bartlett, P. D., In *Perspectives in Organic Chemistry* (edited by A. Todd), Interscience, New York and London 1956, p. 9.
39. Zum Beispiel: Bell, R. P., Gelles, E., und Möller, E., *Proc. Roy. Soc. (London) Ser. A*, 198 (1959) 308.
40. Bücher, T., Luh, W., und Pette, D., In *Handbuch der physiologisch- und pathologisch-chemischen Analyse* (Hoppe-Seyler/Thierfelder), Springer-Verlag Berlin, Göttingen, Heidelberg 1964, Bd VI A, S. 292.
41. Bell, R. P., *The Proton in Chemistry*, Methuen and Co., London 1959.
42. Schwartz, P., und Carter, H. E., *Proc. natl. Acad. Sci. U. S.* 40 (1954) 499.
43. Levy, H. R., Talalay, P., und Vennesland, B., In *Progress in Stereochemistry* (edited by P. B. D. de la Mare and W. Klyne), Butterworths, London 1962, Vol. 3, p. 299.
44. Mislow, K., *Introduction to Stereochemistry*, W. A. Benjamin, New York 1966.
45. Srere, P. A., *Biochim. Biophys. Acta*, 77 (1963) 693.
46. Gillespie, D. C., und Gunsalus, I. C., *Bacteriol. Proc.* 53 (1963) 80.
47. Wheat, R. W., und Ajl, S. J., *J. Biol. Chem.* 217 (1955) 909.
48. Anet, F. A. L., *J. Am. Chem. Soc.* 82 (1960) 994.
49. Gawron, O., Glaid, A. J., und Fondy, T. P., *J. Am. Chem. Soc.* 83 (1961) 3634.
50. Alberty, R. A., In *The Enzymes* (edited by P. D. Boyer, H. Lardy und K. Myrbäck), Academic Press, New York and London 1961, Vol. 5, p. 551.
51. Lynen, F., und Reinwein, D., unveröffentlichte Versuche.
52. Rose, I. A., und Rieder, S. V., *J. Am. Chem. Soc.* 77 (1955) 5764.
53. Rieder, S. V., und Rose, I. A., *Federation Proc.* 15 (1956) 337.
54. Bloom, B., und Y. J. Topper, *Science*, 124 (1956) 982.
55. Topper, Y. J., Mehler, A. H., und Bloom, B., *Science*, 126 (1957) 1287.
56. Horecker, B. L., Rowley, P. T., Grazl, E., Cheng, T., und Tchola, O., *Biochem. Z.* 338 (1963) 36.
57. Eichhorn, G. L., In *The Chemistry of the Coordination Compounds* (edited by C. Bailar and D. H. Busch), Reinhold Publishing Co., New York 1956, p. 698.
58. Smith, E. L., *Proc. natl. Acad. Sci. U. S.* 35 (1949) 80.
59. Johnson, C. K., Gabe, E. J., Taylor, M. R., und Rose, I. A., *J. Am. Chem. Soc.* 87 (1965) 1802.
60. Eigen, M., *Angew. Chem.* 75 (1963) 489; [I. E. 3 (1964) 18].

61. Jencks, W. P., *Ann. Rev. Biochem.* 32 (1963) 639.
62a. Bruice, T. C., In *Brookhaven Symposia in Biology*, Associated Universities, Inc., Brookhaven National Laboratory, Upton, N. Y. 1962, Number 15, p. 52.
62b. Buckel, W., und Eggerer, H., *Abstr., Herbsttagung Ges. Physiol. Chemie*, Marburg/Lahn 1966, S. 18.
63. Lynen, F., und Gehring, U., *Abstracts of the Fed. Europ. Biochem. Soc.* 1964, p. 3.
64. Gehring, U., Promotionsarbeit, Universität München 1964.
65. Stewart, P. R., und Rudney, H., *J. Biol. Chem.* 241 (1966) 1222.
66. Thomas, U., Kalyanpur, M. G., und Stevens, C. M., *Biochemistry*, 5 (1966) 2513.
67. Gottschalk, G., und Barker, H. A., *Biochemistry*, 5 (1966) 1125.
68. Cornforth, J. W., In *Perspectives in Organic Chemistry* (edited by A. Todd) Interscience, New York and London 1956, p. 371.
69. Simon, H. J., und Shemin, D., *J. Am. Chem. Soc.* 75 (1953) 2530.
70. Tabor, H., Mehler, H. A., und Stadtman, E. R., *J. Biol. Chem.* 204 (1953) 127.
71. Warburg, O., und Christian, W., *Biochem. Z.* 310 (1941) 384.
72. Buckel, W., und Eggerer, H., *Biochem. Z.* 343 (1965) 29.
73. Bücher, T., Czok, R., Lamprecht, W., und Latzko, E., In *Methoden der enzymatischen Analyse* (herausgegeben von H. U. Bergmeyer), Verlag Chemie, Weinheim/Bergstraße 1962, S. 253.
74. Hohorst, J., In *Methoden der enzymatischen Analyse* (herausgegeben von H. U. Bergmeyer), Verlag Chemie, Weinheim/Bergstraße 1962, S. 328.
75. Stadtman, E. R., *J. Biol. Chem.* 196 (1952) 535.
76. Wieland, T., und Bokelmann, E., *Angew. Chem.* 64 (1952) 49.
77. Ott, D. G., In *Liquid Scintillation Counting* (edited by C. G. Bell and N. F. Hayes) Pergamon Press, New York 1958, p. 101.
78. Martin, N. O., und Ames, B. N., *J. Biol. Chem.* 236 (1961) 1372.
79a. Layne, E., In *Methods in Enzymology* (edited by S. P. Colowick and N. O. Kaplan), Academic Press, New York 1957, Vol. 3, p. 447.
79b. Numa, S., persönliche Mitteilung.
80. Zollinger, H., *Angew. Chem.* 70 (1958) 204.
81. Kaufmann, A., *Ber. Deut. Chem. Ges.* 42 (1909) 3480.
82. Schweizer, E., Promotionsarbeit, Universität München 1963.
83. Walden, P., *Ber. Deut. Chem. Ges.* 32 (1899) 2707, 2710.
84. Grunert, R., und Phillips, P., *Arch. Biochem. Biophys.* 30 (1951) 217.
85. Lowry, O. H., Rosebrough, N. J., Farr, A. L., und Randall, R. J., *J. Biol. Chem.* 193 (1951) 265.
86. Lynen, F., *J. Cellular Comp. Physiol.* 54 (1959) Suppl. 1, S. 33.
87. Simon, H., und Palm, D., *Angew. Chem.* 78 (1966) 993.

H. Eggerer und A. Klette
Chemisches Laboratorium, Institut für Biochemie der Universität
8 München 2, Karlstraße 23, Deutschland

European J. Biochem. 1 (1967) 476—481

The Importance of SH-Groups for Enzymic Activity

7. The Amino Acid Sequence around the Essential SH-Group of Pig Heart Lactate Dehydrogenase, Isoenzyme I.

J. J. Holbrook, G. Pfleiderer, K. Mella, M. Volz, W. Leskowac, and R. Jeckel

Institut für Biochemie im Institut für Organische Chemie der J. W. Goethe-Universität, Frankfurt am Main

(Received March 20, 1967)

1. The amino acid sequence of a dodecapeptide containing the essential SH-group of pig heart L-lactate dehydrogenase, isoenzyme I, was determined as Val-Ile-Gly-Ser-Gly-Cys-Asn-Leu-Asp-Ser-Ala-Arg. This sequence shows three differences with that of an SH-peptide from chicken LDH-I.

2. It is shown that the possible codons for the amino acids occupying equivalent positions in the heptapeptides around the essential SH-groups of glyceraldehyde 3-phosphate-, yeast alcohol-, liver alcohol- and lactate dehydrogenases can be related to one another by a chain of point mutations.

The NAD-dependent dehydrogenases which have been rigorously studied have been found to be composed of a series of identical subunits of molecular weights from 35,000 to 50,000, each of which appears to carry a single coenzyme-binding site[1]. The catalytic activities of GAPDH, Y-ADH, L-ADH, LDH and of glutamate dehydrogenase are destroyed by the modification of one amino acid residue per subunit[2—6]. Such a residue may be termed essential. The isolation of carboxymethylcysteine from acid hydrolysates of GAPDH and ADH which had been specifically labelled with [^{14}C]iodoacetate was direct chemical demonstration that these enzymes contained essential cysteine residues[4,10]. By allowing N-(N'-acetyl-4-[^{35}S]sulphamoylphenyl)-maleimide to react with pig heart LDH-I Holbrook and Pfleiderer[5] were able to show that each LDH subunit contains one essential SH-group and were able to isolate and determine the composition of a unique dodecapeptide containing the essential SH-group. Gold and Segal[7] and Fondy et al.[8] also isolated tryptic peptides containing modified SH-groups from LDH which had been extracted from a number of species, and presented indirect evidence that the SH-groups concerned were essential in character. The amino acid compositions of the peptides were recognisably similar to, although not in all cases

identical with, that of the essential-SH-containing peptide from pig LDH-I[9]. This was complementary evidence that the peptides isolated by Fondy et al. contained the essential SH-groups of the various enzymes. Harris and his coworkers[2,10,11] had previously shown that sequences of 12 amino acids around the essential SH-groups of GAPDH obtained from 9 species are always identical. As soon as work was commenced on the amino acid sequences of the essential peptide of pig LDH-I it was clear that the sequence was not the same as that of an SH-peptide from chicken LDH-I and special care has thus been taken in establishing the differences. The final results which define the sequence of the peptide containing the essential SH-group of pig heart LDH-I are the subject of this paper.

EXPERIMENTAL PROCEDURE

Materials

Pig heart LDH, [^{35}S]ASPM and most of the other chemicals used in this work were obtained from sources already noted[9]. Aminopeptidase, isolated from pig kidney by a commercial modification of the method of Wachsmuth et al.[12] was a gift from Röhm und Haas GmbH, Darmstadt and was found to have a turn-over of 4.71 μmole leucine p-nitroanilide min⁻¹ mg protein⁻¹ at 22° in 0.02 M phosphate buffer, pH 7.2. The preparation was freed from amino acids by gel-filtration through Sephadex G-25. The α-protease was kindly prepared by G. Reinhard from Crotalus atrox venom (Calbiochem A.G., Lucerne) according to Pfleiderer and Kraus[13] and was found to have an activity of 25,000 Kunitz units/ml[14]. Subtilopeptidase A was donated under the name of Alcalase by Novo Industri A/S, Copenhagen N, Denmark. Dowex 50-X2, Dowex 1-X2

Non-standard abbreviations. N-(N'-acetyl-4-sulphamoylphenyl)maleimide, ASPM; phenylthiohydantoin, PTH; lactate dehydrogenase, LDH; yeast alcohol dehydrogenase, Y-ADH; horse liver alcohol dehydrogenase, L-ADH; glyceraldehyde-3-phosphate dehydrogenase, GAPDH.

Enzymes. Alcohol dehydrogenase (EC 1.1.1.1); glyceraldehyde-3-phosphate dehydrogenase (EC 1.2.1.12); L-glut. amate dehydrogenase (EC 1.4.1.3); pig kidney aminopeptidase (EC 3.4.1. ?); α-protease (EC 3.4.4. ?); L-lactate dehydrogenase, isoenzyme I (EC 1.1.1.27); subtilopeptidase A (EC 3.4.4.16).

and the PTH-amino acids were purchased from Serva Entwicklungslabor, Heidelberg. Aminex-MS resin, blend Q 15, was obtained from Bio-Rad-Laboratories GmbH, Munich. Kieselgel GF_{254} according to Stahl was obtained from Merck AG., Darmstadt.

Methods

The techniques, apparatus, and solvents used for the thin layer chromatography and electrophoresis of amino acids and peptides have been described [9]. Unless otherwise stated, the supporting medium was always a 0.325 mm layer of Kieselgel-starch. Peptides were recovered from the thin layer plates by packing the material into a small column plugged with glass wool, and allowing a few drops of 0.5% ammonia solution to percolate through the bed. If the plate had been sprayed with ninhydrin, this was first eluted with dry acetone.

For amino acid analysis, peptides were hydrolysed at 110° for 48 h in 6 N HCl in evacuated, sealed tubes. The residue after removal of the HCl was analysed for amino acids on the Auto-Analyser (Technicon GmbH, Frankfurt/M.) essentially by the method of Hamilton [15]. The paper strip modification [16] of the method of Edman [17] was employed for the stepwise degradation of peptides. The PTH-derivatives were compared with authentic samples by chromatography on Kieselgel GF_{254} in either $CHCl_3-CH_3OH$ (90:10, v/v) or $CHCl_3-HCOOH$ (100:5, v/v) [18]. PTH-glycine was the only derivative which turned dark orange after prolonged standing or on exposure to NH_3 vapour. The radioactivity eluted from chromatographic columns was continuously recorded by means of flow-through system (NE 5504, Nuclear Enterprises Ltd., Edinburgh, Scotland). The detection efficiency was about 4% for ^{35}S present as solution in either water or the base-acetate solvents used in the ion-exchange chromatography. The eluates from columns were examined for ninhydrin-positive substances by the method of Hirs et al. [19].

Modified Purification of ASPM-T_5. The soluble tryptic peptides (700 mg, prepared from ASPM-LDH as described in ref. [9])were dissolved in 0.2 N pyridine and were adjusted to pH 2.5 with 6 N HCl. The solution was applied to a column of Dowex 50-X2 (85 × 2.25 cm) which had previously been equilibrated with 0.2 N pyridine-acetic acid, pH 3.1 at 40°. The peptides were washed into the column with 100 ml of the same buffer and were eluted with a pH gradient prepared by a linear mixing device containing 2.2 l of the pH 3.1 buffer and 2.2 l of 2N pyridine-acetic acid, pH 5. The radioactivity emerged between pH 3.77 and pH 3.86. The residue after lyophilysation of the radioactive fraction was rechromatographed on a 1 × 100 cm column of Dowex 1 under the conditions described by Schroeder and Robberson [20]. The radioactivity emerged as two barely separated peaks between pH 3.0 and pH 2.7. The material in the peaks appeared homogeneous under the conditions of the standard "fingerprint" [5]. Traces of organic bases were removed from the product obtained after lyophilysation by filtering a solution of the peptide through a 1 × 30 cm column of Sephadex G-25 in water. The modified method yielded 8.35 µmole peptide, that is about 48% of theory. The peptide was designated ASPM-T_5.

Hydrolysis of ASPM-T_5 with α-protease. ASPM-T_5 (2 µmole) was incubated at 25° for 5 h in 0.35 ml of solution which contained 2500 Kunitz units of α-proteolytic activity, and which was $2.5 × 10^{-4}$ M with respect to $CaCl_2$ and 0.065 M in Tris-HCl buffer, pH 7.6. The whole of the incubation mixture was then separated as a 15 cm streak on Kieselgel S by electrophoresis at pH 6.5 at 50 volts/cm for 2 h. Unchanged ASPM-T_5 was still present and accounted for 15% of the radioactivity on the plate. A second radioactive band (85% of the radioactivity), more acidic than ASPM-T_5 was also detected (peptide α-P_1). Spraying guide-strips with t-butylhypochlorite/o-tolidine revealed a third peptide which was neutral and non-radioactive (peptide α-P_2). Peptides α-P_1 and α-P_2 were eluted and the eluates lyophilysed.

Hydrolysis of ASPM-T_5 with Subtilopeptidase. ASPM-T_5 (2 µmole) and subtilopeptidase (0.2 mg) were incubated at 25° in 0.1 N NH_4HCO_3 for 5 h. The lyophilysate was dissolved in 0.2 N pyridine-acetic acid buffer, pH 3.1, and was added to a 0.9 × 15 cm column of Aminex-MS resin which had previously been equilibrated with the same buffer at 40°. The column was developed with a pH-gradient drawn from a linear mixing device containing 250 ml each of pH 3.1 buffer and 2 N pyridine-acetic acid, pH 5 [21]. Ninhydrin-positive material was eluted at pH 3.16 (fraction S_1), pH 3.18 (fraction S_2), pH 3.23 (fraction S_3), and at pH 4.32 (fraction S_4). The material recovered after lyophilysing fraction S_3 was not homogeneous and was further purified by micropreparative electrophoresis in 5% (v/v) HCOOH into an acidic, radioactive peptide ($S_{3.1}$) and a neutral, non-radioactive peptide ($S_{3.2}$).

Aminopeptidase Hydrolysis of ASPM-T_5. ASPM-T_5 (0.6 µmole) was incubated with aminopeptidase (0.1 mg) at 25° in 0.02 M phosphate buffer, pH 7.2 (0.46 ml). At 20, 40, 80, and 240 min samples corresponding to 0.065 µmole of peptide were pipetted into 0.1 N HCl (1 ml) containing 0.1 µmole of norleucine. The samples were stored frozen at −20° until analysed for amino acids.

Hydrazinolysis of a Peptide Containing only Neutral Amino Acids. Peptide S_1 (0.06 µmole) was dissolved in 0.2 ml of anhydrous hydrazine and was heated at 105° in an evacuated, sealed tube for 5 h according to Akabori et al. [22]. The lyophilysate was separated in the first dimension by electrophoresis at pH 6.5 and in the second dimension by chromatography in solvent III. The only neutral, ninhydrin-positive substance detected had an R_F-value the same as that of serine and different to those of glycine, isoleucine and valine.

RESULTS

Edman Degradation of ASPM-T_5. Two different preparations of ASPM-T_5 have been degraded by the Edman method. The order of liberation of PTH-amino acid residues was in both cases the same and indicated that the N-terminal sequence was Val-Ile-Gly-Ser-. A trace of PTH-glycine was also detected after step 4. However, an incomplete reaction often leads to an amino acid being carried over to the following stages. For this reason, although PTH-glycine was detected after the 5th degradation, no conclusion as to the nature of the 5th amino acid was possible. There was no trace of PTH-serine after the 3rd degradation. The amino acid analysis of the residue after 4 stages of degradation was $Ala_{0.9}$, $Asx_{2.21}$, $Gly_{1.11}$, $Leu_{0.9}$, $Ser_{0.88}$, Arg (not determined but assumed as C-terminal),
ASPM
Cys (presence inferred because the residue was still radioactive).

Aminopeptidase Digestion of ASPM-T_5. The N-terminal sequence of ASPM-T_5 can also be deduced from the rate of liberation of amino acids by the relatively non-specific pig kidney aminopeptidase. The results (Table 1) clearly show the liberation of

Table 1. *Kinetics of the aminopeptidase hydrolysis of ASPM-T_5* ASPM-T_5 (0.6 μmole), aminopeptidase (0.11 mg) incubated at 25° in 0.02 M phosphate buffer, pH 7.2. At the times shown aliquots equivalent to 65 nmole peptide were analysed for amino acids

Amino acids	Incubation time (min)			
	20	40	80	240
	nmole amino acid			
Valine	58.5	65.1	57.7	62.9
Isoleucine	8.1	29.8	46.5	70.5
Glycine	5.2	27.1	39.6	82.0
Serine	4.0	18.3	33.9	59.7

valine before isoleucine, isoleucine before glycine, and glycine before serine. It is possible that the high glycine value at 240 min was an indication that glycine was the 5th amino acid. The relative resistance of the peptide bond between amino acid residues 5 and 6 to aminopeptidase hydrolysis was also observed when peptide α-P_1 (which presumably has the
ASPM
sequence Val-Ile-Gly-Ser-Gly-Cys-Asn) was digested with the enzyme. There was a rapid liberation of valine, isoleucine, glycine and serine (detected by electrophoresis in 5% (v/v) HCOOH) whereupon the reaction apparently became very slow and a radio-

active component which coloured yellow with nin-hydrin accumulated. This component yielded aspartic acid, glycine, and radioactive products after acid hydrolysis.

Peptide α-P_2. Amino acid analysis: $Asx_{1.26}$, $Ala_{0.97}$, $Arg_{1.0}$, $Leu_{0.95}$, $Ser_{0.98}$. Total digestion of α-P_2 with aminopeptidase gave products which were identified by electrophoresis as alanine, aspartic acid, arginine, leucine and serine. The neutral nature of the peptide at pH 6.5 was also in accordance with the presence of equal amounts of arginine and aspartic acid residues. Edman degradation: step 1, PTH-leucine; step 2, PTH-aspartic acid. The partial structure is thus: Leu-Asp (Ala, Arg, Ser).

Peptide S_1. Amino acid analysis: $Gly_{1.2}$, $Ile_{1.0}$, $Ser_{0.84}$, and $Val_{0.94}$. Hydrazinolysis yielded serine. The partial structure is thus (Gly, Ile, Val)-Ser.

Peptide S_2. Amino acid analysis: $Asx_{1.85}$, $Gly_{1.03}$,
ASPM
$Leu_{1.10}$, $Cys_{1.02}$ (calculated from radioactivity). Edman degradation: step 1, PTH-Gly; at step 2 the radioactivity was eluted from the paper strip. The
ASPM
partial structure is thus Gly-Cys-(Asx₂, Leu).

Peptide $S_{3.2}$. Amino acid analysis: $Asx_{0.97}$. $Leu_{1.03}$. After aminopeptidase digestion, asparagine and leucine were chromatographically identified. The peptide was neutral at pH 6.5. The partial structure is thus (Asn, Leu).

Peptide S_4. Amino acid analysis: $Ala_{0.96}$, $Arg_{1.04}$. The peptide was basic at pH 6.5. Partial structure (Ala, Arg). As was previously reported [9] carboxypeptidase B liberated only arginine from ASPM-T_5.

These results are summarised in Table 2. They define a unique amino acid sequence for peptide ASPM-T_5. It should be noted that the sequence of residues 3 and 4 is confirmed in 3 independent experiments *viz.* Edman degradation of ASPM-T_5, aminopeptidase digestion of ASPM-T_5, and hydrazinolysis of peptide S_1. The identification of aspartic acid as residue 9 depends upon the neutral character of peptide α-P_2 and the identification of aspartic acid after aminopeptidase digestion. That residue 7 is asparagine depends upon the charge of peptide $S_{3.2}$ and the detection of asparagine after aminopeptidase digestion. In a previous publication some of us reported that ASPM-T_5 prepared by another method did not contain asparagine [9]. Although it is possible that the method of preparation resulted in deamidation, it is now realised that the ochre colour given by ninhydrin and asparagine on thin-layer plates makes it impossible to detect this amino acid under conditions where equivalent amounts of other amino acids are only just revealed.

DISCUSSION

In the following discussion, amino acid residues lying on the carboxyl side of the essential cysteines

Table 2. *Deduction of the amino acid sequence of peptide ASPM-T₅*

The α-P₂ peptide was isolated from an α-protease digest of ASPM-T₅. The peptides S₁₋₄ were isolated from a subtilisin digest of ASPM-T₅. Only when the presence or absence of amide has been established for the peptide under consideration is either Asp or Asn employed instead of the non-commital Asx

Fragment	Partial structure	
Carboxypeptidase B digestion of ASPM-T₅		-Arg
Aminopeptidase digestion of ASPM-T₅	Val-Ile-Gly-Ser-	
	ASPM	
Edman degradation of ASPM-T₅	Val-Ile-Gly-Ser-(Gly,Cys, Asx, Leu,Asx, Ser,Ala,Arg)	
Peptide α-P₂	Leu-Asp(Ser,Ala,Arg)	
Peptide S₁	(Val, Ile, Gly)Ser	
	ASPM	
Peptide S₂	Gly-Cys(Asx,Leu,Asx)	
Peptide S₃.₂	(Asn,Leu)	
Peptide S₄		(Ala,Arg)
	ASPM	
Peptide ASPM-T₅	Val-Ile-Gly-Ser-Gly-Cys- Asn- Leu- Asp-Ser-Ala-Arg	

Table 3. *Comparison of the amino acid sequences around the essential SH-groups of pig and chicken LDH-I*

pig LDH-I	Val-Ile-Gly-Ser-Gly-CYS-Asn-Leu-Asp-Ser-Ala-Arg
chicken LDH-I [9]	Val-Ile-Ser-Gly-Gly-CYS-Asn-Leu-Asp-Thr-Ala-Arg
Number	$-5 -4 -3 -2 -1 \quad 0 \quad +1 +2 +3 +4 +5 +6$

are numbered $+1$, $+2$, etc. and those on the amino side -1, -2, etc.

The sequences of tryptic peptides containing the essential SH-groups of pig and chicken LDH-I will be compared first (see Table 3). There are three exchanges visible. The exchange at position $+4$

$$\left(\text{threonine} \leftrightarrow \text{serine}, \ AC_A^C \leftrightarrow AG_C^U \text{ or } UC_A^C \right)$$

and the exchanges at positions -2 and -3

$$\left(\text{glycine} \leftrightarrow \text{serine}, \ GG_A^C \leftrightarrow AG_C^U \text{ or } UC_A^C \right)$$

can all be accounted for by an alteration of a single base in the messenger RNA codon. The codons shown in brackets above are those proposed by Khorana [23]. Although it has been reported that pig LDH-V [9], chicken LDH-V, beef LDH-I, and frog LDH-V [8] all contain a run of 12 amino acids with the same overall amino acid *composition* as that of the peptide containing the essential SH-group of pig LDH-I, it is clear from Table 3 that they do not necessarily all have the same *sequence*. The amino acid sequences around the essential SH-groups in NAD-dehydrogenases which catalyse the oxidation of ethanol [4], glyceraldehyde 3-phosphate [10] and lactic acid [8] are now known, and are compared in Table 4. In comparing such sequences it is usual to emphasize

that amino acids occupying equivalent positions relative to the reference amino acid residue (in this case cysteine) are functionally related (*e. g.* asparagine and aspartic acid at position $+3$). Thus Harris [4] noted that the exchanges between L-ADH and Y-ADH in the region $+3$ to -3 were between chemically related amino acids. Fondy *et al.* [8] observed that "the functional relationship among the amino acids in the sequences from lactate dehydrogenases and the alcohol dehydrogenases was quite striking" and further that "between the lactate dehydrogenases on the one hand and the triose phosphate dehydrogenases on the other, there is not a readily apparent functional relationship by direct comparison of labelled cysteine residues." It is clear that there is no functional relationship between leucine (GAPDH, $+5$) and histidine (ADH, $+5$) or between serine (L-ADH, $+2$) and leucine (LDH, $+2$). Thus while comparisons based on chemical relationships at equivalent positions are likely to be of value in deducing the chemistry of the catalytic process, the very fact that a single point mutation can cause the exchange of chemically and functionally dissimilar amino acids (*e.g.* CCU $\leftrightarrow$ ACU, proline $\leftrightarrow$ threonine) tends to over-emphasise differences in the polypeptide chains so that possible evolutionary trends may be overlooked. The codons which are thought to code for the amino acids found in the same position relative to the essential cysteine residues are shown in Table 5. The triplet code is up to

Table 4. *A comparison of the RNA base sequences which are believed to code for the amino acids around the essential cysteine residues of three NAD-dehydrogenases*

	−7	−6	−5	−4	−3	−2	−1	0	+1	+2	+3	+4	+5	+6	+7	+8	+9	+10	+11
Glyceraldehyde 3-phosphate dehydrogenase (9 species) [10,11]	Val GU(U/C/A/G) / Lys AA(A/G)	Met AUG / Val GU(U/C/A/G)	Ile AU(U/C)	Ser UC(U/C/A/G)/AG(U/C)	Asn AA(U/C)	Ala GC(U/C/A/G)	Ser UC(U/C/A/G)/AG(U/C)	CYS UG(U/C)	Thr AC(U/C/A/G)	Thr AC(U/C/A/G)	Asn AA(U/C)	Cys UG(U/C)	Leu UUG / CU(U/C/G)	Ala GC(U/C/A/G)	Pro CC(U/C)	Val GU(U/C/A/G)	Ala GC(U/C/G)	Lys AA(A/G)	
Alcohol dehydrogenase — Horse liver [4]		Met AUG	Val GU(U/C/A/G)	Ala GC(U/C/A/G)	Thr AC(U/C/A/G)	Gly GG(U/C/A/G)	Ile AU(U/C)	CYS UG(U/C)	Arg CG(U/C/A/G)/AGA	Ser UC(U/C/A/G)/AG(U/C)	Asp GA(U/C)	Asp GA(U/C)	His CA(U/C)	Val GU(U/C/A/G)	Thr AC(U/C/A/G)	Ser UC(U/C/A/G)/AG(U/C)	Gly GG(U/C/A/G)	Asp GA(U/C)	
Alcohol dehydrogenase — Yeast [4]				Tyr UA(U/C)	Ser UC(U/C/A/G)/AG(U/C)	Gly GG(U/C/A/G)	Val GU(U/C/A/G)	CYS UG(U/C)	His CA(U/C)	Thr AC(U/C/A/G)	Asp GA(U/C)	Leu UUG / CU(U/C/G)	His CA(U/C)	Ala GC(U/C/A/G)	Trp UGG	His CA(U/C)	Gly GG(U/C/A/G)		
Lactate dehydrogenase — Chicken-I [8]					Ser UC(U/C/A/G)/AG(U/C)	Gly GG(U/C/A/G)						Thr AC(U/C/A/G)							
Lactate dehydrogenase — Pig-I (this paper)			Val GU(U/C/A/G)	Ile AU(U/C)	Gly GG(U/C/A/G)	Ser UC(U/C/A/G)/AG(U/C)	Gly GG(U/C/A/G)	CYS UG(U/C)	Asn AA(U/C)	Leu UUG / CU(U/C/G)	Asp GA(U/C)	Ser UC(U/C/A/G)/AG(U/C)	Ala GC(U/C/A/G)	Arg CG(U/C/A/G)/AGA					
Chain of point mutations possible?	−	+	+	−	+	+	+	+	+	+	+	−	−	−	−	−	+	−	

6-fold degenerate and so it is not possible to deduce precisely the triplet which coded for each amino acid. The comparison does, however, show that functionally dissimilar amino acids, such as the leucine—histidine pair mentioned above, can be related by a point mutation (in this case CUC ↔ CAC). The last line in Table 4 indicates the possibility of relating all the amino acids in that column by a chain of point mutations in the codons. For example, a chain of point mutations for position +2 would be UUG (leucine) ↔ UCG (serine) ↔ ACG (threonine). No such chain can be constructed for position +5: UUG

$$\text{or } CU^{C}_{G} \text{ (leucine), } CA^{U}_{C} \text{ (histidine), and } GC^{C}_{A} \text{ (alanine).}$$

At position −4 it would only be possible to construct such a chain if the serine of GAPDH was coded by

$$\text{both } UC^{C}_{A} \text{ and } AG^{U}_{C} \text{ at the same time. It is thus}$$

observed that the codons for the amino acids which occupy an equivalent position in the heptapeptide around the essential cysteine residues of NAD-dehydrogenases of three different substrate specificities can be related by chains of point mutations. At greater distances from the essential cysteine residues such chains may be constructed less frequently. In contrast to a comparison based on a similarity in functions of amino acids, where it was maintained that there is a similarity between GAPDH and ADH's but not apparently between GAPDH and LDH, a comparison based on the possibility that two amino acids in the central heptapeptide region may be related by point mutations, indicates that there is at least as great a similarity between GAPDH and LDH as there is between GAPDH and L-ADH.

In addition to the possible genetic similarity, there are also more general structural similarities between these three dehydrogenases, such as their each being composed of subunits of nearly the same size (36,000) and the fact that each subunit seems to contain one coenzyme binding site [1].

It should be mentioned, that in spite of any possible evolutionary similarity between the NAD-dehydrogenases, there are great differences in the character of the individual essential SH-groups. Those of GAPDH are super-reactive and are alkylated by iodoacetate at assay concentrations and are acylated by 1,3-diphosphoglycerate [24] and acetyl phosphate [2]. The essential SH-groups of LDH are chemically unreactive and, in the native protein, do not react with iodoacetate at all. There are even differences between the reactivities of the essential SH-groups of the two extreme isoenzymes of LDH towards maleimides [9]. There is no evidence that an acyl-enzyme is an intermediate in the LDH reaction. Lowe [25] has recently noted structural similarities in the regions around the active centre SH-groups of the plant proteases and the active centre OH-group of an animal protease and also in the regions around the active centre histidines in plant and animal proteases of different substrate specificities. Lowe was unable to decide whether these similarities were due to a convergent or divergent evolution. Neither is a decision on this point possible with the NAD-dehydrogenases. It is possible that the elucidation of the function, if any, of the essential SH-groups of LDH and ADH will be of use in this respect.

We thank Novo Industri A/S, Copenhagen, for the gift of subtilopeptidase A. Our thanks are due to G. Reinhard for preparing the α-protease. This work was supported by grants from the Deutsche Forschungsgemeinschaft.

REFERENCES

1. Pfleiderer, G., and Auricchio, F., *Biochem. Biophys. Res. Commun.* 16 (1964) 53.
2. Harris, J. I., Meriweather, B. P., and Park, J. H., *Nature*, 197 (1963) 154.
3. Li, T. K., and Vallee, B. L., *Biochemistry*, 3 (1964) 869.
4. Harris, J. I., *Nature*, 203 (1964) 30.
5. Holbrook, J. J., and Pfleiderer, G., *Biochem. Z.* 342 (1965) 111.
6. Pfleiderer, G., Holbrook, J. J., Nowicki, L., and Jeckel, R., *Biochem. Z.* 346 (1966) 297.
7. Gold, A. H., and Segal, H. L., *Biochemistry*, 4 (1965) 1506.
8. Fondy, T. P., Everse, J., Driscoll, G. A., Castillo, F., Stolzenbach, F. E., and Kaplan, N. O., *J. Biol. Chem.* 240 (1965) 4219.
9. Holbrook, J. J., Pfleiderer, G., Schnetger, J., and Diemair, S., *Biochem. Z.* 344 (1966) 1.
10. Perham, R. N., and Harris, J. I., *J. Mol. Biol.* 7 (1963) 316.
11. Allison, W. S., and Harris, J. I., *Abstracts 2nd FEBS Meeting*, (Vienna) 1965, p. 140.
12. Wachsmuth, E. D., Fritze, I., and Pfleiderer, G., *Biochemistry*, 5 (1966) 169.
13. Pfleiderer, G., and Krauss, A., *Biochem. Z.* 342 (1965) 85.
14. Kunitz, M., *J. Gen. Physiol.* 30 (1946) 291.
15. Hamilton, P. B., *Anal. Chem.* 35 (1963) 2055.
16. Schroeder, W. A., Shelton, J. R., Shelton, J. B., Cormick, J., and Jones, R. T., *Biochemistry*, 2 (1963) 992.
17. Edman, P., *Acta Chem. Scand.* 4 (1953) 283 and 288.
18. Brenner, M., Niederwieser, A., and Pataki, G. in *Dünnschicht-Chromatographie* (herausgegeben von E. Stahl) Springer Verlag, Berlin, Göttingen, Heidelberg 1962, p. 403.
19. Hirs, C. H. W., Stein, W. H., and Moore, S., *J. Biol. Chem.* 219 (1956) 623.
20. Schroeder, W. A., and Robberson, B., *Anal. Chem.* 37 (1965) 1585.
21. Jones, R. T., *Cold Spring Harbor Symp. Quant. Biol.* 29 (1964) 297.
22. Akabori, S., Ohno, K., and Narita, K., *Bull. Chem. Soc. Japan*, 25 (1952) 214.
23. Khorana, H. G., *Federation Proc.*, 24 (1965) 1473.
24. Krimsky, I., and Racker, E., *Science*, 122 (1955) 319.
25. Lowe, G., *Nature*, 212 (1966) 1265.

J. J. Holbrook, G. Pfleiderer, K. Mella, M. Volz, W. Leskowac, and R. Jeckel
Institut für Biochemie im Institut für Organische Chemie der J. W. Goethe-Universität, 6 Frankfurt am Main, Robert-Mayer-Straße 7—9, Germany

European J. Biochem. 1 (1967) 482—486

A Quantitative Isotope Method for Regulation Studies
of Aromatic Amino Acid Synthesis under Growth Conditions

R. K. Thauer, K. Jungermann, and K. Decker

Biochemisches Institut, Albert-Ludwigs-Universität Freiburg

(Received April 10, 1967)

A non-oxidative pentose phosphate pathway starting from 3,4-labeled hexoses leads in growing organisms to the formation of ribose with a number of marked positions (n) varying between 1.0 and 1.67. n depends on the percentage ($100 \times b$) of erythrose-4-phosphate that is branched off for the synthesis of aromatic amino acids; it can be calculated using the equation $n = (5-4b)/(3-2b)$.

This method is demonstrated with growing cultures of *Clostridium kluyveri*. Data are presented showing that under normal regulatory conditions about 70% of the erythrose-4-phosphate is utilized for aromatic amino acid synthesis while in the presence of tyrosine or phenylalanine or all three aromatic amino acids respectively about 50%, 40% or almost none of the tetrose is branched off into the synthesis of aromatic compounds.

It is shown by analysis of the cell constituents that in cells growing in media supplemented with aromatic amino acids the decreasing percentage of branched-off tetrose is quantitatively correlated to the decreasing requirement for aromatic amino acids.

The pathway of the biosynthesis of the aromatic compounds phenylalanine, tyrosine, tryptophan, *p*-hydroxybenzoic acid and *p*-aminobenzoic acid branches off from carbohydrate metabolism with the condensation of erythrose-4-phosphate and phosphoenolpyruvate to 7-phospho-2-oxo-3-deoxy-D-arabino-heptonate (PODH) [3—8]. More than 99% of this heptulosonic acid is utilized for the synthesis of the three aromatic amino acids. Therefore *p*-hydroxybenzoate and *p*-aminobenzoate can be neglected in regulation studies of PODH aldolase [11]. It was shown for different microorganisms [9—15] that the three isoenzymes of PODH aldolase are subject to a specific feedback inhibition or repression or both. This is a key step in the regulation of the synthesis of aromatic amino acids. In these studies PODH aldolase activity in cell-free extracts or the accumulation of cyclic intermediates by stationary whole cells of mutants defective of chorismate synthetase were measured. The position of regulation within the pathway of aromatic amino acid synthesis and the type of the regulatory mechanisms were thus clearly established. However, no quantitative statement could be made on the degree of a physiological inhibition of aromatic amino acid synthesis in intact growing cells.

Such a quantitative determination under growth conditions should be possible in organisms which synthesize the pentoses from hexoses via a non-oxidative pentose phosphate pathway, using an isotope method. From hexoses labeled in positions 3 and 4, pentoses will be formed which are expected to have between 1.0 and 1.67 marked positions, depending on the amount of erythrose-4-phosphate that is utilized for the synthesis of the aromatic amino acids [1]. This situation is depicted in Fig. 1.

Three different species of pentoses (I, II, III) having 1, 3, and 1 labeled carbon atoms constitute the ribose pool. The average number of labeled positions(n) in ribose equals the total number of labeled carbon atoms divided by the total number of pentoses formed. n can easily be expressed in terms of the percentage ($100 \times b$) of tetroses being branched off or the percentage ($100 \times a$) remaining in the carbohydrate pathway, $a + b = 1$. If all tetrose is utilized for aromatic amino acid synthesis, *i.e.* $a = 0$, only one pentose (III) is formed carrying one labeled carbon atom, thus the lower limit of n becomes 1.0. If none of the tetrose is utilized for the synthesis of aromatic compounds, *i.e.* $a = 1.0$, all three pentoses are formed in equivalent amounts having together 5 labeled carbon atoms, setting the upper limit of n at $5/3 = 1.67$. In general the total number of pentoses formed depends on a and is $(1 + 2a)$; the pentoses will have $(1 + 4a)$ labeled carbon atoms, n is therefore $(1 + 4a)/(1 + 2a)$. Substituting $(1 - b)$ for a, n equals $(5 - 4b)/(3 - 2b)$. This quantitative relation is given in Fig. 2.

Non-standard abbreviations. Trichloroacetic acid, TCA; 7-phospho-2-oxo-3-deoxy-D-arabino-heptonate, PODH.

Enzymes. 7-phospho-2-oxo-3-deoxy-D-arabino-heptonate aldolase (EC 4.1.2.15); alkaline phosphatase (EC 3.1.3.1); glycerol kinase (EC 2.7.1.30); glycerol-3-phosphate dehydrogenase (EC 1.1.1.8).

The regulation of aromatic amino acid synthesis can thus be studied by determining the number of marked positions, *i.e.* specific radioactivities, in the pentoses.

In the present paper this method is demonstrated with growing cultures of *Clostridium kluyveri*. Data will be presented showing that under normal regulatory conditions approximately 70 % of the erythrose-4-phosphate is utilized for aromatic amino acid synthesis, while in the presence of tyrosine or phenylalanine or all three aromatic amino acids respectively

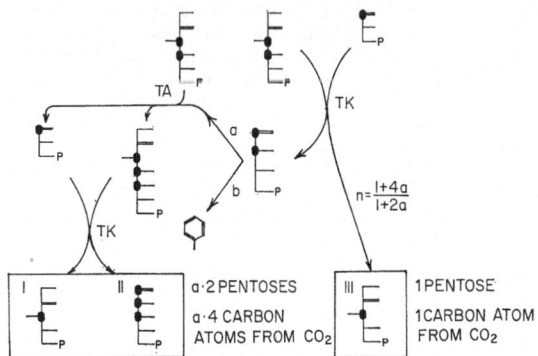

Fig.1. *Scheme of the biosynthesis of pentoses and aromatic amino acids from erythrose-4-phosphate.* ● positions derived from CO_2

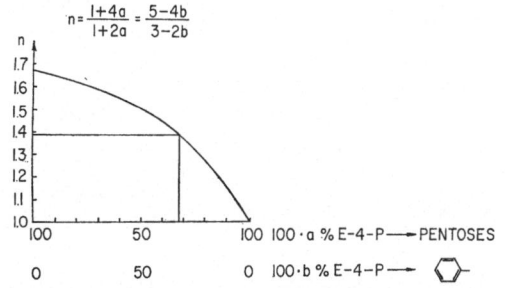

$$n = \frac{1+4a}{1+2a} = \frac{5-4b}{3-2b}$$

Fig.2. *Quantitative correlation between ribose labeling (n) and participation of erythrose-4-phosphate in pentose ($100 \times a$) and aromatic amino acid ($100 \times b$) synthesis*

about 50 %, 40 % or almost none of the tetrose is branched off into the synthesis of aromatic compounds[1].

MATERIALS AND METHODS

All chemicals were reagent grade and were purchased from E. Merck AG, Darmstadt, Germany. Alkaline phosphatase, glycerolkinase and glycerol-

---

[1] Part III of Carbohydrate Synthesis in *Clostridium kluyveri* [1, 2].

phosphate-dehydrogenase were obtained from C. F. Boehringer GmbH, Mannheim, Germany. $Na_2{}^{14}CO_3$ was bought from Radiochamical Centre, Amersham, England.

Radioactivities were determined in a TriCarb Scintillation Spectrometer 314 EX, Packard Instruments Inc., La Grange, Illinois, USA, using the dioxane scintillator[16] for aqueous solutions and the toluene scintillator[17] for $Ba{}^{14}CO_3$ determinations. All measurements were corrected with an internal [${}^{14}C$]toluene standard (Packard). Paperchromatograms were assayed in a radiopaperchromatograph FH 49/407/452 of Frieseke und Hoepfner, Erlangen-Buck, Germany. Amino acid analyses were performed with an automatic amino acid analyzer of Bender and Hobein, Karlsruhe, Germany. Ribose was determined quantitatively with the orcinol reaction [18], and glycerol enzymatically [19].

Clostridium kluyveri cultures were revived from dried crotonate cells [20] (gift of Dr. H.-U. Bergmeyer, C. F. Boehringer Biochemica, Tutzing, Germany). The synthetic medium described by Stadtman and Barker [21] was used; anaerobiosis was secured with a methyleneblue/dithinite indicator. At 33° the cultures reached the end of the logarithmic phase after 60 hours, and were then harvested. A 10 % inoculum of a radioactive subculture was used throughout.

The bacteria were grown in 100 ml volumetric flasks to which a bubble-counter filled with 5 ml 4 N KOH was attached directly. The counter was changed after 25, 42 and 60 hours. CO_2 evolved from the medium, due to bacterial fatty acid excretion, was thus trapped. It was transformed to $BaCO_3$ and its radioactivity was determined as described earlier [1].

The cells were harvested, washed, hydrolyzed, and worked up as reported previously [1]. Ribose and glycerol were isolated from the phosphatasetreated, desalted carbohydrate fraction using de scending paperchromatography on Whatman I with ethylacetate — acetic acid — water (3:1:3). Lipids, DNA, RNA and protein of the dried cells were separated according to Schneider [22]. Lipids and protein were determined gravimetrically, DNA according to Burton [23], RNA according to Mejbaum [24]. Amino acid analysis of the protein fraction was carried out using the method of Spackman, Stein, and Moore[25].

RESULTS AND DISCUSSION

Clostridium kluyveri is a strict anaerobe growing on synthetic media with ethanol, acetate, and bicarbonate as the sole carbon sources [26]. Its carbon assimilation is based on a Ferredoxin-dependent carboxylation of acetyl-CoA to pyruvate [27], which leads to the formation of both amino acids [28,29] and carbohydrates [1,2].

It was demonstrated previously that ribose synthesis proceeds via the following pathways: reductive carboxylation of acetyl-CoA to pyruvate, formation of triose-phosphate from pyruvate, condensation of two triose-phosphates to hexose-phosphate followed by a transaldolase and transketolase catalyzed nonoxidative pentose phosphate pathway. No catabolism nor accumulation of ribose was found [1,2].

Hexoses of cells grown in the presence of ^{14}C-labeled bicarbonate are therefore specifically marked in positions 3 and 4. Ribose formed via the nonoxidative pentose phosphate pathway was found to contain about 1.4 carbon atoms derived from CO_2[1]. *Clostridium kluyveri* is therefore an excellent organism to study the regulation of aromatic amino acid synthesis with the proposed quantitative isotope method.

An exact regulation of the synthesis of the aromatic compounds was to be expected as *Clostridium kluyveri* does not excrete aromatic amino acids or their precursors into the medium [29]. On addition of the three aromatic amino acids to the growth medium a specific control of their synthesis can therefore be predicted. If the proposed theory is correct, the specific radioactivity ($CO_2 = 1$) of ribose should increase from about 1.4 towards 1.67. The results of such experiments are given in Tables 1 and 2.

In the four experiments I_{1-4} and four experiments reported earlier [1] an average of 1.39 with a mean error of 0.021 is obtained in the absence of the aromatic amino acids. Solving the equation $n = (1 + 4a)/(1 + 2a)$ for a, it was calculated that under these conditions $68\% \pm 2.6\%$ of erythrose-4-phosphate is branched off from the carbohydrate pathway into the aromatic amino acid synthesis. Already at a concentration in the growth medium of 0.1 mM of each of the three aromatic amino acids the specific activity of ribose approaches 1.67. No further increase was observed at higher concentrations. In 6 experiments

Table 1. *Incorporation of $^{14}CO_2$ into ribose and glycerol in cells of* Clostridium kluyveri *grown on media with varying concentrations of all three aromatic amino acids*

		Amino acids			Wet cells	Bicarbonate				Ribose			Glycerol		
		DL-Phe	L-Tyr	L-Try		t_{25}	t_{45}	t_{60}	Average						
			mmoles/l		mg/100 ml	decompositions/ min/μmole				μmoles	decompositions/ min/μmole	n	μmoles	decompositions/ min/μmole	n
I	1	—	—	—	110	150,000	—	146,000		0.88	210,000	1.4			
	2	—	—	—	112	149,000	—	151,000		1.09	195,000	1.3	1.42	144,000	0.96
II	1	0.2	0.1	0.1	120	—	147,000	—		1.35	256,000	1.7			
	2				122	—	154,000	—	150,000	0.92	244,000	1.63	1.52	138,000	0.92
III	1	1.0	0.5	0.5	118	—	153,000	—		1.28	236,000	1.57	0.77	142,000	0.95
	2				109	—	152,000	—		1.18	242,000	1.61	—	—	—
IV	1	4.0	2.0	2.0	95	149,000	—	148,500		0.65	248,000	1.66			
	2				123	147,000	—	150,500		1.37	253,000	1.68	1.18	149,500	0.99

Table 2. *Incorporation of $^{14}CO_2$ into ribose and glycerol in cells of* Clostridium kluyveri *grown on media with and without aromatic amino acids*

		Amino acids		Wet cells	Bicarbonate			Ribose			Glycerol		
		DL-Phe	L-Tyr		t_{25}	t_{60}	Average						
		mmoles/l		mg/100 ml	decompositions/ min/μmole			μmoles	decompositions/ min/μmole	n	μmoles	decompositions/ min/μmole	n
I	3	—	—	99	83,500	81,000		0.84	113,000	1.37	0.52	76,500	0.93
	4	—	—	113	—	—		0.65	116,500	1.42	—	—	—
V	1	—		110	81,000	82,000		0.90	123,000	1.50	0.53	75,000	0.91
	2	—	1.0	115	—	—	82,000	0.98	123,000	1.50	—	—	—
	3	—		113	—	—		0.80	121,000	1.47	—	—	—
VI	1	—		114	80,500	83,500		0.55	129,000	1.57	0.47	77,000	0.94
	2	2.0	—	110	—	—		0.62	126,000	1.54	—	—	—
	3	—		117	—	—		0.64	124,000	1.51	—	—	—

an average of 1.64 ± 0.014 was obtained; 89% $\pm 5.5\%$ of the erythrose now remained in the carbohydrate pathway. Addition of tyrosine or phenylalanine alone to the growth medium resulted in a decrease of tetrose utilization for aromatic amino acid synthesis from about 70% to 50% or 40% resp.

The n-values of triose-phosphates as measured by glycerol, are independent of the presence of aromatic amino acids. This is a necessary condition for the applicability of the method.

The utilization pattern of tetroses should be in agreement with the requirement of the cells for pentoses and aromatic amino acids. From an analysis of cell constituents (Table 3) the ratio (r) of pentoses and their derivatives to aromatic compounds

Table 3. *Composition of* Clostridium kluyveri *dried cells*

	Dried cells	Pentoses	Aromatic compounds
	$\%$ (w/w)	μmoles	μmoles
Protein	68		
Phe	3.17		19.2
Tyr	2.74		15.1
Try	1.36	6.7	6.7
His	4.50	29.0	
Nucleic Acids	9		
DNA	1.4	3.75	
RNA	7.6	21.0	
Lipids	3.0		
Cold TCA soluble fraction	20		
Orcinol positive		10.5	
Flavine (ribitol)		0.3	
Pyrimidine ribotides[a]		2.0	
Cell wall (ribitol)[a]		5.0	
		78.25	41.0

[a] Estimated values.

has been determined. From Fig. 1 it can be seen that $(1 + 2a)$ pentoses are formed for each $b = (1 - a)$ aromatic amino acids, $r = (1 + 2a)/(1 - a)$. Therefore a can not only be calculated from the n-values of ribose, obtained from long-time labeling experiments, but also independently from the r-values of cell-constituent analyses. The regulatory effect of added aromatic amino acids can also be calculated from the r-values using the assumption, that added amino acids will not be synthesized by the cells; then r equals the sum of all pentoses divided by the sum of all aromatic amino acids minus the constituent amount of the amino acid added.

In Fig. 3, a as calculated from the n-values is plotted against a as computed from the r-values. Theoretically $a(n)$ should equal $a(r)$. The points obtained from the presented data deviate slightly

from the $45°$ line, indicating that the tracer technique for regulation studies yields reasonable results.

The method, in contrast to the analytical procedures applied to date, allows to describe a physiological state of regulation quantitatively. Such results cannot be obtained with enzyme activity measurements in cell-free extracts as it will remain unknown

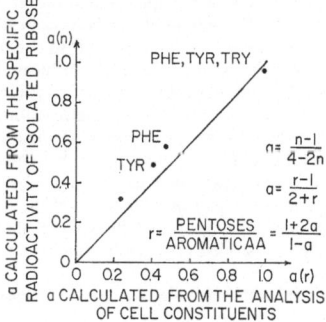

Fig. 3. *Correlation between a calculated from n and a calculated from r*

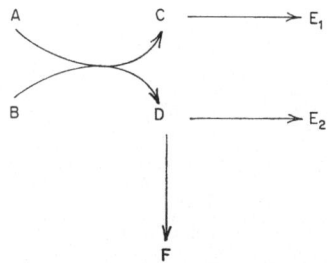

Fig. 4. *Metabolic requirements for the applicability of the isotope method.* E_1 *and* E_2 *yield identical compounds that are differently labeled*

which enzyme activity or level might be limiting *in vivo*. Radioactive dilution experiments with intact cells do not permit quantitative statements on regulation phenomena either, as the exchangeability and the pool sizes are unknown factors.

The accuracy of this method diminishes with increasing values of b, as the slope of the curve (Fig. 2) relating n to b, becomes very small for $b < 0.2$. Therefore even small differences in the single values for n caused by the relative inaccuracy of the ribose determinations [18] lead to diverging results for b and so constitute a disadvantage of the new method.

The applicability of this method is not restricted to Clostridium kluyveri. It can be performed with all growing cells that synthesize the pentoses from hexoses via a non-oxidative pathway and do not have any appreciable pentose catabolism, during growth.

A similar *in vivo* technique may be applied to any regulatory system which meets the requirements outlined in Fig. 4. If E_1 and E_2, having a different labeling pattern, yield identical compounds, the competitive reactions from D to E_2 and F can be studied.

From the presented data it can be seen that the amount of erythrose-4-phosphate used for amino acid synthesis is strictly regulated in *Clostridium kluyveri*. As no ribose is found to be excreted into the medium an exact regulation of pentose synthesis is also indicated. The finding that there is no overproduction of pentose in the presence of aromatic amino acids must be explained with a decrease of the intracellular level of erythrose-4-phosphate. In order to guarantee an appropiate tetrose level in the cells the in-flow reaction (transketolase) as well as the outflow reaction (transaldolase) of erythrose-4-phosphate should be specifically controlled by endproducts or intermediates.

These problems are under further investigation.

This project was supported by grants from the Deutsche Forschungsgemeinschaft, Bad Godesberg.

REFERENCES

1. Decker, K., Thauer, R. K., and Jungermann, K., *Biochem. Z.* 345 (1966) 461.
2. Decker, K., Barth, C., and Metz, H., *Biochem. Z.* 345 (1966) 472.
3. Cohen, G. N., *Ann. Rev. Microbiol.* 19 (1965) 124.
4. Stadtman, E. R., *Advan. Enzymol.* 28 (1966) 62.
5. Srinivasan, P. R., Katagiri, M., and Sprinson, D. B., *J. Am. Chem. Soc.* 77 (1955) 4943.
6. Srinivasan, P. R., Shigeura, H. T., Sprecher, M., Sprinson, D. B., and Davis, B. D., *J. Biol. Chem.* 220 (1956) 477.
7. Srinivasan, P. R., Katagiri, M., and Sprinson, D. B., *J. Biol. Chem.* 234 (1959) 713.
8. Srinivasan, P. R., and Sprinson, D. B., *J. Biol. Chem.* 234 (1959) 716.
9. Fradejas, R. G., Ravel, J. M., and Shive, W., *Biochem. Biophys. Res. Commun.* 5 (1961) 320.
10. Smith, L. C., Ravel, J. M., Lax, S. R., and Shive, W., *J. Biol. Chem.* 237 (1962) 3566.
11. Brown, K. D., and Doy, C. H., *Biochim. Biophys. Acta,* 77 (1963) 170.
12. Smith, L. C., Ravel, J. M., Lax, S. R., and Shive, W., *Arch. Biochem. Biophys.* 105 (1964) 424.
13. Brown, K. D., and Doy, C. H., *Biochim. Biophys. Acta,* 118 (1966) 157.
14. Jensen, R. A., and Nester, E. W., *J. Biol. Chem.* 241 (1966) 3365, 3373.
15. Lingens, F., Goebel, W., and Uessler, H., *Biochem. Z.* 346 (1966) 357.
16. Davidson, J. D., and Feigelson, P., *Int. J. Appl. Radiat.* 2 (1957) 1
17. Kallmann, H., and Furst, M. in *Liquid Scintillation Counting* (edited by C. G. Bell and F. N. Hayes). Pergamon Press New York 1958, p. 56.
18. Ashwell, G., in *Methods in Enzymology* (edited by S. P. Colowick and N. O. Kaplan). Academic Press, New York 1957, Vol. III, p. 88.
19. Wieland, O., in *Methoden der enzymatischen Analyse* (herausgegeben von H. U. Bergmeyer). Verlag Chemie GmbH, Weinheim 1962, p. 211.
20. Bergmeyer, H. U., Holz, G., Klotsch, H., and Lang, G., *Biochem. Z.* 338 (1963) 114.
21. Stadtman, E. R., and Barker, H. A., *J. Biol. Chem.* 180 (1949) 1085.
22. Schneider, W. C., *J. Biol. Chem.* 161 (1945) 293.
23. Burton, K., *Biochem. J.* 62 (1956) 315.
24. Mejbaum, W., *Z. Physiol. Chem.* 258 (1939) 117.
25. Spackman, D. H., Stein, W. H., and Moore, S., *Analyt. Chem.* 30 (1958) 1190.
26. Bornstein, B. T., and Barker, H. A., *J. Biol. Chem.* 172 (1948) 659.
27. Andrew, I. G., and Morris, J. G., *Biochim. Biophys. Acta,* 97 (1965) 176.
28. Tomlinson, N., and Barker, H. A., *J. Biol. Chem.* 209 (1954) 585.
29. Tomlinson, N., *J. Biol. Chem.* 209 (1954) 597, 605.

R. K. Thauer, K. Jungermann and K. Decker
Biochemisches Institut der Universität
78 Freiburg im Breisgau
Hermann Herder-Straße 7, Germany